GRAPHS OF FUNCTIONS

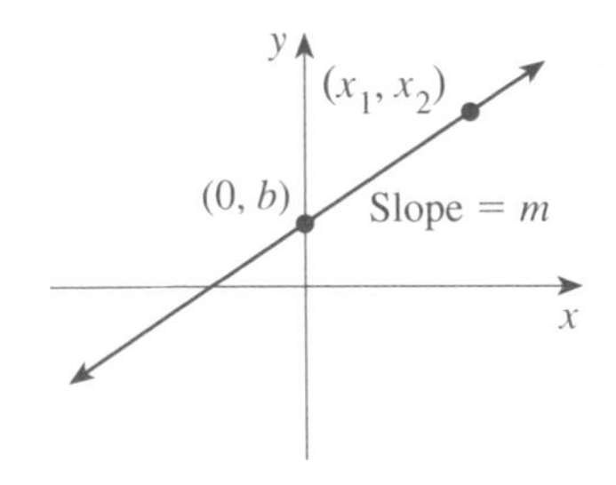

$y = mx + b$

$y - y_1 = m(x - x_1)$

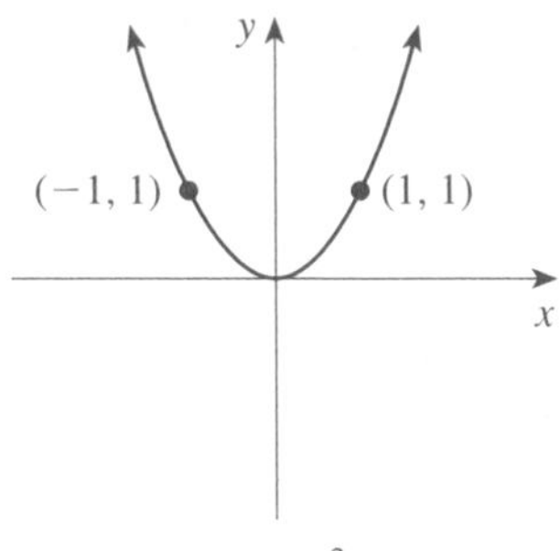

$y = x^2$

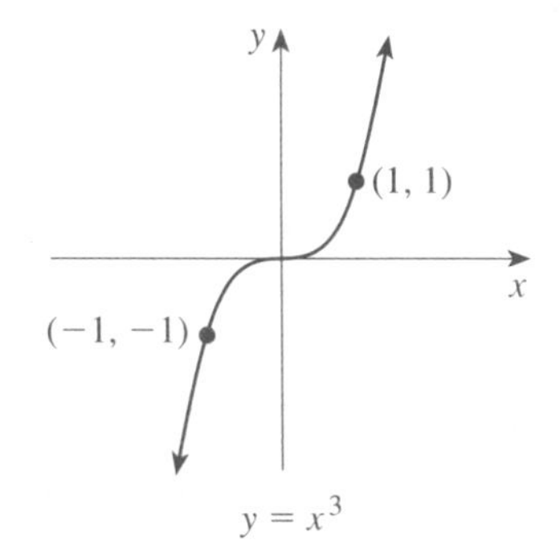

$y = x^3$

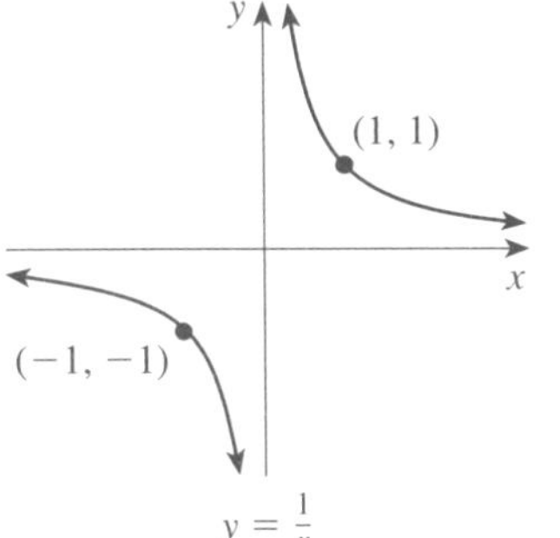
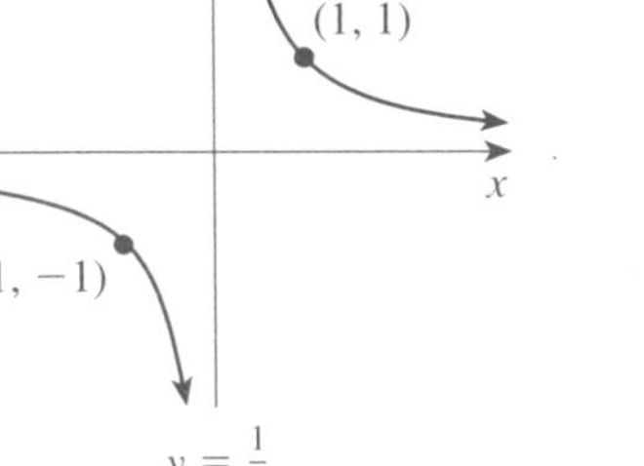

$y = \frac{1}{x}$

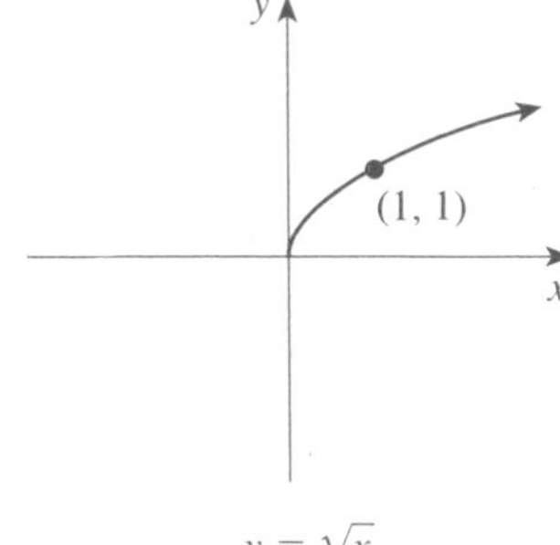

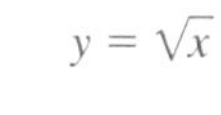

$y = \sqrt{x}$

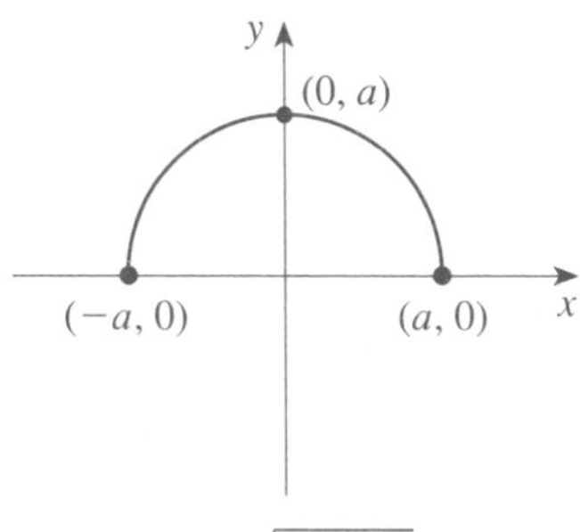

$y = \sqrt{a^2 - x^2}$

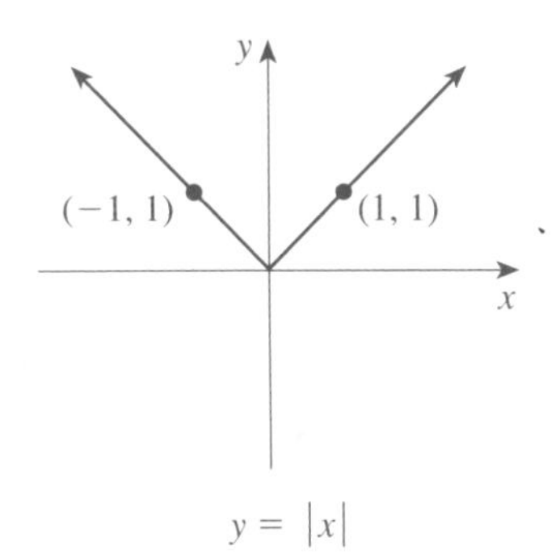

$y = |x|$

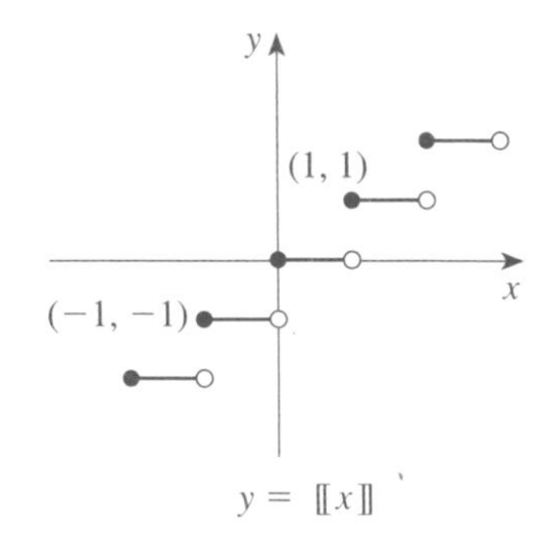

$y = [\![x]\!]$

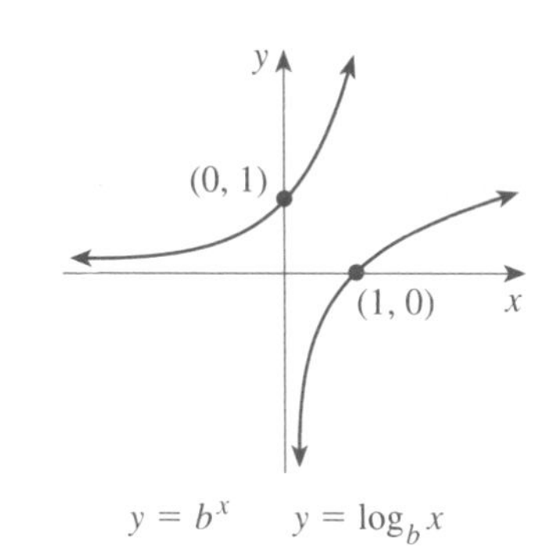

$y = b^x \qquad y = \log_b x$

GRAPHS OF RELATIONS

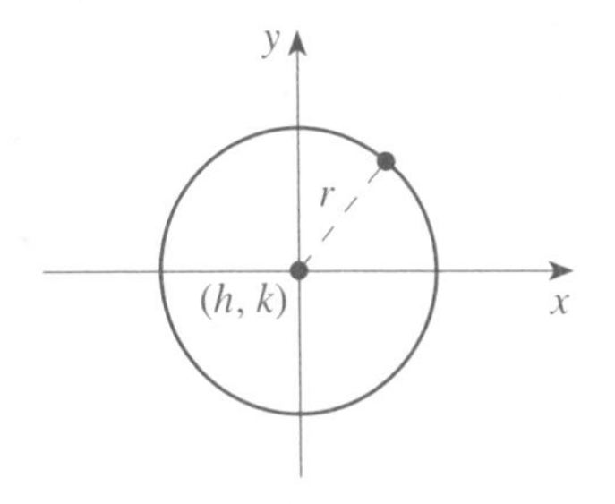

$x^2 + y^2 = r^2$

$(x - h)^2 + (y - k)^2 = r^2$

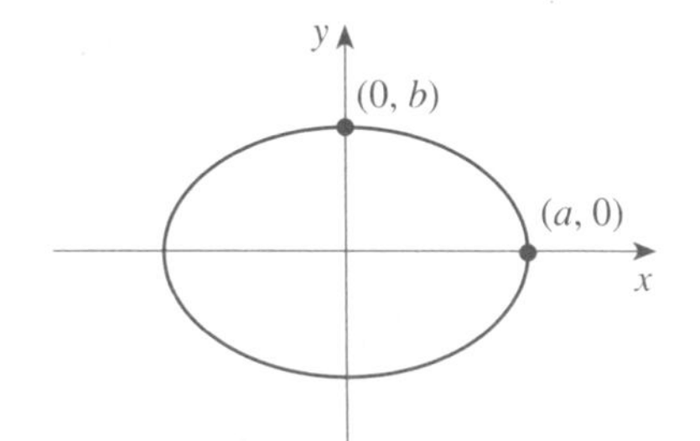

$\frac{x^2}{a^2} + \frac{y^2}{b^2} = 1$

$\frac{(x-h)^2}{a^2} + \frac{(y-k)^2}{b^2} = 1$

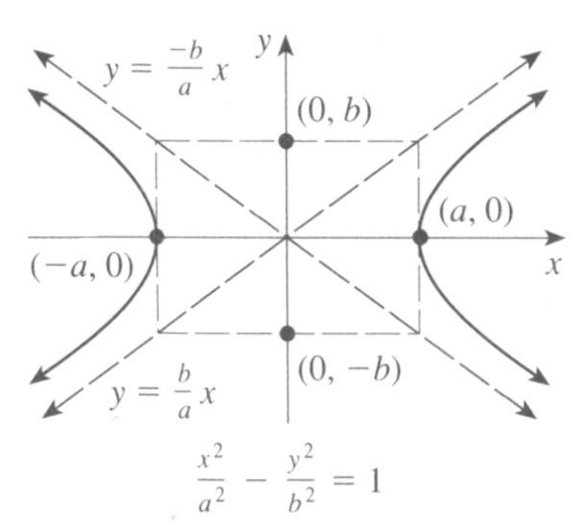

$\frac{x^2}{a^2} - \frac{y^2}{b^2} = 1$

$\frac{(x-h)^2}{a^2} - \frac{(y-k)^2}{b^2} = 1$

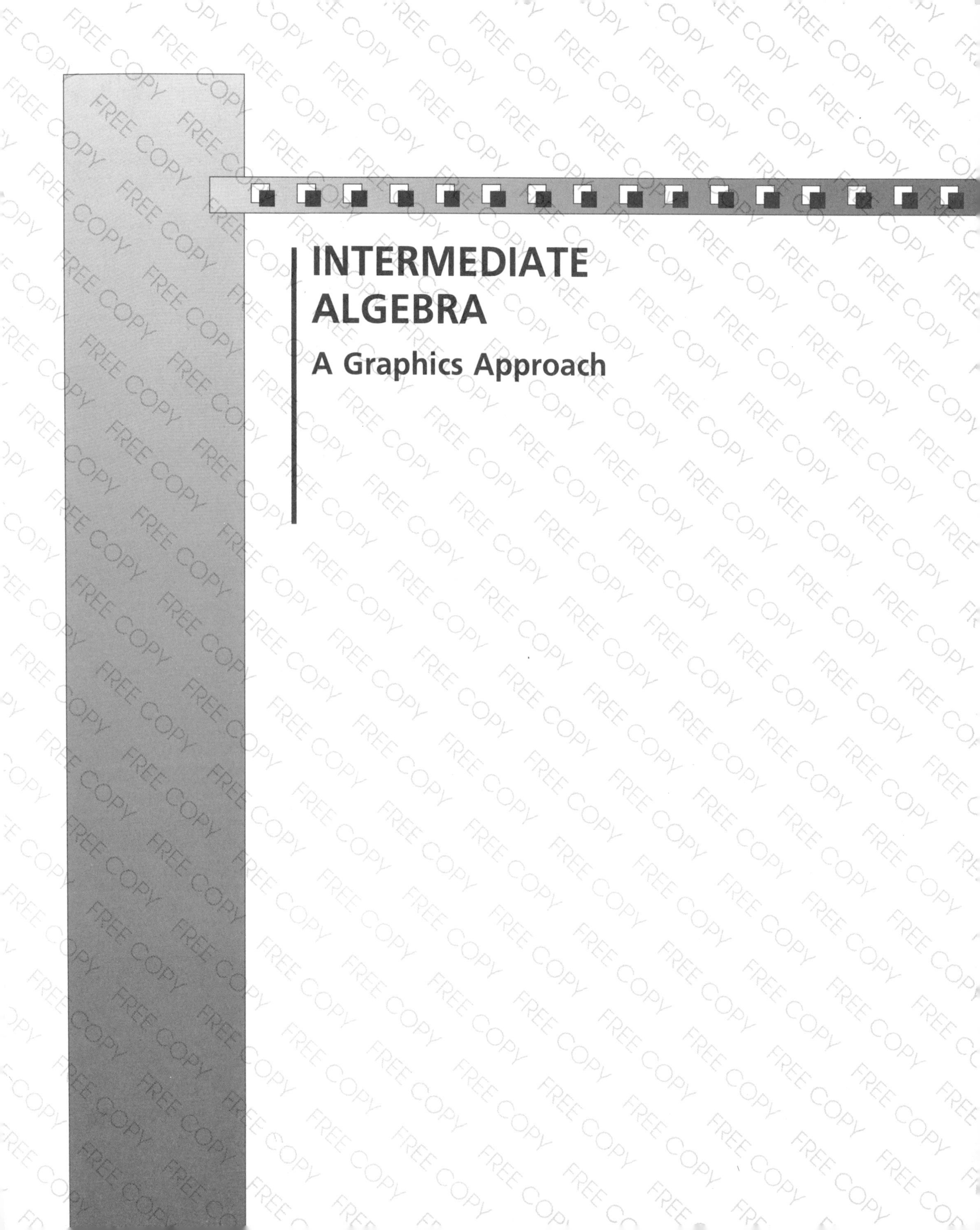

INTERMEDIATE ALGEBRA

A Graphics Approach

INTERMEDIATE ALGEBRA

A Graphics Approach

M. G. Settle
Pensacola Junior College

WEST PUBLISHING COMPANY

Minneapolis/St. Paul
Los Angeles
New York
San Francisco

WEST'S COMMITMENT TO THE ENVIRONMENT

In 1905, West Publishing Company began recycling materials left over from the production of books. This began a tradition of efficient and responsible use of resources. Today, up to 95 percent of our legal books and 70 percent of our college and school texts are printed on recycled, acid-free stock. West also recycles nearly 22 million pounds of scrap paper annually—the equivalent of 181,717 trees. Since the 1960s, West has devised ways to capture and recycle waste inks, solvents, oils, and vapors created in the printing process. We also recycle plastics of all kinds, wood, glass, corrugated cardboard, and batteries, and have eliminated the use of Styrofoam book packaging. We at West are proud of the longevity and the scope of our commitment to the environment.

Printing and binding by West Publishing Company.

Interior design ■ TECH*arts*
Interior illustration ■ TECH*arts*
Composition ■ The Clarinda Company
Cover image ■ David Bishop
Cover design ■ Lois Stanfield, LightSource Images

610 Opperman Drive
P.O. Box 64526
St. Paul, MN 55164-0526

Printed in the United States of America
01 00 99 98 97 96 95 94 8 7 6 5 4 3 2 1 0

Library of Congress Cataloging-in-Publication Data

Settle, M. G.
Intermediate algebra: a graphics approach/M.G. Settle.
p. cm.
Includes index.
ISBN 0-314-02842-0
1. Algebra. I. Title.
QA154.2.S48 1994
512'.9—dc20 93-33044
CIP

To Thelma Mary Dobbs Settle,
for teaching me arithmetic

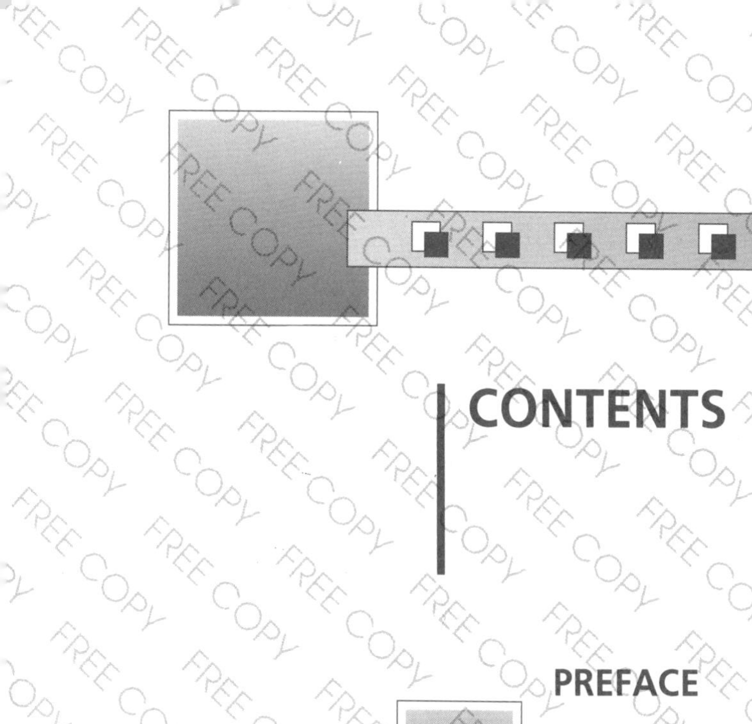

CONTENTS

5 Solving Second-Degree Equations 240

6 Rational Expressions 314

7 Radical Expressions 372

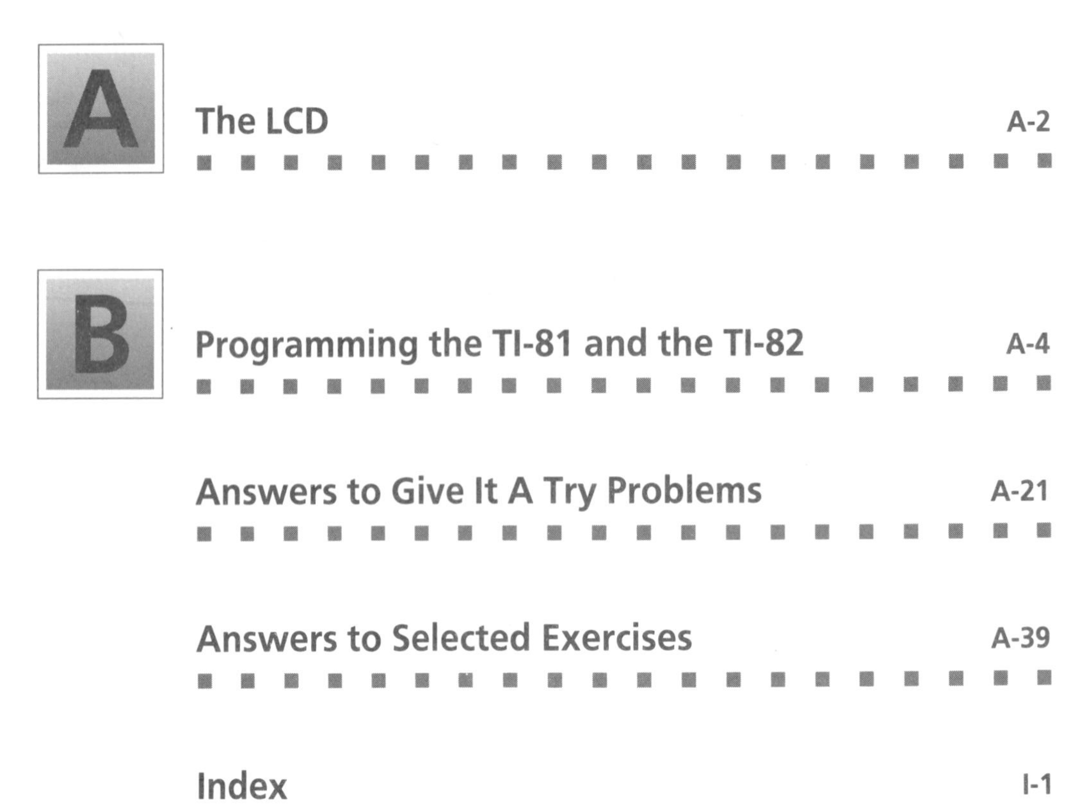

PREFACE

Intermediate Algebra: A Graphics Approach was developed to help students gain a greater understanding of algebra through graphics and to increase their interest and performance in algebra topics through graphics calculator use. The traditional topics of intermediate algebra are covered; however, graphs and functions are introduced early, in Chapter 3. This text is designed for a one-term course in algebra. A background of either an elementary algebra course or an appropriate secondary school course is assumed. No former knowledge of calculators is assumed.

CONTENT FEATURES ■ ■ ■

The introductory material on the graphics calculator is presented at the end of each of the first four sections in Chapter 1 in a guided discovery format. This presentation provides easy access for students (and allows instructors the option of starting the course in Chapter 2). After the introductory material in Chapter 1, the graphics calculator is fully integrated into the course materials. In Chapter 2, the proficiency with the calculator is obtained by testing for errors in work on addition and subtraction of polynomials. These exercises not only give the student a feel for the calculator, but also reinforce the fact that algebraic expressions represent number phrases. Next, the calculator is used to check the solution to an equation in one variable. Here, the graph screen is introduced as a tool for seeing several evaluations of an expression for various replacements of the variable. From the beginning, the use of a *friendly* graph screen is emphasized. After the work in Chapter 2, the student is prepared in both algebra background and graphics calculator skills for the equations in two variables and linear functions of Chapter 3.

Throughout the book, detailed instructions are given for the TI-81 calculator. This calculator was chosen as a baseline graphics calculator because of its features, cost, and user base. Footnotes are provided for the TI-82, the upgrade to the TI-81. These footnotes also provide for the easy use of other graphics calculators, such as the Casio 7700G/8700G and the Sharp 9200/9300EL calculators. These calculators, like the TI-82, work from a 95-pixel screen, using a -4.7-to-4.7 graphics window. Additionally, *Using Graphics Calculators,* a manual written by M. G. Settle and available with the book, provides specific instructions for the Casio and Sharp calculators, keyed to material in the textbook.

The graphics calculator is an integral part of this book—its use is assumed. Although traditional topics are covered, they are greatly enhanced by the use of the graphics calculator. Additionally, by using the graphics calculator, it is natural to move linear equations and functions forward to Chapter 3 from their traditional position later in the course. Function concepts, notation, and graphs are gradually introduced to the student throughout the course.

This book is designed for student use. From the strategically placed *Give It a Try* problems in the content sections (with answers provided) to the *Using the Graphics Calculator* boxes, students are encouraged to read and think about algebra concepts. In many of the exercises, especially the application problems, and in the *Discovery* topics, students are asked to write. This emphasis follows the recommendations of many professional groups to "incorporate writing in mathematics courses."

PEDAGOGICAL FEATURES ■ ■ ■

Examples Several examples are provided to cover each concept presented. These examples are presented in a form that is useful to students. The graphics calculator is fully integrated into these examples. Instructors will find that this feature greatly increases student involvement in the text materials. For students, reading calculator instructions improves critical reading and thinking skills required for success in mathematics (as well as for other subjects).

Using the Graphics Calculator Specific instructions are provided for students on how to use the graphics calculator. These instructions are highlighted and are often accompanied by calculator screens. This feature greatly reduces the burden on instructors and frees valuable classroom time for instruction, discussion, and questions.

Give It a Try Several short sets of problems are placed strategically within each section to provide students with an understanding check. These problems deal with the currently presented concepts and often expand on recently presented examples. Answers plus explanatory material for these problems are conveniently located at the back of the book to provide students with immediate feedback.

Warning and Remember These boxed materials are placed in the margins at critical positions throughout the book. These boxes show common errors to be avoided and important background material for concepts being developed.

Summary At the end of each section, a short summary reviews important concepts and skills developed in the section.

Quick Index A mini-index is presented with each section summary to provide students with page number references for important terms presented within the section.

Exercises At the end of each section, several problems are presented to ensure full development of the concepts and skills presented in the section. Enough problems are presented for review later in the course. Most problems are presented in a matched, odd-even format. Graphics calculator problems are also presented in these exercise sets. At the end of each chapter, a chapter review exercise set is provided, along with a chapter test.

Discovery These sections appear throughout the textbook. Many of the discovery topics are graphics-calculator based, such as finding range settings for friendly graph screens and power regression. Other discovery topics include the Richter scale for measuring earthquake intensity and parametric equations.

Programs An important feature of the graphics calculator is its ability to store a sequence of instructions (a program) in memory. Using this feature, a student can execute programs to do a wide variety of mundane or potentially error-producing procedures. Three such programs are a program for finding the equation of a line, a program for producing the solution to a quadratic equation, and a program for performing synthetic division. All of these programs are given in Appendix B and are optional.

SUPPLEMENTS ■ ■ ■

Instructor's Solutions Manual The *Instructor's Solutions Manual* by Jon D. Weerts contains an answer or solution to every even-numbered exercise and chapter test question in the book. Graphics calculator screens are shown for many of the solutions.

Student's Solutions Manual The *Student's Solutions Manual* by Jon D. Weerts contains complete solutions, including graphics calculator screens, for the odd-numbered exercises and all chapter test questions.

Using Graphics Calculators The *Graphics Calculators Supplement* by M. G. Settle provides instructions matching the introductory materials in Chapter 1 for the Casio 7700G and Sharp 9200/9300EL calculators. (The TI-81 and TI-82 calculators are covered in the textbook.) Also, calculator-specific instructions are provided for other techniques that vary from those presented in the textbook (for example, function evaluation and matrix entry).

Test Bank The *Test Bank* by Charles Heuer contains three versions of a multiple-choice test and two versions of an open-ended test for each chapter in *Intermediate Algebra: A Graphics Approach*.

WestTest© 3.0 (Computer-Generated Testing Programs) Versions of this software are available for both the PC and Macintosh computer systems to qualified adopters of the textbok.

Overhead Transparencies The important figures and tables in the textbook are available as transparency masters to schools adopting *Intermediate Algebra: A Graphics Approach*. **Worksheet masters** for each section provide additional problems that are easy to grade and provide quick feedback to students.

Video Tapes Two video tapes are available to introduce the TI-81, TI-82, Casio 7700G, and Sharp 9200/9300EL graphics calculators.

Graph⁺ Software The *Graph⁺* software by M. G. Settle is a program for PC computers. This program has a user interface similar to a graphics calculator. Students can use the software to produce printed copies of graphs. Introductory materials are provided and the instructions presented in the textbook for the TI-81 calculator can be followed to produce corresponding results on PC computers.

GraphToolZ Software The *GraphToolZ* software by Tom Saxton is a graphing program for the Macintosh family of computers. An introductory package of materials provides students with instructions for getting started with this software.

Computer Tutorials *Natural Language Mathematics Tutorial* software by Mathens, Inc., contains algorithms for generating multiple examples of the basic problem types in *Intermediate Algebra: A Graphics Approach*. These tutorials are available for both the PC and Macintosh computer systems.

Sequences and Series A chapter on the binomial theorem, sequences, series, and probability is available from West Publishing Company. The graphics calculator is used throughout this supplemental chapter.

ACKNOWLEDGMENTS ■ ■ ■

First, I thank my family, my wife Robin and our sons Joshua, Gary, and David. I also thank all my students at Pensacola Junior College, Milton Center, as well as the 1991 fall-term intermediate algebra students of Margaret Greene at Florida Community College—Jacksonville. Their responses and constructive suggestions for improving the text material and the exercises have been very helpful. A special thanks goes to Charles Heuer for his careful work in checking the text and exercises for accuracy. I am especially grateful to Kathi Townes and her staff at TECH*arts* for all their work on this book. I received valuable suggestions and direction from the following reviewers:

Mary Alter ■ University of Maryland–College Park

Rick Armstrong ■ St. Louis Community College—Florissant Valley

Judy Becker ■ Santa Fe Community College

Irene Doo ■ Austin Community College

Dale Green ■ Oregon State University

Margaret Greene ■ Florida Community College—Kent Campus

Thomas Gregory ■ The Ohio State University—Mansfield

William Grimes ■ Central Missouri State

Roseanne Hofmann ■ Montgomery County Community College

Robert Horvath ■ El Camino College

Mary Barr Humphrey ■ Western Kentucky University

Nancy Hyde ■ Broward Community College

Robert Knott ■ University of Evansville
Debra Landre ■ San Joaquin Delta College
Hector Mendez ■ University of Georgia
Ruth Parsons ■ Illinois State University
Jay Sachs ■ Middlesex Community College
Jean Sanders ■ University of Wisconsin—Platteville
Larry Sher ■ Manhattan Community College
Lee Topham ■ Kingwood College
Frances Ventola ■ Brookdale Community College

I also thank Professor Jon D. Weerts, who wrote the supplementary manuals; Kathi Townes, who copyedited the manuscript; Becky Tollerson at West Publishing Company (Atlanta), for handling all the details; and Mary Verrill, for seeing the book through the production process. Finally, I want to thank Peter Marshall, executive editor at West Publishing, for his farsightedness and many helpful suggestions.

I have used a graphics approach to algebra over the past several years—first using the personal computer, and now using hand-held graphics calculators. Such instruction is exciting and enjoyable to give, and is well received by students. I wish both students and instructors all the best in this endeavor.

M. G. Settle
Marquis Bayou, Florida
1994

1

PRELUDE TO ALGEBRA

Algebra is an area of mathematics characterized by the use of variables. A *variable* holds the place for a number and is often denoted by letters of the alphabet, such as x and y. By using variables instead of specific numbers, we can generalize the procedures and techniques of arithmetic. Because the techniques of algebra are generalizations from arithmetic, we will start with a review of arithmetic. We will next consider integers, rational numbers, irrational numbers, radicals, algebraic numbers, and real numbers. We will discuss the operations of addition, subtraction, multiplication, and division on the integers and on rational numbers.

The development of algebra throughout this text will mirror the review in this chapter: we will introduce a collection of objects (such as polynomials or rational expressions) and then discuss the operations of addition, subtraction, multiplication, and division on these objects. For the objects of algebra, we will also discuss applications, such as solving equations, functions, and so on.

Throughout this review chapter we will introduce features of the graphics calculator. When we begin to discuss the objects of algebra, the graphics calculator will be a useful tool.

1.1 ■ ■ ■ SETS AND INTEGERS

Since their introduction over a century ago, sets have been found to be a powerful tool in communicating ideas in mathematics. Thus, we start by introducing some of the basics of sets. We then discuss one of the most used sets in algebra—the set of integers.

Set Notation A **set** is a collection of objects. The objects may be a list of people, or a list of fast-food chains, or a list of dates. In this course, the objects will usually be numbers or pairs of numbers. Sets are denoted by the bracket symbols, { and }, and are named with capital letters of the alphabet. One way to specify the contents of a set is to list them as a **roster.** For example,

$$A = \{4, 5, 6, 10\}$$

is the set containing the four numbers 4, 5, 6, and 10, and

$$B = \{1, 2, 3, \ldots, 20\}$$

is the set containing the twenty numbers 1 through 20. The ellipsis mark, . . . , means to continue the pattern established, in this case, *adding 1 to the previous number*. For set B, we may say that 11 is an *element* of the set and we write $11 \in B$. Note that 11 is not an element of set A. We write this fact as $11 \notin A$. The sets A and B each have a finite number of elements: set A has 4 elements and set B has 20 elements. We say that each of these sets is a *finite set*. An important finite set is the **empty set,** which contains *no* element. We write the empty set as $\varnothing$ or { }.

Some sets are so large that it is impossible to list all the elements individually. We can communicate the idea of an *infinite set* by listing a few elements and then writing an ellipsis: $\{2, 4, 6, \ldots\}$. Two important infinite sets are the **natural numbers**

$$N = \{1, 2, 3, \ldots\}$$

and the **whole numbers**

$$W = \{0, 1, 2, 3, \ldots\}$$

In addition to the roster method of specifying a set, we can also use *set-builder notation*. As an example of using set-builder notation,

$$E = \{x \mid x \text{ is an even number}\}$$

means "the set of x such that x satisfies the condition that *x is an even number*." In *roster notation* we write this as

$$E = \{0, 2, 4, 6, \ldots\}$$

In general, set-builder notation has the following form:

$$A = \{n \mid n \text{ satisfies property } P\}$$

the set of all . . . (points to n) such that (points to $\mid$)

Relations for Sets Two sets, say A and B, are **equal** when they have exactly the same elements. The order of appearance of elements in either set is unimportant, and each element is listed only once in each set:

$$\{2, 3, 4\} = \{4, 2, 3\}$$

> **WARNING**
> Be sure to distinguish between zero and the empty set. For example, $0 \in W$ and $\varnothing \subseteq W$, but $\varnothing \notin W$ and $0 \nsubseteq W$.

When one set, say B, contains another set, say A, every element of the smaller set A is also an element of the larger set B. We say that A is a **subset** of B and write $A \subseteq B$. For the natural numbers and the whole numbers, we can write $N \subseteq W$, because every element in set N is also an element in set W. Notice that W is not a subset of N, because W contains an element (namely, 0) that is not in N. This fact we write as $W \nsubseteq N$. Note that for any set A, $\varnothing \subseteq A$ and $A \subseteq A$.

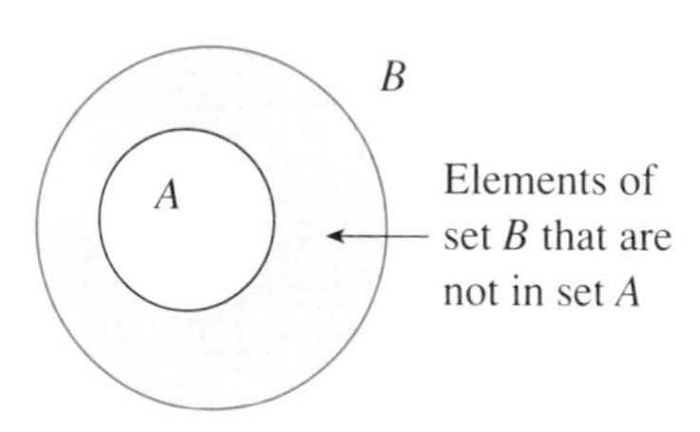

FIGURE 1

Sets are often pictured using a drawing known as a *Venn diagram*. We may picture the subset relationship $A \subseteq B$ as shown in Figure 1. In the language of logic, the subset relationship matches an "IF-THEN" relationship:

$$\text{For } A \subseteq B, \qquad \text{IF } a \in A \quad \text{THEN} \quad a \in B.$$

As an example of some of the concepts we have discussed, work through the following example.

EXAMPLE 1 Set $A = \{1, 2, 3, \ldots, 10\}$ and set $B = \{x \mid x \text{ is an odd number}\}$.
a. Is $A \subseteq B$? b. Is $7 \in A$? c. Is $7 \in B$?

Solution
a. We need to check whether every element in set A is also in set B. We can write B as $\{1, 3, 5, \ldots\}$ because B is the set of numbers where each number is an odd number. Now, $2 \in A$ but $2 \notin B$. Thus, there is an element in A that is not in set B. So A is not a subset of B. We write $A \nsubseteq B$.
b. Set A is $\{1, 2, 3, 4, 5, 6, 7, 8, 9, 10\}$. Thus, $7 \in A$.
c. Since 7 is an odd number, we have $7 \in B$. ■

Operations on Sets So far, we have introduced some objects, namely, sets. We have discussed how sets are named—the roster method and the set-builder method—and two relationships between sets, equal and subset.

The **union** of set A with set B, written $A \cup B$, is a set, and its elements are the elements that are in set A *or* that are in set B. Consider the following examples:

$A = \{2, 4, 6\}$	$B = \{1, 3, 4\}$	$A \cup B = \{1, 2, 3, 4, 6\}$
$V = \{a, e, i, o\}$	$W = \{c, d, f\}$	$V \cup W = \{a, c, d, e, f, i, o\}$

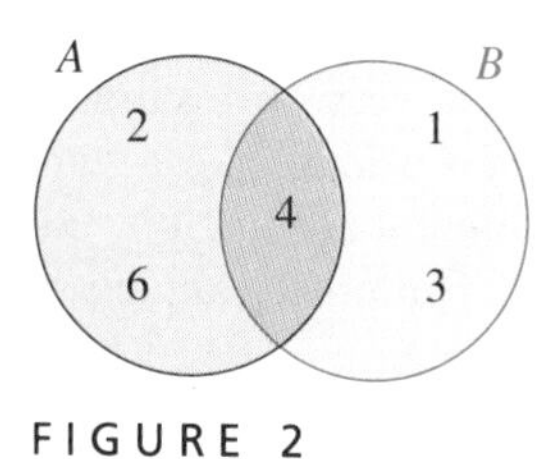

FIGURE 2

In set-builder notation, we can write $A \cup B = \{x \mid x \in A \text{ or } x \in B\}$. A Venn diagram for the union of sets A and B is shown in Figure 2. In the language of logic, the union operation matches an "OR" relationship:

$$a \in (A \cup B) \qquad \text{provided } a \in A \quad \text{OR} \quad a \in B$$

The **intersection** of set A with set B, written $A \cap B$, is a set, and its elements are the elements that are in set A *and* in set B.

Consider the following examples:

$$A = \{2, 4, 6\} \qquad B = \{1, 3, 4\} \qquad A \cap B = \{4\}$$
$$V = \{a, e, i, o\} \qquad W = \{c, d, f\} \qquad V \cap W = \varnothing$$

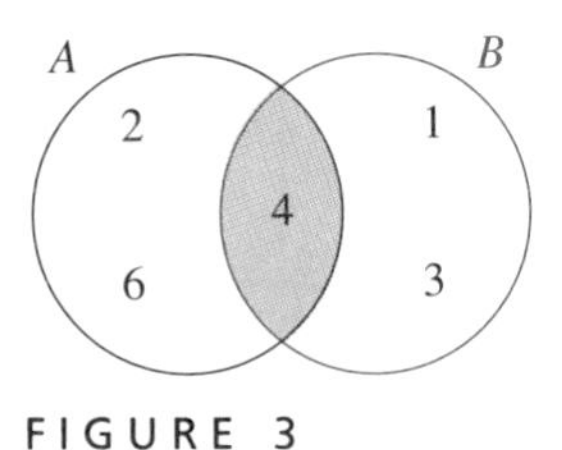

FIGURE 3

Sets V and W have no element in common. Thus, their intersection is empty. We can also write $V \cap W = \{\ \ \}$. In set-builder notation, we can write: $A \cap B = \{x \mid x \in A \textit{ and } x \in B\}$. A Venn diagram diagram for the intersection of sets A and B is shown in Figure 3. In the language of logic, the intersection operation matches an "AND" relationship:

$$a \in (A \cap B) \qquad \text{provided } a \in A \quad \text{AND} \quad a \in B$$

GIVE IT A TRY

Set $A = \{1, 2, \ldots, 6\}$ and set $B = \{x \mid x \text{ is odd and is less than } 10\}$.

1. What is $A \cup B$, and what is $A \cap B$?
2. Is $5 \in A$? Is $5 \in B$?
3. Is $A \subseteq B$? Is $B \subseteq A$?

The Integers We now use set operations to build other sets of numbers, starting with the whole numbers W and the idea of the additive opposite of a whole number. The **additive opposite** (or, the additive inverse) of a whole number, say 3, is the number that is added to the 3 to yield 0. This number (the additive opposite of 3) is not a whole number. The notation for the number added to 3 to yield 0 is -3 (pronounced *negative 3*). The additive opposite of 7 is the number added to 7 to yield 0. We write -7 for the additive opposite of 7. Again, -7 is not a whole number, that is, $-7 \notin W$. The union of the set of the whole numbers with the set of their additive opposites yields the set of numbers known as the **integers.** In set notation we write:

$$I = \{0, 1, 2, 3, \ldots\} \cup \{0, -1, -2, -3, \ldots\}$$
$$= \{\ldots, -3, -2, -1, 0, 1, 2, 3, \ldots\}$$

The natural numbers N are also known as the **positive integers.** The opposites of the natural numbers are known as the **negative integers.** The only number that is its own opposite is 0, because the number that we add to 0 to get 0 is 0: $0 + 0 = 0$.

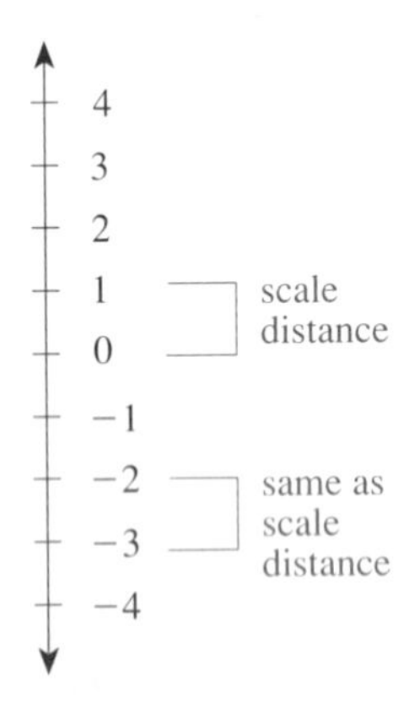

FIGURE 4
Vertical number line, or graph, of the integers

Number Line Graphs We can picture, or graph, the integers on a number line, a one-dimensional graph. This number line can be vertical, like a wall thermometer as shown in Figure 4. The number 0 and the number 1 are placed, and their placement sets the scale, the length of one unit. The positive integers are represented as points on the line above zero, and the negative integers are represented by points below zero.

In this book we will usually use a horizontal number line like the one shown in Figure 5. The positive integers are represented by points to the right of zero, and negative integers are represented by points to the left of zero. We can picture any integer on the number line. For example, Figure 5 shows the *graph,* or *plot,* of -3.

FIGURE 5
Horizontal number line of the integers

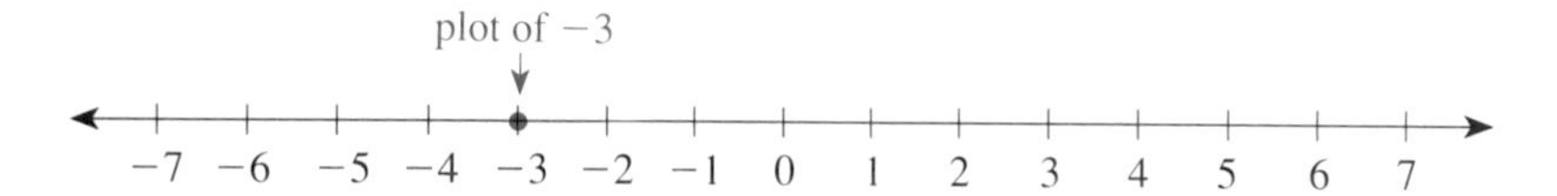

Relations One immediate use of a number line is to make comparisons between integers. Is -3 less than -5, or is -3 greater than -5? By looking at the horizontal number line, we can see that -3 is to the right of -5, so -3 is greater than -5. This statement describes the **relation** between the integers -3 and -5.

> **WARNING**
> As we move to the left on the number line, the numbers become *smaller*, not more negative.

We now introduce a collection of symbols that are used to describe the relation between two integers. If integer a is *less than* integer b (the plot of integer a on the number line is to the left of the plot of integer b), we write $\boldsymbol{a < b}$. If integer a is *greater than* integer b (the plot of integer a on the number line is to the right of the plot of integer b), we write $\boldsymbol{a > b}$. Also, we write $\boldsymbol{a \leq b}$ to mean integer a is *less than or equal to* integer b. We use $\boldsymbol{a \geq b}$ to mean integer a is *greater than or equal to* integer b.

GIVE IT A TRY

Write each relation in words. Draw a number line and plot the numbers. Also, write *true* for a true arithmetic relation or *false* for a false arithmetic relation.

4. $-3 > -1$ **5.** $0 \geq -2$ **6.** $8 < 23$

7. $-8 \leq 2$ **8.** $7 \geq 7$ **9.** $-3 \geq -4$

The Opposite Operation The first operation on the integers we will consider is the **opposite of an integer.** The opposite of 3 is -3, and the opposite of -5 is 5, and the notation for this operation is $-(3)$ and $-(-5)$. The opposite of an integer a is the integer b such that $a + b = 0$. That is, $-5 + -(-5) = 0$ and $5 + -(5) = 0$. The sentence $-5 + -(-5) = 0$ is read *negative 5 plus the opposite of negative 5 is 0*. On a number line graph, we can show the additive opposite as a reflection of a plotted point across the zero point. See Figure 6.

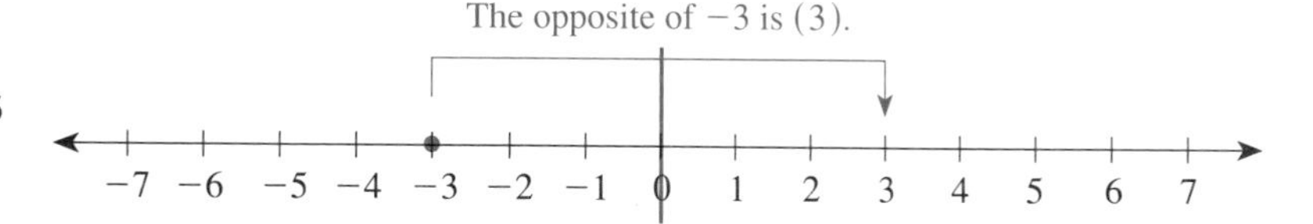

FIGURE 6

Absolute Value An operation closely related to the opposite of an integer is the **absolute value operation.** Graphically, the absolute value of a number is the number's distance from the zero point. Thus, the absolute value of 3, denoted $|3|$, is 3, because the number 3 is 3 units from 0. Also, $|-2| = 2$, because the number -2 is 2 units from 0. See Figure 7.

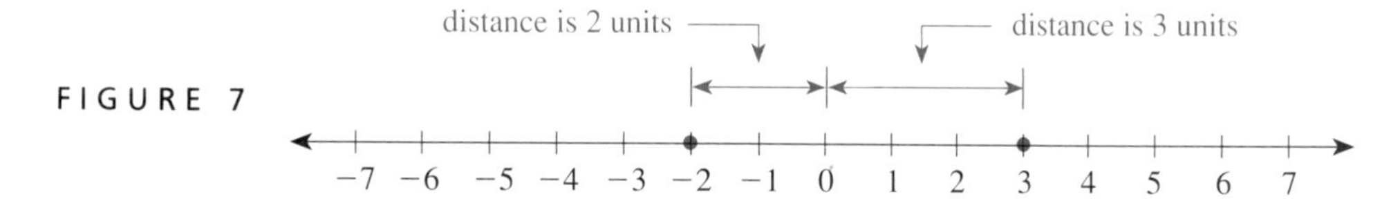

FIGURE 7

The absolute value operation is defined as follows.

DEFINITION OF ABSOLUTE VALUE

The **absolute value** of a is

a if a is a whole number (a positive integer or 0)

or $-(a)$ if a is a negative integer.

The notation for the absolute value of a is $|a|$ (read *the absolute value of a*).

The notation for the definition of absolute value is

$$|a| = \begin{cases} a & \text{if } a \geq 0 \\ -(a) & \text{if } a < 0 \end{cases}$$

Thus, we have $|7| = 7$, because 7 is greater than or equal to zero (that is, 7 is a whole number). By our definition, the absolute value of a whole number is identically that number. Also, we have $|-4| = 4$, because -4 is less than zero (that is, -4 is a negative integer). By our definition, the absolute value of a negative integer is its additive opposite. The opposite of -4, $-(-4)$, is 4. Notice that the absolute value operation on an integer always produces a whole number (the result of the absolute value operation is either 0 or a positive integer).

GIVE IT A TRY

10. Report the opposite of each integer.
a. 6 **b.** -3 **c.** 0 **d.** -147 **e.** 35,678

11. What is the opposite of the opposite of the opposite of -9?

12. Report the absolute value of each expression.
a. -3 **b.** 17 **c.** $-(-4)$ **d.** 0 **e.** $-(-(-5))$

13. Report the result of the indicated operations on each integer.
a. $-(-3)$ **b.** $|8|$ **c.** $-(|-5|)$ **d.** $|-(-(|-4|))|$

14. Is the absolute value of an integer always a positive integer?

Operations The opposite and absolute value are examples of a **unary operation**—an operation performed on a single number. The next collection of operations that we consider are binary operations. A **binary operation** is performed on a pair of numbers. The first binary operation on integers is that of addition. We find the sum of two integers as follows.

ADDITION OF INTEGERS WITH LIKE SIGNS

If two integers have *like signs* (both are positive or both are negative), add the absolute values of the integers, and then assign the common sign to the sum.

Examples:

$$5 + 8 = |5| + |8| = 5 + 8 = 13$$

common sign is negative

$$-5 + (-8) = -(|-5| + |-8|) = -(5 + 8) = -13$$

ADDITION OF INTEGERS WITH UNLIKE SIGNS

If two integers have *unlike signs* (one is positive and one is negative), find the absolute value of the integers. Subtract the smaller absolute value from the larger absolute value. Attach to the difference the sign of the integer with larger absolute value.

Example: For $-8 + 10$, we find the difference $10 - 8$, which is 2. We then attach a positive sign, because the sign of the integer with the larger absolute value, 10, is positive. Thus, $-8 + 10 = 2$. Graphically, we have

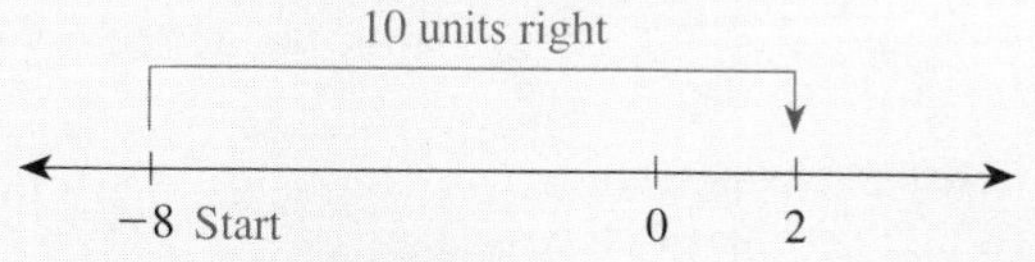

Example: For $5 + (-7)$, we find the difference $7 - 5$, which is 2. We then attach a negative sign, because -7 has the larger absolute value. Thus, $5 + (-7) = -2$. Graphically, we have

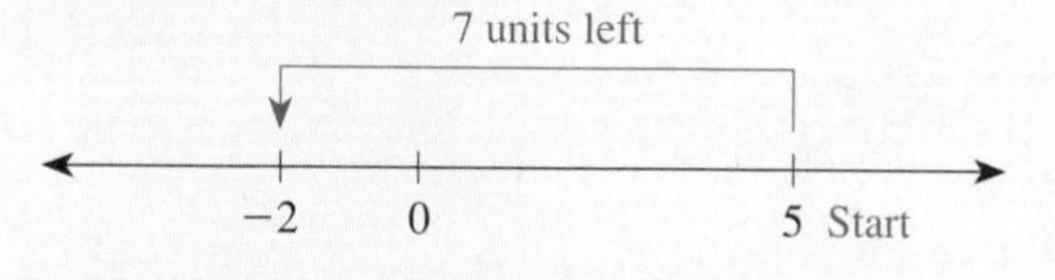

Once the operation of addition is established, we can define the operation of subtraction of integers in terms of addition.

SUBTRACTION OF INTEGERS

For integers a and b,

$$a - b = a + -(b)$$

means that the integer a minus the integer b is the integer a plus the opposite of the integer b.

Examples: $5 - (-8) = 5 + -(-8) = 5 + 8 = 13$ $\qquad -3 - 7 = -5 + -(7) = -5 + (-7) = -12$

We now define multiplication of two integers:

MULTIPLICATION OF INTEGERS WITH LIKE SIGNS

If two integers have the same sign, find the product of their absolute values.

Examples: $(3)(5) = (|3|)(|5|) = (3)(5) = 15$ $\qquad (-3)(-5) = (|-3|)(|-5|) = (3)(5) = 15$

MULTIPLICATION OF INTEGERS WITH UNLIKE SIGNS

If two integers have unlike signs, find the product of their absolute values, and attach a negative sign.

Examples:

$$\begin{aligned}(-3)(5) &= -(|-3|)(|5|)\\ &= -(3)(5)\\ &= -15\end{aligned} \qquad \begin{aligned}(3)(-5) &= -(|3|)(|-5|)\\ &= -(3)(5)\\ &= -15\end{aligned}$$

Finally, we define division of integers:

DIVISION OF INTEGERS

If two integers have the same sign, find the quotient of their absolute values.

Examples:

$$\begin{aligned}(12) \div (3) &= (|12|) \div (|3|)\\ &= (12) \div (3)\\ &= 4\end{aligned}$$

$$\begin{aligned}(-12) \div (-3) &= (|-12|) \div (|-3|)\\ &= (12) \div (3)\\ &= 4\end{aligned}$$

If two integers have unlike signs, find the quotient of their absolute values, and attach a negative sign.

Examples:

$$\begin{aligned}(-12) \div (3) &= -(|-12|) \div (|3|)\\ &= -(12) \div (3)\\ &= -4\end{aligned}$$

$$\begin{aligned}(12)(-3) &= -(|12|) \div (|-3|)\\ &= -(12) \div (3)\\ &= -4\end{aligned}$$

The preceding operation rules yield the following facts for integers a and b:

$$(a)(-b) = (-a)(b) = -(a)(b)$$

and $\quad (a) \div (-b) = (-a) \div (b) = -(a \div b)\quad$ provided $b \neq 0$.

GIVE IT A TRY

Write an integer for each of the following:

15. a. $-13 + (-2)$ **b.** $-13 - (-2)$ **c.** $12 - (-2)$

16. a. $(-5)(-7)$ **b.** $-18 \div (-3)$ **c.** $23 \div (-1)$

Factors and Exponents We next consider the important topic of exponents. In a product, such as $(a)(b)$, we say that a and b are **factors.** If a divides a number, say n, with no remainder (that is, a 0 remainder) then we say a is a *factor* of n. We say that -3 is a factor of -12, because $-12 \div -3 = 4$. Notice that 2 is *not* a factor of 7, because 2 does not divide 7 evenly (that is, $7 \div 2$ does not have a 0 remainder).

Now, consider the product $2 \cdot 2 \cdot 2 \cdot 2$. We can use an exponent to write this product as 2^4. The raised 4 (the superscript) is known as the **exponent** and the number 2 is the **base.** We say that 2^4 is a *power* of 2. In general, we have the following definition.

DEFINITION OF EXPONENT

If a is an integer and n is a natural number, then

$$a^n \text{ is defined as } \underbrace{a \cdot a \cdot a \cdot \cdots \cdot a}_{n \text{ factors of } a}$$

The integer a is the **base** and the natural number n is the **exponent.**

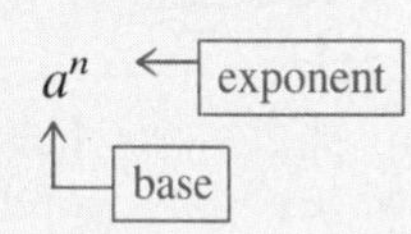

We define $a^0 = 1$, provided $a \neq 0$.

Just as $\frac{a}{0}$ is left undefined, 0^0 is left undefined.

EXAMPLE 2 Compute the following: **a.** $(-3)^4$ **b.** -3^4

Solution **a.** We have:

$$(-3)^4 = (-3)(-3)(-3)(-3) = 81$$

b. We have:

$$-3^4 = (-1)(3)(3)(3)(3) = -81$$

■

In Example 2(b), the exponent applies only to the base, 3. To indicate a negative integer, -3, raised to the fourth power, we must write $(-3)^4$ as in part (a).

GIVE IT A TRY

Compute the following:

17. -5^2 **18.** $(-5)^2$ **19.** -2^3 **20.** $(-2)^3$ **21.** -3^0

Properties of Number Collections All the properties of whole numbers also apply to the integers. We now list these properties and examples for integers.

Commutative Properties $a + b = b + a$ and $a \cdot b = b \cdot a$.

Examples: $-3 + 8 = 8 + (-3)$ $(-3)(8) = (8)(-3)$

Note: The order of appearance of the numbers is altered when the commutative property is used.

Associative Properties $a + (b + c) = (a + b) + c$ and $a \cdot (b \cdot c) = (a \cdot b) \cdot c$.

Examples: $-3 + [8 + 5] = [(-3) + 8] + 5$

$(-3)[(8)(5)] = [(-3)(8)](5)$

Distributive Property $a \cdot (b + c) = a \cdot b + a \cdot c$ and $(b + c) \cdot a = b \cdot a + c \cdot a$.

Example: $(-3)[8 + 5] = (-3)(8) + (-3)(5)$

Identities *Addition* ■ $a + 0 = a$

Multiplication ■ $a \cdot 1 = a$

Examples: $-3 + 0 = -3$ $(-3)(1) = -3$

The next two properties deal with a property of sets known as closure. A set has **closure** with respect to an operation if performing the operation on elements of the set produces another element of the set. The whole numbers have *closure with respect to addition* because adding any two whole numbers produces a whole number. For integers, we have the following closure properties.

Closure for Addition The sum of any two integers is an integer.

Example: $-3 + 8 = 5$, an integer

Closure for Multiplication The product of any two integers is an integer.

Example: $(-3)(8) = -24$, an integer

The integers also have two properties that the whole numbers do not have. First, every integer has an additive inverse. For each integer a there is an integer b such that $a + b = 0$. Of course, the additive inverse of integer a is the opposite of a, denoted $-(a)$. Second, the integers have closure for subtraction. That is, the difference of two integers is always an integer.

Distance between Points A mathematical application of the operations of subtraction of integers and absolute value is finding the distance between two points on the number line. Consider the points 3 and -2, which are plotted on the number line in Figure 8. The distance between these two points is 5 units. (Count the units.) Mathematically, the distance is the difference between the numbers 3 and -2. But distance must be nonnegative. To insure this nonnegativity, we can use the absolute value operation. We make the following definition.

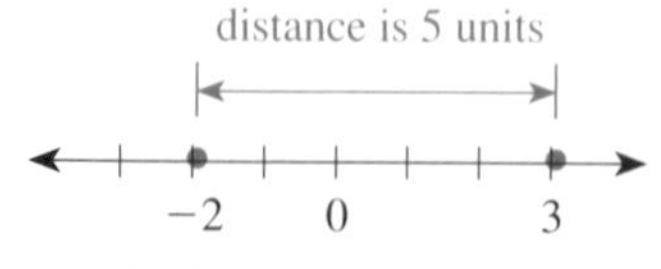

FIGURE 8

DISTANCE BETWEEN POINTS

> The distance d between points a and b is $d = |a - b|$.

For the points 3 and -2 in Figure 8, we have $d = |3 - (-2)| = |3 + 2| = |5| = 5$. To use this definition, it does not matter which point we call a and which point we call b. We could have written $d = |-2 - 3| = |-5| = 5$. Either way, the distance from 3 to -2 is 5 units.

Applications of Integers Before finishing this review of integers, we consider an everyday application of these numbers. Work through the next example.

EXAMPLE 3 Frank buys 300 shares of ABC stock and 200 shares of XYZ stock. The ABC stock gains \$3 per share and the XYZ stock loses \$5 per share. What is Frank's total gain or loss on these stocks?

Solution We can represent Frank's gain as a positive integer and his loss as a negative integer. So, Frank's total gain or loss is:

$$\underbrace{300(3)}_{\text{ABC stock}} + \underbrace{200(-5)}_{\text{XYZ stock}} = 900 + (-1000)$$
$$= -100$$

Thus, Frank has lost \$100 on these stocks. ■

GIVE IT A TRY

22. The distance from -3 to -6 is $|-3 - (-6)|$. Write this distance as an integer.

23. River normally drives from Louisville to Pensacola in 10 hours. On his March 25 trip, he lost 5 hours in a traffic jam in Nashville, but he gained an hour between Birmingham and Montgomery, due to road improvements. How many hours did this trip take?

24. The temperature at 6:00 A.M. was -10°C. The temperature increased 20° during the day, to reach its high. The temperature has now dropped 12° from that high. What is the current temperature?

■ SUMMARY

In this section, we have discussed sets and the operations of union and intersection on sets. One set of numbers essential to the study of algebra is the integers. The set of integers is the collection of whole numbers (0, 1, 2, 3, . . .) unioned with their additive opposites (0, -1, -2, -3, . . .). For the integers, the numbers greater than 0 are known as the positive integers and the numbers less than 0 are known as the negative integers.

For the integers, we discussed two unary operations, opposite and absolute value. The opposite operation [denoted $-(a)$] returns the additive opposite (inverse) of an integer. For example, $-(-3)$ is 3, because $-3 + 3 = 0$. The absolute value of an integer a, denoted $|a|$, is the integer a if $a \geq 0$. If $a < 0$, then the absolute value of a is $-(a)$. For example, $|-5|$ is $-(-5)$ or 5, because -5 is less than 0.

We also discussed binary operations on the integers—addition, subtraction, multiplication, and division. For the binary operations of addition and multiplication, the following properties are valid:

- commutative $[a + b = b + a$ and $(a)(b) = (b)(a)]$
- associative $[a + (b + c) = (a + b) + c$ and $a(bc) = (ab)c]$
- distributive $[a(b + c) = ab + ac]$

Additionally, 0 is the additive identity $[a + 0 = a]$, and 1 is the multiplicative identity $[(a)(1) = a]$. For the integers (unlike the whole numbers), every number has an additive inverse, and the operation of subtraction has closure for the integers.

■ *Quick Index*

1.1 ■ ■ ■ EXERCISES

Insert true *for a true statement, or* false *for a false statement.*

____ 1. Every integer a has an additive inverse, $-(a)$, in the integers.

____ 2. Every whole number a has an additive inverse in the whole numbers.

____ 3. The absolute value of an integer is always an integer.

____ 4. The absolute value of an integer is always positive.

____ 5. For any two integers a and b,

$$|a - b| = |b - a|$$

____ 6. The integers are a subset of the whole numbers.

____ 7. For sets A and B, $A \subseteq (A \cup B)$.

____ 8. The number 0 is a subset of W, the whole numbers.

____ 9. The empty set, $\varnothing$, is a subset of every set.

____ 10. For any integer a, $|a| = |-a|$.

Set $A = \{2, 3, 6, 8\}$ *and set* $B = \{1, 3, 5, \ldots, 11\}$.

11. Is $9 \in B$?
12. Is $A \subseteq B$?
13. What is $A \cup B$?
14. What is $A \cap B$?

Let N be the natural numbers, W the whole numbers, and I the integers.

15. Is $N \subseteq W$?
16. Is $I \subseteq W$?
17. What is $I \cap W$?
18. What is $I \cup N$?

Perform the binary operations to compute each result.

19. $-13 + (-23)$
20. $(-12) + (-8)$
21. $28 + (-3)$
22. $-20 + 5$
23. $-148 + 48$
24. $-432 + (-3)$
25. $-13 - (-23)$
26. $(-12) - (-8)$
27. $-45 - 20$
28. $-36 - 4$
29. $28 - (-3)$
30. $-20 - 5$
31. $-148 - 48$
32. $-432 - (-3)$
33. $-432 - 100$
34. $250 - (-4)$
35. $-3(-23)$
36. $(-12)(-8)$
37. $-45 \cdot 20$
38. $-6 \cdot 4$
39. $(25)(-3)$
40. $-20(5)$
41. $-432 \cdot 0$
42. $(250)(-4)$
43. $-33 \div (-3)$
44. $(-12) \div (-4)$
45. $-40 \div 10$
46. $-36 \div 4$
47. $27 \div (-3)$
48. $-20 \div 5$
49. $-144 \div 12$
50. $-432 \div (-3)$
51. $-400 \div 2$
52. $240 \div (-4)$
53. $-39 + (-3)$
54. $(-12) - (-4)$
55. $-90 \div 5$
56. $-30(-4)$
57. $54 \div (-3)$
58. $-20 + 5$
59. $-144 + 121$
60. $(-432)(-5)$
61. $-800 \div 100$
62. $240 - 10$

63. State the binary operations that have closure on the integers.

64. What is the additive inverse of (-8)?

65. What is the additive identity for the integers?

66. What is the multiplicative identity for the integers?

The integer factors of 6 *are* 1, −1, 2, −2, 3, −3, 6, *and* −6. *Report the integer factors of each integer.*

67. 8
68. −10
69. −15
70. 12
71. 18
72. −20
73. −28
74. 21
75. 13
76. 23

Use the meaning of a^n *to compute the following.*

77. $(-5)^3$
78. -5^3
79. 4^4
80. 4^3
81. $(-10)^3$
82. -10^3
83. $(-6)^2$
84. -6^2
85. -20^3
86. $(-20)^3$
87. $(-5)^0$
88. -5^0
89. 0^3
90. -0^3

Name the property illustrated in each statement.

91. $2(-3 + 6) = 2(6 + (-3))$
92. $(5 + -8) + 3 = 5 + (-8 + 3)$
93. $2(-3 + 6) = 2(-3) + 2(6)$
94. $(-5)[(-8)(3)] = [(-5)(-8)](3)$
95. $(-13)(1) = -13$
96. $3(47) = 3(40) + 3(7)$

Use absolute value to find the distance between the two given points.

97. −8 and 3
98. −13 and 17
99. −51 and −7
100. −78 and −23

101. Susan has 150 shares of Blue Water stock and 220 shares of Desert Sands stock. Her Blue Water stock gains $8 per share and her Desert Sands stock loses $2 a share. What is her total gain or loss on these stocks?

102. Bill is at elevation 2,456 feet when he parachutes into Deep Valley, to elevation 250 feet below sea level. What is the distance of his jump?

D I S C O V E R Y

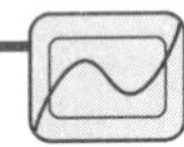

Using the Graphics Calculator

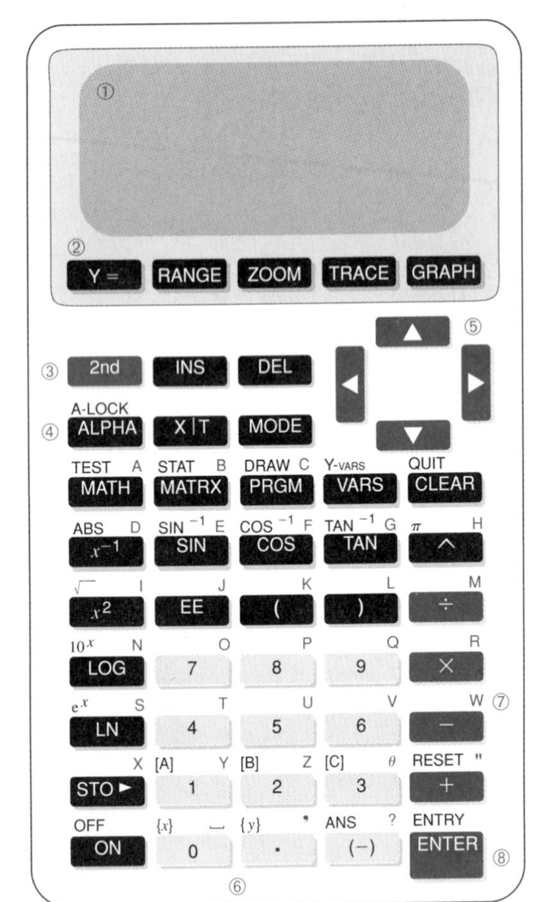

FIGURE 1

Throughout this course, you will find that the graphics calculator can increase your understanding of algebra concepts and decrease the number of computational errors you may make. Additionally, the graphics calculator will be a valuable tool for producing graphs. These features will help you to learn algebra more efficiently and with greater understanding. This introduction is for the TI-81 and TI-82 calculator. If you are using a different calculator, read its user manual to follow this discussion.* Before learning how to use the calculator, let's take a moment to learn its layout and basic features. Refer to Figure 1 as you read the following descriptions.**

The Display Screen At the top of the calculator is an 8-line-by-16-character display screen (1). On this screen, computations are entered and the results are output. Also graphs are displayed.

The Graph Keys Just below the display screen are five keys (2). These keys are used to define, display, manipulate, and interpret the graph of a function.

The 2nd Key The top leftmost key below the graph keys is the 2nd key (3). This key is used to access the functions displayed in blue type, on the left above some keys (such as 10^x and $\sqrt{\ }$).

The ALPHA Key Just below the 2nd key is the ALPHA key (4). This key is used to access the characters displayed in gray type, on the right above some keys (such as A and ?).

Directional Arrow Keys The four blue keys at the top right (5) are used to move the cursor (typically a blinking rectangular character) during editing and to trace along the curve of a graph.

The Function Keys The remaining black keys are used to access various mathematical functions (such as SIN, LOG, x^2), edit entries (the INS and DEL keys), and to display menus (the MATH, MATRX, PRGM, and VARS keys).

The Numeric Keypad The gray keys (6) are used to type numeric values, 0 through 9, the decimal point, and negative sign.

The Operation Keys On the lower right (7) are the four operation keys, +, −, ×, and ÷.

The ENTER Key At the bottom right is the blue ENTER key (8). Pressing this key informs the calculator to execute the command (instruction) you have typed.

TI-82 Note *If you are using the TI-82 calculator, any differences in its operation from that of the TI-81 are documented in these footnotes. If no TI-82 comment is given, you may assume that the instructions work as stated.

**The layout differs slightly—the WINDOW key replaces the RANGE key.

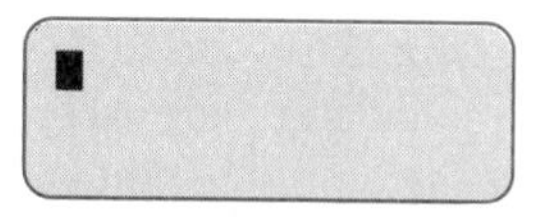

FIGURE 2
The opening screen

Getting Started With the calculator in hand, press the ON key (the leftmost key on the bottom row) to turn on the calculator. The calculator responds by presenting the cursor (a blinking rectangular character). See Figure 2. If you have trouble seeing the screen, try tilting the calculator slightly. If you still can't see the cursor, press the 2nd key and then the down arrow key (the ▼ key) or up arrow key (the ▲ key) to set the contrast. If any material appears on the screen, press the CLEAR key to erase the material.

Doing Calculations with Integers We will now use the calculator to do some computations with integers. The gray (-) key is used for the opposite operation and to indicate a negative integer. The blue − key is used to indicate subtraction.

EXAMPLE 1 Compute: $3 - (-5)$

Solution Press the 3 key, the blue − (subtraction) key, the gray (-) key, and then the 5 key to get 3 − ⁻5. Press the ENTER key. As a result, the subtraction is performed. The result, 8, is displayed to the right of the computation entered. See Figure 3.

FIGURE 3
The subtraction yields 8.

Keystrokes:

■

EXAMPLE 2 Compute: $-3 - 18$

Solution Press the (-) key, the 3 key, the − key, and then type 18. Press the ENTER key. As a result, the computation is performed to yield ⁻21. See Figure 4.

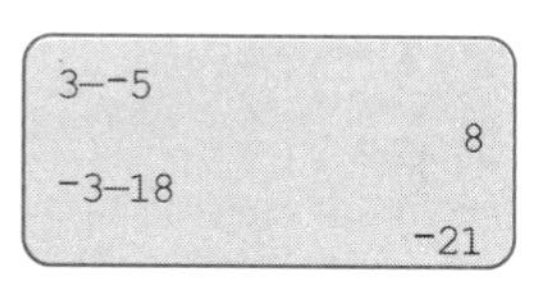

FIGURE 4
The subtraction yields −21.

Keystrokes:

■

ERROR 06 SYNTAX
1:Goto Error
2:Quit

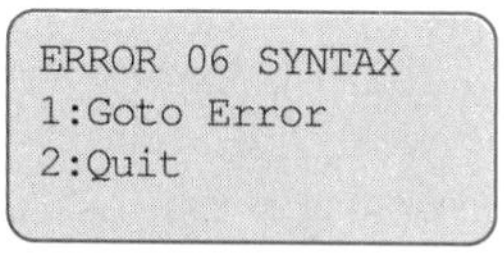

FIGURE 5

Error Handling In the first two examples, our use of the (-) key is important. To see this for yourself, press the 5 key and the + key. Now, press the − key [rather than the (-) key], and then the 3 key. Press the ENTER key. The calculator now enters an error state (see Figure 5).* Press the ENTER key or the 1 key to return to the home screen. The cursor is placed on the error (the − character).

Press the (-) key and then press the ENTER key. The computation is performed and the result, 2, is displayed on the home screen.

TI-82 Note *The ERROR screen differs slightly.

Editing Computations: The INS Key In the previous examples, we used the graphics calculator like a regular calculator. Graphics calculators are very different from most calculators. An important difference is the ability to edit an entry before pressing the `ENTER` key. In the following example, we start by making an error, and then we use the `INS` key to correct it.

EXAMPLE 3 Compute: $-12 - (-8)$

Solution First, we type an erroneous entry, and then we will use the `INS` key to correct the entry. Type `-12(-8)`. See Figure 6. Before pressing the `ENTER` key, press the left arrow key to back the cursor up to the `(` character. Press the `INS` key.* The cursor changes to a blinking underscore character. Press the blue $-$ key and a $-$ character is inserted. See Figure 7. Press the `ENTER` key to have the calculator perform the computation. As Figure 8 shows, `-4` is reported. ■

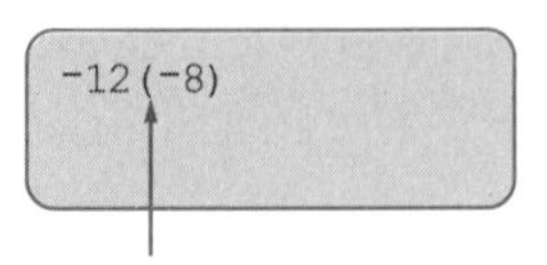

FIGURE 6
Use the left arrow key to move the cursor to the `(` character.

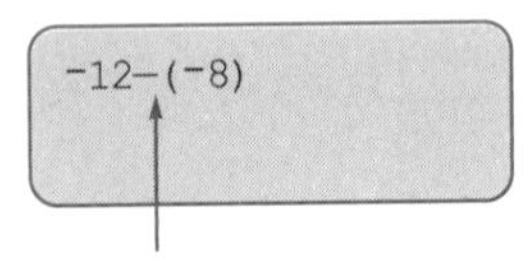

FIGURE 7
Press the `INS` key and press the blue $-$ key to insert the subtraction operation.

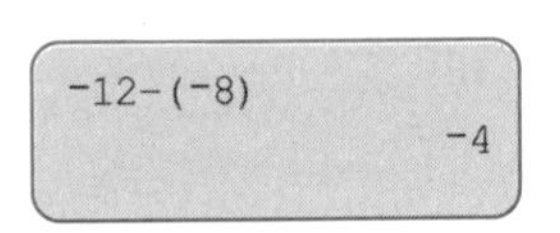

FIGURE 8
The subtraction yields -4.

WARNING
Although parentheses are not always required when entering an expression, it is a good idea to use them if you are in doubt. In Section 1.3 we will review order of operations and the use of parentheses.

Editing Computations: The DEL Key As we demonstrated in Example 3, we can insert text in a computation by using the `INS` key. Text can also be deleted from a computation using the `DEL` key. In the following example, we use the *replay feature* to display the previous computation for editing. Then we use the `DEL` key to delete material from the displayed computation.

EXAMPLE 4 Compute: $-12(-8)$ and $-12 \div (-4)$

Solution To do these computations, we can simply replay the computation from Example 3 and then edit the entry. To do the computation $-12(-8)$, we will need to delete the $-$ operation from the current entry. We use the `DEL` key to make this deletion.

Press the `2nd` key, followed by the `ENTER` key (for the `ENTRY` key), to replay the last entered command, `-12-(-8)`. Use the left-arrow key to move the cursor to the $-$ character following the -12. Press the `DEL` key to remove (delete) the character. Press the `ENTER` key. As you can see in Figure 9, the result, `96`, is written to the home screen.

To do the second computation, $-12 \div (-4)$, press the `2nd` key, then the `ENTER` key (the `ENTRY` key), to replay the last entered command. Move the cursor to the `(` character. Use the `INS` key to enter the insert mode. Press

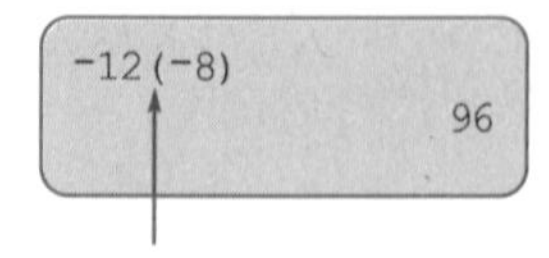

FIGURE 9
Press the `DEL` key to delete the subtraction operation.

TI-82 Note *To use the `INS` key, press the `2nd` key, followed by the `DEL` key.

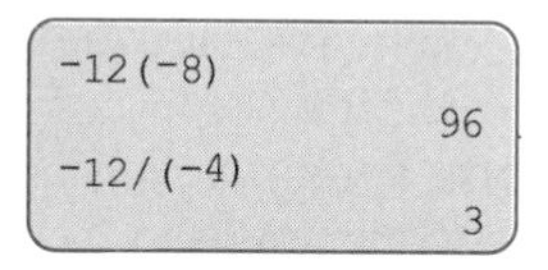

FIGURE 10

To type the * symbol, press the blue × key. To type the / symbol, press the blue ÷ key.

the blue ÷ key and the character / is written to the screen. Now, move the cursor to the 8 character. Notice that the cursor changes back to the (overwrite) cursor. Press the 4 key and the 8 is replaced by a 4. Press the ENTER key. The result, 3, is written to the home screen, as shown Figure 10. ■

GIVE IT A TRY

1. a. Use the graphics calculator to report the value of $-13 + (-5)$.
   b. Replay the command in part (a), and edit it to 13−(⁻5). Press the ENTER key and report the output.
   c. Replay the command in part (b), and edit it to 213+(⁻70). Press the ENTER key and report the output.
2. Use the (−) key to type ⁻(⁻(⁻8)). Press the ENTER key and report the value of the opposite of the opposite of -8.
3. a. Use the graphics calculator to report the value of $-60 \div (-5)$.
   b. Replay the command in part (a), and edit it to 60(⁻5). Press the ENTER key and report the output.

The OFF Key To turn off the calculator, press the 2nd key followed by the ON key (for the OFF key). If no key is pressed within approximately five minutes, the calculator turns off automatically.

The ABS Key So far, we have considered the unary operation of the *opposite,* the gray (−) key. The next unary operation we want to consider is the *absolute value operation*. To find the absolute value of a number using the calculator, we use the ABS key. To learn how to use this key, work through the following example.

EXAMPLE 5 Use the ABS key to compute each result.

a. $|-13|$ b. $|-8 - (-13)|$
c. $|-12| \div (-3)$ d. $-12 \div |-3 - 3|$

Solution a. Press the ABS key (the 2nd key, followed by the x^{-1} key), the characters abs are written to the screen. Press the (−) key, and then type 13. Press the ENTER key. The output is 13. See Figure 11.

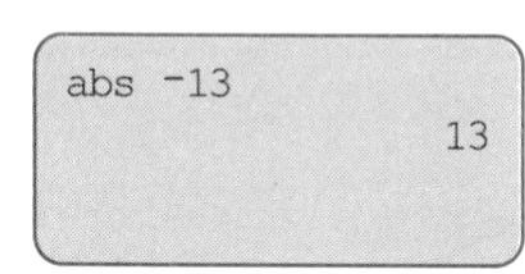

FIGURE 11

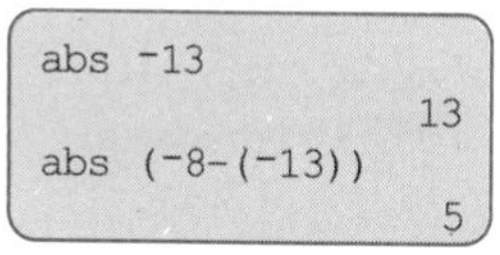

abs ⁻13
13
abs (⁻8-(⁻13))
5

FIGURE 12

b. Press the ABS key, then type (⁻8 − (⁻13)) to get abs(⁻8 − (⁻13)) on the screen. Press the ENTER key. The output is 5. See Figure 12.

```
abs -13
                13
abs (-8-(-13))
                 5
abs (-12)/-3
                -4
```

FIGURE 13

```
abs (-12)/-3
                -4
(-12)/abs (-3-3)
                -2
```

FIGURE 14

c. Type `abs (-12)/-3` and press the `ENTER` key. The number `-4` is output. See Figure 13.

d. Use the `ENTRY` key (the `2nd` key, followed by the `ENTER` key) to replay the last command. The characters `abs (-12)/-3` are written to the screen. Use the left arrow key to move the cursor to the `abs` characters. Press the `DEL` key to delete the `abs` characters. Now, move the cursor to the − character following the / character. Use the `INS` key to enter the insert mode. Now, use the `ABS` key to insert the characters `abs`. Press the `(` key. Move the cursor to the end of the line and type `-3)`. Press the `ENTER` key. The number `-2` is output. See Figure 14. ■

Exponents In addition to the unary operations of opposite and absolute value, we can also view exponentiation (computing a number to an exponent) as a unary operation. Because squaring a number is common in mathematics, this operation is given its own key, the x^2 key. For other whole number exponents, we use the ^ key (the "caret" key is located directly above the ÷ key). Work through the next example.

EXAMPLE 6 Compute: **a.** -4^2 **b.** $(-4)^2$ **c.** -5^4 **d.** $(-6 - 5)^3$

Solution

a. Type `-4` and then press the x^2 key to display `-4²` on the screen. Press the `ENTER` key. The number `-16` is output.

b. Use the parentheses keys to type `(-4)` and then press the x^2 key to display `(-4)²` on the screen. Press the `ENTER` key. The number `16` is output.

c. Type `-5`, press the ^ key and then the `4` key to get `-5^4` on the screen. Press the `ENTER` key. The number `-625` is output on the screen. See Figure 15.

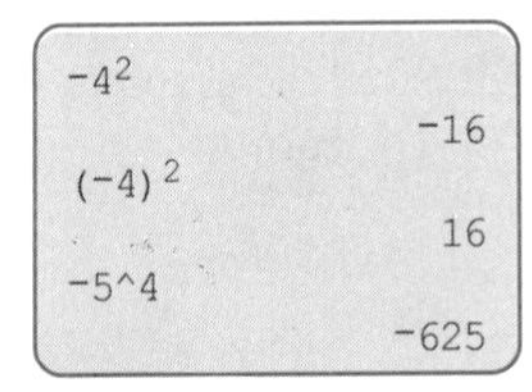

FIGURE 15

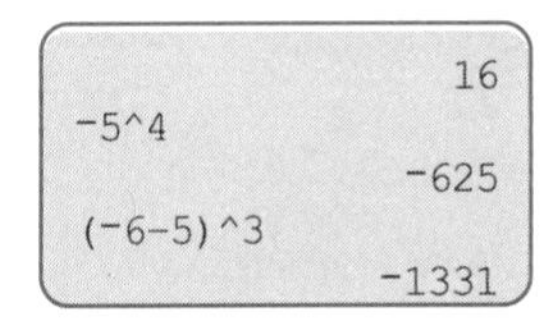

FIGURE 16

d. Use the parentheses keys to type `(-6 − 5)`. Now, press the ^ key followed by the `3` key to display `(-6 − 5)^3` on the screen. Press the `ENTER` key. The number `-1331` is output, as shown in Figure 16. ■

Exercises

1. Use the `ABS` key to compute each result.
   a. $-(37 - |-52 + 278|)$ **b.** $|-7 - |5 - 20| - 8| - 2$

2. Use the x^2 key and the ^ key to compute each result.
   a. -10^2 **b.** $(-10)^2$ **c.** -6^3 **d.** $(-6)^3$ **e.** $(-5 + 7)^4$

3. Return to Exercises 1.1, and use the graphics calculator to check the answers to Problems 19–62, 77–90, and 97–102.

1.2 ■ ■ ■ RATIONAL NUMBERS

In this section, we discuss the next set of numbers that appear in algebra, the rational numbers. We will start by introducing a naming system for these numbers. Next, we review the operations on these numbers.

Rational Numbers To define the integers, we started with the whole numbers of arithmetic. The set of whole numbers unioned with their additive opposites yields the integers. We can define the rational numbers in a similar fashion, we start with the fractions of arithmetic. A fraction of arithmetic is a ratio of whole numbers, where the denominator is not zero. See Figure 1.

FIGURE 1

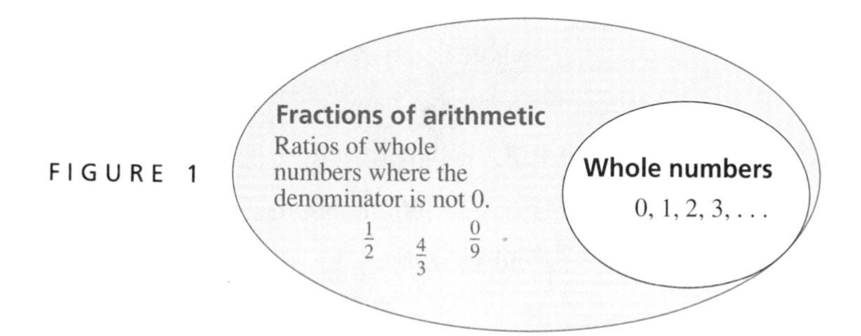

Now, for any fraction $\frac{a}{b}$, its additive opposite (inverse) is the number $\frac{-a}{b}$. For example, the fraction $\frac{2}{3}$ has additive opposite $\frac{-2}{3}$. The additive opposite of $\frac{a}{b}$ is the number we add to $\frac{a}{b}$ to get zero:

Additive opposite

$$\frac{a}{b} + \frac{-a}{b} = 0$$

The set of fractions unioned with the set of their additive opposites yields the rational numbers. Figure 2 shows the relation between integers and rational numbers. Notice the parallel between Figure 1 and Figure 2.

FIGURE 2

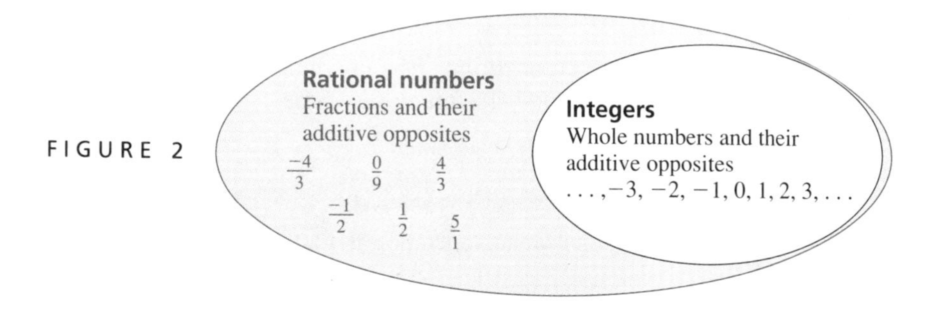

We can use set notation to describe the rational numbers.

THE RATIONAL NUMBERS

$$Q = \left\{\frac{a}{b} \,\middle|\, a \text{ and } b \text{ are integers with } b \neq 0\right\}.$$

For $\frac{a}{b}$, a is the **numerator** and b is the **denominator.**

We can use a number line, such as the one shown in Figure 3, to graph rational numbers.

FIGURE 3

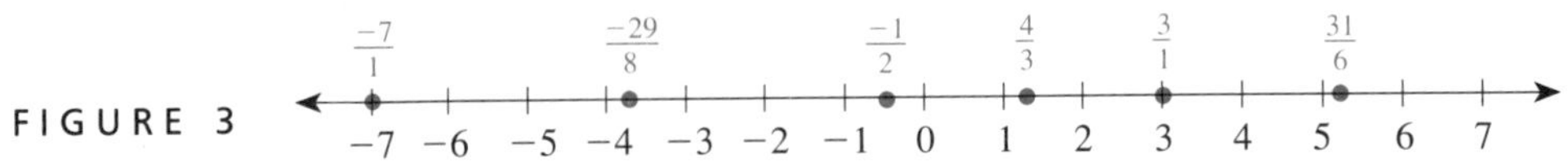

Just as every whole number, say 5, has a fractional name, $\frac{5}{1}$, each integer, say -3, has a rational number name, $\frac{-3}{1}$. Additionally, just as every fraction has an infinite number of fractional names $\left(\text{for } \frac{2}{3} \text{ we have the name } \frac{4}{6}, \frac{6}{9}, \text{ or } \frac{8}{12}\right)$, each rational number has an infinite number of rational names $\left(\text{for } \frac{-3}{5} \text{ we have the names } \frac{-6}{10}, \frac{-9}{15}, \text{ and so on}\right)$.

Naming System Because every rational number has an infinite number of names, we must devise a system for naming rational numbers. *A completely reduced rational number* will mean that the number has the form $\frac{a}{b}$ and that the numerator and denominator have no common factor. To rewrite, or rename, a rational number to this form, we use the following property.

CANCELLATION PROPERTY

The operations of multiplication by a nonzero number, say c, and division by the same nonzero number c, cancel each other. That is

$$\frac{ac}{bc} = \frac{a}{b} \qquad \text{where } c \neq 0 \quad \text{and} \quad b \neq 0$$

The cancellation property is often stated as "multiplication by c and division by c cancel."

Unary Operations The unary operations of opposite and absolute value for rational numbers follow the same patterns as they do for the integers. The opposite of $\frac{-2}{5}$, denoted $-\left(\frac{-2}{5}\right)$, is $\frac{2}{5}$. Likewise, the absolute value of $\frac{-2}{5}$, denoted $\left|\frac{-2}{5}\right|$, is $-\left(\frac{-2}{5}\right)$, because $\frac{-2}{5}$ is less than 0. Thus, $\left|\frac{-2}{5}\right|$ is $\frac{2}{5}$. An additional unary operation is that of reciprocal. The **reciprocal** of a nonzero rational number $\frac{a}{b}$ is the rational number $\frac{b}{a}$. The number $\frac{-2}{5}$ has reciprocal $\frac{-5}{2}$.

Binary Operations Addition, subtraction, multiplication, and division follow the same pattern for rational numbers as they do for fractions:

Definition of Addition $\frac{a}{c} + \frac{b}{c} = \frac{a+b}{c}$ where $c \neq 0$.

Example: Write as a completely reduced rational expression: $\frac{2}{5} + \frac{-1}{3}$

Solution: To use the definition of addition, we must make both denominators the same. So we start by renaming the rational numbers to have the same denominator. To do this, we choose a least common denominator, or LCD, for the two denominators, 3 and 5. Recall that the LCD is the smallest number both denominators will divide. In this case, it is the product (3)(5), or 15. Using the cancellation property, we have

$$\frac{2}{5} = \frac{2(3)}{5(3)} = \frac{6}{15} \quad \text{and} \quad \frac{-1}{3} = \frac{-1(5)}{3(5)} = \frac{-5}{15}$$

Using our definition of addition we have

$$\frac{2}{5} + \frac{-1}{3} = \frac{6}{15} + \frac{-5}{15} = \frac{6+(-5)}{15} = \frac{1}{15}$$

So,
$$\frac{2}{5} + \frac{-1}{3} = \frac{1}{15}$$

For a review of how to find the LCD for two denominators, see Appendix A.

Definition of Subtraction $\frac{a}{c} - \frac{b}{c} = \frac{a}{c} + \left(-\frac{b}{c}\right) = \frac{a}{c} + \frac{-b}{c}$ $c \neq 0$.

The additive opposite of $\frac{b}{c}$

Example: Write as a completely reduced rational expression: $\frac{-1}{6} - \frac{-5}{4}$

Solution: We start by renaming the rational numbers so that their denominators are the same. The LCD for the denominators 6 and 4 is 12. Renaming each rational number to have a denominator of 12, we get

$$\frac{-1}{6} = \frac{-1(2)}{6(2)} = \frac{-2}{12} \quad \text{and} \quad \frac{-5}{4} = \frac{-5(3)}{4(3)} = \frac{-15}{12}$$

We now have

change to addition

opposite of $\frac{-15}{12}$

$$\frac{-2}{12} - \frac{-15}{12} = \frac{-2}{12} + -\left(\frac{-15}{12}\right) = \frac{-2}{12} + \frac{15}{12} = \frac{-2+15}{12} = \frac{13}{12}$$

additive opposite

So, $$\frac{-1}{6} - \frac{-5}{4} = \frac{13}{12}$$

Definition of Multiplication $$\left(\frac{a}{b}\right)\left(\frac{c}{d}\right) = \frac{(a)(c)}{(b)(d)} \quad b \neq 0 \text{ and } d \neq 0$$

Example: Write as a completely reduced rational number: $\left(\frac{2}{5}\right)\left(\frac{-1}{3}\right)$

Solution: $\left(\frac{2}{5}\right)\left(\frac{-1}{3}\right) = \frac{(2)(-1)}{(5)(3)} = \frac{-2}{15}$

Change division to multiplication.

Definition of Division $$\left(\frac{a}{b}\right) \div \left(\frac{c}{d}\right) = \left(\frac{a}{b}\right)\left(\frac{d}{c}\right) = \frac{ad}{bc} \quad b \neq 0,\ c \neq 0,\ d \neq 0$$

Change divisor to reciprocal.

Example: Write as a completely reduced rational number: $\left(\frac{-2}{7}\right) \div \left(\frac{-5}{2}\right)$

Solution: $\left(\frac{-2}{7}\right) \div \left(\frac{-5}{2}\right) = \left(\frac{-2}{7}\right)\left(\frac{-2}{5}\right) = \frac{(-2)(-2)}{(7)(5)} = \frac{4}{35}$

reciprocal of $\frac{-5}{2}$

With these binary operations in place, we make the following observations. To rename a rational number, we simply multiply the number by 1, the rational number $\frac{a}{a}$, where $a \neq 0$. The form of 1 we choose depends on the new denominator desired. The cancellation property, $\frac{a}{b} = \frac{ac}{bc}$, can be viewed as multiplying by 1 in the form $\frac{c}{c}$. To find the additive opposite of a rational number, $\frac{a}{b}$, we multiply by -1:

$$-\left(\frac{a}{b}\right) = \frac{-1}{1} \cdot \frac{a}{b} = \frac{-a}{b}$$

Also, we have the following properties for addition and subtraction.

Addition and Subtraction Properties $$\frac{a}{b} + \frac{c}{d} = \frac{ad}{bd} + \frac{cb}{db} = \frac{ad + cb}{bd} \quad \text{and} \quad \frac{a}{b} - \frac{c}{d} = \frac{ad}{bd} - \frac{cb}{db} = \frac{ad - cb}{bd}$$

The next example demonstrates some of these definitions and properties.

EXAMPLE 1 Write as a completely reduced rational number:

a. $\dfrac{-27}{15}$ b. $\dfrac{-2}{3} - \dfrac{-5}{6}$ c. $\dfrac{-2}{3} \div \dfrac{-2}{15}$

Solution a. We need only factor the numerator and denominator to check for common factors. Now, -27 factors as $(-1)(3)(3)(3)$, and 15 factors as $3(5)$. Thus, we write

$$\frac{-27}{15} = \frac{(-1)(\not{3})(3)(3)}{(\not{3})(5)} = \frac{-9}{5} \qquad \text{the completely reduced name}$$

b. We start by using the subtraction operation to get a rational expression. Next, we proceed as we did in part (a) to get a completely reduced rational expression.

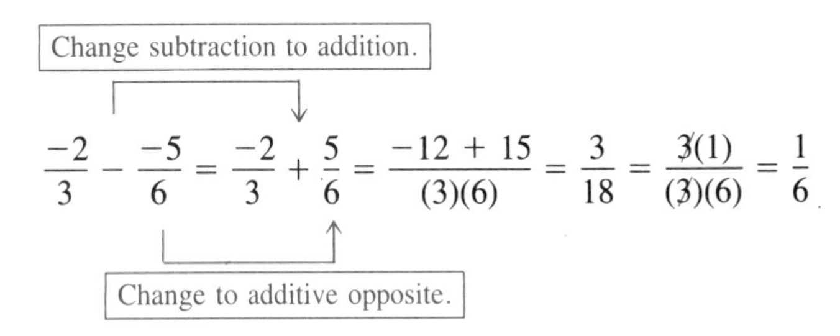

$$\frac{-2}{3} - \frac{-5}{6} = \frac{-2}{3} + \frac{5}{6} = \frac{-12 + 15}{(3)(6)} = \frac{3}{18} = \frac{\not{3}(1)}{(\not{3})(6)} = \frac{1}{6}$$

c. We first do the division, and then take care of the common factors:

Change division to multiplication.

$$\frac{-2}{3} \div \frac{-2}{15} = \frac{-2}{3} \times \frac{-15}{2} = \frac{(-1)(-1)(\not{2})(\not{3})(5)}{(\not{3})(\not{2})(1)} = \frac{5}{1}$$

Change divisor to reciprocal. ■

> **WARNING**
>
> When using the slash mark, /, in working with rational numbers, make sure you cancel only factors! A common error is to write
>
> $$\frac{5 + \not{2}}{6 + \not{2}} = \frac{5}{6}$$
>
> Here, 2 is **not** a factor of the numerator or denominator.

In Example 1, the slash marks mean that multiplication by a number and division by that same number cancel. That is, the slash marks in

$$\frac{(-1)(\not{3})(3)(3)}{(\not{3})(5)}$$

mean that the operation of multiplication by 3 and the operation of division by 3 cancel each other. (It is not the 3s that cancel, it is the operations that cancel.)

GIVE IT A TRY

answers follow Appendix B Write as a completely reduced rational number:

1. $\dfrac{-3}{7} + \dfrac{-2}{5}$ 2. $\dfrac{5}{3} - \dfrac{-5}{6}$ 3. $\dfrac{-2}{9} \times \dfrac{-3}{8}$ 4. $\dfrac{-2}{5} \div \dfrac{4}{15}$

Properties of Rational Numbers All the properties of the integers also hold for rational numbers. Additionally, nonzero rational numbers have multiplicative inverses that are rational numbers. The operation of division has closure for the rational numbers, as do the operations of addition, subtraction, and multiplication. The properties are as follows.

Commutative Properties $a + b = b + a$ and $(a)(b) = (b)(a)$.

Examples: $\frac{5}{3} + \frac{-3}{8} = \frac{-3}{8} + \frac{5}{3}$ $\left(\frac{5}{3}\right)\left(\frac{-3}{8}\right) = \left(\frac{-3}{8}\right)\left(\frac{5}{3}\right)$

Associative Properties $a + (b + c) = (a + b) + c$ and $(a)[(b)(c)] = [(a)(b)](c)$.

Examples: $\frac{7}{8} + \left(\frac{-2}{9} + \frac{5}{7}\right) = \left(\frac{7}{8} + \frac{-2}{9}\right) + \frac{5}{7}$

$$\left(\frac{7}{8}\right)\left[\left(\frac{-2}{9}\right)\left(\frac{5}{7}\right)\right] = \left[\left(\frac{7}{8}\right)\left(\frac{-2}{9}\right)\right]\left(\frac{5}{7}\right)$$

Distributive Property $(a)(b + c) = (a)(b) + (a)(c)$.

Example: $\left(\frac{-6}{11}\right)\left(\frac{4}{7} + \frac{-7}{9}\right) = \left(\frac{-6}{11}\right)\left(\frac{4}{7}\right) + \left(\frac{-6}{11}\right)\left(\frac{-7}{9}\right)$

Identities *Additive* ■ $a + 0 = 0 + a = a$

Example: $\frac{-3}{7} + \frac{0}{1} = \frac{-3}{7}$

Multiplicative ■ $(a)(1) = (1)(a) = a$

Example: $\left(\frac{-3}{7}\right)\left(\frac{1}{1}\right) = \frac{-3}{7}$

Inverses *Additive* ■ For each rational number a, there exists a rational number b, known as the **additive inverse** of a and denoted $-a$, such that $a + b = 0$.

Example: For $\frac{-3}{7}$, $\frac{-3}{7} + \frac{3}{7} = 0$

Thus, $\frac{3}{7}$ is the additive inverse of $\frac{-3}{7}$.

Multiplicative ■ For each nonzero rational number a, there exists a rational number b, known as the **multiplicative inverse** of a and denoted a^{-1} or $\frac{1}{a}$, such that $(a)(b) = 1$.

Example: For $\frac{-3}{7}$, $\left(\frac{-3}{7}\right)\left(\frac{-7}{3}\right) = 1$

Thus, $\frac{-7}{3}$ is the multiplicative inverse of $\frac{-3}{7}$.

Applications of Rational Numbers Before finishing this review, we consider some everyday applications of rational numbers. Work through the next two examples.

EXAMPLE 2 For the next term, Dr. Lopez hopes to limit his drop-out rate to $\frac{1}{5}$ of the number of students enrolled. If he expects 150 students to sign up for his courses, how many students will complete these courses?

Solution In most arithmetic problems dealing with rational numbers, the word *of* translates to multiplication.

Dr. Lopez expects $\frac{1}{5}$ of 150 students to drop his courses. Doing this computation, we get $\left(\frac{1}{5}\right)(150) = 30$. So, 30 students will drop. This means that $150 - 30$, or 120 students will complete these courses. ■

EXAMPLE 3 Joann has recently invested \$20,000 in the stock market. During the first month, she lost $\frac{1}{8}$ of her investment, but in the second month her investment increased by $\frac{1}{5}$. After the second month, what is the value of Joann's investment?

Solution Joann lost $\frac{1}{8}$ of her investment during the first month. So, at the end of the first month,

$$\begin{aligned}\text{Value of investment} &= 20{,}000 \underset{\text{decreased}}{-} \left(\tfrac{1}{8}\right)(20{,}000)\\ &= 20{,}000 - 1{,}500\\ &= 18{,}500\end{aligned}$$

During the second month Joann's investment increased by $\frac{1}{5}$. So, at the end of the second month,

$$\begin{aligned}\text{Value of investment} &= 18{,}500 \underset{\text{increased}}{+} \left(\tfrac{1}{5}\right)(18{,}500)\\ &= 18{,}500 + 3{,}700\\ &= 22{,}200\end{aligned}$$

Thus, at the end of the second month, Joann's investment is worth \$22,200. ■

■ SUMMARY

The set of rational numbers is the set of fractions (numbers of the from $\frac{a}{b}$ where a and b are whole numbers with $b \neq 0$) unioned with the set of their additive opposites (numbers of the form $\frac{-a}{b}$, where a and b are whole numbers with $b \neq 0$). The collection of rational numbers is the set $Q = \left\{\frac{a}{b} \,\middle|\, a \text{ and } b \text{ are integers with } b \neq 0\right\}$. For a rational number in the form $\frac{a}{b}$, a is known as the numerator and b is the denominator.

■ *Quick Index*

Because each rational number has an infinite number of names, we write the number in a proper form known as a completely reduced rational number. A completely reduced rational number is one in which the numerator and denominator have no common factor.

For the rational numbers, we discussed three unary operations: the opposite operation $\left[-\left(\frac{-2}{3}\right) \text{ is } \frac{2}{3}\right]$, the absolute value operation $\left[\left|\frac{-2}{3}\right| \text{ is } \frac{2}{3}\right]$, and the reciprocal operation $\left[\text{the reciprocal of } \frac{-2}{3} \text{ is } \frac{-3}{2}\right]$. The binary operations of addition, subtraction, multiplication, and division follow the same patterns for rational numbers as for fractions. The commutative, associative, and distributive properties also hold. For the rational numbers, $\frac{0}{1}$ is the additive identity and $\frac{1}{1}$ is the multiplicative identity. Each rational number, $\frac{a}{b}$, with $b \neq 0$, has an additive inverse. $\left[\text{The opposite of } \frac{a}{b} \text{ is } \frac{-a}{b}\text{, also denoted } -\left(\frac{a}{b}\right)\right]$. If $a \neq 0$ and $b \neq 0$, then the rational number $\frac{a}{b}$ has a multiplicative inverse $\left(\text{the reciprocal of } \frac{a}{b}\right)$, which is $\frac{b}{a}$.

1.2 ■ ■ ■ EXERCISES

1. State the binary operations that have closure on the rational numbers.
2. State the multiplicative inverse of $\frac{-7}{15}$.
3. What is the additive identity for the rational numbers?
4. What is the multiplicative identity for the rational numbers?

Write each as a completely reduced rational number.

5. $\frac{-12}{36}$ 6. $\frac{-48}{15}$ 7. $\frac{-10}{-18}$ 8. $\frac{-12}{-22}$

9. $\frac{28}{36}$ 10. $\frac{52}{-24}$ 11. $\frac{-90}{-28}$ 12. $\frac{-26}{-50}$

Write each as a completely reduced rational number.

13. $\frac{-8}{36} + \frac{-12}{36}$ 14. $\frac{3}{7} + \frac{-3}{7}$ 15. $\frac{-14}{15} - \frac{-1}{15}$

16. $\frac{-5}{12} - \frac{-2}{12}$ 17. $\frac{-4}{30} - \frac{5}{30}$ 18. $\frac{3}{5} + \frac{17}{5}$

19. $\frac{3}{8} + \frac{3}{4}$ 20. $\frac{3}{4} + \frac{-1}{2}$ 21. $\frac{-1}{2} - \frac{1}{3}$

22. $\frac{1}{5} - \frac{1}{4}$ 23. $\frac{1}{4} - \frac{-3}{5}$ 24. $\frac{-14}{15} - \frac{-5}{12}$

25. $\frac{-13}{20} + \frac{-7}{30}$ 26. $\frac{-17}{30} + \frac{-3}{5}$ 27. $\frac{-5}{12} \cdot \frac{-6}{15}$

28. $\frac{-8}{3} \cdot \frac{6}{15}$ 29. $\frac{-1}{2} \cdot \frac{3}{4}$ 30. $\frac{3}{5} \cdot \frac{3}{8}$

Write each as a completely reduced rational number.

31. $\frac{-2}{5} \div \frac{-6}{5}$ 32. $\frac{-6}{15} \div \frac{-6}{10}$ 33. $\frac{-3}{7} \div (-5)$

34. $\frac{3}{4} \div 6$ 35. $-15 \div \frac{1}{5}$ 36. $-3 \div \frac{-2}{5}$

37. $\left(\frac{-5}{12}\right)^2$ 38. $\left(\frac{-3}{5}\right)^2$ 39. $\left(\frac{3}{4}\right)^3$

40. $\left(\frac{-3}{7}\right)^3$ 41. $\left(\frac{-11}{15}\right)^0$ 42. $\left(\frac{17}{5}\right)^0$

43. $\frac{3}{8} - \frac{-5}{12}$ 44. $\frac{-3}{5} \div \frac{3}{8}$ 45. $\frac{-6}{5} \div \frac{5}{30}$

46. $2\frac{3}{4} + 3\frac{1}{3}$ 47. $\left(\frac{-3}{7}\right)\left(\frac{-14}{12}\right)$ 48. $\frac{17}{5} - \frac{3}{4}$

49. A local telephone exchange has a capacity of 1500 calls per day. It is recommended that the exchange operate at $\frac{4}{5}$ of capacity. How many calls should the exchange handle per day?
50. Ms. Littlefield has traveled $\frac{2}{3}$ the distance from her home to her job. If her daily commute is 23 miles, how many miles has she traveled?
51. When Uncle Joe died, the Brady boys and the Elm Tree Society inherited all of his money. The oldest boy got

$\frac{1}{2}$ of the money, the second oldest got $\frac{1}{2}$ of the amount the oldest boy got, and the youngest boy got $\frac{1}{5}$ of Uncle Joe's money. The remaining money was given to the Elm Tree Society. If Uncle Joe left $100,000, how much money did the Elm Tree Society receive?

52. Professor Burns hopes to cut her drop-out rate by $\frac{2}{3}$ next term. If her drop rate is normally $\frac{1}{5}$ and she expects 150 students to sign up for her courses next term, how many students will complete her courses?

53. A carpenter starts with a board that is 100 inches long. He cuts from the board five pieces, each $8\frac{3}{4}$ inches long. What is the length of the remaining piece of board?

54. Dion has invested $10,000 in the stock market. At the end of the first month, his investment has increased by $\frac{1}{5}$. At the end of the second month, his investment has decreased by $\frac{1}{4}$. What is the value of the investment at the end of the second month?

Name the property illustrated in each statement.

55. $\frac{-3}{7}\left(\frac{-2}{5}+\frac{3}{4}\right)=\frac{-3}{7}\left(\frac{3}{4}+\frac{-2}{5}\right)$

56. $\left(\frac{-6}{5}+\frac{-5}{12}\right)+\frac{-3}{7}=\frac{-6}{5}+\left(\frac{-5}{12}+\frac{-3}{7}\right)$

57. $\frac{-3}{7}\left(\frac{-2}{5}+\frac{3}{4}\right)=\frac{-3}{7}\left(\frac{-2}{5}\right)+\frac{-3}{7}\left(\frac{3}{4}\right)$

58. $\left(\frac{-2}{3}\right)\left(\frac{-3}{2}\right)=1$

59. $\left(\frac{-5}{12}\right)\left[\left(\frac{17}{5}\right)\left(\frac{3}{8}\right)\right]=\left[\left(\frac{-5}{12}\right)\left(\frac{17}{5}\right)\right]\left(\frac{3}{8}\right)$

60. $\frac{3}{4}+\frac{-3}{4}=0$

61. $\left(\frac{4}{7}\right)\left(\frac{-3}{5}+\frac{1}{4}\right)=\left(\frac{4}{7}\right)\left(\frac{1}{4}+\frac{-3}{5}\right)$

62. $\left(\frac{9}{5}\right)(1)=\frac{9}{5}$

$3\left(4\frac{1}{2}\right)=3(4)+3\left(\frac{1}{2}\right)$

$\frac{3}{4}+\frac{2}{9}=\frac{2}{9}+\frac{3}{4}$

D I S C O V E R Y

USING THE GRAPHICS CALCULATOR: Operations on Rational Numbers

Many graphics calculators work with rational numbers in fractional form, but the TI-81 uses only decimal notation. We will now see how the graphics calculator can be used to check the results of operations on rational numbers. Work through the following examples.

EXAMPLE 1 Write $\frac{3}{10}+\frac{2}{5}$ as a completely reduced rational number. Use the graphics calculator to check your work.

Solution To add these rational numbers, we first get a common denominator (in this case, 10). The first number already has this denominator. Renaming the number $\frac{2}{5}$ with a denominator of 10, we get

$$\left(\frac{2}{5}\right)\left(\frac{2}{2}\right)=\frac{4}{10}$$

Now,

$$\frac{3}{10}+\frac{2}{5}=\frac{3}{10}+\frac{4}{10}=\frac{7}{10}$$

To check this work, type the original problem, `(3/10) + (2/5)`, using the ÷ key to type the / symbol. Then press the ENTER key. The number .7 is written to the screen.

Now, type the answer, $\frac{7}{10}$ and press the ENTER key. As Figure 1 indicates, the same number is output. So, we have confidence that $\frac{7}{10}$ is the correct answer. ■

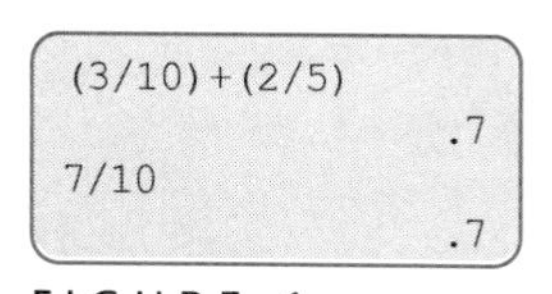

FIGURE 1
To type /, press the ÷ key.

EXAMPLE 2 Write $\left(\frac{4}{11}\right)\left(\frac{7}{8}\right)$ as a completely reduced rational number. Use the graphics calculator to check your work.

Solution To multiply these rational numbers, we multiply the numerators and multiply the denominators to get $\frac{-28}{88}$. Reducing this rational number, we get

$$\frac{-28}{88} = \frac{-(\cancel{2})(\cancel{2})(7)}{(\cancel{2})(\cancel{2})(2)(11)} = \frac{-7}{22}$$

To use the graphics calculator to check this answer, type

(−4/11)∗(7/8),

using the blue multiplication key to type the ∗ symbol. Then press the ENTER key.

Now, type the answer produced, −7/22 and press the ENTER key. As you can see in Figure 2, the same number is output for both computations. So, we have confidence that our work is correct. ■

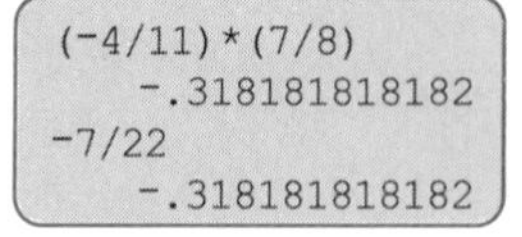

FIGURE 2
To type *, press the blue X key.

Exercises

Return to Exercises 1.2, and use the graphics calculator to check each answer for Problems 13–54.

1.3 ■ ■ ■ NOTATION SYSTEMS AND INTEGER EXPONENTS

Although we have many notation systems, one is commonly used for the integers: a place-value system based on powers of ten (10^0, 10^1, 10^2, and so on). For rational numbers, the most common notation system is fractional notation, $\frac{a}{b}$. In this section we will discuss three other systems—decimal notation, percent notation, and scientific notation. With the scientific notation system, integer exponents play an important role.

Decimal Notation Consider the number −3456. Each digit in this number has a meaning based on its position (place):

$$-3456 \text{ means } (-3)(1000) + (-4)(100) + (-5)(10) + (-6)(1)$$

Thus, each **place value** is a power of ten. We now repeat the definition of a whole number exponent, n.

DEFINITION OF WHOLE NUMBER EXPONENT

For n a natural number,

$$a^n \quad \text{means} \quad \underbrace{a \cdot a \cdot a \cdot \cdots \cdot a}_{n \text{ factors of } a}.$$

If $a \neq 0$, then $a^0 = 1$.

The number a is the **base** and the number n is the **exponent.**

A power of ten has the form 10^n, where 10 is the base and n is the exponent. Some whole number powers of 10 are given in the following table.

Number	*As a product of 10s*	*Power of 10*
1	—	10^0
10	—	10^1
100	(10)(10)	10^2
1,000	(10)(10)(10)	10^3
10,000	(10)(10)(10)(10)	10^4

Using exponents, we can rewrite the number −3456 as

$$(-3)(10^3) + (-4)(10^2) + (-5)(10^1) + (-6)(10^0)$$

Consider the rational number $\frac{3}{5}$. This number can be renamed as $\frac{6}{10}$. Consider the rational number $\frac{13}{20}$. This number can be written as $\frac{6}{10} + \frac{5}{100}$. In the late 1500s, several mathematicians (François Viète of France, Simon Stevin of Belgium, and John Napier of Scotland) suggested the use of decimal notation to represent these kinds of numbers. The number $\frac{3}{5}$ can be written as .6 and the number $\frac{13}{20}$ can be written as .65. Just as the digits to the left of the decimal point represent whole number powers of ten, the digits to the right of the decimal point represent negative integer powers of ten. We now define negative integer exponents, a^{-n}.

DEFINITION OF NEGATIVE INTEGER EXPONENT

For n a positive integer and $a \neq 0$,

$$a^{-n} \quad \text{means} \quad \frac{1}{a^n}$$

Example: $3^{-4} = \frac{1}{3^4}$

Some negative integer powers of ten are shown in the following table.

Number	*As a quotient of 10s*	*Power of 10*
$\frac{1}{10}$	$\frac{1}{10}$	10^{-1}
$\frac{1}{100}$	$\frac{1}{(10)(10)}$	10^{-2}
$\frac{1}{1000}$	$\frac{1}{(10)(10)(10)}$	10^{-3}
$\frac{1}{10000}$	$\frac{1}{(10)(10)(10)(10)}$	10^{-4}

Using these powers, we can write 0.25 as

$$2(10^{-1}) + 5(10^{-2})$$

and we can write 23.4567 as

$$2(10^1) + 3(10^0) + 4(10^{-1}) + 5(10^{-2}) + 6(10^{-3}) + 7(10^{-4})$$

EXAMPLE 1 Give the place value of the underlined digit in each number.

a. $38.08\underline{4}$ **b.** $3\underline{6}7.11235$

Solution **a.** The underlined digit, 4, is in the thousandths place. Its place value is 10^{-3}.

b. The underlined digit, 6, is in the tens place. Its place value is 10^1. ■

Every rational number can be written in decimal notation. For many rational numbers, the decimal name is infinite in length, such as 0.33333. . . for $\frac{1}{3}$. We often use a bar above the repeating part to indicate that the pattern continues: $0.\overline{3}$ for $\frac{1}{3}$ and $0.\overline{12}$ for $\frac{12}{99}$.

Rounding Decimals In business and in the sciences, numbers in decimal notation are often rounded, or approximated. There are two approaches to rounding a number in decimal notation. The first approach is to round the number to a place value (hundreds, tens, ones, tenths, hundredths, and so on). The second approach is to round the number to a selected number of significant digits (such as three significant digits or two significant digits). A digit is one of the symbols 0, 1, 2, 3, 4, 5, 6, 7, 8, or 9. We now state the rule for rounding a number using either of the approaches, place value or significant digits.

RULES FOR ROUNDING A NUMBER

1. If the first digit to the right of the round-off point is 5 or is larger than 5, increase the digit at the round-off point to the next digit.

   increase by 1

   Example: 5.6583 → 5.66

   Round-off point of hundredths

2. If the first digit to the right of the round-off point is less than 5, leave the digit at the round-off point unchanged.

   unchanged

   Example: 5.6583 → 5.658

   Round-off point of thousandths

After using one of the rounding rules, if the round-off point is to the left of the decimal point, replace all digits to the right of the round-off point with a zero. Otherwise, erase all digits to the right of the round-off point.

EXAMPLE 2 Round 0.2376 to hundredths.

Solution The round-off point is as indicated:

Next digit is greater than 5.

0.2376

Round-off point at hundredths

We follow rule 1 to get 0.24 ■

EXAMPLE 3 Round 23,456.456 to thousands.

Solution The round-off point is as indicated:

Next digit is less than 5.

23,456.456

Round-off point at thousandths

Using rule 2, we have 23,000 ■

REMEMBER

0.1234

tenths

hundredths

thousandths

ten-thousandths

Before we round a number to a selected number of *significant digits,* we must first discuss what a significant digits is. Significant digits are used to communicate the accuracy of a measurement. Nonzero digits are always significant. Zero digits between nonzero digits are significant. A zero digit to the right of the decimal point *and* following a nonzero digit is significant. Some examples of significant digits are shown:

all digits are significant	only 5 and 3 are significant	only the last five digits are significant
53000.70	53000	0.00053020
7 significant digits	2 significant digits	5 significant digits

EXAMPLE 4 Round 0.230895 to 4 significant digits.

Solution The round-off point is as indicated:

Next digit is greater than 5.

0.230895

Round-off point at fourth significant digit

We follow rule 1 to get 0.2309 ■

EXAMPLE 5 Round 23,560,300 to 2 significant digits.

Solution The round-off point is as indicated:

Next digit is 5.

23,560,300

Round-off point at second significant digit

We use rule 1 to get 24,000,000 ■

EXAMPLE 6 Round 23.54000 to 3 significant digits.

Solution The round-off point is as indicated:

Next digit is less than 5.

23.54000.

Round-off point at third significant digit

Following rule 2, we have 23.5 ■

GIVE IT A TRY

1. Round each number to the indicated place value.
   a. 0.003456 to thousandths **b.** 132,789.2354 to hundreds
   c. 132,789.2354 to hundredths
2. Round each number to the indicated number of significant digits.
   a. 23,456 to 3 significant digits **b.** 0.2354000 to 2 significant digits
3. Can the number 23.50 be rounded to 5 significant digits? Can 23.50 be rounded to hundredths? To hundreds?

Percent Notation Rational numbers can be written in different notations, including percent notation. First and foremost, *percent* means fractional part of 100. For example, 45% means 45 parts of 100 or, simply, $\frac{45}{100}$. Thus, to find 35% of a number means to find $\frac{35}{100}$ of the number. Consider the following mnemonic evolution of the percent symbol:

$\frac{35}{100}$ became 35/100 became 35 0/0 became 35%.

Percent notation was probably introduced by the Romans. The actual evolution of the percent symbol, %, is more like

35 per cento became 35 P 100 became 35 p c^0
became 35 $^0/$ became 35$^0/_0$ became 35%

The examples that follow deal with percent notation.

EXAMPLE 7 Write 0.3456 and 23.35 in percent notation.

Solution To write a number in percent notation means to rewrite the number so that the denominator is 100.
For 0.3456, we get:

$$\frac{0.3456}{1} \times \frac{100}{100} = \frac{34.56}{100} \quad \text{or} \quad 34.56\%$$

For 23.35, we get:

$$\frac{23.35}{1} \times \frac{100}{100} = \frac{2335}{100} \quad \text{or} \quad 2335\%$$

■

EXAMPLE 8 Write 0.0034% as a decimal and as a completely reduced rational number.

Solution The key to this conversion is knowing what the % symbol means. As a decimal, 0.0034% means

$$\frac{0.0034}{100} \quad \text{or} \quad 0.000034$$

Writing this decimal as a fraction, we get $\frac{34}{1000000}$. Writing this fraction as a completely reduced rational number, we get $\frac{17}{500000}$. ■

EXAMPLE 9 **a.** What is 38% of 80? **b.** What percent of 47 is 10?

Solution **a.** A convenient approach to percent problems such as these is to set up the following form:

$$\_\_\_ \% \text{ of } \_\_\_ = \_\_\_$$

Now, we fill in the blanks with the given information.

$$\underline{38} \% \text{ of } \underline{80} = \_\_\_$$

So we need only do the computation (0.38)(80) to answer the question:

$$38\% \text{ of } 80 \text{ is } \quad (0.38)(80) \quad \text{or} \quad 30.4$$

b. We fill in the form from part (a) to get:

$$\_\_\_ \% \text{ of } \underline{47} = \underline{10}$$

We need the number that multiplied by 47 yields 10. This number is $\frac{10}{47}$. Writing this number as a decimal, we get approximately (rounded to ten-thousandths) 0.2128. As a percent, we get 21.28%. Thus, 10 is approximately 21.28% of 47. ■

30% acid 70% water

FIGURE 1
A 30% acid solution means 30% acid and 70% water.

Using Percents Because percents are used so often in business and in science, many word problems in algebra use percent notation. One common use of percent in algebra is in mixture problems involving acid solutions. An acid solution is a mixture of pure water and an acid. (The type of acid used is unimportant.) We say a "30% acid solution" to mean a mixture of 30% acid and 70% water, as shown in Figure 1. The percent indicates the concentration of the acid in the solution. Consider the following examples.

EXAMPLE 10 If 8 liters of 30% acid solution is mixed with 12 liters of 70% acid solution, find the concentration of the resulting mixture.

Solution To find the concentration of acid in the resulting mixture, we need to know the amount of mixture and the amount of acid in the mixture. Figure 2 shows the solutions and the mixture.

8 liters of 30% acid solution

12 liters of 70% acid solution

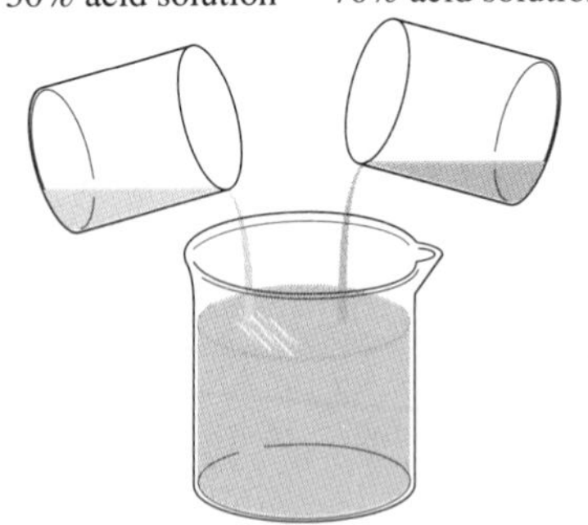

FIGURE 2

$8 + 12 = 20$ liters of solution

30% of $8 = 2.4$ liters of acid in 30% solution
70% of $12 = 8.4$ liters of acid in 70% solution

$2.4 + 8.4 = 10.8$ liters of acid

First, the amount of mixture is $8 + 12$, or 20 liters.
Second, we find the amount of acid. From 8 liters of 30% acid solution, we get

$$30\% \text{ of } 8 = (0.3)(8) = 2.4 \text{ liters of acid}$$

From 12 liters of 70% acid solution, we get

$$70\% \text{ of } 12 = (0.7)(12) = 8.4 \text{ liters of acid}$$

So, the mixture contains $2.4 + 8.4$, or 10.8 liters of acid.

Finally, by completing our percent set up, ____ % of $20 = 10.8$, we get the concentration of acid in the mixture:

$$\frac{10.8}{20} = 0.54 = 54\%$$ ■

EXAMPLE 11 If Lamon drives for 4 hours at 60 mi/h and then increases his speed by 10% and drives another 3 hours, how many miles does he travel?

Solution The basic relationship we need to solve this problem is that distance traveled equals the rate of travel times the time spent traveling. In symbols, we write this relationship as $d = rt$. Now, on the first leg of the trip, Lamon drives at 60 mi/h for 4 hours. So, we have

$$\begin{aligned} d &= (60)(4) \\ &= 240 \text{ miles} \end{aligned}$$

On the second leg, he increases his speed by 10%. Thus, his rate of speed on the second leg is

$$\begin{aligned} 60 + 10\% \text{ of } 60 &= 60 + 0.1(60) \\ &= 60 + 6 = 66 \text{ mi/h} \end{aligned}$$

Thus, for the second leg Lamon drives at 66 mi/h for 3 hours. We have

$$\begin{aligned} d &= (66)(3) \\ &= 198 \text{ miles} \end{aligned}$$

For the two legs of the trip, Lamon travels 240 + 198, or 438 miles. We could also write

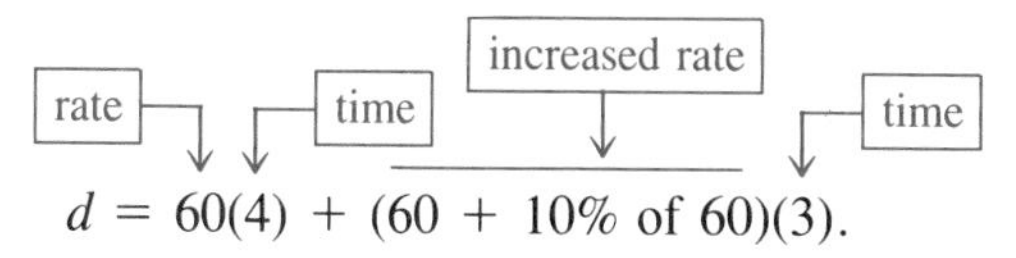

$$d = 60(4) + (60 + 10\% \text{ of } 60)(3).$$

■

GIVE IT A TRY

4. a. Write 3.0045% as a decimal.
   b. Write 234.5% as a completely reduced rational number.
   c. Write 0.002345 in percent notation.
5. If we consider % as a unary operation that means *divide by 100,* what is the decimal name for 23%? For 23%%?
6. a. If 20 liters of 10% acid solution is mixed with 60 liters of 90% acid solution, what is the concentration of the resulting mixture?
   b. Rita drives at 50 mi/h for 3 hours and then increases her speed by 20% and drives another 4 hours. How many miles does she travel?

Scientific Notation Scientific notation was developed to write very large numbers and very small numbers in a compact form. The form for scientific notation is

A number from 1 to 10 including 1 → $\times$ 10 ← An integer power of 10

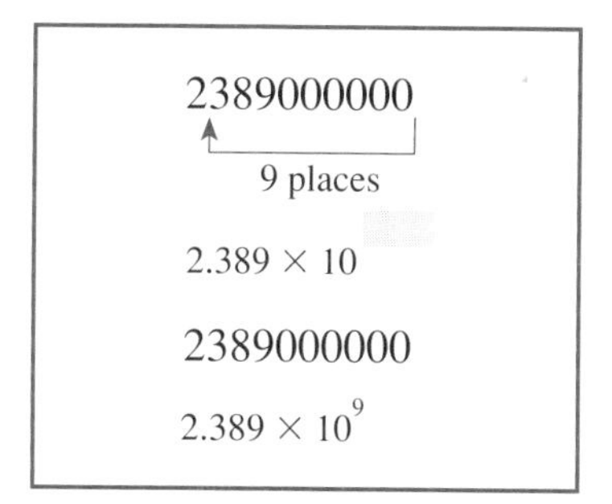

FIGURE 3

To write a number larger than 1 in scientific notation, we follow these rules:

1. Starting at the decimal point, imagine moving the decimal point one place further to the left until we obtain a number from 1 to 10.

   For 2,389,000,000 we get 2.389. See Figure 3.

2. Count the number of times this new number must be *multiplied* by 10 to get the original number. This number is the exponent.

   For 2.389, the number of times to multiply by 10 is 9.
   Thus, 2,389,000,000 written in scientific notation is 2.389×10^9.

0.000002307
6 places
2.307×10
0.000002307
2.307×10^{-6}

FIGURE 4

For a positive number less than 1, follow these rules to write the number in scientific notation:

1. Starting at the decimal point, imagine placing the decimal point one place further to the right until we obtain a number from 1 to 10.

   For 0.000002307 we get 2.307. See Figure 4.

2. Count the number of times this new number must be *divided* by 10 to get the original number. The additive opposite of this number is the exponent.

   For 2.307, the number of times to divide by 10 is 6.
   Thus, 0.000002307 written in scientific notation is 2.307×10^{-6}.

Scientific notation is typically used with positive numbers, but we can write a negative number, such as −34,870, in scientific notation as follows:

$$-34{,}870 = -1(34{,}870) = -1(3.487 \times 10^4) = -3.487 \times 10^4$$

When writing a number in scientific notation, we report all significant digits for the number from 1 to 10. This provides the reader with the *precision* of the number. The integer power of 10 yields the *magnitude* (size) of the number. Thus, a reader can quickly compare any two numbers for magnitude and for precision. For example, in 1798, Sir Henry Cavendish computed the mass of the earth is 6,000,000,000,000,000,000,000,000,000 grams and its weight in tons as 6,600,000,000,000,000,000,000. It is difficult to tell which of these numbers is the largest, but in scientific notation, the numbers are written as 6×10^{27} grams for the mass and 6.6×10^{21} tons for the weight. In this notation we can easily see that the number of grams is greater by several magnitudes than the number of tons ($27 > 21$). Also, we can see that the weight measurement has more precision than the mass measurement (6.6 versus 6). As an example of scientific notation and precision, consider the following.

EXAMPLE 12 Round 23,000,456 to 5 significant digits.

Solution The digit to the right of the round-off point (the thousands) is less than 5:

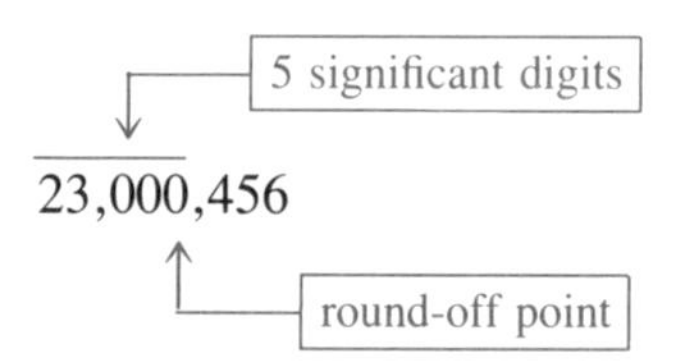

Thus, we round to 23,000,000. This number appears to have only two significant digits. To correct this, we place a dot above a zero digit to indicate that the number has 5 significant digits: $23{,}00\dot{0}{,}000$. In scientific notation, we write this as 2.3000×10^7, and the fact that the number has 5 significant digits (its precision) is evident immediately. ■

Each notation system for the rational numbers is useful for some particular problem or in some particular area of science or business. Fraction notation is useful in many areas of science because rates and ratios can easily be expressed as fractions. For decimal notation, the rules for the operations of addition, subtraction, multiplication, and division are very close to those for the same operations on whole numbers. With decimal notation, a number can be written on a single line (a great advantage for detailed business ledgers). Percent notation is simply a shorthand notation for "divided by 100" or "per 100." The scientific notation system has several advantages in science (and even in business, when discussing the national budget or the salary of a CEO). The major advantage of scientific notation is that we can write very large numbers and very small numbers in a compact manner.

GIVE IT A TRY

7. Write each number in scientific notation.
   a. 23,400,000,000 **b.** 0.000034 **c.** 5 **d.** 10.2×10^3
8. Complete the following table for notational systems.

Fractional	*Decimal*	*Percent*	*Scientific*
$\frac{23}{50}$	____	____	____
____	0.0023	____	____
____	____	23.05%	____
____	____	____	3.24×10^{-3}

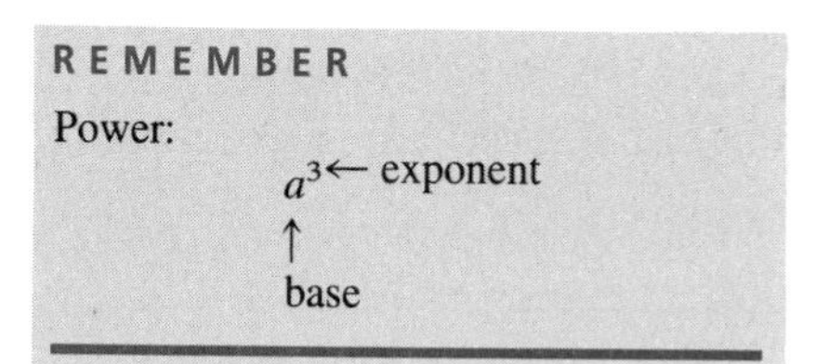

Exponents A positive integer exponent is shorthand notation for a number times itself a specific number of times. For example,

$$4^3 \quad \text{means} \quad (4)(4)(4)$$

A negative exponent is shorthand notation for 1 divided by a number a specific number of times. For example,

$$4^{-3} \quad \text{means} \quad \left(\frac{1}{4^3}\right) \quad \text{or} \quad \left(\frac{1}{4}\right)\left(\frac{1}{4}\right)\left(\frac{1}{4}\right)$$

There are five laws for integer exponents.

LAWS OF EXPONENTS FOR INTEGER EXPONENTS

1. $(a^n)(a^m) = a^{n+m}$ *Example:* $(3^4)(3^{-2}) = 3^{4-2} = 3^2$
2. $(a^n) \div (a^m) = a^{n-m}$ *Example:* $(2^5) \div (2^{-3}) = 2^{5-(-3)} = 2^8$
3. $(a^n)^m = a^{(n)(m)}$ *Example:* $(5^{-2})^3 = 5^{(-2)(3)} = 5^{-6}$
4. $(a \cdot b)^n = (a^n)(b^n)$ *Example:* $(2 \cdot 5)^3 = 2^3 \cdot 5^3$
5. $\left(\frac{a}{b}\right)^n = (a^n)/(b^n)$ *Example:* $\left(\frac{2}{5}\right)^3 = \frac{2^3}{5^3}$

As an example of using the laws of exponents, consider the following operations with numbers in scientific notation.

EXAMPLE 13 Write in scientific notation: $(3.2 \times 10^4)(5 \times 10^4)$

Solution Of course, we can write the numbers as (32,000)(50,000), and perform the computation to get 1,600,000,000. We then write the result in scientific notation, 1.6×10^9.

We can also use the laws of exponents to do this problem. Using the associative and commutative laws, we get

$$\begin{aligned}(3.2 \times 10^4)(5 \times 10^4) &= (3.2 \times 5)\,(10^4 \times 10^4) \\ &= 16 \times 10^8 && (a^n)(a^m) = a^{n+m} \\ &= (1.6 \times 10^1) \times 10^8 && \text{Scientific notation for 16} \\ &= 1.6 \times 10^9 && (a^n)(a^m) = a^{n+m}\end{aligned}$$

■

EXAMPLE 14 Write in scientific notation: $(6.34 \times 10^2) \div (2 \times 10^6)$

Solution Using the associative and commutative laws, we get:

$$
\begin{aligned}
(6.34 \times 10^2) \div (2 \times 10^6) &= \frac{6.34 \times 10^2}{2 \times 10^6} \\
&= \frac{6.34}{2} \times \frac{10^2}{10^6} \\
&= 3.17 \times 10^{2-6} \qquad \frac{a^n}{a^m} = a^{n-m} \\
&= 3.17 \times 10^{-4}
\end{aligned}
$$

■

EXAMPLE 15 Write in scientific notation: $(2.5 \times 10^{-5})^2$

Solution

$$
\begin{aligned}
(2.5 \times 10^{-5})^2 &= (2.5)^2 \times (10^{-5})^2 \qquad (a \times b)^n = (a^n)(b^n) \\
&= 6.25 \times 10^{-10} \qquad (a^n)^m = a^{(n)(m)}
\end{aligned}
$$

■

GIVE IT A TRY

Write each in scientific notation.

9. $3(1.2 \times 10^4)$
10. $1.2 \times 10^3 + 2.3 \times 10^3$
11. $(3.2 \times 10^3) + (-1.2 \times 10^3)$
12. $(3.2 \times 10^3) \times (2 \times 10^4)$
13. $(7.4 \times 10^6) \div (-2 \times 10^4)$
14. $(3 \times 10^5)^3$
15. $(48{,}000) \times (0.0002)^3$

■ SUMMARY

In this section, we have reviewed three notation systems for the rational numbers: decimal notation, percent notation, and scientific notation. The decimal notation system is simply an extension of the place-value system, which we use for whole numbers and integers. Each place to the right of the decimal point is a negative integer power of ten. The first place to the right of the decimal point has value $\frac{1}{10}$ or 10^{-1}, the second place to the right of the decimal point has value $\frac{1}{100}$ or 10^{-2}, and so on. One advantage of the decimal notation system over the fraction (ratio) notation system, $\frac{a}{b}$, which is typically used with rational numbers, is that the processes (or algorithms) for addition, subtraction, multiplication, and division closely resemble those for the whole numbers.

To use the percent notation system we simply write each number so that its denominator is 100. For example, $\frac{4}{5}$ is $\frac{80}{100}$, or 80%, and 3.42 is $\frac{342}{100}$, or 342%. Percent notation grew out of business mathematics as a shorthand for rates—per cento or per one hundred.

Scientific notation was developed to write very large numbers and very small numbers in a way that makes it easy to compare numbers. The form for this system is *a number from 1 to 10 times an integer power of 10* (for example, 2×10^3).

In arithmetic (especially with scientific notation) and in algebra, knowledge of exponents is essential. In this section, we discussed only integer exponents.

The Laws of Exponents consist of the following five generalizations:

1. $(a^n)(a^m) = a^{n+m}$ *Example:* $(5^3)(5^4) = 5^{3+4} = 5^7$
2. $(a^n) \div (a^m) = a^{n-m}$ *Example:* $(7^5) \div (7^2) = 7^{5-2} = 7^3$
3. $(a^n)^m = a^{(n)(m)}$ *Example:* $(2^2)^3 = 2^{(2)(3)} = 2^6$
4. $(a \cdot b)^n = (a^n)(b^n)$ *Example:* $(7 \cdot 4)^3 = 7^3 \cdot 4^3$
5. $\left(\frac{a}{b}\right)^n = \frac{(a^n)}{(b^n)}$ *Example:* $\left(\frac{7}{4}\right)^3 = \frac{7^3}{4^3}$

In this section we used these five generalizations to do computations in scientific notation.

1.3 ■ ■ ■ EXERCISES

Report the place value, such as 10^3 or 10^{-2}, of the digit underlined in each number.

1. $23.\underline{6}78$
2. $23\underline{5}.098$
3. $0.00\underline{4}56$
4. $-0.0045\underline{6}78$
5. $12{,}89\underline{0}.234$
6. $3{,}\underline{9}05{,}006$
7. $-0.00231\underline{0}$
8. $12.00098\underline{0}23$

Report the integer name for the given number.

9. $5(10^4) + 6(10^2)$
10. $3(10^3) + 2(10^0)$
11. $-2(10^2) - 3(10^0)$
12. $-5(10^2) - 7(10^1)$

Report the decimal name for each number.

13. $5(10^{-2}) + 6(10^{-3})$
14. $3(10^{-2}) + 2(10^{-4})$
15. $-2(10^0) - 3(10^{-1})$
16. $-5(10^0) - 7(10^{-2})$
17. $3(10^1) + 2(10^0) + 5(10^{-1}) + 3(10^{-2})$
18. $6(10^1) + 1(10^0) + 8(10^{-2}) + 3(10^{-4})$

Report the number of significant digits in each number.

19. 12.032
20. 23,000
21. 23,001
22. 23,001.2900
23. 0.003400
24. 23%

Round each number to the indicated place value.

25 123,567,890 to thousands
26. 0.00234089 to millionths
27. 999.9 to ones
28. 0.00353 to thousandths
29. 23,890.00456 to hundreds
30. 23,897.0468 to hundredths
31. 0.004567 to hundred-thousandths
32. 0.004567 to ten-thousandths
33. -0.915 to tenths
34. -0.915 to ones

Round each number to the indicated number of significant digits.

35. 15,490 to 2 significant digits
36. 15,490 to 3 significant digits
37. 0.0034569 to 4 significant digits
38. 3.00001 to 2 significant digits
39. 345,009 to 2 significant digits
40. 345,009 to 3 significant digits
41. 5.00363 to 4 significant digits
42. 5.00363 to 5 significant digits
43. 0.0030567 to 4 significant digits
44. 0.0230567 to 1 significant digit

Write each number in percent notation.

45. 0.23
46. 0.0067
47. 5
48. 1.675
49. $\frac{3}{8}$
50. $\frac{17}{40}$

Compute each of the following.

51. 28% of 68
52. 20% of 2.34
53. 200% of 12
54. 15% of 0.26
55. 30% of $\frac{3}{10}$
56. 1.5% of $\frac{32}{5}$
57. 400% of $\frac{35}{17}$
58. 0.5% of $\frac{7}{2}$

59. An acid solution is a combination of acid and pure water. If an acid solution contains 30% acid, and we have 80 liters of the solution, how many liters are acid? How many liters are pure water?

60. The value of Tom's property at Willow Creek increased 12% last year. If the property was worth $52,000 at the start of the year, what was it worth at the end of the year?

61. Silver alloy is a mixture of silver and other metals. If 30 grams of a 30% silver alloy is mixed with 20 grams of 60% silver alloy, the resulting mixture will contain how many grams of silver? The mixture will be what percent silver?

62. Frank travels at 60 mi/h for 5 hours and then decreases his speed by 30% and travels for another 2 hours. How many miles does he travel?

Write each number in scientific notation.

63. 23,000,000,000,000 **64.** 7,002,000

65. 678.2 **66.** 3.0023

67. 0.000000398 **68.** 0.00102

69. 178,000 **70.** 0.000000000456

71. 0.000000000189 **72.** 0.23%

Do the following computations and report each result in scientific notation.

73. $3(1.2 \times 10^3)$ **74.** $(100)(2.67 \times 10^2)$

75. $(0.000000235)^2$ **76.** $(1.2 \times 10^3)^2$

77. $(2.3 \times 10^4) + (5 \times 10^3)$ **78.** $3(7.81 \times 10^{-1})$

79. $12\%/(1.2 \times 10^3)$ **80.** $(3 \times 10^4)(6 \times 10^{-2})$

81. $(9.1 \times 10^4)(1.2 \times 10^3)$ **82.** $\dfrac{36{,}000}{(0.0006)^2}$

83. $(7.2 \times 10^4)/(1.2 \times 10^3)$ **84.** $(0.0005)^2 - (0.01)^3$

85. $(3.25 \times 10^3) - (1.2 \times 10^3)$

86. $(3.25 \times 10^3) - (1.2 \times 10^2)$

87. $(3.25 \times 10^3) - (1.2 \times 10^1)$

88. $(5.17 \times 10^{-2}) + (2.13 \times 10^{-3})$

89. $(9.89 \times 10^{-1}) - (2.67 \times 10^{-2})$

90. $2.3 \times 10^{-2} - 38\% + 0.23$

91. $0.56 - (5.1 \times 10^{-1}) + 80\%$

DISCOVERY

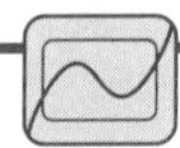

USING THE GRAPHICS CALCULATOR: Scientific Notation

Calculators typically convert a number to scientific notation whenever the absolute value of the output gets large enough or small enough. To see this, we can square the number 23,000,000: type `23000000`, press the x^2 key, and then press the `ENTER` key. The value reported is `5.29E14`. This is the calculator's notation for the value 5.29×10^{14}.

The calculator has a convenient key, the `EE` key, for converting a number from scientific notation to decimal notation, provided the absolute value of the number is between 10^{-4} and 10^{10}. To learn how to use this key, consider the following example.*

EXAMPLE 1 Convert 2.3×10^{-3} to decimal notation.

Solution First, press the `CLEAR` key to erase the home screen. We will use the `EE` key to enter a number in scientific notation. Type `2.3` and then press the `EE` key. The character `E` is written to the screen. Now, press the `(-)` key, followed by the `3` key, to display `2.3E-3`. This is calculator notation for 2.3×10^{-3}. Press the `ENTER` key, and the decimal name is reported, as shown in Figure 1. ■

```
2.3E-3
         .0023
```

FIGURE 1

TI-82 Note *To access the `EE` key, use the `2ND` key, followed by the comma key, `,`.

To do computations using scientific notation, such as $5(2 \times 10^2)$, we use the MODE key to select scientific mode.* To learn how to do this, work through the following example.

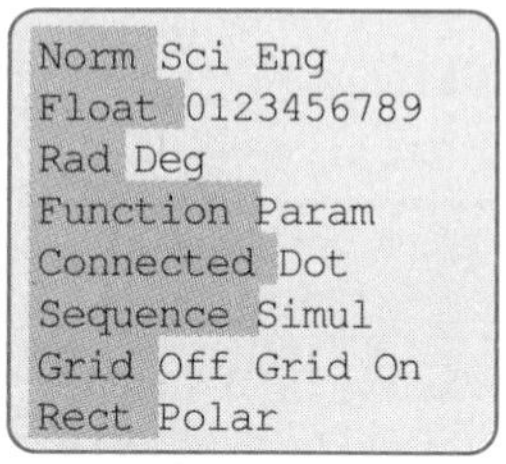

Norm Sci Eng
Float 0123456789
Rad Deg
Function Param
Connected Dot
Sequence Simul
Grid Off Grid On
Rect Polar

FIGURE 2

EXAMPLE 2 Write in scientific notation: $2.3 \times 10^6 - 4.56 \times 10^4$

Solution First, press the CLEAR key to erase the home screen. Now, press the MODE key to display the MODE menu, as shown in Figure 2. With the rectangular highlight on the line Norm Sci Eng, press the right arrow key to move the highlight to the Sci option. Press the ENTER key to select this option. Use the QUIT key to quit editing the mode settings, save the changes, and return to the home screen.

Type 2.3E6−4.56E4 and then press the ENTER key. The value 2.2544E6 is written to the screen. See Figure 3. Thus,

$$2.3 \times 10^6 - 4.56 \times 10^4 = 2.2544 \times 10^6$$ ■

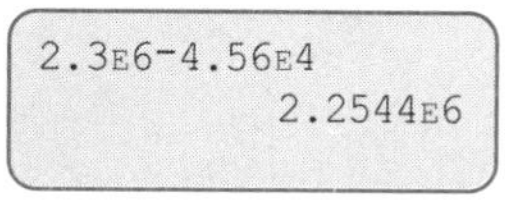

2.3E6-4.56E4
2.2544E6

FIGURE 3

To return to normal mode, press the MODE key, highlight the Norm option, and press the ENTER key. Use the QUIT key to quit editing and return to the home screen.

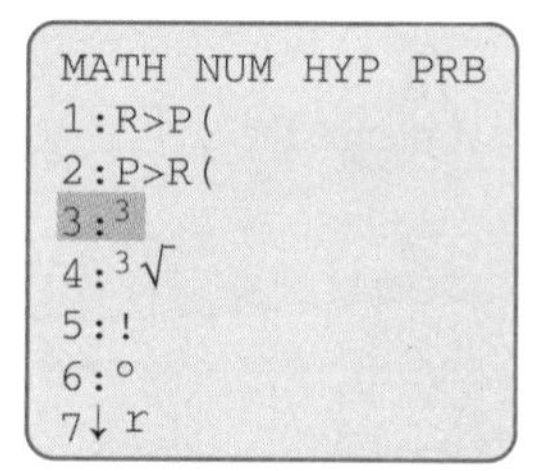

FIGURE 4

Integer Exponents We will now use the calculator to work with the laws of exponents. The calculator provides a key for squaring a number (the x^2 key) and a menu selection for cubing a number. The cubing option, ³, from the MATH menu is shown in Figure 4.* In general, to express a power of a number, we use the ^ key. For example, to compute 2^5, the fifth power of 2, press the 2 key, followed by the ^ key and then the 5 key, to display 2^5. Pressing the ENTER key displays 32. Consider the following example.

EXAMPLE 3 Use the graphics calculator to verify that $(3)^2(3)^3$ equals 3^5.

Solution First, press the CLEAR key to erase the home screen.

Press the 3 key followed by the x^2 key to display 3^2 on the screen. Press the 3 key, then the MATH key, followed by the 3 key to display $3^2 3^3$ on the screen. Press the ENTER key. The value 243 is written to the screen.

Press the 3 key, the ^ key, and then the 5 key to write 3^5 to the screen. Press the ENTER key. The value 243 is written to the screen.

Thus, $(3)^2(3)^3$ equals 3^5. See Figure 5. ■

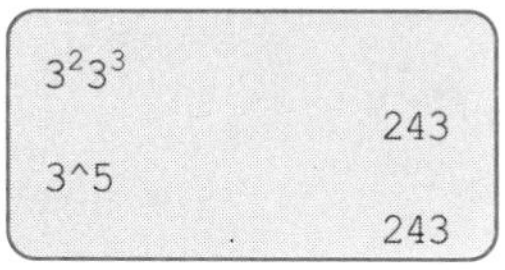

$3^2 3^3$
243
3^5
243

FIGURE 5

EXAMPLE 4 Use the graphics calculator to verify that $20^5 \div 20^8$ equals 20^{-3}.

Solution First, press the CLEAR key to erase the home screen.

Use the following keystrokes to evaluate 20^5/20^8:

[2] [0] [^] [5] [/] [2] [0] [^] [8] [ENTER]

The value output is 1.25E-4.

TI-82 Note *The MODE screen and MATH menu differ slightly.

```
20^5/20^8
          1.25E-4
20^-3
          1.25E-4
```

FIGURE 6

Next, use these keystrokes to evaluate `20^-3`.

The value output is `1.25E-4`.
Thus, $20^5 \div 20^8 = 20^{-3}$. See Figure 6. ■

Exercises

1. Set the mode to `Sci` (select the `Sci` option from the `MODE` menu). Write each number in scientific notation.
   a. $(2.34 \times 10^{-3}) \div (9.08 \times 10^{-4})$ **b.** $(9.02 \times 10^4) \times (3 \times 10^5)$
2. Set the mode to `Normal` (select the `Norm` option from the `MODE` menu). Report the decimal name for each number.
   a. 2.56×10^{-2} **b.** -1.3×10^4 **c.** $5.68 \times 10^5 - 2.35 \times 10^6$
3. From the laws of exponents, we know that
   $$(a)^n(a^m) = a^{n+m}, \qquad \frac{(a)^n}{(a^m)} = a^{n-m}, \qquad \text{and} \qquad ((a)^n)^m = a^{nm}$$
   Use the x^2 key, option `3` (the cubing option) from the `MATH` menu, the x^{-1} key, and the ^ key to verify each statement.
   a. $(5)^2(5)^3 = 5^5$ **b.** $((2)^2)^3 = 2^6$ **c.** $(2)^2 \div (2)^3 = 2^{-1}$
4. From the laws of exponents, we know that
   $$(ab)^n = (a^n)(b^n) \qquad \text{and} \qquad \left(\frac{a}{b}\right)^n = \frac{(a^n)}{(b^n)}$$
   Use the graphics calculator to verify each statement.
   a. $10^3 = (2^3)(5^3)$ **b.** $5^5 = 10^5/2^5$

Return to Exercises 1.3, and use the graphics calculator to check the answers for Problems 63–91.

1.4 ■ ■ ■ ORDER OF OPERATIONS AND FORMULAS

In this section, we review computations that involve more than one operation. We will also consider the topic of formulas.

Precedence Rules Consider the computation $3 - 5 \cdot 2$. Is the result of this computation the number -4, or is is it the number -7? That is, do we subtract first, to get $-2 \cdot 2$, or -4? Or do we do the multiplication first, to get $3 - 10$, or -7? To eliminate this point of possible confusion, mathematians have agreed to use a particular collection of rules. These rules are known as Order of

Operations rules or Precedence rules. Read through Examples 1–3. The rules are illustrated in the solutions.

EXAMPLE 1 Compute $6/2 - 10 \cdot 2$ and $6 \cdot 2 - 10/2$.

Solution *Starting at the left, do all multiplication and division operations (the only preference between these two operations is the order in which they appear).*

For $6/2 - 10 \cdot 2$, the multiplication and division operations are performed to get $3 - 20$, or -17.

For $6 \cdot 2 - 10/2$, the multiplication and division operations are performed to get $12 - 5$, or 7. ■

EXAMPLE 2 Compute $14 - 2 \cdot 3 + 5$ and $14 + 2 \cdot 3 - 5$.

Solution *After all the multiplication and division operations are performed, start at the left and do all the addition and subtraction operations (the only preference between these two operations is the order in which they appear).*

For $14 - 2 \cdot 3 + 5$, the multiplication is performed to get $14 - 6 + 5$. Now, starting at the left, the subtraction is performed to get $8 + 5$, or 13.

For $14 + 2 \cdot 3 - 5$, the multiplication is performed to get $14 + 6 - 5$. Starting at the left, the addition is performed to get $20 - 5$, or 15. ■

EXAMPLE 3 Compute $(12 - 2) \cdot (3 + 5)$ and $12 - (2 \cdot 3 + 5)$.

Solution *To override the first two rules, use parentheses (grouping symbols). Any computation enclosed in parentheses is performed first. If the computation inside the parentheses involves more than one operation, follow the first two rules.*

For $(12 - 2) \cdot (3 + 5)$, the computations enclosed in parentheses are performed first, to get $(10)(8)$, or 80.

For $12 - (2 \cdot 3 + 5)$, the computations in parentheses are performed first, to get $12 - 11$, or 1. ■

We now state the rules for order of operations.

ORDER OF OPERATIONS RULES

1. Clear the parentheses. That is, evaluate computations within grouping symbols, such as parentheses, ( ), brackets, [ ], or braces, { }. For nested groupings (groupings within groupings), start with the innermost grouping. If a grouping contains more than one operation, apply rules 2, 3, and 4.
2. Evaluate all unary operations, such as exponents, *n*th roots, and absolute value.
3. Evaluate all multiplications and divisions, from left to right.
4. Evaluate all additions and subtractions, from left to right.

To learn how to use the Order of Operations Rules, work through the next examples.

EXAMPLE 4 Write each result as a completely reduced rational number.

Solution **a.** 1/2/3 **b.** 1/(2/3)

a. We start at the left: $1/2/3 = \frac{1}{2} \div 3 = \left(\frac{1}{2}\right)\left(\frac{1}{3}\right) = \frac{1}{6}$

b. We start with the parentheses: $1/(2/3) = 1 \div \frac{2}{3} = 1\left(\frac{3}{2}\right) = \frac{3}{2}$ ■

EXAMPLE 5 Write as a completely reduced rational number: $\left(\frac{1}{2} + \frac{2}{3}\right) \div \left(\frac{3}{4}\right)^2$

Solution We start by doing the computation within the parentheses:

$$\frac{1}{2} + \frac{2}{3} = \frac{3}{6} + \frac{4}{6}$$
$$= \frac{7}{6}$$

We now have $$\frac{7}{6} \div \left(\frac{3}{4}\right)^2$$

Next, we evaluate the exponent:

$$\left(\frac{3}{4}\right)^2 = \left(\frac{3}{4}\right)\left(\frac{3}{4}\right) = \frac{9}{16}$$

We then do the remaining operation and reduce:

$$\frac{7}{6} \div \frac{9}{16} = \frac{7}{6} \cdot \frac{16}{9} = \frac{(7)(16)}{(6)(9)}$$
$$= \frac{(7)(\not{2})(2)(2)(2)}{(\not{2})(3)(3)(3)}$$
$$= \frac{56}{27}$$

So, $$\left(\frac{1}{2} + \frac{2}{3}\right) \div \left(\frac{3}{4}\right)^2 = \frac{7}{6} \div \left(\frac{3}{4}\right)^2$$
$$= \frac{7}{6} \div \frac{9}{16}$$
$$= \frac{7}{6} \div \frac{16}{9}$$
$$= \frac{56}{27}$$ ■

GIVE IT A TRY

1. Perform the computations and write each result as an integer.
   a. $-7 - (5 - 3 \cdot 2)$
   b. $-36/2 - 8 \cdot 2$
   c. $-36/(2 - 8) \cdot 2$
   d. $-3^2 - 2 \cdot (5 - 4/2)$

2. Write each result as a completely reduced rational number.

a. $\frac{3}{14} + \frac{5}{7} \div \frac{5}{2}$ **b.** $\left(\frac{3}{8} + \frac{5}{7}\right) \div \left(\frac{2}{5}\right)$ **c.** $\left(\frac{1}{2} + \frac{1}{3}\right)\left(\frac{1}{4} - \frac{1}{9}\right)$ **d.** $\left(\frac{1}{2} + \frac{1}{3}\right)/\left(\frac{1}{4} - \frac{1}{9}\right)$

Complex Fractions One area of arithmetic in which order of operations (precedence) rules are essential is that of complex fractions. These expressions are multiple operations on rational numbers. We will reserve the words *rational number* to mean a number that can be written as a ratio of integers, that is, a number of the form $\frac{a}{b}$, where a and b are integers with $b \neq 0$. We will use the word *fraction* for any expression with a numerator and a denominator. The complex fraction

$$\frac{\frac{1}{4} - \frac{1}{9}}{\frac{1}{3} - \frac{1}{2}}$$

is a division of the differences of rational numbers. The main fraction bar indicates a grouping, acting as parentheses around the numerator and around the denominator. This complex rational number can be written as

$$\left(\frac{1}{4} - \frac{1}{9}\right) \div \left(\frac{1}{3} - \frac{1}{2}\right)$$

Doing the subtractions (clearing the parentheses), we get:

$$\frac{1}{4} - \frac{1}{9} = \frac{9}{36} - \frac{4}{36} = \frac{5}{36} \quad \text{and} \quad \frac{1}{3} - \frac{1}{2} = \frac{2}{6} - \frac{3}{6} = \frac{-1}{6}$$

So,

$$\left(\frac{1}{4} - \frac{1}{9}\right) \div \left(\frac{1}{3} - \frac{1}{2}\right) = \frac{5}{36} \div \frac{-1}{6}$$

Doing the division yields:

$$\frac{5}{36} \cdot \frac{-6}{1} = \frac{(-1)(5)(\not{6})}{(6)(\not{6})} \quad \text{or} \quad \frac{-5}{6}$$

Thus, the complex fraction

$$\frac{\frac{1}{4} - \frac{1}{9}}{\frac{1}{3} - \frac{1}{2}}$$

written as a completely reduced rational number is $\frac{-5}{6}$.

GIVE IT A TRY

Write each expression as a completely reduced rational number.

3. $\dfrac{\frac{1}{25} + \frac{1}{9}}{\frac{1}{5} - \frac{1}{3}}$ **4.** $\frac{-1}{3} + \dfrac{1}{3 - \frac{-1}{2}}$

Nested Parentheses Computations sometimes involve nested parentheses (parentheses within parentheses), such as the expression

$$-3 - (5 - 2(8 - 16))$$

To perform such a computation, we first perform the operation within the innermost parentheses. Thus, we write

$$\begin{aligned} -3 - (5 - 2(8 - 16)) &= -3 - (5 - 2(-8)) \\ &= -3 - (5 - (-16)) \\ &= -3 - (5 + 16) \\ &= -3 - 21 \\ &= -24 \end{aligned}$$

Consider the following example, which involves rational numbers and nested parentheses.

EXAMPLE 6 Write the expression as a completely reduced rational number:

$$\frac{7}{8} - \left(\frac{2}{5} \div \left(\frac{-1}{3} + \frac{1}{2}\right)\right)$$

Solution Starting with the innermost parentheses, we get:

$$\begin{aligned} \frac{-1}{3} + \frac{1}{2} &= \frac{-2}{6} + \frac{3}{6} \\ &= \frac{1}{6} \end{aligned}$$

Thus,

$$\begin{aligned} \frac{7}{8} - \left(\frac{2}{5} \div \left(\frac{-1}{3} + \frac{1}{2}\right)\right) &= \frac{7}{8} - \left(\frac{2}{5} \div \frac{1}{6}\right) \\ &= \frac{7}{8} - \left(\frac{2}{5} \cdot \frac{6}{1}\right) \\ &= \frac{7}{8} - \frac{12}{5} \\ &= \frac{35}{40} - \frac{96}{40} \\ &= \frac{-61}{40} \end{aligned}$$

Change division to multiplication

So, written as a completely reduced rational number,

$$\frac{7}{8} - \left(\frac{2}{5} \div \left(\frac{-1}{3} + \frac{1}{2}\right)\right) \quad \text{is} \quad \frac{-61}{40}$$

■

GIVE IT A TRY

5. Write the expression as a completely reduced rational number.

$$\left(\frac{-3}{4} - \frac{2}{5}\right) - (3)\left(\frac{1}{2} \div \left(\frac{2}{5} + \frac{-9}{5}\right) - 5\right)$$

Scope of an Exponent and $-a^n$ A common source of confusion and error is with an expression such as -3^2. Does this mean $(-3)(-3)$? Or does it mean $-(3)(3)$? In general, an exponent applies *only* to the object immediately to its left. However, this general rule still does not resolve our dilemma for -3^2. Is the object next to exponent the number -3? Or is it the number 3 with the $-$ symbol representing the opposite operation? Here, mathematicians have simply made the decision that -3^2 means $-(3)(3)$, which equals -9. To obtain $(-3)(-3)$, we must write $(-3)^2$. In general, we have $-a^2 = -(a)(a)$.

Formulas A *formula* is a mathematical statement that describes a relationship between variables. Consider the physical situation of an automobile traveling at a selected speed (miles per hour) for a specified amount of time (number of hours), and the distance covered. The relationship between these quantities is: *the distance traveled is the product of rate of travel and the amount of time spent traveling*. Using variables to represent the distance traveled, d, the rate of travel, r, and time spent traveling, t, we get the formula $d = rt$. Using a formula rather than the worded statement of the relationship between the variables can enhance our ability to think about the relationship. For example, if we increase the rate, then the distance traveled will be greater for the same amount of time. Additionally, formulas help to simplify computations involving the relationship. It is with such computations that we are now concerned. Consider the following example.

EXAMPLE 7 The formula for the perimeter of a rectangle is $P = 2L + 2W$, as shown in Figure 1. If $L = 8$ and $W = 5$, find the value for P.

Solution To find the value for P, we replace the variable L by 8 and the variable W by 5 to get

$$P = 2(8) + 2(5)$$

Now we do the computation to get

$$P = 26$$

Thus, the value for P is 26. ■

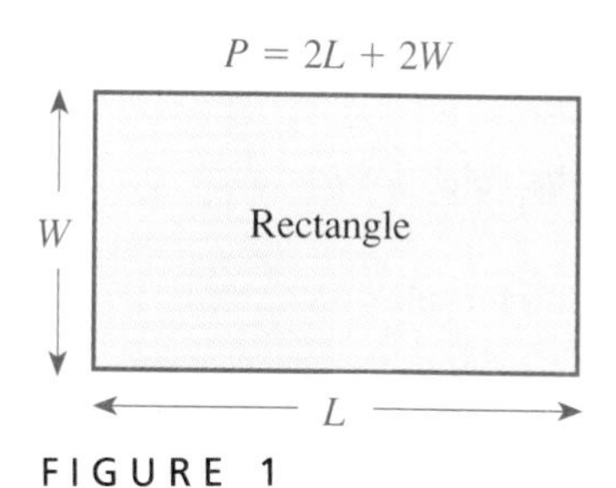

FIGURE 1

From Example 7, we can see that variables such as P, L, and W simply hold the places for numbers. When L is replaced by 8 and W by 5, we get $P = 26$. This means that if P is replaced by 26, a true arithmetic sentence is obtained: $26 = 2(8) + 2(5)$. To increase your understanding of formulas, work through the following examples.

EXAMPLE 8 For the formula $A = \frac{W + C}{R}$ with $W = \frac{2}{3}$, $R = -6$, and $C = \frac{4}{5}$, find the value for A.

Solution We replace W by $\frac{2}{3}$, R by -6, and C by $\frac{4}{5}$ to get

$$A = \frac{\frac{2}{3} + (-6)}{\frac{4}{5}}$$

Working with the complex fraction, we get

$$\begin{aligned} \left(\frac{2}{3} + (-6)\right) \div \left(\frac{4}{5}\right) &= \left(\frac{2}{3} + \frac{-18}{3}\right) \div \left(\frac{4}{5}\right) \\ &= \frac{-16}{3} \div \frac{4}{5} && \text{Clear parentheses.} \\ &= \frac{-16}{3} \cdot \frac{5}{4} && \text{Do division.} \\ &= \frac{-20}{3} && \text{Remove common factors.} \end{aligned}$$

So, for $W = \frac{2}{3}$, $R = -6$, and $C = \frac{4}{5}$, we have $A = \frac{-20}{3}$. ■

EXAMPLE 9 For the formula $y = -x^2 - 2x + 1$ with $x = -2$, find the value for y.

Solution Replacing the x variable with -2, we get

$$\begin{aligned} y &= -(-2)^2 - 2(-2) + 1 \\ &= -4 - (-4) + 1 \\ &= 1 \end{aligned}$$

Thus, if $x = -2$, then $y = 1$. ■

GIVE IT A TRY

6. For the formula $y = \frac{-2}{3}x + 7$ and $x = -3$, find the value for y.
7. For the formula $F = \frac{9}{5}C + 32$ and $C = -12$, find the value for F.
8. For the formula $y = 3 - \frac{2}{x}$ and $x = \frac{-2}{3}$, find the value for y.
9. For the formula $z = \frac{1}{R} + \frac{1}{r}$ with $R = 3$ and $r = 10$, find the value for z.
10. For the formula $A = \frac{|10 - W|}{|SW|}$ with $W = 20$ and $S = -3$, find the value for A.

■ SUMMARY

In this section, we have learned how to do computations that involve more than one operation. To correctly evaluate such computations, we follow the Order of Operations Rules. To evaluate complex fractions (numbers in which the numerator or denominator is a rational number), we must keep in mind that the fraction bar implies a set of parentheses around the numerator and around the denominator. That is, the horizontal fraction bar serves to indicate a division operation and a grouping. We discussed the topic of formula evaluations. A formula is a mathematical statement of the relationship between variables (such as $A = LW$). To evaluate a formula for selected values of the variables, we replace the variables by their respective values and then perform the computation.

1.4 ■ ■ ■ EXERCISES

Use the Order of Operations Rules to do the given computations.

1. $(-4 - 16) \div (4 + -3)$
2. $5 - (7 - 17) \div 5 - 3$
3. $-4 - 16 \div 4 + -3$
4. $5 - (7 - 17) \div (5 - 3)$
5. $(-4 - 16 \div 4) + -3$
6. $(5 - 7 - 17) \div (5 - 3)$
7. $-4 - (16 \div 4 + -3)$
8. $5 - 7 - 17 \div 5 - 3$
9. $-20/5 - 30 \cdot 2 - 5$
10. $-20/5 - 30 \cdot (2 - 5)$
11. $(-20/5 - 30) \cdot 2 - 5$
12. $-20/5 - (30 \cdot 2 - 5)$
13. $12/4/2$
14. $12/(4/2)$
15. $36/(12/3)$
16. $36/12/3$
17. $2/3 + 1/4$
18. $2/(3 + 1)/4$
19. $(2/3 + 1)/4$
20. $2/(3 + 1/4)$
21. $6/2 - 3/5$
22. $6/(2 - 3)/5$

Use the Order of Operations Rules to do the indicated computations. Be sure to do computations within nested parentheses first.

23. $-5 - (16 - (5 + 2 \cdot 10))$
24. $10 - (20 \div (16 - 20))$
25. $-5 - ((16 - 5 + 2) \cdot 10)$
26. $13 + 2(10 - 5(2 - 8))$
27. $13 - 2(10 - 5(2 - 8))$
28. $13 - 2(10 + 5(2 - 8))$
29. $(20 - (16/(6 - 8)) \div -2$
30. $48 \div (10 \div (5 - 3))$
31. $20 - (16/(6 - 8) \div -2)$
32. $20 - 16/((6 - 8) \div -2)$

Write each of the following expressions as a completely reduced rational number.

33. $\left(\frac{3}{4} - \frac{1}{2}\right) \div \left(\frac{1}{3} + \frac{4}{5}\right)$
34. $\left(\frac{1}{4} - \frac{2}{3}\right) \div \left(\frac{3}{4} + \frac{1}{2}\right)$
35. $\frac{34}{5} \div \left(\frac{1}{3} + \frac{4}{5}\right)$
36. $\frac{14}{25} \div \left(\frac{1}{3} - \frac{4}{5}\right)$
37. $\left(\frac{3}{4} - \frac{5}{7}\right) \div \frac{15}{28}$
38. $\left(\frac{3}{4} + \frac{5}{7}\right) \div \frac{41}{28}$
39. $\dfrac{\frac{1}{3} - \frac{4}{5}}{\frac{3}{4} - \frac{1}{2}}$
40. $\dfrac{\frac{1}{4}}{\frac{3}{4} - \frac{1}{2}}$
41. $\dfrac{-2}{3} - \dfrac{3}{\frac{3}{5} + \frac{1}{3}}$
42. $\dfrac{1}{4} - \dfrac{\frac{1}{4} - \frac{3}{2}}{\frac{1}{3}}$
43. $\dfrac{\frac{1}{5} - \frac{4}{7}}{\frac{1}{3}}$
44. $\dfrac{\frac{1}{5} - \frac{4}{7}}{13} + \dfrac{-2}{5}$
45. $\dfrac{1}{\frac{-3}{4} + \frac{1}{2}} - \dfrac{-3}{5}$
46. For the formula $A = 2LW + 2HW$, if $L = 5$, $H = 7$, and $W = 10$, find the value for A.
47. For the formula $y = -3x + 5$, if $x = -5$, find the value for y.
48. For the formula $P = 2L + 2W$, if $L = \frac{17}{4}$ and $W = 2.3$, find the value for P.

49. For the formula $y = -5x + 7$, if $x = \frac{-3}{5}$, find the value for y.

50. For the formula $y = x^2 + x$, if $x = -2$, find the value for y.

51. For the formula $y = -x^2 + x$, if $x = -2$, find the value for y.

52. For the formula $y = \frac{x - 2}{3 - x}$, if $x = -5$, find the value for y.

53. For the formula $y = \frac{3x - 2}{3 - 5x}$, if $x = 9$, find the value for y.

54. For the formula $F = \frac{|E - C|}{-(C + E)}$, if $E = -8$, and $C = -3$, find the value for F.

55. For the formula $y = \frac{x}{3 - x}$, if $x = \frac{-3}{4}$, find the value for y.

56. For the formula $T = \frac{R}{C - R}$, if $R = \frac{-3}{5}$ and $C = \frac{-3}{2}$, find the value for T.

DISCOVERY

USING THE GRAPHICS CALCULATOR: Memory

The graphics calculator uses the same Order of Operations (precedence) Rules that we used in Section 1.4. To see this for yourself, work through the following examples.

EXAMPLE 1 Compute: $3 - 4 \times 5$

Solution Type `3 - 4*5` and then press the ENTER key. The multiplication is performed to yield $3 - 20$, and then the subtraction is performed. The number `-17` is displayed to the right of the computation entered, as shown in Figure 1. ■

FIGURE 1
Order of operations yields -17

To override these rules, we can enter parentheses.

EXAMPLE 2 Compute: $(3 - 4) \times 5$

Solution Type `(3 - 4)*5` and then press the ENTER key. The computation within the parentheses is performed first (subtraction in this case), and then the multiplication is performed to yield `-5`. See Figure 2. ■

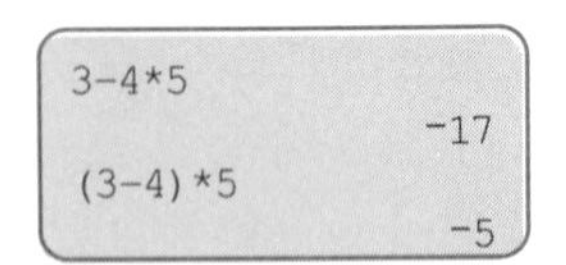

FIGURE 2

Evaluating Formulas and the STO▶ Key Graphic calculators have a collection of memory locations where values (numbers) can be stored. This feature is very useful in evaluating formulas. Consider the following example.

EXAMPLE 3 Evaluate $y = \frac{3x - 2}{x - 5}$ for the values $x = 10$, $x = -2$, $x = -8$, $x = 0$, and $x = 7$.

Solution Press the CLEAR key to erase the home screen. Next, store the value 10 in memory location X. To do this, type `10`, press the STO▶ key, and then press the X | T key to type an X. Press the ENTER key. See Figure 3. To

FIGURE 3
Use the STO▶ key to store the number 10 in memory location X.

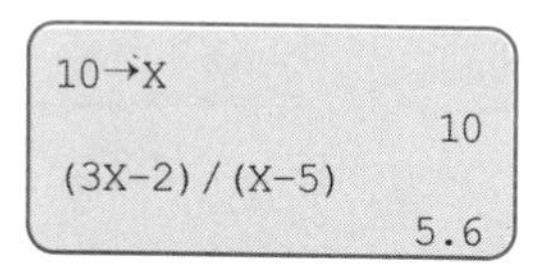

FIGURE 4

```
:Y1=
:Y2=
:Y3=
:Y4=
```

FIGURE 5

```
:Y1=(3X-5)/(X-5)
:Y2=
:Y3=
:Y4=
```

FIGURE 6

```
Y ON OFF
1:Y1
2:Y2
3:Y3
4:Y4
5:X1T
6:Y1T
7↓X2T
```

FIGURE 7

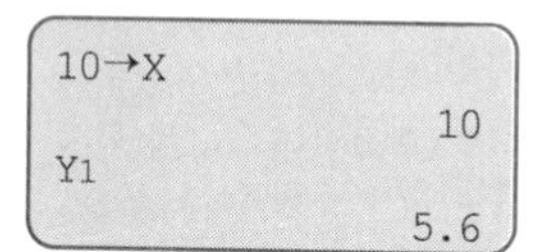

FIGURE 8

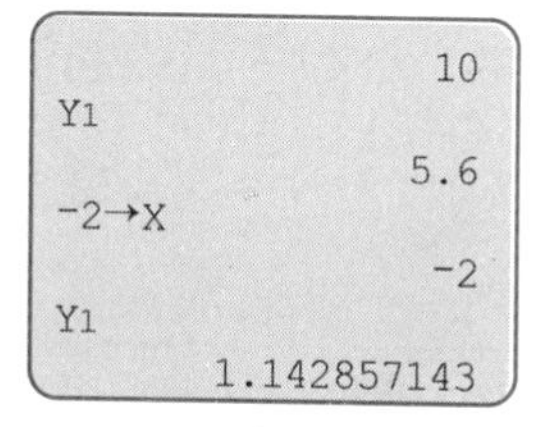

FIGURE 9

evaluate the formula for $x = 10$, type (3X − 2)/(X − 5), and press the ENTER key. The value 5.6 is written to the screen, as shown in Figure 4. Thus, for $x = 10$, the value for y is 5.6. ■

To find the value for y for the other values of x (the values -2, -8, 0, and 7) in Example 2, we could simply repeat the same steps over and over. There is a better way! We can use the Y= key to store the expression (3X − 2)/(X − 5) in a special memory location, Y1. After doing this, we simply change the value in memory location X and then display the value the calculator computes for the expression stored in memory location Y1. We will finish the solution to Example 3 in *Give It a Try* Problems 1–3.

The Y= Key Press the Y= key on the top row, and the Y= screen is displayed. See Figure 5.* If any material is entered for Y1, Y2, Y3, or Y4, press the CLEAR key to erase it. Use the up arrow and down arrow keys to move from line to line. With the cursor to the right of the :Y1= prompt, type (3X − 2)/(X − 5). See Figure 6. Press the ENTER key. Use the QUIT key to return to the home screen.

To display the special memory location Y1 on the screen, we use the calculator menu system. Press the 2nd key, followed by the VARS key, to display the Y-VARS menu. See Figure 7. To select the Y1 option from this menu, press the 1 key.** The characters Y1 are displayed.

Now, press the ENTER key. The calculator now evaluates the expression (3X − 2)/(X − 5) for the value currently stored in memory location X (which is still 10). The value 5.6 is written to the home screen. See Figure 8.

To evaluate the expression for $x = -2$, store the value ⁻2 in memory location X by pressing the (−) key, the 2 key, the STO▶ key, the X | T key, and then the ENTER key.

To display the resulting value stored in memory location Y1, press the 2nd key, followed by the VARS key, the 1 key, and the ENTER key. The value 1.142857143 is written to the home screen, as shown in Figure 9. For $x = -2$, using arithmetic to evaluate the expression, we get $y = \frac{8}{7}$. Type 8/7 and press the ENTER key. The value 1.142857143 is output. This gives us confidence that our work is correct.

TI-82 Note

*The Y= screen differs slightly.

**Use the keystrokes [2nd] [VARS] [1] [1] to type Y1 on the home screen.

GIVE IT A TRY

For the formula $y = \frac{3x - 2}{x - 5}$, use the steps from the preceding paragraph to evaluate the formula for each value of x.

1. $x = -8$ **2.** $x = 0$ **3.** $x = 7$

Formulas and Exponents To use the graphics calculator to evaluate a formula that uses exponents, we follow the same general approach used on page 51. Consider the following example.

EXAMPLE 4 For $y = x^2 - 3x + 2$, if $x = -3$, find the value for y.

Solution For this formula, if x is replaced by -3, we get $y = (-3)^2 - 3(-3) + 2$. We now use the graphics calculator to compute $(-3)^2 - 3(-3) + 2$. First, press the CLEAR key to erase the home screen.
Use the keystrokes

((-) 3) x^2 − 3 ((-) 3) + 2

to type (⁻3)² − 3(⁻3) + 2.

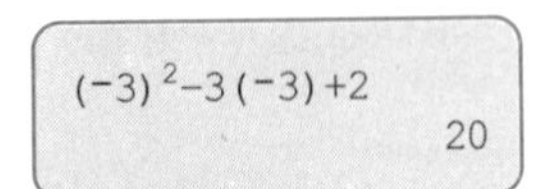

FIGURE 10

Press the ENTER key and the number 20 is output. Thus, for $x = -3$, the value for y is 20. See Figure 10. ■

Suppose we wished to find the value for y in Example 4 for several values of x. We could use arithmetic to complete the following table.

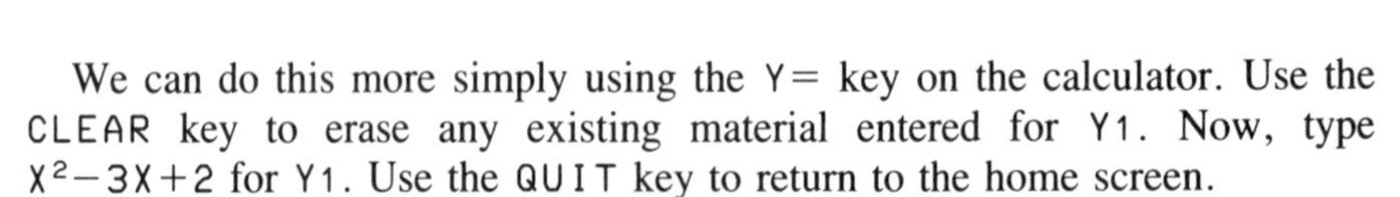

x	−3	−2	−1	0	1	2	3
y	20						

We can do this more simply using the Y= key on the calculator. Use the CLEAR key to erase any existing material entered for Y1. Now, type X²−3X+2 for Y1. Use the QUIT key to return to the home screen.

Store ⁻2 in memory location X. Display the Y1 option from the Y-VARS menu.

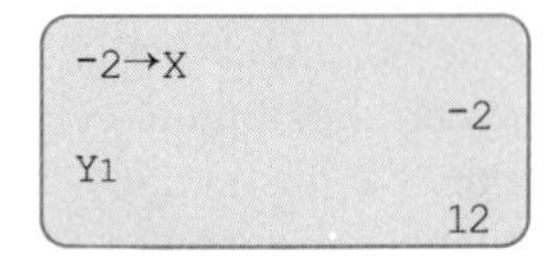

FIGURE 11

Press the ENTER key and the number 12 is output. See Figure 11. Thus, $y = x^2 - 3x + 2$ has the value 12 when x is replaced by -2. ■

GIVE IT A TRY

4. For $y = x^2 - 3x + 2$, use memory locations X and Y1 to complete the following table. Start by storing ⁻1 in memory location X.

x	−3	−2	−1	0	1	2	3
y	20	12					

Finance The formula for an amount of money A produced by money P invested at an interest rate r, compounded periodically, for a number of time periods n is $A = P(1 + r)^n$. As you can see, this formula uses an exponent. The following example shows how to use the calculator to evaluate a formula that involves an exponent.

EXAMPLE 5 If \$100 is invested at 6% interest, and the interest is compounded monthly, what is the amount of money produced after 20 years?

Solution To use the formula $A = P(1 + r)^n$, we must find the interest rate per compound period and number of time periods. Because the money is compounded monthly, the interest rate is 6% ÷ 12 months, or 0.005% per month. The number of time periods is 12 periods per year times 20 years, or 240 periods. Replacing P with 100, r with 0.005, and n with 240, we get:

$$A = 100(1 + 0.005)^{240}$$

```
100(1+0.005)^240
        331.0204476
```

FIGURE 12

To do this computation using the calculator, we use the ^ key to type `100(1+0.005)^240`. Press the ENTER key. The value 331.0204476 is output. See Figure 12.

Thus, the \$100 invested at 6%, and compounded monthly, will be worth \$331.02 after 20 years. ■

Exercises

1. The formula $A = P(1 + r)^n$ works for any quantity (not just money) that is growing at a constant rate. Suppose the population of the world is now 5.2 billion people. If population is increasing at 2% per year, then in 10 years the population will be

$$A = (5.2 \times 10^9)(1 + 0.02)^{10}$$

Use the graphics calculator to find the population in 10 years.

2. The formula for monthly payments on a loan is

$$M = \frac{rL(1 + r)^n}{(1 + r)^n - 1}$$

where M is the monthly payment
L is the loan amount
r is the monthly interest rate (the yearly interest rate divided by 12)
and n is the number of payments for the loan to be repaid

Use the graphics calculator and this formula to find the monthly payments, M, on a car loan for \$10,000 at 12% interest for 5 years. [*Hint:* Find M for $r = 0.01$, $L = 10{,}000$, and $n = 60$.]

3. Return to the problems in Exercises 1.4, and use the graphics calculator to verify each answer.

Graphics Calculator Checklist

You should now know how to use your graphics calculator to carry out the following procedures and tasks. The keys listed below are for the TI-81 and TI-82. Pencil in any differences in key names or procedures for your calculator.

ENTER
1. Use the ENTER key to execute a command that you have typed.

− (−)
2. Use the blue − key for subtraction and the gray (−) key for finding the additive opposite.

3. Recover from the error screen (either go to the error or quit the entry).

2nd ENTER
4. Use the ENTRY key (the 2nd key followed by the ENTER key) to replay the last entry (command).

DEL INS
5. Edit an entry using the DEL key or the INS key.*

ABS
6. Use the ABS key to find the absolute value of a number or expression.

x^2 ^
7. Use the x^2 key and the ^ key to evaluate exponents.

MODE
Norm Sci Eng
8. Execute computations in scientific notation by displaying the MODE screen and setting the mode to Sci. Also, be able to return to the MODE screen and set the mode back to normal (Norm).

MATH 3
9. Use the cubing option ^{3} from the MATH menu.

( )
10. Use parentheses—the (and) keys—to compute an entry in the desired sequence of operations.

STO▶ X|T
11. Use the STO▶ key and the X memory location to store a particular number in variable X for later use in a computation.

Y=
2nd CLEAR
2nd VARS-1
12. Use the Y= key to type an entry for the expression in memory location Y1. Use the QUIT key (the 2nd key followed by the CLEAR key) to return to the home screen.** Use the Y-VARS menu to select Y1 and then display its contents.***

1.5 ■ ■ ■ THE REAL NUMBERS

So far, we have reviewed the integers and the rational numbers. We now expand this collection of numbers to the real numbers. We first consider the idea of the real numbers as the union of rational and irrational numbers. We next consider algebraic numbers of the form $\sqrt[3]{5}$ and $\frac{2 + \sqrt{7}}{2}$. We then consider numbers that

TI-82 Note

*The INS key is obtained by pressing the 2nd key followed by the DEL key.

**The QUIT key is the 2nd key, followed by the MODE key.

***To display Y1 on the screen, use the keystrokes 2nd VARS 1 1 .

transcend algebraic numbers. We start with some background information.

Decimal Notation In Section 1.3 we discussed decimal notation for the rational numbers. We stated that every rational number has a decimal name. Every rational number has a decimal name that is either terminating or nonterminating and repeating. An example of a rational number with a terminating decimal name is $\frac{1}{4}$. Its decimal name is 0.25. An example of a rational number with a nonterminating and repeating decimal name is $\frac{1}{3}$. Its decimal name is 0.3333. . . or $0.\overline{3}$.

We also learned in Section 1.2 that a rational number may be plotted on a number line graph, as shown in Figure 1. Now, each plotted point corresponds to, or can be matched with, its distance from 0. In fact, the absolute value of a number a, denoted $|a|$, is the distance a is from 0.

FIGURE 1

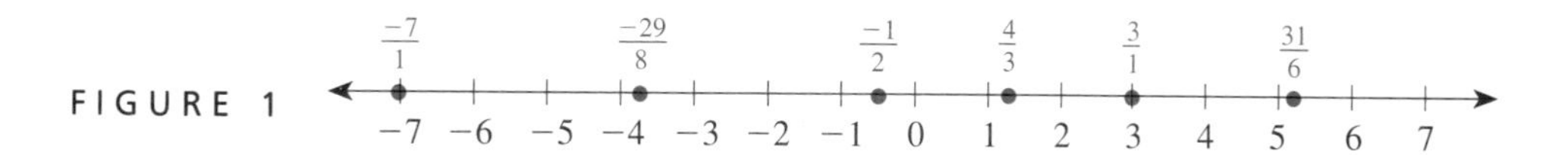

By plotting many numbers on a number line graph, we can see that the rational numbers are dense. By **dense,** we mean that between any two rational numbers, there is another rational number. For rational numbers a and b, the arithmetic average $\frac{a+b}{2}$ is a rational number, and it is between the numbers a and b. For example, consider the rational numbers 0.9 and 0.99. The average is $\frac{0.9 + 0.99}{2}$, or 0.945. We know that $0.9 < 0.945 < 0.99$, so 0.945 is between 0.9 and 0.99. It would seem that the denseness of the rational numbers would fill the number line. However, there are still other numbers on the number line that are not rational numbers, but are between rational numbers!

Consider the number r whose decimal representation (name) is 0.101001000100001. . . . This decimal number is nonterminating *and* nonrepeating. The decimal part has a pattern: the number of zeros increases by one after each successive 1 digit. However, the decimal has no repeating part, as does $\frac{13}{99}$, which is 0.13131313 . . . or $0.\overline{13}$.

Now, the number r, 0.1010010001 . . . , is some distance from 0, and thus has a place on the number line. The number 0.101001 . . . is between 0.1 and 0.2. Also, r is between 0.101 and 0.102, and so forth. We say that the numbers 0.101 and 0.102 *bound* the number r.

A number such as r is known as an **irrational number:** it cannot be written in the form $\frac{a}{b}$, where a and b are integers with $b \neq 0$. There are infinitely many irrational numbers.

We define the **real numbers** to be the union of the set of rational numbers with the set of irrational numbers. We can illustrate the set of real numbers with a Venn diagram as shown in Figure 2.

The irrational numbers are all the real numbers that are *not* rational (they do not have a decimal name that is either terminating or nonterminating and

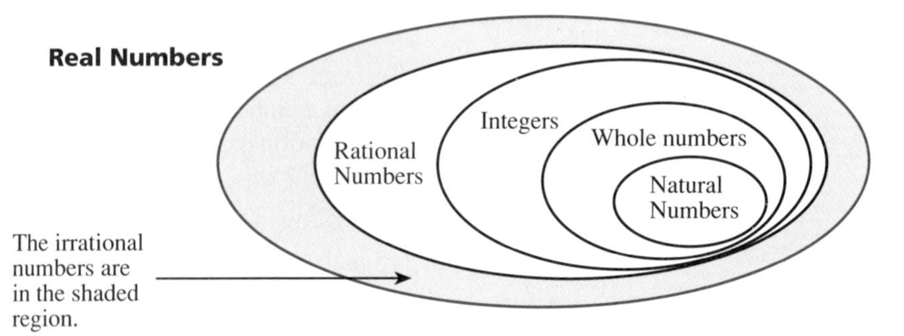

FIGURE 2

repeating). From a geometric viewpoint, the real numbers correspond to the distances of various points on the number line from 0. We will now develop several irrational numbers.

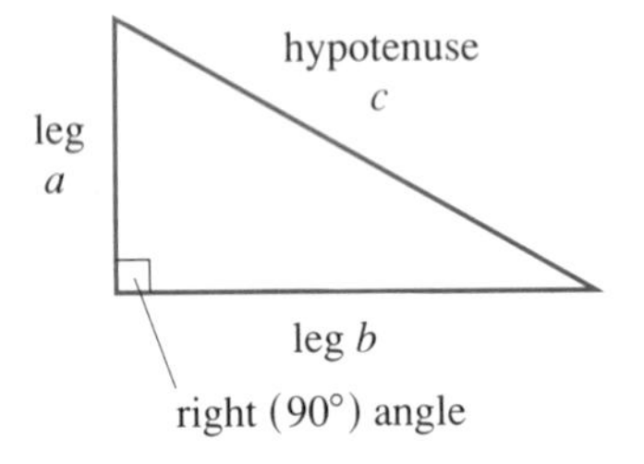

FIGURE 3

Pythagorean Theorem Consider the right triangle shown in Figure 3. The Pythagoreans were a secret fellowship following the teachings of the Greek mathematician Pythagoras (approximately 580–500 B.C.). This group proved that in any right triangle the sum of the squares of the legs equals the square of the hypotenuse, the longest side. In symbols, we write this as $a^2 + b^2 = c^2$. One such set of numbers is 3, 4, and 5. That is, if a right triangle has legs of lengths 3 inches and 4 inches, then the hypotenuse must have length 5 inches, because $3^2 + 4^2 = 5^2$, or $9 + 16 = 25$. The converse of the theorem is also true—if the sides of a triangle satisfy the relationship $a^2 + b^2 = c^2$, then the triangle must be a right triangle.

Now, the Pythagoreans, in their mysterious ways, were devoted to ratios of positive rational numbers (neither zero nor the negative integers had been introduced by 500 B.C.). Imagine their surprise when they summed the sides of a right triangle with legs 1 and 1, like the triangle in Figure 4. Try as they might, they could not find a rational number that was equal to the hypotenuse. Many proofs have been written to show that for a rational number $\frac{a}{b}$, the square of the number, $\left(\frac{a}{b}\right)^2$, cannot equal 2. The positive number that we square to get 2 is named the *square root* of 2, and is denoted $\sqrt{2}$. What the Pythagoreans realized is that the length of the hypotenuse of the right triangle with legs 1 and 1 cannot be a rational number. There must be some nonrational numbers.

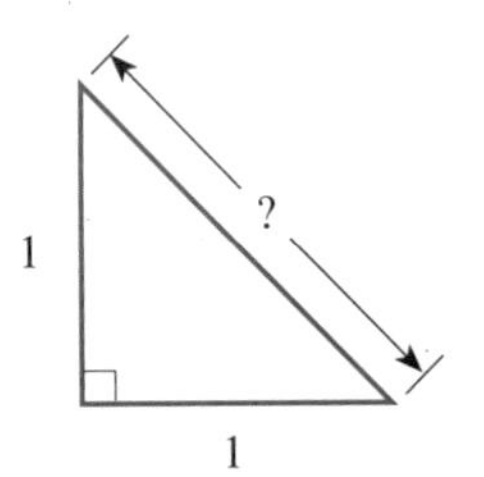

FIGURE 4

The Point $\sqrt{2}$ We can plot a point on a number line graph whose distance from 0 is $\sqrt{2}$ units. To do this, we use a right triangle with sides 1 and 1 and a compass for drawing circles. At the point 1 on the number line, we draw a vertical line segment, 1 unit long, as shown in Figure 5. Next, we draw a line segment from 0 to complete the right triangle. Using the compass, we draw a circle whose radius is the length of the hypotenuse of our right triangle. The intersection of the circle with the positive number line is the point whose distance from 0 is $\sqrt{2}$ units.

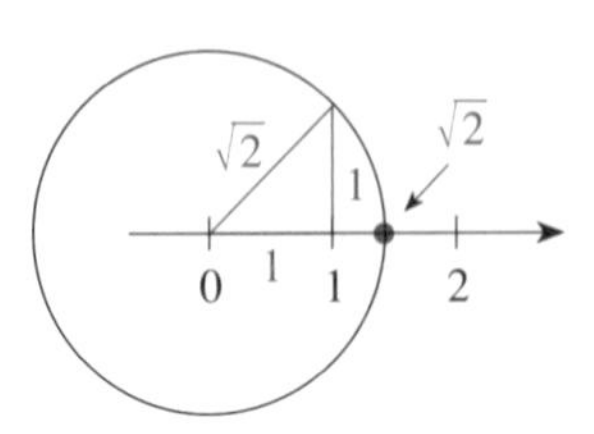

FIGURE 5

We can also plot a point on the number line graph whose distance from 0 is $\sqrt{3}$, as shown in Figure 6. If we repeat this process once more, we plot the number $\sqrt{4}$, which is the same point as 2. See Figure 7.

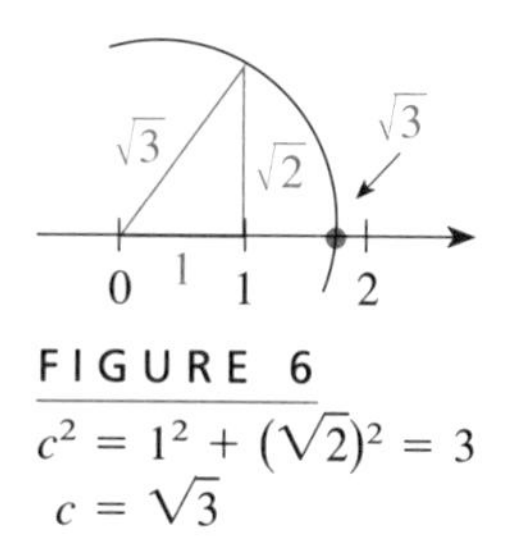

FIGURE 6

$c^2 = 1^2 + (\sqrt{2})^2 = 3$
$c = \sqrt{3}$

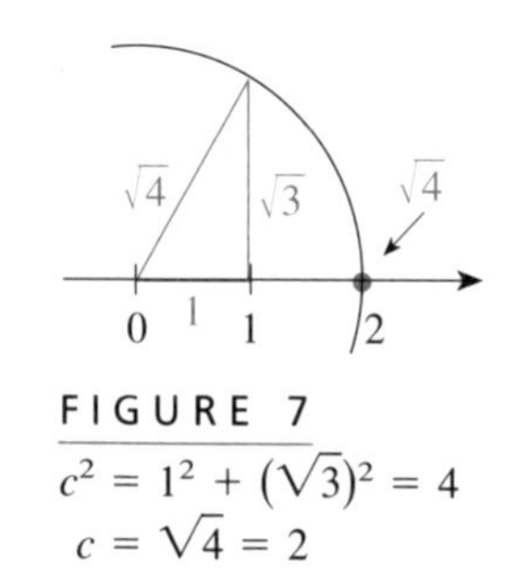

FIGURE 7

$c^2 = 1^2 + (\sqrt{3})^2 = 4$
$c = \sqrt{4} = 2$

The *n*th Root With the whole numbers, we considered the unary operation the *opposite,* and it produced numbers such as -3 and -5, which are negative integers. We now consider the unary operation, *square root,* denoted $\sqrt{a}$ (and pronounced *the square root of a*). The $\sqrt{a}$ is the positive number b such that b times b, or b^2, equals a. Operating on nonnegative rational numbers with the square root operation, we get infinitely many radicals: $\sqrt{1}$, $\sqrt{2}$, $\sqrt{3}$, $\sqrt{4}$, $\sqrt{5}$, $\sqrt{\frac{1}{3}}$, $\sqrt{\frac{1}{4}}$, $\sqrt{\frac{2}{3}}$, and so on. Now, $\sqrt{4}$ is the rational number because $(2)(2) = 4$. So, some radicals, such as $\sqrt{4}$ and $\sqrt{\frac{25}{9}}$, are rational numbers. Most aren't. If we operate on any negative rational number with the square root operation, we do not get a real number. (As we will see in Section 7.5, a nonreal, complex number is produced.)

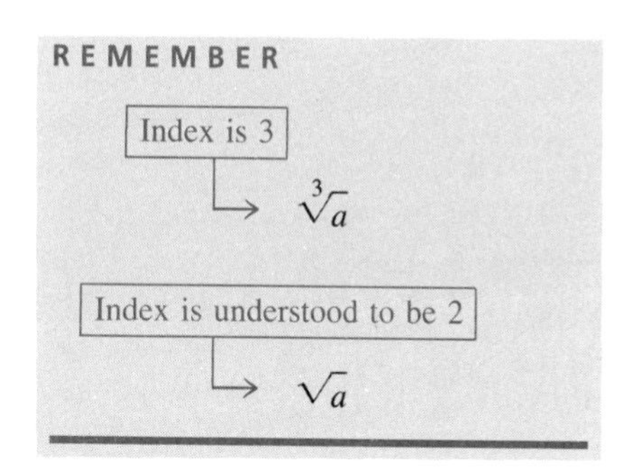

In addition to the square root, we can define the *cube root* operation as $\sqrt[3]{a}$ for a rational number a. The $\sqrt[3]{a}$ is the number b such that b times b times b equals a. Using this operation on rational numbers, we again obtain an infinite number of radicals: $\sqrt[3]{1}$, $\sqrt[3]{-1}$, $\sqrt[3]{2}$, $\sqrt[3]{-2}$, and so on. Note that unlike operating with the $\sqrt{}$, we can operate on a negative number with the $\sqrt[3]{}$. This is possible because a negative number cubed is a negative number. This process can be continued to the nth root operation for any positive integer n. The number n is known as the **index** of the radical and the number being operated on is the **radicand.**

If we operate on rational numbers a finite number of times, using the operations of addition, subtraction, multiplication, division, and nth roots, the numbers produced by these operations are known as **algebraic numbers.** These numbers will be essential in Chapter 8, when we solve general second-degree equations. Other radical and algebraic expressions are discussed in Chapter 7.

The following examples show how we use algebraic numbers.

EXAMPLE 1 For the right triangle shown in Figure 8, what is the value of x?

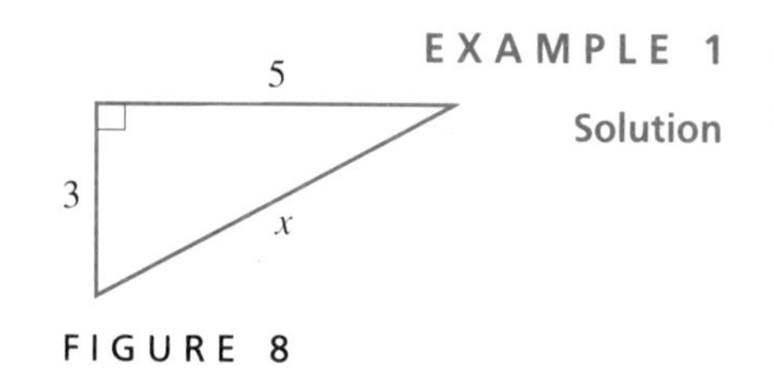

FIGURE 8

Solution We can use the Pythagorean relationship to find the length of side x:

$$x^2 = 3^2 + 5^2$$
$$x^2 = 9 + 25$$
$$x^2 = 34$$

An algebraic number such that it times itself yields 34 is $\sqrt{34}$, so x has a value of $\sqrt{34}$.

The negative number times itself that yields 34 is written $-\sqrt{34}$. Because we are finding the length of the hypotenuse, we ignore the negative number squared that yields 34. ■

EXAMPLE 2 Write each radical as a completely reduced rational number.

a. $\sqrt{64}$ b. $\sqrt[3]{27}$ c. $\sqrt{\dfrac{9}{4}}$ d. $\sqrt{\dfrac{1}{100}}$ e. $\sqrt[3]{-8}$

Solution

a. $\sqrt{64} = 8$ because $(8)(8)$ or $8^2 = 64$.

b. $\sqrt[3]{27} = 3$ because $(3)(3)(3)$ or $3^3 = 27$.

c. $\sqrt{\dfrac{9}{4}} = \dfrac{3}{2}$ because $\left(\dfrac{3}{2}\right)\left(\dfrac{3}{2}\right)$ or $\left(\dfrac{3}{2}\right)^2 = \dfrac{9}{4}$.

d. $\sqrt{\dfrac{1}{100}} = \dfrac{1}{10}$ because $\left(\dfrac{1}{10}\right)\left(\dfrac{1}{10}\right)$ or $\left(\dfrac{1}{10}\right)^2 = \dfrac{1}{100}$

e. $\sqrt[3]{-8} = -2$ because $(-2)(-2)(-2) = -8$ ■

GIVE IT A TRY

1. Name two rational numbers to the nearest thousandth that bound the number 0.2020020002 . . . .
2. Write a rational name for each radical.

   a. $\sqrt{36}$ b. $\sqrt[3]{-27}$ c. $\sqrt{\dfrac{4}{49}}$ d. $\sqrt[3]{\dfrac{1}{8}}$

3. For the right triangle shown, find the value for y.

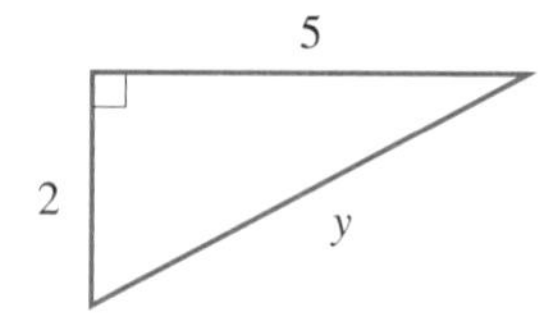

Using the Graphics Calculator: The $\sqrt{\ }$ Key The graphics calculator provides a key for producing a decimal value (often an approximation) to the square root of a number. We will also consider the ANS key. The ENTRY memory location contains the last expression entered, the ANS memory location contains the last number output. To learn how to use these keys, work through the following examples.

EXAMPLE 3 Use the graphics calculator to find a decimal approximation of $\sqrt{2}$.

Solution We start by pressing the CLEAR key to erase the home screen. Press the 2nd key followed by the x^2 key. The $\sqrt{\ }$ character is written to the screen. Now, press the 2 key followed by the ENTER key. The decimal approximation of $\sqrt{2}$ is written to the screen: 1.414213562. See Figure 9. ■

FIGURE 9

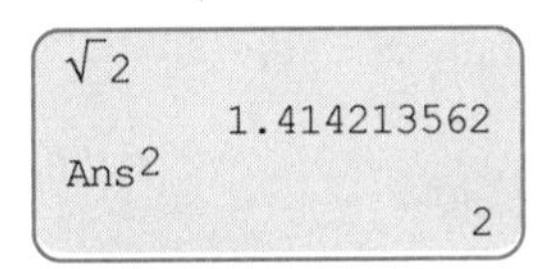

FIGURE 10

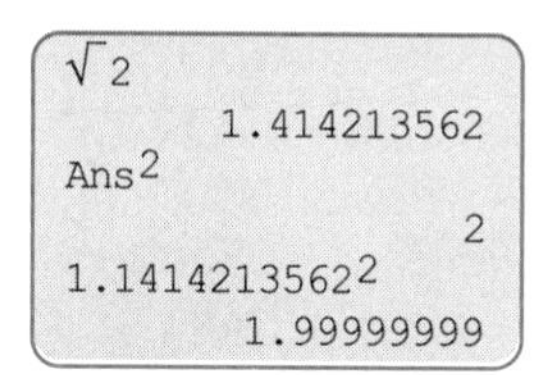

FIGURE 11

Press the x^2 key to display `Ans²` on the screen. Press the `ENTER` key. The calculator squares the number stored in the `Ans` memory location, and the number `2` is written to the screen. See Figure 10. From this output, it appears that 1.414213562 is the number that times itself yields 2. But if this were true, then $\sqrt{2}$ would have a terminating decimal name and thus would be a rational number. Our last result is due to the way the calculator rounds a number. Using paper and pencil to square 1.414213562, we would obtain the number 1.999999998944727844, not 2. Type `1.414213562` and then press the x^2 key. Now press the `ENTER` key. As you can see, the calculator reports `1.999999999`. See Figure 11.

The calculator is very useful in finding a decimal approximation of an algebraic number. Suppose we wished to find the two consecutive integers that bound $\frac{-3-\sqrt{17}}{4}$. Type `(-3-√17)/4` and then press the `ENTER` key. The value reported is `-1.780776406`. Thus, the two consecutive integers that bound $\frac{-3-\sqrt{17}}{4}$ are -2 and -1.

EXAMPLE 4 State the two consecutive integers that bound $\sqrt{50}$.

Solution Because 7^2 is 49 and 8^2 is 64, we have $7 < \sqrt{50} < 8$. So the two consecutive integers that bound $\sqrt{50}$ are 7 and 8.

On the calculator, we press the `2nd` key followed by the x^2 key (for the $\sqrt{\ }$ key). Now, type `50` and then press the `ENTER` key. The number `7.071067812` is output. This result confirms our answer that $\sqrt{50}$ is between 7 and 8. See Figure 12. ■

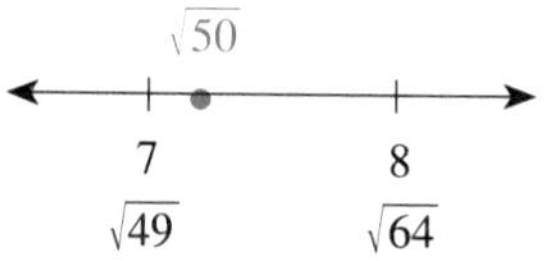

$\sqrt{50}$; 7, 8; $\sqrt{49}$, $\sqrt{64}$

FIGURE 12

EXAMPLE 5 State the two consecutive integers that bound $\frac{-7-\sqrt{5}}{2}$.

Solution Type `(-7-√5)/2` and press the `ENTER` key. The number `-4.618033989` is output. Thus,

$$-5 < \frac{-7-\sqrt{5}}{2} < -4$$

so $\frac{-7-\sqrt{5}}{2}$ is bounded by the integers -5 and -4. See Figure 13. ■

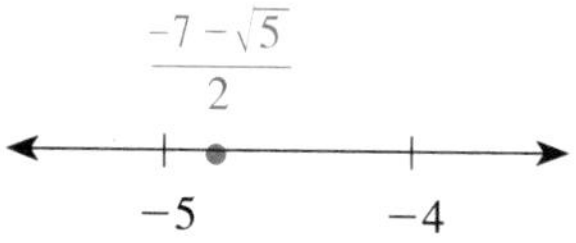

$\frac{-7-\sqrt{5}}{2}$; -5, -4

FIGURE 13

GIVE IT A TRY

4. State the two consecutive integers that bound $\sqrt{150}$.
5. State the two consecutive integers that bound $\frac{3+\sqrt{50}}{2}$.
6. State the two consecutive integers that bound $\frac{-2-\sqrt{21}}{2}$.

Rational Numbers and Radical Expressions Every integer can be written as a rational number. That is, a number such as -2 can be written as $\frac{-2}{1}$. In the same way, every rational number can be written as a radical expression. For example, $\frac{2}{3}$ can be written as $\sqrt{\frac{4}{9}}$. In fact, an infinite number of radical expressions can be used to write the rational number $\frac{2}{3}$. Some other examples are $\sqrt[3]{\frac{8}{27}}$ and $\sqrt[4]{\frac{16}{81}}$.

The Transcendental Numbers The algebraic numbers are the numbers that can be written as a finite combination of rational numbers with the operations of addition, subtraction, multiplication, division, and nth roots. Some real numbers are not algebraic numbers. One such number is π, the ratio of the circumference of a circle to its diameter. A decimal approximation for this number is 3.141592654. Another such number is 0.010010001 . . . . Numbers that are not algebraic are known as **transcendental numbers**. As we have seen, some algebraic numbers are irrational. All transcendental numbers are irrational.

Using the Graphics Calculator: Pi Graphics calculators have built-in approximations for some of the common transcendental numbers. One such real number is π, the ratio of the circumference of a circle to its diameter. To see the calculator's decimal approximation of this number, press the `2nd` key followed by the ^ key. The character π is written to the screen. Press the `ENTER` key. The number `3.141592654` is written to the screen. See Figure 14.

```
π
        3.141592654
```

FIGURE 14

It wasn't until 1767 that the number π was proved to be irrational by the Swiss-German mathematician Johann Heinrich Lambert. In 1981, Kazunori Miyoshi and Kazuhiko Nakayama calculated π to 2,000,000 decimal places without finding a repeating pattern. Since that time, American mathematicians have calculated π to an even greater number of decimal places.

Using the Graphics Calculator: *e* Another real number that is not an algebraic number is e, Euler's constant. This number is the base of the natural logarithms (see Chapter 10). To see the calculator's built-in approximation of this number, we use the e^x key. Press the `2nd` key followed by the `LN` key to access e^x. The characters `e^` are written to the screen. The caret, ^, is a prompt to enter an exponent for the number e. Press the `1` key followed by the `ENTER` key. The decimal approximation of e, `2.718281828`, is written to the screen, as shown in Figure 15.

```
π
        3.141592654
e^1
        2.718281828
```

FIGURE 15

To increase your understanding of real numbers and the absolute value operation, work through the following example.

EXAMPLE 6 Write the expression $|2\pi - e^3|$ without the use of the absolute value symbols.

Solution The absolute value of the real number $2\pi - e^3$ will be either that number or its opposite, depending on whether or not $2\pi - e^3$ is less than 0. To determine this, we can use the graphics calculator. Type `2π−e^3` and then press the `ENTER` key. The number `-13.80235162` is output. We see that $2\pi - e^3$ is less than 0, so its absolute value is the opposite of the number:

$$|2\pi - e^3| = -(2\pi - e^3)$$
$$= e^3 - 2\pi \qquad -(a - b) = b - a$$

■

REMEMBER

$$|a| = \begin{cases} a & \text{if } a \geq 0 \\ -a & \text{if } a < 0 \end{cases}$$

Reals For the real numbers, the commutative, associative, and distributive properties are all true for the operations of addition and multiplication. The number 0 is still the additive identity and the number 1 is still the identity for multiplication. Each number has an additive inverse and each nonzero number has a reciprocal (multiplicative inverse). Each of the binary operations has closure over the real numbers (the operation produces a number in the set of real numbers).

■ SUMMARY

In this section we have considered a collection of numbers that cannot be written as rational numbers, that is, they cannot be written in the form $\frac{a}{b}$, where a and b are integers with $b \neq 0$. Such a number is called an irrational number. One source of these numbers is the square root operation. The square root of a number, written $\sqrt{a}$, is the positive number b such that b times b equals a. When we operate on positive rational numbers with the square root operation, some of the numbers produced are rational numbers $\left(\text{such as } \sqrt{16} \text{ and } \sqrt{\frac{4}{9}}\right)$, but most are not. The $\sqrt{2}$, $\sqrt{3}$, and many other radicals are irrational numbers. The cube root, the fourth root, and even the nth root can be defined in the same manner as the square root operation. Any number that can be written as a finite collection of rational numbers with the operations of addition, subtraction, multiplication, division, and nth roots is known as an algebraic number.

Finally, we discussed a collection of numbers that are not algebraic numbers. These numbers, known as the transcendental numbers, include numbers like π, e, and 0.1010010001 . . . . All transcendental numbers are irrational numbers. The union of the set of rational numbers with the set of irrational numbers produces the set of real numbers.

■ *Quick Index*

1.5 ■ ■ ■ EXERCISES

Insert true *for a true statement, or* false *for a false statement.*

____ 1. The number $\sqrt{2}$ is a rational number and can be written in the form $\frac{a}{b}$ where a and b are integers with $b \neq 0$.

____ 2. For the number $\sqrt{7}$, the radicand is 7 and the index is 1.

____ 3. The number $\sqrt{5}$ means the positive number, say b, such that b^2 is 5.

____ 4. The $\sqrt[3]{-125}$ is -5, because $(-5)^3 = -125$.

____ 5. The number $\sqrt{36}$ is irrational.

____ 6. The number e is a real number.

____ 7. The numbers π and 3.14 are equal.

____ 8. For any triangle with sides a, b, and c, we have $c^2 = a^2 + b^2$.

Every rational number can be written as a radical number, just as every integer can be written as a rational number. To write $\frac{2}{7}$ as a radical number using the square root radical, we write $\sqrt{\frac{4}{49}}$. Write each fraction as a square root radical.

9. $\frac{3}{5}$

10. $\frac{5}{6}$

11. $\frac{7}{8}$

12. $\frac{2}{11}$

Use the Pythagorean Theorem $a^2 + b^2 = c^2$, for Problems 13–15.

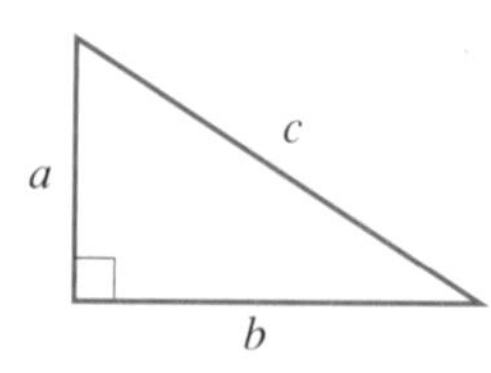

13. For the following triangle, what is the value for x?

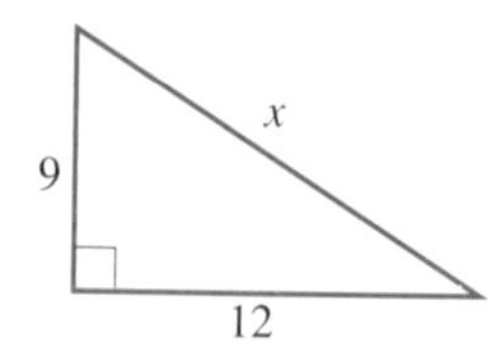

14. For the triangle shown, what is the value for z?

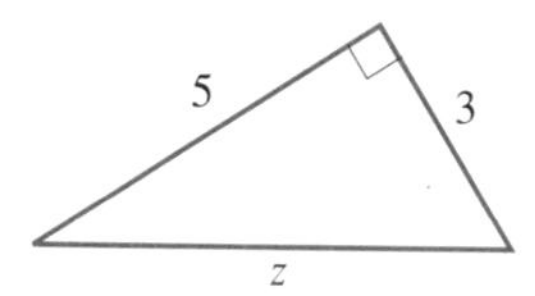

15. For the following triangle, what is the value for c?

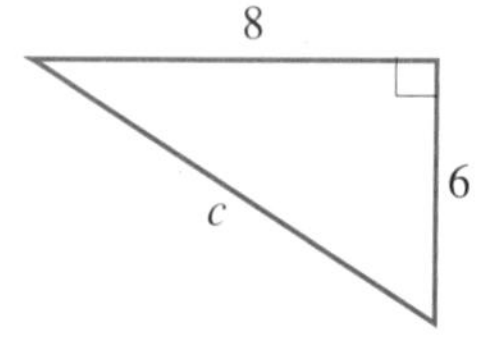

16. Use the fact that the rational numbers are dense to report a rational number between $\frac{2}{3}$ and $\frac{4}{5}$.

17. Use the fact that the rational numbers are dense to report a rational number between $\frac{-3}{7}$ and $\frac{-4}{7}$.

18. Report two rational numbers, accurate to thousandths, that bound the irrational number 0.030030003 . . . .

Use the graphics calculator to find two integers that bound the given real number.

19. $\sqrt{17}$ **20.** $\sqrt{23}$

21. $\sqrt{26}$ **22.** $\sqrt{47}$

23. $\sqrt{91}$ **24.** $\sqrt{143}$

25. $\sqrt{199}$ **26.** $\sqrt{291}$

27. $\sqrt{300}$ **28.** $\sqrt{141}$

29. $\sqrt{523}$ **30.** $\sqrt{1003}$

31. $\frac{11 - \sqrt{51}}{4}$ **32.** $\frac{13 - \sqrt{21}}{6}$

33. $\frac{5 - \sqrt{37}}{4}$ **34.** $\frac{-5 - \sqrt{80}}{6}$

35. $\frac{-14 - \sqrt{73}}{10}$ **36.** $\frac{23 + \sqrt{123}}{12}$

37. $\frac{-13 - \sqrt{26}}{9}$ **38.** $\frac{35 - \sqrt{125}}{15}$

39. $\frac{15 - \sqrt{613}}{40}$ **40.** $\frac{-50 - \sqrt{1100}}{60}$

Write each number or expression as a completely reduced rational number.

41. $\sqrt{64}$ **42.** $-8 - \sqrt{25}$

43. $35 - \sqrt{121}$ **44.** $17 - \sqrt{81}$

45. $\frac{13 - \sqrt{81}}{4}$ **46.** $\frac{-3 - \sqrt{25}}{16}$

47. $\frac{-6 - \sqrt{100}}{8}$ **48.** $\frac{16 + \sqrt{121}}{6}$

49. $\frac{7 + \sqrt{1}}{6}$ **50.** $\frac{-7 + \sqrt{49}}{5}$

Use the graphics calculator to write each number without the absolute value symbols.

Example: $|6 - 3\pi| = -(6 - 3\pi)$ *because* $6 - 3\pi < 0$
$= 3\pi - 6$

51. $|\pi - 5|$ **52.** $|3e - 2|$

53. $|3\pi - 2e|$ **54.** $|\pi^2 - e^2|$

55. $|\sqrt{19} - 5\pi|$ **56.** $|\sqrt{\pi} - 2|$

57. $|\pi^2 - \sqrt{21}|$ **58.** $|\sqrt{17} - \pi^2|$

CHAPTER 1 REVIEW EXERCISES

1. The following is a collection of real numbers:

$0,\quad 13,\quad -3,\quad -1,\quad \frac{-3}{7},\quad \frac{-\sqrt{4}}{2},\quad \sqrt{16},\quad -\sqrt{3},\quad \pi$

  a. List the whole numbers in this collection.
  b. List the integers in this collection.
  c. List the rational numbers in this collection.
  d. List the irrational numbers is this collection.

2. Identify the property illustrated by each statement.

  a. $(-2 + 5)(3) = (3)(-2 + 5)$
  b. $(-2 + 5)(3) = (-2)(3) + (5)(3)$
  c. $-3 + (5 + 7) = (-3 + 5) + 7$
  d. $-3 + (5 + 7) = -3 + (7 + 5)$
  e. $(-3)(1) = -3$

Carry out the following computations.

3. $-13 + 8$
4. $-5(-9)$
5. $-20 \div -5$
6. $20 - (-9)$
7. $|-3|$
8. $5 - |8|$
9. $|-(-3)|$
10. $-|-17|$
11. $3 - |5 - (-6)|$
12. $-|-2 - 9|$
13. $12 - (5 - 9)$
14. $3(-5 + 8)$
15. $24 \div (6 - 3)$
16. $24 \div 6 - 3$
17. $24 \div 8 \div 2$
18. $24 \div (8 \div 2)$
19. $(20 - 10)(8 - 13)$
20. $38 - (10 \div -5)$
21. $24 - 12 \div (-3)(6)$
22. $120 \div (40 + (-20))$

Write each fraction as a completely reduced rational expression.

23. $\frac{-30}{21}$
24. $\frac{-42}{-24}$
25. $\frac{180}{12}$
26. $\frac{12 - 18}{10}$
27. $\frac{10 - 28}{12 - 6}$
28. $\frac{15 - 28}{32 - 6}$

Write each expression as a completely reduced rational expression.

29. $\frac{-3}{7} + \frac{17}{7}$
30. $\frac{2}{5} - \frac{4}{9}$
31. $\frac{2}{3} - \frac{7}{2}$
32. $\frac{3}{5} \div \frac{-30}{7}$
33. $\left(\frac{2}{3}\right)\left(\frac{-3}{7}\right)$
34. $\left(\frac{7}{8}\right)\left(\frac{1}{3} - \frac{2}{5}\right)$
35. $1 \div \frac{-3}{2}$
36. $1 \div \left(\frac{-3}{10} - \frac{2}{5}\right)$
37. $\frac{3}{14} \div \left(\frac{-30}{7}\right)$
38. $\frac{2}{5} \div \left(1 \div \frac{2}{3}\right)$

Round each number as indicated.

39. 23.0089 to thousandths
40. 0.003456 to 2 significant digits
41. 23,780 to 3 significant digits
42. 23,785.6 to tens
43. 23,785.6 to hundreds
44. 0.003452 to 3 significant digits

Write each number in percent notation.

45. 0.035
46. 2.3
47. $\frac{3}{12}$
48. 0.000126
49. 5
50. $\frac{47}{20}$

Report the value of the given percent expression.

51. 37% of 90
52. 30% of 15
53. 200% of 0.3
54. 0.5% of 0.5
55. 0.01% of 7
56. 500% of 0.08

57. Kelly earns $300 a week plus a 3% commission on her total sales. If she had sales worth $20,000 last week, what was her pay for the week?

58. The sales tax on a new car is 5% of the price. For a car priced at $13,500, how much is the sales tax?

59. If 30 liters of 60% acid solution is mixed with 40 liters of 90% acid solution, what is the acid concentration of the resulting mixture?

60. Bill drives 3 hours at 50 mi/h, then increases his speed by 30% and drives another 4 hours. How many miles does he travel?

61. Joan's paycheck for last week was $560. If 20% of this money is a special bonus, how much is Joan's regular pay?

62. Ms. Wu has $20,000 invested in an account paying 8% interest, compounded yearly. She also has $30,000 invested in a city bond that pays 5% interest, compounded yearly. How much will Ms. Wu earn in interest from these investments this year?

Write each number or expression in scientific notation.

63. 23,000,000
64. 156,000
65. 0.0004056
66. 0.02013
67. 58
68. 3.2001
69. $(3.0 \times 10^2)^3$
70. $(2 \times 10^{-2})^{-1}$
71. $(4.56 \times 10^4)(2 \times 10^3)$
72. $(3.8 \times 10^{-2}) \div (2 \times 10^{-3})$

Round each number as indicated.

73. 345.34 to 2 significant digits
74. 0.23589 to hundredths
75. 0.0005672 to 3 significant digits
76. 34,560,786 to thousands
77. Draw a number line graph and plot these numbers.

 a. -2 b. $\sqrt{5}$ c. $\dfrac{-3\pi}{7}$ d. $\dfrac{-\sqrt{14}}{2}$

78. Report the two consecutive integers that bound the given number.

 a. $-\sqrt{17}$ b. $\dfrac{-2+\sqrt{13}}{2}$

Write each expression as a completely reduced rational number.

79. $-5 - \dfrac{\frac{-3}{5}}{\frac{1}{2} - 2}$
80. $\dfrac{3}{\frac{1}{2} + \frac{-3}{5}} - 1$
81. $\dfrac{\frac{1}{3} + \frac{-1}{4}}{\frac{5}{6} - \frac{1}{2}}$

Write a completely reduced rational number for each expression.

82. $\sqrt{36}$
83. $5 - \sqrt{100}$
84. $\sqrt{81}$
85. $-\sqrt{16}$
86. $\dfrac{10 - \sqrt{49}}{2}$
87. $\dfrac{8 - \sqrt{64}}{6}$
88. $\dfrac{-4 + \sqrt{81}}{2}$
89. $\dfrac{18 - \sqrt{100}}{24}$
90. $\dfrac{-2 + \sqrt{36}}{2}$
91. $\dfrac{10 + \sqrt{64}}{8}$

CHAPTER 1 TEST

1. The following is a collection of real numbers:

$$-3, \quad 0, \quad \sqrt{36}, \quad e, \quad \frac{\sqrt{5}}{3}, \quad \frac{8-\sqrt{25}}{6}, \quad \pi$$

 a. List the whole numbers in this collection.
 b. List the integers in this collection.
 c. List the rational numbers in this collection.
 d. List the irrational numbers in this collection.

2. Identify the property illustrated by each statement.

 a. $(3)(9-5) = (3)(9) - (3)(5)$
 b. $-12(5 \cdot 3) = (-12 \cdot 5)(3)$

Carry out the following computations.

3. $-13 - 8 + 9$
4. $-5 - (-9)(-6)$
5. $-20 \div (-5 + 12 \div 4)$
6. $-|20 - (-9)|$

Write each expression as a completely reduced rational number.

7. $\dfrac{5}{3} + \dfrac{-7}{4}$
8. $\left(\dfrac{1}{2}\right)\left(\dfrac{-5}{9} - \dfrac{5}{3}\right)$
9. $3 - \left(2 \div \dfrac{-4}{11}\right)$
10. $\dfrac{\frac{-1}{4} + 2}{\frac{-1}{3} - 2}$

11. If 20 liters of 30% acid solution is mixed with 10 liters 80% acid solution, what is the concentration of the resulting mixture?

Write each number or expression in scientific notation.

12. 2,378,000
13. 0.0000456
14. $(8.1 \times 10^4)^2$
15. $(2.3 \times 10^{-2}) + (5.1 \times 10^{-3})$
16. $(3.6 \times 10^4) \div (2 \times 10^{-1})$
17. $(5.9 \times 10^3) - (6 \times 10^4)$
18. Round 0.00345 to 2 significant digits.
19. Report two consecutive integers that bound the given number.

 a. $5 - \sqrt{20}$ b. $\dfrac{-\sqrt{36}}{5}$ c. $\dfrac{5 - \sqrt{13}}{2}$

20. Write each number or expression as a completely reduced rational number.

 a. $\sqrt{81}$ b. $\dfrac{51 - \sqrt{121}}{20}$ c. $\dfrac{15 - \sqrt{49}}{6}$

2

SOLVING FIRST-DEGREE EQUATIONS IN ONE VARIABLE

Solving equations is an important skill in algebra. In this chapter, we will learn how to solve a first-degree equation in one variable, such as $3x - 5 = 16 - x$. We start with the main skill required for solving first-degree equations—addition and subtraction of polynomials. Next, we discuss solving first-degree equations. We then consider direct applications of first-degree equations in one variable: solving first-degree inequalities in one variable, solving word problems by creating mathematical models, and rewriting formulas.

2.1 ■ ■ ■ ADDITION AND SUBTRACTION OF POLYNOMIALS

Most of our work in algebra will involve the numbers of arithmetic (the real numbers) and variables. A **variable** is simply a placeholder for a number or for a collection of numbers. Variables are often denoted by symbols such as x or y, but any character or shape (such as $\square$) could be used. When we operate on variables and numbers with a finite number of the operations of addition, subtraction, multiplication, division, and nth roots (such as square roots), we form *algebraic expressions,* like these:

$$2x^2 - 2x + 5 \qquad \frac{3x + 1}{x^2 - 2} \qquad \sqrt{x^3 - 5x - 1}$$

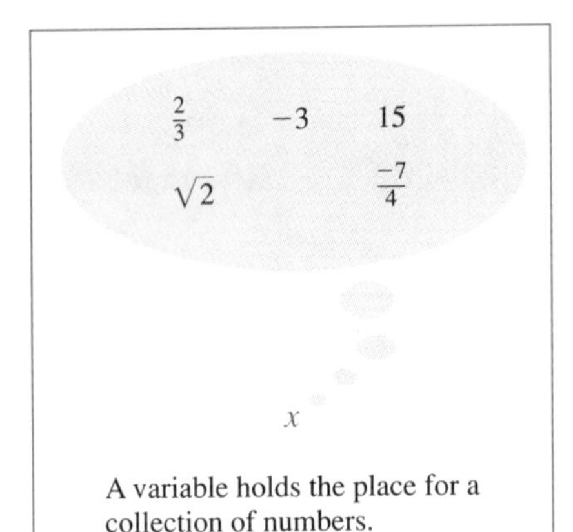

A variable holds the place for a collection of numbers.

The expression 2^x is *not* an algebraic expression because the operation of raising a number to a variable exponent is not one of the operations that forms algebraic expressions.

Terms and Polynomials To define a polynomial, we start by defining its terms. A **term of a polynomial** is a number or a product of a number with one or more variables. For example, $3x^2y^4$ is a term. Notice that $3x^2y^4$ is a shorthand way of writing $3xxyyyy$. In this expanded form we can see that $3x^2y^4$ is a product of a number and two variable factors, x and y. The exponent for each of the variables must be a whole number. The number factor of a term is known as the **coefficient** of the term (for example, the 3 in $3x^2y^4$ is called the coefficient).

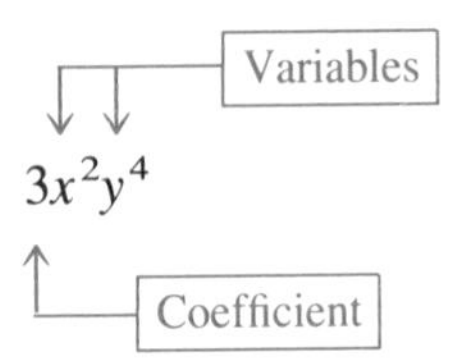

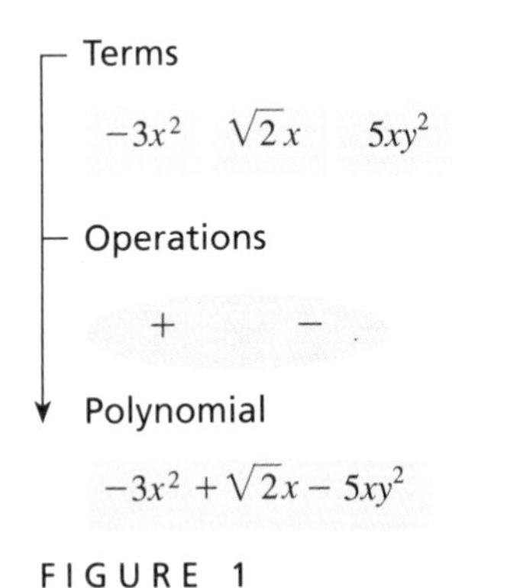

FIGURE 1

Other examples of a term are $3x^2$, $5x^3y^2$, $\sqrt{3}\,x$, and $\frac{1}{5}x$. A term that has no variable factor and is only a number is known as a **constant term**.

A **polynomial** is either a single term, or is a finite sum or difference of several terms. (The prefix *poly-* means many; a polynomial is many terms.) For example, in Figure 1, we have combined the terms $-3x^2$, $\sqrt{2}\,x$, and $5xy^2$ using addition and subtraction to produce the polynomial $-3x^2 + \sqrt{2}\,x - 5xy^2$. Other examples of polynomials are

$$3x^3 - 5x + 2 \qquad 5xy - x^2 + y^3 - 2$$

$$w + z - 2xy \qquad \sqrt{2}\,x - 1$$

Examples of expressions that are *not* polynomials are $\frac{1}{x}$ and $\sqrt{xy}$. Each of these is an algebraic expression, but not a polynomial, because neither is a term of a polynomial. Neither division by a variable nor the square root or nth root of a variable is allowed in a term of a polynomial. Only the multiplication operation is allowed.

$(-3)^2 - 2(-3)$

$4^2 - 2(4)$

$x^2 - 2x$

A polynomial holds the place for a collection of number phrases.

Evaluating a Polynomial Just as a variable holds the place for a collection of numbers, a polynomial holds the place for a collection of number phases. For instance, the polynomial $x^2 - 3x + 5$ represents all number phrases such as these:

$$(-2)^2 - 3(-2) + 5 \qquad \left(\frac{2}{3}\right)^2 - 3\left(\frac{2}{3}\right) + 5 \qquad \left(\sqrt{5}\right)^2 - 3\left(\sqrt{5}\right) + 5$$

To **evaluate** a polynomial such as $3xy - y^2 + x^2$ for $x = 2$ and $y = -3$ means to create a number phrase by replacing the x with 2 and the y with -3, and then compute the resulting number phrase.

EXAMPLE 1 Evaluate $x^2 - 2x - y + 1$ for $x = -1$ and $y = 3$.

Solution We replace the x with -1 and the y with 3 to form the number phrase $(-1)^2 - 2(-1) - 3 + 1$. Now we perform the operations, and get 1. Thus, for $x = -1$ and $y = 3$, the polynomial $x^2 - 2x - y + 1$ evaluates to the number 1. ■

Using the Graphics Calculator

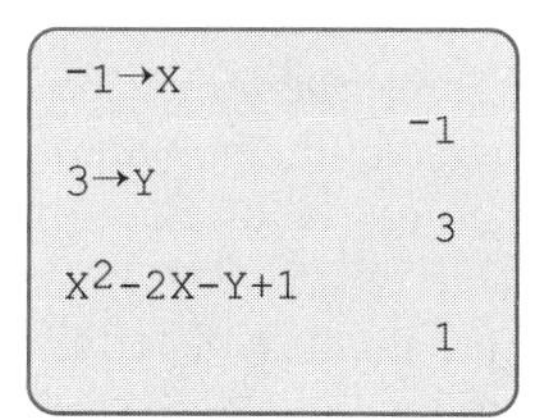

FIGURE 2

To check our work in Example 1, store -1 in memory location X. Type -1, press the STO▶ key, press the X | T key, and then press the ENTER key. Store 3 in memory location Y. Pressing the STO▶ key sets the calculator in ALPHA mode.* So, to type Y in 3→Y, we simply type 3, then press the STO▶ key, followed by the 1 key (for Y). Press the ENTER key to store the value.

To evaluate $x^2 - 2x - y + 1$, type X²−2X−Y+1. To type the Y in X²−2X−Y+1, press the ALPHA key followed by the 1 key. Check the expression, and correct any errors by using the INS and DEL keys. Then press the ENTER key. The value 1 is output. See Figure 2.

GIVE IT A TRY

Evaluate each polynomial for the indicated replacement of the variable(s).

1. $2x^2 - 3x - 5$ for $x = -2$
2. $3x - 5y$ for $x = 4$ and $y = -3$
3. $x^2 - y^2 - 3y$ for $x = 5$ and $y = 3$
4. $2w^3 - 3w - 8$ for $w = 10$
5. Use the graphics calculator to check your work in Problems 1–4.

Nomenclature The naming system for polynomials involves the number of terms, the number of variables, and the degree. The **degree of a term** is the sum of the exponents that appear on the variables in the term. For example, the degree of $3x^2y^5$ is 7, because 2 + 5 equals 7:

The degree of term is sum of exponents.

$3x^2y^5$ has degree 7

The degree of the term $3x$ is 1, because $3x$ is thought of as $3x^1$. The degree of a constant term like -5 is 0, because -5 is shorthand for $-5x^0$. Although the number 0 is a term (and thus a polynomial), its degree is undefined.

TI-82 Note *To type a letter of the alphabet, first press the ALPHA key. To type 3→Y, use the keystrokes

The **degree of a polynomial** is the maximum (largest) of the degrees of all of the terms in the polynomial:

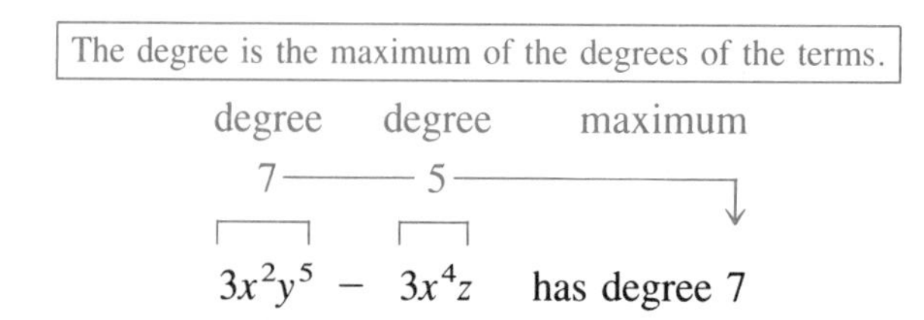

The degree of $3x^3 - 5x^2 - x + 2$ is 3, because the degrees of the terms are 3, 2, 1, and 0, and the largest of these numbers is 3. The degree of the polynomial $3x^2y^3 - y^2 + x^6 + 1$ is 6, because the maximum of the degrees of the terms 5, 2, 6, and 0 is 6.

In addition to its degree, a polynomial is named by the number of terms it contains. A one-term polynomial is known as a **monomial** (*mono-* meaning one), a two-term polynomial is a **binomial** (*bi-* meaning two), a three term polynomial is a **trinomial** (*tri-* meaning three), and a four term polynomial is a **quadrinomial** (*quadri-* meaning four). Some examples of polynomials and their names are shown here:

Polynomial	*Number of terms*	*Degree*	*Number of variables*	*Name*
$3x^2 - 3x + 1$	3	2	1	trinomial second degree in one variable
$2xy^2 + 5x^4$	2	4	2	binomial fourth degree in two variables
$5x^3yz$	1	5	3	monomial fifth degree in three variables

Standard Form of a Polynomial Although $3x - 2x^2 + 1$ is a polynomial, it is not written in proper form. (This is much like the rational number $\frac{8}{6}$, which has the completely reduced name $\frac{4}{3}$.) Standard form for this polynomial is $-2x^2 + 3x + 1$. For a **standard polynomial** in one variable, the terms appear in descending order of degree—the highest degree term, then the next highest degree term, and so forth, ending with the lowest degree term (often the constant term or zero degree term).

Addition of Polynomials Once we have established a naming system, we can discuss operations on polynomials. The first operation on polynomials is that of addition. Because a variable holds the place for a number, all the properties of numbers are valid for variables: commutative properties, associative properties, and distributive properties. The next example shows how these properties are used.

EXAMPLE 2 Write as a standard polynomial: $-2x^2 + 5x + x^2 + 1$

Solution

$$-2x^2 + 5x + x^2 + 1 = -2x^2 + x^2 + 5x + 1 \quad \text{Commutative property of addition}$$
$$= (-2 + 1)x^2 + 5x + 1 \quad \text{Distributive property}$$
$$= -1x^2 + 5x + 1 \quad \text{Addition fact: } -2 + 1 = -1$$

Thus, the standard polynomial for $-2x^2 + 5x + x^2 + 1$ is $-1x^2 + 5x + 1$. ■

Fortunately, we do not usually have to proceed step-by-step, as shown in Example 2, in order to add polynomials. The method we used is summarized by the phrase *combine like terms*. Terms are **like terms** provided they have the same variables to the same exponents (only their coefficients can be different). For example, $3xy^2$ and $-5xy^2$ are like terms, but $3xy^2$ and $-5x^2y$ are *not* like terms: they have the same variables, but those variables have different exponents.

To write the polynomial $-2x^2 + 5x + x^2 + 1$ as a standard polynomial, we combine the like terms:

Combine the x^2 terms using the distributive property.

$$-2x^2 + 5x + x^2 + 1 = -1x^2 + 5x + 1$$

The meaning of this equality is twofold. First, anywhere the polynomial $-2x^2 + 5x + x^2 + 1$ appears, it can be replaced by the polynomial $-1x^2 + 5x + 1$. This meaning will be important in Section 2.2, when we solve equations involving polynomials. Second, the statement means that for any replacement of x by a number, say -3, the polynomial $-2x^2 + 5x + x^2 + 1$ will yield, or evaluate to, the same number as the polynomial $-1x^2 + 5x + 1$. Using this last meaning and the graphics calculator, we can test our work for errors.

```
:Y1=-2X²+5X+X²+1
:Y2=⁻1X²+5X+1
:Y3=
:Y4=
```

FIGURE 3

```
-3→X
                -3
Y1
               -23
Y2
               ⁻23
```

FIGURE 4

To check our work in Example 2, we can evaluate the original polynomial and the standard polynomial for the same value. Press the Y= key, and erase any expression entered for Y1, Y2, Y3, and Y4. For Y1 type -2X²+5X+X²+1, and for Y2 type ⁻1X²+5X+1. See Figure 3. Return (quit) to the home screen. Store a number, say ⁻3, in memory location X. Use the Y-VARS menu to display Y1 on the screen. Press the ENTER key. The value ⁻23 is displayed. Use the Y-VARS menu to display Y2 on the screen. Press the ENTER key. The value ⁻23 is displayed. Figure 4 displays the results—replacing the variable x by -3 yields -23 for both the original polynomial $-2x^2 + 5x + x^2 + 1$ and the standard polynomial, $-1x^2 + 5x + 1$. This should give us confidence that our work is correct.

The screen in Figure 4 shows that the two polynomials evaluate to the same number when x is replaced by -3. Try storing a different number in `X`, say 7, and evaluating the two polynomials to see that they yield the same number. If these two expressions do not evaluate to the same number for a particular replacement of the variable, then an error is present. Either an incorrect character has been typed, or we have not combined like terms correctly to write the polynomial in standard form.

The next example involves the addition operation on polynomials.

EXAMPLE 3 Write as a standard polynomial:

$$(-3x^2 - 5x - 2) + (-x^3 - 2x^2 + 3x - 6)$$

Solution To write this as a standard polynomial, we must follow the form

$$\square x^3 + \square x^2 + \square x + \square$$

We first combine the like terms, then arrange the terms in decreasing order, based on the degree of each term.

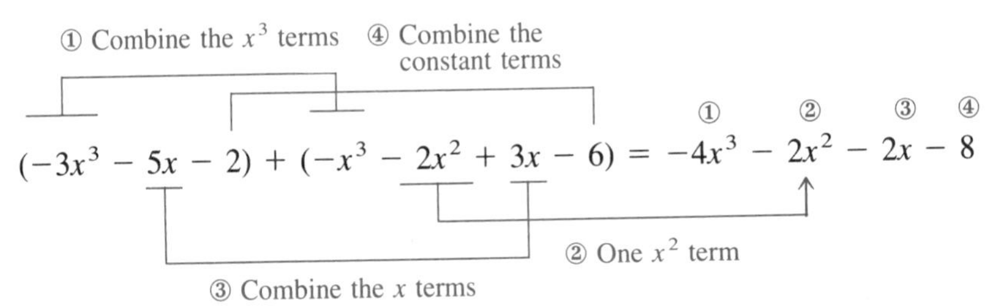

Thus, for the polynomial

$$(-3x^3 - 5x - 2) + (-x^3 - 2x^2 + 3x - 6)$$

the standard polynomial is

$$-4x^3 - 2x^2 - 2x - 8$$ ■

Using the Graphics Calculator

```
:Y1=(-3X3-5X-2)+
(-X3-2X2+3X-6)
:Y2=-4X3-2X2-2X-
8
```

FIGURE 5

We can use the graphics calculator to test our work in Example 3. Press the `Y=` key, and type `(-3X³-5X-2)+(-X³-2X²+3X-6)` for `Y1`. (To type the 3 exponent, either press the `MATH` key followed by the 3 key, or use the ^ key and type `X^3` for x^3.)

For `Y2`, type `-4X³-2X²-2X-8`. See Figure 5. Return to the home screen, and then store a number, say 10, in memory location `X`.

Use the `Y-VARS` menu to display `Y1`. Press the `ENTER` key. The value `-4228` is displayed. Use the `Y-VARS` menu to display `Y2`. Press the `ENTER` key. The value `-4228` is displayed. See Figure 6.

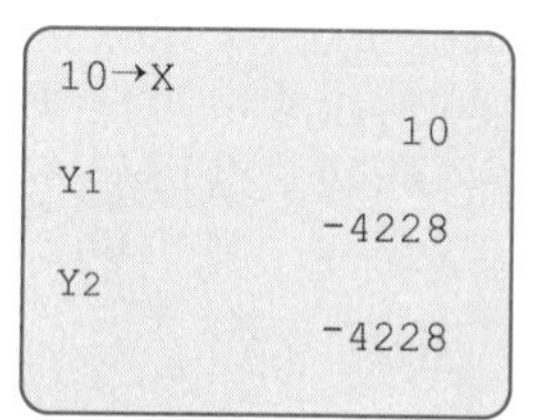

FIGURE 6

Since replacing the variable x by 10 yields -4228 for both

$$(-3x^3 - 5x - 2) + (-x^3 - 2x^2 + 3x - 6)$$

and

$$-4x^3 - 2x^2 - 2x - 8$$

We have confidence that our work is correct.

The advantage of using the Y1 and Y2 memory locations is that if an error has been made, and the outputs are not the same, we can enter a correction for Y2, and run the test again.

GIVE IT A TRY

6. A standard polynomial in one variable has *no* like terms (they have been combined) and is written in decreasing order of the terms. A standard polynomial in two or more variables is also written without any like terms and in decreasing order of the terms. Write the sum as a standard polynomial:

$$(x^2 + 3xy - 2y^2) + (-xy + 3x^2 + y)$$

7. Write each expression as a standard polynomial.

a. $(-x^3 - 5x - 2) + (-x^3 - 2x^2 + x + 6)$

b. $(-5z^2 - 5xz - 2) + (-x^3 - 2z^2 + 3x - 6xz)$

8. Use the graphics calculator to confirm your work in Problems 6 and 7. For Problem 6, store -5 in memory location X and 3 in memory location Y. Report the values produced by the given expression and by the standard polynomial. [*Hint:* Use ALPHA 2 to type the Z character.]

Subtraction of Polynomials To perform subtraction of polynomials, we use the definition of subtraction. That is, we combine finding the opposite of a polynomial with addition of polynomials.

The definition of subtraction is $a - b = a + (-b)$.

Example: $7 - 5$ means $7 + (-5)$

To find the opposite of a number, we can multiply the number by -1. That is,

$$a - b \text{ means } a + (-1)b$$

Example: $7 - 5$ means $7 + (-1)(5)$ or $7 + (-5)$

Since variables hold the place for numbers and polynomials hold the place for number phrases, we can use the definition of subtraction with polynomials as we did with numbers. We start by finding the opposite of a polynomial.

EXAMPLE 4 Find the opposite of the polynomial $-2x^2 + 6x - 1$.

Solution To find the opposite of $-2x^2 + 6x - 1$, we multiply the polynomial by -1. Using the distributive property, we get:

$$-1(-2x^2 + 6x - 1) \quad \text{or} \quad 2x^2 - 6x + 1$$

Thus, the opposite of $-2x^2 + 6x - 1$ is $2x^2 - 6x + 1$. ■

In our first example of subtraction of polynomials, we will show our work in a column format so that the process is easy to follow. Example 5 uses the our work from Example 4 as the first step of the subtraction operation, finding the opposite of the polynomial.

EXAMPLE 5 Write as a standard polynomial: $(-3x^2 + 5x - 2) - (-2x^2 + 6x - 1)$

Solution We start by setting up the problem in column format:

$$\begin{array}{rr} & -3x^2 + 5x - 2 \\ - & -2x^2 + 6x - 1 \\ \hline \end{array} \qquad \begin{array}{rr} & -3x^2 + 5x - 2 \\ + & 2x^2 - 6x + 1 \\ \hline & -x^2 - x - 1 \end{array}$$

Change subtraction to addition; Change to opposite; Do addition.

We can also write this work in linear form:

$$(-3x^2 + 5x - 2) - (-2x^2 + 6x - 1) = -3x^2 + 5x - 2 + 2x^2 - 6x + 1$$

Change subtraction to addition; Change to opposite

$$= -x^2 - x - 1 \qquad \text{Do addition.}$$

Thus, $(-3x^2 + 5x - 2) - (-2x^2 + 6x - 1)$ as a standard polynomial is $-x^2 - x - 1$. ■

WARNING
A common error is to evaluate $-x^2$ for $x = -3$ as 9. However, $-x^2$ means $-1xx$, and replacing x by -3 yields $-1(-3)(-3)$, or -9. If you have trouble with this error, it is helpful to write $-x^2$ as $-1x^2$.

Using the Graphics Calculator

To test our work in Example 4, type `-(-2X²+6X-1)` for `Y1`. Now, type `2X²-6X+1` for `Y2`. Store a number, say `-2`, in memory location `X`. Display the value for `Y1`. The value `21` is output. Display the value for `Y2`. Again, the value `21` is output.

To test our work in Example 5, type

`(-3X²+5X-2)-(-2X²+6X-1)`

for `Y1`. Now, type `-X²-X-1` for `Y2`. Store a number, say `7`, in memory location `X`. Display the value for `Y1`. The value `-57` is output. Display the value for `Y2`. Again, the value `-57` is output. This result confirms that our work is correct for $x = 7$. You may want to run the test again for $x = -3$.

The preceding approach to the subtraction operation is very useful in working with more advanced situations. Consider the next example.

EXAMPLE 6 Write as a standard polynomial: $-3x^2 - (5 - (2x^2 - 3x + 7) - 3x)$

Solution We first find the opposite of the expression in the innermost parentheses:

$$-3x^2 - (5 \underline{- 2x^2 + 3x - 7} - 3x)$$

Opposite of $2x^2 - 3x + 7$

Now, we find the opposite of the expression in the remaining parentheses:

$$-3x^2 - 5 + 2x^2 - 3x + 7 + 3x$$

Finally, combining like terms yields

$$-x^2 + 2$$

Thus, the polynomial $-3x^2 - (5 - (2x^2 - 3x + 7) - 3x)$ written as a standard polynomial is $-x^2 + 2$. ■

We can reverse the procedure we used in Example 6, first combining like terms and then clearing parentheses. Example 7 uses this method on the polynomial in Example 6.

EXAMPLE 7 Write as a standard polynomial: $-3x^2 - (5 - (2x^2 - 3x + 7) - 3x)$

Solution Starting with the expression in the innermost parentheses, we find its opposite:

$$-3x^2 - (5 - 2x^2 + 3x - 7 - 3x)$$

Combining like terms within the remaining parentheses, we get

$$-3x^2 - (-2x^2 - 2)$$

Clearing the last set of parentheses, we get

$$-3x^2 + 2x^2 + 2$$

Finally, combining like terms yields $-x^2 + 2$. ■

Using the Graphics Calculator

To test our work in Examples 6 and 7, type

```
-3X²-(5-(2X²-3X+7)-3X)
```

for `Y1` and type `-X²+2` for `Y2`. Store a number, say -6, in memory location `X`. Display the value, for `Y1` and for `Y2`. Is the same number output for both? If not, an error is present. If an error is found when you check your work, make any necessary corrections and run the test again.

As Examples 6 and 7 show, there is usually more than one way to approach a given problem. The following examples suggest more than one way to approach each problem.

EXAMPLE 8 Evaluate $-3x^2 - (5 - (2x^2 - 3x + 7) - 3x)$ for $x = 10$.

Solution We could replace the x by 10 and do the computation

$$-3(10)^2 - (5 - (2(10)^2 - 3(10) + 7) - 3(10))$$

Instead, we can evaluate the standard form of the polynomial. From Example 7, we know that the polynomial $-3x^2 - (5 - (2x^2 - 3x + 7) - 3x)$ in standard form is $-x^2 + 2$.

Now, we evaluate $-x^2 + 2$ for $x = 10$ to get $-100 + 2$, or -98.

Thus, $-3x^2 - (5 - (2x^2 - 3x + 7) - 3x)$ evaluated for $x = 10$ is -98. ■

EXAMPLE 9 Evaluate $-3x^2 - (3y - (2x^2 - 5y + x))$ for $x = -2$ and $y = 5$.

Solution We can replace x by -2 and y by 5, and do the computation:

$$-3(-2)^2 - (3(5) - (2(-2)^2 - 5(5) + (-2)))$$

Or, we can write the polynomial $-3x^2 - (3y - (2x^2 - 5y + x))$ in standard form, and then evaluate:

$$\begin{aligned} -3x^2 - (3y - (2x^2 - 5y + x)) &= -3x^2 - (3y - 2x^2 + 5y - x) \\ &= -3x^2 - 3y + 2x^2 - 5y + x \\ &= -x^2 + x - 8y^2 \end{aligned}$$

Now, we evaluate $-x^2 + x - 8y$ for $x = -2$ and $y = 5$ to get:

$$-4 + (-2) - 8(5) \qquad \text{or} \qquad -46$$

Thus, $-3x^2 - (3y - (2x^2 - 5y + x))$ evaluated for $x = -2$ and $y = 5$ is -46. ■

GIVE IT A TRY

9. Find the opposite of $-6x^2 + 2x - 3$

10. Write as a standard polynomial: $3x - 5 - (-6x^2 + 2x - 3)$

11. Write as a standard polynomial.
 a. $(3x^2 - 5x + 2) - (4x^2 - 2x - 3)$
 b. $-3x - (8 - (3x^2 - x + 1) - 3x^2)$

12. a. Evaluate $-3x - (8 - (3x^2 - x + 1) - 3x^2)$ for $x = -1$.
 b. Evaluate $-z - (3y - (3z - (5 + y^2)))$ for $y = -4$ and $z = 9$.

MEANING, MEMORY, AND AVOIDING ERRORS

In this section we have learned two important concepts:

1. How to add and subtract polynomials.
2. What it means to say the sum or difference of two polynomials is some third polynomial.

These concepts are equally important. In elementary arithmetic, so much time is spent learning how to do an operation, such as division, that it is easy to forget what it means to divide two numbers. Algebra presents the same pitfall—after spending so much time learning *how* to add and subtract polynomials, it is easy to forget what the operations mean. Doing addition and subtraction to write a polynomial, such as $3x - (5 - (7x - 3) - x)$, in standard form (as $11x - 8$) makes it easier to think about and work with the polynomial. This is especially important in solving equations, the topic of the next section.

A second pitfall to avoid in the study of algebra is memorization. Although some level of memory is certainly required, trying to memorize how to do each individual task is not possible for most students. Algebra is a building type of subject: you learn the next topic by understanding and developing skill with previous topics. Imagine memorizing how to spell words and the rules of grammar in a foreign language. Without knowing the meaning of any of the words, do you think you could use them to write a paper?

The greatest impediment to learning later topics in algebra is probably making errors in addition and subtraction of polynomials. Beyond this section, you may sometimes arrive at an incorrect answer for a problem. At this point, you naturally may assume that you do not understand the concept being presented; however, you may have understood the new concept, but simply made an error while doing a known skill, such as addition or subtraction. It is important to learn to work as error-free as possible. Many of the techniques shown in this and future sections are included to help you learn how to avoid making errors and how to check your work to detect the presence of such errors. Using the approach we have introduced usually produces written material that you check back over for errors. Using the graphics calculator to test your work for errors will usually identify the presence of an error. Once you are aware of its presence, you can check back over your work to find and correct the error. Be sure to test your work again after correcting the error, since another error may still be present.

■ SUMMARY

■ *Quick Index*

In this section, we have discussed polynomials, which are algebraic expressions formed by adding terms and subtracting terms. A term is a number or a product of a number (the coefficient) with one or more variables. Just as a variable holds the place for a collection of numbers, a polynomial holds the place for a collection of number phrases. A polynomial is named, or classified, by its degree, its number of terms (monomial, binomial, trinomial) and the number of variables. Thus, we speak of a first-degree binomial in two variables or a fourth-degree trinomial in one variable. A polynomial should be written as a standard polynomial, which means that like terms are combined and the polynomial is written in descending order, based on the degree of its terms.

We also introduced the operations of addition and subtraction on polynomials. The operation of addition is achieved by using commutative, associative, and distributive properties. The procedure for adding polynomials is often described as "combining like terms" (such as x^2 terms, x^2y terms, or xy^2 terms). Once the like terms are combined, the polynomial is written so that the terms appear in descending order, based on their degree. The operation of subtraction is performed by combining a concept and a skill: the meaning of subtraction [$a - b$ means $a + (-b)$] and the skill of adding polynomials.

To say that the sum of polynomials $x^2 + 3x$ and $-2x^2 + 5x$ is the polynomial $-x^2 + 8x$ means that any place the sum $(x^2 + 3x) + (-2x^2 + 5x)$ appears, it can be replaced by the polynomial $-x^2 + 8x$. It also means that the polynomials $(x^2 + 3x) + (-2x^2 + 5x)$ and $-x^2 + 8x$ evaluate to the same number for any replacement of the variable x. Based on this last meaning, we can use the graphics calculator to test for errors in work involving addition or subtraction of polynomials.

2.1 ■ ■ ■ EXERCISES

Insert true *for a true statement, or* false *for a false statement.*

____ 1. A term of a polynomial is a polynomial.

____ 2. The term $-3x^2y$ has coefficient -3.

____ 3. A variable holds the place for a collection of numbers.

____ 4. The polynomial $x^2 - 3x + 2$ is a second-degree binomial in one variable.

____ 5. The polynomial $5 - 3x$ is a standard polynomial.

____ 6. A term of a polynomial is a monomial.

____ 7. The monomial $3x^2y^3$ has degree 6.

____ 8. The polynomial $-x^2$ evaluates to 1 for $x = -1$.

9. Which of the following is a first-degree polynomial in one variable?
   i. $3x + 2$ ii. $x^2 - 3x + 1$
   iii. $3x - 5y$ iv. 23

10. Which of the following is a binomial?
   i. $3x + 5y$ ii. $-2x^2$
   iii. $5x^2y^3$ iv. $x^2 - 3xy + 2$

11. When $-x^2$ is evaluated for $x = -3$, the result is:
   i. 9 ii. -9 iii. 6 iv. -6

Name each polynomial, giving its degree, classification by number of terms (monomial, binomial, trinomial, or quadrinomial), and the number of variables. For example, $x^3 - 3xy$ is a third-degree binomial in two variables.

12. $x^2y - 3xy + y^4$
13. $3 - x^2 - 4x^3$
14. $3x^2y^4z$
15. -13
16. $2x^2 - 3xy^3$
17. $w^5 - w^3x^4 - t^2 - 1$
18. Give an example of a third-degree trinomial in two variables.
19. Give an example of two second-degree polynomials in one variable whose *sum* produces a zero-degree polynomial.
20. When two polynomials are added, can the degree of their sum be larger than the degree of either of the original two polynomials?
21. Is it possible to have a zero-degree binomial?
22. What is the smallest degree a trinomial can have?

Evaluate each polynomial for the given value(s) of the variable(s). Use the graphics calculator to test your work for errors.

23. $-2x^2 - x - 5$ for $x = -3$
24. $-3xy^2 - y^2$ for $x = 10$ and $y = -1$
25. $-x^2y - xy - 5y^2$ for $x = -1$ and $y = -3$
26. $3w^2 - w - 5$ for $w = -10$
27. $x^2y - 3xy + y^4$ for $x = -10$ and $y = 10$
28. $3x^2y^4z$ for $x = 17$, $y = 0$, and $z = -32$

Add the given polynomials and write the sum as a standard polynomial. Use the graphics calculator to test your work for errors.

29. $(-7x^2 + 2x + 5) + (-2x^2 - 3x + 1)$
30. $(-7x^2 + 2x + 5) + (2x^2 + 3x - 1)$
31. $(3w^2 - w - 5) + (-w^2 - 3w + 2)$
32. $(3w^2 - w - 5) + (w^2 + 3w - 2)$
33. $(5x^3 + 4x^2 - 7) + (x^2 - 3x - 1)$
34. $(-x^3 - 4x^2 + 7) + (-3x^2 - 3x + 1)$
35. $3p + (5 + (4p + 5) + 3p^2)$
36. $3x + 2(5 - 4x) + 3x^2$
37. $-3x^2 + 5 + (3x^2 + 2x + 1)$
38. $(-2r^2 + r + 2) + 3(-2r^2 - 2r + 1)$

Perform the subtraction and write the difference as a standard polynomial. Use the graphics calculator to test your work for errors.

39. $(-3x^2 + 4x + 1) - (-2x^2 - 3x + 1)$
40. $(3w^2 - w - 5) - (-w^2 - 3w + 2)$
41. $(5x^3 + 4x^2 - 7) - (x^2 - 3x - 1)$
42. $(-x^3 - 4x^2 + 7) - (-3x^3 - 3x^2 + 1)$
43. $(3s^2 - s - 3) - (-s^2 - 3s + 6)$

44. $\left(\frac{2}{3}x + \frac{3}{5}\right) - \left(\frac{1}{2}x - \frac{2}{3}\right)$

45. $\frac{2x + 3}{-5} - \frac{3x + 8}{-5}$

46. $2(3x^2 - 2x + 3) - 3(2x^2 - x - 5)$

Write each expression as a standard polynomial. Use the graphics calculator to test your work for errors.

47. $3x - (5 - (4x + 5) + 3x^2)$

48. $y^2 - [3xy - (x^2 - 2xy - y^2)]$

49. $-3(2m - 6) - 3(m - 2) - 5[7 - (3m - m^2) + 1]$

50. $3p - [5 - (4p - 5) - 3p^2]$

51. $3x - [2x^2 - (7 - 2x - x^2) - 2(x - x^2)] - 1$

Evaluate each polynomial for the indicated value(s) of the variable(s). Use the graphics calculator to test your work for errors.

52. $-3x^2 - [5 - (3x^2 + 2x + 1) + 2x^3]$ for $x = -2$

53. $3y - [2xy - 3(x + 2y)]$ for $x = -3$ and $y = 10$

54. $3x - 7 - (x^2 - 7x - 1) - 4x^2$ for $x = -3$

55. $pt - (3p - t^2) - 2(pt - p)$ for $p = 2$ and $t = 2$

56. $-3x^3 - [5 - (x^2 - 2x + 1) - x^3]$ for $x = 10$

D I S C O V E R Y

USING THE GRAPHICS CALCULATOR: The Graph Screen

You should now be able to use the graphics calculator to confirm your work by testing for errors when doing addition and subtraction of polynomials. This procedure has four steps:

1. Store a number in one or more of the memory locations, such as X, Y, Z, A, or B.
2. Type the original expression and press the ENTER key to evaluate the expression.
3. Type the standard polynomial produced and press the ENTER key to evaluate it.
4. Observe the numbers output from steps 2 and 3 and determine if an error is present.

For step 2, you can press the Y= key and store an expression in one of the special memory locations: Y1, Y2, Y3, or Y4. Display the value of Y1, Y2, Y3, or Y4 by selecting it from the Y-VARS menu.

In this section we will introduce the GRAPH key, the TRACE key, and the RANGE key. We now look at a procedure that will display the value of a polynomial in one variable, x, for various values of the variable. This procedure involves the graph screen. Press the Y= key and erase any material entered for Y1, Y2, Y3, and Y4. With the cursor at the :Y1= prompt, type 3X−2. This polynomial represents number phrases such as $3(-5) - 2$ or $3(\sqrt{2}) - 2$. Press the ZOOM key to display the ZOOM menu (see Figure 1). Press the 6 key to select the Standard option. The graph screen is presented. See Figure 2. As you can see, a horizontal line (known as the x-axis) is drawn, a vertical line (known as the y-axis) is drawn, and a third line is drawn. It is on this third line that we focus our interest.

```
ZOOM
1:Box
2:Zoom In
3:Zoom Out
4:Set Factors
5:Square
6:Standard
7↓Trig
```

FIGURE 1

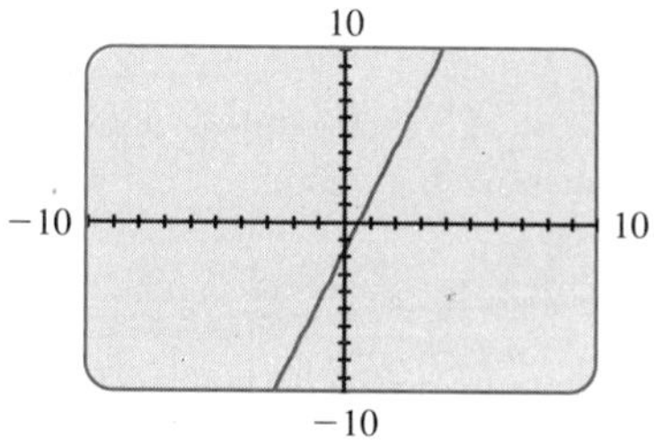

FIGURE 2

Press the TRACE key. A flashing square appears on the third line. Notice the message presented at the bottom of the screen. Press the right arrow key. The message changes, and the numbers presented are rather awkward. We will now use the RANGE key to set values that produce a more friendly graph screen, that is, the TRACE feature will display convenient values for

```
RANGE
Xmin=-10
Xmax=10
Xscl=1
Ymin=-10
Ymax=10
Yscl=1
Xres=1
```

FIGURE 3

```
RANGE
Xmin=-9.6
Xmax=9.4
Xscl=1
Ymin=-6.4
Ymax=6.2
Yscl=1
Xres=1
```

FIGURE 4

```
RANGE
Xmin=-4.8
Xmax=4.7
Xscl=1
Ymin=-3.2
Ymax=3.1
Yscl=1
Xres=1
```

FIGURE 5

X, such as 1.2 and 2.6. Press the RANGE key to display the setting for the graph screen.* See Figure 3. With the cursor on the (-) character of the prompt Xmin=-10, type -9.6 and press the ENTER key. At the Xmax prompt, type 9.4, and press the ENTER key. For Xscl, just press the ENTER key. At the Ymin prompt, type -6.4 and at the Ymax prompt, type 6.2. See Figure 4.** Press the GRAPH key to return to the graph screen. Press the TRACE key. As you can see, the message X=0 Y=-2 is presented. Press the right arrow key until the message reads X=1 Y=1. This indicates that when the x-variable is replaced by 1, the polynomial $3x - 2$ evaluates to 1.

Exercises

1. Use paper and pencil to evaluate the polynomial $3x - 2$ when x is replaced by each value.
   a. 1.5 **b.** 3 **c.** -0.4 **d.** 4
2. With the polynomial $3x - 2$ entered for Y1 and the range settings from Figure 4, use the TRACE key and evaluate the polynomial in Problem 1 for the values of x in parts (a)–(d).
3. Use paper and pencil to evaluate the polynomial $x^2 - 3x - 2$ when x is replaced by each value.
   a. 0 **b.** -1 **c.** 1.4 **d.** 4
4. Type X²−3X−2 for Y1. Use the range settings in Figure 4, and the TRACE key to evaluate the polynomial in Problem 3 for the values of x in parts (a)–(d).
5. Suppose you want to evaluate the polynomial $x^2 - 3x - 2$ for $x = 1.1$ using the graph screen. With the range settings in Figure 4, the x-value changes by 0.2 each time the right or left arrow key is pressed. Use the range settings shown in Figure 5,*** and the graph screen to evaluate the polynomial $x^2 - 3x - 2$ for the following values of x.
   a. 0.5 **b.** 1.1 **c.** -1.3 **d.** -2.1
   (To learn more about these range settings and programming the graphics calculator, see Appendix B.)
6. In Example 5 in Section 2.1, we stated that $(-3x^2 + 5x - 2) - (-2x^2 + 6x - 1)$ can be written in standard form as the polynomial $-x^2 - x - 1$. We found that for every replacement of the x-variable with a number, the polynomial $(-3x^2 + 5x - 2) - (-2x^2 + 6x - 1)$ and the standard polynomial $-x^2 - x - 1$ evaluate to the same number.
   Type (-3X²+5X−2)−(-2X²+6X−1) for Y1. Type -X²−X−1 for Y2. Use the range settings in Figure 5. From the graph screen, press the TRACE key. The values displayed are evaluations of the polynomial $(-3x^2 + 5x - 2) - (-2x^2 + 6x - 1)$ for various values of x. Press the down arrow key. This switches the trace cursor to the graph of Y2, the standard polynomial. Notice that the displayed values are the same for both graphs. That is, the expressions for Y1 and Y2 produce the same curve.
   Change the entry for Y2 to -2X²+6X+1. How is the graph screen changed?

TI-82 Note

*Press the WINDOW key followed by the down arrow key, to display the settings for the graph screen.
**Type -9.4 for Xmin and -6.2 for Ymin.
***Type -4.7 for Xmin and -3.1 for Ymin.

2.2 ■ ■ ■ SOLVING EQUATIONS

$3(7) + 5 = 10$

$3(\sqrt{2}) + 5 = 10$

$3\left(\frac{5}{7}\right) + 5 = 10$

$3x + 5 = 10$

An equation holds the place for a collection of arithmetic sentences.

Consider the arithmetic sentences $3(-2) + 5 = 10$ and $-7 + 8 = 1$. Both of these sentences have a truth value. The first one, $3(-2) + 5 = 10$, is a false arithmetic sentence, and the second one, $-7 + 8 = 1$, is a true arithmetic sentence. Every arithmetic sentence has a truth value (true or false). Now, consider the equation, $3x + 5 = 10$. This sentence has no truth value. Once the variable is replaced by a value, an arithmetic sentence results. Just as a variable holds the place for a number, and as a polynomial holds the place for a number phrase, an equation holds the place for various arithmetic sentences.

Background An **equation** is a relationship between two variable expressions using the equal relation. A polynomial equation is an equation in which each of the variable expressions is a polynomial. We have classified polynomials by degree, number of terms, and number of variables. Similarly, we classify a polynomial equation by its degree and by the number of its variables. The following table gives examples of first-degree and second-degree equations.

Degree of Equation	*Number of Variables*	*Example*
First-degree equations	1	$3x - 5 = 16 - x$
	2	$3x - 5y = 18$
	3	$x + 2y - z = 5$
Second-degree equations	1	$x^2 - 3x = 6$
	2	$x^2 + y^2 = 16$
	3	$z = x^2 - y^2$

In a first-degree polynomial equation, the highest degree of any term in the equation is 1. Likewise, the highest degree of any of the terms in a second-degree equation is 2. Note that $xy = 4$ is a second-degree equation in two variables.

To solve an equation in one variable means to find the collection of numbers that produce a true arithmetic sentence when any of the numbers from the collection replaces the variable in the equation. The variable in an equation can be replaced by any real number. (In this case, the set of real numbers is known as the *replacement set*.) Some of the numbers replacing the variable produce a true arithmetic sentence; some of the numbers produce a false arithmetic sentence. The collection of numbers that produce a true arithmetic sentence is known as the **solution set**. This collection (set) of numbers is known as the **solution** to the equation. We say that a number in the solution set solves the equation or satisfies the equation. We have several processes (algorithms) for solving an equation, just as we have many processes for solving general problems.

A common algorithm for solving equations in one variable is *prior knowledge*. That is, we simply know the solution from earlier experiences with numbers. Consider the equation $x + 5 = 12$. Knowledge of addition facts from arithmetic lets us state that the solution to this equation is the number 7. We do not do anything other than remember a fact—use prior knowledge.

Another method of solving equations in one variable is *guessing*. Consider the equation $3x - 5 = 18 + x$. We might start by guessing 10. Now, replacing the variable by 10 produces the arithmetic sentence $3(10) - 5 = 18 + 10$, which is a false arithmetic sentence. We can continue guessing until we find a number that produces a true arithmetic sentence. Now, based on our guess of 10 (which made the left-hand side too small), we then guess 11. Again, we get a false arithmetic sentence, with the left-hand side still too small. Now, guessing 12 produces a false arithmetic sentence, with the left-hand side too large. Next we try 11.5, which produces a true arithmetic sentence.

Solve: $3x - 5 = 18 + x$

Guess: 10 $3(10) - 5 = 18 + 10$

$25 = 28$ a false arithmetic sentence

Guess: 11 $3(11) - 5 = 18 + 11$

$28 = 29$ a false arithmetic sentence

Guess: 12 $3(12) - 5 = 18 + 12$

$31 = 30$ a false arithmetic sentence

Guess: 11.5 $3(11.5) - 5 = 18 + 11.5$

$29.5 = 29.5$ a true arithmetic sentence

Thus, the solution to the equation $3x - 5 = 18 + x$ is 11.5, because replacing the variable by 11.5 yields the true arithmetic sentence

$$3(11.5) - 5 = 18 + 11.5$$

With some refinements, this approach to equation solving might be great for a computer, since the computer can do all the computations very rapidly to check the guesses. This approach is not very useful to us at the moment.

The Algebraic Approach The algebraic approach to solving equations uses arithmetic and the following properties of equalities.

PROPERTIES OF EQUALITY

1. Reflexive Property: $a = a$
2. Symmetric Property: If $a = b$, then $b = a$.
3. Transitive Property: If $a = b$ and $b = c$, then $a = c$.
4. Addition Property: If $a = b$, then $a + c = b + c$.
5. Multiplication Property: If $a = b$ and $c \neq 0$, then $ac = bc$.

These properties provide a systematic approach to solving equations—finding the numbers that replace the variable to yield a true arithmetic sentence. To demonstrate again how algebra develops step by step, consider the following sequence of equations.

Solving Equations of Level I We start with equations of the form $ax = b$, for which a and b are real numbers. We call an equation of this form a level I equation. To solve this type of equation, we use the following property.

MULTIPLICATION PROPERTY OF EQUATION SOLVING

> Multiplying each side of an equation by a nonzero number never alters the solution to the equation.

Consider the trivial equation $x = 10$. The solution is 10.

We can multiply both sides by -3 to get $-3x = -30$. The solution to this new equation is still 10, because $-3(10) = -30$ is a true arithmetic sentence. Although the equation has been altered, the solution is *not* altered.

To see why the multiplier must be nonzero, multiply each side of the equation $x = 10$ by 0. We get $0x = 0$. For this equation, every number is in the solution, because 0 times a number is always 0. Thus, multiplying each side of an equation by zero *can* alter the solution. To obtain an equation whose solution is obvious, we must multiply each side by the reciprocal of the variable coefficient, provided the reciprocal exists.

EXAMPLE 1 Solve: $-3x = 17$

Solution We start by multiplying both sides by the number $\frac{-1}{3}$, which is the reciprocal of -3:

$$\left(\frac{-1}{3}\right)(-3x) = \left(\frac{-1}{3}\right)(17)$$

$$x = \frac{-17}{3}$$

Thus, the solution to $-3x = 17$ is $\frac{-17}{3}$. ■

EXAMPLE 2 Solve: $3x = \sqrt{17}$

Solution From the preceding discussion, we know that the equation is in the form $ax = b$. So, we multiply each side by $\frac{1}{3}$:

$$\left(\frac{1}{3}\right)(3x) = \left(\frac{1}{3}\right)(\sqrt{17})$$

$$x = \frac{\sqrt{17}}{3}$$

Thus, the equation $3x = \sqrt{17}$ has solution $\frac{\sqrt{17}}{3}$. ■

EXAMPLE 3 Solve: $\frac{-7}{5} = \frac{2}{3}x$

Solution Here, we multiply each side by $\frac{3}{2}$, the reciprocal of the x-coefficient:

$$\frac{3}{2} \cdot \frac{-7}{5} = \frac{3}{2} \cdot \left(\frac{2}{3}x\right)$$

$$\frac{-21}{10} = x$$

Thus, the equation has solution $\frac{-21}{10}$. ■

In general, the solution to an equation in the form $ax = b$ has the form $\frac{b}{a}$ provided $a \neq 0$. If $a = 0$ and $b \neq 0$, then the equation has no solution, and if $a = 0$ and $b = 0$, then every number solves the equation.

Solving Equations of Level II The next level of first-degree equations in one variable includes equations such as $3x + 6 = 34$. Any equation of the form $ax + b = c$ is a level II equation. To solve level II equations, we use another basic property of equations.

ADDITION PROPERTY OF EQUATION SOLVING

For any equation, adding a polynomial to each side never alters the solution.

To demonstrate this property, consider the equation $x = 10$, which has the obvious solution 10. Now, we will add the polynomial $x + 5$ to each side of the equation. (We could add any polynomial; we have selected $x + 5$ for demonstration only.)

$$x = 10 \qquad \text{Solution: } 10$$
$$\underline{x + 5} \quad \underline{x + 5} \qquad \text{Add polynomial to each side.}$$
$$2x + 5 = x + 15 \qquad \text{Solution: } 10$$

Notice that $2(10) + 5 = (10) + 15$ is a true arithmetic sentence. Adding a polynomial to each side of the equation has altered the equation, but *not* the solution. Now, we use this property to solve the equation $3x + 6 = 34$. Our strategy is to alter this level II equation to a level I equation, $ax = b$, which we know how to solve.

EXAMPLE 4 Solve: $3x + 6 = 34$

Solution

$$3x + 6 = 34$$

Add -6 to each side

$$\underline{-6} \quad \underline{-6}$$

to get:

$$3x = 28 \qquad \text{Level I equation}$$

Multiply each side by $\frac{1}{3}$ to get:

$$x = \frac{28}{3}$$

Thus, the solution to $3x + 6 = 34$ is $\frac{28}{3}$. ■

We can add any number (or polynomial) to each side of an equation without altering the solution. We choose a number that when added produces a new equation (a Level I equation of the form $ax = b$).

Using the Graphics Calculator

We can use the graphics calculator to see that $\frac{28}{3}$ really does solve the equation, $3x + 6 = 34$. Store `28/3` in memory location `X`. Now, type `3X+6` and press the `ENTER` key. The number `34` is output. See Figure 1. Thus, $\frac{28}{3}$ is the number that replaces the x-variable to produce a true arithmetic sentence, $3\left(\frac{28}{3}\right) + 6 = 34$.

```
28/3→X
3X+6
            34
```

FIGURE 1

```
RANGE
Xmin=-9.6
Xmax=9.4
Xscl=1
Ymin=-6.4
Ymax=6.2
Yscl=1
Xres=1
```

FIGURE 2

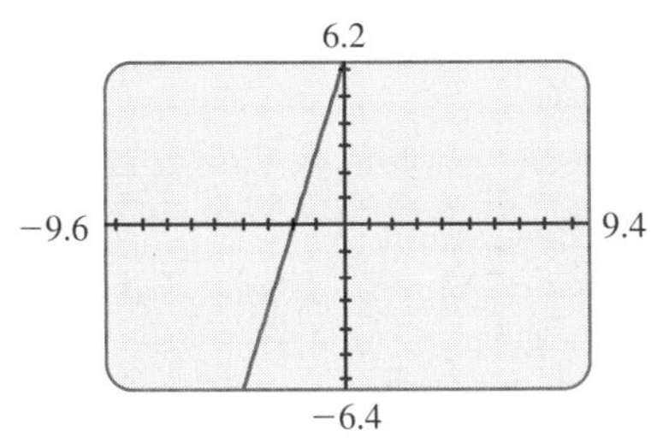

FIGURE 3

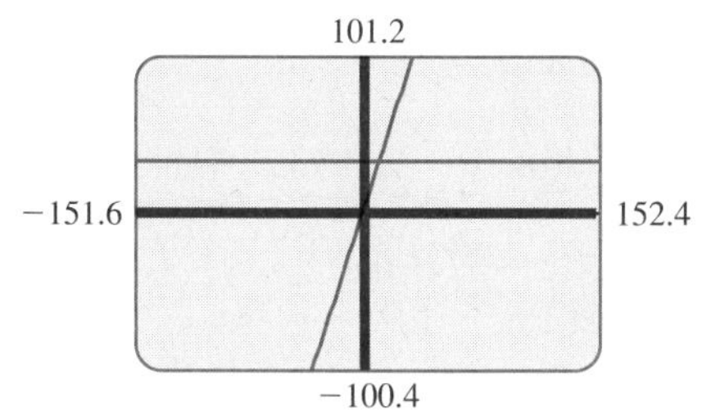

FIGURE 4

We can use the graph screen to increase our understanding of solving equations. For `Y1`, type `3X+6`. Press the `RANGE` key and enter the values in Figure 2.* The graph screen displays evaluations of the polynomial $3x + 6$ for various values of the x-variable. See Figure 3. Press the `TRACE` key. The message at the bottom of the screen means that when the x-variable is replaced by the value listed for `X`, the polynomial $3x + 6$ evaluates to the number listed for `Y`. To solve the equation $3x + 6 = 34$, we seek a replacement value for the x-variable that results in the polynomial $3x + 6$ evaluating to 34. Press the right arrow key several times. As you can see, we need a larger view to see where $3x + 6$ evaluates to a number as larger as 34.

For `Y2`, type `34`. Press the `ZOOM` key, followed by the `4` key, to set the Zoom factors.** Check that both factors listed are set to 4. If not, type `4` for each value. Quit to the home screen. Press the `GRAPH` key to view the graph screen. Press the `ZOOM` key, followed by the `3` key, to select the `Zoom Out` option. Press the `ENTER` key. We need an even larger view. Zoom out again. Now a horizontal line is drawn for `Y2` (the constant value 34). See Figure 4. Press the `TRACE` key, and then the right arrow key, to move the cursor to the intersection point of the two lines. The replacement for the x-variable that results in the polynomial $3x + 6$ evaluating to 34 is between 8 and 12. From our earlier work in Example 4, we know that the number is exactly $\frac{28}{3}$, or $9\frac{1}{3}$.

GIVE IT A TRY

Solve each first-degree equation in one variable.

1. $-3x = 11$ **2.** $\frac{-2}{5}x = \frac{-12}{7}$ **3.** $-2x + 5 = 12$ **4.** $-1 = 5 - \frac{3}{4}x$

5. Use the graphics calculator to check that the solution you found in Problem 1 does indeed produce a true arithmetic sentence when it replaces the variable. Repeat this procedure for Problems 2–4.

Solving Equations of Level III The next level of first-degree equations in one variable is a binomial equal to a binomial. A level III equation is any equation of the form $ax + b = cx + d$. Again, we will use the properties of equality to alter the equation to the form $1x = a$, which has an obvious solution. Work through the following example.

TI-82 Note *Type `-9.4` for `Xmin` and `-6.2` for `Ymin`.
**Press the `ZOOM` key, highlight `MEMORY`, select option `4`, `SetFactors`.

EXAMPLE 5 Solve: $3x - 17 = 18 - x$

Solution First, we write the binomials as standard polynomials:

$$3x - 17 = -1x + 18$$

Now, add $1x$ to each side $\quad 1x \qquad 1x$

to get $\quad 4x - 17 = 18 \quad$ Level II equation

Next, add 17 to each side $\quad 17 \qquad 17$

to get: $\quad 4x = 35 \quad$ Level I equation

Finally, multiply by $\frac{1}{4}$ to get: $\quad x = \frac{35}{4}$

This last equation has solution $\frac{35}{4}$.

Thus, the original equation $3x - 17 = 18 - x$ has solution $\frac{35}{4}$. ■

Using the Graphics Calculator

To check our work in Example 5, for Y1 type 3X−17, and for Y2 type 18−X. Store the number 35/4 in memory location X. Display the resulting values stored in Y1 and Y2. Does this x-value result in the polynomials $3x - 17$ and $18 - x$ evaluating to the same number? If not, an error is present!

To gain insight into the solution to the equation in Example 5, we start with 3X−17 entered for Y1 and 18−X entered for Y2. View the graph screen using the Standard option (press the ZOOM key followed by the 6 key). Two lines are drawn, as shown in Figure 5. To see more of the lines, zoom out (press the ZOOM key, the 3 key, and then the ENTER key). See Figure 6. The first line drawn represents the evaluations of $3x - 17$ for various replacements of the x-variable. The second line drawn represents the evaluations of $18 - x$ for various replacements of the x-variable. Press the TRACE key to move the cursor to the intersection point. Is the x-value approximately $\frac{35}{4}$, or $8\frac{3}{4}$?

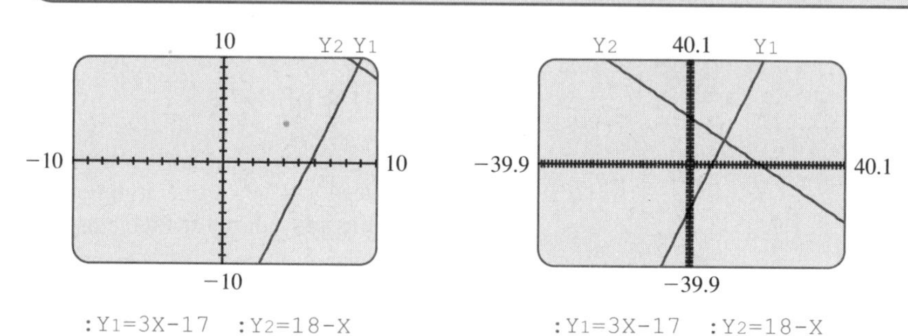

FIGURE 5

FIGURE 6

GIVE IT A TRY

Solve each equation.

6. $3x = -147$

7. $-3x + 6 = 3x + 23$

8. $15 - x = 20 - 5x$

9. $2x - 1 = 4x + 2$

10. Use the graphics calculator to check your solutions to the equations in Problems 6–9.

Rational Numbers in Equations We can always add a polynomial to each side of an equation without altering the solution. We can also multiply each side of an equation by a nonzero number without altering the solution. Although the equation is altered, the solution is *not* altered. The multiplication property is very useful in solving equations that involve rational numbers. Consider the following examples.

EXAMPLE 6 Solve: $\frac{2}{3}x - 3 = \frac{1}{2} + 5x$

Solution We start by multiplying each side of the equation by 6, a common denominator for the rational numbers $\frac{2}{3}$ and $\frac{1}{2}$:

$$6\left(\frac{2}{3}x - 3\right) = 6\left(\frac{1}{2} + 5x\right)$$

$$\frac{12}{3}x - 18 = \frac{6}{2} + 30x$$

$$4x - 18 = 3 + 30x$$

$$-26x - 18 = 3 \qquad \text{Add } -30x \text{ to each side.}$$

$$-26x = 21 \qquad \text{Add 18 to each side.}$$

$$x = \frac{-21}{26} \qquad \text{Multiply each side by } \frac{-1}{26}.$$

Thus, $\frac{2}{3}x - 3 = \frac{1}{2} + 5x$ has solution $\frac{-21}{26}$. ■

EXAMPLE 7 Solve: $\frac{3x + 1}{5} = \frac{6 - x}{3}$

Solution

We chose 15 because both 5 and 3 are factors of 15.

Multiply each side by 15 to get:	$\frac{15(3x + 1)}{5} = \frac{15(6 - x)}{3}$	
Do division to get:	$3(3x + 1) = 5(6 - x)$	
Clear parentheses:	$9x + 3 = -5x + 30$	Level III equation
Add $5x$ to each side to get:	$\underline{5x} \quad \underline{5x}$ $14x + 3 = 30$	Level II equation
Add -3 to each side to get:	$\underline{-3} \quad \underline{-3}$ $14x = 27$	Level I equation
Multiply each side by $\frac{1}{14}$ to get:	$x = \frac{27}{14}$	

This last equation has solution $\frac{27}{14}$. Thus, our original equation,

$$\frac{3x + 1}{2} = \frac{6 - x}{3},$$

has solution $\frac{27}{14}$. ■

For the equation in Example 7, we could write each side as a standard polynomial to get:

$$\frac{3}{5}x + \frac{1}{5} = \frac{-1}{3}x + 2$$

Now, multiply each side by 15 to get: $9x + 3 = -5x + 30$

Add $5x$ to each side to get: $14x + 3 = 30$

Add -3 to each side to get: $14x = 27$

Multiply each side by $\frac{1}{14}$: $x = \frac{27}{14}$

This last equation has solution $\frac{27}{14}$. Thus, the original equation has solution $\frac{27}{14}$.

Using the Graphics Calculator

To check our work in Example 6, for `Y1` type `(2/3)X−3`, and for `Y2` type `(1/2)+5X`. Store the number `-21/26` in memory location `X`. Display the resulting value stored in `Y1` and `Y2`. Does this x-value result in the polynomials `(2/3)X−3` and `(1/2)+5X` evaluating to the same number? If not, there is an error!

To check our work in Example 7, for `Y1` type `(3X+1)/5`, and for `Y2` type `(6−X)/3`. Store the number `27/14` in memory location `X`. Display the resulting value stored in `Y1` and `Y2`. Does this x-value result in the polynomials `(3X+1)/5` and `(6−X)/3` evaluating to the same number? If not, there is an error!

Equation Solving and Standard Polynomials The most difficult first-degree equation in one variable is a binomial equal to a binomial, for example, $-3x + 5 = 7x + 2$. Of course, the original equation can appear to be more difficult, but after writing each side as a standard polynomial, the equation will become (at most) a binomial equal to a binomial, a Level III equation. Consider the following examples.

EXAMPLE 8 Solve: $-3x - (5 - (7x + 2)) = 3 - x$

Solution Writing the left-hand side as a standard polynomial produces

$$\begin{aligned} -3x - (5 - (7x + 2)) &= -3x - (5 - 7x - 2) \\ &= -3x - 5 + 7x + 2 \\ &= 4x - 3 \end{aligned}$$

Writing the right-hand side as a standard polynomial yields $-1x + 3$. Thus, the equation becomes

$$4x - 3 = -1x + 3$$

Add $1x$ to each side to get:

$$\begin{array}{rcl} \underline{1x\quad\;\;} & & \underline{\;\;1x\quad\;\;} \\ 5x - 3 & = & 3 \end{array}$$

Add 3 to each side to get:

$$\begin{array}{rcl} \underline{\quad\;\;3} & & \underline{3\quad\;\;} \\ 5x & = & 6 \end{array}$$

Multiply each side by $\frac{1}{5}$ to get: $\quad x = \dfrac{6}{5}$

This last equation has solution $\frac{6}{5}$. Thus, the original equation

$$-3x - (5 - (7x + 2)) = 3 - x$$

has solution $\frac{6}{5}$. ■

EXAMPLE 9 Solve: $-2(2x - 1) = 3x - 5(x + 2)$

Solution Writing the left-hand side as a standard polynomial (using the distributive property), we get $-4x + 2$. Writing the right-hand side as a standard polynomial, we get $3x - 5(x + 2) = 3x - 5x - 10 = -2x - 10$. Thus, the equation becomes:

$$-4x + 2 = -2x - 10$$

Add $2x$ to each side to get:

$$\begin{array}{r} 2x \quad \quad 2x \\ \hline -2x + 2 = -10 \end{array}$$

Add -2 to each side to get:

$$\begin{array}{r} -2 \quad \quad -2 \\ \hline -2x = -12 \end{array}$$

Multiply each side by $\frac{-1}{2}$ to get: $\quad x = 6$

This last equation has solution 6. Thus, $-2(2x - 1) = 3x - 5(x + 2)$ has solution 6. ■

Using the Graphics Calculator

```
RANGE
Xmin=-9.6
Xmax=9.4
Xscl=1
Ymin=-6.4
Ymax=6.2
Yscl=1
Xres=1
```

FIGURE 7

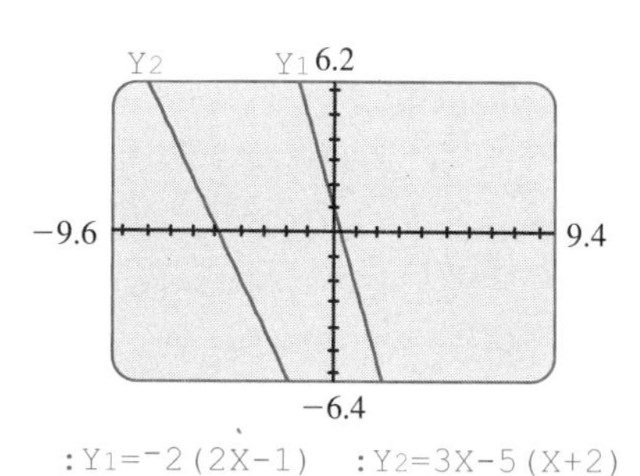

FIGURE 8

To check our work in Example 8, for `Y1` type `⁻3X−(5−(7X+2))`, and for `Y2` type `3−X`. Store the number `6/5` in memory location `X`. Display the resulting values stored in `Y1` and `Y2`. Does this x-value result in the polynomials `⁻3X−(5−(7X+2))` and `3−X` evaluating to the same number? If not, an error is present!

To check the work in Example 9, for `Y1`, type the left-hand side of the equation, `⁻2(2X−1)`. For `Y2`, type the right-hand side of the equation, `3X−5(X+2)`. Store the number `6` in memory location `X`. Display the resulting value for `Y1` and the resulting value for `Y2`. Is the same number output for both expressions? If not, an error is present!

We can increase our understanding of the solution in Example 9 by pressing the `RANGE` key and entering the values in Figure 7.* The graph screen displays two lines: one line shows evaluations of the polynomial for the left-hand side of the equation. A second line shows evaluations of the polynomial for the right-hand side. See Figure 8. To see the x-value that results in the polynomials evaluating to the same number, press the `ZOOM` key and the `3` key (to select the `Zoom Out` option), and then the `ENTER` key. The screen is redrawn showing an expanded view of the lines. Press the `TRACE` key. Press the right

TI-82 Note *Type `⁻9.4` for `Xmin` and `⁻6.2` for `Ymin`.

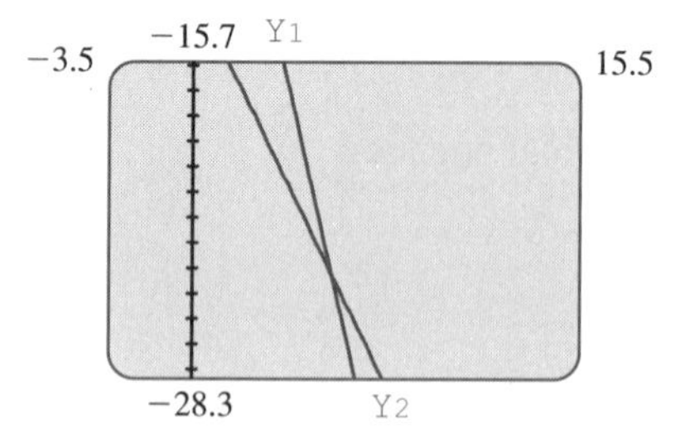

:Y1=⁻2(2X−1) :Y2=3X−5(X+2)

FIGURE 9

arrow key to move the cursor to the intersection of the two lines. Press the ZOOM key and then the 2 key for the Zoom In option. Press the ENTER key. The screen is redrawn with an enlarged view of the intersection point, as shown in Figure 9. Press the TRACE key and move the cursor to the intersection point. As you can see, this x-value is approximately 6, the solution to the equation.

The sequence of examples we have worked in this section is typical of the algebra topics throughout this text. To solve an equation, we must combine skill in operations on polynomials with skill in solving first-degree equations in one variable. In the next section we will use equation-solving skills to solve first-degree inequalities in one variable.

Special Cases of First-Degree Equations in One Variable The solution to a first-degree equation in one variable typically consists of a single number, such as 143, or −5, or 0. However, the solution can be the empty set, which means that no number solves the equation. Consider the equation

$$x + 5 = x + 7$$

It should be obvious that no number solves this equation. Adding 5 to a number (represented by x) always yields a different number than adding 7 to that number. If we tried to solve the equation using the algebraic approach we discussed earlier, we would write the following:

Solve:	$x + 5 = x + 7$
Add $-1x$ to each side	$\underline{-1x \qquad\quad -1x}$
to get:	$5 = 7$

```
RANGE
Xmin=⁻9.6
Xmax=9.4
Xscl=1
Ymin=⁻6.4
Ymax=6.2
Yscl=1
Xres=1
```

FIGURE 10

We say that the "variable has dropped out" and that an arithmetic sentence is produced. When a polynomial is added to each side of the equation and a false arithmetic sentence is produced, the solution to the original equation is the empty set of numbers, since no number solves the equation. For the solution, we write *none* or the symbol $\varnothing$.

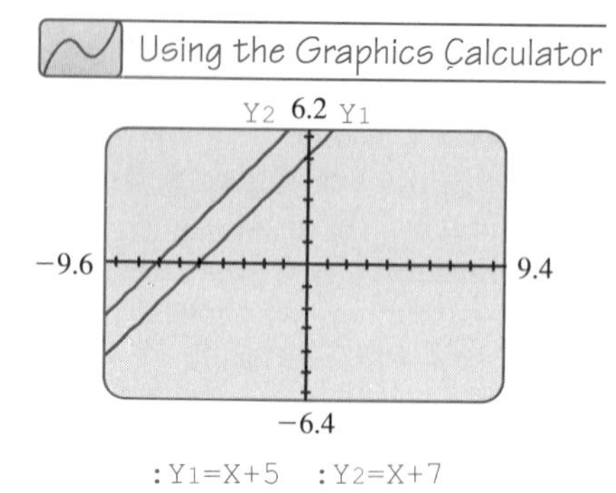

FIGURE 11

We can use the graphics calculator to confirm the solution we found for $x + 5 = x + 7$. Type X+5 for Y1 and X+7 for Y2. Press the RANGE key and type the values shown in Figure 10.* Viewing the graph screen shown in Figure 11, we see evaluations of the polynomials $x + 5$ and $x + 7$ for various replacement of the x-variable. Press the TRACE key. Use the down arrow key to switch from one line to the other. The polynomials $x + 5$ and $x + 7$ always evaluate to numbers that are 2 units apart; never do they evaluate to the same number. Thus, there is no number that solves the equation $x + 5 = x + 7$.

TI-82 Note *Type ⁻9.4 for Xmin and ⁻6.2 for Ymin.

In the other special case for first-degree equations in one variable, all the real numbers solve the equation. Consider the equation

$$2x + 6 = 2(x + 3)$$

This special case is an example of the distributive property: If x is replaced by any number, a true arithmetic sentence is produced. Such an equation is known as an **identity**. The solution to an identity is all real numbers. If we tried to solve an identity using our algebraic approach, we would write the following:

Solve:	$2x + 6 = 2x + 6$
Add $-2x$ to each side	$-2x \qquad -2x$
to get:	$6 = 6$

The variable drops out and an arithmetic sentence is produced. When a polynomial is added to each side of the equation and a true arithmetic sentence is produced, the solution to the original equation is the set of all numbers, which means that every number solves the equation. For the solution, we write *all* or the symbol $\mathbb{R}$ (for the set of real numbers).

Using the Graphics Calculator

We can use the graphics calculator to check our work for the solution we found for $2x + 6 = 2(x + 3)$. Type 2X+6 for Y1, and type 2(X+3) for Y2. Keep the RANGE settings from Figure 10. Viewing the graph screen (see Figure 12) we see evaluations of the polynomials $2x + 6$ and $2(x + 3)$ for various replacement of the x-variable. Press the TRACE key. Use the down arrow key to switch from one line to the other. For each replacement of the x-variable, the evaluation is the same for both polynomials. Thus, the solution is all real numbers.

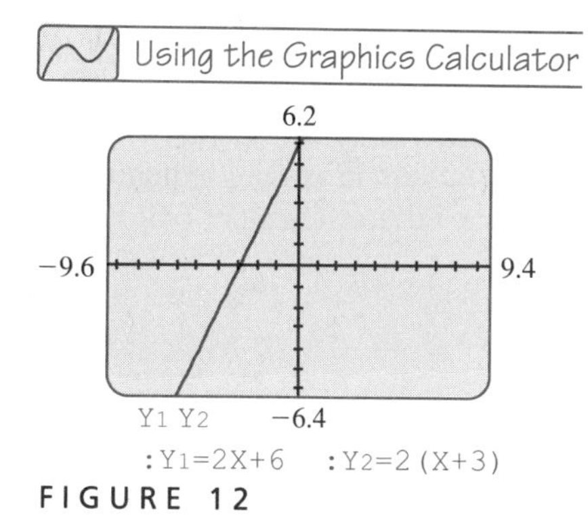

FIGURE 12

GIVE IT A TRY

Solve each equation.

11. $3x - (5 - 12x) = 17 - 2x$

12. $5 - (3x - (2x + 3)) = 3$

13. $3(x - 2) = 5x - (4 + 2x)$

14. $3(x - 2) = 5x - (6 + 2x)$

15. Use the graphics calculator to check the solutions to Problems 11–14.

STRUCTURE

The first two sections of this chapter have opened the gates of algebra. In Section 2.1 we learned what polynomials do—they hold the place for number phrases—and we learned how to carry out the operations of addition and subtraction on polynomials. Section 2.2 introduced equations, which hold the

TI-82 Note *Type −9.4 for Xmin and −6.2 for Ymin.

place for number (arithmetic) sentences. To solve an equation is to find the number, or numbers, that replace the variable to produce a true arithmetic sentence.

We have learned that a polynomial equation is one in which both sides are polynomials. A first-degree equation in one variable is a polynomial equation in which both polynomials have a single variable and the degree of each term is at most 1. An algebraic approach to solving first-degree equations in one variable is to reduce the equation to the form $x = a$. This is accomplished by adding a polynomial to each side of the equation or by multiplying each side of the equation by a nonzero number, using each procedure as necessary.

To solve first-degree equations in one variable requires that we be able to add and subtract polynomials. This skill is a key to learning algebra at this level. To be successful at solving first-degree equations in one variable, we must know a previous topic—addition and subtraction of polynomials. In the next sections, situations are presented that require solving first-degree equations in one variable. To be successful with those topics, we must thoroughly understand solving first-degree equations in one variable. In fact, throughout the remainder of this course, the skills from Sections 2.1 and 2.2 will be used in other topics.

In Chapter 5, for example, we will discuss the operations of multiplication and factoring of polynomials. These operations will then be used in solving second-degree equations. Chapter 6 introduces operations on rational expressions (ratios of polynomials) and uses these operations in solving equations that involve rational expressions. All these topics in later chapters build on the topics of Sections 2.1 and 2.2.

■ SUMMARY

■ ***Quick Index***

In this section, we have learned to solve first-degree equations in one variable. To solve an equation in one variable, we must find the numbers that replace the variable to produce a true arithmetic sentence. For first-degree equations in one variable, we find the solution by using two properties of equations:

1. Any polynomial can be added to each side of an equation, and the solution is not altered.
2. Each side of an equation can be multiplied by a nonzero number, and the solution will not be altered.

To solve some first-degree equations in one variable, the left-hand and right-hand sides of the equation must each be written as standard polynomials (a skill from Section 2.1). Once this is accomplished, we use the two properties of equations to alter the equation to the form $x = a$, and then write the solution.

Using these properties, we can usually alter a first-degree equation in one variable (say, the variable x) to the form $x = a$. With this form, the solution can be arrived at by observation (in this case, a). If the equation is altered to the form $a = b$ (the variable drops out), and $a = b$ is a true arithmetic sentence, then all numbers solve the equation. If the equation is altered to the form $a = b$ (the variable drops out), and $a = b$ is a false arithmetic sentence, then no number solves the equation. Thus, the solution to a first-degree equation in one variable will be either a single number, all the numbers, or the empty set (no number solves the equation).

2.2 ■ ■ ■ EXERCISES

Insert true *for a true statement, or* false *for a false statement.*

____ **1.** The solution to an equation in one variable is a collection of numbers (possibly the empty set).

____ **2.** The solution to the equation $-x = 5$ is the number 5.

____ **3.** If a polynomial is added to each side of a first-degree equation in one variable, the solution is never altered.

____ **4.** If each side of a first-degree equation in one variable is multiplied by a number, the solution is not altered.

____ **5.** If a first-degree equation in one variable is altered to produce the arithmetic sentence $7 = 8$, then the solution to the equation is all numbers.

____ **6.** A first-degree equation in one variable can have a solution consisting of two numbers, such as -4 and 5.

____ **7.** No number solves the equation $3x = 0$.

____ **8.** Every number solves the equation $0x = 3$.

____ **9.** The solution to $2 = 5x$ is the number $\frac{5}{2}$.

10. Complete steps (a)–(d) to solve the equation

$$3x - 17 = 7x - 6$$

a. Report the new equation produced when $-7x$ is added to each side.

b. Report the new equation produced when 17 is added to each side of the equation reported in step (a).

c. Report the new equation produced when each side of the equation reported in step (b) is multiplied by $\frac{-1}{4}$.

d. Report the solution to the original equation $3x - 17 = 7x - 6$.

11. Complete steps (a)–(d) to solve the equation

$$3x - (5 - (7 - 5x)) = 5 - (x + 3)$$

a. Report the new equation produced when both sides are written as standard polynomials.

b. Report the new equation produced when x is added to each side of the equation reported in step (a).

c. Report the new equation produced when -2 is added to each side of the equation reported in step (b).

d. Report the new equation produced when each side of the equation reported in step (c) is multiplied by -1.

e. Report the solution to the original equation $3x - (5 - (7 - 5x)) = 5 - (x + 3)$.

12. Consider solving the equation

$$\frac{2}{3}x - \frac{1}{5} = 5 - \frac{1}{5}x$$

a. Report the new equation produced by multiplying each side by the number 15.

b. Report the solution to the equation produced in part (a).

c. Report the solution to the equation $\frac{2}{3}x - \frac{1}{5} = 5 - \frac{1}{5}x$.

Solve each equation. Use the graphics calculator to store the answer in memory location X *and then check the solution. Always use* X *for the variable on the calculator.*

13. $-3x = 129$

14. $-47 = -5y$

15. $16 = 7w$

16. $-5x = 200$

17. $\frac{2}{3}z = 5$

18. $\frac{-5}{6}x = -3$

19. $\frac{4}{7} = \frac{-3}{5}x$

20. $\frac{3}{10}p = 1.5$

21. $-0.02x = 4.32$

22. $1.2R = 5.6$

23. $-5x = 12.7$

24. 30% of $x = 8$

25. $3B = 10\%$ of 9

26. 15% of $4x = 10$

27. $3w = \sqrt{25}$

28. $-5x = \sqrt{81}$

29. $-3T = -\sqrt{121}$

30. $3x = -\sqrt{36}$

31. $-10x = \sqrt{5}$

32. $0.2x = \sqrt{7}$

33. $5x + 8 = 6$

34. $-3x - 2 = 20$

35. $16 - 2x = 18$

36. $20 - 5x = 50$

37. $32 = 18 - 5x$

38. $7 = 12 - 8x$

39. $12 - x = -2$

40. $15 = -2 - x$

41. $3x - 8 = -7x + 3$

42. $2W - 5 = 5W + 1$

43. $12 - R = 8 - 7R$

44. $12 - 5x = 2x - 7$

45. $15 - x = 7x$

46. $25 + 3x = x - 8$

47. $12x - 5 = 7 + 6x$

48. $-5 + 2x = 6 - 9x$

49. $2(5x - 1) = -3(x + 4)$

50. $-2(5 - x) = 6x$

51. $\frac{2}{5}x + 3 = \frac{1}{2}$

52. $6 - \frac{3}{5}x = \frac{7}{2}x$

53. $\frac{2}{3}p + 3 = \frac{1}{2} - 2p$

54. $y = \frac{-7}{5}y - \frac{4}{3}$

55. $\frac{4}{9}x + \frac{3}{5} = \frac{1}{5} - \frac{2}{9}x$

56. $\frac{3}{7} - \frac{1}{2}n = \frac{6}{7}n + \frac{1}{2}$

57. $0.3x - 2 = 7 - 2.1x$

58. $1.3 - x = 2 - 0.5x$

59. $0.3x + 0.8(20) = 0.4(x + 20)$

60. $0.2x + 0.8(70) = 0.4(x + 70)$

61. $2 - (3x - (5 - 4x)) = x + 2$

62. $5r - (6 - r) = 12 - (5 - 3r)$

63. $8x = 13 - (2x - (2x - (5 + 3x) - 1))$

64. $5(w + 1) - 2 = w - 3(6 - w)$

65. $17 - (3x - 2) = x - 3(5 - 2x)$

66. $13 - (2x - 3) = 2x - 3(7 - x)$

67. $\frac{2x + 5}{3} = \frac{16 - 2x}{4}$

68. $\frac{7x - 2}{2} = \frac{5x - 2}{5}$

69. $\frac{2 - 5x}{2} = \frac{6 - 5x}{7}$

70. $\frac{7 - 6x}{8} = \frac{x - 1}{3}$

■ **BONUS PROBLEMS**

Solve each equation, and use the graphics calculator to check the solution.

71. $5x + 3 = \sqrt{16}$

72. $2x - 5 = \sqrt{64}$

73. $6m = 5 - \sqrt{3}$

74. $8x - 7 = \sqrt{31}$

75. $3x - \sqrt{1} = \sqrt{4}$

76. $\sqrt{25} - 2x = 3$

77. $\sqrt{5} + 3x = \sqrt{9}$

78. $5x - 6 = \sqrt{71}$

DISCOVERY

USING THE GRAPHICS CALCULATOR: Setting Up A Friendly Screen

You should now be able to use the graphics calculator and its memory locations, such as X, Y1, Y2, to confirm your work when writing standard polynomials and when solving equations. Also, you should be able to use the graph screen to view evaluations of a polynomial in x for various replacements of the x-variable. The graph screen, along with the TRACE key, can be used to display a message that gives the evaluation for an expression stored in either Y1, Y2, Y3, or Y4 for successive x-values. The x-values displayed with the TRACE key depend on the values Xmin and Xmax shown on the RANGE screen. To understand this relationship, we must discuss picture elements, or *pixels*.

```
RANGE
Xmin=-9.6
Xmax=9.4
Xscl=1
Ymin=-6.4
Ymax=6.2
Yscl=1
Xres=1
```

FIGURE 1

```
RANGE
Xmin=-4.8
Xmax=4.7
Xscl=1
Ymin=-3.2
Ymax=3.1
Yscl=1
Xres=1
```

FIGURE 2

The TI-81 Screen The screen for a graphic calculator is made up of picture elements or pixels. The TI-81 screen is 96 pixels wide. When the calculator evaluates an expression from the Y= screen, it computes 95 evaluations of the expression. These 95 evaluations are for x-values equally spaced from Xmin to Xmax. As we have seen, if the range values are those shown in Figure 1, the x-values are the numbers: $-9.6, -9.4, -9.2, \ldots, 9.4$. Note that Xmax − Xmin = 19. In TRACE mode, each press of the right or left arrow key moves the cursor 0.2 of a unit.

Press the Y= key and for Y1, type 0.5X+1. Erase (or clear) any other entries on the Y= screen. Press the RANGE key and type the values shown in Figure 1. Press the TRACE key and verify that as the cursor is moved right or left, the x-value does increase or decrease by 0.2 each time. Now, press the RANGE key and divide the entry for Xmin, Xmax, Ymin, and Ymax by 2 to get the values shown in Figure 2. Press the TRACE key and verify that as the cursor is moved, the x-value does increase or decrease by 0.1 each time. Here, Xmax − Xmin is 9.5, or one-half of 19. Each of these graph screens is known as a *friendly screen,* because the x-increment is an easily recognized value, such as 0.1 or 0.2.

```
RANGE
Xmin=-48
Xmax=47
Xscl=10
Ymin=-32
Ymax=31
Yscl=1
Xres=1
```

FIGURE 3

An easy way to display a friendly graph screen is to use the Integer option on the ZOOM menu. Press the ZOOM key and then the 8 key to select the Integer option. Move the cursor to the origin, X=0 Y=0. Press the ENTER key. Press the RANGE key and observe the values, as shown in Figure 3. Press the TRACE key and verify that as the cursor is moved, the x-value does increase or decrease by 1 each time. Here, Xmax − Xmin is 95. Press the RANGE key and divide each entry by 10. We get the values shown in Figure 2. Now, by doubling the values for Xmin, Xmax, Ymin, and Ymax, we get the values shown in Figure 1.

If Xmax − Xmin is 95 or a multiple of 95, such as twice 95 or one-tenth of 95, we get a friendly graph screen. Suppose we want to see evaluations of $0.5x + 1$ for x-values from -10 to 120 on a friendly graph screen. The difference, Xmax − Xmin, is 130. Divide this number by 95 to get 1.4, rounded to the nearest one-tenth. Press the RANGE key and type -10 for Xmin. For Xmax type 123, which is the value of $(1.4)(95) - 10$. For Xscl, type 10. The Ymin and Ymax values do not have to be as precise as Xmin and Xmax. We can evaluate the Y1 expression at Xmin and Xmax to get the approximate values to type for Ymin and Ymax. For $x = -10$, the expression $0.5x + 1$ evaluates to -4 and for $x = 123$, it evaluates to 62.5. So, for Ymin we can type -5 and for Ymax we can type 70. For Yscl, we type 10. View the graph screen and use the TRACE key to see evaluations of $0.5x + 1$ for x-values from -10 to 120. What is the x-increment?

```
RANGE
Xmin=-4.7
Xmax=4.7
Xscl=1
Ymin=-3.1
Ymax=3.1
Yscl=1
Xres=1
```

FIGURE 4

The TI-82 Screen To display a friendly graph screen on the TI-82, we start with the Decimal option on the ZOOM menu. For Y1, type 0.5X+1. Press the ZOOM key, followed by the 4 key, for the ZDecimal option, and then press the ENTER key. Press the WINDOW key and observe the values shown in Figure 4. The screen for the TI-82 is 95 pixels wide. The TI-82 makes 94 calculations from Xmin to Xmax. Notice that Xmax − Xmin is 9.4, or one-tenth of 94.

If Xmax − Xmin is 94 or a multiple of 94, such as twice 94 or one-tenth of 94, we get a friendly graph screen. Suppose we want to see evaluations of $0.5x + 1$ for x-values from -10 to 120 on a friendly graph screen. The difference, Xmax − Xmin, is 130. Divide this number by 94 to get 1.4, rounded to the nearest one-tenth. Press the WINDOW key and type -10 for Xmin. For Xmax, type 121.6, the value of $(1.4)(94) - 10$. For Xscl, type 10. The Ymin and Ymax values do not have to be as precise as Xmin and Xmax. We can evaluate the Y1 expression at Xmin and Xmax to get the approximate values to type for Ymin and Ymax. For $x = -10$, the expression $0.5x + 1$ evaluates to -4 and for $x = 121.6$, it evaluates to 61.8. So, for Ymin we can type -5 and for Ymax we can type 70. For Yscl, we type 10. View the graph screen and use the TRACE key to see evaluations of $0.5x + 1$ for x-values from -10 to 120. See Figure 5.

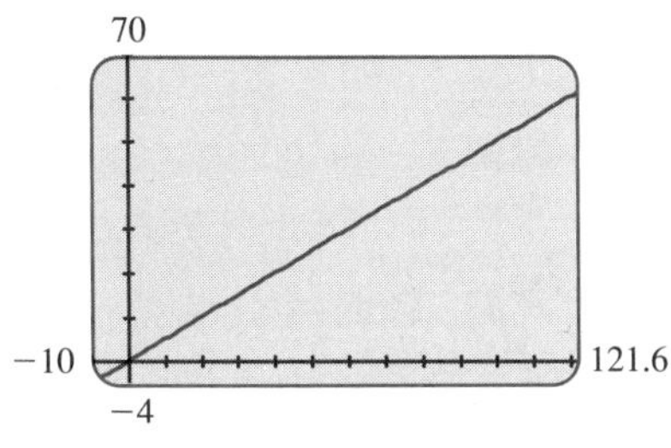

FIGURE 5

Exercises

1. Display a friendly graph screen (using `-4.8` to `4.7` for x-values on the TI-81 or use the `ZDecimal` option on the TI-82) for the evaluations of $-3x + 2$. Use the `TRACE` key to complete the following table.

x-value	-2.1	-1.4	0.7	3.2	4.5
Evaluation of $-3x + 2$					

2. Display a friendly graph screen of the evaluation of $-3x + 2$ for x-values from 0 to 130. Use the `TRACE` key to complete the following table.

x-value	56	98	105	112	126
Evaluation of $-3x + 2$					

2.3 ■ ■ ■ SOLVING FIRST-DEGREE INEQUALITIES IN ONE VARIABLE

In this section, we will apply our knowledge of solving first-degree equations in one variable to solving first-degree inequalities in one variable, such as

$$3x - 5 \geq 15$$

To solve the inequality we must identify the numbers that replace the variable to produce a true arithmetic sentence.

$3(-2) - 5 \leq 3$

$3(\sqrt{2}) - 5 \leq 3$

$3(9.3) - 5 \leq 3$

$3x - 5 \leq 3$

A first-degree inequality in one variable holds the place for arithmetic sentences.

Inequalities An inequality is much like an equation in that it holds the place for arithmetic sentences. However, the solution to a first-degree inequality in one variable is typically an infinite collection of numbers, rather than a single number. Remember, that a major objective of algebra at this level is learning to solve equations. Inequalities are a direct application of solving equations, and thus are quite important.

A typical first-degree inequality in one variable is $3x - 5 \leq 3$. This inequality holds the place for arithmetic sentences such as

$$3(-5) - 5 \leq 3 \qquad 3(5) - 5 \leq 3$$

$$3\left(\frac{2}{3}\right) - 5 \leq 3 \qquad 3(3.1) - 5 \leq 3$$

$$3(\sqrt{5}) - 5 \leq 3 \qquad 3(4) - 5 \leq 3$$

Some of these arithmetic sentences are true (the three in the left-hand column), and some are false (the three on the right). Some of the numbers in the solution to $3x - 5 \leq 3$ are -5, $\frac{2}{3}$, and $\sqrt{5}$.

Using the Graphics Calculator

We can use the graphics calculator to quickly judge each of the arithmetic statements we have listed for the inequality $3x - 5 \leq 3$. For Y1, type 3X−5. Store ⁻5 in memory location X. Display the resulting output value in Y1. The output value is ⁻20, a number less than 3. So, ⁻5 solves the inequality. Next, store 2/3 in memory location X, and display the resulting value in Y1. Again, a number less than 3 is output. Repeat these steps to check the rest of the numbers that we have used to replace x in the inequality $3x - 5 \leq 3$.

We now plot the six replacements for x: $-5, \frac{2}{3}, \sqrt{5}, 3.1, 4,$ and 5. We plot a solid dot for each number that produces a true arithmetic sentence and an open dot for each number that produces a false sentence. We get the graph shown in Figure 1. It appears that each number to the left of some particular value solves the inequality, but each number to the right of this *key point* fails to solve the inequality.

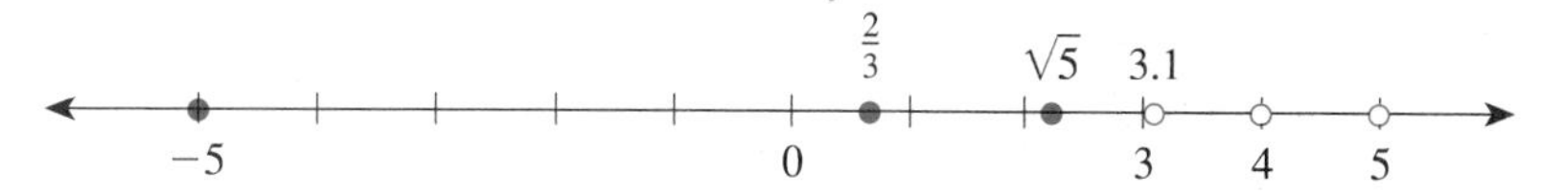

FIGURE 1

Using the Graphics Calculator

We can use the graphics calculator to approximate the key point for the inequality $3x - 5 \leq 3$. Press the RANGE key and enter the values shown in Figure 2.* Press the TRACE key to display the screen shown in Figure 3, and then press the right arrow key. The polynomial 3X−5 evaluates to a number less than 3 for x-values up to 2.6. When x is 2.8, the polynomial is greater than 3. So the *key point* must be between 2.6 and 2.8. As we will see, the key point is the solution to the equation $3x - 5 = 3$.

```
RANGE
Xmin=-9.6
Xmax=9.4
Xscl=1
Ymin=-6.4
Ymax=6.2
Yscl=1
Xres=1
```

FIGURE 2

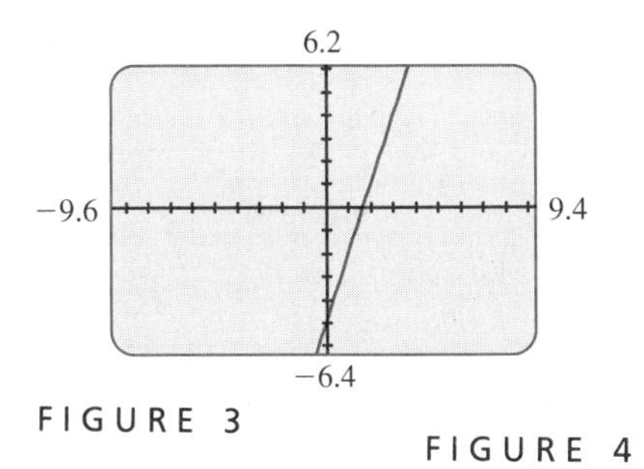

FIGURE 3

The solution to $3x - 5 \leq 3$ contains an infinite number of values. So, we cannot list each value as we can for most first-degree equations. To communicate the solution to an inequality, we will use a number line graph. We now present an organized, algebraic approach to solving inequalities.

Key Points The graph of the solution to a first-degree inequality in one variable is an interval on the number line. An **interval** is a continuous, uninterrupted collection of points on a number line. See Figure 4 for examples of intervals.

FIGURE 4

The leftmost or rightmost point in such a collection is an endpoint of the interval. Notice from Figure 4 that an interval can have either one or two

TI-82 Note *Type ⁻9.4 for Xmin and ⁻6.2 for Ymin.

endpoints or no endpoint. For a first-degree inequality in one variable, the solution to the related first-degree equation determines the endpoints of the interval. An endpoint is also known as a key point in the solution. For the inequality $3x - 5 \leq 3$, the solution to $3x - 5 = 3$ is the key point, the endpoint for the interval. This key point divides the number line into two intervals, one of which represents the solution to the inequality. The following property is true for inequalities.

KEY POINT PROPERTY FOR INEQUALITIES

> If one number in an interval (with an endpoint that is a key point) solves an inequality, then every number in that interval solves the inequality. If one number in such an interval fails to solve the inequality, then every number in that interval fails to solve the inequality.

The strategy we use to graph the solution to a first-degree inequality is to solve the related first-degree equation. The solution to this equation yields the key point or endpoint for the interval that is the graph of the solution to the inequality. We then test a number from each of the intervals determined by the key point. If the test number produces a true arithmetic sentence, or satisfies the inequality, then all the numbers represented by the interval solve the inequality.

EXAMPLE 1 Graph the solution to $3x - 5 \leq 3$.

Solution **Step 1** Solve the equation:

$$3x - 5 = 3$$

$$3x = 8 \qquad \text{Add 5 to each side.}$$

$$x = \frac{8}{3} \qquad \text{Multiply each side by } \tfrac{1}{3}.$$

The solution is $\frac{8}{3}$. Thus, $\frac{8}{3}$ is the key point in the solution to the inequality.

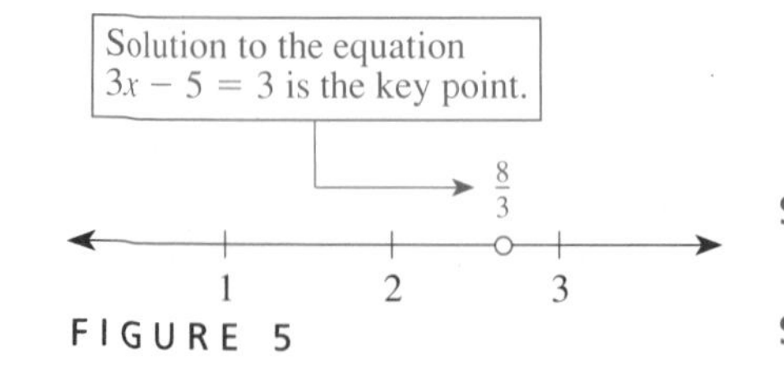

FIGURE 5

Step 2 Plot the number $\frac{8}{3}$ on a number line by placing a small circle on $\frac{8}{3}$. Label the point, as shown in Figure 5.

Step 3 Because $\frac{8}{3}$ solves the inequality $\left[3\left(\frac{8}{3}\right) - 5 \leq 3\right.$ is a true arithmetic sentence$\left.\right]$, we place a solid dot at $\frac{8}{3}$. See Figure 6. (If the number $\frac{8}{3}$ did not solve the inequality, we would leave the circle open.)

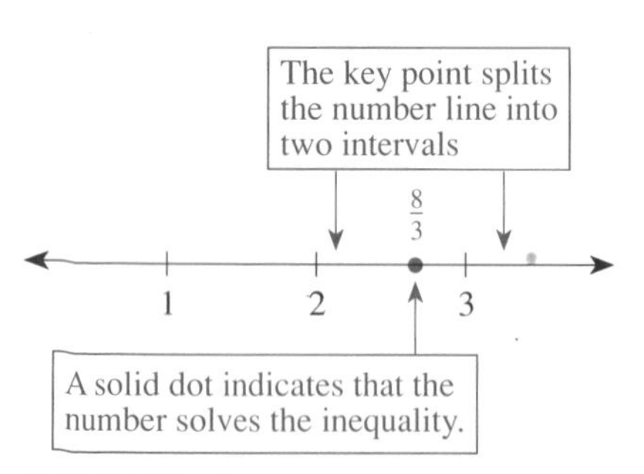

FIGURE 6

Step 4 Try a number in the left-hand interval, a number less than $\frac{8}{3}$.
Try 0: $3(0) - 5 \leq 3$ is a true arithmetic sentence.
If one number in this interval solves the inequality, then all the numbers in the interval solve the inequality. Shade the left-hand interval. See Figure 7.

Try a number in the right interval, a number greater than $\frac{8}{3}$.
Try 10: $3(10) - 5 \leq 3$ is a false arithmetic sentence.
If one number in an interval does not solve the inequality, then none of the values in that interval solves the inequality. Leave the right-hand interval unshaded.

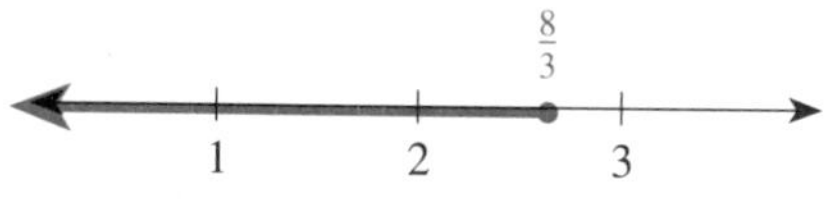

FIGURE 7

The points in this interval represent the numbers that solve the inequality.

Thus, the solution to the inequality $3x - 5 \le 3$ is $\frac{8}{3}$ and all the numbers to the left of $\frac{8}{3}$. This solution is pictured in the number line graph in Figure 7. ■

This approach to solving inequalities will be productive in all the examples in this text. We now use the approach in a more involved example.

EXAMPLE 2 Graph the solution to $3x - [5 - (12 - x)] > 3 - x$.

Solution

Step 1 Solve the equation $3x - [5 - (12 - x)] = 3 - x$.
To do this, we write the left-hand side as a standard polynomial:

$$\begin{aligned} 3x - (5 - (12 - x)) &= 3x - (5 - 12 + x) \\ &= 3x - 5 + 12 - x \\ &= 2x + 7 \end{aligned}$$

and the right-hand side as a standard polynomial: $-1x + 3$

Thus, the equation becomes: $2x + 7 = -1x + 3$

Add $1x$ to each side to get: $3x + 7 = 3$

Add -7 to each side to get: $3x = -4$

Multiply by $\frac{1}{3}$ to get: $x = \dfrac{-4}{3}$

The equation $3x - (5 - (12 - x)) = 3 - x$ has solution $\frac{-4}{3}$, so the key point for the solution to the inequality is $\frac{-4}{3}$.

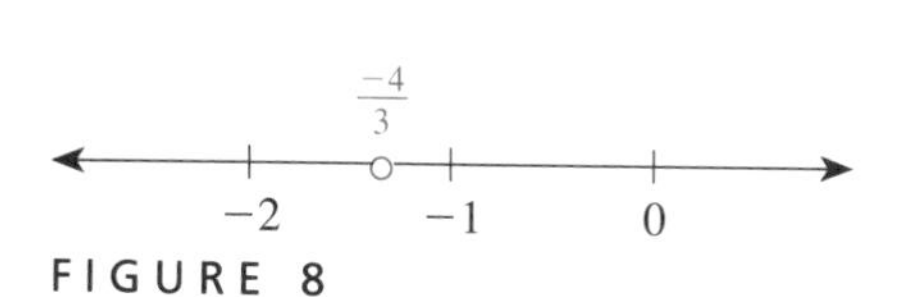

FIGURE 8

Step 2 Plot and label the point $\frac{-4}{3}$ as in Figure 8.

Step 3 We leave the circle for the point open because $\frac{-4}{3}$ does not solve the inequality:

$$3\left(\frac{-4}{3}\right) - \left[5 - \left(12 - \left(\frac{-4}{3}\right)\right)\right] > 3 - \frac{-4}{3}$$

Note the strictly greater than sign.

is a false arithmetic sentence.

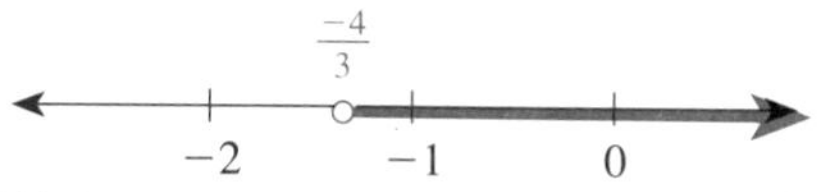

Step 4 Try 0 (a number in the right-hand interval):

$$0 - (5 - (12 - 0)) > 3 - 0$$

is a true arithmetic sentence. Thus, all the numbers to the right of the key point solve the inequality. Shade the right-hand interval, as shown in Figure 9.

FIGURE 9

Try -2 (a number in the left interval):

$$3(-2) - [5 - (12 - (-2))] > 3 - (-2)$$

is a false arithmetic sentence. Thus, -2 is not in the solution, nor is any number left of the key point $\frac{-4}{3}$. Leave the interval unshaded. ■

When drawing the graph of the solution to a first-degree inequality in one variable, if the key point is not an integer always label the key point and the two consecutive integers that bound the key point (see Figure 9).

In the first two examples, the solution to the inequality runs in the same direction as the inequality symbol. For example, the inequality $3x - 5 \leq 3$ has a solution interval that points to the left, like the symbol $\leq$. The next example shows that this is *not* always the case.

EXAMPLE 3 Graph the solution to $3x + 5 \geq 5x + 8$.

Solution Solving the equation $3x + 5 = 5x + 8$, we get:

$$\begin{aligned} 3x + 5 &= 5x + 8 \\ -5x \quad &= -5x \\ \hline -2x + 5 &= 8 \\ -5 &= -5 \\ \hline -2x &= 3 \\ x &= \frac{-3}{2} \end{aligned}$$

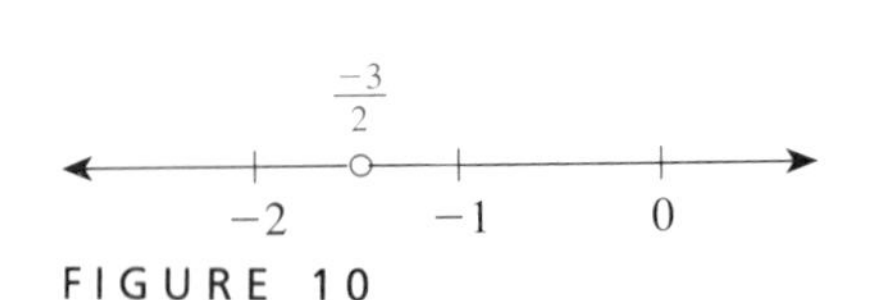

FIGURE 10

Thus, the key point in the solution to the inequality is $\frac{-3}{2}$. In Figure 10 we have plotted and labeled this point. The integer bounds are also labeled.

The number $\frac{-3}{2}$ solves the inequality, so we place a solid dot at $\frac{-3}{2}$. See Figure 11.

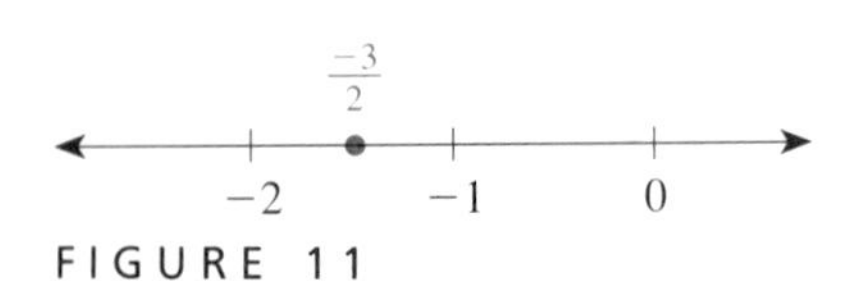

FIGURE 11

Now, we try a number in the right-hand interval, say 0:

$$3(0) + 5 \geq 5(0) + 8$$

is a false arithmetic sentence. Thus, 0 nor any of the numbers greater than $\frac{-3}{2}$ solves the inequality. We leave the interval unshaded.

Finally, we try a number in the left-hand interval, say −2:

$$3(-2) + 5 \geq 5(-2) + 8$$

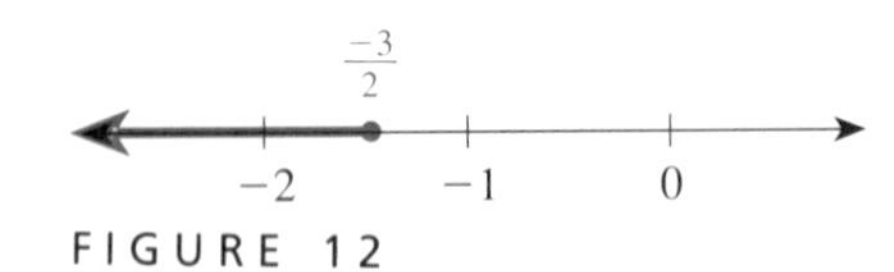

FIGURE 12

is a true arithmetic sentence. Thus, −2 and every number in the left interval solves the inequality, so we shade this interval. See Figure 12. ■

As the graph for Example 3 shows, the interval can run in the direction opposite that of the inequality symbol.

GIVE IT A TRY

Use a number line to graph the solution to each inequality.

1. $3x - 2 > 7$ **2.** $5 - 2x \leq x + 2$ **3.** $4 - (2x - 3) > 7 - x$

Interval Notation Another method for reporting the solution to a first-degree inequality in one variable is the use of interval notation. With interval notation, we use square brackets, [ ], to indicate that the key point is included in the solution (as we use a solid dot for the graph). We use parentheses, ( ), to indicate that the key point is not included in the solution (as we use an open dot for the graph). We use the symbol $-\infty$ to indicate that the solution continues to the left (as we use a left arrow for the graph), and use $+\infty$ to indicate that the solution continues to the right (as we use a right arrow for the graph). The symbol $+\infty$ is read *positive infinity*. It indicates the concept of a number being as large as we wish. Likewise, $-\infty$, or *negative infinity*, is the concept of a number being as small as we wish. The following table shows parallels between graphs and interval notation.

Inequality	*Number line graph*	*Interval notation*
$x \geq a$		$[a, +\infty)$
$x > a$		$(a, +\infty)$
$x < a$		$(-\infty, a)$
$x \leq a$		$(-\infty, a]$

To see how interval notation is used to report the solution to a first-degree inequality, consider the next example.

EXAMPLE 4 Solve $3x - 5 \leq 4$ and report the solution in interval notation.

Solution We can sketch a number line graph and then convert the graph to interval notation.

FIGURE 13

Solving the equation $3x - 5 = 4$, we get a solution of 3. Thus, the key point is 3.

Next, we plot and label the key point. Because the key point solves the inequality, we use a solid dot, as plotted in Figure 13.

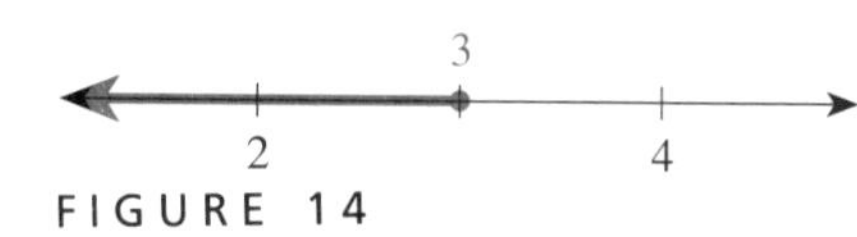

FIGURE 14

Since 0, a number in the left interval, solves the inequality, we shade the left interval. Since 4, a number in the right interval does not solve the inequality, we have the right interval unshaded. See Figure 14.

When we convert the number line graph to interval notation, the arrow pointing to the left becomes the symbol $-\infty$, and the solid dot at 3 becomes a bracket,].

Thus, we write the solution as the interval $(-\infty, 3]$. ■

GIVE IT A TRY

Use interval notation to report the solution to each inequality.

4. $3x - 2 > 7$ **5.** $5 - 2x \leq x + 2$ **6.** $4 - (2x - 3) > 7 - x$

The next example shows how to use the graphics calculator to check the solution to an inequality.

EXAMPLE 5 Use the graphics calculator to check that the solution to the inequality $-3x - 10 \geq 6 + 5x$ is the interval $[-2, +\infty)$.

Solution We start by storing ⁻2 in memory location X.

Now, type the left-hand side of the inequality, ⁻3X−10, and press the ENTER key. The number ⁻4 is output. Next, type the right member, 6+5X, and press the ENTER key. Again, the number ⁻4 is output.

Thus, the value stored in memory location X, ⁻2, is the key point because it solves the equation $-3x - 10 = 6 + 5x$.

```
TEST
1:=
2:≠
3:>
4:≥
5:<
6:≤
```

FIGURE 15

The graphics calculator is useful for identifying errors made in operations on polynomials and solving first-degree equations in one variable. To check the solution to a first-degree inequality in one variable, we use the TEST menu to judge the truth of an arithmetic sentence. The calculator outputs 1 for TRUE or 0 for FALSE.

The interval given, $[-2, +\infty)$, indicates that every number greater than -2 solves the inequality. Store a number greater than -2, say -1, in memory location X.

Type ⁻3X−10 and then press the 2nd key, followed by the MATH key, for the TEST menu shown in Figure 15. Press the 4 key to select the ≥ option. You are returned to the home screen. Complete the command to read ⁻3X−10≥6+5X. Press the ENTER key. The number 0 is output. The 0 indicates a false arithmetic sentence. So, -1 does not solve the inequality $-3x - 10 \geq 6 + 5x$. An error is present!

The correct solution is $(-\infty, -2]$. To verify this, store a number less than -2, say ⁻3, in memory location X. Use the TEST menu to type ⁻3X−10≥6+5X. Press the ENTER key. The number 1 is output. The 1 indicates a true arithmetic sentence. That is, -3 is a number that solves the inequality. ■

Using the Graphics Calculator

```
RANGE
Xmin=⁻4.8
Xmax=4.7
Xscl=1
Ymin=⁻5
Ymax=10
Yscl=1
Xres=1
```

FIGURE 16

We can use the graph screen on a graphics calculator to gain insight into the solution to a first-degree inequality in one variable. Consider the inequality $-2x + 4 \leq 5$. The solution to this inequality is the collection of numbers (replacement values for x) that result in the polynomial $-2x + 4$ evaluating to a value smaller than or equal to 5. To see these values, for Y1 type ⁻2X+4 and for Y2, type 5. Press the RANGE key and enter the values shown in Figure 16.* Figure 17 shows the graph screen. Press the TRACE key. When x is replaced by 0, the polynomial $-2x + 4$ evaluates to 4, which is less than 5. Thus, 0 is in the solution. Press the left arrow key. When x is replaced by -0.5, the polynomial $-2x + 4$ equals 5. So, -0.5 is the key point. For numbers greater than -0.5, the polynomial $-2x + 4$ evaluates to a number less than 5. For the numbers less than -0.5, the polynomial $-2x + 4$ evaluates to a number greater than 5. See the number line graph of the solution shown in Figure 18. In interval notation, we write the solution as $[-0.5, +\infty)$.

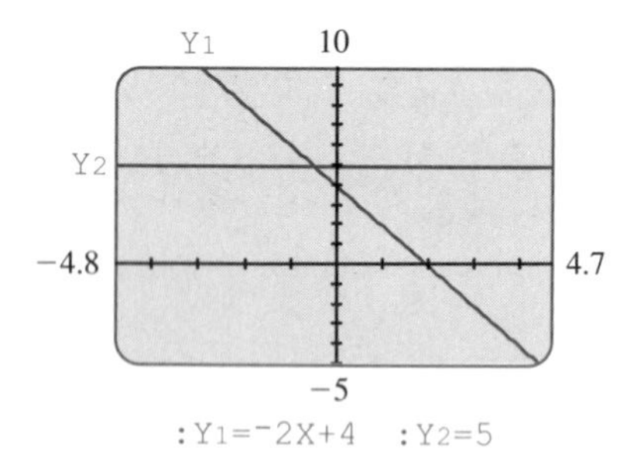

FIGURE 17

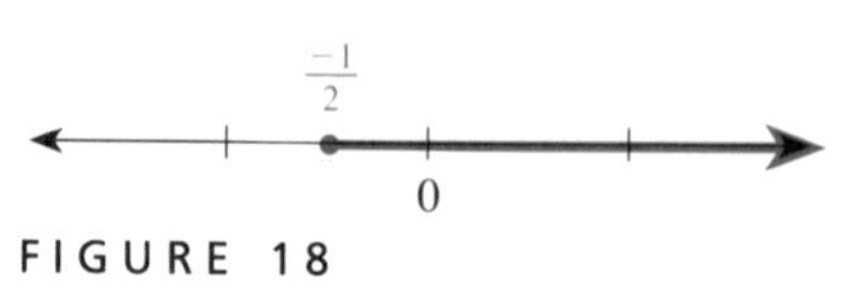

FIGURE 18

GIVE IT A TRY

7. Use the graphics calculator to check that the solution to the inequality $-2x + 10 \geq 3 + 5x$ is the interval $[1, +\infty)$.

8. Solve each inequality, and use the graphics calculator and its graph screen to check the answer.
   a. $2x - 3 \leq -3$ **b.** $-2 \geq 3 - 2x$

TI-82 Note *Type ⁻4.7 for Xmin.

Special Cases of Inequalities Just as there are special cases of solutions for first-degree equations in one variable (a solution of *all* numbers and a solution of *none*), we have special cases of solutions for inequalities. Consider the next example.

EXAMPLE 6 Graph the solution to $x + 3 > x$.

Solution Perhaps from just looking at this inequality, you can see that the solution is all the numbers. (If you add three to a number, the results will be larger than the original number.)

> **WARNING**
> The solution to the related equation can be *none,* yet the solution to the inequality is *all*.

Using the key-point approach to produce a number line graph, we first solve the equation $x + 3 = x$.

Solve:	$x + 3 = x$
Add $-1x$ to each side	$-1x \quad -1x$
to get:	$3 = 0$

This is a false arithmetic sentence. The equation has solution *none*. Thus, there is no key point in the solution to the inequality $x + 3 > x$. So, there is no number that divides the number line into intervals. There is only one interval (the entire number line), and the solution is either *none* or *all*.

Try a number in this interval, say 0: $0 + 3 > 0$ is a true arithmetic sentence. Thus, every number in the interval solves the inequality, so the solution is *all*. Figure 19 shows that the interval is the entire number line. ■

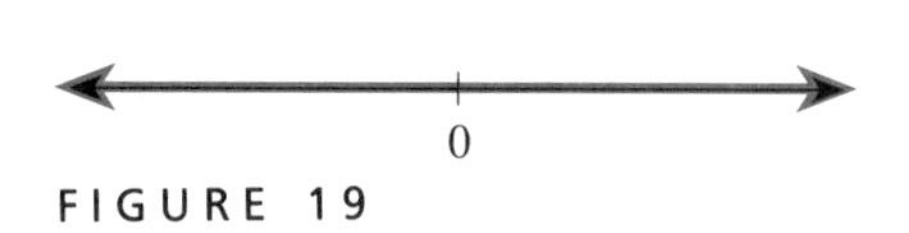

FIGURE 19

In interval notation, we report the solution to the inequality in Example 6 as $(-\infty, +\infty)$. The next example shows the other special case, the empty set, as the solution.

EXAMPLE 7 Graph the solution to $x + 1 < 1 + x$.

Solution Looking at the inequality reveals that the solution is *none*. For every replacement of x, the polynomial $x + 1$ is never less than $1 + x$ (they are always equal). Using our key-point approach to solving inequalities, we first solve the equation $x + 1 = 1 + x$.

Solve:	$x + 1 = 1 + x$	
Add $-1x$ to each side	$-1x \quad -1x$	
to get:	$1 = 1$	A true arithmetic sentence

Thus, the solution to the equation is *all* numbers. So, each number is a key point. Of course, we cannot place an open circle on each point on the number line, but, we can imagine doing so. Because none of these key points produce a true arithmetic sentence, none of the points is shaded. Figure 20 shows the number line graph. ■

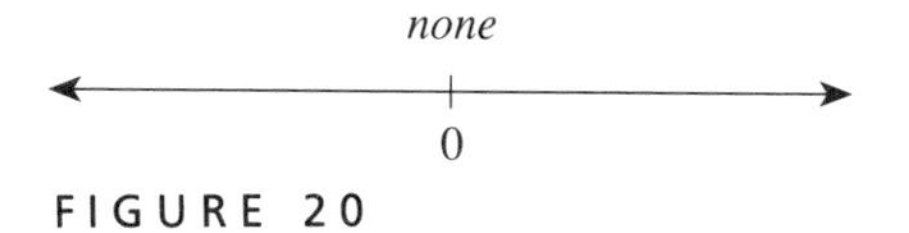

FIGURE 20

To report the solution to the inequality in Example 7 using interval notation, we report $\varnothing$, which indicates the interval that contains no number, the empty interval.

As a last example of a special case, work through the next example.

EXAMPLE 8 Graph the solution to $x + 1 < 3 + x$.

Solution Again, looking at the inequality reveals that the solution is *all*. For every replacement of x, the polynomial $x + 1$ is always less than $3 + x$. Using the key-point approach to solving inequalities, we first solve the related equation:

Solve:	$x + 1 = 3 + x$	
Add $-1x$ to each side	$-1x \quad -1x$	
to get:	$1 = 3$	A false arithmetic sentence

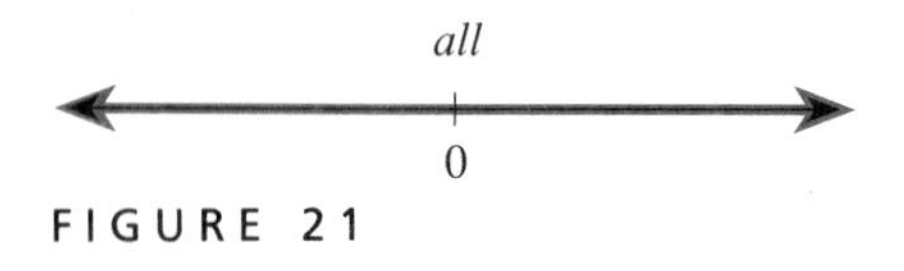

FIGURE 21

Thus, the solution to the equation is *none*, so no number is a key point. The solution will be either the entire number line or *none*. Try a number from the one interval (the entire number line), say 0: $0 + 1 < 3 + 0$ is a true arithmetic sentence. So, the entire number line is shaded. In interval notation we report a solution of $(-\infty, +\infty)$. See Figure 21. ■

Using the Graphics Calculator

To check the solution for the inequality in Example 8, type X+1 for Y1 and 3+X for Y2. View the graph screen using the RANGE settings in Figure 22.* The first line drawn represents evaluations of $x + 1$ for various replacements of the x-variable. The second line drawn represents evaluations of $3 + x$ for replacements of the x-variable. Figure 23 shows the graph screen. For any given replacement of the x-variable, the evaluation of $x + 1$ is less than the evaluation for $3 + x$.

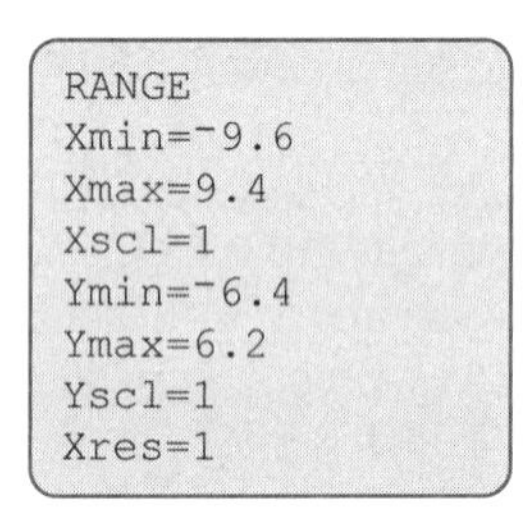

FIGURE 22

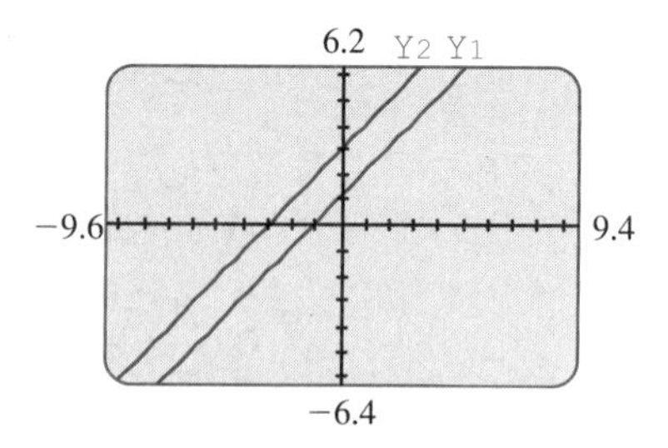

FIGURE 23

GIVE IT A TRY

Graph the solution to each inequality, and report the solution in interval notation.

9. $-3x + 5 > 6 - (x + 2)$

10. $-3x + 5 \leq -6 - 5x$

11. $-3x + 5 \leq 5 - 3x$

12. $5 - x < 2 - x$

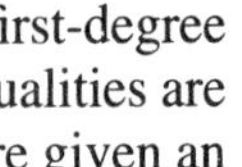

Equivalent Inequalities An alternative approach to solving a first-degree inequality in one variable is that of *equivalent inequalities*. Two inequalities are equivalent inequalities provided they have the same solution. If we are given an inequality, we can create an equivalent inequality using any of the following properties.

TI-82 Note *Type ⁻9.4 for Xmin and ⁻6.2 for Ymin.

PROPERTIES OF EQUIVALENT INEQUALITIES

To produce an equivalent inequality,

1. add a polynomial to the left- and right-hand sides of the inequality.
   Example: If $x < 5$, then $x + (-8) < 5 + (-8)$
   $x - 8 < -3$
2. multiply both the left- and right-hand sides by a positive number.
   Example: If $x < 5$, then $3x < 3(5)$
   $3x < 15$
3. multiply both the left- and right-hand sides by a negative number, and then reverse the inequality (replace $<$ with $>$, or $\geq$ with $\leq$, and so on).
   Example: If $x < 5$, then $-3x > -3(5)$
   $-3x > -15$

To solve an inequality by this approach, we create equivalent inequalities until we have one for which the solution is obvious. Consider the following examples.

EXAMPLE 9 Solve $-3x + 2 \geq 7$, and give the solution in interval notation.

Solution We begin by working directly with the inequality:

$$-3x + 2 \geq 7$$

Add -2 to each side to get:

$$\frac{\quad -2 \quad -2}{-3x \geq 5}$$

Multiply each side by $\frac{-1}{3}$, and reverse the inequality sign:

$$x \leq \frac{-5}{3}$$

The solution to $x \leq \frac{-5}{3}$ is $\left(-\infty, \frac{-5}{3}\right]$.

Because $-3x + 2 \geq 7$ is equivalent to the inequality $x \leq \frac{-5}{3}$, it also has solution $\left(-\infty, \frac{-5}{3}\right]$. ■

EXAMPLE 10 Solve $5x + 2 < 7 - 2x$, and give the solution in interval notation.

Solution Again, we begin by working directly with the inequality.

$$5x + 2 < 7 - 2x$$

Add $2x$ to each side to get:

$$\frac{2x \qquad\qquad 2x}{7x + 2 < 7}$$

Add -2 to each side to get:

$$\frac{\quad -2 \quad -2}{7x < 5}$$

Multiply each side by $\frac{1}{7}$:

$$x < \frac{5}{7}$$

The solution to $x < \frac{5}{7}$ is $\left(-\infty; \frac{5}{7}\right)$.

Because $5x + 2 < 7 - 2x$ is equivalent to the inequality $x < \frac{5}{7}$, it also has solution $\left(-\infty, \frac{5}{7}\right)$. ■

GIVE IT A TRY

Solve each inequality using the equivalent inequalities approach. Report the solution in interval notation.

13. $-3x - 5 \geq 2x + 7$

14. $6 - 3x \geq -3$

15. $2x - (6 - x) < 2 + 2x$

16. $2 - \frac{1}{2}x > \frac{-1}{3}x + 1$

The equivalent inequalities approach to solving inequalities is very efficient at this level of algebra. However, as inequalities become more complicated (such as $x^2 - 3x < 1$), the equivalent inequalities approach becomes more difficult to use. A strong point of the key-point approach is that as the inequalities get more involved, only the number of key points increases. The basic approach (the steps used) never varies.

■ SUMMARY

In this section, we have used our knowledge and skill in solving first-degree equations in one variable to solve first-degree inequalities in one variable. A first-degree inequality in one variable typically has a solution consisting of an infinite collection of numbers. We can describe this set of numbers on a number line graph. The solution is usually an interval starting at a particular number—the key point—and extending indefinitely to the left or to the right. Special cases of the graph of the solution are the entire number line and no number on the line.

To solve a first-degree inequality in one variable and graph the solution on a number line graph, we can use the key-point approach:

1. **Solve the equation** to find the key point for the graph of the solution.
2. **Plot and label the key point** found in Step 1.
3. **See if the key point solves the inequality**. Place a solid dot on the key point if it solves the inequality. Otherwise place an open dot on the key point.
4. **Test the intervals formed**. Check a number in the interval to the left of the key point to see if it solves the inequality. If a number from the left-hand interval solves the inequality, then every number in this interval solves the inequality, and the interval is shaded on the graph. If not, none of the numbers in this interval solves the inequality, and the interval remains unshaded. Similarly, check a number in the right interval, if a number from this interval solves the inequality, then all the numbers in the interval solve the inequality, and it is shaded on the graph.

Instead of drawing a number line graph, we can report the solution to a first-degree inequality in one variable in interval notation. An interval such as $[-1, +\infty)$ indicates that -1, indicated by the symbol [, and all the numbers larger than -1, indicated by the symbol $+\infty$, solve the inequality. Reporting an interval such as $(-\infty, 8)$ indicates that all the numbers less than 8 (indicated by $-\infty$) solve the inequality. The symbol) indicates that the number 8 is not included in the interval. To indicate that all numbers solve the inequality, we write the interval $(-\infty, +\infty)$. To indicate that no number solves the inequality, we write $\varnothing$.

An alternative to the graphical key-point approach is the use of equivalent inequalities. An equivalent inequality is produced by adding a polynomial to each side of the inequality, or by multiplying each side of the inequality by a positive number, or by multiplying each side of the inequality by a negative number and then reversing the inequality sign ($\geq$ becomes $\leq$). Using this approach, equivalent inequalities are written until the solution becomes obvious (such as $x > 2$).

2.3 ■ ■ ■ EXERCISES

Report true *for a true statement, or* false *for a false statement.*

____ 1. The solution to a first-degree inequality in one variable is usually a single number.

____ 2. In solving a first-degree inequality in one variable, we can solve the related equation to find the key point for graphing the solution to the inequality.

____ 3. The graph of the solution to a first-degree inequality in one variable is drawn on a number line graph.

____ 4. The solution to a first-degree inequality in one variable can be the empty interval (no number solves the inequality).

____ 5. The solution to the inequality $2 + x > x$ is all the real numbers.

____ 6. Interval notation for all the numbers to the right of -2 is $(-2, +\infty)$.

____ 7. Interval notation for all the numbers to the left of 5 is $(-\infty, 5]$.

____ 8. The graph of the solution to a first-degree inequality in one variable may have two key points, say -3 and 7.

Describe each number line graph in interval notation.

9.

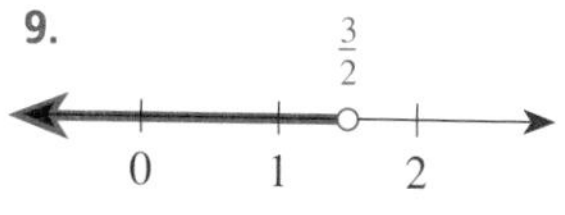

10.

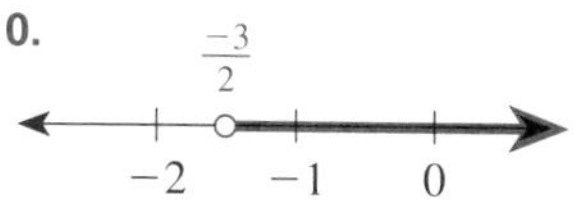

11.

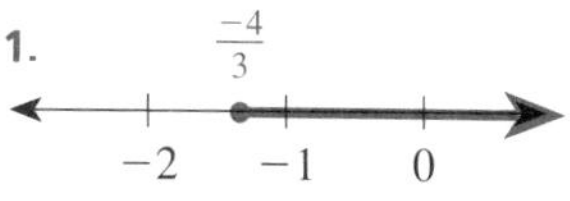

12.

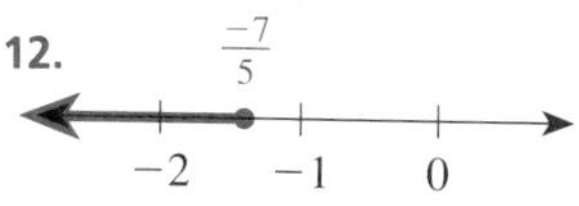

Draw a number line graph for each interval.

13. $(-2, +\infty)$

14. $(-\infty, 1)$

15. $\left(-\infty, \dfrac{-9}{5}\right]$

16. $\left[\dfrac{-7}{2}, +\infty\right)$

Graph the solution to the given first-degree inequality in one variable. Also report the solution in interval notation. Use the graphics calculator to check the solution.

17. $3x > -12$

18. $9x > 27$

19. $-3x \geq 12$

20. $-x \leq -5$

21. $\dfrac{2}{5}x \leq \dfrac{3}{4}$

22. $-2 \leq \dfrac{-3}{7}x$

23. $5x - 8 > 14$

24. $6 + 4x \geq 8$

25. $6 - 3p \leq -2$

26. $12 - 5k \leq 2$

27. $5x + 6 \geq 6$

28. $-3x - 2 \leq 8$

29. $15 < 3 - 2x$

30. $15 \geq -2 - x$

31. $-21 \leq 5w + 4$

32. $2 \geq 3a + 5$

33. $\dfrac{1}{2}x - 2 > \dfrac{3}{4}$

34. $\dfrac{2}{5}x + \dfrac{1}{3} \leq 2$

35. $\dfrac{-5}{7} - \dfrac{2}{5}x \leq -2$

36. $3 - \dfrac{3}{4}x < \dfrac{7}{6}$

37. $3r + 8 \geq 5 - 2r$

38. $5 - 5y < 2y + 6$

39. $2 - x < 3 + 5x$

40. $16 - 5x \geq 3x$

41. $2c > 2c - 5$

42. $5m \geq 5m - 5$

43. $-2(3 - 2x) \leq x$

44. $16 - x \geq 2x + 1$

45. $16 - 3x \geq 2(3 - x)$

46. $1.2x < -0.2(3 - x)$

47. $3x - (5 - x) \leq 5(x - 1)$

48. $5x - 2 > 7x + 4$

49. $5x - 2 \geq 7x + 1$

50. $3x - 3 \leq x - (x - 1)$

51. $2x - (3 - 2x) > 3$

52. $2x - (3 - 2x) < 3$

53. $5x + 2 \leq 3x - 2(2 + x)$

54. $3 - [6 - (2x - 5)] > 5 - x$

55. $2x - 3 \leq x - [2 - (2x - 1)]$

56. $0.2(x - 1) - 1 < 0.1x$

57. $\dfrac{2x+5}{3} > \dfrac{16-2x}{4}$

58. $\dfrac{7x-2}{2} < \dfrac{5x-2}{5}$

59. $\dfrac{2-5x}{2} \le \dfrac{6-5x}{7}$

60. $\dfrac{7-6x}{8} \ge \dfrac{x-1}{3}$

Solve each inequality, and use the graphics calculator and its graph screen to check the solution. If no number solves the inequality, write none. *If all numbers solve the inequality, write* all.

61. $2x - 3 \le 2x$

62. $-2x \ge 3 - 2x$

63. $5x - 2 \ge 3 + 5x$

64. $-3 - x < -1 - x$

65. $-3x + 1 \ge -3x$

66. $2x - 1 \ge 2x - 2$

67. $5x - 3 > 3$

68. $2x + 1 > 1 + 2x$

Solve each inequality using the equivalent inequalities approach. Produce a number line graph of the solution and also report the solution in interval notation. Use the graphics calculator to check the solution.

69. $2x - 3 \le x + 5$

70. $3x \ge 3 - 2x$

71. $2x - 1 \le 7x + 4$

72. $2x \ge 3 + 4x$

73. $x - 3 > 3x + 5$

74. $x - 5 \le 3 - 5x$

75. $6 - 3x < x - 6$

76. $1 - x > 3 + 2x$

77. $2x - 3 \ge 2x + 5$

78. $x \ge 3 - 2(x - 5)$

2.4 ■ ■ ■ COMPOUND INEQUALITIES

In this section we continue our work with inequalities by learning how to solve compound sentences that involve first-degree inequalities. This new concept will depend on our ability to solve first-degree inequalities in one variable. We will use the graphics calculator to check and to gain greater understanding of the solution.

Compound Inequalities In Section 2.3 we learned to solve first-degree inequalities in one variable. An extension of this topic involves inequalities in the form $a \le x \le b$. This type of inequality is known as a **compound inequality**. An example of a compound inequality is

$$-3 \le 2x + 5 \le 8$$

The solution to this inequality is the set of x-values that result in the binomial $2x + 5$ evaluating to any number *between* -3 and 8, inclusive.

```
RANGE
Xmin=-9.6
Xmax=9.4
Xscl=1
Ymin=-6
Ymax=10
Yscl=1
Xres=1
```

FIGURE 1

Using the Graphics Calculator

The graphics calculator can quickly provide a *feel* for the numbers that solve a compound inequality such as $-3 \le 2x + 5 \le 8$. For Y1 type 2X+5, for Y2 type ⁻3, and for Y3 type 8. View the graph screen using the RANGE settings shown in Figure 1. See Figure 2. Press the TRACE key and then use the left arrow key to move to the intersection of the line for $2x + 5$ and the horizontal line for -3. The x-value is -4. Now, press the right arrow key. Observe that for x-values between -4 and 1.4, the evaluations of $2x + 5$ are between -3 and 8. So, the solution appears to have two key points, -4 and approximately 1.4.

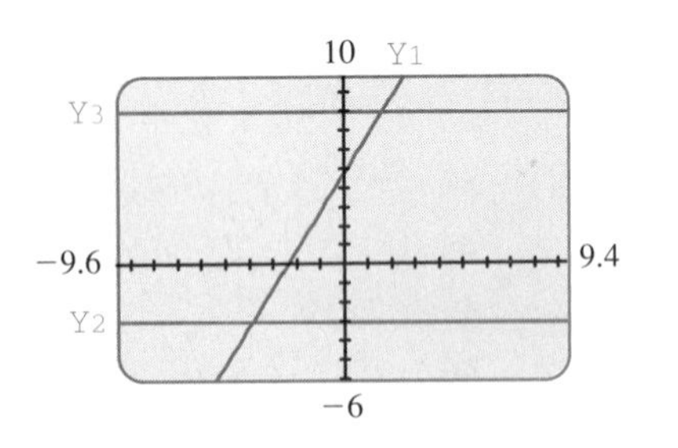

FIGURE 2

We will now solve the inequality $-3 \le 2x + 5 \le 8$ using an algebraic approach. This inequality translates to the compound sentence

$$-3 \le 2x + 5 \quad and \quad 2x + 5 \le 8$$

TI-82 Note *Type ⁻9.4 for Xmin and ⁻6.2 for Ymin.

A number that is a solution of the inequality $-3 \leq 2x + 5 \leq 8$ must be a solution of *each* of the inequalities $-3 \leq 2x + 5$ and $2x + 5 \leq 8$. We can solve the inequality $-3 \leq 2x + 5 \leq 8$ by first finding the key points from the equations $-3 = 2x + 5$ and $2x + 5 = 8$. Each of these equations produces one key point. Plotting these key points on a number line creates three intervals. We test a number in each of the three intervals to determine the solution interval. In the next example we show this approach.

EXAMPLE 1 Graph the solution to $-3 \leq 2x + 5 \leq 8$.

Solution The solution is all the numbers that result in a true arithmetic sentence. Such a number will result in both $-3 \leq 2x + 5$ yielding a true arithmetic sentence and $2x + 5 \leq 8$ yielding a true arithmetic sentence. To find the key points, we solve the related equations.

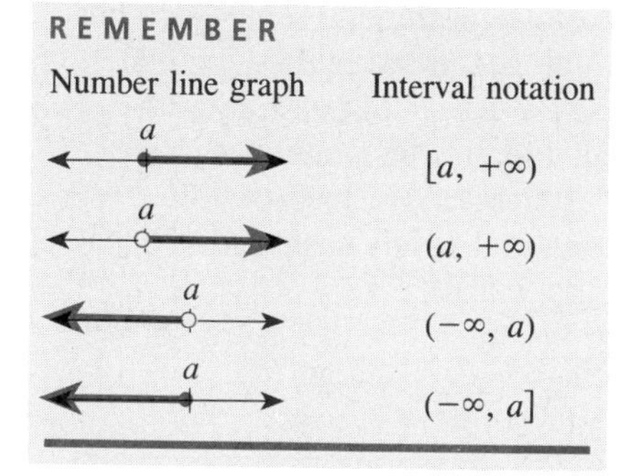

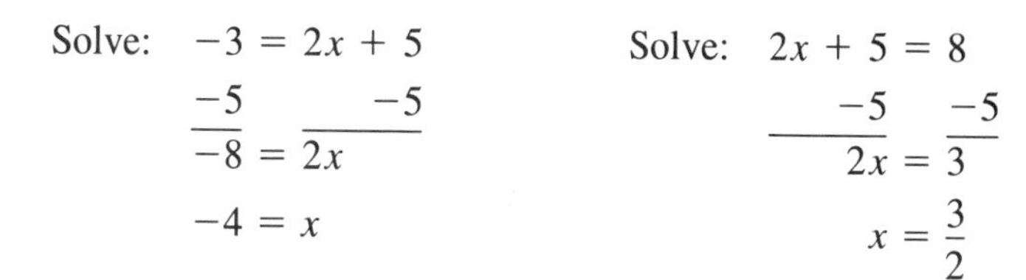

Solve: $-3 = 2x + 5$; $-5 \quad -5$; $-8 = 2x$; $-4 = x$

Solve: $2x + 5 = 8$; $-5 \quad -5$; $2x = 3$; $x = \frac{3}{2}$

Thus, the key points are -4 and $\frac{3}{2}$. We now plot and label these key points, as shown in Figure 3.

FIGURE 3

A solid dot is used at each key point because the number -4 and the number $\frac{3}{2}$ each solves the inequality $-3 \leq 2x + 5 \leq 8$. See Figure 4.

FIGURE 4

Now we test a number in each interval. We try a number in the leftmost interval, say -5. Replacing the variable by -5 yields

$$-3 \leq 2(-5) + 5 \leq 8 \qquad \text{or} \qquad -3 \leq -5 \leq 8$$

a false arithmetic sentence. Thus, the leftmost interval is left unshaded.

Trying a number in the middle interval, say 0, yields $-3 \leq 2(0) + 5 \leq 8$, a true arithmetic sentence. Thus, we shade the middle interval. See Figure 5.

Trying a number in the rightmost interval, say 2, yields

$$-3 \leq 2(2) + 5 \leq 8 \qquad \text{or} \qquad -3 \leq 9 \leq 8$$

a false arithmetic sentence. Thus, the rightmost interval is left unshaded.

The graph of the solution is as shown in Figure 5. ■

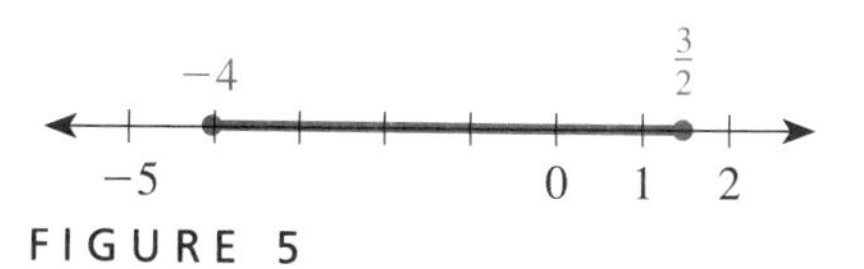

FIGURE 5

The solution to the inequality in Example 1 is written in interval notation as $\left[-4, \frac{3}{2}\right]$.

As a second example of solving compound inequalities, work through the following.

EXAMPLE 2 Graph the solution to $2 > -3x + 5 \geq -6$.

Solution We are seeking the numbers that result in a true arithmetic sentence. Such a number will result in both $2 > -3x + 5$ yielding a true arithmetic sentence *and* $-3x + 5 \geq -6$ yielding a true arithmetic sentence. To find these numbers, we solve the equations $2 = -3x + 5$ and $-3x + 5 = -6$ to get the key points:

Solve:
$$\begin{aligned} 2 &= -3x + 5 \\ -5 \quad & \quad -5 \\ \hline -3 &= -3x \\ 1 &= x \end{aligned}$$

Solve:
$$\begin{aligned} -3x + 5 &= -6 \\ -5 \quad & \quad -5 \\ \hline -3x &= -11 \\ x &= \frac{11}{3} \end{aligned}$$

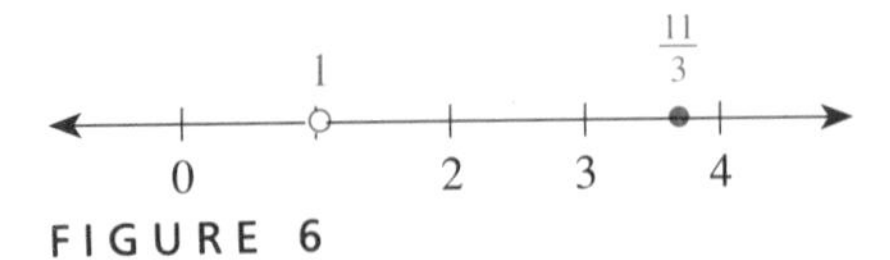

FIGURE 6

Thus, the key points are 1 and $\frac{11}{3}$. Plot and label these key points as shown in Figure 6. The dot at $\frac{11}{3}$ is solid, because $\frac{11}{3}$ solves the inequality

$$2 > -3x + 5 \geq -6$$

The dot at 1 is left open, because 1 does not solve the inequality.

Test the intervals. Try a number in the leftmost interval, say 0. Now, $2 > -3(0) + 5 \geq -6$, or $2 > 5 \geq -6$, is a false arithmetic sentence. Thus, the leftmost interval is left unshaded. Try a number in the middle interval, say 2. Because $2 > -3(2) + 5 \geq -6$, or $2 > -1 \geq -6$, is a true arithmetic sentence, the middle interval is shaded. Finally, try a number in the rightmost interval, say 4. Because $2 > -3(4) + 5 \geq -6$, or $2 > -7 \geq -6$, is a false arithmetic sentence, the rightmost interval is left unshaded. The solution to the inequality is as shown in Figure 7.

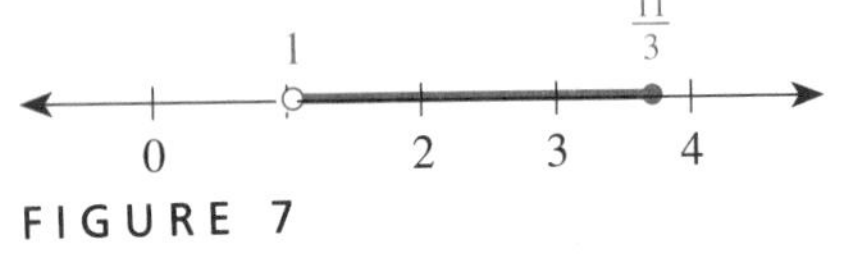

FIGURE 7

Using interval notation to report the solution to $2 > -3x + 5 \geq -6$, we write $\left(1, \frac{11}{3}\right]$. ■

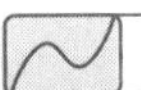
Using the Graphics Calculator

REMEMBER

The TI-81 and T1-82 evaluates a relation, such as $3 \geq 5$, and outputs a 1 for TRUE and a 0 for FALSE.

To check our work in Example 2, start by storing a number in the leftmost interval, say 0, in memory location X. Now, type 2>-3X+5 (using the TEST menu to type the > character), and press the ENTER key. The value 0 is output, indicating a false arithmetic sentence. For an AND statement to be true, both parts must be true. So, we need not test the other inequality, $-3x + 5 \geq -6$ to verify that the leftmost interval should be unshaded.

Store a number from the middle interval, say 2, in memory location X. Type 2>−3X+5 and press the ENTER key. The number 1 is output, indicating a true arithmetic sentence. Type -3X+5≥-6 and press the ENTER key. The number 1 is output, indicating a true arithmetic sentence. See Figure 8. Because both inequalities are satisfied, the number in memory location X, the number 2, satisfies the compound inequality. So the interval containing 2 should be shaded.

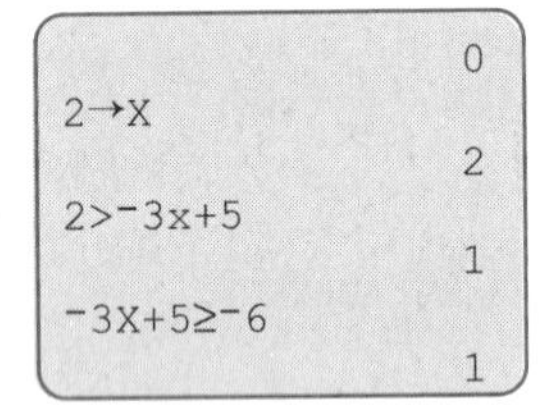

FIGURE 8

Finally, store a number from the rightmost interval, say 5, in memory location X. Type 2>-3X+5 and press the ENTER key. A 1 is output, indicating a true arithmetic sentence. Testing the other inequality, type -3X+5≥-6 and press the ENTER key. A 0 is output, indicating a false arithmetic sentence. See Figure 9. Thus, the number in memory location X, the number 5, does not satisfy the compound inequality (it did not yield a true arithmetic for *both* inequalities). The rightmost interval is left unshaded. Figure 7 shows a number line graph of the solution.

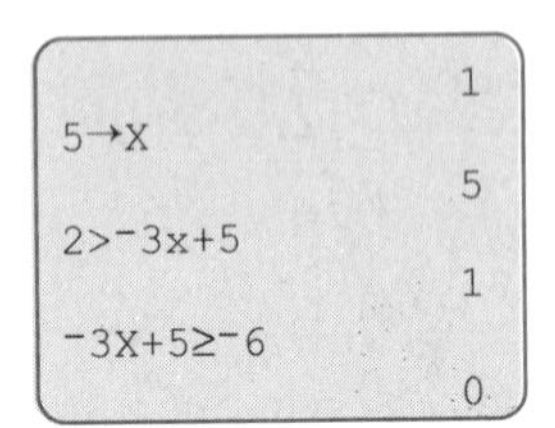

FIGURE 9

GIVE IT A TRY

Solve each compound inequality. Draw a number line graph for the solution and also report the solution in interval notation.

1. $-3 \le 2x - 3 \le 2$
2. $3 \ge 2x - 3 \ge -2$
3. $-5 < 2x - 3 < 2$
4. $-5 < -2x + 3 \le 2$
5. $2 \ge -3x + 5 \ge -2$
6. $-2 < -2 - 3x \le 5$
7. Use the graphics calculator to check the solutions to Problems 1–6.

Using the Graphics Calculator

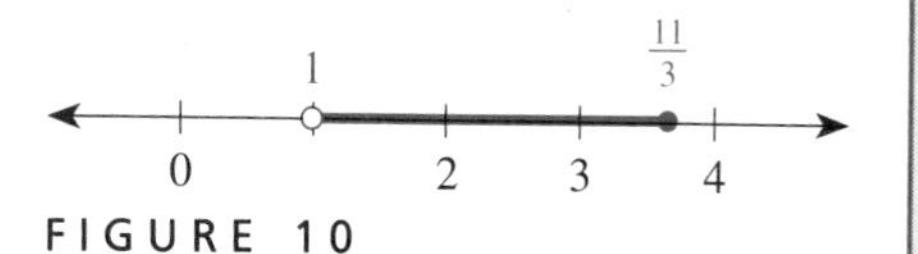

FIGURE 10

```
RANGE
Xmin=-9.6
Xmax=9.4
Xscl=1
Ymin=-10
Ymax=4
Yscl=1
Xres=1
```
FIGURE 11

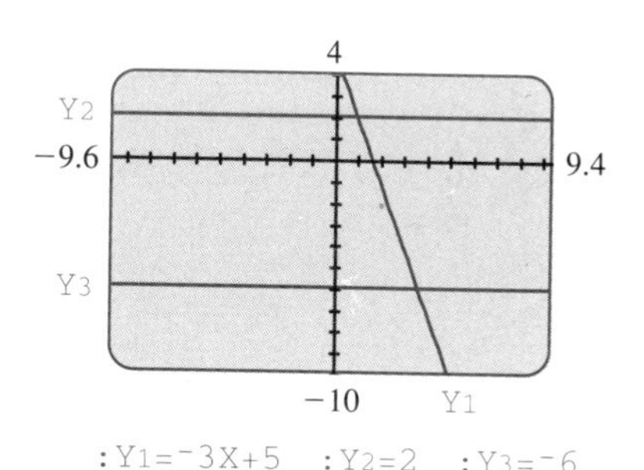

FIGURE 12

We can use the graphics calculator to increase our understanding of the solution to a compound inequality. Consider the compound inequality $2 > -3x + 5 \ge -6$ from Example 2. We reported a solution of $\left(1, \frac{11}{3}\right]$. The number line graph is shown in Figure 10. For `Y1`, type the polynomial `-3X+5`. For `Y2`, type `2`, and for `Y3` type `-6`. The solution to the compound inequality is all the replacements for the x-variable that result in the polynomial $-3x + 5$ evaluating to a number that is less than 2, but larger than -6, and the number -6.

View the graph screen using the settings in Figure 11. A line is drawn for the evaluations of the polynomial $-3x + 5$. Also, a horizontal line is drawn for 2, and a horizontal line is drawn for -6, as shown in Figure 12. We will now trace along the line and use evaluations that we can compute mentally. Press the `TRACE` key. When x is replaced by 0, the polynomial $-3x + 5$ evaluates to 5, a number outside the band of 2 to -6. Thus, 0 is not in the solution to the inequality. Press the right arrow key. When x is replaced by 1, the polynomial $-3x + 5$ evaluates to 2. Thus, 1 is a key point. Press the right arrow key. As you can see, for these x-values larger than 1, the polynomial evaluates to a number less than 2 and larger than -6. These x-values are in the solution to the inequality (graphically, they result in the polynomial evaluating to a number in the band from 2 down to -6). Continue to press the right arrow key. When the x-value is larger than $3\frac{2}{3}$ $\left(\text{or } \frac{11}{3}\right)$, the polynomial $-3x + 5$ evaluates to a number less than -6. The x-values larger than $3\frac{2}{3}$ do not solve the compound inequality.

Equivalent Compound Inequalities An alternative approach to solving a compound inequality in one variable is that of *equivalent inequalities*. Two inequalities are equivalent provided they have the same solution. Given a compound inequality, we can write an equivalent compound inequality using the properties of inequalities listed in Section 2.3. Here, we must apply the property to each *member* of the inequality.

1. Add a polynomial to each member of the inequality.

   Example: If $-3 < x < 5$, then $-3 + 3 < x + 3 < 5 + 3$
   or $0 < x + 3 < 8$

2. Multiply each member by a positive number.

   Example: If $-3 < x < 5$, then $3(-3) < 3x < 3(5)$
   or $-9 < 3x < 15$

3. Multiply each member by a negative number, and then reverse each inequality (replace $<$ with $>$ or $\geq$ with $\leq$ and so on).

   Example: If $-3 < x < 5$ then $-3(-3) > -3x > -3(5)$
   or $-9 > -3x > -15$

To solve a compound inequality by this approach, we write equivalent inequalities until the solution is obvious. Consider the following examples.

EXAMPLE 3 Solve $-2 \leq -3x + 2 \leq 7$, and report the solution in interval notation.

Solution Here, we work directly with the inequality.

$$-2 \leq -3x + 2 \leq 7$$

Add -2 to each member to get:

$$\begin{array}{rcccl} -2 & & -2 & & -2 \\ \hline -4 \leq & & -3x & & \leq 5 \end{array}$$

Multiply by $\frac{-1}{3}$ and reverse the inequality:

$$\frac{4}{3} \geq x \geq \frac{-5}{3}$$

The solution to $\frac{4}{3} \geq x \geq \frac{-5}{3}$ is $\left[\frac{-5}{3}, \frac{4}{3}\right]$.

Because $-2 \leq -3x + 2 \leq 7$ is an equivalent inequality for $\frac{4}{3} \geq x \geq \frac{-5}{3}$, it also has solution $\left[\frac{-5}{3}, \frac{4}{3}\right]$. ■

WARNING
It is improper to write the solution interval for Example 3 as $\left[\frac{4}{3}, \frac{-5}{3}\right]$. Instead, write the smaller endpoint first, $\left[\frac{-5}{3}, \frac{4}{3}\right]$.

GIVE IT A TRY

Solve each inequality using the equivalent inequality approach. Report the solution using interval notation.

8. $0 < -3x - 5 < 7$
9. $8 \geq 6 - 3x \geq -3$

Advantages of the Key Point Approach Although the equivalent inequalities approach to solving inequalities is important and is often the most efficient approach, we now consider writing equivalent inequalities for the compound inequality:

$$9 \leq 2x + 3 < 4x + 1$$

Add -3 to each member:

$$\begin{array}{rcl} -3 & -3 & -3 \\ \hline 6 \leq & 2x & < 4x - 2 \end{array}$$

Multiply by $\frac{1}{2}$:

$$3 \leq x < 2x - 1$$

Notice that this last inequality does not immediately yield an interval. That is, writing equivalent inequalities does not produce an inequality whose solution interval is obvious. In the next example, we show how to solve this inequality using the key point approach.

EXAMPLE 4 Solve: $9 \leq 2x + 3 < 4x + 1$

Solution The solution is all the numbers that result in a true arithmetic sentence. Such a number will result in both $9 \leq 2x + 3$ yielding a true arithmetic sentence and $2x + 3 < 4x + 1$ yielding true arithmetic sentence. To find the key points, we solve the related equations.

Solve:
$$\begin{aligned} 9 &= 2x + 3 \\ -3 & \quad\ -3 \\ \hline 6 &= 2x \\ 3 &= x \end{aligned}$$

Solve:
$$\begin{aligned} 2x + 3 &= 4x + 1 \\ -4x \quad & \quad -4x \\ \hline -2x + 3 &= 1 \\ -3 & \quad -3 \\ \hline x &= 1 \end{aligned}$$

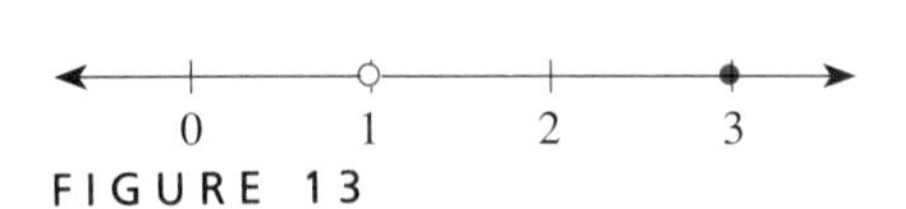

FIGURE 13

Thus, the key points are 1 and 3. We now plot and label these key points. A solid dot is used for 3 and an open dot is used for 1. See Figure 13.

Test a number in the leftmost interval, say 0: $9 \le 2(0) + 3 < 4(0) + 1$, or $9 \le 3 < 1$, is a false arithmetic sentence. So, we leave the interval unshaded.

Test a number in the middle interval, say 2: $9 \le 2(2) + 3 < 4(2) + 1$, or $9 \le 7 < 9$, is a false arithmetic sentence. So, the interval is left unshaded.

FIGURE 14

Test a number in the rightmost interval, say 4: $9 \le 2(4) + 3 < 4(4) + 1$, or $9 \le 11 < 17$, is a true arithmetic sentence. So, we shade the interval. See Figure 14.

In interval notation, we report the solution as $[3, +\infty)$. ■

GIVE IT A TRY

Solve each inequality. Report the solution using interval notation.

10. $2 \le 2x - (6 - x) < 4 + 2x$ **11.** $x - 1 < 2(3 - x) \le x + 1$

■ SUMMARY

In this section we have applied the general procedure, or algorithm, for solving inequalities to compound inequalities:

1. Solve the related equation(s) to find the key point(s).
2. Plot and label the key points on a number line graph.
3. Determine whether each key point should be shown as an open dot or a solid dot.
4. Test the intervals created by the key points.

A compound inequality, such as $-2 \le 3x - 5 \le 4$, has two related equations, $-2 = 3x - 5$ and $3x - 5 = 4$. These equations are solved to get two key points, which divide the number line into three intervals. A number in each interval is tested. If the test number solves the original inequality, then each number in the interval solves the inequality and the interval is shaded. If the test number fails, each number in that interval fails to solve the inequality and the interval is left unshaded.

2.4 ■ ■ ■ EXERCISES

Write the interval notation for the interval shown on each number line graph.

1.

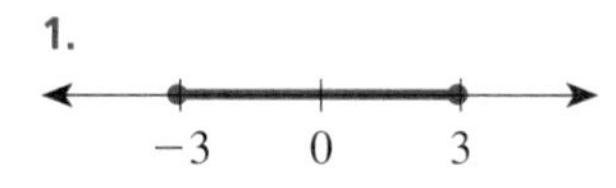

2.

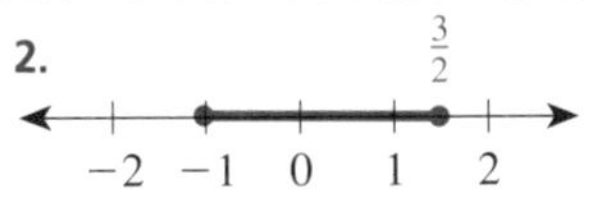

3.

4.

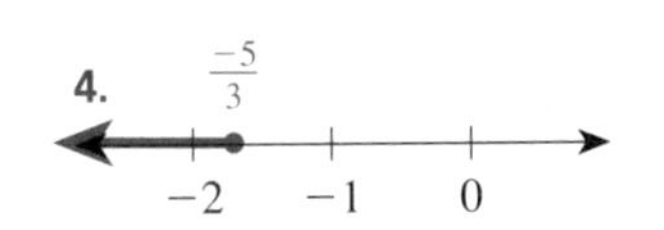

Draw a number line graph for each interval.

5. $[-3, 5)$ **6.** $(2, 6]$

7. $(2, +\infty)$ **8.** $[4, +\infty)$

9. $\left[\frac{-5}{4}, \frac{-1}{3}\right]$ **10.** $\left(\frac{-9}{5}, \frac{9}{2}\right)$

Solve each compound inequality. Draw a number line graph of the solution and report the solution in interval notation. Use the graphics calculator to check your work.

11. $-2 \le x - 1 \le 3$ **12.** $5 > 3x + 1 > -1$

13. $6 > 3x > -3$ **14.** $-4 \le 2x \le 12$

15. $-4 \le -2x < 10$ **16.** $4 > -3x > 0$

17. $8 \le 3x + 5 \le 13$ **18.** $-3 \le 5x + 1 < 6$

19. $-10 < 3 - 4x < 2$ **20.** $6 \ge 3 - 2x \ge -2$

21. $\frac{1}{3} < 3x + \frac{5}{3} < 6$ **22.** $3 \le \frac{5}{4}x + 1 \le \frac{9}{2}$

23. $-2 \le \frac{x+2}{3} \le 2$ **24.** $1 \le \frac{3x+1}{2} < 5$

25. $\frac{5}{2} > 3 - 2x > \frac{-2}{3}$ **26.** $\frac{5}{4} \ge -2x + 1 \ge \frac{-1}{2}$

27. $-1 \le \frac{2x-1}{2} < 4$ **28.** $2 < \frac{5-3x}{2} < 5$

29. $2 > -3(2 - x) > -5$ **30.** $6 \le 5(2x - 3) \le 10$

31. $x > -3x + 5 > x - 2$ **32.** $-3 < 3 - 2x < 4x$

2.5 ■ ■ ■ ABSOLUTE VALUE EQUATIONS AND INEQUALITIES

In this section, we continue working with inequalities, learning to solve inequalities that contain absolute values. Our work with these inequalities depends on the ability to solve first-degree inequalities in one variable. We will continue using the graphics calculator to check the solutions to inequalities in order to gain greater understanding of such solutions.

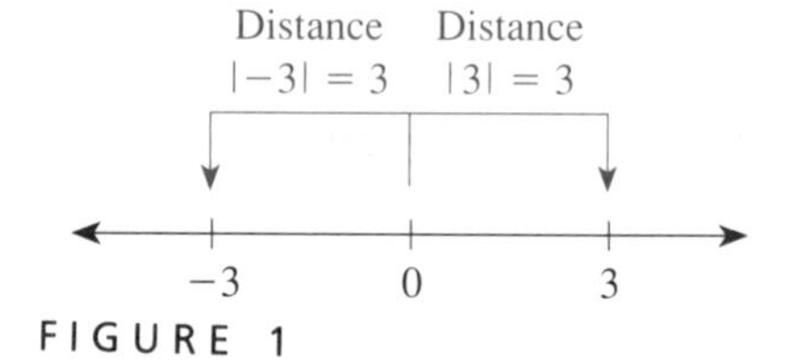

FIGURE 1

Absolute Value Equations Before we can solve inequalities involving absolute values, we need to learn how to solve equations that involve absolute values. In Section 1.1, we stated that the absolute value of a real number is its distance from 0. For instance, $|-3| = 3$, because -3 is 3 units from 0. Likewise, $|3| = 3$, because 3 is 3 units from 0. See Figure 1.

Now, consider the equation $|x| = 5$. The solution to this absolute value equation is the collection of numbers that replace the x-variable to yield a true arithmetic sentence. There are two such numbers, -5 and 5. Thus, the solution to the equation $|x| = 5$ is the numbers -5 and 5. A procedure to arrive at these numbers algebraically uses the following definition of the absolute value operation on a real number a.

If $a \ge 0$, then $|a| = a$.
If $a < 0$, then $|a| = -a$.

or

$$|a| = \begin{cases} a & \text{if } a \ge 0 \\ -a & \text{if } a < 0 \end{cases}$$

Thus, when the absolute value operates on a number, the output will be either the number or the opposite of the number. For the equation $|x| = 5$, we get two equations:

Absolute value produces x or $-x$.

$$x = 5 \quad \textit{or} \quad -x = 5$$

The first equation has solution 5. The second equation has solution -5. Thus, the absolute value equation, $|x| = 5$ has solution 5 and -5. Checking, we see that both 5 and -5 solve the original equation.

Extraneous Solutions When we obtain solutions by this algebraic process, we must always check that these numbers do in fact solve the original equation. Consider the absolute value equation $|x| = -2$. Using this approach, we get

$$x = -2 \qquad or \qquad -x = -2$$

These equations produce a solution of -2 and 2. Checking these numbers in the original, we find that neither number solves the original: $|-2| = -2$ is a false arithmetic sentence, as is $|2| = -2$. The equation $|x| = -2$ has no solution (because the absolute value always produces a nonnegative number). The numbers -2 and 2 are **extraneous** (extra) **solutions** to the equation $|x| = -2$.

Using the Graphics Calculator

```
:Y1=abs (X)
:Y2=-2
```

FIGURE 2

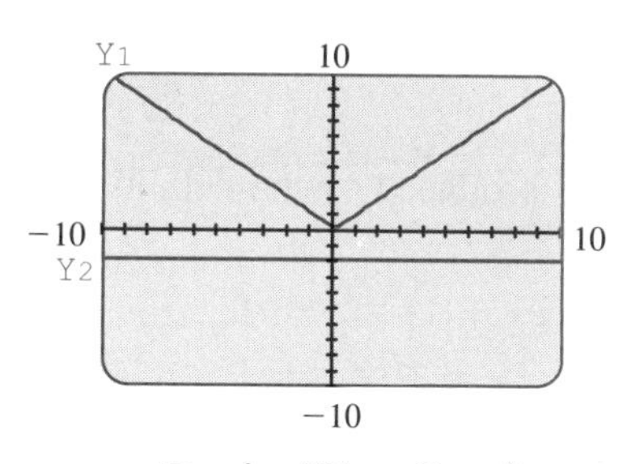

FIGURE 3

```
:Y1=abs (3X-5)
:Y2=2
```

FIGURE 4

```
RANGE
Xmin=-9.6
Xmax=9.4
Xscl=1
Ymin=-6.4
Ymax=6.2
Yscl=1
Xres=1
```

FIGURE 5

The graphics calculator can provide insight about extraneous solutions. For Y1, type abs (X) (Use the 2nd key followed by the x^{-1} for the ABS key.) For Y2, type -2. See Figure 2. View the graph screen using the Standard option. A V-shaped curve is drawn for $|x|$ and a horizontal line is drawn for -2, as displayed in Figure 3. Press the TRACE key and then use the right and left arrow keys to explore the evaluations of the absolute value of x. As you can see, these evaluations are always 0 or larger. Press the down arrow key to move the cursor to the horizontal line for -2. Here, the evaluations are always -2. Notice that there is no x-value for which the evaluation of the absolute value of x is -2. So, the equation $|x| = -2$ has no solution.

We can look at another equation, $|3x - 5| = 2$, to consider absolute value operations in general. For Y1 type abs (3X−5), and for Y2, type 2. See Figure 4. View the graph screen using the settings in Figure 5.* Figure 6 shows the V-shaped curve that is drawn for $|3x - 5|$ and the horizontal line drawn for 2. The graph indicates that two x-values result in $|3x - 5|$ evaluating to the number 2. Press the TRACE key, and then the right arrow key, until the cursor reaches the intersection point. When x is replaced by 1, the expression $|3x - 5|$ evaluates to 2. Continue to press the right arrow key until the rightmost intersection point is reached. The expression $|3x - 5|$ also evaluates to 2 for an x-value between 2 and 3, approximately 2.33.

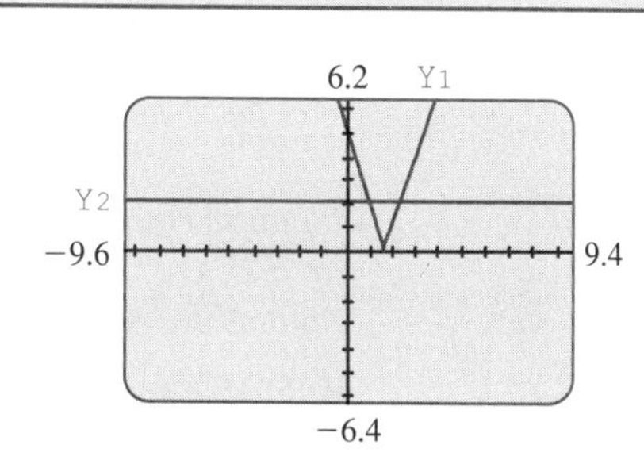

FIGURE 6

TI-82 Note *Type -9.4 for Xmin and -6.2 for Ymin.

The next two examples show an algebraic approach to solving absolute value equations.

EXAMPLE 1 Solve: $|3x - 5| = 2$

Solution Using the definition of absolute value, we conclude that $3x - 5$ must equal 2 or that $-(3x - 5)$ must equal 2. We get two equations:

$$3x - 5 = 2 \qquad or \qquad -(3x - 5) = 2$$

Now, we can solve these first-degree equations in one variable.

Solve:	Solve:
$3x - 5 = 2$	$-(3x - 5) = 2$
$\underline{\quad 5 \quad 5}$	$-3x + 5 = 2$
$3x = 7$	$\underline{\quad -5 \quad -5}$
$x = \frac{7}{3}$	$-3x = -3$
	$x = 1$
Solution: $\frac{7}{3}$	Solution: 1

Thus, $|3x - 5| = 2$ has solutions $\frac{7}{3}$ and 1. ■

Compare the numbers in the solution for Example 1 to the numbers we found using the graphics calculator.

EXAMPLE 2 Solve: $|2x + 1| = x + 1$

Solution Again the we solve the two first-degree equations produced by the definition of absolute value.

Solve:	Solve:
$2x + 1 = x + 1$	$-(2x + 1) = x + 1$
$\underline{-x \quad -x}$	$-2x - 1 = x + 1$
$x + 1 = 1$	$\underline{-x \quad -x}$
$\underline{-1 \quad -1}$	$-3x - 1 = 1$
$x = 0$	$\underline{1 \quad 1}$
	$-3x = 2$
	$x = \frac{-2}{3}$
Solution: 0	Solution: $\frac{-2}{3}$

Thus, the solution to $|2x + 1| = x + 1$ is 0 and $\frac{-2}{3}$. ■

Using the Graphics Calculator

To check our solution in Example 2, for `Y1`, type `abs (2X+1)` and for `Y2`, type `X+1`. Store `0` in memory location `X`. Display the resulting evaluations stored in `Y1` and `Y2`. Is the output the same in both cases? If not, there is an error! Next, store $\frac{-2}{3}$ in memory location `X`. Again, display the resulting evaluations stored in `Y1` and `Y2`. Is the output the same in both cases? If not, there is an error!

GIVE IT A TRY

Solve each absolute value equation.

1. $|2x - 5| = 1$ **2.** $\left|5 - \frac{1}{3}x\right| = \frac{2}{5}$ **3.** $|2x + 3| = -2$

4. Use the graphics calculator to check the solutions to Problems 1–3.

Absolute Value Inequalities Now that we have experience solving equations involving the absolute value operation, we can discuss solving absolute value inequalities.

Using the Graphics Calculator

```
RANGE
Xmin=-9.6
Xmax=9.4
Xscl=1
Ymin=-6.4
Ymax=6.2
Yscl=1
Xres=1
```

FIGURE 7

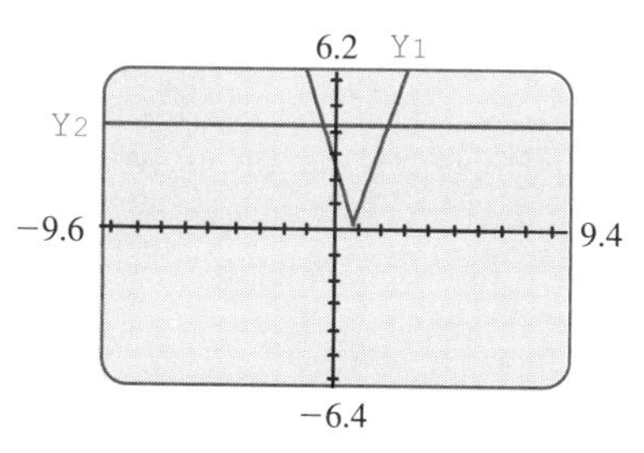

:Y1=abs(3X−2) :Y2=4

FIGURE 8

Before using algebra, we will use the graphics calculator to develop a feel for the solution to an absolute value inequality, say $|3x - 2| > 4$. For Y1, type abs(3X−2) and for Y2, type 4. View the graph screen using the settings in Figure 7.* A V-shaped curve is drawn for the evaluations of $|3x - 2|$ and a horizontal line is drawn for the constant 4, as displayed in Figure 8. The solution to the inequality $|3x - 2| > 4$ is all the x-values that result in the expression $|3x - 2|$ evaluating to a number larger than 4. The graphs in Figure 8 show two such intervals of x-values, one interval on the left branch and one interval on the right. Press the TRACE key. Use the right arrow key to move the cursor to the right intersection point. The expression $|3x - 2|$ evaluates to a number larger than 4 for x-values of $x = 2$ and all values to the right of 2. Continue to press the right arrow key to convince yourself of this. Press the left arrow to move the cursor to the left intersection point. Press the left arrow key again. As we will find in the next example, the key point is $\frac{-2}{3}$. The expression $|3x - 2|$ evaluates to a number larger than 4 for x-values of $x = \frac{-2}{3}$ and all values to the left of $\frac{-2}{3}$. Continue to press the left arrow key to convince yourself of this.

REMEMBER

Number line graph	Interval notation
solid dot at a, shaded to the right	$[a, +\infty)$
open dot at a, shaded to the right	$(a, +\infty)$
open dot at a, shaded to the left	$(-\infty, a)$
solid dot at a, shaded to the left	$(-\infty, a]$

We now solve the inequality $|3x - 2| > 4$ using algebra, following the same algorithm (procedure) we used in Sections 2.3 and 2.4:

1. Solve the related equation(s) to find the key point(s).
2. Plot and label the key points on a number line.
3. Determine whether each key point should be shown as an open or a solid dot.
4. Test the intervals created by the key points to see which interval(s) should be shaded.

The next example demonstrates this procedure.

TI-82 Note *Type −9.6 for Xmin and −6.2 for Ymin.

EXAMPLE 3 Solve $|3x - 2| > 4$, and report the solution in interval notation.

Solution We start by solving the equation $|3x - 2| = 4$.

Solve:
$$\begin{aligned} 3x - 2 &= 4 \\ 2 \quad & \quad 2 \\ \hline 3x &= 6 \\ x &= 2 \end{aligned}$$

Solution: 2

Solve:
$$\begin{aligned} -(3x - 2) &= 4 \\ -3x + 2 &= 4 \\ -2 \quad & \quad -2 \\ \hline -3x &= 2 \\ x &= \frac{-2}{3} \end{aligned}$$

Solution: $\frac{-2}{3}$

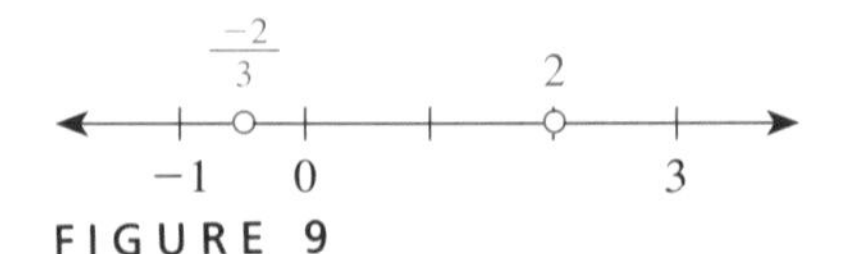

FIGURE 9

FIGURE 10

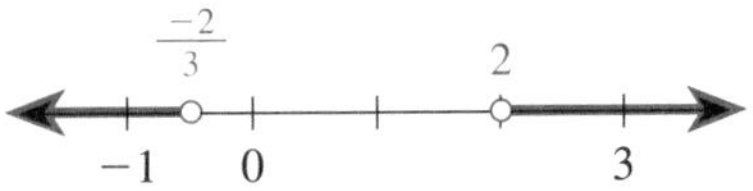

FIGURE 11

Thus, the solution of the equation $|3x - 2| = 4$ is 2 and $\frac{-2}{3}$. The key points for the inequality $|3x - 2| > 4$ are 2 and $\frac{-2}{3}$.

Next, we plot and label the key points. Each key point is drawn as an open circle, because neither solves the original inequality (these numbers result in equality). See Figure 9.

Try a number in the leftmost interval, say -1: $|3(-1) - 2| > 4$ is a true arithmetic sentence, so the leftmost interval is shaded. See Figure 10.

Try a number in the middle interval, say 0: $|3(0) - 2| > 4$ is a false arithmetic sentence. Thus, the middle interval is left unshaded.

Try a number in the rightmost interval, say 3: $|3(3) - 2| > 4$ is a true arithmetic sentence, so the rightmost interval is shaded. See Figure 11.

In interval notation we must use the union (join) symbol, $\cup$, to write the solution. We write $\left(-\infty, \frac{-2}{3}\right) \cup (2, +\infty)$. ■

To increase your understanding of solving absolute value inequalities, work through the following example.

EXAMPLE 4 Solve: $3 - |5 - 2x| \geq x$

Solution As usual, we start by solving the related equation, $3 - |5 - 2x| = x$. Using the meaning of absolute value, we get two equations.

> **REMEMBER**
>
> To write $3 - [-(5 - 2x)]$ as a standard polynomial, we clear the innermost parentheses to get
>
> $$3 - [-5 + 2x]$$
>
> Then clearing the brackets, we get
>
> $$3 + 5 - 2x$$
>
> Combining like terms, we get
>
> $$-2x + 8$$

Solve:
$$\begin{aligned} 3 - (5 - 2x) &= x \\ 2x - 2 &= x \\ -1x \quad & \quad -1x \\ \hline x - 2 &= 0 \\ 2 \quad & \quad 2 \\ \hline x &= 2 \end{aligned}$$

Solution: 2

Solve:
$$\begin{aligned} 3 - [-(5 - 2x)] &= x \\ -2x + 8 &= x \\ -1x \quad & \quad -1x \\ \hline -3x + 8 &= 0 \\ -8 \quad & \quad -8 \\ \hline -3x &= -8 \\ x &= \frac{8}{3} \end{aligned}$$

Solution: $\frac{8}{3}$

The solution to $3 - |5 - 2x| = x$ is 2 and $\frac{8}{3}$, and so the key points for the inequality $3 - |5 - 2x| \geq x$ are also 2 and $\frac{8}{3}$.

We plot and label the key points on a number line. A solid dot is used for each point because each number solves the inequality. See Figure 12.

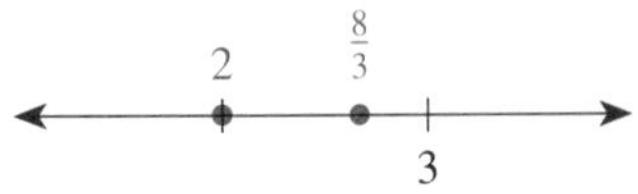

FIGURE 12

Now test the three intervals created by the key points:

Try a number in the leftmost interval, say 0: $3 - |5 - 2(0)| \geq 0$, or $3 - |5| \geq 0$, is a false arithmetic sentence, so we leave the leftmost interval unshaded.

Try a number in the middle interval, say 2.1: $3 - |5 - 2(2.1)| \geq 2.1$, or $3 - |0.8| \geq 2.1$, is a true arithmetic sentence. Shade the middle interval, as shown in Figure 13.

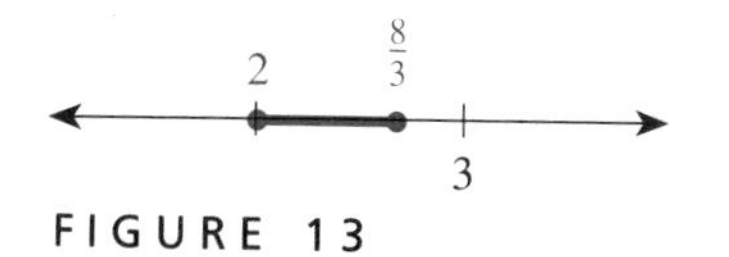

FIGURE 13

Try a number in the rightmost interval, say 3: $3 - |5 - 2(3)| \geq 3$, or $3 - |-1| \geq 3$, is a false arithmetic sentence, so we leave the rightmost interval unshaded.

In interval notation, the solution is $\left[2, \frac{8}{3}\right]$. ■

Using the Graphics Calculator

We can use the graphics calculator to check that the solution to $3 - |5 - 2x| \geq x$ is the interval $\left[2, \frac{8}{3}\right]$. Type `3-abs (5-2X)` for `Y1` and type `X` for `Y2`. Store the number 2 in memory location `X`. Display the resulting values in `Y1` and `Y2`. As the screen in Figure 14 shows, the output values are the same, so, 2 is a key point. Next, store $\frac{8}{3}$ in memory location `X`. Display the resulting values in `Y1` and `Y2`. As Figure 15 shows, the outputs are the same, so, $\frac{8}{3}$ is a key point.

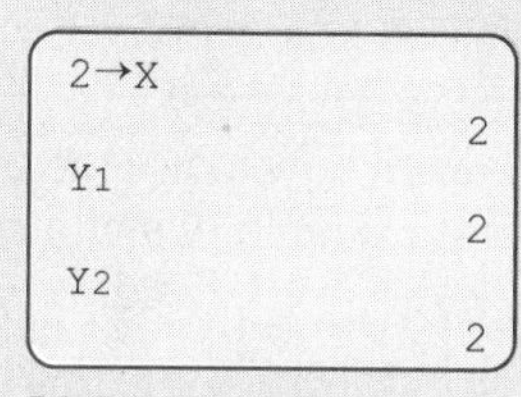

FIGURE 14

```
8/3→X
        2.666666667
Y1
        2.666666667
Y2
        2.666666667
```

FIGURE 15

To test the interval, store a number from the interval $\left[2, \frac{8}{3}\right]$, say 2.1, in memory location `X`. Display the resulting values in `Y1` and `Y2`. Figure 16 shows that the value output for `Y1` is greater than the value output for `Y2`. This means that the value of the left-hand side of the inequality, $3 - |5 - 2x|$, is greater than that of the right-hand side, x. So the number 2.1, along with all the other numbers in the interval $\left[2, \frac{8}{3}\right]$, solves the inequality.

```
2.1→X
        2.1
Y1
        2.2
Y2
        2.1
```

FIGURE 16

To be certain that we have identified the solution interval, store a number from outside the interval, say 0, in memory location `X`. Display the resulting values in `Y1` and `Y2`. (See Figure 17.) We can see that the value for `Y1` is *not* greater than or equal to the value for `Y2`, so, the number 0 fails to solve the inequality. This gives us confidence that the solution is in fact the interval $\left[2, \frac{8}{3}\right]$.

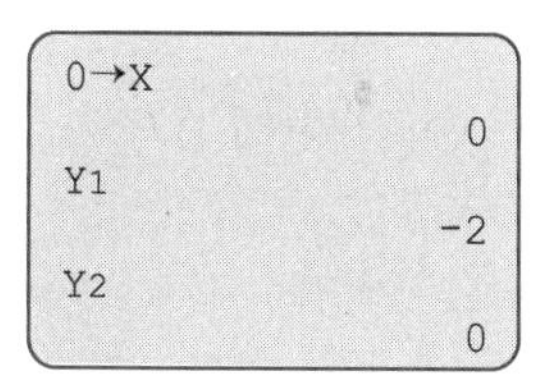

FIGURE 17

GIVE IT A TRY

Solve each absolute value inequality. Use the graphics calculator to check the solution.

5. $|2x - 5| \leq 1$ **6.** $|5 - 3x| > 2$ **7.** $|2x + 3| \geq x + 1$ 

```
RANGE
Xmin=-9.6
Xmax=9.4
Xscl=1
Ymin=-6.4
Ymax=6.2
Yscl=1
Xres=1
```

FIGURE 18

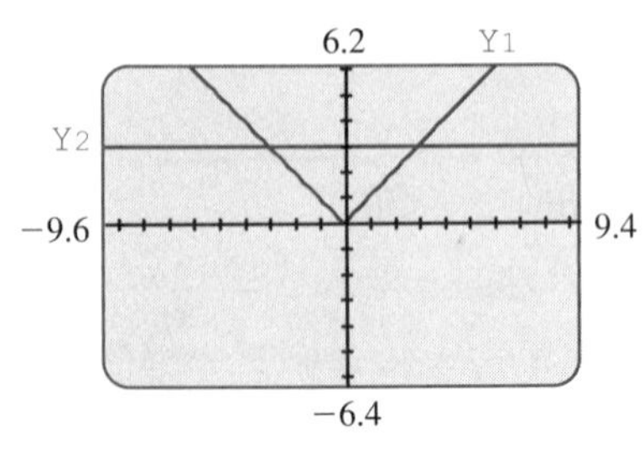

:Y1=abs(X) :Y2=3

FIGURE 19

Solving $|x| > a$ and $|x| < a$ for $a > 0$ Many inequalities involving absolute values have the form $|x| > a$ or the form $|x| < a$ for $a > 0$. For example, consider the absolute inequalities $|2x - 3| > 3$ and $|2x - 3| < 3$. If $a < 0$, then $|x| > a$ has all real numbers for a solution. If $a < 0$, then $|x| < a$ has no number as its solution. With the graphics calculator, we can see a generalization for these types of absolute value inequalities. We start with the simplest cases, $|x| > 3$ and $|x| < 3$.

On the graphics calculator, type `abs (X)` for `Y1` and `3` for `Y2`. Press the `RANGE` key and enter the values in Figure 18* for a friendly viewing screen. The graph screen appears as shown in Figure 19. The x-values where the evaluation of $|x|$ is greater than 3 are those x-values such that $x > 3$ or those x-values such that $x < -3$. In general, we have the following.

> For $|x| > a$ with $a > 0$, then $x < -a$ OR $x > a$.

From the graph screen in Figure 19 we can also see that the x-values where the evaluation of $|x|$ are less than 3 are those x-values such that $x < 3$ *and* those x-values such that $x > -3$. In general we have the following generalization:

> For $|x| < a$ with $a > 0$, then $x > -a$ AND $x < a$.
> That is, x is *between* $-a$ and a, $-a < x < a$.

To learn how to use these generalizations to solve inequalities, work through the next example, which uses the equivalent inequalities approach (page 103).

EXAMPLE 5 Solve $|2x - 3| > 3$, and report the solution in interval notation.

Solution This absolute value inequality has the form $|x| > a$, so its solution is the collection of numbers that solve either $2x - 3 < -3$ or $2x - 3 > 3$.

$$\begin{array}{lrcr} \text{Solve:} & 2x - 3 < -3 & \quad \textit{or} \quad & 2x - 3 > 3 \\ & \underline{\phantom{2x - {}}3 \quad \underline{3}} & & \underline{\phantom{2x - {}}3 \quad \underline{3}} \\ & 2x < 0 & & 2x > 6 \\ & x < 0 & & x > 3 \end{array}$$

The solution is $(-\infty, 0) \cup (3, +\infty)$. ■

EXAMPLE 6 Solve $|2x - 3| < 3$, and report the solution in interval notation.

Solution This absolute value inequality has the form $|x| < a$, so its solution is the collection of numbers that solve both $2x - 3 > -3$ and $2x - 3 < 3$.

$$\begin{array}{rcr} 2x - 3 > -3 & \quad \textit{and} \quad & 2x - 3 < 3 \\ \underline{\phantom{2x - {}}3 \quad \underline{3}} & & \underline{\phantom{2x - {}}3 \quad \underline{3}} \\ 2x > 0 & & 2x < 6 \\ x > 0 & & x < 3 \end{array}$$

So, the solution is $(0, 3)$. ■

TI-82 Note *Type `-9.4` for `Xmin` and `-6.2` for `Ymin`.

GIVE IT A TRY

Use the equivalent equalities approach to solve each inequality.

8. $|3x - 1| > 8$ **9.** $|2x + 3| < 7$ **10.** $|5 - 4x| \geq 3$

11. Does the equivalent inequalities approach work for the inequality $|2x - 3| < x + 1$? What is the solution to this inequality?

12. Use the graphics calculator to verify each solution in Problems 8–11.

■ SUMMARY

In this section we have applied the general algorithm (procedure) for solving inequalities to absolute value inequalities:

1. Solve the related equation(s) to find the key points.
2. Plot and label the key points on a number line graph.
3. Determine whether each key point should be shown as an open or a solid dot.
4. Test the intervals created by the key points.

The skill of solving absolute value inequalities uses skill we have previously acquired—solving first-degree equations and inequalities.

To solve an absolute value inequality, we first solve the related absolute value equation, such as $|5x - 2| = 4$. From the meaning of absolute value, two first-degree equations are produced, $5x - 2 = 4$ and $-(5x - 2) = 4$. After solving the absolute value equation, we must check that the numbers obtained actually do solve the original equation. Once we have solved the absolute value equations, we can use the solutions as key points to find the solution interval(s) of the absolute value inequality, such as $|5x - 2| > 4$.

2.5 ■ ■ ■ EXERCISES

Report the shaded interval(s) on each number line graph in interval notation.

1. −2 0 3

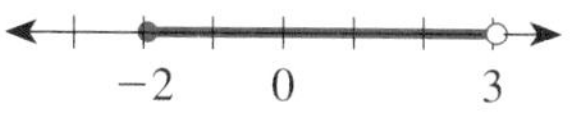

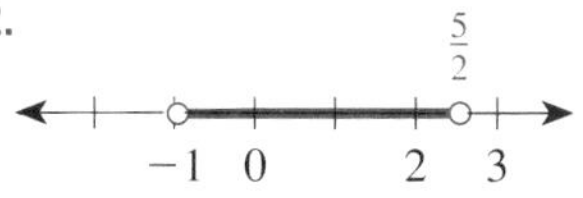

2. $\frac{5}{2}$ −1 0 2 3

3. −1 0 2

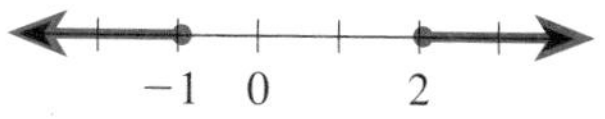

4. $\frac{-5}{3}$ −2 0 2

Draw a number line graph for each.

5. $(-4, -1]$

6. $[-2, 3)$

7. $(-\infty, -1] \cup (2, +\infty)$

8. $(-\infty, 2) \cup [4, +\infty)$

9. $\left[\frac{-5}{4}, \frac{-1}{4}\right]$

10. $\left(\frac{-9}{2}, \frac{9}{5}\right]$

Solve the absolute value equation. Use the graphics calculator to check the solution.

11. $|2x| = 8$

12. $|-2x| = 4$

13. $|-3x| = -3$

14. $|5x| = -10$

15. $|x + 5| = 8$

16. $|x - 2| = 1$

17. $|5 - 2x| = 0$

18. $|5 - 3x| = 6$

19. $|5(x - 3)| = -1$

20. $-|3 - 2x| = -2$

21. $|3x - 1| = 5$

22. $|7 - 5x| = 3$

23. $-3|2x + 7| = -18$

24. $-2|3 - 2x| = 5$

25. $\left|\frac{-5}{4}x + \frac{1}{2}\right| = \frac{9}{2}$

26. $\left|\frac{-1}{3}x - 3\right| = \frac{1}{2}$

27. $\left|\frac{-2x - 1}{5}\right| = 3$

28. $\left|\frac{3x + 1}{2}\right| = 5$

29. $3 + |5 - 2x| = 6$

30. $5 - |4x - 3| = -1$

31. $|2 - 4x| = 2x + 1$

32. $|x| = x - 5$

33. $\left|\dfrac{5 - 3x}{2}\right| = x$

34. $\left|\dfrac{2 + 3x}{5}\right| = -7$

Solve the absolute value inequality. Draw a number line graph of the solution, and report the solution in interval notation. Use the graphics calculator to check the solution.

35. $|2x| < 6$

36. $|-2x| \geq 4$

37. $\left|\dfrac{x}{2}\right| \geq 3$

38. $\left|\dfrac{2x}{5}\right| < 1$

39. $|3 - x| \leq 5$

40. $|x - 1| < 2$

41. $|2x - 5| > 3$

42. $|5 - 3x| \geq 3$

43. $|10 - 2x| \leq -2$

44. $|7 + 3x| > -1$

45. $|2x| - 3 \geq 5$

46. $6 - |3x| < 10$

47. $2 - 3|x| < 6$

48. $-3 + 2|-x| > -1$

49. $\left|\dfrac{2x - 1}{5}\right| \leq 3$

50. $\left|\dfrac{2 - 3x}{4}\right| > 4$

51. $-2|3 - 2x| > -5$

52. $3 + 2|5 - 4x| \geq 5$

53. $\left|\dfrac{7 - 5x}{6}\right| \geq \dfrac{1}{2}$

54. $|2x - 5| \geq x - 2$

2.6 ■ ■ ■ APPLICATION: WORD PROBLEMS

In this section, we use knowledge of solving first-degree equations in one variable to solve word problems. The ability to solve first-degree equations in one variable lets us model some real-world situations and solve them as word problems. We start by considering translations from word phrases to polynomials. Next, we consider translations from sentences to equations, and then we solve word problems.

Phrases to Polynomials In Section 2.1 we stated that polynomials held the place for number phrases. As a prelude to solving word problems, we now translate a phrase to a polynomial. The following table shows various wordings of the basic arithmetic operations.

Operation	*Words*	*Example*	*Polynomial*
Addition	sum of	*sum of* a number and 5	$x + 5$
	added to	-3 *added to* a number	$w + (-3)$
	increased by	a number *increased by* 8	$y + 8$
	more than	5 *more than* a number	$n + 5$
Subtraction	minus	a number *minus* 5	$x - 5$
	difference from	*difference* of a number *from* 5	$5 - x$
	decreased by	a number *decreased by* 10	$n - 10$
	less than	7 *less than* a number	$x - 7$
Multiplication	of	three-fourths *of* a number	$\frac{3}{4}y$
		30% *of* a number	$0.3x$
	times	five *times* a number	$5n$
	product of	*product of* a number and 8	$8x$
Division	divided by	a number *divided by* 7	$\dfrac{x}{7}$
	quotient of	the *quotient of* a number and 3	$\dfrac{n}{3}$
	ratio of	the *ratio of* a number to 5	$n{:}5$ or $\dfrac{n}{5}$

We now consider translations from number phrases to polynomial expressions. In most cases, a variable is not stated explicitly to represent the unknown quantity, and we must select a variable. We can start with a phrase such as

Let x represent the unknown. Or Let $x =$ the number.

to state the variable explicitly. Then we use this variable for the translation, such as $x + 5$ or $n:5$. The variable we use is unimportant, as long as we use it to represent the number throughout the translation.

EXAMPLE 1 Translate to a polynomial expression: Twice the sum of a number and 6

Solution Let $x =$ the number. Then the translation of this phrase is

twice

$$2(x + 6)$$

Sum of x and 6 ■

EXAMPLE 2 Translate: The sum of a number and 30% of the number

Solution Let $n =$ the number. We translate the phrase as

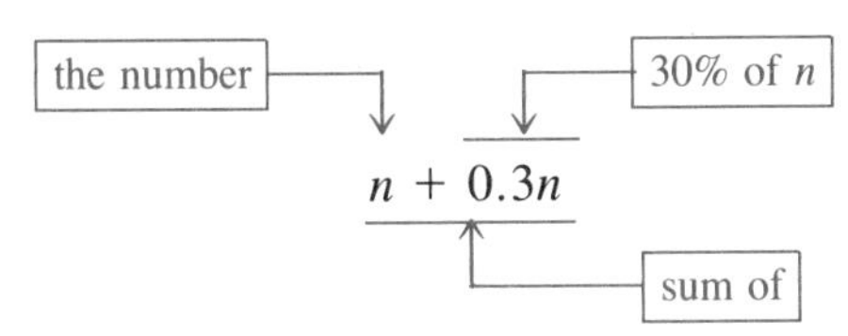

■

EXAMPLE 3 Translate: 6 more than twice a number decreased by 3

Solution Let $y =$ the number.

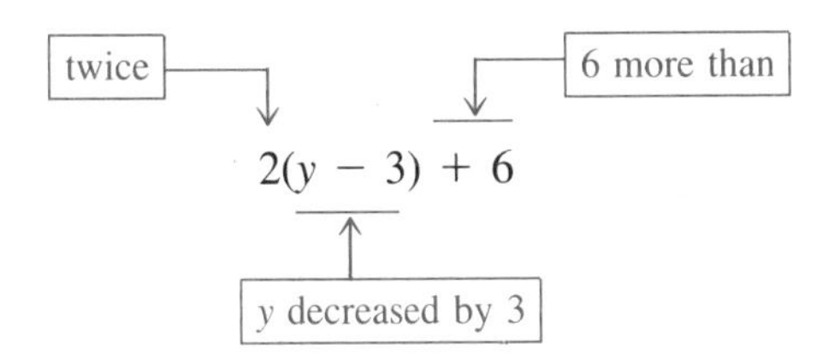

■

Many consider the word phrase given in Example 3 to be vague, since it could also be translated as $2y - 3 + 6$. For this reason we will try to avoid such phrases in the future! A more accurate phrase is *6 more than twice the difference of 3 from a number*.

GIVE IT A TRY

Translate each number phrase to a polynomial.

1. Twelve less than the sum of twice a number and 8.
2. Three times the difference of 7 from a number.
3. A number decreased by 30% of the number.

Added Information Often in solving word problems, we are expected to provide information not stated explicitly in the problem, such as the formula for the perimeter of a triangle or rectangle. The following examples demonstrate translations that involve such "outside" information. Study these examples carefully, since the outside information may be useful for solving later word problems.

EXAMPLE 4 The length of a rectangle is 3 more than twice its width. Let x represent the width. Write a polynomial for the length and a polynomial for the perimeter of the rectangle.

Solution Let x = width of the rectangle. Then the length is

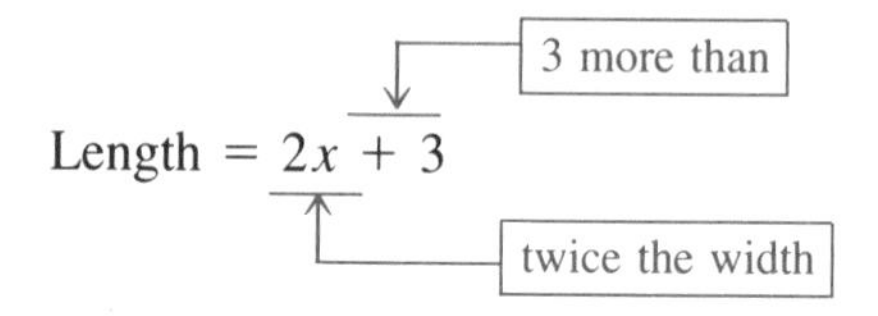

We use the formula $P = 2W + 2L$ for the perimeter:

$$\text{Perimeter} = 2x + 2(2x + 3) \qquad \text{Use } x \text{ for } W \text{ and } (2x + 3) \text{ for } L.$$
$$= 6x + 6 \qquad \text{Write as a standard polynomial.}$$ ■

EXAMPLE 5 The number of adult tickets sold for a concert is 40% more than the number of student tickets. The price of an adult ticket is \$15 and the price of a student ticket is \$6. Let n represent the number of student tickets sold. Write a polynomial for the total income from the tickets.

Solution Let n = number of student tickets sold.
We know that the number of adult tickets sold is

40% of student tickets

$$\text{Adult tickets} = n + 0.4n$$

We multiply the number of tickets sold by the price of each ticket to find the income:

Income from student tickets — Income from adult tickets

$$\text{Total income} = 6n + 15(n + 0.4n)$$
$$= 27n \qquad \text{Write as a standard polynomial.}$$ ■

To write the polynomial in Example 5, we had to use outside information:

Income = price per ticket *times* number of tickets sold

EXAMPLE 6 Tom travels 4 hours at a fixed speed. He then increases his speed by 15 mi/h and travels another 3 hours. Let x represent Tom's original speed. Write a polynomial for the total distance Tom has traveled.

Solution Let x = Tom's original speed. So Tom's increased speed = $x + 15$. Using the formula $d = rt$, we get:

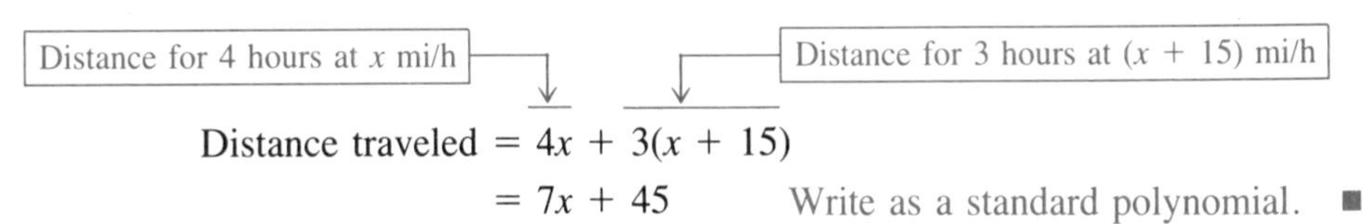

$$\text{Distance traveled} = 4x + 3(x + 15)$$
$$= 7x + 45 \qquad \text{Write as a standard polynomial.}$$ ■

GIVE IT A TRY

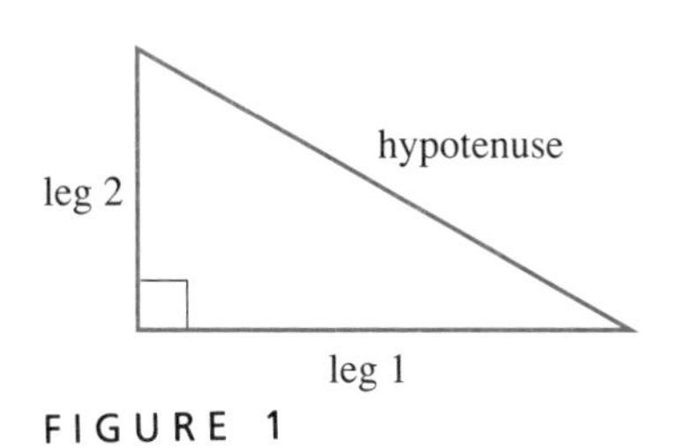

FIGURE 1

4. Write a polynomial for the perimeter of the right triangle shown in Figure 1. The length of the longest side (the hypotenuse) is twice the length of one leg and 6 more than the length of the other leg.

Word Problems We now discuss a *real-world* situation, such as an automobile traveling at different speeds, and produce an equation that describes this situation. This topic commonly has the title *word problems*. The rest of the problems (*real-world* situations) discussed in this section have as a mathematical model a first degree-equation in one variable. Additionally, all the problems request one value. (Later, when two values are requested, we will use as a mathematical model a system of equations with two variables.) To introduce this application, we start with a word problem about numbers.

EXAMPLE 7 Find a number such that five less than twice the number equals twenty percent of the number.

Solution We start in a fashion similar to writing a polynomial for a phrase.
Let x = the number.
Five less than twice the number = $2x - 5$.
Twenty percent of the number = $0.2x$.
To write the equation, we equate the two polynomials:

$$2x - 5 = 0.2x$$

5 less than twice the number (→ $2x - 5$) 20% of the number (→ $0.2x$)

To find the number, we solve this first-degree equation in one variable.

Solve:	$2x - 5 = 0.2x$
Multiply each side by 10:	$20x - 50 = 2x$
Add $-2x$ to each side	$-2x \qquad -2x$
to get:	$18x - 50 = 0$
Add 50 to each side	$50 \qquad 50$
to get:	$18x = 50$
Multiply by $\frac{1}{18}$:	$x = \frac{50}{18}$

The solution to this equation is $\frac{50}{18}$, or $\frac{25}{9}$ (reduced), or $2\frac{7}{9}$. The number requested is $2\frac{7}{9}$. ■

Using the Graphics Calculator

```
RANGE
Xmin=⁻9.6
Xmax=9.4
Xscl=1
Ymin=⁻6.4
Ymax=6.2
Yscl=1
Xres=1
```

FIGURE 2

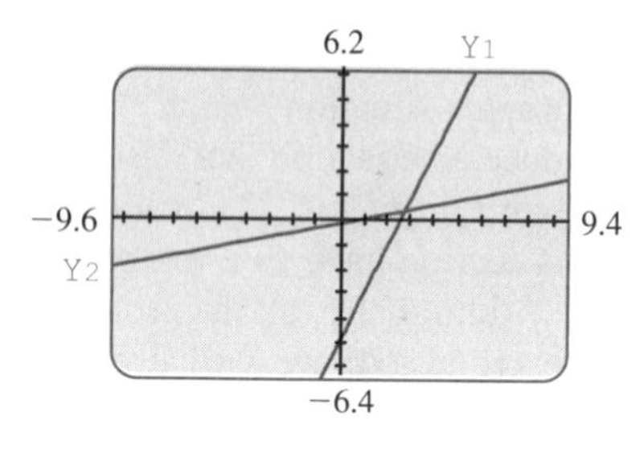

FIGURE 3

We can verify the number we found for Example 7. Return to the worded version of the problem. Read the problem. For *five less than twice the number,* type `2X−5` at the `:Y1` prompt. For *twenty percent of the number,* type `.2X` for `Y2`. Store the number $\frac{25}{9}$ in memory location `X`. Display the resulting values for `Y1` and `Y2`. Is the same number output in both cases? If not, there is an error!

To increase our understanding of the problem, we can view the graph screen using the `RANGE` settings in Figure 2.* The graph is shown in Figure 3. The first line drawn represents 5 less than twice a number. The second line drawn represents 20% of that number. Press the `TRACE` key and move the cursor until the message reads `X=2  Y=⁻1`. Five less than twice the number 2 is the number −1. Press the down arrow key to move to the other line. The message reads `X=2  Y=.4`. Twenty percent of 2 is 0.4.

Move the cursor to the intersection point of the two lines. Zoom in on the point. Press the `TRACE` key and move the cursor to the intersection point. The x-value, which is approximately 2.76, is *the* number such that 5 less than twice the number equals 20% of the number.

A general approach to solving word problems in algebra involves following the steps given in the next box. These steps provide a general guide that is productive in most situations which involve word problems.

GUIDE TO SOLVING WORD PROBLEMS

1. Read through the problem, and then decide what the problem is requesting—a particular number, the length of a rectangle, how fast someone is traveling, or some other quantity.
2. Select a variable (such as x or n) and write down what the variable represents (often the quantity being requested in the word problem).
3. Make a sketch or diagram of the situation. This is often helpful because you must read the problem carefully and record the information given in order to make a sketch.
4. Write polynomials for other quantities given in the word problem, using the variable from step 2.
5. Reread the word problem and write an equation using the polynomials from step 4.
6. Solve the equation.
7. Use the solution to the equation to report the quantity requested by the word problem. Report the units with the number, such as 4 cm. Check to make sure that your answer is reasonable!

To demonstrate this approach to solving word problems, work through the following example.

TI-82 Note *Type `⁻9.4` for `Xmin` and `⁻6.2` for `Ymin`.

EXAMPLE 8 The perimeter of a rectangle is 200 meters. The length is 5 meters more than twice the width. Find the width of the rectangle.

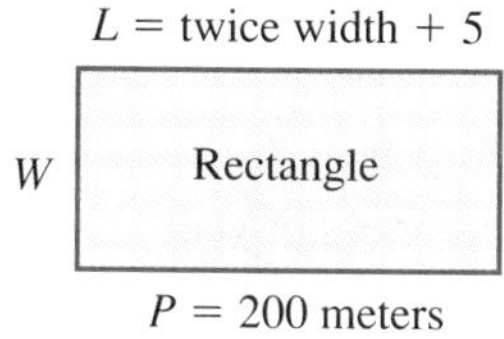

FIGURE 4

Solution

Step 1 By reading the problem, we decide that the quantity we must find is the width of the rectangle.

Step 2 Let w = width of the rectangle.

Step 3 Sketching a picture of a rectangle and inserting the given information produces the drawing shown in Figure 4.

Step 4 The other quantities can be represented as

$$\begin{aligned} \text{Length} &= 2w + 5 \\ \text{Perimeter} &= 2w + 2(2w + 5) \qquad \text{Use } P = 2W + 2L. \end{aligned}$$

Step 5 From the problem statement, and from step 3, we see that the perimeter is 200. Thus, we get the equation

$$200 = 2w + 2(2w + 5)$$

Step 6 Solving this equation, we get

$$\begin{aligned} 2w + 2(2w + 5) &= 200 \\ 2w + 4w + 10 &= 200 \\ 6w + 10 &= 200 \\ -10 \quad & \quad -10 \\ \hline 6w &= 190 \\ w &= \frac{190}{6} = \frac{95}{3} \end{aligned}$$

The solution is $\frac{95}{3}$, or $31\frac{2}{3}$.

The width of the rectangle is $31\frac{2}{3}$ meters. ■

Using the Graphics Calculator

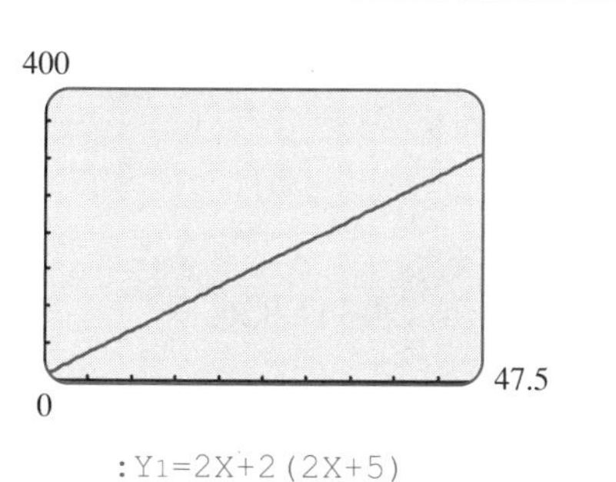

FIGURE 5

We can use the graphics calculator to verify the number we found in Example 8. Type `2X+2(2X+5)` for `Y1`. This expression represents the perimeter for various values of the width, x. Press the `RANGE` key and type `0` for `Xmin`, `47.5` for `Xmax`*, `5` for `Xscl`, `0` for `Ymin`, `400` for `Ymax`, and `50` for `Yscl`. Press the `GRAPH` key to view the graph screen using these range settings. See Figure 5. The line drawn represents the perimeter for different widths (the x-value). Press the `TRACE` key and move the cursor until the message reads `X=30 Y=190`. When the width of the rectangle is 30, the perimeter is 190. Press the `Y=` key and type `200` (the desired perimeter) for `Y2`. Press the `TRACE` key and move the cursor to the intersection point of the two lines. The x-value of this point is approximately 31.5, which is *the* width such that the perimeter is 200.

TI-82 Note *Type `47` for `Xmax`.

Percent Problems Many problems in algebra contain percent notation. These problems are actually a continuation of the basic percent problems from arithmetic (see Section 1.3). In arithmetic, we have worked problems based on the form

$$\square\ \%\ \text{of}\ \square = \square$$

For example, retail stores buy items at "cost" and then sell these items at a higher price. The "markup" on an item is the difference between these amounts. A coat costs a store \$80 and sells for \$110. The business school advisor from a local college recommends the store have at least a 35% markup on all items. What is the percent of the markup for the coat? Filling in our percent form with the appropriate numbers, we get

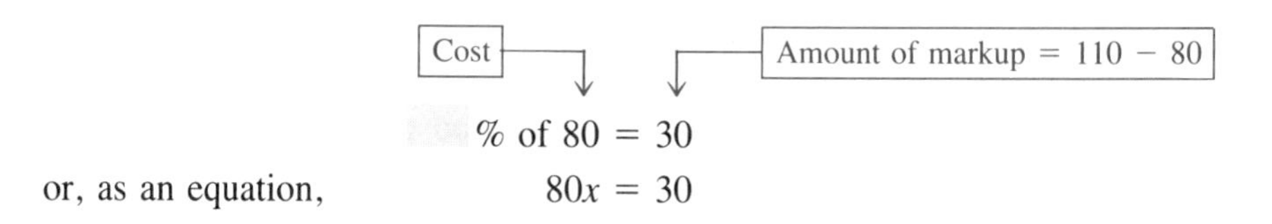

$$\square\ \%\ \text{of}\ 80 = 30$$

or, as an equation, $$80x = 30$$

The solution to this equation is $\frac{3}{8}$. Writing this number in percent notation, we get 37.5%. Thus the percent markup is 37.5% and is within the recommendation of the business school advisor.

The following examples show percent problems with solutions that use skills from algebra.

EXAMPLE 9 Sandra earned 15% more income this year than she did last year. If her income this year is \$28,000, what was her income last year?

Solution Let x = Sandra's income last year.
This year's income $= x + (15\% \text{ of } x) = x + .15x = 1.15x$

This year's income

Equation: $$1.15x = 28000$$

Multiply by 100: $$115x = 2800000$$

Multiply by $\frac{1}{115}$: $$x = \frac{2800000}{115} = \frac{560000}{23} \approx 24{,}347.826$$

To answer the question, we report that Sandra's income last year was \$24,347.83, to the nearest cent. ■

EXAMPLE 10 Joel is paid a base salary of \$400 a week, plus 10% commission on all his sales. If his salary last week was \$600, what was his amount of sales?

Solution Let y = Joel's amount of sales last week.
Joel's salary $= 400 + (10\% \text{ of } y) = 400 + 0.1y$

Last week's salary

Equation:	$400 + 0.1y = 600$
Multiply by 10:	$4000 + y = 6000$
Add -4000:	$-4000 \quad -4000$
	$y = 2000$

To answer the question, Joel had sales of \$2,000 last week. ■

■ SUMMARY

In this section, we have learned to translate a phrase—such as *three more than twice a number*—to a polynomial for the given phrase, such as $2x + 3$. This skill is a requirement for solving word problems. To solve a word problem, read the story, and then write an equation that models the physical situation (such as $2x + 3 = 10$). Once an equation is obtained, we can solve it and then state the answer to the word problem.

■ *Quick Index*

2.6 ■ ■ ■ EXERCISES

For Problems 1–10, write a polynomial for the phrase.

1. The sum of three times a number and 6
2. Twenty less than twice a number
3. The difference of 3 from half a number
4. The product of 3 more than a number and 6 less than the number
5. A number increased by sixty percent of the number
6. Twice a number decreased by half the sum of the number and 3
7. The perimeter of a rectangle, where the width is x and the length is 20 more than the width
8. The distance traveled, where the rate of travel is x miles per hour and the time in hours spent traveling is one-fifth of the rate
9. The sum of two consecutive integers
10. The difference of twice a number from 60% of the number

Write a word phrase for each polynomial.

11. $3x - 5$
12. $3(x - 5)$
13. $6 - \frac{2}{3}x$
14. $x + 0.4x$
15. $\frac{5}{7}y + \frac{2}{5}y$
16. $50(x + 2) - 3$
17. Let b represent the showroom sale price of a bed. This price is discounted 30% to get the store's purchase price. The sales tax on the purchase price is 5%. Write a polynomial for the purchase price plus tax.
18. Let n represent Joan's salary last year. This year, Joan's salary decreased by 12%. Write a polynomial for Joan's salary this year.
19. The AA Car Rental Company charges \$45 a day plus \$0.15 per mile driven. Let m represent the number of miles driven on a particular day. Write a polynomial for the total rental charge for that day.
20. Sara McMurphy earns a base salary plus a 30% bonus. Let S represent her base salary. Write a polynomial for her base salary plus the bonus.

Solve each word problem. Report the following: (a) the variable and what it represents, (b) polynomials for other quantities in the problem, (c) the equation, and (d) the answer to the problem.

21. Find a number such that twice the number equals twenty decreased by the number.
22. Find a number such that the number increased by 30% of itself equals 50.

23. Joey is twice as old as his brother today. In five years, Joey's brother's age will be 70% of Joey's age at that time. How old is Joey today?

24. The width of a rectangle is 6 meters less than the length. If the perimeter of the rectangle is 73 meters, what is the length of the rectangle?

25. The perimeter of a triangle is 172 meters. If one side is 42 meters and the other two sides are such that one has length 30% of the length of the other, what is the length of the shortest side?

26. Joan earned 20% more this year than she did last year. If her salary this year is $40,000, what was her salary last year?

27. Pete's monthly salary is $2,000 plus a 12% commission on his sales for the month. If Pete's salary in June was $2,800, what was his amount of sales that month?

28. Tom travels 4 hours at a fixed speed. He then increases his speed by 15 mi/h and travels another 3 hours. If Tom has traveled 395 miles, what was Tom's original speed? [*Hint*: See Example 6.]

29. The number of adult tickets sold at a concert is 40% more than the number of student tickets sold. The price of an adult ticket is $15 and the price of a student ticket is $6. If the total income from the tickets is $2700, how many student tickets were sold? [*Hint*: See Example 5.]

30. Bill has test grades of 70, 65, 85, and 90. What score must he make on the fifth test to get an average of 80 on the five tests?

31. The price of a new CD player is $154, including tax. If the tax rate is 5%, what is the sticker price of the CD player?

32. Twice a number minus 30% of the sum of the number and 8 equals 34. What is the number?

33. If 30% of the difference of 8 from a number equals two-thirds of the number, what is the number?

34. Jan's salary decreased by 12% from last year's salary. Her salary this year is $25,000. What was her salary last year?

35. The AA Car Rental Company charges $45 a day plus $0.15 per mile driven. If Jose's rental charge for one day was $90, how many miles did he drive the rental car?

36. LoTech, Inc. pays a flat rate of $200 for the first 500 local telephone calls made in a month. For each local call over 500, the charge is $0.20 per call. If LoTech's telephone bill for local calls last month is $420, how many local calls were made?

37. Today Bonnie is two-fifths as old as her sister. In 10 years, Bonnie's age will be 70% of her sister's age. How old is Bonnie today?

38. Two fractions have the same numerator and they have denominators of 2 and 7. If their sum is $\frac{99}{14}$, what is the smaller fraction?

39. The ratio of three more than a number to 7 equals the ratio of the number to 2. What is the number?

40. The sticker price of a coat equals the cost of the coat plus the percent markup. If the percent markup is set at 30%, what must the cost of the coat be in order to set the sticker price at $130?

41. Juan is planning to sell 100 T-shirts at a local seafood festival. The shirts cost him $4 each, and he must pay $200 for a booth from which to sell the shirts. Given that his profit equals sales revenue minus all his costs, at what price must he sell the 100 T-shirts to earn a 30% profit on his investment?

42. A local computer store advertises a computer sale price of $1400. They claim that this price is a 30% reduction from the original. What is the original price?

2.7 ■ ■ ■ MIXTURE AND UNIFORM RATE WORD PROBLEMS

In this section we continue solving word problems, using first-degree equations in one variable to model some real-world situations. The first set of word problems are known as *mixture problems*—the physical situation involves mixing two different quantities. The second set of word problems are known as

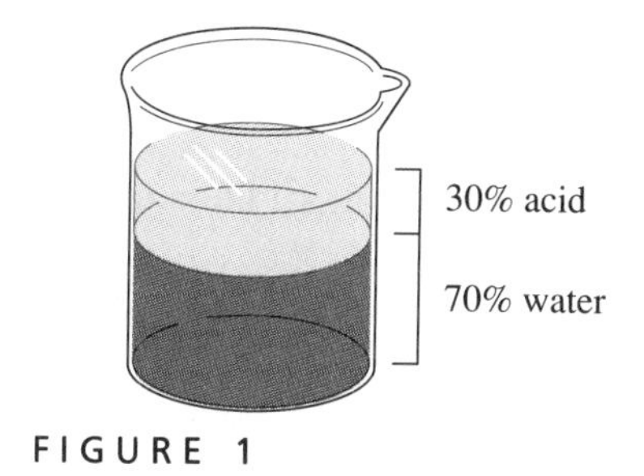

FIGURE 1

uniform rate problems—the physical situation involves rates, such as miles per hour and gallons per hour. As you work through the examples, note the similarity of the situations.

Acid Problems One type of mixture problem deals with acid solutions. As with most problems in this area, we must know the basics for the situation: An acid solution is a mixture of acid and water. For example, a 30% acid solution means the solution is 30% acid and 70% water. See Figure 1. We will start with an arithmetic problem to introduce some background knowledge of acid solutions.

EXAMPLE 1 Suppose 20 liters of 30% acid solution is mixed with 60 liters of 70% acid solution. See Figure 2.

a. How many liters of solution are in the mixture?
b. How many liters of acid are in the mixture?
c. What is the acid concentration of the mixture?

Solution a. The mixture contains $20 + 60 = 80$ liters of solution.

b. The liters of acid in the mixture is

$$\begin{aligned} 30\% \text{ of } 20 + 70\% \text{ of } 60 &= \\ 0.3(20) \quad + \quad 0.7(60) &= 6 + 2 \\ &= 48 \end{aligned}$$

The mixture contains 48 liters of acid.

c. The acid concentration of the mixture is

$$\begin{aligned} \square\% \text{ of } 80 &= 48 \\ 80x &= 48 \\ x &= \frac{48}{80} \quad \text{or} \quad 60\% \text{ acid} \end{aligned}$$

The mixture is a 60% acid solution. ■

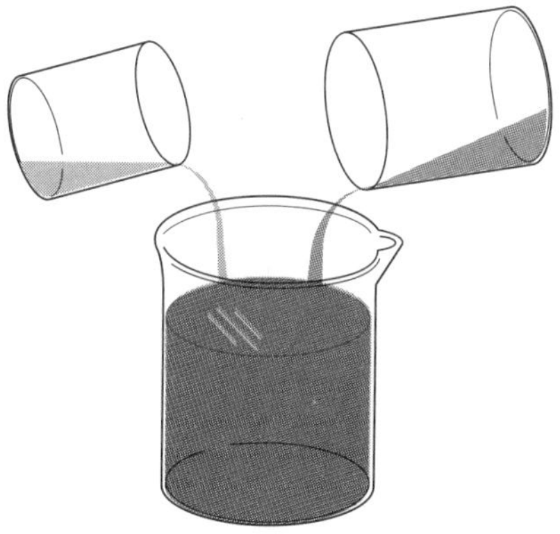

20 liters of 30% acid solution

60 liters of 70% acid solution

Mixture

FIGURE 2

When working algebra problems involving mixture problems, it is often helpful to create a table, such as the following one, to record given information along with the requested information.

WARNING
The acid concentration of the mixture is *not* the average of 30% and 70%; that is, it is *not* 50%!

Solution type	*Solution amount*	*Acid amount*	*Acid concentration*
30% acid	20 liters	6 liters	$\frac{6}{20} = 30\%$ acid
70% acid	60 liters	42 liters	$\frac{42}{60} = 70\%$ acid
Mixture	80 liters	48 liters	$\frac{48}{80} = 60\%$ acid

Now, consider the following algebra problem dealing with mixtures. We will follow our general approach to solving word problems, stated on page 124.

These steps are general guidelines that are productive in most situations that involve word problems.

EXAMPLE 2 How many liters of a 30% acid solution must be mixed with 10 liters of an 80% acid solution to produce a 40% acid solution?

Solution Read the problem; the quantity requested is the amount of 30% acid solution. Figure 3 illustrates the physical situation.

Let x represent the liters of 30% acid solution. Then other quantities are:

Acid in x liters of 30% acid solution $= 0.3x$
Acid in 10 liters of 80% acid solution $= 0.8(10)$, or 8
Acid in $x + 10$ liters of 40% acid solution $= 0.4(x + 10)$

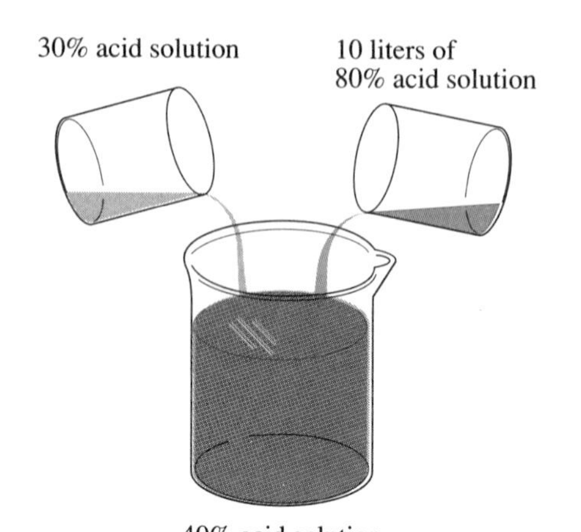

FIGURE 3

We can organize the information we have so far in a table:

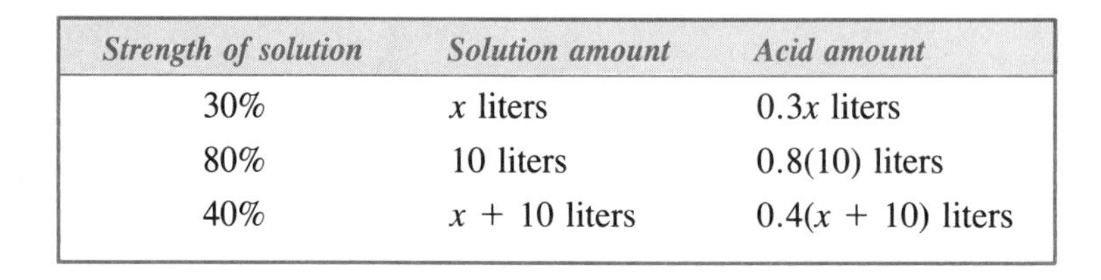

Strength of solution	*Solution amount*	*Acid amount*
30%	x liters	$0.3x$ liters
80%	10 liters	$0.8(10)$ liters
40%	$x + 10$ liters	$0.4(x + 10)$ liters

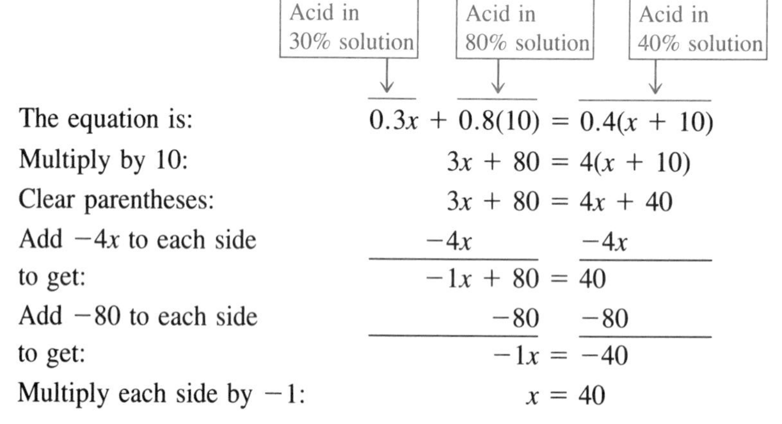

The equation is: $0.3x + 0.8(10) = 0.4(x + 10)$

Multiply by 10: $3x + 80 = 4(x + 10)$

Clear parentheses: $3x + 80 = 4x + 40$

Add $-4x$ to each side: $-4x \quad -4x$

to get: $-1x + 80 = 40$

Add -80 to each side: $-80 \quad -80$

to get: $-1x = -40$

Multiply each side by -1: $x = 40$

The solution to this equation is 40. So, 40 liters of 30% acid solution should be used. ■

Using the Graphics Calculator

To check our work in Example 2, store 40 in memory location X. Now type the left-hand side of the equation, `.3X+.8(10)`, and press the ENTER key. This is the acid in x liters of 30% acid solution plus the acid in 10 liters of 80% acid solution. Next, type the right-hand side, `.4(X+10)`, and press the ENTER key. This is the acid in the $(x + 10)$ liters of 40% mixture. Is the same number output in each case? If not, an error is present!

As a second example of a mixture problem, we consider two people working at different speeds. If the two work together, their combined "rate" is much like the new acid solution we found in Example 2.

EXAMPLE 3 Frank can type 6 pages per hour and Ruth can type 4 pages per hour. How long will it take Frank and Ruth together to type a 200-page report?

Solution Reading the problem, we see that the quantity requested is the time required to type 200 pages.

Let x = time required (in hours) to type the report. The other quantities are then

$$\text{Pages typed by Frank in } x \text{ hours} = 6x$$
$$\text{Pages typed by Ruth in } x \text{ hours} = 4x$$
$$\text{Pages typed by both in } x \text{ hours} = 6x + 4x$$

Typist	*Typing rate*	*Time typing*	*Pages typed*
Frank	6 pg/h	x	$6x$
Ruth	4 pg/h	x	$4x$
Together	10 pg/h	x	$10x$

We can create a table, as shown at the left, to organize these facts.

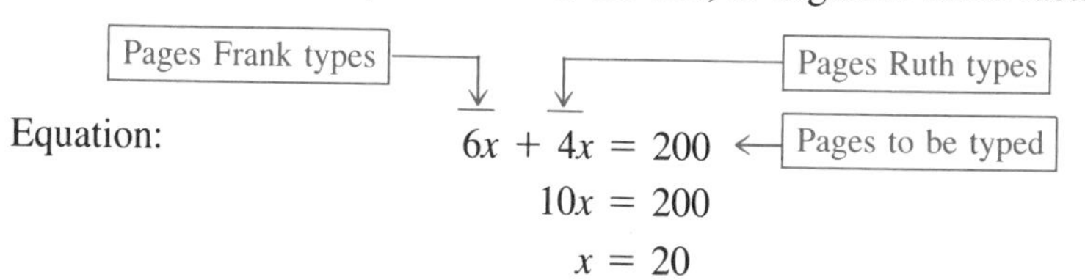

Equation:

$$6x + 4x = 200$$
$$10x = 200$$
$$x = 20$$

Thus, it will take Frank and Ruth 20 hours each to type the 200-page report. ■

In the next mixture problem, we are combining nuts that have different costs per pound. Combining the nuts into a mixture is much like creating a new acid solution.

EXAMPLE 4 Walnuts cost $3.50 per pound and pecans cost $5.00 per pound. How many pounds of walnuts should be mixed with 20 pounds of pecans to get a mixture that costs $4.00 a pound to produce?

Solution Rereading the problem, the quantity requested is the number of pounds of walnuts in the mixture.

Let x = number of pounds of walnuts.

$$\text{Value of } x \text{ pounds of walnuts} = 3.5x$$
$$\text{Value of 20 pounds of pecans} = 5(20)$$
$$\text{Value of } (x + 20) \text{ pounds of the mixture} = 4(x + 20)$$

The table shows our work so far:

Type of nut	*Price per pound*	*Number of pounds*	*Value of pounds of nuts*
Walnut	$3.50	x	$3.5x$
Pecan	$5.00	20	$5(20)$
Mixture	$4.00	$x + 20$	$4(x + 20)$

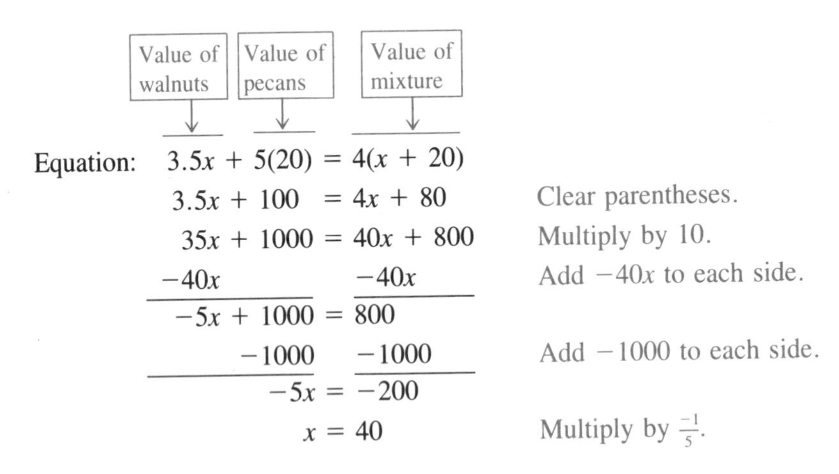

Equation:

$$
\begin{aligned}
3.5x + 5(20) &= 4(x + 20) \\
3.5x + 100 &= 4x + 80 && \text{Clear parentheses.} \\
35x + 1000 &= 40x + 800 && \text{Multiply by 10.} \\
-40x \qquad &\quad\; -40x && \text{Add } -40x \text{ to each side.} \\
-5x + 1000 &= 800 \\
-1000 &\quad\; -1000 && \text{Add } -1000 \text{ to each side.} \\
-5x &= -200 \\
x &= 40 && \text{Multiply by } \tfrac{-1}{5}.
\end{aligned}
$$

Thus, add 40 pounds of walnuts to the 20 pounds of pecans. ■

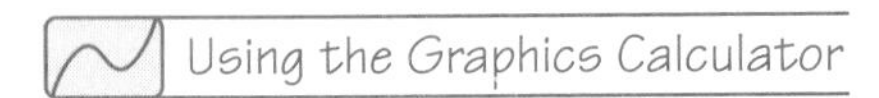

To check the answer to Example 4, store 40 in memory location `X`. Now, type `3.5X+5(20)`, for the value of the walnuts plus the value of the pecans, and press the `ENTER` key. Next, type `4(X+20)` for the value of mixture, and press the `ENTER` key. Are the outputs the same in both cases? If not, there is an error!

GIVE IT A TRY

1. Rework Example 2 to produce a 70% acid solution.
2. Rework Example 3 if Ruth can type 8 pages per hour.
3. Rework Example 4 if the cost of pecans is \$6.00 per pound.

Uniform Rate Problems The next type of word problem deals with *uniform rates,* such as an automobile traveling for 5 hours at 60 miles per hour. Uniform rate problems are actually just mixture problems, much like the typing rates we combined in Example 3.

EXAMPLE 5 Robin drives 4 hours at a fixed speed. She then increases her speed by 15 miles per hour and drives another 5 hours. If the distance traveled is 400 miles, what was Robin's original speed, to the nearest one-tenth mile per hour?

Solution From reading the problem, we see that the quantity requested is the original speed Robin was traveling. Figure 4 is a sketch of the physical situation.

Let x = Robin's original speed.

Distance traveled driving 4 hours at x miles per hour $= 4x$
Robin's new speed $= x + 15$
Distance traveled driving 5 hours at
$(x + 15)$ miles per hour $= 5(x + 15)$
Total distance traveled $= 4x + 5(x + 15)$

Time: 4 hours 5 hours
Rate: x miles/hour $(x + 15)$ miles/hour
START END
Distance: $4x$ miles $5(x + 15)$ miles
400 miles

FIGURE 4

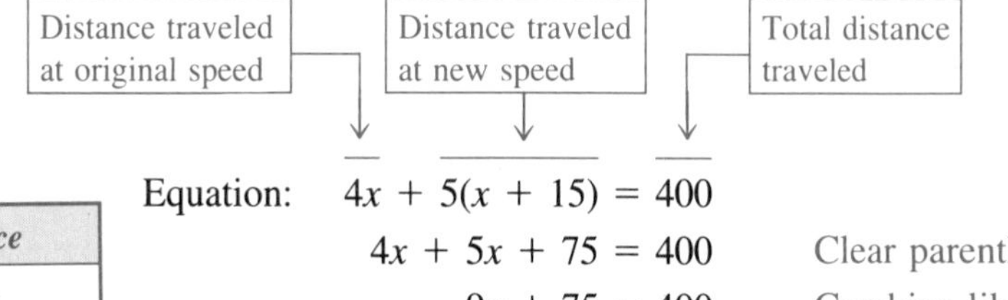

Rate	*Time*	*Distance*
x	4	$4x$
$x + 15$	5	$5(x + 15)$

Equation: $4x + 5(x + 15) = 400$

$$4x + 5x + 75 = 400 \quad \text{Clear parentheses.}$$
$$9x + 75 = 400 \quad \text{Combine like terms.}$$
$$\underline{-75 \quad -75} \quad \text{Add } -75 \text{ to each side.}$$
$$9x = 325$$
$$x = \frac{325}{9} \quad \text{Multiply each side by } \tfrac{1}{9}.$$

The solution to the equation is $\frac{325}{9}$. Robin's original speed (to nearest one-tenth mile) is 36.1 miles per hour. ■

EXAMPLE 6 Ted leaves Tallahassee at 8:00 A.M., driving east toward Jacksonville. Bill leaves Jacksonville at 8:30 A.M., driving west toward Tallahassee. Ted drives at 60 mi/h and Bill drives at 65 mi/h. Jacksonville and Tallahassee are 150 miles apart. How far from Jacksonville will Bill and Ted meet?

Solution Reading the problem, we determine that the quantity requested is the number of miles from Jacksonville to the point where the drivers meet.

Let x = distance of meeting point from Jacksonville.

Distance traveled by Bill = x
Distance traveled by Ted = $150 - x$

Time for Bill to travel x miles at 65 mi/h = $\frac{x}{65}$

Time for Ted to travel $(150 - x)$ miles at 60 mi/h = $\frac{150 - x}{60}$

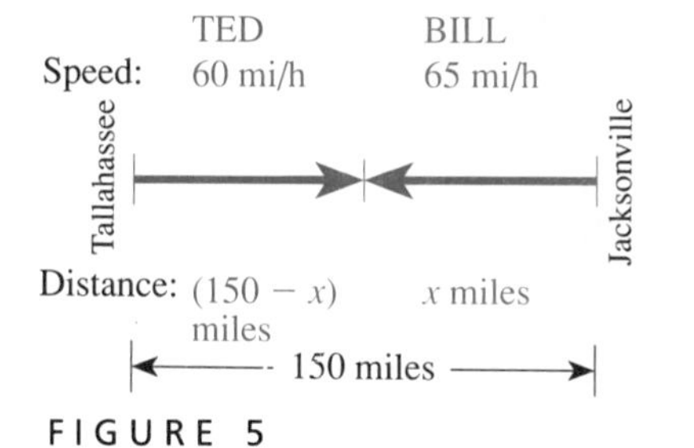

FIGURE 5

Figure 5 is a sketch of these facts.

Time Ted travels — Time Bill travels

Equation: $\frac{150 - x}{60} = \frac{x}{65} + \frac{1}{2}$ ← Bill left $\frac{1}{2}$ hour later. So, add $\frac{1}{2}$ to Bill's time to make the two times equal.

$$\frac{(60)(65)(2)(150 - x)}{60} = \frac{(60)(65)(2)x}{65} + \frac{(60)(65)(2)}{2} \quad \text{Multiply by a nonzero number.}$$
$$130(150 - x) = 120x + 3900 \quad \text{Remove common factors.}$$
$$19500 - 130x = 120x + 3900 \quad \text{Clear parentheses.}$$
$$\underline{-120x \quad -120x} \quad \text{Add } -120x \text{ to each side.}$$
$$19500 - 250x = 3900$$
$$\underline{-19500 \quad -19500} \quad \text{Add } -19500 \text{ to each side.}$$
$$-250x = -15600$$
$$x = \frac{-15600}{-250} \quad \text{Multiply each side by } \tfrac{-1}{250}.$$

The solution is $\frac{15600}{250} = 62.4$.

Thus, Bill and Ted will meet 62.4 miles from Jacksonville. ■

GIVE IT A TRY

4. Rework Example 5 if Robin drives 2 hours at her original speed.
5. Use the formulas and approach from Example 6 to solve the following word problem:

   Train A leaves Centerville at 6:00 A.M, traveling at 50 mi/h. Train B leaves Centerville at 9:00 A.M. on a parallel track, traveling at 80 mi/h. How far from Centerville will Train A pass Train B? [*Hint*: The distance traveled by each train is the same.]

Our last example is another kind of uniform rate problem. Here, two pumps work at different speeds, like the typists in Example 3.

EXAMPLE 7 Pump A can drain a 40,000-gallon tank in five hours. Pump B requires eight hours to empty the same tank. If the tank is full, and then both pumps are turned on at once, how long does it take to drain the tank?

Solution Reading the problem, we see that the quantity is the number of hours for both pumps working together to empty the tank. The situation is illustrated in Figure 6.

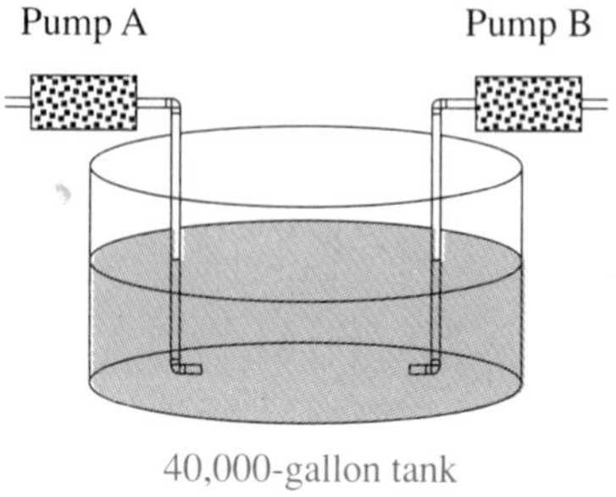

Pump A Pump B

40,000-gallon tank

FIGURE 6

Let x = number of hours both pumps operate.

Number of gallons drained by Pump A = $8000x$

$\frac{40{,}000 \text{ gal}}{5 \text{ h}} = 8000 \text{ gal/h}$

Number of gallons drained by Pump B = $5000x$

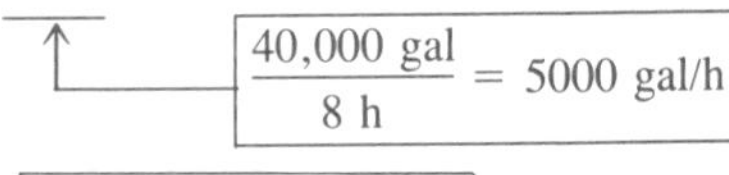

Gallons pumped by A | Gallons pumped by B

$$\text{Equation:} \quad 8000x + 5000x = 40000 \quad \leftarrow \text{Total gallons pumped}$$

$$13000x = 40000$$

$$x = \frac{40000}{13000}$$

Solution: $\frac{40}{13}$

Thus, it takes $3\frac{1}{13}$ hours to drain the tank. ■

■ SUMMARY

In this section we have expanded the general idea of solving word problems, particularly we solved mixture problems and uniform rate problems. Our approach to solving such problems is to write a first-degree equation in one variable as a mathematical model of the physical situation. Solving the equation leads to an answer for the question posed in the word problem. All the word problems in this chapter request one value (thus a first-degree equation in one variable is the appropriate model for the problem). Later, when we consider similar problems that request two values, we will use a system of first-degree equations in two variables to model the situation.

■ ***Quick Index***

2.7 ■ ■ ■ EXERCISES

Solve the following arithmetic problems as a warm-up for the algebra word problems that follow.

1. Suppose 20 liters of 60% acid solution is mixed with 40 liters of 90% acid solution.
   a. How many liters are in the mixture?
   b. How many liters of acid are in the mixture?
   c. What is the acid concentration of the mixture?
2. To make a special blend, 45 pounds of coffee worth $1.20 per pound are mixed with 50 pounds of coffee worth $1.80 per pound.
   a. How many pounds are in the mixture?
   b. How many dollars is the mixture worth?
   c. How much per pound is the mixture worth?
3. Freda earns $8 per hour and her husband Ben earns $7. If Freda worked 30 hours last week and Ben worked 38 hours, how much did the couple earn?
4. Lawyer drives 3 hours at 55 mi/h and then drives 4 hours at 65 mi/h.
   a. How many hours did he drive?
   b. How many miles did he drive?
   c. For the trip, what was the average speed? [*Hint*: Divide miles traveled by 7 hours.]

Solve the following algebra word problems.

5. a. Let x represent the number of liters of 10% acid solution. Write a polynomial for the acid in x liters of 10% acid solution and 70 liters of 40% acid solution.
   b. Write a polynomial for the acid in $(x + 70)$ liters of 30% acid solution.
   c. Create an equation by equating the polynomial from part (a) to the polynomial from part (b). Solve the equation. What is the solution?
   d. How many liters of 10% acid solution must be mixed with 70 liters of 40% acid solution to get a 30% acid solution?
6. a. Let b represent the number of hours Jan and Dan spend typing. Write a polynomial for the number of pages typed in b hours if Dan types 6 pages per hour and Jan types 5 pages per hour.
   b. Using the polynomial from part (a), write an equation for Jan and Dan, working together, to type 165 pages in b hours. Solve the equation, and report the solution.
   c. If Dan types 6 pages per hour and Jan types 5 pages per hour, how long will it take them working together to type a 165-page report?
7. a. Let n represent Mike's speed in miles per hour. Write a polynomial for the distance Mike will travel in 3 hours.
   b. If Mike increases his speed in part (a) by 20% and drives 4 more hours, write a polynomial for the total distance traveled in 7 hours.
   c. Using the polynomial from part (b), write an equation for Mike traveling a total of 500 miles at the two speeds. Solve the equation, and report the solution.
   d. Mike drives 3 hours at one speed, then increases his speed by 20% and drives another 4 hours. The total distance traveled is 500 miles. What is Mike's original speed?
8. How many liters of 40% acid solution must be mixed with 90 liters of 70% acid solution to produce a 60% acid solution?
9. Sally can paint 20 toy trucks per day and Jill can paint 15 toy trucks per day. If they work together, how many days will it take Sally and Jill to paint 200 toy trucks?
10. Della drives for 2 hours at one speed, then increases her speed by 20 mi/h and drives another 3 hours. If she travels 200 miles, what is her original speed?
11. Cashews are worth $4.50 per pound and almonds are worth $3.20 per pound. How many pounds of cashews must be mixed with 10 pounds of almonds to get a mixture worth $4.20 per pound?
12. Senator Woo leaves Wichita, heading toward Topeka, driving at 60 mi/h. Senator Thompson leaves Topeka, heading for Wichita on the same road as Senator Woo, driving at 50 mi/h. If Wichita and Topeka are 100 miles apart, how far from Topeka will they meet?
13. Carol has two acid solutions, one is 30% acid and the other is 70% acid. If she mixes 5 liters of the 30% acid solution with 8 liters of the 70% acid solution, what is the concentration of the resulting mixture?
14. Tom must make a 60% acid solution from a 10% acid solution and a 90% acid solution. If he has 3 liters of the 10% acid solution, how many liters of the 90% acid solution should he add?
15. A tank holds 400 gallons of water. Pipe A can fill it in 4 hours, and Pipe B can fill it in 5 hours. If the valves of both pipes are opened, how long will it take to fill the empty tank?
16. A 400-gallon water heater can be filled in 2 hours from the water main. The kitchen faucet, fully opened, drains the water heater in 5 hours. The bathroom faucet, fully opened, drains the water heater in 4 hours. If the water

heater is half full when the valves on the water main, kitchen faucet, and the bathroom faucet are all opened, how long will it take to refill the water heater?

17. Paul beats Mark in 70% of the Scowl games they play. He beats Wanda in 40% of the Scowl games they play. If Paul plays Wanda 35 games, how many games must he play Mark in order to win 60% of the Scowl games he plays?

18. **a.** Debbie jogs at 8 mi/h and bikes at 12 mi/h. If she jogs for 2 hours, how long must she bike to travel 30 miles?

    b. Debbie starts at her house and jogs to a friend's house to pick up her bike, and then bikes back home. If the trip took 3 hours, how far away does her friend live?

19. Joshua is driving on the interstate at 65 mi/h. Shawn is driving a car 3 miles in front of Joshua and is traveling at 60 mi/h. How long will it take Joshua to catch up to Shawn?

20. The AZT Company adds 30% to (marks up) the cost of all items that it sells. If the AZT Company has a coat sticker-priced at $125, what was the cost of the coat to the AZT Company?

21. How many liters of pure water (0% acid solution) must be mixed with 20 liters of 70% acid solution to produce a 30% acid solution?

22. Ms. Gilly can get a 9% return on her investment in Stock A and a 12% return on her investment in Stock B. If she invests $10,000 in Stock A, how much should she invest in Stock B to earn a total return of $2,000 on these investments?

23. Pump A can fill an 80,000-gallon tank in 8 hours. A drain can empty the same tank in 10 hours. The tank contains 10,000 gallons at 8:00 A.M., when pump A is turned on and the drain is set to half-open. At what time will the tank be filled up to 65,000 gallons?

24. Silver costs $18 per ounce. How many ounces of silver must be mixed with 40 ounces of an alloy costing $3 per ounce to obtain a silver alloy costing $15 per ounce?

■ BONUS PROBLEM

25. The road over Nob Hill is a one-mile incline up the hill, followed by a one-mile decline down the hill. Bobbie drives up the hill at 30 mi/h. How fast must she drive down the other side to average 60 mi/h for the two miles?

2.8 ■ ■ ■ REWRITING FORMULAS

In this section, we apply knowledge of solving first-degree equations in one variable to rewriting formulas, such as $3x + 5y = 18$ or $F = \frac{9}{5}C + 32$. This topic is a direct application of solving first-degree equations in one variable.

Formulas The importance of using formulas in algebra, and the ability to work with formulas, cannot be overstated. Formulas appear in all the sciences and in business. In Chapter 1, we learned how to evaluate a formula—replacing the variables of the formula with values, then performing the computation. In this section, we will learn to rearrange, or rewrite, a formula for a particular variable. To accomplish this, we use solving first-degree equations as a model.

Background Consider the formula relating temperature in degrees Fahrenheit, F, to degrees Celsius, C:

$$F = \frac{9}{5}C + 32$$

We say this formula has two variables, F and C, and that the formula is **written for** F. Notice that F appears on the left-hand side of the equal sign and is the only variable (or term) on the left side. Also, F appears *only* on the left-hand side of the equal sign. Next, consider the formula

$$s = gt^2 + ut$$

This formula has four variables, s, g, t, and u, and is written for the variable s. Finally, consider the formula

$$3x + 5y = 12$$

W	P	V
30	6.1	?
35	6.1	?
40	6.1	?
50	6.1	?
30	6.8	?
35	6.8	?
40	6.8	?
50	6.8	?

This formula has two variables, x and y, and it is not written for either variable.

Suppose we have the formula $W = P(V + 4)$ and wish to find the value for V when $W = 30$ and $P = 6$. One way to approach this problem is to replace W with 30 and P with 6, and then solve the equation $30 = 6(V + 4)$. Now, suppose we wish to find the value for V for all the values W and P shown in the table. Here, we would have to solve an equation similar to $30 = 6(V + 4)$, eight times. An alternative would be to rewrite the formula $W = P(V + 4)$ for V, and then just do computations.

To rewrite the formula, we can solve the equation $30 = 6(V + 4)$ as a guide. This is known as *reasoning by analogy*: using a simple, or known, concept to help explain a more difficult, or unknown, concept. Consider the following.

EXAMPLE 1 Rewrite $W = P(V + 4)$ for V.

Solution

Analogy (for $W = 30$ and $P = 6$)		*Formula*	
Equation:	$30 = 6(V + 4)$		$W = P(V + 4)$
	$6(V + 4) = 30$		$P(V + 4) = W$
Clear parentheses:	$6V + 24 = 30$		$PV + 4P = W$
Add -24 to each side:	$-24 \quad -24$	Add $-4P$ to each side:	$-4P \quad -4P$
Multiply each side by $\frac{1}{6}$:	$6V = 6$	Multiply each side by $\frac{1}{P}$:	$PV = W - 4P$
	$V = \frac{6}{6}$		$V = \frac{W - 4P}{P}$
Solution: 1		Formula for V:	$V = \frac{W - 4P}{P}$

The formula for V could also be written as

$$V = \frac{W}{P} - \frac{4P}{P}$$

The operations of multiplication by a number and division by the same number cancel each other.

or $$V = \frac{W}{P} - 4$$ ■

The first step in solving the equation and in rewriting the formula in Example 1 uses the reflexive property of equality: *If $a = b$, then $b = a$.* To create an equation for the analogy, we replace each variable (other than the one we are rewriting for) with a number (any number). This activity is often referred to as *solving for V* or *solving for y*. For the moment, we will use the word *solve* for the activity of finding numbers that replace the variable to yield a true arithmetic sentence. Here we are simply altering the form of an equation.

As a second example of rewriting a formula for a particular variable, consider the following problem.

EXAMPLE 2 Rewrite $3x + 5y = 12$ for y.

Solution

Analogy (for $x = 7$)			*Formula*
Equation:	$3(7) + 5y = 12$		$3x + 5y = 12$
	$21 + 5y = 12$		$3x + 5y = 12$
Add -21 to each side:	$-21 \qquad -21$	Add $-3x$:	$-3x \qquad -3x$
	$5y = -9$		$5y = -3x + 12$
Multiply each side by $\frac{1}{5}$:	$y = \frac{-9}{5}$	Multiply by $\frac{1}{5}$:	$y = \frac{-3x + 12}{5}$
Solution: $\frac{-9}{5}$		Formula for y:	$y = \frac{-3x + 12}{5}$

The formula for y can also be written as $y = \frac{-3}{5}x + \frac{12}{5}$. ■

GIVE IT A TRY

1. Rewrite $R(T + 2) = C - 3$ for T.
2. Rewrite $5x - 2y = 13$ for y.

Special Notation in Formulas Many formulas in mathematics, science, and business use special notation, such as letters from the Greek alphabet: pi (π), alpha (α), delta (Δ), and sigma (Σ). Formulas may also include one or more subscripts. A **subscript** is a number or letter placed to the right of and slightly below a variable. For example, the 2 in x_2 is a subscript. Subscripts are used to expand the collection of characters available for variable representation. Usually we use them when two or more quantities in a formula are closely related, such as two velocities, v_1 and v_2, or two values of x (x_1 and x_2). The following formulas show the use of subscripts and Greek letters.

Formula	*Area*	*Meaning*
$A = \pi r^2$	Mathematics	The Greek letter pi, π, represents the ratio of the circumference to the diameter of a circle (the number π is ≈ 3.14).
$S = r\theta$	Mathematics	The Greek letter theta, θ, is used to represent the length of an arc.
$v_c = R\omega$	Physics	The subscript c represents the center of mass, and the Greek letter omega, ω, represents angular velocity.
$m = \frac{\Delta y}{\Delta x}$	Mathematics	The Greek letter delta, Δ, is used to represent the difference, or change.
$m = \frac{y_2 - y_1}{x_2 - x_1}$	Mathematics	The subscripts are used for a first pair of numbers, x_1 and y_1, and a second pair of numbers, x_2 and y_2.

The next example shows how we can rewrite a formula with special notation. Consider the following.

EXAMPLE 3 Rewrite the formula $m = \dfrac{y - y_1}{x - x_1}$ for y.

Solution We start by creating an analogy.

Analogy (for $y_1 = 5$, $x = 7$, $x_1 = 3$)		*Formula*	
Equation:	$3 = \dfrac{y - 5}{7 - 3}$		$m = \dfrac{y - y_1}{x - x_1}$
	$3 = \dfrac{y - 5}{4}$		$m = \dfrac{y - y_1}{x - x_1}$
Multiply by 4:	$12 = \dfrac{\not{4}(y - 5)}{\not{4}}$	Multiply by $x - x_1$:	$(x - x_1)m = \dfrac{\cancel{(x - x_1)}(y - y_1)}{\cancel{x - x_1}}$
Remove common factor:	$12 = y - 5$	Remove common factor:	$(x - x_1)m = y - y_1$
Add 5:	$\underline{5 \quad 5}$	Add y_1:	$\underline{y_1 \quad y_1}$
	$17 = y$		$(x - x_1)m + y_1 = y$
Solution: 17		Formula for y:	$y = m(x - x_1) + y_1$

■

To create the equation for the analogy, replace each variable (other than the one we are rewriting for) with a number (any number). As a last example of rewriting formulas for a particular variable consider the following.

EXAMPLE 4 Rewrite $RW + 5 = 7W + T$ for W.

Solution Again, we start by creating an analogy.

Analogy (for $R = 10$ and $T = 8$)		*Formula*
Equation:	$10W + 5 = 7W + 8$	$RW + 5 = 7W + T$
	$\underline{-7W \qquad -7W}$	$\underline{-7W \qquad -7W}$
	$3W + 5 = 8$	$(R - 7)W + 5 = T$
	$\underline{-5 \quad -5}$	$\underline{-5 \quad -5}$
	$3W = 3$	$(R - 7)W = T - 5$
	$W = \dfrac{-3}{3}$	$W = \dfrac{T - 5}{R - 7}$
Solution:	-1	Formula for W: $W = \dfrac{T - 5}{R - 7}$ $R \neq 7$

■

In Example 4, for the first line we subtract $7W$ from RW by using the distributive property, $RW - 7W = (R - 7)W$. Notice the analogy with the equation $10W - 7W = (10 - 7)W = 3W$.

GIVE IT A TRY

3. Rewrite $s = \dfrac{x - \mu}{\sigma}$ for x.

4. Rewrite $V(R + 3) = TR - 2$ for R.

■ SUMMARY

In this section, we have used knowledge of, and skill with solving, a first-degree equation in one variable to rewrite a formula for a particular variable. That is, we learned how to rewrite a formula such as

$$F = \frac{9}{5}C + 32$$

for C to get

$$C = \frac{5F - 160}{9} \quad \text{or} \quad C = \frac{5}{9}F - \frac{160}{9}$$

To accomplish this, we use the technique of *reasoning by analogy*.

2.8 ■ ■ ■ EXERCISES

1. For the formula $P = 3(W + T)$, if $P = 100$ and $W = 30$, then what is the value for T?
2. For the formula $3x + 5y = 18$,
   a. if $y = -3$, then what is the value for x?
   b. if $x = 2$, then what is the value for y?
3. For the formula $m = \dfrac{y - y_1}{x - x_1}$, if $m = \dfrac{-2}{3}$, $x = 6$, $x_1 = 5$, and $y_1 = -3$, what is the value for y?
4. **a.** Solve $3(7) - 5y = 9$. Report the solution.
   b. Rewrite $3x - 5y = 9$ for y.
5. **a.** Solve $3T - 8 = 2T + 5$. Report the solution.
   b. Rewrite $3T - R = CT + 5$ for T.
6. **a.** Solve $5 = \dfrac{T - 5}{9 - 7}$. Report the solution.
   b. Rewrite $C = \dfrac{T - 5}{M - 7}$ for T.

Rewrite each formula for the indicated variable.

7. $3x + 7y = 21$ for y
8. $2x - 5y = 12$ for y
9. $2x - 5y = 12$ for x
10. $P = 2L + 2W$ for W
11. $PV + R = 10 - R$ for V
12. $PV + R = 10 - R$ for R
13. $W(T + 3) = CW$ for W
14. $F = \dfrac{9}{5}C + 32$ for C
15. $A = \dfrac{1}{2}h(B + b)$ for B
16. $AB(p + 1) = 24f$ for p
17. $s = s_0 + gt^2$ for g
18. $3(x - 2) + 3(y - 4) = 5 - y$ for y
19. $CF - 2 = RF + T$ for F
20. $S = 2\pi rh + 2\pi r^2$ for h
21. $3x - y = 5 + y$ for y
22. $y = \dfrac{-2}{3}x + \dfrac{1}{6}$ for x
23. $P = \dfrac{3T + 2}{3}$ for T
24. $z = \dfrac{x - \mu}{\sigma}$ for μ
25. $y = \dfrac{3 - 2x}{5}$ for x
26. $A = \dfrac{h(b_2 + b_1)}{2}$ for b_1
27. $A = \dfrac{h(b_2 + b_1)}{2}$ for h
28. $F = \dfrac{km_1m_2}{d^2}$ for m_1

CHAPTER 2 REVIEW EXERCISES

Report the degree and the number of variables in each polynomial. Identify the polynomial as a monomial, binomial, trinomial, or quadrinomial.

1. $3x^2y - y^5$
2. $x^2 - 3x + x^2z$
3. $-5w^4p^2t^2$
4. $x^4 - x^2 - xy + y$
5. $x^5 - x^2 - 2x + 1$
6. ab
7. $2x - 3yz$
8. $y^2 - 3xy + x^2$
9. $-7x^2 + 2y + y^3$
10. $x^4 - y^4$
11. Which of the following expressions is a first-degree polynomial in one variable?

    $3x - 2y \quad 3x - 8 \quad 2xy - 1 \quad x^2 - 3x + 2$
12. Which of the following expressions is *not* a second-degree binomial?

    $3x^2 - y \quad 3xy - x \quad 2x^2 - x \quad 3x - 2y$

Write each polynomial in standard form.

13. $(x^2 - 3x + 5) + (-3x^2 - 5x - 8)$
14. $(x^2 - 3xy + y^2) + (2x^2y - 5xy - y^2)$
15. $(p^2 - 3p - 5) - (2p^2 + 5p - 8)$
16. $w^3 - (2w - 6) - (w^2 - 3w^3)$
17. $3x^2 - (x^2 - 6x + 2) - (3x - 2)$
18. $2y - [y^2 - (2y^2 - 5y - 2) - 4y]$
19. $-(2y - (3 - y^2) - 5)$
20. $\left(\frac{3x + 2}{3}\right) - \left(\frac{x + 5}{3}\right)$
21. $3y^2 - [5 - (6 - 2y) - y^2]$
22. $r^3 - (r^2 - 3r + 2) - (r^3 - r^2 + r)$

Evaluate the polynomial for the given values.

23. $3x^2 - 2x - (5 - 4x - x^2)$ for $x = -3$
24. $2x - [5 - (2 - x - x^2) - 5x]$ for $x = 5$
25. $\frac{-1}{2}t^2 - 2t - 16$ for $t = 10$
26. $3xy - [(y^2 - 2xy - x^2) - 2y^2]$ for $x = 3, y = -1$

Solve the equation.

27. $-3x = 135$
28. $13 = -5x$
29. $\frac{1}{6} = -5x$
30. $\frac{3}{4}x = \frac{-2}{7}$
31. $2x + 7 = 19$
32. $-3x + 5 = 23$
33. $16 - 5x = -4$
34. $2 - 6x = 7$
35. $-4x - 8 = 2$
36. $-5x + 7 = 32$
37. $\frac{1}{2}x + \frac{3}{5} = 3$
38. $\frac{3}{8} - \frac{1}{4}x = -1$
39. $\frac{3x - 1}{5} = \frac{3}{10}$
40. $\frac{3 - 2x}{6} = \frac{-1}{3}$
41. $3x - 6 = 7 - 8x$
42. $y - 6 = 14 - 3y$
43. $-5 + 8c = 3 - c$
44. $w + 5 = 6w - 9$
45. $16 - [3x - (5 - 2x)] = 8 - x$
46. $1 - (4 - 3x) = 14x - (6 + 12x)$

Solve each inequality. Draw a number line graph of the solution, and report the solution in interval notation.

47. $-3x \geq 9$
48. $-2x < 8$
49. $5 + 3x > 5$
50. $5 - 3x > 12$
51. $16 - 2x \leq x + 12$
52. $2x + 7 \geq 5x + 7$
53. $3 - (2x - 4) > x - 1$
54. $5x < 6 + 5x$
55. $5 - 2x \leq -(2x - 5)$
56. $5 - x > 3x - (2 - x)$
57. $\frac{3x - 1}{4} \geq -2$
58. $\frac{6x - 7}{5} \leq x$
59. $-2 \leq 3x + 4 \leq 5$
60. $6 \geq 3x - 5 \geq 0$
61. $1 \geq 3 - 2x \geq -4$
62. $-5 \leq 2 - 4x \leq -1$
63. $x \leq 3x + 7 < x + 3$
64. $x + 1 < 2x \leq 6$
65. $\frac{3}{4} < \frac{3x - 5}{2} < \frac{3}{2}$
66. $-1 \leq \frac{3 - 4x}{5} \leq \frac{5}{2}$

Solve each absolute value equation.

67. $|2x - 3| = 10$
68. $|3 - 2x| = 12$
69. $|5 - 2x| = -1$
70. $|3x + 6| = 0$
71. $|4 - 2x| = x$
72. $|1 - x| = 2x + 3$

Solve each absolute value inequality.

73. $|2x - 3| \leq 10$
74. $|3 - 2x| < 12$
75. $|6 - 2x| > 3$
76. $|3x - 5| \geq 7$
77. $|4 - 2x| \leq x$
78. $|1 - x| \geq 2x + 3$

Write a polynomial for each phrase.

79. Three less than twice a number.
80. The distance traveled by a train going 60 mi/h for t hours.
81. The value of a mixture of p pounds of walnuts worth \$2 per pound and 12 pounds of cashews worth \$5 per pound.
82. The amount of acid in y liters of 30% acid solution and 7 liters of 90% acid solution.

83. The distance traveled by Goldie driving 5 hours at x miles per hour, then increasing her speed by 20% and driving 3 more hours.

84. The interest from investing \$20,000 at an interest rate of 9% and d dollars at an interest rate of 15%.

Solve each word problems. Report (a) the variable used and what it represents, (b) polynomials for the quantities in the problem, (c) the equation (mathematical model) to be solved and its solution, and (d) the answer to the problem.

85. Fifteen more than 3 times a number is 38. Find the number.

86. Today Cathy is three times as old as her sister. In 10 years, her sister will be 70% of Cathy's age in 10 years. How old is Cathy today?

87. A number decreased by 8 equals 10% of the sum of the number and 8. What is the number?

88. The perimeter of an isosceles triangle (two sides are equal) is 120 meters. The two equal sides each have a length 10 meters greater than the nonequal side. What is the length of the nonequal side?

89. Ms. Mellot wants to enclose a rectangular garden plot with fencing that costs \$3 a foot. She has \$300 available for the project. If the width of the plot must be 10 feet, what length can she make the plot?

90. Joe has test scores of 98, 85, and 70. What score must he make on the fourth test to average 82 on the four tests?

Solve each problem. Report (a) the variable used and what it represents, (b) a polynomial for each quantity given in the problem, (c) the equation (mathematical model) to be solved and its solution, and (d) the answer to the problem.

91. How many liters of a 6% alcohol solution must be mixed with 3 liters of a 10% alcohol solution to produce a 9% alcohol solution?

92. How many pounds of a tea blend worth \$1.30 per pound must be combined with 3 pounds of a tea blend worth \$2.50 per pound to get a tea blend worth \$1.80 per pound?

93. Wilbur drives for 5 hours at a fixed speed and then decreases his speed by 15 mi/h and drives another hour. If he travels 300 miles, what is his original speed?

94. Lyle drives 3 hours at 50 mi/h. How fast must he drive for the next 2 hours to average 55 mi/h for the trip?

95. Becky can assemble care packages at a rate of 7 per hour. Holly can assemble 10 of the care packages per hour. If Becky and Holly work together, how long will it take them to assemble 170 care packages?

96. Denny can paint 12 toy soldiers per day. Cal paints 30% more toy soldiers per day than Denny. If Denny and Cal work together, how many days will it take them to paint 200 toy soldiers?

97. Pump A can empty a 20,000-gallon tank in 8 hours. Pump B can empty the same tank in 5 hours. If both pumps are turned on, how long will it take to empty the tank?

98. John can ride his bike 20 mi/h and he can walk 10 mi/h. John bikes from his house to a friend's house and then walks home. If the entire trip takes an hour, how far away does his friend live?

99. Train A leaves Seatown, traveling at 80 mi/h. If Train B leaves Seatown 2 hours later, traveling at 100 mi/h, how far from Seatown will Train B overtake Train A?

100. How many liters of 10% acid solution must be mixed with 50 liters of 70% acid solution to produce a 20% acid solution?

Rewrite each formula for the indicated variable.

101. $2x - 9y = 18$ for y

102. $5x + 6y = 20$ for y

103. $-2x + 8 = 5x + 2y$ for x

104. $12 - (3x + 2y) = 5y - 8$ for y

105. $\frac{2x}{3} + \frac{y}{5} = 1$ for y

106. $\frac{x + 2}{5} = \frac{t - 1}{2}$ for x

107. $PV - W = C$ for V

108. $R - 3T = 6R + 2$ for T

109. $C(3 - D) = T$ for D

110. $P = 2L + 2W + 2H$ for W

111. $C = 2\pi r$ for r

112. $c = \frac{b + 2a}{7}$ for a

113. $3W - P = 5W + C$ for W

114. $3W - P = C(W + 1)$ for W

115. $RT - 5 = 3 - 2R$ for R

116. $C = \frac{5}{9}(F - 32)$ for F

117. $m = \frac{y - y_1}{x - x_1}$ for y_1

118. $F = \frac{km_1m_2}{d^2}$ for k

CHAPTER 2 TEST

Use the graphics calculator to check your work.

1. State the degree of the polynomial $-3x^2 - 5xy^2 - y$, and evaluate for $x = 2$ and $y = -3$.
2. Write each polynomial in standard form.
   a. $(-x^2 + 2x - 5) + (3x^2 - 3x - 5)$
   b. $(-x^2 + 2x - 5) - (3x^2 - 3x - 5)$
   c. $3p - (5 - (p^2 - 6p - 1) + 2p)$
3. Solve each first-degree equation in one variable.
   a. $\frac{3}{5}x = \frac{-1}{9}$
   b. $3x - 8 = 17 - 7x$
   c. $5 - (2x - 8) = 17 - x$
4. Solve each inequality. Draw a number line graph of the solution, and report the solution in interval notation.
   a. $2x - 5 \geq 4x - 15$
   b. $-3 \leq 5 - 3x < 6$
   c. $2 > \frac{3x - 4}{5} > -1$
5. Solve the absolute value equation $|2x - 9| = 3$.
6. Solve the absolute value inequality $|2x - 9| > 3$. Draw a number line graph of the solution, and report the solution in interval notation.

For Problems 7 and 8, report (a) the variable used and what it represents, (b) a polynomial for each quantity given in the problem, (c) the equation (mathematical model) to be solved and its solution, and (d) the answer to the problem.

7. How many liters of 20% alcohol solution must be mixed with 15 liters of 60% alcohol solution to produce a 30% alcohol solution?
8. Judy drives for 5 hours at a fixed speed, and then decreases her speed by 10 mi/h and drives another 2 hours. If she travels 400 miles, what is her original speed?
9. Rewrite $3x - 8y = 20$ for y.
10. Rewrite $PV = R - 3V$ for V.

3

EQUATIONS IN TWO VARIABLES AND LINEAR FUNCTIONS

In this chapter, we continue working with first-degree equations. (Remember that solving equations is a major objective of algebra at this level.) We now consider equations that have two variables, say x and y. Solving such equations requires the skill of solving first-degree equations in one variable. The major difference between a first-degree equation in one variable, such as $2x = 8$, and a first-degree equation in two variables, such as $3x + 5y = 12$, is the solution. A collection of ordered pairs of numbers is always the solution of a first-degree equation in two variables. When graphed, this solution appears as a line. For this reason, these equations are known as linear equations.

In this chapter we will also study first-degree inequalities in two variables, linear relationships, and we introduce function notation.

3.1 ■ ■ ■ SOLVING LINEAR EQUATIONS: DRAWING THE LINE

The solution to a first-degree equation in one variable, such as $2x + 4 = 12$ is typically one number (see Figure 1). The solution to a first-degree inequality in one variable, such as $2x + 4 < 12$, is typically an infinite collection of numbers. To show this infinite set of numbers, we use a number line as shown

Solve: $2x + 4 = 12$

$$\begin{aligned} 2x + 4 &= 12 \\ -4 \quad & \quad -4 \\ \hline 2x &= 8 \\ x &= 4 \end{aligned}$$

Solution: 4

FIGURE 1

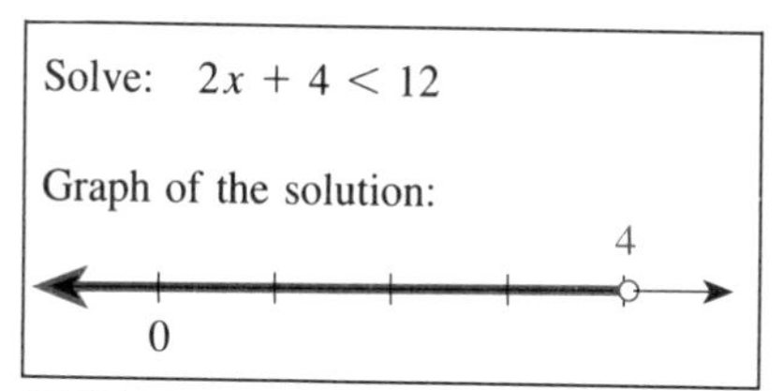

FIGURE 2

in Figure 2. The solution to a *first-degree equation in two variables,* such as $2x + 4y = 12$, is a collection of pairs of numbers—each pair has a number for the x-variable and a number for the y-variable. In fact the pairs of numbers are ordered (the x-value first and the y-value second). Thus, we say the solution to an equation, such as $2x + 4y = 12$, is an infinite set of **ordered pairs,** such as $\{(6, 0), \left(3, \frac{3}{2}\right), (-6, 6), (4, 1), \ldots\}$. Remember, equations with one variable have a solution of numbers, and equations with two variables have a solution of ordered pairs of numbers.

An ordered pair that solves the equation $2x + 4y = 12$ is $(4, 1)$, because replacing the x with 4 and the y with 1 yields $2(4) + 4(1) = 12$, a true arithmetic sentence. An ordered pair that does *not* solve the equation is $(1, 4)$, because $2(1) + 4(4) = 12$ is a false arithmetic sentence. As we can see from this example, order is very important! Suppose we wish to find the ordered pair that solves $2x + 4y = 12$ if the x-value (known as the ***x*-coordinate**) is 1. To accomplish this, we replace the x with 1 to get first-degree equation in one variable, $2 + 4y = 12$. Solving this equation, we get

$$\begin{aligned} 2 + 4y &= 12 \\ -2 \quad & \quad -2 \\ \hline 4y &= 10 \\ y &= \frac{10}{4} = \frac{5}{2} \end{aligned}$$

This equation has solution $\frac{5}{2}$. Thus, the ordered pair in the solution with x-coordinate 1 is $\left(1, \frac{5}{2}\right)$. The use of parentheses is critical, since $\left(1, \frac{5}{2}\right) \neq 1, \frac{5}{2}$.

Suppose we wish to find the ordered pair that solves $2x + 4y = 12$ if the y-value (the ***y*-coordinate**) is 1. To accomplish this, we replace the y with 1 to get a first-degree equation in one variable, $2x + 4 = 12$. Solving this equation, we get

$$\begin{aligned} 2x + 4 &= 12 \\ -4 \quad & \quad -4 \\ \hline 2x &= 8 \\ x &= 4 \end{aligned}$$

This equation has solution 4. Thus, in the solution to $2x + 4y = 12$ the ordered pair with y-coordinate 1 is $(4, 1)$. Work through the following example.

EXAMPLE 1 For the first-degree equation in two variables, $3x - 5y = 12$, find the ordered pair in the solution with the given coordinate.

a. x-coordinate 2 **b.** x-coordinate 0

c. y-coordinate 0 **d.** y-coordinate -2

Solution **a.** We replace x with 2 to get the equation $6 - 5y = 12$. Solving this equation, we get

$$\begin{aligned} 6 - 5y &= 12 \\ -6 \quad & \quad -6 \\ \hline -5y &= 6 \\ y &= \frac{-6}{5} \end{aligned}$$

Thus, the ordered pair is $\left(2, \frac{-6}{5}\right)$.

b. Replace x with 0 in $3x - 5y = 12$ to get the equation $-5y = 12$, which has the solution $\frac{-12}{5}$. Thus, the ordered pair is $\left(0, \frac{-12}{5}\right)$.

c. Replace y with 0 to get the equation $3x = 12$, which has the solution 4. Thus, the ordered pair is $(4, 0)$.

d. Replace y with -2 to get the equation $3x + 10 = 12$, which has the solution $\frac{2}{3}$. Thus, the ordered pair is $\left(\frac{2}{3}, -2\right)$. ■

Ordered pairs that solve
$\mathbf{3x - 5y = 12}$

x	y	*Ordered pair*
2	$\frac{-6}{5}$	$\left(2, \frac{-6}{5}\right)$
0	$\frac{-12}{5}$	$\left(0, \frac{-12}{5}\right)$
4	0	$(4, 0)$
$\frac{2}{3}$	-2	$\left(\frac{2}{3}, -2\right)$

As Example 1 shows, we could continue forever, generating ordered pairs that solve the equation $3x - 5y = 12$. Simply select a number for x, replace the variable x with the number, and solve the resulting first-degree equation to get the y-coordinate. We can record the results of our work in a table. For instance, a table for our work in Example 1 is shown in the margin. ■

GIVE IT A TRY

1. For the first-degree equation in two variables, $3x - 5y = 12$, find the ordered pair that solves the equation with the given coordinate.
   a. x-coordinate -1 **b.** y-coordinate 3 **c.** y-coordinate -3
2. For the equation $-5x + 2y = 10$, complete the following table.

x	-2	-1	0	____	____
y	____	____	____	1	2

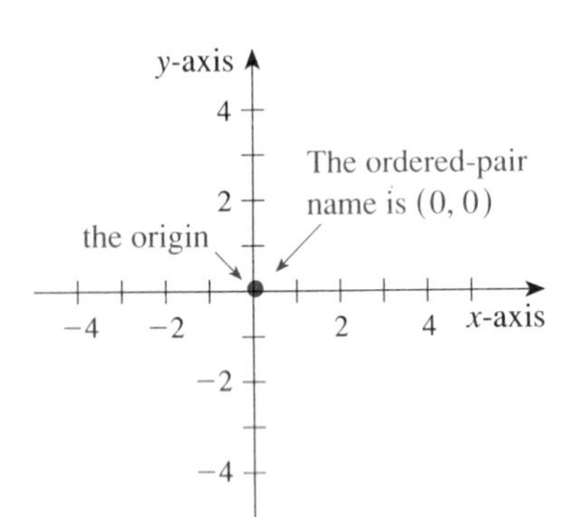

FIGURE 3

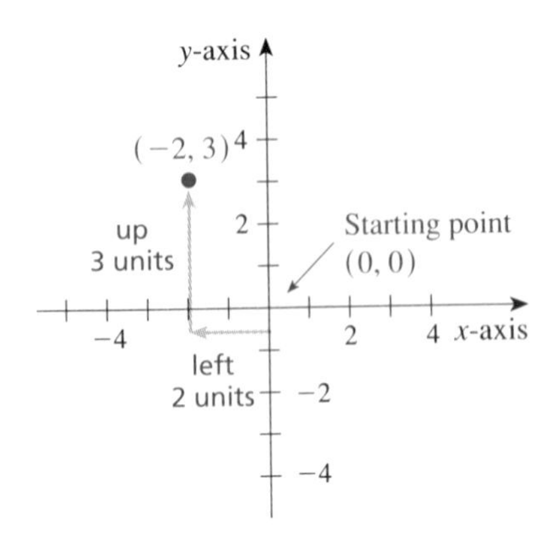

FIGURE 4

The Coordinate System Just as we can graph the solution of a first-degree inequality in one variable as an interval on a number line, we can show on a graph the infinite collection of ordered pairs that solves a first-degree equation in two variables. But rather than a one-dimensional number-line graph, we use a two-dimensional graph, which is simply two number lines set at right angles to each other. The horizontal line is known as the ***x*-axis,** and the vertical line is the ***y*-axis.** Arrows are placed on each axis to indicate the positive direction. The intersection point of the horizontal number line with the vertical number line is called the **origin.** The set of axes is shown in Figure 3. Every ordered pair of numbers can be plotted on this set of axes by moving left or right from the origin the x-value amount, and then moving up or down the y-value amount. For example, we plot the ordered pair $(-2, 3)$ by starting at the origin and moving 2 units in the negative x direction (to the left of the origin), and then 3 units in the positive y direction (up). Figure 4 shows the plot.

The Quadrants The axes divide the graph into four sections, or quadrants. The points to the right of and above the origin make up quadrant I. We say that the point $(2, 5)$ is in quadrant I, because it is 2 units to the right of and 5 units above the origin. The point $(-2, 3)$ is said to be in quadrant II, which includes all the points to the left of and above the origin. The point $(-2, -4)$ is in quadrant III, as are all the points to the left of and below the origin. Finally, the point to the right of and below the origin make up quadrant IV. The quadrants are labeled on the set of axes shown in Figure 5. The points *on* the axes, those with x-coordinate 0 or y-coordinate 0, are not in any quadrant.

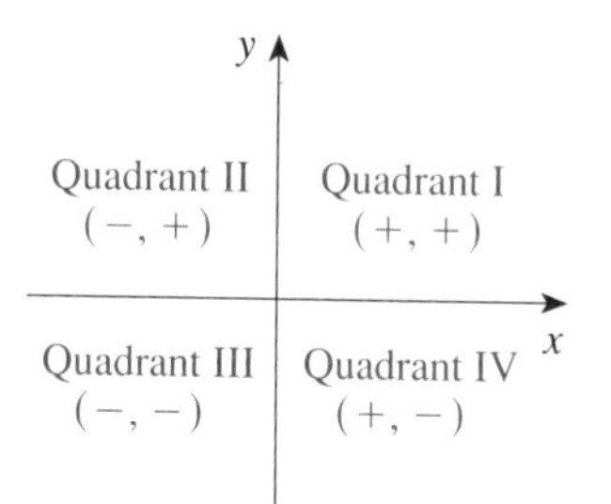

FIGURE 5

Using the Graphics Calculator

```
RANGE
Xmin=-4.8
Xmax=4.7
Xscl=1
Ymin=-3.2
Ymax=3.1
Yscl=1
Xres=1
```

FIGURE 6

We can use the graphics calculator to achieve greater understanding of the coordinate system. Press the `Y=` key and use the `CLEAR` key to erase any existing entries. Press the `RANGE` key and enter the values shown in Figure 6* to create a friendly view screen. These settings will display the x-axis (the horizontal number line) from -4.8 to 4.7 and the y-axis (the vertical number line) from -3.2 to 3.1. Press the `GRAPH` key to display the graph screen and then press the right arrow key. The message at the bottom of the screen, `X=.1 Y=0`, means that the point has an ordered-pair name of (0.1, 0). Continue to press the right arrow key until the message reads `X=2 Y=0`. All the points displayed have y-coordinate 0. Any point on the x-axis has y-coordinate 0 (the points are 0 unit up or down). Press the up arrow key until the message reads `X=2 Y=3`. This point is in quadrant I. Now, press the left arrow key until the message reads `X=-1 Y=3`. This point, $(-1, 3)$, is in quadrant II. Notice that every point in quadrant II has a negative x-coordinate and a positive y-coordinate. Press the down arrow key until the message reads `X=-1 Y=-2`. The point $(-1, -2)$ is in quadrant III. Finally, press the right arrow key until the message reads `X=2 Y=-2`. This point $(2, -2)$ is in quadrant IV.

We now return to the equation of Example 1, $3x - 5y = 12$, and the ordered pairs we found in its solution (see the following table). To plot a point on the set of axes, we start at the origin, (0, 0), and then move the number of units in each direction indicated by the x- and y-coordinates:

$$\underset{\text{right or left}}{\overset{\uparrow}{(x,}}\ \underset{\text{up or down}}{\overset{\uparrow}{y)}} \qquad \underset{\text{right 2}}{\overset{\uparrow}{\Big(2,}}\ \underset{\text{down } \frac{6}{5}}{\overset{\uparrow}{\frac{-6}{5}\Big)}}$$

A plot of the ordered pairs given in the table is shown in Figure 7.

Ordered pairs that solve
$3x - 5y = 12$

x	y	*Ordered pair*
0	$\frac{-12}{5}$	$\left(0, \frac{-12}{5}\right)$
$\frac{2}{3}$	-2	$\left(\frac{2}{3}, -2\right)$
2	$\frac{-6}{5}$	$\left(2, \frac{-6}{5}\right)$
4	0	(4, 0)

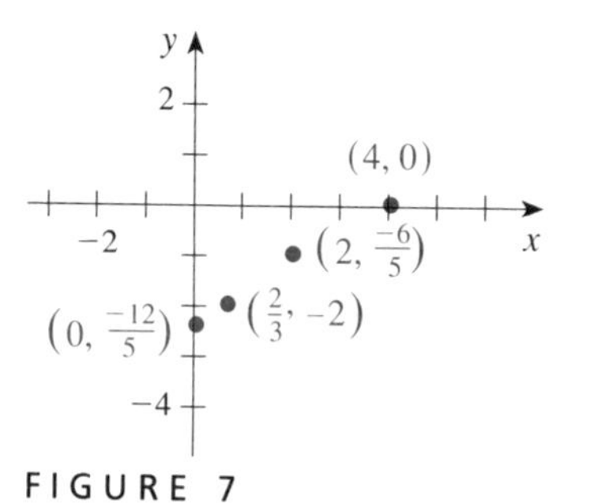

FIGURE 7

TI-82 Note *Type `-4.7` for `Xmin` and `-3.1` for `Ymin` (or use `ZOOM-option 4` for `ZDecimal`).

If we continue plotting ordered pairs of numbers that solve the equation $3x - 5y = 12$, we find that they all appear to lie on a straight line. To show the solution to a first-degree equation in two variables, we need the following property.

LINEAR EQUATION PROPERTY

> The solution to a first-degree equation in two variables is a collection of ordered pairs. When the ordered pairs are plotted on a set of axes, their graph is a straight line. The plot of every ordered pair that solves the equation will appear on the line. Likewise, the ordered-pair name of any point on the line will solve the equation.

Because of the appearance of its graph, a first-degree equation in two variables is known as a **linear equation.**

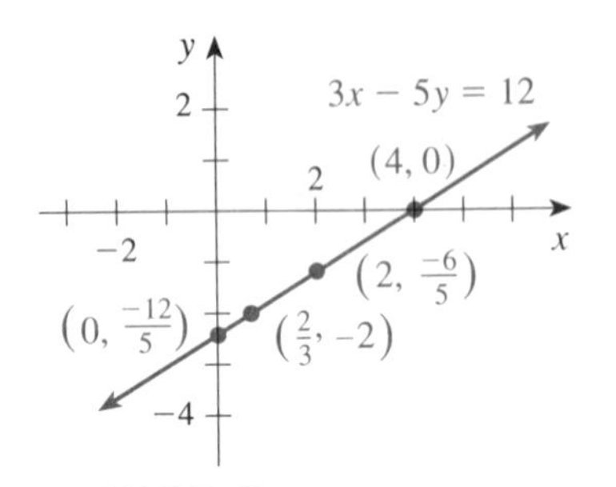

FIGURE 8

Use a straightedge, such as a ruler or the cover of your calculator, to draw the straight line through the points.

Drawing a line through the points plotted in Figure 7 produces the graph shown in Figure 8. We place an arrowhead on each end of the line to show that the line extends indefinitely in the indicated directions. The point where the line crosses the x-axis is known as the ***x*-intercept**. The x-intercept is always 0 unit up or down, and so its y-coordinate is always 0. To find the name for the x-intercept, we replace the y-variable with 0 and solve the resulting equation, $3x = 12$. (The solution is 4.) Thus, the x-intercept is $(4, 0)$.

The ***y*-intercept** of a line is the point where the line meets, or *intercepts*, the y-axis. The y-intercept is always 0 unit right or left, and so its x-coordinate is 0. To find the name of the y-intercept, we replace the x-variable with 0 and solve the resulting equation, $-5y = 12$. $\left(\text{The solution is } \frac{-12}{5}.\right)$ Thus, the y-intercept is $\left(0, \frac{-12}{5}\right)$. See Figure 8. In general, we have the following definition.

***x*- AND *y*-INTERCEPTS**

> An ***x*-intercept** of a graph has the form $(a, 0)$ and a ***y*-intercept** of a graph has the form $(0, b)$.

Constructing a Graph We will now use the x-intercept and y-intercept to draw a graph. Work through the following examples.

EXAMPLE 2 Graph the solution to $5x - 2y = 9$.

Solution We could generate any two ordered pairs, plot them, and then draw a line through the two points. Here we will use the x-intercept and the y-intercept.

To find the x-intercept replace y with 0 to get: $\quad 5x = 9$

This equation has solution $\frac{9}{5}$. The x-intercept is $\left(\frac{9}{5}, 0\right)$. $\quad x = \frac{9}{5}$

To find the y-intercept, replace x with 0 to get: $\quad -2y = 9$

This equation has solution $\frac{-9}{2}$, so the y-intercept is $\left(0, \frac{-9}{2}\right)$. $\quad y = \frac{-9}{2}$

Now we plot and label these two points. See Figure 9.

Finally, we draw a line through the points and add an arrowhead at each end of the line, as shown in Figure 10.

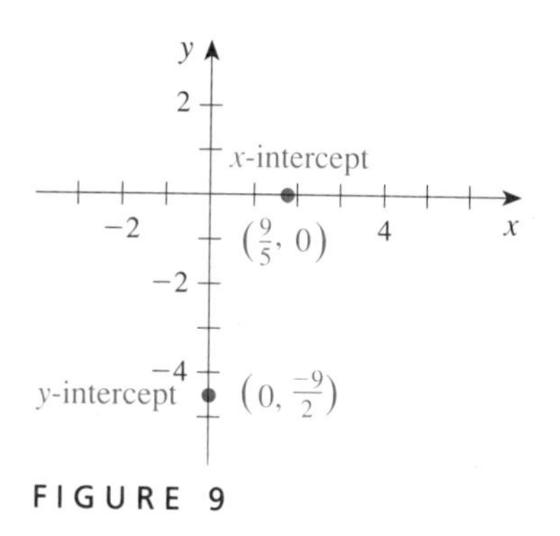

FIGURE 9

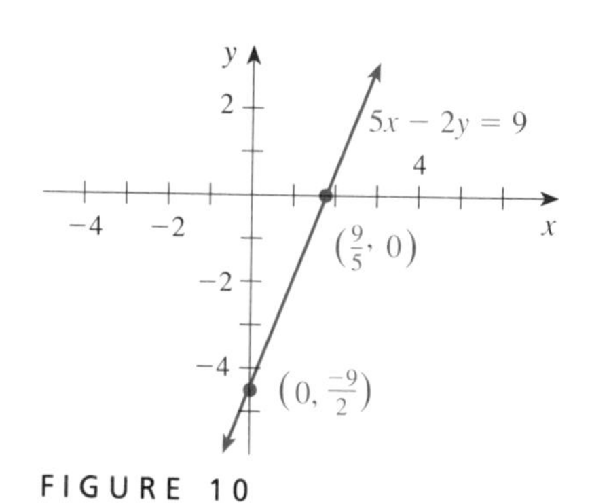

FIGURE 10

GIVE IT A TRY

3. For the equation $5x + 2y = 10$, complete each statement.
   a. The ordered pair in the solution with x-coordiante 0 is (____, ____).
   b. The ordered pair in the solution with y-coordinate 0 is (____, ____).
   c. The ordered pair in the solution with x-coordinate 3 is (____, ____).
   d. Draw a set of x- and y-axes. Plot the three points in parts (a)–(c). Draw a straight line through the points. Label the line with its equation and label the x- and y-intercepts with their ordered-pair names.

4. For the equation $3x - 5y = 8$, what is the x-intercept? What is the y-intercept? Draw a set of x- and y-axes. Plot the two points. Draw a straight line through the points. Label the line with its equation and label the x- and y-intercepts with their ordered-pair names.

Using the Graphics Calculator

```
RANGE
Xmin=-9.6
Xmax=9.4
Xscl=1
Ymin=-6.4
Ymax=6.2
Yscl=1
Xres=1
```

FIGURE 11

```
DRAW
1:ClrDraw
2:Line(
3:PT-On(
4:PT-Off(
5:PT-Chg(
6:DrawF
7 Shade(
```

FIGURE 12

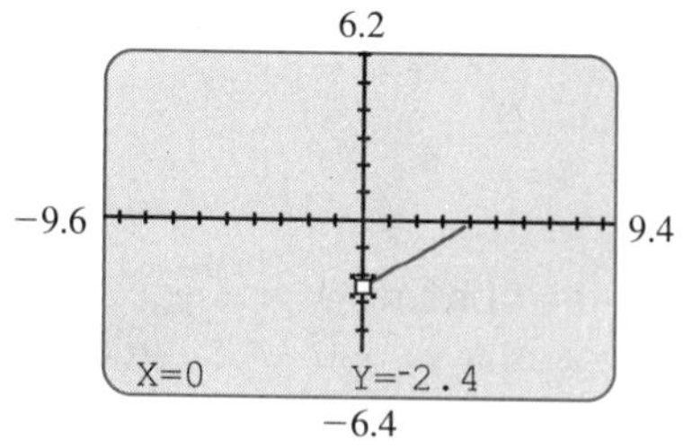

FIGURE 13

We can use the graphics calculator to increase our understanding of the graph of the solution to a first-degree equation in two variables. Press the Y= key and use the CLEAR key to erase any existing entries. Press the RANGE key and enter the settings shown in Figure 11* for a friendly view screen. Press the GRAPH key to view the graph screen, where only the x-axis and the y-axis are currently drawn.

We will now plot two points that solve the first-degree equation $3x - 5y = 12$. Two such ordered pairs are (4, 0) and (0, −2.4). We start by displaying the DRAW menu: press the 2nd key followed by the PRGM key. Figure 12 shows the DRAW menu. Press the 1 key to select the ClrDraw option. Press the ENTER key to execute the command; the word Done is written to the screen. The graph screen has been cleared of any graphs drawn on the screen. Press the GRAPH key. From the graph screen, display the DRAW menu and select the Line(option. Move the cursor to the point (4, 0) and press the ENTER key. Move the cursor to the point (0, −2.4) and press the ENTER key. A *line segment* is displayed as in Figure 13. This segment represents part of the graph of the equation $3x - 5y = 12$. Move the cursor until the message reads X=2 Y=⁻1.2. Is this point on the line? To find out, check to see if (2, −1.2) solves the equation $3x - 5y = 12$, that is, determine whether $3(2) - 5(-1.2) = 12$ is a true arithmetic sentence.

TI-82 Note *Type ⁻9.4 for Xmin and ⁻6.2 for Ymin.

The set of points currently plotted on the graph screen is a line segment with *endpoints* at (4, 0) and $\left(0, \frac{-12}{5}\right)$. This line segment represents only a part of the entire solution to the equation $3x - 5y = 12$. We can use the Y= screen to get a better idea of all the ordered pairs that solve $3x - 5y = 12$. First, we must rewrite $3x - 5y = 12$ for y:

$$\begin{aligned} 3x - 5y &= 12 \\ \underline{-3x} \quad\quad &\quad \underline{-3x} \\ -5y &= -3x + 12 \\ y &= \frac{3}{5}x - \frac{12}{5} \end{aligned}$$

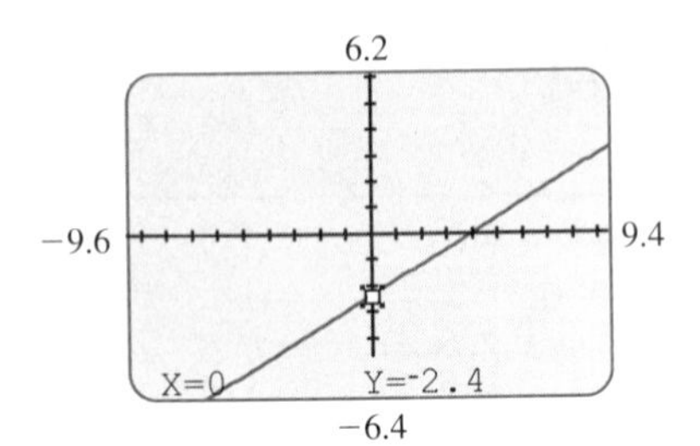

FIGURE 14

Press the Y= key, and for Y1 type (3/5)X-(12/5). Press the GRAPH key. The graph screen now displays a larger portion of the line representing the solution to $3x - 5y = 12$. Press the TRACE key. The message X=0 Y=⁻2.4 means that the y-intercept is (0, −2.4). See Figure 14. Press the right arrow key until the message reads X=2 Y=⁻1.2. As we have shown already, the ordered pair (2, −1.2) solves the equation $3x - 5y = 12$. Press the right arrow key until the ordered-pair name of the x-intercept is displayed. What is the ordered-pair name of the x-intercept? Continue to press the right arrow key until the message reads X=9.4 Y=3.24. Press the right arrow key again. The screen scrolls to the right displaying even more points on the line. Use the arrow keys to find the ordered pair with x-coordinate 12 that solves the equation $3x - 5y = 12$.

GIVE IT A TRY

Use the graphics calculator for Problems 5–8.

5. Display the solution to $3x - 5y = 12$ on the graph screen (see Figure 14). Use the TRACE key and the left arrow key to find and report the ordered pair with x-coordinate −2 that solves the equation. What is the ordered pair with y-coordinate −3 that solves the equation?
6. Rewrite the equation $3x + 5y = 12$ for y. Press the Y= key, and for Y1 enter the binomial that appears on the right-hand side of the rewritten equation. Press the RANGE key and enter the values shown in Figure 15*. Use the TRACE feature to report the ordered-pair names of the y-intercept, the x-intercept, and the point with y-coordinate −2.

```
RANGE
Xmin=-4.8
Xmax=4.7
Xscl=1
Ymin=-3.2
Ymax=3.1
Yscl=1
Xres=1
```

FIGURE 15

7. Report the x- and y-intercepts of the graph of the solution to $y = -2x + 5$.
8. Graph the solution to $5x - 2y = 8$. Use the TRACE feature to report the ordered-pair names of the x-intercept and the y-intercept.

■ SUMMARY

In this section, we learned how to graph the solution of a first-degree equation in two variables. The solution to this type of equation is an infinite set of ordered pairs. The ordered pairs of numbers that solve the equation are those such that when the first number replaces the x-variable and the second number replaces the y-variable, a true arithmetic sentence results.

TI-82 Note *Use ZOOM-option 4 for ZDecimal.

■ ***Quick Index***

To show a collection of ordered pairs, we use a set of axes—a horizontal number line (the x-axis) and a vertical number line (the y-axis). The point where the x- and y-axes cross, or intersect, is known as the origin; it represents the ordered pair (0, 0). From the origin, all other ordered pairs can be located. The first number in the ordered pair (the x-coordinate) indicates a move right or left from the origin. The second number in the ordered pair (the y-coordinate) indicates a move up or down. For example, the ordered pair $(5, -2)$ is located, or plotted, by starting at the origin, moving right 5 units, and then moving down 2 units.

If all of the ordered pairs that solve a first-degree equation in two variables (such as $3x + 2y = -5$) are plotted on a set of axes, the result is a straight line. For this reason, such equations are known as linear equations. Two important features of a line are its intercepts. The x-intercept is the point at which the line crosses the x-axis. Its ordered-pair name has the form $(a, 0)$, where a is any real number. The y-intercept is the point where the line crosses the y-axis. Its ordered-pair name has the form $(0, b)$, where b is any real number.

3.1 ■ ■ ■ EXERCISES

Insert true *for a true statement, or* false *for a false statement.*

____ 1. The solution to a first-degree equation in two variables is typically a single ordered pair.

____ 2. The ordered pair $(-1, 2)$ solves the equation $3x - y = -5$.

____ 3. An ordered pair that does *not* solve $x + 2y = 6$ is $(6, -1)$.

____ 4. The ordered pair $(-3, 5)$ has y-coordinate -3.

____ 5. When $(-1, -2)$ is plotted on a set of axes, the point appears to the left of and below the origin.

____ 6. When plotted on a set of axes, the ordered pairs that solve the equation $3x - 5y = 8$ form a circle.

____ 7. The graph of the solution to $y = 3x$ has no x-intercept.

____ 8. For the line determined by $3x - y = -2$, the y-intercept is $(0, -2)$.

____ 9. The plot of the ordered pair $(-3, 0)$ is in quadrant III.

10. Which of the following is a first-degree polynomial in one variable?

 i. $3x - 5$ ii. $3x - 5y \leq 10$
 iii. $3x - 5 = 2$ iv. $y = 3x$

11. Which of the following is a first-degree equation in two variables?

 i. $3x - 5y$ ii. $3x - 5y \leq 10$
 iii. $3x - 5 = 2$ iv. $y = 3x$

12. At which of the given points does the graph of the solution to $3x - 2y = 8$ cross the y-axis?

 i. $(0, -4)$ ii. $\left(\frac{8}{3}, 0\right)$ iii. $(0, 4)$ iv. $\frac{3}{2}$

Match the equations given in Problems 13–16 with graphs A–D.

A.

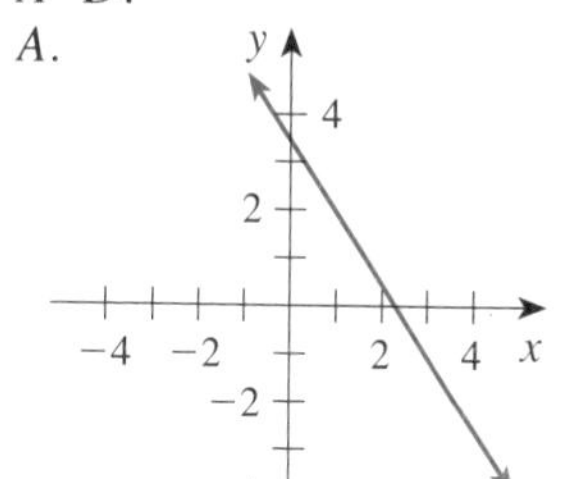

B.

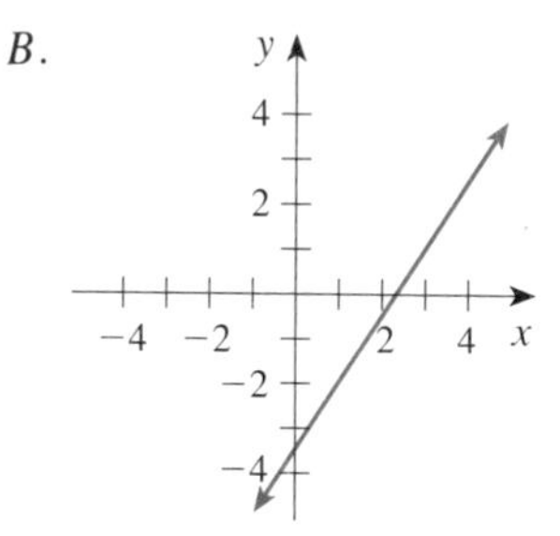

C.

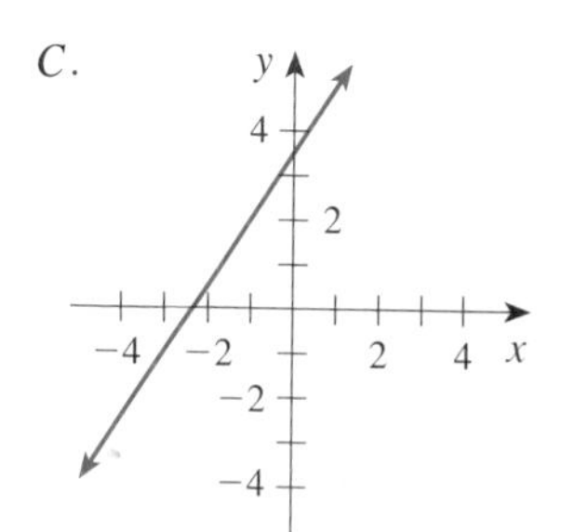

D.

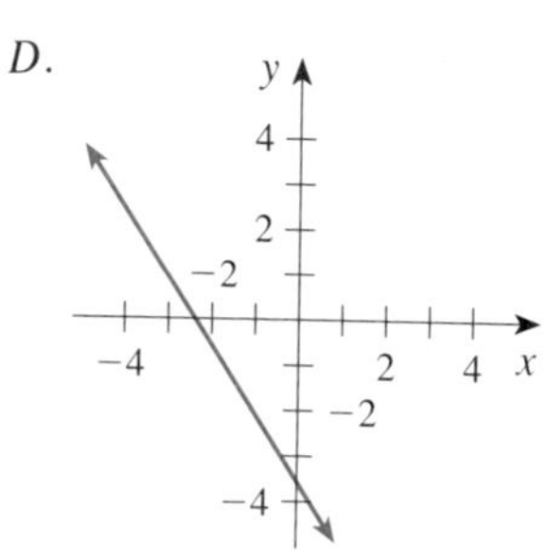

13. $3x - 2y = 7$

14. $3x + 2y = 7$

15. $-3x + 2y = 7$

16. $-3x - 2y = 7$

For the given pairs of points, draw a set of axes, plot the two points, and then draw a line through the points.

17. $(-2, 4)$ and $(1, 5)$

18. $(1, -1)$ and $(4, 1)$

19. $(5, 2)$ and $(1, -3)$

20. $(-2, 6)$ and $(1, -3)$

21. $(-6, 2)$ and $(1, 1)$

22. $(-3, -3)$ and $(-1, -1)$

23. For the equation $4x - 5y = 12$, complete the following table. Use the graphics calculator to check your work.

x	-2	0	1	3	4
y					

24. For the equation $3x + 5y = 12$, complete the following table. Use the graphics calculator to check your work.

x	-4	-1	0	2	3
y					

25. Draw a set of axes, then plot the ordered pairs produced in Problem 23. Draw a straight line through the points.

26. Draw a set of axes, then plot the ordered pairs produced in Problem 24. Draw a straight line through the points.

Rewrite each equation for y. Use the graphics calculator to graph the solution to the equation using the `RANGE` *values* `Xmin=-9.6`, `Xmax=9.4`, `Xscl=1`, `Ymin=-6.4`, `Ymax=6.2`, *and* `Yscl=1`.* *Use the* `Trace` *feature to find and report the calculator decimal approximations for the x- and y-intercepts. Use algebra to report the actual intercepts.*

27. $-3x - y = 5$

28. $-3x + y = 5$

29. $2x + 3y = 7$

30. $-2x + 3y = 7$

31. $6x - 7y = 15$

32. $3x + 7y = 14$

33. $-5x - 2y = 10$

34. $5x + 2y = 12$

35. $7x - 3y = 13$

36. $2x - 9y = -15$

37. $2(x - 3) - (5 - 2y) = 2 - 3x$

38. $-3(y - 4) + 2(3 - 2x) = 0$

39. $\frac{x}{5} + \frac{y}{2} = 1$

40. $\frac{x}{3} - \frac{y}{4} = 1$

For each first-degree equation in two variables, report the y-intercept and the x-intercept. Then draw the graph of the solution. Use the graphics calculator to check your work.

41. $3x - y = -2$

42. $y = \frac{4}{5}x + -3$

43. $-3x + 2y = 8$

44. $y = -2x + 3$

45. $y = -2x + 5$

46. $3y - 2x = -1$

47. $5y - 2x = -4$

48. $10x + 30y = 40$

49. $-50x - 60y = 60$

50. $15x - 5y = 20$

51. $y = \frac{-3}{2}x - \frac{7}{2}$

52. $y = \frac{7}{5}x - \frac{13}{5}$

53. $\frac{y + 2}{3} = \frac{x - 1}{5}$

54. $\frac{x - 3}{5} = \frac{y}{2}$

3.2 ■ ■ ■ THE SLOPE OF A LINE

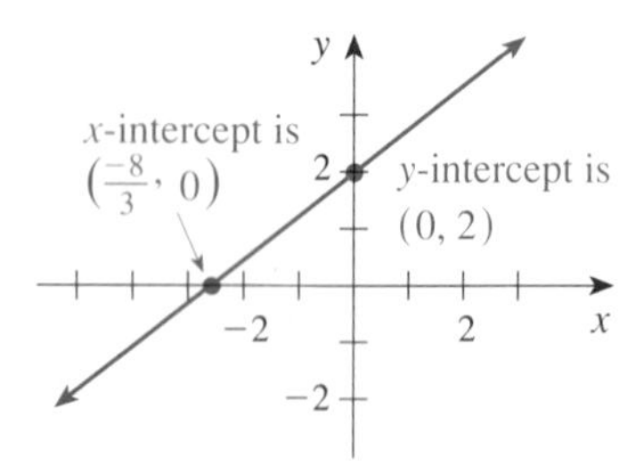

FIGURE 1
Graph of the solution to $-3x + 4y = 8$

The solution to a first-degree equation in two variables, such as $-3x + 4y = 8$, is an infinite collection of *ordered pairs*. In Section 3.1 we learned to graph these ordered pairs and to identify the x- and y-intercepts of a line, as shown in Figure 1. Although the x- and y-intercepts are important, the defining characteristic of a straight line is its slope. The *slope* of a line is a description of its steepness, or direction, or incline. Read carefully through the following discussion.

Background For the point $(3, 2)$, an infinite number of lines can be drawn through the point, First, consider the **horizontal line** through the point. See

TI-82 Note *Type `-9.4` for `Xmin` and `-6.2` for `Ymin`.

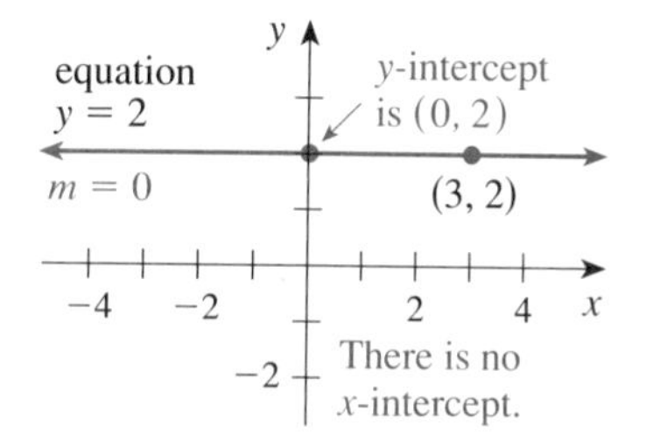

FIGURE 2
A horizontal line has slope zero.

Figure 2. This line is parallel to the x-axis. It has slope measure zero. We write $m = 0$, using m for slope measure. The line is flat—it has no inclination. The equation of the line is $0x + 1y = 2$ or, simply, $y = 2$. For this line, the y-coordinate of its points is always 2. (The x-coordinate can be any number.) Notice that this line has no x-intercept.

A line through the point (3, 2) that is **increasing** (or *going up,* from left to right) has slope of some positive number, say $\frac{2}{3}$. The greater the increase of the line, the larger the slope measure. Figure 3 illustrates two lines with positive slope, $m = \frac{2}{3}$ and $m = 2$.

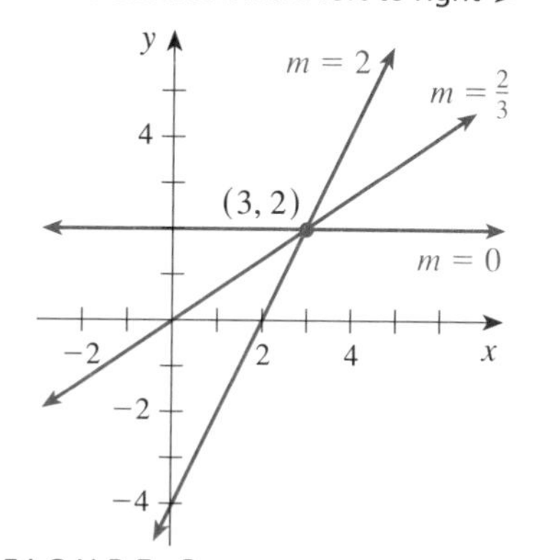

FIGURE 3
Positive slope: line goes *up*.

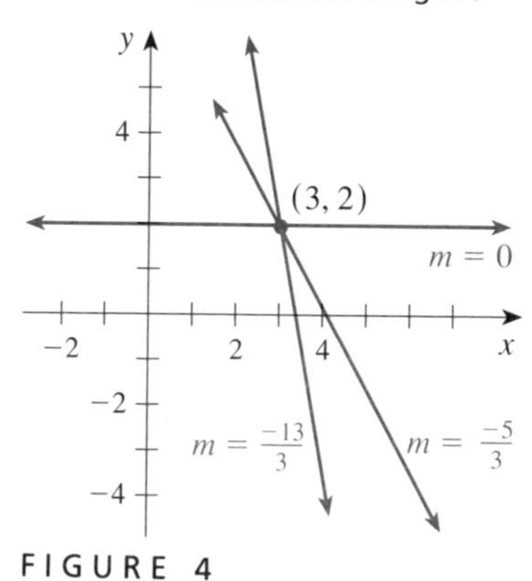

FIGURE 4
Negative slope: line goes *down*.

If a line through the point is **decreasing** (or *going down,* from left to right) it has a negative slope measure, say $\frac{-5}{3}$. Figure 4 shows some lines with negative slopes. The greater the decrease, the smaller the slope measure.

What about a vertical line through the point (3, 2)? The equation of this line is $0y + 1x = 3$, or simply $x = 3$. A **vertical line** is the collection of points that have the same x-coordinate, and the line is parallel to the y-axis.

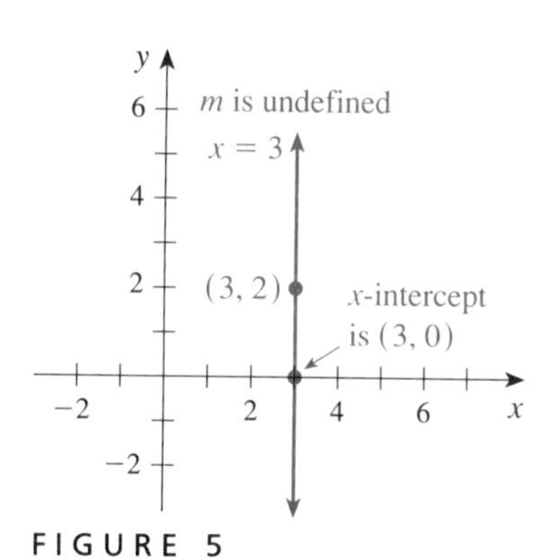

FIGURE 5
A vertical line has no slope measure.

For the line $x = 3$, every point has x-coordinate 3. What is the slope of this line? Is it a very large positive value, such as 100 or 10^8, or is it a very small negative value, such as -300 or -7^9? The slope of any vertical line is undefined, and we say that a vertical line has no slope measure. See Figure 5. Notice that this line has no y-intercept.

Before we actually compute the slope of a line, consider the following situation. The formula for the revenue y from selling x sun visors at \$2 each is given by the equation $y = 2x$. If you sell 0 sun visors, your revenue is \$0. Notice that the ordered pair (0, 0) solves the equation $y = 2x$. If you sell 1 sun visor, your revenue is $y = 2(1)$, or \$2. Again, the ordered pair (1, 2) solves the equation $y = 2x$. In general, for each additional sun visor that you sell, your revenue increases by \$2. The graph of the ordered pairs that solve $y = 2x$ is a straight line, and that line has slope measure $\frac{2}{1}$. See the line in Figure 6. For each increase of 1 unit in x (the number of sun visors sold), there is an increase of 2 units in y (the revenue). Notice that the line is going up (increasing, when

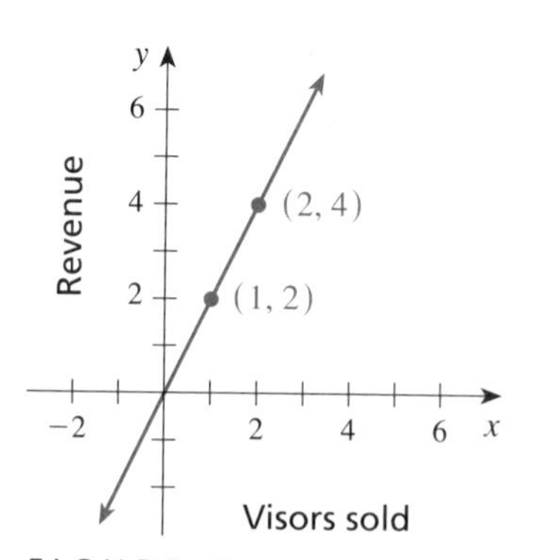

FIGURE 6

viewed from left to right), and the slope measure is 2. Here slope represents the revenue earned per sun visor sold.

Using the Graphics Calculator

```
RANGE
Xmin=-9.6
Xmax=9.4
Xscl=1
Ymin=-6.4
Ymax=6.2
Yscl=1
Xres=1
```

FIGURE 7

We can use the graphics calculator to check our work in the preceding revenue situation. For Y1, type 2X. Enter the range settings shown in Figure 7.* View the graph screen. A line is displayed for the equation $y = 2x$. Notice that the line is increasing (going up). Press the right arrow key to move the cursor 1 unit right to (1, 0). Now, move the cursor up to the point (1, 2). Notice that for an increase of 1 units in x, there is an increase of 2 units in y. Again, move the cursor 1 unit right, to the point (2, 2), then 2 units up, to the point (2, 4). For the preceding revenue example, each additional sun visor sale (an increase of 1 unit in x) results in a revenue increase of $2 (an increase of 2 units in y).

WARNING

A large percentage of errors made by students in finding the slope measure involves the sign on the slope measure. By sketching a graph and observing whether the line is increasing or decreasing, you can avoid such errors.

Computing the Slope Measure We now discuss how to find the slope measure of a line. Before we start, remember that if the line is *increasing* (viewed from left to right), then its slope measure is positive; if the line is *decreasing*, then its slope measure is negative. Given the ordered-pair names of two points on a line, we can compute the slope of the line. We define the slope measure as follows.

DEFINITION OF SLOPE MEASURE

The **slope measure** m is the ratio of the *change in y* (Δy) to the *change in x* (Δx), or

$$m = \frac{\Delta y}{\Delta x}$$

Consider the graph of the solution to the equation $3x - 5y = 12$. Some of the ordered pairs that solve this equation are given in the table at the left. The coordinates of two points on this line are $(-1, -3)$ and $(4, 0)$. Now, we know that the slope measure is positive, because in the graph shown in Figure 8, the line is increasing.

Ordered pairs that solve $3x - 5y = 12$

x	y	*Ordered pair*
-1	-3	$(-1, -3)$
0	$\frac{-12}{5}$	$\left(0, \frac{-12}{5}\right)$
2	$\frac{-6}{5}$	$\left(2, \frac{-6}{5}\right)$
4	0	$(4, 0)$

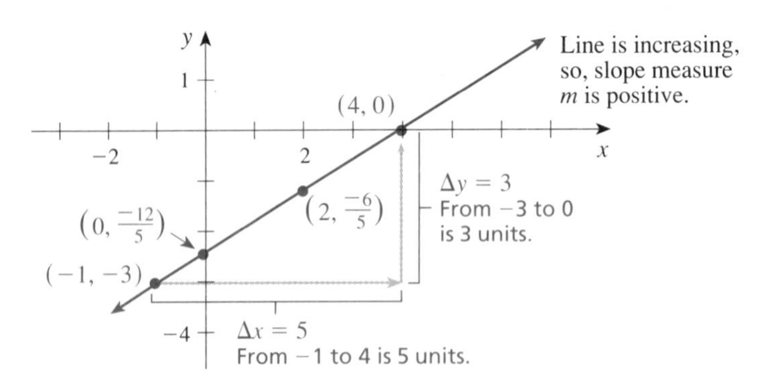

FIGURE 8

Computing the slope measure of the graph of the solution to $3x - 5y = 12$

TI-82 Note *Type -9.4 for Xmin and -6.2 for Ymin.

Next, from the point $(-1, -3)$ to $(4, 0)$, the x-value changes from -1 to 4, or 5 units. (Count the units on the graph.) That is, Δx is 5. Between these two points, the y-value changes from -3 to 0, or 3 units (count the units on the graph in Figure 8). Thus, Δy is 3. So, we have

$$m = \frac{\Delta y}{\Delta x} = \frac{3}{5}$$

We say this line has a slope measure of $\frac{3}{5}$.

GIVE IT A TRY

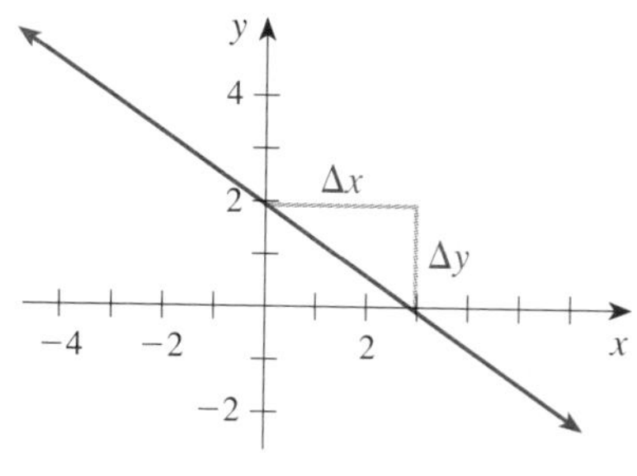

FIGURE 9

1. For the equation $3x - 5y = 12$, we can compute the slope measure using the points $\left(0, \frac{-12}{5}\right)$ and $(4, 0)$. For these two points, report each quantity.
   a. Δy b. Δx c. the slope measure, m
2. Consider the equation $2x + 3y = 6$.
   a. Report the y-intercept and the x-intercept.
   b. The graph of this equation is shown in Figure 9. What is the sign for the slope measure of the line?
   c. Using the two points from part (a), report Δy, Δx, and the slope measure, m.
3. Draw a set of axes. Plot the points $(-2, 5)$ and $(1, -1)$, and draw the line through these points. What is the slope measure of the line?

A Formula for the Slope Measure, m Before we discuss a formula for m, we review the process for finding the slope measure of a line.

1. Plot two points on the line.
2. Draw the line through the points.
3. Observe the line to see if it is increasing (going up) or decreasing (going down). Determine the sign for slope (positive or negative).
4. Compute the change in x (Δx).
5. Compute the change in y (Δy).
6. Compute the ratio of the change in y (Δy) to the change in x (Δx). Then assign the sign found in step 3 to this ratio.

These steps are summarized in the following formula.

THE SLOPE FORMULA

For two points on a line, (x_1, y_1) and (x_2, y_2), the slope measure of the line is

$$m = \frac{\Delta y}{\Delta x} = \frac{y_2 - y_1}{x_2 - x_1}$$

Using the formula for the slope measure, we can compute the slope between two points, say $(-3, 5)$ and $(1, -2)$:

$(-3, 5)$ $(1, -2)$

$$m = \frac{5 - (-2)}{-3 - 1} = \frac{7}{-4} = \frac{-7}{4}$$

The graph shown in Figure 10 confirms that the slope is negative.

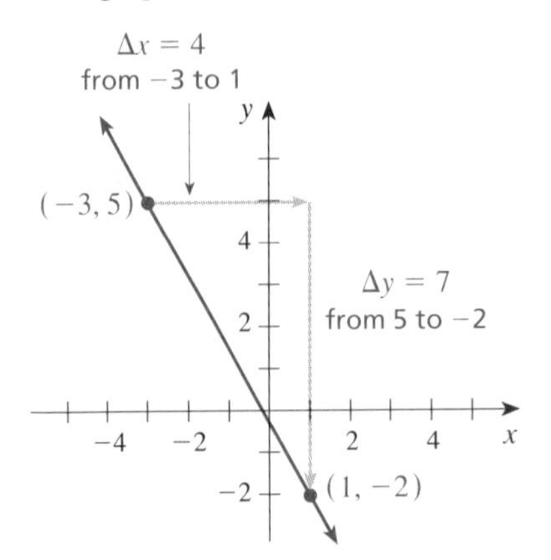

FIGURE 10

$$m = \frac{\Delta y}{\Delta x}$$

$$= \frac{-7}{4}$$

Line is decreasing, so sign is negative.

Also, counting units for Δx and Δy, we confirm that the slope is $\frac{-7}{4}$.

EXAMPLE 1 Plot the ordered pairs $(-1, -2)$ and $(4, 1)$. Draw the line through the two points. Find the slope measure of the line.

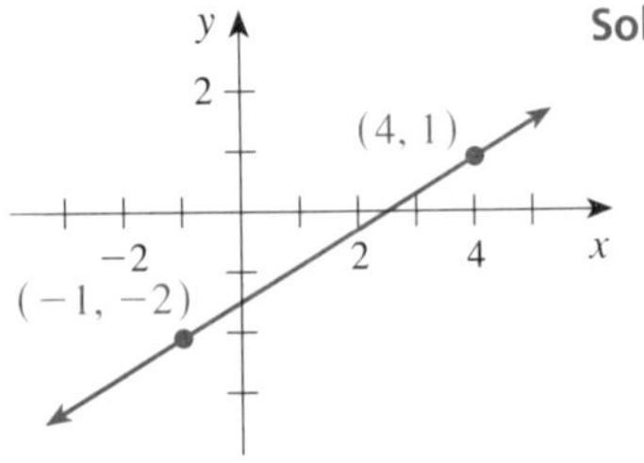

Solution A plot of the two points and the resulting line is shown in Figure 11. Notice that the line is increasing (going up). So, the slope measure is positive. To compute the slope, we use the slope formula:

$$m = \frac{-2 - 1}{-1 - 4} = \frac{-3}{-5} = \frac{3}{5}$$

or

$$m = \frac{1 - (-2)}{4 - (-1)} = \frac{3}{5}$$ ■

FIGURE 11

GIVE IT A TRY

Use the slope formula to find the slope measure of the line through the given ordered pairs. Plot the points, and draw the line through the two points.

4. $(-3, -2)$ and $(4, 3)$ **5.** $(0, 5)$ and $(3, 0)$ **6.** $(-1, 3)$ and $(3, 1)$

Slope, *y*-Intercept of a Linear Equation Consider the first-degree equation in two variables $3x + 5y = 12$. This form, $Ax + By = C$, where A, B, and C are real numbers, is known as **standard form**. Rewriting this equation for y, we get

$$\begin{aligned} 3x + 5y &= 12 \\ -3x \quad &\quad -3x \\ 5y &= -3x + 12 \\ y &= \frac{-3}{5}x + \frac{12}{5} \end{aligned}$$

This last form is known as the **slope, *y*-intercept form** of the equation. The x-coefficient $\frac{-3}{5}$ is the slope measure of the graph. Also, the number term $\frac{12}{5}$ yields the y-intercept, $\left(0, \frac{12}{5}\right)$. In general, we have the following property.

SLOPE, y-INTERCEPT PROPERTY

For an equation in the form $y = mx + b$, the x-coefficient m is the slope measure and the constant term b determines y-intercept $(0, b)$ of the graph. For example,

$$y = \frac{-3}{5}x + \frac{12}{5}$$

$m = -\frac{3}{5}$ ⟵   ⟶ y-intercept is $\left(0, \frac{12}{5}\right)$

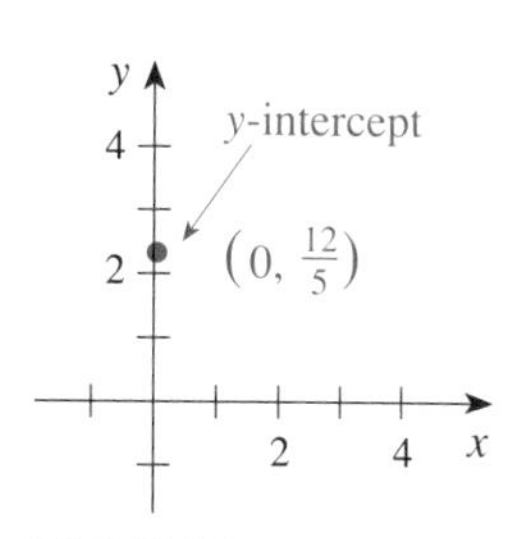

FIGURE 12

Using this property, we can construct the graph of the solution to

$$3x + 5y = 12 \quad \text{or} \quad y = \frac{-3}{5}x + \frac{12}{5}$$

To draw the graph, we first plot the y-intercept, $\left(0, \frac{12}{5}\right)$. See Figure 12. Now, we apply the slope measure $\frac{-3}{5}$ to the point. That is, from the y-intercept, move right 5 units (Δx) and then move down 3 units (Δy). See Figure 13. The name of this new point is $\left(5, \frac{-3}{5}\right)$.

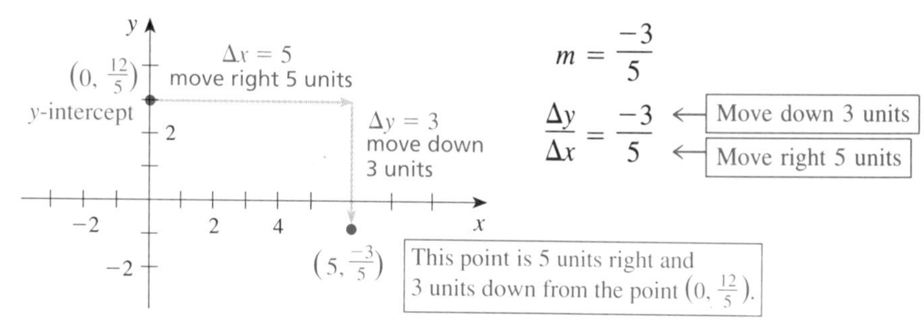

FIGURE 13

The ordered pair $\left(5, \frac{-3}{5}\right)$ solves the equation $3x + 5y = 12$ because

$$3(5) + 5\left(\frac{-3}{5}\right) = 12 \text{ is a true arithmetic sentence.}$$

Check it! Finally, we draw the line through the two points. See Figure 14.

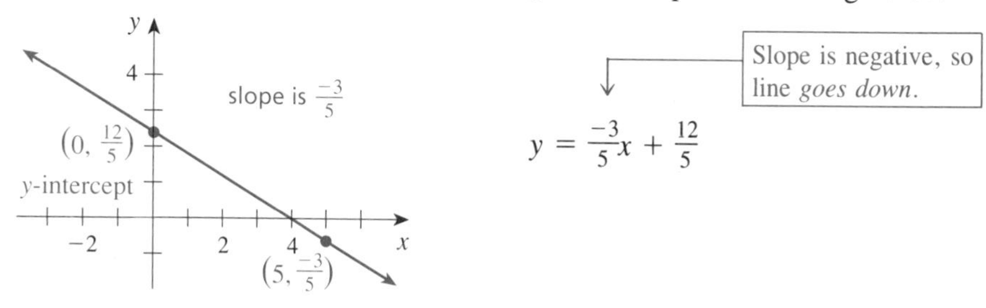

FIGURE 14

This approach to constructing the graph of the solution to a first-degree equation is more than just an alternative to the point-plotting approach. The approach is essential to finding the equation of a line, which we will discuss in Chapter 4.

EXAMPLE 2 For the equation $4x - 3y = 6$, identify the slope and y-intercept of its graph. Apply the slope to the y-intercept in order to generate another point on the line determined by $4x - 3y = 6$. Graph the line.

Solution To quickly identify the slope and y-intercept, we can rewrite the equation for y:

$$\begin{aligned} 4x - 3y &= 6 \\ -4x \qquad &\quad -4x \\ -3y &= -4x + 6 \\ y &= \frac{4}{3}x - 2 \end{aligned}$$

The x-coefficient is the slope measure, so $m = \frac{4}{3}$. We get the y-intercept from the number term of $y = \frac{4}{3}x - 2$. The y-intercept is $(0, -2)$. Plotting the y-intercept and applying the slope, we get the point $(3, 2)$. See Figure 15.

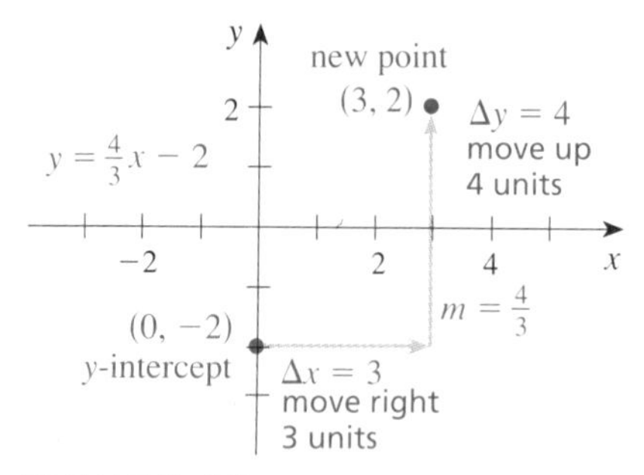

FIGURE 15

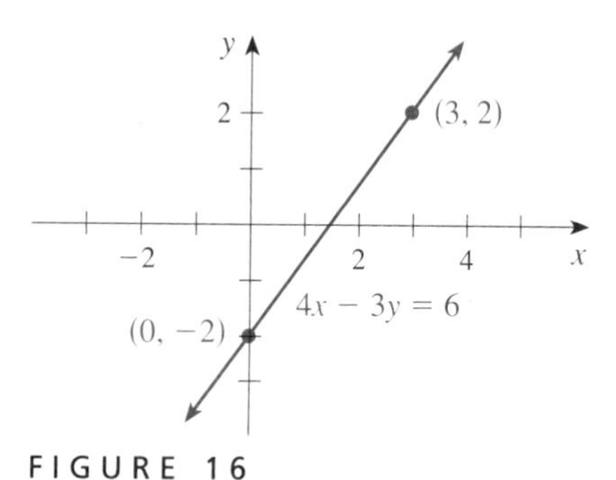

FIGURE 16

Drawing the line through the two points $(0, -2)$ and $(3, 2)$ produces the graph of the solution to $4x - 3y = 6$. See Figure 16. ■

GIVE IT A TRY

7. **a.** Rewrite the equation $2x + 3y = 9$ for y. Report the new equation. Report the slope measure and the y-intercept.
   b. Plot the y-intercept, and then apply the slope measure to the y-intercept. Report the ordered-pair name of the new point.
   c. Use the information found in parts (a) and (b) to draw the graph of the solution to the equation $2x + 3y = 9$.
8. **a.** For the line through $(-2, -1)$ with slope measure 3, report the ordered-pair name of another point on the line. [*Hint*: Many answers are possible.]
   b. Draw the line through the point $(-2, -1)$ with slope 3.

Using the Graphics Calculator

We can use the graphics calculator to quickly illustrate many of the concepts we have discussed for the slope of a line. Press the `Y=` key and use the `CLEAR` key to erase any existing entries. For `Y1` key, then type 3. Use the `RANGE` key to enter the settings shown in Figure 17.* View

TI-82 Note *Type `-9.4` for `Xmin` and `-6.2` for `Ymin`.

```
RANGE
Xmin=-9.6
Xmax=9.4
Xscl=1
Ymin=-6.4
Ymax=6.2
Yscl=1
Xres=1
```

FIGURE 17

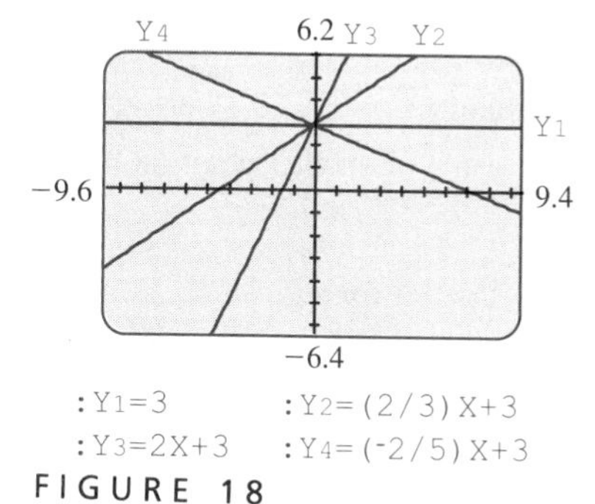

FIGURE 18

the graph screen. As you can see, a horizontal line is drawn for `Y1`. This line has no inclination, so its slope measure is 0. Its y-intercept is (0, 3). Press the `Y=` key, move the cursor to `Y2`, and type `(2/3)X+3`. Press the `GRAPH` key. This line also has y-intercept (0, 3), but its slope measure is $\frac{2}{3}$. Press the up arrow key to move the cursor to the y-intercept (the message reads `X=0  Y=3`). Now, press the right arrow key and move right 3 units (Δx is 3). Next, move up 2 units (Δy is 2). Is the cursor on the line? What is the ordered-pair name of the point? (Check the message at the bottom of the screen.) Does this ordered pair solve the equation?

Press the `Y=` key, move the cursor to `Y3`, and type `2X+3`. Press the `GRAPH` key. As you can see, this new line also has y-intercept (0, 3), but its slope measure is 2. Its incline (rate of increase) is greater than that of the graph of $y = \frac{2}{3}x + 3$. Press the `Y=` key and for `Y4`, type `(-2/5)X+3`. Press the `GRAPH` key. Again, we get a line whose y-intercept is (0, 3). However, this line has a negative slope (since the line is decreasing). The slope measure is $\frac{-2}{5}$. See Figure 18.

■ SUMMARY

When we plot the ordered pairs that solve a first-degree equation in two variables (such as $3x + 2y = -5$) on a set of axes, the result is a straight line. The important features of a line are its intercepts (discussed in Section 3.1) and its slope. The slope measure m of a straight line is obtained by taking the ratio of the change in y (denoted Δy) to the change in x (denoted Δx) for *any* two points on the line. This measure is positive if the line is increasing (*going up*, observed from left to right), or negative if the line is decreasing (*going down*). A horizontal line has slope measure 0, and the slope measure of a vertical line is undefined.

An equation such as $3x + y = 2$ is in standard form, $Ax + By = C$. An equation such as $y = -3x + 2$ is in slope, y-intercept form, $y = mx + b$. When an equation is rewritten for y (in the form $y = mx + b$), the x-coefficient m is the slope measure and the number term b yields the y-intercept, $(0, b)$. The slope, y-intercept form also provides a way to graph the line: plot the y-intercept, apply the slope to generate a new point, and then draw a line through the y-intercept and the new point.

■ ***Quick Index***

3.2 ■ ■ ■ EXERCISES

Insert true *for a true statement or* false *for a false statement.*

____ 1. The slope of the line determined by $y = 2x - 3$ is $\frac{2}{1}$.

____ 2. The vertical line $x = -2$ has slope measure 0.

____ 3. A line that passes through the point $(-2, 3)$ and is decreasing (from left to right) has a negative slope measure.

____ 4. For a line with slope -3, an increase of 1 unit in the x-variable corresponds to an increase of 3 units in the y-variable.

____ 5. For the horizontal line $y = 3$, the slope is undefined.

____ 6. If a line through $(-1, 2)$ has slope measure $\frac{2}{3}$, then the point $(2, 4)$ is also on the line.

____ 7. The graph of $y = \frac{2}{3}x - \frac{7}{3}$ has y-intercept $\left(0, \frac{7}{3}\right)$.

____ 8. The line through $(-2, 3)$ and $(4, 3)$ has slope measure 0.

9. Which of the following equations determines a line with a slope measure of zero?

   i. $y = 3x - 5$ ii. $5y = 10$
   iii. $3x - 5 = 2y$ iv. $x = 3$

10. Which of the following is the slope measure of an increasing line?

   i. 0 ii. $\dfrac{-2}{3}$ iii. 2 iv. -4

11. Which of the following is the slope measure of the graph of the solution to $3x - 2y = 8$?

   i. 3 ii. -2 iii. 8 iv. $\dfrac{3}{2}$

12. Which of the following describes the slope measure of the graph of the solution to $2x + 6y = 9$?

   i. positive ii. negative
   iii. zero iv. undefined

13. a. Rewrite $3x + 5y = 13$ for y.
   b. What is the x-coefficient of the equation rewritten for y in part (a)?
   c. Report the slope measure for the graph of the solution to $3x + 5y = 13$.

For the lines shown in the graph, list their slope measures in order from largest to smallest.

14.

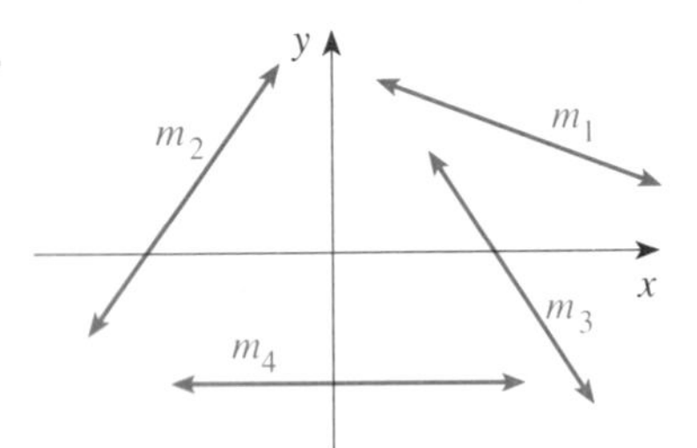

15.

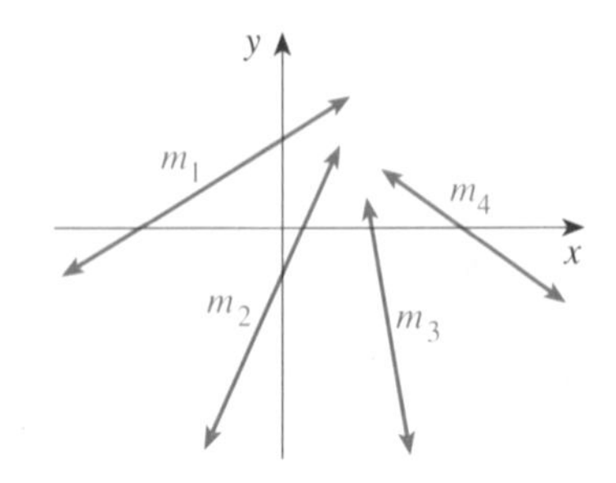

16. Consider the points $(-3, 5)$ and $(4, 2)$.
   a. Report Δx, the change in x.
   b. Report Δy, the change in y.
   c. Plot the two points, and draw a line through them. Is the line increasing or decreasing?
   d. Is the slope of the line in part (c) positive or negative?
   e. Report the slope measure m of the line in part (c).

17. For the line through the point $(-2, 3)$ with slope measure $\frac{-2}{5}$, report the ordered-pair name of another point on the line. Draw the line.

18. For the line through the point $(2, -3)$ with slope measure $\frac{3}{4}$, report the ordered-pair name of another point on the line. Draw the line.

For the given ordered pairs, plot the corresponding points on a set of axes. Report Δx, the change in x, and Δy, the change in y. Report the slope measure. Use the slope formula on page 155 to check your work.

19. $(-2, 4)$ and $(1, 5)$
20. $(1, -1)$ and $(4, 1)$
21. $(5, 2)$ and $(1, -3)$
22. $(-2, 6)$ and $(1, -3)$
23. $(-6, 2)$ and $(1, 1)$
24. $(-3, -3)$ and $(-1, -1)$
25. $\left(\dfrac{-3}{2}, \dfrac{9}{2}\right)$ and $\left(-2, \dfrac{7}{2}\right)$
26. $\left(\dfrac{9}{5}, \dfrac{-3}{4}\right)$ and $\left(\dfrac{9}{2}, \dfrac{-9}{4}\right)$

Rewrite each equation for y to get the slope, y-intercept form. Identify the slope and y-intercept. Draw a graph of the solution to the equation. Use the graphics calculator to check your work.

27. $3x - 5y = 20$
28. $2x - y = 5$
29. $x - 2y = -3$
30. $3x - 2y = -1$
31. $5x - 4y = 8$
32. $5x + 4y = 9$

For the given first-degree equation in two variables, report the slope measure and y-intercept. Draw the graph of the solution. Use the graphics calculator to check your work.

33. $2x - y = -3$
34. $y = \dfrac{4}{5}x - 1$
35. $2x - 3y = 8$
36. $y = -x + 3$
37. $y = -x - 1$
38. $3y - 2x = -3$
39. $5y - 2x = -4$
40. $10x + 30y = 40$
41. $-50x - 60y = 60$
42. $15x - 5y = 20$
43. $\dfrac{y + 2}{3} = \dfrac{x - 1}{5}$
44. $\dfrac{x - 3}{5} = \dfrac{y}{2}$

DISCOVERY

Points of View

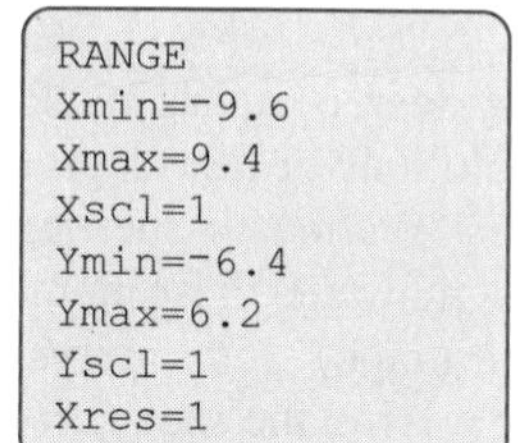

FIGURE 1

At the beginning of Chapter 2, we discussed polynomials, such as $2x - 3$. This polynomial is a first-degree binomial in one variable: It holds the place for (or represents) number phrases of the form

$$2(5) - 3 \qquad 2\left(\frac{-2}{3}\right) - 3 \qquad 2(\sqrt{3}) - 3$$

and so forth. Using the graphics calculator, we can see several evaluations of this polynomial for different replacements of the variable, x. Type 2X−3 for Y1, then press the RANGE key and enter the values shown in Figure 1.* View the graph screen. The line drawn represents the evaluations of the polynomial $2x - 3$ for various replacements of the x-variable. Press the TRACE key, and then press the right arrow key until the message reads X=1 Y=⁻1. (See Figure 2.) This message means that replacing x by 1 results in the polynomial $2x - 3$ evaluating to the number -1. Use the arrow keys to complete the following table for evaluations of the polynomial $2x - 3$ for various replacements of the x-variable.

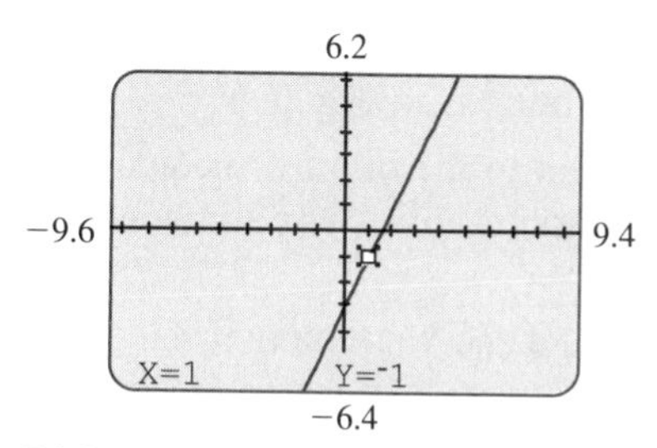

FIGURE 2

x	-1	0	1	2	3	4
$2x - 3$			-1			

In Chapter 2, we also solved first-degree equations in one variable, such as $2x - 3 = 1$. To solve this equation, we find a number to replace the x-variable in order to produce a true arithmetic sentence. One way of viewing this problem is to say that we want an x-value which results in the polynomial $2x - 3$ evaluating to the number 1. Using the graphics calculator, press the Y= key and type 1 for Y2. View the graph screen. Use the TRACE key and the arrow keys to move the cursor to the point of intersection of the line representing the evaluations of the polynomial $2x - 3$ and the horizontal line representing the number 1. The message reads X=2 Y=1, as shown below the graph in Figure 3. Thus, when the x-variable is replaced by 2, the polynomial $2x - 3$ evaluates to 1; that is, the number 2 solves the equation $2x - 3 = 1$.

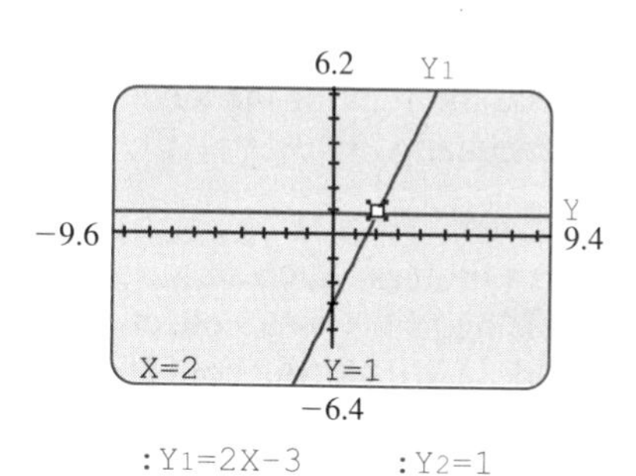

FIGURE 3

In Section 3.1 we considered first-degree equations in two variables. An example of such an equation is $y = 2x - 3$. The solution to this equation is an infinite set of ordered pairs of numbers. On the graphics calculator, make sure 2X−3 is still entered for Y1. (Erase any entries for Y2, Y3, and Y4). View the graph screen. Press the TRACE key. As Figure 4 shows, the message X=0 Y=⁻3 appears. This message means the ordered pair $(0, -3)$ solves the equation $y = 2x - 3$. In fact, the ordered-pair name of every point on the line drawn solves the equation $y = 2x - 3$. Use the TRACE key to find the ordered pair that solves the equation with y-coordinate 1.

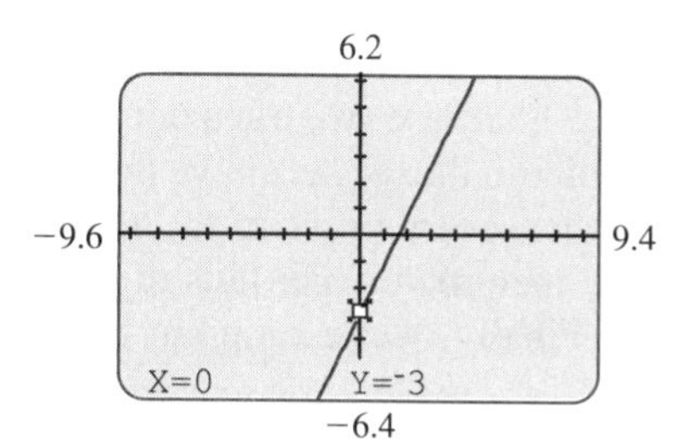

FIGURE 4

TI-82 Note *Type ⁻9.4 for Xmin and ⁻6.2 for Ymin.

3.3 ■ ■ ■ LINEAR INEQUALITIES

In the first two sections we learned how to solve, or graph the solution to, a linear equation (a first-degree equation in two variables). In this section we apply this skill of graphing the solution to a linear equation to graphing the solution to a linear inequality. Throughout this discussion we will review and reinforce the skills we used in graphing the solutions to linear equations.

Linear Inequalities In Chapter 2, after solving first-degree equations in one variable, such as $3x + 7 = 13$, we discussed the solutions of first-degree inequalities in one variable, such as $3x + 7 \leq 13$. The solution to such an inequality is an infinite collection of numbers. To communicate these numbers we use a number-line graph (a one-dimensional graph) and interval notation. For review, we can list the basic technique for solving the inequality $3x + 7 \leq 13$:

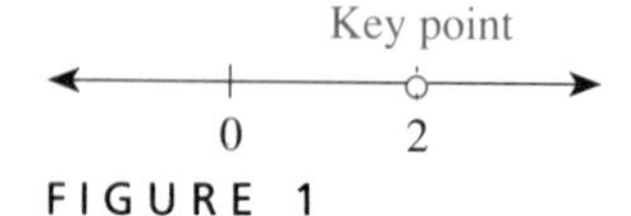

FIGURE 1

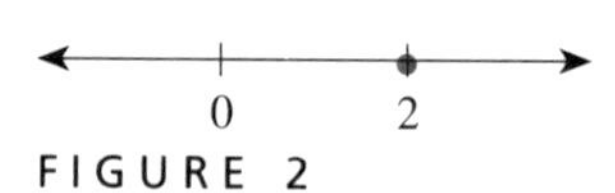

FIGURE 2

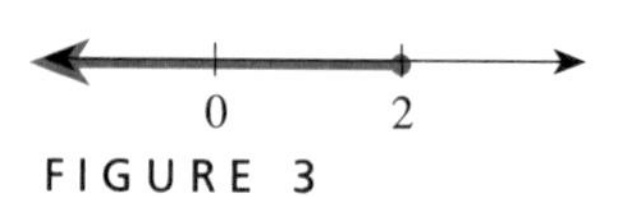

FIGURE 3

1. Solve the related equation to obtain a key point (the solution to the equation), such as the number 2 for the equation $3x + 7 = 13$.
2. Plot and label the key point on a number line, as in Figure 1.
3. Determine if the key point is in the solution to the original inequality. If so, draw a solid dot for the plotted point; otherwise, draw an open dot. See Figure 2.
4. Try a number from the interval to the left of the key point. If the number solves the original inequality, then every number in the interval solves the inequality, and we shade the interval. If the number fails to solve the inequality, then every number in the interval fails, and the interval is left unshaded.
5. Repeat step 4 for a number in the interval to the right of the key point. Figure 3 shows the solution to the inequality $3x + 7 \leq 13$. In interval notation we have $(-\infty, 2]$.

We now solve a first-degree inequality in two variables. Such an inequality is $2x - 5y \leq 12$. The solution for this type of inequality is a collection of ordered pairs of numbers—there are two variables. In fact, the collection of ordered pairs that solves the inequality forms a region. This region is determined by the graph of the solution to the related equation. To solve an inequality in two variables, we will follow the same basic steps that we used for inequalities in one variable. However, rather than getting a key point, we will find a **key line**, as shown in Figure 4. If the ordered pairs on the key line solve the original inequality, then we draw a solid line. Otherwise, we use a dotted or dashed line. Next, we test an ordered pair for a point in the region above the key line. If the ordered pair solves the inequality, then every point in the region solves the inequality, and we shade the region. If the ordered pair fails to solve the inequality, then the region is left unshaded. Finally, we test a point in the region below the key line.

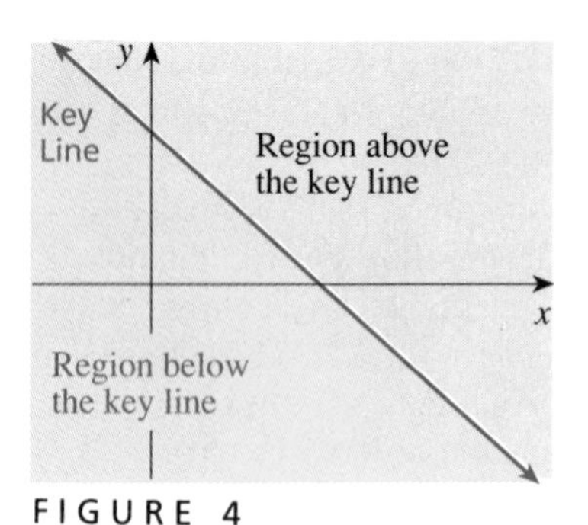

FIGURE 4

The Technique We now demonstrate this process to solve a first-degree inequality in two variables: $3x + 5y \leq 10$. Work through the following example.

EXAMPLE 1 Graph the solution to $3x + 5y \le 10$.

Solution

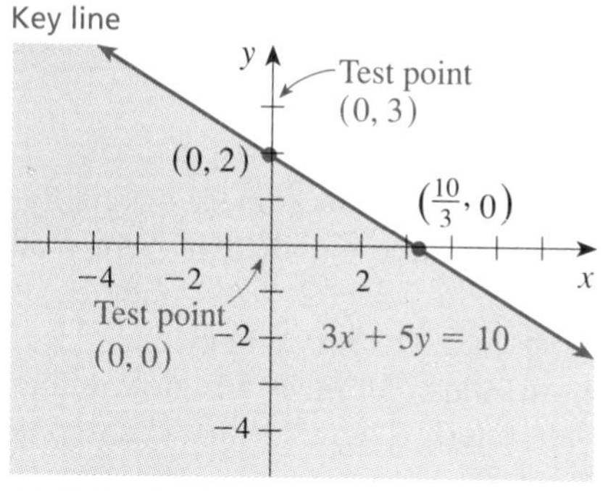

FIGURE 5
The graph of the solution to $3x + 5y \ge 10$.

Step 1 Start by graphing the solution to the related equation, $3x + 5y = 10$. The graph crosses the y-axis at $(0, 2)$. It crosses the x-axis at $\left(\frac{10}{3}, 0\right)$.

Step 2 Do the ordered pairs plotted on the key line solve the inequality? Yes. So, the key line is drawn solid rather than dotted (or dashed).

Step 3 Test an ordered pair for a point above the key line, say $(0, 3)$. Now, $3(0) + 5(3) \le 10$ is a false arithmetic sentence, so we leave the region unshaded.

Step 4 Test an ordered pair for a point in the region below the key line, say $(0, 0)$. Since $3(0) + 5(0) \le 10$ is a true arithmetic sentence, we shade the region below the key line. See Figure 5.

Graphically, the solution to the inequality $3x + 5y \le 10$ is the line $3x + 5y = 10$ and the region below this line. ■

As a second demonstration of graphing the solution to a first-degree inequality in two variables, consider the following example.

EXAMPLE 2 Graph the solution to $y > \frac{2}{5}x - 1$.

Solution

Step 1 Start by graphing the solution to the equation $y = \frac{2}{5}x - 1$. The y-intercept is $(0, -1)$. Applying the slope measure, $\frac{2}{5}$, to this point yields the point 5 units right and 2 units up. This point's ordered-pair name is $(5, 1)$. We plot the two points and draw the line through them.

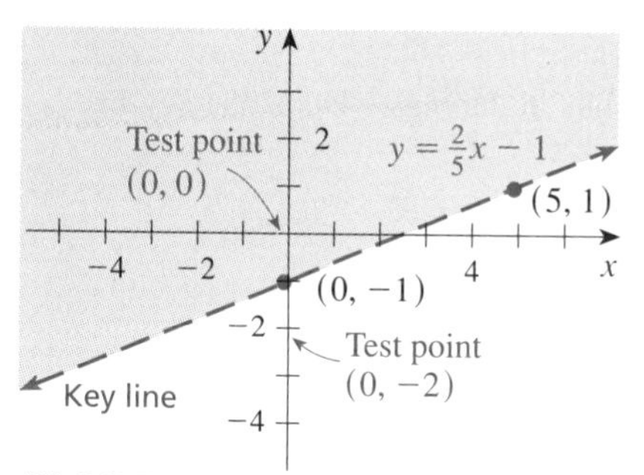

FIGURE 6
The graph of the solution to $y > \frac{2}{5}x - 1$

Step 2 Test an ordered pair on the line, say $(0, -1)$. The resulting inequality $-1 > \left(\frac{-2}{5}\right)(0) - 1$ is a false arithmetic sentence. Thus, the line is shown as a dashed (or dotted) line to indicate that the ordered pairs on the line do *not* solve the inequality.

Step 3 Now, test a point in the region above the key line, say $(0, 0)$. The inequality $0 > \left(\frac{2}{5}\right)(0) - 1$ is a true arithmetic sentence. So, we shade the region above the key line, as shown in Figure 6.

Step 4 Test a point in the region below the key line, say $(0, -2)$. The inequality $-2 > \left(\frac{2}{5}\right)(0) - 1$ is a false arithmetic sentence, so we leave the region unshaded.

Graphically, the solution to the inequality $y > \frac{2}{5}x - 1$ is the region above the line. ■

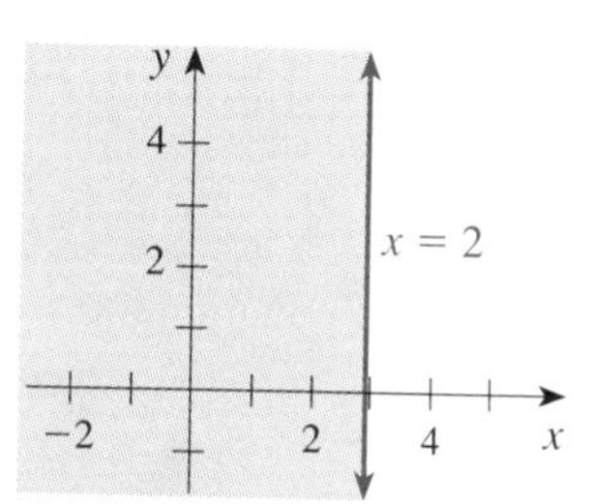

FIGURE 7
The graph of the solution to $x \le 2$

In Example 1 we used the x- and y-intercepts to graph the solution to the linear equation because the equation was in standard form. In Example 2 we used the y-intercept and the slope to graph the solution to the linear equation because the equation was in slope, y-intercept form.

If the key line is a vertical line, such as $x = 2$, we must consider the region to the right or the left of the key line, like the one illustrated in Figure 7 for the inequality $x \le 2$.

GIVE IT A TRY

1. Graph the solution to the inequality $3x - 2y > 7$.
2. Graph the solution to the inequality $y \le -3x + 4$.

■ SUMMARY

In this section we have learned how to graph the solution to a linear inequality, such as $3x - 2y \ge 8$. The technique we use is the same one that we have used for first-degree inequalities in one variable, such as $3x - 2 > 8$. With linear inequalities, rather than determining a key point, we find a key line. Also, instead of reporting the solution as an interval, the solution is a region, either above or below the line. Of course, if the key line is a vertical line, the region is to the left or right of the line.

3.3 ■ ■ ■ EXERCISES

1. **a.** For $3x - 2y > 10$, report the y-intercept, the x-intercept, and the slope measure of the key line.
   b. Draw a set of axes and sketch the key line from part (a). Plot and label the intercepts.
   c. Name three points in the region above the key line with y-coordinate 2.
   d. Name three points in the region below the key line with x-coordinate -1.
   e. Graph the solution to $3x - 2y > 10$.
2. **a.** For $y \le \frac{-2}{3}x + 4$, report the y-intercept and the slope measure of the key line.
   b. Draw a set of axes and sketch the key line from part (a). Plot and label the y-intercept and another point on the line.
   c. Name three points in the region above the key line with x-coordinate 1.
   d. Name three points in the region below the key line with y-coordinate -2.
   e. Graph the solution to $y \le \frac{-2}{3}x + 4$.
3. **a.** For $y > \frac{3}{5}x - 3$, report the y-intercept and the slope measure of the key line.
   b. Draw a set of axes and sketch the key line from part (a). Plot and label the y-intercept and another point on the line.
   c. Name three points in the region above the key line with y-coordinate 3.
   d. Name three points in the region below the key line with x-coordinate -2.
   e. Graph the solution to $y > \frac{3}{5}x - 3$.
4. **a.** For $4x + 5y \le 8$, report the y-intercept, the x-intercept, and the slope of the key line.
   b. Draw a set of axes and sketch the key line from part (a). Plot and label the intercepts.
   c. Name three points in the region above the key line with x-coordinate -3.
   d. Name three points in the region below the key line with y-coordinate 4.
   e. Graph the solution to $4x + 5y \le 8$.

Graph the solution to each linear inequality. On the key line, plot and label the intercepts. Use the graphics calculator to check your work.

5. $5x - 2y \ge 4$
6. $2y - 6x < 9$
7. $y > \frac{2}{5}x - 3$
8. $y \le \frac{-1}{3}x + 7$
9. $6(3 - x) \ge 2(y - 1)$
10. $13 \le 2(1 - y) + x$
11. $y \le -2x + 1$
12. $y > 2x + 5$
13. $y < \frac{x - 1}{3}$
14. $y \ge \frac{5 - 2x}{4}$
15. $\frac{y - 2}{3} \le \frac{2x}{3}$
16. $\frac{5y}{4} < \frac{3x - 2}{2}$
17. $x > 3y - x + 5$
18. $x - 1 \le y - 1$
19. $y - 3 > 0$
20. $x < 2x + 3$

3.4 ■ ■ ■ LINEAR RELATIONS

In the sciences and in business, we often discuss the relationship between two variables. For example, we may say that the tax rate is related to business activity (we may not know *precisely* how these two are related, but only that they are related). Likewise, health specialists know that smoking cigarettes is related to having health problems (again, they may not know how the two are related). The relations we discuss now are known as *linear relations*. We know from earlier work that the revenue earned from selling an item which costs $5 is related linearly to the number of items sold. That is, if we plot the ordered pairs of the form *(items sold, revenue earned)*, the graph is a straight line.

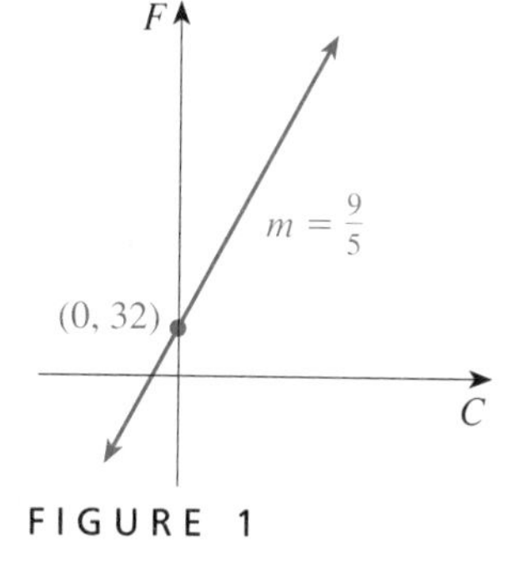

FIGURE 1

Relations A **relation** is any set of ordered pairs. Thus, the solution to a first-degree equation in two variables is a relation. If a plot of the ordered pairs of a relation produces a straight line, then the relation is known as a linear relation. We say that the two variables are **linearly related**, or that a **linear relationship** exists between the two variables. Two variables that are linearly related are Fahrenheit temperature F and Centigrade temperature C. Notice that $F = \frac{9}{5}C + 32$ is a first-degree equation in two variables. If we plot the ordered pairs (C, F) that solve the equation, the graph is the straight line shown in Figure 1. We say that *F is linearly related to C.*

Consider the set of ordered pairs $\{(-1, 1), (0, 2), (1, 1), (2, 3), (3, -1)\}$. This set is a relation, a set of ordered pairs. It is a linear relation? No. If we plot these points on a set of axes, we get the graph shown in Figure 2. These points do not lie on a straight line. Consider the solution to the equation $3x - y = 3$. To plot the ordered pairs that solve this equation, we use the facts that the x-intercept of the graph is $(1, 0)$ and the y-intercept is $(0, -3)$. This set of ordered pairs is a relation, and Figure 3 shows that this relation is a linear relation (the graph *is* a straight line).

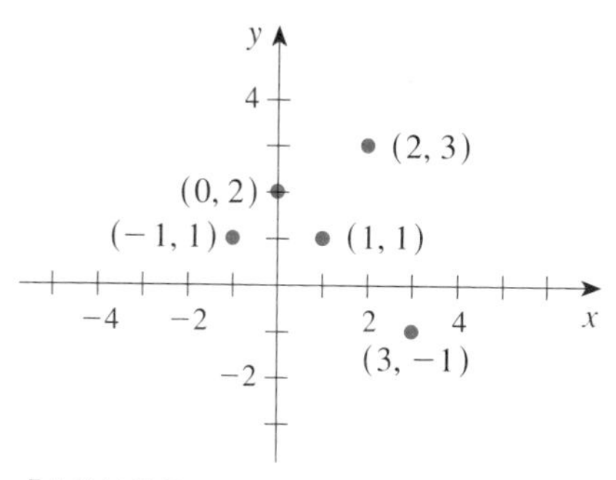

FIGURE 2

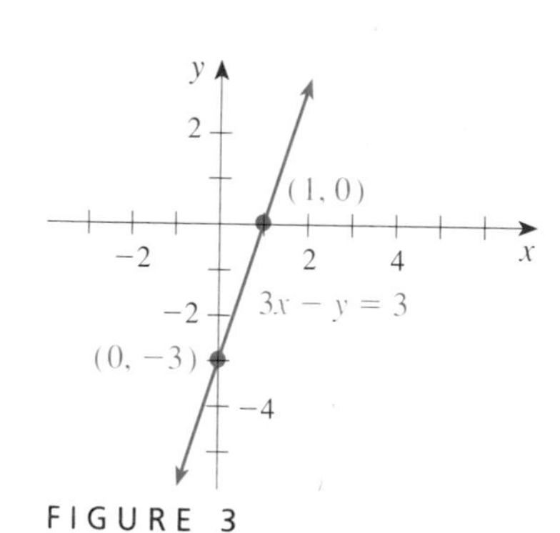

FIGURE 3

GIVE IT A TRY

Determine whether each of the following is a relation. Which of the relations are linear?

1. $\{2, 3, 6, 8, 9\}$
2. $\{(-3, 2), (0, 1), (3, 4)\}$
3. the solution to $3x - 5y = 8$
4. the solution to $|3x - 2| = -5$
5. the solution to $|3x - 2| > 2$

Using the Graphics Calculator

```
RANGE
Xmin=-10
Xmax=110
Xscl=10
Ymin=0
Ymax=240
Yscl=20
Xres=1
```

FIGURE 4

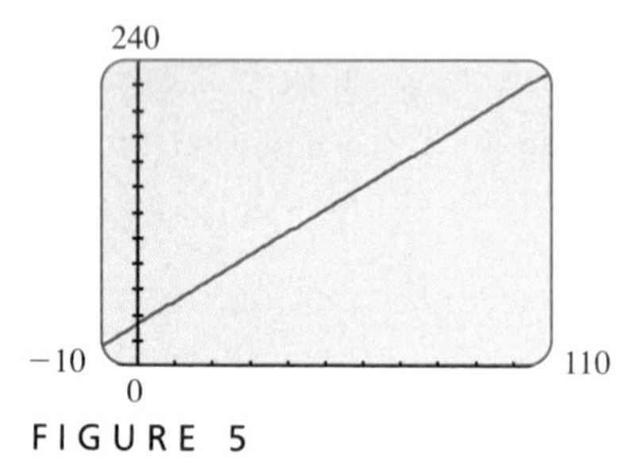

FIGURE 5

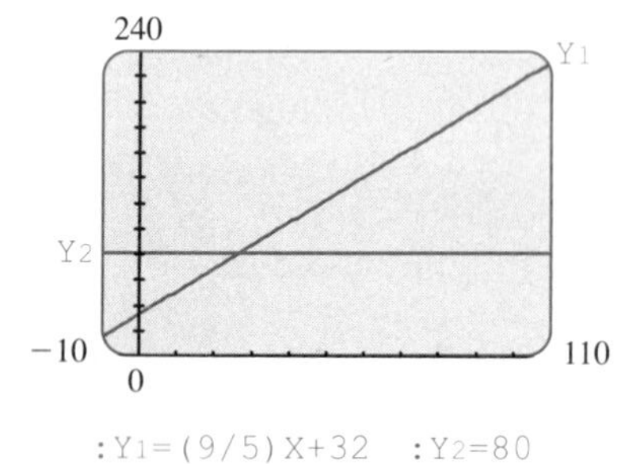

FIGURE 6

To graph the linear relationship determined by $F = \frac{9}{5}C + 32$, we must first decide which variable to represent by x and which variable to represent by y. Because of the way the equation is currently written, it is natural to let x represent C and y represent F. For `Y1`, type `(9/5)X+32`. We now need to decide what portion of the graph we wish to view. Let's suppose we're interested in Centigrade temperatures from -10 to 110. The corresponding Fahrenheit values will be from approximately 0 to 240. (If we guess incorrectly, we can adjust the y-range values, and try again.) Press the `RANGE` key and enter the values shown in Figure 4. View the graph screen: A line is drawn as shown in Figure 5. Press the `TRACE` key. The message `X=50.631579 Y=123.13684` means that when the Centigrade temperature is about 50, the corresponding Fahrenheit temperature is about 123.*

The slope measure of the line in Figure 5, $\frac{9}{5}$ means that, as the Centigrade temperature increases 5 degrees, the Fahrenheit temperature increases 9 degrees. The rate of increase is $\frac{9}{5}$. To find the Fahrenheit temperature when the Centigrade temperature is 100, press the right arrow key until the message reads `X=99.894737 Y=211.81053`. Press the right arrow key again. The message reads `X=101.15789 Y=214.08421`. To find the exact value, we can replace C by 100 and do the computation: $F = \left(\frac{9}{5}\right)(100) + 32$. The value for F is exactly 212. To redraw the graph at an enlarged size, press the `ZOOM` key and select the `Zoom In` option. Press the `ENTER` key. The location of the cursor determines the center of the graph at the new view. Press the `TRACE` key. Move the cursor until the message reads `X=100.05263 Y=212.09474`. This ordered pair is an even better approximation of the actual ordered pair, (100, 212). We could repeat this zoom-in process to get even better approximations of the actual point.

To find the Centigrade temperature when the Fahrenheit temperature is 80, press the `Y=` key and type `80` for `Y2`. Use the range settings in Figure 4 to view the graph screen, as shown in Figure 6. Press the `TRACE` key and move the cursor to the point where the two lines intersect. The ordered-pair name of that point is close to (26.6, 79.9). That is, when the Centigrade temperature is about 26.6, the Fahrenheit temperature is close to 80. To find the exact value, we solve the equation $80 = \frac{9}{5}C + 32$.

$$80 = \frac{9}{5}C + 32$$

$$400 = 9C + 160 \qquad \text{Multiply by 5.}$$

$$240 = 9C \qquad \text{Add } -160 \text{ to each side.}$$

$$\frac{240}{9} = C \qquad \text{Multiply by } \tfrac{1}{9}.$$

TI-82 Note *The TI-82 values are slightly different.

This equation has solution $\frac{240}{9}$, or $26\frac{2}{3}$. Thus, when the Centigrade temperature is $26\frac{2}{3}$, the Fahrenheit temperature is exactly 80. On the graphics calculator, we can get an improved approximation of the actual ordered-pair name of the point $\left(26\frac{2}{3}, 80\right)$ by using the ZOOM key and the Zoom In option. Try it.

The next example of working with a linear relation involves depreciation. It illustrates one of several methods of computing equipment depreciation, the loss of its value over the years of its useful life. This method is often called *straight-line depreciation*.

EXAMPLE 1 The value of a piece of equipment is linearly related to the number of years since its purchase. Suppose this relationship is given by

$$V = -5000t + 30000$$

where the value V is measured in dollars and the time t is given in years.

a. Graph this linear relationship.
b. Report the value of the equipment at 3.6 years.
c. When is the equipment worth $18,000?

Solution

a. We can use the graphics calculator to graph the relationship. Erase any entries on the Y= screen. For Y1, type -5000X+30000. The x-variable represents time t and the y-variable represents value V. We are interested in time values from 0, when the machine was new, to 6, when the machine has no value [the solution to the equation for zero value, $0 = -5000(t) + 30000$]. The machine value varies from $30,000 to $0. Press the RANGE key and type 0 for Xmin, 6 for Xmax, 1 for Xscl, 0 for Ymin, 30000 for Ymax, and 5000 for Yscl. View the graph (see Figure 7). Notice that the scale for the x-axis is 1 unit, but the scale for the y-axis is 5,000 units. Such a difference in scales is common in graphs for applications.

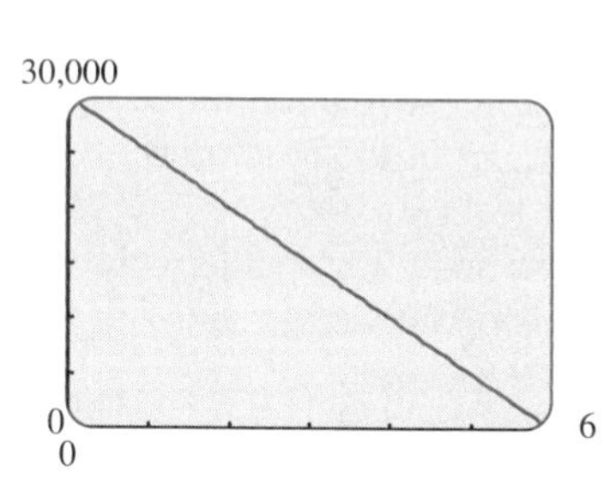

FIGURE 7

b. Press the TRACE key to find the value of the equipment at 3.6 years. Move the cursor as close as possible to an x-value of 3.6. Thus, at time $t = 3.6$ years, its value is $12,000. Using arithmetic, we can replace t by 3.6 and do the computation: $V = -5000(3.6) + 30000$.
c. Press the Y= key and type 18000 for Y2. View the graph screen. Press the TRACE key and move to the intersection of the two lines. At time 2.4 years (or 2 years 4.8 months) the equipment is worth $18,000. Algebraically, we replace V by 18,000 and solve the equation

$$18000 = -5000t + 30000$$

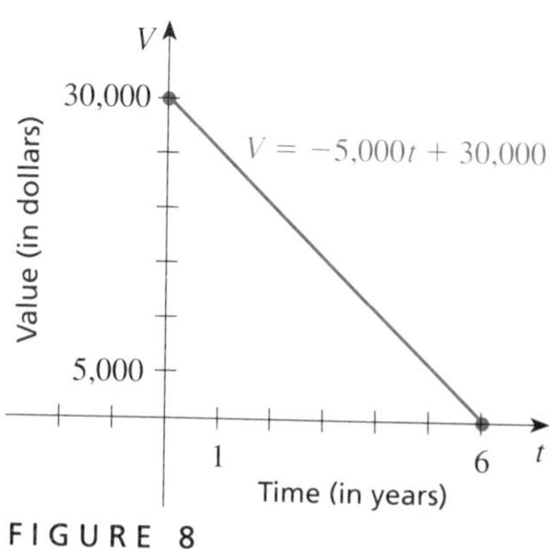

FIGURE 8

To transfer the graph in Figure 7 to paper, we first draw a set of axes. Mark off 6 units on the horizontal axis and 6 units on the vertical axis. Plot the ordered-pair names of two points, say (0, 30000) and (6, 0). Draw the line. See Figure 8. ■

GIVE IT A TRY

6. The cost C of producing n geography games is given by $C = 1.2n + 10000$. Use the graphics calculator to graph this linear relationship from $n = 0$ to $n = 5{,}000$ (using x for n and y for C). How many games can be produced for a cost of $14,000?

7. The world record time T for running a mile is linearly related to the year, x. The x-values decrease from John Paul Jones' record in 1911, of 4.26 minutes, to Jim Ryun's record in 1966, of 3.86 minutes. This linear relationship is $T = -0.0069x + 17.53$. Use the graphics calculator to construct a graph of this linear relationship (using x for the year and y for the time T). Use this graph to predict what year the world record for the mile will be 3.5 minutes (3 minutes 30 seconds). Is this a realistic prediction?

Ordered pairs that solve $y = |x - 3| - 1$

x	y	*Ordered pair*
0	2	(0, 2)
1	1	(1, 1)
2	0	(2, 0)

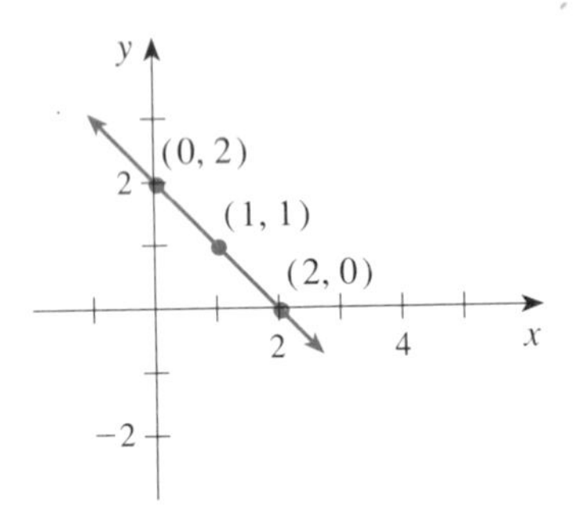

A Nonlinear Relationship To improve our understanding of linear relationships, we now consider a nonlinear relationship involving the absolute value operation. Consider the equation $y = |x - 3| - 1$. This equation has two variables, so its solution is a set of ordered pairs—a relation. To generate some of these ordered pairs, we can use a table like the one in the margin.

By plotting these points as we have shown in Figure 9, we can see that we do have a linear relation, at least up to this point! The equation of the line is $y = -x + 2$. [Notice that for the x-values in the table, the expression $x - 3$ is less than 0, so its absolute value is $-(x - 3)$. The relationship for these values is given by $y = -(x - 3) - 1$, or $y = -x + 2$.]

If we continue to generate ordered pairs that solve $y = |x - 3| - 1$, we see a change in the ordered pairs. Plotting the points generated for x values from 3 to 7, as shown in the second table, we also have a linear relation to the right of the x-value 3. See Figure 10. The equation of this part of the relation is given by $y = x - 4$.

Overall, the graph of the solution to the equation $y = |x - 3| - 1$ is a V-shaped line, not a straight line. Thus, the relationship between x and y is *not* a linear relationship.

Ordered pairs that solve $y = |x - 3| - 1$

x	y	*Ordered pair*
0	2	(0, 2)
1	1	(1, 1)
2	0	(2, 0)
3	−1	(3, −1)
4	0	(4, 0)
5	1	(5, 1)
6	2	(6, 2)
7	3	(7, 3)

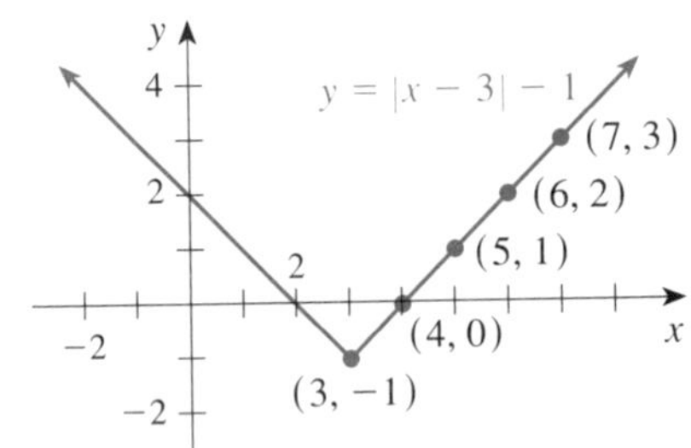

FIGURE 10

GIVE IT A TRY

8. **a.** Graph the ordered pairs that solve the equation $y = -|x| + 1$.
   b. Report the y-intercept of the graph in part (a).
   c. Report the x-intercept(s).

9. **a.** Graph the ordered pairs that solve $y = |x + 2|$.
   b. Report the y-intercept of the graph in part (a).
   c. Report the x-intercept(s).

Using the Graphics Calculator

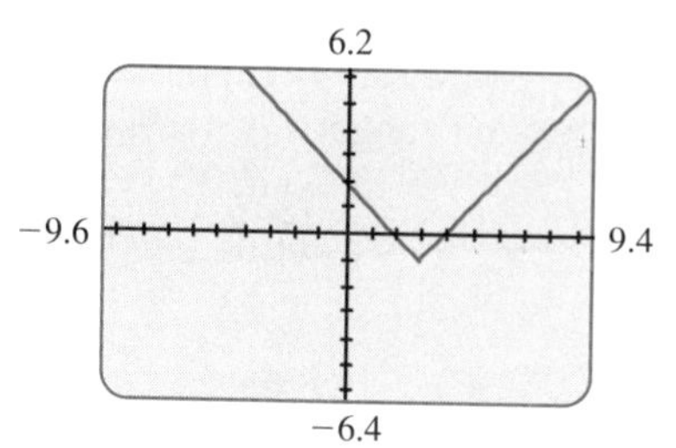

FIGURE 11

To see the graph of the solution to $y = |x - 3| - 1$, erase any existing entries on the Y= screen. For Y1, type abs (X−3)−1. On the RANGE screen, enter ⁻9.6 for Xmin, 9.4 for Xmax, 1 for Xscl, ⁻6.4 for Ymin, 6.2 for Ymax, and 1 for Yscl*. View the graph screen, as in Figure 11. Press the TRACE key. As we can see, the y-intercept is (0, 2). Use the right arrow key and move the cursor until the message reads X=2 Y=0. Thus, one x-intercept is (2, 0). Continue to press the right arrow key until the message X=3 Y=⁻1 appears. This is a **turning point** of the graph. To the left of this point, the graph is *decreasing* (the slope is negative). To the right of this point, the graph is *increasing* (its slope is positive). This point is also known as a **vertex**. Continue to press the right arrow key until the message reads X=4 Y=0. A second x-intercept is (4, 0).

GIVE IT A TRY

```
RANGE
Xmin=-4.8
Xmax=4.7
Xscl=1
Ymin=-3.2
Ymax=3.1
Yscl=1
Xres=1
```

FIGURE 12

10. Using the graphics calculator, enter the expression ⁻abs (X) + 1 for Y1. Press the RANGE key and enter the values, given in Figure 12.** Is the set of ordered pairs that solve $y = -|x| + 1$ a linear relation? What is the y-intercept of the graph? What are the x-intercepts?

11. Using the graphics calculator, enter the polynomial X²−2X−3 for Y1. View the graph using the values shown in Figure 12. Is the set of ordered pairs that solve $y = x^2 - 2x - 3$ a linear relation? What is the y-intercept of the graph? What are the x-intercept(s)?

Domain and Range As our last topic on relations, we now discuss some basic ideas about relations and their graphs. We start with the definition of domain and range.

DEFINITION OF DOMAIN AND RANGE

For a relation where the ordered pairs are (x, y), the collection of numbers x that compose the first member of the ordered pairs is the **domain** of the relation. The set of numbers y that compose the second member of the ordered pairs is the **range** of the relation.

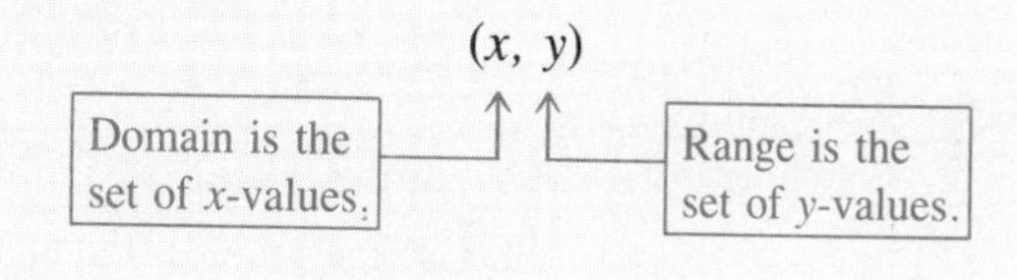

For a linear relationship, the domain is usually the entire set of real numbers. In some situations the domain is limited, such as the linear relationship we discussed in Example 1, $V = -5000t + 30000$. For this linear relationship, the domain $= \{t \mid 0 \le t \le 6\}$, or "the set of numbers t such that t is greater than or equal to zero and t is less than or equal to 6" (the bar | is read *such that*).

TI-82 Note

*Type ⁻9.4 for Xmin and ⁻6.2 for Ymin.
**Use ZOOM=4 for ZDecimal.

In interval notation, we have $[0, 6]$. This domain is restricted because the piece of equipment has no value to the company either prior to purchase (at $t = 0$) or after $t = 6$ years.

For most linear relations, the range is usually the entire set of real numbers. As with the domain, some situations result in a limited range. For the linear relationship $V = -5000t + 30000$, the range is $\{V \mid 0 \leq V \leq 30000\}$. The value of the piece of equipment cannot be less than zero or more than its purchase price, \$30,000.

EXAMPLE 2 Identify the domain and the range of the given graph.

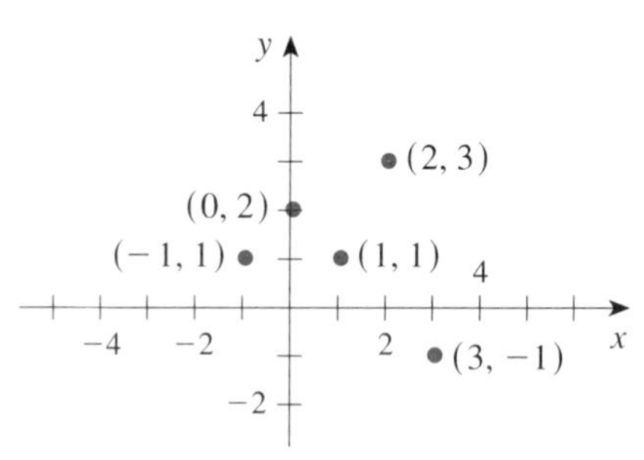

Solution To identify the domain, we observe the x-coordinates of the ordered pairs to get the set $\{-1, 0, 1, 2, 3\}$. Observing the y-coordinates of the ordered pairs, we get the range—the set $\{1, 2, 3, -1\}$. Notice that the domain and the range are each a set of numbers. ■

EXAMPLE 3 For the solution to $y = |x - 3| - 1$, identify the domain and the range.

Solution Viewing the graph of the relation, shown in Figure 13, we see that the first element of the ordered pairs can be any real number. Thus, domain $= \{x \mid x \text{ is real}\}$. In interval notation, we have $(-\infty, +\infty)$.

Viewing the graph of the relation, we see that the second element of the ordered pairs is never less than -1. Thus, the range $= \{y \mid y \geq -1\}$. In interval notation, we have $[-1, +\infty)$.

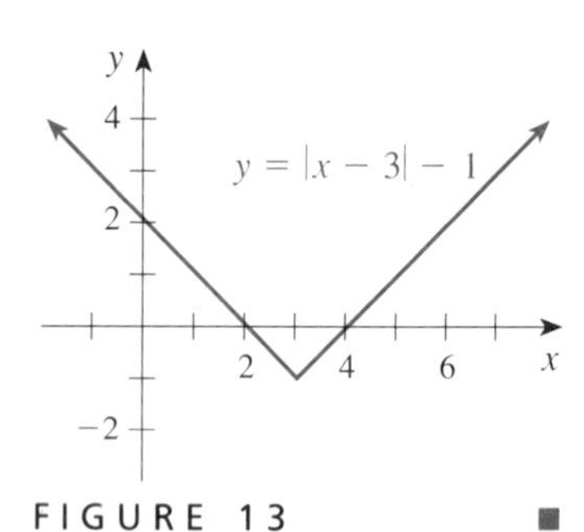

FIGURE 13 ■

GIVE IT A TRY

12. For the relation determined by $2x - 5y = 8$, identify the domain and range.
13. Use the graphics calculator to graph the relation determined by $y = -|3 - x| + 2$. Identify the domain and range.

■ ***Quick Index***

■ SUMMARY

In this section, we discussed linear relationships between variables. This topic finds use in the sciences and in the business world. We also considered several nonlinear relationships, such as the solution to $y = |x - 2|$. Finally, we defined the domain and range of a relation. The domain and the range are each a set of numbers.

3.4 ■ ■ ■ EXERCISES

1. The daily rental cost C for an automobile from AAZ Rentals is linearly related to the number of miles driven, x. The relation is determined by $C = 0.15x + 23$.
   a. Graph this linear relation for x values from 0 to 500.
   b. If 100 miles are driven in one day, what is the total daily cost?
   c. If the daily cost C is \$92, how many miles have been driven?
2. Leona has a weekly salary S that is linearly related to the number of hours h over 40 that she works each week. The relation is determined by $S = 14h + 280$.
   a. Graph this linear relation for h values from 0 to 30.
   b. If Leona works 50 hours ($h = 10$), how much is her salary?
   c. If Leona's salary last week was \$320, how many hours over 40 did she work?
3. The perimeter of a rectangle of width 5 meters is linearly related to the length of the rectangle by $P = 2L + 10$.
   a. Graph this linear relation for L values from 0 to 30.
   b. If the length is 10 meters, what is the perimeter?
   c. If the perimeter is 54 meters, what is the length?
4. The consumption C in billions of dollars, is linearly related to disposable income x, in billions of dollars, by $C = 0.7x + 10$.
   a. Graph this linear relation for x values from 0 to 50.
   b. If the amount of disposable income is \$30 billion, what is the consumption?
   c. If consumption is \$20 billion, what is the disposable income?
5. Use the graphics calculator to graph the relation determined by $y = \sqrt{(x-2)}$. By viewing the graph, decide whether this is a linear relation.
6. Use the graphics calculator to graph the relation determined by $y = 1/(x + 2)$. By viewing the graph determine whether this is a linear relation. Use the TRACE key. The y value does not exist when the x value is -2. Why is this? What is the domain of this relation?

For the graph of the relation shown in the figure, identify the domain and the range.

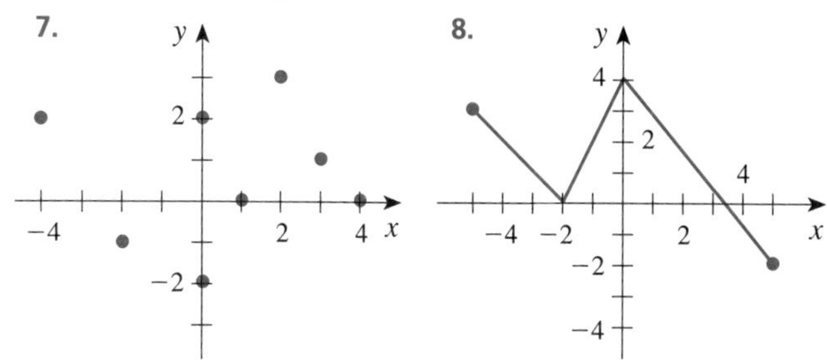

Use the graphics calculator to graph each relation. Use the TRACE *key to approximate each of the following, if it exists:*

a. *the* y-*intercept* **b.** *the* x-*intercept(s)* **c.** *the turning point*

9. $y = |x| - 3$
10. $y = |x| + 2$
11. $y = |x + 3| - 2$
12. $y = 3 - |x + 1|$
13. $y = -|x - 2| + 1$
14. $y = -|2x - 3| + 3$
15. $y = x^2 - 4x + 3$
16. $y = x^2 - 4x$
17. $y = 4 - x^2$
18. $y = 3 - 2x - x^2$

Use the graphics calculator to help graph each relation. Report the domain and the range of each graph using interval notation.

19. $y = 2x - 5$
20. $y = x^2$
21. $y = |2x - 3|$
22. $y = \sqrt{(x - 2)}$
23. $y = 3 - |x|$
24. $3x - 5y = 9$

DISCOVERY

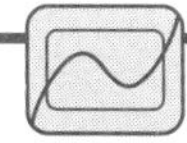

Absolute Value Equations in Two Variables

A very important and exciting facet of learning is *exploration*—working with known information and then trying to discover additional information or make a generalization. Here, we will start with knowledge that we have developed in earlier work and then attempt to discover a generalization about the graph of an absolute-value equation in two variables.

```
RANGE
Xmin=-4.8
Xmax=4.7
Xscl=1
Ymin=-3.2
Ymax=3.1
Yscl=1
Xres=1
```

FIGURE 1

We can graph the solution to the equation $y = |x|$ using the graphics calculator by typing `abs (X)` for `Y1`. Type the values shown in Figure 1.* View the graph, as shown in Figure 2. What is the effect on this graph of adding a number term to the right-hand side of this equation? That is, what is the graph $y = |x| + A$, where A is any real number? For `Y2`, type `abs (X)+2` and view the graph screen (Figure 3). Do you have an idea about the effect of adding a number to the right-hand side? For `Y3`, type `abs (X)+5`. What is the effect on the graph of adding a positive number to the right-hand side of $y = |x|$? What do you think the graph will look like if the value of A is negative? To obtain more evidence, use the graphics calculator to graph the solution to an equation such as $y = |x| - 3$. For `Y4`, type `abs (X)-3`. View the graph screen, as shown in Figure 4.

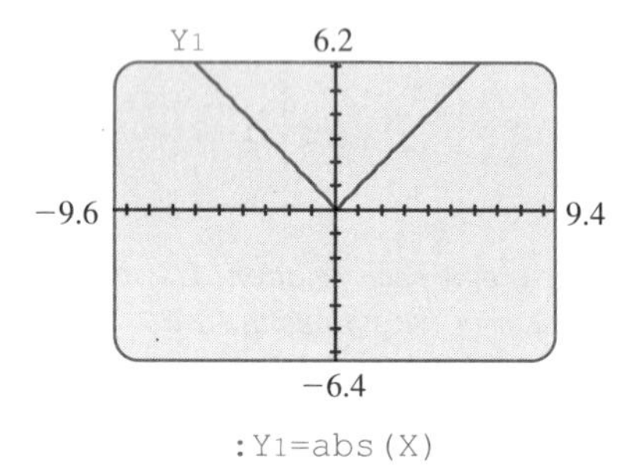

FIGURE 2

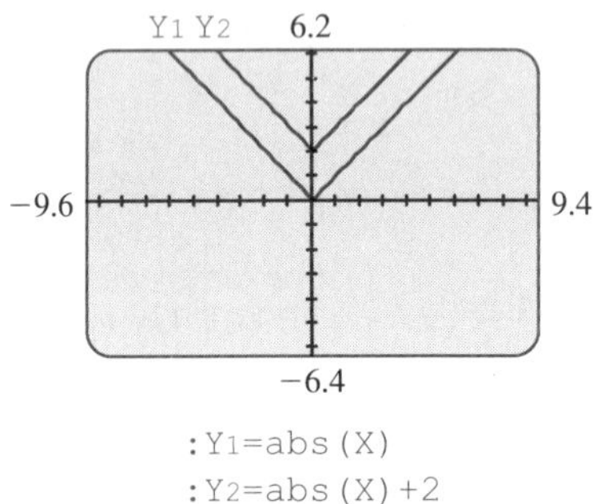

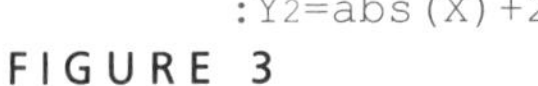

FIGURE 3

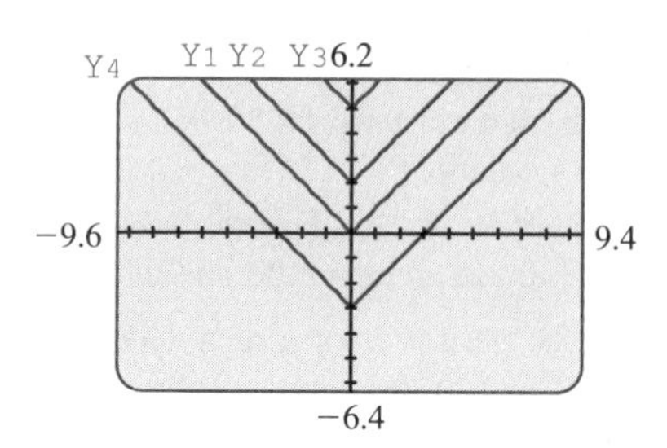

:Y1=abs(X :Y2=abs(X)+2
:Y3=abs(X)+5 :Y4=abs(X)-3

FIGURE 4

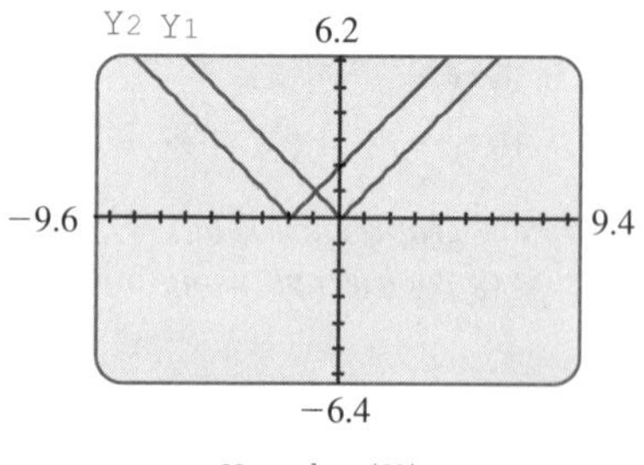

:Y1=abs(X)
:Y2=abs(X+2)

FIGURE 5

From this activity we have discovered that adding a constant, say A, to the right-hand side of $y = |x|$ alters the graph by moving the V-shaped curve A units, up or down. Now, consider how the graph of $y = |x|$ is related to the graph of $y = |x + B|$, where B is any real number. On the graphics calculator, erase the existing entries for `Y2`, `Y3`, and `Y4`. For `Y2`, type `abs (X+3)`. View the graph screen (Figure 5). Do you have an idea of the effect of B on the graph? Confirm your hypothesis by testing it with $y = |x + 5|$. For `Y3`, type `abs (X+5)`. View the graph screen. What happens to the graph if B is negative? That is, what is the relationship of the graph of $y = |x - 2|$ to the graph of $y = |x|$? In your own words, write this relationship.

Exercises

1. Using the graphics calculator, display the graph of $y = x^2$. Report how the graph of $y = x^2 + A$, where A is any real number, would differ from the graph of $y = x^2$.
2. Report how the graph of $y = (x + B)^2$, where B is any real number, would differ from the graph of $y = x^2$.

TI-82 Note *Type `-9.4` for `Xmin` and `-6.2` for `Ymin`.

3.5 ■ ■ ■ LINEAR FUNCTIONS

In Section 3.4 we learned about linear relations, such as the solution to a first-degree equation in two variables. We now introduce the closely related topic of linear functions. First, the general idea of a function and the notation for functions will be presented. Next, we will graph linear functions.

Background The concept of functions is used widely in mathematics, in the sciences, and in business. In mathematics, we often speak of linear functions, quadratic functions, polynomial functions, rational functions, exponential functions, and so forth. For example, the circumference of a circle is a linear function of its radius. In the sciences, we say that the distance traveled is a function of the speed and the time. In business, we say the cost of production is a function of the number of items produced. Because the idea of function is so widely used, it is an important concept to master.

Earlier, we considered equations such as $10x - y = 15$. When such an equation is rewritten (or rearranged) for y, we get the slope, y-intercept form of the equation, $y = 2x - 3$. In this form, we say y is a function of x; that is, y is dependent on x. The y-variable is known as the **dependent variable** and the x-variable is the **independent variable**. Given any value for x, we know exactly what the y-value is—we apply the rule $2x - 3$. For a particular x-value, say $x = 5$, the y-value is $2(5) - 3$, or 7.

In function notation, we write $y = f(x) = 2x - 3$ or, simply, $f(x) = 2x - 3$. We say that $f(x)$, or the **function of x,** is given by the rule $2x - 3$. Also we say that $f(5)$ is 7 or, simply, $f(5) = 7$. We read $f(5)$ as *f of 5* or *the function f evaluated at 5*.

For the function $f(x) = 2x - 3$, the value of x is known as the **input value**. The function value $f(x)$ is the **output value**. See Figure 1. For an input value of 5, we apply the function rule to get an output value, $f(5)$, or 7. See Figure 2.

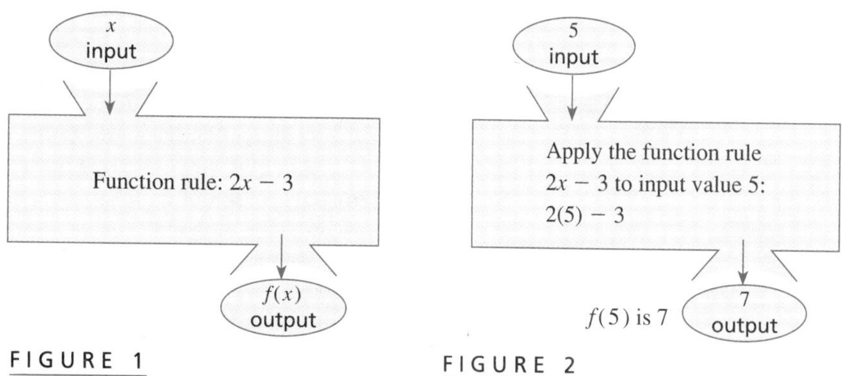

FIGURE 1

Input-Process-Output for the function $f(x) = 2x - 3$

FIGURE 2

Input-Process-Output for an input of 5

For the function $f(x) = 3x - 6$, the function rule is $3x - 6$. For each input value x, we find the function value $f(x)$ by multiplying by 3 and then subtracting 6. For an input of 0, the function rule $3x - 6$ will output -6; that is, the output value is $f(0) = 2(0) - 6 = -6$.

GIVE IT A TRY

1. For the function $f(x) = -2x + 5$, for an input of 3 the output is $-2(3) + 5$, or -1. We write $f(3) = -1$.
   a. What is $f(-3)$? b. For an input of 5, what is the output?
2. Consider the function given by $g(x) = 4x + 2$.
   a. What is the function rule? b. Find $g(0)$.

Using the Graphics Calculator

Input	-5→X	
	-5	
Process	2X-3	
	-13	Output

FIGURE 3

We can use the graphics calculator to improve our understanding of the IPO (Input-Process-Output) concept of functions. Consider the function whose rule is the polynomial $2x - 3$. In function notation, we have $f(x) = 2x - 3$. To simulate an input of -5 to this function, store -5 in memory location X. Now, to simulate processing this input value, type 2X−3 and press the ENTER key. The value -5 is processed and the number -13 is output. See Figure 3. In function notation, we write $f(-5) = -13$.

Functions and Ordered Pairs If we view a function as a rule that determines an output value for each input value, then we can think of the function as a collection of input values and output values, that is, a collection of ordered pairs, where the first member is the input value and the second member is the output value. We can then say that a function is the collection of ordered pairs $(x, f(x))$. One restriction must be placed on this ordered pair idea:

For every input value x, there is exactly one output value $f(x)$.

That is, we *cannot* have a function which outputs $f(2) = 3$ *and* $f(2) = 8$. A function is a special case of a relation, which is any set of ordered pairs. We can now define a function.

DEFINITION OF A FUNCTION

A **function** is a set of ordered pairs, denoted $(x, f(x))$, such that for every first element, the input value x, there is *exactly* one second element, the output value $f(x)$. A function is also thought of as a rule that assigns each element of a set A to exactly one element of a set B.

In Section 3.4, we stated that a relation is a set of ordered pairs. The solution to any equation in two variables is a relation. Some of these relations are also functions. The solution to a linear equation, such as $3x - 5y = 10$, is a linear function. There is a parallel between linear functions and the solution to a first-degree equation in two variables:

Linear equation:	$y \;=\; 3x - 2$	$(x, \;\; y)$
	$\downarrow \qquad \downarrow$	$\downarrow \quad \downarrow$
Linear function:	$f(x) = 3x - 2$	$(x, f(x))$
		(input, output)

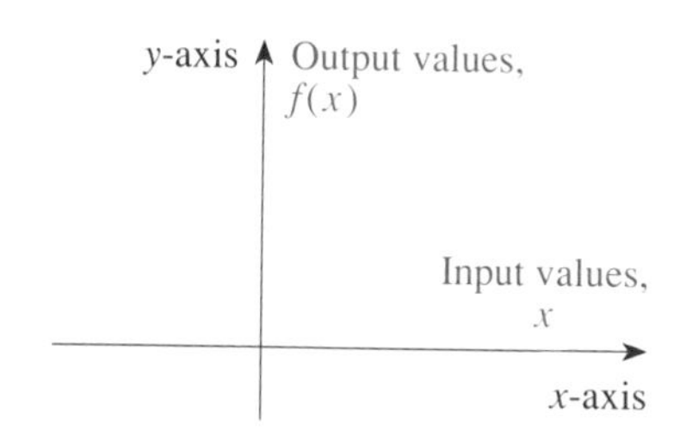

FIGURE 4

Every linear relation is a function, except the one determined by an equation of the form $x = a$. However, some relations are not functions. For example, the ordered pairs that solve the equation $y^2 = x$ are a relation, but not a function. Two ordered pairs in the solution to this equation are (4, 2) and (4, −2). This violates the definition of a function because every first element must be matched with exactly one second element.

Graphs If we view a function as a collection of ordered pairs, then we can graph it on a set of axes. As Figure 4 shows, the horizontal axis, or x-axis, represents the input values, and the vertical axis represents the output values produced, $f(x)$. The following example shows how to graph a function.

EXAMPLE 1 Graph the function $f(x) = 3x - 2$.

Solution For this function, $f(0) = 3(0) - 2 = -2$ and $f(2) = 3(2) - 2 = 4$. We can record this data in a table like the one shown. Plotting this information, we get the points in Figure 5.

x	$f(x)$
0	−2
2	4

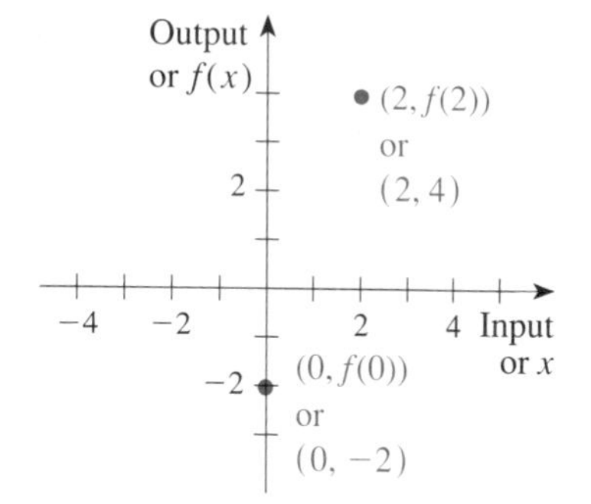

FIGURE 5

■

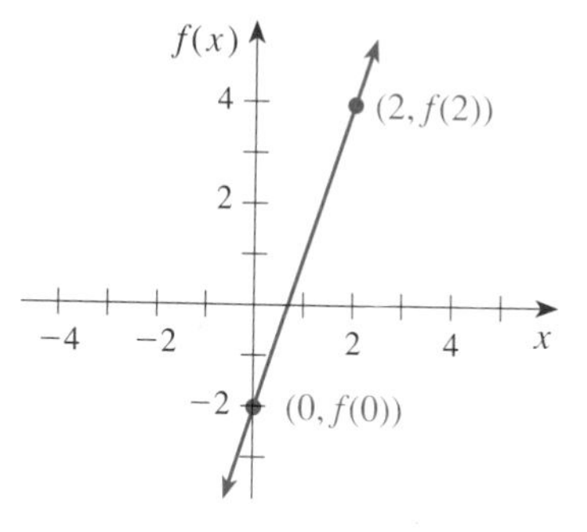

FIGURE 6

If we continue to generate ordered pairs from the function and plot each pair, we get the straight line, shown in Figure 6. If a function rule is a first-degree polynomial in one variable, then the graph of the function is a straight line. For this reason, such a function is called a **linear function**. The graph of the linear function $f(x) = 3x - 2$ is shown in Figure 6.

A **constant function** is a function whose rule is a number. (This special case of a linear function—the rule has degree zero.) For example, $f(x) = 2$ is a constant function. Here, for every input value, x, the output value is 2: $f(-8) = 2$, $f(100) = 2$, and so on. The graph of the constant function $f(x) = 2$ is a horizontal line, as shown in Figure 7.

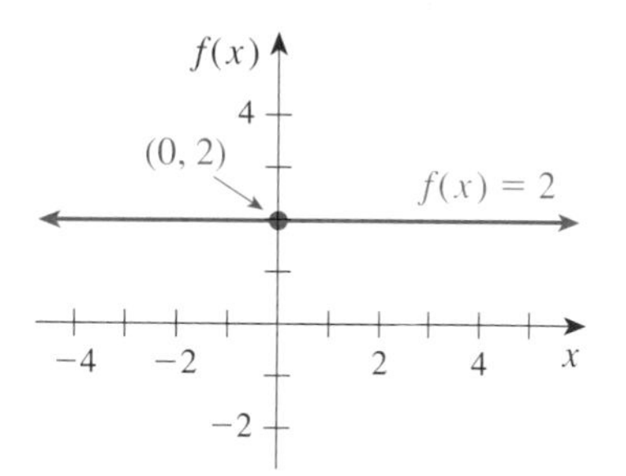

FIGURE 7

EXAMPLE 2 Consider the linear function $g(x) = \frac{2}{3}x + 2$.

a. Find $g(3)$.
b. For an input of 2, what is the output?
c. What input value generates an output of 5?
d. Graph the linear function.

Solution a. Because the function rule is $\frac{2}{3}x + 2$, we have

$$g(3) = \left(\frac{2}{3}\right)(3) + 2$$

(input 3)

Thus, $g(3) = 4$.

b. An input of 2 means the x-value will be 2, so the output value is

$$g(2) = \left(\frac{2}{3}\right)(2) + 2 = \frac{10}{3}$$

The output is $\frac{10}{3}$.

c. To get an output of 5 means that $5 = \frac{2}{3}x + 2$. We must solve this first-degree equation in one variable:

$$5 = \frac{2}{3}x + 2$$

Multiply by 3 to get: $15 = 2x + 6$

Add -6 to each side to get: $\frac{-6 \qquad -6}{9 = 2x}$

Multiply by $\frac{1}{2}$: $\frac{9}{2} = x$

So, to generate an output of 5, the input must be $\frac{9}{2}$.

d. The graph is shown in Figure 8. ■

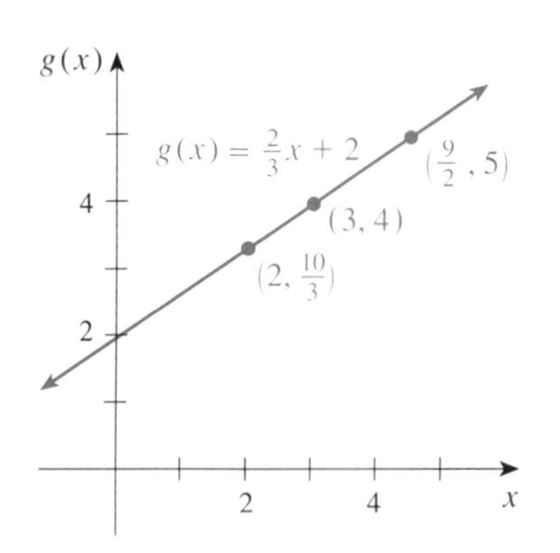

FIGURE 8

Using the Graphics Calculator

We now use the graphics calculator to rework Example 2. Erase any existing entries on the Y= screen. For Y1, type (2/3)X+2. Press the RANGE key and enter the values shown in Figure 9.* View the graph screen. A line is drawn similar to the one in Figure 8. Press the TRACE key. The message X=0 Y=2 means that "for an input of 0, the output is 2."

a. *Find $g(3)$.* Use the right arrow key to move the cursor until the message reads X=3 Y=4. Thus, for an input of 3, the output is 4, or $g(3) = 4$.

b. *For an input of 2, what is the output?* Move the cursor until the message reads X=2 Y=3.3333333. For an input of 2, the graphics calculator approximates the exact output value, $\frac{10}{3}$.

c. *What input value generates an output of 5?* Press the Y= key and type 5 for Y2. View the graph screen, press the TRACE key, and then move the cursor to the intersection point of the two lines. The

```
RANGE
Xmin=-9.6
Xmax=9.4
Xscl=1
Ymin=-6.4
Ymax=6.2
Yscl=1
Xres=1
```

FIGURE 9

TI-82 Note *Type ⁻9.4 for Xmin and ⁻6.2 for Ymin.

message reads `X=4.4  Y=4.9333333`. Press the right arrow key again. The message reads `X=4.6  Y=5.0666667`. Thus, to produce an output of 5, the input must be between 4.4 and 4.6. From algebra, we know that this input value is exactly 4.5. Press the `ZOOM` key and select the `Zoom In` option. Press the `ENTER` key. Use the `TRACE` key and move to the point of intersection.

Graph the linear function. A portion of the graph is displayed on the screen. Press the `ZOOM` key and select the `Zoom Out` option to display a more complete view of the graph. (The horizontal line `Y2=5` is still displayed.)

Applications Functions are used in many fields of study. Throughout mathematics, the function concept is used for quadratic functions, exponential functions, logarithmic functions, and so forth. Function concepts are also used in the sciences and in business. In biology, the population of a particular bacteria is said to be a function of time. In chemistry, the volume of a gas is said to be a function of temperature. In business, supply is described as a function of demand.

The next example demonstrates a business use of functions.

EXAMPLE 3 The cost C (in dollars) of producing a selected number x of Z83 computer chips is given by the linear function $C(x) = 3x + 437$.

a. What is the cost of producing ten Z83 computer chips?
b. How many Z83 computer chips can be produced for a cost of $700?
c. Draw a graph plotting the cost as a function of the number of Z83 computer chips produced.

Solution a. Because the cost function is $C(x) = 3x + 437$, the cost of producing ten Z83 computer chips is the output value $C(10)$:

$$\begin{aligned} C(10) &= 3(10) + 437 \\ &= 467 \end{aligned}$$

Thus, the cost of producing ten Z83 computer chips is $467.

b. To find how many chips can be produced for $700, we want to find the input x that generates an output of 700. So, we solve the equation $700 = 3x + 437$:

$$\begin{aligned} 700 &= 3x + 437 \\ -437 & \quad\;\; -437 \qquad \text{Add } -437 \text{ to each side.} \\ 263 &= 3x \qquad \text{Multiply each side by } \tfrac{1}{3}. \\ \frac{263}{3} &= x \end{aligned}$$

This equation has solution $\frac{263}{3}$, or $87\frac{2}{3}$. Thus, 87 computer chips can be produced for $700.

c. The graph of the function is shown in Figure 10. ■

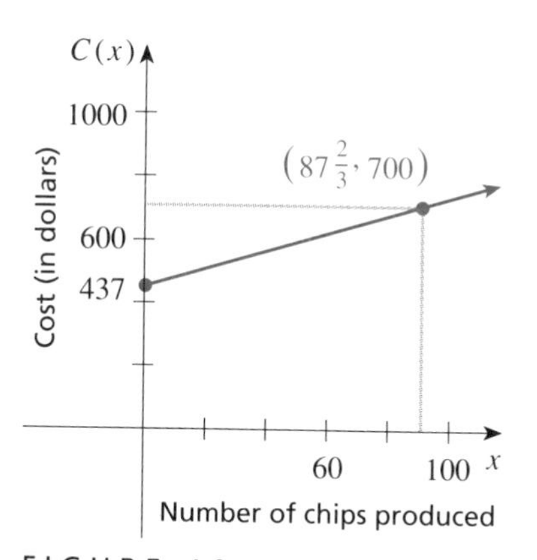

FIGURE 10

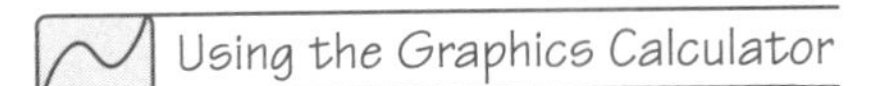

To check our work in Example 3, for Y1 type 3X+437. Use the RANGE key and enter the values 0 for Xmin, 100 for Xmax, 10 for Xscl, 0 for Ymin, 1000 for Ymax, and 100 for Yscl. View the graph screen. (The graph should look similar to the line in Figure 10.) Use the TRACE key to check that when the input value X is near 10, then the output value Y1 is near 467. To check that an output of 700 is achieved when the input value is $87\frac{2}{3}$, type 700 for Y2. View the graph, press the TRACE key, and then move the cursor to the intersection point. The x-value should be approximately 87.

GIVE IT A TRY

3. Consider the linear function $f(x) = \frac{1}{3}x - \frac{5}{3}$.
a. Find $f(-2)$. **b.** Find the input value x such that $f(x) = -3$.

4. The temperature in Fahrenheit F is a linear function of the temperature in Celsius C. The function is $F(C) = \frac{9}{5}C + 32$.
a. Find $F(0)$.
b. Find the Fahrenheit temperature for a Celsius temperature of 30.
c. What Celsius temperature is needed to obtain a Fahrenheit temperature of 120?
d. Graph this function.

The Domain and the Range Like a relation, every function has a domain and a range. The definition of these terms follows.

DOMAIN AND RANGE OF A FUNCTION

For a function, the set of numbers that are allowed as input values, say x, is the **domain** of the function. The resulting set of numbers that are the output values, $f(x)$, is the **range** of the function.

$$(x, f(x))$$

domain → x; range → $f(x)$

Every linear function has the understood domain of the entire set of real numbers, because it is possible to input any of these values. This domain produces a range that is all the real numbers. Functions may have limited domains or ranges. Consider the following examples of a linear and a nonlinear function:

$$f(x) = \frac{2}{3}x + 5 \qquad \text{Domain: } = \{x \mid x \text{ is real}\} \quad \text{Range} = \{y \mid y \text{ is real}\}$$

$$f(x) = \sqrt{x} \qquad \text{Domain} = \{x \mid x \geq 0\} \quad \text{Range} = \{y \mid y \geq 0\}$$

The domain of each function is the set of allowable input values, and the range is the set of output values produced from the domain. For the linear function we used in Example 3 to describe, or model, a real-world situation, the domain is restricted to positive real numbers. We write this as

$$\text{domain} = \{x \mid x \geq 0\}$$

For this domain, the range $= \{y \mid y \geq 437\}$, because these are the values output by the function.

Using the Graphics Calculator

Consider the nonlinear function whose rule is $f(x) = |x - 3| - 1$. Using the graphics calculator, type `abs (X-3)-1` for `Y1`. Type the values shown in Figure 11.* View the graph screen (see Figure 12). The domain for this function is all real numbers, so we write domain $= \{x \mid x \text{ is real}\}$. The numbers output are all -1 or greater. The range includes only the real numbers greater than or equal to -1, so we write range $= \{y \mid y \geq -1\}$.

```
RANGE
Xmin=-9.6
Xmax=9.4
Xscl=1
Ymin=-6.4
Ymax=6.2
Yscl=1
Xres=1
```

FIGURE 11

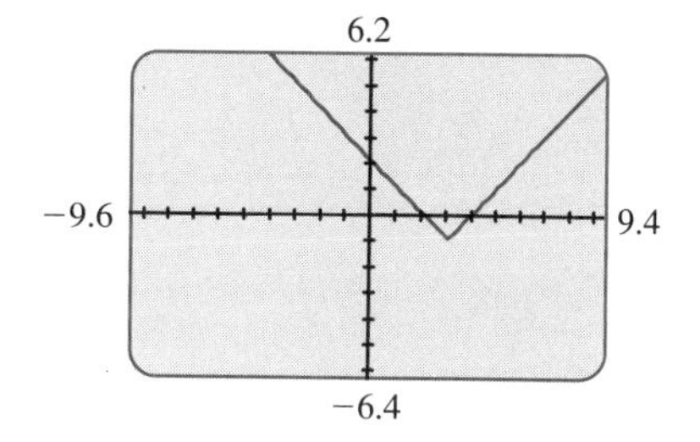

FIGURE 12

GIVE IT A TRY

For each function, use the graphics calculator to view the graph. Report the domain and the range.

5. $h(x) = 2 - |3 - x|$ **6.** $g(x) = \sqrt{(3 - x)}$ **7.** $r(x) = |x| - 3$

■ SUMMARY

In this section we have learned about a topic closely related to solving first-degree equations in two variables—linear functions. A function is a rule that assigns to each input value exactly one output value. The notation for a function appears as $f(x) = 3x + 1$, where x represents the input value and $f(x)$ represents the output value. When the rule for a function is a first-degree polynomial [or a zero-degree polynomial for the special case of a constant function, such as $f(x) = 3$], then the function is known as a linear function. By considering the input values and output values as ordered pairs of numbers, we can draw a graph for the function.

After introducing linear functions and their graphs, we discussed the domain and range of a function. For any function, the domain is the set of allowable input values. For a linear function, it is usually understood that all real numbers are allowed in the domain. In some special situations, such as using a linear function as a model for a physical situation, the domain is limited (often to the nonnegative real numbers). The range of a function is the set of output values produced by the domain values.

■ *Quick Index*

TI-82 Note *Type `-9.4` for `Xmin` and `-6.2` for `Ymin`.

3.5 ■ ■ ■ EXERCISES

1. For the function $f(x) = -3x + 5$, find each output value:

   a. $f(0)$ **b.** $f(-3)$ **c.** $f(4)$ **d.** $f\left(\frac{-2}{3}\right)$

Use the function $f(x) = \frac{2}{3}x - 1$ to answer Problems 2–6.

2. Find $f(6)$.
3. Find the output for an input of -6.
4. Find the input value for an output of 0.
5. Find the input value for an output of 2.
6. Graph the function.

Use the function $f(x) = \frac{-7}{2}x + 5$ to answer Problems 7–12.

7. Find $f(-1)$.
8. Find the output for an input of 4.
9. Find the input value for an output of 0.
10. Find the input value for an output of $\frac{2}{3}$.
11. Graph the function.
12. Use the graphics calculator to graph the function. Then use the TRACE key to find $f(-2)$.

Graph each linear function, and report the x-intercept and the y-intercept. Also report the slope measure of the line. Use the graphics calculator to check your work.

13. $f(x) = -3x - 1$
14. $g(x) = 2x + 1$
15. $h(x) = 5x - 4$
16. $d(x) = -4x + 7$
17. $n(x) = \frac{-3}{5}x + 2$
18. $r(x) = \frac{7}{2}x - 1$
19. $f(x) = \frac{7}{4}x - \frac{11}{4}$
20. $g(x) = \frac{-5}{4}x$
21. $r(x) = \frac{2x - 5}{3}$
22. $h(x) = \frac{7 - 3x}{5}$
23. $g(x) = 0.5x - 2.3$
24. $b(x) = 6 - 1.2x$
25. $f(x) = \frac{17}{8}x$
26. $k(x) = 1.2x + 350$
27. $j(x) = \frac{8 - 5x}{6}$
28. $f(x) = \frac{-3x - 2}{4}$
29. $v(x) = -4$
30. $c(x) = -x$
31. The temperature in Celsius C is a function of the temperature in Fahrenheit F. The function is $C(F) = \frac{5}{9}F - \frac{160}{9}$.
    a. Find $C(0)$.
    b. Report the Celsius temperature for a Fahrenheit temperature of 32.
    c. What Fahrenheit temperature is needed to obtain a Celsius temperature of 20?
    d. Graph this function.
32. The cost of a car rental for one day, C (in dollars), is a function of the miles driven, x. The function is $f(x) = 0.2x + 39$.
    a. What is the cost for renting the car for one day, but driving 0 miles?
    b. What is the cost for driving 100 miles?
    c. If the cost is \$100, how many miles were driven?
    d. Graph this function.
33. The daily profit of a fish-cake company, P (in thousands of dollars), is a function of the number of fish cakes sold, x (in thousands). The function is $P(x) = 0.8x - 12$.
    a. What is the daily profit when 10,000 fish cakes are sold? [*Hint*: Input 10 for x.]
    b. To get a daily profit of \$20,000, how many fish cakes must be sold for the day?
    c. Graph the function.
34. The number of full-time instructors N at a local college is a function of the number of credit hours h that students signed up for in the previous year. The function is $N(h) = 0.0005h + 60$.
    a. If students signed up for 300,000 credit hours last year, how many full-time instructors should be employed at the college for this year?
    b. If the college has 300 full-time instructors this year, how many credit hours were taken the previous year?
    c. Graph this function.

CHAPTER 3 REVIEW EXERCISES

1. Which of the following is a first-degree equation in two variables?

   i. $3x - 7 = 10$ **ii.** $xy = 9$
   iii. $5x - 2y > 8$ **iv.** $y = -2x + 5$

2. Which of the following is a first-degree inequality in two variables?

   i. $|3x - 7| \leq 10$ **ii.** $x^2 - 2x = 9$
   iii. $5x - 2y = 8$ **iv.** $y > -2x + 5$

3. Which of the following is a *false* statement?
   i. The solution to a first-degree equation in one variable is a collection of numbers.
   ii. The solution to a first-degree equation in two variables is a collection of ordered pairs of numbers.
   iii. For the graph of any first-degree equation in two variables, the y-intercept has y-coordinate 0.
   iv. For the graph of a first-degree equation in two variables, the slope measure can be negative.
4. Which of the following is a *false* statement?
   i. The graph of the solution to a first-degree inequality in one variable is an interval.
   ii. The graph of the solution to a first-degree equation in two variables is a straight line.
   iii. A linear relation is a set of numbers.
   iv. The graph of a linear inequality is a region.

Use the equation $3x - 2y = 9$ to answer Problems 5–9.

5. Report the ordered pair in the solution with x-coordinate -3.
6. Report the ordered pair in the solution with y-coordinate 4.
7. Complete the following table to generate ordered pairs that solve the equation.

x	-1	0	2	3.1	___	___	___
y	___	___	___	___	-2	0	1

8. Report the y-intercept and the x-intercept.
9. Rewrite the equation for y.

Use the equation $5x - 3y = 12$ to answer Problems 10–14.

10. Report the ordered pair in the solution with x-coordinate 2.
11. Report the ordered pair in the solution with y-coordinate -1.
12. Complete the following table to generate ordered pairs that solve the equation.

x	-1	0	1	2.3	___	___	___
y	___	___	___	___	-2	0	1

13. Report the y-intercept and the x-intercept.
14. Rewrite the equation for y.

For each pair of ordered pairs, plot the points, draw a line through the plotted points, and report the slope of the line.

15. $(-4, -1)$ and $(6, 2)$
16. $(5, -1)$ and $(3, 4)$
17. $(-4, 1)$ and $(-1, -3)$
18. $(-3, -3)$ and $(2, -3)$
19. $(3, 6)$ and $(3, -2)$
20. $(1, 0.5)$ and $(3.2, 2)$
21. Plot the point $(-3, 5)$ and draw a line through the point with slope measure $\frac{-2}{5}$. Report the ordered-pair name of another point on the line.
22. Plot the point $(2, -1)$ and draw the line through the point with slope measure $\frac{2}{3}$. Report the ordered-pair name of another point on the line.
23. Plot the point $(1, -5)$ and draw the line through the point with slope measure $\frac{7}{3}$. Report the ordered-pair name of another point on the line.
24. Plot the point $(-2, -3)$ and draw the line through the point with slope measure 0. Report the ordered-pair name of another point on the line.

Use graphs A–D to answer Problems 25–28.

A.

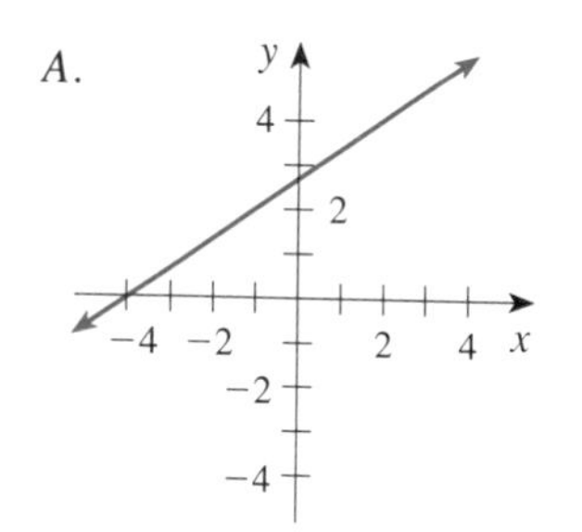

B.

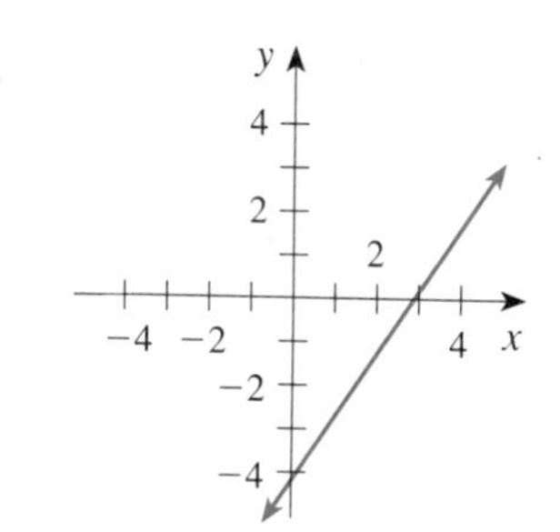

C.

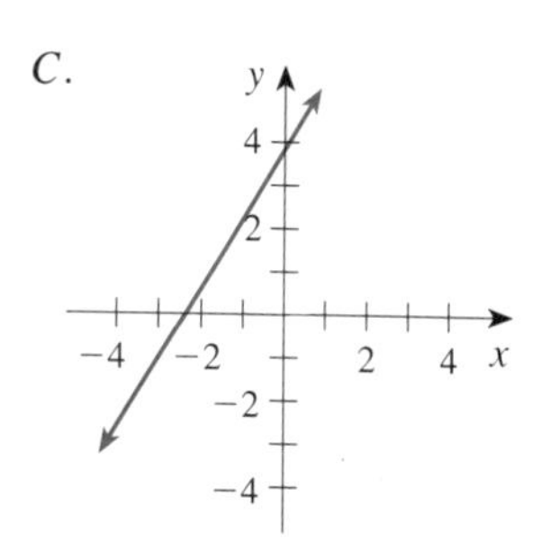

D.

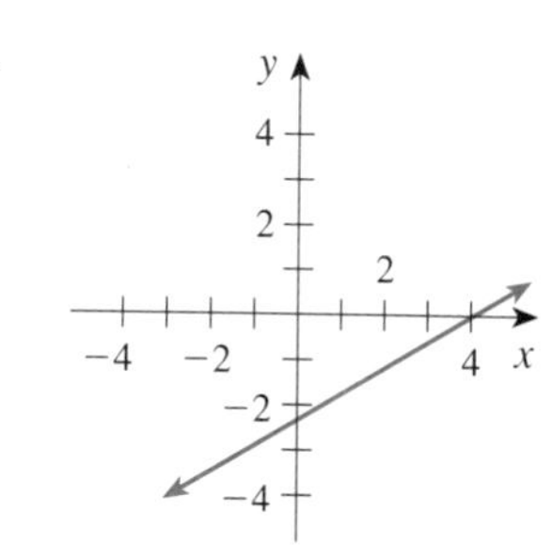

25. Which is the graph of $2x - 3y = 8$?
26. Which is the graph of $3x - 2y = 8$?
27. Which is the graph of $-3x + 2y = 8$?
28. Which is the graph of $-2x + 3y = 8$?
29. A vertical line has a slope measure that is
   i. undefined
   ii. negative
   iii. positive
   iv. 0

30. The graph of the solution to the equation $3x - 5y = 12$ has the following y-intercept:

i. $(4, 0)$ **ii.** $(0, 4)$

iii. $\left(0, \frac{-12}{5}\right)$ **iv.** $\left(\frac{-5}{12}, 0\right)$

31. What is the slope of the straight line through the points $(-2, 6)$ and $(3, -1)$?

32. What is the slope of the straight line through the points $(5, -1)$ and $(-3, -1)$?

Graph the solution to each first-degree equation in two variables. Report the y-intercept, the x-intercept, and the slope measure.

33. $2x - 5y = -4$ $3x + 2y = -6$

35. $y = -3x + 6$ $y = 5x - 6$

37. $y = \frac{2}{5}x - \frac{7}{5}$ $\frac{5 - 3x}{2} = \frac{y - 3}{4}$

39. Rewrite the equation $3x - 2y = 5y + 1$ for y. What is the slope measure of the graph of the solution to the equation? What is the y-intercept for the graph?

40. Rewrite the equation $7x + 3y = 12 + x$ for y. What is the slope measure of the graph of the solution to the equation? What is the y-intercept for the graph?

Use graphs A–D to answer Problems 41–44.

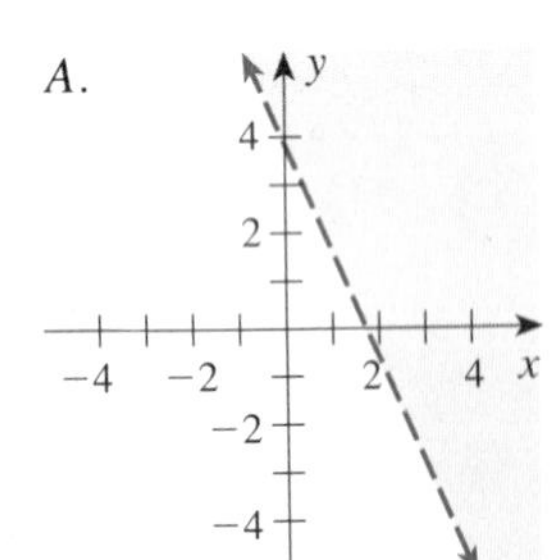

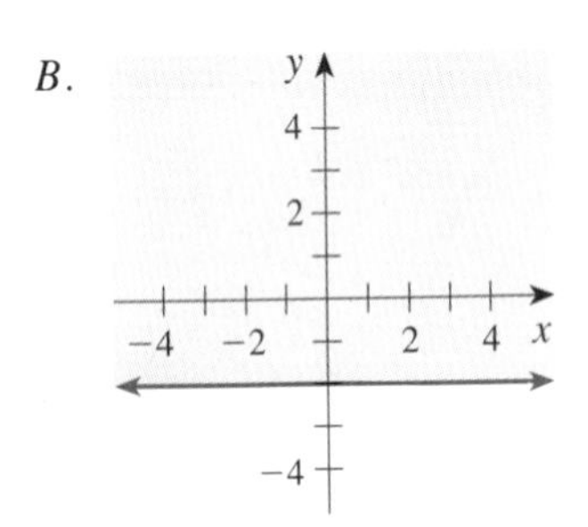

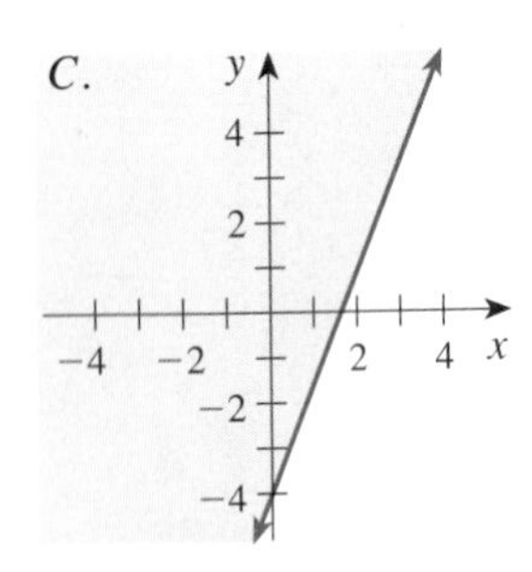

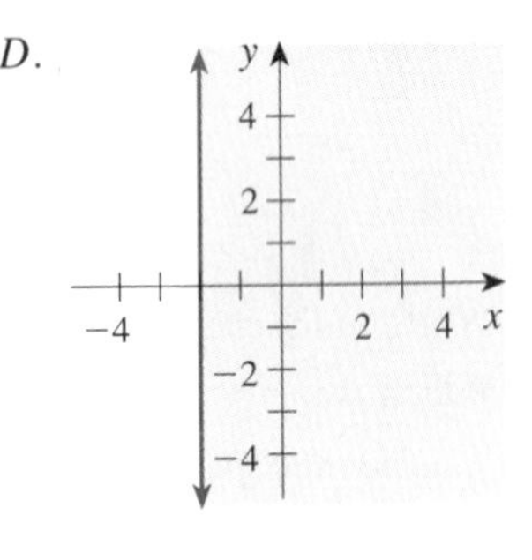

41. Which is the graph of $5x + 2y > 8$?

42. Which is the graph of $5x - 2y \leq 8$?

43. Which is the graph of $y \geq -2$?

44. Which is the graph of $x \geq -2$?

Graph the solution to each first-degree inequality in two variables.

45. $3x - 2y \leq -6$ **46.** $5x - 2y > 8$

47. $y > \frac{2}{3}x + 3$ **48.** $y \leq \frac{5 - 2x}{3}$

49. The graph of the solution to an equation in two variables is a straight line that crosses the y-axis at $(0, -2)$ and has slope measure 3. The equation is

i. $y = -2x + 3$ **ii.** $y - 3x = -2$

iii. $3x + y = 2$ **iv.** $-3x - y = 2$

50. The graph of the solution to an equation in two variables is a straight line that crosses the y-axis at $(0, 4)$ and has slope measure -3. The equation is

i. $y = 4x + 3$ **ii.** $3y - x = -12$

iii. $9x + 3y = 12$ **iv.** $3x - y = 12$

For the lines shown in the graph, list their slope measures in order from largest to smallest.

51. **52.**

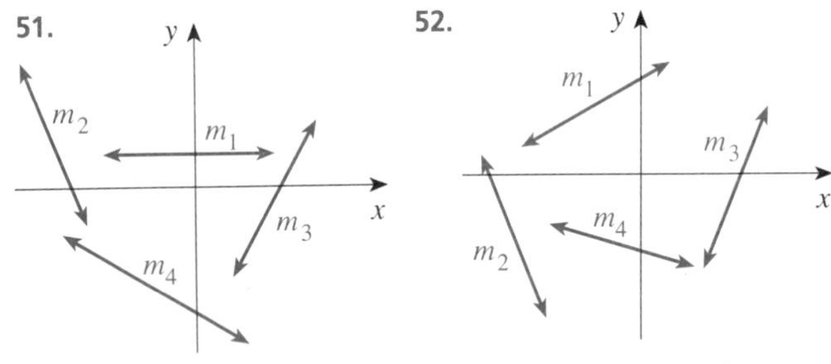

53. The cost of a bicycle rental is linearly related to the number of hours the bicycle is used. The linear relationship is given by $C = 6x + 10$.

a. Graph this linear relationship.

b. What is the cost of renting the bicycle for 5 hours?

c. If the cost of the rental is \$19, for how many hours was the bicycle used?

d. If the domain of this relation is the set of x-values from 0 to 14, what is the range of the relation?

54. Use the graphics calculator to graph each relation. View the graph, then report the domain and the range of the relation.

a. $y = |x - 2| + 1$ **b.** $y = \sqrt{(x + 2)}$

55. For the linear function $f(x) = 3x - 5$, find each output value.

a. $f(-4)$ **b.** $f(3)$ **c.** $f(0)$ **d.** $f\left(\frac{1}{3}\right)$

56. For the linear function $f(x) = -1.2x + 6$, find each output value.
   a. $f(-2)$
   b. $f(5)$
   c. $f\left(\frac{-1}{4}\right)$
   d. $f(0)$

57. For the linear function $g(x) = \frac{-2}{3}x + 2$,
   a. An input of -3 produces what output value?
   b. What input value produces an output of -1?
   c. Graph the function.

58. For the linear function $h(x) = 2x - 1$,
   a. An input of 3 produces what output value?
   b. What input value produces an output of 5?
   c. Graph the function.

59. The price (in dollars) of a box of cat food as a linear function of the weight (in ounces) of the box is $P(x) = 0.1x + 2.9$.
   a. What is the price of a 48-ounce box?
   b. If the price of a box is $5, what is the weight of the box?
   c. Graph this linear function for x-values from 0 to 100 ounces.

60. A fast-food chain finds that its selling price P (in dollars) for a 16-ounce burger is a linear function of the amount of filler, x (in ounces) and is given by $P(x) = -0.5x + 5$.
   a. If 6 ounces of filler is used, what is the selling price of a burger?
   b. If the selling price of a burger is $3, how many ounces of filler are used?
   c. Graph this linear function for a domain of [0, 9].

CHAPTER 3 TEST

Use the graphics calculator to check your work.

1. a. Graph the solution to $5x - 2y = 7$.
   b. The y-intercept is (____, ____).
   c. The x-intercept is (____, ____).
   d. The slope of the line is $m =$ ____.

2. a. Rewrite $-2x + 3y = 9$ for y.
   b. Use the graphics calculator with the range settings* given in the figure to graph the solution to the equation in part (a).
   c. Use the TRACE feature to complete the table:

```
RANGE
Xmin=-9.6
Xmax=9.4
Xscl=1
Ymin=-6.4
Ymax=6.2
Yscl=1
Xres=1
```

x	-4.8	-1	2	5.2
y				

   d. Draw a set of axes and plot the ordered pairs from the table in part (c). Draw a straight line through the points.

3. What is the slope of the line through the points $(-2, 5)$ and $(4, -3)$?

4. a. Draw a set of axes and plot the point $(-1, 2)$.
   b. Draw the line through the point in part (a) that has slope measure $\frac{2}{5}$.
   c. What is the ordered-pair name of another point on the line drawn in part (b)?

5. Graph the solution $3x - 2y \leq 5$. Plot and label the x- and y-intercepts of the key line for the graph.

6. The position P of a particle on the number line is linearly related to time t. The relation is defined by $P = -3t + 5$.
   a. Where is the particle at time $t = 2$?
   b. When is the particle at $P = 2$?
   c. For a domain of $\{t \mid 0 \leq t \leq 10\}$, report the range of the relation.

7. Consider the relation $\{(-2, 3), (0, 1), (2, -1), (3, 4)\}$.
   a. Report the domain.
   b. Report the range.

8. Consider the function $f(x) = \frac{5}{2}x - \frac{7}{2}$.
   a. Find $f(-4)$.
   b. What input value produces an output of 10?

9. The daily cost C of a car rental is a linear function of the number of miles driven, x. The function is defined by $C(x) = 0.3x + 10$.
   a. Find $C(100)$.
   b. If the cost is $100, how many miles were driven?

10. Use the graphics calculator to view the graph of $g(x) = x^2 - 1$.
   a. Is this a linear function?
   b. Report the range of the function.

TI-82 Note *Type `-9.4` for `Xmin` and `-6.2` for `Ymax`.

4

APPLICATIONS OF LINEAR EQUATIONS

In Chapter 3, we learned about linear equations and how to graph the solution to such an equation. In this chapter we reverse the process—that is, we find a linear equation when given information about its solution. Additionally, we will use graphs to learn about coordinate geometry. Finally, we will study systems of linear equations and applications of systems.

4.1 ■ ■ ■ FINDING THE EQUATION OF A LINE

So far, we have learned how to graph a linear relation, such as the one determined by $3x - 5y = 10$. The graph is always a straight line. The slope is the defining characteristic of a line. Also of importance in graphing a line are the x-intercept and the y-intercept. In fact, you have learned that when a first-degree equation in two variables is written for y, such as $y = 3x + 2$, the x-coefficient is the slope measure and the constant term yields the y-intercept. These ideas will be used in this section as we will learn to find the equation of a line. That is, given some information about a line, we will find the equation whose graph produces the line. This concept will then be applied to finding linear functions. In Section 4.2, we will apply this skill to topics in coordinate geometry.

Background When we learned to solve a first-degree equation in one variable (see Section 2.2), we had no corresponding reverse process like the following.

Problem Find the first-degree equation in one variable whose solution is 5.

One reason for this lack of procedure is that the solution to the problem is very simple: one such equation is $x = 5$. This is not the case with first-degree equations in two variables. Consider the following problem.

Problem Find the first-degree equation in two variables such that the ordered pairs $(-3, 4)$ and $(2, 1)$ solve the equation.

WARNING

The key to solving these problems is finding the correct slope measure. Sketch a graph, and observe whether the line is increasing or decreasing to find the correct sign for the slope.

This is not a trivial problem, but it is not overly difficult (although it does require some concentration). A starting point in solving this problem is a graph. The first-degree equation in two variables that has the ordered pairs $(-3, 4)$ and $(2, 1)$ in its solution has as a graph the line passing through the points $(-3, 4)$ and $(2, 1)$. That is, the equation we seek is the equation of the line through the two points. This line is shown in Figure 1.

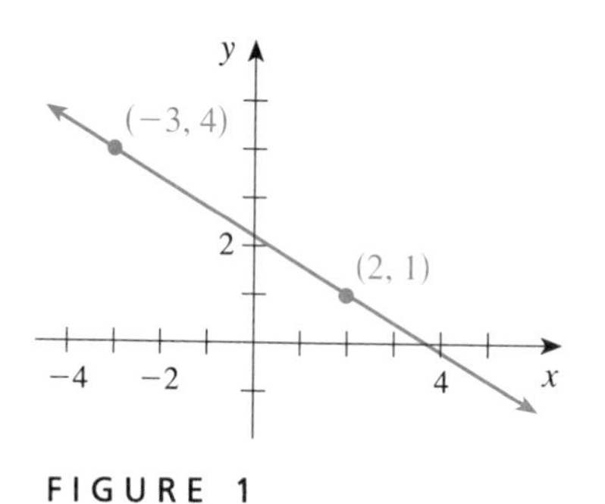

FIGURE 1

To find the equation of this line, or the first-degree equation in two variables whose solution contains the ordered pairs $(-3, 4)$ and $(2, 1)$, we must remember that the equation of a nonvertical line has the form

$$y = mx + b$$

In this slope, y-intercept form, the x-coefficient m is the slope measure and the constant term b yields the y-intercept $(0, b)$. Given the two ordered pairs in the solution, we can determine the slope measure of the line. First, we know that the slope measure is negative (observe the graph of the line shown in Figure 2), because the line is decreasing. Next, the change in x, or Δx, is 5 and the change in y, that is, Δy, is 3. Thus, $m = \frac{-3}{5}$.

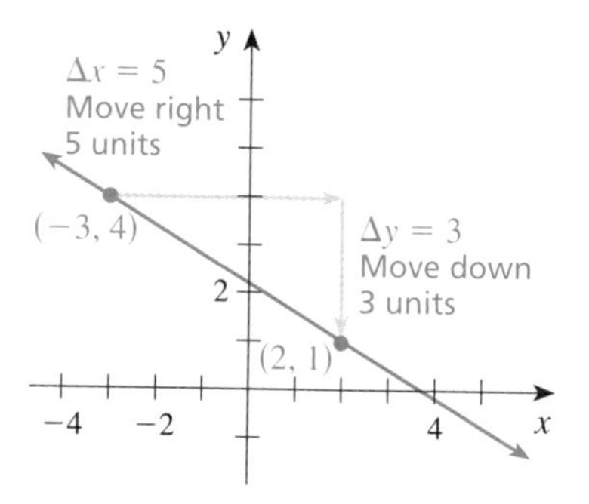

FIGURE 2

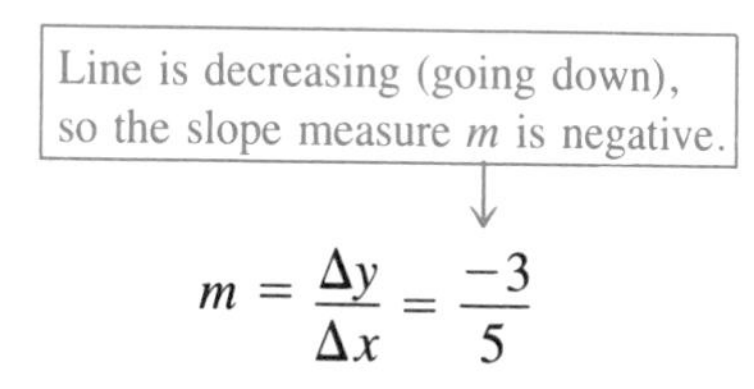

$$m = \frac{\Delta y}{\Delta x} = \frac{-3}{5}$$

We could also find this number using the formula for the slope:

$$m = \frac{y_2 - y_1}{x_2 - x_1} = \frac{4 - 1}{-3 - 2} = \frac{3}{-5} = \frac{-3}{5}$$

So far, we have $y = \frac{-3}{5}x + b$. Once we have found the value for b, we will have the equation of the line. It appears from the graph that b is a little more than 2. To find the exact value for the variable b, we use the fact that the ordered pair $(2, 1)$ must solve the equation. That is, whatever value replaces b in the equation $y = \frac{-2}{3}x + b$,

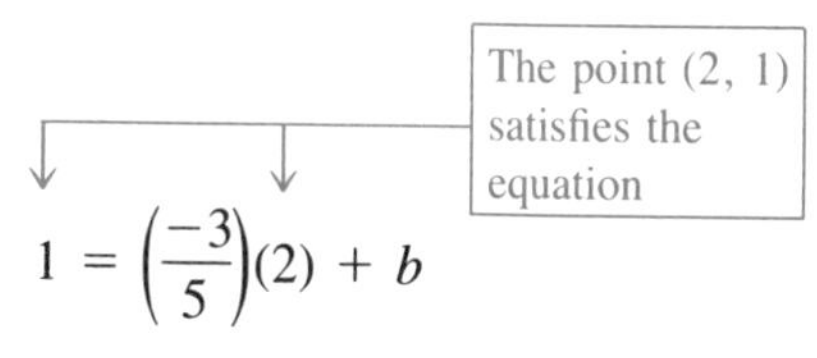

$$1 = \left(\frac{-3}{5}\right)(2) + b$$

must yield a true arithmetic sentence. Solving this equation, we get:

$$1 = \left(\frac{-3}{5}\right)(2) + b$$

Do arithmetic: $1 = \frac{-6}{5} + b$

Multiply each side by 5: $5 = -6 + 5b$

Add 6 to each side: $\underline{6 \qquad 6}$

Multiply each side by $\frac{1}{5}$: $11 = 5b$

$$\frac{11}{5} = b$$

Thus, the value for b must be $\frac{11}{5}$, and the equation of the line is

$$y = \frac{-3}{5}x + \frac{11}{5}$$

Using the ordered pair $(-3, 4)$ rather than the ordered pair $(2, 1)$ to find b produces the same value for b. That is, $4 = \left(\frac{-3}{5}\right)(-3) + b$ also has solution $\frac{11}{5}$.

Check The equation $y = \frac{-3}{5}x + \frac{11}{5}$ can be rewritten to standard form, $Ax + By = C$ as follows:

$$y = \frac{-3}{5}x + \frac{11}{5}$$

Multiply each side by 5 to get: $5y = -3x + 11$

Add $3x$ to each side to get: $\underline{3x \qquad 3x}$ $\quad 3x + 5y = 11$

In this form, it is easier to see that we do have the desired equation.

For $(-3, 4)$: $3(-3) + 5(4) = 11$ is a true arithmetic sentence.
For $(2, 1)$: $3(2) + 5(1) = 11$ is a true arithmetic sentence.

We can also find the equation of a line if we are given just one point and the slope measure. The following example shows how we do this.

EXAMPLE 1 Find the equation of the line through the point $(-3, -1)$ with slope measure $\frac{3}{7}$.

Solution The place to start is by drawing a graph. Consider the one shown in Figure 3. The equation of this line has the form

$$y = mx + b$$

Knowing that $m = \frac{3}{7}$ yields the equation

$$y = \frac{3}{7}x + b$$

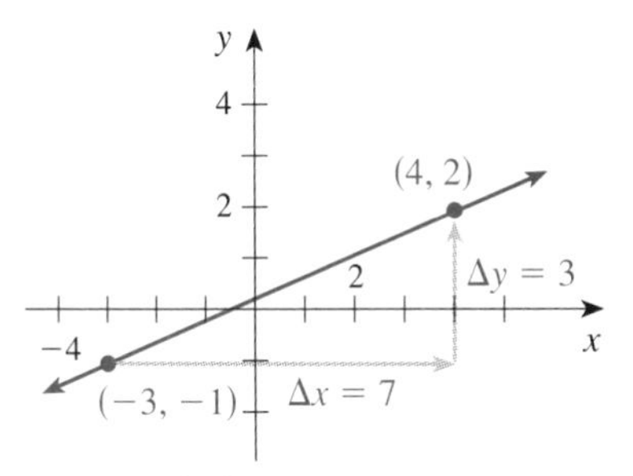

FIGURE 3

Slope measure is $\frac{3}{7}$, so the line is increasing.

Thus, we need to find only the value of b. That is, we need to find the value where the line crosses (intercepts) the y-axis. By observing the graph, we can see that this number appears to between 0 and 1. To find the exact value, we use the fact that the ordered pair $(-3, -1)$ must solve the equation. That is, whatever the value replacing b *in* $y = \frac{-3}{7}x + b$, the equation $-1 = \left(\frac{-3}{7}\right)(-3) + b$ must be a true arithmetic sentence.

Solve:	$-1 = \left(\frac{3}{7}\right)(-3) + b$
Do arithmetic:	$-1 = \frac{-9}{7} + b$
Multiply each side by 7:	$-7 = -9 + 7b$
Add 9 to each side	$\underline{9 \qquad 9}$
to get:	$2 = 7b$
Multiply each side by $\frac{1}{7}$	$\frac{2}{7} = b$

Thus, the equation of the line is $y = \frac{3}{7}x + \frac{2}{7}$.

Check Does this equation produce a line with slope $\frac{3}{7}$? Yes, the x-coefficient is $\frac{3}{7}$.

Does this line pass through the point $(-3, -1)$? Yes. Replacing x with -3 and y with -1 yields:

$$-1 = \left(\frac{3}{7}\right)(-3) + \frac{2}{7} \quad \text{a true arithmetic sentence.}$$

■

Using the Graphics Calculator

We can quickly check our work in Example 1 by using the graph screen. Erase any existing entries on the `Y=` screen. For `Y1`, type `(3/7)X+(2/7)`. Press the `RANGE` key, enter the values given in Figure 4*, and view the graph screen as shown in Figure 5. We know that the line has slope measure $\frac{3}{7}$ (from the x-coefficient). Use the `TRACE` key to verify that the line passes through the point $(-3, -1)$.

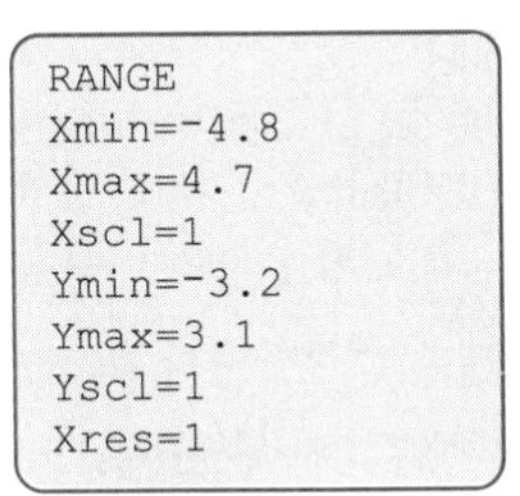

FIGURE 4

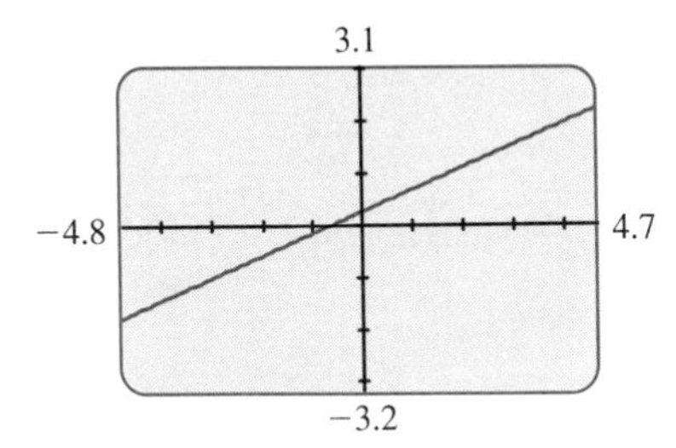

FIGURE 5

TI-82 Note *Use `ZOOM-option 4` for `ZDecimal`.

Consider another example of finding the equation of a line.

EXAMPLE 2 Find the equation of the line through the point $(0, -2)$ with slope $\frac{3}{2}$.

Solution We start by drawing a graph, using the given information. The graph is shown in Figure 6.

Now, the equation of a line has the form $y = mx + b$. Because the slope is $\frac{3}{2}$, we have

$$y = \frac{3}{2}x + b$$

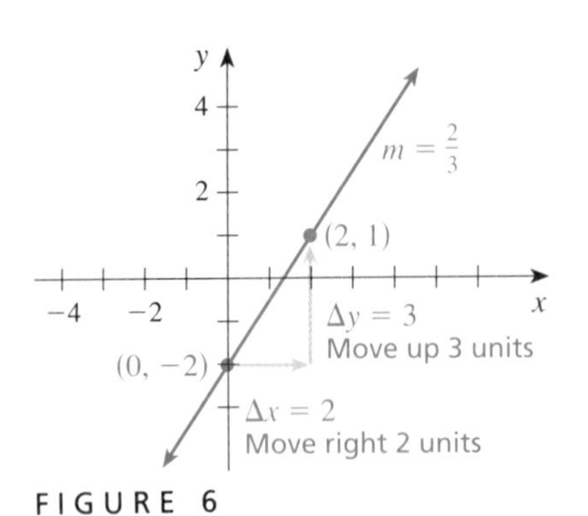

FIGURE 6

Now, the value for b yields the y-intercept. But in this problem, we already know that the y-intercept is $(0, -2)$. Thus, b is -2. So, the equation of the line is

$$y = \frac{3}{2}x - 2$$

■

In Example 2, we are given exactly the information needed to find the equation—the slope and the y-intercept.

GIVE IT A TRY

1. A line passes through the points $(-2, 5)$ and $(3, -1)$.
   - **a.** Plot these points, and draw the line through them.
   - **b.** Observe the graph from part (a). Is the line increasing or decreasing? Is the slope measure positive or negative?
   - **c.** What is the slope measure of the line?
   - **d.** The equation of every nonvertical line has the form $y = mx + b$. Knowing the slope measure of the line in part (c) gives us $y = \frac{-6}{5}x + b$. Now, the ordered pair $(-2, 5)$ must solve the equation. Solve the equation $5 = \left(\frac{-6}{5}\right)(-2) + b$, and report the solution.
   - **e.** Using the solution to the equation $5 = \left(\frac{-6}{5}\right)(-2) + b$ as the value for b [see part (d)], report the equation of the line through $(-2, 5)$ and $(3, -1)$.
   - **f.** Use the graphics calculator to check that the equation found in part (e) does indeed pass through the points $(-2, 5)$ and $(3, -1)$.
2. **a.** Graph the line through the point $(-2, 3)$ with slope 2. What is the equation of the line?
   - **b.** On the set of axes used in part (a), graph the line through $(-2, 3)$ with slope 0. What is the equation of this line?
3. Graph the line through $(0, 3)$ with slope -2. What is the equation of this line?

Finding Linear Functions We now learn how to find the rule for a linear function given information (data) about the function. The following example shows how we can do this.

EXAMPLE 3 Find the linear function such that $f(-2) = 4$ and $f(3) = 1$.

Solution Recall that the notation $f(-2) = 4$ means that for an input of -2, the output is 4. As an ordered pair, $f(-2) = 4$ yields $(-2, 4)$. Likewise, $f(3) = 1$ yields the ordered pair $(3, 1)$. Graphing the desired linear function produces the graph in Figure 7. Now, the linear function will have the form $f(x) = mx + b$, where m is the slope measure of the line and b yields the y-intercept $(0, b)$.

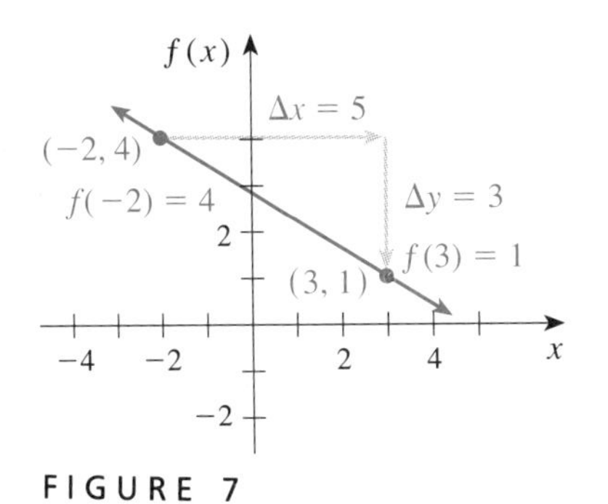

FIGURE 7

Looking at the graph, we see that the line is decreasing, so the slope measure is negative. For the two ordered pairs we have, Δy is 3 and Δx is 5. Thus, the slope measure, m, is $\frac{-3}{5}$:

$$m = \frac{\Delta y}{\Delta x} = \frac{y_2 - y_1}{x_2 - x_1}$$

$$= \frac{1 - 4}{3 - (-2)} = \frac{-3}{5}$$

At this point, we know the linear function is $f(x) = \frac{-3}{5}x + b$. To find the value for b, we use the data from one of the function outputs, $f(3) = 1$. That is, we need a value for b such that

$$1 = \left(\frac{-3}{5}\right)(3) + b$$

Solving this equation yields:

$$\begin{aligned} 1 &= \frac{-9}{5} + b & \\ 5 &= -9 + 5b & \text{Multiply by 5.} \\ \underline{9} & \quad \underline{\;9\qquad\qquad} & \\ 14 &= 5b & \text{Add 9 to each side.} \\ b &= \frac{14}{5} & \text{Multiply by } \tfrac{1}{5}. \end{aligned}$$

Finally, we write the linear function:

$$f(x) = \frac{-3}{5}x + \frac{14}{5}$$

■

Using the Graphics Calculator

We can check our work in Example 3 by typing `(-3/5)X+(14/5)` for `Y1`. Enter the `RANGE` values shown in Figure 8*, and view the graph screen. Use the `TRACE` key to confirm that $(-2, 4)$ and $(3, 1)$ are on this line.

```
RANGE
Xmin=-9.6
Xmax=9.4
Xscl=1
Ymin=-6.4
Ymax=6.2
Yscl=1
Xres=1
```

FIGURE 8

GIVE IT A TRY

4. Find the linear function such that $f(-3) = 5$ and $f(4) = 1$.

TI-82 Note *Type `-9.4` for `Xmin` and `-6.2` for `Ymin`.

Applications We now consider a "word problem" that involves finding the equation of a line. Here, we will be working with a linear relationship between two variables. To solve this type of problem, we first read the story to find the needed data—two ordered pairs, or a point and the slope. Work through the following example.

EXAMPLE 4 The daily cost C of a car rental is linearly related to the number of miles driven, x. The daily cost of the rental is \$33 when 50 miles are driven and the daily cost is \$43 when 100 miles are driven. Find the linear relationship between C and x.

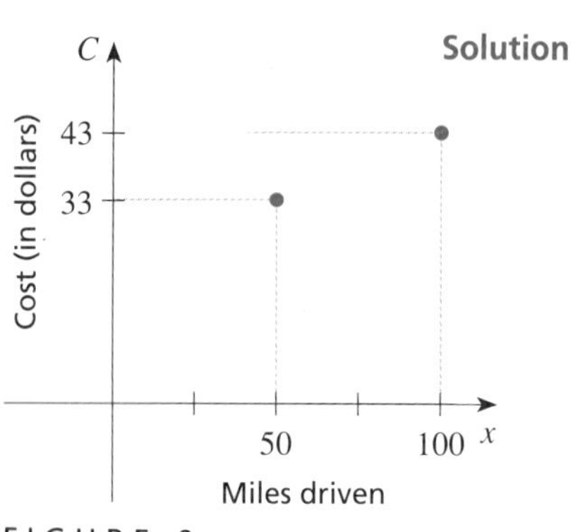

FIGURE 9

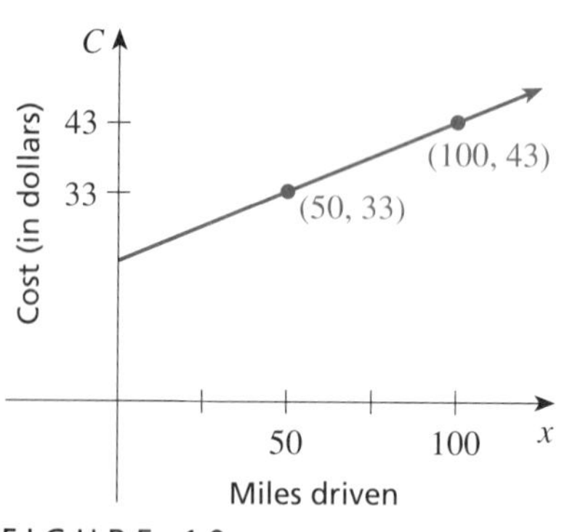

FIGURE 10

Solution As usual, we start by sketching a graph. It does not matter which variable is represented by which axis, so we will let the horizontal axis represent x, the number of miles driven. The vertical axis will be used for the other variable, cost C. Once we have made this decision, we must use this order for all our following work. Plotting the data on the set of axes, as in Figure 9, we get the ordered pairs (50, 33) and (100, 43). Because the relationship between C and x is linear, we draw in the straight line as in Figure 10. The slope measure of this line is

$$m = \frac{43 - 33}{100 - 50} = \frac{1}{5}$$

Using $m = \frac{1}{5}$, we have the equation $\quad C = \frac{1}{5}x + b$

Using (50, 33) to find the value of b, we get $\quad 33 = \frac{1}{5}(50) + b$

Solving this equation, we get $\quad b = 23$

So, the linear relationship between C and x is $C = \frac{1}{5}x + 23$. ■

In the equation we found in Example 4, $C = 0.2x + 23$, the x-coefficient 0.2 represents the *cost per mile driven*. The constant term 23 is the *cost per day*. This is the **fixed cost**—if no miles are driven ($x = 0$), the daily rental cost is \$23.

To use the skill of finding the equation of a line to produce a linear function in another business situation, work through the following example.

EXAMPLE 5 A shipping company earns a revenue of \$15 from shipping a five-pound package and a revenue of \$23 for shipping a ten-pound package.

a. Find the revenue R as a linear function of the weight x of the package shipped.
b. What is the company's revenue from shipping a 40-pound package?

Solution a. We are seeking a linear function such that $R(5) = 15$ and $R(10) = 23$. Figure 11 shows the two points generated by these function values, along with the straight line drawn through the points.

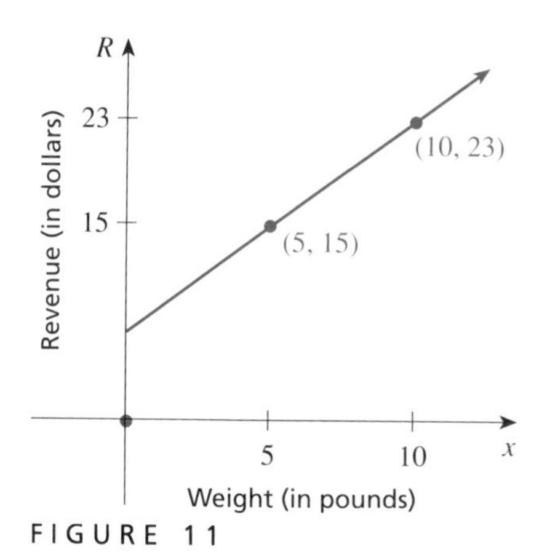

FIGURE 11

Now, we use the procedure for finding the equation of a line to produce the linear function. The function must have the form $R(x) = mx + b$.

The slope measure m for the line is

$$m = \frac{23 - 15}{10 - 5} \quad \text{or} \quad \frac{8}{5} \quad \text{or} \quad 1.6$$

In this situation, the slope measure represents the revenue per pound shipped, that is, the revenue earned is \$1.60 per pound. So far, we have

$$R(x) = 1.6x + b$$

To find b, or the intercept of the vertical axis, we use one of the data values, $R(5) = 15$. That is, b must be a value such that $15 = 1.6(5) + b$ is a true arithmetic sentence. Solving this equation yields:

$$\begin{aligned} 15 &= 1.6(5) + b \\ 15 &= \ \ 8 + b \\ -8 & \quad -8 \\ \hline 7 &= b \end{aligned}$$

Thus, the value for b is 7. So the revenue function is $R(x) = 1.6x + 7$. In this situation, the vertical axis intercept represents the ''flat'' shipping charge. That is, a flat rate of \$7 is charged for each item shipped, and an additional charge is calculated on the weight of the item.

b. The revenue earned from shipping a 40-pound package is $R(40)$. For this function,

$$\begin{aligned} R(40) &= 1.6(40) + 7 \\ &= 71 \end{aligned}$$

Thus, the revenue earned from shipping a 40-pound package is \$71. ■

Using the Graphics Calculator

We can check our work in Example 5 by typing `1.6X+7` for `Y1`. Press the `RANGE` key and enter the values shown in Figure 12. View the graph screen. Press the `TRACE` key and move the cursor as close as possible to an x-value of 5. Is the output value approximately 15? Move the cursor as close as possible to an x-value of 10. Is the output value approximately 23? Move the cursor as close as possible to an x-value of 40. Is the output value approximately 71?

```
RANGE
Xmin=0
Xmax=50
Xscl=10
Ymin=0
Ymax=100
Yscl=10
Xres=1
```

FIGURE 12

As we have seen in the preceding examples, finding a linear function from data given about the function is essentially the same process as finding the equation of a line when given two ordered pairs. This process is repeated over and over in business and the sciences. In these areas, the data is usually obtained from measurements, and then a linear function is produced. Of course, rather than using two ordered pairs, hundreds or even thousands of ordered pairs are used. The process is altered slightly (and is known as *linear regression*), but the result is the same—a linear function that describes the relationship between two variables.

GIVE IT A TRY

5. Sally's weekly take-home pay P is linearly related to the amount of her gross wages A. Her take-home pay is $188 when her gross wages are $200, and her take-home pay is $228 when her gross wages are $250. Graph the linear relationship between A and P (let the horizontal axis represent A). State the linear relationship as a function.
6. The temperature T of air is a linear function of the height above sea level, h. For a height of 3000 feet, the temperature is 56°F. For a height of 8000 feet, the temperature is 36°F.
   a. Graph the linear relationship.
   b. Find the linear function, that is, $T(h) = \quad h + \quad$
   c. At a height of 4500 feet, what is the temperature?
   d. At what height is the temperature 38°F?

Point-Slope Form of an Equation So far, we have seen that knowing how to use the form $y = mx + b$ is very productive in finding the linear equation that satisfies selected conditions. We now develop a convenient form for finding the equation of a line when the slope and a point are known. Suppose we have slope m and some point, say (x_1, y_1). Knowing that some general point on the line, say (x, y), must satisfy the formula for the slope (see page 155), we can find the slope from these two points:

$$m = \frac{y - y_1}{x - x_1} \qquad \text{or} \qquad \frac{y - y_1}{x - x_1} = m$$

Now, altering this formula by multiplying each side by the expression $(x - x_1)$, we get

$$\frac{\cancel{(x - x_1)}(y - y_1)}{\cancel{x - x_1}} = m(x - x_1) \quad \text{for } x \neq x_1$$

Removing the common factor, we get:

$$y - y_1 = m(x - x_1) \quad \text{for } x \neq x_1$$

This last formula is known as the **point-slope formula**. We now state this formally.

THE POINT-SLOPE FORMULA

> Given a point (x_1, y_1) and a slope measure m, the equation of the line through the point with the given slope measure is
> $$y - y_1 = m(x - x_1)$$

The next example demonstrates the convenience of the point-slope formula.

EXAMPLE 6 Find the equation of the line through the point $(-3, 2)$ with slope measure -2.

Solution We use the point-slope formula with $x_1 = -3$, $y_1 = 2$, and $m = -2$:

$$y - y_1 = m(x - x_1) \quad \text{point } (-3, 2)$$

$$y - 2 = -2[x - (-3)] \quad \text{slope } -2$$

$$y - 2 = -2(x + 3)$$

$$y - 2 = -2x - 6$$

$$\underline{\quad 2 \quad} \quad \underline{\quad\quad 2}$$

The equation is: $y = -2x - 4$ ■

Using the Graphics Calculator

To check our work in Example 6, type `-2X-4` for `Y1`. Press the `RANGE` key and type the values shown in Figure 13.* Use the `TRACE` key to confirm that the line does indeed pass through the point $(-3, 2)$.

```
RANGE
Xmin=-4.8
Xmax=4.7
Xscl=1
Ymin=-3.2
Ymax=3.1
Yscl=1
Xres=1
```

FIGURE 13

Collinear Points All our work in this section has used the fact that any two points in the xy-plane are *collinear,* that is, they lie on a line. What about three points? How can we determine if three points in the plane are collinear? Basically, we can find the equation of the line containing two of the points, and then check to see if the ordered-pair name of the third point satisfies this equation. Work through the following example.

EXAMPLE 7 Are the points $(-2, 3)$, $(1, -1)$, and $(3, -4)$ collinear? If so, find the equation of the line containing the three points.

Solution We start by drawing a graph of the situation. (This graph may save us a lot of work, since it may show immediately that the three points don't lie on a single line.) From the graph in Figure 14, it appears that the points might all be on a single line. Using the first two points, $(-2, 3)$ and $(1, -1)$, we find the slope measure:

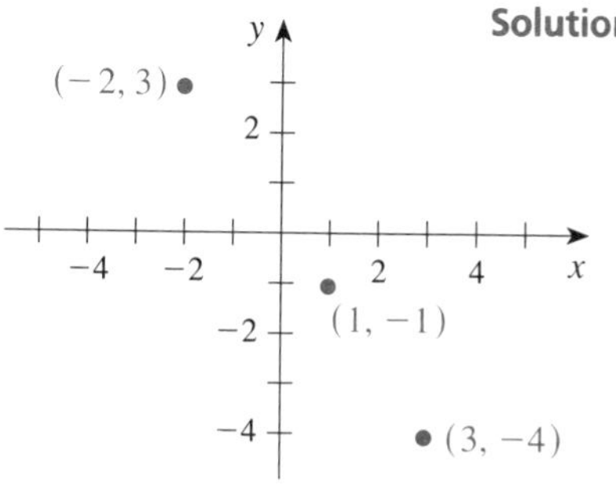

$$m = \frac{3 - (-1)}{-2 - 1}$$

$$= \frac{4}{-3} = \frac{-4}{3}$$

FIGURE 14

TI-82 Note *Use `ZOOM`-option `4` for `ZDecimal`.

Using the slope $\frac{-4}{3}$, the point $(-2, 3)$, and the point-slope form of an equation, we get

$$y - 3 = \left(\frac{-4}{3}\right)[x - (-2)]$$

$$y - 3 = \frac{-4}{3}x - \frac{8}{3}$$

$$y = \frac{-4}{3}x + \frac{1}{3} \qquad \text{Slope, } y\text{-intercept form}$$

$$3y = -4x + 1$$

$$4x + 3y = 1 \qquad \text{Standard form}$$

Now, we test the third point, $(3, -4)$, to see if it lies on this line. That is, does it satisfy the equation? We get $4(3) + 3(-4) = 1$, or $0 = 1$, a false arithmetic sentence. So, the point $(3, -4)$ does not lie on the line through the other two points. The three points are not collinear. ■

GIVE IT A TRY

7. Are the points $(-1, -3)$, $(2, 1)$, and $(5, 5)$ collinear? If so, find the equation of the line through the three points.

■ SUMMARY

■ ***Quick Index***

In this section, we have learned the basic process of finding the equation of a line. That is, given information about the line—such as two ordered pairs, or a point and the slope—we can derive the equation of the line. This is accomplished by knowing that the equation of a line can be written in either of these forms:

Slope, *y*-Intercept Form $\quad y = mx + b,$

where m is the slope measure and b yields the y-intercept, $(0, b)$

or

Point-slope Form $\quad y - y_1 = m(x - x_1)$

where m is the slope measure and x_1 and y_1 are the coordinates of some point (x_1, y_1) on the line.

Remember that in addition to these two forms of an equation, we can use the standard form of an equation, $Ax + By = C$.

The process of finding the equation of a line is also used to find the linear relationship between two variables or to find the linear function when given data concerning the variables. We also used this process to determine whether three points are collinear, that is, we determined if all three points lie on a straight line.

4.1 ■ ■ ■ EXERCISES

1. **a.** Graph the line determined by the first-degree equation whose solution contains the ordered pairs $(-3, 1)$ and $(5, 3)$.
   b. Is the line drawn in part (a) increasing or decreasing? Is the slope of the line positive or negative?
   c. For the line drawn in part (a), find Δy and Δx. What is the slope measure?
   d. Find the equation of the line drawn in part (a).
   e. Report the ordered pair in the solution to the equation found in part (d) with x-coordinate 3.
   f. Report the ordered pair in the solution to the equation found in part (d) with y-coordinate 3.
2. Draw a graph of the line through the points $(-5, 4)$ and $(2, -1)$. What is the equation of the line? For this line, report the name of the point with x-coordinate 0.
3. Draw a graph of the line through the point $(-2, 5)$ with slope measure $\frac{2}{3}$. What is the equation of the line? For this line, find the name of the point with x-coordinate 1.
4. Draw a graph of the line through the point $(-2, 1)$ with slope measure $\frac{-5}{3}$. What is the equation of the line? For this line, report the name of the point with y-coordinate 0.
5. Draw a graph of the line through the origin with slope measure 1. What is the equation of the line? For this line, find the name of the point with x-coordinate -3.
6. Find the equation of the line through $(-1, 5)$ and $(3, -4)$. Using the graphics calculator, graph the solution to the equation. Use the TRACE and arrow keys to confirm that the line does indeed pass through the points.

Find the equation of the line that satisfies the given conditions. Use the graphics calculator to check your work.

7. slope $= -3$ and line passes through $(-2, 5)$
8. slope $= 2$ and line passes through $(0, 6)$
9. slope $= \frac{-2}{3}$ and line passes through $(2, -1)$
10. slope $= \frac{4}{5}$ and line passes through $(-5, -2)$
11. slope $= \frac{-1}{2}$ and line passes through $(-1, -4)$
12. slope $= \frac{-5}{2}$ and line passes through $(4, -2)$
13. line passes through $(4, 1)$ and $(-1, 3)$
14. line passes through $(-2, 5)$ and $(2, 0)$
15. line passes through $(-4, 5)$ and $(3, 1)$
16. line passes through $(-3, -2)$ and $(5, 1)$
17. **a.** Graph the linear function such that $f(0) = 5$ and $f(4) = -1$.
    b. What is the linear function in part (a)?
    c. For this function, find $f(-2)$.
    d. For this function, an input value of 2 produces what output value?
    e. For this function, to get an output of 2.5, what must be the input value?
18. **a.** Graph the linear function such that $f(-2) = -2$ and $f(4) = 5$.
    b. What is the linear function in part (a)?
    c. For this function, find $f(0)$.
    d. For this function, an input value of 3 produces what output value?
    e. For this function, to get an output of -1, what must be the input value?

Graph the linear function that satisfies the conditions given in Problems 19–26. Find the rule for the function. Use the graphics calculator to check your work.

19. $f(-4) = 8$ and $f(-1) = 2$
20. $h(0) = 3$ and $h(5) = 1$
21. $g(-4) = -3$ and $g(5) = 2$
22. $n(4) = 8$ and $n(-1) = -2$
23. An input of 5 produces an output of 6, and an input of 1 produces an output of 2.
24. An input of -2 produces an output of -5, and an input of 4 produces an output of -1.
25. An input of -7 produces an output of 6, and an input of 3 produces an output of 6.
26. An input of -2 produces an output of -7, and an input of 1 produces an output of 4.
27. The cost C of producing plastic tubs is linearly related to the number of tubs produced, n. The production cost is $178 for 28 tubs and $193 for 33 tubs.
    a. Graph the given data, using n as the horizontal axis.
    b. State the linear relationship as a function.
28. The monthly income M of an executive is linearly related to x, the sales for the month. The income is $3,000 when sales are $50,000, and the income is $4,100 when sales are $70,000.
    a. Graph this data, using the horizontal axis for sales, x.
    b. State the linear relationship as a function.

29. The position x of a particle moving along the x-axis is linearly related to the time t. The particle is at $x = 3$ at time $t = 0$, and its position is $x = -2$ at time $t = 4$.
 a. Graph this data, using the horizontal axis for time t.
 b. Report the linear relationship as a function.

30. Julia has the following data on the Cordoba store she opened in 1982 (year 1): In 1983 (year 2), her profit was $-\$10,000$. In 1992 (year 11), her profit was \$83,000. From her books, she has determined that her profit P on the store is linearly related to the number of years n the store has been open.
 a. Graph the linear relation, using the horizontal axis for n.
 b. State the linear relationship as a function.
 c. Based on the equation found in part (b), predict Julia's profit for 1996.
 [*Hint:* The year 1996 is year 15.]

31. The daily cost of a car rental is \$78 when 150 miles are driven, and the daily cost of the rental is \$98 if 250 miles are driven. The daily cost C is a linear function of the miles driven x.
 a. Graph the linear relationship.
 b. What is the linear function?
 c. Find $C(200)$.
 d. If 80 miles are driven, what is the daily rental cost?

32. a. Given that 0°C is the same temperature as 32°F, and that 100°C is 212°F, graph the linear function for C as a function F, that is, $C(F)$.
 b. Find $C(78)$.
 c. What is the input value that results in an output of 50?

33. A local furniture outlet finds that when \$1,000 is spent on newspaper advertisements, their sales are \$90,500, and when \$1,500 is spent on newspaper advertisements, their sales are \$95,800. If sales S is a linear function of the dollars x spent of advertising, what is the function? Using this linear function, what sales would be expected if \$1,200 is spent on newspaper advertisements?

34. The cost C per unit produced for a whiz-bat is a linear function of the number of whiz-bats produced, x. When 100 whiz-bats are produced the cost per unit is \$5.80, and when 500 are produced the cost per unit is \$2.20. Report C as a linear function of x.

In Problems 35–37, use the definition of collinear points—three points are collinear if all three lie on the same line.

35. Find the equation of the line through $(-3, 5)$ and $(4, 2)$. Is the point $(11, -1)$ on this line? Are the points $(-3, 5)$, $(4, 2)$, and $(11, -1)$ collinear?

36. Are the points $(-3, 6)$, $(2, 3)$, and $(5, -2)$ collinear? If so, report the equation of the line through the three points. If not, give the equation of the line containing the first two points, and give the equation of the line containing the last two points.

37. Are the points $(-3, -4)$, $(0, -2)$, and $(5, 1)$ collinear? If so, give the equation of the line containing the points. If not, give the equation of the line containing the first two points and give the equation of the line containing the last two points.

DISCOVERY

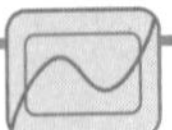

Parallel and Perpendicular Lines

We will now use exploration and the graphics calculator to discover a generalization about the graphs of first-degree equations in two variables.

```
RANGE
Xmin=-9.6
Xmax=9.4
Xscl=1
Ymin=-6.4
Ymax=6.2
Yscl=1
Xres=1
```

FIGURE 1

Parallel Lines We start by graphing the solution to the equation $y = 2x$. Type `2X` for `Y1`. Then view the graph using the settings shown in Figure 1.* The graph is shown in Figure 2. We now want to see the effect on this graph of adding a constant number to the right-hand side of the equation. That is, we want to graph $y = 2x + A$ if A is any real number. For `Y2`, type `2X+3` and view the graph screen (see Figure 3). Do you have an idea about the

TI-82 Note *Type `-9.4` for `Xmin` and `-6.2` for `Ymin`.

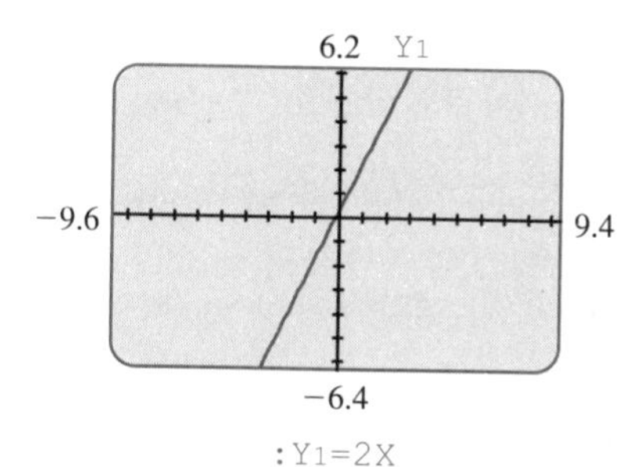

FIGURE 2

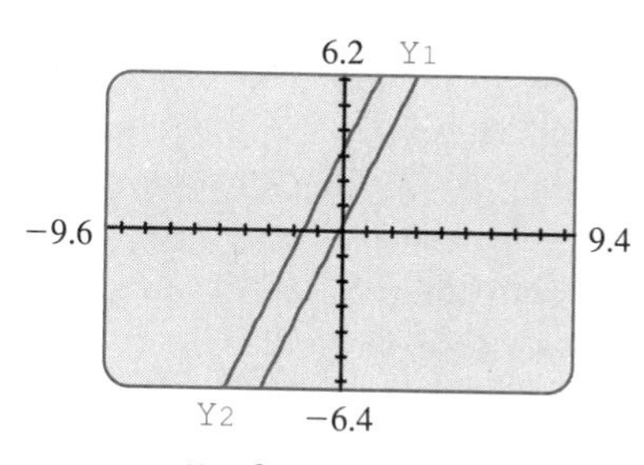

FIGURE 3

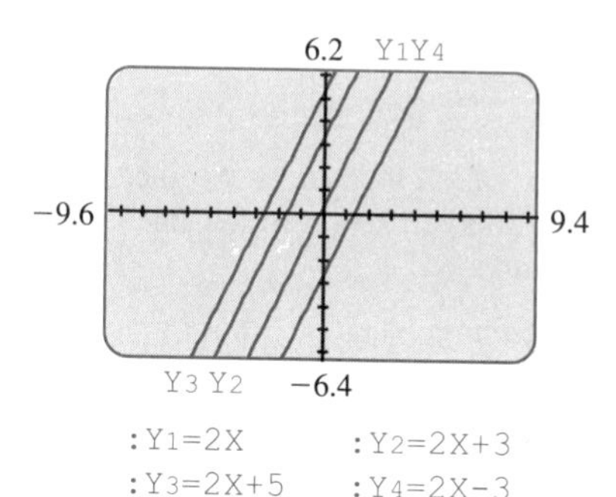

FIGURE 4

effect of adding a number to the right-hand side of the equation? For Y3, type 2X+5, and view the graph. Does this graph confirm your idea?

What is the effect on the graph when a negative number is added to the right-hand side of $y = 2x$? To see if you are right, graph the solution to an equation such as $y = 2x - 3$: For Y4, type 2X−3. View the graph screen, as shown in Figure 4. These lines are known as *parallel lines*. They are always the same distance apart. That is, adding a number to the right-hand side of the equation $y = 2x$ simply moves the line up or down that many units. Notice that the slope of each line is $\frac{2}{1}$.

From this activity, we have discovered that adding a constant, say A, to the right-hand side of $y = 2x$ alters the graph by moving the line A units, up or down. That is, a line parallel to the original line is produced. How is the graph of $y = -3x$ related to the graph of $y = -3x + B$, where B is any real number? Will each value of B produce a parallel line? Use the graphics calculator to graph the solutions to $y = -3x$, $y = -3x + 2$, and $y = -3x - 5$. What is the slope measure of each of these three lines?

In your own words, write down the condition for which the line determined by $y = mx + b$, where m and b are real numbers, is parallel to the line determined by $y = cx + d$, where c and d are real numbers.

Perpendicular Lines We next consider the line determined by $y = 3x$ and a line perpendicular to this line. Two lines are *perpendicular* if they intersect and form a right, or 90°, angle. The slope of the line determined by $y = 3x$ is $\frac{3}{1}$. Now, a line perpendicular to a line with a positive slope should have a negative slope. Consider the line determined by $y = -3x$. Do the lines $y = 3x$ and $y = -3x$ intersect at a right angle? On the graphics calculator, erase the existing entries on the Y= screen. For Y1 type 3X, and for Y2 type ⁻3X. View the graph screen using the current range settings. See Figure 5.

The lines in Figure 5 do not appear to be perpendicular. They aren't. Press the Y= key and type ⁻(1/3)X for Y3. View the graph screen. See Figure 6. Do the graphs for Y1 and Y3 appear to be perpendicular? They are!

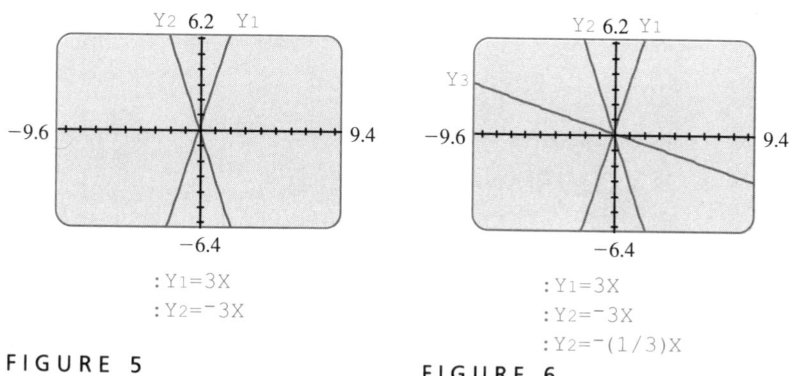

FIGURE 5

FIGURE 6

Erase the Y= entries. For Y1, type 2X+1. For Y2, type ⁻(1/2)X+1. Do you think these lines will be perpendicular? View the graph screen using the current range settings. For Y3, type ⁻(1/2)X−5. View the graph screen, as shown in Figure 7. What is the relationship between the line for Y1 and the line for Y3? What is the relationship for the line for Y2 and the line for Y3?

Do you think you know what condition determines whether two lines are perpendicular? Erase the entries on the Y= screen. For Y1, type (2/3)X+3. For Y2, type -(3/2)X−1. View the graph screen, (see Figure 8). Again, the lines are perpendicular.

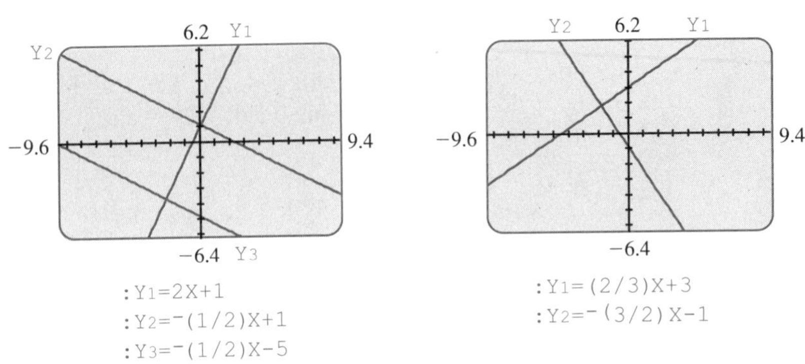

FIGURE 7

FIGURE 8

Exercises

1. In your own words, write down the condition for which the line determined by $y = mx + b$, where m and b are real numbers, is perpendicular to the line determined by $y = cx + d$, where c and d are real numbers.
2. For the line determined by $y = \frac{-3}{4}x + 2$, write a linear equation whose graph will be a line perpendicular to the given line. Use the graphics calculator to check your answer.
3. For the line determined by $y = \frac{-3}{5}x + 3$, write a linear equation whose graph will be a line parallel to the given line. Use the graphics calculator to check your answer.

4.2 ■ ■ ■ COORDINATE GEOMETRY

So far, we have learned that the solution to a first-degree equation in two variables is a collection of ordered pairs. When these ordered pairs are plotted on a set of axes, the graph is a line. We have also learned to reverse the process in order to find the equation of a line when given information about the line, such as its slope measure and a point on the line. In this section, we learn to use the coordinate system to accomplish various tasks in geometry—finding the length of a line segment and finding the midpoint of a line segment. Also, we will discuss the relationships between the slope and parallel lines and the slope and perpendicular lines. Finally, we will find the equation of the perpendicular bisector of a line segment.

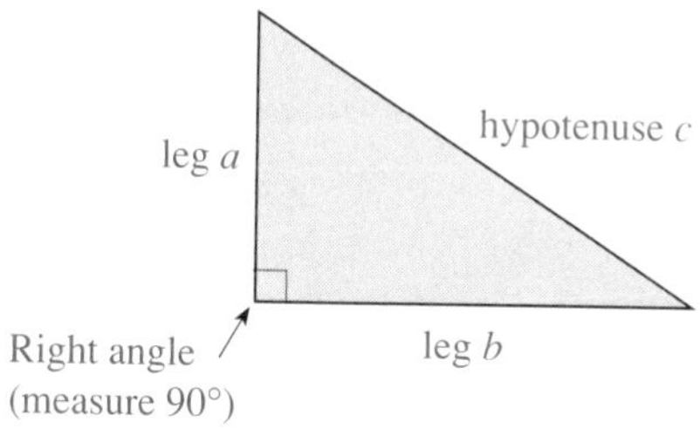

Line Segments A line through two points extends indefinitely and so, has no length measure (or, we can say it has infinite length measure). A **line segment** is the collection of points between two points, say A and B. We say that A and B are the **endpoints** of the line segment. The segment is denoted as $\overline{AB}$, whereas the line through points A and B is often denoted $\overleftrightarrow{AB}$. We can find the length of a line segment if given the coordinates of the endpoints. To do this, we use the Pythagorean theorem: in a right triangle, the side opposite the right angle (the hypotenuse) squared is equal to the sum of the squares of the legs. This is often written as $c^2 = a^2 + b^2$, where c represents the hypotenuse and a and b represent the legs. Figure 1 illustrates a right triangle with the parts labeled.

FIGURE 1
Pythagorean theorem: $a^2 + b^2 = c^2$

Distance between Two Points Finding the length of a line segment, or finding the distance between two points, is shown in the following example.

EXAMPLE 1 Given the points $A = (-3, 4)$ and $B = (2, 1)$, find the length of $\overline{AB}$.

Solution As is true for most problems dealing with ordered pairs, the place to start to solve this problem is to draw a graph. Figure 2 shows the graph of line segment AB.

Next, we draw a vertical line through point A and a horizontal line through point B. Figure 3 shows these lines.

Because a vertical line and a horizontal line form a right angle—the lines are perpendicular—a right triangle is formed, as shown in Figure 3. Line segment AB is the hypotenuse of this right triangle. One leg of the triangle has length Δy, and the other leg has length Δx.

Now, for the points $(-3, 4)$ and $(2, 1)$, the change in y, or Δy, is 3 and the change in x, or Δx, is 5. Redrawing the right triangle without the axes yields the right triangle shown in Figure 4.

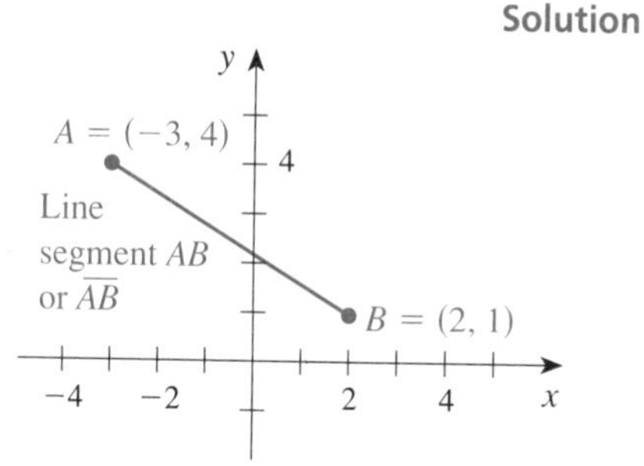

FIGURE 2

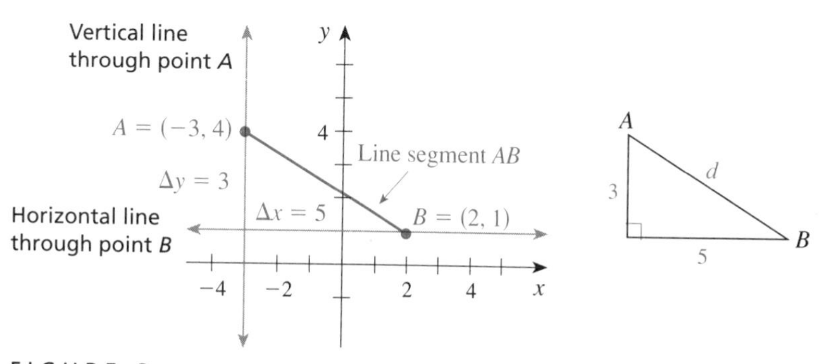

FIGURE 3

FIGURE 4

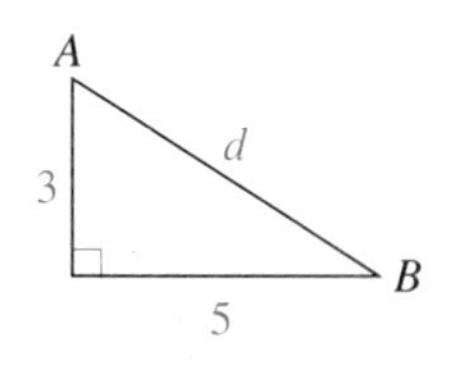

Using the Pythagorean theorem, we can find the length, d, of the hypotenuse or the length of line segment AB:

$$d^2 = 3^2 + 5^2$$
$$d^2 = 34$$

This equation has solutions $\sqrt{34}$ and $-\sqrt{34}$, the numbers that yield 34 when squared. Because d represents a linear measurement, we use only the positive value, $\sqrt{34}$. Thus, the length of $\overline{AB}$ is $\sqrt{34}$. ■

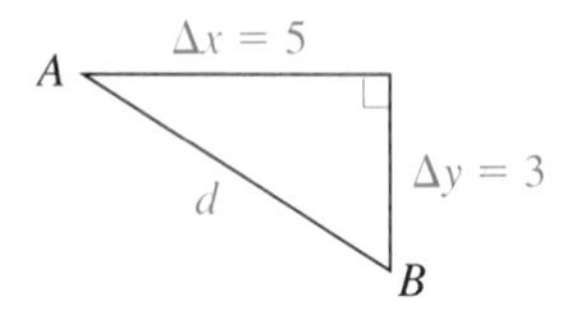

FIGURE 5

In Example 1, we could have drawn a horizontal line through point A and a vertical line through point B. The right triangle formed would still have the hypotenuse $\overline{AB}$. Figure 5 shows a comparison of these triangles. The length of this segment is still $\sqrt{34}$.

We now state the formula for finding the distance between two points, or finding the length of a line segment.

LENGTH OF A LINE SEGMENT
DISTANCE BETWEEN TWO POINTS

Given two points, (x_1, y_1) and (x_2, y_2), the distance between these points is

$$d = \sqrt{(\Delta x)^2 + (\Delta y)^2}$$
$$= \sqrt{(x_2 - x_1)^2 + (y_2 - y_1)^2}$$

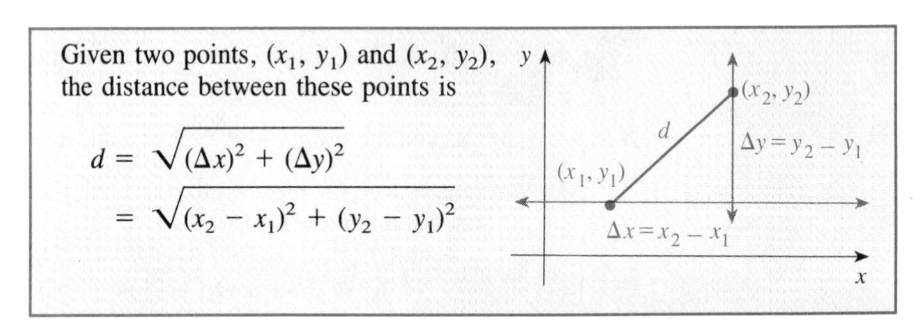

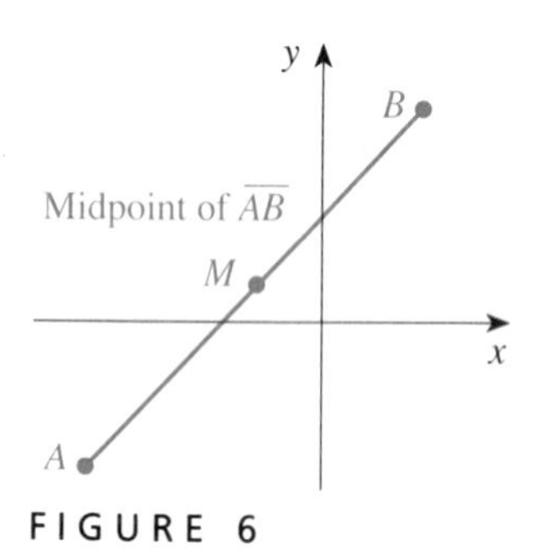

FIGURE 6

Midpoints In addition to finding the length of a line segment, we can find the midpoint of the segment. The **midpoint** of a line segment with endpoints A and B is a point M between A and B such that the distance from A to M equals the distance from M to B. Figure 6 shows the midpoint M of $\overline{AB}$. Given the ordered-pair names of endpoints A and B, we can find the ordered-pair name of the midpoint. To see how this is done, work through the following example.

EXAMPLE 2 For points $A = (-3, 4)$ and $B = (2, 1)$, find the ordered-pair name of the midpoint of $\overline{AB}$.

Solution Again, our solution starts with a graph (see Figure 7). Now, the midpoint is simply the average of the x-coordinates,

$$\frac{-3 + 2}{2} \quad \text{or} \quad \frac{-1}{2}$$

and the average of the y-coordinates

$$\frac{4 + 1}{2} \quad \text{or} \quad \frac{5}{2}$$

That is, the ordered-pair name of the midpoint is $\left(\frac{-1}{2}, \frac{5}{2}\right)$.

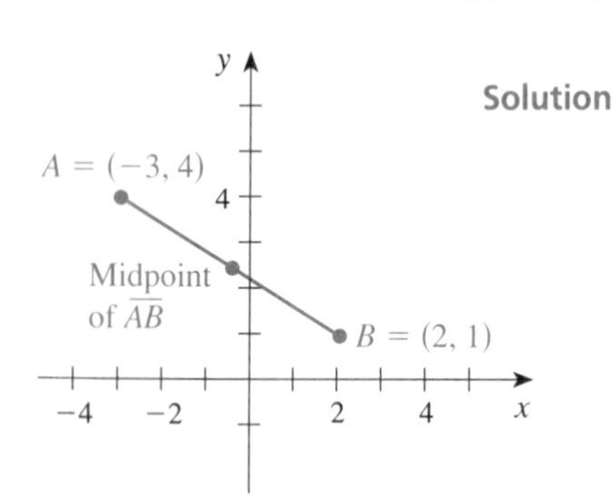

FIGURE 7

To check this fact, we compute the distance from the endpoint $(-3, 4)$ to the midpoint $\left(\frac{-1}{2}, \frac{5}{2}\right)$, and the distance from $\left(\frac{-1}{2}, \frac{5}{2}\right)$ to the endpoint $(2, 1)$, to see if the distances are equal. We can set up triangles to find the distances as shown in Figure 8, and we find that the two distances are equal:

$$\text{Distance from } A \text{ to } M\text{:}\quad d = \sqrt{\left(\frac{5}{2}\right)^2 + \left(\frac{3}{2}\right)^2} = \sqrt{\frac{34}{4}}$$

$$\text{Distance from } M \text{ to } B\text{:}\quad d = \sqrt{\left(\frac{5}{2}\right)^2 + \left(\frac{3}{2}\right)^2} = \sqrt{\frac{34}{4}}$$

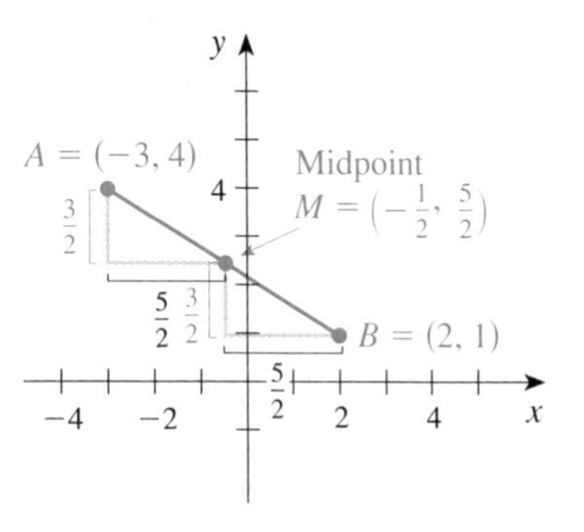

FIGURE 8

■

Notice that the graph in Figure 8 suggests that the midpoint of $\overline{AB}$ is the point halfway over and halfway down from point A to point B. We now state the general formula used to find the midpoint of a line segment.

MIDPOINT OF A LINE SEGMENT

Given a line segment with endpoints (x_1, y_1) and (x_2, y_2), the ordered-pair name of the midpoint is

$$\left(\frac{x_1 + x_2}{2}, \frac{y_1 + y_2}{2}\right)$$

Average of x-coordinates | Average of y-coordinates

y, (x_2, y_2), $\left(\frac{x_1 + x_2}{2}, \frac{y_1 + y_2}{2}\right)$, Midpoint, (x_1, y_1), x

GIVE IT A TRY

1. Consider the points $A = (2, -1)$ and $B = (4, 5)$.
   a. Find the length of line segment AB.
   b. Report the ordered-pair name of the midpoint of $\overline{AB}$.
2. Consider the points $A = (-3, 1)$ and $B = (2, -3)$.
   a. Find the length of line segment AB.
   b. Report the ordered-pair name of the midpoint of $\overline{AB}$.

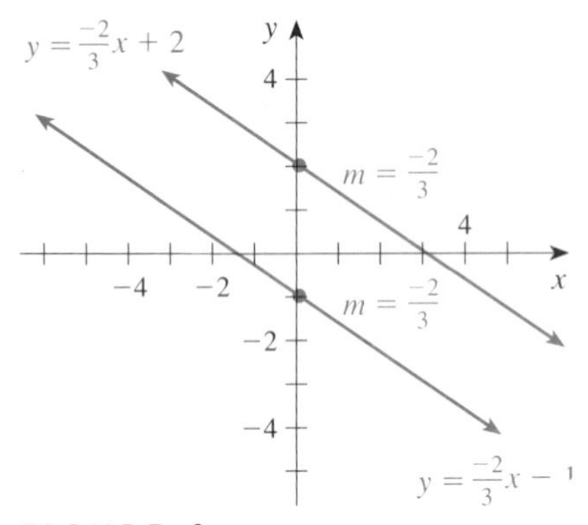

FIGURE 9
Parallel lines have equal slope measures.

As we learned in Chapter 3, the defining characteristic of a line is its slope. We now use the slope measure to determine when two lines are parallel and when two lines are perpendicular.

Parallel Lines Two lines in the same plane are **parallel** if they are nonintersecting—they have no point in common. In terms of slope measure, two lines are parallel if their slope measures are equal. Consider the lines defined by the equations.

$$y = \frac{-2}{3}x + 2 \qquad \text{and} \qquad y = \frac{-2}{3}x - 1$$

The graphs of the solutions to these equations are shown in Figure 9. The slope measure (the x-coefficient) determines that the two lines are parallel, and the constant term simply determines where the line crosses (intersects) the y-axis. Work through the following example.

EXAMPLE 3 Find the equation of the line through the point (2, 4) that is parallel to the graph of the solution to $3x + 5y = 8$.

Solution As usual, we start with a graph like the one in Figure 10.

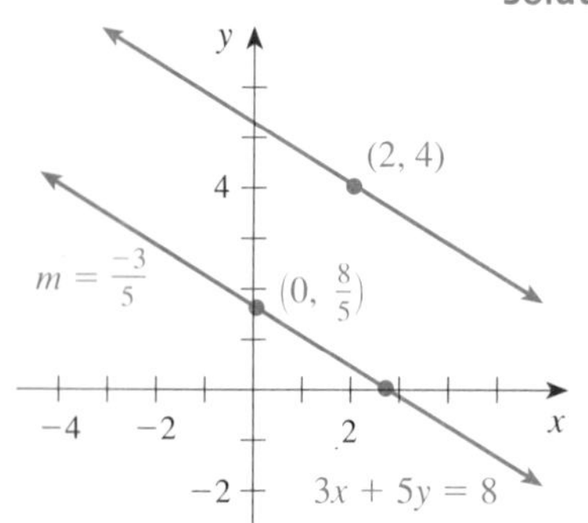

FIGURE 10
The line that passes through the point (2, 4) and is parallel to the line determined by $3x + 5y = 8$.

We then rewrite $3x + 5y = 8$ for y:

$$\begin{aligned} 3x + 5y &= 8 \\ -3x \qquad &\quad -3x \\ \hline 5y &= -3x + 8 \\ y &= \frac{-3}{5}x + \frac{8}{5} \end{aligned}$$

The slope measure of the line determined by this equation is $\frac{-3}{5}$. We want to find the equation of a line parallel to the line determined by $3x + 5y = 8$. This line must also have slope $\frac{-3}{5}$. Additionally, the line must pass through the point (2, 4). So, we want to find the equation of the line through the point (2, 4) with slope measure $\frac{-3}{5}$. You should recognize this problem from the beginning of Section 4.1. In algebra and other problem-solving situations, this approach is very useful—making a new problem look like an old one that you know how to solve!

We now find the equation of the line through the point (2, 4) with slope measure $\frac{-3}{5}$:

The equation of line has the form: $\qquad y = mx + b$

Because $m = \frac{-3}{5}$, we have: $\qquad y = \frac{-3}{5}x + b$

Because (2, 4) must solve the equation, we have:

$$4 = \frac{-3}{5}(2) + b$$

$$4 = \frac{-6}{5} + b$$

$$20 = -6 + 5b$$

$$\underline{6} \qquad \underline{6 \qquad\qquad}$$

$$26 = 5b$$

$$\frac{26}{5} = b$$

Thus, b is $\frac{26}{5}$ and the equation of the line through (2, 4) parallel to the graph of the solution to $3x + 5y = 8$ is

$$y = \frac{-3}{5}x + \frac{26}{5}$$

In standard form, the equation is $3x + 5y = 26$. ■

Using the Graphics Calculator

```
RANGE
Xmin=-9.6
Xmax=9.4
Xscl=1
Ymin=-6.4
Ymax=6.2
Yscl=1
Xres=1
```

FIGURE 11

To check our work in Example 3, we use the original line $3x + 5y = 8$. For `Y1`, type `(-3/5)X+(8/5)`. For `Y2`, we use the equation we have found—type `(-3/5)X+(26/5)`. View the graph screen using the values in Figure 11.* Do the lines appear parallel? Yes. Press the `TRACE` key and then the down arrow key to switch to the line for `Y2`. Move the cursor until the message reads `X=2  Y=4`. Thus, the new line does pass through the point (2, 4).

Perpendicular Lines A vertical line and a horizontal line are **perpendicular lines**—they intersect to form a right angle. In general, any two lines that intersect to form a right angle are perpendicular lines. In Figure 12, line L_1 is perpendicular to line L_2.

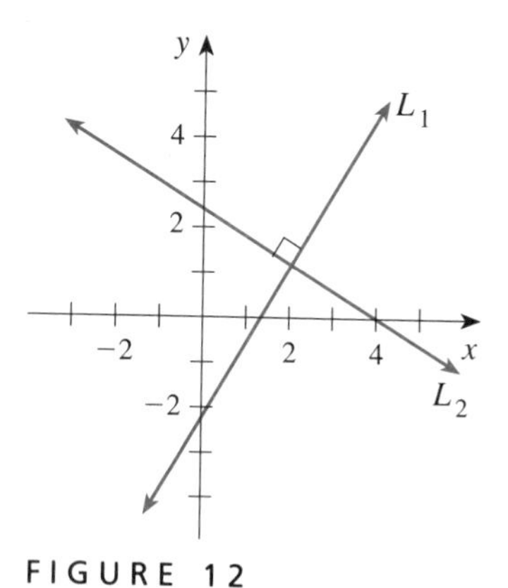

FIGURE 12

In terms of slope measure, two lines are perpendicular if their slope measures are **negative reciprocals**: If the slope measure of line L_1 is $\frac{a}{b}$, then the slope measure of a perpendicular line L_2 is $\frac{-b}{a}$. The following examples show how to find a negative reciprocal:

Number	*Reciprocal*	*Negative reciprocal*
$\frac{2}{3}$	$\frac{3}{2}$	$\frac{-3}{2}$
-3	$\frac{-1}{3}$	$\frac{1}{3}$

TI-82 Note *Type `-9.4` for `Xmin` and `-6.2` for `Ymin`.

To learn how to work with perpendicular lines, work through the following example.

EXAMPLE 4 Find the equation of the line through the point (2, 4) that is perpendicular to the line determined by $3x + 5y = 8$.

Solution A graph of the situation is shown in Figure 13.

We next rewrite $3x + 5y = 8$ for y:

$$\begin{aligned} 3x + 5y &= 8 \\ -3x \qquad &\quad -3x \\ 5y &= -3x + 8 \\ y &= \frac{-3}{5}x + \frac{8}{5} \end{aligned}$$

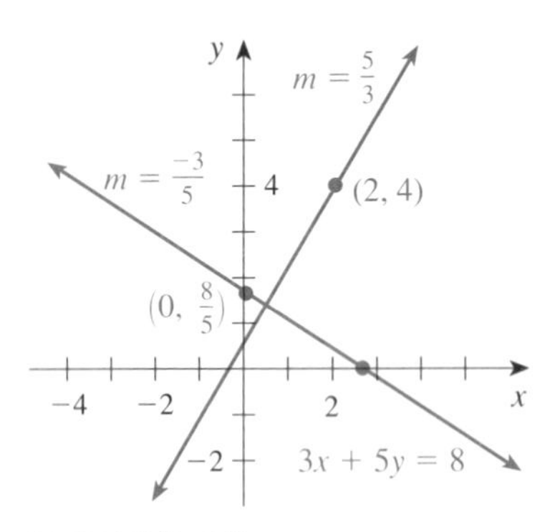

FIGURE 13
The line that passes through the point (2, 4) and is perpendicular to the graph of the solution to $3x + 5y = 8$

So, the slope of the line determined by $3x + 5y = 8$ is $\frac{-3}{5}$. The slope of a line perpendicular to this line must have slope measure $\frac{5}{3}$ $\left(\text{the negative reciprocal of } \frac{-3}{5}\right)$. Thus, we seek the equation of the line through the point (2, 4) with slope measure $\frac{5}{3}$.

Using the point-slope form, with the point (2, 4) and slope $\frac{5}{3}$, we get the equation

$$\begin{aligned} y - 4 &= \left(\frac{5}{3}\right)(x - 2) \\ y - 4 &= \frac{5}{3}x - \frac{10}{3} \\ y &= \frac{5}{3}x + \frac{2}{3} \end{aligned}$$

■

Using the Graphics Calculator

We can check our work in Example 4 as follows. Press the Y= key. Type `(-3/5)X+(8/5)` for Y1 and type `(5/3)X+(2/3)` for Y2. View the graph screen using the values in Figure 14.* Do the lines appear to be perpendicular? Press the TRACE key and then the down arrow key to move the cursor to the second line. Then move the cursor until the message reads X=2 Y=4. Thus, this perpendicular line, the one determined by $y = \frac{5}{3}x + \frac{2}{3}$, does pass through the point (2, 4).

```
RANGE
Xmin=-9.6
Xmax=9.4
Xscl=1
Ymin=-6.4
Ymax=6.2
Yscl=1
Xres=1
```

FIGURE 14

In general, we have the following properties.

PARALLEL LINES AND PERPENDICULAR LINES

Two lines are **parallel** if and only if their slope measures are equal.

Two lines are **perpendicular** if and only if their slope measures are negative reciprocals of each other.

TI-82 Note *Type -9.4 for Xmin and -6.2 for Ymin.

GIVE IT A TRY

3. Graph the line determined by $y = 3x - 2$. On the same set of axes, graph the line through the point $(-3, -1)$ that is parallel to the original line. Find the equation of the parallel line.

4. Graph the line determined by $y = 3x$. On the same set of axes, graph the line through the point $(-3, -1)$ that is perpendicular to the original line. Find the equation of the perpendicular line.

The Perpendicular Bisector To conclude this introduction to coordinate geometry, we now find the equation of a perpendicular bisector of a line segment. This task combines several of the concepts and skills from this and previous sections.

A **bisector** of a line segment is a line that passes through the midpoint of the line segment. The bisector divides the line segment into two equal pieces. A given line segment has an infinite number of bisectors, because any line passing through the midpoint of the line segment is a bisector of the segment. Figure 15 shows several bisectors of line segment AB.

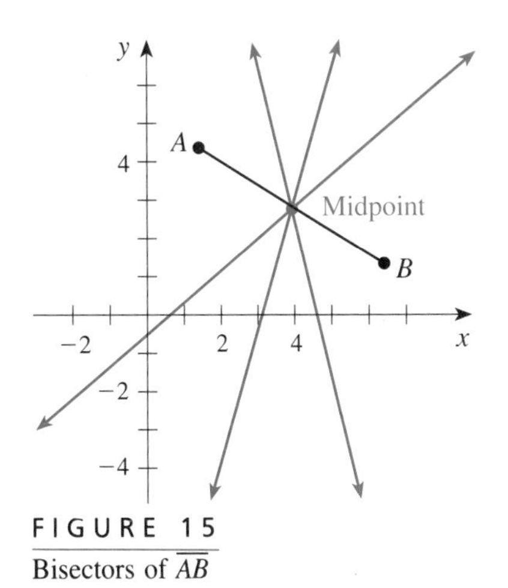

FIGURE 15
Bisectors of $\overline{AB}$

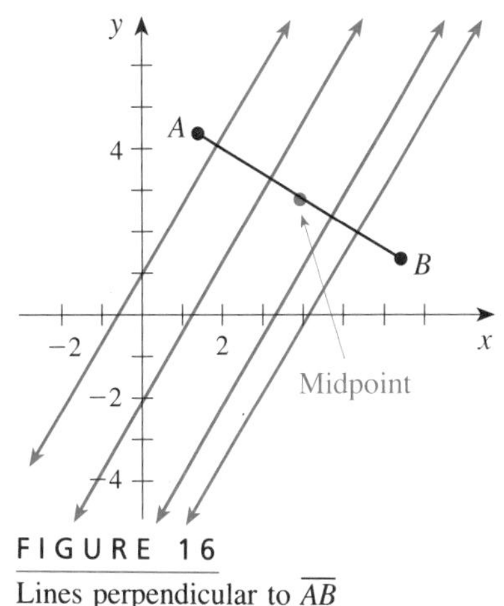

FIGURE 16
Lines perpendicular to $\overline{AB}$

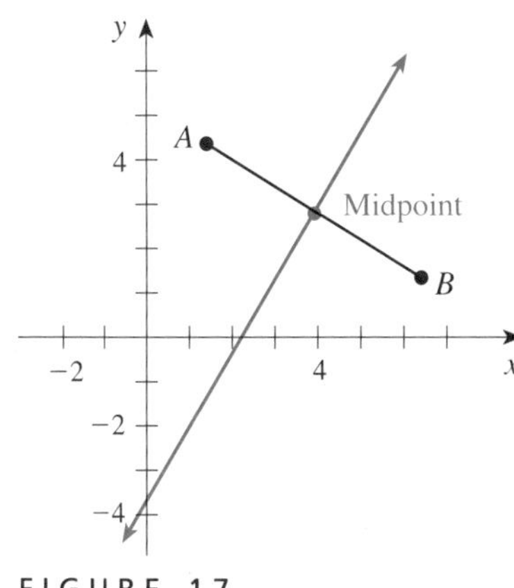

FIGURE 17
The perpendicular bisector of $\overline{AB}$

An infinite number of lines can be drawn perpendicular to a line segment. Figure 16 shows several lines perpendicular to the line segment AB.

A **perpendicular bisector** of a line segment is a bisector that is also perpendicular to the line which passes through the endpoints of the line segment. The perpendicular bisector of a line segment is unique; there is one and only one. Figure 17 shows the perpendicular bisector of the line segment AB.

Work through the following example to find the equation of the perpendicular bisector of a line segment.

EXAMPLE 5 Find the equation of the perpendicular bisector of the line segment with endpoints $A = (-3, 2)$ and $B = (1, 5)$.

Solution A graph of $\overline{AB}$ is given in Figure 18.

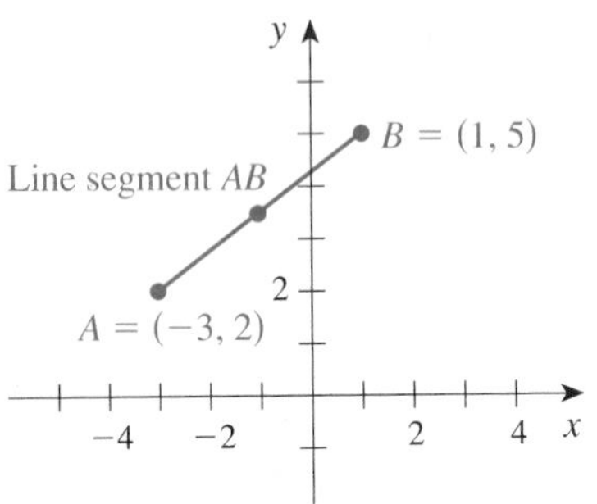

FIGURE 18

First, the midpoint of line segment AB is

$$\left(\frac{-3+1}{2}, \frac{2+5}{2}\right) \quad \text{or} \quad \left(-1, \frac{7}{2}\right)$$

From the graph, we observe that the line containing points A and B is increasing (going up), and its slope measure is a positive number. Figure 19 shows that for points A and B, the Δy is 3 and the Δx is 4. So, the line through points A and B has slope measure $\frac{3}{4}$.

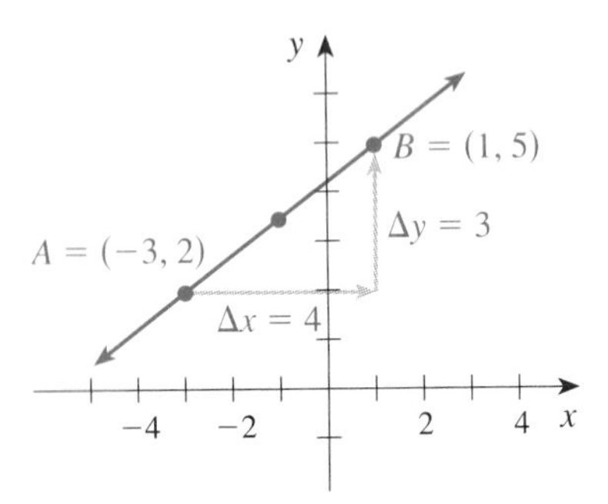

FIGURE 19

Line is increasing (going up), so the slope measure m is positive.

$$m = \frac{\Delta y}{\Delta x} = \frac{3}{4}$$

or

$$m = \frac{y_2 - y_1}{x_2 - x_1} = \frac{5 - 2}{1 - (-3)} = \frac{3}{4}$$

A line perpendicular to the line with slope $\frac{3}{4}$ must have slope measure $\frac{-4}{3}$, the negative reciprocal of $\frac{3}{4}$. The equation of the perpendicular bisector is the equation of the line through the midpoint $\left(-1, \frac{7}{2}\right)$ with slope measure $\frac{-4}{3}$. The perpendicular bisector is drawn in Figure 20.

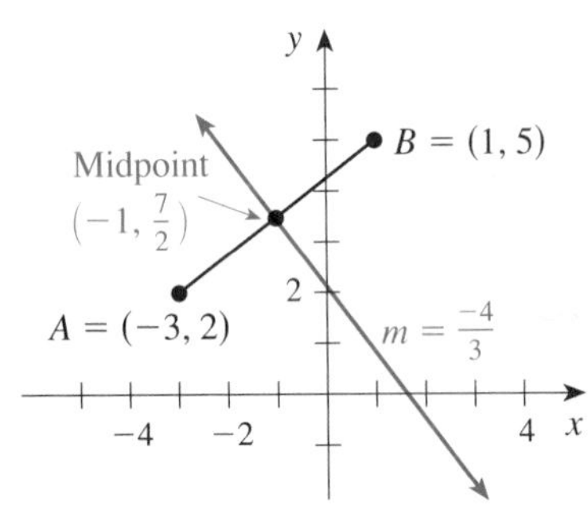

FIGURE 20

To find the equation of this line, we use the slope-intercept form:

$$y = mx + b$$

Because the slope is $\frac{-4}{3}$, we have

$$y = \frac{-4}{3}x + b$$

Since the line passes through the midpoint $\left(-1, \frac{7}{2}\right)$, we can solve for b:

$$\frac{7}{2} = \left(\frac{-4}{3}\right)(-1) + b$$

Solving this equation yields:

$$\frac{7}{2} = \frac{4}{3} + b$$

Multiplying each side by 6, we get:

$$21 = 8 + 6b$$

$$\underline{-8 \quad -8}$$

$$13 = 6b$$

$$\frac{13}{6} = b$$

Thus, b is $\frac{13}{6}$, and the equation of the perpendicular bisector is

$$y = \frac{-4}{3}x + \frac{13}{6}$$

■

Using the Graphics Calculator

```
RANGE
Xmin=-4.8
Xmax=4.7
Xscl=1
Ymin=-1.1
Ymax=5.2
Yscl=1
Xres=1
```

FIGURE 21

To check the equation in Example 5, type `(-4/3)X+(13/6)` for `Y1`. Press the `RANGE` key and enter the values shown in Figure 21.* View the graph screen. Press the `2nd` key followed by the `PRGM` key to select the `DRAW` menu. Press the `2` key for the `Line(` option. To draw a line segment AB, move the cursor to the point $(-3, 2)$ and press the `ENTER` key. Then move the cursor to the point $(1, 5)$ and press the `ENTER` key. Press the `TRACE` key, and move the cursor until the message reads `X=-1   Y=3.5`. As you can see, the line passes through the midpoint of $\overline{AB}$.

The problem in Example 5 is typical of many problems, not only those in mathematics. Solving it requires knowledge of midpoints, the relationship between perpendicular lines, and perpendicular bisectors. Additionally, this knowledge must be combined with a previous skill—finding the equation of a line. As you continue to study algebra and mathematics, this process will be repeated over and over: A basic skill will be developed (finding the equation of a line, in this case), then new concepts will be introduced (midpoints, perpendicular lines, and bisectors, in this case) and combined with a previous basic skill in order to solve a new problem.

GIVE IT A TRY

Use the points $A = (1, 4)$ and $B = (3, -2)$ to answer Problems 5–12.

5. Graph line segment AB.

6. Find the length of line segment AB.

7. Find the midpoint of line segment AB.

8. Plot the midpoint on the graph from Problem 5.

9. Find the slope of the line through points A and B.

10. Find the slope of a line perpendicular to line AB.

11. Find the equation of the perpendicular bisector of $\overline{AB}$.

12. Graph the perpendicular bisector on the graph from Problem 8.

TI-82 Note *Type `-4.7` for `Xmin` and `-1` for `Ymin`.

■ ***Quick Index***

■ SUMMARY

In this section we have introduced several basic concepts from coordinate geometry: the length of a line segment, the midpoint of a line segment, the relationship between the slope and parallel lines, and the relationship between slope and perpendicular lines. We combined these concepts with the idea of a coordinate system—a set of axes. With the coordinate system, ordered-pair names can be given for any point, such as the endpoints of a line segment.

We used these concepts and the Pythagorean theorem, $c^2 = a^2 + b^2$, to find the length of a line segment. Additionally, we found the ordered-pair name of the midpoint of a line segment by taking the average of the x-coordinates and the average of the y-coordinates.

If we know the slope of a line, we can state the slope of a parallel line or a perpendicular line. Parallel lines have equal slope measures, and perpendicular lines have slope measures that are negative reciprocals. Finally, we used this knowledge of perpendicular lines to find the equation of a perpendicular bisector of a line segment—a line passing through the midpoint, perpendicular to the line containing the line segment.

4.2 ■ ■ ■ EXERCISES

Insert true *for a true statement, or* false *for a false statement.*

____ 1. If the length of a line segment AB is $\sqrt{23}$, then the midpoint of line segment AB is $\frac{\sqrt{23}}{2}$ units from point A.

____ 2. For two points A and B, the x-coordinate of the midpoint is $\frac{\Delta x}{2}$ and the y-coordinate is $\frac{\Delta y}{2}$.

____ 3. Two lines are parallel if their slope measures are equal.

____ 4. Perpendicular lines are nonintersecting lines.

____ 5. Two lines are perpendicular if their slope measures are reciprocals of each other.

____ 6. A bisector of a line segment passes through the midpoint of the line segment.

____ 7. A line perpendicular to a line segment always passes through the midpoint of the line segment.

____ 8. A perpendicular bisector of a line segment passes through the midpoint of the line segment and is perpendicular to the line through the endpoints of the line segment.

9. For the points $A = (-2, 5)$ and $B = (4, 1)$, the midpoint of $\overline{AB}$ is

   i. $(2, 6)$ **ii.** $(3, 2)$ **iii.** $(1, 3)$ **iv.** $\left(\frac{3}{2}, 3\right)$

10. For the points $A = (-2, 5)$ and $B = (4, 1)$, the length of $\overline{AB}$ is

    i. 8 **ii.** $\sqrt{10}$ **iii.** $2\sqrt{10}$ **iv.** $\sqrt{52}$

11. The slope of a line parallel to the line determined by $y = \frac{2}{3}x + 3$ is

    i. $\frac{2}{3}$ **ii.** 3 **iii.** $(0, 3)$ **iv.** $\frac{-3}{2}$

12. The slope of a line perpendicular to the line determined by $y = 2x + 1$ is

    i. 2 **ii.** 1 **iii.** $\frac{1}{2}$ **iv.** $\frac{-1}{2}$

Use the points $A = (-1, -2)$ *and* $B = (3, 4)$ *to answer Problems 13–16.*

13. Graph line segment AB.
14. Find the length of line segment AB.
15. Report the ordered-pair name of the midpoint of $\overline{AB}$.
16. Plot and label the midpoint of $\overline{AB}$ on the graph in Problem 13.

Use the equation $3x - 5y = 10$ *to answer Problems 17–19.*

17. Graph the solution to the equation.
18. Graph the line through the point $(-3, 4)$ that is parallel to the line in Problem 17.
19. What is the equation of the line graphed in Problem 18?

Use the equation $y = \frac{2}{3}x + 1$ to answer Problems 20–22.

20. Graph the solution to the equation.

21. Graph the line through the point $(-3, 4)$ that is perpendicular to the line in Problem 20.

22. What is the equation of the line graphed in Problem 21?

Use the points $A = (-5, 1)$ and $B = (3, -1)$ to answer Problems 23–25.

23. Graph line AB and the line parallel to line AB through the point $(0, 4)$.

24. Find the equation of line AB.

25. Find the equation of the line through $(0, 4)$ that is parallel to line AB.

Use the points $A = (-3, -2)$ and $B = (3, 4)$ to answer Problems 26–28.

26. Graph line AB and the line perpendicular to line AB through the point $(0, -2)$.

27. Find the equation of line AB.

28. Find the equation of the line through $(0, -2)$ that is perpendicular to line AB.

Use the points $A = (2, -4)$ and $B = (4, 6)$ to answer Problems 29–32.

29. Graph line segment AB.

30. Plot the midpoint of line segment AB.

31. Find the equation of the perpendicular bisector of line segment AB.

32. Graph the perpendicular bisector of line segment AB.

Use the points $A = (5, 1)$ and $B = (-3, -1)$ to answer Problems 33–35.

33. Find the midpoint of line segment AB.

34. Graph line segment AB and its midpoint.

35. Find the length of line segment AB.

Use the points $A = (-2, 5)$, $B = (-2, -1)$, and $C = (4, -1)$ to answer Problems 36–42.

36. Graph the triangle with vertices A, B, and C.

37. What is perimeter of the triangle in Problem 36?

38. Find the equation of the line containing points A and B.

39. Find the equation of the line containing points A and C.

40. Find the equation of the line containing points B and C.

41. Is the triangle in Problem 36 a right triangle? If so, why? If not, why not?

42. What is the area of the triangle in Problem 36?

Find the midpoint and length of the line segment with the given endpoints.

43. $(-2, 5)$ and $(4, 1)$

44. $(-3, 1)$ and $(3, -3)$

45. $(5, 1)$ and $(1, -2)$

46. $(-1, -5)$ and $(3, 3)$

47. $(-4, 1)$ and $(3, 2)$

48. $(3, -6)$ and $(-2, 1)$

49. $\left(\frac{2}{3}, -2\right)$ and $\left(-1, \frac{8}{3}\right)$

50. $\left(4, \frac{-5}{2}\right)$ and $\left(\frac{-3}{2}, \frac{3}{2}\right)$

■ BONUS PROBLEMS

51. a. For a circle centered at the origin and passing through the point $(-2, 3)$, what is the equation of the line containing the radius to this point?

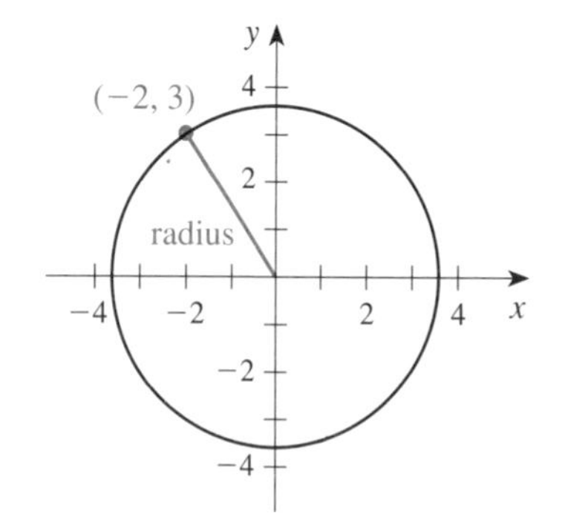

b. The tangent line to a circle at a point on the circle is always perpendicular to the radius of the circle drawn to that point. What is the equation of the tangent line to the circle at the point $(-2, 3)$?

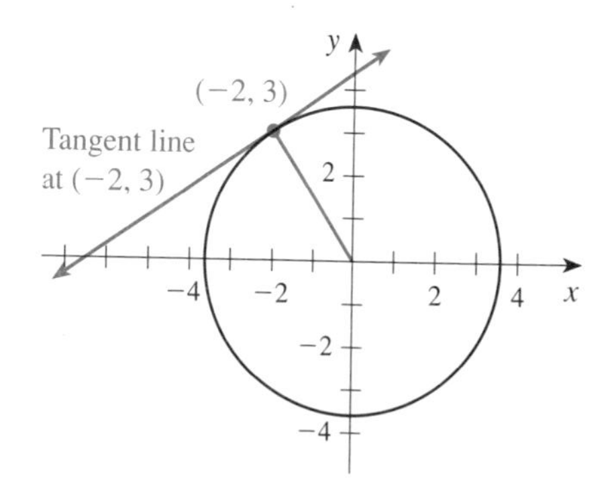

52. Using Problem 51, for the circle centered at the origin and passing through the point $(3, 4)$, find the equation of the tangent line at the point $(3, 4)$.

DISCOVERY

Intersection Points of Lines

For this discovery section, we will start with knowledge that has been developed up to this point and attempt to discover a generalization about the intersection point of two lines.

We can use the graphics calculator to graph the input-output ordered pairs for a linear function such as $f(x) = 2x - 1$. The graph of these ordered pairs is a straight line. We will use this information to work with a real-world application.

Problem The AA Auto Rental Company charges a daily rental fee of \$32 plus a fee of \$0.25 per mile driven. The Econo Rental Company charges \$26 a day, plus \$0.30 per mile driven. How many miles per day must a car be driven for the two rental fees to be equal?

To solve this problem, we know that we can write a linear function describing the rental fee from AA Auto, and we can write a linear function describing the rental fee from Econo. Our plan is to write these linear functions, graph the functions, then use the graphs to find the number of miles driven that results in the fees being the same.

We want to write a linear "cost" function—that is, cost as a function of the number of miles driven. From our earlier work, we know that a linear cost function is based on a cost per unit and a fixed cost.

For AA Auto, the cost per mile is \$0.25. That is, the cost per unit is \$0.25. The fixed cost is \$32.

So, a cost function for AA Rental is $C(x) = 0.25x + 32$.

For Econo, the cost per mile is \$0.30. That is, the cost per unit is \$0.30. The fixed cost is \$26.

So, a cost function for Econo is $C(x) = 0.3x + 26$.

Using the Graphics Calculator

For Y1, type 0.25X+32. Press the RANGE key and enter the settings shown in Figure 1.* (These settings are a trial, so we may have to adjust them later.) View the graph screen, as shown in Figure 2. Press the TRACE key, and move the cursor until the message reads X=80 Y=52. This message means that for an input of 80, the output

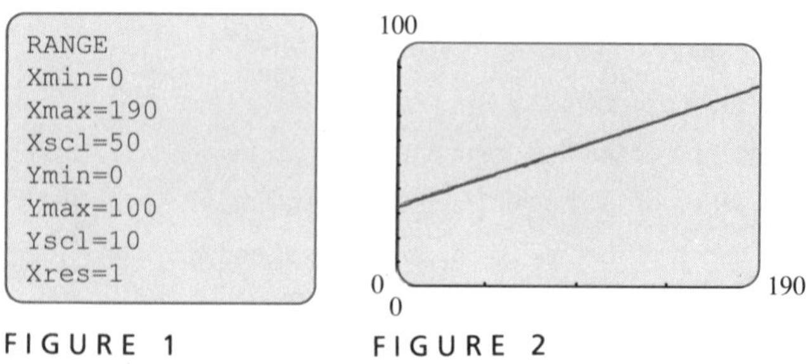

FIGURE 1 FIGURE 2

TI-82 Note *Type 188 for Xmax.

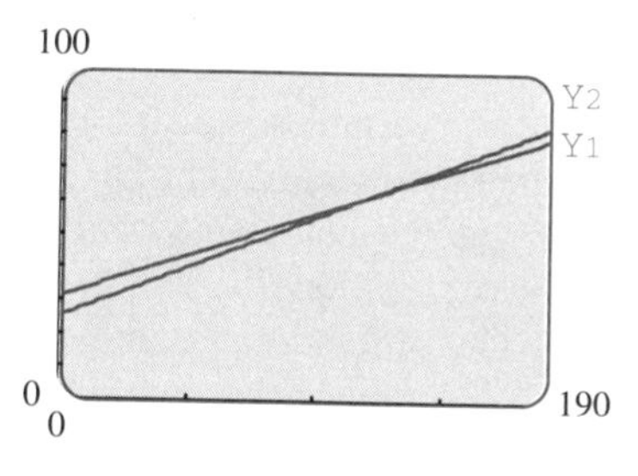

FIGURE 3

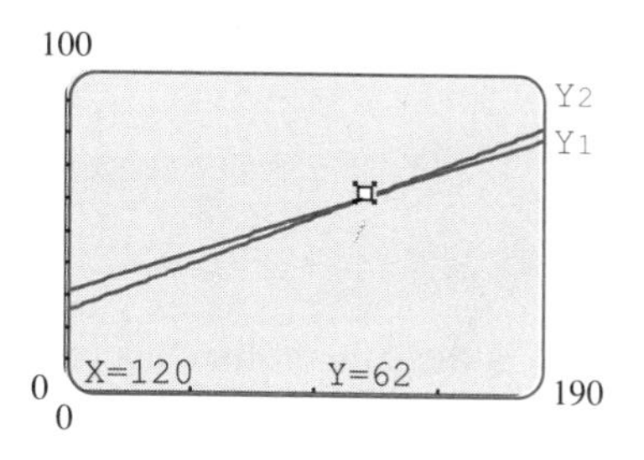

FIGURE 4

is 52; that is, when the car is driven 80 miles in a day, the AA Auto rental fee is \$52. Notice that the slope of this line is 0.25; that is, for each increase of 1 mile driven, the charge increases by \$0.25.

For Y2, type 0.3X+26. View the graph screen, as shown in Figure 3. Press the TRACE key, and then the down arrow key to move the cursor to the Y2 line. Move the cursor on this line until the message reads X=80 Y=50. This message indicates that for an input of 80, the output is 50; that is, when the car is driven 80 miles in a day, the Econo auto rental fee is \$50. Notice that the slope of this line is 0.3; that is, for each increase of 1 mile driven, the charge increases by \$0.30. Press the down arrow key to move the cursor to the AA Auto line (their rental fee is \$52 for 80 miles). Move the cursor to the point of intersection of the two lines (the message now reads X=120 Y=62). Press the down arrow key to move the cursor to the other line, and confirm that this point is on both lines. The intersection point is shown in Figure 4. So, when 120 miles are driven (for an input of 120), the rental fee for both AA Auto and Econo is \$62 (both linear functions output 62). In the next section we will say that the solution to this system of linear equations (or functions)

$$y = 0.25x + 32 \qquad \text{and} \qquad y = 0.3x + 26$$

is the ordered pair (120, 62).

Exercises

1. Delux Car Rental charges a rental fee of \$50 per day plus \$0.45 per mile driven. El Cheapo Car Rental charges \$40 per day plus \$0.50 per mile. Find the daily cost function for each of these rental companies. Use the graphics calculator to graph these linear functions. Use the TRACE and Zoom In features to find how many miles (to the nearest one-tenth mile) must be driven in a day for the two car rental companies to have the same charge.

2. Della has two job offers—one pays \$250 per week plus 10% commission on sales: For sales of \$4,000 in a week, she will earn \$250 + 0.1(\$4,000), or \$650. The second offer pays \$150 per week plus 20% commission on sales: For sales of \$4,000 in a week, she will earn \$150 + 0.2(\$4000), or \$950. Write a linear function for her earnings y as a function of sales amount x for each of the two offers. Use the graphics calculator to graph these linear functions. Use the TRACE and Zoom In features to find how many dollars in sales (to the nearest ten dollars) Della must have to earn the same pay from the two jobs.

3. The Celsius temperature y as a function of Fahrenheit temperature x is given by $y = \frac{5}{9}x - \frac{160}{9}$. The Fahrenheit temperature y as a function of the Celsius temperature x is given by $y = \frac{9}{5}x + 32$. Use the graphics calculator to find the input value at which the Celsius temperature is equal to the Fahrenheit temperature. [*Hint*: Try different range values.]

4.3 ■ ■ ■ SYSTEMS OF LINEAR EQUATIONS

The graph of the solution to a first-degree equation in two variables is a straight line. For two such equations with their solutions graphed on the same set of axes as shown in Figure 1, we might ask, "What is the ordered-pair name of the intersection point of the lines determined by the two equations?" The two equations are known as a *system of linear equations*. The ordered-pair name of the intersection point of the two lines is known as the *solution to the system of linear equations*. That is, the solution to a system is the set of ordered pairs that solve the equations simultaneously (at the same time). In this section, we will solve systems of equations. In the next section, we will model some real-world situations using systems of linear equations.

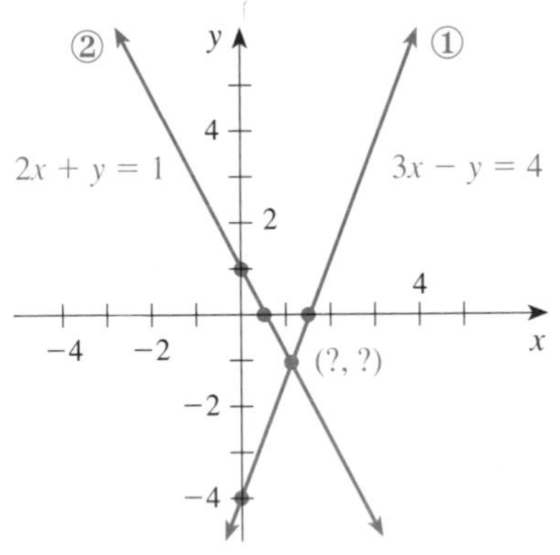

FIGURE 1

$$\begin{cases} 3x - y = 4 \\ 2x + y = 1 \end{cases}$$

The Solution to a System of Linear Equations Consider a system of two equations:

$$\begin{cases} 3x - y = 4 \\ 2x + y = 1 \end{cases}$$

The graph of the solution to the equation $3x - y = 4$ is the straight line labeled ① in Figure 1. The graph of the solution to the equation $2x + y = 1$ is the straight line labeled ② in the graph shown in the figure. The solution to this system of equations is the ordered-pair name of the point of intersection of the two lines. In general, we have the following definition.

DEFINITIONS: A SYSTEM OF EQUATIONS AND THE SOLUTION TO A SYSTEM

A collection of two or more equations is known as a **system of equations**. When all the equations are first-degree equations, the system is known as a **system of linear equations.**

The **solution to a system** of equations in two variables is the collection of ordered pairs that simultaneously solve every equation in the system.

Using the Graphics Calculator

```
RANGE
Xmin=-9.6
Xmax=9.4
Xscl=1
Ymin=-6.4
Ymax=6.2
Yscl=1
Xres=1
```

FIGURE 2

We can display the graphs of the solutions to the preceding two equations, and then use the TRACE key to identify (or approximate) the ordered-pair name of the point of intersection of the graphs. First, we rewrite the equations for y. The equation $3x - y = 4$ yields $y = 3x - 4$, and $2x + y = 1$ yields $y = -2x + 1$. Press the Y= key and erase any existing entries. For Y1 type 3X−4, and for Y2 type ⁻2X+1. Use the settings in Figure 2* to view the graph screen shown in Figure 3. Press the TRACE key. The y-intercept for the first equation, $3x - y = 4$, is $(0, -4)$. Press the down arrow key to move the cursor to the other graph. The y-intercept for $2x + y = 1$ is $(0, 1)$. Move the cursor to the point of intersection, then zoom in on this point. The view screen will look like Figure 4. Use the down arrow key to switch the cursor back and forth between the lines in order to see how far apart the y-values

TI-82 Note *Type ⁻9.4 for Xmin and ⁻6.2 for Ymin.

are for various x-values. It appears the ordered-pair name of the point of intersection is $(1, -1)$. That is, we claim the solution to the system is the ordered pair $(1, -1)$. This ordered pair solves both equations in the system, since $3(1) - (-1) = 4$ and $2(1) + (-1) = 1$ are both true arithmetic sentences. The *lines* shown in Figure 4 represent the collections of ordered pairs that solve "$2x + y = 1$ *or* $3x - y = 4$." The *point of intersection* represents the solution to "$2x + y = 1$ *and* $3x - y = 4$." We will soon introduce an algebraic approach that will yield the single ordered pair which solves a linear system.

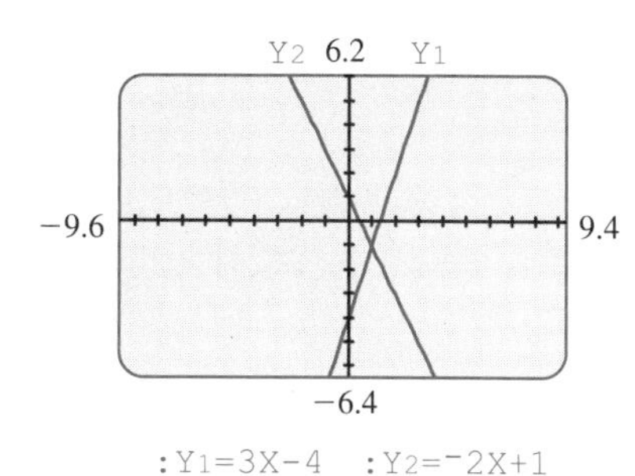

FIGURE 3

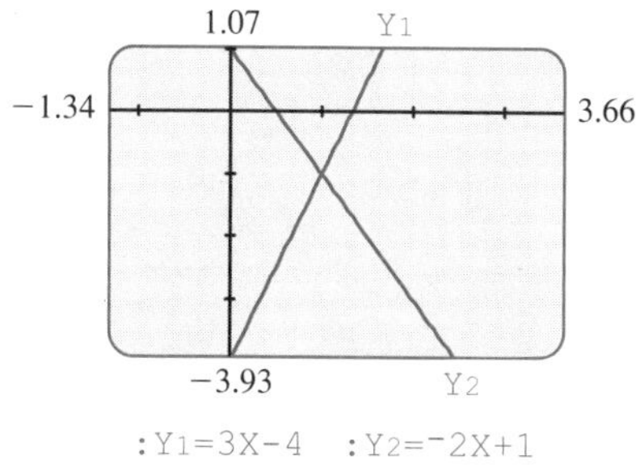

1.07 Y1
−1.34
3.66
−3.93 Y2
:Y1=3X-4 :Y2=-2X+1

FIGURE 4

GIVE IT A TRY

Rewrite each pair of equations for y. Then use the graphics calculator to graph the system. Use the TRACE and the ZOOM features to approximate the solution to the system.

1. $\begin{cases} 3x + 2y = 8 \\ 3x - 2y = 4 \end{cases}$

2. $\begin{cases} 3x - 5y = 9 \\ 2x + 3y = -9 \end{cases}$

Properties For a system of linear equations in two variables, exactly one of the following cases is true.

PROPERTIES OF THE SOLUTION TO A LINEAR SYSTEM

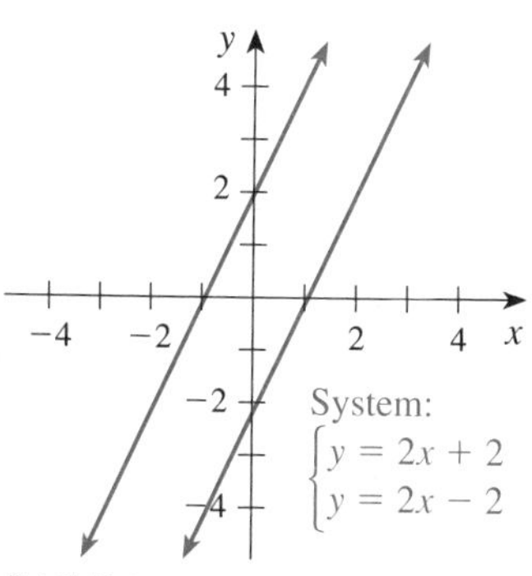

FIGURE 5

1. The lines determined by the equations in the system intersect in exactly one point. In this case, the solution to the system is a single ordered pair. The system is said to be **independent and consistent**. Such a system is graphed in Figure 4.
2. The lines determined by the equations in the system are parallel. In this case, the solution to the system is *none*, or empty, and the system is said to be **inconsistent**. Such a system is graphed in Figure 5.
3. The lines determined by the equations in the system are the same line. That is, the equations in the system are *equivalent equations*. In this case, the solution to the system is the infinite set of ordered pairs that solve either of the equations. Such a system is said to be **dependent**. Its graph is a single line.

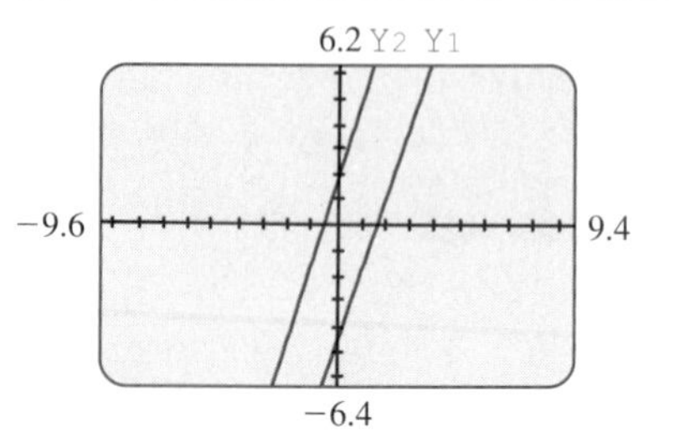

FIGURE 6

We can graph an inconsistent system of equations (case 2) on the graphics calculator: type 3X−4 for Y1, and type 3X+2 for Y2. View the graph screen using the settings given in Figure 2 (on page 212). As Figure 6 shows, the lines are parallel (both have slope measure 3), so the graphs have no intersection point. The solution to the system

$$\begin{cases} y = 3x - 4 \\ y = 3x + 2 \end{cases}$$

is *none*. This linear system is inconsistent.

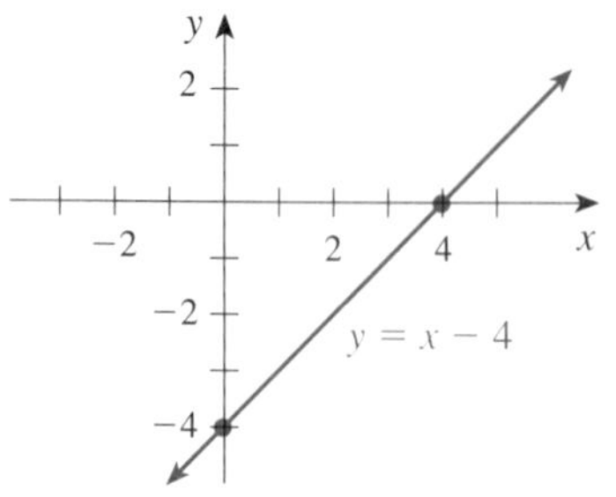

FIGURE 7

Graph of the solution to $\begin{cases} x - y = 4 \\ 3y = 3x - 12 \end{cases}$

An example of a dependent linear system (case 3) is

$$\begin{cases} x - y = 4 \\ 3y = 3x - 12 \end{cases}$$

When these equations are rewritten for y, we get:

$$\begin{cases} y = x - 4 \\ y = x - 4 \end{cases}$$

It is now obvious that the two equations are identical. The solution is the infinite collection of ordered pairs that solve the equation $y = x - 4$. The graph of the solution is the single line drawn in Figure 7. Often the solution is reported by reporting the equation $y = x - 4$, or by reporting the solution using set notation: $\{(x, y) \mid y = x - 4\}$. The ordered pairs that solve the system all have the form $(a, a - 4)$, where a is a real number.

The Sum Equation For a system of two linear equations, the **sum equation** is the equation whose left-hand member is the sum of the left members of the two equations in the system, and the right-hand member is the sum of the right members of the two equations in the system. Thus, the sum equation of the first linear system we considered at the start of the section is

$$\text{System:} \quad \begin{cases} 3x - y = 4 \\ 2x + y = 1 \end{cases}$$

$$\text{Sum equation:} \quad 5x + 0y = 5 \quad \text{or} \quad x = 1$$

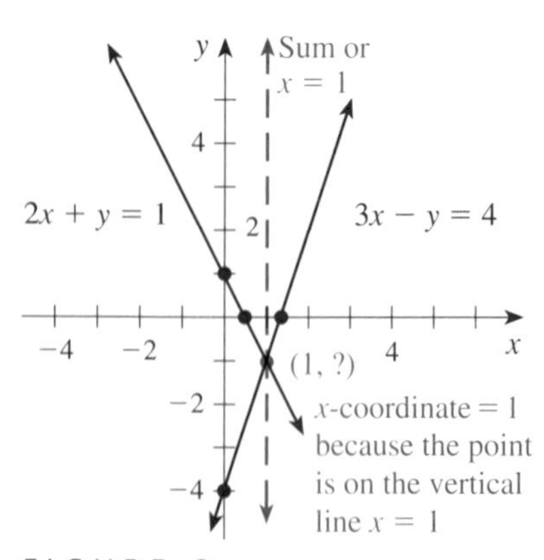

FIGURE 8

Now, if we graph the solution to this sum equation, its line will pass through the intersection point of the two lines determined by the system (see Figure 8). The line determined by this sum equation is the vertical line $x = 1$: every point on this vertical line has x-coordinate 1. Thus, the x-coordinate of the point of intersection is 1.

To find the y-coordinate of the point of intersection, we can replace x by 1 in $3x - y = 4$, and solve $3 - y = 4$. (We could also replace x by 1 in the other equation, $2x + y = 1$, and solve $2 + y = 1$.) The solution to this equation in one variable is -1.

So the name of the intersection point is $(1, -1)$, and the solution to the linear system is $(1, -1)$. We write

$$\begin{cases} 3x - y = 4 \\ 2x + y = 1 \end{cases} \qquad \text{Solution:} \quad (1, -1)$$

Check

The ordered pair $(1, -1)$ solves $3x - y = 4$, because $3(1) - (-1) = 4$ is a true arithmetic sentence.

The ordered pair $(1, -1)$ solves $2x + y = 1$, because $2(1) + (-1) = 1$ is a true arithmetic sentence.

Thus, $(1, -1)$ is the solution to the linear system.

Solving a System of Equations by the Addition Method The solution to a system of two equations in two variables is the collection of ordered pairs that solves the first equation *and* the second equation. For a system of linear equations the solution is usually a single ordered pair. We solved the preceding system of linear equations using the **addition method**. This method is based on the following property.

PROPERTY OF THE SUM EQUATION

> For a system of two equations in two variables, the graph of the solution to the sum equation passes through the intersection point of the graphs of the equations in the system.

For a more complete description of the addition method, work through the following example.

EXAMPLE 1 Solve the system:

$$\begin{cases} 2x - 3y = 8 \\ 3x - y = -2 \end{cases}$$

Solution We start by drawing a graph such as the one in Figure 9, where the solution to $2x - 3y = 8$ is labeled as ① and the solution to $3x - y = -2$ is labeled as ②. The solution to the system will be the ordered-pair name of the point of interesection of the two lines. To find this ordered pair, we use the property of the sum equation. The sum equation is $5x - 4y = 6$. Graphing the solution to the sum equation ($5x - 4y = 6$ in this case) produces another line that passes through the intersection point. This line is graphed in Figure 10, but an infinite number of such lines exist.

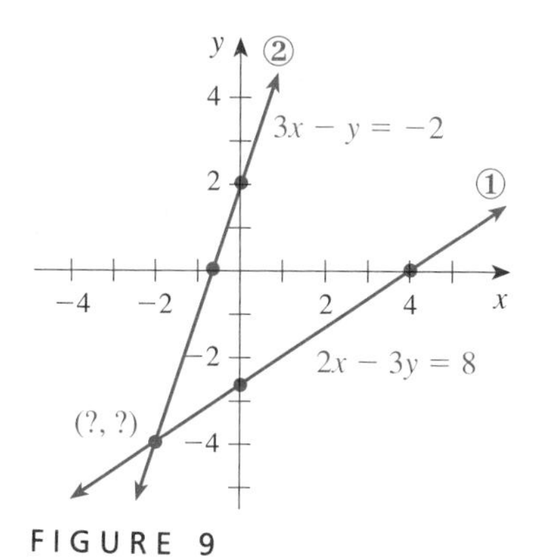

FIGURE 9

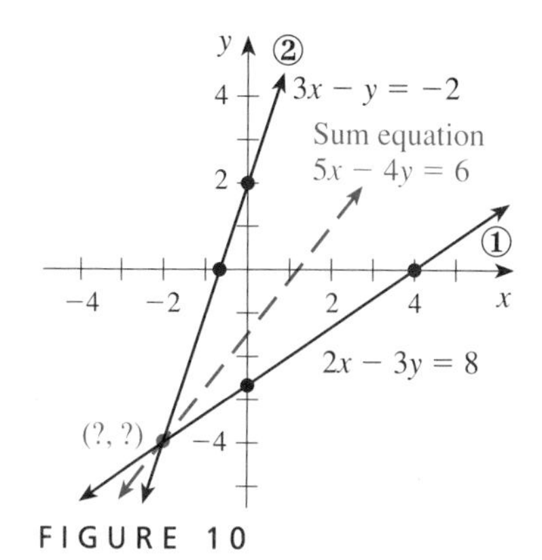

FIGURE 10

We are looking for a vertical or horizontal line through the point of intersection. From one of these lines we would know the x-coordinate or the y-coordinate of the intersection point. To find one of these lines, we alter the system by multiplying each equation by a nonzero number. Remember that multiplying an equation by a nonzero number never alters the solution to the equation. For a system in the form

$$\begin{cases} ax + by = c \\ dx + ey = f \end{cases}$$

we can eliminate x by multiplying the first equation by d and the second equation by $-a$. For the system at hand, we use 3 and -2:

$$\begin{matrix} \text{Multiply by 3:} \\ \text{Multiply by } -2\text{:} \end{matrix} \quad \begin{cases} 2x - 3y = 8 \\ 3x - y = -2 \end{cases} \quad \text{to get} \quad \begin{cases} 6x - 9y = 24 \\ -6x + 2y = 4 \end{cases}$$

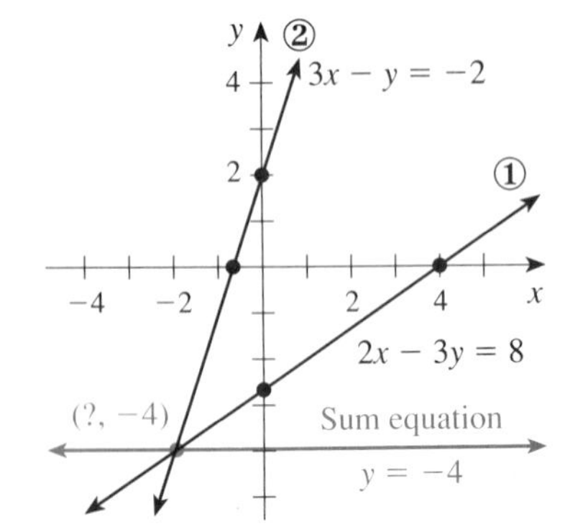

FIGURE 11

The graph of the solution to the equation $6x - 9y = 24$ is still line ①, and the graph of the solution to $-6x + 2y = 4$ is still line ②. Now, the sum of the two equations in this altered system is

$$\begin{array}{rr} & 6x - 9y = 24 \\ & -6x + 2y = 4 \\ \hline \text{Sum equation:} & 0x - 7y = 28 \quad \text{or} \quad y = -4 \end{array}$$

The equation $y = -4$ represents the horizontal line through the intersection point. Thus, the y-coordinate of that point is -4 (see Figure 11).

To find the x-coordinate of the intersection point, we replace the y-variable in $2x - 3y = 8$ with -4, and solve the equation $2x - 3(-4) = 8$. This equation has solution -2. Thus, the solution to the linear system is the ordered pair $(-2, -4)$. ■

Using the Graphics Calculator

To check our work in Example 1, we first rewrite $2x - 3y = 8$ for y to get $y = \frac{2}{3}x - \frac{8}{3}$, and we rewrite $3x - y = -2$ for y to get $y = 3x + 2$. Now, press the Y= key and erase any existing entries. For Y1 type (2/3)X-(8/3), and for Y2 type 3X+2. View the graph using the settings in Figure 12.* With the graph of the system displayed (as in Figure 13) press the TRACE key and move to the point of intersection.

```
RANGE
Xmin=-9.6
Xmax=9.4
Xscl=1
Ymin=-6.4
Ymax=6.2
Yscl=1
Xres=1
```

FIGURE 12

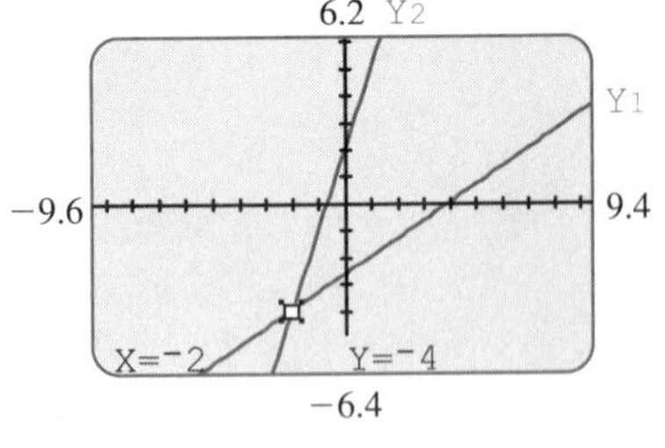

:Y1=(2/3)X-(8/3) .:Y2=3X+2

FIGURE 13

TI-82 Note *Type -9.4 for Xmin and -6.2 for Ymin.

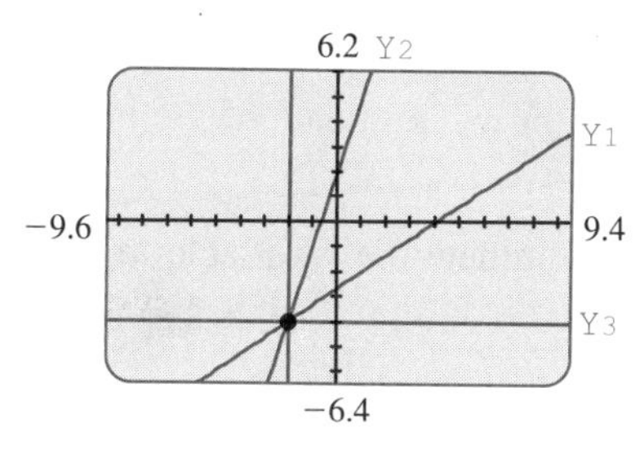

:Y1=(2/3)X-(8/3)
:Y2=3X+2
:Y3=-4

FIGURE 14

Use the down arrow key to verify that this point is on both lines. Press the Y= key and type -4 for Y3. Press the GRAPH key. The horizontal line $y = -4$ does pass through the point of intersection. Return to the home screen (press the CLEAR key), and select the Line(command from the DRAW menu. Complete the command to read Line(-2,10,-2,-10). To type a comma, press the ALPHA key, followed by the period key. Press the ENTER key. As Figure 14 shows, the vertical line $x = -2$ also passes through the point of intersection.

To check that $(-2, -4)$ does in fact produce two true arithmetic sentences, return to the home screen. Store -2 in memory location X and -4 in memory location Y. Type 2X-3Y. (Remember that to type Y, press the ALPHA key, followed by the 1 key.) Press the ENTER key. The number 8 is output. So, the ordered pair $(-2, -4)$ solves $2x - 3y = 8$. Type 3X-Y, and press the ENTER key. The number -2 is output. So, $(-2, -4)$ also solves $3x - y = -2$. We have checked that the ordered pair $(-2, -4)$ solves both equations simultaneously.

As another demonstration of the addition approach to solving a system of linear equations, consider the following example.

EXAMPLE 2 Solve the system:

$$\begin{cases} 5x - (5 - 3y) = 5 + x \\ 5y = 12 - (3x - 2) \end{cases}$$

Solution To solve this system using the addition approach, we first rewrite each equation in standard form:

$$\begin{cases} 4x + 3y = 10 \\ 3x + 5y = 14 \end{cases}$$

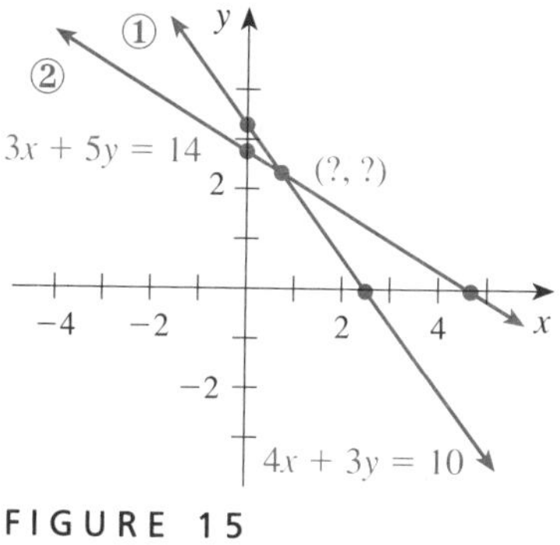

FIGURE 15

The graphs of the solutions to these first-degree equations are shown in Figure 15. The solution to the system is the ordered-pair name of the point of intersection of these lines. The sum equation for this system is the equation $7x + 8y = 24$, and its graph is simply another of the infinite number of lines through the intersection point. Thus, we alter the system in order to produce a sum equation whose graph is the vertical or horizontal line through the intersection point. To get the horizontal line, we multiply the first equation by 3 (the x-coefficient of the second equation). We multiply the second equation by -4 (the additive opposite of the x-coefficient for the first equation) This will eliminate the x-variable:

Multiply by 3: Multiply by −4: $\begin{cases} 4x + 3y = 10 \\ 3x + 5y = 14 \end{cases}$ to get: $\begin{cases} 12x + 9y = 30 \\ -12x - 20y = -56 \end{cases}$

Sum equation: $0x - 11y = -26$

or $y = \dfrac{26}{11}$

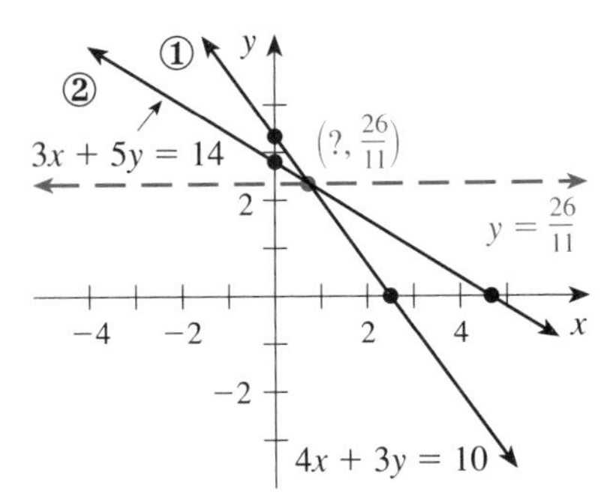

FIGURE 16

So, $y = \frac{26}{11}$ is the equation of the horizontal line through the intersection point. This line is graphed in Figure 16. Thus, the y-coordinate of the intersection point is $\frac{26}{11}$.

Now we return to the original system, and eliminate the y-variable to get the equation of the vertical line through the intersection point. For a system in the form

$$\begin{cases} ax + by = c \\ dx + ey = f \end{cases}$$

we can eliminate y, by multiplying the first equation by e and the second equation by $-b$. For the system at hand, $e = 5$ and $-b = -3$:

$$\begin{array}{ll} \text{Multiply by 5:} & \left\{ \begin{array}{rcr} 4x + 3y &=& 10 \\ 3x + 5y &=& 14 \end{array} \right. \\ \text{Multiply by } -3: & \\ & \\ \text{to get:} & \left\{ \begin{array}{rcr} 20x + 15y &=& 50 \\ -9x - 15y &=& -42 \end{array} \right. \\ \text{with sum:} & \quad 11x + 0y = 8 \quad \text{or} \quad x = \dfrac{8}{11} \end{array}$$

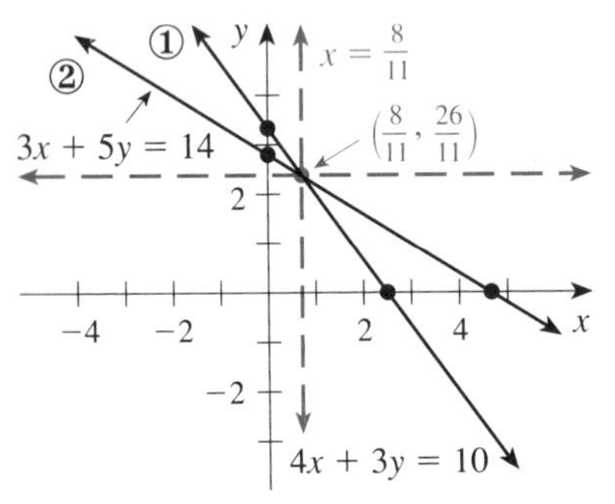

FIGURE 17

So, $x = \frac{8}{11}$ is the equation of the vertical line through the intersection point. We have added this line in Figure 17. Thus, the x-coordinate of the intersection point is $\frac{8}{11}$.

The solution to the system is $\left(\frac{8}{11}, \frac{26}{11}\right)$. ■

Using the Graphics Calculator

```
RANGE
Xmin=-9.6
Xmax=9.4
Xscl=1
Ymin=-6.4
Ymax=6.2
Yscl=1
Xres=1
```

FIGURE 18

Y2 Y1 6.2

−9.4 9.6

−6.4

:Y1=(⁻4/3)X+10/3
:Y2=(⁻3/5)X+14/5

FIGURE 19

When checking the solution to a system, always return to the original equations. Store `8/11` in memory location `X` and `26/11` in memory location `Y`. Type `5X−(5−3Y)`, and then press the `ENTER` key. Type `5+X` and then press the `ENTER` key. In both cases the number `5.727272727` is output. Thus, $\left(\frac{8}{11}, \frac{26}{11}\right)$ solves the equation

$$5x - (5 - 3y) = 5 + x$$

Now, type `5Y` and press the `ENTER` key. Type `12−(3X−2)` and press the `ENTER` key. In both cases the number `11.81818182` is output. Thus, the ordered pair $\left(\frac{8}{11}, \frac{26}{11}\right)$ also solve the equation $5y = 12 - (3x - 2)$.

To increase our understanding of the system in Example 2, we can graph the system and graphically check the intersection point. To do so, we must rewrite each equation for y. For `Y1` type `(⁻4/3)X+10/3`, and for `Y2` type `(⁻3/5)X+14/5`. View the graph screen using the settings in Figure 18.* The graph is shown in Figure 19. For `Y3` type `26/11`. View the graph. As you can see, the horizontal line $y = \frac{26}{11}$

TI-82 Note *Type `⁻9.4` for `Xmin` and `⁻6.2` for `Ymin`.

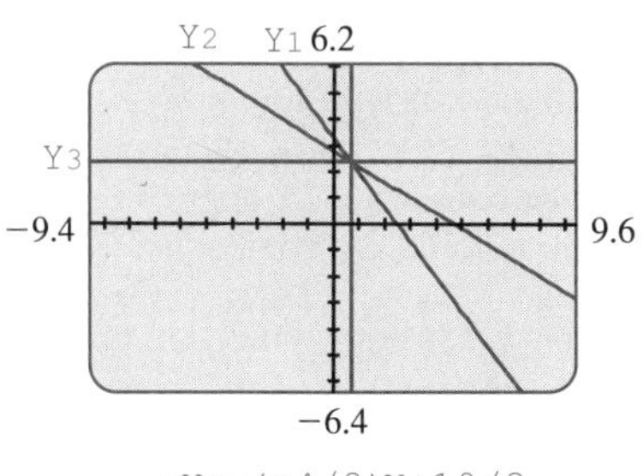


```
:Y1=(-4/3)X+10/3
:Y2=(-3/5)X+14/5
:Y3=26/11
```

FIGURE 20

passes through the point of intersection. Return to the home screen and select the `Line(` command from the `DRAW` menu. Complete the command to read `Line(8/11,9,8/11,-9)` and then press the `ENTER` key.

Figure 20 shows that the new horizontal line `Y3` and the new vertical line $x = \frac{8}{11}$ passes through the intersection point.

For an improved view, select the `Zoom In` option from the `ZOOM` menu, move the cursor to the point of intersection, and then press the `ENTER` key. After each Zoom command, you must redraw the vertical line $x = \frac{8}{11}$.

GIVE IT A TRY

3. Consider the system

$$\begin{cases} 3x - y = 8 \\ 2x + y = 3 \end{cases}$$

Graph the solution to the equation $3x - y = 8$, and label the line as ①. On the same set of axes, graph the solution to $2x + y = 3$, and label this line as ②. Add the two equations together, graph the solution to this sum equation, and label it *sum*. What is the x-coordinate of the intersection point? Replace the x-variable in $3x - y = 8$ with the x-coordinate of the intersection point, and solve the resulting equation in one variable. What is its solution? What is the solution to the system?

4. Use the following system to answer parts (a)–(f):

$$\begin{cases} 3x + 2y = 9 \\ 4x - 3y = 6 \end{cases}$$

a. Graph the solution to the equation $3x + 2y = 9$, and label the line as ①. On the same set of axes, graph the solution to $4x - 3y = 6$, and label the line as ②.

b. Multiply the first equation by 4 and the second equation by -3. Report the altered system.

c. The graph of the solution to the equations in the altered system in part (b) is the same as the graph of solution to the equations in the original system. For the altered system, find the sum equation and graph it on the axes used in part (a). What is the y-coordinate of the intersection point?

d. Multiply the first equation in the original system by 3 and the second equation in the original system by 2. Report the altered system.

e. For the altered system in part (d), find the sum equation and graph it on the same axes as used in part (a). What is the x-coordinate of the intersection point?

f. What is the solution to the original system?

```
RANGE
Xmin=-4.8
Xmax=4.7
Xscl=1
Ymin=-3.2
Ymax=3.1
Yscl=1
Xres=1
```

FIGURE 21

5. Using the graphics calculator, rewrite the equation $3x + 2y = 9$ for y, and type the resulting right-hand member for `Y1`. Rewrite the equation $4x - 3y = 6$ for y, and type the resulting right-hand member for `Y2`. Press the `RANGE` key and enter the values shown in Figure 21.* View the graph. Use the arrow keys to

TI-82 Note *Use the `ZDecimal` option.

move the cursor to the point (2.3, 1.05). Press the ZOOM key and select the Zoom In option (2). Press the ENTER key to zoom in with the new view centered on the point (2.3, 1.05). Move the cursor to the point of intersection. Give an approximation of the point of intersection. How does the approximate value compare with the exact solution to the system found in Problem 4?

Special Cases What happens if we attempt to use the addition method to solve an inconsistent system of linear equations (a system whose solution is *none*, because the lines are parallel)? Consider the following system and its sum equation:

$$\begin{cases} 3x - 2y = 8 \\ 3x - 2y = 2 \end{cases}$$

Multiply $3x - 2y = 8$ by 1:
Multiply $3x - 2y = 2$ by -1:
The sum equation is:

$$\begin{cases} 3x - 2y = 8 \\ -3x + 2y = -2 \end{cases}$$
$$0 = 6$$

When a system produces a sum equation that is a false arithmetic sentence, the system is inconsistent. The solution is *none*.

What happens if we attempt to solve a dependent system of linear equations (a system whose solution is an infinite collection of ordered pairs, because the lines are identical)? Consider the following system and its sum equation:

$$\begin{cases} 2x - 3y = 8 \\ 4x - 6y = 16 \end{cases}$$

Multiply $2x - 3y = 8$ by -2:
Multiply $4x - 6y = 16$ by 1:
The sum equation is:

$$\begin{cases} -4x + 6y = -16 \\ 4x - 6y = 16 \end{cases}$$
$$0 = 0$$

When a system produces a sum equation that is a true arithmetic sentence, the system is a dependent system. The two equations are equivalent equations (they have the same solution). We can report the solution as an equation by writing

$$\text{Solution:} \quad y = \frac{2}{3}x - \frac{8}{3}$$

or we can report the solution in set notation by writing

$$\text{Solution:} \quad \left\{(x, y) \mid y = \frac{2}{3}x - \frac{8}{3}\right\}$$

If the variable is eliminated from an equation in *one* variable, and a true arithmetic sentence is produced, then the solution is *all* numbers. When solving a system of linear equations in *two* variables, if both variables are eliminated and a true arithmetic sentence is produced, the solution is *not* all ordered pairs. However, the solution is an infinite set of ordered pairs—the ordered pairs that satisfy either equation in the system.

The Substitution Approach Because systems of linear equations are so important in mathematics and in areas related to mathematics, several different approaches have been developed for solving a system of linear equations. Consider solving the following system of linear equations:

$$\begin{cases} y = 3x - 2 \\ y = -2x + 3 \end{cases}$$

We could easily rewrite these equations in standard form and use the addition method. However, we want to introduce another approach to solving a system of equations—the **substitution method**.

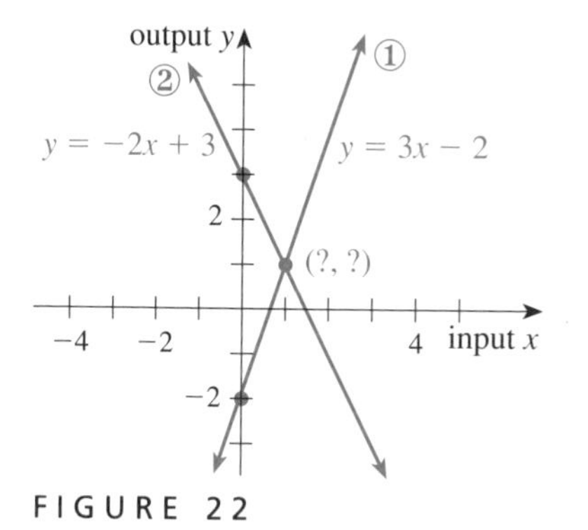

FIGURE 22

The equations in this system are in ''function'' form (think of y as a function of x). For the equation $y = 3x - 2$, we think of inputting an x-value, and then getting an output of a y-value (the function rule is: multiply the input value by 3, then subtract 2). Following this rule with all the x-values possible produces line ① shown in Figure 22. For the equation $y = -2x + 3$, we think of an input of an x-value, and then get an output of the y-value (the function rule is: multiply the input value by -2, and then add 3). Following this rule for all the x-values possible produces line ② in Figure 22. The point of intersection of the two lines can be thought of as the point where, for the given input value x, the two rules, $3x - 2$ and $-2x + 3$, produce the same output value y. For this input value, we have:

Solving this equation, we get:

$$\begin{array}{rcl} 3x - 2 & = & -2x + 3 \\ \underline{2x \qquad} & & \underline{\qquad 2x} \\ 5x - 2 & = & 3 \\ \underline{\qquad 2} & & \underline{2 \qquad} \\ 5x & = & 5 \\ x & = & 1 \end{array}$$

So, the solution is the number 1. When the number 1 is input for x, the rule $3x - 2$ outputs the y-value 1. When 1 is input for x, then the rule $-2x + 3$ also outputs the y-value 1. Thus, the solution to the system is the ordered pair (1, 1).

We now will use the substitution approach to solve the following system of linear equations.

EXAMPLE 3 Solve the system, using the substitution approach:

$$\begin{cases} 5x - (5 - 3y) = 5 + x \\ 5y = 12 - (3x - 2) \end{cases}$$

Solution For this system, we start by rewriting each of the linear equations for y (in function form). The altered system is

$$\begin{cases} y = \dfrac{-4}{3}x + \dfrac{10}{3} \\[2ex] y = \dfrac{-3}{5}x + \dfrac{14}{5} \end{cases}$$

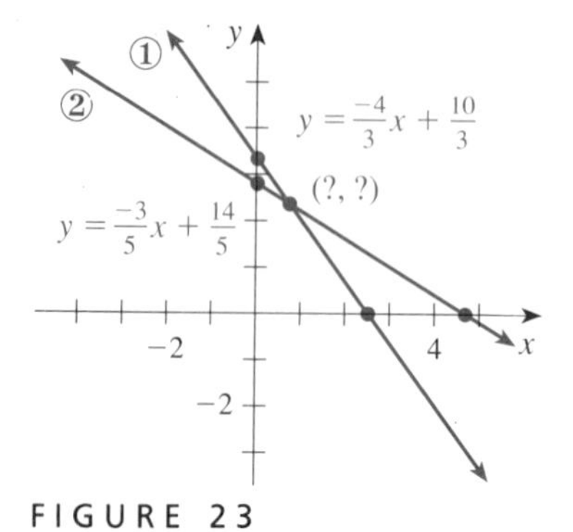

FIGURE 23
The graph of the system
$$\begin{cases} 5x - (5 - 3y) = 5 + x \\ 5y = 12 - (3x - 2) \end{cases}$$

The graphs of the lines determined by these equations are shown in Figure 23. The intersection point is where the function outputs are equal:

$$\frac{-4}{3}x + \frac{10}{3} = \frac{-3}{5}x + \frac{14}{5}$$

We solve this equation in one variable:

$$\frac{-4}{3}x + \frac{10}{3} = \frac{-3}{5}x + \frac{14}{5}$$

	$\frac{-4}{3}x + \frac{10}{3} = \frac{-3}{5}x + \frac{14}{5}$
Multiply each side by 15 to get:	$-20x + 50 = -9x + 42$
Add $9x$ to each side	$9x \qquad 9x$
to get:	$-11x + 50 = 42$
Add -50 to each side	$-50 \qquad -50$
to get:	$-11x = -8$
Multiply by $\frac{-1}{11}$:	$x = \frac{8}{11}$

The solution to this equation is $\frac{8}{11}$. Thus, the input value x that results in $\frac{-4}{3}x + \frac{10}{3}$ and $\frac{-3}{5}x + \frac{14}{5}$ producing the same output is $\frac{8}{11}$. So, the x-coordinate of the intersection point is $\frac{8}{11}$.

To find the y-coordinate (the output value), we substitute (or input) $\frac{8}{11}$ for the x-variable in either of the equations. Selecting the first equation yields

$$y = \frac{-4}{3}\left(\frac{8}{11}\right) + \frac{10}{3}$$

or

$$y = \frac{26}{11}$$

Thus, the solution to the system is $\left(\frac{8}{11}, \frac{26}{11}\right)$. ■

You may want to compare the substitution approach to the addition approach shown in Example 2 for this same system. Although the approaches are different, the result is the same—the solution to the system is the ordered pair $\left(\frac{8}{11}, \frac{26}{11}\right)$. Which approach is best, the addition approach or the substitution approach? There is no *best* approach—both methods are valid and are used frequently. Often the addition approach is used when the equations in the system are given in standard form, $Ax + By = C$. The substitution approach is typically used when the equations in the system are given in slope, y-intercept form, $y = mx + b$.

The next example shows a form of the substitution approach used on a linear system whose equations are given in standard form. *Although a graph of the system is not required to solve the system, drawing one can provide a quick visual check for your work.* Producing graphs repeatedly (like solving first-degree equations in one variable) provides the practice needed to be certain that you understand this process.

EXAMPLE 4 Use the substitution method to solve the system:

$$\begin{cases} 3x + 2y = 6 \\ 2x - 5y = 8 \end{cases}$$

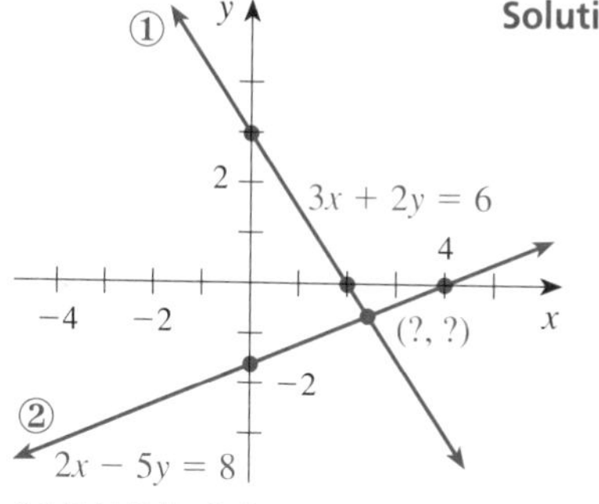

FIGURE 24

Solution As usual, we start with a graph of the system as shown in Figure 24.

Next we rewrite the first equation for y to get: $y = \frac{-3}{2}x + 3$

This expression for y is substituted in the second equation: $2x - 5\left(\frac{-3}{2}x + 3\right) = 8$

This produces a first-degree equation in one variable, which we can solve.

$$\begin{aligned} 2x + \frac{15}{2}x - 15 &= 8 && \text{Clear parentheses.} \\ 4x + 15x - 30 &= 16 && \text{Multiply each side by 2.} \\ 19x - 30 &= 16 && \text{Combine like terms.} \\ 30 &\quad 30 && \text{Add 30 to each side.} \\ 19x &= 46 && \text{Multiply by } \tfrac{1}{19}. \\ x &= \frac{46}{19} \end{aligned}$$

So, the x-coordinate of the ordered pair that solves the system is $\frac{46}{19}$ (check this against the graph in Figure 24).

Replacing x with this value in the first equation rewritten for y, we get:

$$\begin{aligned} y &= \left(\frac{-3}{2}\right)\left(\frac{46}{19}\right) + 3 \\ y &= \frac{-69}{19} + \frac{57}{19} \\ y &= \frac{-12}{19} \end{aligned}$$

So, the y-coordinate is $\frac{-12}{19}$ (check this against the graph).

The solution to the system is $\left(\frac{46}{19}, \frac{-12}{19}\right)$. ■

Using the Graphics Calculator

To check the solution $\left(\frac{46}{19}, \frac{-12}{19}\right)$ to the system in Example 4, store `46/19` in memory location `X`. Store `-12/19` in memory location `Y`. Type `3X+2Y` and press the `ENTER` key. Is `6` output? Type `2X-5Y` and press the `ENTER` key. Is `8` output? If so, our solution is correct.

GIVE IT A TRY

Solve each system using the substitution method.

6. $\begin{cases} y = 3x + 3 \\ y = 3x + 1 \end{cases}$ 7. $\begin{cases} y = -3x + 3 \\ y = \frac{1}{2}x - 1 \end{cases}$ 8. $\begin{cases} -3x + 2y = 3 \\ 2x + 6y = -10 \end{cases}$

■ SUMMARY

■ ***Quick Index***

In this section, we have learned to solve a system of first-degree equations in two variables. To solve such a system, we must find the ordered pairs that satisfy (solve) both equations. The solution to a first-degree equation in two variables is a straight line when graphed. Two straight lines can intersect in at most one point, so the solution to a system of first-degree equations in two variables is typically a single ordered pair. Such a system is independent and consistent. If the lines are parallel, then the system has solution *none,* and we say that the system is inconsistent. Finally, if the lines determined are the same line, then the solution is the infinite collection of ordered pairs for that line. The system is dependent.

Two methods of solving a system of linear equations have been presented in this section—the addition method and the substitution method. In different situations, one method may be preferred over the other. For example, if the equations are in the form $Ax + By = C$, then the addition method is preferred. If the equations are in the form $y = mx + b$, then the substitution method is preferred. No matter which of the two methods of solving the linear system is used, the solution produced is the same.

4.3 ■ ■ ■ EXERCISES

For each linear system, graph the solution to the first equation and label it A. Graph the solution to the second equation and label it B. Using the graphics calculator, its TRACE *key, and the* Zoom In *option, find the intersection point and label the graph with its ordered-pair name (approximated to the nearest 0.01).*

1. $\begin{cases} 3x - y = 8 \\ 2x + y = -2 \end{cases}$
2. $\begin{cases} 2x + 5y = 12 \\ -2x + 3y = 8 \end{cases}$
3. $\begin{cases} 4x + 2y = 4 \\ x + 4y = -2 \end{cases}$
4. $\begin{cases} 4x - y = 6 \\ -2x + 3y = 8 \end{cases}$
5. $\begin{cases} x - y = 4 \\ x + y = -3 \end{cases}$
6. $\begin{cases} 12x + 2y = 6 \\ 6x + y = 4 \end{cases}$
7. $\begin{cases} 3x - y = 2 \\ 2x + y = 3 \end{cases}$
8. $\begin{cases} 2x - 4y = -4 \\ x - y = -3 \end{cases}$
9. $\begin{cases} x - 2y = -2 \\ 3x + 10y = -2 \end{cases}$
10. $\begin{cases} 8x + y = 2 \\ 2x - y = 4 \end{cases}$

Solve each linear system. Use the graphics calculator to check your work.

11. $\begin{cases} 4x - 5y = 9 \\ 5x + 2y = 8 \end{cases}$
12. $\begin{cases} -3x + 2y = 9 \\ x - 4y = 6 \end{cases}$
13. $\begin{cases} -3x + 7y = 12 \\ -2x + 5y = 9 \end{cases}$
14. $\begin{cases} y = -3x + 8 \\ 2(x + y) = 6 - y \end{cases}$
15. $\begin{cases} 3(x - 2) = 2(y - 5) \\ 3x = y - 5(2 - x) \end{cases}$
16. $\begin{cases} 2x - y = 6 - 3y \\ 3(x - 1) + 3y = 7 \end{cases}$
17. $\begin{cases} 11x - 9y = 15 \\ 4x - 10y = -8 \end{cases}$
18. $\begin{cases} 6x - 10y = -25 \\ 3x - 5y = 5 \end{cases}$
19. $\begin{cases} 6y - 3x = 16 \\ 7x + 19y = -188 \end{cases}$
20. $\begin{cases} x - 3y = 10 \\ 2x - y = 6 \end{cases}$
21. $\begin{cases} y = 3x - 5 \\ y = -2x + 3 \end{cases}$
22. $\begin{cases} y = 3x + 1 \\ y = \frac{1}{2}x - 2 \end{cases}$

23. $\begin{cases} \frac{2}{3}x + \frac{1}{2}y = 3 \\ \frac{2}{5}x - \frac{3}{2}y = -1 \end{cases}$

24. $\begin{cases} \frac{1}{5}x + \frac{7}{2}y = -2 \\ \frac{7}{3}x + \frac{4}{5}y = 3 \end{cases}$

25. $\begin{cases} x + y = 18 \\ 0.3x + 0.2y = 5 \end{cases}$

26. $\begin{cases} x - y = 10 \\ 0.6x - 0.1y = 2 \end{cases}$

27. Graph the line determined by $y = 2x - 3$. On the same set of axes, graph the line through the point (2, 3) that is perpendicular to the line determined by $y = 2x - 3$. What is the ordered-pair name of the intersection point of the two lines?

28. Graph the line determined by $y = -3x + 5$. On the same set of axes, graph the line through the point $(-2, -1)$ that is perpendicular to the line determined by the equation $y = -3x + 5$. What is the ordered-pair name of the intersection point of the two lines?

Label each system as consistent, inconsistent, *or* dependent. *(Do not solve the system.)*

29. $\begin{cases} 2x + 3y = 3 \\ 10x - 15y = 6 \end{cases}$

30. $\begin{cases} 2x + 3y = 3 \\ 10x + 15y = 6 \end{cases}$

31. $\begin{cases} y = 4x + 7 \\ 4x - y = 2 \end{cases}$

32. $\begin{cases} 5x = 2(y - 3) \\ 2y = 5(x + 1) \end{cases}$

Find the input value x that results in the given linear functions producing the same output value.

33. $\begin{cases} f(x) = -2x + 5 \\ h(x) = 3x - 4 \end{cases}$

34. $\begin{cases} g(x) = 3x - 2 \\ d(x) = -4x + 5 \end{cases}$

DISCOVERY

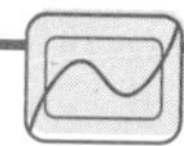

Revenue and Cost

In this discovery section, we explore cost functions and revenue functions. Our goal is to find where the functions are equal (produce the same output).

The revenue received for a product is a function of the number of units of the product sold. Typically, a revenue function appears as $R(x) = 3x$, where x is the number of units sold. The x-coefficient 3 is the slope of the line. It represents the price of each unit. That is, if no units are sold, then the revenue is \$0. If 1 unit is sold, then the revenue is \$3. For each increase of 1 unit sold, the revenue increases by \$3. Suppose 100 units are sold. The revenue is then $R(100) = 3(100) = 300$, that is, the revenue is \$300.

```
RANGE
Xmin=0
Xmax=950
Xscl=100
Ymin=0
Ymax=2850
Yscl=500
Xres=1
```

FIGURE 1

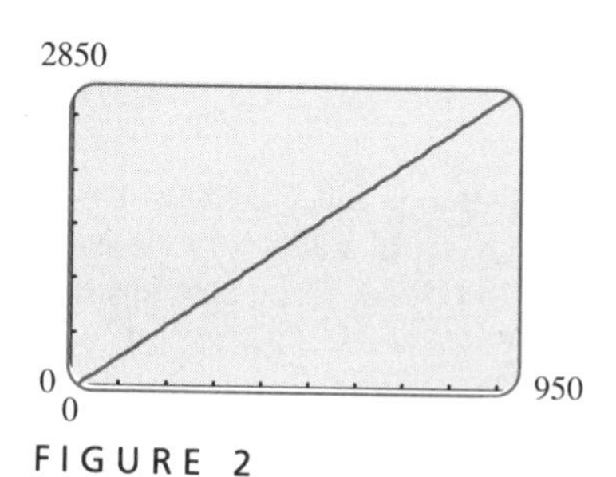

FIGURE 2

Using the Graphics Calculator For `Y1`, type `3X`. Suppose we are interested in a domain from 0 to 950 (units). The range for these domain values is all values from $R(0)$ to $R(950)$. So, the range is from 0 to 2850 (dollars). Press the `RANGE` key and enter the settings shown in Figure 1.* (These settings are a trial, so we may have to adjust them later.) View the graph screen, as shown in Figure 2. Press the `TRACE` key. Move the cursor until the message reads `X=400 Y=1200`. This message means that when 400 units are sold, the revenue is \$1,200. Press the right arrow key. The message reads `X=410 Y=1230`. This message indicates that an increase in units sold of 10 units results in an increase in revenue of \$30.

TI-82 Note *Type `940` for `Xmax`.

Suppose that to produce this product we have a start-up cost of \$1,000, and that it costs \$1.50 for each unit made available for sale. The cost function will be

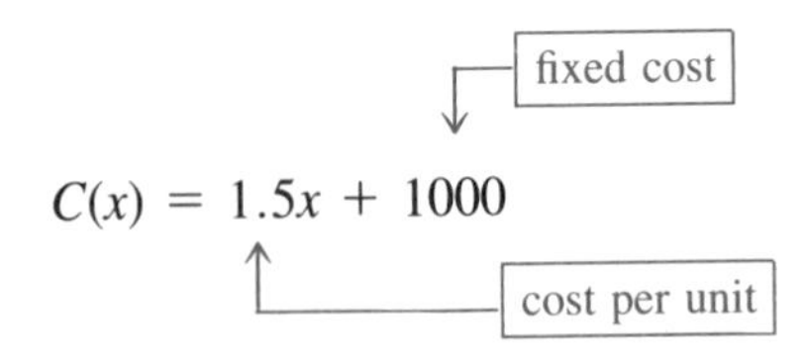

$$C(x) = 1.5x + 1000$$

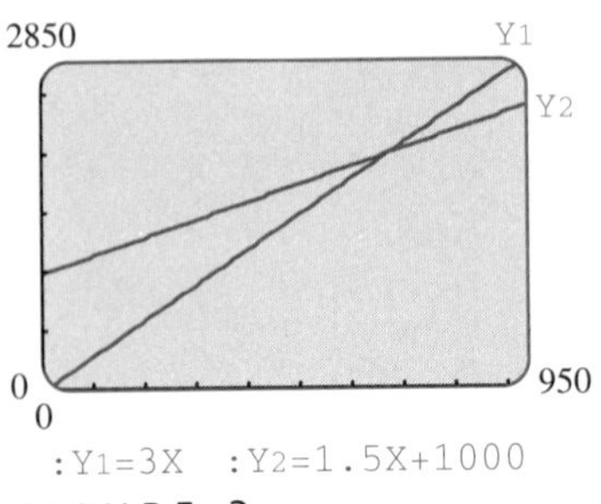

FIGURE 3

The \$1,000 start-up cost is the fixed cost (the constant term). The cost of \$1.50 per unit is the x-coefficient (the slope). We now want to know how many units must be sold to break even, that is, how many units must be sold for revenue received to equal the production cost. The revenue earned from all units sold after this *break-even point* will result in a profit.

For Y2, type 1.5X+1000. View the graph screen, as shown in Figure 3. Press the TRACE key, and then the down arrow key, to move the cursor to the cost function. Move the cursor until the message reads X=400 Y=1600. This message means that the cost for making 400 units available for sale is \$1,600. This amount, \$1,600, includes the \$1,000 fixed cost plus \$600, the cost of producing 400 units at \$1.50 each [400(1.5) is 600].

Move the cursor to the intersection point of the two lines. Zoom in on the point. Press the TRACE key and move the cursor to the intersection point. It appears that when about 667 units are sold, the revenue equals the cost. Press the CLEAR key to return to the home screen. Store 667 in memory location X. Display the value in Y1 (the revenue). That value is 2001. Display the value in Y2 (the cost). That value is 2000.5. For 667 units the revenue is greater than the cost, so the profit is \$0.50. Store 666 in memory location X. Display the revenue (the Y1 value). That value is 1998. Display the cost (the Y2 value). That value is 1999. So, when 666 units are sold the revenue is less than the cost, and a loss of \$1 is incurred. We may say that the break-even point is approximately (667, 2001): when 667 units are sold, the revenue is approximately equal to the cost. If fewer units are sold, we have a loss. If 667 units or more than 667 units are sold, a profit is made.

Exercises

1. The Kid Books Company has a start-up cost of \$3,000 to produce a coloring book of famous works of art (especially those by Monet). In addition to these start-up costs, the cost of production per book is \$2.00. What is the cost function for the coloring book?
2. If the Kids Books Company sells the coloring book for \$5.00 per copy, what is the revenue function for the product?
3. Use the graphics calculator to graph the cost and revenue functions. The domain should start at 0 (that is Xmin and Ymin on the RANGE screen should be 0).

Experiment with various settings for Xmax and Ymax so that the intersection point is at the center of the screen. Transfer the graph to paper. Use the TRACE key and the Zoom In option to approximate the coordinates of the break-even point.

4. From the home screen, use the X, Y1, and Y2 memory locations to report the cost for producing 1,200 coloring books and the revenue from selling 1,200 coloring books. If all the books are sold, is there a loss or a profit? How much?

4.4 ■ ■ ■ APPLICATIONS OF SYSTEMS

In the previous section we learned that the solution to a system of two first-degree equations in two variables is the collection of ordered pairs that solve the first equation *and* the second equation. Typically, a single ordered pair solves such a system. In this section, we will learn that a system of first-degree equations in two variables provides a mathematical model for some kinds of real-world problems (word problems that request two values).

Mathematical Models A system of linear equations can serve as a mathematical model for many situations where two unknown values are requested. Consider the following problem:

Problem Find two numbers such that their sum is 58 and their difference is 14.

Unlike the word problems in Chapter 2, here we are looking for *two* numbers. Thus, rather than using a single equation in one variable as a mathematical model, we will use a system of two equations in two variables. To solve this problem we must read it, produce a system of equations, and then solve the system. We will follow the same steps listed on page 124. Thus, we start by determining what quantities we wish to find, and then listing a variable to represent each of these quantities.

Let x = one of the numbers

and y = the other number.

Next, we list relevant information in terms of these variables:

The sum of the numbers is 58: $x + y = 58$
The difference of two numbers is 14: $x - y = 14$

Using the Graphics Calculator

Before we solve the preceding system of equations, we can use the calculator to visualize the situation. First, we rewrite the equations for y. The equation $x + y = 58$ yields $y = -x + 58$, and the equation $x - y = 14$ yields $y = x - 14$. For Y1 type -X+58, then press the

```
RANGE
Xmin=0
Xmax=95
Xscl=10
Ymin=0
Ymax=60
Yscl=10
Xres=1
```

FIGURE 1

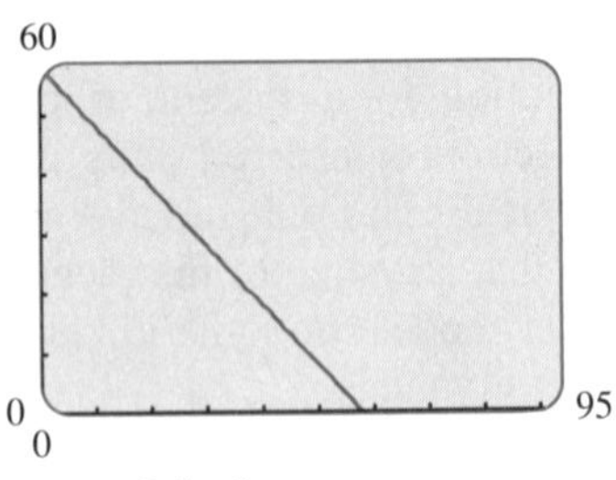

FIGURE 2

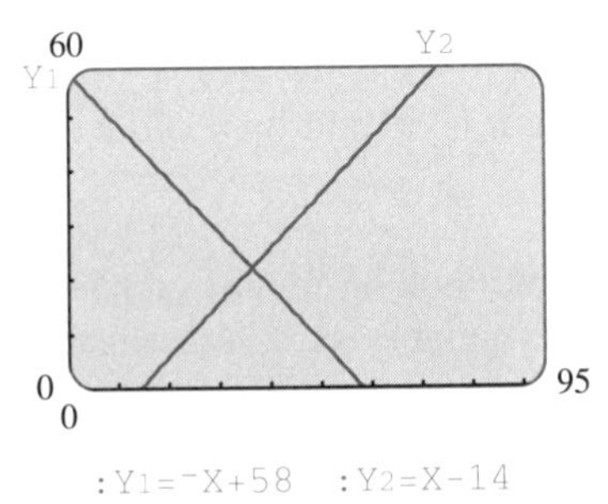

FIGURE 3

RANGE key and type the values shown in Figure 1.* View the graph screen, as shown in Figure 2. This line represents the pairs of numbers that add up to 58. Press the TRACE key, and move the cursor until the message reads X=24 Y=34. Notice that the sum of 24 and 34 is 58. As we can see, the graph indicates an infinite number of pairs of numbers whose sum is 58 (they are all represented by the line drawn). For Y2 type X−14, and then view the graph screen. Figure 3 shows this screen. This new line represents the pairs of numbers for which the x-coordinate minus the y-coordinate is 14. Press the TRACE key, and then the down arrow key to move the cursor to this second line. Move the cursor until the message reads X=24 Y=10. Notice that $24 - 10$ is 14; however, the sum of these two numbers is not 58. Again, there is an infinite number of pairs whose difference is 14. The intersection point of these two lines represents a pair of numbers whose sum is 58 *and* whose difference is 14. Move the cursor to the intersection point of the two lines. The message reads X=36 Y=22. These two numbers have a sum of 58 *and* a difference of 14.

Although we have found the answer graphically, we will continue showing the algebraic approach. Our next step is to solve the system of equations.

$$\begin{cases} x + y = 58 \\ x - y = 14 \end{cases}$$

Using the addition method, we find the sum equation: $2x = 72$ or $x = 36$

Thus, the x-coordinate of the ordered pair is 36. Using this information we get

$$\begin{aligned} 36 + y &= 58 \\ -36 \quad & \quad -36 \\ \hline y &= 22 \end{aligned}$$

Thus, the ordered pair that solves the system is (36, 22).

We can now answer the question posed by the problem: One number is 36, the other 22.

Check Is the sum 58? Yes: $36 + 22 = 58$.
Is the difference 14? Yes: $36 - 22 = 14$.

We now consider a word problem that is very similar to the number problem we just solved.

EXAMPLE 1 An airplane flies with the wind at a speed of 600 miles per hour (mi/h). When flying against the wind, the plane's speed is 500 mi/h. What is the speed of the plane, and what is the speed of the wind?

Solution We are seeking two values, thus a system of equations is selected to model the situation. We can identify two variables:

TI-82 Note *Type 94 for Xmax.

Let x = speed of the airplane and y = speed of the wind.

Reading the problem, we get the equation

$$x + y = 600$$

because when flying with the wind, the speed of the airplane plus the speed of the wind is 600 mi/h. Also, we get the equation

$$x - y = 500$$

because when flying against the wind, the speed of the airplane minus the speed of the wind is 500 mi/h. Thus, we have the system

$$\begin{cases} x + y = 600 \\ x - y = 500 \end{cases}$$

Solving this system,
we find the sum equation: $2x = 1100$ or $x = 550$

Replacing the x-variable in $x + y = 600$ with 550, we get:

$$\begin{array}{rcr} 550 + y & = & 600 \\ -550 & & -550 \\ \hline y & = & 50 \end{array}$$

So, the solution to the system is the ordered pair (550, 50).
Using our list for the variables, we answer the question:

The airplane's speed is 550 mi/h.
The wind's speed is 50 mi/h.

Check The speed of the plane with the wind is 550 + 50, or 600 mi/h.
The speed of the plane against the wind is 550 − 50, or 500 mi/h. ■

```
RANGE
Xmin=0
Xmax=950
Xscl=50
Ymin=0
Ymax=610
Yscl=50
Xres=1
```

FIGURE 4

Using the Graphics Calculator

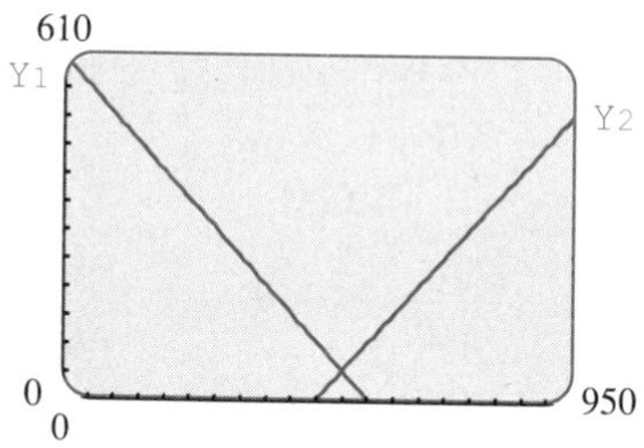

FIGURE 5

To check our work in Example 1, we first rewrite the equations for y: $y = -x + 600$ *and* $y = x - 500$. Type ⁻X+600 for Y1 and X−500 for Y2. Press the RANGE key and enter the values shown in Figure 4.* View the graph screen, as shown in Figure 5. Press the TRACE key. The points on this line represent the possibilities for flying *with the wind* to get a speed of 600 mi/h. (The x-coordinate is the speed of the plane, and the y-coordinate is the wind speed.) Press the down arrow key. The points on this line represent the possibilities of flying *against the wind* to get a speed of 500 mi/h. Move the cursor to the point of intersection. We see that it is (550, 50).

TI-82 Note *Type 940 for Xmax.

We can also use a system of equations to model a mixture problem, as in the following example.

EXAMPLE 2 The L Club recently ran a bake sale, selling cupcakes and cups of coffee to earn money for charity. The coffee was sold for \$0.50 a cup and the cupcakes for \$0.70 each. Although the records were lost, a member does recall that 120 items were sold and that the total receipts were \$68. How many cups of coffee and how many cupcakes did the club sell?

Solution Two values are requested. Thus, we define the variables as follows:

Let x = number of cups of coffee sold

and y = number of cupcakes sold.

The facts given in the problem are recorded in the table shown in the margin. Because the total number of items sold is 120, we have the following equation:

Item	*Price*	*Number sold*	*Value*
Coffee	\$0.50	x	$0.5x$
Cupcake	\$0.70	y	$0.7y$
Total		120	\$68

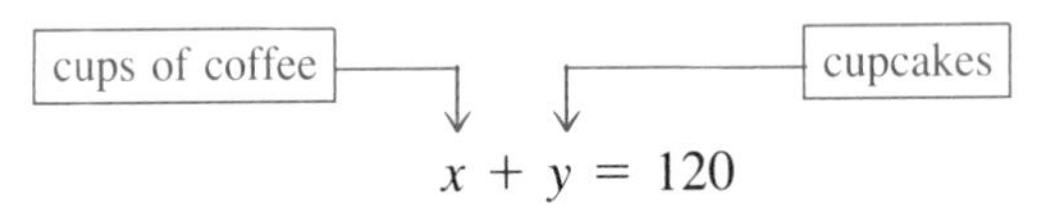

$$x + y = 120$$

Because the total receipts are \$68, we have:

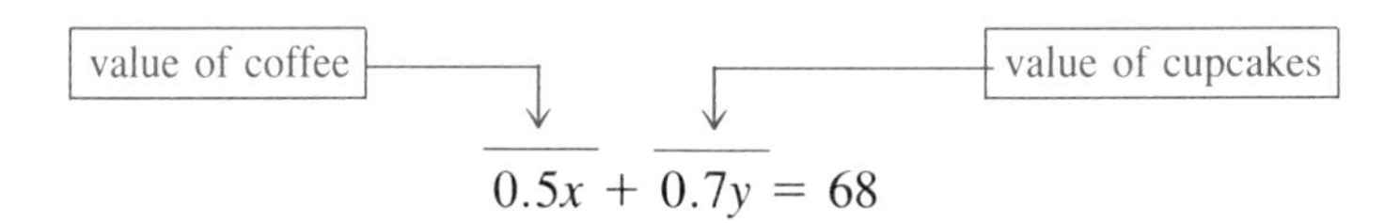

$$0.5x + 0.7y = 68$$

Thus, the system of equations is:

$$\begin{cases} x + y = 120 \\ 0.5x + 0.7y = 68 \end{cases}$$

To solve this system using the substitution method, we first rewrite the equations for y:

$$\begin{cases} y = -x + 120 \\ y = \dfrac{-5}{7}x + \dfrac{680}{7} \end{cases}$$

Setting the rules equal, we get: $-x + 120 = \dfrac{-5}{7}x + \dfrac{680}{7}$

Multiply by 7: $-7x + 840 = -5x + 680$

Add $5x$ to each side to get: $-2x + 840 = 680$

Add -840 to each side to get: $-2x = -160$

Multiply by $\frac{-1}{2}$: $x = 80$

Substituting 80 for x in the first equation, we get $y = -80 + 120$, or $y = 40$.

WARNING
Without the definition of the variables, x and y, it is difficult to answer the question posed correctly.

Thus, the solution to the system is the ordered pair (80, 40). Using our definition of the variables, we answer the question: The club sold 80 cups of coffee and 40 cupcakes.

Check The number of items sold, 80 + 40, is 120.
The value of these 120 items is

$$0.5(80) + 0.7(40) = 40 + 28 = \$68$$ ■

The next example is similar to the acid solution word problem we solved in Section 2.7. In that problem, we were looking for one unknown value and a first-degree equation in one variable served as the mathematical model. Here, two unknown values are requested, and a *system* of first-degree equations in two variables serves as the mathematical model.

EXAMPLE 3 How many liters of 30% acid solution and how many liters of 90% acid solution must be mixed to produce 200 liters of 80% acid solution?

Solution We are seeking two values. So, a system of two equations in two variables is a natural model. We start by specifying the variables and what they represent.

Percent acid	*Amount of solution*	*Amount of acid*
30%	x	$0.3x$
90%	y	$0.9y$
80%	200	0.8(200)

Let x represent the amount of 30% acid solution.

Let y represent the amount of 90% acid solution.

Using the facts from the problem (see the table in the margin and the sketch in Figure 6), we write the equation

$$x + y = 200 \qquad \text{Amount of solution}$$

because the two solutions produce 200 liters of the new solution. We also get the equation

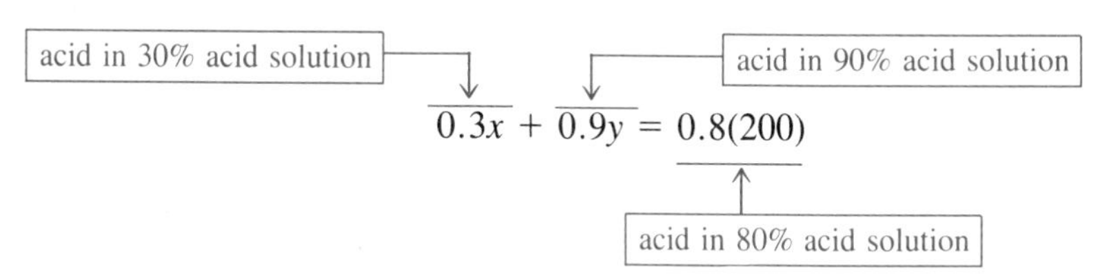

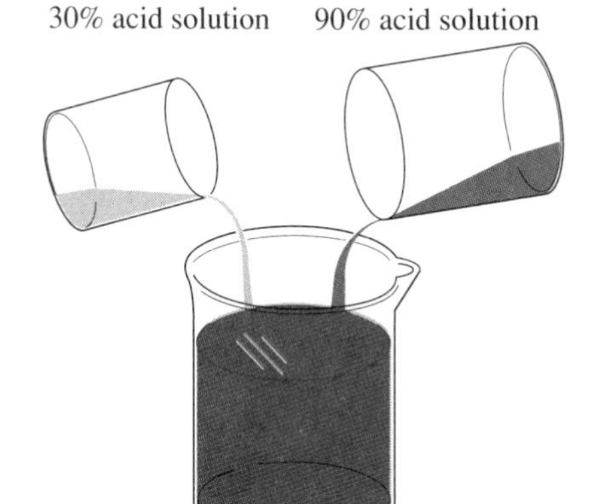

200 liters of 80% acid solution
FIGURE 6

because the acid in the 30% solution plus the acid in the 90% solution must equal the acid in 200 liters of the 80% solution.

Thus, the system produced is

$$\begin{cases} x + y = 200 \\ 0.3x + 0.9y = 0.8(200) \end{cases}$$

We first multiply the second equation by 10 to get

$$\begin{cases} x + y = 200 \\ 3x + 9y = 1600 \end{cases}$$

We solve this system of equations using the addition method. To eliminate x, we multiply the first equation by -3 to get

$$\begin{cases} -3x - 3y = -600 \\ 3x + 9y = 1600 \end{cases}$$

Sum equation: $$6y = 1000$$

$$y = \frac{1000}{6} = \frac{500}{3}$$

Thus, the y-coordinate is $\frac{500}{3}$. Using this value in the original system, we solve the first-degree equation in one variable:

$$\begin{aligned} x + \frac{500}{3} &= 200 \\ 3x + 500 &= 600 && \text{Multiply by 3.} \\ -500 &= -500 && \text{Add } -500. \\ 3x &= 100 \\ x &= \frac{100}{3} && \text{Multiply by } \tfrac{1}{3}. \end{aligned}$$

The solution to the system is $\left(\frac{100}{3}, \frac{500}{3}\right)$.

So, we should mix $33\frac{1}{3}$ liters of 30% acid solution with $166\frac{2}{3}$ liters of 90% acid solution to produce 200 liters of 80% acid solution. ■

Using the Graphics Calculator

```
RANGE
Xmin=0
Xmax=190
Xscl=50
Ymin=0
Ymax=210
Yscl=50
Xres=1
```

FIGURE 7

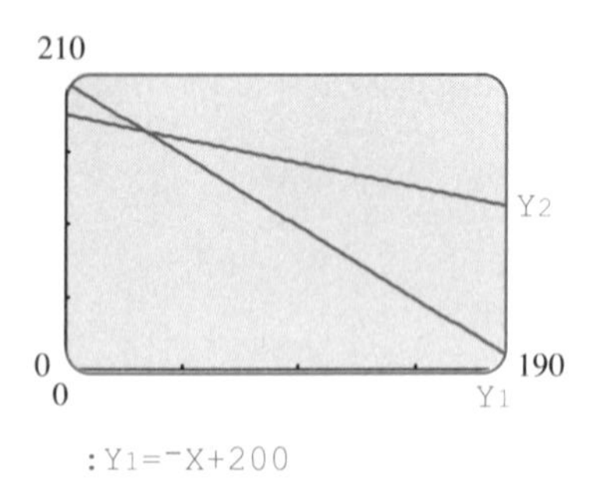

:Y1=⁻X+200
:Y2=⁻(1/3)X+(1600/9)

FIGURE 8

We can check the solution to Example 3 on the graphics calculator. Store `100/3` in memory location `X` and `500/3` in memory location `Y`. Type `X+Y` and press the `ENTER` key. The output is `200`. So, we do have 200 liters of solution. Type `.3X+.9Y`, and press the `ENTER` key. The output is `160`. This mixture contains 160 liters of acid. Divide 160 by the amount of solution, 200 liters, to get 0.8. That is, this mixture is an 80% acid solution.

To visualize the problem in Example 3, rewrite each equation for y. For `Y1`, type `⁻X+200`. Press the `RANGE` key and enter the values shown in Figure 7.* View the graph screen. The points on the line represent the different amounts that can be mixed together to yield a 200 liters of solution. For `Y2`, type `⁻(1/3)X+(1600/9)`. View the graph screen (Figure 8). Press the `TRACE` key, and then the down arrow key to move the cursor to the `Y2` line. The points on this line represents the different amounts of solutions that yield an 80% acid solution. Move the cursor until the message reads `X=84 Y=149.77778`. This means that if 84 liters of 30% acid solution is mixed with $149\frac{7}{9}$ liters of 90% acid solution, the mixture will be an 80% acid solution (however, the mixture will contain $233\frac{7}{9}$ liters of solution, not the desired 200 liters). Move the cursor to the intersection point of the two lines. Zoom in on the point. Are the coordinates of this point approximately $\left(33\frac{1}{3}, 166\frac{2}{3}\right)$? If so, then our solution is correct.

TI-82 Note *Type `188` for `Xmax`.

GIVE IT A TRY

1. In the problem given in Example 3, change the liters of 80% acid solution to be produced to 150, and solve this new problem.
2. Judy drives from Jacksonville to New Orleans, a distance of 630 miles, in 10 hours. She travels at only two speeds, 55 mi/h and 65 mi/h. How many hours does she travel at each speed?

Systems of Linear Functions As a second application of solving a system of first-degree equations in two variables, we will consider a problem from business. This application will yield a system of two linear functions. To graph such a system, the graphs of the individual functions are overlaid to produce the graph of the system.

EXAMPLE 4 The McCoots Company knows that its revenue (in dollars) from producing x duck decoys is given by $R(x) = 35x$. The company also knows that its cost (in dollars) for producing x duck decoys is given by $C(x) = 15x + 1500$.

a. How many duck decoys must be produced for the McCoots Company to break even (revenue equals cost)?
b. If the company produces 100 duck decoys will it make a profit? If so, how much? If not, how much will the company lose?

Solution A graph of the revenue function is shown in Figure 9, and a graph of the cost function is shown in Figure 10. The two linear functions are graphed on the same set of axes in Figure 11.

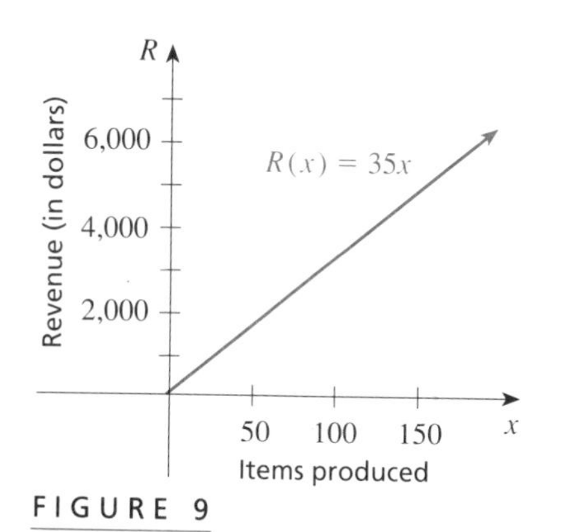

FIGURE 9
Revenue function

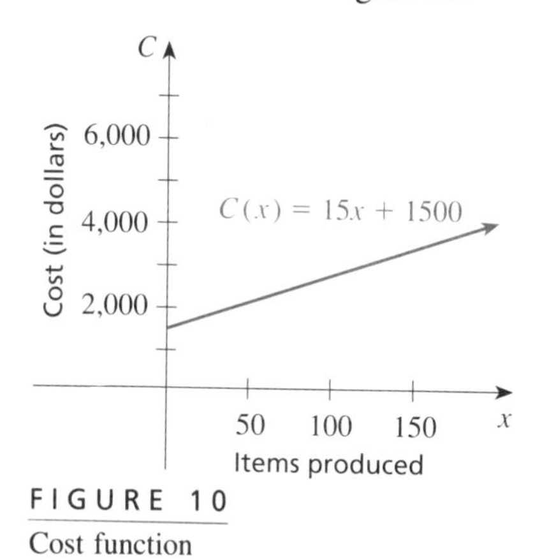

FIGURE 10
Cost function

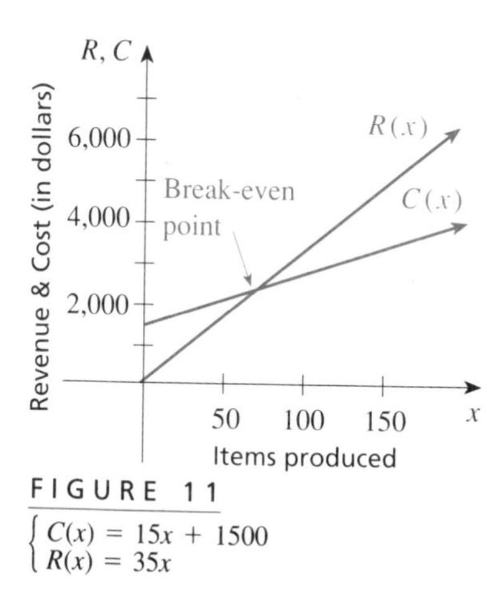

FIGURE 11

$$\begin{cases} C(x) = 15x + 1500 \\ R(x) = 35x \end{cases}$$

a. The x-value at which revenue equals cost (the break-even point) is the x-coordinate of the intersection point. To find this value, we solve the system of revenue and cost functions:

$$\begin{cases} R(x) = 35x \\ C(x) = 15x + 1500 \end{cases}$$

To find the x-value at which $R(x) = C(x)$, we solve the equation

$$35x = 15x + 1500$$

The solution to this equation is 75. Thus, the x-coordinate of the intersection point is 75. To find the y-coordinate, we may find either $R(75)$ or $C(75)$. They both produce (output) the value 2625. Thus, the ordered-pair name of the intersection point is (75, 2625). This point is shown in Figure 12.

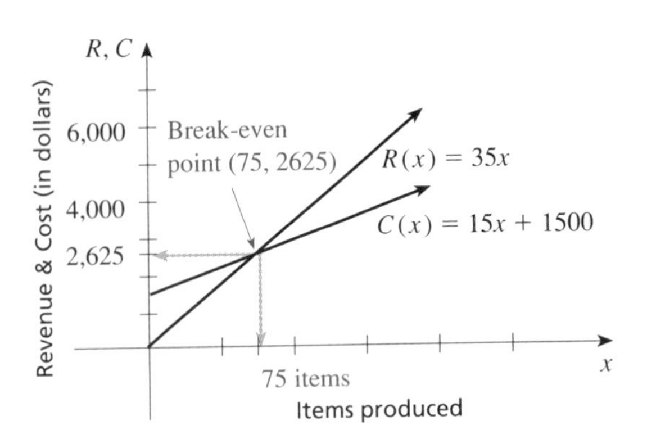

FIGURE 12

So, McCoots must produce 75 duck decoys for revenue to equal cost. When 75 duck decoys are produced, the cost is \$2625 and the revenue is \$2625, so the profit, (or revenue − cost) is 0. Thus, the company will break even when 75 duck decoys are produced.

b. If 100 duck decoys are produced, then $R(100)$ is \$3500 and $C(100)$ is \$3000. So, the company will make a profit. The amount of profit will be revenue minus cost, which is \$3500 − \$3000, or \$500. ■

As another example of solving a system of linear functions, consider the following.

EXAMPLE 5 The ABC Car Rental Company charges a rental fee of \$35 per day plus \$0.20 per mile driven. The XYZ Car Rental Company charges \$20 per day and \$0.40 per mile driven. Assume a rental of one day.

a. How many miles must be driven for the cost of a rental to be the same for both companies?

b. If 200 miles are driven, which rental fee is less expensive? By how much?

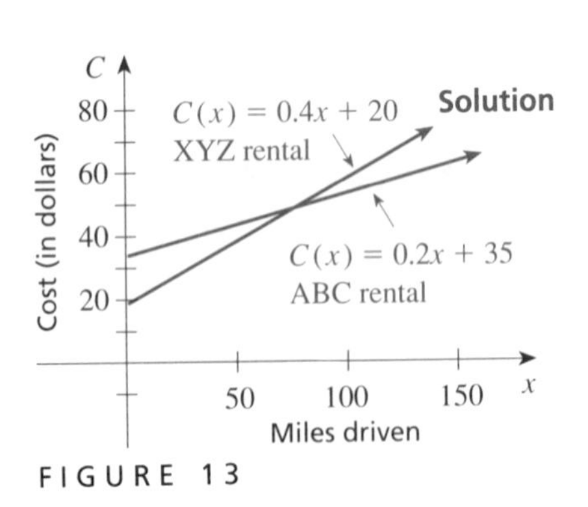

FIGURE 13

Solution In both cases, the cost of the car rental is a linear function of the miles driven. For the ABC Company the linear function is $C(x) = 0.2x + 35$, and for the XYZ Company the linear function is $C(x) = 0.4x + 20$. The two functions are graphed in Figure 13.

a. To find the number of miles driven for the cost of the ABC rental to equal the cost of the XYZ rental, we can solve the following system of linear functions:

$$\begin{cases} C(x) = 0.2x + 35 \\ C(x) = 0.4x + 20 \end{cases}$$

At the point of intersection, the function rules $0.2x + 35$ and $0.4x + 20$ must output the same cost value. Thus, we solve the following first-degree equation in one variable:

$$\begin{aligned} 0.2x + 35 &= 0.4x + 20 \\ 2x + 350 &= 4x + 200 \qquad \text{Multiply by 10.} \\ -4x \quad & \quad -4x \\ -2x + 350 &= 200 \\ -350 \quad & \quad -350 \\ -2x &= -150 \\ x &= 75 \end{aligned}$$

Thus, when 75 miles are driven, the cost for the ABC rental is

$$C(75) = 0.2(75) + 35 \qquad \text{or} \qquad \$50$$

and this is equal to the cost of the XYZ rental

$$C(75) = 0.4(75) + 20 \qquad \text{or} \qquad \$50$$

The intersection point is shown in Figure 14.

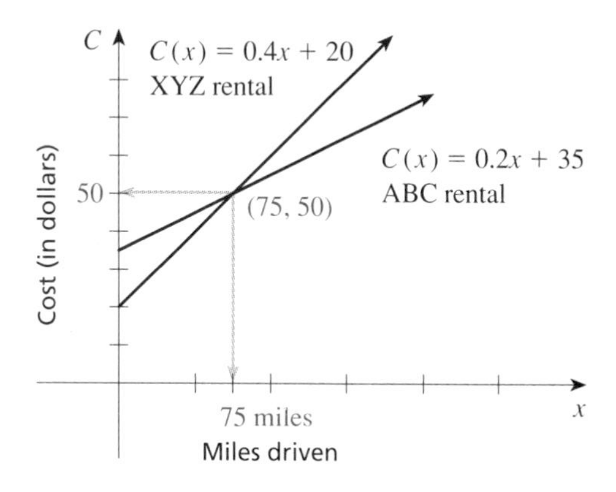

FIGURE 14

b. If 200 miles are driven, which rental is the least expensive? By how much? To answer these questions, we compute $C(200)$ for the ABC rental obtaining \$75, and $C(200)$ for the XYZ rental, which is \$100. Thus, when 200 miles is driven ABC provides the less expensive rental, cheaper by \$25 (compute $100 - 75$). ■

GIVE IT A TRY

3. For the car rental in Example 5, suppose the ABC Car Rental Company reduces its daily charge to \$30. How many miles must be driven for the cost of a rental to be the same for both companies?

4. The price P of an item (in dollars) can be given as a linear function of the number of units x, by either the number of units demanded by customers or the number of units supplied by manufacturers. When expressed in terms of the number of units demanded, the function is given by $P(x) = 100 - 3x$. When given in terms of the units supplied, the function is given by $P(x) = 2x$. Graph these two functions on the same set of axes. At what price will the number of units demanded equal the number of units supplied?

■ SUMMARY

In this section, we have used linear systems to model selected physical situations. A linear system can provide a mathematical model for a real-world situation (word problem) for which two unknown values are requested. Once the system of linear equations is found, we can solve it and state the answer to the word problem.

Many business and science applications can be solved by finding the point at which two linear functions with the same input variable produce the same output. We can accomplish this by viewing the two linear functions as a system and then solving the system. One common use of this approach is finding the production level at which revenue equals cost (the break-even point).

4.4 ■ ■ ■ EXERCISES

For each problem, report the following:

a. *The variables used, and what they represent.*

b. *The system of equations or functions produced.*

c. *The solution to the system and the answer to the problem.*

1. The sum of two numbers is 120, and their difference is 34. Find the two numbers.
2. For two numbers, the larger is 5 more than twice the smaller number. Their sum is -8. Find the two numbers.
3. A rectangle has a width that is 7 feet less than the length of the rectangle. If the perimeter of the rectangle is 86 feet, what is the length and the width?
4. An isosceles triangle has two sides of equal length. The base of a particular triangle is 28 cm more than either of the two equal sides. The perimeter of the triangle is 142 cm. Report the lengths of the three sides.
5. An airplane flies with the wind at a speed of 300 miles per hour (mi/h). Flying against the wind, its speed is 240 mi/hr. Find the speed of the plane and the speed of the wind.
6. A boat going downstream has a speed of 60 mi/hr. Running against the current, its speed is 46 mi/h. Find the speed of the boat and the speed of the current.
7. The Nuts-R-Us Company pays \$1.20 per pound for peanuts and \$4.50 for almonds. How many pounds of each kind should Nut-R-Us mix to make 10 pounds of a mixture that costs them \$2.00 per pound to make?
8. Cashews cost \$4.50 per pound and almonds cost \$3.20 per pound. How many pounds of each must be mixed to make 20 pounds of mixture that costs \$4.20 per pound to make?
9. The price of an adult ticket is \$15 and the price of a student ticket is \$6. If the total income from the tickets is \$2700, and 300 tickets are sold, how many of each kind are sold?
10. A farmer grows and sells tomatoes and carrots. She earns \$2 per pound for tomatoes and \$3 per pound for carrots. If a total of 60 pounds are sold on a given day and the receipts total \$130, how many pounds of each are sold?
11. Carol has two acid solutions—one is 30% acid and the other is 70% acid. How many liters of each must she mix to get 120 liters of 60% acid solution?
12. Tom must make a 300 liters of 60% acid solution from a 10% acid solution and a 90% acid solution. How many liters of each type of acid solution must he use?

13. How many liters of 10% acid solution and how many liters of 60% acid solution must be mixed to produce 70 liters of 20% acid solution?

14. Ms. Alverez, a jeweler, wants to combine a 30% silver alloy with a 60% silver alloy to produce 300 grams of a 40% silver alloy. How many grams of the 30% silver alloy and how many grams of the 60% silver alloy should she mix?

15. Tom travels 8 hours at only two speeds, 50 mi/h and 65 mi/h. If Tom travels 495 miles in 8 hours, how many hours does he travel at 50 mi/h and how many hours does he travel at 65 mi/h?

16. Elmore drove from Jacksonville to Mobile, a distance of 400 miles. He traveled at only two speeds, 55 mi/h and 65 mi/h. If he made the trip in 7 hours, how long did he travel at each of these speeds?

17. Senator Woo leaves Wichita, traveling toward Topeka. Senator Tom leaves Topeka, headed for Wichita on the same road as Woo, at the same time, and driving 10 mi/h faster than Woo. If Wichita and Topeka are 100 miles apart, and the senators meet after 40 minutes of driving, find the rate at which each is driving.

18. Pete's monthly salary is based on a commission on his sales for the month. If Pete's salary is $2,800 when his amount of sales is $30,000, and his salary is $3,500 when his amount of sales if $70,000, find his base salary and commission rate.

19. Michel invests $20,000 in stocks and bonds. The stocks return 15% a year and the bonds return 10% a year. If the total return on the investment is $2,000, how much did Michel invest in stocks and how much did he invest in bonds?

20. Paul beats Mark in 70% of the Scowl games they play. He beats Wanda in 30% of their Scowl games. If Paul plays 50 games and wins 28 of the games, how many games did he play against Mark, and how many games did he play against Wanda?

21. Sally and Jill have painted 150 toy trucks in 8 hours. If Sally painted 20 trucks per hour and Jill painted 15 trucks per hour, how many hours has Sally worked and how many hours has Jill worked?

22. A tank holds 400 gallons of water. Pump A can fill the tank in 4 hours, while pump B can fill it in 10 hours. The tank can also be filled using both pumps at once. If it took 8 hours to fill the tank yesterday, how many hours was pump A used and how many hours was pump B used?

23. For $y = mx + b$, if $(-3, 6)$ and $(5, -1)$ satisfy the equation, find the value for m and the value for b.

24. For $Ax + By = 16$, if $(4, 5)$ and $(-2, -1)$ satisfy the equation, find the values for A and for B.

25. Graph the linear function $f(x) = -3x + 6$ and the linear function $g(x) = 2x - 3$ on the same set of axes. What input value results in both functions producing the same output? What is that output?

26. Root² Inc. knows that its revenue (in dollars) from producing x mobiles is given by $R(x) = 15x$. The company also knows that its cost (in dollars) for producing x mobiles is given by $C(x) = 9x + 2000$.
    a. How many mobiles must be produced for Root² to break even (revenue is equal to cost)?
    b. If the company produces 80 mobiles, will it make a profit? If so, how much? If not, how much money will the company lose?

27. The cost of a car rental at El Cheapo is $30 per day plus $0.30 per mile driven. The cost of a car rental at Delux is $20 per day and $0.50 per mile. How many miles must be driven for the cost of the car rental to be the same for both El Cheapo and Delux?

28. The Wedlock Group produces wedding cakes. The revenue function for this term is $R(x) = 120x$ and the cost function is $C(x) = 20x + 800$, where $R(x)$ and $C(x)$ are given in dollars and x is the number of cakes produced.
    a. How many cakes must be produced for the Wedlock Group to break even (revenue equals cost)?
    b. If 15 cakes are produced, will the Wedlock Group make a profit? If so, how much? If not, how much will the group lose?

29. The *depreciation* of an item, or its loss of value through use, is a function of time. One method of computing depreciation is known as the *straight-line method*. By this method, the value of the item V (in dollars) is reported as a linear function of time t (in years). For item A, the depreciation function is $V(t) = 600 - 100t$. For Item B, the depreciation function is $V(t) = 800 - 150t$. If the items are purchased at the same time, how many years will it take for the value of item A to equal the value of item B?

CHAPTER 4 REVIEW EXERCISES

1. Find the equation of the line that passes through the points $(-2, 5)$ and $(-5, 1)$.
2. Find the equation of the line that passes through the point $(2, -3)$ and has slope measure $\frac{-5}{2}$.
3. Find the linear function such that $f(-3) = 6$ and $f(2) = -1$.
4. Find the linear function such that $f(0) = -1$ and $f(-2) = 3$.
5. The temperature C of a metal rod is a linear function of time. At time $t = 3$ minutes the temperature is 10°C, and at time $t = 5$ minutes the temperature is 23°C.
   a. Using this data, report the linear function.
   b. When is the temperature 19°C?
   c. At time $t = 4.1$ minutes, what is the temperature?
6. The cost C of producing Glitz Boxes is a linear function of the number of boxes produced, x. For $x = 18$ boxes, the cost is \$215, and for 38 boxes, the cost is \$228.
   a. Using this data, report the linear function.
   b. When the cost is \$432, how many boxes are produced?
   c. What is the cost to produce 42 boxes?

For a line segment with the given endpoints, report the length of the line segment and the ordered-pair name of the midpoint of the line segment.

7. $(-3, 4)$ and $(5, 2)$
8. $(1, 5)$ and $(3, -3)$
9. $(-6, -2)$ and $(3, -4)$
10. $(0, -2)$ and $(-1, 1)$
11. Find the equation of the line through $(-3, -2)$ that is parallel to the graph of $4x - 3y = 10$.
12. Find the equation of the line through $(-3, 2)$ that is perpendicular to the graph of $4x - 3y = 10$.

For a line segment with the given endpoints, find the equation of the perpendicular bisector of the line segment.

13. $(-3, 1)$ and $(5, 3)$
14. $(1, -5)$ and $(3, -3)$
15. For the equations $3x - 5y = 9$ and $x + 5y = 1$, what is the sum equation? What is the x-coordinate of the point of intersection of the lines determined by these equations? What is the y-coordinate of the intersection point?
16. For the equations $2x + 3y = 5$ and $-2x + y = 3$, what is the sum equation? What is the y-coordinate of the point of intersection of the lines determined by these equations? What is the x-coordinate of the intersection point?

Solve each system of linear equations or functions.

17. $\begin{cases} 3x - 5y = 10 \\ 4x + 5y = -3 \end{cases}$
18. $\begin{cases} 2x + y = 8 \\ 6x - y = -6 \end{cases}$
19. $\begin{cases} 5x - 3y = 8 \\ 3x + 2y = -6 \end{cases}$
20. $\begin{cases} x + 3y = -7 \\ 3x - 4y = 13 \end{cases}$
21. $\begin{cases} y = \frac{-5}{3}x + 5 \\ y = -2x + \frac{11}{2} \end{cases}$
22. $\begin{cases} f(x) = 23x + 10 \\ h(x) = -12x + 30 \end{cases}$
23. The sum of two numbers is 48 and their difference is 60. Find the two numbers.
24. One length of a rectangular field borders a river, and the remaining three sides of the field are to be fenced. If the length of the field is 20 yards more than its width, and the amount of fencing to be used is 500 yards, find the dimensions of the field.

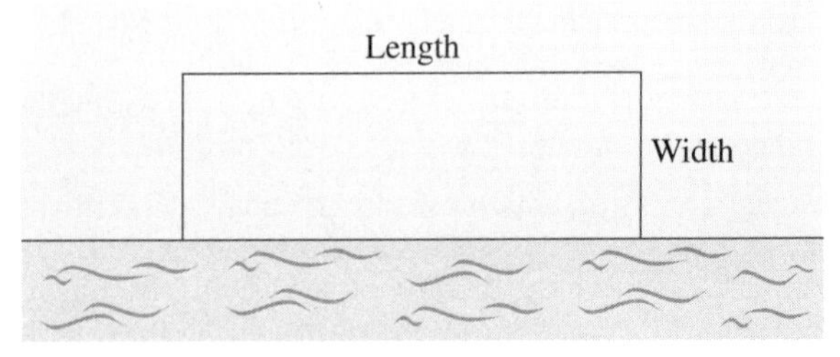

25. How many liters of 10% alcohol solution and how many liters of 25% alcohol solution must be mixed to produce 50 liters of a 20% alcohol solution?
26. Bonnie drives 350 miles in 6 hours, traveling only at speeds of 55 mi/h and 60 mi/h. How many hours does she drive at each speed?
27. Aplex produces hot tubs. The company knows that its revenue is given by $R(x) = 120x$ and its cost is given by $C(x) = 70x + 2000$, where x is the number of tubs produced. What is the break-even point for the hot tubs? If 25 tubs are produced, will Aplex make a profit? If so, how much? If not, how much will the company lose?

CHAPTER 4 TEST

Use the graphics calculator to check your work.

1. **a.** Find the equation of the line through the point $(3, -2)$ with slope measure $\frac{1}{2}$.

   b. For the line in part (a), what is the x-intercept?

2. A small company can produce 15 backpacks for $230, and it can produce 20 backpacks for $250.

   a. Graph the linear relationship between backpacks produced, x, and cost of production, C.

   b. State the linear relationship graphed in part (a).

   c. How many backpacks can be produced for $310?

3. Are the points $(-1, 5)$, $(2, -1)$, and $(5, -7)$ collinear? If so, report the equation of the line through the three points. If not, give the equation of the line through the first two points and the equation of the line through the last two points.

4. A line segment has endpoints $A(2, 5)$ and $B(6, -1)$.

   a. Find the length of $\overline{AB}$.

   b. Report the ordered-pair name of the midpoint of $\overline{AB}$.

   c. Find the equation of line AB.

   d. Report the slope of a line perpendicular to line AB.

   e. Find the equation of the perpendicular bisector of $\overline{AB}$.

   f. Graph $\overline{AB}$ and its perpendicular bisector.

Solve each system of equations.

5. $\begin{cases} 3x - 4y = 13 \\ -2x + 5y = 10 \end{cases}$

6. $\begin{cases} y = 2x - 3 \\ y = \frac{-2}{3}x + 1 \end{cases}$

7. How many liters of 20% acid solution and how many liters of 70% acid solution must be mixed together to produce 160 liters of 30% acid solution?

8. Mary's farm has only pigs and chickens. She counts 69 heads and 222 feet (not counting her own). How many chickens and how many pigs does Mary have?

5

SOLVING SECOND-DEGREE EQUATIONS

In this chapter we have two learning objectives—factoring polynomials and using factoring to solve second-degree equations, such as $x^2 - 3x = 4$. Factoring is the act of reversing multiplication of polynomials in order to write a polynomial as a product of polynomials. Because factoring is dependent on multiplication of polynomials, we start this chapter with a discussion of multiplication of polynomials.

After discussing multiplication and factoring of polynomials, we will apply this knowledge and skill to solving second-degree equations in one variable. Recalling our work with linear equations in Chapters 2, 3, and 4, we will begin by solving second-degree equations in one variable, then solving second-degree equations in two variables. These equations in two variables have solutions that are ordered pairs of numbers. When graphed, the ordered pairs form a shape known as a *parabola*.

At the end of this chapter we consider the binary operation of division of polynomials. This operation completes our discussion of the four basic operations on polynomials.

5.1 ■ ■ ■ MULTIPLICATION OF POLYNOMIALS

In Chapters 2, 3, and 4, we solved first-degree equations using the operations of addition and subtraction of polynomials and the following two equation-solving properties:

1. Any polynomial can be added to each side of an equation without altering the solution.
2. Each side of an equation can be multiplied by a nonzero number without altering the solution.

To solve a second-degree equation, such as $x^2 + 5x = 6$, we will need more than these two properties. In fact, we must be able to factor polynomials. Since the key to factoring polynomials is multiplication of polynomials, we begin this chapter with a discussion of this topic.

Background To multiply polynomials, we use the following laws of exponents (from Section 1.3):

For positive integers m and n: $(a^n)(a^m) = a^{n+m}$

For positive integers m and n: $(a^n)^m = a^{nm}$

For a positive integer n: $(ab)^n = a^n b^n$

We will also use the commutative, associative, and distributive properties of real numbers. Finally, remember that a *standard polynomial* is a polynomial that has no like terms and is written in decreasing order of its terms.

Monomials The simplest case of multiplication of polynomials is a monomial times a monomial. Work through the following example.

EXAMPLE 1 Write each product as a standard polynomial:

a. $(-3x^2y^5)(5x^3y^4)$ b. $(2x^3y^2)^4$

Solution a. We use the associative and commutative properties:

$$\begin{aligned}(-3x^2y^5)(5x^3y^4) &= (-3)(5)(x^2x^3)(y^5y^4) \\ &= -15x^5y^9 \qquad (a^n)(a^m) = a^{n+m}\end{aligned}$$

b.

$$\begin{aligned}(2x^3y^2)^4 &= (2^4)(x^3)^4(y^2)^4 \qquad (ab)^n = a^nb^n \\ &= 16x^{12}y^8 \qquad (a^n)^m = a^{nm}\end{aligned}$$

■

Just as addition or subtraction of polynomials has two meanings, an equality involving a polynomial product, such as $(-3x^2y^5)(5x^3y^4) = -15x^5y^9$, can be used in two ways:

a. anywhere the polynomial product $(-3x^2y^5)(5x^3y^4)$ appears, it can be replaced by the polynomial $-15x^5y^9$, and
b. for any replacement of x and y by numbers, the polynomial $-15x^5y^9$ evaluates to the same number as $(-3x^2y^5)(5x^3y^4)$.

Using the Graphics Calculator

We can use the description in part (b) to test our work in Example 1(a) for an error. Store a number, say `-6`, in memory location `X` and a number, say `7`, in memory location `Y`. Type `(-3X²Y^5)(5X³Y^4)` and press the `ENTER` key. Next, type `-15X^5Y^9` and press the `ENTER` key. If the same number is *not* output in both cases, an error is present!

The Distributive Property The next level of multiplication of polynomials is a monomial times a binomial. To do this multiplication, we use the distributive properties of multiplication over addition:

$$a(b + c) = ab + ac \qquad \text{and} \qquad (b + c)a = ba + ca$$

Work through the following example.

EXAMPLE 2 Write as a standard polynomial: $(3x^2)(2x^2 - 3)$

Solution We use the distributive property of multiplication over addition:

distribute

$$(3x^2)(2x^2 - 3) = (3x^2)(2x^2) - (3x^2)(3)$$

Multiplying the monomials, we get: $\qquad = 6x^4 - 9x^2$

So, $(3x^2)(2x^2 - 3)$ written as a standard polynomial is $6x^4 - 9x^2$. ■

EXAMPLE 3 Write as a standard polynomial: $(-2x^2)(x^2 - x - 3)$

Solution Using the distributive property of multiplication over addition, we get:

$$(-2x^2)(x^2) - (-2x^2)(x) - (-2x^2)(3)$$

Multiplying the monomials, we get:

$$-2x^4 - (-2x^3) - (-6x^2) \qquad \text{or} \qquad -2x^4 + 2x^3 + 6x^2$$

So, $(-2x^2)(x^2 - x - 3)$ written as a standard polynomial is

$$-2x^4 + 2x^3 + 6x^2$$

■

GIVE IT A TRY

Write each product as a standard polynomial.

1. $(-5x^2)(4xy^3)^2$ **2.** $(3p^2)(-2p^3 - p^2 + 3)$ **3.** $(-x^2)(x^3 - 2x^2 - 1)$

Binomials At this level of algebra the most common multiplication of polynomials is multiplication by a binomial. Work through the following example.

EXAMPLE 4 Write each product as a standard polynomial:

a. $(3x^2 - 4)(2x^2 - 3)$ b. $(3d^2 - 4)(9d^4 - 12d^2 + 16)$

Solution a. We use the distributive property:

$$(3x^2 - 4)(2x^2 - 3) = (3x^2 - 4)(2x^2) - (3x^2 - 4)(3)$$

(distribute; distribute; distribute)

Next, using the distributive property again, we get:
$$= (3x^2)(2x^2) - (4)(2x^2) - (3x^2)(3) + (4)(3)$$

Using multiplication of monomials, we get:
$$= 6x^4 - 8x^2 - 9x^2 + 12$$

Note the sign change.

Finally, using addition of polynomials to combine like terms, we get:
$$= 6x^4 - 17x^2 + 12$$

Thus, the product of the binomials $3x^2 - 4$ and $2x^2 - 3$ is the polynomial $6x^4 - 17x^2 + 12$.

b. We can show the multiplication in column format, as we often do in arithmetic, in order to align like terms (in arithmetic, powers of ten are aligned).

$$\begin{array}{rrrrl}
 & 9d^4 & -12d^2 & +16 & \\
 & & 3d^2 & -4 & \\
\hline
 & -36d^4 & +48d^2 & -64 & \text{Multiply } -4 \text{ times } 9d^4 - 12d^2 + 16. \\
27d^6 & -36d^4 & +48d^2 & & \text{Multiply } 3d^2 \text{ times } 9d^4 - 12d^2 + 16. \\
\hline
27d^6 & -72d^4 & +96d^2 & -64 & \text{Combine like terms.}
\end{array}$$

Thus, $(3d^2 - 4)(9d^4 - 12d^2 + 16)$ written as a standard polynomial is $27d^6 - 72d^4 + 96d^2 - 64$. ■

FOIL Method In algebra, finding the product of two binomials is a very common multiplication (much like multiplication by 10 is very common in arithmetic). For this reason, a *fast* method is used to find the product of binomials. This method is known as the **FOIL method**. Consider our work from Example 4(a):

$$(3x^2 - 4)(2x^2 - 3) = 6x^4 - 8x^2 - 9x^2 + 12$$
$$= 6x^4 - 17x^2 + 12$$

F: First times First: $(3x^2)(2x^2)$

L: Last times Last: $(-4)(-3)$

O + I: Outer + Inner: $(3x^2)(-3) + (-4)(2x^2)$

The first term of the product is the *first times the first* (**F**). The middle term of the product is the *sum of the product of the outer* (**O**) *terms and inner terms* (**I**). The last term is the *last times the last* (**L**). This method of multiplying binomials is more than just a shortcut, it is essential for our work with factoring trinomials in Section 5.3. For another example of the FOIL method of multiplying binomials, consider the following example.

EXAMPLE 5 Write as a standard polynomial: $(3x^2 - 3)(2x + 5)$

Solution Using the FOIL method, we get:

$$(3x^2 - 3)(2x + 5) = \overset{\text{F}}{6x^3} + \overset{\text{O}}{15x^2} - \overset{\text{I}}{6x} - \overset{\text{L}}{15}$$

The sum of the product of outer terms ($3x^2$ and 5) and the product of the inner terms (-3 and $2x$) produces $15x^2 + (-6x)$. Because these two terms are not like terms, the product of these two binomials is a quadrinomial, a four-term polynomial. ■

GIVE IT A TRY

Write each product as a standard polynomial.

4. $(-3x^2)(-7x^5y^2)$ **5.** $(2x - 5)(-3x^2)$ **6.** $(x + 8)(x + 7)$

7. $(2x - 3)(3x + 5)$ **8.** $(5x + 8)(x - 1)$ **9.** $(x^2 - 2)(x - 3)$

The Square of a Binomial A special product of binomials is the square of a binomial, that is, a binomial multiplied by itself. Consider squaring the binomial $2x + 3$:

$$(2x + 3)^2 = (2x + 3)(2x + 3)$$

$$= \overset{\text{F}}{4x^2} + \overset{\text{O}}{6x} + \overset{\text{I}}{6x} + \overset{\text{L}}{9}$$

$$= 4x^2 + 12x + 9$$

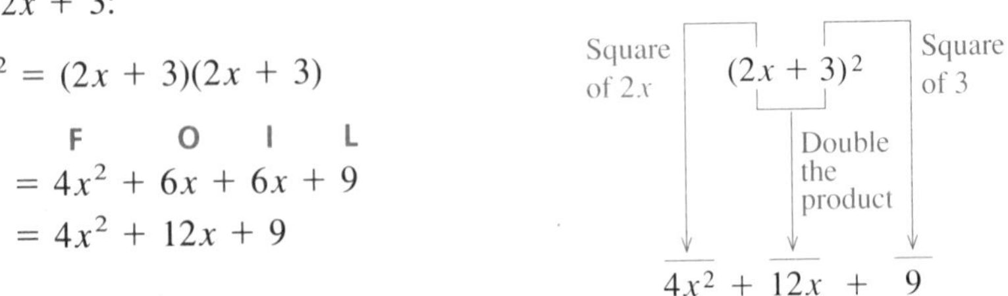

We can compare the product, $4x^2 + 12x + 9$, to the original binomial, $2x + 3$:

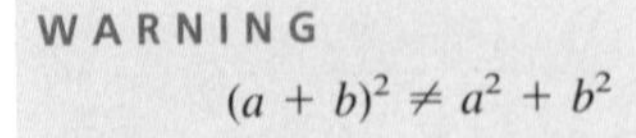

- the first term of the product, $4x^2$, is the square of the first term of the original binomial, $2x$.
- the second term of the product, $12x$, is twice the product of the two terms of the original binomial, $2x$ and 3;
- the third term of the product, 9, is the square of the last term of the original binomial, 3.

These comparison can be stated as a general rule.

THE SQUARE OF A BINOMIAL

The square of a binomial is the square of the first term, plus twice the product of the two terms, plus the square of the last term.

$$(a + b)^2$$

$$a^2 + 2ab + b^2$$

We write the product as

$$(a + b)^2 = a^2 + 2ab + b^2$$

We can also square a binomial that is the difference of two terms:

$$(a - b)^2 = a^2 - 2ab + b^2$$

EXAMPLE 6 Write as a standard polynomial: $(-2x^3 - 1)^2$

Solution We use the special product $(a - b)^2 = a^2 - 2ab + b^2$ with $a = -2x^3$ and $b = 1$.

$$\begin{aligned}(-2x^3 - 1)^2 &= (-2x^3)^2 - 2(-2x^3)(1) + (1)^2 \\ &= 4x^6 + 4x^3 + 1\end{aligned}$$ ■

Sum and Difference A monomial times a monomial always yields a monomial. The product of two binomials typically yields a trinomial, but, as Example 5 shows, the product of two binomials can be a quadrinomial. A third possibility for the product of two binomials is a binomial.

$$\begin{aligned}& \qquad\quad\ \ \text{F} \qquad\ \ \text{O} \qquad\ \ \text{I} \qquad \text{L} \\ (4x^2 + 5)(4x^2 - 5) &= 16x^4 - 20x^2 + 20x^2 - 25 \\ &= 16x^4 - 25\end{aligned}$$

Here, the product of the outer terms, $-20x^2$, added to the product of the inner terms, $20x^2$, is 0. Thus, a binomial results. In fact, the result is a special binomial—the difference of two perfect squares, $16x^4$ and 25.

THE SUM TIMES THE DIFFERENCE

The sum of two terms times the difference of the same two terms is the square of the first term minus the square of the second term. We write the products as

$$(a + b)(a - b) = a^2 - b^2$$

The preceding multiplications are known as *special products*. We now summarize their formulas.

SPECIAL PRODUCTS

$(a + b)^2 = a^2 + 2ab + b^2$	Square of a sum
$(a - b)^2 = a^2 - 2ab + b^2$	Square of a difference
$(a + b)(a - b) = a^2 - b^2$	Sum times a difference

EXAMPLE 7 For each expression, identify the monomials a and b. Then use a special product to write each as a standard polynomial.

a. $(-2x + 3)^2$ **b.** $(-2x + 3)(-2x - 3)$

Solution **a.** We have $a = -2x$ and $b = 3$. Using the special product for the square of a sum, $(a + b)^2$, we get:

$$(-2x + 3)^2 = (-2x)^2 + 2(-2x)(3) + 3^2$$
$$= 4x^2 - 12x + 9$$

b. We have $a = -2x$ and $b = 3$. Using the special product for a sum times a difference, $(a + b)(a - b)$, we get:

$$(-2x + 3)(-2x - 3) = (-2x)^2 - (3)^2$$
$$= 4x^2 - 9$$

■

GIVE IT A TRY

Write each product as a standard polynomial. Use a special product if possible, and identify the monomials a and b.

10. $(3x + 2)(3x - 2)$ **11.** $(3x^2 - 2)(3x^2 + 2)$ **12.** $(3x + 2)(3x + 2)$

13. $(3x^2 - 2)(3x^2 - 2)$ **14.** $(3x - 2)^2$ **15.** $(3x^2 + 2)^2$

More than One Operation For operations on polynomial expressions that involve more than one operation, we use the order of operations rules. Most errors made in performing these operations involve errors in subtraction: be sure to change the sign of each term when clearing a set of parentheses that is preceeded by a subtraction operation. Consider the following example.

EXAMPLE 8 Write as a standard polynomial: $5x - [3 - (x + 2)(x - 3) - (x - 1)^2]$

Solution

First, we do the multiplications to get:	$5x - [3 - (x^2 - x - 6) - (x^2 - 2x + 1)]$
Now, clear the parentheses, starting with the innermost parentheses:	$5x - (3 - x^2 + x + 6 - x^2 + 2x - 1)$
Clearing the outer parentheses, we get:	$5x - 3 + x^2 - x - 6 + x^2 - 2x + 1$
Combining like terms yields the standard polynomial:	$2x^2 + 2x - 8$

Thus, $5x - [3 - (x + 2)(x - 3) - (x - 1)^2]$ written as a standard polynomial is $2x^2 + 2x - 8$.

■

The result of Example 8 has two meanings. Anywhere the polynomial

$$5x - (3 - (x + 2)(x - 3) - (x - 1)^2)$$

appears, it can be replaced by the standard polynomial $2x^2 + 2x - 8$. Also, for any replacement value of x, such as 10 or $\frac{-2}{3}$, both the polynomials

$$5x - (3 - (x + 2)(x - 3) - (x - 1)^2) \quad \text{and} \quad 2x^2 + 2x - 8$$

must evaluate to the same number.

GIVE IT A TRY

Write each polynomial in standard form.

16. $3x - (2x - 1)(x - 2)$

17. $3x - (2x - 5x(x - 3))$

18. $3x - (2x - 1)^2$

19. $3x - (5 - (2x - 1)^2 - (2x - 1)(x - 2))$

20. $2 - (x - 1)(x^2 + x + 1) - (x - 1)(x + 1)$

Using the Graphics Calculator

We can test our work for errors in Example 8. Store a number, say 7, in memory location X. Type 5X−(3−(X+2)(X−3)−(X−1)²), and then press the ENTER key. Observe the output. Next, type 2X²+2X−8 and then press the ENTER key. Observe the output. If our work is correct, the two outputs are the same. Use the graphics calculator to check your work in *Give It a Try* Problems 16–20.

■ SUMMARY

■ ***Quick Index***

In this section, we have discussed multiplication of polynomials. As with most topics in algebra, multiplication of polynomials builds from simple beginnings. We started with the product of monomials. Next, we considered a monomial times a binomial, and then the product of binomials. Much like multiplication by 10 is important in arithmetic, multiplication of binomials is very important in algebra. We use the FOIL process to quickly find the product of two binomials. In multiplying binomials, the result is typically a trinomial; however, the product may be a quadrinomial (four-term polynomial) or a binomial.

In addition to the FOIL method of multiplying binomials, we have considered some special products of binomials. The sum of two terms times the difference of the same two terms, $(a + b)(a - b)$, produces the square of the first term minus the square of the second term, $a^2 - b^2$. A second special product is a binomial squared, $(a + b)^2$. Here, the standard polynomial produced is the square of the first term, plus twice the product of the two terms, plus the square of the second term, $a^2 + 2ab + b^2$. Likewise, $(a - b)^2$ yields $a^2 - 2ab + b^2$.

The last topic we have considered in this section is writing an expression that involves more than one operation as a standard polynomial. Following order of operations rules, all multiplication operations (including powers) are performed first. Then all addition and subtraction operations are performed.

5.1 ■ ■ ■ EXERCISES

Insert true *for a true statement, or* false *for a false statement.*

____ 1. The product of two monomials is always a monomial.

____ 2. The product of two binomials is always a binomial.

____ 3. The product of two binomials can be a monomial.

____ 4. If two binomials do not have the same degree, their product will be a quadrinomial (a four-term polynomial).

____ 5. The square of $(2x + 3)$ is $4x^2 + 9$.

____ 6. The sum of $2x$ and 3 times the difference of $2x$ and 3 is $4x^2 - 9$.

____ 7. The square of a monomial is a binomial.

____ 8. The product $(2^3x^4)(2^5x^3)$ is 4^8x^7.

9. Use the FOIL approach to multiply $(3x + 2)(x - 5)$. Identify F, O + I, and L. Report the standard polynomial for the product $(3x + 2)(x - 5)$.

10. Use the graphics calculator to check that $(2x - 3)(x^2 - 3x + 4)$ as a standard polynomial is $2x^3 - 9x^2 + 17x - 12$: Store a number, say $^-5$, in memory location X. Report the calculator evaluation of $(2x - 3)(x^2 - 3x + 4)$ and its evaluation of $2x^3 - 9x^2 + 17x - 12$. Do you think an error is present?

Use multiplication by monomials to write each product as a standard polynomial. Use the graphics calculator to check your work.

11. $(3x^2y^3)(-5x^2y^4)$
12. $(-6n^2m^3)(2n^2p^4)$
13. $(-x^2z)(2yz)^3$
14. $(-a^2b)^3(2ab^2)$
15. $(2x)(3x - 5)$
16. $(-3r - 8)(\frac{2}{3}r)$
17. $(2x^2)(3x - 5)$
18. $(-x^2 - 2)(-3x)$
19. $(-3)(x^2 - 3x + 5)$
20. $(-3x)(x^2 - 3x + 5)$
21. $(6w)(3 - w - 2w^2)$
22. $9x - 3x(2x)$

Use the FOIL approach to multiplication of binomials to write each product as a standard polynomial. Use the graphics calculator to check your work.

23. $(x + 3)(x + 7)$
24. $(2x - 3)(2x + 3)$
25. $(x - 4)(x + 1)$
26. $(5x + 3)(5x - 3)$
27. $(a + b)(2a - b)$
28. $(c - 3d)(2c + d)$
29. $(2x^3 - 3)(2x^3 + 3)$
30. $(-x^2 + 3)(-x^2 - 3)$
31. $(2x + 1)(x - 4)$
32. $(3 - 2p)(-2 + p)$
33. $(x^2 - 7)(x^2 - 7)$
34. $(c^2 + b)(c^2 + b)$
35. $(3p - 4)(2p + 5)$
36. $(2x^2 - 9)(x + 1)$
37. $(3 - 2y)(2 - y)$
38. $(3x - 4)(5x - 3)$
39. $(x^2 - 3)(x + 3)$
40. $(2x^2 - 3)(2x^3 - 3)$

Use special products to write each product as a standard polynomial. Use the graphics calculator to check your work.

41. $(x - 5)^2$
42. $(2x + 3)^2$
43. $(3t + 5)^2$
44. $(3 - 2r)^2$
45. $(2y - 1)^2$
46. $(5z - 2)^2$
47. $(x + 3)(x - 3)$
48. $(5x - 1)(5x + 1)$
49. $(3x - 4)(3x + 4)$
50. $(5x + 3)(5x + 3)$
51. $(2a + b)(2a - b)$
52. $(2c - 3d)(2c + 3d)$
53. $(2x^3 - 3)(2x^3 - 3)$
54. $(-x^2 + 3)(-x^2 + 3)$

Follow order of operations to write the following as a standard polynomial. Use the graphics calculator to test your work for errors.

55. $x^4 - (x - 1)(x^2 + x + 1)$
56. $x^2 - (3x - (x - 1)(x + 1) - (x + 2)^2)$
57. $3x - [(x - 3)(x^2 + 3x + 6) - (x + 2)(x - 3)]$
58. $(3x - 2y)(2x - 5y)$
59. $(3x^2 - 2y)(2x - 5y^2)$
60. $3xy - [(x - y)^2 - (x + y)(x - y)]$
61. $\left(\frac{3}{5}x + \frac{2}{3}\right)\left(\frac{4}{5}x - \frac{4}{3}\right)$
62. $\left(\frac{1}{2}x - \frac{1}{3}\right)\left(\frac{3}{5}x - \frac{1}{2}\right)$
63. Evaluate the following for $x = -2$ and $y = 3$.
$$2x - 3y(2x - y) - (2xy - (x - y)^2 - (2x - y)^2)$$
64. Evaluate the following for $x = -1$ and $y = -2$:
$$2y - 2y(2x - 3) - (xy - (x - 2y)^2 - (2x - y)^2)$$

■ BONUS PROBLEM

65. Consider the following pattern:

$(a)^2 = a^2$ — 1 term
$(a + b)^2 = a^2 + 2ab + b^2$ — 3 terms
$(a + b + c)^2 = a^2 + b^2 + c^2 + 2ab + 2ac + 2bc$ — 6 terms

If the entire alphabet is squared,
$$(a + b + c + \cdots + z)^2$$
how many terms will be in the product?

DISCOVERY

Cubes

In Section 5.1 we found that the product of the *sum* of two monomials and the *difference* of the same two monomials produces a binomial (the difference of two perfect square monomials). Now, consider the following special product of a binomial and a trinomial:

distribute

$$
\begin{aligned}
(2x + 3)(4x^2 - 6x + 9) &= (2x + 3)(4x^2) - (2x + 3)(6x) + (2x + 3)(9) \\
&= 8x^3 + 12x^2 - 12x^2 - 18x + 18x + 27 \\
&= 8x^3 + 27
\end{aligned}
$$

This product is a binomial. In fact, it is the cube of $2x$ plus the cube of 3. Work through the following examples.

$$
\begin{aligned}
(3x + 5)(9x^2 - 15x + 25) &= (3x + 5)(9x^2) - (3x + 5)(15x) + (3x + 5)(25) \\
&= 27x^3 + 45x^2 - 45x^2 - 75x + 75x + 125 \\
&= 27x^3 + 125
\end{aligned}
$$

$$
\begin{aligned}
(x + 2)(x^2 - 2x + 4) &= (x + 2)(x^2) - (x + 2)(2x) + (x + 2)(4) \\
&= x^3 + 2x^2 - 2x^2 - 4x + 4x + 8 \\
&= x^3 + 8
\end{aligned}
$$

Each product consists of a cube plus a cube. In fact, the cubes produced are based on the lead binomial in the product. If the binomial is $a + b$, then the binomial produced is $a^3 + b^3$. The lead binomial is also related to the trinomial factor. For $a + b$, the trinomial factor is $a^2 - ab + b^2$. Identify these relationships in the products produced earlier:

$$
\begin{aligned}
(2x + 3)(4x^2 - 6x + 9) &= 8x^3 + 27 \\
(3x + 5)(9x^2 - 15x + 25) &= 27x^3 + 125 \\
(x + 2)(x^2 - 2x + 4) &= x^3 + 8
\end{aligned}
$$

As you can see, these products follow the general formula

$$(a + b)(a^2 - ab + b^2) = a^3 + b^3$$

Next, consider what happens if the lead binomial is $a - b$. We can write $a - b$ as $a + (-b)$, and then use the same special product generalization:

$$(a + (-b))(a^2 - a(-b) + (-b)^2) = a^3 + (-b)^3$$

or

$$(a - b)(a^2 + ab + b^2) = a^3 - b^3$$

We can use this special product, $(a - b)(a^2 + ab + b^2) = a^3 - b^3$, to quickly find the products of selected polynomials. For example,

$$(2x - 3)(4x^2 + 6x + 9) = 8x^3 - 27$$
$$(3x - 5)(9x^2 + 15x + 25) = 27x^3 - 125$$
$$(x - 2)(x^2 + 2x + 4) = x^3 - 8$$

We can now add these two generalizations to the list of special products we developed in Section 5.1.

SPECIAL PRODUCTS

$$(a + b)^2 = a^2 + 2ab + b^2$$
$$(a - b)^2 = a^2 - 2ab + b^2$$
$$(a + b)(a - b) = a^2 - b^2$$
$$(a + b)(a^2 - ab + b^2) = a^3 + b^3$$
$$(a - b)(a^2 + ab + b^2) = a^3 - b^3$$

Exercises

Write each product as a standard polynomial. Use the graphics calculator to test your work for errors.

1. $(3x + 1)(9x^2 - 3x + 1)$
2. $(3x - 1)(9x^2 + 3x + 1)$
3. $(3x - 1)(9x^2 - 3x + 1)$
4. $(x + 3)(x^2 - 3x + 9)$
5. $(3n - 2)(9n^2 + 6n + 4)$
6. $(2x + 5)(4x^2 - 10x + 25)$
7. $x^4 - 2(x^2 - 2)(x^2 + 2)$
8. $u^3 - (u - 1)(u^3 + u + 1)$

5.2 ■ ■ ■ FACTORING POLYNOMIALS, PART I

In Section 5.1 we discussed multiplication of polynomials. In this section, we discuss factoring of polynomials. *Factoring* is the reverse process of multiplication. Success with factoring requires knowledge of multiplication of polynomials—not only how to carry out the operation, but also the generalizations. This is much like arithmetic—we know 2 is factor of 348 because 348 is an even number. When factoring a trinomial, we look for binomial factors because, from multiplication, we known that the product of binomials can be a trinomial. We start with the meaning of factoring and the first basic rule of factoring—look for a common factor. We then consider some additional principles of factoring.

The Meaning of Factoring If $ab = c$, then a is a *factor* of c and b is a factor of c. At this level of algebra, we consider only integer factors and variable factors. Consider factoring the number 12. What does this mean? A correct response would be ''Write 12 as a product of whole numbers.'' Thus, to factor 12, we can write (2)(6). We are often asked to *completely* (or *prime*) *factor* a number. This means that each factor must be prime. Remember that a prime number is a whole number larger than 1, whose only factors are itself and 1. (Some primes are 2, 3, 5, 7, 11, 13, 17, 19, 23.) Thus, for 12 we write (2)(2)(3): We have written 12 as a product, and each factor in the product is a prime number. That is, we have completely factored 12.

To **factor** a polynomial, we write the polynomial as a product of polynomials. In a **completely (prime) factored** polynomial, each factor in the product is a prime polynomial, a polynomial that has only 1 and itself as factors—like $3x + 2$. As usual, we will start with the simplest case.

Factoring Monomials Consider the monomial $12x^3y^4$. This polynomial is already factored, because it is written as a product. We could also write $(2)(2)(3)xxxyyyy$. Usually, we leave monomials as they are written. However, some situations, such as finding the *greatest common factor,* we factor monomials to obtain a common factor.

The GCF A **common monomial factor** of a standard polynomial is a monomial whose factors are factors of each of the terms of the polynomial. The **greatest common monomial factor**, or GCF, is either one or the common factor with the largest number of prime factors. The GCF of a collection of natural numbers is the largest number that evenly divides the numbers. For example, the GCF of 12, 18, and 30 is 6, because 6 is the largest integer that evenly divides 12, 18, and 30. Prime factoring these numbers makes this fact more apparent:

$$12 = (2)(2)(3) \qquad 18 = (2)(3)(3) \qquad 30 = (2)(3)(5)$$

For a collection of powers of a variable, such as x^2, x^5, x^6, and x^7, the GCF is the lowest power in the collection. For these powers, the GCF is x^2. For the terms y, y^3, and y^7, the GCF is y. If the terms contain more than one variable, say x^2y^3, x^4y, and x^3y^3z, we find the GCF by treating the variables individually. For the powers of x we get x^2, and for the powers of y we get y. For the powers of z we get 1, because the first two terms have *no* z-factor (or, we have z^0 in the first two terms). So the GCF for x^2y^3, x^4y, and x^3y^3z is x^2y.

We now combine the ideas for a GCF for integers and powers of variables to find the GCF for a polynomial.

EXAMPLE 1 Find the GCF of the polynomial $3x^2 - 9x$.

Solution Consider the terms of the polynomial: $3x^2$ and $9x$.
For the numbers 3 and 9, the GCF is 3.
For the x-powers, x^2 and x, the GCF is x.
So, the GCF is $3x$. ■

EXAMPLE 2 Find the GCF of the polynomial $4rs - 12rs^2 + 10r^3s$.

Solution Consider the terms of the polynomial: $4rs$, $12rs^2$, and $10r^3s$.
For the numbers 4, 12, and 10, the GCF is 2.
For the r-powers, r and r^3, the GCF is r.
For the s-powers, s and s^2, the GCF is s.
So, we have a GCF of $2rs$. ■

GIVE IT A TRY

Find the GCF for each polynomial.

1. $3x^3 - 2x^2 - 6x$
2. $x^3 - x^2 - 5$
3. $20x^4y^2 - 10x^3y^2$

Looking for a Common Factor A basic rule of factoring is to *look for a common (monomial) factor*. If the polynomial is to be factored completely, the GCF must appear as a factor! Thus, it is best to look for this greatest common factor, or GCF, at the start. To *factor out* the GCF, we use the distributive property, $ab + ac = a(b + c)$. Consider the following examples.

EXAMPLE 3 Factor out the GCF from the polynomial $3x^2 - 9x$.

Solution The GCF for the terms of this polynomial is $3x$ (see Example 1). So, we can rewrite the polynomial as

$$\begin{aligned} 3x^2 - 9x &= (3x)(x) - (3x)(3) \\ &= (3x)(x - 3) \end{aligned}$$

■

EXAMPLE 4 Factor out the GCF from the polynomial $4rs - 12rs^2 + 10r^3s$.

Solution The GCF for the terms of this polynomial is $2rs$ (see Example 2). Thus, we factor the polynomial as

$$\begin{aligned} 4rs - 12rs^2 + 10r^3s &= (2rs)(2) - (2rs)(6s) + (2rs)(5r^2) \\ &= (2rs)(2 - 6s + 5r^2) \end{aligned}$$

To write the factors of this factored form as standard polynomials, we write $(2rs)(5r^2 - 6s + 2)$. It is very important to be able to recognize factored form, that is, the final result is a *product of polynomials*! Additionally, each polynomial factor should be in standard form.

Because of the commutative property of multiplication, we can report the answer as either $(2rs)(5r^2 - 6s + 2)$ or $(5r^2 - 6s + 2)(2rs)$.

Finally, to check our work, we multiply the resulting polynomial factors to see that the product is the original polynomial. ■

GIVE IT A TRY

Factor out the GCF from each polynomial. If the GCF is 1, write prime for your answer.

4. $30x^3 - 20x^2$
5. $-12w^2t - 15wt^2$
6. $xy^3 - xt + 3$
7. $5x^4 + 15x^3 - 10x^2$

Grouping We can use the idea of factoring a monomial from each term of a polynomial to factor other polynomials from an expression. Consider the following expression.

common factor

$$3x(x - 2) + 2(x - 2)$$

We can write this expression in factored form by using the distributive property, $ac + bc = (a + b)(c)$:

$$(3x + 2)(x - 2)$$

The following example shows how to use grouping to factor a polynomial.

EXAMPLE 5 Factor completely: $x(x - 5) + 5(x - 5)$

Solution We recognize a common factor of $x - 5$. Using the distributive property, we write $(x + 5)(x - 5)$. ■

Now, consider completely factoring $ax + ay + 2x + 2y$. The GCF is 1 (the terms have no common factor other than 1). If we group the first two terms, we do have a common factor of a. Likewise, grouping the last two terms yields a common factor of 2. Thus, we can write:

common factor $x + y$

$$ax + ay + 2x + 2y = a(x + y) + 2(x + y)$$

common factor a common factor 2

Now, we factor out $x + y$ to get $(a + 2)(x + y)$.

To learn how to use groupings in factoring, work through the following examples.

EXAMPLE 6 Factor completely: $x^3 - 2x^2 + 3x - 6$

Solution First, all the terms have no common factor, so the GCF is 1. Next, we try grouping.

group group

$$\begin{aligned} x^3 - 2x^2 + 3x - 6 &= x^2(x - 2) + 3(x - 2) \\ &= (x^2 + 3)(x - 2) \end{aligned}$$

■

EXAMPLE 7 Factor completely: $ay - y - a + 1$

Solution The polynomial has a GCF of 1, so we try grouping.

group group Note the sign change.

$$\begin{aligned} ay - y - a + 1 &= y(a - 1) - (a - 1) \\ &= y(a - 1) - 1(a - 1) \\ &= (y - 1)(a - 1) \end{aligned}$$

■

We now state a general procedure for using grouping to factor a quadrinomial. The procedure assumes the GCF for all the terms has been factored out.

FACTORING QUADRINOMIALS USING GROUPING

1. Arrange the four terms into two groups of two terms each, such that each group of terms has a GCF other than 1.
2. Factor out the GCF from each group of terms.
3. If the remaining groupings have a common binomial factor, factor out the common binomial from the two groupings.

In this procedure, if any of the steps fail, grouping cannot be used to factor the polynomial. For example, suppose we have the polynomial $x^2 - 3x - 3a + a^2$. Here, we can carry out step 1 to get $(x^2 - 3x) - (3a - a^2)$. We can also carry out step 2 to get $x(x - 3) - a(3 - a)$. But we cannot carry out step 3, because the binomial factors are not the same. So, we cannot factor this polynomial using grouping! We do *not* report part of the work, $x(x - 3) - a(3 - a)$, because this is not factored form! We simply write *prime,* or *cannot factor,* for the given polynomial.

EXAMPLE 8 Prime factor: $x^3 + x^2 - 3x - 3$

Solution First, the terms have no common factor.
Now, we group the first two terms and the last two terms to get:

Note the sign change.

$$(x^3 + x^2) - (3x + 3)$$

Next, we factor each of these groupings to get:

$$x^2(x + 1) - 3(x + 1)$$

Finally, we factor out $(x + 1)$ from the groupings to get:

$$(x^2 - 3)(x + 1)$$

■

EXAMPLE 9 Prime factor: $2x^4 + 2x^3 + 6x^2 + 6x$

Solution First, the common factor in each of the terms is $2x$. Factoring out $2x$, we get

$$(2x)(x^3 + x^2 + 3x + 3)$$

Now, we group the first two terms and the last two terms to get:

$$(2x)[(x^3 + x^2) + (3x + 3)]$$

Next, we factor each of these groupings to get:

$$(2x)[x^2(x + 1) + 3(x + 1)]$$

Finally, we factor out $(x + 1)$ from the groupings to get:

$$(2x)(x^2 + 3)(x + 1)$$

■

GIVE IT A TRY

Completely (prime) factor each polynomial.

8. $x^3 + x^2 + 5x + 5$
9. $x^3 + x^2 - 5x - 5$
10. $ac + bc - 2a - 2b$
11. $2y^4 + 6y^3 + 2y^2 + 6y$

■ ***Quick Index***

■ SUMMARY

In this section, we have discussed the skill of factoring polynomials. To factor a polynomial means to write it as a product of polynomials. The keys to this skill are knowledge of multiplication of polynomials and an organized approach. Some basic rules are very helpful with factoring polynomials:

1. Look for a common monomial factor in each of the terms of the polynomial. Factor out this common factor, the GCF, and then try to factor the remaining polynomial.
2. For a four-term polynomial—a quadrinomial—use grouping and a binomial factor.

If the polynomial will not factor, we report the answer as *prime*.

5.2 ■ ■ ■ EXERCISES

For each polynomial, report the GCF *(the greatest common factor).*

1. $12x^2 - 8x$
2. $20x^3 + 15x^2$
3. $4x^4 - 2x^2 - 8x$
4. $25x^5 - 10x^3 + 15x^2$
5. $20 - 15n - 35n^2$
6. $15pm - 10p^3m + 5pm^2$
7. $45a^{10} - 30a^7$
8. $9cv^3 - 27cv^2 + 36c^2v^2$
9. $36 - x^4$
10. $2cd^5 - 10c^3d^5 + 18d^2$

Factor each polynomial by factoring out the GCF.

11. $22x^2 - 33x$
12. $4hx^3 + 16hx^2$
13. $6x^2 - 3x - 12$
14. $12x^2 - 3x + 6$
15. $20 - 10x$
16. $48c + 16hc^2$
17. $7x^2 - 14x + 28$
18. $6x^3 + 18x^2 + 28x$
19. $10m^5x^2 - 4m^3x$
20. $48nc^3 + 18n^2c^2 + 28nc$
21. $6x^2 - 3x - 15$
22. $12x^2 + 3x + 6$
23. $6x - 3xy - 12y$
24. $9gk + 18g^2k^3 - 27g$
25. $6x^2 - 3xy - 12y^2$
26. $19mcx + 38m^2c^3x$

Completely (prime) factor each quadrinomial. If the polynomial does not factor, report prime.

27. $x^3 - x^2 - 3x + 3$
28. $x^3 + x^2 + 3x + 3$
29. $x^3 - 2x^2 - 5x + 10$
30. $x^3 + x^2 - 3x - 3$
31. $x^3 - x^2 + 5x - 5$
32. $x^3 + 2x^2 + 3x + 6$
33. $x^5 + x^3 + 2x^2 + 2$
34. $6p^3 - 12p^2 + 6p - 12$
35. $x^3 - 2x^2 + x - 2$
36. $x^4 + 2x^3 + 8x + 16$
37. $ax + bx + a + b$
38. $c^2 - 2ac + bc - 2ab$
39. $a^2 + ab + a + b$
40. $x^3 + x^2 + 2x + 2$
41. $2x^3 - 4x^2 + x - 2$
42. $x^4 + 3x^3 - 5x - 15$
43. $3x^5 + 6x^3 - x^2 - 2$
44. $6p^3 + 9p^2 - 2p - 3$
45. $ax + bx - a - b$
46. $a^2 + ab - a - b$
47. $a^2 - ab - a + b$
48. $x^3 + x^2 - 2x - 2$
49. $x^3 - x^2 - 3x - 3$
50. $ax - bx - a + b$
51. $a^2 + ab + a - b$
52. $x^3 + x^2 - 2x + 2$
53. $x^5 - 4x^3 - 8x^2 + 2$
54. $y^3 - 2y^2 - 4y + 8$
55. $3p^3 + 6p^2 - 9p - 18$

5.3 ■ ■ ■ FACTORING POLYNOMIALS, PART II

In the preceding section, we considered two of the fundamental rules of factoring. The first of these is to factor out a common factor (the GCF). The second fundamental rule is to use grouping to factor a quadrinomial. In this section we focus on trinomials. A key feature of the factoring technique for trinomials is the FOIL approach to multiplication. (You may want to review the FOIL approach to multiplication in Section 5.1.)

Sign of Last Term Consider the products

$$(x + 3)(x + 5) = x^2 + 8x + 15 \quad \text{and} \quad (x - 3)(x - 5) = x^2 - 8x + 15$$

These products demonstrate that when multiplying binomials that are sums, such as $x + 3$ and $x + 5$, the last term of the product is positive. Also, when multiplying binomials that are differences, such as $x - 3$ and $x - 5$, the last term of the product is positive. Next, consider the products

$$(x + 3)(x - 5) = x^2 - 2x - 15 \quad \text{and} \quad (x - 3)(x + 5) = x^2 + 2x - 15$$

These products show that when multiplying binomials that are a sum, like $x + 3$, and a difference, like $x - 5$, the last term of the product is negative. We summarize these observations in the following property.

SIGN OF LAST TERM PROPERTY

> For $x^2 + nx + m$, if the polynomial factors and $m > 0$, then both factors are sums (if $n > 0$) or both factors are differences (if $n < 0$).
>
> For $x^2 + nx + m$, if the polynomial factors and $m < 0$, then the factors are a sum and a difference.

Using FOIL in Factoring We now study completely (prime) factoring of second-degree trinomials. From our earlier work with multiplication of polynomials, we have the following generalization.

FACTORS OF A SECOND-DEGREE TRINOMIAL

> A second-degree trinomial (without a common factor for the terms) factors as the product of first-degree binomials.
>
> Otherwise the trinomial is prime (it will not factor).

Work through the following example.

EXAMPLE 1 Completely factor: $x^2 - x + 1$

Solution First, we look for a common factor. The terms have none—the GCF is 1. Next, to factor $x^2 - x + 1$, we must be able to write it as a binomial times a binomial. From the FOIL method of multiplication, we know that the first term comes from the first term times the first term (**F**). Thus, we have:

$$x^2 - x + 1 = (x + ??)(x + ??)$$

Also, from FOIL we know that the last term comes from the last term times the last term (**L**). Thus, we have two possibilities:

$$x^2 - x + 1 = (x + 1)(x + 1) \quad or \quad x^2 - x + 1 = (x - 1)(x - 1)$$

Finally, the middle term comes from the sum of the product of the outer terms and the product of the inner terms (**O + I**). We use trial and error to check each of the two possibilities:

$(x + 1)(x + 1)$ — O: $1x$, I: $1x$; O + I: $2x$ *or* $(x - 1)(x - 1)$ — O: $-1x$, I: $-1x$; O + I: $-2x$

So, neither possibility produces the desired middle term, $-x$. Thus, $x^2 - x + 1$ is a prime polynomial. ■

A general approach to factoring a second-degree trinomial $x^2 + bx + c$, based on FOIL, is given next. These steps assume the GCF has been factored out.

FACTORING $x^2 + bx + c$ USING FOIL

1. Look for two binomial factors. Factor the first term of the trinomial for the first terms of the binomials.
2. Factor the last term of the trinomial for the last terms of the binomials.
3. Use trial and error to produce middle term.

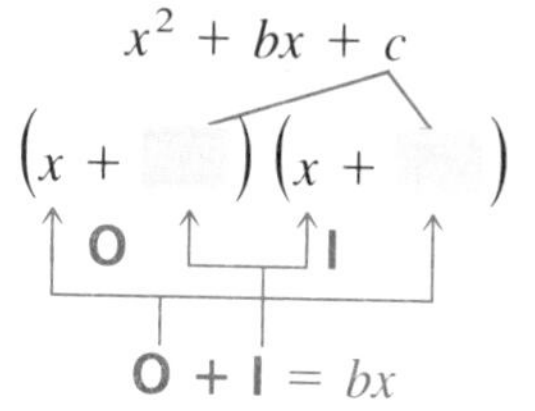

$$x^2 + bx + c$$

$$(x + \square)(x + \square)$$

$$\mathbf{O} + \mathbf{I} = bx$$

If the middle term cannot be produced, the trinomial is prime.

To study this approach, work through the following examples.

EXAMPLE 2 Prime factor: $x^2 - 5x + 6$

Solution First, the terms have no common factor.
Now, we look for two binomial factors. The first term x^2 yields x and x. Because the last term is positive and the middle term is negative, 6 factors as either -1 and -6 or -2 and -3. So, we start by trying -1 and -6:

$(x - 1)(x - 6)$ — inner: $-1x$, outer: $-6x$; sum: $-7x$

This product produces a middle term of $-7x$. This doesn't match our original term, $-5x$, in $x^2 - 5x + 6$, so we try the second possibility, -2 and -3:

$$(x - 2)(x - 3)$$

$$\begin{array}{r} -2x \\ \underline{-3x} \\ -5x \end{array}$$

This factoring produces a middle term of $-5x$. Success.
We report that $x^2 - 5x + 6$ factors as $(x - 2)(x - 3)$. ■

EXAMPLE 3 Prime factor: $3x^2 + 24x + 45$

Solution First, the GCF is 3. Factoring out this GCF, we get

$$3(x^2 + 8x + 15)$$

Now, to factor the trinomial $x^2 + 8x + 15$, we look for two binomials. Because the first term is x^2, the first terms of the binomial factors are x and x. Because the last and the middle terms are positive, 15 factors as either 1 and 15 or 3 and 5. So, we start by trying 1 and 15:

$$(x + 1)(x + 15)$$

$$\begin{array}{r} 1x \\ \underline{15x} \\ 16x \end{array}$$

This product produces a middle term of $16x$. We need a middle term of $8x$, so we try the second possibility, 3 and 5:

$$(x + 3)(x + 5)$$

$$\begin{array}{r} 3x \\ \underline{5x} \\ 8x \end{array}$$

This factoring produces a middle term of $8x$. Success. So the factored form for $3x^2 + 24x + 45$ is $(3)(x + 3)(x + 5)$. ■

EXAMPLE 4 Prime factor: $x^2 + 2x - 15$

Solution First, the terms have no common factor. Next, we look for two binomials. Because the first term is x^2, the first terms of the binomial factors are x and x. Because the last term is negative, we factor -15 as 1 and -15, as -1 and 15, as 3 and -5, or as -3 and 5. We start by trying 1 and -15:

$$(x + 1)(x - 15)$$

$$\begin{array}{r} 1x \\ \underline{-15x} \\ -14x \end{array}$$

This product produces a middle term of $-14x$, which does not match our term, $2x$. Likewise, -1 and 15 does not yield the desired middle term, so we try

$$(x - 3)(x + 5)$$

$$-3x$$

$$\underline{5x}$$

$$2x$$

This factoring produces a middle term of $2x$. Success. So the factored form for $x^2 + 2x - 15$ is $(x - 3)(x + 5)$. ■

Suppose that the polynomial in Example 4 is $15 - 2x - x^2$. First, we write it as a standard polynomial, $-x^2 - 2x + 15$. Next, we factor out -1 to get $-1(x^2 + 2x - 15)$. Now, we factor $x^2 + 2x - 15$ as we did in Example 4. Thus, $15 - 2x - x^2$ factors as $(-1)(x - 3)(x + 5)$. Compare this with the result of Example 4.

GIVE IT A TRY

Prime factor each trinomial. Use multiplication to check your work.

1. $x^2 - 5x - 6$ **2.** $4x^3 - 20x^2 + 24x$ **3.** $-x^2 - 11x - 18$

Factoring $ax^2 + bx + c$ We now use the FOIL technique to factor a trinomial whose x^2 coefficient is not 1. Such a trinomial has the general form $ax^2 + bx + c$.

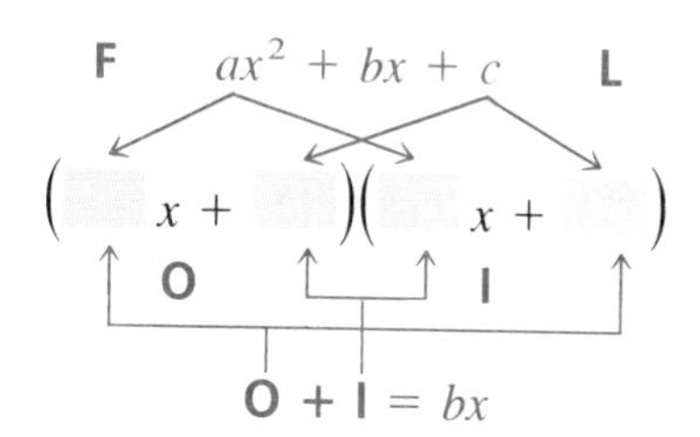

In general, for polynomials of the form $ax^2 + bx + c$, we factor out any common factors, then try two binomials. To get the binomials, we factor the first term, factor the last term, then use trial and error until we produce the middle term. Remember to keep the first term and the last term factored correctly. Resist getting so involved in trying to achieve the middle term, that you lose the correct factoring for the first term and the last term. Consider the next example.

EXAMPLE 5 Prime factor: $3x^2 + 11x + 6$

Solution First, the terms have no common factor. Now, we look for two binomials: the first term, $3x^2$, yields $3x$ and x, the last term, 6, yields either 1 and 6 or 3 and 2. We know to use positive factors because the constant term as well as the x-term is positive.

So, we start to factor $3x^2 + 11x + 6$ by trying 1 and 6:

$$(3x + 1)(x + 6)$$

$1x$

$18x$

$19x$

This product produces a middle term of $19x$, which does not match our middle term, $11x$. We know *not* to try $(3x + 6)(x + 1)$ because the binomial factor $3x + 6$ has a common factor of 3. (Remember, $3x^2 + 11x + 6$ has no common factor.) Next, we try 2 and 3:

$$(3x + 2)(x + 3)$$

$2x$

$9x$

$11x$

This factoring produces a middle term of $11x$. So we have prime factored $3x^2 + 11x + 6$ as $(3x + 2)(x + 3)$. ■

Factoring trinomials is always a straightforward process. However, when the first term and last term have a large number of factors, many trials may be necessary to find the desired middle term. For this reason, we often construct a table to show the factors of the first term, the last term, and the sum of the factors (the middle term).

In general, for $(ax + b)(cx + d)$ we have the product

$$(ac)x^2 + (ad)x + (bc)x + bd \qquad \text{or} \qquad (ac)x^2 + (ad + bc)x + bd$$

So, in our table, the middle term will be $(ad + bc)$. Such a table for the polynomial in Example 5, $3x^2 + 11x + 6$, is:

First	Last	Middle (**O** + **I**)
3 and 1	6 and 1	$(3)(1) + (1)(6) = 9$
3 and 1	1 and 6	$(3)(6) + (1)(1) = 19$
3 and 1	3 and 2	$(3)(2) + (1)(3) = 9$
3 and 1	2 and 3	$(3)(3) + (1)(2) = 11$ Success!

The lines of the table correspond to trials of various binomials:

$$(3x + 6)(1x + 1) = 3x^2 + 9x + 6$$
$$(3x + 1)(1x + 6) = 3x^2 + 19x + 6$$
$$(3x + 3)(1x + 2) = 3x^2 + 9x + 6$$
$$(3x + 2)(1x + 3) = 3x^2 + 11x + 6 \quad \text{Success!}$$

EXAMPLE 6 Prime factor: $3x^2 - 17x + 10$

Solution The terms have no common factor (the GCF is 1). So, we look for two binomials. The first terms of the binomials will be $3x$ and x (the factors of $3x^2$). The last terms will be either -10 and -1 or -2 and -5. We know the signs will be negative because the last term is positive and the middle term is negative.

We can construct the table:

First	Last	Binomials	Middle (**O** + **I**)
3 and 1	-10 and -1	$(3x - 10)(x - 1)$	$(3)(-1) + (1)(-10) = -13$
3 and 1	-1 and -10	$(3x - 1)(x - 10)$	$(3)(-10) + (1)(-1) = -31$
3 and 1	-5 and -2	$(3x - 5)(x - 2)$	$(3)(-2) + (1)(-5) = -11$
3 and 1	-2 and -5	$(3x - 2)(x - 5)$	$(3)(-5) + (1)(-2) = -17$ Success!

So, we write $3x^2 - 17x + 10 = (3x - 2)(x - 5)$. ■

The next example of factoring a trinomial shows a slight variation from factoring $ax^2 + bx + c$. However, the technique used is the same.

EXAMPLE 7 Completely factor: $4x^2 - 5xy - 6y^2$

Solution Again, the terms have no common factor.

Searching for two binomials, we consider the form:

$$(\quad x + \quad y)(\quad x + \quad y).$$

Because the last term is negative, we know we must use a sum and a difference for the binomials. Building a table of possibilities, we get:

First	Last	Binomials	Middle (**O** + **I**)
4 and 1	-6 and 1	$(4x - 6y)(x + y)$	omit, common factor
4 and 1	6 and -1	$(4x + 6y)(x - y)$	omit, common factor
4 and 1	3 and -2	$(4x + 3y)(x - 2y)$	$(4)(-2) + (1)(3) = -5$ Success!

The factors of $4x^2 - 5xy - 6y^2$ are $(4x + 3y)(x - 2y)$. ■

In Example 7, we found the correct factors early in the table listing of the factors. If the coefficients of the trinomial have many different factorings, this approach can lead to a large table of values. In the *Discovery* following Exercises 5.3, we give a more detailed approach for determining whether the trinomial will factor and for reducing the number of possibilities for the factors. The preceding approach, using the FOIL technique and possibly a table, is very productive for most factoring in algebra at this level.

GIVE IT A TRY

Completely factor each trinomial. Use multiplication to check your work.

4. $x^2 - 9x + 14$ **5.** $5x^2 - 2x - 3$

6. $x^2 + xy - 2y^2$ **7.** $18x^2 + 6x - 4$

Beyond $ax^2 + bx + c$ Although the factoring approach we have shown is designed for second-degree trinomials, it will often work for trinomials of degree greater than 2. Work through the next example.

EXAMPLE 8 Prime factor: $4x^5 - 10x^3 - 6x$.

Solution First, the terms have a common factor of $2x$, so we get $(2x)(2x^4 - 5x^2 - 3)$. Now, to factor $2x^4 - 5x^2 - 3$, we look for two binomials:

$2x^4$ yields $2x^2$ and x^2 -3 yields -3 and 1 or 3 and -1

We start by trying -3 and 1:

$$(2x)(2x^2 - 3)(x^2 + 1)$$

$$\begin{array}{r} -3x^2 \\ 2x^2 \\ \hline -1x^2 \end{array}$$

This product produces a middle term of $-1x^2$. Since we are looking for a middle term of $-5x^2$, we try the other possibility for -3 and 1:

$$(2x)(2x^2 + 1)(x^2 - 3)$$

$$\begin{array}{r} 1x^2 \\ -6x^2 \\ \hline -5x^2 \end{array}$$

This factoring produces a middle term of $-5x^2$. Success. The factors of $4x^5 - 10x^3 - 6x$ are $(2x)(2x^2 + 1)(x^2 - 3)$. ■

GIVE IT A TRY

Completely factor each trinomial. Use multiplication to check your work.

8. $x^4 - 9x^2 + 14$ **9.** $x^4 - 2x^3 - 3x^2$

10. $x^4 + 3x^2y^2 + 2y^4$ **11.** $9x^4 + 3x^2 - 2$

■ SUMMARY

In this section, we have developed the skill of factoring trinomials. To factor a polynomial means to write it as a product of polynomials. The key to this skill is knowledge of multiplication of polynomials and an organized approach.

Some basic rules are very helpful with factoring polynomials:

1. Look for a common monomial factor in each of the terms of the polynomial. Factor out this common monomial factor, and then try to factor the remaining polynomial.
2. For a second-degree trinomial (with no common monomial factor) to factor, it must factor as two binomials. Reverse the FOIL approach to multiplication to:
   a. factor the first term of the trinomial to get the first terms of the binomials;
   b. factor the last term of the trinomial to get the last terms of the binomials;
   c. check to see if the outer and inner products for the binomials produce the middle term of the trinomial. If so, the trinomial is factored. If not, try other factorings of the first and last terms of the trinomial. If all the possibilities fail to produce the middle term, report the polynomial as prime.

This approach can also be used to factor some higher-degree trinomials.

5.3 ■ ■ ■ EXERCISES

Insert true *for a true statement, or* false *for a false statement.*

____ 1. To factor a polynomial means to write it as a product of polynomials.

____ 2. The polynomial $3x(x + 2) - 1$ is in factored form.

____ 3. To factor a polynomial, first look for a common factor for each of the terms.

____ 4. The greatest common factor for $2x^3 - 4x^2 + 12$ is $2x$.

____ 5. The GCF for $6x^4 - 3x^3 - 12x^2$ is x^2.

____ 6. Every second-degree trinomial will factor.

____ 7. The trinomial $x^2 + x + 3$ is a prime polynomial.

____ 8. The product $(4x)(x^2 - 2x)$ is the prime factorization of the polynomial $4x^3 - 8x^2$.

Completely factor each trinomial. If the polynomial will not factor, report it as prime.

9. $x^2 - 6x + 8$
10. $x^2 - 3x - 10$
11. $m^2 - 3m + 2$
12. $a^2 - 7a + 6$
13. $m^2 - m - 2$
14. $3w^2 + 12wd + 6d^2$
15. $x^2 - 9x + 8$
16. $6x^2 - 12x + 18$
17. $x^2 - 5x - 14$
18. $14y^2 + 28y - 21$
19. $d^2 - 3d - 40$
20. $x^2 - 6x - 16$
21. $2t^2 - 6t - 8$
22. $x^2 + 10x + 16$
23. $t^2 - 6t - 3$
24. $5x^2 - 10x + 20$
25. $7p^2 - 28p + 21$
26. $x^2 - 10x + 16$
27. $2t^2 + 6t - 8$
28. $n^2 - 15n + 56$
29. $y^2 - 2y - 63$
30. $n^2 - 16n + 63$

Completely (prime) factor each trinomial. If the polynomial does not factor, report it as prime.

31. $2x^2 - 3x + 1$
32. $3a^2 - 7a + 2$
33. $2n^2 - 7n + 6$
34. $2x^2 + x - 6$
35. $2x^2 - 11x - 6$
36. $3p^2 - p - 2$
37. $14y^2 + y - 3$
38. $5 + 3t - 2t^2$
39. $12x^2 - 17x - 6$
40. $12x^2 - 18x + 6$
41. $2t^2 - 9t - 5$
42. $2c^2 - 7c + 5$
43. $4x^2 + 12x + 9$
44. $4x^2 - 12x + 9$
45. $3w^2 + 7wd + 2d^2$
46. $6x^2 - 5x + 1$
47. $10x^2 + 17x + 3$
48. $15r^2t + 74rt - 5t$
49. $3x^2 - 5x + 2$
50. $4a^2 - 11a - 3$

Completely (prime) factor each polynomial. If the polynomial does not factor, report it as prime.

51. $x^2 - 6x + 8$
52. $2x^2 - 3x - 1$
53. $3x^2 - 9x$
54. $5mn - 25m^2n^2$
55. $20m - 27n$
56. $20m^2 - 28$
57. $3t^3 - 6t^2 - 45t$
58. $w^4 - 5w^3 - 24w^2$
59. $x^3 - 7x^2 + 10x$
60. $x^4 - 3x^3 - 10x^2$
61. $y^3 - 11y^2 + 30y$
62. $n^4 - 7n^3 - 30n^2$
63. $v^2 - 5v - 24$
64. $d^5 - 10d^4 - 24d^3$
65. $2x^4 - 3x^2 - 1$
66. $4b^4 - 11b^2 - 3$
67. $2x^3 + 16x^2 + 4x$
68. $3s^2t - 6st - 9t$
69. $36t^5 + 18xt^3$
70. $4x^2y^4 - 8x^3y^2 - 12x^2y^2$
71. $3x^2 - 13x - 10$
72. $3x^3 - 15x^2 + 2x - 10$
73. $2x^3 + 2x^2 - 40x$
74. $x^3 + 5x^2 + 7x + 35$
75. $6x^4 - x^2 - 2$

DISCOVERY

The *ac* Approach to Factoring $ax^2 + bx + c$

In factoring a polynomial of the form $ax^2 + bx + c$, where the terms have no common factor, we look for two binomials of the form

$$(\quad x + \quad)(\quad x + \quad)$$

As we found in Section 5.3, when a and c have several factors, using the FOIL approach and a table can become quite a tedious process. If the polynomial is prime, we must exhaust every possibility before stating that the polynomial will not factor—that it is prime. It would be convenient to have a method that would quickly determine whether or not the polynomial will factor. Such an approach exists and is known as the *ac approach.*

THE *ac* TEST FOR FACTORING $ax^2 + bx + c$

> For $ax^2 + bx + c$, with no common factor, if there exist integers, say p and q, such that pq equals ac and $p + q = b$, then the polynomial factors over the integers. If no such integers p and q exist, then the polynomial is a prime polynomial over the integers.

Consider the polynomial $x^2 - 3x - 5$. We have $a = 1$, $b = -3$, and $c = -5$. The product ac is -5. Now, if integers p and q exist such that $pq = -5$ and $p + q = -3$, then the polynomial will factor. The only integer factors of -5 either are 5 and -1 or -5 or 1. Neither of these pairs has a sum of the b-value, -3. Thus, we conclude $x^2 - 3x - 5$ is a prime polynomial.

Consider the polynomial $3x^2 - 17x + 10$ from Example 6 in Section 5.3. Remember that this polynomial factors as $(3x - 2)(x - 5)$. Here, $a = 3$, $b = -17$, and $c = 10$. The product ac is $(3)(10)$, or 30. Now, if integers p and q exist such that $pq = 30$ and $p + q = -17$, then the polynomial will factor. Using the factors -15 and -2 for 30 gives us the required integers. That is, for $p = -15$ and $q = -2$, we have $pq = 30$ and $p + q = -17$. Thus, we conclude that the polynomial will factor.

Using grouping (from Section 5.2), we can use the *ac* approach to factor the polynomial $3x^2 - 17x + 10$. We write the polynomial as follows:

$$3x^2 - 17x + 10 = 3x^2 \underbrace{- 15x - 2x}_{p \text{ and } q} + 10$$

Now, using grouping, we get $3x(x - 5) - 2(x - 5) = (3x - 2)(x - 5)$. It does not matter how p and q are used in the grouping. If we write

$$3x^2 - 17x + 10 = 3x^2 - 2x - 15x + 10$$

we factor as $x(3x - 2) - 5(3x - 2) = (x - 5)(3x - 2)$.

To learn how to use the *ac* approach, work through the following examples.

EXAMPLE 1 Completely factor: $8x^2 - 51x + 18$

Solution First, the terms have no common factor.

Next, $a = 8$, $b = -51$, and $c = 18$. The product ac is $(8)(18) = 144$. So, we need to find integers p and q such $pq = 144$ and $p + q = -51$:

p	q	$p + q$	
-1	-144	$-1 + (-144) = -145$	
-2	-72	$-2 + (-72) = -74$	
-3	-48	$-3 + (-48) = -51$	Success!

The polynomial $8x^2 - 51x + 18$ does factor. To find the factors, we use groupings:

$$8x^2 \underbrace{- 3x - 48x}_{p \text{ and } q} + 18$$
$$x(8x - 3) - 6(8x - 3)$$
$$(x - 6)(8x - 3)$$

Thus, $8x^2 - 51x + 18$ factors as $(x - 6)(8x - 3)$. ■

We can also use this technique on a polynomial in two variables, such as $24x^2 + 32xy - 6y^2$.

EXAMPLE 2 Completely factor $24x^2 + 32xy - 6y^2$.

Solution Here, we have a common factor of 2, so we write $2(12x^2 + 16xy - 3y^2)$. Now, $a = 12$, $b = 16$, and $c = -3$, the product ac is $(12)(-3) = -36$. We need to find integers p and q such that $pq = -36$ and $p + q = 16$:

p	q	$p + q$	
-1	36	$-1 + 36 = 35$	
-2	18	$-2 + 18 = 16$	Success!

The polynomial $12x^2 + 16xy - 3y^2$ does factor. To find the factors, we use groupings:

p and q

$$12x^2 - 2xy + 18xy - 3y^2$$
$$2x(6x - y) + 3y(6x - y)$$
$$(2x + 3y)(6x - y)$$

Thus, $24x^2 + 32xy - 6y^2$ factors as $2(2x + 3y)(6x - y)$. ■

Exercises

Use the *ac* method to completely factor each polynomial. Use the graphics calculator to test your work for errors.

1. $x^2 - 5x - 84$
2. $6x^2 - 17x - 10$
3. $30y^2 - 117y + 42$
4. $8x^3 - 62x^2 - 16x$
5. $x^2 - 12x - 17$
6. $2x^2 + 7x + 5$
7. $9x^2 - 15xy + 16y^2$
8. $12x^2 - 36xy + 15y^2$
9. $12z^2 + 16z - 3$
10. $40x^2 - 130x + 30$

5.4 ■ ■ ■ FACTORING POLYNOMIALS, PART III

In the last two sections, we have learned to factor polynomials by first looking for a common factor, the GCF. We have also learned to factor trinomials of the form $ax^2 + bx + c$ and some quadrinomials. The keys to these factorings are the FOIL technique of multiplication and groupings. In this section, we will concentrate on factoring binomials and some special trinomials (perfect squares). We start with binomials.

Factoring Binomials Beyond finding a common factor, binomials factor based on the special products from multiplication of polynomials. These special products are as follows.

SPECIAL PRODUCTS THAT PRODUCE A BINOMIAL

$$(A + B)(A - B) = A^2 - B^2$$
$$(A + B)(A^2 - AB + B^2) = A^3 + B^3$$
$$(A - B)(A^2 + AB + B^2) = A^3 - B^3$$

If you do not yet know these products, it will be advantageous to commit them to memory now. These products imply that binomials which are the difference of two perfect squares will factor. Also, binomials that are the sum or the difference of perfect cubes will factor.

Difference of Squares The sum of perfect squares, such as $x^2 + 4$, never factor (over the integers). That is, a binomial of the form $A^2 + B^2$ will not factor over the integers (in fact, it will not factor over the reals). Also, the binomial $x^2 - 3$ is prime. It is a binomial, has no common factor, but is not the difference of two perfect squares. Thus, it is prime. Remember, the product of the sum of two monomials, $(a + b)$, and the difference of the same two monomials, $(a - b)$, is $a^2 - b^2$. Thus, any binomial of the form $a^2 - b^2$ will factor into the form $(a + b)(a - b)$. Consider the following examples of factoring such binomials.

EXAMPLE 1 Completely factor: $4x^2 - 16$

Solution First, we factor out the common factor to get $4(x^2 - 4)$.

Next, $x^2 - 4$ is the difference of squares, namely, x and 2. We can write the binomial as

$$4(x^2 - 2^2)$$

Thus, it factors as:

$$4(x + 2)\ (x - 2)$$

sum difference ■

EXAMPLE 2 Completely factor: $32x^2 - 2$

Solution Here, we have a common factor of 2, so we write $32x^2 - 2 = 2(16x^2 - 1)$. Now, we write $16x^2 - 1$ as

$$(4x)^2 - 1^2$$

$$(4x + 1)(4x - 1)$$

Thus, $32x^2 - 2$ factors as $2(4x + 1)(4x - 1)$. ■

The difference-of-squares techniques can also work for higher-degree binomials, provided they are still the difference of two squares.

EXAMPLE 3 Completely factor: $25x^4 - 16$.

Solution Here, we have no common factor (the GCF is 1).

Next, $25x^4 - 16$ is the difference of squares, namely, $5x^2$ and 4. We can write $25x^4 - 16$ as

$$(5x^2)^2 - 4^2$$

$$(5x^2 + 4)(5x^2 - 4)$$

Thus, $25x^4 - 16$ factors as $(5x^2 + 4)(5x^2 - 4)$. ■

GIVE IT A TRY

Completely factor each binomial. Use multiplication to check your work.

1. $x^2 - 49$ **2.** $x^2 + 25$ **3.** $x^3 - 16x$

4. $49x^2 - 100$ **5.** $36x^2 - 9$ **6.** $12x^3 - 27x$

Cubes The techniques we have used to factor a binomial that is the difference of two squares can be modified to factor a binomial that is the sum of cubes or the difference of cubes. You may wish to review the *Discovery* following Exercises 5.1, which deals with multiplication of polynomials that produce these binomials.

EXAMPLE 4 Completely factor: $8x^3 - 27$

Solution Here, we have no common factor. The GCF is 1.
Next, $8x^3 - 27$ is the difference of cubes, using the general form $a^3 - b^3 = (a - b)(a^2 + ab + b^2)$ with $a = 2x$ and $b = 3$, we write

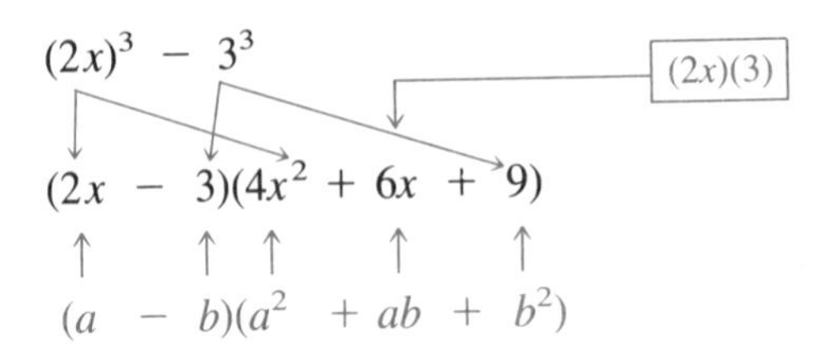

So, $8x^3 - 27 = (2x - 3)(4x^2 + 6x + 9)$. This is factored completely because the trinomial $4x^2 + 6x + 9$ is a prime polynomial. ■

EXAMPLE 5 Completely factor: $2x^4 + 16x$

Solution Here, we have the common factor of $2x$. Thus, we have $2x(x^3 + 8)$.
Next, $x^3 + 8$ is the sum of cubes, $x^3 + 2^3$.
Using the general form $a^3 + b^3 = (a + b)(a^2 - ab + b^2)$ with $a = x$ and $b = 2$, we write

(x)(2)

$$(x + 2)(x^2 - 2x + 4)$$

Thus, $2x^4 + 16x$ factors as $(2x)(x + 2)(x^2 - 2x + 4)$. ■

The technique for factoring cubes also works for a higher-degree binomial, provided it is the sum of two cubes or the difference of two cubes.

EXAMPLE 6 Completely factor: $x^6 - 1$

Solution Here, we have no common factor (the GCF is 1).
Next, we can think of $x^6 - 1$ as the difference of two squares, $(x^3)^2 - 1^2$. Thus, we write

$$(x^3)^2 - 1^2$$

$$(x^3 + 1)(x^3 - 1)$$

Now, $x^3 + 1$ is the sum of cubes and $x^3 - 1$ is the difference of cubes.

So, we write

$$(x^3 + 1)(x^3 - 1)$$

$$(x + 1)(x^2 - x + 1) \qquad (x - 1)(x^2 + x + 1)$$

So, $x^6 - 1$ factors as $(x + 1)(x^2 - x + 1)(x - 1)(x^2 + x + 1)$. ■

GIVE IT A TRY

Completely factor each polynomial. Use multiplication to check your work.

7. $x^3 - 8$ **8.** $8x^3 + 1$ **9.** $64y^6 + 1$

10. $8x^2 - 125$ **11.** $3x^4 - 81x$ **12.** $y^6 - 64$

Perfect Square Trinomials In Section 5.3 we factored trinomials using the FOIL approach. Here, we consider some special trinomials—those that are the square of a binomial. These factorings are based on the special products below.

SPECIAL PRODUCTS THAT PRODUCE A TRINOMIAL

$$(A + B)^2 = A^2 + 2AB + B^2$$
$$(A - B)^2 = A^2 - 2AB + B^2$$

In the following examples, we use these special products to factor perfect square trinomials.

EXAMPLE 7 Prime factor: $x^2 + 6x + 9$

Solution Using the special product $(A + B)^2 = A^2 + 2AB + B^2$, we must recognize that the form is present with $A = x$ and $B = 3$, that is,

$$x^2 + 6x + 9 = x^2 + 2(3)(x) + 3^2$$

Thus, we get $x^2 + 6x + 9 = (x + 3)(x + 3)$ or $(x + 3)^2$ ■

EXAMPLE 8 Prime factor: $36x^2 - 60x + 25$

Solution Using the special product $(A - B)^2 = A^2 - 2AB + B^2$, we must recognize that the form is present with $A = 6x$ and $B = 5$, that is,

$$36x^2 - 60x + 25 = (6x)^2 - 2(6x)(5) + 5^2$$

Thus, we get $36x^2 - 60x + 25 = (6x - 5)(6x - 5)$ or $(6x - 5)^2$ ■

This special product technique also works for a higher-degree trinomial, provided it is the perfect square of a binomial.

EXAMPLE 9 Prime factor: $x^4 - 8x^2 + 16$

Solution Using the special product $(A - B)^2 = A^2 - 2AB + B^2$, we must recognize that the form is present with $A = x^2$ and $B = 4$, that is,

$$x^4 - 8x^2 + 16 = (x^2)^2 - 2(x^2)(4) + 4^2$$

Thus, we get $x^4 - 8x^2 + 16 = (x^2 - 4)(x^2 - 4)$

Now, each factor in $(x^2 - 4)(x^2 - 4)$ is the difference of squares, so we get:

$$(x + 2)(x - 2)(x + 2)(x - 2)$$

or

$$(x + 2)^2(x - 2)^2$$

So, $x^4 - 8x^2 + 16$ factors as $(x + 2)^2(x - 2)^2$. ■

GIVE IT A TRY

Completely factor each trinomial. Use multiplication to check your work.

13. $4w^2 - 12w + 9$ **14.** $49y^2 + 28y + 4$ **15.** $x^6 - 2x^3 + 1$

Using Substitution We can expand the collection of polynomials that can be factored by using substitution in addition to special products. Consider factoring the polynomial $(x - 2)x - (x - 2)(3)$. Returning to our original idea of factoring out common factors, we can factor out the binomial $(x - 2)$ to obtain:

common factor

$$\underline{(x - 2)}x - \underline{(x - 2)}(3) = (x - 2)(x - 3)$$

Here, we have used the form $Ax - 3A = A(x - 3)$ with $(x - 2)$ substituted for A. Likewise, for $y^2 - (x - 3)^2$ we can use the form $A^2 - B^2$ with y substituted for A and $(x - 3)$ substituted for B:

$$\begin{aligned} A^2 - B^2 &= (A + B)\ (A - B) \\ y^2 - (x - 3)^2 &= [y + (x - 3)][y - (x - 3)] \\ &= (y + x - 3)(y - x + 3) \end{aligned}$$

Note the sign change.

We now use this substitution approach to factor polynomials.

EXAMPLE 10 Prime factor: $(x - 2)^2 - 3(x - 2) - 4$

Solution First, the polynomial has no common factor.
Now, substituting A for $(x - 2)$, we have the form $A^2 - 3A - 4$. This form factors as $(A - 4)(A + 1)$. Thus,

$$\begin{aligned} (x - 2)^2 - 3(x - 2) - 4 &= [(x - 2) - 4][(x - 2) + 1] \\ &= (x - 6)(x - 1) \end{aligned}$$

We have factored $(x - 2)^2 - 3(x - 2) - 4$, because we have written it as a product of the polynomials $(x - 6)$ and $(x - 1)$. ■

EXAMPLE 11 Prime factor: $(x - 1)^3 + 8$

Solution First, the polynomial has no common factor.
Now, substituting A for $(x - 1)$, we have the form $A^3 + 8$. This form factors as $(A + 2)(A^2 - 2A + 4)$. Thus,

$$\begin{aligned} & (A + 2)\ (A^2 - 2A + 2^2) \\ (x - 1)^3 + 8 &= [(x - 1) + 2][(x - 1)^2 - 2(x - 1) + 4] \\ &= (x + 1)(x^2 - 2x + 1 - 2x + 2 + 4) \\ &= (x + 1)(x^2 - 4x + 7) \end{aligned}$$

■

Using the Graphics Calculator

Checking our work in Example 11 by using multiplication requires several multiplications (which can themselves produce errors) in order to show that the polynomials are equivalent. To use the graphics calculator to test that $(x - 1)^3 + 8$ and $(x + 1)(x^2 - 4x + 7)$ are equivalent polynomials, type `(X−1)³+8` for `Y1` and type `(X+1)(X²−4X+7)` for `Y2`. Return to the home screen and store a number, say `⁻4`, in memory location `X`. Display the values in `Y1` and `Y2`. Are the outputs the same? If not, the polynomials are *not* equivalent, and an error is present!

GIVE IT A TRY

Prime factor each polynomial.

16. $(x + 3)^2 - 4(x + 3) + 3$ **17.** $(2x - 3)^2 - 25$ **18.** $(x + 1)^3 + 1$

19. Use the graphics calculator to test your work for errors in Problems 16–18.

Quadrinomials and Special Products In Section 5.2, we considered factoring of quadrinomials. Now, we return to this topic and use the factoring of special products discussed in this section. Work through the following examples.

EXAMPLE 12 Completely factor: $2x^3 - x^2 - 8x + 4$

Solution First, the polynomial has no common factor.
Here, we can factor x^2 from the first two terms and 4 from the second two terms to get:

Note the sign change.

$$x^2(2x - 1) - 4(2x - 1)$$

Now, we recognize that $2x - 1$ is a common factor. Thus, we rewrite this last expression as

$$(x^2 - 4)(2x - 1)$$

Finally, factoring $x^2 - 4$, we get $(x + 2)(x - 2)(2x - 1)$

So, $2x^3 - x^2 - 8x + 4$ factors as $(x + 2)(x - 2)(2x - 1)$. ■

EXAMPLE 13 Prime factor: $x^5 - x^3 - 8x^2 + 8$

Solution First, the polynomial has no common factor.
Now, we group the first two terms and the last two terms to get:

$$(x^5 - x^3) - (8x^2 - 8)$$

Note the sign change.

Next, we factor each of these groupings in $(x^5 - x^3) - (8x^2 - 8)$ to get:

$$x^3(x^2 - 1) - 8(x^2 - 1)$$

Finally, we factor out $(x^2 - 1)$ from the groupings to get:

$$(x^3 - 8)(x^2 - 1)$$

The factor $x^3 - 8$ factors as

$$(x - 2)(x^2 + 2x + 4)$$

The factor $x^2 - 1$ factors as

$$(x + 1)(x - 1)$$

So, $x^5 - x^3 - 8x^2 + 8$ factors as

$$(x - 2)(x^2 + 2x + 4)(x + 1)(x - 1)$$ ■

The quadrinomial in Example 12 is the product of three binomials. The quadrinomial in Example 13 is the product of three binomials and a trinomial.

GIVE IT A TRY

Completely factor each quadrinomial.

20. $x^3 + x^2 - 4x - 4$

21. $ac^2 + bc^2 - 9a - 9b$

22. $x^4 - x^3 + 8x - 8$

23. $2y^4 + 6y^3 - 2y^2 - 6y$

24. Use the graphics calculator to test your work for errors in Problems 20–23.

■ SUMMARY

■ ***Quick Index***

In this section we have continued developing the skill of factoring polynomials. (To factor a polynomial means to write it as a product of polynomials.) Skill in the area of factoring is essential to continued success in algebra. In this section we learned to factor binomials and perfect square trinomials. Factoring such polynomials requires the following special products from multiplication:

$$(A + B)(A - B) = A^2 - B^2$$
$$(A + B)(A^2 - AB + B^2) = A^3 + B^3$$
$$(A - B)(A^2 + AB + B^2) = A^3 - B^3$$
$$(A + B)^2 = A^2 + 2AB + B^2$$
$$(A - B)^2 = A^2 - 2AB + B^2$$

Using these special products and the technique of substitution expands the collection of polynomials that we can factor. Finally, we combined factoring these special binomials and trinomials with our earlier work of factoring quadrinomials.

5.4 ■ ■ ■ EXERCISES

Completely (prime) factor each binomial. If the binomial does not factor, report prime.

1. $y^2 - 81$
2. $x^3 - 9x$
3. $m^3 - m$
4. $x^3 - 64$
5. $16x^2 - 1$
6. $8x^3 + 27$
7. $4x^2 + 25$
8. $z^6 - x^6$
9. $4x^2 - 16$
10. $h^3r^3 - 8$
11. $25 - 9a^2$
12. $z^6 + x^6$
13. $x^3 + 125$
14. $100x^4 - 900$
15. $x^6 - 1$
16. $64x^3 - 27$
17. $100x^4 - 9$
18. $x^2 - 8$
19. $x^4 - 16$
20. $9x^4 - 16$
21. $4x^2 - 49$
22. $81x^4 - 16$
23. $5x^2 - 10$
24. $x^6 - x^3$
25. $5x^2 - 20$
26. $x^6 - 1$
27. $16x^4 - 9$
28. $x^5 - 25x^3$
29. $16x^4 - 81$
30. $x^6 - 16$

Completely (prime) factor each trinomial. (Look for perfect squares.) If the polynomial does not factor, report prime.

31. $y^2 - 18y + 81$
32. $x^2 + 10x + 25$
33. $m^2 + 16m + 64$
34. $x^2 - 22x + 121$
35. $25x^2 - 10x + 1$
36. $9x^2 + 24x + 16$
37. $x^4 - 6x^2 + 9$
38. $x^4 + 8x^2 + 16$
39. $x^4 - 8x^2 + 16$
40. $9p^3 - 30p^2 + 25p$
41. $100x^2 - 60x + 9$
42. $25r^2t + 20rt + 4t$

Completely (prime) factor each polynomial. (Use substitution.) If the polynomial does not factor, report prime.

43. $(x - 3)^2 - 5(x - 3)$
44. $(x + 2)^2 - 2(x + 2) - 3$
45. $(2x + 1)^2 - 16$
46. $(3 - 2x)^3 - 8$
47. $(3 - 2x)^3 + 27$
48. $(2x^2)^2 - 3(2x^2) + 2$
49. $9x^4 - 3x^2 - 2$
50. $(x - 1)^4 - 16$

Completely (prime) factors each quadrinomial (using grouping). If the polynomial does not factor, report prime.

51. $ax^2 + bx^2 - a - b$
52. $c^2 + 4cb - ac - 4ab$
53. $a^3 + a^2b + a + b$
54. $x^3 + x^2 + 4x + 4$
55. $x^3 - x^2 - 3x - 3$
56. $ax^2 + bx^2 + a + b$
57. $c^2 - 4cb + 4b^2 - a^2$
   [*Hint*: $c^2 - 4cb + 4b^2 = (c - 2b)^2$]
58. $a^3 + a^2b - a - b$
59. $x^3 + x^2 - 4x - 4$
60. $x^5 - 4x^3 - 8x^2 + 32$
61. $y^3 - 2y^2 - 4y + 8$
62. $4p^3 + 8p^2 - 9p - 18$

Completely (prime) factor each polynomial. If the polynomial does not factor, report prime.

63. $9x^2 - 25$
64. $16x^2 - 49$
65. $27y^3 + 125$
66. $125w^3 + 1$
67. $8d^3 - 1$
68. $64a^3 - 27$
69. $4x^2 - 10xy + 25y^2$
70. $9w^2 + 12wp + 4p^2$
71. $3(x + 5)^2 - 2(x + 5)$
72. $2(x - 4)^2 + 5(x - 4)$
73. $(x - 1)^2 - 36$
74. $(x + 3)^3 - 27$
75. $x^3 - 3x^2 + 2x - 6$
76. $6x^3 + 4x^2 - 3x - 6$
77. $y^2 - 3xy + xy - 3x^2$
78. $s^3 + 3s^2 + 2s + 6$
79. $x^2 + 16x + 28$
80. $15x^2 - 14x - 8$
81. $64d^2 - 25t^2$
82. $36y^4 - 1$
83. $b^4 - 81c^4$
84. $z^2 + z + 2$
85. $25x^2 - x^4$
86. $a^2 - b^2 - 2b - 1$
87. $8x^3 - 2x^2 - 12x + 3$
88. $4x^3y^2 - 18x^2y + 20xy^2$
89. $9x^4 + 42x^3 - 15x^2$
90. $3x^3 - x^2 - 12x + 4$

Completely (prime) factor each polynomial. If the polynomial does not factor, report prime.

91. $9x^2 - 18x$
92. $16x^2 - 4x + 8$
93. $y^3 - 16y$
94. $25w^3 + w$
95. $8d^3 - 16d$
96. $64a^3 - 16a - 8$
97. $4x^2 - 10x + 4$
98. $9w^2 - 12wp + 4p^2$
99. $3(x - 5)^2 - 2(x - 5) - 1$
100. $2(x - 4)^2 + 5(x - 4) + 2$
101. $4(x - 1)^2 - 36$
102. $3(x - 3)^3 - 81$
103. $x^3 + 3x^2 - 2x - 6$
104. $6x^3 - 4x^2 + 3x - 6$
105. $y^2 - 3xy - xy + 3x^2$
106. $s^3 - 3s^2 - 2s + 6$
107. $x^2 - 16x + 28$
108. $15x^2 + 14x - 8$
109. $64d^4 - 25t^2$
110. $36y^4 - 81$
111. $4b^4 - 81c^4$
112. $2z^2 + z + 2$
113. $25x^2 - 16x^4$
114. $a^2 - 3b^2 - 2b - 3$

D I S C O V E R Y

Picturing Factors

```
RANGE
Xmin=-9.6
Xmax=9.4
Xscl=1
Ymin=-6.4
Ymax=6.2
Yscl=1
Xres=1
```

FIGURE 1

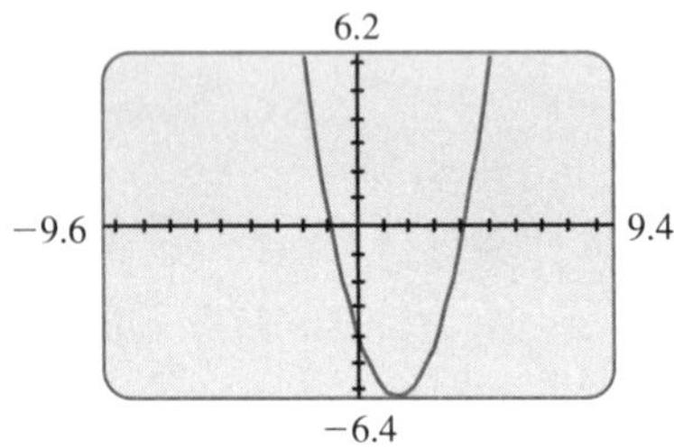

FIGURE 2

We can use the graphics calculator and its graph screen to gain more understanding of factoring polynomials in one variable. In this session, we will investigate a relationship between evaluations of a polynomial, such as $x^2 - 3x - 4$, and its factors, $x - 4$ and $x + 1$.

Type the polynomial X²−3X−4 for Y1. Press the RANGE key and type in the values shown in Figure 1.* Press the GRAPH key. The curve displayed in Figure 2 represents the evaluations of the polynomial $x^2 - 3x - 4$ for various x-values. Press the TRACE key. The message X=0 Y=-4 means that when the x-variable is replaced by 0, the polynomial $x^2 - 3x - 4$ evaluates to -4. Now, use the left arrow key to move the cursor until the message reads X=-1 Y=0. When the x-variable is replaced by -1, the polynomial $x^2 - 3x - 4$ evaluates to 0. Use the right arrow key and move the cursor until the message reads X=4 Y=0. When the x-variable is replaced by 4, the polynomial $x^2 - 3x - 4$ evaluates to 0.

As we mentioned at the beginning of this section, the polynomial $x^2 - 3x - 4$ factors as $(x + 1)(x - 4)$. Press the Y= key and type (X+1)(X−4) for Y2. View the graph screen—the evaluations are the same for the polynomial as they are for its factors. This is what is meant by equivalent expressions. This factored form, $(x + 1)(x - 4)$, shows why the polynomial evaluates to 0 when the x-variable is replaced by -1 and by 4.

The polynomial $x^2 - x - 2$ factors as $(x - 2)(x + 1)$. Press the Y= key and type (X−2)(X+1) for Y1. Before viewing the graph screen, predict for which x-values the polynomial $x^2 - x - 2$ will evaluate to 0. View the graph screen as shown in Figure 3. Use the TRACE feature and the arrow keys to verify your prediction of the x-values that result in the polynomial evaluating to 0. What are those x-values? Suggest how these x-values are related to the factors of $x^2 - x - 2$.

Press the Y= key and type X²+2X−3 for Y1. Before viewing the graph screen, factor the polynomial $x^2 + 2x - 3$. Use the factors to predict the x-values where the polynomial $x^2 + 2x - 3$ evaluates to 0. Now, view the graph screen (see Figure 4). Use the TRACE feature and the arrow keys to verify your prediction.

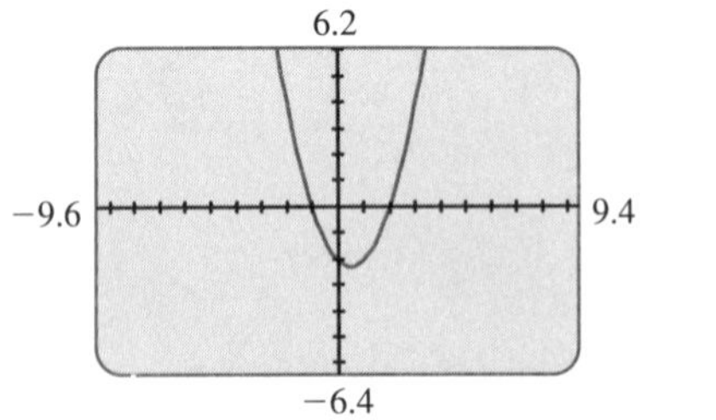

FIGURE 3

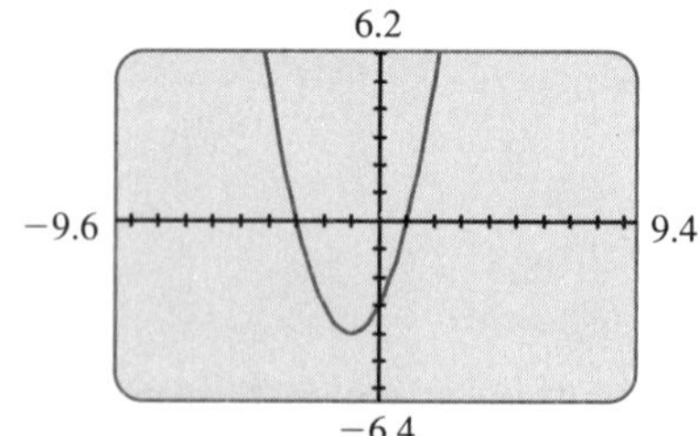

FIGURE 4

TI-82 Note *Type -9.4 for Xmin and -6.2 for Ymin.

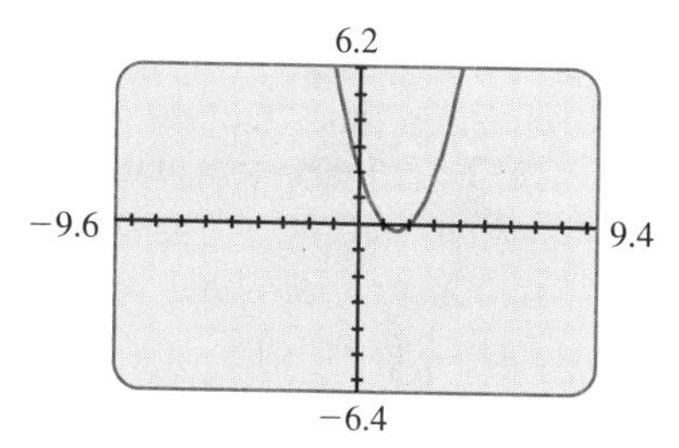

FIGURE 5

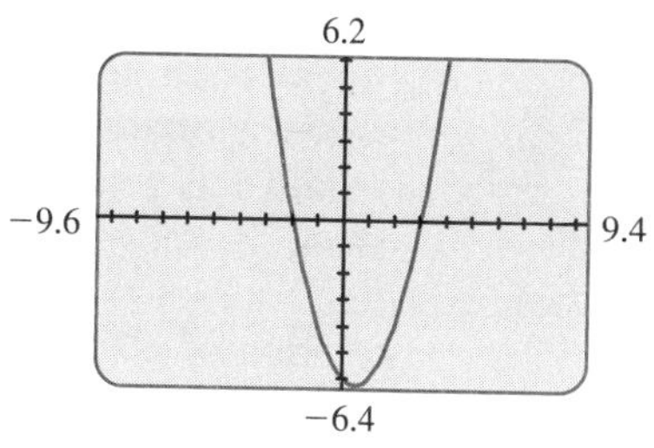

FIGURE 6

Do you think we could use the graph screen showing evaluations of a polynomial in one variable to help factor such a polynomial? Press the Y= key and type X²−3X+2 for Y1. Figure 5 shows the graph screen. Use the TRACE feature and the arrow keys to identify the x-values that result in the polynomial evaluating to 0. What are those x-values? Use these values to find the factors of $x^2 - 3x + 2$.

Press the Y= key and type X²−X−6 for Y1. View the graph screen, as shown on Figure 6. Use the TRACE feature and the arrow keys to identify the x-values that result in the polynomial evaluating to 0. What are those x-values? Use these values to factor the polynomial $x^2 - x - 6$.

In Section 5.4, we stated that the sum of two squares, such as $x^2 + 4$, does not factor over the integers. Press the Y= key and type X²+4 for Y1. View the graph screen shown in Figure 7. As you can see, no x-values result in the polynomial $x^2 + 4$ evaluating to 0. In fact, $x^2 + a^2$ always evaluates to a number greater than 0 for every replacement of the x-variable, provided $a \neq 0$.

We also stated in Section 5.4 that the binomial $x^2 - 3$ is prime over the integers. Type X²−3 for Y1 and then view the graph screen (see Figure 8). Two x-values result in the binomial $x^2 - 3$ evaluating to 0; however, these x-values are not integers, nor are they rational numbers. The x-values that replace the x-variable and result in the polynomial $x^2 - 3$ evaluating to 0 are the irrational numbers $\sqrt{3}$ and $-\sqrt{3}$. Because these numbers are irrational, we conclude that the polynomial $x^2 - 3$ is prime over the integers.

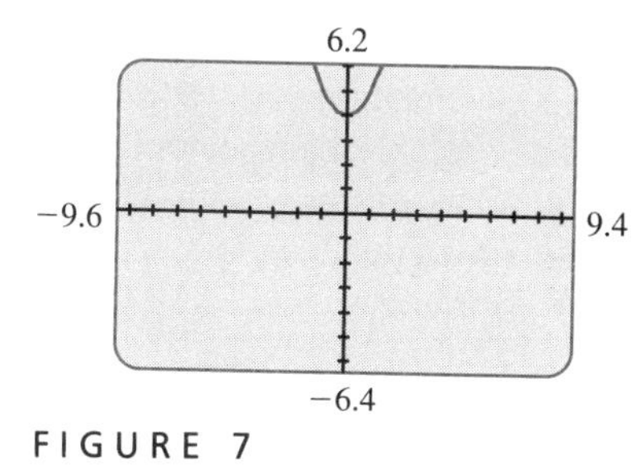

FIGURE 7

FIGURE 8

Exercises

Use the graphics calculator and its graph screen to factor each polynomial. *If* the polynomial does not factor over the integers, report it as *prime*.

1. $x^2 - 3x - 4$
2. $x^2 + 3x - 4$
3. $x^2 - 5x + 4$
4. $x^2 + 5x + 4$
5. $x^2 - 5x - 6$
6. $x^2 + 5x - 6$
7. $x^2 - 1$
8. $x^2 - 9$
9. $x^2 + 1$
10. $x^2 - 2$

5.5 ■ ■ ■ SOLVING QUADRATIC EQUATIONS USING FACTORING

Throughout this chapter we have studied multiplication and factoring of polynomials. We now make use of these skills to solve second-degree equations in one variable, such as $x^2 - 3x = 4$. These equations are also known as *quadratic equations*.

Background The equation-solving properties we discussed earlier are still valid and are required for solving quadratic equations:

1. Any polynomial can be added to each side of an equation without altering the solution.
2. Any *nonzero* polynomial can be multiplied times each side of an equation without altering the solution to the equation.

A first-degree equation, such as $2x + 4 = 0$, is known as a **linear equation**. A second-degree equation, such as $x^2 - 3x = 7$, is known as a **quadratic equation**. A third-degree equation, such as $x^3 - x = 5$, is known as a **cubic equation**. A fourth-degree equation, such as $x^4 - 3x^2 = 4$, is known as **quartic**, or **biquadratic equation**. In general, these equations are known as **polynomial equations**. Linear equations typically have a solution of a single number. Quadratic equations typically have a solution of two numbers.

Factoring From arithmetic, we know that if the product of two numbers is zero, then one of the numbers must be zero. We state this fact as a property.

ZERO PROPERTY OF MULTIPLICATION

If $ab = 0$, then $a = 0$ or $b = 0$.

To solve the equation $xy = 0$, we simply replace one of the variables with zero (the other variable can be replaced by any number). We will now use this property of zero along with factoring polynomials, to solve some selected second-degree equations in one variable. We start by writing the equation in standard form.

STANDARD FORM

For a quadratic equation in one variable, *standard form* is

$$ax^2 + bx + c = 0$$

where a, b, and c are real numbers and $a \neq 0$. The x^2-coefficient is a, the x-coefficient is b, and c is the constant term.

Before working through the following examples, take a moment to clarify the differences between a polynomial and a polynomial equation. Now, consider the following examples.

EXAMPLE 1 Solve: $3x^2 = 18x$

Solution First and foremost, the solution to this equation is the collection of numbers that result in a true arithmetic sentence when the variable is replaced by one of the numbers.

To arrive at these numbers, we start by writing the equation in standard form (as a standard polynomial equal to zero):

Add $-18x$ to each side	$3x^2 = 18x$ $\underline{-18x \quad -18x}$
to get standard form:	$3x^2 - 18x = 0$
Factor the left-hand side:	$(3x)(x - 6) = 0$
Using the zero property, we get:	$3x = 0$ or $x - 6 = 0$
Solving these equations, we get:	$x = 0$ or $x = 6$

The number that results in the factor $3x$ evaluating to 0 is 0, and the number resulting in the factor $x - 6$ evaluating to 0 is 6.

Thus, the solution to the equation $3x^2 = 18x$ is 0 and 6. ■

We have reduced the equation $3x^2 = 18x$ to an *equivalent equation*, an equation with the same solution: $3x^2 - 18x = 0$. Using factoring and the zero property of multiplication, we produced the equations, $3x = 0$ and $x - 6 = 0$. The solution to each of these equations is obvious.

EXAMPLE 2 Solve: $x^2 - 2x = 8$

Solution

First, we write the equation in	$x^2 + 2x = 8$ $\underline{-8 \quad -8}$
standard form:	$x^2 - 2x - 8 = 0$
Now, we write the left-hand side in factored form:	$(x - 4)(x + 2) = 0$
Using the zero property, we get the equations:	$x - 4 = 0$ or $x + 2 = 0$
	$\underline{4 \quad 4}$ $\quad$ $\underline{-2 \quad -2}$
	$x = 4$ $\quad$ $x = -2$

Thus, the solution to $x^2 - 2x = 8$ is 4 and -2. ■

In Example 2, we could write $x(x - 2) = 8$. However, this factoring is of *no* value. To solve the equation, we must obtain a product equal to 0. This provides information about the factors namely, that one must be 0. Having a product equal to 8 provides no useful information about the factors!

EXAMPLE 3 Solve: $2x^2 - 3x = -1$

Solution Again, we start by writing the equation in standard form:

$$\begin{aligned} 2x^2 - 3x &= -1 \\ \underline{ 1} &\quad \underline{1} \\ 2x^2 - 3x + 1 &= 0 \end{aligned}$$

Now, writing the left-hand side, $2x^2 - 3x + 1$, in factored form, we get

$$(2x - 1)(x - 1) = 0$$

Using the zero property, we get two equations:

$$\begin{array}{rclcrcl} 2x - 1 &=& 0 & \text{or} & x - 1 &=& 0 \\ 1 && 1 & & 1 && 1 \\ \hline 2x &=& 1 & & x &=& 1 \\ x &=& \frac{1}{2} & & & & \end{array}$$

Thus, the solution to $2x^2 - 3x = -1$ is $\frac{1}{2}$ and 1. ■

In Example 3, we are trying to reason out which numbers we can *double the square of, then subtract three times the number, to obtain the number −1*. We are turning a difficult mental problem into a simpler problem—which number results in the factor $2x - 1$ equaling 0, and which number results in the factor $x - 1$ equaling 0. The key to the factoring approach to solving second-degree equations in one variable is getting a product equal to zero. When such an equation is found, a number in the solution can be found by finding a number that makes each factor zero.

Notice that the solution is a collection of numbers. If an equation has one variable, the solution to the equation is always a collection of numbers. For equations in two variables, like $3x + 2y = 9$, the solution is always a collection of ordered pairs of numbers. The solution to a first-degree equation in one variable, such as $2x - 7 = 9$, is typically a single number. A second-degree equation in one variable, such as $2x^2 - 3x = -1$, usually has a solution of two numbers.

GIVE IT A TRY

Use factoring and the zero product property to solve each equation.

1. $x^2 - 6x = 0$

2. $2x^2 - x - 3 = 0$

3. $x^2 - 5x = 6$

4. $x^2 = 16$

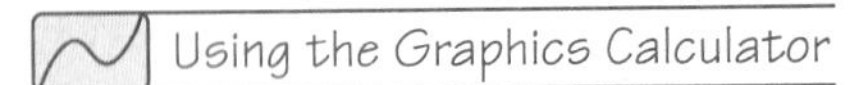

To check our work in Example 3, press the Y= key and type 2X²−3X for Y1. Return to the home screen. Store the number 1/2 in memory location X. Display the resulting value in Y1. Is the number ⁻1 output? Store the number 1 in memory location X. Display the resulting value in Y1. Is the number ⁻1 output? As you can see, $\frac{1}{2}$ and 1 are the x-values that result in the polynomial $2x^2 - 3x$ evaluating to the number -1. Use this approach to check your work in *Give It a Try* Problems 1–4.

We can improve our understanding of the solution to a second-degree equation in one variable by using the graph screen. Consider the equation from Example 2, $x^2 - 2x = 8$. Although we could enter

```
RANGE
Xmin=-9.6
Xmax=9.4
Xscl=1
Ymin=-6.4
Ymax=6.2
Yscl=1
Xres=1
```

FIGURE 1

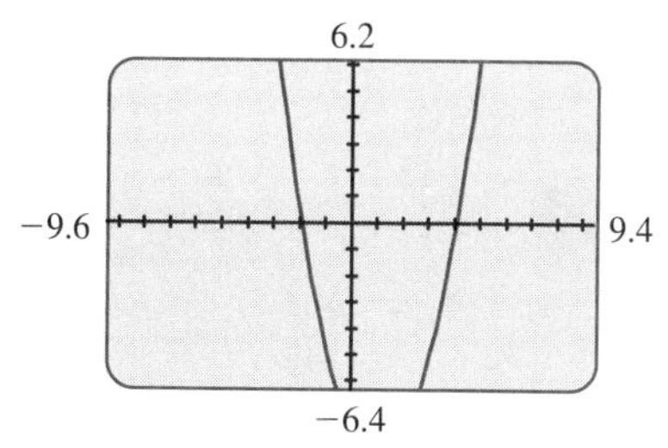

FIGURE 2

```
RANGE
Xmin=-4.8
Xmax=4.7
Xscl=1
Ymin=-3.2
Ymax=3.1
Yscl=1
Xres=1
```

FIGURE 3

$x^2 - 2x$ for Y1, and observe the x-values where this polynomial evaluates to 8, it is easier to enter $x^2 - 2x - 8$ for Y1, and observe where this polynomial evaluates to 0. Actually, this is the meaning of the step $x^2 - 2x - 8 = 0$ in the solution to the equation. Press the Y= key and type X²−2X−8 for Y1. Press the RANGE key and enter the values in Figure 1.* View the graph screen, as shown in Figure 2. Press the TRACE key. The message X=0 Y=-8 means that when the x-variable is replaced by 0, the polynomial $x^2 - 2x - 8$ evaluates to -8. We want to find the x-values that result in the polynomial evaluating to 0. Press the right arrow key until the message reads X=4 Y=0. Thus, when the x-value is 4, the polynomial $x^2 - 2x - 8$ evaluates to 0. That is, 4 solves the equation $x^2 - 2x - 8 = 0$. Use the left arrow key to move the cursor until the message reads X=-2 Y=0. Thus, when the x-value is -2, the polynomial $x^2 - 2x - 8$ evaluates to 0. That is, -2 also solves the equation $x^2 - 2x - 8 = 0$.

To check the equation in Example 3, $2x^2 - 3x = -1$, we could observe the x-values where the polynomial $2x^2 - 3x$ evaluates to -1. We would find two such numbers, $\frac{1}{2}$ and 1. However, it is easier to observe the x-values where the polynomial $2x^2 - 3x + 1$ is 0. Press the Y= key and type 2X²−3X+1 for Y1. Press the RANGE key, enter the values in Figure 3.** View the graph screen (see Figure 4). Press the TRACE key and then the right arrow key until the message reads X=.5 Y=0. Thus, when the x-value is 0.5, the polynomial $2x^2 - 3x + 1$ evaluates to 0; that is, 0.5 solves the equation $2x^2 - 3x + 1 = 0$. Use the right arrow key to move the cursor until the message reads X=1 Y=0. Thus, when the x-value is 1 the polynomial $2x^2 - 3x + 1$ evaluates to 0; that is, 1 also solves the equation $2x^2 - 3x + 1 = 0$. Type X²−2X+1 for Y1. View the graph screen, as shown in Figure 5. How many numbers solve $x^2 - 2x + 1 = 0$?

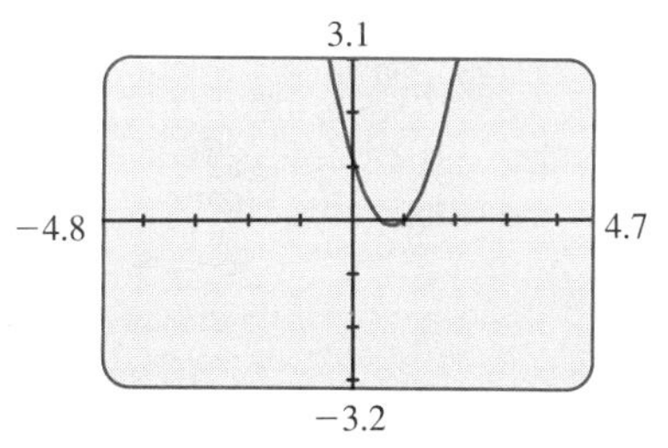

FIGURE 4

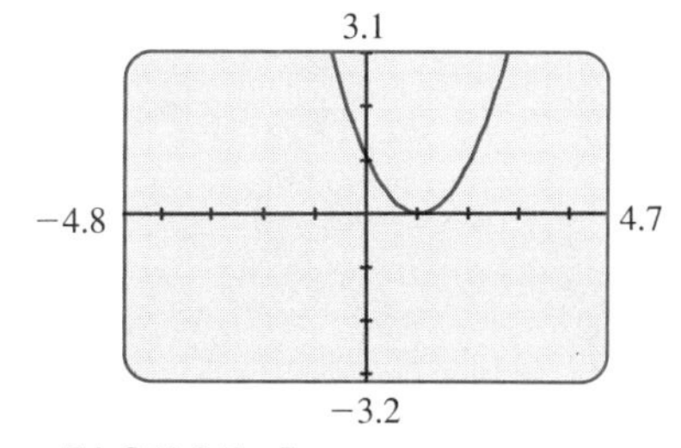

FIGURE 5

TI-82 Note *Type -9.4 for Xmin and -6.2 for Ymin.
**Use the ZDecimal option.

Additional Examples We now consider some additional examples of using factoring to solve a second-degree equation in one variable.

EXAMPLE 4 Solve: $(x - 4)(x + 2) = 7$

Solution

We start by writing the equation in standard form: $x^2 - 2x - 8 = 7$

Add -7 to each side: $-7 \quad -7$

to get: $x^2 - 2x - 15 = 0$

Write in factored form: $(x - 5)(x + 3) = 0$

Using the zero property, we get two equations

$$\begin{array}{rrcrr} x - 5 = & 0 & \text{or} & x + 3 = & 0 \\ 5 & 5 & & -3 & -3 \\ \hline x = & 5 & & x = & -3 \end{array}$$

So, the solution to $(x - 4)(x + 2) = 7$ is 5 and -3. ■

EXAMPLE 5 Solve: $-x^2 - 3x + 10 = 0$

Solution This equation is in standard form, but before we attempt to factor the polynomial on the left-hand side, we multiply each side by -1 in order to make the x^2 term positive.

> **REMEMBER**
> We can multiply both sides of any equation by a nonzero number without altering the solution to the equation.

Multiply by -1 to get: $x^2 + 3x - 10 = 0$

Factoring, we get: $(x + 5)(x - 2) = 0$

Using the zero property, we get:

$$\begin{array}{rrcrr} x + 5 = & 0 & \text{or} & x - 2 = & 0 \\ -5 & -5 & & 2 & 2 \\ \hline x = & -5 & & x = & 2 \end{array}$$

The solution to $-x^2 - 3x + 10 = 0$ is -5 and 2. ■

Using the Graphics Calculator

To check the solution to Example 4, type `(X−4)(X+2)` for `Y1`. Return to the home screen and store `5` in memory location `X`. Display the resulting value in `Y1`. Is the number 7 output? Store `-3` in memory location `X`. Display the resulting value in `Y1`. Is the number 7 output? Use this approach to check the solution to Example 5.

GIVE IT A TRY

Solve each equation using factoring.

5. $2x^2 - 5x = 0$ **6.** $-2x^2 + 5x + 3 = 0$ **7.** $2x^2 + 5x = 3$

8. $x^2 - 6x = 7$ **9.** $(x - 3)(x + 2) = x + 2$ **10.** $x^2 = (2x + 1)^2$

11. Use the graphics calculator to check your work for Problems 5–10.

Solving Other Polynomial Equations The factoring approach for quadratic equations has limitations. If the polynomial does not factor, the method fails. The equation may have a solution, but we can't find it by factoring. We will consider second-degree equations such as these in Chapter 8. At this time, we wish to consider using factoring and the zero product rule to solve polynomial equations of degree greater than 2. First, the zero product property can be expanded to three factors: If $(a)(b)(c) = 0$, then $a = 0$ or $b = 0$ or $c = 0$. Now, consider the following example.

EXAMPLE 6 Solve the polynomial equation $x^3 + x^2 = 4x + 4$.

Solution We start by rewriting the equation in standard form:

$$\begin{array}{rcl} x^3 + x^2 & = & 4x + 4 \\ -4x & & -4x \\ \hline x^3 + x^2 - 4x & = & 4 \\ -4 & & -4 \\ \hline x^3 + x^2 - 4x - 4 & = & 0 \end{array}$$

Next, we write the resulting quadrinomial in factored form:

$$(x^2 - 4)(x + 1) = 0$$
$$(x + 2)(x - 2)(x + 1) = 0$$

Using the zero product property, we get:

$$\begin{array}{rcl} x + 2 & = & 0 \\ -2 & & -2 \\ \hline x & = & -2 \end{array} \quad \text{or} \quad \begin{array}{rcl} x - 2 & = & 0 \\ 2 & & 2 \\ \hline x & = & 2 \end{array} \quad \text{or} \quad \begin{array}{rcl} x + 1 & = & 0 \\ -1 & & -1 \\ \hline x & = & -1 \end{array}$$

Thus, the equation $x^3 + x^2 = 4x + 4$ has solution -2, -1, and 2. ■

GIVE IT A TRY

Solve each polynomial equation.

12. $3x - 17 = 2$

13. $x^2 = 5x + 6$

14. $x^3 - 2x^2 = 3x$

15. $x^4 - 16 = 0$

Applications We now consider some physical situations for which a second-degree equation in one variable serves as a mathematical model. In these word problems, the equation produced is a second-degree equation in one variable. We use the same process as in Section 2.6 for first-degree equations in one variable. Work through the following examples.

EXAMPLE 7 The area measure of a rectangle is 40 square inches. If the length of the rectangle is 2 inches greater than twice the width of the rectangle, what is the rectangle's width?

Solution The facts are shown in Figure 6.

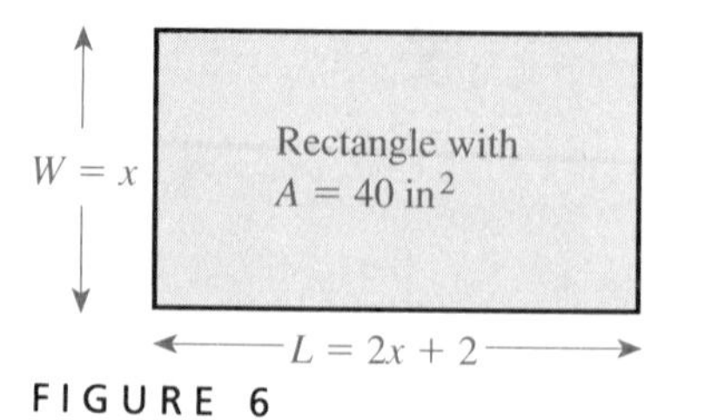

FIGURE 6

Step 1 Identify the requested quantity, the width.

Step 2 Define what the variable represents:

$$\text{Let } x = \text{rectangle width}$$

Step 3 Write expressions (polynomials) for the other quantities given in the story.

$$2x + 2 = \text{rectangle's length}$$

and $$x(2x + 2) = \text{rectangle's area measure [Use } A = LW\text{].}$$

Step 4 Write an equation.

LW is area measure of the rectangle. → $x(2x + 2)$

Area measure of rectangle → 40

$$x(2x + 2) = 40$$

$$2x^2 + 2x = 40 \qquad \text{Clear parentheses.}$$

$$x^2 + x = 20 \qquad \text{Multiply by } \tfrac{1}{2}.$$

$$x^2 + x - 20 = 0 \qquad \text{Standard form}$$

$$(x - 4)(x + 5) = 0 \qquad \text{Factored form}$$

$$x - 4 = 0 \quad \text{or} \quad x + 5 = 0$$

The solution is 4 and -5.

Step 5 Answer the question:
The width is 4 inches (the -5 is discarded because a linear measurement must be a positive value). ■

EXAMPLE 8 A football is thrown by a quarterback, as shown in Figure 7. The height of the football above the ground, t seconds after leaving the quarterback's hand, is given by the formula $h = -16t^2 + 32t + 8$. The height at time $t = 0$ is 8. When is the height of the football again 8 feet above the ground?

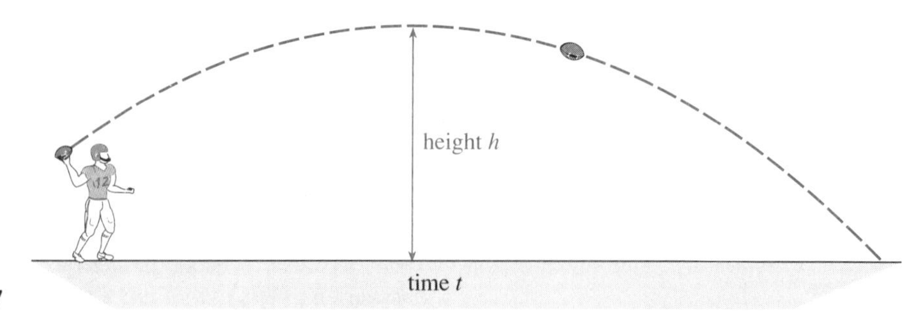

FIGURE 7

Solution In this problem the relationship (equation) is given by the formula. We simply want to find the value of t when $h = 8$. Thus, we solve the equation:

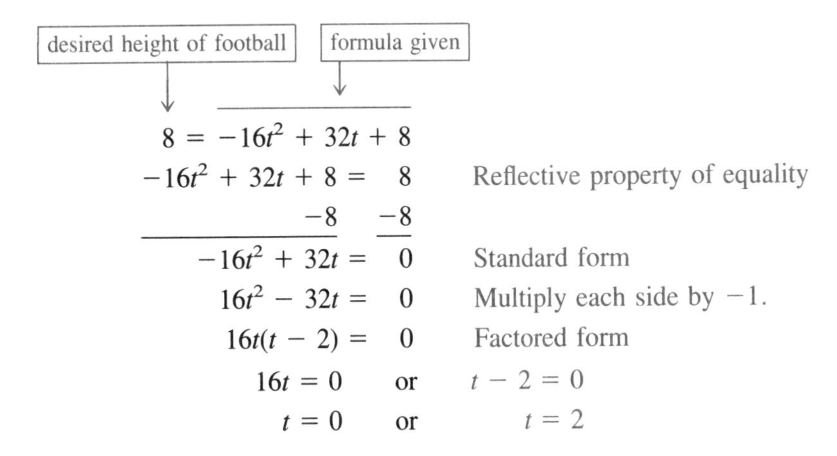

$$\begin{aligned}
8 &= -16t^2 + 32t + 8 \\
-16t^2 + 32t + 8 &= 8 && \text{Reflective property of equality} \\
-8 \quad & \quad -8 \\
-16t^2 + 32t &= 0 && \text{Standard form} \\
16t^2 - 32t &= 0 && \text{Multiply each side by } -1. \\
16t(t - 2) &= 0 && \text{Factored form} \\
16t = 0 \quad &\text{or} \quad t - 2 = 0 \\
t = 0 \quad &\text{or} \quad t = 2
\end{aligned}$$

Solution to the equation $8 = -16t^2 + 32t + 8$ is 0 and 2.

The number 8 is the value of h at time $t = 0$. (When the football leaves the quarterback's hand, it has a height of 8 feet.) The answer to the question is time $t = 2$. That is, 2 seconds after the ball leaves the quarterback's hand, it has a height of 8 feet again. ■

Using the Graphics Calculator

Type `-16X²+32X+8` for `Y1`. Press the `RANGE` key and type the values in Figure 8. View the graph screen, as shown in Figure 9.* Press the `TRACE` key. The message `X=.5  Y=20` means that $\frac{1}{2}$ second after the ball leaves the quarterback's hand, its height is 20 feet. Use the right arrow key to identify when the height is 8 feet off the ground. If the ball is not caught, approximately how long will it be in the air?

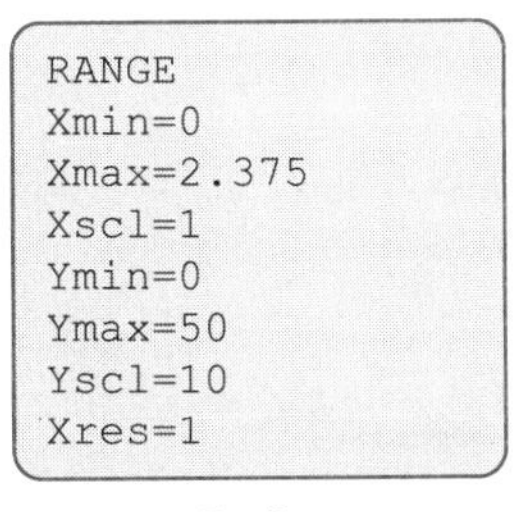

FIGURE 8

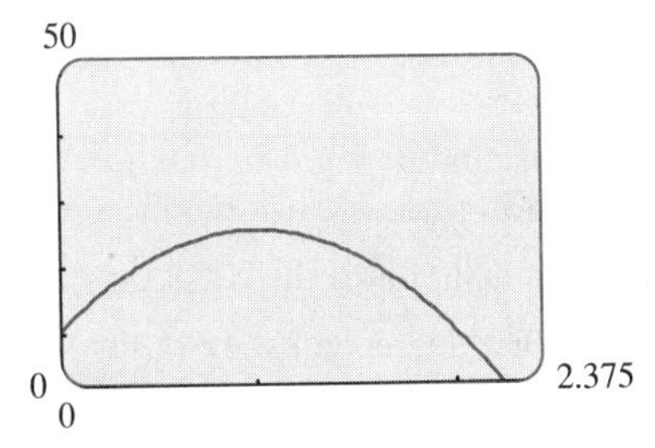

FIGURE 9

TI-82 Note *Type `2.35` for `Xmax`.

■ SUMMARY

In this section, we have learned to apply the ability of factoring polynomials to solving second-degree (quadratic) equations in one variable. The solution to such an equation is typically two numbers. Along with factoring, we used the zero product property to arrive at these two numbers. Essentially, the process has three steps:

1. Write the equation in standard form (a standard polynomial equal to 0).
2. Factor the polynomial.
3. Identify the numbers that result in each of the factors evaluating to zero.

5.5 ■ ■ ■ EXERCISES

Match equations A–J with statements 1–10:

A. $3x - 5 = 9$ **B.** $3x - 2y = 9$

C. $3x - 5 \geq 9$ **D.** $3x - 2y \leq 6$

E. $\begin{cases} 3x - y = 8 \\ x + y = -2 \end{cases}$ **F.** $x^2 - 3x = 21$

G. $(x - 2)(x - 5) = 0$ **H.** Zero Product Property

I. $ax^2 + bx + c = 0$ **J.** -3

____ 1. Standard form for a second-degree equation in one variable

____ 2. A second-degree equation whose solution is the numbers 2 and 5

____ 3. A first-degree inequality in two variables

____ 4. A linear system; its solution is typically a single ordered pair of numbers.

____ 5. A first-degree inequality in one variable

____ 6. If $(a)(b) = 0$, then $a = 0$ or $b = 0$.

____ 7. A first-degree equation in two variables

____ 8. A second-degree equation in one variable; its solution is typically two numbers.

____ 9. The solution to the equation $x^2 - 3x = x^2 + 9$.

____ 10. A first-degree equation in one variable

Solve each equation using the zero product property and factoring.

11. $(x - 3)(x - 5) = 0$
12. $(5 - x)(x + 3) = 0$
13. $(3x - 2)(x + 1) = 0$
14. $(5x + 7)(2x - 3) = 0$
15. $(x - 2)(x + 3)(x + 5) = 0$
16. $(2x - 1)(x - 5)(x + 2) = 0$
17. $(3x)(x - 7) = 0$
18. $(2x^2)(x - 4) = 0$
19. $(x^2)(2x + 3) = 0$
20. $x^2 = 0$
21. $x^2 - 5x = 0$
22. $3x^2 + 6x = 0$
23. $x^2 - 3x = 0$
24. $x^2 - 5x + 6 = 0$
25. $x^2 = 13x$
26. $2x^2 - 3x - 5 = 0$
27. $x^2 + x - 4 = 8$
28. $x^2 - 5x = 24$
29. $x^2 + 5x = 24$
30. $2y^2 - 3y = 20$
31. $2y^2 - 6y = 20$
32. $3t^2 = t + 4$
33. $3t^2 = -4t + 4$
34. $4x^2 = 3 - 11x$

Solve each equation using factoring and the zero product property. Use the graphics calculator to check your work.

35. $x^2 - 3x = 40$
36. $x^2 - 3x + 6 = 10$
37. $4x^2 = 13x$
38. $2x^2 - 3x - 4 = x^2$
39. $(2r + 1)(r - 2) + 3 = 0$
40. $x^2 + x - 4 = 8$
41. $6d^2 - 5d = 50$
42. $6p^2 + 5p = 50$
43. $v^2 = 9v - 14$
44. $12s^2 = 1 - 4s$
45. $(x - 3)^2 = 1$
46. $(x + 4)^2 = 25$
47. $(2x + 3)^2 = 9$
48. $(3x - 2)^2 = 4$
49. $6 - x - x^2 = 0$
50. $10 - 3x - x^2 = 0$
51. $(3 - x)(x + 2) = 6x$
52. $x(3 - x) = 2$
53. $(x - 1)^2 = x^2$
54. $(5 - x)^2 = x(x - 1)$
55. $(x + 3)^2 = x^2 + 9$
56. $(2x - 1)^2 = 4x^2 + 1$
57. $x^3 = x(2x + 3)$
58. $x^3 + 2x^2 = x + 2$
59. $(x - 3)^2 = (x - 2)(x + 2)$
60. $(2x - 3)(x + 1) = x + 1$

61. $(2x + 3)(x + 1) = x^2 + 9$

62. $(x - 3)(x + 2) = x^2 - x$

63. $x(x - 2) = (x - 4)(x + 2)$

64. $6 - 3x^2 - x^4 = 0$

65. $x^3 + 3x^2 = 4x + 12$

66. $2x^3 + 1 = x^2 + 2x$

67. Find a positive number such that the sum of the square of the number and five times the number is 24.

68. Find all the numbers such that the square of the sum of the number and 3 equals 6 squared.

69. Use the Pythagorean theorem (see page 56) to find the value of x in the triangle:

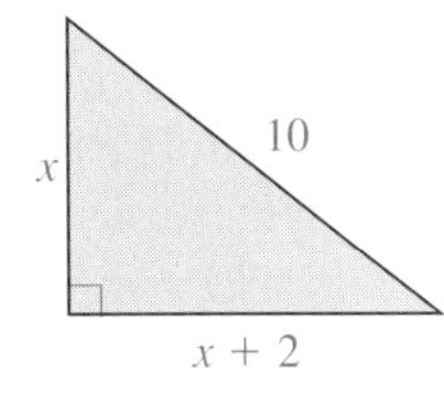

70. Find two consecutive integers (such as 6 and 7) whose product is 156.

71. If the radius of a circle is increased by 3 cm, then the area measure of the circle is increased by 39π cm^2. What is the radius of the circle?

72. The height of a triangle is 2 inches more than the base to which the height is drawn. The area measure of the triangle is 4 square inches. Find the height of the triangle.

73. The area measure of a rectangle is 60 square meters. If the length is 14 meters less than twice the width, what is the width of the rectangle?

74. The length of a rectangle is 15 inches more than twice the width. If the area measure is 50 square inches, what is the perimeter of the rectangle?

75. For a right triangle, one leg is 2 cm less than the hypotenuse and the other leg is 1 cm less than the hypotenuse. What is the length of the hypotenuse?

76. The hypotenuse of a right triangle is 20 feet. If one leg is 4 feet greater than the other leg, what is the perimeter of the triangle?

77. If the side of a square is increased by 2 cm, the area measure is 8 cm^2 more than twice the original area measure. What is the length of the side of the square?

78. Freda and Eddie leave a stoplight at the same time, one heading east and the other north. If Freda is traveling at 40 mi/h and Eddie is driving 30 mi/h, how far apart will they be in 2 hours?

79. A 17-foot ladder is leaning against a vertical wall. The height reached by the top of the ladder is 7 feet more than the distance from the base of the ladder to the wall. How far is the base of the ladder from the wall?

80. The height h (in feet) reached by a model rocket t seconds after liftoff is given by the formula

$$h = -16t^2 + 320t$$

At what time is the height 1200 feet?

81. The height h (in feet) reached by a ball t seconds after being thrown upward is given by the formula

$$h = -16t^2 + 256t$$

At what time is the height 1024 feet?

DISCOVERY

With 3.57 You Get Chaos

Often in ecology an equation such as $x_{n+1} = kx_n(1 - x_n)$ is used as a mathematical model to predict population growth of a species, such as the growth rate of a certain fish in a lake. In this model, x_{n+1} is the growth rate of the population at the next time step (say, a month later), x_n is the current population growth rate, and k is constant value (known as a parameter). Let's investigate this type of equation for $k = 2$ and $x_0 = 0.2$. The parameter k is determined by various environmental factors (food supply or acid rain) and x_0 is the initial (starting) population growth rate, which is found by research to be 20%. Now, we get $x_1 = 2(0.2)(1 - 0.2) = (0.4)(0.8) = 0.32$. That is, for month 1, the growth rate will be 32%.

```
          .4352
2(Ans)(1-Ans)
      .49160192
2(Ans)(1-Ans)
    .4998589445
2(Ans)(1-Ans)
    .4999999602
```

FIGURE 1

```
     .5130189944
3.2(Ans)(1-Ans)
     .7994576185
3.2(Ans)(1-Ans)
     .5130404311
3.2(Ans)(1-Ans)
     .7994558309
```

FIGURE 2

```
           .5712
3.57(Ans)(1-Ans)
     .8744020992
3.57(Ans)(1-Ans)
     .3920683532
3.57(Ans)(1-Ans)
     .8509122118
```

FIGURE 3

On the graphics calculator, type `2(0.2)(1−0.2)` and press the `ENTER` key. Use the `ANS` key to display the characters `Ans`. Press the `ENTER` key. The last computed value is output, `0.32` in this case. Now, type `2(Ans)(1−Ans)`. Press the `ENTER` key. The value `0.4352` is output. For the next month (month 2) the population is growing at 43.52%. Replay the last command. Press the `ENTER` key. The value `0.49160192` is output. For the next month—month 3—the population is growing at approximately 49.16%. Repeat these steps to find that the population growth for the next month, month 4, is approximately 49.99%. Repeating these steps over and over (see Figure 1) indicates that population growth for this situation is 50% after many months.

Once a value for k is selected (based on knowledge of the situation) and a value for x_0 is found from experimental data, the pattern of growth is determined—it is deterministic. Suppose $k = 3.2$ and x_0 is 0.2. Repeating the above process on the graphics calculator, something a little strange occurs. Type `3.2(0.2)(1−0.2)` and press the `ENTER` key. Now, type `3.2(Ans)(1−Ans)` and press the `ENTER` key. Replay the last command and then press the `ENTER` key. Repeating this replay over and over shows that population growth rate jumps between two values, approximately 51.3% and 79.9%. See Figure 2.

Suppose 3.57 is selected for k and x_0 is 0.2. Repeat the above steps for these values. Here, the value produced jumps about wildly—the growth rate is chaotic! See Figure 3.

Use 0.3 for x_0 and repeat the above experiment. For $k = 2$, does population growth reach 50% in the long run? For $k = 3.2$, does population growth oscillate between 51.3% and 79.9%? For $k = 3.57$, does population growth rate become chaotic? The k-value is determined from environmental factors. In your own words, explain what this discovery indicates for the population growth rate.

5.6 ■ ■ ■ SECOND-DEGREE INEQUALITIES

Now that we can solve some second-degree equations in one variable, we will put that knowledge and skill to work. In this section, we solve second-degree inequalities in one variable. This development parallels that of solving first-degree inequalities in one variable (see Section 2.3).

Inequalities Once we know how to solve a second-degree equation in one variable, we can apply that knowledge to solving a second-degree inequality in one variable. A typical example of one of these inequalities is $x^2 \leq 4$. This inequality holds the place for an infinite collection of arithmetic sentences. Some examples are:

$$(-3)^2 \leq 4 \qquad (-1)^2 \leq 4 \qquad 0^2 \leq 4 \qquad 2^2 \leq 4 \qquad 4^2 \leq 4$$

FIGURE 1

Some of these sentences are true arithmetic sentences, others are false arithmetic sentences (the first and last are false arithmetic sentences). The numbers that produce true arithmetic sentences are the ones in the solution to the inequality. We can plot these replacements for x on a number line, as in Figure 1. (We use a solid dot for the ones in the solution and an open dot for those not in the solution.)

As with first-degree inequalities in one variable, the solution is usually an infinite collection of numbers. To communicate this collection, a one-dimensional graph (a number line) is used. From the number line, it appears that there will be two key points (rather than one key point, as is common with first-degree inequalities).

Using the Graphics Calculator

```
RANGE
Xmin=-9.6
Xmax=9.4
Xscl=1
Ymin=-6.4
Ymax=6.2
Yscl=1
Xres=1
```

FIGURE 2

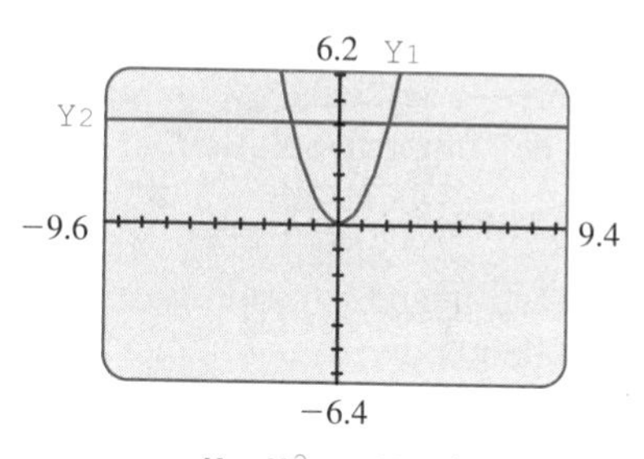

FIGURE 3

-4 -2 0 2 4

FIGURE 4

Press the RANGE key and type in the values shown in Figure 2.* For Y1 type X². View the graph screen. The curve drawn represents evaluations of the polynomial x^2 for various replacements of the x-variable. Press the TRACE key. When x is replaced by 0, the polynomial x^2 evaluates to 0. Use the arrow keys and explore some of the evaluations of x^2. For Y2 type 4 and then view the graph screen. A horizontal line is drawn for 4, as shown in Figure 3. Now, what are the x-values resulting in x^2 evaluating to a number less than or equal to 4? Press the TRACE key and use the arrow key. These first x-values result in x^2 evaluating to a number less than 4. When the message reads X=2 Y=4, we have a key point. To the right of the x-value 2, the polynomial x^2 evaluates to a number larger than 4. Press the left arrow key until the message reads X=-2 Y=4. To the left of -2, the polynomial x^2 also evaluates to a number larger than 4. The solutions to the inequality $x^2 \leq 4$ are the numbers between -2 and 2, including the numbers -2 and 2. A number-line graph of the solution is shown in Figure 4. In interval notation, we write $[-2, 2]$.

For an organized approach to finding the collection of numbers that solves a second-degree inequality, we will use the same process that we developed for first degree inequalities:

1. Solve the related equation to find the key points.
2. Plot these points. (Draw a number line, and circle the numbers representing the key points.) Then label the key points.
3. Determine whether a particular key point solves the inequality. If so, place a solid dot on that point. If not, leave the circle open.
4. The key points divide the number line into several intervals. Test a number in each interval. If the number yields a true arithmetic sentence, shade the interval. If the test number in the interval fails (yields a false arithmetic sentence), leave the interval unshaded.

TI-82 Note *Type -9.4 for Xmin and -6.2 for Ymin.

To use this process to solve a second-degree inequality in one variable, work through the following example.

EXAMPLE 1 Solve the inequality (graph the solution to) $x^2 - 2x \geq 15$.

Solution **Step 1** Solve the equation $x^2 - 2x = 15$.

$$x^2 - 2x - 15 = 0 \quad \text{Standard form}$$
$$(x - 5)(x + 3) = 0 \quad \text{Factored form}$$
$$x - 5 = 0 \quad \text{or} \quad x + 3 = 0$$
$$x = 5 \qquad x = -3$$

Solution: 5 and -3.
Thus, the key points are 5 and -3.

Step 2 Plot and label these key points. See Figure 5.

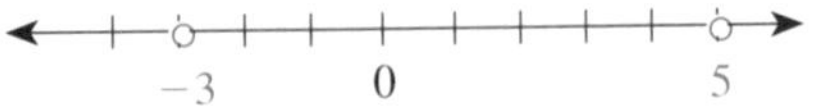

Step 3 Both dots are solid, because these numbers result in true arithmetic sentences. Figure 6 shows the dots.

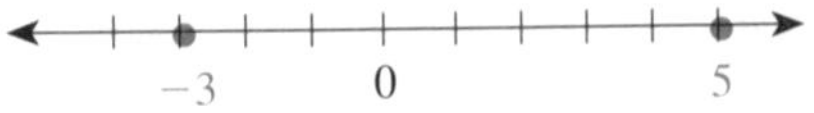

Step 4 Test a number in the leftmost interval, say -4:
$(-4)^2 - 2(-4) \geq 15$.
We get $16 + 8 \geq 15$, which is a true arithmetic sentence, so we shade the leftmost interval. Figure 7 shows the interval.

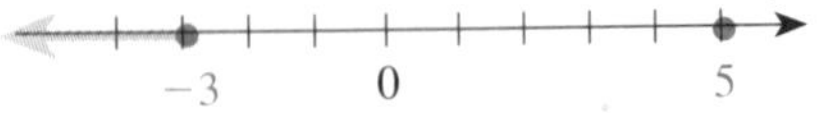

Test a number in the middle interval, say 0: $0^2 - 2(0) \geq 15$. This is a false arithmetic sentence. The interval is left unshaded.

Test a number in the rightmost interval, say 6: $6^2 - 2(6) \geq 16$. We get $36 - 12 \geq 15$, which is a true arithmetic sentence, so we shade the rightmost interval. Figure 8 shows the complete solution.

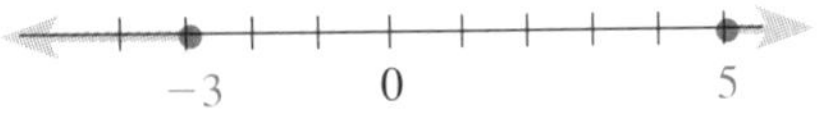

In internal notation we write $(-\infty, -3] \cup [5, +\infty)$. ■

-3 0 5

FIGURE 5

-3 0 5

FIGURE 6

-3 0 5

FIGURE 7

-3 0 5

FIGURE 8

As the solution to Example 1 shows, a second-degree inequality in one variable typically has two key points. These key points divide the number line into three intervals. In Example 1 the points outside the key points (those points in the leftmost and rightmost intervals) represent the solution. Of course, if the inequality sign has been $\leq$, then the points between the key points (the middle interval) would have represented the solution. Try it.

```
RANGE
Xmin=-9.6
Xmax=9.4
Xscl=1
Ymin=-2
Ymax=20
Yscl=5
Xres=1
```

FIGURE 9

Using the Graphics Calculator

We can get an alternative view of the solution to the inequality $x^2 - 2x \geq 15$ by using the graph screen. For Y1 type X²−2X. For Y2 type 15. View the graph screen using the settings in Figure 9.* As Figure 10 indicates, the x-values where the $x^2 - 2x$ is equal to 15 are -3 and 5. We want the x-values where Y1 (the polynomial $x^2 - 2x$) is greater than or equal to Y2 (the constant 15). Use the TRACE key and the arrow keys to confirm that these x-values are the values to the left of -3 and the values to the right of 5.

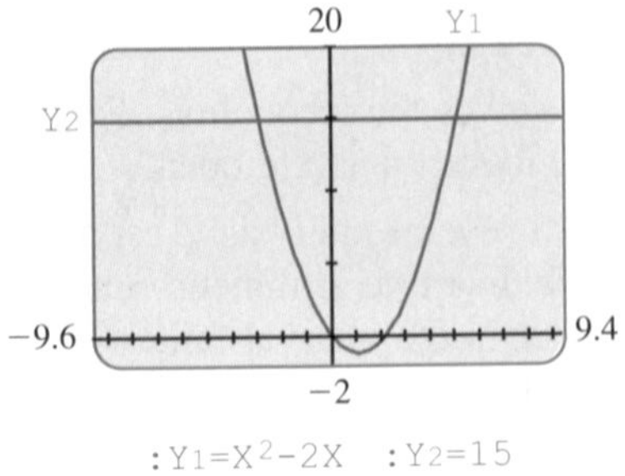

20 Y1
Y2
-9.6 9.4
-2

:Y1=X²-2X :Y2=15

FIGURE 10

TI-82 Note *Type −9.4 for Xmin.

GIVE IT A TRY

1. Solve $x^2 - 3x < -2$. Display a number-line graph of the solution, and report the solution in interval notation.
2. Solve $5 \le x^2 - 4x$. Display a number-line graph of the solution, and report the solution in interval notation.

■ SUMMARY

In this section we have learned to apply knowledge of (and skills with) solving a second-degree equation in one variable. The solution to a second-degree inequality in one variable is typically an infinite collection of numbers. To communicate these numbers, a one-dimensional graph, or number line, is used. To arrive at the numbers that result in a true arithmetic sentence when each number replaces the variable, we used the same procedure (process) as for solving first-degree inequalities.

This procedure starts by solving the equation (a second-degree equation in this case) to obtain the key points. Typically, second-degree inequalities have two key points. These points divide the number line into intervals. If one number in a particular interval solves the inequality, then every number in that interval solves the inequality. A number from each of the intervals is tested in the inequality. If the number from an interval produces a true arithmetic sentence, the interval is shaded, otherwise, the interval is left unshaded. Typically, the solution will be the points outside the key points or the points between the key points.

5.6 ■ ■ ■ EXERCISES

Draw a number-line graph of the solution to each quadratic inequality. Also, report the interval notation for the solution. Check your work using the graphics calculator.

1. $x^2 - 5x \ge 0$
2. $x^2 - 2x - 3 \le 0$
3. $(x - 2)(x + 3) > 0$
4. $(x + 4)(x - 1) < 0$
5. $(3x - 2)(2x + 3) \le 0$
6. $4x^2 - 9 \ge 0$
7. $(x - 2)^2 > 9$
8. $(x - 2)^2 \le 1$
9. $(x - 2)(x + 3) > 6$
10. $x(3 - x) \le 5$
11. $4x^2 > 13x$
12. $2x^2 - 3x - 4 < x^2$
13. $(2r + 1)(r - 2) + 3 \le 0$
14. $x^2 + x - 4 \ge 8$
15. $6d^2 - 5d \ge 50$
16. $6p^2 + 5p \le 50$
17. $v^2 \le 9v - 14$
18. $12s^2 > 1 - 4s$
19. $(x - 3)^2 > 1$
20. $(x + 4)^2 < 25$
21. $(2x + 3)^2 < 9$
22. $(3x - 2)^2 > 4$
23. $6 - x - x^2 \ge 0$
24. $10 - 3x - x^2 < 0$
25. $x^2 < -5$
26. $x^2 > -2$
27. The sum of the square of a number and five times the number is less than 24. Display a number-line graph of the possible values for the number.
28. Display a number-line graph of all the numbers such that the square of the sum of 3 and the number is less than 36.
29. The length of a rectangle is 14 meters less than twice the width. What values for the width insure that the area is less than 16 square meters?
30. The length of a rectangle is 15 inches more than twice the width. If the area measure is less than 50 square inches, what are the possible values for the perimeter of the rectangle?

5.7 ■ ■ ■ SECOND-DEGREE EQUATIONS IN TWO VARIABLES

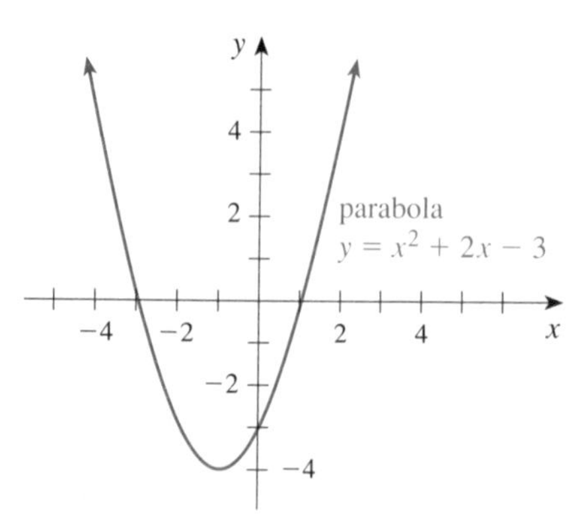

FIGURE 1

Consider the equation $y = x^2 + 2x - 3$. This equation has two variables, x and y. Thus, its solution will be a collection of ordered pairs of numbers. If we plot the ordered pairs that solve this second-degree equation in two variables on an xy-coordinate system, the graph is a curve known as a *parabola,* as shown in Figure 1. Compare this with plotting the ordered pairs that solve $y = 2x + 1$. With this first-degree equation in two variables, the graph is that of a line, which is shown in Figure 2.

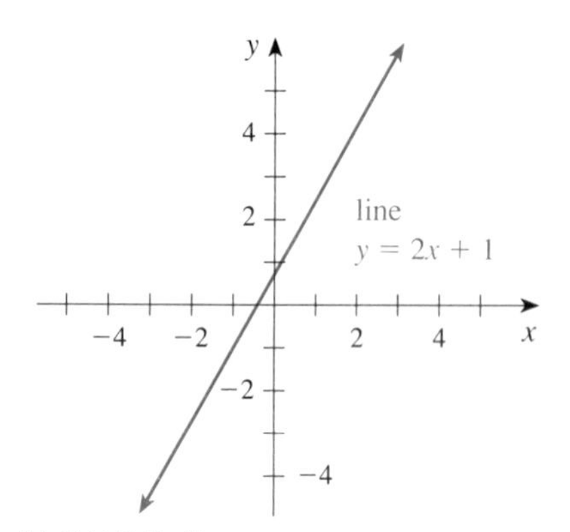

FIGURE 2

In this section, we will learn the characteristic features of a parabola—the line of symmetry and the vertex—and how to graph a parabola. The discussion will be restricted to equations that are second-degree in only one of the two variables. The more general case is covered in Chapter 12. Additionally, we will work with quadratic functions.

Solving Second-Degree Equations in Two Variables Once we know how to solve a second-degree equation in one variable, we can apply that knowledge to solving a second-degree equation in two variables. The solution to a second-degree equation in two variables, such as $y = x^2 - 4$, is a collection of ordered pairs of numbers that result in a true arithmetic sentence when the x-variable is replaced by the first number in the ordered pair and the y-variable is replaced by the second number in the ordered pair. For example, the ordered pair (2, 0) solves the equation $y = x^2 - 4$, because $0 = 2^2 - 4$ is a true arithmetic sentence. However, (0, 2) does not solve $y = x^2 - 4$, because $2 = 0^2 - 4$ is a false arithmetic sentence. In the table shown in the margin, some of the infinite collection of ordered pairs that solve $y = x^2 - 4$ are listed. Plotting these ordered pairs produces the picture shown in Figure 3.

First, it should be obvious from the points plotted in Figure 3 that the graph of the ordered pairs which solve $y = x^2 - 4$ is *not* a straight line. Sketching in a curve for the plotted points yields the curve in Figure 4, which is known as a *parabola*. Second, notice that each point, except the point $(0, -4)$, has a "mate." That is, if (a, b) is on the graph, then $(-a, b)$ is also on the graph. This means the graph is **symmetrical** about the y-axis. If we fold the graph paper along the y-axis, one side of the graph will exactly match the other side of the graph. Such a line is known as a *line of symmetry*. The unmated point, $(0, -4)$ in this case, is the *vertex* of the parabola.

Some ordered pairs that solve $y = x^2 - 4$

x	y	(x, y)
0	−4	(0, −4)
1	−3	(1, −3)
−1	−3	(−1, −3)
2	0	(2, 0)
−2	0	(−2, 0)
$\sqrt{2}$	−2	$(\sqrt{2}, -2)$
$-\sqrt{2}$	−2	$(-\sqrt{2}, -2)$
3	5	(3, 5)
−3	5	(−3, 5)

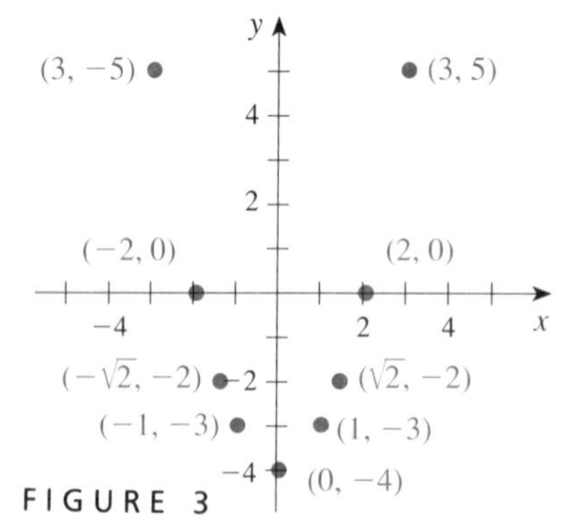

FIGURE 3

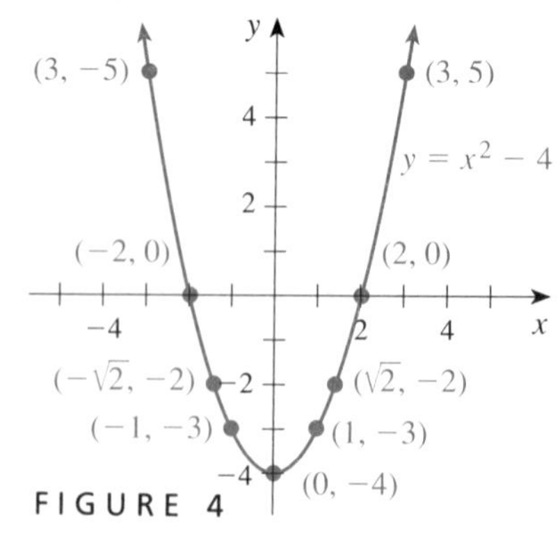

FIGURE 4

Using the Graphics Calculator

```
RANGE
Xmin=-9.6
Xmax=9.4
Xscl=1
Ymin=-6.4
Ymax=6.2
Yscl=1
Xres=1
```

FIGURE 5

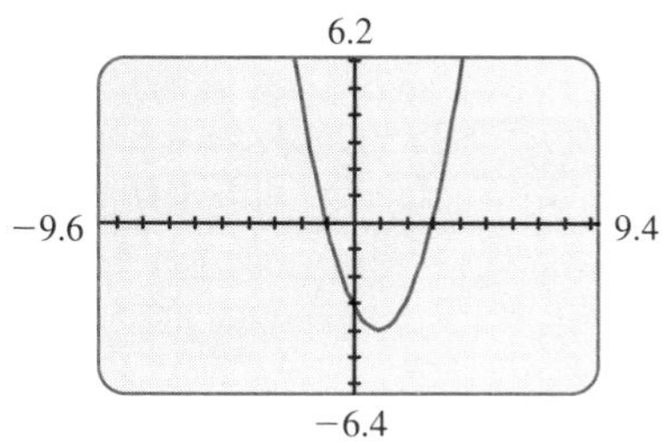

FIGURE 6

We can use the graph screen to quickly see the graph of the solution to a second-degree equation in two variables. Type the rule X²−2X−3 for Y1. Press the RANGE key and enter the values shown in Figure 5.* View the graph screen. The graph of the solution to $y = x^2 - 2x - 3$ is displayed in Figure 6. This curve is known as a parabola. Press the TRACE key. As you can see from the message at the bottom of the screen, the y-intercept is $(0, -3)$. Also, notice that the graph has two x-intercepts, $(-1, 0)$ and $(3, 0)$. Move the cursor to the vertex, the lowest point of the curve. What is the ordered-pair name of this point? We will now use algebraic techniques to confirm this analysis of the curve.

Features of a Parabola The major feature of a straight line is its slope. Additionally, the x-intercept and y-intercept are important. For a parabola, the x- and y-intercepts are still important (as they will be on any graph we study). The major feature of the parabola determined by $y = ax^2 + bx + c$, where $a \neq 0$, is its symmetry about a vertical line through its vertex.

LINE OF SYMMETRY AND VERTEX

For the parabola determined by $y = ax^2 + bx + c$ with $a \neq 0$, the **vertex** is the point on the graph whose x-coordinate is $\frac{-b}{2a}$.

The **line of symmetry** is the vertical line determined by $x = \frac{-b}{2a}$.

We will establish this property in Chapter 8. For most parabolas, if we know the y-intercept, the x-intercepts, the line of symmetry, and the vertex, we can sketch a graph of the parabola. To see that this is the case, work through the following example.

EXAMPLE 1 Consider the equation $y = x^2 - 2x - 3$.

a. Identify the y-intercept.
b. Identify the two x-intercepts.
c. Identify the vertex.
d. What is the equation of the line of symmetry?
e. Graph the parabola determined.

Solution a. To find the y-intercept, we replace x with 0 to get

$$y = 0^2 - 2(0) - 3 = -3$$

Thus, the y-intercept is $(0, -3)$.

TI-82 Note *Type ⁻9.4 for Xmin and ⁻6.2 for Ymin.

b. To find the x-intercepts for the graph of $y = x^2 - 2x - 3$, we replace y with 0 to get $0 = x^2 - 2x - 3$, and then solve this equation:

$$x^2 - 2x - 3 = 0$$
$$(x - 3)(x + 1) = 0$$
$$x - 3 = 0 \quad \text{or} \quad x + 1 = 0$$
$$x = 3 \qquad\qquad x = -1$$

The solution is 3 and -1. Thus, the x-intercepts are (3, 0) and $(-1, 0)$.

c. The x-coordinate of the vertex has the form $\frac{-b}{2a}$.

For this equation, $a = 1$ and $b = -2$.

So, the x-coordinate of the vertex is $\frac{-(-2)}{2(1)}$, or 1.

So, we have (1, ?). To find the y-coordinate, we replace x with 1 in the original equation to get $y = 1^2 - 2(1) - 3 = -4$.
Thus, the vertex is $(1, -4)$.

d. The line of symmetry is a vertical line through the vertex. Because the x-coordinate of the vertex is 1, the equation of line of symmetry is $x = 1$.

e. To draw the parabola determined by $y = x^2 - 2x - 3$, we plot the y-intercept, the x-intercepts, the vertex, and then draw a dotted or dashed line for the line of symmetry (this line is *not* part of the parabola, it is only a reference line). Finally, we sketch in the parabola. Figure 7 shows this graph. ■

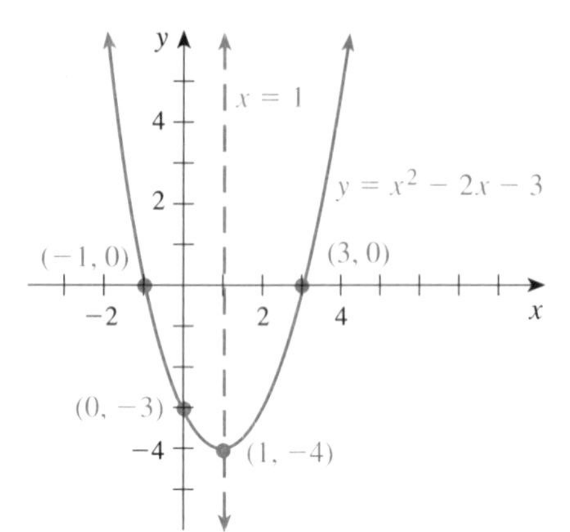

FIGURE 7

With the rule X^2−2X−3 entered for Y1, we can display the line of symmetry by using the Line option from the DRAW menu. Display the current graph screen (see Figure 6). Return to the home screen and press the 2nd key followed by the PRGM key (for the DRAW key). Select the Line option. The characters Line(are written to the home screen. To draw the vertical line $x = 1$, the x-coordinate of the points must be 1. The y-coordinate can be any value—let's choose 10 and -10. Complete the command to read Line(1,10,1,⁻10) and then press the ENTER key. As you can see, a vertical line is drawn, and this vertical lines passes through the vertex of the parabola.

GIVE IT A TRY

Report the requested information for each second-degree equation in two variables:

a. the ordered-pair name of the y-intercept,
b. the ordered-pair name for each of the two x-intercepts,
c. the ordered-pair name of the vertex, and
d. the equation of the vertical line that is the line of symmetry.

Use the information from parts (a)–(d) to graph the solution.

1. $y = x^2 - 5x - 6$ **2.** $y = (3 - x)(2 - x)$ **3.** $y = x^2 - 3x + 2$

A Down Turn So far, we have considered parabolas where the vertex is a *minimum point*—the vertex is the lowest point on the graph—but this is not always the case. The vertex can be the *maximum point* of the graph. That is, the vertex is the highest point on the graph.

Using the Graphics Calculator

```
RANGE
Xmin=-9.6
Xmax=9.4
Xscl=1
Ymin=-6.4
Ymax=6.2
Yscl=1
Xres=1
```

FIGURE 8

For `Y1` type `-X²+3X+4`. View the graph screen using the values in Figure 8.* Viewing the solution to the equation $y = -x^2 + 3x + 4$, we can see that the curve is still a parabola. As Figure 9 shows, the curve turns down, so the vertex is a maximum point (the highest point on the graph). If we alter the polynomial for `Y1` to $x^2 - 3x - 4$, the parabola turns up (the vertex is a minimum point). Press the `Y=` key and type `X²-3X-4` for `Y1`. Press the `GRAPH` key to display the graph shown in Figure 10. Compare this graph with the one in Figure 9.

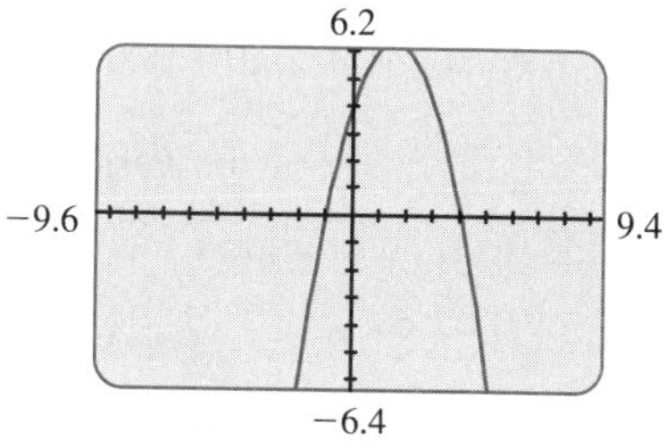

FIGURE 9

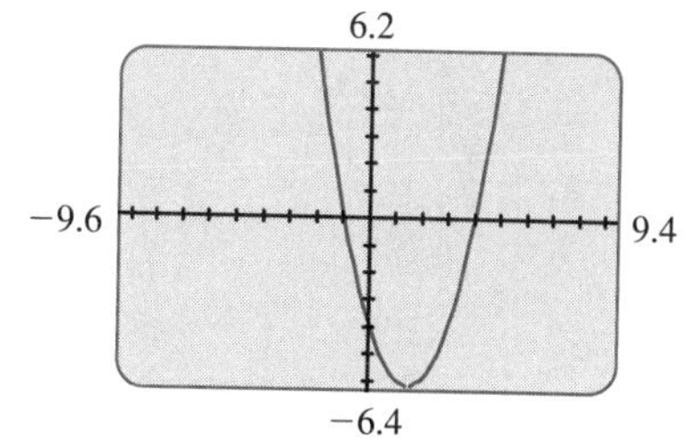

FIGURE 10

In general, $y = ax^2 + bx + c$ determines a parabola whose vertex is a minimum point if the x^2-coefficient is greater than 0. We say the parabola *turns upward*. If the x^2-coefficient is less than 0, the vertex is a maximum point, and we say the parabola *turns downward*. We summarize this information as follows.

PARABOLAS AND THE x^2-COEFFICIENT

For $y = ax^2 + bx + c$,

if $a > 0$, then the vertex is a *minimum point* (the parabola turns upward)

and

if $a < 0$, then the vertex is a *maximum point* (the parabola turns downward).

TI-82 Note *Type `-9.4` for `Xmin` and `-6.2` for `Ymin`.

EXAMPLE 2 Consider the equation $y = -x^2 + 3x + 4$.

a. Identify the y-intercept.
b. Identify the two x-intercepts.
c. Identify the vertex.
d. What is the equation of the line of symmetry?
e. Graph the parabola determined by this equation.

Solution **a.** To find the y-intercept, replace x with 0 in the equation. The y-intercept is (0, 4).

b. To find the x-intercepts, we replace y with 0 and solve the equation $0 = -x^2 + 3x + 4$:

$$-x^2 + 3x + 4 = 0 \qquad \text{Standard form}$$
$$x^2 - 3x - 4 = 0 \qquad \text{Multiply by } -1.$$
$$(x - 4)(x + 1) = 0$$
$$x - 4 = 0 \quad \text{or} \quad x + 1 = 0$$
$$x = 4 \qquad\qquad x = -1$$

The solution is 4 and -1. Thus, the x-intercepts are $(-1, 0)$ and $(4, 0)$.

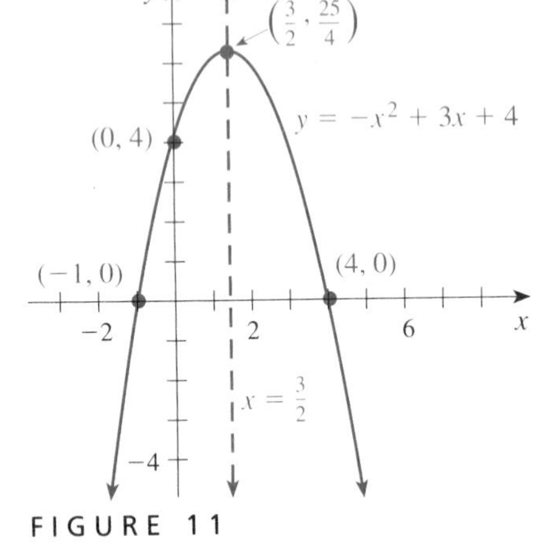

FIGURE 11

c. The x-coordinate of the vertex is $\frac{-b}{2a} = \frac{-3}{2(-1)} = \frac{3}{2}$. Notice that we return to the original equation for the a and b values!

Replacing x by $\frac{3}{2}$, we get $y = -\left(\frac{3}{2}\right)^2 + 3\left(\frac{3}{2}\right) + 4$. Doing the arithmetic, we find the y-coordinate, $\frac{25}{4}$.

The ordered-pair name for the vertex is $\left(\frac{3}{2}, \frac{25}{4}\right)$.

d. The x-coordinate of the vertex is $\frac{3}{2}$, so the line of symmetry is $x = \frac{3}{2}$.

e. Plotting this information and then drawing the parabola yields the graph in Figure 11. ■

GIVE IT A TRY

Report the requested information for each second-degree equation in two variables:

a. the ordered-pair name of the y-intercept,
b. the ordered-pair name for each of the two x-intercepts,
c. the ordered-pair name of the vertex, and
d. the equation of the vertical line that is the line of symmetry.

Use the information from parts (a)–(d) to graph the solution.

4. $y = -x^2 + 5x + 6$ **5.** $y = (3 - x)(2 + x)$

Quadratic Functions In Section 3.5, we discussed linear functions, such as $f(x) = x + 3$. We now want to consider quadratic functions, such as

$$f(x) = x^2 - 3x - 4$$

The *rule* for this function is $x^2 - 3x - 4$, that is, for each input value x, the output value $f(x)$ is determined by the rule $x^2 - 3x - 4$. Thus, $f(0)$ is -4 because $0^2 - 3(0) - 4$ is -4. We say that for an input of 0, the output is -4, and we write $f(0) = -4$. Likewise, $f(2) = -6$ because $2^2 - 3(2) - 4$ is -6.

If we plot the input values x against the output values $f(x)$ for the quadratic function $f(x) = x^2 - 3x - 4$, the curve produced will be a parabola. We approach the graph of quadratic functions the same way we graph the solution to a second-degree equation in two variables. Consider the following example.

EXAMPLE 3 Consider the quadratic function $f(x) = x^2 - 3x - 4$.

a. Find $f(-3)$.
b. For an input of 3, what is the output?
c. To get an output of -6, what value is input?
d. Graph the quadratic function.
e. What is the domain and range of the function?

Solution a. To find $f(-3)$, we input -3 for x (we replace x by -3). Thus,

$$f(-3) = (-3)^2 - 3(-3) - 4$$

Doing the arithmetic, we get $f(-3) = 14$.

b. To find the output for an input of 3 means to find $f(3)$. Now,

$$f(3) = 3^2 - 3(3) - 4$$

Thus, $f(3) = -4$. So, for an input of 3, the output is -4.

c. To find the input value that yields an output of -6, we replace the output value $f(x)$ with -6, and we get $-6 = x^2 - 3x - 4$. We then solve this equation.

$$x^2 - 3x + 2 = 0 \qquad \text{Standard form}$$
$$(x - 2)(x - 1) = 0 \qquad \text{Factored form}$$
$$x - 2 = 0 \quad \text{or} \quad x - 1 = 0$$
$$x = 2 \qquad\qquad x = 1$$

The solution is 2 and 1.

Thus, we have a choice of two input values to produce an output of -6—the input can be 2, or the input value can be 1.

d. The vertical axis intercept (the y-intercept) is the ordered pair $(0, -4)$. The x-intercepts are the values for which the function outputs 0. Thus, we solve $0 = x^2 - 3x - 4$.

$$0 = x^2 - 3x - 4$$
$$x^2 - 3x - 4 = 0$$
$$(x - 4)(x + 1) = 0$$
$$x - 4 = 0 \quad \text{or} \quad x + 1 = 0$$
$$x = 4 \qquad\qquad x = -1$$

The solution is 4 and -1. So, the x-intercepts are $(4, 0)$ and $(-1, 0)$.

The x-coordinate of the vertex is $\frac{-b}{2a} = \frac{-(-3)}{2(1)} = \frac{3}{2}$. To find the y-coordinate of the vertex, we find $f\left(\frac{3}{2}\right)$:

$$f\left(\frac{3}{2}\right) = \left(\frac{3}{2}\right)^2 - 3\left(\frac{3}{2}\right) - 4$$
$$= \frac{-25}{4}$$

The vertex is $\left(\frac{3}{2}, \frac{-25}{4}\right)$. The line of symmetry is $x = \frac{3}{2}$. See Figure 12 for the graph.

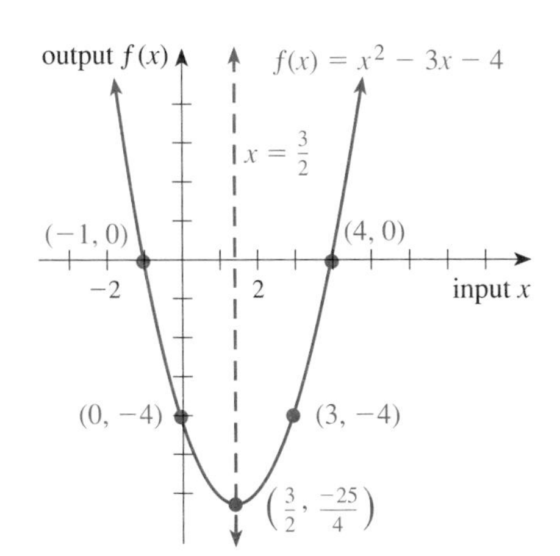

FIGURE 12

e. The domain is the allowable replacement values for x. For this function, the x-variable can be replaced by any real number. So, the domain is $\{x \mid x \text{ is real}\}$. The range is the set of output values produced. For this function, only the real numbers larger than or equal to $\frac{-25}{4}$ are output. So, the range $= \left\{y \mid y \geq \frac{-25}{4}\right\}$. ■

GIVE IT A TRY

6. Use the quadratic function $f(x) = 15 + 2x - x^2$ to answer each question.
   a. What is $f(-3)$?
   b. What two input values output the number 7?
   c. What are the x-intercepts of the graph of the function?
   d. What is the domain? What is the range?

Application As with linear functions, quadratic functions are useful in the sciences and in business. Consider the following example.

EXAMPLE 4 The profit P (in dollars) from producing a selected number x of Z8086 computer chips is given by the quadratic function $P(x) = 100x - x^2$.

a. What is the profit from producing 40 Z8086 computer chips?
b. What is the profit from producing 51 Z8086 computer chips?
c. To earn a profit of \$2100, how many Z8086 computer chips must be produced?
d. How many Z8086 computer chips must be produced for the profit to be at its maximum?
e. If 101 Z8086 computer chips are produced, what is the profit? What does this mean?

Solution We start by producing a graph of the quadratic function $P(x) = 100x - x^2$. The vertical axis intercept is (0, 0) because $P(0) = 0$.

The x-intercepts are found by solving $0 = 100x - x^2$:

$$-x^2 + 100x = 0 \quad \text{Standard form}$$
$$x^2 - 100x = 0 \quad \text{Multiply by } -1.$$
$$x(x - 100) = 0$$
$$x = 0 \quad \text{or} \quad x - 100 = 0$$

The equation $0 = 100x - x^2$ has a solution of 0 and 100. So, the x-intercepts are (0, 0) and (100, 0).

The x-coordinate of the vertex is

$$\frac{-b}{2a} = \frac{-100}{2(-1)} = 50$$

The y-coordinate of the vertex is $P(50) = 100(50) - (50)^2 = 2500$. Thus, the ordered-pair name of the vertex is (50, 2500).

The axis of symmetry is the vertical line through the vertex. Thus, we have $x = 50$.

Using this information, the graph appears as shown in Figure 13.

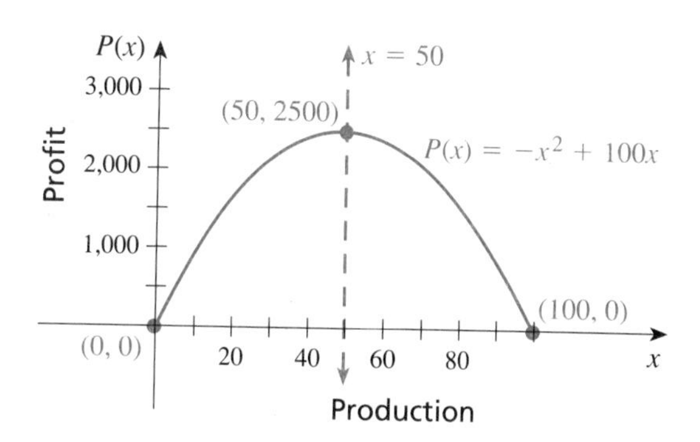

FIGURE 13

a. The profit from producing 40 computer chips is $P(40)$. Applying the function rule to 40 yields

$$\begin{aligned} P(40) &= 100(40) - (40)^2 \\ &= 2400 \end{aligned}$$

Thus the profit from producing 40 Z8086 computer chips is \$2400.

b. The profit from producing 51 computer chips is

$$\begin{aligned} P(51) &= 100(51) - (51)^2 \\ &= 2499 \end{aligned}$$

So, the profit for producing 51 Z8086 chips is \$2499.

c. To find the production level that yields a profit of \$2100, we solve the equation $2100 = 100x - x^2$:

$$\begin{aligned} -x^2 + 100x - 2100 &= 0 && \text{Standard form} \\ x^2 - 100x + 2100 &= 0 && \text{Multiply by } -1. \\ (x - 30)(x - 70) &= 0 \end{aligned}$$

$$x - 30 = 0 \quad \text{or} \quad x - 70 = 0$$

$$x = 30 \qquad\qquad x = 70$$

The solution is 30 and 70.

Thus, if 30 Z8086 chips are produced, the profit is \$2100. If 70 Z8086 chips are produced, the profit is also \$2100.

d. The production level x that produces the maximum profit occurs at the x-coordinate of the vertex for this function. We know that the x-coordinate of the vertex is 50 (see Figure 13). Thus, if 50 Z8086 computer chips are produced, the profit is at a maximum, \$2500.

e. If 101 Z8086 computer chips are produced, the profit, $P(101)$, is

$$\begin{aligned} P(101) &= 100(101) - 101^2 \\ &= -101 \end{aligned}$$

A loss of \$101 is incurred. ■

■ SUMMARY

In this section, we found the solutions to second-degree equations in two variables (second degree in x-only, such as $y = x^2 - 3x - 4$). The solution produces a graph whose shape is known as a parabola. Just as the major feature of a line is its slope, the major features of a parabola are its line of symmetry and its vertex. We also outlined a general procedure for graphing the solution to a second-degree equation for two variables (in standard form, $y = ax^2 + bx + c$):

1. Replace the x-variable by 0 to find the y-intercept.
2. Replace the y-variable by 0 to find the x-intercepts.
3. Find the x-coordinate of the vertex using $\frac{-b}{2a}$. Find the y-coordinate using arithmetic.
4. Find the equation of the line of symmetry: $x = \frac{-b}{2a}$.

In general, this information is sufficient to sketch in the parabola.

In addition to this general technique for graphing a parabola, we can predict the direction of the parabola determined by the equation in standard form, $y = ax^2 + bx + c$. If $a > 0$, the parabola turns upward, and the vertex is a minimum point. If $a < 0$, the parabola turns downward, and the vertex is a maximum point.

A quadratic function is a function whose rule is a second-degree polynomial in one variable [for example, $f(x) = x^2 - 3x - 4$]. If the input values for the function are plotted against the output values, a parabola is produced. As with linear functions, quadratic functions are used frequently to model situations in the sciences and in business.

■ ***Quick Index***

5.7 ■ ■ ■ EXERCISES

Insert true *for a true statement, or* false *for a false statement.*

____ 1. The graph of the solution to $y = x^2$ is a straight line.

____ 2. The parabola determined by $y = x^2 - 2x - 3$ has y-intercept $(0, -3)$.

____ 3. The graph of a quadratic function always produces a parabola.

____ 4. The parabola determined by $f(x) = ax^2 + bx + c$ has a minimum point if $a > 0$.

____ 5. The line of symmetry for the graph of the solution to $y = x^2 - 4x + 3$ is the vertical line $x = 2$.

____ 6. The graph of the solution to $y = (x - 3)(x + 2)$ has x-intercepts at $(-3, 0)$ and $(2, 0)$.

____ 7. The graph of the solution to $y = x^2 + 3$ does not have a y-intercept.

____ 8. For the vertex of a parabola to be a maximum point, the x^2-coefficient a must be less than 0.

____ 9. The range for $f(x) = x^2$ is all real numbers.

10. For the equation $y = (x - 3)(x + 2)$, complete each statement.
    a. The y-intercept is (____, ____).
    b. The x-intercepts are (____, ____) and (____, ____).
    c. The vertex is (____, ____).
    d. The line of symmetry is $x =$ ____.
    e. Use the information reported in parts (a)–(d) to graph the solution to $y = (x - 3)(x + 2)$.

For each second-degree equation, in two variables, report the requested information:

a. *the y-intercept* **b.** *the x-intercepts*
c. *the vertex* **d.** *the line of symmetry*

Use this information to graph the solution to the equation. Use the graphics calculator to check your work.

11. $y = (x - 2)(x + 3)$
12. $y = (x + 1)(x - 2)$
13. $y = x^2 + 2x - 3$
14. $y = x^2 - 2x - 3$

15. $y = (2x - 2)(x + 3)$
16. $y = (3x - 4)(x + 2)$
17. $y = 2x^2 + 5x - 3$
18. $y = 2x^2 + x - 1$
19. $y = -x^2 + 2x + 3$
20. $y = -x^2 + 4x + 5$
21. $y = x^2 + 4x - 5$
22. $y = x^2 + 4x - 5$
23. $y = -2x^2 + 3x + 5$
24. $y = -2x^2 + 3x + 9$
25. $y = x^2 + 2x - 8$
26. $y = -x^2 + 2x + 3$
27. $y = (x - 1)^2$
28. $y = (x + 2)^2$
29. $y = -(x + 1)^2$
30. $y = -(2x - 3)^2$
31. $y = (x - 1)^2 - 4$
32. $y = (x + 1)^2 - 4$
33. $y = -(x - 1)^2 + 4$
34. $y = -(x - 1)^2 + 9$
35. $y = 8x - x^2$
36. $y = -6x - x^2$

Graph each quadratic function. Label the vertical-axis intercept and the x*-axis intercept(s), if any exist, with their ordered-pair names. Label the line of symmetry with its equation. Label the vertex with its ordered-pair name. Report the domain and the range for the function.*

37. $f(x) = x^2 - 3x + 2$
38. $g(x) = x^2 - 4x - 5$
39. $h(x) = 2x^2 - 3x - 2$
40. $d(x) = 3x^2 - 4x + 1$
41. $s(t) = -8t - t^2$
42. $h(t) = 16 - t^2$
43. $R(a) = 80 - 16a - a^2$
44. $P(n) = 16n - n^2$
45. For the Z8086 chips discussed in Example 4, find the production levels that produce a profit of \$2400.
46. Maxwell makes caps that are sold at local festivals (a mullet-head cap is the biggest seller). He has found that his cost per cap C is a function of the number of caps produced, x. That function is the quadratic function $C(x) = x^2 - 8x + 30$.
    a. What is the cost per cap when 10 caps are produced?
    b. Graph the cost-per-cap function.
    c. How many caps must be produced for the cost per cap to be minimum? What is the minimum cost per cap?
47. A model rocket is launched straight upward. The height h (in feet) reached by the rocket t seconds after it is launched is given by the quadratic function

$$h(t) = -10t^2 + 90t$$

    a. Graph the quadratic function.
    b. How high is the rocket 2 seconds after launch?
    c. At what time does the rocket reach its maximum height? What is the maximum height?
    d. How long does the rocket remain aloft (its height is greater than 0)?
48. Of two numbers, one is 6 more than the other. Their product P is a function of the smaller number, x. Report the function rule for function P. At what value of x is the product minimum? What is the value of the minimum product?
49. For two numbers, one is 8 more than twice the other. Their product P is a function of the smaller number, x. Report the function rule for function P. At what value of x is the product minimum? What is the value of the minimum product?
50. The height h (in feet) of a ball t seconds after being dropped from a tall building is given by

$$h(t) = -16t^2 + 256$$

At what time will the ball be 192 feet high? At what time will the ball hit the ground (reach height zero)?
51. The perimeter of rectangle is 80 meters. The area measure A of the rectangle is a function of the length of the rectangle, x. Report the rule for function A. At what value of x is the area measure maximum?

5.8 ■ ■ ■ DIVISION OF POLYNOMIALS

We have discussed some operations on polynomials: addition and subtraction of polynomials in Section 2.1 and multiplication, and the closely related topic of factoring, of polynomials in Sections 5.1–5.4. We now discuss the last of the basic binary operations—division of polynomials. This operation leads to the next major object in algebra, the rational expression, (which we will introduce in Chapter 6).

A Division Algorithm An *algorithm* is a specific set of steps that can be used to solve selected problems. Consider the following division of polynomials:

$$\underbrace{(x^2 - 3x + 2)}_{\text{Dividend}} \div \underbrace{x}_{\text{Divisor}}$$

Arithmetic Analogy: $47 \div 2$

We set up this problem much like dividing whole numbers, that is, we write

$$x\overline{)x^2 - 3x + 2}$$

See Analogy 1.

1. $2\overline{)47}$

Now, $x^2 \div x = x$ (we say that x goes into x^2, x times, because x times x is x^2). Thus, we write

$$\begin{array}{r} x \\ x\overline{)x^2 - 3x + 2} \end{array}$$

See Analogy 2.

2. $\begin{array}{r} 2 \\ 2\overline{)47} \end{array}$

Multiplying x times the divisor, x, we get:

$$\begin{array}{r} x \\ x\overline{)x^2 - 3x + 2} \\ x^2 \end{array}$$

See Analogy 3.

3. $\begin{array}{r} 2 \\ 2\overline{)47} \\ 4 \end{array}$

Subtracting x^2 from the dividend, $x^2 - 3x + 2$, we get:

$$\begin{array}{r} x \\ x\overline{)x^2 - 3x + 2} \\ \underline{x^2 } \\ - 3x + 2 \end{array}$$

See Analogy 4.

Subtract

4. $\begin{array}{r} 2 \\ 2\overline{)47} \\ \underline{4} \\ 7 \end{array}$

Now, we repeat the process. First, $-3x \div x = -3$ (we say x goes into $-3x$, -3 times, because x times -3 is $-3x$). Thus, we write:

$$\begin{array}{r} x - 3 \\ x\overline{)x^2 - 3x + 2} \\ \underline{x^2 } \\ - 3x + 2 \\ \underline{- 3x } \\ 2 \end{array}$$

See Analogy 5.

Multiply -3 times x
Subtract

5. $\begin{array}{r} 23 \\ 2\overline{)47} \\ \underline{4} \\ 7 \\ \underline{6} \\ 1 \end{array}$

Once the subtraction produces a polynomial of degree less than the divisor, the process stops. So, $(x^2 - 3x + 2) \div x$ is $x - 3$ with a remainder of 2. As with division of whole numbers,

$(x^2 - 3x + 2) \div x$ is $x - 3$ with a remainder of 2

means the product of x and $x - 3$ plus the remainder 2 yields the dividend $x^2 - 3x + 2$.

$$(x^2 - 3x + 2) \div x = x - 3 + \frac{2}{x}$$

dividend: $x^2 - 3x + 2$; divisor: x; quotient: $x - 3$; remainder: 2

The various parts of the division are shown in the margin.

We write

$$(x^2 - 3x + 2) \div x = x - 3 + \frac{2}{x}$$

This last expression, $\frac{2}{x}$, is not a polynomial because of the division by a variable. The expression $\frac{2}{x}$ is a *rational expression,* a ratio of polynomials. We will study rational expressions in detail in Chapter 6. As a second example of dividing polynomials, consider the following example.

EXAMPLE 1 Find the quotient and remainder for $(2x^2 - x + 1) \div (x - 2)$.

Solution We set up this problem as follows:

$$x - 2 \overline{\smash{)}\, 2x^3 - x + 1}$$

Now, we say x goes into $2x^3$, $2x^2$ times. So, we write

$$\begin{array}{rl} & 2x^2 \qquad \leftarrow (2x^3) \div x = 2x^2 \\ x - 2 & \overline{\smash{)}\, 2x^3 - x + 1} \\ & 2x^3 - 4x^2 \qquad \text{Multiply } 2x^2 \text{ times } x - 2. \end{array}$$

Next, we subtract $2x^3 - 4x^2$ from $2x^3 - x + 1$ to get $4x^2 - x + 1$. Thus,

$$\begin{array}{rl} & 2x^2 \\ x - 2 & \overline{\smash{)}\, 2x^3 - x + 1} \\ & 2x^3 - 4x^2 \\ & \overline{\qquad 4x^2 - x + 1} \qquad \text{Subtract} \end{array}$$

WARNING

Most of the errors made with the division algorithm occur in the subtraction operation. An error-reducing approach to subtraction is to add the opposite.

We now repeat the process. We say x goes into $4x^2$, $4x$ times, and we write:

$$\begin{array}{rl} & \qquad (4x^2) \div x = 4x \\ & 2x^2 + 4x \\ x - 2 & \overline{\smash{)}\, 2x^3 - x + 1} \\ & 2x^3 - 4x^2 \\ & \overline{\qquad 4x^2 - x + 1} \\ & \qquad 4x^2 - 8x \qquad \text{Multiply } 4x \text{ times } x - 2. \\ & \qquad \overline{\qquad 7x + 1} \qquad \text{Subtract} \end{array}$$

We start the process again. We say x goes into $7x$, 7 times, and write:

$$\begin{array}{rl} & 2x^2 + 4x + 7 \qquad \leftarrow (7x) \div x = 7 \\ x - 2 & \overline{\smash{)}\, 2x^3 - x + 1} \\ & 2x^3 - 4x^2 \\ & \overline{\qquad 4x^2 - x + 1} \\ & \qquad 4x^2 - 8x \\ & \qquad \overline{\qquad 7x + 1} \\ & \qquad \qquad 7x - 14 \qquad \text{Multiply 7 times } x - 2. \\ & \qquad \qquad \overline{\qquad 15} \qquad \text{Subtract} \end{array}$$

Because the result of the last subtraction has degree less than that of the divisor, the process ends. We write

$$(2x^3 - x + 1) \div (x - 2) = 2x^2 + 4x + 7 + \frac{15}{x - 2}$$

This means the product of $2x^2 + 4x + 7$ and $x - 2$ plus the remainder 15 yields $2x^3 - x + 1$. ■

To check our work in Example 1, we can multiply $x - 2$ times $2x^2 + 14x + 7$, then add the remainder, 15, to the product. If the result is not the quotient, $2x^3 - x + 1$, then an error is present. That is, we must have:

$$(x - 2)(2x^2 + 14x + 7) + 15 = 2x^3 - x + 1$$

Using the Graphics Calculator

To test our work in Example 1 for errors, we start by storing a number, say `10`, in memory location `X`. For `Y1`, use option `3` from the `MATH` menu to type `(2X³−X+1)/(X−2)`. For `Y2`, type the quotient, `2X²+4X+7+15/(X−2)`. Return to the home screen. Display the value in `Y1` and the value in `Y2`. These two expressions must output the same value. If not, an error is present.

GIVE IT A TRY

Report the quotient and the remainder for each division.

1. $(2x^3 - x^2 + 2) \div (2x)$
2. $(x^4 - x^2 + 3) \div (x^2 - 1)$
3. $(x^3 - 1) \div (x + 1)$
4. Use the graphics calculator to check your work in Problems 1–3.

Polynomial Division and Factoring One application of division of polynomials is determining whether one polynomial is a factor of another polynomial. For example, is the polynomial $2x + 3$ a factor of the polynomial $8x^3 + 27$? This is similar to the question "Is the number 7 a factor of the number 987?" To answer the number question, we divide 987 by 7 and see if the remainder is 0. If it is, then 7 is a factor 987. Try it. To answer the algebra problem, we divide $8x^3 + 27$ by $2x + 3$. If the remainder is 0, then $2x + 3$ is a factor. Doing this division, we get

$$
\begin{array}{r|l}
 & 4x^2 - 6x + 9 \\
2x + 3 & 8x^3 + 27 \\
 & 8x^3 + 12x^2 \\
 & -12x^2 + 27 \\
 & -12x^2 - 18x \\
 & 18x + 27 \\
 & 18x + 27 \\
 & 0
\end{array}
$$

Multiply $4x^2$ times $2x + 3$.
Subtract $8x^3 + 12x^2$ from $8x^3 + 27$.
Multiply $-6x$ times $2x + 3$
Subtract $-12x^2 - 18x$ from $-12x^2 + 27$.
Multiply 9 times $2x + 3$.
Subtract $18x + 27$ from $18x + 27$.

Because the division produced a 0 remainder, we have shown that $2x + 3$ is a factor of $8x^3 + 27$. In fact, the other factor of $8x^3 + 27$ is $4x^2 - 6x + 9$. Remember from factoring polynomials in Section 5.4 that $8x^3 + 27$ factors as $(2x + 3)(4x^2 - 6x + 9)$.

Consider the following example, which also deals with factoring and division of polynomials.

EXAMPLE 2 If one factor of $x^5 + 3x^3 - 5x^2 - 15$ is $x^2 + 3$, what is the other factor?

Solution To find the other factor, we use the long division algorithm:

$$
\begin{array}{r|l}
 & x^3 - 5 \\
x^2 + 3 & x^5 + 3x^3 - 5x^2 - 15 \\
 & x^5 + 3x^3 \qquad \text{Multiply } x^3 \text{ times } x^2 + 3. \\
 & -5x^2 - 15 \qquad \text{Subtract} \\
 & -5x^2 - 15 \qquad \text{Multiply } -5 \text{ times } x^2 + 3 \\
 & 0 \qquad \text{Subtract}
\end{array}
$$

Thus, the other factor is $x^3 - 5$. That is, the product of $x^2 + 3$ and $x^3 - 5$ yields the polynomial $x^5 + 3x^3 - 5x^2 - 15$. ■

GIVE IT A TRY

5. If one factor of $x^4 + x^2 + 1$ is $x^2 + x + 1$, what is the other factor?
6. If one factor of $4x^4 + 1$ is $2x^2 + 2x + 1$, what is the other factor?
7. Given that $x^2 - x + 1$ is a factor of $x^6 - 1$, completely factor $x^6 - 1$.

Division and Equation Solving As a second application we will see how division of polynomials can be used in solving equations. So far, we have looked at solving first- and second-degree equations in one variable. We can always arrive at a solution to these types of equations by following an organized procedure (in Chapter 8 we will introduce a procedure, the quadratic formula, for solving *any* second-degree equation). For third- and fourth-degree equations, such as $x^3 - 3x^2 + 2x = 1$ and $x^4 - x = 2$, definite procedures exist, but they are beyond the scope of this course. For higher-degree polynomial equations in one variable (fifth-degree and greater), Paolo Ruffini (Italian) and Niels Abel (Norwegian) proved in the 1800s that it is impossible to find a general solution (formula). We can use some ingenuity and knowledge of basic algebra to *solve* some of these equations.

Using the Graphics Calculator

Consider the equation $x^3 + 3x^2 - x = 3$. If we write the equation in standard form, we get $x^3 + 3x^2 - x - 3 = 0$. Now, we look at the graph of the solution to $y = x^3 + 3x^2 - x - 3$. For `Y1`, type `X³+3X²−X−3`. View the graph screen using the settings shown in

Figure 1. The graph is given in Figure 2. Press the TRACE key. It appears that when the x-variable is replaced by -1, the y-variable is 0. That is, it appears that the number -1 solves the equation $x^3 + 3x^2 - x = 3$. If we replace the x-variable by -1, we get $(-1)^3 + 3(-1)^2 - (-1) = 3$, a true arithmetic sentence. Thus, -1 solves the equation. Saying -1 is in the solution to $x^3 + 3x^2 - x = 3$ is equivalent to saying that $(-1, 0)$ is an x-intercept of the graph of $y = x^3 + 3x^2 - x - 3$. We will now show that saying that $(-1, 0)$ is an x-intercept of the graph of $y = x^3 + 3x^2 - x - 3$ is equivalent to saying that $x + 1$ is a factor of the polynomial $x^3 + 3x^2 - x - 3$.

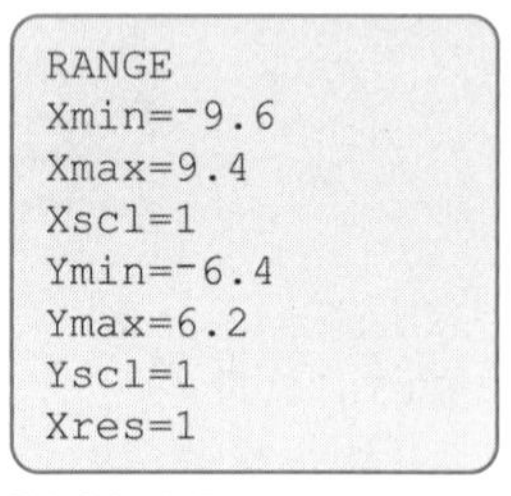

FIGURE 1

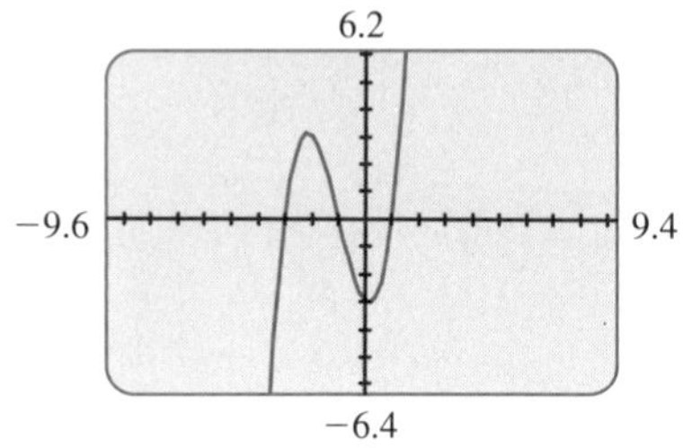

FIGURE 2

EXAMPLE 3 Suppose we have found that -1 solves the equation $x^3 + 3x^2 - x = 3$. Find the other numbers in the solution.

Solution Rewriting the equation to standard form yields $x^3 + 3x^2 - x - 3 = 0$. Knowing that -1 solves the equation means that we have

$$(x + 1)(\textit{some polynomial}) = 0$$

↑

-1 makes this factor 0.

That is, $x + 1$ is a factor of $x^3 + 3x^2 - x - 3$. We now use division of polynomials to find the other factor. We get:

$$\begin{array}{rll}
 & x^2 + 2x - 3 & \\
x + 1 \,\big) & x^3 + 3x^2 - x - 3 & \\
 & x^3 + x^2 & \text{Multiply } x^2 \text{ times } x + 1. \\
\hline
 & 2x^2 - x - 3 & \text{Subtract} \\
 & 2x^2 + 2x & \text{Multiply } 2x \text{ times } x + 1. \\
\hline
 & -3x - 3 & \text{Subtract} \\
 & -3x - 3 & \text{Multiply } -3 \text{ times } x + 1. \\
\hline
 & 0 & \text{Subtract}
\end{array}$$

TI-82 Note *Type ⁻9.4 for Xmin and ⁻6.2 for Ymin.

Thus, we have $(x + 1)(x^2 + 2x - 3) = 0$. The number that results in the factor $x + 1$ equaling 0 is -1.

Now, we factor $x^2 + 2x - 3$ to get $(x + 3)(x - 1)$. So, we have

$$x^3 + 3x^2 - x - 3 = 0$$
$$(x + 1)(x + 3)(x - 1) = 0$$
$$x + 1 = 0 \quad \text{or} \quad x + 3 = 0 \quad \text{or} \quad x - 1 = 0$$
$$x = -1 \qquad x = -3 \qquad x = 1$$

The solution to the equation $x^3 + 3x^2 - x = 3$ is -1, -3, and 1. ■

As a second example of using division of polynomials to help find the solution to an equation, consider the following.

EXAMPLE 4 Given that $\frac{2}{3}$ is in the solution to the equation $3x^4 - 5x^3 = 4x^2 - 4x$, find the other numbers in the solution.

Solution We start by writing the equation in standard form,

$$3x^4 - 5x^3 - 4x^2 + 4x = 0$$

Now, x is a common factor of the left-hand side. So, we factor it out:

$$x(3x^3 - 5x^2 - 4x + 4) = 0$$

Next, knowing that $\frac{2}{3}$ solves the equation means that $(3x - 2)$ is a factor $\left(\text{because } 3x - 2 = 0 \text{ is equivalent to } x = \frac{2}{3}\right)$. We get:

$$x(3x - 2)(\textit{some polynomial}) = 0$$

To find the third factor, we use long division of polynomials:

$$\begin{array}{r|l} & x^2 - x - 2 \\ \hline 3x - 2 & 3x^3 - 5x^2 - 4x + 4 \\ & 3x^3 - 2x^2 \\ \hline & \quad - 3x^2 - 4x + 4 \\ & \quad - 3x^2 + 2x \\ \hline & \qquad\quad - 6x + 4 \\ & \qquad\quad - 6x + 4 \\ \hline & \qquad\qquad\quad 0 \end{array}$$

Thus, we have

$$x(3x - 2)(x^2 - x - 2) = 0$$
$$x(3x - 2)(x - 2)(x + 1) = 0$$
$$x = 0 \quad \text{or} \quad 3x - 2 = 0 \quad \text{or} \quad x - 2 = 0 \quad \text{or} \quad x + 1 = 0$$
$$x = 0 \qquad x = \frac{2}{3} \qquad x = 2 \qquad x = -1$$

The solution to $3x^4 - 5x^3 = 4x^2 - 4x$ is 0, $\frac{2}{3}$, 2, and -1. ■

Using the Graphics Calculator

For `Y1`, type `3X^4−5X³−4X²+4X`. Use the settings in Figure 3* to view the graph screen, as shown in Figure 4. Now, use the `TRACE` key. The message `X=0 Y=0` means that when the x-variable is replaced by 0, the polynomial $3x^4 - 5x^3 - 4x^2 + 4x$ evaluates to 0. That is, the number 0 solves $3x^4 - 5x^3 = 4x^2 - 4x$. Move the cursor until the message reads `X=-1 Y=0`. This means that -1 solves the equation $3x^4 - 5x^3 = 4x^2 - 4x$. Using the right arrow key, we can see that a number between 0.6 and 0.7 also solves the equation. From the algebra work in Example 4, we know that this number is $\frac{2}{3}$. Find the fourth number that solves the equation $3x^4 - 5x^3 = 4x^2 - 4x$.

```
RANGE
Xmin=-4.8
Xmax=4.7
Xscl=1
Ymin=-3.2
Ymax=3.1
Yscl=1
Xres=1
```

FIGURE 3

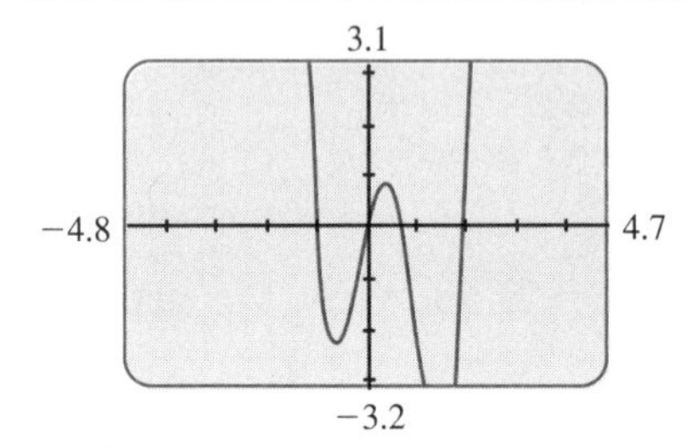

FIGURE 4

GIVE IT A TRY

```
RANGE
Xmin=-9.6
Xmax=9.4
Xscl=1
Ymin=-6.4
Ymax=6.2
Yscl=1
Xres=1
```

FIGURE 5

8. Given that 3 solves $x^3 - 2x^2 - 5x + 6 = 0$, find the other two numbers in the solution.

9. Given that -2 solves $x^4 + x^3 - 6x^2 - 4x = -8$, find the other numbers in the solution.

10. Use the graphics calculator with the range settings shown in Figure 5,** to find the x-intercepts of the function $g(x) = x^4 - x^3 - 7x^2 + x + 6$. Use your answer to factor the polynomial $x^4 - x^3 - 7x^2 + x + 6$.

Synthetic Division As we have seen from the application of solving equations, polynomial division often involves a first degree binomial such as $x - 2$, as the divisor. Dividing by a first-degree binomial is so routine that an algorithm has been developed that uses only the coefficients of the polynomials (omitting the variables). This algorithm, known as *synthetic division,* works only for division by a first-degree binomial. Consider the following division problem.

$$\begin{array}{r|l} & 2x^2 + 4x \;\; + 10 \\ \hline x - 2 & 2x^3 + 0x^2 + \;\; 2x + \;\; 1 \\ & 2x^3 - 4x^2 \\ \hline & \qquad 4x^2 + \;\; 2x + \;\; 1 \\ & \qquad 4x^2 - \;\; 8x \\ \hline & \qquad\qquad 10x + \;\; 1 \\ & \qquad\qquad 10x - 20 \\ \hline & \qquad\qquad\qquad 21 \end{array}
\qquad
\begin{array}{r|l} & 2 + 4 + 10 \\ \hline 1 - 2 & 2 \quad 0 \quad 2 \quad 1 \\ & 2 - 4 \\ \hline & \quad 4 + \;2 + \;1 \\ & \quad 4 - \;8 \\ \hline & \qquad 10 + \;1 \\ & \qquad 10 - 20 \\ \hline & \qquad\qquad 21 \end{array}
\quad \textit{(with variables omitted)}$$

TI-82 Note *Use the `ZDecimal` option.
**Type `-9.4` for `Xmin` and `-6.2` for `Ymin`.

The synthetic division algorithm abstracts the long division process by omitting the variables. Once the problem is set up, the process can be carried out without reference to the variables. Work through the following steps. Compare each step in the synthetic division algorithm with the preceding division.

1. Write down the coefficients of the dividend (using a 0 for any missing terms): $2x^3 + 0x^2 + 2x + 1$

$$\underline{\quad}\,|\;2 \quad 0 \quad 2 \quad 1$$

2. Write down the divisor, a, for the dividing binomial, $x - a$. In this case, $a = 2$.

$$\underline{\mathbf{2}}\,|\;2 \quad 0 \quad 2 \quad 1$$

This takes care of the set up. We omit the x-coefficient of the divisor, $x - 2$, because it will always be a 1. Now, the division process starts.

3. "Bring down" the lead number of the dividend.

$$\begin{array}{r|rrrr} \mathbf{2} & 2 & 0 & 2 & 1 \\ & \downarrow & & & \\ \hline & 2 & & & \end{array}$$

4. Multiply the divisor times the bottom number, and then place the product under the next place to the right:

$$\begin{array}{r|rrrr} \mathbf{2} & 2 & 0 & 2 & 1 \\ & & 4 & & \\ \hline & 2 & & & \end{array}$$

multiply

5. "Bring down" the sum of the next column.

$$\begin{array}{r|rrrr} \mathbf{2} & 2 & 0 & 2 & 1 \\ & & 4 & & \\ \hline & 2 & 4 & & \end{array}$$

sum

6. Multiply the divisor times the bottom number, and then place the product under the next place to the right.

$$\begin{array}{r|rrrr} \mathbf{2} & 2 & 0 & 2 & 1 \\ & & 4 & 8 & \\ \hline & 2 & 4 & & \end{array}$$

multiply

7. "Bring down" the sum of the next column.

$$\begin{array}{r|rrrr} \mathbf{2} & 2 & 0 & 2 & 1 \\ & & 4 & 8 & \\ \hline & 2 & 4 & 10 & \end{array}$$

sum

8. Multiply the divisor times the bottom number, and then place the product under the next place to the right:

$$\begin{array}{r|rrrr} 2 & 2 & 0 & 2 & 1 \\ & & 4 & 8 & 20 \\ \hline & 2 & 4 & 10 & \end{array}$$

multiply

9. ''Bring down'' the sum of the next column.

$$\begin{array}{r|rrrr} 2 & 2 & 0 & 2 & 1 \\ & & 4 & 8 & 20 \\ \hline & 2 & 4 & 10 & 21 \end{array}$$

sum

At this point, the division process ends. Notice that the first three numbers in the bottom row match the coefficients of the dividend found by long division:

$$(2x^3 + 0x^2 + 2x + 1) \div (x - 2) = 2x^2 + 4x + 10 + \frac{21}{x - 2}$$

Because the divisor is always a first-degree binomial, the quotient always has degree one less than the dividend, $2x^3 + 2x + 1$. So, in this case, the quotient starts with an x^2 term. The rightmost number on the bottom row is the remainder. We read the quotient and remainder from the bottom row:

$$\begin{array}{r|rrrr} 2 & 2 & 0 & 2 & 1 \\ & & 4 & 8 & 20 \\ \hline & 2 & 4 & 10 & 21 \\ & \downarrow & \downarrow & \downarrow & \downarrow \end{array}$$

$$2x^2 + 4x + 10 + \frac{21}{x - 2}$$

EXAMPLE 5 Write the quotient for $(3x^3 - 2x^2 - 4) \div (x - 3)$.

Solution Because the divisor is a first-degree binomial, we can use synthetic division. Setting up the synthetic division, we get:

$$\begin{array}{r|rrrr} 3 & 3 & -2 & 0 & -4 \end{array}$$

Next, we carry out the synthetic division process:

$$\begin{array}{r|rrrr} 3 & 3 & -2 & 0 & -4 \\ & & 9 & 21 & 63 \\ \hline & 3 & 7 & 21 & 59 \end{array}$$

Thus,

$$(3x^3 - 2x^2 - 4) \div (x - 3) = 3x^2 + 7x + 21 + \frac{59}{x - 3}$$

■

EXAMPLE 6 Find the quotient for $(x^6 - 1) \div (x + 1)$.

Solution Setting up the problem, we get:

$$\underline{-1}| \; 1 \quad 0 \quad 0 \quad 0 \quad 0 \quad 0 \quad -1$$

WARNING

Synthetic division expects $x - a$ for the divisor. If the divisor has the form $x + a$, rewrite as $x - (-a)$.

Now, the process of synthetic division appears as follows:

$$\begin{array}{r|rrrrrrr} -1 & 1 & 0 & 0 & 0 & 0 & 0 & -1 \\ & & -1 & 1 & -1 & 1 & -1 & 1 \\ \hline & 1 & -1 & 1 & -1 & 1 & -1 & 0 \end{array}$$

Thus,

$$(x^6 - 1) \div (x + 1) = x^5 - x^4 + x^3 - x^2 + x - 1$$ ■

Using the Graphics Calculator

To test our work in Example 5 for errors, type `(3X³−2X²−4)/(X−3)` for `Y1`. Type `3X²+7X+21+59/(X−3)` for `Y2`. Return to the home screen. Store a number, say `5`, in memory location `X`. Display the resulting values stored in `Y1` and `Y2`. Is the same number output in both cases? If not, an error is present!

To test our work in Example 6 for errors, type `(X^6−1)/(X+1)` for `Y1`. Type `X^5−X^4+X³−X²+X−1` for `Y2`. Return to the home screen. Store a number, say `7`, in memory location `X`. Display the resulting values stored in `Y1` and `Y2`. Is the same number output in both cases? If not, an error is present!

GIVE IT A TRY

Use synthetic division to report the quotient for each division.

11. $(2x^3 - x - 4) \div (x + 3)$ **12.** $(x^4 - 1) \div (x - 1)$ **13.** $(x^3 + 1) \div (x - 1)$

14. Use the graphics calculator to check your work in Problems 11–13.

■ SUMMARY

In this section we have learned how to do the binary operation of division on polynomials. The algorithm for division of polynomials resembles closely the arithmetic algorithm for the division of whole numbers. To divide a polynomial (the dividend) by a polynomial (the divisor), we use the long division algorithm. The process is basically a repetition of a three-step process: (1) divide a monomial (the first term of the divisor) into a monomial (the first term of the dividend); (2) multiply a monomial times the divisor, and (3) subtract the polynomials. The process is repeated until the subtraction step produces a polynomial of degree less than the divisor. This last polynomial is known as the remainder. The divisor times the quotient plus the remainder must equal the dividend.

■ ***Quick Index***

Two applications of division of polynomials were discussed—its use in factoring and in solving equations. If we know one factor of a polynomial, division will yield the other factor. Likewise, if we know one number, say a, that solves a polynomial equation, then the binomial $(x - a)$ is a factor of the left-hand side of the equation provided the equation is in standard form, a standard polynomial equal to 0. We can use division to find the other factor. Finally, we introduced synthetic division, a shortcut to the long division algorithm. Synthetic division works only for first-degree binomial divisors.

5.8 ■ ■ ■ EXERCISES

1. To check that the division
$$(3x^3 - 4x^2 + 2x - 1) \div (x - 3)$$
yields $\quad 3x^2 + 5x + 17 + \dfrac{40}{x - 3}$
find the product of -3 (the number term of the divisor) and 17 (the number term of the quotient), then add the remainder, 40. The result should be the number term of the dividend, -1. Use this procedure to show that the above division is incorrect. What is the correct remainder?

2. To use the graphics calculator to check that the division
$$(3x^3 - 4x^2 + 2x - 1) \div (x - 3)$$
yields $\quad 3x^2 + 5x + 17 + \dfrac{40}{x - 3}$
type `(3X³-4X²+2X-1)/(X-3)` for `Y1`. Type the quotient, `3X²+5X+17+40/(X-3)` for `Y2`. Store a number say, `13`, in memory `X`. Report the number output for `Y1`. Report the number output for `Y2`. Is the correct quotient and remainder listed? Why or why not?

Report the quotient and the remainder for each division. Use the graphics calculator to check your work.

3. $(2x^3 - x^2 + 2) \div (2x)$
4. $(x^4 - x^2 + 3) \div (x^2 - 1)$
5. $(x^3 - 1) \div (x - 1)$
6. $(x^4 - 3x^2 - 3) \div (x^2 - 2)$
7. $(2x^3 - x - 5) \div x$
8. $(-3x^5 - x + 2) \div x^3$
9. $(-3x^5 - x + 2) \div (x^3 - 1)$
10. $(27x^3 - 8) \div (3x - 2)$
11. $(x^3 - 2x + 5) \div (x + 5)$

For the given factor, report the other *factor of each polynomial.*

	Polynomial	*Given factor*
12.	$x^3 - 4x^2 + 9x - 10$	$x - 2$
13.	$x^3 - 5x - 2$	$x^2 - 2x - 1$
14.	$x^4 - 81$	$x - 3$
15.	$x^6 + 5x^3 - 6$	$x^3 - 1$

Using the given factor, completely factor each polynomial.

	Polynomial	*Given factor*
16.	$x^3 - 8x^2 + 19x - 12$	$x - 3$
17.	$x^3 + 3x^2 - 5x - 4$	$x + 4$
18.	$3x^3 - 6x^2 + 2x - 4$	$x - 2$
19.	$4x^4 - 25x^2 + 36$	$2x + 3$

20. Given that 2 is in the solution to the equation
$$x^3 - 3x^2 - 4x + 12 = 0$$
then $x - 2$ is a factor of the polynomial
$$x^3 - 3x^2 - 4x + 12$$
   a. Find the other binomial factors of
$$x^3 - 3x^2 - 4x + 12$$
   b. Report the three numbers that solve
$$x^3 - 3x^2 - 4x + 12 = 0$$

Using the given number(s) in the solution, solve each equation.

	Equation	*Solution contains:*
21.	$x^3 = 5x^2 - 2x - 8$	2
22.	$x^3 + 3x^2 = 6x + 8$	-4
23.	$x^4 + x^3 = 11x^2 + 9x - 18$	1 and -3

Use synthetic division to find each quotient.

24. $(x^3 - 3x^2 + 5) \div (x - 2)$
25. $(x^2 - 2x - 1) \div (x + 5)$
26. $(x^4 - 3x^2) \div (x + 1)$
27. $(x^3 - 3x^2 + 2x - 1) \div (x - 1)$
28. $(x^6 + x^4 + 2x^2 - 1) \div (x - 3)$

CHAPTER 5 REVIEW EXERCISES

Insert true *for a true statement, or* false *for a false statement.*

____ 1. The product of two polynomials is always a polynomial.

____ 2. The product of $(a + b)$ and $(a - b)$ is $a^2 - b^2$.

____ 3. The product of two binomials can be a trinomial.

____ 4. If two binomials have the same degree, their product will be a quadrinomial, a four-term polynomial.

____ 5. The square of $(2x - 3)$ is $4x^2 - 9$.

____ 6. The GCF of $18x^2 + 6xy + 12xy^2$ is $2xy$.

____ 7. If the GCF of a polynomial is 1, then the polynomial is prime.

____ 8. The solution to a second-degree equation in one variable is typically two numbers.

____ 9. The solution to $(x - 3)^2 = 0$ is a single number, -3.

____ 10. The solution to $(x - 3)(x + 2) = 0$ is the numbers -3 and 2.

11. Which of the following is a first-degree equation in one variable?
    i. $3x - 5 = 17$ ii. $3x - 5y = 17$
    iii. $3x^2 - x = 2$ iv. $|2x - 7| = x - 1$

12. Which of the following is a second-degree equation in one variable?
    i. $3x - 5 = 17$ ii. $3x - 5y = 17$
    iii. $3x^2 - x = 2$ iv. $|2x - 7| = x - 1$

13. Which of the following is a first-degree equation in two variables?
    i. $3x - 5 = 17$ ii. $3x - 5y = 17$
    iii. $3x^2 - x = 2$ iv. $|2x - 7| = x - 1$

14. Which of the following is *not* a polynomial equation?
    i. $3x - 5 = 17$ ii. $3x - 5y = 17$
    iii. $3x^2 - x = 2$ iv. $|2x - 7| = x - 1$

Write each product as a standard polynomial. Use the graphics calculator to check your work.

15. $\left(\frac{-2}{3}x^3y^2\right)^2$
16. $\left(\frac{-1}{2}s^2r^4\right)^3$
17. $(3x^2)(x^2 - 3x - 2)$
18. $(-x)(2x^3 - 3x^2 + 5)$
19. $(3x^2 + 1)(2x - 3)$
20. $(-2x + 5)(x^2 - 1)$
21. $(4x^2 + 5)(2x^2 - 1)$
22. $(-2x^3 + 1)(x^3 + 3)$
23. $(x + 5)(x - 1)$
24. $(x - 5)(x - 1)$
25. $(3x - 5)(x - 1)$
26. $(3x + 5)(x - 1)$
27. $(2x + 3)(2x - 3)$
28. $(7x - 5)(7x + 5)$
29. $(x^2 + 5)(x^2 - 5)$
30. $(x^2 - 6)(x^2 + 6)$
31. $(2x - 5)^2$
32. $(-3u + 1)^2$
33. $(3x + 1)^2$
34. $(2y + 4)^2$
35. $(x + 3)(x^2 - 3x + 9)$
36. $(x - 3)(x^2 + 3x + 9)$
37. $3x - [5 - (x - 2)(x + 2) - (2x - 3)^2]$
38. $x^2 - [2x - (3 - x)^2 + (2x + 3)(2x - 3)]$
39. $(x - 1)^2 - [(3 - x)^2 + (x + 3)(x - 3)]$
40. $(z + 2)^2 - [(z - 2)^2 + (2z + 1)(2z - 1)]$

Use the GCF to completely (prime) factor each polynomial. Use multiplication or the graphics calculator to check your work.

41. $3x^2 + 9x^3$
42. $-6x^3 + 15x^2 + x$
43. $5xy^2 - 20x^2y$
44. $r^2t^2 - rt^2 + r^2t$
45. $8m^4x^5 - 12m^2x^3$
46. $28n^2z^3 + 12n^2z^2$
47. $6x^3 - 30x^2 - 15x$
48. $2x^4 + 2x^3 + 20x$
49. $3y^3x^2 - 8y^2x^3$
50. $4ab^3 - 8a^2b - 8b$

Completely (prime) factor each quadrinomial. If the polynomial does not factor, report prime.

51. $x^3 - x^2 - 5x + 5$
52. $x^3 + x^2 + 5x + 5$
53. $x^3 - 6x^2 - 2x + 12$
54. $x^3 + 4x^2 - 2x - 8$
55. $x^3 - 3x^2 + 3x - 9$
56. $x^3 + 2x^2 - 3x - 6$

57. $x^4 + x^3 + x^2 + 1$

58. $3p^3 - 6p^2 + 3p - 6$

59. $x^4 - x^3 - x^2 + 1$

60. $p^3 - p^2 + p - 6$

Completely (prime) factor each polynomial. If polynomial does not factor, report prime.

61. $x^2 - 3x - 10$

62. $x^2 - 3x - 18$

63. $6 - 5x - x^2$

64. $7 - 6x - x^2$

65. $5x^2 - 16x + 3$

66. $4x^2 - 7x - 2$

67. $3x^2 - 8x - 3$

68. $2x^2 - 3x - 5$

69. $14y^2 - y - 3$

70. $5 - 3t - 2t^2$

71. $12x^2 - 21x - 6$

72. $12x^2 - 22x + 6$

73. $2t^2 - 13t - 7$

74. $2c^2 - 11c + 5$

75. $49x^2 + 28x + 4$

76. $36x^2 - 60x + 25$

77. $81x^2 - 18x + 1$

78. $100x^2 + 60x + 9$

79. $3x^2 - x + 7$

80. $81x^2 + 9$

81. $16x^2 - 81$

82. $16x^2 - 4$

83. $8z^3 - 27$

84. $27c^3 + 125$

85. $(x - 3)^2 - 2(x - 3) - 3$

86. $(2x - 1)^2 - 25$

87. $x^4 - 6x^2 - 7$

88. $5x^4 - 18x^2 + 9$

89. $x^3 - 2x^2 - 4x + 8$

90. $x^4 + x^3 - 8x - 8$

Solve each equation.

91. $(x - 3)(x + 7) = 0$

92. $(2x - 5)(x + 1) = 0$

93. $(3x)(x - 4) = 0$

94. $(-5x)(2x + 7) = 0$

95. $3x^2 - 6x = 0$

96. $-2x^2 - x = 0$

97. $3x = 2x^2$

98. $5x^2 = 10x$

99. $x^2 - 3x - 10 = 0$

100. $7 - 6x - x^2 = 0$

101. $x^2 - 7x = 18$

102. $x^2 + 15x = x - 13$

103. $(x - 5)(x + 7) = 13$

104. $(3x + 5)(x - 1) = 3$

105. $2x^2 = 15 - 7x$

106. $3x^2 = 10 - 13x$

107. $x^3 + 3x^2 = 4x + 12$

108. $x^4 = 3x^2 + 4$

109. Use the Pythagorean theorem (see page 56) to find the value of x in the triangle:

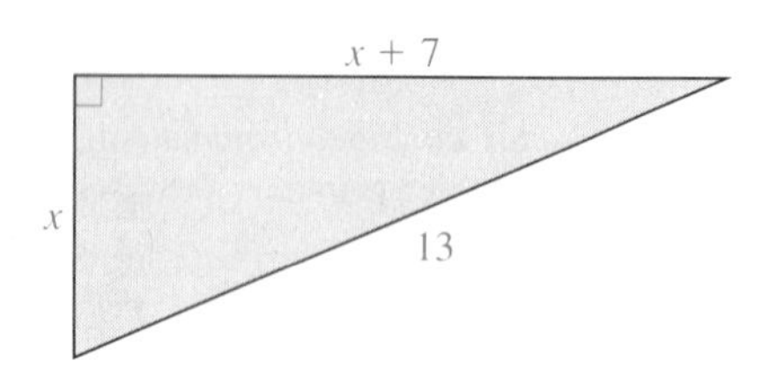

110. The length of a rectangular box is 5 cm more than twice the width of the box. If the area measure is 18 cm^2, what is the width of the box?

111. The square of the sum of a number and 2 is 9. Report all the real numbers that satisfy this condition.

Solve the given quadratic inequality. Draw a number-line graph of the solution, and report the solution in interval notation.

112. $(x - 3)(x + 2) \geq 0$

113. $(2x + 1)^2 \leq 9$

114. $x^2 - 4x > 12$

115. $2x^2 \leq x + 3$

116. The base of a triangle is 3 inches more than the altitude drawn to it. If the area measure is 35 square inches, what is the altitude of the triangle?

117. The height h (in feet) reached by a model rocket t seconds after it is launched upward is given by the formula $h = -16t^2 + 640t$. When is the height of the rocket 1500 feet?

Graph the solution to each equation. Report x-intercept(s), the y-intercept, the line of symmetry, and the vertex.

118. $y = (x - 1)(x + 3)$

119. $y = x^2 - 4x - 5$

120. $y = x^2 - 3x + 2$

121. $y = 2x^2 - x - 1$

122. $y = 4 - (x + 1)^2$

123. $y = 8 - 2x - x^2$

Graph the given quadratic function, and report the y-intercept, x-intercept(s), line of symmetry, and the vertex of the graph.

124. $g(x) = x^2 - 4x - 5$

125. $h(t) = 16 - t^2$

126. $r(n) = 5 - 4n - n^2$

127. $f(x) = 8 + 2x - x^2$

Report the quotient in each division.

128. $(x^3 - x^2 + 3) \div (x^2 - 3)$

129. $(x^4 - x^2 - 2) \div (x + 3)$

130. $(4x^5 - 2x^2 + x) \div (2x - 3)$

131. If one factor of $x^5 - 2x^3 + 8x^2 - 16$ is $x^2 - 2$, what are the other factors?

132. If 3 is one of the numbers that solves

$$x^3 - 9x = x^2 - 9$$

what are the other numbers in the solution?

Use synthetic division to find each quotient.

133. $(x^3 - 3x + 5) \div (x - 2)$

134. $(x^4 - 16) \div (x + 2)$

135. $(x^5 - 3x^2 - 2) \div (x - 4)$

CHAPTER 5 TEST

1. Write each as a standard polynomial.
   a. $\left(\frac{-1}{2}x^2y^3\right)^2$
   b. $(3x^2 - 2)(2x^2 + 1)$
   c. $(3x - 5)(3x + 5)$
   d. $(2x - 3)(4x^2 + 6x + 9)$
   e. $3x - (5 - (x + 2)(x - 3) - (x + 1)^2)$
2. Completely (prime) factor each polynomial.
   a. $2x(x - 5) - 7(x - 5)$
   b. $8x^3 + 16x^2 + 3x + 6$
   c. $2y^3 - y^2 - 15y$
   d. $4x^4 - 81$
   e. $27s^3 - 125$
   f. $6x^2 - 7x - 5$
   g. $x^3 + 3x^2 - 4x - 12$
3. Use factoring to solve each equation.
   a. $4x^2 = 20x$
   b. $(3x - 5)(x - 2) = 0$
   c. $x^2 = 5x - 6$
   d. $2x^2 = 6 - x$
   e. $x^3 + 3x^2 = 4x + 12$
4. A storage bin has a height of 48 feet. As shown in the diagram, the length of a ramp to the top of the bin is 60 feet. How far is the base of the ramp from the base of the bin?

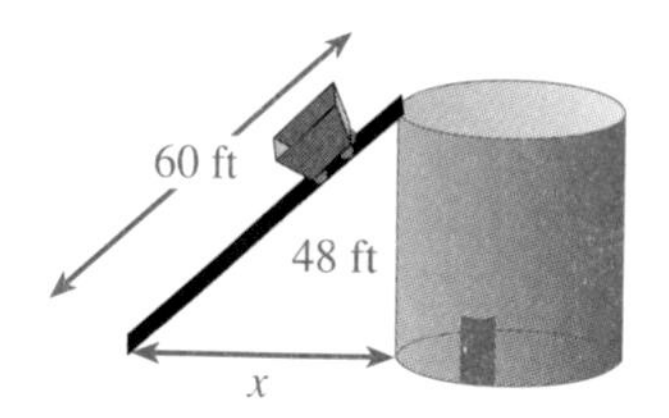

5. For $y = (2x - 5)(x + 3)$, report:
   a. the y-intercept,
   b. the x-intercept(s),
   c. the line of symmetry,
   d. the vertex, and
   e. draw a graph of the solution.
6. Consider the quadratic function $f(x) = -4x^2 + 2x + 5$.
   a. Find $f(-2)$.
   b. What two input values produce an output of 5?
7. Report the quotient of $(2x^3 - x^2 + 3) \div (x - 2)$.

6

RATIONAL EXPRESSIONS

Just as the whole numbers are the heart of arithmetic, polynomials are the heart of algebra. So far we have used the four binary operations—addition, subtraction, multiplication, and division—on polynomials. Using these operations and some basic properties of equations, we have solved polynomial equations, in both one and two variables. We now consider rational expressions—ratios of polynomials. After learning to name rational expressions, we will add, subtract, multiply, and divide rational expressions. Using these binary operations, we can solve rational expression equations in one variable, and then apply this equation-solving skill to real-world situations. In the last section of this chapter we discuss integer exponents.

6.1 ■ ■ ■ COMPLETELY REDUCED RATIONAL EXPRESSIONS

A polynomial is an expression that either is a term or is the sum or difference of a finite number of terms. For example, $3x^2 - 5x + 2$ and $5x^2y^3$ are polynomials. A **rational expression** is an expression of the form $\frac{P}{Q}$, where P and Q are polynomials with $Q \neq 0$. Examples of rational expressions are listed here. Note that the numerators and denominators are all standard polynomials:

$$\frac{-6}{x+1} \qquad \frac{1}{2w} \qquad \frac{x^2-3x+6}{x^3+1} \qquad \text{and} \qquad \frac{x^5-3y^2}{2x-y}$$

Using 1 for Q, every polynomial can be written as a rational expression. For example, the polynomial $x^2 - 2$ can be written as $\frac{x^2-2}{1}$. Thus, the set of rational expressions contains the set of polynomials (much like the set of rational numbers contains the set of integers). Another parallel from arithmetic is the operation of division. Just as division of whole numbers may result in a fraction, division of polynomials may result in a rational expression:

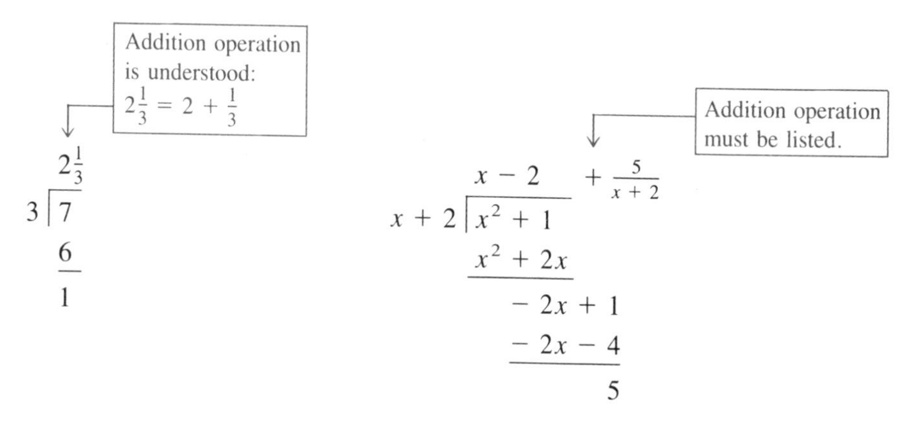

For $(x^2 + 1) \div (x + 2)$ the divisor, $x + 2$, is known as the **denominator**, and the dividend, $x^2 + 1$, is the **numerator**:

$$\frac{x^2+1}{x+2} \quad \begin{matrix} \leftarrow \text{numerator} \\ \leftarrow \text{denominator} \end{matrix}$$

Domain of Definition One important difference between polynomials and rational expressions concerns the replacement of the variable by a number. When the variables in a polynomial are replaced by numbers, the result is a number. This is not always the case with rational expressions. Consider the rational expression $\frac{x}{x-3}$. If the variable x is replaced by 3, then the result is $\frac{3}{0}$, which is undefined (meaningless). Thus, we often list a *restriction set* for rational expressions, that is we write

$$\frac{x}{x-3} \quad \text{for } x \neq 3$$

This means that the variable can be replaced by any number *except* 3. We say the **domain of definition** is $\{x \mid x \neq 3\}$. Restriction sets are important in equation-solving, and they will be essential when we discuss rational expression inequalities and rational functions. To learn how to find the domain of definition, work through the following example.

WARNING

For $\dfrac{x}{x-3}$ replacing x with 3 yields $\frac{3}{0}$, a meaningless expression, but replacing x with 0 yields $\frac{0}{-3}$, which is 0.

EXAMPLE 1 Report the domain of definition for the given rational expression.

a. $\dfrac{7x - 4}{2x - 5}$ b. $\dfrac{x - 4}{x^2 - 3x - 4}$

Solution We set the denominator equal to 0, and then solve the resulting equation.

a.

$$\begin{aligned} 2x - 5 &= 0 && \text{Set denominator equal to 0.} \\ \underline{\quad 5} &\quad \underline{5} && \text{Add 5 to each side.} \\ 2x &= 5 && \text{Multiply by } \tfrac{1}{2}. \\ x &= \frac{5}{2} \end{aligned}$$

Thus, the domain of definition for $\frac{7x - 4}{2x - 5}$ is $\left\{x \mid x \neq \frac{5}{2}\right\}$.

b.

$$\begin{aligned} x^2 - 3x - 4 &= 0 && \text{Set denominator equal to 0.} \\ (x - 4)(x + 1) &= 0 && \text{Factored form} \end{aligned}$$

$$x - 4 = 0 \quad \text{or} \quad x + 1 = 0$$
$$x = 4 \qquad\qquad x = -1$$

The solution is 4 and -1. Thus, the domain of definition for $\frac{x - 4}{x^2 - 3x - 4}$ is $\{x \mid x \neq 4, x \neq -1\}$. ■

Using the Graphics Calculator

```
RANGE
Xmin=-9.6
Xmax=9.4
Xscl=1
Ymin=-6.4
Ymax=6.2
Yscl=1
Xres=1
```

FIGURE 1

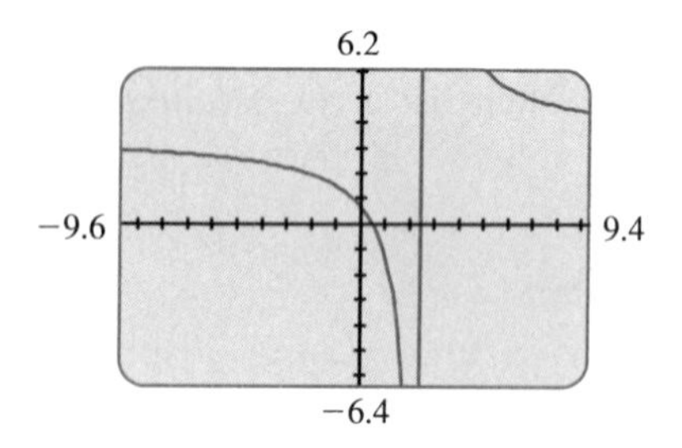

FIGURE 2

To visualize the domain of definition for the rational expression in part (a) of Example 1, type `(7X−4)/(2X−5)` for `Y1`. Using the values in Figure 1,* view the graph screen (see Figure 2). From this point on we will refer to the `RANGE` settings in Figure 1 as the Window 9.4 range settings. Remember that, like a polynomial, a rational expression holds the place for number phrases. The curve displayed (known as a *hyperbola*) shows the evaluations of $\frac{7x - 4}{2x - 5}$ for various values of the x-variable. Press the `TRACE` key, then move the cursor until the message reads `X=2 Y=-10`. When the x-variable is replaced by 2, the rational expression evaluates to -10. Press the right arrow key twice. When the x-variable is 2.4, the rational expression evaluates to -64. Press the right arrow key again. When the x-variable is 2.6, the rational expression evaluates to 71. Something strange happens between x-values 2.4 and 2.6. The evaluations jump from -64 to 71. This strange jump occurs because the rational expression $\frac{7x - 4}{2x - 5}$ is undefined for an x-value of 2.5. The graphics calculator simply connects the dots and, by doing so, creates the line at `X=2.5` that appears vertical. Press the `MODE` key and select the `Dot` option from the `Connected Dot` line. Quit to the home screen. Now display the graph screen, the vertical line at `X=2.5` is no longer displayed.

TI-82 Note *Type `-9.4` for `Xmin` an `-6.2` for `Ymin`.

```
RANGE
Xmin=0
Xmax=4.75
Xscl=1
Ymin=-100
Ymax=100
Yscl=20
Xres=1
```

FIGURE 3

Press the RANGE key and enter the values shown in Figure 3.* Press the TRACE key and move the cursor until the message reads X=2.5 Y= . When the x-variable is replaced by 2.5, there is no y-value. Return to the home screen. Store 2.5 in memory location X. Display the resulting value in Y1. As you can see, the ERROR screen is presented. The reason, of course, is that the rational expression is undefined when the x-variable is $\frac{5}{2}$. Press the 2 key to leave the ERROR screen and return to the home screen.

GIVE IT A TRY

Report the domain of definition of each rational expression.

1. $\dfrac{x^2 - 4}{2x + 8}$ **2.** $\dfrac{2x - 1}{x^2 - 6x + 5}$ **3.** $\dfrac{2x - 1}{x^2 - 3x + 2}$ 

An Infinite Number of Names Each rational expression has an infinite number of names. In arithmetic, $\frac{1}{2}$ can be named $\frac{3}{6}, \frac{5}{10}$, and so on. In algebra, a rational expression, such as $\frac{3}{x}$, can be named in an unlimited number of ways by using the following property.

FUNDAMENTAL PROPERTY OF RATIONAL EXPRESSIONS

For $\frac{a}{b}$, where $b \neq 0$, if $c \neq 0$, then $\quad \dfrac{a}{b} = \dfrac{ac}{bc}$.

If we multiply numerator and denominator by a nonzero polynomial, then we obtain an equivalent rational expression. For example,

The multiplier c is 2.

$$\frac{3}{x} = \frac{(3)(2)}{(x)(2)} = \frac{6}{2x}$$

Thus, $\frac{6}{2x}$ is another name for $\frac{3}{x}$. Continuing this process of multiplying the numerator and denominator by $c \neq 0$, and selecting a different expression for c each time, produces an infinite number of names for a rational expression. This is useful because we often want to name a rational expression with a particular polynomial as the denominator. Consider the following example.

TI-82 Note *Type 4.7 for Xmax.

EXAMPLE 2 Rename $\frac{x-5}{x+1}$ so that the denominator is $x^2 - 3x - 4$.

Solution The key to this problem is selecting the appropriate polynomial for c. To do this, we can use factoring or division. The problem is set up as follows:

Some polynomial selected for c

$$\frac{x-5}{x+1} = \frac{(x-5)(\quad)}{(x+1)(\quad)} = \frac{\quad}{x^2 - 3x - 4}$$

The product of $x + 1$ and the polynomial selected for c

To find the polynomial to use for c, we can factor $x^2 - 3x - 4$ or we can divide it by $x + 1$. Factoring, we get $(x - 4)(x + 1)$. So, multiplying $x + 1$ by $x - 4$ yields the desired denominator. Thus, the polynomial to use for c is $x - 4$. We get:

The required polynomial for c — The product of $(x - 5)$ and $(x - 4)$

$$\frac{x-5}{x+1} = \frac{(x-5)(x-4)}{(x+1)(x-4)} = \frac{x^2 - 9x + 20}{x^2 - 3x - 4}$$

product yields the desired denominator.

We have found *one* of the infinite number of names for $\frac{x-5}{x+1}$. In particular, we have found the name whose denominator is $x^2 - 3x - 4$. ■

Using the Graphics Calculator

```
7→X
Y1
                .25
Y2
                .25
```

FIGURE 4

To say that two rational expressions, such as $\frac{x-5}{x+1}$ and $\frac{x^2 - 9x + 20}{x^2 - 3x - 4}$, both name the same object means that both of these expressions evaluate to the same number when the variable (x in this case) is replaced by a number (provided, of course, that x is not replaced by -1 or 4). Type `(X²-9X+20)/(X²-3X-4)` for `Y1` and `(X-5)/(X+1)` for `Y2`. Return to the home screen. Store a number, say `7`, in memory location `X`. Display the resulting values in `Y1` and `Y2`. See Figure 4. If these output values are not the same, then an error is present. If the results are the same, we have *not* proved that the two expressions are equivalent, but we can be more confident that our work is correct. Store several other numbers (such as `⁻8` and `5.23`) in memory location `X` and compare the values in `Y1` and `Y2` for each of these numbers. What are two numbers that produce an error when `Y1` and `Y2` are evaluated?

GIVE IT A TRY

Rename the rational expression to have the given denominator.

4. $\frac{x+3}{x} = \frac{\quad}{2x^2}$ **5.** $\frac{3x}{2x+3} = \frac{\quad}{2x^2+3x}$ **6.** $\frac{x^2-2}{x+5} = \frac{\quad}{x^2-25}$

7. Use the graphics calculator to test your work in Problems 4–6 for errors. Report the value stored in X, and the evaluations for the original rational expression and for the renamed rational expression.

A Proper Naming System As our work so far has shown, a rational expression has an infinite number of names. As with rational numbers, a proper naming system has been developed for naming rational expressions. This system requires that the rational expression be named so that the numerator and denominator have *no* common factor. Naturally, factoring is involved in properly naming a rational expression. The instruction "Write as a completely reduced rational expression." means that a rational expression (a ratio of polynomials) should be written so that the numerator and denominator have no common factor. This naming system is also based on the fundamental property of rational expressions, but here we reverse the process.

As usual, to demonstrate the naming system, we first reduce a simple rational expression, and then work with more involved rational expressions. We can restate fundamental property as "The operation of *multiplying by a polynomial* and the operation of *dividing by that same polynomial* cancel."

EXAMPLE 3 Write as a completely reduced rational expression: $\frac{15m^3y^4}{12x^2m^2}$

Solution Because both the numerator and denominator are monomials, we can quickly write the numerator and denominator in factored form:

$$\frac{(3)(5)(m^2m)y^4}{(3)(4)x^2(m^2)}$$

Using the fundamental property, we can cancel the common factors:

$$\frac{\cancel{(3)}(5)(\cancel{m^2}m)y^4}{\cancel{(3)}(4)x^2(\cancel{m^2})} = \frac{5my^4}{4x^2}$$

Thus, as a completely reduced rational expression,

$$\frac{15m^3y^4}{12x^2m^2} \quad \text{is} \quad \frac{5my^4}{4x^2} \qquad \text{Restrictions: } x \neq 0 \quad m \neq 0$$

■

The result of Example 3 means that any place the expression $\frac{15m^3y^4}{12x^2m^2}$ appears, it can be replaced by $\frac{5my^4}{4x^2}$. Also, for any numeric replacement of the variables, except $x = 0$ and $m = 0$, the expressions $\frac{15m^3y^4}{12x^2m^2}$ and $\frac{5my^4}{4x^2}$ evaluate to the same number. Note that replacing m by 0, x by 3, and y by 4 yields $\frac{0}{0} = \frac{0}{36}$. This is *not* a true arithmetic sentence! The expression $\frac{0}{36}$ is 0, but the expression $\frac{0}{0}$ has no meaning. Thus, we cannot know whether or not the expression $\frac{0}{0}$ is the same as $\frac{0}{36}$.

To increase our skill with writing rational expressions in completely reduced form, work through the following examples.

EXAMPLE 4 Write as a completely reduced rational expression: $\dfrac{x-3}{x^2-5x+6}$

Solution Again, we start by factoring:

$$\frac{x-3}{x^2-5x+6} = \frac{\cancel{x-3}}{(\cancel{x-3})(x-2)} = \frac{1}{x-2} \qquad \text{Restrictions: } x \neq 3 \quad x \neq 2$$

We include $x \neq 3$ in the restriction set, since we are excluding *any* number that results in *any* denominator evaluating to 0. ■

EXAMPLE 5 Write as a completely reduced rational expression: $\dfrac{x^2-9}{x^3-27}$

Solution Again, we start by factoring:

$$\frac{x^2-9}{x^3-27} = \frac{(x+3)(\cancel{x-3})}{(\cancel{x-3})(x^2+3x+9)} = \frac{x+3}{x^2+3x+9} \qquad \text{Restriction: } x \neq 3$$

This last expression is a rational expression (a ratio of polynomials). It is completely reduced because the numerator and denominator (the polynomials $x + 3$ and $x^2 + 3x + 9$) have no common factor. ■

WARNING

In the expression $\frac{\cancel{3}(x+3)}{\cancel{3}x}$ the slash marks mean that *multiplication by 3* and division by 3 cancel. That is, the *operations* of multiplying by 3 and dividing by 3 cancel out each other. The 3s do not cancel. A common error is to cancel a number or variable within a polynomial factor: $\frac{x-\cancel{3}}{y+\cancel{3}}$ to get $\frac{x}{y}$. Before drawing a slash mark, make sure that you have selected a factor!

Using the Graphics Calculator

We can test the results of Examples 4 and 5—the original rational expression and the completely reduced rational expression are equivalent expressions—by replacing the variable(s) with a number. For Example 4, store 5 in memory location X. For Y1 type (X−3)/(X²−5X+6) and for Y2 type 1/(X−2). From the home screen, display the resulting values in Y1 and Y2. If these values are not the same, an error is present. Now, run a similar test for Example 5. This type of test can reveal an error in our work, but it cannot indicate whether the final expression is completely reduced (numerator and denominator have no common factor).

The $a - b$ Factor The next example demonstrates that a binomial factor such as $5 - x$ can be written as $-1(x - 5)$. Because of the communative property of addition, $a + b$ can be written as $b + a$, but $a - b$ cannot be written as $b - a$ (that is, $a - b \neq b - a$). However, $a - b$ *is* equal to $-1(b - a)$.

EXAMPLE 6 Write as a completely reduced rational expression: $\dfrac{x^2 - 5x}{25 - x^2}$

Solution Factoring the numerator and denominator, we get

$$\frac{x^2 - 5x}{25 - x^2} = \frac{x(x-5)}{(5+x)(5-x)} = \frac{x\cancel{(x-5)}}{(-1)(x+5)\cancel{(x-5)}}$$

$5 - x = -1(x - 5)$

$$= \frac{x}{-1(x+5)} = \frac{-x}{x+5}$$

Restrictions: $x \neq 5 \quad x \neq -5$

We could leave the factor -1 in the denominator and report the expression as

$$\frac{x}{-1(x+5)} \quad \text{or} \quad \frac{x}{-x-5}$$

However, we usually place the factor -1 in the numerator, $\frac{-x}{x+5}$, by multiplying numerator and denominator by -1. This last expression is a rational expression (a ratio of polynomials) and is completely reduced because the numerator and denominator have no common factor. ■

REMEMBER

For $\frac{a}{b}$, with $b \neq 0$,

$$\frac{a}{-b} = \frac{-a}{b} = -\frac{a}{b}$$

GIVE IT A TRY

Write each expression as a completely reduced rational expression.

8. $\dfrac{6w^3y^4}{12x^2w^2}$

9. $\dfrac{49 - x^2}{x^2 - 6x - 7}$

10. $\dfrac{8x^3 - 1}{2x^2 - x}$

11. $\dfrac{2x^2 - 10x^3}{25x^2 - 1}$

■ SUMMARY

In this section we have introduced some important topics. A rational expression is a ratio of polynomials. For example, $\frac{3}{x+5}$ is a rational expression—the ratio of the polynomial 3, the numerator, to the polynomial $x + 5$, the denominator. Just as fractions result naturally from the division of whole numbers, rational expressions result from the division of polynomials.

Unlike polynomials, which always evaluate to a number when the variables are replaced with numbers, a rational expression can result in a undefined expression when the variables are replaced with numbers. Because division by 0 is undefined, the domain must exclude any number that results in the

■ ***Quick Index***

denominator of a rational expression evaluating to zero. For $\frac{2x}{x+2}$, the restriction is $x \neq -2$, which indicates that the variable cannot be replaced by the number -2. We say that the domain of definition is $\{x \mid x \neq -2\}$.

Each rational expression has an infinite number of names. A new name can be created by multiplying the numerator and denominator by the same polynomial. We find this polynomial multiplier either by factoring or by using long division. In our later work with the operations of addition and subtraction, we will often rename a rational expression to have a particular denominator.

A proper (completely reduced) naming system for rational expressions requires that the rational expression be written so that the numerator and denominator have no common factor. This is accomplished by using the fact that the operations of multiplication by a polynomial and division by the same polynomial cancel. The polynomials do not cancel, it is the operations that cancel.

The rational expression $\frac{a-b}{b-a}$ reduces to $\frac{-1}{1}$, or -1. The factor $b - a$ can be written as $-1(a - b)$. Thus,

$$\frac{a-b}{b-a} = \frac{\cancel{a-b}}{-1(\cancel{a-b})} = \frac{1}{-1} = \frac{-1}{1} = -1 \qquad \text{Restriction: } a \neq b$$

6.1 ■ ■ ■ EXERCISES

Report the domain of definition for the given rational expression.

1. $\frac{x-3}{x}$
2. $\frac{y^2}{y-2}$
3. $\frac{s}{s^2-4}$
4. $\frac{2x^2+3x-2}{x^2+5x+6}$
5. $\frac{r^2-5r+14}{r^2+14r+49}$
6. $\frac{x^2-10x+16}{x^2-6x+9}$
7. $\frac{a^2-6-a}{a^2-9}$
8. $\frac{25-y^2}{y^3-125}$
9. $\frac{3x^2+2x-8}{8-10x+3x^2}$
10. $\frac{x^2-3x-4}{5-4x-x^2}$
11. $\frac{V^2-D^2}{V^2-V}$
12. $\frac{P^2-T^2}{PM-PM^2}$

Give three different names for each rational expression.

13. $\frac{x}{y+x}$
14. $\frac{x+1}{x}$
15. $\frac{3}{y^2-x^2}$
16. $\frac{2x}{4-x^2}$

Use the graphics calculator to test that the given expressions are equivalent rational expressions. Report the number stored in memory location X, and the numbers output.

17. $\frac{3x^2+2x-8}{8-10x+3x^2}$ and $\frac{4x+4}{3x-4}$
18. $\frac{x^3+5x}{x^2-25x}$ and $\frac{x+5}{1+x}$

Write a rational expression with the indicated denominator.

19. $\frac{x-3}{x} = \frac{\quad}{2x^2-4x}$
20. $\frac{r+5}{3r} = \frac{\quad}{6r^2-9r}$
21. $\frac{y^2}{y+2} = \frac{\quad}{y^2-2y-8}$
22. $\frac{d^2}{d-7} = \frac{\quad}{d^2-5d-14}$
23. $\frac{2x+3}{x-2} = \frac{\quad}{x^2-4}$
24. $\frac{x+5}{x+3} = \frac{\quad}{x^2+3x}$
25. $\frac{x+1}{x-4} = \frac{\quad}{16-x^2}$
26. $\frac{b+1}{b-1} = \frac{\quad}{b^3-1}$
27. $\frac{x^2-3x+5}{x^2+x+1} = \frac{\quad}{x^3-1}$

28. $\frac{2}{x+3} = \frac{}{x^5 + 3x^4 - 3x^3 - 9x^2 + 2x + 6}$

[*Hint:* Use division algorithm.]

Write the given expression as a completely reduced rational expression. Use the graphics calculator to test your work for errors.

29. $\frac{5m^3}{10m}$

30. $\frac{-12s^6}{18s^2}$

31. $\frac{-28dx^4}{14d^3x}$

32. $\frac{5mx^2}{15m^2x}$

33. $\frac{6x+3}{3x}$

34. $\frac{5x^2 + 10x}{15xy}$

35. $\frac{12 - 4h}{16h^2}$

36. $\frac{x^2 + 16}{16 - x^2}$

37. $\frac{8s^2 + 12s}{24su}$

38. $\frac{r^2 + 5r - 14}{r^2 + 14r + 49}$

39. $\frac{3x^2 + 2x - 8}{8 - 10x + 3x^2}$

40. $\frac{4x^2 - 11x - 3}{2 + 7x - 4x^2}$

41. $\frac{4 - x^2}{x^2 - 4x - 5}$

42. $\frac{x^2y^2(x^2 - y^2)}{yx^2 - y^3}$

43. $\frac{x^2y^2(x - y)}{xy(x^2 - y^2)}$

44. $\frac{8 - x^3}{2x - 4}$

6.2 ■ ■ ■ MULTIPLICATION AND DIVISION OF RATIONAL EXPRESSIONS

In the last section, we learned how to write a rational expression as a completely reduced rational expression. In this section we will see the binary operations of multiplication and division with rational expressions. With these two operations, the algorithms (procedures) are straightforward. The difficulty with multiplication (or division) is in achieving completely reduced form (removing common factors from the numerator and denominator).

Multiplication The model for multiplication of rational expressions is multiplication of fractions (or rational numbers) from arithmetic. Using this model, we have the following definition.

MULTIPLICATION OF RATIONAL EXPRESSIONS

For $\frac{a}{b}$ and $\frac{c}{d}$, with $b \neq 0$, $d \neq 0$,

$$\left(\frac{a}{b}\right)\left(\frac{c}{d}\right) = \frac{ac}{bd}$$

In arithmetic, we write multiplication of fractions as

$$\left(\frac{2}{3}\right)\left(\frac{5}{8}\right) = \frac{(2)(5)}{(3)(8)}$$

The product

$$\frac{(2)(5)}{(3)(8)} \quad \text{reduces to} \quad \frac{(\not{2})(5)}{(3)(\not{2})(4)}$$

Thus, $\left(\frac{2}{3}\right)\left(\frac{5}{8}\right) = \frac{5}{12}$.

The operation of multiplication is simple—multiply the numerators, multiply the denominators. The difficulty is in removing the common factors. Consider the following examples.

EXAMPLE 1 Write as a completely reduced rational expression: $\frac{5m^2y}{2x} \cdot \frac{10x^3y}{15m^2}$

Solution We multiply the numerators, multiply the denominators, and then remove the common factor.

$$\frac{5m^2y}{2x} \cdot \frac{10x^3y}{15m^2} = \frac{50m^2x^3y^2}{30m^2x} \qquad \left(\frac{a}{b}\right)\left(\frac{c}{d}\right) = \frac{ac}{bd}$$

$$= \frac{5\cancel{(10)}\cancel{m^2}\cancel{(x)}(x^2)y^2}{3\cancel{(10)}\cancel{m^2}\cancel{x}} \qquad \text{Remove common factors.}$$

$$= \frac{5x^2y^2}{3} \qquad \text{Restrictions: } x \neq 0 \quad m \neq 0$$

The result, $\frac{5x^2y^2}{3}$, can be written as the polynomial $\frac{5}{3}x^2y^2$. ■

EXAMPLE 2 Write as a completely reduced rational expression:

$$\frac{x^2 - 4}{x^2 + 3x + 2} \cdot \frac{x^2 - 5x + 4}{x^2 - 6x + 8}$$

Solution Rather than multiplying the numerators (and denominators), we can factor each numerator (denominator), and write the indicated products.

> **REMEMBER**
> A completely reduced rational expression is a ratio (quotient) of polynomials in which the numerator and denominator have no common factor.

$$\frac{x^2 - 4}{x^2 + 3x + 2} \cdot \frac{x^2 - 5x + 4}{x^2 - 6x + 8} = \frac{\cancel{(x-2)}\cancel{(x+2)}\cancel{(x-4)}(x-1)}{\cancel{(x+2)}(x+1)\cancel{(x-4)}\cancel{(x-2)}}$$

$$= \frac{x-1}{x+1} \qquad \text{Restrictions: } x \neq -2,\ x \neq -1,\ x \neq 4,\ x \neq 2$$ ■

The result of Example 2 has two meanings:

1. Any place the expression

$$\frac{x^2 - 4}{x^2 + 3x + 2} \cdot \frac{x^2 - 5x + 4}{x^2 - 6x + 8}$$

appears, it can be replaced by $\frac{x-1}{x+1}$.

2. For any replacement of x (except -2, -1, 4, or 2), the expressions

$$\frac{x^2 - 4}{x^2 + 3x + 2} \cdot \frac{x^2 - 5x + 4}{x^2 - 6x + 8} \quad \text{and} \quad \frac{x-1}{x+1}$$

evaluate to the same number.

If we replace x by a number, say 3, we can check the result. If our result is correct, then

$$\frac{9-4}{9+9+2}\cdot\frac{9-15+4}{9-18+8} \quad \text{evaluates to} \quad \frac{3-1}{3+1}$$

Try it.

Using the Graphics Calculator

To test our work in Example 2 for errors, type

```
(X²−4)/(X²+3X+2)*(X²−5X+4)/(X²−6X+8)
```

for `Y1`. Type `(X−1)/(X+1)` for `Y2`. Store a number, say 3, in memory location `X`. Display the resulting values stored in `Y1` and `Y2`. If these values are not the same, then an error is present.

GIVE IT A TRY

Write each expression as a completely reduced rational expression.

1. $\dfrac{-3c^2y}{12d}\cdot\dfrac{16d^3y}{c^3}$
2. $\dfrac{x-5}{x+2}\cdot\dfrac{x^2+2x}{x+5}$
3. $\dfrac{5x}{x^2-25}\cdot\dfrac{5-x}{30x^2}$
4. $\dfrac{x^2-36}{x^2-25}\cdot\dfrac{x^2-6x+5}{x^2-5x-6}$

Division The division of rational expressions is also modeled after the division of fractions (rational numbers). In arithmetic, we write

Change ÷ operation to multiplication.

$$\frac{2}{3}\div\frac{4}{5}=\left(\frac{2}{3}\right)\left(\frac{5}{4}\right)$$

Change divisor to its reciprocal.

Now, doing the multiplication, we get:

$$\frac{2}{3}\div\frac{4}{5}=\left(\frac{2}{3}\right)\left(\frac{5}{4}\right)=\frac{(\cancel{2})(5)}{(3)(2)(\cancel{2})}=\frac{5}{6}$$

The division operation is reworded as multiplication. Division is defined as follows.

DIVISION OF RATIONAL EXPRESSIONS

For $\frac{a}{b}$ and $\frac{c}{d}$, with $b \neq 0$, $c \neq 0$, and $d \neq 0$,

$$\frac{a}{b}\div\frac{c}{d}=\left(\frac{a}{b}\right)\left(\frac{d}{c}\right)$$

Division has its parallel in subtraction. Subtraction is defined in terms of addition, that is, $a - b$ means: $a +$ (opposite of b), or $a + (-b)$. In division, $\left(\frac{a}{b}\right) \div \left(\frac{c}{d}\right)$ means

$$\left(\frac{a}{b}\right) \cdot \left(\text{reciprocal of } \frac{c}{d}\right) \quad \text{or} \quad \left(\frac{a}{b}\right) \cdot \left(\frac{d}{c}\right)$$

Consider the following examples.

EXAMPLE 3 Write as a completely reduced rational expression: $\dfrac{6m^3y}{5x} \div \dfrac{8my^3}{10x^2}$

Solution Change ÷ operation to multiplication.

$$\frac{6m^3y}{5x} \div \frac{8my^3}{10x^2} = \frac{6m^3y}{5x} \cdot \frac{10x^2}{8my^3}$$

Change divisor to its reciprocal.

$$= \frac{60m^3x^2y}{40mxy^3} \quad \text{Perform multplications.}$$

$$= \frac{\cancel{2} \cdot \cancel{2} \cdot 3 \cdot \cancel{5}\ \cancel{m}mm\cancel{x}x\cancel{y}}{\cancel{2} \cdot \cancel{2} \cdot 2 \cdot \cancel{5}\ \cancel{m}\cancel{x}\cancel{y}yy}$$

$$= \frac{3m^2x}{2y^2} \quad \text{Restrictions: } x \neq 0,\ y \neq 0,\ m \neq 0$$

■

REMEMBER
In algebra, both the dot · and parentheses are used to indicate multiplication. That is,

$$\begin{aligned} ab &= a \cdot b \\ &= (a)(b) \\ &= (a) \cdot (b) \end{aligned}$$

After we have rewritten the divisor as its reciprocal and changed the operation to multiplication, the problem becomes a multiplication of rational expressions problem, much like those in Examples 1 and 2.

EXAMPLE 4 Write as a completely reduced rational expression: $\dfrac{x^2 - 1}{x^2 - 3x} \div \dfrac{x^3 - 1}{2x}$

Solution

$$\frac{x^2 - 1}{x^2 - 3x} \div \frac{x^3 - 1}{2x} = \frac{x^2 - 1}{x^2 - 3x} \cdot \frac{2x}{x^3 - 1} \qquad \frac{a}{b} \div \frac{c}{d} = \frac{a}{b} \cdot \frac{d}{c}$$

$$= \frac{(x + 1)\cancel{(x - 1)}(2)\cancel{(x)}}{\cancel{(x)}(x - 3)\cancel{(x - 1)}(x^2 + x + 1)} \qquad \frac{a}{b} \cdot \frac{d}{c} = \frac{ad}{bc}$$

$$= \frac{2x + 2}{(x - 3)(x^2 + x + 1)} \qquad \text{Restrictions: } x \neq 0,\ x \neq 3,\ x \neq 1$$

■

In Example 4, we could write the denominator

$$(x - 3)(x^2 + x + 1) \quad \text{as} \quad x^3 - 2x^2 - 2x - 3$$

Traditionally, the denominator is left in factored form so that the domain of definition can be quickly observed (in most cases).

GIVE IT A TRY

Write each as a completely reduced rational expression.

5. $\dfrac{8my^3}{10x^2} \div \dfrac{6m^3y}{5x}$ **6.** $\dfrac{x^3 - 1}{2x} \div \dfrac{x^2 - 1}{x^2 - 3x}$

7. $\dfrac{x^2 - 1}{x^2 + 3x} \div \dfrac{2x - 2}{(x^2 + 4x + 3)}$ **8.** $\dfrac{x^2 - 4x - 5}{x^2 - 5x - 6} \div \dfrac{x^2 - 25}{x^2 - 36}$

Multiplication by 1 In Section 6.1, we renamed rational expressions using the fundamental property of rational expressions. Now that we know how to multiply rational expressions, we can multiply by 1 to rename rational expressions.

FUNDAMENTAL PROPERTY OF RATIONAL EXPRESSIONS AND MULTIPLICATION BY 1

For $\dfrac{a}{b}$, with $b \neq 0$,

$$\frac{a}{b} = \left(\frac{a}{b}\right)\left(\frac{c}{c}\right) = \frac{ac}{bc} \quad \text{provided } c \neq 0.$$

We can use this property to change the name of a rational expression by multiplying by a form of 1, $\frac{c}{c}$ with $c \neq 0$. Consider the following example.

EXAMPLE 5 Rename the given rational expression with the indicated denominator:

$$\frac{x + 3}{x - 2} = \frac{\quad}{x^2 - 4}$$

Solution Rather than multiplying the numerator and denominator by a polynomial (as we did in Section 6.1), we will multiply by a form of 1. We select the form of 1 by factoring the desired denominator, $x^2 - 4$. This polynomial factors as $(x + 2)(x - 2)$. Thus, we use $\frac{x + 2}{x + 2}$ as the form of 1, and get:

a form of 1

$$\frac{x + 3}{x - 2} = \frac{x + 3}{x - 2} \cdot \frac{x + 2}{x + 2} = \frac{x^2 + 5x + 6}{x^2 - 4}$$

■

GIVE IT A TRY

Multiply by a form of 1 to rename the given rational expression to a rational expression with the indicated denominator.

9. $\dfrac{-1}{x - 2} = \dfrac{\quad}{x^2 - 4x + 4}$ **10.** $\dfrac{x}{2x - 3} = \dfrac{\quad}{8x^3 - 27}$

We can gain greater understanding of rational expressions by using the graph screen. To check our work in Example 5, type `(X+3)/(X−2)` for `Y1`. Enter the Window 9.4 `RANGE` settings, and view the graph screen, as shown in Figure 1. Evaluations of the expression $\frac{x+3}{x-2}$ for various values of the x-variable are displayed. Press the `TRACE` key. The message `X=0 Y=-1.5` means that when the x-variable is replaced by `0`, the rational expression evaluates to `-1.5`. Press the right arrow key until the message reads `X=1.8 Y=-24`. For replacements of the x-variables with numbers close to 2 (but still less than 2), the value of the rational expression $\frac{x+3}{x-2}$ gets smaller and smaller.

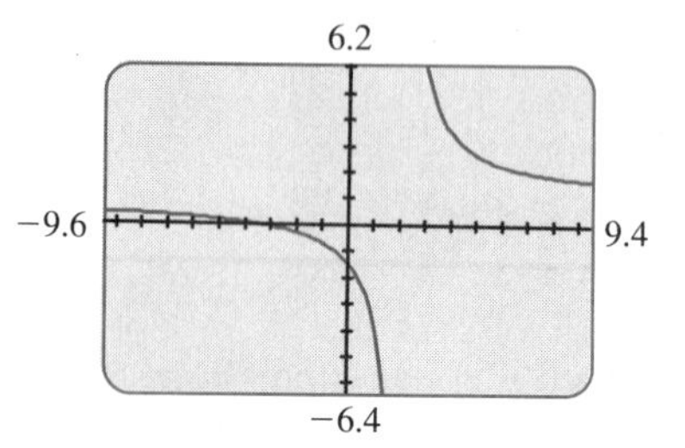

FIGURE 1

Press the `RANGE` key and enter the values shown in Figure 2.* View the graph screen, as shown in Figure 3. Press the `TRACE` key and move the cursor until the message reads `X=2 Y=` . When the x-variable is replaced by `2`, the calculator does not know how to compute a value for the rational expression. That is, the rational expression is undefined for $x = 2$. The domain of definition for this rational expression is $\{x \mid x \neq 2\}$. Press the right arrow key. The message now reads `X=2.01 Y=501`. For values of x close to 2 (but still greater than 2), the rational expression evaluates to a larger and larger number. See the table of values in the margin. The vertical line $x = 2$ is known as a *vertical asymptote*.

```
RANGE
Xmin=1.5
Xmax=2.45
Xscl=1
Ymin=-100
Ymax=100
Yscl=10
Xres=1
```

FIGURE 2

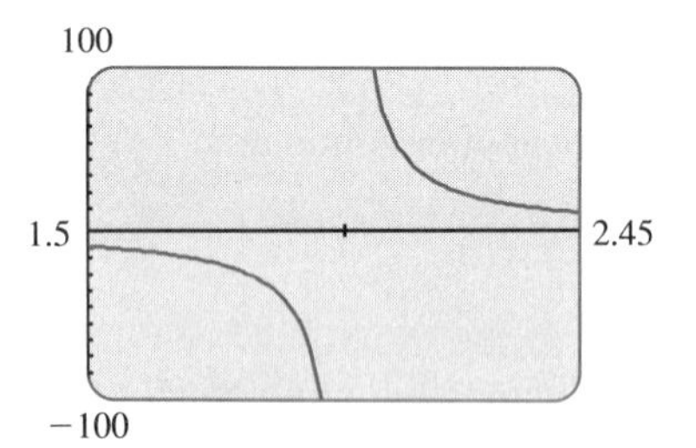

FIGURE 3

x-value	value of $\frac{x+3}{x-2}$
1.8	−24
1.9	−49
1.99	−499
2	undefined
2.01	501
2.1	51
2.2	26

Press the `Y=` key and, for `Y2`, type the equivalent rational expression `(X²+5X+6)/(X²−4)`. This is the rational expression we obtained in Example 5. Using the Window 9.4 range settings, and view the graph screen. The graph for `Y1`, `(X+3)/(X−2)`, is drawn. Next, the graph for `Y2`, `(X²+5X+6)/(X²−4)`, is drawn. Although we cannot see any difference between the two curves drawn, they are different. These two expressions are equivalent—they evaluate to the same value for each replacement of the x-variable except for 2 and -2. To see that -2 is also in the restriction set, press the `TRACE` key, and then the left arrow key, until the message reads `X=-2 Y-.25`. When x is replaced by -2, the `Y1` expression evaluates to -0.25. Press the down arrow key to move the cursor to the curve for the `Y2` expression. The message now reads `X=2 Y=` . When the x-variable is replaced by -2, the expression $\frac{x^2+5x+6}{x^2-4}$ is undefined. Thus we write

$$\frac{x+3}{x-2} = \frac{x^2+5x+6}{x^2-4} \qquad x \neq 2 \quad x \neq -2$$

TI-82 Note *Type `2.44` for `Xmax`.

■ SUMMARY

In this section, we have learned how to multiply and divide rational expressions. The product of two rational expressions is the product of the numerators divided by the product of the denominators. To achieve completely reduced form, we must remove the common factors from the product. The quotient of two rational expressions is found by converting the division operation to multiplication, and then converting the divisor to its reciprocal. Next, perform the multiplication operation. Also, we have learned that we can multiply by a form of 1 to rename a rational expression to have a particular denominator.

6.2 ■ ■ ■ EXERCISES

Perform the multiplication, and write as a completely reduced rational expression. Use the graphics calculator to test your work for errors.

1. $\frac{3}{x+1} \cdot \frac{x^2-1}{6}$

2. $\frac{3y^2}{4-x} \cdot \frac{x^2-16}{9y}$

3. $\frac{3x}{4m^2} \cdot \frac{5m}{15x^3}$

4. $\frac{x-y}{xy} \cdot \frac{x^2y^2}{x^2-y^2}$

5. $\frac{3m^2y}{2ab} \cdot \frac{4a^2}{21m^3y^2}$

6. $\frac{y+2}{x} \cdot \frac{x-3}{y}$

7. $\frac{4x+2}{x-2} \cdot \frac{x-2}{4x+2}$

8. $\frac{x^2+3x+2}{x+4} \cdot \frac{x^2+5x+4}{x+2}$

9. $(x^2-4) \cdot \frac{3}{x-2}$

10. $\frac{x^2+5x+6}{3x+1} \cdot \frac{3x^2+7x+2}{x^2-36}$

11. $\frac{x^2-5x-6}{x^3+1} \cdot \frac{x^2-x+1}{x^2-6x}$

12. $\frac{x^2-3x+2}{x^2-4} \cdot \frac{x^2+2x}{x^3-x}$

13. $\frac{6x^3+x^2-x}{2x^2-x+1} \cdot \frac{1-4x^2}{3x^3-4x^2+x}$

14. $\frac{x^2+2x-15}{2x-6} \cdot \frac{3-x}{x^2+7x+10}$

15. $\frac{d^2-c^2}{d^2+2d+1} \cdot \frac{-4d-4}{2c-2d}$

16. $\frac{x^2-3x+2}{x^2-1} \cdot (x^2+3x-4)$

17. $\frac{10x^2-17x+3}{15x^2-8x+1} \cdot \frac{6x^2-11x+3}{4x^3+4x^2-9x-9}$

18. $\frac{x^2-16}{x^2-3x} \cdot \frac{x^3-3x^2-x+3}{x^2+5x+4}$

Perform the division and write as a completely reduced rational expression. Use the graphics calculator to test your work for errors.

19. $\frac{2}{3} \div \frac{4m^2}{9x^3}$

20. $\frac{y-x}{xy} \div \frac{y^2-x^2}{x^2y^2}$

21. $\frac{x}{y} \div \frac{x^2+x}{y^2-y}$

22. $\frac{4-x}{4x} \div \frac{16-x^2}{16x^2}$

23. $\frac{x^2+x}{y^2-y} \div \frac{x}{y}$

24. $\frac{5}{x^3+x^2} \div \frac{10x}{x+1}$

25. $\frac{2x}{5m^2} \div \frac{6xy}{15m^2y^2}$

26. $\frac{x^2-9}{2x+3} \div \frac{x+3}{2x^2+5x+3}$

27. $(x^2-4) \div \frac{3x^2+6x}{x^2-4x+4}$

28. $\frac{x^4-4}{x^3+2x^2} \div \frac{x^3-3x^2+2x-6}{x+2}$

29. $\frac{3x^3-12x}{x^2-4x+4} \div \frac{6x^3-12x^2}{2x^2-18}$

30. $\dfrac{x^2 + x}{y^2 - y} \div \dfrac{y^2 + 2y - 3}{2x^2 + 2x}$

Multiply the rational expressions by a form of 1 to rename it with the indicated denominator.

31. $\dfrac{x - y}{xy} = \dfrac{\quad}{x^2y + xy^2}$

32. $\dfrac{2x - 1}{3x + 1} = \dfrac{\quad}{9x^2 - 1}$

33. $\dfrac{x + 2}{x - 2} = \dfrac{\quad}{2x^2 + x - 10}$

34. $\dfrac{x^2 - 2x + 4}{x^2 - 3} = \dfrac{\quad}{x^3 + 2x^2 - 3x - 6}$

6.3 ■ ■ ■ ADDITION AND SUBTRACTION OF RATIONAL EXPRESSIONS

In the first two sections we learned that a rational expression is a ratio of polynomials, and the phrase *completely reduced* means that the common factors of the numerator and denominator have been cancelled (removed). Also, we have learned to carry out the operations of multiplication and division in order to obtain a completely reduced rational expression. In this section, we perform the binary operations of addition and subtraction on rational expressions. The model for this work is addition and subtraction of fractions (rational numbers).

Level I—Same Denominators At the first level of addition and subtraction of rational expressions, the denominators are the same polynomial.

ADDITION AND SUBTRACTION OF RATIONAL EXPRESSIONS

For $\frac{a}{b}$ and $\frac{c}{b}$, where $b \neq 0$,

$$\frac{a}{b} + \frac{c}{b} = \frac{a + c}{b} \quad \text{and} \quad \frac{a}{b} - \frac{c}{b} = \frac{a - c}{b}$$

When the denominators are the same, we simply add (or subtract) the numerators. As will be the case with all the binary operations, we write the result as *a completely reduced rational expression,* a ratio of polynomials where the numerator and denominator have no common factor. Consider the following example.

EXAMPLE 1 Write as a completely reduced rational expression: $\dfrac{x^2}{x + 1} - \dfrac{3x + 4}{x + 1}$

Solution The first step in doing addition (or subtraction) of rational expressions is to *observe the denominators*. Here, they are the same. Thus, we subtract the numerators:

MODEL FROM ARITHMETIC

$$\frac{3}{7} - \frac{2}{7} = \frac{3 - 2}{7} = \frac{1}{7}$$

$$\frac{x^2}{x + 1} - \frac{3x + 4}{x + 1} = \frac{x^2 - (3x + 4)}{x + 1} = \frac{x^2 - 3x - 4}{x + 1}$$

To finish the problem we must check that $\frac{x^2 - 3x - 4}{x + 1}$ is completely reduced. Factoring the numerator, we get $\frac{(x - 4)(x + 1)}{x + 1}$.

Removing the common factors we get:

$$\frac{(x - 4)\cancel{(x + 1)}}{\cancel{x + 1}} \quad \text{or} \quad x - 4$$

Thus, as a completely reduced rational expression,

$$\frac{x^2}{x + 1} - \frac{3x + 4}{x + 1} = x - 4 \qquad \text{Restriction: } x \neq -1$$

■

As Example 1 shows, the difference of two rational expressions can yield a polynomial. Of course, the polynomial could be written as the rational expression $\frac{x + 4}{1}$, but it is best to write the result as a polynomial when possible.

In the next example of subtracting rational expressions with the same denominators, we will make use of the fact that $a - b = -1(b - a)$.

EXAMPLE 2 Write as a completely reduced rational expression: $\frac{5}{x^2 - 25} - \frac{x}{x^2 - 25}$

Solution Again, we start by observing the denominators—they are the same. Thus, we subtract the numerators:

$$\frac{5}{x^2 - 25} - \frac{x}{x^2 - 25} = \frac{5 - x}{x^2 - 25}$$

Next, we check to make sure the result is completely reduced. Factoring, we get

$$\frac{(5 - x)}{(x + 5)(x - 5)}$$

Using $a - b = -1(b - a)$ we can identify the common factor:

$$\frac{-1\cancel{(x - 5)}}{(x + 5)\cancel{(x - 5)}} \quad \text{or} \quad \frac{-1}{x + 5}$$

Thus, as a completely reduced rational expression,

$$\frac{5}{x^2 - 25} - \frac{x}{x^2 - 25} = \frac{-1}{x + 5} \qquad \text{Restrictions: } x \neq 5 \quad x \neq -5$$

The meaning of this statement is that any place the expression

$$\frac{5}{x^2 - 25} - \frac{x}{x^2 - 25}$$

appears, it can be replaced by $\frac{-1}{x+5}$. It also means that for any replacement of the variable, except 5 and -5, the expressions

$$\frac{5}{x^2-25} - \frac{x}{x^2-25} \quad \text{and} \quad \frac{-1}{x+5}$$

evaluate to the same number. ■

Using the Graphics Calculator

To check our work in Example 2, type `5/(X²−25)−X/(X²−25)` for `Y1` and type `⁻1/(X+5)` for `Y2`. Store a number, say `10`, in memory location `X`. Display the resulting values in `Y1` and `Y2`. If these values are not the same, an error is present.

GIVE IT A TRY

Perform the operation and write the result as a completely reduced rational expression.

1. $\frac{2x-5}{x-1} + \frac{x^2+2}{x-1}$
2. $\frac{x^2-5x-2}{4-x} - \frac{2x^2-x-2}{4-x}$
3. $\frac{4x}{x^2-25} + \frac{x^2-5}{x^2-25}$
4. $\frac{x+1}{x^3+1} - \frac{2x-x^2}{x^3+1}$

Level II—Unlike Denominators At the next level of addition and subtraction of rational expressions, the denominators for the rational expressions are *not* the same. In this situation, we rename the rational expressions (see Example 5 in Section 6.2) so that the denominators are the same. Thus, we make this new problem look like an old problem—adding rational expressions with the same denominator. Consider the following example.

EXAMPLE 3 Write as a completely reduced rational expression:

$$\frac{x-3}{x+3} + \frac{x}{x-2}$$

Solution We observe that the denominators are *not* the same. So, we rename the rational expressions with denominator $(x+3)(x-2)$, the product of the denominators. The denominator $(x+3)(x-2)$ is known as the *common denominator* for the denominators $x+3$ and $x-2$.

a form of 1 ↓ (applied to $\frac{x-2}{x-2}$) a form of 1 ↓ (applied to $\frac{x+3}{x+3}$)

$$\frac{x-3}{x+3} \cdot \frac{x-2}{x-2} = \frac{x^2-5x+6}{(x+3)(x-2)} \qquad \frac{x}{x-2} \cdot \frac{x+3}{x+3} = \frac{x^2+3x}{(x+3)(x-2)}$$

> **MODEL FROM ARITHMETIC**
>
> This example is much like
>
> $$\frac{2}{3}+\frac{4}{5}$$
>
> Here the fractions are renamed as 15s—the common denominator is (3)(5):
>
> $$\frac{2}{3}+\frac{4}{5}=\frac{10}{15}+\frac{12}{15}$$
>
> $$=\frac{22}{15}$$

Now, $\dfrac{x-3}{x+3}+\dfrac{x}{x-2}$ becomes

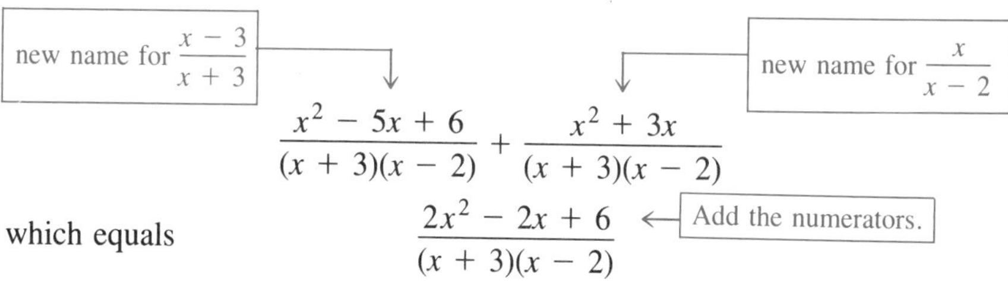

$$\frac{x^2-5x+6}{(x+3)(x-2)}+\frac{x^2+3x}{(x+3)(x-2)}$$

which equals

$$\frac{2x^2-2x+6}{(x+3)(x-2)}$$

Thus,

$$\frac{x-3}{x+3}+\frac{x}{x-2}=\frac{2x^2-2x+6}{(x+3)(x-2)}$$

Restrictions: $x \neq -3 \quad x \neq 2$

Our last step is to make certain that the rational expression is in completely reduced form. The numerator factors as $2(x^2 - x + 3)$. So, the numerator and denominator have no common factor. We can write the denominator as $(x + 3)(x - 2)$ or as $x^2 + x - 6$. Traditionally, it is left in factored form. Remember, the reason for leaving the denominator in factored form is so that the restricted values can be quickly observed (in most cases). ■

GIVE IT A TRY

Write each as a completely reduced rational expression.

5. $\dfrac{x-3}{x+3}-\dfrac{x}{x-2}$ **6.** $\dfrac{x}{x-1}+\dfrac{x-3}{x^2+6}$ **7.** $\dfrac{x}{x-y}-\dfrac{y}{x+y}$

8. Use the graphics calculator to test your work in Problems 5–7 for errors.

A Least Common Denominator (LCD) The next situation for addition and subtraction of rational expressions also involves unlike denominators. In this situation, rather than taking the product to get the common denominator, we find a **least common denominator** (LCD). The least common denominator (LCD) is "least" in the sense that it has fewer factors than any other common denominator. Consider the following example.

EXAMPLE 4 Write as a completely reduced rational expression: $\dfrac{x-3}{x^2}+\dfrac{x}{2x}$

Solution In this case we have unlike denominators. We could find a common denominator by taking the product, $2x^3$, but the resulting sum would have a numerator and denominator with a common factor. (We would have to reduce the new expression.) We can find a least common denominator by forming the product that uses each factor the greatest number of times that it appears in either denominator. Here, the least common denominator is $2x^2$. Now we rename each rational expression to have this denominator:

a form of 1 a form of 1

$$\frac{x-3}{x^2}\cdot\frac{2}{2}=\frac{2x-6}{2x^2} \quad \text{and} \quad \frac{x}{2x}\cdot\frac{x}{x}=\frac{x^2}{2x^2}$$

new names

Thus, $$\frac{x-3}{x^2}+\frac{x}{2x}=\frac{2x-6}{2x^2}+\frac{x^2}{2x^2}$$
$$=\frac{x^2+2x-6}{2x^2} \qquad \text{Add the numerators.}$$

So, as a completely reduced rational expression,

$$\frac{x-3}{x^2}+\frac{x}{2x} \quad \text{is} \quad \frac{x^2+2x-6}{2x^2} \qquad \text{Restriction: } x \neq 0$$ ■

In the next example we work with a more complicated situation involving addition and subtraction of rational expressions using an LCD.

EXAMPLE 5 Write as a completely reduced rational expression: $\frac{x-1}{x^2-4}-\frac{x+3}{x^2-4x+4}$

Solution A clue for using an LCD is the fact that each denominator can be factored:

$$x^2-4 \quad \text{factors as} \quad (x+2)(x-2)$$
$$x^2-4x+4 \quad \text{factors as} \quad (x-2)(x-2)$$

The factor $(x+2)$ appears at most once in either factoring.
The factor $(x-2)$ appears at most twice in either factoring.
The LCD is

Factor $(x+2)$ appears at most once.

$$(x+2)(x-2)(x-2)$$

Factor $(x-2)$ appears at most twice.

Now, we rename each rational expression to have the new denominator:

a form of 1

$$\frac{x-1}{x^2-4}\cdot\frac{x-2}{x-2}=\frac{x^2-3x+2}{(x+2)(x-2)(x-2)}$$

a form of 1

$$\frac{x+3}{x^2-4x+4}\cdot\frac{x+2}{x+2}=\frac{x^2+5x+6}{(x+2)(x-2)(x-2)}$$

Thus,

new names

$$\frac{x-1}{x^2-4}-\frac{x+3}{x^2-4x+4}=\frac{x^2-3x+2}{(x+2)(x-2)(x-2)}-\frac{x^2+5x+6}{(x+2)(x-2)(x-2)}$$
$$=\frac{-8x-4}{(x+2)(x-2)(x-2)} \qquad \text{Restrictions: } x \neq -2 \quad x \neq 2$$ ■

To find the LCD for two or more rational expressions, we can use the following steps.

ALGORITHM FOR THE LCD FOR RATIONAL EXPRESSIONS

1. Factor each polynomial denominator.
2. From the factorings in step 1, write each of the factors the *greatest number of times* that it appears in any one of the individual factorings.

In Example 5 we could have selected the product $(x^2 - 4)(x^2 - 4x + 4)$ for the denominator. But, the rational expression resulting from taking the difference would be

$$\frac{-8x^2 + 12x + 8}{(x + 2)(x - 2)(x - 2)(x - 2)} \qquad \text{Restrictions: } x \neq -2 \quad x \neq 2$$

and this rational expression must be reduced:

$$\frac{-8x^2 + 12x + 8}{(x + 2)(x - 2)(x - 2)(x - 2)} = \frac{-4(2x + 1)\cancel{(x - 2)}}{(x + 2)(x - 2)(x - 2)\cancel{(x - 2)}}$$
$$= \frac{-8x - 4}{(x + 2)(x - 2)(x - 2)}$$

Notice that we obtain the same result as in Example 5.

When a problem involves this much work, be sure to lay out the problem in a neat, orderly fashion so that you can easily check back through the steps for errors. It is also a good idea to replace the variable with a number, say 5, and evaluate the original expression as well as the completely reduced expression. The graphics calculator is very useful for this check.

Many skills are involved and much concentration is required to add and subtract rational expressions. A common problem is forgetting what you were doing (or trying to accomplish)! You can overcome this problem through the understanding of what you are doing, as well as much practice.

GIVE IT A TRY

Write each as a completely reduced rational expression.

9. $\dfrac{x}{x^2 - 4} + \dfrac{2}{x^2 + 3x + 2}$ **10.** $\dfrac{5m}{6x^2y} - \dfrac{2}{3xy^2}$ **11.** $\dfrac{1}{x^2 - 4x - 5} + \dfrac{x}{(x - 5)^2}$

12. Use the graphics calculator to test your work in Problems 9–11 for errors.

The Slash Division Symbol The division operation is displayed on some graphics calculators, (such as the TI-81) as a slash, /. This is also the case with computers—in spreadsheet and other mathematics-related programs, the / symbol appears for division. For these reasons, experience with this symbol will be useful for further work in mathematics. The following examples also provide practice with the basic operations on rational expressions and with order of operations.

EXAMPLE 6 Write $3 - x + 1/x$ as a completely reduced rational expression.

Solution Using order of operations, $3 - x + 1/x$ means $3 - x + \frac{1}{x}$. To produce a completely reduced rational expression, we find the LCD:

$$3 - x + \frac{1}{x} = \frac{3 - x}{1} + \frac{1}{x}$$
$$= \frac{3x - x^2}{x} + \frac{1}{x}$$
$$= \frac{-x^2 + 3x + 1}{x} \qquad \text{Restriction: } x \neq 0$$

■

WARNING

The graphics calculator interprets $1/2x$ as $\frac{1}{2x}$, whereas, by order of operations, $1/2x$ is $\frac{1}{2}x$. When using the graphics calculator, type `(1/2)X` or `1/2 * X` for $1/2x$.

GIVE IT A TRY

Using order of operations, and your skills with addition and subtraction, write each as a completely reduced rational expression. Use the graphics calculator to test your work for errors.

13. $x^2 - 1/x - 1/x^2$

14. $(x^2 - 1)/x - 1/x^2$

15. $3/(x - 3) - 2/x$

16. $3/x - 3 - 2/x$

■ SUMMARY

■ ***Quick Index***

In this section, we have performed the operations of addition and subtraction on rational expressions. These two operations can be considered in levels. At the first level, the denominators are the same, and we simply add (or subtract) the numerators:

$$\text{Level I (same denominator):} \quad \frac{3}{x + 1} + \frac{5}{x + 1} = \boxed{\frac{8}{x + 1}}$$

At the second level, the denominators are unlike. Here, we first rename the rational expressions so that they have the same denominator (known as a common denominator). The common denominator is the product of the individual denominators. Once this is accomplished, we add (subtract) the (possibly altered) numerators.

Level II (unlike denominators):

$$\frac{3}{x + 1} - \frac{5}{x + 2} = \frac{x + 2}{x + 2} \cdot \frac{3}{x + 1} - \frac{5}{x + 2} \cdot \frac{x + 1}{x + 1}$$
$$= \frac{3x + 6}{(x + 2)(x + 1)} - \frac{5x + 5}{(x + 2)(x + 1)}$$
$$= \frac{3x + 6 - (5x + 5)}{(x + 2)(x + 1)}$$
$$= \boxed{\frac{-2x + 1}{(x + 2)(x + 1)}}$$

At this second level, for unlike denominators, a least common denominator (LCD) is selected. That is, we can often select a common denominator that has fewer factors than the product of the individual denominators. To determine whether a denominator with fewer factors than the product of the denominators may be possible, observe the denominators and see that they factor. (If the individual denominators do *not* factor, then the product of the unlike denominators is the smallest common denominator possible.)

$$\text{Level II (LCD):}\quad \frac{3}{2x^2}-\frac{3}{4x}=\frac{2}{2}\cdot\frac{3}{2x^2}-\frac{3}{4x}\cdot\frac{x}{x}$$
$$=\frac{6}{4x^2}-\frac{3x}{4x^2}$$
$$=\boxed{\frac{-3x+6}{4x^2}}$$

6.3 ■ ■ ■ EXERCISES

1. Which of the following is *not* a polynomial?
   i. $3x^2 - 2x + 1$ ii. 17 iii. $1/x$ iv. x^2
2. Which of the following is *not* a rational expression?
   i. $x^2 - 2x$ ii. $1/(x + 2)$ iii. $\sqrt{7}/x$ iv. $\sqrt{x} + 1$

For the given rational expression with like denominators, report a completely reduced *rational expression. Use the graphics calculator to test your work for errors.*

3. $\dfrac{2x+5}{x-3}+\dfrac{x-5}{x-3}$
4. $\dfrac{6}{x^2-9}-\dfrac{2x}{x^2-9}$
5. $\dfrac{2x+3}{x^2-1}-\dfrac{x+2}{x^2-1}$
6. $\dfrac{x^2-3x+2}{3x+1}+\dfrac{2x^2-5x-5}{3x+1}$

For the given rational expressions with unlike denominators, report a completely reduced *rational expression. Use the graphics calculator to test your work for errors.*

7. $\dfrac{2x+3}{2}+\dfrac{x-5}{4}$
8. $\dfrac{3}{a}-\dfrac{5}{b}$
9. $\dfrac{3-x}{1}-\dfrac{1}{x}$
10. $(x+3)+\dfrac{x+3}{x-3}$
11. $(2x-1)-\dfrac{5}{2x+1}$
12. $\dfrac{x-1}{x+2}+\dfrac{2}{x-3}$
13. $\dfrac{x^2}{x+1}+\dfrac{2}{x-3}$
14. $\dfrac{x}{x-y}-\dfrac{y}{x+y}$

For each rational expression with unlike denominators, use the LCD, *to report a* completely reduced *rational expression. Use the graphics calculator to test your work for errors.*

15. $\dfrac{x+1}{2x^2}-\dfrac{5}{6x}$
16. $\dfrac{x+2}{x^2-3x}+\dfrac{x-1}{x^2-6x+9}$
17. $\dfrac{3}{x^2-4}+\dfrac{x}{2x+4}$
18. $\dfrac{x}{2x-1}-\dfrac{x+1}{4x^2-1}$
19. $\dfrac{x-1}{x^2-3x+2}-\dfrac{2}{x^2-x-2}$
20. $\dfrac{5}{x^3-8}+\dfrac{x}{x^2-4}$
21. $\dfrac{7}{x^3-x^2}-\dfrac{5}{x^2-2x+1}$
22. $\dfrac{3y}{y^2+y}-\dfrac{3}{y+2}$
23. $\dfrac{x+1}{x^2-2x-3}+\dfrac{2x}{x^2-6x+9}$
24. $\dfrac{y+2}{4mx^2}+\dfrac{y-1}{6m^2x}$
25. $\dfrac{x+1}{x^3-8}-\dfrac{1}{2x^2-4x}$
26. $\dfrac{5}{x^2+x+3}-\dfrac{1}{x^2-1}$
27. $\dfrac{y}{x-x^2}+\dfrac{x}{xy-y^2}$
28. $\dfrac{3}{x}-\dfrac{5}{x}$
29. $\dfrac{x+1}{x+3}-\dfrac{x+3}{x+5}$
30. $\dfrac{r+2}{r+3}+\dfrac{r-2}{r}$
31. $\dfrac{s}{s^2-9}-\dfrac{s}{s-2}$
32. $\dfrac{3y}{y^2-9}-\dfrac{3}{y+1}$

Using order of operations and your skills with addition and subtraction, write each result as a completely reduced rational expression. Use the graphics calculator to test your work for errors.

33. $1/(x-1) - x/(x+1)$
34. $1/x + 1/(2x) + 1/(2x)$
35. $1/x + (1/2)x + (1/2)x$
36. $3/x - 5/x$
37. $1/(x+1) + 1/x + 1$
38. $1/x + 1 + 1/x + 1$
39. $1/(x+1) + 1/(x+1)$
40. $x/(x-y) - y/(x+y)$
41. $y/x^2 + 2/(3x)$
42. $y/x^2 + (2/3)x$

6.4 ■ ■ ■ COMPLEX FRACTIONS

In the last two sections, we have learned how to add, subtract, multiply, and divide rational expressions. We now learn to work with more than one operation, using the order of operations (precedence) rules on rational expressions. Also, in this section complex fractions are introduced. A complex fraction is any expression in which the numerator or denominator contains a rational expression. Frequently, the numerator and denominator may contain sums and differences of rational expressions. We will use two different approaches to writing a complex fraction as a completely reduced rational expression.

Order of Operations We start by considering order of operations (precedence) rules with rational expressions. That is, we consider situations involving more than one operation. To be successful with this topic, you must

1. know how to add, subtract, multiply, and divide rational expressions.
2. know the order of operations rules (see Section 1.4).
3. concentrate.

Consider the following examples.

EXAMPLE 1 Write the result as a completely reduced rational expression:

$$\left(\frac{1}{x+2}\right) - \left(\frac{x^2-9}{x^2-4}\right) \div \left(\frac{x-3}{x+2}\right)$$

Solution We start by doing the division:

$$\left(\frac{1}{x+2}\right) - \left(\frac{x^2-9}{x^2-4}\right) \div \left(\frac{x-3}{x+2}\right) = \left(\frac{1}{x+2}\right) - \left(\frac{x^2-9}{x^2-4}\right)\left(\frac{x+2}{x-3}\right)$$

$$= \left(\frac{1}{x+2}\right) - \frac{(x+3)\cancel{(x-3)}\cancel{(x+2)}}{\cancel{(x+2)}(x-2)\cancel{(x-3)}}$$

$$= \frac{1}{x+2} - \frac{x+3}{x-2}$$

Next, we do the subtraction:

$$= \frac{1(x-2)}{(x+2)(x-2)} - \frac{(x+3)(x+2)}{(x-2)(x+2)}$$

$$= \frac{(x-2)-(x^2+5x+6)}{(x+2)(x-2)}$$

$$= \frac{-x^2-4x-8}{(x+2)(x-2)}$$

Restrictions: $x \neq 2$, $x \neq -2$, $x \neq 3$

■

Using the Graphics Calculator

To test our work in Example 1 for errors, store a number, say 5, in memory location X. For Y1 type

```
(1/(X+2))-((X²-9)/(X²-4))/((X-3)/(X+2))
```

and for Y2 type `(-X²-4X-8)/((X+2)(X-2))`

Use parentheses to make sure that the graphics calculator carries out the computations as you intend. From the home screen, display the resulting values stored in Y1 and Y2. Is the same number output in both cases? If not, an error is present.

EXAMPLE 2 Write as a completely reduced rational expression: $\left(\frac{x}{x-1}\right)^2 - \left(\frac{x-2}{x-1}\right)$

Solution Applying order of operations, we first do the multiplication (clear the exponent):

$$\left(\frac{x}{x-1}\right)^2 - \left(\frac{x-2}{x-1}\right) = \frac{x^2}{(x-1)^2} - \frac{x-2}{x-1}$$

Rename to make denominator the LCD: $= \dfrac{x^2}{(x-1)^2} - \dfrac{(x-2)(x-1)}{(x-1)(x-1)}$

Subtract numerators: $= \dfrac{x^2 - (x^2 - 3x + 2)}{(x-1)^2}$

$= \dfrac{3x-2}{(x-1)^2}$ Restrictions: $x \neq 1$ ■

EXAMPLE 3 Write as a completely reduced rational expression: $1 + \left(1 \div \left(1 + \frac{1}{x-2}\right)\right)$

Solution First, we clear the innermost parentheses: [rename 1]

$$1 + \left(1 \div \left(1 + \frac{1}{x-2}\right)\right) = 1 + \left(1 \div \left(\frac{x-2}{x-2} + \frac{1}{x-2}\right)\right)$$

$$= 1 + \left(1 \div \frac{x-1}{x-2}\right)$$

Clear the remaining parentheses: $= 1 + \left(1 \cdot \dfrac{x-2}{x-1}\right)$

Now, we have $= 1 + \dfrac{x-2}{x-1}$

[rename 1]

Rename 1, using LCD: $= \dfrac{x-1}{x-1} + \dfrac{x-2}{x-1}$

Add numerators: $= \dfrac{x-1+x-2}{x-1}$

$= \dfrac{2x-3}{x-1}$ Restrictions: $x \neq 2 \quad x \neq 1$ ■

GIVE IT A TRY

Write each result as a completely reduced rational expression.

1. $\left(\frac{x+1}{x+3}\right) + \left(\frac{x}{x+3}\right) \div \left(\frac{x}{x^2-9}\right)$

2. $1 - \left(1 \div \left(1 - \frac{1}{x-2}\right)\right)$

Complex Notation The expression $1 + \left(1 \div \left(1 + \frac{x-1}{x-2}\right)\right)$ from Example 3 often appears in complex notation (fraction form) as

$$1 + \frac{1}{1 + \frac{x-1}{x-2}}$$

The fraction bar serves a dual purpose—it indicates *division* and it represents a *set of parentheses*:

$$1 + \frac{1}{1 + \frac{x-1}{x-2}}$$

The fraction bar implies
1. a division and
2. a set of parentheses.

$$1 + \left(1 \div \left(1 + \frac{x-1}{x-2}\right)\right)$$

We now consider writing a completely reduced rational expression for expressions that appear as complex fractions. We start with an example from arithmetic.

EXAMPLE 4 Write as a completely reduced rational number:

$$\frac{\frac{1}{3} - \frac{1}{2}}{\frac{1}{9} - \frac{1}{4}}$$

Solution Doing the operation in the numerator, we get

$$\frac{1}{3} - \frac{1}{2} = \frac{2}{6} - \frac{3}{6} = \frac{-1}{6}$$

Doing the operation in the denominator, we get

$$\frac{1}{9} - \frac{1}{4} = \frac{4}{36} - \frac{9}{36} = \frac{-5}{36}$$

Finally, we do the implied division:

$$\frac{-1}{6} \div \frac{-5}{36} = \left(\frac{-1}{6}\right)\left(\frac{-36}{5}\right) = \frac{36}{(6)(5)} = \frac{\cancel{(6)}(6)}{\cancel{(6)}(5)} = \frac{6}{5}$$ ■

EXAMPLE 5 Write as a completely reduced rational expression:

$$\frac{\dfrac{1}{x} - \dfrac{1}{2}}{\dfrac{1}{x^2} - \dfrac{1}{4}}$$

Solution First, we rewrite to division notation:

$$\left(\frac{1}{x} - \frac{1}{2}\right) \div \left(\frac{1}{x^2} - \frac{1}{4}\right)$$

Next, we clear the parentheses:

$$\frac{1}{x} - \frac{1}{2} = \frac{2}{2x} - \frac{x}{2x} = \frac{2 - x}{2x} \quad \text{and} \quad \frac{1}{x^2} - \frac{1}{4} = \frac{4}{4x^2} - \frac{x^2}{4x^2} = \frac{4 - x^2}{4x^2}$$

Now we do the division:

$$\frac{2 - x}{2x} \div \frac{4 - x^2}{4x^2} = \left(\frac{2 - x}{2x}\right)\left(\frac{4x^2}{4 - x^2}\right) = \frac{(2 - x)(2)(2)(x)(x)}{2x(2 + x)(2 - x)}$$

$$= \frac{2x}{x + 2} \qquad \text{Restrictions: } x \neq 0 \quad x \neq 2 \quad x \neq -2$$

■

EXAMPLE 6 Write as a completely reduced rational expression:

$$\frac{\dfrac{x + 2}{x - 3}}{\dfrac{x^2 - 4}{x^2 - 4x + 3} + \dfrac{1}{x - 1}}$$

Solution Rewriting the expression using division notation, we get

$$\frac{x + 2}{x - 3} \div \left(\frac{x^2 - 4}{x^2 - 4x + 3} + \frac{1}{x - 1}\right)$$

Applying order of operations, we next clear the parentheses:

$$\frac{x + 2}{x - 3} \div \left(\frac{x^2 - 4}{x^2 - 4x + 3} + \frac{1}{x - 1}\right) = \frac{x + 2}{x - 3} \div \left(\frac{x^2 - 4}{(x - 3)(x - 1)} + \frac{1(x - 3)}{(x - 1)(x - 3)}\right)$$

$$= \frac{x + 2}{x - 3} \div \frac{x^2 + x - 7}{(x - 3)(x - 1)}$$

$$= \left(\frac{x + 2}{x - 3}\right)\left(\frac{(x - 3)(x - 1)}{x^2 + x - 7}\right)$$

$$= \frac{(x + 2)(x - 3)(x - 1)}{(x - 3)(x^2 + x - 7)}$$

$$= \frac{x^2 + x - 2}{x^2 + x - 7}$$

So,

$$\frac{\dfrac{x+2}{x-3}}{\dfrac{x^2-4}{x^2-4x+3}+\dfrac{1}{x-1}} = \frac{x^2+x-2}{x^2+x-7}$$

■

Using the Graphics Calculator

To use the graphics calculator to test our work in Example 5 for errors, we must use parentheses carefully. A rule of thumb is: If in doubt, insert a set of parentheses. Often it is helpful to break the entry into parts. For Y1, type the numerator, (X+2)/(X−3). For Y2, type the denominator, (X²−4)/(X²−4X+3)+1/(X−1). For Y3, type Y1/Y2. Now type (X²+X−2)/(X²+X−7) for Y4. Return to the home screen. Store a number, say 7, in memory location X. Display the resulting values in Y3 and Y4. If the same number is *not* output in both cases, an error is present.

GIVE IT A TRY

Write each as a completely reduced rational expression.

3. $\left(\dfrac{1}{x-2}\right)^2 - \left(\dfrac{1}{x}\right)^2$ **4.** $\dfrac{\dfrac{1}{x}-\dfrac{1}{3}}{x^2-9}$ **5.** $\dfrac{\dfrac{1}{x}}{\dfrac{1}{x}+\dfrac{1}{y}}$

The LCD Approach So far, our work with complex rational expressions has been based on knowledge of order of operations and our ability to add, subtract, multiply, and divide rational expressions. We may now consider a more sophisticated approach. This approach is based on multiplying numerator and denominator by the least common denominator, the LCD, for all the denominators in the expression. That is, we will multiply by a form of 1. To learn how to do this, work through the following examples.

EXAMPLE 7 Write as a completely reduced rational expression:

$$\frac{\dfrac{1}{x}-\dfrac{1}{2}}{\dfrac{1}{x^2}-\dfrac{1}{4}}$$

Solution First, the LCD for the denominators 2, 4, x, and x^2 is $4x^2$. Thus, we multiply the fraction by a form of 1 that uses the LCD, $\frac{4x^2}{4x^2}$.

$$\frac{\frac{1}{x}-\frac{1}{2}}{\frac{1}{x^2}-\frac{1}{4}} = \underbrace{\frac{4x^2}{4x^2}}_{\text{a form of 1}} \cdot \frac{\frac{1}{x}-\frac{1}{2}}{\frac{1}{x^2}-\frac{1}{4}} = \frac{4x^2\left(\frac{1}{x}\right) - 4x^2\left(\frac{1}{2}\right)}{4x^2\left(\frac{1}{x^2}\right) - 4x^2\left(\frac{1}{4}\right)}$$

$$= \frac{\frac{4x^2}{x} - \frac{4x^2}{2}}{\frac{4x^2}{x^2} - \frac{4x^2}{4}}$$

Remove common factors:

$$= \frac{4x - 2x^2}{4 - x^2}$$

Remove common factors:

$$= \frac{-2x\cancel{(x-2)}}{-1(x+2)\cancel{(x-2)}}$$

$$= \frac{2x}{(x+2)}$$

■

GIVE IT A TRY

Use the LCD approach to write the given expression as a completely reduced rational expression.

6. $\dfrac{y - \frac{1}{x}}{x - \frac{1}{y}}$

7. $\dfrac{\frac{1}{y} + \frac{1}{x}}{x^2 - y^2}$

A Comparison The LCD approach for the complex rational expression in Example 7 is probably simpler than the division approach used in Example 5. You may wish to compare the two approaches. The completely reduced rational expression produced is the same, regardless of which approach is used. For the complex rational expressions in Example 6, the LCD approach becomes more complicated. Give it a try—the LCD for the denominators is $(x - 3)(x - 1)$.

We now work the same example, using each approach.

EXAMPLE 8 Write as a completely reduced rational expression:

$$\frac{\frac{1}{x}}{\frac{x+3}{x+1} + \frac{x+5}{x}}$$

Solution We start by rewriting the problem using division notation:

$$\frac{1}{x} \div \left(\frac{x+3}{x+1} + \frac{x+5}{x}\right)$$

We next clear the parentheses:

$$\frac{1}{x} \div \left(\frac{x+3}{x+1} + \frac{x+5}{x}\right) = \frac{1}{x} \div \left(\frac{x^2+3x}{x(x+1)} + \frac{x^2+6x+5}{x(x+1)}\right)$$

$$= \frac{1}{x} \div \frac{2x^2+9x+5}{x(x+1)}$$

Now, we do the division:

$$= \frac{1}{x} \cdot \frac{x(x+1)}{2x^2+9x+5}$$

$$= \frac{\cancel{x}(x+1)}{\cancel{x}(2x^2+9x+5)}$$

$$= \frac{x+1}{2x^2+9x+5}$$

So, we have

$$\frac{\frac{1}{x}}{\frac{x+3}{x+1} + \frac{x+5}{x}} = \frac{x+1}{2x^2+9x+5}$$

We now use the LCD approach. The LCD is $x(x + 1)$.

$$\frac{\frac{1}{x}}{\frac{x+3}{x+1} + \frac{x+5}{x}} \cdot \underbrace{\frac{x(x+1)}{x(x+1)}}_{\text{a form of 1}} = \frac{\frac{\cancel{x}(x+1)}{\cancel{x}}}{\frac{(x+3)(x)\cancel{(x+1)}}{\cancel{x+1}} + \frac{(x+5)(\cancel{x})(x+1)}{\cancel{x}}}$$

$$= \frac{x+1}{(x+3)(x) + (x+5)(x+1)}$$

$$= \frac{x+1}{2x^2+9x+5}$$

■

GIVE IT A TRY

8. Write as a completely reduced rational expression:

$$\frac{3}{3 + \frac{x-3}{x-2}}$$

9. The expressions $2/3x$ and $2/(3x)$ are not equivalent rational expressions. Using order of operations, $2/3x$ is $\frac{2}{3}x$ and $2/(3x)$ is $\frac{2}{3x}$. Write each rational expression in completely reduced form.

a. $3/x/y$ **b.** $3/x/y/6$ **c.** $3/(x/y)/6$

■ SUMMARY

Complex rational expressions are actually an application of the basic operations on rational expressions and the order of operations rules. That is, a complex rational expression is an expression involving more than one operation on rational expressions. So, working with complex rational expressions requires knowledge of order of operations on rational expressions. To write such an expression in completely reduced form, one approach is to rewrite the expression using division and parentheses. An alternate technique is to multiply by a form of 1, where the numerator and denominator for 1 are the LCD for all the denominators in the expression.

6.4 ■ ■ ■ EXERCISES

Write each as a completely reduced rational expression. Use the graphics calculator to test your work for errors.

1. $\frac{1}{x} \cdot \frac{1}{x-1} + \frac{1}{x}$

2. $\left(\frac{1}{x}\right)\left(\frac{1}{x-1} + \frac{1}{x}\right)$

3. $\frac{1}{x} \div \frac{1}{x-1} + \frac{1}{x}$

4. $\frac{1}{x} \div \left(\frac{1}{x-1} + \frac{1}{x}\right)$

5. $\frac{1}{x-3} \div \frac{1}{x^2-9} + \frac{1}{x+2}$

6. $\frac{x^2-1}{x-2} \div \frac{x+1}{x^2-4} + \frac{1}{x+2}$

7. $\frac{1}{x-3} \div \frac{1}{x-3} - \frac{1}{x+2}$

8. $\frac{1}{x-3} \div \left(\frac{1}{x-3} - \frac{1}{x+2}\right)$

9. $1 + \frac{x+1}{2x^2-3}$

10. $3 - \frac{2x+3}{x-2}$

11. $\frac{x^2-1}{x^2+5x+6} \cdot \frac{x+3}{x+1} \div \frac{x^2-4}{(x-1)^2}$

12. $\left(\frac{3}{x-2} - \frac{1}{x+1}\right) \div \frac{1}{x^2-25}$

13. $\frac{3}{x-2} - \frac{1}{x+1} \div \frac{1}{x^2-25}$

14. $\frac{1}{h} \cdot \left(\frac{-3}{x-h} - \frac{3}{h}\right)$

15. $\frac{1}{h} \cdot \left(\frac{1}{x+2+h} - \frac{1}{x+2}\right)$

16. $\left(x - \frac{x-1}{3}\right) \div \left(\frac{2}{3} - \frac{1}{6x}\right)$

17. $\left(\frac{r+3}{r} - \frac{4}{r-1}\right) \div \left(\frac{r}{r-1} + \frac{1}{r}\right)$

18. $\frac{r+3}{r} - \frac{4}{r-1} \div \frac{r}{r-1} + \frac{1}{r}$

19. $\frac{r+3}{r} - \left(\frac{4}{r-1} \div \frac{r}{r-1} + \frac{1}{r}\right)$

20. $\frac{r+3}{r} - \frac{4}{r-1} \div \left(\frac{r}{r-1} + \frac{1}{r}\right)$

21. $\dfrac{1}{1 - \dfrac{1}{x-1}}$

22. $\dfrac{1}{1 - \dfrac{1}{x^2+1}}$

23. $\dfrac{1}{\dfrac{1}{s} - \dfrac{1}{s-1}}$

24. $(p + 3) \div (1/p^2 - 1/9)$

25. $1 + (1 + (1 \div (a + 1)))$

26. $(1/x + 1/y) \div (x^2 - y^2)$

27. $1 - 1/(n - 1)/n^2$

28. $\left[x - 1 \div \left(1 + \frac{1}{x}\right)\right] \div \left[x - 1 \div \left(1 - \frac{1}{x}\right)\right]$

Write the complex fraction as a completely reduced rational expression. Use the graphics calculator to test your work for errors.

29. $\frac{1}{x} + \dfrac{\dfrac{1}{x-1}}{\dfrac{1}{x}}$

30. $\dfrac{\dfrac{1}{x} - \dfrac{1}{x-1}}{\dfrac{1}{x}}$

31. $\frac{1}{x-3} \div \dfrac{\dfrac{1}{x^2-9}}{\dfrac{1}{x+3}}$

32. $\dfrac{\dfrac{1}{x-3} \div \dfrac{1}{x+2}}{\dfrac{1}{x-3} + \dfrac{1}{x+2}}$

33. $1 \div \dfrac{1}{1 - \dfrac{1}{x-1}}$

34. $1 + \dfrac{1}{1 - \dfrac{1}{x^2+1}}$

35. $3 - \dfrac{\dfrac{1}{x}}{\dfrac{1}{x} - \dfrac{1}{5}}$

36. $\dfrac{\dfrac{1}{x^2}}{\dfrac{1}{x^2} - \dfrac{1}{25}}$

37. $1 - \dfrac{1}{\dfrac{1}{x} - \dfrac{1}{x-1}}$

38. $\dfrac{1}{2 - \dfrac{1}{x+1}} + \dfrac{x-1}{x-2}$

39. $\dfrac{1 - \dfrac{2}{x} - \dfrac{3}{x^2}}{1 + \dfrac{3}{x} + \dfrac{2}{x^2}}$

40. $\dfrac{\dfrac{1}{p-3} - \dfrac{p}{p-3}}{\dfrac{p}{p+1} - \dfrac{1}{p-3}}$

41. $\dfrac{1}{x} - \dfrac{\dfrac{12x+8}{3x-2}}{\dfrac{7x-6}{3x-2}}$

42. $x - \dfrac{1}{\dfrac{y}{x} - \dfrac{1}{y - \dfrac{1}{y}}}$

43. $1 - \dfrac{1}{\dfrac{1}{1 + \dfrac{1}{x+1}}}$

44. $x^2 - \dfrac{1}{1 - \dfrac{1}{x^2}}$

45. $\dfrac{\dfrac{1}{(xy)^2}}{\dfrac{1}{x^2} - \dfrac{1}{y^2}}$

46. $\dfrac{\dfrac{(x-y)^2}{xy}}{\dfrac{1}{x^2} - \dfrac{1}{y^2}}$

6.5 ■ ■ ■ SOLVING RATIONAL EXPRESSION EQUATIONS IN ONE VARIABLE

Now that we know what rational expressions are and how to perform the basic binary operations on these objects, we can apply this knowledge to solving equations. As with polynomial equations, we begin with the simplest level and work to more complex equations. The knowledge and skill we gain in this section will be used in Section 6.6, when we discuss applications.

Background An example of a rational expression equation in one variable is

$$\frac{3}{x-2} = \frac{-2}{x+1}$$

To solve any equation in one variable means that we find the number(s) that yield a true arithmetic sentence when one of the numbers replaces the variable. For example, is 7 in the solution to this equation? To find out, we replace the variable with 7 to get

$$\frac{3}{7-2} = \frac{-2}{7+1}$$

This is a false arithmetic sentence because $\frac{3}{5}$ is not equal to $\frac{-2}{8}$. Thus, 7 is *not* in the solution to the equation.

Using the Graphics Calculator

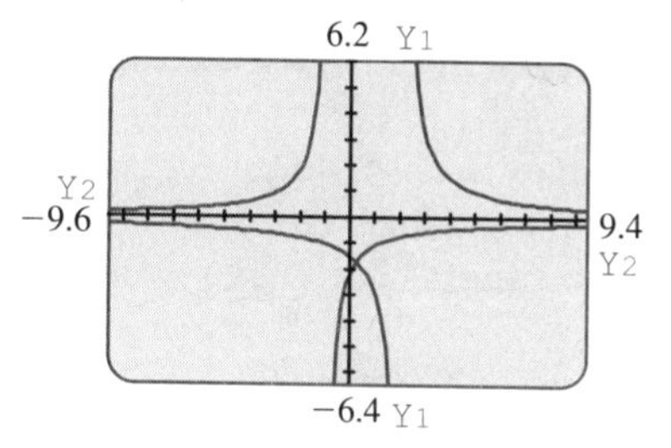

:Y1=3/(X-2) :Y2=-2/(X+1)

We know from earlier work with equations that we can use the graph screen to see evaluations of an expression for various replacements of the x-variable. A number in the solution to any rational expression equation will be an x-value that results in both the left-hand member of the equation and the right-hand member of the equation evaluating to the same number. To find a number in the solution of the equation $\frac{3}{x-2} = \frac{-2}{x+1}$, we will view the evaluations of $\frac{3}{x-2}$ and the evaluations of $\frac{-2}{x+1}$ on the same graph screen.

For Y1 enter the expression 3/(X−2). For Y2 enter the expression -2/(X+1). Use the Window 9.4 range settings to view the graph screen, which displays the evaluations of the two expressions (see Figure 1). The x-value where the two expressions evaluate to the same number is near zero. Press the TRACE key and then move the cursor to that intersection point. Does the x-value solve the equation $\frac{3}{x-2} = \frac{-2}{x+1}$? We will now use an algebraic approach to arrive at the number that solves the above rational expression equation.

The Algebraic Approach To solve a rational expression equation, we use an expansion of the multiplication property of equation solving.

GENERAL MULTIPLICATION PROPERTY OF EQUATION SOLVING

> Multiplying each side of an equation by an expression produces an equation that *contains* the solution to the original equation.

We can multiply each side of an equation by an expression. If the expression represents a nonzero number, the solution is not altered (see Section 2.2). The following example shows why we must have the containment clause.

EXAMPLE 1 Solve: $\frac{x^2}{x+2} = \frac{4}{x+2}$

Solution The expression we will multiply by is $x + 2$. Multiplying each side by the $x + 2$, we get:

$$\frac{(x+2)(x^2)}{x+2} = \frac{(x+2)(4)}{x+2}$$

$$x^2 = 4 \quad \text{Remove common factors.}$$

$$x^2 - 4 = 0 \quad \text{Standard form}$$

$$(x+2)(x-2) = 0 \quad \text{Factored form}$$

$$x + 2 = 0 \quad \text{or} \quad x - 2 = 0$$

$$x = -2 \qquad\qquad x = 2$$

The second-degree equation has solution 2 and -2. However, only the number 2 solves the original rational expression equation. Notice that replacing the variable in the original equation with -2 results in a denominator of 0. The number -2 is known as an *extraneous solution* and is discarded. ■

The possibility of an extraneous solution makes it absolutely necessary to *check your work*. You should always check your work for possible errors, but when solving rational expression equations, you *must* check your work for extraneous solutions.

In general, the expression we will choose as a multiplier is the LCD for the denominators in the equation. This will produce a new equation that we *can* solve (such as a first- or second-degree equation). In the next example, we solve algebraically the rational expression equation whose solution we found at the beginning of this section using the graphics calculator.

EXAMPLE 2 Solve: $\dfrac{3}{x-2} = \dfrac{-2}{x+1}$

Solution We start by selecting the LCD for the denominators. Here, the LCD is the product of the denominators, $(x + 1)(x - 2)$.

We multiply each side by $(x + 1)(x - 2)$ to get:

$$\frac{(x-2)(x+1)(3)}{x-2} = \frac{(x-2)(x+1)(-2)}{x+1}$$

Reduce the rational expressions: $3(x + 1) = -2(x - 2)$

Clear the parentheses: $3x + 3 = -2x + 4$

Add $2x$ to each side to get: $5x + 3 = 4$

Add -3 to each side to get: $5x = 1$

Multiply by $\frac{1}{5}$: $x = \dfrac{1}{5}$

Thus, $\dfrac{3}{x-2} = \dfrac{-2}{x+1}$ has solution $\frac{1}{5}$. ■

Using the Graphics Calculator

To check our work in Example 2, check that the left-hand member of the equation, `3/(X−2)`, is entered for `Y1` and the right-hand member, `-2/(X+1)`, is entered `Y2`. Return to the home screen and store `1/5` in memory location `X`. Display the evaluations for `Y1` and `Y2`. As the outputs in Figure 2 show, the number 1/5 does result in the left-hand and right-hand members of the equation evaluating to the same number. Thus, the x-value `1/5` produces a true arithmetic sentence when it replaces the variable.

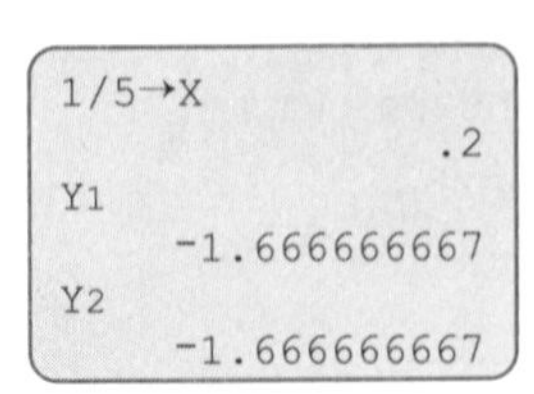

FIGURE 2

The next example shows that a rational expression equation may produce a second-degree equation. Again, we get an equation we know how to solve—a second-degree equation in one variable (see Section 5.5).

EXAMPLE 3 Solve: $\dfrac{x+2}{2} = \dfrac{2}{x-1}$

Solution We start by selecting the LCD for the denominators. Here, the LCD is the product of the denominators, $(2)(x - 1)$.

We multiply each side by $(2)(x + 1)$ to get: $\dfrac{2(x-1)(x+2)}{2} = \dfrac{2(x-1)(2)}{x-1}$

Reduce the rational expressions: $(x - 1)(x + 2) = 4$

Write in standard form: $x^2 + x - 6 = 0$

Write in factored form: $(x + 3)(x - 2) = 0$

$x + 3 = 0$ or $x - 2 = 0$

$x = -3$ $\qquad$ $x = 2$

Thus, $\dfrac{x+2}{2} = \dfrac{2}{x-1}$ has solution -3 and 2. ■

WARNING

Check that none of the numbers listed in the solution results in a denominator evaluating to zero.

Using the Graphics Calculator

To increase our understanding of the equation in Example 3, type `(X+2)/2` for `Y1`. For `Y2` type `2/(X-1)`. View the graph screen using the settings in Figure 3.* From this point on, we will refer to these `RANGE` settings as the Window 4.7 range settings. As the graph in Figure 4 shows, the evaluations of `(X+2)/2` are displayed (the line), and then the evaluations of `2/(X-1)` are displayed. There are two x-values where these evaluations are the same. Use the `TRACE` feature to move the cursor to the rightmost intersection point. The message is `X=2  Y=2`. As you can see, the x-value is 2. That is, replacing the x-variable with 2 results in both sides evaluating to the same number (the evaluation is 2 in this case). Move the cursor to the leftmost intersection point. The message reads `X=-3  Y=-.5`. Another x-value where the two expressions evaluate to the same number is -3.

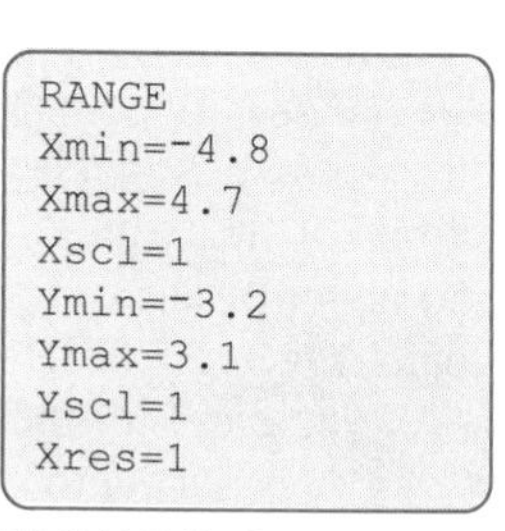

FIGURE 3

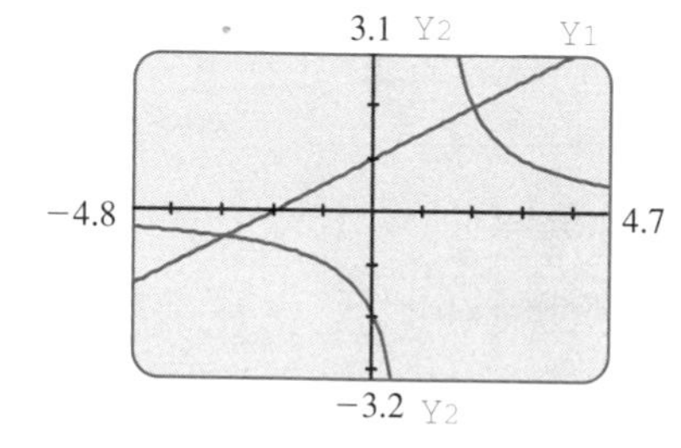

FIGURE 4

TI-82 Note *Use the `ZDecimal` option.

GIVE IT A TRY

Solve each rational expression equation.

1. $\dfrac{3}{x+2} = \dfrac{2}{3}$

2. $\dfrac{3}{x+2} = \dfrac{x}{1}$

3. $\dfrac{2}{2x+5} = \dfrac{1}{2x+5}$

4. $\dfrac{2}{x-2} = \dfrac{x}{x+6}$

Level II The LCD approach works for any rational expression equation at this level of algebra. The following examples show how to solve more involved rational expression equations.

EXAMPLE 4 Solve: $\dfrac{x}{x+2} + \dfrac{2}{x} = 1$

Solution The LCD for the denominators is $x(x+2)$. Multiplying each side by the LCD, we get:

$$\frac{x(x+2)(x)}{x+2} + \frac{x(x+2)(2)}{x} = x(x+2)(1)$$

Remove common factors: $x(x) + (x+2)(2) = x^2 + 2x$

Clear the parentheses: $x^2 + 2x + 4 = x^2 + 2x$

Add $-x^2 - 2x$ to each side to get:

$$\begin{array}{r} x^2 + 2x + 4 = x^2 + 2x \\ \underline{-x^2 - 2x \quad\quad -x^2 - 2x} \\ 4 = 0 \end{array}$$

↑ a false arithmetic sentence

Solution: *None* ■

Using the Graphics Calculator

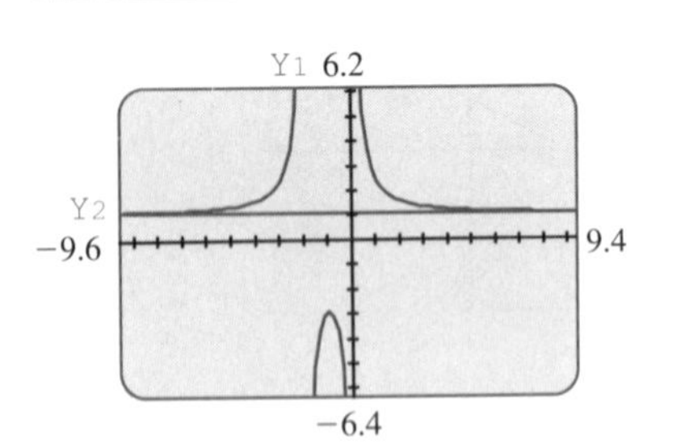

FIGURE 5

To see why the equation in Example 4 has no solution, press the `Y=` key and type `X/(X+2)+2/X` for `Y1`. For `Y2`, type `1`. Use the Window 9.4 range settings to view the graph screen, as shown in Figure 5. Press the `TRACE` key. Use the right and left arrow keys to explore the evaluations of the expression typed for `Y1`. The evaluations of this expression never produce the number 1. Move the cursor until the message reads `X=9.2 Y=1.038819`. Zoom in on the point. Zoom in again. As you can see, the curve representing the evaluations of the left-hand member of the equation, $\dfrac{x}{x+2} + \dfrac{2}{x}$, is above the horizontal line representing the right-hand member, 1.

EXAMPLE 5 Solve: $\dfrac{3}{2x+1} + 1 = \dfrac{x}{x+2}$

Solution The LCD for the denominators, $2x + 1$ and $x + 2$, is $(2x + 1)(x + 2)$. Multiplying both sides by the LCD, we get:

$$\frac{(2x+1)(x+2)(3)}{2x+1} + (2x+1)(x+2) = \frac{(2x+1)(x+2)(x)}{x+2}$$

Remove common factors: $(x + 2)(3) + (2x + 1)(x + 2) = (2x + 1)(x)$

Clear the parentheses: $3x + 6 + 2x^2 + 5x + 2 = 2x^2 + x$

Combine like terms: $2x^2 + 8x + 8 = 2x^2 + x$

Add $-2x^2 - x$ to each side $\quad -2x^2 - x \quad -2x^2 - x$

to get:

$$7x + 8 = 0$$
$$7x = -8$$
$$x = \frac{-8}{7}$$

The solution is $\frac{-8}{7}$. ■

Using the Graphics Calculator

To check our work in Example 5, for Y1 type the right-hand member of the equation, 3/(2X+1)+1. For Y2 type the left-hand member of the equation X/(X+2). Return to the home screen and store -8/7 in memory location X. Display the resulting values in Y1 and Y2. Is the same number output in both cases? If not there is an error. The advantage of using this approach to checking is that if you find an error, you can correct it, then store the correct solution in memory location X (without having to reenter the expressions for Y1 and Y2).

GIVE IT A TRY

Solve each rational expression equation.

5. $\dfrac{2}{x+3} + \dfrac{1}{x-3} = \dfrac{5}{x^2-9}$

6. $\dfrac{x^2}{x+2} = \dfrac{4}{x+2}$

Rewriting Formulas One application of solving rational expression equations in one variable is rewriting formulas. As with earlier examples (see Section 2.8), we will use an analogy as an aid. To rewrite a formula for a particular variable, say P, means that we want P to appear as the left-hand member of the equation, and that the left-hand side is the only place P appears in the formula. Consider the following example.

EXAMPLE 6 Rewrite the formula $\dfrac{1}{P} = \dfrac{R+C}{P+2}$ for P.

Solution To create an analogy (a rational expression equation), we replace each variable except P with a number. Solving this equation provides a guide for working with the formula.

	Analogy (for R = 7 and C = 8)	*Formula*
	$\frac{1}{P} = \frac{7+8}{P+2}$	$\frac{1}{P} = \frac{R+C}{P+2}$
Multiply by the LCD, $P(P+2)$:	$\frac{P(P+2)}{P} = \frac{P(P+2)(15)}{P+2}$	$\frac{P(P+2)}{P} = \frac{P(P+2)(R+C)}{P+2}$
Reduce:	$P + 2 = (15)P$	$P + 2 = P(R + C)$
	$(15)P = P + 2$	$P(R + C) = P + 2$
	$15P = P + 2$	$PR + PC = P + 2$
	$-P \quad -P$	$-P \quad -P$
	$14P = 2$	$(R + C - 1)P = 2$
	$P = \frac{2}{14} = \frac{1}{7}$	$P = \frac{2}{R + C - 1}$

So, the formula rewritten for P is $P = \frac{2}{R+C-1}$. ■

As a check to Example 6 be sure to answer the following questions:

- Does P appear on the left-hand side of the equal sign, by itself?
- Is the left-hand side the only place P appears in the formula?

Your answer should be *yes* to both questions.

Using the Graphics Calculator

To check our work on Example 6, store 7 in memory location R and store 8 in memory location C (the numbers used to create the analogy). Next, type `2/(R+C−1)` and then press the `ENTER` key. Type `1/7` (the number from the analogy), and press the `ENTER` key. The numbers output should be the same. If not, an error is present.

EXAMPLE 7 Rewrite the formula $\frac{1}{x} = \frac{1}{a} + \frac{1}{c}$ for a.

Solution

	Analogy (for x = 5 and c = 3)	*Formula*
	$\frac{1}{5} = \frac{1}{a} + \frac{1}{3}$	$\frac{1}{x} = \frac{1}{a} + \frac{1}{c}$
Multiply by LCD:	$\frac{15a}{5} = \frac{15a}{a} + \frac{15a}{3}$	$\frac{acx}{x} = \frac{acx}{a} + \frac{acx}{c}$
Reduce:	$3a = 15 + 5a$	$ac = cx + ax$
	$-5a \quad -5a$	$-ax \quad -ax$
	$-2a = 15$	$(c - x)a = cx$
	$a = \frac{-15}{2}$	$a = \frac{cx}{c - x}$

So, the formula rewritten for a is $a = \frac{cx}{c-x}$. ■

EXAMPLE 8 Rewrite $A = \dfrac{rLF}{F - 1}$ for F.

Solution

	Analogy (for A = 10, r = 3, and L = 2)	*Formula*
	$\dfrac{10}{1} = \dfrac{(3)(2)F}{F - 1}$	$\dfrac{A}{1} = \dfrac{rLF}{F - 1}$
Multiply by LCD:	$\dfrac{(F-1)(10)}{1} = \dfrac{(F-1)(3)(2)F}{F - 1}$	$\dfrac{(F-1)A}{1} = \dfrac{(F-1)rLF}{F - 1}$
Reduce:	$(F - 1)(10) = 6F$	$(F - 1)A = rLF$
	$10F - 10 = 6F$	$AF - A = rLF$
	$-6F \qquad -6F$	$-rLF \qquad -rLF$
	$4F - 10 = 0$	$(A - rL)F - A = 0$
	$10 \qquad 10$	$A \qquad A$
	$4F = 10$	$(A - rL)F = A$
	$F = \dfrac{10}{4} = \dfrac{5}{2}$	$F = \dfrac{A}{A - rL}$

So, the formula rewritten for F is $F = \dfrac{A}{A - rL}$. ■

GIVE IT A TRY

7. Rewrite $A = \dfrac{rLF}{F - 1}$ for L.

8. Rewrite $y = \dfrac{x + 2}{x - 1}$ for x.

9. Rewrite $\dfrac{a}{b} = \dfrac{1}{c} - a$ for a.

■ SUMMARY

We started this section by solving rational expression equations. The key to this process is to multiply each side of the equation by the LCD for the denominators in the equation. Because the LCD is often a variable expression and the multiplication property of equation solving allows multiplication only by a nonzero number, using this LCD can generate extraneous solutions. You must check that *none* of the numbers listed in the solution result in a denominator evaluating to 0. If such a number is found, it must be removed from the numbers listed for the solution. An application of solving rational expression equations is rewriting formulas. To rewrite a formula, an analogy of a rational expression equation can be created to serve as a guide. The steps involved in solving the equation are mirrored in rewriting the formula.

6.5 ■ ■ ■ EXERCISES

1. Which one of the following expressions is *not* a polynomial?

   i. $x^2 - 3x + 2$ ii. $\frac{2}{3}x$ iii. $\frac{2}{3x}$ iv. $\frac{2}{3}$

2. Which one of the following expressions is *not* a rational expression?

   i. $x^2 - 3x$ ii. $\sqrt{2}x$ iii. $\frac{\sqrt{2}}{x}$ iv. $\sqrt{2x}$

3. Which one of the following is a first degree-equation in one variable?

   i. $3x + 2y = 8$ ii. $\frac{1}{x} = \frac{2}{3}$

   iii. $x^2 - 3x = 7$ iv. $7 - 3x = \frac{2}{3}$

4. Which one of the following is a rational expression equation in one variable?

   i. $\frac{3}{x + 1} = 2$ ii. $3(x + 1) = y$

   iii. $|2x - 1| = 5$ iv. $\sqrt{x + 1} = 2$

5. (true/false) The solution to $\frac{2x + 5}{x - 3} = \frac{x}{x - 3}$ and $(x - 3)(2x + 5) = x(x - 3)$ are the same.

6. (true/false) The solution to a rational expression equation in one variable is always a single number.

7. Explain in your own words, the difference between problems A and B:

   A. Write as a completely reduced rational expression:

   $$\frac{x}{x - 3} + \frac{1}{2}$$

   B. Solve: $\frac{x}{x - 3} + \frac{1}{2} = \frac{1}{x}$

8. As a completely reduced rational expression, $\frac{x}{x - 3} + \frac{1}{2}$ is $\frac{3x - 3}{2(x - 3)}$. Report the solution to the equation.

   $$\frac{x}{x - 3} + \frac{1}{2} = \frac{3x - 3}{2(x - 3)}$$

Solve each rational expression equation. Use the graphics calculator to check your work.

9. $\frac{1}{x} = \frac{1}{2}$
10. $\frac{1}{x} = \frac{5}{x + 2}$
11. $\frac{3}{x + 1} = \frac{2}{x - 1}$
12. $\frac{x + 2}{x - 3} = \frac{2}{3}$
13. $\frac{2}{3} = \frac{2x + 5}{x - 3}$
14. $\frac{x + 1}{5} = \frac{2x + 3}{2}$
15. $\frac{3}{x - 2} = \frac{x + 3}{2}$
16. $\frac{1}{x - 2} = \frac{x + 3}{6}$
17. $\frac{1}{x + 2} = \frac{x + 3}{12}$
18. $\frac{2x + 3}{2} = \frac{1}{x}$
19. $\frac{x + 2}{x - 3} = \frac{x}{x - 1}$
20. $\frac{x - 1}{x - 3} = \frac{x + 3}{4}$
21. $\frac{1}{x - 2} + \frac{2x + 1}{x - 2} = \frac{-1}{4}$
22. $\frac{x + 3}{2x} + \frac{1}{2x} = \frac{1}{2}$
23. $\frac{1}{x - 2} + \frac{1}{x + 1} = \frac{2}{x}$
24. $\frac{1}{x - 2} = \frac{1}{x + 1}$

Rewrite each formula for the indicated variable.

25. $y = \frac{5}{x - 2}$ for x
26. $y = \frac{x}{x - 2}$ for x
27. $A = \frac{C + 2}{B - C}$ for C
28. $A = \frac{C + 2}{B - C}$ for B
29. $\frac{y}{x - 2} = \frac{5}{2}$ for x
30. $\frac{y}{x - 2} = \frac{5}{2}$ for y
31. $\frac{1}{R_1} = \frac{R_2 - R}{RR_2}$ for R
32. $\frac{1}{R_1} + \frac{1}{R_2} = \frac{1}{R}$ for R
33. $\frac{1}{y} = \frac{5}{x - 2} - \frac{1}{5}$ for y

6.6 ■ ■ ■ APPLICATIONS OF RATIONAL EXPRESSIONS

In the last section, we learned how to solve rational expression equations. We will now apply that knowledge to creating mathematical models (solving word problems) that are based on rational expression equations.

Word Problems Involving Work Rational expression equations often serve as mathematical models for physical situations that involve people or machines working at a constant rate. Suppose Katie can mow a lawn in 4 hours. Assuming Katie works at a uniform rate, we reason that she can do $\frac{1}{4}$ of the job in 1 hour. Now, suppose Bill mows the same lawn in 5 hours. (In one hour, Bill will do $\frac{1}{5}$ of the job.) We want to know how long it will take Katie and Bill to mow the lawn if they work together. We assume the part of the job done by Katie plus the part of the job done by Bill will equal one job done. Using the problem-solving procedure discussed in Section 2.6, we have these quantities.

Worker	Rate of work	Time taken	Fraction done
Katie	$\frac{1}{4}$	x	$\frac{x}{4}$
Bill	$\frac{1}{5}$	x	$\frac{x}{5}$

Let x = the number of hours for Katie and Bill, working together, to mow the lawn.

$\frac{x}{4}$ = fraction of the job Katie does. $\frac{x}{5}$ = fraction of the job Bill does.

We can also set up this information as shown in the table.

Fraction of job done by Katie ↓ Fraction of job done by Bill ↓

Equation:

$$\frac{x}{4} + \frac{x}{5} = 1 \quad \leftarrow \text{1 job done}$$

$$\frac{20x}{4} + \frac{20x}{5} = 20 \quad \text{Multiply by LCD, 20.}$$

$$4x + 5x = 20 \quad \text{Remove common factor.}$$

$$9x = 20$$

$$x = \frac{20}{9}$$

So, Katie and Bill can do the job in $\frac{20}{9}$ hours, or $2\frac{2}{9}$ hours.

The next example illustrates a word problem that produces a rational expression equation as a mathematical model.

EXAMPLE 1 Machine A and machine B, working together, can do a job in 30 minutes. If machine A can complete the task working alone in 50 minutes, how long does it take machine B working alone to complete the task?

Solution Let x = time it takes machine B to do the task.

Using the table shown below, we get:

$\frac{30}{50}$ = fraction of job done by machine A $\qquad$ $\frac{30}{x}$ = fraction of job done by machine B

Machine	Rate of work per minute	Time	Fraction done
A	$\frac{1}{50}$	30	$\frac{30}{50}$
B	$\frac{1}{x}$	30	$\frac{30}{x}$

Fraction of job done by A ↓ Fraction of job done by B ↓

Equation $\dfrac{30}{50} + \dfrac{30}{x} = 1$ ← 1 job done

$$\begin{aligned}
\frac{50x(30)}{50} + \frac{50x(30)}{x} &= 50x && \text{Multiply by LCD, } 50x.\\
30x + 1500 &= 50x && \text{Remove common factors.}\\
-50x \qquad\quad & \quad -50x && \text{Add } -50x \text{ to each side.}\\
\hline
-20x + 1500 &= 0 \\
-1500 & \quad -1500 && \text{Add } -1500 \text{ to each side.}\\
\hline
-20x &= -1500 \\
x &= \frac{-1500}{-20} && \text{Multiply by } \tfrac{-1}{20}.
\end{aligned}$$

Now, $\frac{-1500}{-20} = 75$. Thus, Machine B can do the job in 75 minutes. ■

Using the Graphics Calculator

```
RANGE
Xmin=0
Xmax=95
Xscl=10
Ymin=0
Ymax=2
Yscl=1
Xres=1
```

FIGURE 1

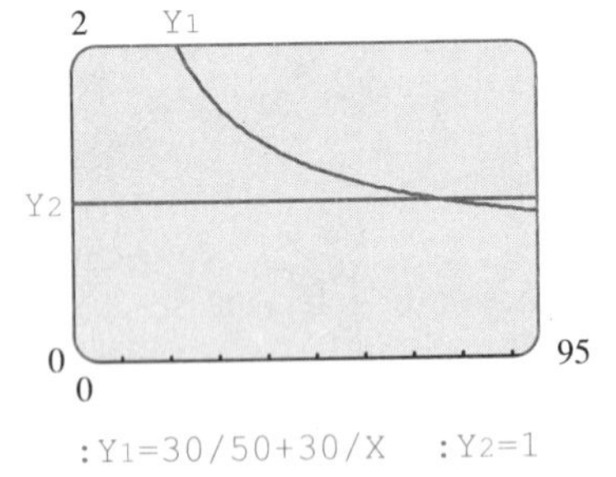

FIGURE 2

To increase our understanding of Example 1, for `Y1` type the left-hand member of the equation, `30/50+30/X`. For `Y2`, type the right-hand member of the equation, `1`. To find an appropriate view, we reason that we are interested only in positive x-values (the times taken for machine B). Also, we wish to see where the left-hand member of the equation equals the right-hand member, 1—the full job completed. From the solution to Example 1, we know the x-value is 75. Press the `RANGE` key and type the values shown in Figure 1.* View the graph screen, as shown in Figure 2. Press the `TRACE` key and then the left arrow key until the message reads `X=40 Y=1.35`. This means that if machine B takes 40 minutes to do the job working alone, the two machines working together for 30 minutes will do the job 1.35 times. That is, they will have done the complete job once and will have completed about one-third of the job again. Notice that moving to the left means machine B can do the job faster (in a shorter amount of time). The time, the x-value, never reaches 0 (machine B can't do the job instantly). Move the cursor to the right until the message reads `X=80 Y=.975`. This means that if machine B takes 80 minutes to do the job alone, the two machines working together will not quite complete the job in 30 minutes (they will complete only 97.5% of the job). Of course, when the x-value is 75, then the part of the job done in 30 minutes is 1, or 100%.

TI-82 Note *Type `94` for `Xmax`.

GIVE IT A TRY

For each situation, report the quantity represented by the variable, the equation produced, its solution, and the answer to the word problem.

1. A farmer can plow a field in 6 hours, and his daughter can plow the same field in 10 hours. Working together, how many hours will it take to plow the field?
2. Computer A can process CRTs payroll in 6 hours. If computer A and computer B working together can process this payroll in 4 hours, how long does it take computer B to process the payroll, working alone?
3. A painter can paint a house in 10 hours. Her helper can paint the same house in 15 hours. If the painter works on the house for 4 hours, and then her helper finishes the job, how many hours will the helper have to paint? [*Hint*: Let $x =$ the number of hours the helper paints.] If the painter is paid $15 an hour, and her helper gets $8 per hour, how much should they bill for the job?

Uniform Motion Another type of word problem that often leads to a rational expression equation is that of uniform motion. The basic relationship between distance traveled, the rate of travel, and the time spent traveling is $d = rt$. We can rewrite this formula for t as $t = \frac{d}{r}$. We now use this relationship to solve the following problems.

EXAMPLE 2 An express bus travels 350 miles in the same time that it takes a local bus to travel 300 miles. If the local bus travels at a rate 10 mi/h less than the express bus, what is the rate of the express bus?

Solution Let $x =$ rate of the express bus.
Filling in a chart for this situation produces the table shown in the margin. Using the information from this table, we have

$$\frac{350}{x} = \text{time for express bus} \quad \text{and} \quad \frac{300}{x-10} = \text{time for the regular bus}$$

Bus	d	r	t
Express	350	x	$\frac{350}{x}$
Regular	300	$x - 10$	$\frac{300}{x-10}$

The time of travel is the same for both busses, so we solve the following:

Time for express bus → $\frac{350}{x} = \frac{300}{x-10}$ ← Time for regular bus

$$\frac{350x(x-10)}{x} = \frac{300x(x-10)}{x-10} \qquad \text{Multiply by LCD, } x(x-10).$$

$$350(x-10) = 300x \qquad \text{Reduce}$$

$$350x - 3500 = 300x \qquad \text{Clear parentheses.}$$

$$\underline{-300x \qquad\qquad -300x} \qquad \text{Add } -300x \text{ to each side.}$$

$$50x - 3500 = 0$$

$$\underline{3500 \qquad 3500} \qquad \text{Add 3500 to each side.}$$

$$50x = 3500$$

$$x = \frac{3500}{50} = 70 \qquad \text{Multiply by } \tfrac{1}{50}.$$

The solution is 70. Thus, the rate for the express bus is 70 mi/h. ■

As a final example of word problems with rational expression equations, work through the following uniform motion problem.

EXAMPLE 3 Gil drives from his home to work at 50 mi/h. On the return trip, traffic is heavy, and he can travel at only 40 mi/h. If his total driving time for the round trip is 4 hours, how far does Gil live from his job?

Solution The quantity requested is the distance from Gil's home to his job.
Let x = distance from home to job.
Next, we fill in the chart shown in the margin. Using the information from the chart,

Trip	*d*	*r*	*t*
Going	x	50	$\frac{x}{50}$
Returning	x	40	$\frac{x}{40}$

$$\frac{x}{50} = \text{time to travel to work} \qquad \text{and} \qquad \frac{x}{40} = \text{time to travel home}$$

Time to travel to work ↓ ; Time to travel home ↓ ; Total travel time ↓

$$\frac{x}{50} + \frac{x}{40} = 4$$

$$\frac{200x}{50} + \frac{200x}{40} = 4(200) \qquad \text{Multiply by LCD.}$$

$$4x + 5x = 800$$

$$9x = 800$$

$$x = \frac{800}{9}$$

The solution is $\frac{800}{9}$, or $88\frac{8}{9}$. To answer the question, Gil lives $88\frac{8}{9}$ miles from his job. ■

GIVE IT A TRY

For each problem, report the variable and what it represents, the equation that models the problem, and the answer to the problem.

4. Jill drove 300 miles at two different speeds, 65 mi/h and 50 mi/h. If the total time for the trip was 5 hours, how many miles did Jill travel at 50 mi/h?

5. Dante can fly his ultralight plane against the wind 200 miles in the same time it takes him to fly 250 miles with the wind. If the wind is blowing at 10 mi/h, how fast would the plane fly in calm air (no wind)?

■ SUMMARY

In this section, we have learned that rational expression equations in one variable often serve as mathematical models for situations involving work rates and uniform rates of travel. The process (procedure) we use to solve these problems is the same as the one we use with first-degree equations in one variable (Section 2.6) and with second-degree equations in one variable (Section 5.5). The major difference is that the equation produced is a rational expression equation in one variable.

6.6 ■ ■ ■ EXERCISES

For each word problem, report the quantity represented by the variable, the equation produced, its solution, and the answer to the problem.

1. A farmer can plow a field in 5 hours, and his son can plow the same field in 12 hours. If they both work on plowing the field, how many hours will it take for them to plow the field?
2. A painter can paint a house in 8 hours. Her helper can paint the same house in 12 hours. The painter is paid $17 an hour, and her helper earns $9 per hour. If the painter works on the house for 5 hours, and then her helper finishes the job, how much is the bill for the job? [*Hint*: Let $x =$ the number of hours the helper works.]
3. Pump A can empty a tank in 6 hours. If pump A and pump B working together can empty the tank in 4 hours, how long does it take pump B to empty the tank working alone?
4. Drainage pipe A can fill a holding pond in 6 weeks, and drainage pipe B can fill the same pond in 8 weeks. Evaporation and seepage will empty the pond in 16 weeks. If the pond is empty, and then both pipes A and B are opened, how long will it take to fill the holding pond?
5. Fredda can paint 30 toy soldiers in a day and Sal can paint 40 toy soldiers in a day. How long will it take Fredda and Sal working together to paint 300 toy soldiers?
6. Computer A can process a job in 3 hours. Computer B can process the same job in 5 hours. If computer B has been processing the job for 2 hours when computer A starts working on the job, how long will it take the two computers to finish processing the job?
7. Kelly and Walt can clean their apartment in 2 hours when they work together. If it takes Kelly 3 hours to clean the apartment when she does it by herself, how long will it take Walt to clean the apartment by himself?
8. If both the hot and cold water taps are opened, the bathtub fills in 13 minutes. With only the cold water tap open, it takes 30 minutes to fill the tub. How long will it take to fill the tub if only the hot water tap is opened?
9. Bosworth can mow a lawn in 4 hours. If Bosworth has been working for an hour when his friend Ashley joins him, and they are finished in 2 hours, how long will it take Ashley to mow the lawn working alone?
10. Pipe A can fill a tank in 9 hours. Pipe B fills the same tank in 12 hours. The drain on the tank empties the tank in 10 hours. If both pipes and the drain are opened, how long will it take to fill the tank?
11. An express bus travels 400 miles in the same time that it takes a local bus to travel 290 miles. If the local bus travels at a rate 20 mi/h less than the express bus, what is the rate of the express bus?
12. Gil drives from his home to work at 40 mi/h. On the return trip, traffic is heavy and he can travel at only 30 mi/h. If his total driving time for the round-trip is 3 hours, how far does Gil live from his job?
13. Dante can fly his ultralight plane against the wind 150 miles in the same time it takes him to fly 190 miles with the wind. If the wind is blowing at 20 mi/h, how fast would the plane fly in calm air?
14. Jill drove 400 miles at two different speeds, 60 mi/h and 50 mi/h. If the total time for the trip was 7 hours, how many miles did Jill travel at 60 mi/h?
15. Thelma leaves home, traveling at 50 mi/h. An hour later, her husband leaves home, traveling at 60 mi/h. How long will it take Thelma's husband to catch her?
16. A charter bus leaves St. Paul for a summer campsite at the same time a car leaves on the same trip. The bus averages 70 mi/h and the car averages 50 mi/h. If the bus arrives 2 hours ahead of the car, how far is the summer campsite from St. Paul?
17. A boat running against the current travels 5 miles in the same time that it takes the boat to travel 8 miles running with the current. If the current is 4 mi/h, what is the speed of the boat in calm water?
18. Gary's power boat will cruise at 40 mi/h in calm water. He can run his boat 50 miles downstream in 75% of the time it takes him to run the boat 50 miles upriver. Find the speed of the current.
19. John can ride his bike 10 mi/h faster than he can jog. If he can bike 10 miles in the same amount of time that it takes him to jog 4 miles, how fast can John ride his bike?
20. The Smelts drove to a party down the coast from Jacksonville on I-95 at 60 mi/h. On the return trip they took the scenic route along US 1 at 30 mi/h. If the total driving time for the trip was 3 hours, how far was the party from Jacksonville?
21. Train A leaves A-town traveling at 80 mi/h. for C-ville, 300 miles away. Train B leaves B-burgh at the same time, traveling at 70 mi/h, headed for C-ville. If Train A reaches C-ville an hour earlier than train B, how far is B-burgh from C-ville?

22. Frank can buy 38 cans of beans for the same amount of money that he can buy 50 cans of corn. If the beans cost 7¢ more per can than the corn, what is the price of a can of beans?

23. The sum of a positive integer and 2 decreased by 8 times the reciprocal of the integer is 9. What is the positive integer?

24. Linda purchased $4,000 worth of T-shirts at two prices, $8 per shirt and $12 per shirt. If the total number of T-shirts purchased was 400, how many T-shirts did she buy for $12 per shirt?

25. A runner and a biker left Gull Point at the same time, on a trail that is 20 miles long. If the biker traveled 5 times faster than the runner and the biker finished the trail 3 hours earlier than the runner, what was the speed of the runner?

6.7 ■ ■ ■ PROPORTION AND VARIATION

In this section, we will again apply our knowledge of solving a rational expression equation to real-world situations. We will first discuss proportion, then variation.

Proportions A common mathematical model used for many situations in the sciences and in business is a proportion. You already know that a ratio is a comparison of two numbers by division (typically, the ratio of 3 to 5 is written as $3 : 5$). The ratio $3 : 5$ means $\frac{3}{5}$. A *proportion* is a statement that two ratios are equal. Consider the following example, in which an equation is written in proportion notation.

EXAMPLE 1 Solve the proportion: $x : (x + 3) = 5 : 7$

Solution The key to solving this equation is knowing what the symbol : means. The ratio of x to $x + 3$, written $x : (x + 3)$, means $\dfrac{x}{x + 3}$.

Thus, we solve the equation

$$\frac{x}{x+3} = \frac{5}{7}$$

$$\frac{7(x+3)(x)}{x+3} = \frac{7(x+3)(5)}{7} \qquad \text{Multiply by the LCD.}$$

$$7x = 5(x+3)$$

$$7x = 5x + 15$$

$$\underline{-5x \quad -5x}$$

$$2x = 15$$

$$x = \frac{15}{2}$$

The equation (proportion) $x : x + 3 = 5 : 7$ has the solution $\frac{15}{2}$. ■

GIVE IT A TRY

1. Solve: $3 : x = 5 : (x - 2)$

2. Write $\dfrac{3x}{x-1} = \dfrac{2}{5}$ as a proportion.

Variation A closely related idea to proportions that we often use to describe relationships between variables is that of variation. To say that a quantity, say y, *varies directly* with another quantity, say x, means $y = kx$, where k is the *constant of variation*. We also say that y is *directly proportional* to x and that k is the *constant of proportionality*. The value of k is determined by the data for a particular situation. Of course, we would also say that y is a linear function of x. To say that a quantity, say y, *varies directly with the square* of another quantity means that $y = kx^2$, where k is the constant of variation. In this case, we would say that y is a quadratic function of x. Consider the following example.

EXAMPLE 2 Property tax varies directly with the value of the property. If a piece of land is valued at \$20,000 and has a property tax of \$200, what is the constant of variation? What is the property tax on a piece of property valued at \$68,000?

Solution

Knowing that the property tax T varies directly with the value of the property V yields:	$T = kV$
To find the constant of variation k, we use the data given (taxes are \$200 on property valued at \$20,000) to get:	$200 = k(20000)$
Solving this equation, we get	$k = 0.01$
Thus, the formula is	$T = 0.01V$
To find the tax on a property with a value of \$68,000, we use the formula	$T = 0.01(68{,}000) = 680$

The tax is \$680. ■

Using the Graphics Calculator

To increase our understanding of the work in Example 2, type `0.01X` for `Y1`. Use the settings in Figure 1* to view the graph (see Figure 2). Use the `TRACE` feature and move the cursor to an x-value of `68000`. Is the y-value 680?

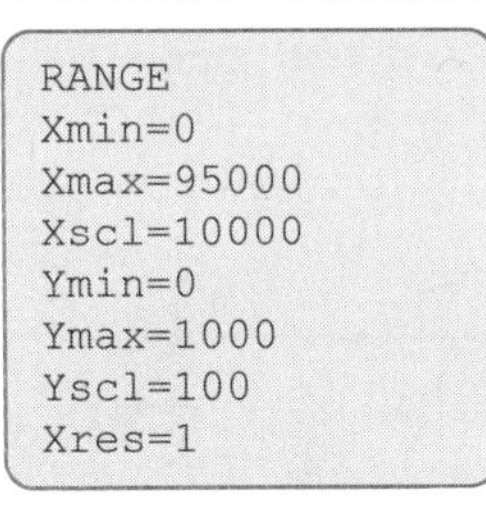

FIGURE 1

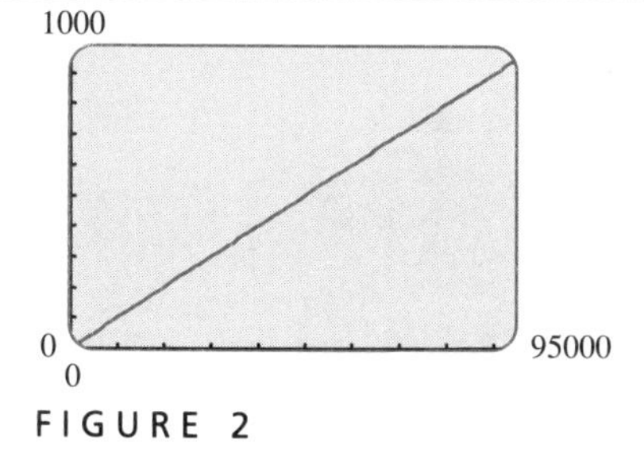

FIGURE 2

To say that a quantity, say y, *varies inversely* with another quantity, say x, means that $y = \frac{k}{x}$. Again, k is the constant of variation. To say that a quantity, say y, *varies inversely with the square* of another quantity, say x, means

TI-82 Note *Type `94000` for `Xmax`.

$y = \frac{k}{x^2}$, where k is the constant of variation. In both of these cases, we would say that y is a rational function of x. Consider the following example.

EXAMPLE 3 The intensity of light I varies inversely with the square of the distance d the measuring device is positioned from the source of the light (see Figure 3). If a light source produces 16 foot-candles at a distance of 15 feet, what is the constant of variation? If the intensity is measured at 9 foot-candles, how far is the measuring device from the light source?

Solution Knowing that I varies inversely with the square of d yields the equation

$$I = \frac{k}{d^2}$$

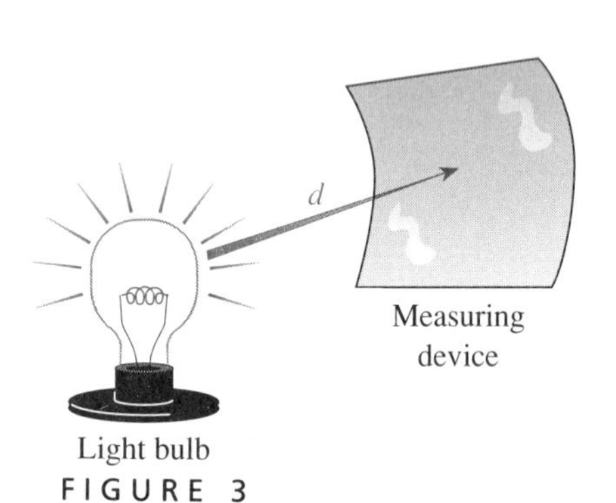

FIGURE 3

Using the data, we have: $16 = \frac{k}{(15)^2}$ or $k = 3600$

So, the constant of variation is 3600. Thus, the formula is: $I = \frac{3600}{d^2}$

To find the distance for an intensity of 9, we solve:

$$9 = \frac{3600}{d^2}$$
$$9d^2 = 3600$$
$$d^2 = 400$$
$$d^2 - 400 = 0$$
$$(d + 20)(d - 20) = 0$$
$$d + 20 = 0 \quad \text{or} \quad d - 20 = 0$$
$$d = -20 \qquad\qquad d = 20$$

The solution is 20 and -20. Because we are looking for a distance, we discard the negative solution, -20. Thus, the distance is 20 feet. ■

Using the Graphics Calculator

To increase our understanding of the work in Example 3, type `3600/X²` for `Y1`. For `Y2` type `9`. Use the settings in Figure 4* to view the graph, as shown in Figure 5. Use the `TRACE` a feature to estimate the x-value at the intersection point of the curve and the horizontal line.

```
RANGE
Xmin=0
Xmax=47.5
Xscl=5
Ymin=0
Ymax=20
Yscl=5
Xres=1
```

FIGURE 4

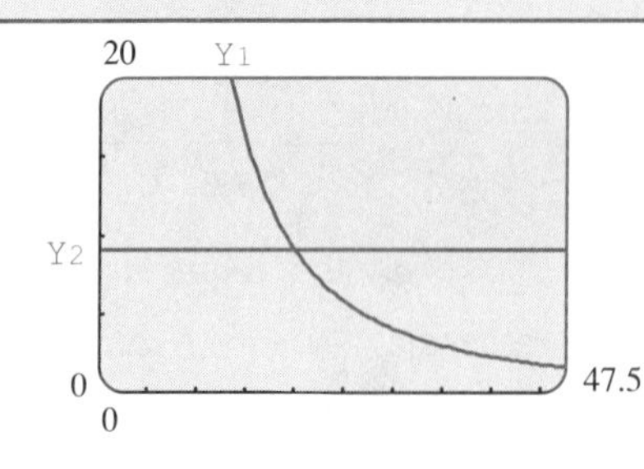

:Y1=3600/X²
:Y2=9

FIGURE 5

TI-82 Note *Type `47` for `Xmax`.

To say that a quantity, say y, *varies jointly* with two other quantities, say x and z, means that $y = kxz$, where k is the constant of variation. Here, y is a function of two variables. (We have considered only functions of a single variable so far.) In the language of variation, the terms *directly, inversely,* and *jointly* can be mixed. Consider Newton's famous statement of the relationship of the gravitational force F between particles of mass m_1 and m_2 to the distance d that the particles are apart:

$$F = \frac{km_1m_2}{d^2}$$

Using the language of variation, we state this relationship as

> Force F varies jointly with m_1, the mass of particle 1, and m_2, the mass of particle 2, and inversely with the square of distance d between the two particles.

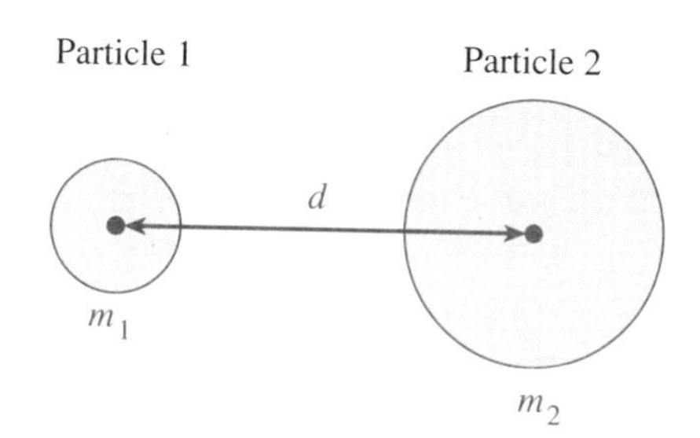

FIGURE 6

Figure 6 illustrates the situation. Using functions, we state this relationship as:

Force F is a rational function of three variables, m_1, m_2, and d, given by the rule $\frac{km_1m_2}{d^2}$. That is,

$$F(m_1, m_2, d) = \frac{km_1m_2}{d^2}$$

Which language is better, variation or function? It really doesn't matter, since both are used in the sciences and in business.

GIVE IT A TRY

3. The volume V of a gas varies directly with temperature T and inversely with the pressure P. If the volume of a gas is 30 in^3 (cubic inches) when the temperature is 60°F and the pressure is 50 psi (pounds per square inch), what is the constant of variation? What is the formula for V? What is the volume of this gas if the temperature is raised to 70°F and the pressure is held at 50 psi?

4. Use variation to describe the relationship $R = \frac{2T^2}{V}$.

■ SUMMARY

In this section, we have learned to solve a proportion (a statement that two ratios are equal). We also considered the language of variation. Direct variation means that one variable equals a constant times another variable, or a constant times the square of another variable. From a mathematical viewpoint, direct variation produces either a linear or quadratic function. Inverse variation means that one variable equals a constant divided by another variable, or a constant divided by the square of another variable. Mathematically, inverse variation produces a rational function. Joint variation and combinations of direct, inverse, and joint variation are used to describe relationships between variables in formulas.

■ ***Quick Index***

The following is a summary of variation.

Relationship	*Formula*
y varies directly with x	$y = kx$
y varies directly with the square of x	$y = kx^2$
y varies inversely with x	$y = \frac{k}{x}$
y varies inversely with the square of x	$y = \frac{k}{x^2}$
y varies jointly with x and z	$y = kxz$

6.7 ■ ■ ■ EXERCISES

Solve each proportion.

1. $3 : x = 5 : 16$
2. $5 : 7 = 3 : x$
3. $17 : 8 = (x + 3) : 2$
4. $(x - 2) : 5 = 2 : 8$
5. $(x + 1) : 2 = x : 7$
6. $5 : (x + 3) = 7 : (2x)$
7. $(2x - 5) : x = 4 : 9$
8. $3 : 2 = x : (3x + 4)$
9. $1 : (x - 2) = (x + 2) : 5$
10. $x : 5 = 2 : (x - 3)$
11. The distance a spring stretches varies directly with the force applied. If a force of 5 pounds stretches the spring 6 inches, what is the constant of variation? How far will a force of 8 pounds stretch the spring?

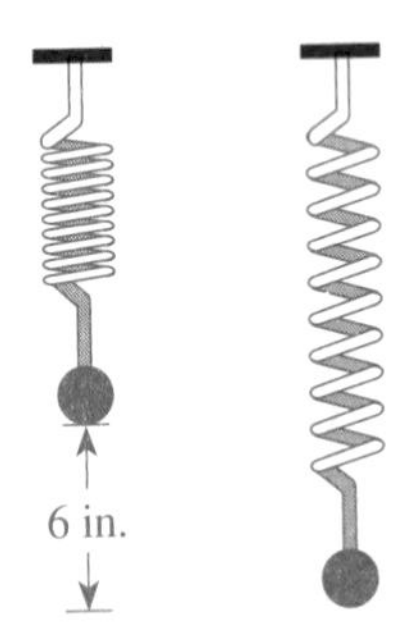

12. The area of a triangle varies directly with the height h of the triangle and the base b (the side to which the altitude is drawn), as shown in the diagram. What is the constant of variation?

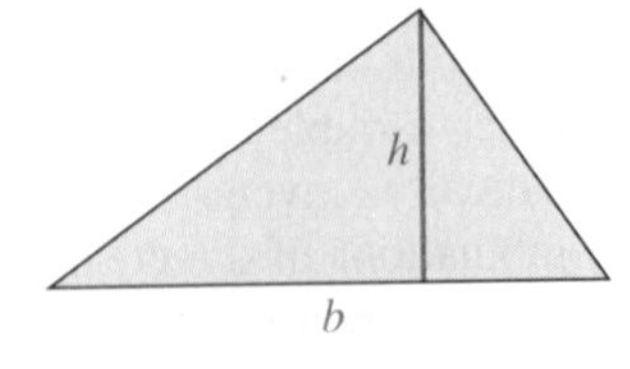

13. The area of the circle sketched in the diagram varies directly with the square of the radius. What is the constant of variation?

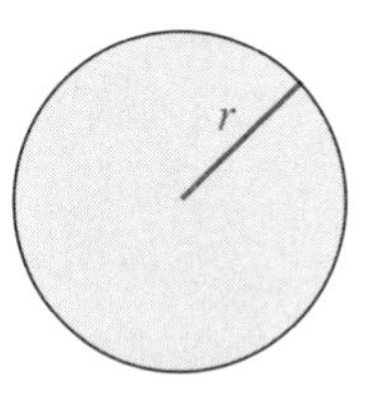

14. The volume of a right circular cone varies directly with the height h of the cone and the square of the radius r (see the sketch). The constant of variation is $\frac{\pi}{3}$. What is the volume for a right circular cone of height 4 feet and a base with raduis 2 feet?

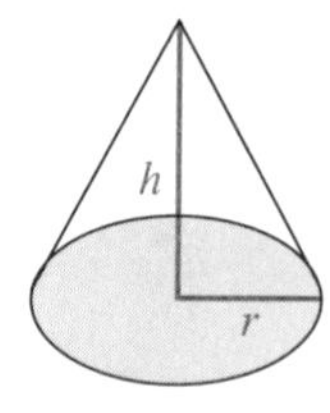

15. The cost of developing a new product at C-Zoom varies directly with time required and inversely with the number of researchers assigned to the project. If the projected cost is \$4.2 million for 5 researchers to develop the ScanT product in 6 months, what is the constant of variation? Find the projected cost for 8 researchers to develop the ScanW product in 5 months.

16. The volume of a gas varies directly with the temperature and inversely with the pressure. For a particular gas, the volume is 6 L (liters) at a temperature of 360°K (degrees Kelvin) with a pressure of 75 kg/cm^2 (kilograms per square centimeter). What is the constant of variation? What is the temperature for this gas when the volume is 5 L and the pressure is 75 kg/cm^2?

Write each formula in variation language.

17. $V = \pi r^2 L$

18. $F = ma$

19. $N = \dfrac{2.01}{d}$

20. $K = \dfrac{2\pi}{F}$

21. $K = \dfrac{1}{2}mv^2$

22. $L = \dfrac{4M}{5T}$

23. $I = 0.03Pt$

24. $T_2 = \dfrac{V_2}{V_1}T_1$

25. $v = \dfrac{d}{t}$

26. $T_2 = \dfrac{P_2V_2}{nR}$

6.8 ■ ■ ■ INTEGER EXPONENTS

In previous chapters we have used only whole number exponents (0, 1, 2, 3, . . .) when working with polynomials. In the rational expressions discussed in this chapter, we have used ratios of polynomials and only whole number exponents. In this section, we will expand the topic of exponents to include the negative integers (. . . , −3, −2, −1). In Chapter 7 we will expand exponents to include any rational number, such as $\frac{-2}{3}$ or $\frac{4}{5}$. Later, when working with exponential and logarithmic expressions, we will expand exponents to any real number, such as $\sqrt{2}$ or π. Exponents will be discussed for the next several chapters.

Zero and Negative One Earlier we stated that any number (except 0) with the exponent 0 is 1. That is, if $a \neq 0$, then $a^0 = 1$. We also stated that if n is a natural number, then x^n means $xx \cdot \cdot \cdot x$, for n factors of x. Finally, we have used the basic laws of exponents, which we list again here.

LAW OF EXPONENTS

For m and n integers [and no 0 denominator]:

1. $a^n a^m = a^{n+m}$ — *Example:* $(x^3)(x^2) = x^{3+2} = x^5$
2. $a^n \div a^m = a^{n-m}$ — *Example:* $(x^3) \div (x^2) = x^{3-2} = x^1$
3. $(a^n)^m = a^{nm}$ — *Example:* $(x^3)^2 = x^{(3)(2)} = x^6$
4. $(ab)^n = a^n b^m$ — *Example:* $(xy)^3 = x^3y^3$
5. $\left(\dfrac{a}{b}\right)^n = \dfrac{a^n}{b^n}$ — *Example:* $\left(\dfrac{x}{y}\right)^3 = \dfrac{x^3}{y^3}$

Worded phrases for these laws are often helpful:

1. Multiplying like bases, add the exponents.
2. Dividing like bases, subtract the exponents.
3. A power to a power, multiply the exponents.
4. A product to a power is each factor to the power.
5. A quotient to a power is the numerator and denominator to the power.

Let's take a moment to see that these definitions agree with the work we've done with rational expressions.

With rational expressions, we write: $\frac{\cancel{x^2}}{\cancel{x^2}} = 1$

With the laws of exponents, we write: $\frac{x^2}{x^2} = x^{2-2} = x^0$

So, we have $x^0 = 1$.

With rational expressions, we write: $\frac{x^2}{x^3} = \frac{\cancel{x^2}}{\cancel{x^2}x} = \frac{1}{x}$

With the laws of exponents, we write: $\frac{x^2}{x^3} = x^{2-3} = x^{-1}$

So, we have $x^{-1} = \frac{1}{x}$.

DEFINITION OF A NEGATIVE EXPONENT

> In general, for n a positive integer x^{-n} is $\frac{1}{x^n}$, provided $x \neq 0$.

REMEMBER

x^2	means	$(x)(x)$
x^1	means	x
x^0	means	1
x^{-1}	means	$\frac{1}{x^1}$
x^{-2}	means	$\frac{1}{x^2}$

Most of the work we've done in this chapter can be reworded using integer exponents. This is much like making the transition from fractional notation (in arithmetic) to decimal notation. That is, we aren't learning any new concepts or skills, we are simply gaining experience with a notation system. Recall that a completely reduced rational expression is a ratio of polynomials in which the numerator and denominator have no common factor. Thus, a completely reduced rational expression *never* contains a negative exponent. Work through the following example.

EXAMPLE 1 Write as a completely reduced rational expression: $2(x + 1)^{-1} + (2x)^{-1}$

Solution Using the meaning of the negative one exponent, we can write

$$2\left(\frac{1}{x+1}\right) + \frac{1}{2x}$$

WARNING
As with $2x^2$ and $(2x)^2$, for $2x^{-1}$ and $(2x)^{-1}$ the scope of the exponent is extended by the use of parentheses.

Multiplying the first rational expression by 2 to get $\frac{2}{x+1}$, and then adding the rational expressions, we get:

$$\frac{2}{x+1} + \frac{1}{2x} = \frac{2(2x)}{2x(x+1)} + \frac{1(x+1)}{2x(x+1)} = \frac{4x}{2x(x+1)} + \frac{x+1}{2x(x+1)}$$

$$= \frac{5x+1}{2x(x+1)} \qquad \text{Restrictions: } x \neq 0 \quad x \neq -1$$

Thus, as a completely reduced rational expression,

$$2(x+1)^{-1} + (2x)^{-1} \quad \text{is} \quad \frac{5x+1}{2x(x+1)} \qquad \text{Restrictions: } x \neq 0 \quad x \neq -1$$

■

Next, we consider an example using the multiplication and division operations.

EXAMPLE 2 Write as a completely reduced rational expression: $(3x^2y^{-4}) \div (6x^{-2}y^3)^{-1}$

Solution First, $3x^2y^{-1}$ means $3x^2\left(\frac{1}{y}\right)$, or $\frac{3x^2}{y}$.
Next, $(6x^{-2}y^3)^{-1}$ means

$$\frac{1}{6\left(\frac{1}{x^2}\right)y^3} \quad \text{or} \quad 1 \div \frac{6y^3}{x^2} \quad \text{or} \quad \frac{x^2}{6y^3}$$

Thus,

$$\begin{aligned}(3x^2y^{-1}) \div (6x^{-2}y^3)^{-1} &= \frac{3x^2}{y} \div \frac{x^2}{6y^3}\\ &= \frac{3x^2}{y} \cdot \frac{6y^3}{x^2}\\ &= 18y^2 \qquad \text{Restrictions: } x \neq 0 \quad y \neq 0\end{aligned}$$

■

The technique shown in Example 2 is based strictly on the meaning of negative exponents. We may also work this problem using the laws of exponents.

$$\begin{aligned}(3x^2y^{-1}) \div (6x^{-2}y^3)^{-1} &= \left(\frac{3x^2}{y}\right) \div (6^{-1}(x^{-2})^{-1}(y^3)^{-1}) \qquad (ab)^m = a^mb^m\\ &= \frac{3x^2}{y} \div \left[\left(\frac{1}{6}\right)(x^2)\left(\frac{1}{y^3}\right)\right] \qquad (a^n)^m = a^{nm}\\ &= \frac{3x^2}{y} \div \frac{x^2}{6y^3}\\ &= \frac{3x^2}{y} \cdot \frac{6y^3}{x^2}\\ &= 18y^2 \qquad \text{Restrictions: } x \neq 0 \quad y \neq 0\end{aligned}$$

We can also use the laws of exponents as follows:

$$\begin{aligned}(3x^2y^{-1}) \div (6x^{-2}y^3)^{-1} &= (3x^2y^{-1})[(6x^{-2}y^3)^{-1}]^{-1} && a \div b = ab^{-1}\\ &= (3x^2y^{-1})(6x^{-2}y^3) && (a^n)^m = a^{nm}\\ &= (3)(6)(x^2)(x^{-2})(y^{-1})(y^3)\\ &= 18x^{2-2}y^{-1+3} && (a^n)(a^m) = a^{n+m}\\ &= 18x^0y^2\\ &= 18y^2 && a^0 = 1\end{aligned}$$

Using the Graphics Calculator

To check our work in Example 1, for Y1 type 2(X+1)⁻¹+(2X)⁻¹, and for Y2 type (5X+1)/(2X(X+1)). Return to the home screen and store a value, say 7, in memory location X. Display the resulting values in Y1 and Y2. The two outputs should be the same. If not, an error is present.

To check our work in Example 2, for Y1 type

(3X²Y⁻¹)/(6X^(⁻2)Y³)⁻¹

and for Y2 type 18Y². Return to the home screen and store a value, say 7, in memory location X and a value, say 11, in memory location Y. Display the resulting values in Y1 and Y2. The two outputs should be the same. If not, an error is present.

GIVE IT A TRY

Write each as a completely reduced rational expression.

1. $3x^{-2} - (3x)^{-2}$ **2.** $(3x^{-2}y^5)^{-2}$

To continue our review of rational expressions (using negative exponents), consider the following examples of solving rational expression equations.

EXAMPLE 3 Solve: $-3x^{-2} + 2x^{-1} + 1 = 0$

Solution Using the meaning of the negative exponents, we can rewrite the equation as

$$\frac{-3}{x^2} + \frac{2}{x} + 1 = 0$$

$$\frac{-3x^2}{x^2} + \frac{2x^2}{x} + x^2 = 0 \qquad \text{Multiply by LCD, } x^2.$$

$$-3 + 2x + x^2 = 0 \qquad \text{Remove common factors.}$$

$$x^2 + 2x - 3 = 0 \qquad \text{Standard form}$$

$$(x + 3)(x - 1) = 0 \qquad \text{Factored form}$$

$$x + 3 = 0 \quad \text{or} \quad x - 1 = 0$$

$$x = -3 \qquad\qquad x = 1$$

The solution is -3 and 1 (after checking that neither number results in a denominator evaluating to 0). ■

Using the Graphics Calculator

To check our work in Example 3, type ⁻3X^(⁻2)+2X⁻¹+1 for Y1. Store the number ⁻3 in memory location X, then display the resulting value stored in Y1. Is the number 0 displayed? No. However, the value displayed, 1E⁻13 is a number very close to 0. So, we conclude that ⁻3 does solve the equation. Store the number 1 in memory location X. Display the resulting value stored in Y1. Is the number 0 displayed? If not, an error is present.

EXAMPLE 4 Solve: $2 = (x - 1)(x - 2)^{-1}$

Solution Using the meaning of the negative exponents, we get:

$$2 = (x - 1)\left(\frac{1}{x - 2}\right)$$

$$2 = \frac{x - 1}{x - 2}$$

$$2(x - 2) = x - 1 \qquad \text{Multiply by the LCD, } x - 2.$$

$$2x - 4 = x - 1$$

$$\underline{-x \qquad -x}$$

$$x - 4 = -1$$

$$x = 3$$

The solution is 3 (after checking that 3 is not an extraneous solution). ■

GIVE IT A TRY

Solve each equation.

3. $3(x + 1)^{-1} = x(x + 1)^{-1}$ **4.** $3x^{-1} = 2(x - 3)^{-1}$

■ SUMMARY

In this section we introduced an alternate notation for rational expressions. Typically, a rational expression is written in fractional notation—as a ratio of two polynomials. Using the facts that if $a \neq 0$ and n is a positive integer, a^{-1} means $\frac{1}{a}$ and a^{-n} means $\frac{1}{a^n}$, we can write rational expressions in a linear notation. That is, the rational expression $\frac{2x + 1}{x - 1}$ can be written as

$$(2x + 1)(x - 1)^{-1}$$

This is much like writing a fraction (ratio of whole numbers) in decimal notation: $\frac{2}{5}$ is written as 0.4 [because 0.4 means $4(10^{-1})$].

In this section we also reviewed the work we've done with rational expressions, using the laws of exponents to write rational expressions in completely reduced form. The laws of exponents will be important in Chapter 7, on radical expressions, and in Chapter 11, on exponential and logarithmic expressions.

6.8 ■ ■ ■ EXERCISES

Write each as a completely reduced rational expression.

1. $(6x^2y^3)(12x^{-3}y^4)^{-1}$
2. $(6x^2y^3) \div (12x^{-3}y^4)^{-1}$
3. $(3x^{-2}) \div 2x^{-1}$
4. $(5m^2y^{-3})(10x^{-3}y^2)^{-1}$
5. $(-h^2)^{-1} \div (h^{-3})^3$
6. $(2y^{-1})^{-2}(3yz^{-2})^{-2}$
7. $2(c^3c^4)^{-1}[3(2c)^{-2}]$
8. $(2^{-1}y^3x^{-2})^{-2} \div (4^{-1}y^{-2}x)^{-1}$
9. $(2^{-1}y^3x^{-2})^{-2}(4^{-1}y^{-2}x)^{-1}$
10. $\dfrac{(3x)^{-2}}{2x^{-1}}$

11. $\dfrac{(a^3)^{-2}}{2a^{-1}}$

12. $\left(\dfrac{(3b^3)^{-2}}{2b^2}\right)^{-2}$

13. $\left(\dfrac{-3x^3y^4}{7x^2y^{-5}}\right)^0$

14. $\left(\dfrac{(3b^3)^{-2}}{2b^2}\right)^0$

15. $\dfrac{(16r)^{-2}}{36r^{-3}}$

16. $5(x + 2)^{-1} + (x + 3)^{-1}$

17. $3(x - 2)^{-1} + (x + 1)^{-1}$

18. $3x^{-2} + 2x^{-1}$

19. $(3x)^{-2} + 2x^{-1}$

20. $(x + 3)(x^2 + 4x + 5)^{-1} \div [(x^2 - 9)(x^2 - 1)^{-1}]$

21. $(x + 2)(x^2 - 4x - 5)^{-1} \div [(x^2 - 4)(x^2 - 25)^{-1}]$

22. $(a^{-2} - b^{-2})(a^{-1} + b^{-1})^{-1}$

Solve each equation in one variable.

23. $3x^{-2} - x^{-1} - 2 = 0$

24. $2p^{-2} - 3p^{-1} - 2 = 0$

25. $3(r + 1)^{-1} = 2(r - 2)^{-1}$

26. $3(x - 3)^{-1} = 2(2x + 1)^{-1}$

27. $(x + 2)(x - 3)^{-1} = 2 \cdot 3^{-1}$

28. $(w - 2)(w - 4)^{-1} = 4^{-1}$

29. $(x + 2)x^{-1} + 3(x + 1)^{-1} = 1$

30. $(x - 2)x^{-1} - 3(x - 1)^{-1} = 1$

CHAPTER 6 REVIEW EXERCISES

Report the domain of definition (by finding the restricted values) for each rational expression.

1. $\dfrac{2}{3x - 9}$

2. $\dfrac{5x^2 + 10x}{15x + 20}$

3. $\dfrac{x^2 - 4x - 8}{4x - x^2}$

4. $\dfrac{2}{x^2 + 5x + 6}$

5. $\dfrac{2x}{2x^2 + 3x - 2}$

6. $\dfrac{2}{x^2 + 5x - 6}$

Write the fraction as a completely reduced rational expression. (Remove common factors from the numerator and denominator.)

7. $\dfrac{-12w^6}{24w^3}$

8. $\dfrac{x^2 - 5x + 4}{16 - x^2}$

9. $\dfrac{T^2 - S^2}{S^2 - ST}$

10. $\dfrac{2x^2 - 3x - 2}{x^2 - 5x + 6}$

11. $\dfrac{y^2 - 5y}{25 - y^2}$

12. $\dfrac{27 - x^3}{3x^2 - 9x}$

Perform the indicated operation, and report the result as a completely reduced rational expression.

13. $\dfrac{7m^4y}{8a^3b} \cdot \dfrac{4a^2}{21m^3y^2}$

14. $\dfrac{4x + 2}{x - 2} \cdot \dfrac{x - 2}{4x + 2}$

15. $\dfrac{x^2 + x}{y^2 - y} \div \dfrac{y^2}{x^2}$

16. $\dfrac{3y^2 - 9y}{2x^2} \div \dfrac{6y^2}{x^3}$

17. $\dfrac{x^2 + 3x + 2}{x + 4} \cdot \dfrac{x^2 + 5x + 4}{x + 2}$

18. $\dfrac{x^2 - 9}{2x + 3} \div \dfrac{x + 3}{2x^2 + 5x + 3}$

19. $\dfrac{x - 1}{x + 1} + \dfrac{2}{x - 3}$

20. $\dfrac{3}{4x^2 - 4} + \dfrac{x}{2x + 2}$

21. $\dfrac{3}{y^3 - y^2} - \dfrac{2}{y^2 - 2y + 1}$

22. $\dfrac{s}{s^2 - 9} - \dfrac{s}{s + 3}$

Write each as a completely reduced rational expression.

23. $\dfrac{1}{x} + \dfrac{1}{x - 1} \div \dfrac{1}{x}$

24. $\left(\dfrac{3}{2x - 2} - \dfrac{1}{x + 1}\right)\left(\dfrac{1}{x^2 - 25}\right)$

25. $\dfrac{\dfrac{9}{2x}}{\dfrac{6y^2}{4x^2}}$

26. $\dfrac{\dfrac{x^2 - 4x - 5}{4x - x^2}}{\dfrac{2x - 2}{x^2 - 3x - 4}}$

27. $\dfrac{1}{1 - \dfrac{1}{x - 1}}$

28. $\dfrac{\dfrac{1}{x} - \dfrac{1}{y}}{y^2 - x^2}$

Solve each rational expression equation in one variable. Check for extraneous solutions.

29. $\dfrac{x - 1}{x + 2} = \dfrac{1}{3}$

30. $\dfrac{x + 1}{3x - 2} = \dfrac{-2}{5}$

31. $\dfrac{2}{x+1} = \dfrac{5}{2x-3}$

32. $\dfrac{-2}{2x+3} = \dfrac{1}{x}$

33. $\dfrac{x}{2x+3} = \dfrac{1}{x}$

34. $\dfrac{3}{x+3} = \dfrac{x+1}{2x+6}$

35. $\dfrac{2}{x-3} + \dfrac{1}{x} = \dfrac{1}{x}$

36. $\dfrac{1}{1 - \dfrac{1}{x-1}} = \dfrac{1}{2}$

37. Pump A can empty a tank in 8 hours. If pump A and pump B working together can empty the tank in 5 hours, how long does it take pump B to empty the tank, working alone?

38. A boat running against the current travels 7 miles in the same time it can travel 11 miles running with the current. If the current is 5 mi/h, what is the speed of the boat in calm water?

Rewrite each formula for the indicated variable.

39. $A = \dfrac{R+2}{R-S}$ for S

40. $\dfrac{1}{y} = \dfrac{5}{x-2}$ for x

41. $\dfrac{1}{b} + \dfrac{c}{a} = \dfrac{1}{c}$ for a

42. $z = \dfrac{x-u}{\sigma}$ for u

43. The distance a spring stretches varies directly with the force applied. If a force of 15 pounds stretches the spring 3 inches, what is the constant of variation? How far will a force of 28 pounds stretch the spring?

44. The circumference of a circle varies directly with the radius. What is the constant of variation?

Write each formula in variation language.

45. $V = \pi r^2 h$

46. $F = \dfrac{5x}{y^2}$

Write the result as a completely reduced rational expression. (Your answer should have no negative exponent.)

47. $(3x)^{-2} \div 2x^{-1}$

48. $(5m^2y^{-3})(10x^{-3}y^2)^{-1}$

49. $3(x-2)^{-1} + (x+1)^{-1}$

50. $3x^{-2} + 2x^{-1}$

51. $(x^{-1} - 2^{-1})(x^{-2} - 2^{-2})^{-1}$

52. $2 - (1 - (2 + x^{-1})^{-1})^{-1}$

Solve each equation.

53. $3(x-2)^{-1} = 2^{-2}$

54. $x^{-2} - 3x^{-1} = 4$

55. $(3x+1)^{-1} = (2x)^{-2}$

56. $3^{-2} = (x)(2x-3)^{-1}$

CHAPTER 6 TEST

1. Report the domain of definition for $\dfrac{x+3}{3x-5}$.

Write each as a completely reduced rational expression.

2. $\dfrac{12x^3y^6}{3xy^2}$

3. $\dfrac{x^2-9}{x^2+x-6}$

4. $\dfrac{2x^2+3x-2}{x^2+x-6} \cdot \dfrac{x^2-x}{4x^2-1}$

5. $\dfrac{x+3}{3x-5} \div \dfrac{x^2-9}{9x^2-25}$

6. $\dfrac{x}{2x-5} + \dfrac{1}{x-6}$

7. $\dfrac{(x-1)^2}{\dfrac{1}{x^2} - 1}$

Solve each equation.

8. $x^{-2} - 3x^{-1} = 4$

9. $\dfrac{-2}{x-2} = \dfrac{2}{x+2}$

10. Becky and Barbara working together can copyedit a manuscript in 35 hours. Barbara, working by herself, could copyedit the same manuscript in 50 hours. How long would it take Becky, working alone, to copyedit this manuscript?

11. The height of a building varies directly with the length of its shadow. If a building that is 100 meters tall has a shadow of 21 meters, what is the height of a nearby building whose shadow has length 17 meters?

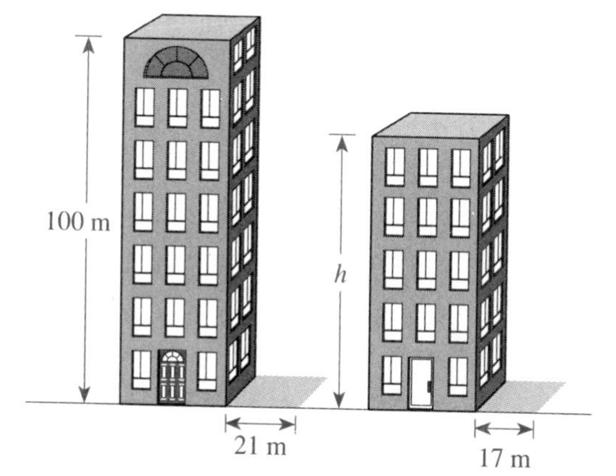

12. Write the formula $H = \dfrac{3w^2}{2p}$ using the language of variation. What is the constant of variation?

7

RADICAL EXPRESSIONS

Having completed our discussion of polynomials and rational expressions, we now consider radical expressions. As we did with polynomials and rational expressions, we will develop a naming system for a completely reduced radical expression. We will learn how to add, subtract, multiply, and divide radical expressions. Then, using another equation-solving property, we will solve equations in one variable that contain radical expressions. We start with a discussion of roots and rational number exponents.

7.1 ■ ■ ■ ROOTS AND RATIONAL NUMBER EXPONENTS

The **radical expressions** we are concerned with at this level of algebra involve square roots, cube roots, or other roots, applied to polynomials and rational expressions. Some examples of radical expressions are shown here:

$$\sqrt{x} + 2, \qquad \sqrt{x^2 - 2y + t^3}, \qquad \sqrt[3]{2y^2}$$

$$\sqrt[3]{\frac{1}{2w}}, \qquad \text{and} \qquad \sqrt{\frac{x^2 - 3x + 6}{x^3 + 1}}$$

Before discussing completely reduced radical expressions, we will formalize the meaning of roots, discuss some of the properties of radicals, and introduce rational number exponents. We start by reviewing and extending our work with radical expressions (from Chapter 1).

Roots When we write $\sqrt{x}$, we mean *the square root of x,* and the symbol $\sqrt{}$ is called the **radical sign**. The number or expression under the radical sign is known as the **radicand**. When we write $\sqrt[3]{x}$, we mean *the cube root of x,* and the 3 is known as the **index**. For $\sqrt{x}$, the index is understood to be 2.

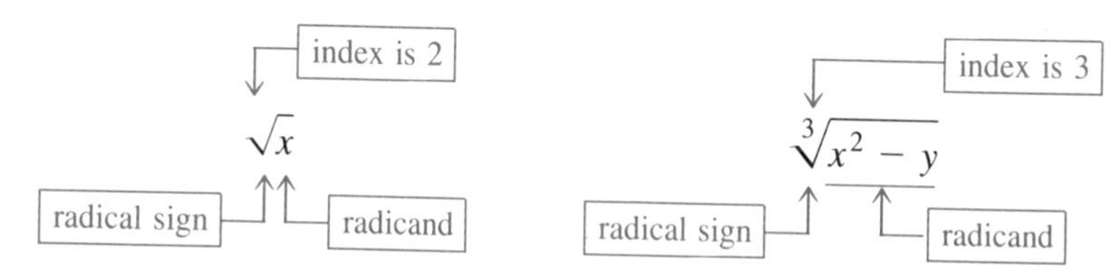

The square root of a nonnegative number a is a number, say b, such that $b^2 = a$. Each positive number a has two square roots. For example, one square root of 25 is 5, because $5^2 = 25$. Another square root of 25 is -5, because $(-5)^2 = 25$. For a positive number a, the positive square root of a, known as the **principal square root**, is denoted $\sqrt{a}$. To indicate the negative square root of a, we write $-\sqrt{a}$.

The square root of a negative number is not a real number. Consider $\sqrt{-25}$. There is *no* real number b such that b^2 is -25. The square of any real number b is either 0 or a number greater than 0. Thus, $\sqrt{-25}$ is not a real number. We study such numbers in Section 7.5.

The cube root of *any* real number a is a number b such that $b^3 = a$. For example, the cube root of 8 is 2, because $2^3 = 8$. We write $\sqrt[3]{8} = 2$. Likewise, the cube root of -8 is -2, because $(-2)^3 = -8$. We write $\sqrt[3]{-8} = -2$.

When the index is an even number, such as 2, 4, 6, . . . , we say that the radical is an **even root**. When the index is an odd number, such as 3, 5, 7, . . . , we say we have an **odd root**. For an nth root, we consider two cases: n is an even number, and n is an odd number. We have the following.

DEFINITION OF *n*TH ROOT FOR *n* EVEN

> An ***n*th root** of a, for n an even number and $a > 0$, is a number, say b, such that $b^n = a$. The **principal *n*th root**, denoted $\sqrt[n]{a}$, is the positive number b such that $b^n = a$.

With any even root of a positive real number a, there are two real roots, one positive and one negative. When we write $\sqrt[n]{x}$ for n an even number, it is understood that we are referring to the principal root (the positive root). Some examples of the definition of even nth roots are:

$\sqrt{16} = 4$ because 4 is a positive number and $4^2 = 16$. Note: $-\sqrt{16} = -4$
$\sqrt[4]{16} = 2$ because 2 is a positive number and $2^4 = 16$. Note: $-\sqrt[4]{16} = -2$
$\sqrt[6]{1} = 1$ because 1 is a positive number and $1^6 = 1$. Note: $-\sqrt[6]{1} = -1$

Next, we consider the nth root for n an odd number.

DEFINITION OF nTH ROOT FOR n ODD

> The nth root of a, denoted $\sqrt[n]{a}$, for n an odd number and a any real number, is the number, say b, such that $b^n = a$.

With any odd root of a real number a, there is only one real number root. Some examples of the definition with n odd are given here:

$$\sqrt[3]{8} = 2 \quad \text{because } 2^3 = 8.$$
$$\sqrt[3]{-8} = -2 \quad \text{because } (-2)^3 = -8.$$
$$\sqrt[5]{-243} = -3 \quad \text{because } (-3)^5 = -243.$$

In general, we have said that for an even index, the radicand must be a nonnegative number for the nth root to be a real number, and the principal nth root is always a nonnegative number. For an odd index, the radicand can be any real number. If the radicand is positive, then the odd nth root is positive; and if the radicand is negative, then the odd nth root is negative. Also, we have the following definition.

DEFINITION OF $\sqrt[n]{0}$

> For n an integer greater than 1, the nth root of 0 is 0, that is, $\sqrt[n]{0} = 0$.

GIVE IT A TRY

If the radical is a real number, rewrite it as an integer. Otherwise, report as *not a real number*.

1. $\sqrt{36}$ **2.** $\sqrt[3]{-27}$ **3.** $\sqrt[4]{-16}$ **4.** $\sqrt[4]{16}$

Domain of Definition Just as polynomials and rational expressions hold the place for number phrases, so do radical expressions. That is, a radical expression such as $\sqrt{x + 3}$ holds the place for number phrases such as $\sqrt{2 + 3}$ or $\sqrt{5 + 3}$. With polynomials, such as $3x - 7$, the variable can be replaced by any real number, and a real number is produced. With rational expressions, such as $\dfrac{x + 1}{x - 3}$, the variable is restricted. For this expression the variable cannot be replaced by the number 3. With rational expressions, the restricted values—usually one or two numbers—are those that result in a denominator evaluating to 0. With radical expressions, the restrictions are more involved. For a radical expression with an even index, such as a square root, the radicand must evaluate to a number greater than or equal to 0. For the radical expression $\sqrt{x + 3}$, we must have $x + 3 \geq 0$, or $x \geq -3$. Thus, the variable is restricted to the numbers greater than or equal to -3. We say the *domain of definition* is $\{x \mid x \geq -3\}$. The following example shows how to find the domain of definition.

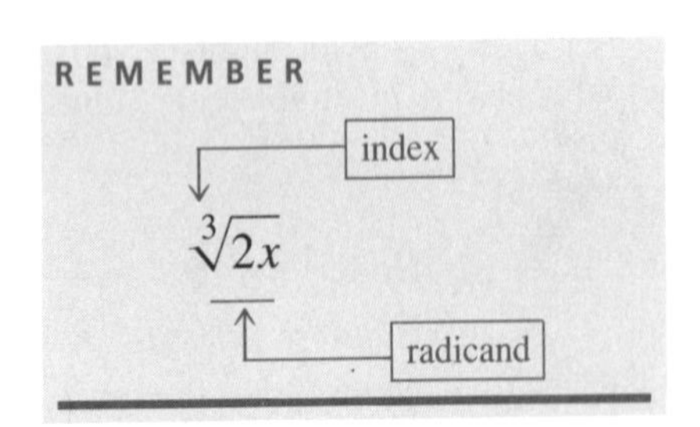

EXAMPLE 1 Report the domain of definition for $\sqrt{x^2 + 3x}$.

Solution We must have a radicand that is greater than or equal to 0. So, we solve the inequality

$$x^2 + 3x \geq 0$$

by solving the related equation:

$$x^2 + 3x = 0$$
$$x(x + 3) = 0$$
$$x = 0 \quad \text{or} \quad x + 3 = 0$$
$$x = -3$$

Thus, we get two key points, 0 and -3. We plot these key points, and then test the intervals determined.

Try -4, a number in the leftmost interval: $(-4)^2 + 3(-4) \geq 0$ is a true arithmetic sentence, so the interval is shaded.

Try a number in the middle interval, say -1: $(-1)^2 + 3(-1) \geq 0$ is a false arithmetic sentence, so the interval is left unshaded.

Try a number in the rightmost interval: $1^2 + 3(1) \geq 0$ is a true arithmetic sentence, so we shade the interval. Using these results, we get the number-line graph shown in Figure 1.

The domain of definition is $\{x \mid x \leq -3\} \cup \{x \mid x \geq 0\}$. ■

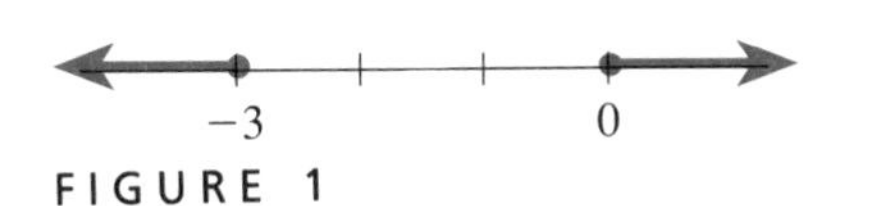

FIGURE 1

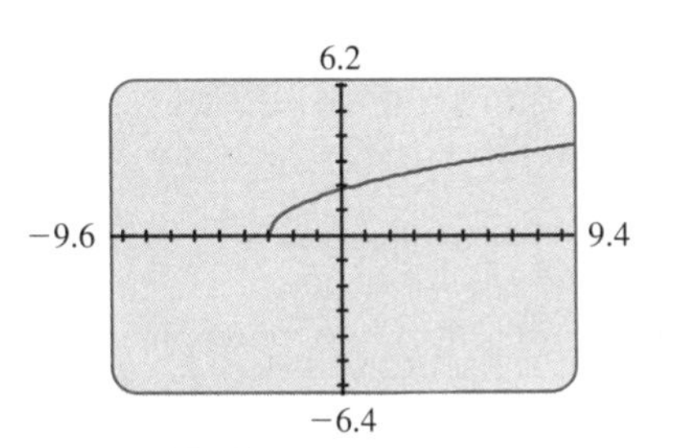

FIGURE 2

We can use the graph screen to increase our understanding of radical expressions and the domain of definition. For Y1, type √(X+3). Press the RANGE key and enter the Window 9.4 range settings. View the graph screen, as shown in Figure 2. The curve drawn is the upper half of a horizontal parabola (which we will discuss in Chapter 12). Press the TRACE key. The message, X=0 Y=1.7320508, means that replacing the x-value with 0 results in the expression $\sqrt{x + 3}$ evaluating to $\sqrt{3} \approx 1.7320508$. As we discussed in Chapter 1, the number $\sqrt{3}$ is an irrational number. Press the left arrow key until the message reads X=-3 Y=0. When x is replaced by -3 the radical expression evaluates to 0. That is, $\sqrt{0}$ is 0. Press the left arrow key to move the cursor to the x-value -3.2. The message X=-3.2 Y= means that when x is replaced by a number less than -3, the expression $\sqrt{x + 3}$ does not evaluate to a real number. The domain of definition is the set of x-values that are -3 and greater. Press the right arrow key and move the cursor to the x-value 6. (Notice that most of the evaluations are irrational numbers, such as $\sqrt{2}$, $\sqrt{3}$, or $\sqrt{5}$). The message now reads X=6 Y=3. When the x-variable is replaced by 6, the expression $\sqrt{x + 3}$ evaluates to 3. Notice that for the x-values in the domain of definition, the evaluations are all greater than or equal to 0. Remember, $\sqrt{x}$ is nonnegative.

Press the Y= key and type √(X²+3X) for Y1. View the graph screen. The evaluations of $\sqrt{x^2 + 3x}$ for various replacements of the x-variable are displayed. The curve displayed is the upper half of a *hyperbola* (which we will discuss in Chapter 12). The curve is shown

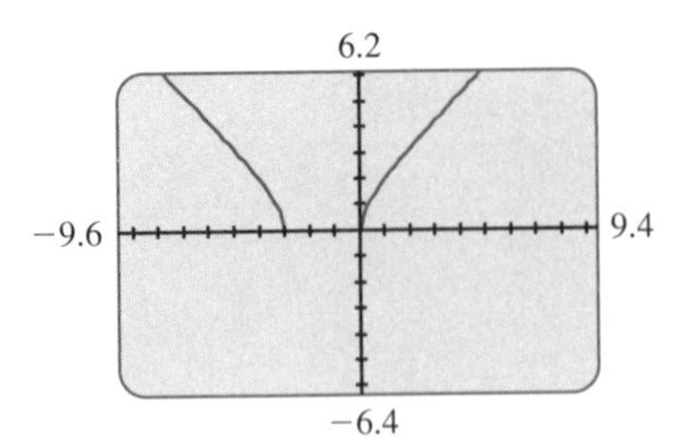

FIGURE 3

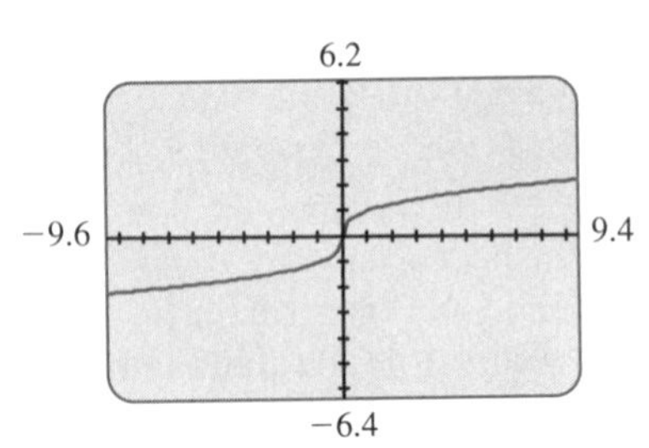

FIGURE 4

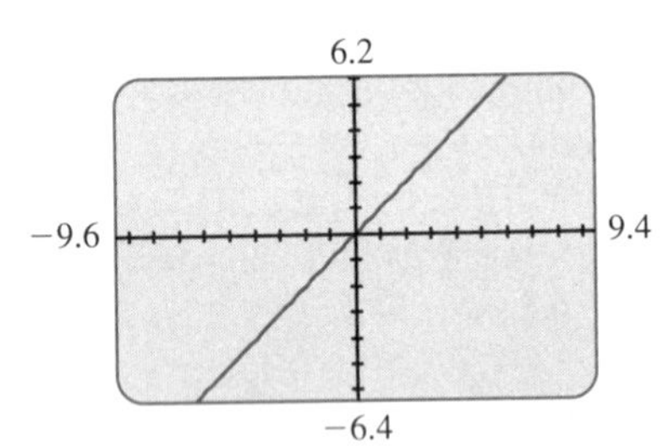

FIGURE 5

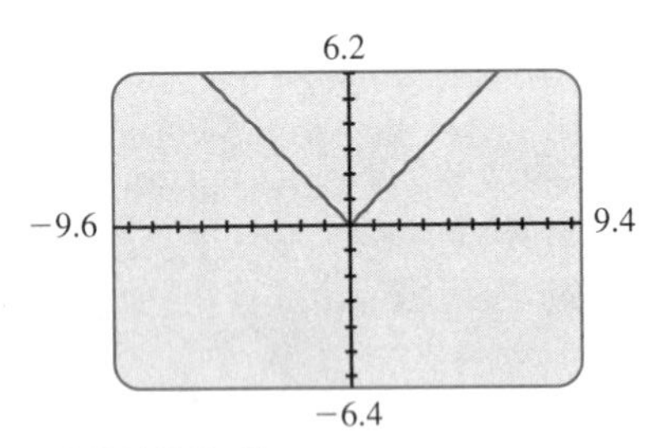

FIGURE 6

in Figure 3. Notice that $\sqrt{x^2 + 3x}$ is undefined for the x-values between -3 and 0. Does this agree with our result from Example 1?

Let's determine the domain of definition for some other radical expressions. For `Y1`, type `³√X`. [Use option `4` from the `MATH` menu to type `³√` or type `X^(1/3)`.]. View the graph screen, as shown in Figure 4. As you can see, evaluations of `³√X` are displayed for various replacements of `X`. Here, the domain of definition is all real numbers. That is, the x-variable can be replaced by any real number. Press the `TRACE` key and use the arrow keys to explore some of the evaluations.

Now, press the `Y=` key and type `³√X³` for `Y1`. (Use option `3` from the `MATH` menu for the `3` exponent.) View the graph screen (see Figure 5). A straight line is drawn. Press the `TRACE` key and use the arrow keys to explore the evaluations of $\sqrt[3]{x^3}$. For each replacement of x, the evaluation of the radical expression is that x-value. We say that $\sqrt[3]{x^3} = x$. Now, type `√X²` for `Y1`. View the graph screen, as shown in Figure 6. This time something different happens. For x-values greater than or equal to 0, the evaluation of $\sqrt{x^2}$ is that x-value. That is, for $x \geq 0$, $\sqrt{x^2} = x$. For x-values less than 0, the evaluation of $\sqrt{x^2}$ is not the same as the x-value. That is, for $x < 0$, we have $\sqrt{x^2} = -x$. We've seen this expression before—it is the absolute value of x. So, $\sqrt{x^2} = |x|$. To be sure that this is the case, type `abs (X)` for `Y2`. (Use the `2nd` key, followed by the x^{-1} key, for the absolute value). View the graph screen. The evaluations for $\sqrt{x^2}$ are displayed. The evaluations for $|x|$ are displayed. As you can see, they are identical. We now discuss this result from an algebraic viewpoint.

The $\sqrt[n]{a^n}$ What is the value of $\sqrt{(-5)^2}$? We often think of square and square root as inverse operations (like *addition of n* and *subtraction of n* or *multiplication by n* and *division by n*). This is not strictly true in this case. First, $\sqrt{(-5)^2} = \sqrt{25}$, and $\sqrt{25}$ is the positive number 5. So, for $a = -5$, we have $\sqrt{a^2} \neq a$. Here is the relationship between the square and square root operations: $\sqrt{a^2} = |a|$. That is, $\sqrt{a^2}$ is the absolute value of a. In our example, we have $\sqrt{(-5)^2} = |-5|$. This relationship also holds for other even nth roots as well. Odd nth roots are the inverse operation of their powers. That is, $\sqrt[3]{(-3)^3} = \sqrt[3]{-27} = -3$. We summarize this information as follows.

THE $\sqrt[n]{a^n}$ PROPERTY

For n an even positive integer and a any real number, $\sqrt[n]{a^n} = |a|$.

For n an odd positive integer and a any real number, $\sqrt[n]{a^n} = a$.

If $a \geq 0$, the preceding property states that $\sqrt[n]{a^n} = a$. *Throughout this chapter, we will often assume that variable expressions in a radicand represent nonnegative values*. That is, we will write $\sqrt{4x^2} = 2x$. This is valid as long as the replacements for the x-variable are greater than or equal to 0. In general, we must know that $\sqrt{4x^2} = 2|x|$.

Using the Graphics Calculator

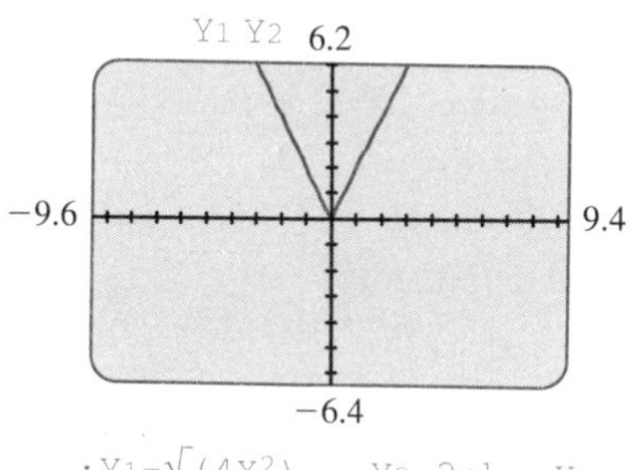

FIGURE 7

Type √(4X²) for Y1. Use the Window 9.4 range settings to view the graph screen. For Y2 type 2abs X. View the graph screen, as shown in Figure 7. As you can see, these two expressions, $\sqrt{4x^2}$ and $2|x|$, have the same evaluation for each replacement of x.

A number such as 1, 4, 9, 16, 25, . . . is known as a **perfect square**, because each one can be written as a square: 1^2, 2^2, 3^2, 4^2, 5^2, ... Likewise, a monomial such as $4x^2y^4$ is a perfect square, since it can be written as $(2xy^2)^2$. A number such as 1, 8, 27, 64, 125, . . . is known as a **perfect cube**, because it can be written as a cube: 1^3, 2^3, 3^3, 4^3, 5^3, ... Similarly, the monomial $-8x^3y^6$ is a perfect cube, since it can be written as $(-2xy^2)^3$. We can extend the idea to a **perfect *n*th power**. As you can see by reviewing the $\sqrt[n]{a^n}$ property, when the nth root is applied to a perfect nth power, the result does not contain a radical.

Rational Number Exponents With polynomials, the only exponents we have used are whole number exponents: 0, 1, 2, ... Remember that for $x \neq 0$, we have $x^0 = 1$. With rational expressions, we expanded exponents to include the integers (. . . , −3, −2, −1, 0, 1, 2, 3, . . .). Remember that for $x \neq 0$, we have $x^{-1} = \frac{1}{x}$. We now wish to expand exponents to rational numbers. That is, we want to give meaning to expressions such as $x^{1/3}$ and $x^{3/2}$. We want the meaning of these exponents to be consistent with our earlier work. That is, we want the following laws of exponents to be valid for rational number exponents.

THE LAWS OF EXPONENTS

Law	Example
1. $a^n a^m = a^{n+m}$	*Example:* $(x^{-3})(x^2) = x^{-3+2} = x^{-1}$
2. $a^n \div a^m = a^{n-m}$	*Example:* $(x^3) \div (x^{-2}) = x^{3-(-2)} = x^5$
3. $(a^n)^m = a^{nm}$	*Example:* $(x^3)^{-2} = x^{(3)(-2)} = x^{-6}$
4. $(ab)^n = a^n b^n$	*Example:* $(xy)^3 = x^3y^3$
5. $\left(\frac{a}{b}\right)^n = \frac{a^n}{b^n}$	*Example:* $\left(\frac{x}{y}\right)^{-3} = \frac{x^{-3}}{y^{-3}}$

Consider $5^{1/2}$. Now, using law 3, we have $(5^{1/2})^2 = 5^1$. That is, $5^{1/2}$ is a number whose square is 5. We already have a name for this number: $\sqrt{5}$.

So, we define $5^{1/2}$ to be $\sqrt{5}$. Now, consider $7^{1/3}$. Again using law 3, we have $(7^{1/3})^3 = 7^1$. That is, $7^{1/3}$ is a number whose cube is 7. We have a name for this number, $\sqrt[3]{7}$. So, we define $7^{1/3}$ to be $\sqrt[3]{7}$.

We now define $x^{1/n}$, where n is a positive integer. For $n = 1$, we know that $x^{1/n} = x^1 = x$. For $n > 1$, we have the following definition.

DEFINITION OF $x^{1/n}$

> For n an integer greater than 1, $x^{1/n} = \sqrt[n]{x}$.

This definition means that

$$x^{1/2} = \sqrt{x} \qquad x^{1/3} = \sqrt[3]{x} \qquad x^{1/4} = \sqrt[4]{x}$$

and so forth. Work through the following example.

EXAMPLE 2 Write each expression using only whole number exponents.

a. $(-8x^3)^{1/3}$ **b.** $(-1)^{1/3}$ **c.** $(32)^{1/5}$ **d.** $(25x^2)^{1/2}$

Solution

a. $(-8x^3)^{1/3} = \sqrt[3]{-8x^3} = -2x$ $\qquad -8x^3 = (-2x)^3$

b. $(-1)^{1/3} = \sqrt[3]{-1} = -1$ $\qquad -1 = (-1)^3$

c. $(32)^{1/5} = \sqrt[5]{32} = 2$ $\qquad 32 = 2^5$

d. $(25x^2)^{1/2} = \sqrt{25x^2} = 5\,|x|$ $\qquad 25x^2 = (5x)^2$ or $25x^2 = (-5x)^2$ ■

Next, we want to give meaning to an exponent such as $\frac{2}{3}$. Again, we want agreement with the laws of exponents. Now, $\frac{2}{3} = 2\left(\frac{1}{3}\right)$. So, for $8^{2/3}$, we have $8^{(2)(1/3)}$ and, by law 3, we have $(8^2)^{1/3}$, or $64^{1/3}$. Now, using our new meaning for the $\frac{1}{3}$ power, we can write this as $\sqrt[3]{64}$, which is 4. Thus, $8^{2/3} = 4$. We define $x^{m/n}$ as follows.

DEFINITION OF A RATIONAL NUMBER EXPONENT, $x^{m/n}$

> For integers m and n with $n \geq 0$, m/n a completely reduced rational number, and $a^{1/n}$ a real number,
>
> $$a^{m/n} = \left(\sqrt[n]{a}\right)^m = \sqrt[n]{(a^m)}$$

For experience with rational number exponents, work through the following examples.

EXAMPLE 3 Write each expression as an integer.

a. $16^{3/4}$ **b.** $(-8)^{2/3}$ **c.** $(-25)^{3/2}$

Solution

a. $16^{3/4}$ means $(16^{1/4})^3$. Now, $16^{1/4}$ is $\sqrt[4]{16}$, or 2.

So, we have

$$\begin{aligned} 16^{3/4} &= (16^{1/4})^3 \\ &= 2^3 \\ &= 8 \end{aligned}$$

b. $(-8)^{2/3}$ means $[(-8)^{1/3}]^2$. Now, $(-8)^{1/3}$ is $\sqrt[3]{-8}$, or -2.
So, we have

$$\begin{aligned}(-8)^{2/3} &= [(-8)^{1/3}]^2\\ &= (-2)^2\\ &= 4\end{aligned}$$

c. $(-25)^{3/2}$ means $[(-25)^{1/2}]^3$. Now, $(-25)^{1/2}$ is $\sqrt{-25}$, which is not a real number. So, $(-25)^{3/2}$ is not a real number. ■

EXAMPLE 4 Write each power as a completely reduced rational number.

a. $16^{-3/4}$ b. $(-1)^{2/5}$ c. $25^{-5/2}$

Solution a. $16^{-3/4}$ means $(16^{1/4})^{-3}$. Now, $16^{1/4}$ is $\sqrt[4]{16}$, or 2.
So, we have

$$\begin{aligned}16^{-3/4} &= (16^{1/4})^{-3}\\ &= 2^{-3}\\ &= \frac{1}{8}\end{aligned}$$

b. $(-1)^{2/5}$ means $[(-1)^{1/5}]^2$. Now, $(-1)^{1/5}$ is $\sqrt[5]{-1}$, or -1.
So, we have

$$\begin{aligned}(-1)^{2/5} &= [(-1)^{1/5}]^2\\ &= (-1)^2\\ &= 1\end{aligned}$$

c. $25^{-5/2}$ means $(25^{1/2})^{-5}$. Now, $25^{1/2}$ is $\sqrt{25}$, or 5.
So, we have

$$\begin{aligned}(25)^{-5/2} &= (25^{1/2})^{-5}\\ &= 5^{-5}\\ &= \frac{1}{3125}\end{aligned}$$

■

EXAMPLE 5 Use the laws of exponents and the meaning of rational number exponents to write the expression using only whole number exponents. Assume $x > 0$.

a. $(16x^4)^{-3/4}$ b. $x^{2/5}x^{2/5}$ c. $x^{1/2} \div x^{3/2}$

Solution a. $(16x^4)^{-3/4} = [(16x^4)^{1/4}]^{-3}$ The meaning of the exponent $\frac{-3}{4}$

With $x > 0$, we have $(16x^4)^{1/4} = \sqrt[4]{16x^4} = 2x$ Because $(2x)^4 = 16x^4$
So,

$$\begin{aligned}(16x^4)^{-3/4} &= [(16x^4)^{1/4}]^{-3}\\ &= (2x)^{-3}\\ &= \frac{1}{(2x)^3}\\ &= \frac{1}{8x^3}\end{aligned}$$

As an alternative method, we can use the laws of exponents and the fact that $16 = 2^4$:

$$\begin{aligned}(16x^4)^{-3/4} &= (2^4x^4)^{-3/4}\\ &= (2^4)^{-3/4}(x^4)^{-3/4}\\ &= 2^{-3}x^{-3} \qquad \text{Because} \quad (4)\left(\frac{-3}{4}\right) = -3\end{aligned}$$

Now, $$2^{-3}x^{-3} = \left(\frac{1}{2^3}\right)\left(\frac{1}{x^3}\right) = \frac{1}{8x^3}$$

REMEMBER

For $x^{m/n}$, the denominator n yields the index for the radical.

b. For $x^{2/5}x^{2/5}$, we use law 1 to write

$$x^{2/5}x^{2/5} = x^{2/5+2/5} = x^{4/5} \quad \text{Because } \frac{2}{5} + \frac{2}{5} = \frac{4}{5}$$

Using the meaning of the exponent $\frac{4}{5}$, we write $\sqrt[5]{x^4}$.

c. For $x^{1/2} \div x^{3/2}$, we use law 2 to write

$$x^{1/3} \div x^{3/2} = x^{1/2-3/2} = x^{-2/2} = x^{-1} = \frac{1}{x}$$ ■

GIVE IT A TRY

Write the expression using whole number exponents only. Assume $x > 0$.

5. $(-27)^{-2/3}$ **6.** $(16x^2)^{1/2}$ **7.** $(-125x^3)^{-1/3}$

We can use the graphics calculator to increase our understanding of rational number exponents. To compute $16^{1/4}$, type `16^(1/4)` and then press the `ENTER` key. The number `2` is output. Now, consider part (c) of Example 3: $(-25)^{3/2}$. We found that $(-25)^{3/2}$ is not a real number. On the graphics calculator type, `(-25)^(3/2)` and press the `ENTER` key. The `ERROR` screen is displayed. For $a^{m/n}$, where $\frac{m}{n}$ is a completely reduced rational number and n is an even number, the base a must be nonnegative. To see the reason that $\frac{m}{n}$ must be completely reduced, consider $(-32)^{1/5}$. This number is $\sqrt[5]{-32}$, which is -2, because $(-2)^5 = -32$. Now, consider $(-32)^{2/10}$. Without writing $\frac{2}{10}$ as a completely reduced rational number, we would have $\left(\sqrt[10]{-32}\right)^2$, but $\sqrt[10]{-32}$ is not a real number. So, we have a conflict! To avoid this situation, we require that $\frac{m}{n}$ be a completely reduced rational number, $\frac{2}{10} = \frac{1}{5}$.

From Example 3(b), we know that $(-8)^{2/3}$ is the number 4. On the graphics calculator, type `(-8)^(2/3)` and press the `ENTER` key. The `ERROR` screen is presented. This particular error is due to the way calculator is programmed, not due to mathematics. If we type `((-8)^(1/3))²` or `((-8)²)^(1/3)` and press the `ENTER` key, the correct value, `4`, is output.

From multiplication of polynomials, we know that

$$(a + b)^2 = a^2 + 2ab + b^2$$

Now, just as $(a + b)^2 \neq a^2 + b^2$, the same generalization can be

WARNING
Avoid these errors:

$$(a^{1/2} + b^{1/2})^2 \neq a + b$$

$$(a^2 + b^2)^{1/2} \neq a + b$$

$$\sqrt{9 + 4} \neq 3 + 2$$

stated for roots: $(a + b)^{1/2} \neq a^{1/2} + b^{1/2}$. In radicals, we write this fact as $\sqrt{x + y} \neq \sqrt{x} + \sqrt{y}$. Using the graphics calculator, type `(16+9)^(1/2)` and press the `ENTER` key. The number `5` is output. Now, type `16^(1/2)+9^(1/2)` and press the `ENTER` key. The number 7 is output. So, we have $(a + b)^{1/2} \neq a^{1/2} + b^{1/2}$.

Another common error with rational number exponents is to write $(a^2 + b^2)^{1/2} = a + b$. To see that $(a^2 + b^2)^{1/2} \neq a + b$, type `(3²+4²)^(1/2)` and press the `ENTER` key. The number `5` is output. Type `3+4` and press the `ENTER` key. The number `7` is output. As you can see, for $a = 3$ and $b = 4$, we have $(a^2 + b^2)^{1/2} \neq a + b$.

To see that the laws of exponents hold for rational number exponents, consider the first law: $a^n a^m = a^{n+m}$. Based on this law, we should have that $(5^{1/2})(5^{1/3}) = 5^{1/2+1/3} = 5^{5/6}$. Type `(5^(1/2))(5^(1/3))` and press the `ENTER` key. Type `5^(5/6)` and press the `ENTER` key. As you can see, the same number is output in both cases. Use this kind of test to check the other laws of exponents for rational number exponents.

■ SUMMARY

■ ***Quick Index***

In this section, we have been introduced to roots—the square root, the cube root, and so on. In general, the nth root of a real number a is a number, say b, such that $b^n = a$. The nth root is denoted $\sqrt[n]{a}$, where n is the index and a is the radicand.

We have learned that for n an even number, the root is known as an even root, and the radicand a must be nonnegative for the nth root to be a real number. Also, for radicand $a > 0$ and n even, there are two nth roots of a, one negative and one positive. The positive root is the principle root. When we write radicals such as $\sqrt{a}$ or $\sqrt[4]{a}$, we are referring to the principle root. For n an odd number, the nth root of a or $\sqrt[n]{a}$, is a single real number.

A radical expression is an expression in which an nth root has been applied to a polynomial or a rational expression. For a radical expression with an odd root the radicand can be any real number. For an even root, the radicand must evaluate to a number greater than or equal to 0. Solving an inequality, we typically find that the domain of definition is an interval of real numbers such as $\{x \mid x \geq 2\}$.

An important relation for nth roots involves $\sqrt[n]{x^n}$. For an odd root, $\sqrt[n]{x^n} = x$. For an even root, $\sqrt[n]{x^n} = |x|$, that is, the evaluation will always be nonnegative.

We defined rational number exponents such that the laws of exponents hold for rational number exponents. First, $x^{1/n}$ is defined as $\sqrt[n]{x}$. Next, for m/n a completely reduced rational number and $\sqrt[n]{x}$ a real number, $x^{m/n}$ is defined as $\left(\sqrt[n]{x}\right)^m$ or $\sqrt[n]{x^m}$. The denominator, n, for the exponent m/n yields the index of the radical.

7.1 ■ ■ ■ EXERCISES

1. For $\sqrt{2x}$, what is the radicand? What is the index?
2. For $\sqrt[3]{3ab + 5}$, what is the radicand? What is the index?
3. What are the two real number square roots of 49? Which one is the principal square root?
4. **a.** For a real number a and an odd number n greater than 1, how many real number nth roots does a have?
   b. If n is an even number and a is less than 0, how many real number nth roots does a have?
5. For any integer n greater than 1, what is the nth root of 0?
6. **a.** For every real number a, is $\sqrt{a^2}$ a real number?
   b. Is $\sqrt{a^2} = a$ for every real number a?

If the radical expression is a real number, write it as an integer. Otherwise, report not a real number.

7. $\sqrt{100}$
8. $\sqrt[3]{-64}$
9. $\sqrt[4]{-36}$
10. $\sqrt[4]{-25}$
11. $\sqrt[5]{-1}$
12. $\sqrt{400}$
13. $\sqrt[7]{0}$
14. $\sqrt[6]{-1}$

Report the domain of definition for the given expression. Use set notation to report the answer. A typical domain is $\{x \mid x \text{ is real}\}$, $\{x \mid x \neq -3\}$, *or* $\{x \mid x \geq 2\}$.

15. $x^2 - 3x - 2$
16. $2x - 3$
17. $\dfrac{x + 5}{x + 3}$
18. $\dfrac{x + 5}{x^2 + 3x - 4}$
19. $\sqrt{3x - 5}$
20. $\sqrt{6 - 2x}$
21. $\sqrt[3]{x + 5}$
22. $\sqrt[5]{x^2 - 3x + 2}$
23. $\sqrt{x^2 - 5x}$
24. $\sqrt{x^2 - x - 6}$

If the expression is a real number, write it as an integer. Otherwise report not a real number.

25. $(36)^{3/2}$
26. $(64)^{3/2}$
27. $(-64)^{2/3}$
28. $(-64)^{3/2}$
29. $(100)^{3/2}$
30. $(81)^{5/4}$
31. $(0)^{7/8}$
32. $(-1)^{4/7}$

Write the expression as a completely reduced rational number.

33. $64^{-3/2}$
34. $(-1)^{-3/5}$
35. $(4)^{-5/2}$
36. $(-27)^{-2/3}$
37. $(125)^{-1/3}$
38. $(400)^{-3/2}$
39. $(0)^{-5/2}$
40. $(16)^{-3/4}$

Use the laws of exponents and the meaning of rational number exponents to write the expression using only whole number exponents. Assume $x > 0$.

41. $(16x^4)^{-3/2}$
42. $x^{3/5}x^{2/5}$
43. $x^{5/2} \div x^{3/2}$
44. $x^{2/3}x^{4/3}$
45. $(x^{5/2})^4$
46. $(x^{2/3})^{9/2}$
47. $(16x^2)^{1/2}$
48. $(36x^6)^{1/2}$
49. $x^{5/4} \div x^{1/4}$
50. $x^{2/3} \div x^{5/3}$
51. $(25x^{-2})^{1/2}$
52. $(-27x^6)^{1/3}$
53. $(25 \div x^{-2})^{1/2}$
54. $(-27 \div x^6)^{1/3}$
55. $(x^{1/4})(x^{7/4})$
56. $(x^{2/3})(x^{1/3})$
57. $(2x^{2/3})(x^{4/3})$
58. $(3x^{1/5})(2x^{4/5})$

7.2 ■ ■ ■ COMPLETELY REDUCED RADICAL EXPRESSIONS

The radical expressions we are concerned with at this level of algebra are expressions made up of polynomials, rational expressions, and nth roots. In Section 7.1, we learned that $\sqrt[n]{x^n} = x$, provided that $x \geq 0$. Also, we learned that rational number exponents can be defined in terms of the nth root: $x^{1/n} = \sqrt[n]{x}$. That is, for m/n a completely reduced rational number and $x^{1/n}$ a real number, $x^{m/n} = \left(\sqrt[n]{x}\right)^m = \sqrt[n]{x^m}$. Of great importance to this section is the fact that the laws of exponents are valid for rational number exponents.

Here, we will write a radical expression as a completely reduced radical expression. This is much like writing a standard polynomial or writing a rational expression as a completely reduced rational expression (removing common factors from the numerator and denominator). Writing a radical expression in completely reduced form is also closely related to the operations of addition, subtraction, multiplication, and division of radical expression, which we will discuss in the next section.

The Product Rule We start by considering nth roots of polynomials—expressions such as $\sqrt{2x^3y}$. First, consider these roots of numbers:

$$\sqrt{9}\sqrt{4} = (3)(2) = 6 \qquad \sqrt{(9)(4)} = \sqrt{36} = 6$$

and

$$\sqrt[3]{8}\,\sqrt[3]{125} = (2)(5) = 10 \qquad \sqrt[3]{(8)(125)} = \sqrt[3]{1000} = 10.$$

These examples suggest the following product rule for nth roots.

THE PRODUCT RULE

> For real numbers $\sqrt[n]{a}$ and $\sqrt[n]{b}$, $\sqrt[n]{a}\sqrt[n]{b} = \sqrt[n]{ab}$.

This property is identical to law 1 of the laws of exponents with $\sqrt[n]{a} = a^{1/n}$. That is, $a^{1/n}b^{1/n} = (ab)^{1/n}$. Secondly, in the product rule, the restriction that $\sqrt[n]{a}$ and $\sqrt[n]{b}$ must be real numbers eliminates nonreal nth roots such as $\sqrt{-25}$ and $\sqrt[4]{-1}$.

Completely Reduced Radical Expressions We are now ready to discuss a naming system for radical expressions. By a radical expression we mean any expression that contains an nth root. At this level of algebra, these expressions are typically nth roots of polynomials and nth roots of rational expressions. We now state the first rule for a completely reduced radical expression.

RULE 1 FOR A COMPLETELY REDUCED RADICAL EXPRESSION

> The radicand for an nth root radical contains no perfect nth power factor.

This means that the radicand for a square root contains no perfect square factor, and the radicand for a cube root contains no perfect cube factor. (For a completely reduced rational expression, the similar rule is that the numerator and denominator contain no common factor.) If the radicand contains an exponent that is greater than or equal to the index of the radical, the radical expression is *not* completely reduced. Consider the following examples, where all variable replacements are assumed to be greater than or equal to zero.

EXAMPLE 1 Write as a completely reduced radical expression: $\sqrt{54x^2y^3z^5}$

Solution The radicand includes factors such as x^2, which have exponents greater than or equal to the index, 2. Thus, this radical expression is not completely reduced. We factor the radicand as a product of (at most) squares:

Now, using $\sqrt{ab} = \sqrt{a}\sqrt{b}$, we get

$$\begin{aligned}
\sqrt{54x^2y^3z^5} &= \sqrt{(2)(3^2)(3)(x^2)(y^2)(y)(z^2)(z^2)(z)} \\
&= \left(\sqrt{2}\right)\left(\sqrt{3^2}\right)\left(\sqrt{3}\right)\left(\sqrt{x^2}\right)\left(\sqrt{y^2}\right)\left(\sqrt{y}\right)\left(\sqrt{z^2}\right)\left(\sqrt{z^2}\right)\left(\sqrt{z}\right) \\
&= \left(\sqrt{2}\right)(3)\left(\sqrt{3}\right)(x)(y)\left(\sqrt{y}\right)(z)(z)\left(\sqrt{z}\right) && \sqrt{a^2} = a \\
&= 3xyz^2\left(\sqrt{2}\right)\left(\sqrt{3}\right)\left(\sqrt{y}\right)\left(\sqrt{z}\right) && \text{Commutative property} \\
&= 3xyz^2\sqrt{6yz} && \sqrt{a}\sqrt{b} = \sqrt{ab}
\end{aligned}$$

■

> **WARNING**
> Avoid these errors:
>
> $\sqrt[n]{a+b} \neq \sqrt[n]{a} + \sqrt[n]{b}$
>
> $\sqrt{a^2+b^2} \neq a + b$

This last radical expression is completely reduced—the exponent on each of the factors in the radicand is 1, which is less than the index, 2. In future work we will omit these steps. We often denote these steps by the phrase *perfect square factors* (or *perfect nth power factors*) *pull out from under the radical sign*. For a radical involving a polynominal other than a monomial, such as $\sqrt{x^2 + 6x + 9}$, the polynomial must factor as a perfect square (or cube, fourth power, nth power). Thus, $\sqrt{x^2 + 6x + 9} = \sqrt{(x+3)^2} = x + 3$. The radical expression $\sqrt{x^2 + 4}$ is completely reduced.

EXAMPLE 2 Write as a completely reduced radical expression: $\sqrt[3]{54x^2y^3z^5}$

Solution Since the radicand contains factors such as y^3, which have exponents greater than or equal to the index, 3, this radical expression is not completely reduced. We factor the radicand as a product of (at most) cubes:

$$\begin{aligned}
\sqrt[3]{54x^2y^3z^5} &= \sqrt[3]{(2)(3^3)(x^2)(y^3)(z^3)(z^2)} \\
&= 3yz\sqrt[3]{2x^2z^2}
\end{aligned}$$

This last radical expression is completely reduced, because the exponent on each of the factors in the radicand is a number less than the index, 3. ■

Using the Graphics Calculator

To check our work for Example 2, type `³√(54X²Y³Z^5)` for `Y1`. (Use the `MATH` menu to type `³√`.) Type the completely reduced radical, `3YZ(³√(2X²Z²))`, for `Y2`. Return to the home screen and store a number in memory locations `X`, `Y`, and `Z`, say `5` in `X`, `3` in `Y`, and `11` in `Z`. Display the resulting values for `Y1` and `Y2`. If the two outputs are not the same, an error is present.

EXAMPLE 3 Write as a completely reduced radical expression: $\sqrt[4]{54x^2y^3z^5}$

Solution Since the radicand includes factors such as z^5, which have exponents greater than or equal to the index, 4, this radical expression is not completely reduced. We factor the monomial as a product of (at most) fourth powers:

$$\begin{aligned}\sqrt[4]{54x^2y^3z^5} &= \sqrt[4]{(54)(x^2)(y^3)(z^4)(z)}\\ &= z\sqrt[4]{54x^2y^3z}\end{aligned}$$

■

EXAMPLE 4 Write as a completely reduced radical expression: $\sqrt[5]{54x^2y^3z^5}$

Solution Since the radicand contains a factor, z^5, which has an exponent greater than or equal to the index, 5, this radical expression is not completely reduced. We factor the monomial as a product of (at most) fifth powers:

$$\begin{aligned}\sqrt[5]{54x^2y^3z^5} &= \sqrt[5]{(54)(x^2)(y^3)(z^5)}\\ &= z\sqrt[5]{54x^2y^3}\end{aligned}$$

■

EXAMPLE 5 Write as a completely reduced radical expression: $\sqrt[6]{54x^2y^3z^5}$

Solution Since no factor in the radicand has an exponent greater than or equal to the index, 6, this radical expression is completely reduced. ■

In Chapter 8 we will need to reduce expressions such as $\frac{10 + \sqrt{24}}{12}$. Work through the next two examples to learn how to write such a number as a completely reduced radical expression.

EXAMPLE 6 Write as a completely reduced radical expression: $\frac{10 + \sqrt{24}}{12}$

Solution

$$\frac{10 + \sqrt{24}}{12} = \frac{10 + 2\sqrt{6}}{12} = \frac{\cancel{2}(5 + \sqrt{6})}{\cancel{2}(6)} = \frac{5 + \sqrt{6}}{6}$$

Reduce the radical; Factor 2 from numerator and denominator; Cancel common factors

■

EXAMPLE 7 Write as a completely reduced radical expression: $\frac{10 + \sqrt{27}}{12}$

Solution

$$\frac{10 + \sqrt{27}}{12} = \frac{10 + 3\sqrt{3}}{12} \qquad \text{Reduce the radical.}$$

■

WARNING

When using the slash to write a division such as $\frac{5 + \sqrt{6}}{6}$, use parentheses to write

$(5 + \sqrt{6})/6$

Writing $5 + \sqrt{6}/6$ means $5 + \frac{\sqrt{6}}{6}$.

Notice that in Example 6, the numerator and denominator have a common factor of 2. In Example 7, the numerator and denominator have no common factor. When we say that *common factors cancel,* we mean the operation of *multiplication by the factor* and the operation of *division by the same factor* cancel. The operations cancel each other's effect, but the numbers do not cancel!

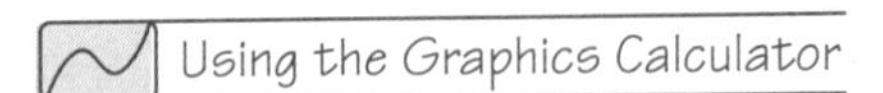

Using the Graphics Calculator

We can check our work in Examples 6 and 7 by using the √ key. For Example 6, type `(10+√24)/12` and press the ENTER key. Now, type `(5+√6)/6` and press the ENTER key. Is the same number output in both cases? If not, an error is present! This level of checking will not insure that we have found the completely reduced form of the number. For example, if we write $2\sqrt{18}$ for the completely reduced form of $\sqrt{72}$, the error would *not* show up in this level of checking. Why not?

GIVE IT A TRY

Write each expression as a completely reduced radical expression.

1. $\sqrt{24x^3y^2}$ **2.** $\sqrt[3]{-16x^5y^3}$ **3.** $\sqrt[4]{25w^6n^5}$ **4.** $\frac{8+\sqrt{28}}{6}$ 

WARNING

When writing an expression such as $3xy\sqrt[3]{2x^2z^2}$, be careful to write the index in the correct position. Otherwise, it may be confused with y^3 times the square root, $3xy^3\sqrt{2x^2z^2}$.

The Quotient Rule Next, we discuss expressions such as $\sqrt{\frac{x}{6}}$. First, consider the following numeric examples:

$$\sqrt{\frac{36}{25}} = \frac{6}{5} \quad \text{and} \quad \frac{\sqrt{36}}{\sqrt{25}} = \frac{6}{5} \qquad \sqrt[3]{\frac{27}{8}} = \frac{3}{2} \quad \text{and} \quad \frac{\sqrt[3]{27}}{\sqrt[3]{8}} = \frac{3}{2}$$

These examples suggest the following quotient rule for nth roots.

QUOTIENT RULE

For real numbers $\sqrt[n]{a}$ and $\sqrt[n]{b}$ with $b \neq 0$, $\frac{\sqrt[n]{a}}{\sqrt[n]{b}} = \sqrt[n]{\frac{a}{b}}$.

This property is identical to law 5 of the laws of exponents with $\sqrt[n]{a} = a^{1/n}$, that is,

$$\frac{a^{1/n}}{b^{1/n}} = \left(\frac{a}{b}\right)^{1/n}$$

We now state the second rule for a completely reduced radical expression.

RULE 2 FOR A COMPLETELY REDUCED RADICAL EXPRESSION

The denominator for an nth root radical contains no radical.

This means that the denominator will be a polynomial. To accomplish this, we multiply the expression by a form of 1. The following examples show how to select the form of 1 to be used.

EXAMPLE 8 Write as a completely reduced radical expression: $\dfrac{2x}{\sqrt{3x}}$

Solution Since the denominator includes a radical, this expression is not completely reduced. To remove the radical from the denominator, we multiply by a form of 1:

A form of 1

$$\frac{2x}{\sqrt{3x}} \cdot \frac{\sqrt{3x}}{\sqrt{3x}} = \frac{2x\sqrt{3x}}{\sqrt{9x^2}}$$ Multiply numerators, multiply denominators.

$$= \frac{2\cancel{x}\sqrt{3x}}{3\cancel{x}}$$ Reduce the radical $\sqrt{9x^2}$.

$$= \frac{2\sqrt{3x}}{3}$$ Remove the common factor. ■

Now we consider a radical expression that contains an index of 3. Again, we multiply by a form of 1 selected to eliminate the radical from the denominator. The form of 1 used is the form that yields a perfect cube under the radical sign.

EXAMPLE 9 Write as a completely reduced radical expression: $\dfrac{2x}{\sqrt[3]{3x}}$

Solution To remove the radical from the denominator, we multiply by a form of 1:

A form of 1

$$\frac{2x}{\sqrt[3]{3x}} = \frac{2x}{\sqrt[3]{3x}} \cdot \frac{\sqrt[3]{9x^2}}{\sqrt[3]{9x^2}}$$

$$= \frac{2x\sqrt[3]{9x^2}}{\sqrt[3]{27x^3}}$$

$$= \frac{2\cancel{x}\sqrt[3]{9x^2}}{3\cancel{x}}$$

$$= \frac{2\sqrt[3]{9x^2}}{3}$$ ■

We could also work Example 9 as follows:

$$\frac{2x}{\sqrt[3]{3x}} = \frac{2x}{\sqrt[3]{3x}} \cdot \frac{\sqrt[3]{3x}}{\sqrt[3]{3x}} \cdot \frac{\sqrt[3]{3x}}{\sqrt[3]{3x}}$$

$$= \frac{2x\sqrt[3]{9x^2}}{\sqrt[3]{27x^3}}$$

$$= \frac{2\cancel{x}\sqrt[3]{9x^2}}{3\cancel{x}} = \frac{2\sqrt[3]{9x^2}}{3}$$

GIVE IT A TRY

Write each expression as a completely reduced radical expression.

5. $\dfrac{2x}{\sqrt{6xy}}$ **6.** $\dfrac{x^2}{\sqrt{2xy}}$ **7.** $\dfrac{2x}{\sqrt[3]{9xy}}$ **8.** $\dfrac{x^2}{\sqrt[3]{2x^2y}}$

Radicand Rule We now state the third rule for a completely reduced radical expression. This rule involves nth roots of rational expressions.

RULE 3 FOR A COMPLETELY REDUCED RADICAL EXPRESSION

> Each radicand contains no rational expression.

Consider the radical expression $\sqrt{\frac{1}{2x}}$. Its radicand is a rational expression. Thus, the radical expression is not completely reduced. We start by using the property $\sqrt{\frac{a}{b}} = \frac{\sqrt{a}}{\sqrt{b}}$, to get

$$\sqrt{\frac{1}{2x}} = \frac{\sqrt{1}}{\sqrt{2x}}$$

$$= \frac{1}{\sqrt{2x}} \cdot \frac{\sqrt{2x}}{\sqrt{2x}} \quad \text{(A form of 1)}$$

$$= \frac{\sqrt{2x}}{2x}$$

EXAMPLE 10 Write as a completely reduced radical expression: $\sqrt[3]{\dfrac{3}{2x^2}}$

Solution Since the radicand is a rational expression, the radical expression is not completely reduced. To remove the rational expression in the radicand, we use the property $\sqrt[n]{\frac{a}{b}} = \frac{\sqrt[n]{a}}{\sqrt[n]{b}}$, to get

$$\sqrt[3]{\frac{3}{2x^2}} = \frac{\sqrt[3]{3}}{\sqrt[3]{2x^2}}$$

$$= \frac{\sqrt[3]{3}}{\sqrt[3]{2x^2}} \cdot \frac{\sqrt[3]{4x}}{\sqrt[3]{4x}} \quad \text{(A form of 1)}$$

$$= \frac{\sqrt[3]{12x}}{\sqrt[3]{8x^3}}$$

$$= \frac{\sqrt[3]{12x}}{2x}$$

■

GIVE IT A TRY

Write each expression as a completely reduced radical expression.

9. $\sqrt{\dfrac{6y}{2x}}$ **10.** $\sqrt{\dfrac{9}{x^2}}$ **11.** $\sqrt[3]{\dfrac{6y}{2x}}$ **12.** $\sqrt[3]{\dfrac{9}{x^2}}$ 

Using the Graphics Calculator

To test our work in Example 10, for Y1 type the original problem, ³√(3/(2X²)), and for Y2 type the completely reduced radical expression ³√(12X)/(2X). Return to the home screen and store a number, say 9, in memory location X. Display the resulting value in Y1 and the value in Y2. If these two values are not the same, an error is present. Return to *Give It a Try* Problems 9–12, and use the graphics calculator check your work. As usual, be careful with your use of parentheses!

Index Rule The last rule for a completely reduced rational expression concerns radicals in which the radicand is an nth power. For example, $\sqrt[6]{x^4}$ is *not* a completely reduced radical expression, because it violates the following rule.

RULE 4 FOR A COMPLETELY REDUCED RADICAL EXPRESSION

> For a radical expression in which the radicand is an nth power, the exponent for the radicand and the index of the radical have no common factor.

Using rational number exponents, we can write $\sqrt[6]{x^4}$ as $(x^4)^{1/6}$. Now, using law 3 of the laws of exponents, $(a^n)^m = a^{nm}$, we get $x^{4/6}$. Reducing the fraction $\frac{4}{6}$ to $\frac{2}{3}$, we have $x^{2/3}$. As a radical, we write $\sqrt[6]{x^4} = \sqrt[3]{x^2}$. For $\sqrt[3]{x^2}$, the exponent on the radicand and the index of the radical have no common factor. Thus, this radical is written in completely reduced form.

The radical $\sqrt[6]{2x^4}$ is a completely reduced radical expression. Here, the radicand is *not* a perfect fourth power—only the factor x is raised to the fourth power.

■ SUMMARY

In this section, we have learned how to write a radical expression in completely reduced form. To accomplish this, we use the laws of exponents and the fact that $\sqrt[n]{a} = a^{1/n}$ in order to write the following generalizations:

For real numbers $\sqrt[n]{a}$ and $\sqrt[n]{b}$,

$$\sqrt[n]{a}\sqrt[n]{b} = \sqrt[n]{ab}$$

For real numbers $\sqrt[n]{a}$ and $\sqrt[n]{b}$ with $b \neq 0$,

$$\frac{\sqrt[n]{a}}{\sqrt[n]{b}} = \sqrt[n]{\frac{a}{b}}$$

Using these generalizations and multiplying by a particular form of 1, we can rename any radical expression to meet the following requirements, that is, to write the expression as a completely reduced radical expression:

1. The radicand for an nth root radical contains no perfect nth power factor.
2. The denominator for an nth root radical contains no radical.
3. Each radicand contains no rational expression.
4. For a radical expression where the radicand is an nth power, the exponent for the radicand and the index of the radical contain no common factor.

In addition to these four requirements, the numerator and denominator must contain no common factor.

7.2 ■ ■ ■ EXERCISES

Insert true *for a true statement, or* false *for a false statement.*

____ 1. For every replacement of x, $\sqrt{x^2} = x$.

____ 2. For every replacement of x, $\sqrt[3]{x^3} = x$.

____ 3. For every replacement of x, $\sqrt{x^2} = |x|$.

____ 4. For $x \geq 0$ and an integer $n > 1$, $\left(\sqrt[n]{x}\right)^n = x$.

____ 5. As a completely reduced radical expression, $\sqrt{8x^3}$ is $2x\sqrt{x}$, provided $x \geq 0$.

____ 6. As a completely reduced radical expression, $\left(\sqrt{x} + \sqrt{2}\right)^2$ is $x + 2$, provided $x \geq 0$.

____ 7. As a completely reduced radical expression, $\sqrt[4]{x^6} = x\sqrt[4]{x^2}$, provided $x \geq 0$.

____ 8. To completely reduce $\dfrac{1}{\sqrt{x}}$, first multiply by 1 in the form $\dfrac{\sqrt{x}}{\sqrt{x}}$.

____ 9. To completely reduce $\dfrac{1}{\sqrt[3]{x}}$, first multiply by 1 in the form $\dfrac{\sqrt[3]{x^2}}{\sqrt[3]{x^2}}$.

10. Using the graphics calculator, type `2/√X` for `Y1`. View the graph screen using the Window 9.4 range settings. Evaluations of the expression are displayed for various replacements of `X`. What is the domain of definition? For `Y2`, type the completely reduced radical expression name for $\dfrac{2}{\sqrt{x}}$. What expression did you type for `Y2`? View the graph screen. Use the `TRACE` key, are the evaluations for these two expressions the same for every replacement of the x-variable in the domain of definition?

Write each expression as a completely reduced radical expression. Assume all variable replacements are greater than zero. Use the graphics calculator to check your work.

11. $\sqrt{200xy^2}$
12. $\sqrt{27x^3y^2}$
13. $\sqrt[3]{27x^3y^2}$
14. $\sqrt[3]{40r^5s^4}$
15. $\sqrt[4]{81m^7n^5}$
16. $\sqrt[4]{48x^8y^6}$
17. $\sqrt{18x^5y^3}$
18. $\sqrt{125d^5t^3z^2}$
19. $\sqrt[5]{96d^5x^3y^6}$
20. $\sqrt[5]{-a^2b^7c^8}$
21. $\sqrt[3]{-81m^7n^5}$
22. $\sqrt[3]{-16x^4y^3}$

23. $\sqrt{150a^4b^5}$

24. $\sqrt{72x^2y^4}$

25. $\sqrt[3]{48x^2y^5}$

26. $\sqrt[3]{-x^3y^6}$

27. $\dfrac{5x}{\sqrt{x}}$

28. $\dfrac{3ab}{\sqrt{a}}$

29. $\dfrac{6x}{\sqrt{2x}}$

30. $\dfrac{3y}{\sqrt{12x}}$

31. $\dfrac{2x}{\sqrt{x^3}}$

32. $\dfrac{2a^2}{\sqrt{a^3}}$

33. $\dfrac{3ab}{\sqrt[3]{a}}$

34. $\dfrac{8x}{\sqrt[3]{2x}}$

35. $\dfrac{2x}{\sqrt[3]{9x}}$

36. $\dfrac{2x}{\sqrt[3]{2x^3}}$

37. $\sqrt{\dfrac{5}{x}}$

38. $\sqrt{\dfrac{9x}{2ax}}$

39. $\sqrt{\dfrac{y}{3a}}$

40. $\sqrt{\dfrac{2}{5x^3}}$

41. $\sqrt[3]{\dfrac{2}{x}}$

42. $\sqrt[3]{\dfrac{10}{3b}}$

43. $\sqrt[3]{\dfrac{-1}{9a^2}}$

44. $\sqrt[3]{\dfrac{3}{2y^2}}$

45. $\sqrt[4]{y^2}$

46. $\sqrt[6]{8y^3}$

7.3 ■ ■ ■ OPERATIONS ON RADICAL EXPRESSIONS

Just as we learned to add, subtract, multiply, and divide polynomials and rational expressions, we now learn how to carry out these operations on radical expressions. In this process we will need to write radical expressions in completely reduced form, according to the rules we developed in the previous section.

WARNING

Assume all variable expressions represent nonnegative values when the index of the radical is an even number.

Addition and Subtraction In addition and subtraction of polynomials, we combine like terms. Like terms of polynomials are terms that have exactly the same variables, and those variables have identically the same exponents. For example, $3x^3y$ and $5x^3y$ are like terms for polynomials. To add radical expressions, we write each radical expression in completely reduced form, and then combine the like radicals. **Like radicals** are radicals with exactly the same index and exactly the same radicand. As with polynomials, we combine like radicals by use of the distributive property. That is,

$$3\sqrt{2x} + 5\sqrt{2x} = (3 + 5)\sqrt{2x} = 8\sqrt{2x}$$

Likewise, for $8\sqrt{5x} - 2\sqrt{2x} - 3\sqrt{5x}$, we use the commutative property to write:

$$\begin{aligned} 8\sqrt{5x} - 2\sqrt{2x} - 3\sqrt{5x} &= 8\sqrt{5x} - 3\sqrt{5x} - 2\sqrt{2x} \\ &= (8 - 3)\sqrt{5x} - 2\sqrt{2x} \\ &= 5\sqrt{5x} - 2\sqrt{2x} \end{aligned}$$

Notice that $\sqrt{5x}$ and $\sqrt{2x}$ are not like radicals. They have the same index, 2, but different radicands.

Consider the following example.

EXAMPLE 1 Write as a completely reduced radical expression: $\sqrt{50x} - \sqrt{18x} + \sqrt{12x}$

Solution We start by reducing each radical expression to get:

$$\sqrt{5^2(2x)} - \sqrt{3^2(2x)} + \sqrt{2^2(3x)} = 5\sqrt{2x} - 3\sqrt{2x} + 2\sqrt{3x}$$

Combining like radicals, we get: $= 2\sqrt{2x} + 2\sqrt{3x}$

Thus, as a completely reduced radical expression, $\sqrt{50x} - \sqrt{18x} + \sqrt{12x}$ is written as $2\sqrt{2x} + 2\sqrt{3x}$. ■

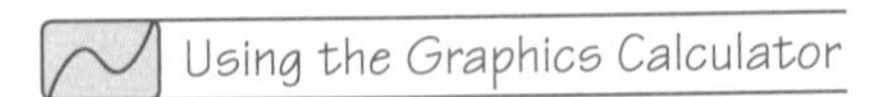

We test the work in Example 1 for errors as follows. Type the original expression, `√(50X)−√(18X)+√(12X)` for `Y1`. For `Y2` type the completely reduced radical expression, `2√(2X)+2√(3X)`. Return to the home screen and store a number, say `5`, in memory location `X`. Display the resulting values from `Y1` and `Y2`. If these values are not the same, an error is present.

The next example of addition and subtraction of radical expressions involves rational expressions in the radicands.

EXAMPLE 2 Write as a completely reduced radical expression: $\sqrt{\frac{2}{x}} - \sqrt{\frac{3}{2x}}$ Assume $x > 0$.

Solution We start by writing each radical expression in completely reduced form.

$$\sqrt{\frac{2}{x}} = \frac{\sqrt{2}}{\sqrt{x}} = \frac{\sqrt{2}}{\sqrt{x}} \cdot \frac{\sqrt{x}}{\sqrt{x}} = \frac{\sqrt{2x}}{\sqrt{x^2}} = \frac{\sqrt{2x}}{x}$$

$$\sqrt{\frac{3}{2x}} = \frac{\sqrt{3}}{\sqrt{2x}} = \frac{\sqrt{3}}{\sqrt{2x}} \cdot \frac{\sqrt{2x}}{\sqrt{2x}} = \frac{\sqrt{6x}}{\sqrt{4x^2}} = \frac{\sqrt{6x}}{2x}$$

So, we have: $\sqrt{\frac{2}{x}} - \sqrt{\frac{3}{2x}} = \frac{\sqrt{2x}}{x} - \frac{\sqrt{6x}}{2x}$

Next, we write the expressions with a common denominator: $= \frac{2\sqrt{2x}}{2x} - \frac{\sqrt{6x}}{2x}$

Finally, subtracting the numerators, we get the completely reduced radical expression: $= \frac{2\sqrt{2x} - \sqrt{6x}}{2x}$ ■

GIVE IT A TRY

Write each expression as a completely reduced radical expression, assuming $x > 0$.

1. $\sqrt{50x^3} - 2\sqrt{12x^3} + 3\sqrt{18x^3}$
2. $\dfrac{2}{\sqrt{3x}} + \dfrac{1}{\sqrt{3x}}$
3. $\sqrt{9x} - 3\sqrt{36y} + \sqrt{16x}$
4. $\dfrac{2}{\sqrt{3x}} - \dfrac{1}{\sqrt{2x}}$

Multiplication The operation of multiplication is based on the property $\sqrt[n]{a^n}\sqrt[n]{b} = \sqrt[n]{ab}$, the distributive property, and the ability to write a radical in completely reduced form. Work through the following examples.

EXAMPLE 3 Write as a completely reduced radical expression: $\left(\sqrt{3xy}\right)\left(\sqrt{6x}\right)$

Solution

$$\left(\sqrt{3xy}\right)\left(\sqrt{6x}\right) = \sqrt{18x^2y} \qquad \sqrt[n]{a}\sqrt[n]{b} = \sqrt[n]{ab}$$
$$= 3x\sqrt{2y} \qquad \text{Reduce the radical expression } \sqrt{2(3^2)x^2y}.$$
■

EXAMPLE 4 Write as a completely reduced radical expression: $\sqrt{2x}\left(\sqrt{3y} - \sqrt{10x}\right)$

Solution

$$\sqrt{2x}\left(\sqrt{3y} - \sqrt{10x}\right) = \sqrt{2x}\sqrt{3y} - \sqrt{2x}\sqrt{10x} \qquad \text{Distributive property}$$
$$= \sqrt{6xy} - \sqrt{20x^2} \qquad \sqrt[n]{a}\sqrt[n]{b} = \sqrt[n]{ab}$$
$$= \sqrt{6xy} - 2x\sqrt{5}$$
■

EXAMPLE 5 Write as a completely reduced radical expression: $\left(\sqrt{2x} + 1\right)\left(\sqrt{3x} - 2\right)$

Solution

Distributive property

$$\left(\sqrt{2x} + 1\right)\left(\sqrt{3x} - 2\right) = \left(\sqrt{2x} + 1\right)\left(\sqrt{3x}\right) - \left(\sqrt{2x} + 1\right)(2)$$

Distributive property   Distributive property

$$= \sqrt{6x^2} + \sqrt{3x} - 2\sqrt{2x} - 2$$
$$= x\sqrt{6} + \sqrt{3x} - 2\sqrt{2x} - 2$$
■

Compare the work in Example 5 with the binomial multiplication

$$(2x^2 + 1)(3x - 2) = 6x^3 - 4x^2 + 3x - 2$$

Just as we use the FOIL approach to multiplying binomials, this approach can be used to multiply radical expressions that are written in "binomial" form.

EXAMPLE 6 Write as a completely reduced radical expression: $\left(\sqrt{3x}+1\right)\left(\sqrt{3x}+2\right)$

Solution Using FOIL, we get:

$$\left(\sqrt{3x}+1\right)\left(\sqrt{3x}+2\right) = \overset{\text{F}}{\sqrt{9x^2}} + \overset{\text{O}}{2\sqrt{3x}} + \overset{\text{I}}{\sqrt{3x}} + \overset{\text{L}}{2}$$

Reduce $\sqrt{9x^2}$, and combine like radicals: $= 3x + 3\sqrt{3x} + 2$ ■

EXAMPLE 7 Write as a completely reduced radical expression: $\left(\sqrt{3x}-2\right)\left(\sqrt{3x}+2\right)$

Solution Using FOIL, we get:

$$\left(\sqrt{3x}-2\right)\left(\sqrt{3x}+2\right) = \overset{\text{F}}{\sqrt{9x^2}} + \overset{\text{O}}{2\sqrt{3x}} - \overset{\text{I}}{2\sqrt{3x}} - \overset{\text{L}}{4}$$

Reduce $\sqrt{9x^2}$, and combine like radicals: $= 3x - 4$ ■

The special products, such as $(a+b)^2 = a^2 + 2ab + b^2$, used with multiplication of polynomials also hold for radical expressions. Rework Example 7 using $(a+b)(a-b) = a^2 - b^2$. Now, work through Example 8.

EXAMPLE 8 Write as a completely reduced radical expression: $\left(\sqrt[3]{3x}+\sqrt[3]{2}\right)^2$

Solution Using the special product $(a+b)^2 = a^2 + 2ab + b^2$ with $a = \sqrt[3]{3x}$ and $b = \sqrt[3]{2}$, we get

$$\overset{(a\ +\ b)^2}{\left(\sqrt[3]{3x}+\sqrt[3]{2}\right)^2} = \overset{a^2}{\sqrt[3]{9x^2}} + \overset{2ab}{2\sqrt[3]{6x}} + \overset{b^2}{\sqrt[3]{4}}$$ ■

GIVE IT A TRY

Write each product as a completely reduced radical expression.

5. $\left(\sqrt[3]{4xy^2}\right)\left(2\sqrt[3]{2xy}\right)$ **6.** $\left(\sqrt{x}+2\right)\left(\sqrt{x}-\sqrt{3}\right)$ **7.** $\left(\sqrt{x}-\sqrt{2}\right)^2$

8. Use the graphics calculator to check your work in Problems 5–7.

Algebraic Numbers In Chapter 1, we stated that an algebraic number is a finite collection of rational numbers and the operations of addition, subtraction, multiplication, division, and nth roots. Some examples of algebraic numbers are

$$\sqrt{3} \qquad \sqrt[3]{17} \qquad \sqrt[5]{12} \qquad \frac{-3+\sqrt{13}}{2} \qquad \frac{\sqrt[3]{13}}{5}$$

In Chapter 8 we will solve second-degree equations in one variable that cannot be solved using factoring (for example, $x^2 - 3x = 1$). The solution to such an equation is typically two algebraic numbers involving square roots. As a prelude to this topic and as an example of operations on radical expressions, work through the following examples.

EXAMPLE 9 Write as a completely reduced radical expression: $\left(\frac{3+\sqrt{2}}{2}\right)^2$

Solution

$$\left(\frac{3+\sqrt{2}}{2}\right)^2 = \frac{3+\sqrt{2}}{2}\cdot\frac{3+\sqrt{2}}{2}$$

$$= \frac{(3+\sqrt{2})(3+\sqrt{2})}{(2)(2)} \qquad \frac{a}{b}\cdot\frac{c}{d} = \frac{ac}{bd}$$

$$= \frac{9+3\sqrt{2}+3\sqrt{2}+\sqrt{4}}{4}$$

$$= \frac{9+6\sqrt{2}+\sqrt{4}}{4} = \frac{11+6\sqrt{2}}{4}$$

■

Algebraic numbers—such as the one in Example 9—are the type of numbers that typically solve quadratic equations. To see that this is the case, work through the next example.

EXAMPLE 10 Show that the number $\frac{3-\sqrt{17}}{2}$ solves the equation $x^2 - 3x = 2$.

Solution To show that $\frac{3-\sqrt{17}}{2}$ solves the equation, we need only show that

$$\left(\frac{3-\sqrt{17}}{2}\right)^2 - 3\left(\frac{3-\sqrt{17}}{2}\right)$$

as a completely reduced radical expression is 2. First, we do the multiplications:

$$\left(\frac{3-\sqrt{17}}{2}\right)^2 - 3\left(\frac{3-\sqrt{17}}{2}\right) = \left(\frac{3-\sqrt{17}}{2}\right)\left(\frac{3-\sqrt{17}}{2}\right) - \frac{3}{1}\left(\frac{3-\sqrt{17}}{2}\right)$$

$$= \frac{9-6\sqrt{17}+\sqrt{17^2}}{4} - \frac{9-3\sqrt{17}}{2}$$

$$= \frac{26-6\sqrt{17}}{4} - \frac{9-3\sqrt{17}}{2}$$

$$= \frac{26-6\sqrt{17}}{4} - \frac{18-6\sqrt{17}}{4}$$

$$= \frac{26-18-6\sqrt{17}+6\sqrt{17}}{4}$$

$$= \frac{8}{4}$$

$$= 2$$

So, $\frac{3-\sqrt{17}}{2}$ is a solution to the given equation. ■

Using the Graphics Calculator

To check our work in Example 10, store the number `(3−√17)/2` in memory location `X`. The decimal approximation of the number, `-.5615528128`, is written to the screen. Now, type `X²−3X` and press the `ENTER` key. Is the number 2 output? If so, the number $\frac{3-\sqrt{17}}{2}$ is in the solution to the equation $x^2 - 3x = 2$.

GIVE IT A TRY

9. Write $\left(2-\sqrt{3}\right)^2 - 3\left(2-\sqrt{3}\right)$ as a completely reduced radical expressi

10. Show that the number $2+\sqrt{3}$ is in the solution to the equation $x^2 - 4x = -1$.

11. Write $\left(\frac{-1+\sqrt{7}}{2}\right)^2 + \left(\frac{-1+\sqrt{7}}{2}\right)$ as a completely reduced radical expression.

Division The operation of division has already been introduced in our naming system (see Section 7.2). That is, we already know how to write expressions like $3 \div \sqrt{2x}$ and $\sqrt[3]{2x^2} \div \sqrt[3]{4x}$ in completely reduced form. For more involved problems using the division operation, we use the following fact from multiplication:

$$\left(\sqrt{a}+\sqrt{b}\right)\left(\sqrt{a}-\sqrt{b}\right) = a - b \qquad \text{See Example 7.}$$

For $\sqrt{a}+\sqrt{b}$, the radical expression $\sqrt{a}-\sqrt{b}$ is known as its **conjugate**. Likewise, the conjugate of $\sqrt{a}-\sqrt{b}$ is $\sqrt{a}+\sqrt{b}$. In general, two expressions of the form $a+b$ and $a-b$ are known as **conjugate pairs**. Work through the following examples.

EXAMPLE 11 Write as a completely reduced radical expression: $\frac{\sqrt{2}}{3-\sqrt{5}}$

Solution First, the denominator of this number contains a radical, so the expression is not completely reduced. To eliminate the radical from the denominator, we multiply by a form of 1 that uses the conjugate of the denominator.

$$\frac{\sqrt{2}}{3-\sqrt{5}} = \frac{\sqrt{2}}{3-\sqrt{5}} \cdot \frac{3+\sqrt{5}}{3+\sqrt{5}} \qquad \text{(a form of 1)}$$

$3+\sqrt{5}$ is the conjugate of $3-\sqrt{5}$.

$$= \frac{\left(\sqrt{2}\right)\left(3+\sqrt{5}\right)}{9+3\sqrt{5}-3\sqrt{5}-\sqrt{25}}$$

$$= \frac{3\sqrt{2}+\sqrt{10}}{9-5}$$

$$= \frac{3\sqrt{2}+\sqrt{10}}{4}$$

■

Using the Graphics Calculator

To check our work in Example 11, type √2/(3−√5) and press the ENTER key, then type (3√2+√10)/4 and press the ENTER key. If the same number is not output in both cases, an error is present.

EXAMPLE 12 Write as a completely reduced radical expression: $\dfrac{1-\sqrt{2}}{3+\sqrt{5}}$

Solution Again, we multiply by a form of 1 (the conjugate of the denominator, divided by itself) to remove the radical from the denominator.

$$\begin{aligned}\frac{1-\sqrt{2}}{3+\sqrt{5}} &= \frac{1-\sqrt{2}}{3+\sqrt{5}}\cdot\frac{3-\sqrt{5}}{3-\sqrt{5}}\\ &= \frac{(1-\sqrt{2})(3-\sqrt{5})}{9-3\sqrt{5}+3\sqrt{5}-\sqrt{25}}\\ &= \frac{3-\sqrt{5}-3\sqrt{2}-\sqrt{10}}{9-5}\\ &= \frac{3-\sqrt{5}-3\sqrt{2}-\sqrt{10}}{4}\end{aligned}$$

■

EXAMPLE 13 Write as a completely reduced radical expression: $\left(\sqrt{3x}+1\right)\div\left(\sqrt{x}+\sqrt{2}\right)$

Solution First, we have $\left(\sqrt{3x}+1\right)\div\left(\sqrt{x}+\sqrt{2}\right)=\dfrac{\sqrt{3x}+1}{\sqrt{x}+\sqrt{2}}$

Now,

a form of 1

$$\begin{aligned}\frac{\sqrt{3x}+1}{\sqrt{x}+\sqrt{2}} &= \frac{\sqrt{3x}+1}{\sqrt{x}+\sqrt{2}}\cdot\frac{\sqrt{x}-\sqrt{2}}{\sqrt{x}-\sqrt{2}}\\ &= \frac{\sqrt{3x^2}+\sqrt{x}-\sqrt{6x}-\sqrt{2}}{\sqrt{x^2}+\sqrt{2x}-\sqrt{2x}-\sqrt{4}}\\ &= \frac{x\sqrt{3}+\sqrt{x}-\sqrt{6x}-\sqrt{2}}{x-2}\end{aligned}$$

$\sqrt{x}-\sqrt{2}$ is conjugate of $\sqrt{x}+\sqrt{2}$.

■

EXAMPLE 14 Write as a completely reduced radical expression: $\left(\sqrt{3x}+1\right)\div\left(\sqrt{x+2}\right)$

a form of 1

Solution

$$\begin{aligned}\left(\sqrt{3x}+1\right)\div\left(\sqrt{x+2}\right) &= \frac{\sqrt{3x}+1}{\sqrt{x+2}}\cdot\frac{\sqrt{x+2}}{\sqrt{x+2}}\\ &= \frac{\sqrt{3x^2+6x}+\sqrt{x+2}}{x+2}\end{aligned}$$

■

GIVE IT A TRY

Write each as a completely reduced radical expression.

12. $\left(\sqrt{x} - \sqrt{2}\right) \div \left(\sqrt{x + 2}\right)$
13. $\left(\sqrt{x} - \sqrt{2}\right) \div \left(\sqrt{x} + \sqrt{2}\right)$
14. $\dfrac{\sqrt{3}}{-2 + \sqrt{3}}$
15. $\dfrac{2 - \sqrt{5}}{1 - \sqrt{5}}$
16. $\left(3 - \sqrt{x}\right) \div \left(2 + \sqrt{x}\right)$
17. $\sqrt{y} \div \left(\sqrt{y} + 3\right)$

■ SUMMARY

In this section, we have learned how to add, subtract, multiply, and divide radical expressions. The operations of addition and subtraction are accomplished by writing the radicals in completely reduced form, and then combining like radicals. Like radicals have exactly the same index and the same radicand, and we combine them by using the distributive property. For example, $\sqrt{x} + 2\sqrt{x} = (1 + 2)\sqrt{x} = 3\sqrt{x}$.

The operation of multiplication is accomplished by using the laws of radicals, the distributive property, and the ability to do addition and subtraction of radical expressions. We found that the FOIL technique (used earlier to multiply binomials) also works with radical expressions in binomial form. For example, $\left(3 - \sqrt{x}\right)\left(2 + 2\sqrt{x}\right)$. We also found that special products developed for polynomials, such as $(a + b)^2 = a^2 + 2ab + b^2$, may be used to perform multiplications like $\left(\sqrt{2x} + \sqrt{3}\right)^2$.

Finally, the operation of division of radicals is accomplished by writing the division in fraction form. The main task is to remove any resulting radical expression from a denominator. This is accomplished by multiplying by a particular form of 1—for example, $\dfrac{\sqrt{x} + 2}{\sqrt{x} + 2}$. The form of 1 selected for a denominator of the form $a + b$ is its conjugate, $a - b$. That is, if the denominator is $2 + \sqrt{x}$, the form of 1 selected is $\dfrac{2 - \sqrt{x}}{2 - \sqrt{x}}$, because the conjugate of $2 + \sqrt{x}$ is $2 - \sqrt{x}$.

■ ***Quick Index***

7.3 ■ ■ ■ EXERCISES

Perform the indicated operations, and then write the result as a completely reduced radical expression. Use the graphics calculator to test your work for errors.

1. $\sqrt{27x^3} + \sqrt{12x^3}$
2. $\sqrt{2y} - \sqrt{8y} + 3\sqrt{50y}$
3. $\sqrt[3]{16x^3} + \sqrt[3]{54x^3}$
4. $\sqrt[3]{81x^5} - \sqrt[3]{54x^5} - \sqrt[3]{32x^3}$
5. $\sqrt{8x^3} - \sqrt{50x^3}$
6. $3\sqrt{12w} - 2\sqrt{18w} + 5\sqrt{9w}$
7. $\sqrt{12x} - \sqrt{18x} - \sqrt{12x}$
8. $\sqrt[3]{40x^2} + \sqrt[3]{250x^2}$
9. $\sqrt{2x}\left(\sqrt{3x} - \sqrt{2y}\right)$
10. $\left(\sqrt{x} - 3\right)^2$
11. $\sqrt[3]{x}\left(\sqrt[3]{x^3} - \sqrt[3]{x^2}\right)$
12. $\left(\sqrt{x} + 3\right)^2$
13. $\left(\sqrt{x} - 3\right)\left(\sqrt{x} + 5\right)$
14. $\left(\sqrt{2x} - \sqrt{3}\right)\left(\sqrt{3x} + \sqrt{10}\right)$
15. $\left(\sqrt{x} - \sqrt{3}\right)\left(\sqrt{x} + \sqrt{5}\right)$
16. $\left(\sqrt[3]{x} + 1\right)\left(\sqrt[3]{x^2} - 1\right)$

17. $\left(\sqrt{x} - 3\right)\left(\sqrt{x} + 3\right)$
18. $\left(\sqrt[3]{x} + 2\right)\left(\sqrt[3]{x^2} - 2\sqrt[3]{x} + 4\right)$
19. $\left(3 - \sqrt{2}\right)^2 - 6\left(3 - \sqrt{2}\right)$
20. $\left(-1 + \sqrt{7}\right)^2 - 2\left(-1 + \sqrt{7}\right)$
21. $\left(-2 + \sqrt{11}\right)^2 + 4\left(-2 + \sqrt{11}\right)$
22. $\left(-2 - \sqrt{11}\right)^2 + 4\left(-2 - \sqrt{11}\right)$
23. $\left(\frac{1 - \sqrt{2}}{2}\right)^2 - \left(\frac{1 - \sqrt{2}}{2}\right)$
24. $\left(\frac{2 + \sqrt{5}}{2}\right)^2 - 2\left(\frac{2 + \sqrt{5}}{2}\right)$
25. $\left(\frac{1 + \sqrt{5}}{6}\right)^2 + 3\left(\frac{1 + \sqrt{5}}{6}\right)$
26. $\left(\frac{-2 + \sqrt{3}}{3}\right)^2 + 5\left(\frac{-2 + \sqrt{3}}{3}\right)$
27. Show that the number $1 - \sqrt{2}$ solves the second-degree equation $x^2 - 2x = 1$.
28. Show that the number $1 + \sqrt{2}$ solves the second-degree equation $x^2 - 2x = 1$.
29. Show that the number $-3 + \sqrt{5}$ solves the second-degree equation $x^2 + 6x = -4$.
30. Show that the number $-3 - \sqrt{5}$ solves the second-degree equation $x^2 + 6x = -4$.

Perform the indicated operations, and then write the result as a completely reduced radical expression. Use the graphics calculator to test your work for errors.

31. $1 \div \sqrt{2xy}$
32. $2 \div \sqrt{3a}$
33. $x \div \sqrt[3]{27x^3}$
34. $\sqrt{x} \div \sqrt{27x^3y^2}$
35. $x \div \sqrt{2x + 1}$
36. $\sqrt{x} \div \sqrt{x - 2}$
37. $x \div \left(\sqrt{2x} + 1\right)$
38. $\sqrt{x} \div \left(\sqrt{x} - 2\right)$
39. $\left(\sqrt{x} + 2\right) \div \left(\sqrt{x - 2}\right)$
40. $\left(\sqrt{a} - 3\right) \div \left(\sqrt{a + 5}\right)$
41. $\left(\sqrt{2x} + \sqrt{3x}\right) \div \left(\sqrt{3x} - \sqrt{2x}\right)$
42. $\left(\sqrt{x} + \sqrt{2}\right) \div \left(\sqrt{2x} - \sqrt{2}\right)$
43. $\frac{\sqrt{2x}}{\sqrt{x} + \sqrt{3}}$
44. $\frac{\sqrt{3x}}{\sqrt{x} + \sqrt{2}}$
45. $\frac{\sqrt{3x - 1}}{\sqrt{x} - \sqrt{2}}$
46. $\frac{\sqrt{3x - 1}}{\sqrt{x} - \sqrt{3}}$
47. $\frac{\sqrt{x} + 1}{\sqrt{x} - 1}$
48. $\frac{\sqrt{x} - 1}{\sqrt{x} + 1}$
49. $\frac{\sqrt{x + 1}}{\sqrt{x} - 1}$
50. $\frac{\sqrt{x} - 1}{\sqrt{x + 1}}$
51. $\frac{\sqrt{x} + 1}{\sqrt{2x} + 3}$
52. $\frac{\sqrt{x} - 1}{\sqrt{3x} - 2}$

■ BONUS PROBLEMS

53. For a radical expression with a denominator of $\sqrt{a} + \sqrt{b}$, we multiply by 1 in the form

$$\frac{\sqrt{a} - \sqrt{b}}{\sqrt{a} - \sqrt{b}}$$

This is based on the multiplication fact

$$(a + b)(a - b) = a^2 - b^2$$

We also know from multiplication that

$$(a + b)(a^2 - ab + b^2) = a^3 + b^3$$

and that $\quad (a - b)(a^2 + ab + b^2) = a^3 - b^3$

Use this information to write each fraction as a completely reduced radical expression.

a. $\frac{1}{\sqrt[3]{x} + 2}$ **b.** $\frac{1}{\sqrt[3]{x} - 3}$

54. Often in algebra, we try an action that doesn't produce the desired result, but does make *progress* toward the desired result. For example,

$$\frac{1}{\sqrt{x} + \sqrt{2} + \sqrt{3}} \cdot \frac{\sqrt{x} - \left(\sqrt{2} + \sqrt{3}\right)}{\sqrt{x} - \left(\sqrt{2} + \sqrt{3}\right)} = \frac{\sqrt{x} - \left(\sqrt{2} + \sqrt{3}\right)}{x - 5 - 2\sqrt{6}}$$

doesn't yield the completely reduced form (because the denominator still contains a radical). But, now we have only one radical in the denominator. Finish the process to write $\frac{1}{\sqrt{x} + \sqrt{2} + \sqrt{3}}$ as a completely reduced radical expression.

7.4 ■ ■ ■ SOLVING RADICAL EXPRESSION EQUATIONS

Now that we know how to write radical expressions in completely reduced form and how to perform the basic binary operations on these objects, we can apply this knowledge to solving equations. As with polynomial equations and rational expression equations, we start with the simplest level, and work to the more complex levels.

Meaning An example of a rational expression equation in one variable is

$$\sqrt{2x - 1} = 3$$

To solve any equation in one variable means that we want to find the numbers that yield a true arithmetic sentence when one of the numbers replaces the variable. For example, is 7 in the solution to the equation $\sqrt{2x - 1} = 3$? To find out, we replace the variable with 7 to get

$$\sqrt{2(7) - 1} = 3$$

This is a false arithmetic sentence, since $\sqrt{13}$ is not equal to 3 (because $3^2 \neq 13$). Thus, 7 is not in the solution to the equation.

Using the Graphics Calculator

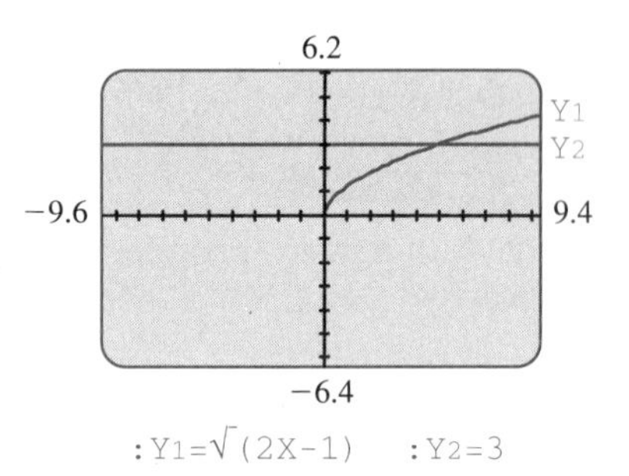

FIGURE 1

We can visualize the number that solves $\sqrt{2x - 1} = 3$ by using the graph screen to observe the evaluations of $\sqrt{2x - 1}$ for various replacements of the x-variable. Type √(2X−1) for Y1 and 3 as the rule for Y2. Use the Window 9.4 range settings to view the graph (see Figure 1). Press the TRACE key. The message displayed means that $\sqrt{2x - 1}$ is not defined when the variable is replaced by the number 0. The domain of definition for $\sqrt{2x - 1}$ is $\left\{x \mid x \geq \frac{1}{2}\right\}$. Why?

Press the right arrow key and move the cursor to the intersection with the horizontal line. What number do you think solves the equation $\sqrt{2x - 1} = 3$?

An Algebraic Approach To solve a radical expression equation using algebra, we use the following equation-solving property.

THE nth POWER PROPERTY OF EQUATIONS

> The solution to an equation in one variable is contained in (is a subset of) the solution to the equation obtained by raising each side of the original equation to the nth power (such as squaring or cubing).

The nth power property of equations says that if c is a number which solves the equation $a = b$, then c must also solve the equation $a^2 = b^2$.

Consider the equation $x = 10$. The solution is 10. If we square both sides, we get $x^2 = 100$. The solution to this equation is 10 and -10.

Unlike adding a polynomial to each side of an equation, squaring each side *can alter* the solution—we do not always get an equivalent equation. However, the solution is altered in a convenient way—the solution to the new equation contains the solution to the original equation. The extra number, -10, in the solution to the new equation, $x^2 = 100$, is an **extraneous root** to the original equation, $x = 10$. Remember from our work with rational expression equations that multiplying each side of an equation by an expression can also produce an extraneous root (see Section 6.5).

REMEMBER

In equation solving, a number that solves an equation is known as a *root* of the equation. An extraneous solution and an extraneous root are synonymous terms.

In Section 7.1, we discussed the facts that $\sqrt[n]{x^n} = |x|$ if n is an even number and that $\sqrt[n]{x^n} = x$ if n is an odd number. With radical expression equations, if we raise each side to the nth power, we often get an expression such as $\left(\sqrt[n]{x}\right)^n$. We have the following property.

THE nth POWER OF a^n

For n an integer with $n > 1$ and $\sqrt[n]{a}$ a real number, $\left(\sqrt[n]{a}\right)^n = a$.

Our strategy for solving radical expression equations in one variable is to square (or raise to a higher power) each side of the equation in order to arrive at a new equation that we *can* solve (such as a first- or second-degree equation, or a rational expression equation). However, we must check each of the numbers in the solution to make sure that each one does indeed solve the original equation (they may be extraneous roots). Consider the following example.

EXAMPLE 1 Solve: $\sqrt{2x - 1} = 3$

Solution Squaring each side of the equation, we get

$$\left(\sqrt{2x-1}\right)^2 = 3^2$$

or

$$\begin{aligned} 2x - 1 &= 9 \\ 1 &\quad 1 \\ \hline 2x &= 10 \\ x &= 5 \end{aligned}$$

The first-degree equation in one variable, $2x - 1 = 9$, has solution 5.

Check To check for an extraneous root, we replace the variable in the original equation with the number, 5, and get $\sqrt{2(5) - 1} = 3$, a true arithmetic sentence $\left(\sqrt{9} = 3\right)$. Thus, the solution to $\sqrt{2x - 1} = 3$ is 5. ■

We now consider a situation similar to Example 1, except this equation has no solution.

EXAMPLE 2 Solve: $\sqrt{2x - 1} = -3$

Solution From the definition of square root, we know that a square root cannot be a negative number, such as -3. This equation has *no* solution.

If we proceed as in Example 1, we would arrive at the same answer. We square each side to get $2x - 1 = 9$. This equation has solution 5.

Check Checking to see that 5 does indeed solve the original equation, we get

$\sqrt{2(5) - 1} = -3$, a false arithmetic sentence $\left(\sqrt{9} \neq -3\right)$.

Because 5 does not solve the equation, the equation has no solution. (Remember, the solution to the new equation obtained by squaring each side *contains* the solution to the original equation). ■

To increase our understanding of the result in Example 2, make sure that √(2X−1) is still entered for Y1, and enter ⁻3 for Y2. Use the Window 9.4 range settings to view the graph. As Figure 2 shows, no x-value results in the radical expression $\sqrt{2x - 1}$ evaluating to -3.

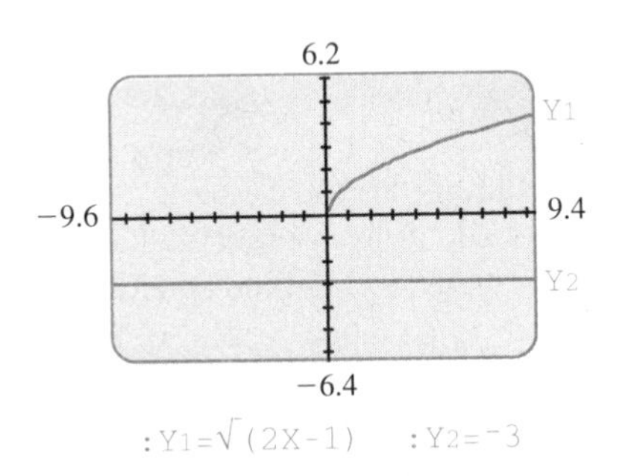

Notice that the solution to the equation $-\sqrt{2x - 1} = -3$ is 5, because $-\sqrt{2(5) - 1}$ is -3.

We can also use this algebraic approach to solve radical expression equations involving other nth roots. For example, the following radical expression equation involves a cube root radical.

EXAMPLE 3 Solve: $\sqrt[3]{5 - 3x} = -4$

Solution Rather than squaring each side, we cube each side to get:

$$\begin{aligned} \left(\sqrt[3]{5 - 3x}\right)^3 &= (-4)^3 \\ 5 - 3x &= -64 \\ -5 \quad &\quad -5 \\ \hline -3x &= -69 \\ x &= 23 \end{aligned}$$

The solution is the number 23.

Check Although with an odd nth root we do not have to check for an extraneous root, we should still check our work. Checking 23 in the original equation, we get $\sqrt[3]{5 - 3(23)} = -4$, or $\sqrt[3]{-64} = -4$, a true arithmetic sentence. Thus, the solution to $\sqrt[3]{5 - 3x} = -4$ is 23. ■

Using the Graphics Calculator

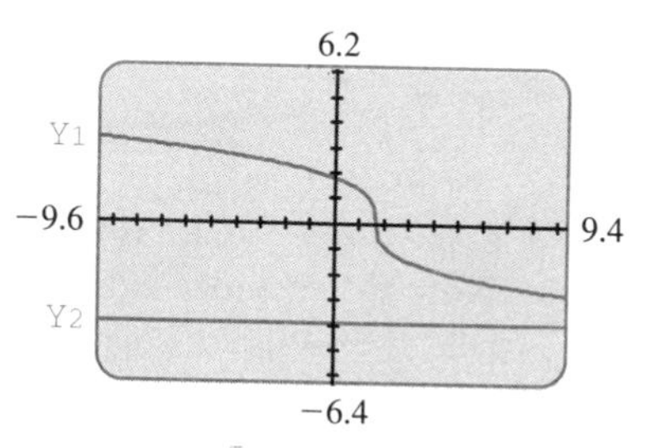

:Y1=³√(5−3X) :Y2=−4

FIGURE 3

To increase our understanding of the solution found in Example 3, type ³√(5−3X) for Y1 and −4 for Y2. (Use the MATH menu to type ³√.) Use the Window 9.4 range settings to view the graph (see Figure 3). Zoom out to view the intersection point. Use the TRACE feature and the arrow key to move the cursor to the intersection point. Zoom in on the point of intersection until you think you know the x-value that results in the radical expression ³√(5−3X) evaluating to −4. Does it agree with the number we found in Example 3?

In the next example, squaring each side produces a second-degree equation in one variable.

EXAMPLE 4 Solve: $\sqrt{x+2} = x$

Solution We use our usual approach:

$$\begin{aligned}
\left(\sqrt{x+2}\right)^2 &= x^2 && \text{Square each side}\\
x + 2 &= x^2 &&\\
x^2 - x - 2 &= 0 && \text{Standard form}\\
(x-2)(x+1) &= 0 && \text{Factoring}
\end{aligned}$$

$$x - 2 = 0 \quad \text{or} \quad x + 1 = 0$$

$$x = 2 \qquad\qquad x = -1$$

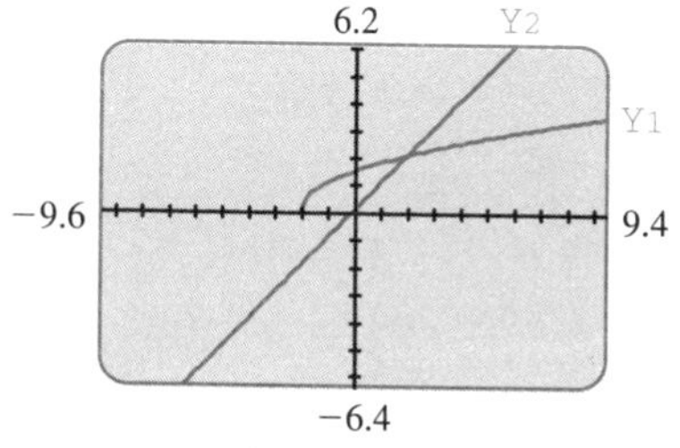

:Y1=√(X+2) :Y2=X

FIGURE 4

The solution is 2 and −1.

Check

Substitute 2: $\sqrt{2+2} = 2$ a true arithmetic sentence.

Substitute −1: $\sqrt{-1+2} = -1$ a false arithmetic sentence.

Thus, $\sqrt{x+2} = x$ has solution 2. (The number −1 is extraneous.) ■

Using the Graphics Calculator

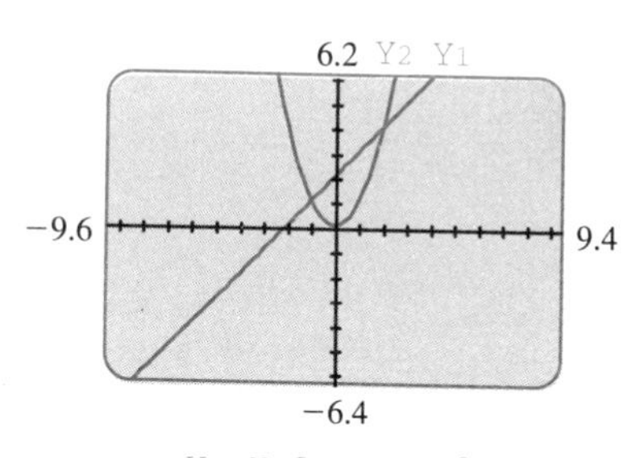

:Y1=X+2 :Y2=X²

FIGURE 5

To increase our understanding of the solution found in Example 4, type √(X+2) for Y1 and X for Y2. Use the Window 9.4 range settings to view the graph, as shown in Figure 4. Use the TRACE feature and the arrow key to move the cursor to the intersection point. The x-value that results in the radical expression √(X+2) and the polynomial X evaluating to the same value is 2. Do you see that −1 does not solve the equation $\sqrt{x+2} = x$? Now, from the second line in the solution to Example 4, type X+2 for Y1 and type X² for Y2. View the graph screen as shown in Figure 5. The equation $x + 2 = x^2$ has two x-values, −1 and 2, that result in the two polynomials, $x + 2$ and x^2, evaluating to the same number.

GIVE IT A TRY

Solve each radical expression equation, and check for extraneous roots.

1. $\sqrt{2-3x} = 5$ 2. $\sqrt[3]{2x - 4} = 2$ 3. $\sqrt{x + 8} = x + 2$

Equations Involving Two Radicals All radical expression equations we have solved up to this point involve a single radical. Now we consider radical expression equations with two radicals. Here, we expand the equation-solving property regarding adding an expression to each side. Any radical expression can be added to each side of an equation without altering the solution. We will use this property to *isolate* a radical, that is, we alter the equation to an equivalent form in which either the left-hand or right-hand member of the equation is a single radical.

EXAMPLE 5 Solve: $\sqrt{x + 2} = \sqrt{x} - 1$

Solution Again, we start by squaring each side:

$$\left(\sqrt{x + 2}\right)^2 = \left(\sqrt{x} - 1\right)^2$$
$$x + 2 = x - 2\sqrt{x} + 1$$

> **WARNING**
> A common error is to square $\sqrt{x} - 1$ as $x + 1$. Remember, $(a + b)^2 \neq a^2 + b^2$.

Although we did not eliminate all the radicals, we did arrive at an equation involving only one radical. Now, we can isolate this radical, $-2\sqrt{x}$.

	$x + 2 = x - 2\sqrt{x} + 1$
Add $-x$ to each side	$-x \qquad -x$
to get:	$2 = -2\sqrt{x} + 1$
Add -1 to each side	$-1 \qquad -1$
to isolate $-2\sqrt{x}$:	$1 = -2\sqrt{x}$
Square each side	$(1)^2 = \left(-2\sqrt{x}\right)^2$
to get:	$1 = 4x$
Multiply by $\frac{1}{4}$:	$\frac{1}{4} = x$

Check Replacing x with $\frac{1}{4}$, we get

$$\sqrt{\frac{1}{4} + 2} = \sqrt{\frac{1}{4}} - 1$$
$$\sqrt{\frac{9}{4}} = \frac{1}{2} - 1$$
$$\frac{3}{2} = \frac{-1}{2} \quad \text{a false arithmetic sentence}$$

Thus, the solution to $\sqrt{x + 2} = \sqrt{x} - 1$ is *none*. ■

EXAMPLE 6 Solve: $\sqrt{x+1} + \sqrt{x} = 2$

Solution Although we could start by squaring each side, it is more convenient to first isolate one of the radicals. We can do this by adding $-\sqrt{x}$ to each side of the equation:

$$\sqrt{x+1} = -\sqrt{x} + 2$$

Now, we square each side

$$\left(\sqrt{x+1}\right)^2 = \left(-\sqrt{x} + 2\right)^2$$

to get:

$$x + 1 = x - 4\sqrt{x} + 4$$

Next, we isolate the radical $-4\sqrt{x}$:

$$\begin{array}{rcl} -x - 4 & & -x \qquad -4 \\ \hline -3 & = & -4\sqrt{x} \end{array}$$

Square each side:

$$9 = 16x$$

Multiply by $\frac{1}{16}$:

$$\frac{9}{16} = x$$

Check Replacing x with $\frac{9}{16}$, we get

$$\sqrt{\frac{9}{16} + 1} + \sqrt{\frac{9}{16}} = 2$$

$$\sqrt{\frac{25}{16}} + \sqrt{\frac{9}{16}} = 2$$

$$\frac{5}{4} + \frac{3}{4} = 2$$

$$\frac{8}{4} = 2 \qquad \text{a true arithmetic sentence}$$

Thus, the equation $\sqrt{x+1} + \sqrt{x} = 2$ has solution $\frac{9}{16}$. ■

Using the Graphics Calculator

To increase our understanding of Example 5, type √(X+2) for Y1 and √X−1 for Y2. Use the Window 9.4 range settings to view the graph. As Figure 6 shows, at no x-value do these two expressions evaluate to the same number. Does this agree with our result in Example 5?

To check our solution in Example 6, type √(X+1)+√X for Y1 and 2 for Y2, then view the graph. As Figure 7 shows; at one x-value

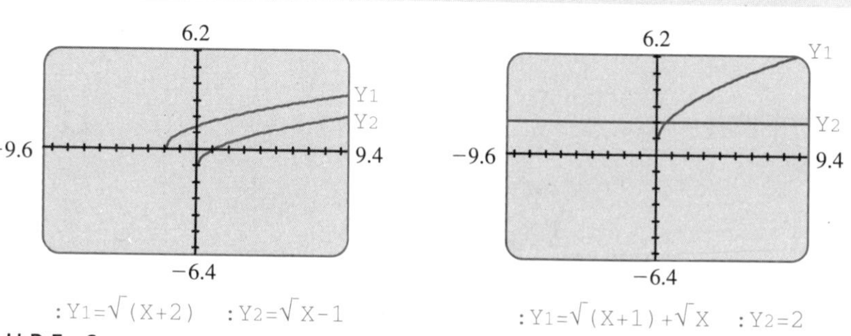

FIGURE 6 FIGURE 7

the radical expression evaluates to the number 2. Press the TRACE key and move to the point of intersection. Does the x-value agree with the solution we found in Example 6?

GIVE IT A TRY

Solve each radical expression equation, and check for extraneous roots.

4. $\sqrt{x-3} - \sqrt{x} = 2$ **5.** $\sqrt{x+2} + \sqrt{x} = 2$

Application: The Pythagorean Theorem We now consider briefly an application of solving radical expression equations in one variable. This application uses a radical expression equation as a mathematical model and involves the Pythagorean theorem (see Section 1.5).

EXAMPLE 7 Two automobiles leave an intersection at the same time. One automobile is traveling northbound at 50 mi/h, while the other is heading eastward at 60 mi/h.

a. After two hours, how far apart are the automobiles?

b. If the automobiles leave the intersection at 10 A.M., at what time are they 100 miles apart?

Solution a. We use arithmetic. From the sketch in Figure 8, we see that $d = \sqrt{x^2+y^2}$ (the Pythagorean theorem).

The automobile traveling north at 50 mi/h for 2 hours travels a distance of $y = 50(2)$, or 100 miles.

The automobile traveling east at 60 mi/h for 2 hours travels a distance of $x = 60(2)$, or 120 miles.

We can add these facts to our sketch. Using Figure 9, we write the expression to be computed:

$$\begin{aligned} d &= \sqrt{120^2+100^2} \\ &= \sqrt{24400} \\ &= 20\sqrt{61} \qquad \approx 156.20 \text{ miles apart.} \end{aligned}$$

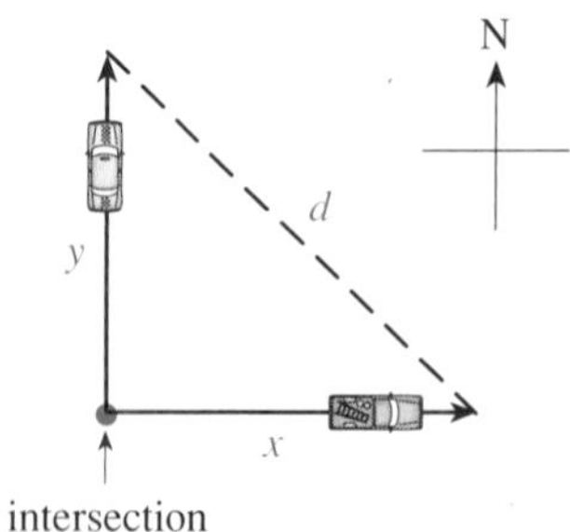

FIGURE 8

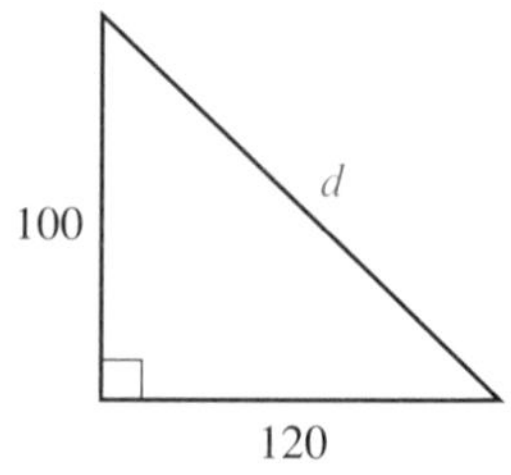

FIGURE 9

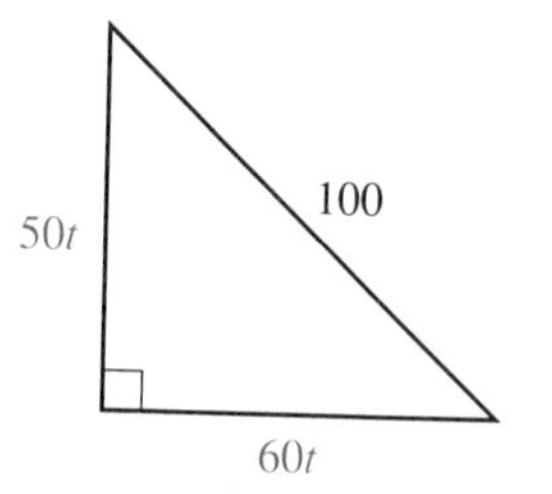

FIGURE 10

	rate	time	distance
North	50	t	$50t$
East	60	t	$60t$

After two hours, the automobiles are approximately 156.20 miles apart.

b. We use algebra to find the time at which the automobiles are 100 miles apart. A sketch of the situation is shown in Figure 10. The value requested is the time. So we let the variable represent the number of hours after leaving the intersection.

Let t = number hours after autos leave the intersection
$50t$ = number of miles traveled by northbound auto
$60t$ = number of miles traveled by eastbound auto

Using the table shown in the margin, we write the equation to be solved:

$$
\begin{aligned}
100 &= \sqrt{(60t)^2 + (50t)^2} \\
10000 &= (60t)^2 + (50t)^2 \qquad \text{Square each side.} \\
10000 &= 3600t^2 + 2500t^2 \\
10000 &= 6100t^2 \\
\frac{100}{61} &= t^2
\end{aligned}
$$

There are two numbers whose square is $\frac{100}{61}$: $\sqrt{\frac{100}{61}}$ and $-\sqrt{\frac{100}{61}}$.

Reduced, we have $\frac{10\sqrt{61}}{61}$ and $\frac{-10\sqrt{61}}{61}$

Of these two, we discard the negative value. (Why?)

To answer the question, we approximate $\frac{10\sqrt{61}}{61}$ to get 1.28. Thus, the two automobiles are 100 miles apart approximately 1.28 hours after 10 A.M. Now, 0.28 hour is about 17 minutes. So, we report an answer of 11:17 A.M. ■

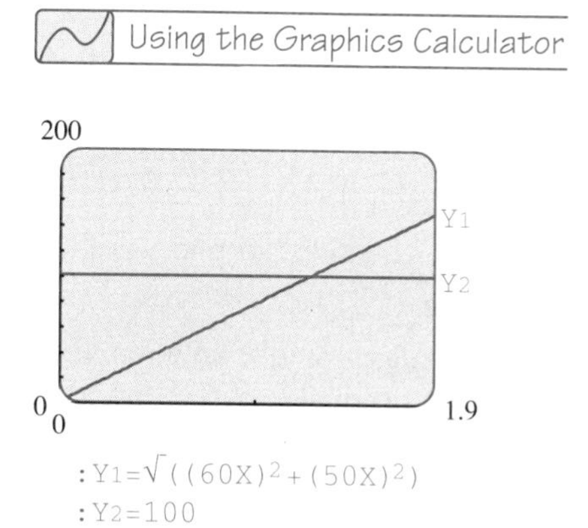

FIGURE 11

To increase our understanding of the physical situation in Example 7, type `√((60X)²+(50X)²)` for `Y1` and `100` for `Y2`. We can use the algebra work from the example to select the range settings, shown on the graph screen in Figure 11.* Use the `TRACE` feature to move the cursor left until the message reads `X=.4 Y=31.240999`. At 0.4 hour (24 minutes) after leaving the intersection, the two automobiles are a little more than 31 miles apart. Now, move the cursor to the right. Reading the messages, you can see that as time (the x-value) increases, the distance between the automobiles also increases. When the evaluation of the radical expression is 100, the x-value is approximately 1.28. From the algebra, we know that the x-value is exactly $\frac{10\sqrt{61}}{61}$.

TI-82 Note *Type `1.88` for `Xmax`.

Rewriting Formulas Another application we will consider is that of rewriting formulas which involve radical expressions. Again, we create an analogy to use as a guide. Consider the following example.

EXAMPLE 8 Rewrite $v = \sqrt{\frac{RT}{m}}$ for m.

Solution

	Analogy (for $v = 7$, $R = 2$, and $T = 3$)	*Formula*
	$7 = \sqrt{\frac{(2)(3)}{m}}$	$v = \sqrt{\frac{RT}{m}}$
Do arithmetic.	$7 = \sqrt{\frac{6}{m}}$	$v = \sqrt{\frac{RT}{m}}$
Square each side.	$(7)^2 = \left(\sqrt{\frac{6}{m}}\right)^2$	$v^2 = \left(\sqrt{\frac{RT}{m}}\right)^2$
	$49 = \frac{6}{m}$	$v^2 = \frac{RT}{m}$
Multiply by m.	$49m = 6$	$mv^2 = RT$
Multiply by $\frac{1}{49}$.	$m = \frac{6}{49}$	$m = \frac{RT}{v^2}$
Solution: $\frac{6}{49}$		Formula: $m = \frac{RT}{v^2}$

■

GIVE IT A TRY

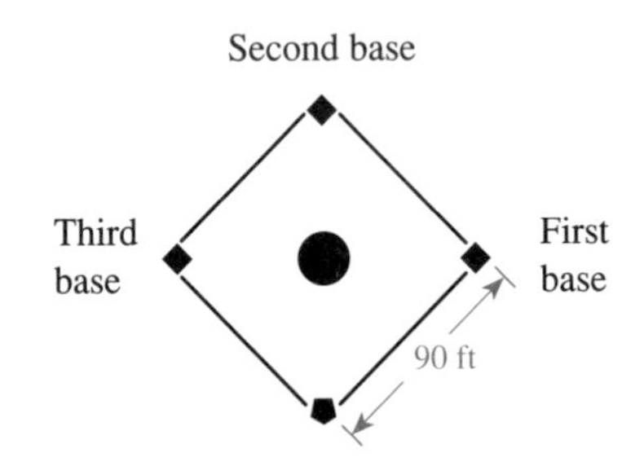

FIGURE 12

6. A baseball diamond is a square. Each side—the distance between the bases—measures 90 feet (see Figure 12). What is the shortest distance from first base to third base?

7. Rewrite the formula $T = \frac{\pi\sqrt{L}}{16}$ for L.

■ SUMMARY

In this section, we have solved radical expression equations in one variable, using the nth power property:

> The solution to an equation is contained in (is a subset of) the solution to the equation obtained by raising each side of the original equation to the nth power (such as squaring or cubing).

If the original equation involves square root radicals, each side is squared. If the original equation contains cube root radicals, each side is cubed. Again, the objective is to obtain an equation we know how to solve (such as a first-degree equation, a second-degree equation, or a rational expression equation).

The *n*th power property simply promises containment. Each number obtained from solving the altered equation must be checked to make sure it solves the original equation (because extraneous roots can be generated by raising each side of the equation to a power). If the numbers that solve the altered equation do not solve the original equation, or the altered equation has no solution, then the solution to the original equation is *none*.

In addition to solving radical expression equations in one variable, we discussed mathematical models. A radical expression equation can serve as a mathematical model for many situations, especially those described by a right triangle. Finally, we discussed rewriting formulas that contain radicals. To rewrite such a formula, we can use a radical expression equation as an analogy (model) of the formula.

7.4 ■ ■ ■ EXERCISES

1. Which one of the following is a first-degree equation in one variable?
   i. $\sqrt{3x - 1} = -3$ ii. $2w - 5 = 7 - w$
   iii. $\dfrac{x}{x + 2} = 2$ iv. $x^2 - 3x = 2$
2. Which one of the following is a second-degree equation in two variables?
   i. $y = 3(y - 2)$ ii. $y = x^2 - 3x$
   iii. $y = \sqrt{2x - 5}$ iv. $y = \dfrac{3}{x + 2}$
3. Which one of the following is a rational expression equation in one variable?
   i. $\sqrt{3x - 1} = -3$ ii. $2w - 5 = 7 - \sqrt{w}$
   iii. $\dfrac{x}{x + 2} = 2$ iv. $x^2 - 3x = y$
4. Which one of the following is a radical expression equation in one variable?
   i. $\sqrt[3]{3x - 1} = -3$ ii. $2w - 5 = 7 - y$
   iii. $\dfrac{x}{x + 2} = y$ iv. $x^2 - 3x = y$

Solve each equation. Use the graphics calculator to check your work.

5. $\sqrt{2x - 3} = 7$
6. $\sqrt{5 - 2x} = 3$
7. $5 = \sqrt{3x + 2}$
8. $1 = 5 - \sqrt{2x}$
9. $\sqrt{5 - 2x} = 3$
10. $\sqrt{3 - 4x} = 1$
11. $\sqrt{x + 2} = x$
12. $\sqrt{2x + 3} = x$
13. $-\sqrt{2x - 1} = -2$
14. $3 - \sqrt{2x} = -2$
15. $\sqrt[3]{x - 5} = 3$
16. $\sqrt[3]{2x - 5} = -1$
17. $\sqrt[3]{2 - 5x} = -2$
18. $1 - \sqrt[3]{x - 2} = 0$
19. $\sqrt[3]{2x - 1} = \sqrt[3]{5 - 4x}$
20. $\sqrt{x + 3} = \sqrt{x^2 - 1}$
21. $\sqrt{x + 2} + \sqrt{x} = 4$
22. $\sqrt{r + 2} + \sqrt{r} = 4$
23. $\sqrt{y - 1} = \sqrt{y} - 3$
24. $\sqrt{x - 2} - \sqrt{x + 1} = 2$
25. $\sqrt{x - 2} + \sqrt{x + 1} = 2$
26. $\sqrt{\dfrac{3}{x}} = 1$
27. $\sqrt{\dfrac{2}{x - 1}} = \dfrac{1}{3}$
28. $\sqrt{\dfrac{1}{x} + \dfrac{1}{x}} = 2$
29. $\sqrt{2x - 5} = \sqrt{4x - 3} - 2$
30. $1 - \sqrt{s + 2} = \sqrt{s + 1}$
31. Rework Example 7 for the automobile traveling north at 40 mi/h. When are the automobiles 300 miles apart?
32. Rework Example 7 for the automobile traveling east at 70 mi/h. When are the automobiles 300 miles apart?

33. Jill has attached a rope to the top of a flag pole. She knows that the rope is 100 feet long and that she is standing 30 feet from the base of the flag pole (see the sketch). Find the height of the flag pole.

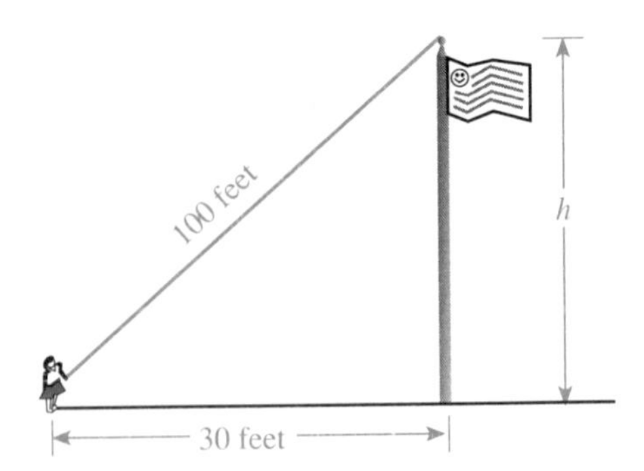

34. A right triangle has one leg 3 inches longer than the other leg. If the hypotenuse is 15 inches, then how long is the shortest leg in the triangle?

35. Blake is flying a kite and has released 500 yards of string. As the diagram shows, his kite is directly over a building which he knows is 100 yards from where he is standing. Find the height of the kite.

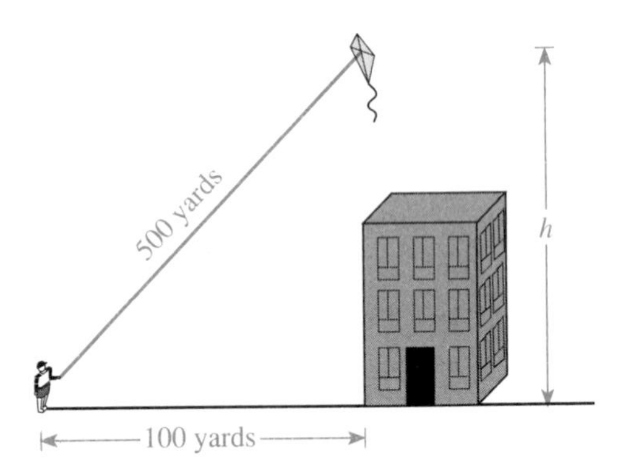

36. The *period* of a pendulum is the length of time required for the pendulum to make a complete swing, back and forth, as shown in the diagram.

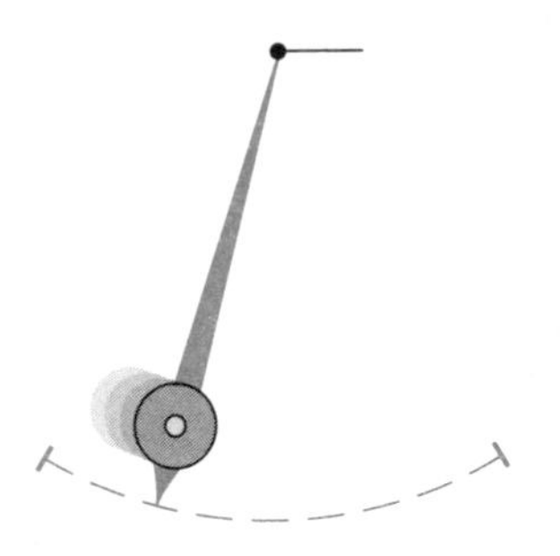

For a pendulum of length L (in feet), the period T (in seconds) is given by the formula

$$T = 2\pi\sqrt{\frac{L}{32}}$$

If the period is 2 seconds, what is the length of the pendulum?

37. The radius r of a circle is given by the formula $r = \sqrt{\frac{A}{\pi}}$, where A is the area measure of the circle. For a circle with radius 3 inches, what is the area measure of the circle?

Rewrite the given formula for the indicated variable.

38. $y = \sqrt{16 - 2x}$ for x

39. $R = 3 - \sqrt{T}$ for T

40. $f = \sqrt{\frac{1}{2t}}$ for t

41. $z = \sqrt{\frac{3}{x} - \frac{1}{x}}$ for x

42. $\sqrt[3]{y} = \sqrt[3]{2x - 3}$ for x

43. Use the fact that $|x| = \sqrt{(x^2)}$ to rewrite the equation $y = |x| - 3$ for x.

7.5 ■ ■ ■ COMPLEX NUMBERS

So far, we have discussed the integers, the rational numbers, and the real numbers. In this section we discuss a set of numbers that is the "superset" for all of these sets—the complex numbers. These numbers extend the real numbers so that equations such as $x^2 = -1$, which have no solutions in the reals, do have a solution in this superset.

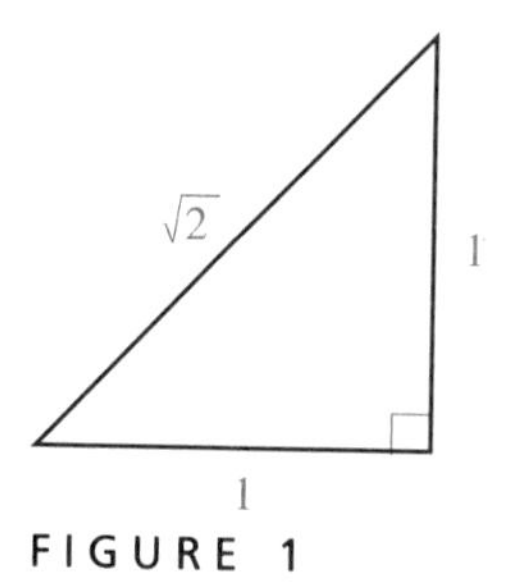

FIGURE 1

An Imaginary Number Trying to find the hypotenuse of a right triangle with legs of length 1 and 1 produces the irrational number $\sqrt{2}$, as shown in Figure 1. This is much like trying to find a number whose square is -1. Just as there is no rational number whose square is 2, there is no real number whose square is -1. For a real number a, where $a \geq 0$, the square of a is greater than or equal to zero. For a real number a, where $a < 0$, the square of a is greater than zero. Thus, any real number squared must be zero or larger. Because there is no real number that can be squared to yield -1, we must consider the numbers in a larger set, the complex numbers.

Just as mathematicians throughout the ages made up numbers such as 1, 2, 3, . . . , and later, $\frac{2}{3}$ and 0, mathematicians also created a number whose square is -1.

DEFINITION OF i

> The number i, known as the **imaginary unit**, is the number such that $i^2 = -1$.

With i defined, we can now define the square root of a negative real number.

DEFINITION OF $\sqrt{-a}$

> For a real number $a > 0$, $\sqrt{-a} = i\sqrt{a}$.

With this definition, we can write $\sqrt{-3} = i\sqrt{3}$, and for $\sqrt{-4}$ we write $i\sqrt{4}$, or $i(2)$, or $2i$.

Two warnings are in order: First, we do *not* write $\sqrt{-3}$ as $\sqrt{3}i$, because of possible confusion with $\sqrt{3i}$. Instead, we write $i\sqrt{3}$. Second, for

$$\sqrt{-4}\sqrt{-9}, \quad \text{we write} \quad \left(i\sqrt{4}\right)\left(i\sqrt{9}\right) \quad \text{or} \quad (2i)(3i) = 6i^2 = 6(-1) = -6$$

A common error is to write $\sqrt{-4}\sqrt{-9} = \sqrt{(-4)(-9)} = \sqrt{36} = 6$. The error here is assuming that the generalization $\sqrt{a}\sqrt{b} = \sqrt{ab}$ holds for negative values. As you can see, for $\sqrt{a}\sqrt{b} = \sqrt{ab}$, we must have $a \geq 0$ and $b \geq 0$.

We now define complex numbers.

DEFINITION OF COMPLEX NUMBERS

> For real numbers a and b, a **complex number** is any number that can be written in the form $a + bi$, where $i = \sqrt{-1}$. The form $a + bi$ is **standard form** for a complex number.

For the complex number $a + bi$, the number a is known as the **real part** and bi is the **imaginary part**. If $b = 0$, then $a + bi$ is a real number. (This is much like rational numbers—for integers a and b, $\frac{a}{b}$ is an integer when $b = 1$). Just as the set of rational numbers contains the set of integers as a subset, and the set of real numbers contains the set of rational numbers as a subset, the set of complex numbers contains the set of real numbers as a subset. These sets are shown in Figure 2.

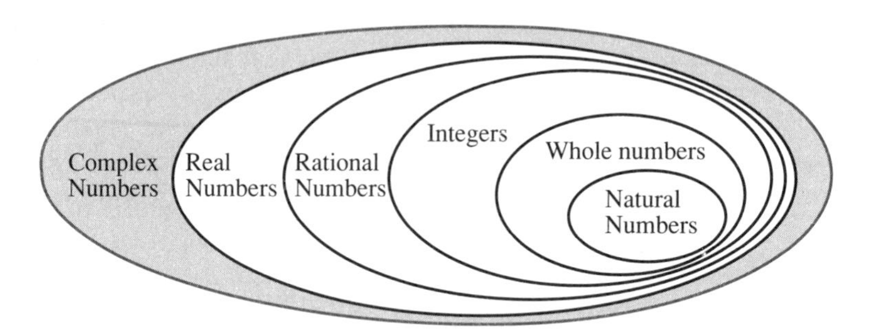

FIGURE 2

Two complex numbers, $a + bi$ and $c + di$, are equal provided $a = c$ and $b = d$. The commutative, associative, and distributive properties are still valid for complex numbers. Likewise, the additive identity is still 0, and the multiplicative identity is still 1 for the set of complex numbers.

Standard form for complex numbers follows the same general rules we used with radical expressions in Section 7.2. That is, we write a complex number such as $\sqrt{-12}$ as

$$i\sqrt{12} = i\sqrt{(2^2)(3)} = i(2)\sqrt{3} = 2i\sqrt{3}$$

Additionally, if the denominator contains a radical, we remove it by multiplying by a form of 1, such as

$$\frac{i\sqrt{3}}{i\sqrt{3}} \quad \text{or} \quad \frac{i}{i}.$$

Consider the following examples.

EXAMPLE 1 Write in standard $a + bi$ form: $\sqrt{45} + \sqrt{-12}$

Solution First, we write $\sqrt{45} + i\sqrt{12}$. Next, we reduce the radicals:

$$3\sqrt{5} + i\left(2\sqrt{3}\right) = 3\sqrt{5} + 2i\sqrt{3}$$ ■

In Example 1, we have $a = 3\sqrt{5}$ and $b = 2\sqrt{3}$.

EXAMPLE 2 Write in standard $a + bi$ form: $\dfrac{1}{\sqrt{-3}}$

Solution First, we write $\dfrac{1}{\sqrt{-3}} = \dfrac{1}{i\sqrt{3}}$. Now, we multiply by a form of 1, namely, $\dfrac{i\sqrt{3}}{i\sqrt{3}}$.

$$\frac{1}{i\sqrt{3}} \cdot \frac{i\sqrt{3}}{i\sqrt{3}} = \frac{i\sqrt{3}}{(i^2)(3)}$$
$$= \frac{i\sqrt{3}}{-3}$$

In $a + bi$ form, we write $\frac{-\sqrt{3}}{3}i$. ■

In Example 2, we have $a = 0$ and $b = \frac{-\sqrt{3}}{3}$.

GIVE IT A TRY

Write each complex number in standard $a + bi$ form.

1. $\sqrt{-18} - \sqrt{20}$ **2.** $\sqrt{25} - \sqrt{-36}$ **3.** $\frac{1}{\sqrt{-5}}$ **4.** $\frac{\sqrt{-12}}{\sqrt{-3}}$

Operations The binary operations—addition, subtraction, multiplication, and division—follow the same general rules for complex numbers as for radical expressions (see Section 7.3). Remember, i^2 is -1 by definition (i *is* the number whose square is -1). We define **addition of complex numbers** as follows.

DEFINITION OF ADDITION

For $a + bi$ and $c + di$, the sum

$$(a + bi) + (c + di) \quad \text{is} \quad (a + c) + (b + d)i$$

Notice that, just as we did when adding polynomials, to add complex numbers, we combine *like terms*. Here, we combine the real parts of all the numbers, then the imaginary parts. Work through the following examples.

EXAMPLE 3 Write in standard $a + bi$ form: $(-3 + 5i) + (7 + 6i)$

Solution To add complex numbers, we use the commutative and associative laws to combine the real parts and to combine the imaginary parts:

$$(-3 + 5i) + (7 + 6i) = (-3 + 7) + (5i + 6i)$$
$$= 4 + 11i$$
■

The **additive opposite of $a + bi$** is $-a - bi$ because

$$(a + bi) + (-a - bi) = 0$$

We define **subtraction of complex numbers** just as we have defined subtraction for real numbers: $a - b = a + (-b)$, where $-b$ is the additive opposite of b. Work through the next example.

EXAMPLE 4 Write in standard $a + bi$ form: $(-3 + 5i) - (-7 + 6i)$

Solution Using the additive opposite, we get:

Change to addition. — additive opposite of $-7 + 6i$

$$\begin{aligned}(-3 + 5i) - (-7 + 6i) &= (-3 + 5i) + (7 - 6i)\\ &= (-3 + 7) + (5i - 6i)\\ &= 4 - i\end{aligned}$$

■

For multiplication of complex numbers, we multiply the complex numbers as if they were binomials:

$$\begin{aligned}(a + bi)(c + di) &= ac + adi + bci + bdi^2 && \text{Use FOIL technique.}\\ &= ac + (ad + bc)i + bd(-1) && i^2 = -1\\ &= (ac - bd) + (ad + bc)i && \text{Combine like terms.}\end{aligned}$$

To learn how to multiply complex numbers, work through the following examples.

EXAMPLE 5 Write in standard $a + bi$ form: $(-3 + 5i)(-7 + 6i)$

Solution To multiply complex numbers, we use the FOIL technique.

$$\begin{aligned}(-3 + 5i)(-7 + 6i) &= \overset{\text{F}}{21} - \overset{\text{O}}{18i} - \overset{\text{I}}{35i} + \overset{\text{L}}{30i^2}\\ &= 21 - 18i - 35i - 30 && i^2 = -1\\ &= -9 - 53i && \text{Combine like terms.}\end{aligned}$$

■

EXAMPLE 6 Write in standard $a + bi$ form: $(-3 + 5i)(-3 - 5i)$

Solution Notice that one factor has the form $a + bi$, while the other factor has the form $a - bi$.

$$\begin{aligned}(-3 + 5i)(-3 - 5i) &= \overset{\text{F}}{9} + \overset{\text{O}}{15i} - \overset{\text{I}}{15i} - \overset{\text{L}}{25i^2}\\ &= 9 + 25\\ &= 34\end{aligned}$$

This product produces a real number. ■

GIVE IT A TRY

Write the result of each operation in standard $a + bi$ form.

5. $(-2 + 5i) + (7 - 6i)$

6. $(-2 + 5i) - (7 - 6i)$

7. $(-2 + 5i)(7 - 6i)$

8. $(-1 + 3i)^2$

9. $(-1 + 3i)(-1 - 3i)$

10. $\left(3 - \sqrt{-4}\right)\left(2 + \sqrt{-9}\right)$

The Conjugate For the complex number $a + bi$, the number $a - bi$ is its **complex conjugate**. Likewise, $a - bi$ is the complex conjugate of $a + bi$. As you can see in Example 6 and in *Give It a Try* Problem 9, a complex number $a + bi$ times its conjugate $a - bi$ always yields a real number, $a^2 + b^2$. Conjugates are used in the process of finding the multiplicative inverse of a complex number.

EXAMPLE 7 Find the multiplicative inverse of $2 + 3i$.

Solution The number $2 + 3i$ has multiplicative inverse $(2 + 3i)^{-1}$, or $\frac{1}{2 + 3i}$. To write this complex number in standard $a + bi$ form, we multiply by a form of 1. We select the form $\frac{2 - 3i}{2 - 3i}$, the conjugate of the denominator divided by itself. This will produce a real number for the denominator.

$$\frac{1}{2 + 3i} \cdot \frac{2 - 3i}{2 - 3i} = \frac{2 - 3i}{4 - 9i^2} = \frac{2 - 3i}{13}$$

(a form of 1)

In standard $a + bi$ form, we write $(2 + 3i)^{-1}$ as $\frac{2}{13} + \frac{-3}{13}i$. ■

To define the division operation, we use the multiplicative inverse. Thus, **division of complex numbers** a and b is defined as $a \div b = (a)(b^{-1})$. That is, a divided by b is a times the multiplicative inverse of b. However, in actual practice, rather than finding the multiplicative inverse, as we have in Example 7, it is much easier to work with fractions and conjugates. To learn how to do this, work through the following examples.

EXAMPLE 8 Write in standard $a + bi$ form: $(-3 + 5i) \div (-7 + 6i)$

Solution First, we write the quotient in fraction form, $\frac{-3 + 5i}{-7 + 6i}$.

Next, we rename the number so that the denominator is a real number. To accomplish this, we multiply by a form of 1. From earlier examples, we know that if we multiply $a + bi$ times $a - bi$, we get $a^2 + b^2$, a real number. Thus, for a denominator of $(-7 + 6i)$, we will multiply by 1 in the form $\frac{-7 - 6i}{-7 - 6i}$.

(a form of 1)

$$\frac{-3 + 5i}{-7 + 6i} = \frac{-3 + 5i}{-7 + 6i} \cdot \frac{-7 - 6i}{-7 - 6i} = \frac{21 + 18i - 35i - 30i^2}{49 - 36i^2} = \frac{51 - 17i}{85}$$

In standard $a + bi$ form, we write $\frac{51}{85} + \frac{-17}{85}i$, or $\frac{3}{5} + \frac{-1}{5}i$. ■

GIVE IT A TRY

11. Write $(2 - 3i) \div (1 - i)$ in standard $a + bi$ form.

Equations Complex numbers have many applications, especially in electricity and in electronics. We are studying these numbers now because they often occur in the solution to second-degree equations. In fact, we define i as a number that solves the equation $x^2 = -1$. Consider the second-degree equation $x^2 - 4x = -13$. A number that solves this equation is $2 + 3i$. To see that this is true, work through the next example.

EXAMPLE 9 Show that $2 + 3i$ is in the solution to the equation $x^2 - 4x = -13$.

Solution To show that $2 + 3i$ is in the solution, we first write the following expression in standard $a + bi$ form:

$$(2 + 3i)^2 - 4(2 + 3i)$$

Squaring the number $2 + 3i$ yields

$$\begin{aligned}(2 + 3i)(2 + 3i) &= 4 + 12i + 9i^2 \\ &= 4 + 12i - 9 \\ &= -5 + 12i\end{aligned}$$

We now replace the number $(2 + 3i)^2$ with $-5 + 12i$ in the original expression, and perform the subtraction:

$$\begin{aligned}(2 + 3i)^2 - 4(2 + 3i) &= \underbrace{-5 + 12i}_{(2+3i)^2} + (-4)(2 + 3i) \\ &= -5 + 12i + (-8) + (-12i) \\ &= -13 + 0i\end{aligned}$$

Since we have shown that $(2 + 3i)^2 - 4(2 + 3i)$ is -13, we can say that $2 + 3i$ is in the solution to the equation $x^2 - 4x = -13$. ■

GIVE IT A TRY

12. Show that the complex number $2 + i$ is in the solution to the equation $x^2 - 4x + 5 = 0$.

Algebraic Numbers Algebraic numbers are often characterized as the numbers that solve polynomial equations having integer coefficients. That is, the algebraic number $2 - \sqrt{3}$ solves the equation $x^2 - 4x + 1 = 0$. From Example 9 we have learned that $2 + 3i$ solves the polynomial equation $x^2 - 4x + 13 = 0$. Thus, some algebraic numbers are real numbers, such as $2 - \sqrt{3}$, and other algebraic numbers, such as $2 + 3i$, are nonreal.

Given an algebraic number, we can find a polynomial equation with integer coefficients such that the algebraic number solves the equation. As a simple example, consider the algebraic number $\frac{2}{3}$. A polynomial equation with integer coefficients whose solution contains this number is $3x = 2$, or $3x - 2 = 0$. Work through the following examples to see how we can find such a polynomial equation.

EXAMPLE 10 Find a polynomial equation with integer coefficients whose solution contains $3 + \sqrt{2}$.

Solution We can easily write an equation whose solution is $3 + \sqrt{2}$, that is, $x = 3 + \sqrt{2}$, but this equation does not have integer coefficients $\left(\sqrt{2}\text{ is not an integer}\right)$. To get an equation with integer coefficients, we write the equation in standard form, as follows:

$$\begin{aligned} x &= 3 + \sqrt{2} \\ -3 &\quad -3 \\ \hline x - 3 &= \sqrt{2} \\ (x-3)^2 &= \left(\sqrt{2}\right)^2 && \text{Square each side.} \\ x^2 - 6x + 9 &= 2 \\ -2 &\quad -2 \\ \hline x^2 - 6x + 7 &= 0 && \text{Standard form for the polynomial equation} \end{aligned}$$

So, a polynomial equation with integer coefficients that has $3 + \sqrt{2}$ in its solution is $x^2 - 6x + 7 = 0$. ■

EXAMPLE 11 Find a polynomial equation with integer coefficients whose solution contains $1 + 2i$.

Solution First we can write $x = 1 + 2i$. This equation does not have integer coefficients. To get an equation with integer coefficients, we write the equation in standard form, as follows:

$$\begin{aligned} x &= 1 + 2i \\ -1 &\quad -1 \\ \hline x - 1 &= 2i \\ (x-1)^2 &= (2i)^2 && \text{Square each side.} \\ x^2 - 2x + 1 &= -4 \\ 4 &\quad 4 \\ \hline x^2 - 2x + 5 &= 0 && \text{Standard form for the polynomial equation} \end{aligned}$$

So, a polynomial equation with integer coefficients that has $1 + 2i$ in its solution is $x^2 - 2x + 5 = 0$. ■

GIVE IT A TRY

13. Find a polynomial equation with integer coefficients whose solution contains $3 - \sqrt{2}$.

14. Find a polynomial equation with integer coefficients whose solution contains $1 - 2i$.

■ SUMMARY

In this section, we have learned to work with complex numbers. These numbers all have the form $a + bi$, where a and b are real numbers and i is $\sqrt{-1}$. The set of complex numbers is a superset for the real numbers. We learned how to write these numbers in standard $a + bi$ form and how to use the four basic operations (addition, subtraction, multiplication, and division) with these numbers. We will use these numbers in later chapters, when we solve quadratic equations.

■ ***Quick Index***

7.5 ■ ■ ■ EXERCISES

Write the result in standard $a + bi$ *form.*

1. $\sqrt{-12}$
2. $(-i)^2$
3. $3i - \sqrt{-27} + 2\sqrt{-18}$
4. $\sqrt{-16} - \sqrt{-25} + 2i$
5. $(-3 + 5i) + (-2 - 7i)$
6. $(15 - 3i)(15 - 3i)$
7. $5(3 + 9i)$
8. $(-6 + 2i)(-6 - 2i)$
9. $(5 + 2i)(3 + 9i)$
10. $(2 + 3i)^2$
11. $(-2 + i)(3 - i)$
12. $(2 + 3i)^2 - 3(2 + 3i)$
13. $(1 + i)(1 - i)$
14. $(5 - 2i)^2 - 3(5 - 2i)$
15. $(3 - 7i)(3 + 7i)$
16. $(-2 + 5i)(-2 - 5i)$
17. $(3 - 7i)^2$
18. $(-2 + 5i)^2$
19. $(-2 + 3i)^2 - (2i)^2$
20. $(-4 - 5i)^2 - (5i)^2$
21. $(2 + 3i)^{-1}$
22. $5 \div (2 + 3i)$
23. $(-2 + i) \div (3 - i)$
24. $(-3 + 4i) \div i$
25. $\dfrac{-1 + 2i}{2 - 3i}$
26. $\dfrac{5 + i}{2 + 3i}$

27. Show that $2 + 2i$ solves the equation $x^2 - 4x + 8 = 0$.

28. Show that $3 - i$ solves the equation $x^2 - 6x + 10 = 0$.

29. Show that both $1 + i$ and $1 - i$ solve the equation $x^2 - 2x + 2 = 0$.

30. Show that both $-1 + 2i$ and $-1 - 2i$ solve the equation $x^2 + 2x + 5 = 0$.

Find a polynomial equation with integer coefficients whose solution contains the given number.

31. $1 - \sqrt{2}$
32. $1 + \sqrt{2}$
33. $-1 + 3i$
34. $-1 - 3i$
35. $1 + \sqrt[3]{2}$

DISCOVERY

Conjugate Pairs

In the preceding section, we have shown that the number $3 + \sqrt{2}$ is in the solution to the equation $x^2 - 6x + 7 = 0$, and we have shown that the number $2 + 3i$ is in the solution to the equation $x^2 - 4x + 13 = 0$. In both cases, we replace the variable by the number and show that a true arithmetic sentence is produced. That is, we show that $(3 + \sqrt{2})^2 - 6(3 + \sqrt{2}) + 7$ is the number 0. Likewise, we show that $(2 + 3i)^2 - 4(2 + 3i) + 13$ is the number 0, or (in standard $a + bi$ form) $0 + 0i$.

From Chapter 5, we know that the solution to a second-degree equation in one variable (a quadratic equation) is usually two numbers. Knowing that $3 + \sqrt{2}$ is in the solution to $x^2 - 6x + 7 = 0$, what do you think the other number in the solution is? Try $3 - \sqrt{2}$.

We evaluate $(3 - \sqrt{2})^2 - 6(3 - \sqrt{2}) + 7$. First, we square $3 - \sqrt{2}$:

$$\begin{aligned}(3 - \sqrt{2})^2 &= (3 - \sqrt{2})(3 - \sqrt{2}) \\ &= 9 - 3\sqrt{2} - 3\sqrt{2} + \sqrt{4} \\ &= 9 - 6\sqrt{2} + 2 \\ &= 11 - 6\sqrt{2}\end{aligned}$$

Now, we perform the addition and subtractions:

$$\begin{aligned}(3 - \sqrt{2})^2 - 6(3 - \sqrt{2}) + 7 &= 11 - 6\sqrt{2} - 18 + 6\sqrt{2} + 7 \\ &= 11 - 18 + 7 - 6\sqrt{2} + 6\sqrt{2} \\ &= -7 + 7 \\ &= 0\end{aligned}$$

So, $3 - \sqrt{2}$ is the other number in the solution to the equation

$$x^2 - 6x + 7 = 0.$$

What is the relationship between $3 + \sqrt{2}$ and $3 - \sqrt{2}$? They are conjugates.

Given that $2 + 3i$ is one number in the solution to the quadratic equation $x^2 - 4x + 13 = 0$, what do you think the other number in the solution is? The correct answer is $2 - 3i$. What is the relationship between the numbers $2 + 3i$ and $2 - 3i$?

Exercises

1. In your own words, write a generalization about solutions to second-degree equations in one variable, based on our work in this Discovery.
2. Given that $5 - 2i$ solves the equation $x^2 - 10x + 29 = 0$, what is the other number that solves this equation?

CHAPTER 7 REVIEW EXERCISES

Write each expression as a completely reduced radical expression.

1. $\sqrt{54x^4y^3}$
2. $\sqrt{80x^3y^2}$
3. $\dfrac{2x}{\sqrt[3]{9x}}$
4. $\dfrac{2x}{\sqrt[3]{2x^3}}$
5. $\sqrt{27x^3y^2}$
6. $\sqrt{64x^5y^2}$
7. $\sqrt{\dfrac{5}{x}}$
8. $\sqrt{\dfrac{5}{2x}}$
9. $\sqrt[3]{18x^5y^3}$
10. $\sqrt[3]{24x^2y^5}$
11. $\sqrt[3]{\dfrac{5}{3x}}$
12. $\sqrt[3]{\dfrac{5}{2y}}$
13. $(18x^3y^5)^{1/2}$
14. $(81r^3s^8t^2)^{1/3}$
15. $(4x)^{-1/2}$
16. $(-16x^2)^{-1/3}$
17. $(2x^{-2})^{-1/2}$
18. $(3a^3y^{-2})^{-1/3}$

Perform the indicated operations, and report the result as a completely reduced radical expression. Use the graphics calculator to test your work for errors.

19. $\sqrt{18x^3} + \sqrt{8x^3}$
20. $\sqrt{3y} - \sqrt{27y}$
21. $\sqrt[3]{8x^3} + \sqrt[3]{27x^3}$
22. $\sqrt[3]{54x^5} - \sqrt[3]{16x^5} - \sqrt[3]{2x^3}$
23. $\sqrt{16x^3} - \sqrt{25x^3}$
24. $\sqrt{3x}\left(\sqrt{5x} - \sqrt{3y}\right)$
25. $\left(\sqrt{x} - 4\right)^2$
26. $\sqrt[3]{y}\left(\sqrt[3]{y^2} - \sqrt[3]{y^4}\right)$
27. $\left(\sqrt{x - 3}\right)^2$
28. $\left(\sqrt{x} - 2\right)\left(\sqrt{x} + 1\right)$
29. $\left(\sqrt{3x} - \sqrt{3}\right)\left(\sqrt{2x} + \sqrt{6}\right)$
30. $\left(\sqrt{2x} - \sqrt{3}\right)\left(\sqrt{2x} + \sqrt{5}\right)$
31. $\left(\sqrt[3]{x} + 3\right)\left(\sqrt[3]{x^2} - 3\right)$
32. $\left(\sqrt{x} - 2\right)\left(\sqrt{x} + 2\right)$
33. $2 \div \sqrt{6xy}$
34. $\left(\sqrt{x} - 2\right) \div \left(\sqrt{x + 2}\right)$
35. $3x \div \sqrt[3]{27x^3}$
36. $\left(\sqrt{2x} + \sqrt{3x}\right) \div \left(\sqrt{3x} - \sqrt{2x}\right)$

Solve each radical equation in one variable. Check for extraneous roots.

37. $\sqrt{3x - 2} = x$
38. $\sqrt{-2x + 3} = x$
39. $-\sqrt{2x - 1} = 2$
40. $3 + \sqrt{2x} = -2$
41. $\sqrt[3]{x - 5} = -3$
42. $\sqrt[3]{2x - 5} = 1$
43. $\sqrt{x - 2} = \sqrt{x} + 3$
44. $\sqrt{r} + \sqrt{r - 2} = 3$
45. $\sqrt{3y - 2} = \sqrt{y} + 2$
46. $\sqrt{x - 2} - \sqrt{x + 1} = 3$
47. $\sqrt{x + 1} - \sqrt{x - 2} = 1$
48. $\sqrt{\dfrac{2}{x}} = 1$
49. $\sqrt{\dfrac{2}{x + 1}} = \dfrac{1}{3}$
50. $\sqrt{\dfrac{1}{x} - \dfrac{1}{x}} = -2$
51. $\sqrt{3x + 4} = \sqrt{2x + 1} + 1$
52. $\sqrt{s + 2} - 1 = \sqrt{s + 1}$
53. The period of a pendulum is the length of time required for the pendulum to make a complete swing, back and forth. For a pendulum of length L (in feet), the period T (in seconds) is given by the formula

$$T = 2\pi\sqrt{\frac{L}{32}}$$

If the period is 3 seconds, what is the length of the pendulum?

54. The diagonal of a rectangle with length 4 inches more than its width w is given by the formula $d = \sqrt{w^2 + (w + 4)^2}$. If the diagonal is 20 inches, what is the width w?

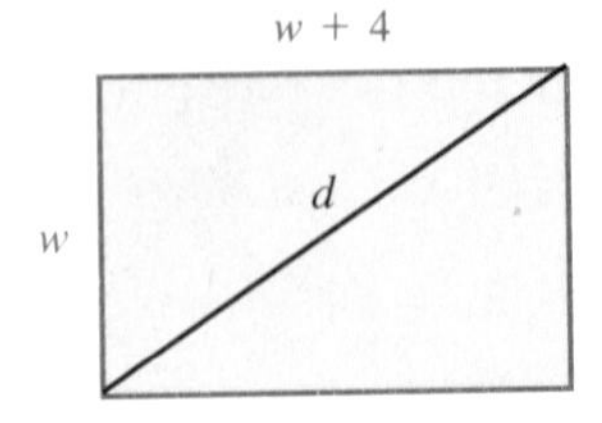

Rewrite the formula for the indicated variable.

55. $y = \sqrt{3x-5}$ for x

56. $A = 3 - \sqrt{C+1}$ for C

57. $g = \sqrt{\dfrac{3}{d}}$ for d

58. $\sqrt{y+2} = \sqrt{x} - 3$ for y

Write the result in standard a + bi *form.*

59. $\sqrt{12} - 3\sqrt{-20}$

60. $2\sqrt{18} - \sqrt{-36}$

61. $(3 - 2i) + (5 + 2i)$

62. $(1 + i) + (3 - 2i)$

63. $(2 - 5i) - (7 + 2i)$

64. $(3i) - (6 - 5i)$

65. $(3 + 5i)(-2 + i)$

66. $(-2 - 4i)(3 - i)$

67. $(5 + 3i)(5 - 3i)$

68. $(-3 + 2i)(-3 - 2i)$

69. $(-2 + 5i)^2$

70. $(4 - 3i)^2$

71. $(1 - 4i)^2$

72. $(-2 + 5i)^2$

73. $(3 - 2i) \div (2i)$

74. $(6 + i) \div (-i)$

75. $(4 - 2i) \div (1 - i)$

76. $(3 + i) \div (2 + i)$

77. $\dfrac{-2 + 3i}{3 - 2i}$

78. $\dfrac{2 - 3i}{3 + i}$

79. Show that $1 - 2i$ solves the equation $x^2 - 2x + 5 = 0$.

80. Show that $3 + 2i$ solves the equation $x^2 - 6x + 13 = 0$.

81. Show that both $2 + i$ and $2 - i$ solve the equation $x^2 - 4x + 5 = 0$.

82. Show that both $-1 + 2i$ and $-1 - 2i$ solve the equation $x^2 + 2x + 5 = 0$.

83. Find a polynomial equation with integer coefficients whose solution contains $3 + \sqrt{2}$.

84. Find a polynomial equation with integer coefficients whose solution contains $5 - i\sqrt{2}$.

CHAPTER 7 TEST

Write each expression as a completely reduced radical expression.

1. $\sqrt{72x^3y^4z^5}$

2. $(2x^2)^{-2/3}$

3. $\left(\sqrt{3x} - 5\right)\left(\sqrt{2x} + 5\right)$

4. $\sqrt{x} \div \left(\sqrt{x} - \sqrt{2}\right)$

5. $(x^{1/2} - y^{1/2})^{-1}$

6. Solve: $\sqrt[3]{2x-5} = -4$

7. Solve: $\sqrt{x-2} - \sqrt{x} = 3$

8. Write each number in standard $a + bi$ form.
   - **a.** $(-3 + 5i) - (6 + 2i)$
   - **b.** $(-3 + 5i)(2 + 3i)$
   - **c.** $(-2 + 7i)^2$
   - **d.** $(7 - 3i)(7 + 3i)$
   - **e.** $(3 + 7i) \div (2 - 5i)$

9. Rewrite $\sqrt{y-1} = \sqrt{x} + 2$ for y.

10. Two cars leave an intersection at the same time. One car is traveling south at 40 mi/h, and the other car is traveling westbound at 50 mi/h (see the sketch). How long after leaving are the cars 50 miles apart?

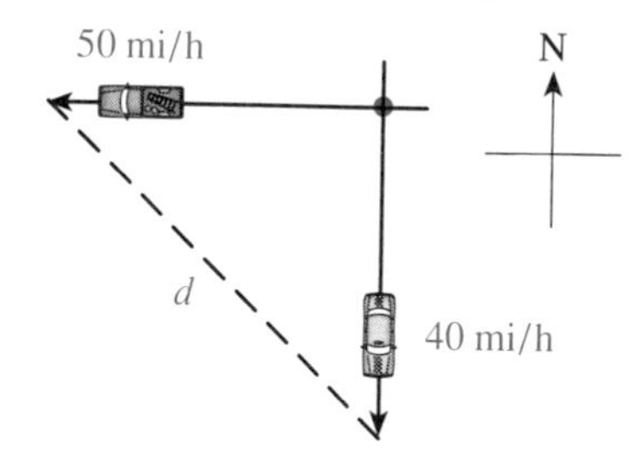

11. Show that both $4 - 3i$ and $4 + 3i$ solve the equation $x^2 - 8x + 25 = 0$.

8

QUADRATIC EQUATIONS AND FUNCTIONS

In this chapter, we will learn a general procedure for solving second-degree equations in one variable, such as $x^2 - 3x = 5$. This procedure is known as the quadratic formula. Once we have developed the ability to solve second-degree equations in one variable, this knowledge and skill is applied to graphing the solution to

1. second-degree inequalities in one variable, such as $x^2 - 3x \leq 5$, and
2. second-degree equations in two variables and quadratic functions, such as $f(x) = x^2 - 3x - 5$.

We will also discuss the graphs of other polynomial functions, rational functions, and radical functions.

8.1 ■ ■ ■ COMPLETING THE SQUARE

In Chapter 5 we solved second-degree (quadratic) equations in one variable by using factoring and the zero product property. Not all second-degree equations can be solved by factoring. To solve nonfactorable quadratic equations, we will use a method known as *completing the square*. The solutions to such equations often include numbers that contain radicals, such as $5 - \sqrt{3}$.

Irrational Numbers Because not all second-degree trinomials factor, not all second-degree equations in one variable can be solved by the factoring approach. Consider solving the equation

$$x^2 - x - 3 = 0$$

The trinomial $x^2 - x - 3$ is a prime polynomial—it does not factor over the integers. Thus, this equation *cannot* be solved by the factoring approach. However, it does have real numbers in its solution, namely,

$$\frac{1 + \sqrt{13}}{2} \quad \text{and} \quad \frac{1 - \sqrt{13}}{2}$$

As you can see, these are irrational numbers. If a second-degree equation can be solved by the factoring approach, the numbers in its solution will always be rational numbers. If the equation cannot be solved by factoring over the integers, the real numbers in its solution will be irrational numbers.

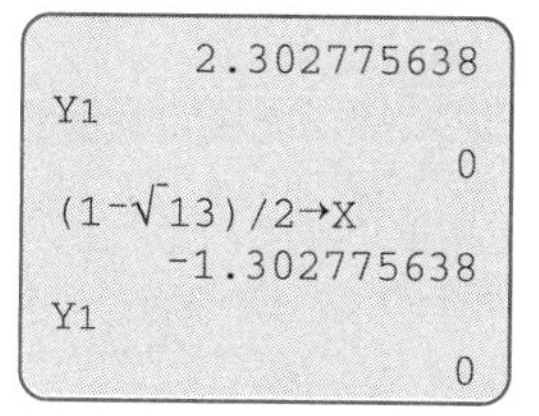

FIGURE 1

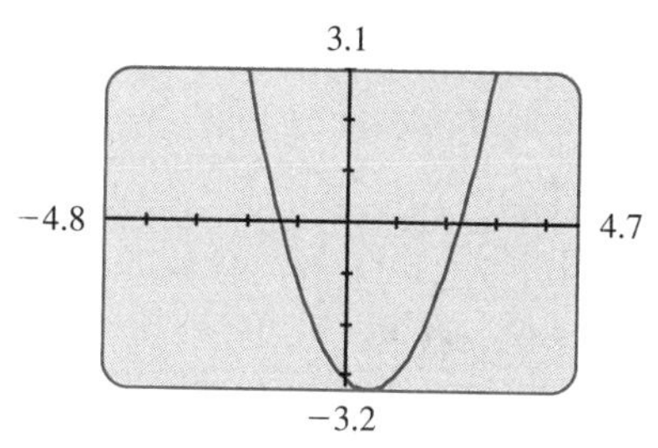

FIGURE 2

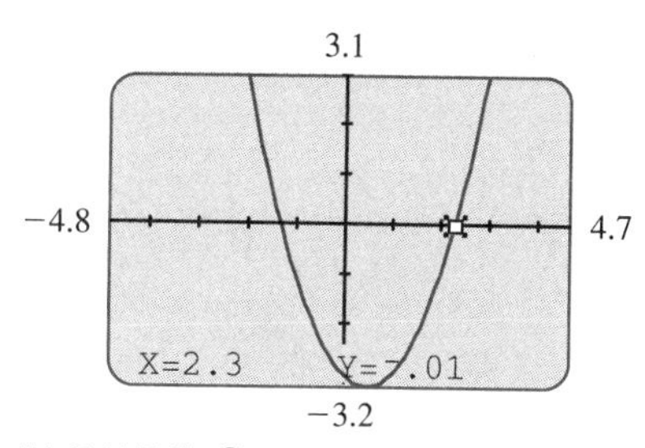

FIGURE 3

To see that the two numbers $\frac{1+\sqrt{13}}{2}$ and $\frac{1-\sqrt{13}}{2}$ do indeed solve the equation $x^2 - x - 3 = 0$, type `X²−X−3` for `Y1`. From the home screen, store the number `(1+√13)/2` in memory location `X`. The number `2.302775638` is displayed, as shown in Figure 1. This number is an approximation of the irrational number $\frac{1+\sqrt{13}}{2}$. Display the resulting value stored in `Y1`. As we expected, the number 0 is output. Now store the number `(1−√13)/2` in memory location `X`. Display the resulting value stored in `Y1`. Again, the polynomial $x^2 - x - 3$ evaluates to 0. Thus, $\frac{1-\sqrt{13}}{2}$ is also in the solution to $x^2 - x - 3 = 0$.

To get a graphical view of the solution to the equation $x^2 - x - 3 = 0$, use the Window 4.7 range settings to view the graph. The graph screen in Figure 2 shows the evaluations of the polynomial $x^2 - x - 3$ for various replacements of the x-variable. Two such replacements for x result in the polynomial evaluating to 0. These two numbers are the ones that solve the quadratic equation $x^2 - x - 3 = 0$.

Use the `TRACE` feature to move the cursor until the message reads `X=2.3  Y=-.01`. See Figure 3. When the x-variable is replaced by 2.3, the polynomial evaluates to -0.01. Press the right arrow key. The message now reads `X=2.4  Y=.36`. When x is replaced by 2.4, the polynomial $x^2 - x - 3$ evaluates to 0.36. The number that results in the polynomial evaluating to 0 is between 2.3 and 2.4. The exact number is $\frac{1+\sqrt{13}}{2}$.

Move the cursor until the message reads `X=-1.3  Y=-.01`. The other number that results in the polynomial $x^2 - x - 3$ evaluating to 0 is near -1.3. That number is exactly $\frac{1-\sqrt{13}}{2}$.

Perfect Square Binomials Before developing an approach that will produce numbers such as the ones we have just discussed, we consider squaring a binomial. From our earlier work (in Section 5.1), we know the following products result from squaring a binomial.

THE SQUARE OF A BINOMIAL

$$(a + b)^2 = a^2 + 2ab + b^2 \quad \text{and} \quad (a - b)^2 = a^2 - 2ab + b^2$$

From these results, we have $(x + a)^2 = x^2 + 2ax + a^2$, where a is a real number. Now, consider the relationship between the x-coefficient, $2a$, and the constant term, a^2. Observe the squares of several binomials:

Binomial squared	$2a$	a^2
$(x - 3)^2 = x^2 - 6x + 9$	-6	9
$(x + 8)^2 = x^2 + 16x + 64$	16	64
$(x - 12)^2 = x^2 - 24x + 144$	-24	144

The relationship between $2a$ and a^2 is *the square of one-half the x-coefficient, 2a, equals the constant term, a^2*. That is,

$$\left[\frac{1}{2}(2a)\right]^2 = a^2$$

x-coefficient (points to $2a$) constant term (points to a^2)

For future work, we will want to add a number to a polynomial such as $x^2 + 6x$ in order to obtain a perfect square. The number we add to this polynomial is *the square of one-half of* 6, or 3^2. This will yield the polynomial $x^2 + 6x + 9$, which is the perfect square $(x + 3)^2$.

EXAMPLE 1 Add a number to the polynomial $x^2 + 8x$ to produce a perfect square binomial. Report the factored form of the perfect square.

Solution The x-coefficient is 8. So, we add the number

$$\left[\frac{1}{2}(8)\right]^2 = 4^2 = 16$$

to the polynomial to get $x^2 + 8x + 16$, which is $(x + 4)^2$ in factored form. ■

In this chapter we will often work with polynomials that have rational number coefficients. Work through the next example.

EXAMPLE 2 Add a number to the polynomial $x^2 - 3x$ to produce a perfect square binomial. Report the factored form of the perfect square.

Solution The x-coefficient is -3. So, we add the number

$$\left[\frac{1}{2}(-3)\right]^2 = \left(\frac{-3}{2}\right)^2 = \frac{9}{4}$$

to get $x^2 - 3x + \frac{9}{4}$, which is $\left(x - \frac{3}{2}\right)^2$ in factored form. ■

In Example 2 we have extended, for the moment, our idea of factoring polynomials to include rational number coefficients.

GIVE IT A TRY

Add a number to each polynomial to produce a perfect square binomial. Report the factored form of the perfect square binomial produced.

1. $x^2 - 10x$ **2.** $x^2 + 2x$ **3.** $x^2 + 5x$ 

The Square Root Property The next idea we must discuss is the square root property. From our earlier work with factoring, we know how to find the solution to an equation such as $x^2 = 16$. Also, we know that $\sqrt{x^2} = |x|$, and we know how to solve absolute value equations. In each of the next two examples, compare the two approaches presented.

EXAMPLE 3 Solve: $x^2 = 16$

Solution

Standard form: $x^2 - 16 = 0$	Take square root $\sqrt{x^2} = \sqrt{16}$
Factored form: $(x + 4)(x - 4) = 0$	to get: $\|x\| = \sqrt{16}$
$x + 4 = 0$ or $x - 4 = 0$	Now, $x = \sqrt{16}$ or $x = -\sqrt{16}$
Solution: -4 and 4	Solution: 4 and -4

■

EXAMPLE 4 Solve: $x^2 = 100$

Solution

Standard form: $x^2 - 100 = 0$	Take square root $\sqrt{x^2} = \sqrt{100}$
Factored form: $(x + 10)(x - 10) = 0$	to get: $\|x\| = \sqrt{100}$
$x + 10 = 0$ or $x - 10 = 0$	Now, $x = \sqrt{100}$ or $x = -\sqrt{100}$
Solution: -10 and 10	Solution: 10 and -10

■

In general, we have the following equation-solving property.

SQUARE ROOT PROPERTY

> The equation $x^2 = a$, where $a \geq 0$, is equivalent to
>
> $$x = \sqrt{a} \quad \text{or} \quad x = -\sqrt{a}$$

We now develop the *completing-the-square approach* to solving second-degree equations in one variable. As usual, we start with simple cases and build up to the general method. Work through the following examples.

EXAMPLE 5 Solve: $x^2 = 25$

Solution We use the square root property to get $x = \sqrt{25}$ or $x = -\sqrt{25}$.
The solution is $\sqrt{25}$ and $-\sqrt{25}$.
Of course, these numbers reduce to 5 and -5. ■

EXAMPLE 6 Solve: $x^2 = 17$

Solution We use the square root property to get $x = \sqrt{17}$ or $x = -\sqrt{17}$.
The solution is $\sqrt{17}$ and $-\sqrt{17}$.
These numbers are completely reduced. ■

EXAMPLE 7 Solve: $(x + 3)^2 = 16$

Solution We use the square root property to get

$$\begin{array}{rlcrl} x + 3 = & \sqrt{16} & \text{or} & x + 3 = & -\sqrt{16} \\ x + 3 = & 4 & & x + 3 = & -4 \\ -3 & -3 & & -3 & -3 \\ \hline x = & 1 & & x = & -7 \end{array}$$

The solution to $(x + 3)^2 = 16$ is 1 and -7. ■

EXAMPLE 8 Solve: $(x + 3)^2 = 17$

Solution We use the square root property:

$$\begin{array}{rlcrl} x + 3 = & \sqrt{17} & \text{or} & x + 3 = & -\sqrt{17} \\ -3 & -3 & & -3 & -3 \\ \hline x = & -3 + \sqrt{17} & & x = & -3 - \sqrt{17} \end{array}$$

The solution to $(x + 3)^2 = 17$ is $-3 + \sqrt{17}$ and $-3 - \sqrt{17}$. ■

GIVE IT A TRY

Solve each equation. Report each number in completely reduced form.

4. $(x - 3)^2 = 36$ **5.** $(x + 5)^2 = 81$ **6.** $(x - 3)^2 = 23$

Completing the Square Every second-degree equation in one variable, such as $x^2 - 3x = 7$, can be altered to the form $(x + b)^2 = a$, where a and b are real numbers. We produce this form by completing the square. Once we have this form, we can use the square root property to write the solution to the equation. Work through the following examples.

EXAMPLE 9 Solve: $x^2 - 6x = 7$

Solution We start by adding a number to each side of the equation in order to get a perfect square binomial on the left-hand side.

$$x^2 - 6x = 7$$

Add $\left(\frac{1}{2} \text{ of } -6\right)^2$ to each side: $\quad 9 \quad 9$

to get: $\quad x^2 - 6x + 9 = 16$

Write left-hand side as a square: $\quad (x - 3)^2 = 16$

Use the square root property to get:

$$x - 3 = \sqrt{16} \quad \text{or} \quad x - 3 = -\sqrt{16}$$
$$x - 3 = 4 \qquad\qquad x - 3 = -4$$
$$\underline{3 \quad 3} \qquad\qquad \underline{3 \quad 3}$$
$$x = 7 \qquad\qquad x = -1$$

Solution: 7 and -1 ■

The fact that the numbers in the solution are rational numbers suggests that the equation in Example 9 could have been solved by factoring. Now, consider an example that cannot be solved by factoring.

EXAMPLE 10 Solve: $x^2 - 6x = 3$

Solution We start by adding a number to each side. $x^2 - 6x = 3$

Add $\left(\frac{1}{2} \text{ of } -6\right)^2$ to each side $\underline{9 \quad 9}$

to get: $x^2 - 6x + 9 = 12$

Write left-hand side as a square: $(x - 3)^2 = 12$

Use the square root property to get:

$$x - 3 = \sqrt{12} \quad \text{or} \quad x - 3 = -\sqrt{12}$$
$$x - 3 = 2\sqrt{3} \qquad\qquad x - 3 = -2\sqrt{3}$$
$$\underline{3 \quad 3} \qquad\qquad \underline{3 \quad 3}$$
$$x = 3 + 2\sqrt{3} \qquad\qquad x = 3 - 2\sqrt{3}$$

Solution: $3 + 2\sqrt{3}$ and $3 - 2\sqrt{3}$

Check Is $\left(3 + 2\sqrt{3}\right)^2 - 6\left(3 + 2\sqrt{3}\right) = 3$ a true arithmetic sentence?

$$9 + 12\sqrt{3} + 12 - \left(18 + 12\sqrt{3}\right) = 21 + 12\sqrt{3} - 18 - 12\sqrt{3} = 3$$

Is $\left(3 - 2\sqrt{3}\right)^2 - 6\left(3 - 2\sqrt{3}\right) = 3$ a true arithmetic sentence?

$$9 - 12\sqrt{3} + 12 - \left(18 - 12\sqrt{3}\right) = 21 - 12\sqrt{3} - 18 + 12\sqrt{3} = 3$$ ■

GIVE IT A TRY

Solve each quadratic equation using the completing-the-square approach. Report the numbers in completely reduced form.

7. $x^2 - 6x = 16$ **8.** $x^2 - 6x = 8$ **9.** $x^2 + 4x = 7$ 

Coefficients In the next two examples, we demonstrate a constraint of this approach to solving second-degree equations in one variable—when rational numbers (fractions) appear as coefficients.

EXAMPLE 11 Solve: $x^2 - 3x - 5 = 0$

Solution

$$x^2 - 3x - 5 = 0$$

Add 5 to each side

$$\underline{\qquad\qquad 5 \quad 5}$$

to get:

$$x^2 - 3x = 5$$

Add $\left(\frac{1}{2} \text{ of } -3\right)^2$ to each side

$$\underline{\qquad\qquad \frac{9}{4} \quad \frac{9}{4}}$$

to get:

$$x^2 - 3x + \frac{9}{4} = \frac{29}{4}$$

Rewrite the left-hand side:

$$\left(x - \frac{3}{2}\right)^2 = \frac{29}{4}$$

Use the square root property:

$$x - \frac{3}{2} = \frac{\sqrt{29}}{2} \qquad \text{or} \qquad x - \frac{3}{2} = \frac{-\sqrt{29}}{2}$$

$$\underline{\quad \frac{3}{2} \quad \frac{3}{2}} \qquad\qquad \underline{\quad \frac{3}{2} \quad \frac{3}{2}}$$

$$x = \frac{3 + \sqrt{29}}{2} \qquad\qquad x = \frac{3 - \sqrt{29}}{2}$$

Solution: $\dfrac{3 + \sqrt{29}}{2}$ and $\dfrac{3 - \sqrt{29}}{2}$ ■

The completing-the-square step where $\left(\frac{1}{2} \text{ of the } x\text{-coefficient}\right)^2$ is added to each side depends on the x^2-coefficient being the number 1. If this is not the case, we can multiply each side of the equation by a number to achieve this situation.

EXAMPLE 12 Solve: $3x^2 - 2x = 4$

Solution

$$3x^2 - 2x = 4$$

Multiply by $\frac{1}{3}$ to get:

$$x^2 - \frac{2}{3}x = \frac{4}{3}$$

Add $\left(\frac{1}{2} \text{ of } \frac{-2}{3}\right)^2$ to each side

$$\underline{\qquad\qquad \frac{1}{9} \quad \frac{1}{9}}$$

to get:

$$x^2 - \frac{2}{3}x + \frac{1}{9} = \frac{13}{9}$$

Rewrite the left-hand side:

$$\left(x - \frac{1}{3}\right)^2 = \frac{13}{9}$$

Use the square root property:

$$x - \frac{1}{3} = \frac{\sqrt{13}}{3} \qquad \text{or} \qquad x - \frac{1}{3} = \frac{-\sqrt{13}}{3}$$

$$\underline{\quad \frac{1}{3} \quad \frac{1}{3}} \qquad\qquad \underline{\quad \frac{1}{3} \quad \frac{1}{3}}$$

$$x = \frac{1 + \sqrt{13}}{3} \qquad\qquad x = \frac{1 - \sqrt{13}}{3}$$

Solution: $\dfrac{1+\sqrt{13}}{3}$ and $\dfrac{1-\sqrt{13}}{3}$ ■

As you can see, fractions do not affect the algebra—the process is the same—but they do affect the difficulty of the arithmetic. Like factoring, the skills and concepts we have developed in the completing-the-square approach to second-degree equations will be applied in other areas of algebra (such as working with circles, in Chapter 12).

Using the Graphics Calculator

To check our work in Example 11, we show that the number $\frac{3+\sqrt{29}}{2}$ does solve the equation $x^2 - 3x - 5 = 0$. Store `(3+√29)/2` in memory location `X`. Type `X²−3X−5` and press the `ENTER` key. Is 0 output? If not, an error is present! In a similar fashion we can check that the number $\frac{3-\sqrt{29}}{2}$ solves the equation. Use this approach to check that the two numbers found in Example 12 solve the equation $3x^2 - 2x = 4$.

GIVE IT A TRY

Solve each quadratic equation. Report the numbers in completely reduced form.

10. $x^2 - 5x = 2$ **11.** $2x^2 - 3x = 5$ **12.** $3x^2 - 2x = x + 6$

13. Use the graphics calculator to check the solution to Problems 10–12.

Special Cases With first-degree equations in one variable, the solution is typically a single number. However, for some special cases the solution is either empty (such as $x = x + 1$) or is all numbers (such as $3x + 2 = 2 + 3x$). With second-degree equations in one variable, the solution is typically two numbers. The following examples illustrate special cases of second-degree equations in one variable.

EXAMPLE 13 Solve: $x^2 = x^2 + 1$

Solution By inspection, no number solves this equation (adding 1 always alters a number squared). Attempting to place the equation in standard form yields a false arithmetic sentence:

$$\begin{aligned} x^2 &= x^2 + 1 \\ -x^2 &\quad -x^2 \\ \hline 0 &= 1 \end{aligned}$$

Add $-x^2$ to each side to get: ■

When an equation reduces to a false arithmetic sentence, the solution to that equation is *none* (no number solves the equation).

EXAMPLE 14 Solve: $x(x + 3) = x^2 + 3x$

Solution By inspection, all numbers solve this equation, because it is simply an example of the distributive property. Placing the equation in standard form yields a true arithmetic sentence:

$$x(x + 3) = x^2 + 3x$$

Clear parentheses: $x^2 + 3x = x^2 + 3x$

Add $-x^2 - 3x$ to each side to get:

$$\begin{array}{r} x^2 + 3x = x^2 + 3x \\ \underline{-x^2 - 3x} \quad \underline{-x^2 - 3x} \\ 0 = 0 \end{array}$$

■

When an equation reduces to a true arithmetic sentence, the solution is *all* (every number solves the equation).

EXAMPLE 15 Solve: $(x - 3)^2 = 0$

Solution We already have a binomial squared equal to a number. We use the square root property:

$$\begin{array}{rclcrcl} x - 3 &=& \sqrt{0} & \text{or} & x - 3 &=& -\sqrt{0} \\ 3 && 3 && 3 && 3 \\ x &=& 3 && x &=& 3 \end{array}$$

Solution: 3 ■

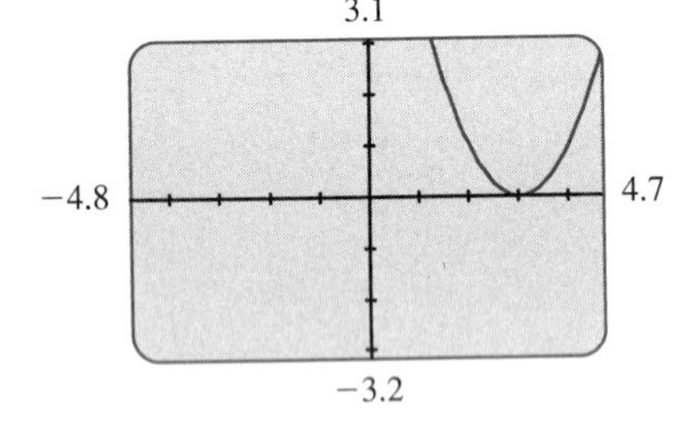

FIGURE 4

Example 15 shows a quadratic equation whose solution is a single number, namely, 3. Such a number is referred to as a *root of order 2,* because it is viewed as occurring twice. A **root** of an equation is a number that solves the equation. We could say the solution is 3 and 3. Contrast this with the numbers that solve $(x - 2)(x + 3) = 0$. These numbers, 2 and -3, are referred to as *roots of order 1,* because each occurs only once.

Using the Graphics Calculator

Type `(X−3)²` for `Y1`. Use the Window 4.7 range settings to view the graph screen, as shown in Figure 4. Press the `TRACE` key and move the cursor to the right until the message reads `X=3  Y=0`. As you can see, the expression $(x - 3)^2$ evaluates to 0 at one single number, namely, 3.

EXAMPLE 16 Solve: $x^2 = -5$

Solution We have a square equal to a number. So we use the square root property:

$$\begin{array}{rclcrcl} x &=& \sqrt{-5} & \text{or} & x &=& -\sqrt{-5} \\ x &=& i\sqrt{5} && x &=& -i\sqrt{5} \end{array}$$

The solution to this equation is *no real number*. The square of a real number must be zero or larger. Thus, no matter which real number the variable is replaced by, its square cannot be -5. The solution to this equation is two nonreal complex numbers, namely, $i\sqrt{5}$ and $-i\sqrt{5}$. ■

For a first-degree equation in one variable, the solution is either *none,* a single number, or *all.* With second-degree equations in one variable, the solution is either *none,* a single number, two numbers, or *all.*

GIVE IT A TRY

Solve each quadratic equation.

14. $x^2 - 5x = x(x - 1)$

15. $x^2 = 5 + x^2$

16. $x^2 - 2x = (2 - x)(-x)$

17. $x^2 = -3$

■ SUMMARY

■ ***Quick Index***

In this section, we have learned a general approach to solving second-degree equations in one variable. The factoring approach to solving these equations has one shortcoming—prime polynomials. The completing-the-square approach to solving second-degree equations in one variable always produces the numbers that solve the equation. It is based on the fact that every second-degree equation in one variable can be written in the form of a binomial squared equal to a number, $(x + b)^2 = a$. Using the square root property, the solution to the equation is produced. The two numbers that solve the equation have the form $-b + \sqrt{a}$ and $-b - \sqrt{a}$.

To solve a second-degree equation in one variable by the completing-the-square approach, we use the following steps:

1. Write the equation with the variable terms on the left-hand side and the number term on the right-hand side.
2. Make sure the x^2 coefficient is 1. If not, multiply each side by the reciprocal of the x^2 coefficient.
3. Add $\left(\frac{1}{2} \text{ of the } x\text{-coefficient}\right)^2$ to each side of the equation.
4. Write the left-hand side as a binomial squared. The equation now has the form $(x + b)^2 = a$.
5. Use the square root property to write the solution, $-b + \sqrt{a}$ and $-b - \sqrt{a}$. If $a > 0$, the solution is two real numbers. If $a < 0$, then the solution is *no real number* (two nonreal complex numbers).

The solution to a second-degree equation in one variable is typically two numbers. Special cases are equations whose solution is *none* (such as $x^2 = x^2 + 1$), a single number [such as $(x - 2)^2 = 0$]; or *all* [such as the equation $x(x + 1) = x^2 + x$]. When the solution is two numbers, we have the special case of a solution that is two nonreal complex numbers.

8.1 ■ ■ ■ EXERCISES

1. Which one of the following equations can be solved by the factoring approach?
   a. $x^2 + x - 5 = 0$ b. $x^2 + 3x + 4 = 0$
   c. $x^2 - 3x = 4$ d. $x^2 - 3x = 7$
2. Which one of the following polynomials is a perfect square binomial?
   a. $x^2 - 2x + 3$ b. $2x^2 + 12x + 6$
   c. $x^2 - 6x + 9$ d. $x^2 - \frac{1}{2}x + \frac{1}{4}$
3. Which one of the following equations has a solution of two rational numbers?
   a. $x^2 = 7$ b. $x^2 - 3x - 1 = 0$
   c. $(2x - 3)(3x - 1) = 0$ d. $x^2 - 4x - 7 = 0$
4. Which one of the following has a solution of two nonreal complex numbers?
   a. $(x - 3)^2 = 7$ b. $2x^2 - x = 5$
   c. $x^2 = -9$ d. $(x - 3)^2 = 0$
5. Which one of the following has a solution of *none*?
   a. $x^2 - 3x = 7$ b. $x^2 = x^2 - 4$
   c. $x^2 = -9$ d. $(x - 3)^2 = x^2 + 9$
6. Which one of the following has exactly one number in its solution?
   a. $x^2 - 3x = 7$ b. $x^2 = x^2 - 4$
   c. $(x - 2)^2 = (2 - x)^2$ d. $(x - 3)^2 = x^2 + 9$

Report the number that when added to the given polynomial produces a perfect square binomial? Report the factored form of the perfect square polynomial produced.

7. $x^2 - 8x$
8. $x^2 - 12x$
9. $x^2 + 10x$
10. $x^2 + 14x$
11. $x^2 - 7x$
12. $x^2 + 9x$
13. $x^2 + \frac{2}{3}x$
14. $x^2 - \frac{3}{5}x$

Solve each equation. Report the numbers in the solution in completely reduced form.

15. $(x - 4)^2 = 25$
16. $(x - 3)^2 = 100$
17. $(x + 1)^2 = 16$
18. $(x + 9)^2 = 64$
19. $(x - 7)^2 = 19$
20. $(x - 3)^2 = 45$

$(x + 2)^2 = 20$

$(x + 13)^2 = 125$

Solve each equation and report the numbers in the solution in completely reduced form. Use the graphics calculator to check your work.

23. $\left(x - \frac{1}{2}\right)^2 = \frac{7}{4}$
24. $\left(x - \frac{3}{2}\right)^2 = \frac{3}{4}$
25. $\left(x + \frac{1}{4}\right)^2 = \frac{25}{16}$
26. $\left(x + \frac{1}{4}\right)^2 = \frac{81}{16}$
27. $\left(x - \frac{7}{2}\right)^2 = \frac{27}{4}$
28. $\left(x + \frac{5}{2}\right)^2 = \frac{45}{4}$

Solve each quadratic equation using the completing-the-square approach. Use the graphics calculator to check your work.

29. $x^2 - 6x = 4$
30. $x^2 - 8x = 9$
31. $(x - 2)(x - 3) = 8$
32. $(x - 1)(x + 5) = 1$
33. $x^2 - 4x = 31$
34. $x^2 - 8x = 41$
35. $x^2 + 10x = 13$
36. $x^2 + 12x = 9$
37. $2x^2 - 2x = x^2 + 5$
38. $2x^2 = (x - 1)^2$
39. $5 - 2x - x^2 = 1$
40. $3 - 6x - x^2 = 2$

Solve each quadratic equation, using either the factoring approach or the completing-the-square approach. Use the graphics calculator to check your work.

41. $x^2 - 3x = 5$
42. $x^2 - 2x = 7$
43. $2x^2 - 4x = 1$
44. $2x^2 - 10x = 3$
45. $2x^2 - 12x = 3$
46. $3x^2 - 6x = 7$
47. $3x^2 - 9x = 13$
48. $x^2 - x = 15$
49. $(x - 2)^2 - x = 1$
50. $(x + 1)^2 - x = 2$
51. $(2 - 3x)x = 5$
52. $(3 - x)x = 1$
53. $2x^2 = (x - 3)^2$
54. $3x^2 = (x + 1)^2$
55. $x^2 - 3x = x(x - 3)$
56. $x^2 = x(x + 7) - 7x$
57. $(x - 5)^2 = x^2 - 25$
58. $(2 + x)^2 = (x + 2)^2$
59. $x^2 = -16$
60. $x^2 = -36$
61. $2x^2 = 2(1 + x^2)$
62. $(2 - x)^2 = 4x^2 - 4x$

8.2 ■ ■ ■ THE QUADRATIC FORMULA

So far, we have seen two approaches to solving second-degree equations in one variable. The first approach, factoring, is easy to use, but is limited in scope. Only *some* second-degree equations in one variable can be solved by this approach (those whose solutions are rational numbers). In the last section, we used the second approach, completing the square. This approach will solve *all* second-degree equations in one variable. However, this approach often introduces greater complexity in the arithmetic produced. In this section, we will generalize the completing-the-square approach to produce a formula that can be used to write the solution to any second-degree equation in one variable. This formula is known as the **quadratic formula**.

> **REMEMBER 1**
>
> $\frac{1}{2}$ of $\frac{-b}{a}$ is $\frac{-b}{2a}$ and $\left(\frac{b}{2a}\right)^2$ is $\frac{b^2}{4a^2}$
>
> **REMEMBER 2**
>
> $$\frac{-c}{a} + \frac{b^2}{4a^2} = \frac{-4ac}{4a^2} + \frac{b^2}{4a^2} = \frac{b^2 - 4ac}{4a^2}$$

Developing the Formula We will now use an analogy to develop a formula for writing the solution to a second-degree equation in one variable. (We have used analogies to rewrite formulas in Sections 2.8, 6.5, and 7.4.) That is, we will use the completing-the-square approach as a model (analogy) for a general quadratic equation in one variable, namely, $ax^2 + bx + c = 0$. We assume that $a > 0$. If this is not the case, we will multiply by -1 to achieve this form.

Analogy (for $a = 2$, $b = 5$, and $c = 1$)		*Quadratic formula*	
Solve:	$2x^2 + 5x + 1 = 0$	Rewrite for x:	$ax^2 + bx + c = 0 \quad a > 0$
Multiply each side by $\frac{1}{2}$ to get:	$x^2 + \frac{5}{2}x + \frac{1}{2} = 0$	Multiply each side by $\frac{1}{a}$:	$x^2 + \frac{b}{a}x + \frac{c}{a} = 0$
Add $\frac{-1}{2}$ to each side to get:	$x^2 + \frac{5}{2}x = \frac{-1}{2}$	Add $\frac{-c}{a}$ to each side to get:	$x^2 + \frac{b}{a}x = \frac{-c}{a}$
Add $\left(\frac{1}{2} \text{ of } \frac{5}{2}\right)^2$ to each side to get:	$x^2 + \frac{5}{2}x + \frac{25}{16} = \frac{17}{16}$	Add $\left(\frac{1}{2} \text{ of } \frac{b}{a}\right)^2$ to each side to get: (See Remember 1 and 2.)	$x^2 + \frac{b}{a}x + \frac{b^2}{4a^2} = \frac{b^2 - 4ac}{4a^2}$
Rewrite the left-hand side to get:	$\left(x + \frac{5}{4}\right)^2 = \frac{17}{16}$	Rewrite the left-hand side:	$\left(x + \frac{b}{2a}\right)^2 = \frac{b^2 - 4ac}{4a^2}$
Take the square root of each side to get:		Take square root of each side to get:	
$x + \frac{5}{4} = \sqrt{\frac{17}{16}}$	or $x + \frac{5}{4} = -\sqrt{\frac{17}{16}}$	$x + \frac{b}{2a} = \sqrt{\frac{b^2 - 4ac}{4a^2}}$	or $x + \frac{b}{2a} = -\sqrt{\frac{b^2 - 4ac}{4a^2}}$
$x + \frac{5}{4} = \frac{\sqrt{17}}{4}$	$x + \frac{5}{4} = \frac{-\sqrt{17}}{4}$	$x + \frac{b}{2a} = \frac{\sqrt{b^2 - 4ac}}{2a}$	$x + \frac{b}{2a} = -\frac{\sqrt{b^2 - 4ac}}{2a}$
$\frac{-5}{4} \quad \frac{-5}{4}$	$\frac{-5}{4} \quad \frac{-5}{4}$	$\frac{-b}{2a} \quad \frac{-b}{2a}$	$\frac{-b}{2a} \quad \frac{-b}{2a}$
$x = \frac{-5 + \sqrt{17}}{4}$	$x = \frac{-5 - \sqrt{17}}{4}$	$x = \frac{-b + \sqrt{b^2 - 4ac}}{2a}$	$x = \frac{-b - \sqrt{b^2 - 4ac}}{2a}$

Using the Formula The quadratic formula

$$x = \frac{-b + \sqrt{b^2 - 4ac}}{2a} \quad \text{or} \quad x = \frac{-b + \sqrt{b^2 - 4ac}}{2a}$$

gives the form of the numbers that solve an equation of the form

$$ax^2 + bx + c = 0$$

Although we could simply identify the coefficients a, b, and c, and then insert these values into the formula, an organized approach is preferable. This approach reduces the possibility of errors and provides insight to the solution. Consider the following example.

EXAMPLE 1 Solve: $x^2 - 3x = 7$

Solution **Step 1** Rewrite the equation to standard form to get:

$$\begin{aligned} x^2 - 3x &= 7 \\ -7 &\quad -7 \\ \hline x^2 - 3x - 7 &= 0 \end{aligned}$$

Step 2 Identify a, b, and c:

$$\underline{\ \ }x^2 - \underline{3}x - \underline{7} = 0$$

$$\downarrow \qquad \downarrow \qquad \downarrow$$

$$a = 1 \qquad b = -3 \qquad c = -7$$

Step 3 Compute the radicand, $b^2 - 4ac$:

$$\begin{aligned} b^2 - 4ac &= (-3)^2 - 4(1)(-7) \\ &= 9 + 28 \\ &= 37 \end{aligned}$$

Step 4 Write the solution:

opposite of b ↓ ; $b^2 - 4ac$ →

$$\frac{3 + \sqrt{37}}{2} \text{ and } \frac{3 - \sqrt{37}}{2}$$

↑ $2a$

■

In the quadratic formula, the radicand—the number under the radical sign, the expression $b^2 - 4ac$—is known as the **discriminant**. We can think of it as "discriminating" (telling the difference) between the two numbers in the solution. One number in the solution adds the square root of the discriminant, while the other number in the solution subtracts the square root of the discriminant.

Using the Graphics Calculator

To check that $\frac{3 + \sqrt{37}}{2}$ solves the equation $x^2 - 3x = 7$, store (3+√37)/2 in memory location X. Type X²−3X and then press the ENTER key. Is 7 output?

Run a similar check for the number $\frac{3 - \sqrt{37}}{2}$.

With the development of the quadratic formula, we have two approaches for solving quadratic equations that cannot be solved by factoring:

1. Solve the second-degree equation in one variable by using the completing-the-square approach (with its inherent arithmetic complexities and errors).
2. Use the quadratic formula.

Although the quadratic formula effectively eliminates the need for using the factoring approach to solving quadratic equations, factoring is still a valid technique. Because factoring is a much simpler approach, it should be used whenever possible (whenever the polynomial factors easily). The quadratic formula works with all second-degree equations in one variable. Consider the equation $x^2 - 5x + 6 = 0$. Using factoring, this equation can be written as $(x - 2)(x - 3) = 0$. Now, by inspection, we know the solution is the numbers 2 and 3. The following example shows that the quadratic formula produces these same two numbers. It does not matter which technique is used to find the solution. Here, we simply want to show that the quadratic formula produces the same result as does factoring.

EXAMPLE 2 Solve: $(x - 3)(x + 2) = 0$

Solution We already know that this equation has solution 3 and -2. Observe what happens when the quadratic formula is used to solve this equation.

Step 1 Write in standard form: $x^2 - x - 6 = 0$

Step 2 Identify a, b, and c: $a = 1, b = -1, c = -6$

Step 3 Compute $b^2 - 4ac$: $(-1)^2 - 4(1)(-6) = 1 + 24 = 25$
As you can see, the discriminant is a *perfect square*. This implies that the equation can be solved by factoring.

Step 4 Write the solution:

$$\frac{1 + \sqrt{25}}{2} \quad \text{and} \quad \frac{1 - \sqrt{25}}{2}$$

Reducing these numbers, we get 3 and -2. ■

Example 2 displays one of the *best* practice techniques for learning to use the quadratic formula—use an equation whose solution we know, such as

$$(2x - 1)(x + 2) = 0$$

Solve the equation using the quadratic formula, to see if we get the correct values, $\frac{1}{2}$ and -2.

Consider solving the equation, $(x - 2)^2 = -4$. Here, the solution is two nonreal complex numbers. No matter which *real* number replaces x, subtracting 2 and then squaring will produce a nonnegative number (not -4). Using the quadratic formula, we find that the discriminant (the number placed under the radical) is a number less than 0. At that point, we can report the solution as *no real number* or we can write the two nonreal complex numbers that solve the equation. Your instructor will tell you which reporting method is preferred.

EXAMPLE 3 Solve: $(x - 2)^2 = -4$

Solution **Step 1** Write in standard form: $x^2 - 4x + 8 = 0$

Step 2 Identify a, b, and c: $a = 1, b = -4, c = 8$

Step 3 Compute $b^2 - 4ac$: $(-4)^2 - 4(1)(8) = 16 - 32 = -16$
As you can see, the discriminant is less than 0.

Step 4 Write the solution: *no real number* or

$$\frac{4 + \sqrt{-16}}{2} \quad \text{and} \quad \frac{4 - \sqrt{-16}}{2}$$

Reducing these numbers, $\frac{4 + 4i}{2}$ and $\frac{4 - 4i}{2}$

we get: $2 + 2i$ and $2 - 2i$ ■

Finally, consider how the quadratic formula approach appears where the solution is a single number.

EXAMPLE 4 Solve: $(x - 2)^2 = 0$

Solution **Step 1** Write in standard form: $x^2 - 4x + 4 = 0$

Step 2 Identify a, b, and c: $a = 1, b = -4, c = 4$

Step 3 Compute $b^2 - 4ac$: $(-4)^2 - 4(1)(4) = 16 - 16 = 0$

Step 4 Write the solution:

$$\frac{4 + \sqrt{0}}{2} \quad \text{and} \quad \frac{4 - \sqrt{0}}{2}$$

Both of these numbers reduce to the number 2. ■

The step-by-step approach used in these examples can provide insight to the solution: The discriminant is the key—we can use it to determine the type of solution. The following generalizations provide useful information about the solutions of second-degree equations.

PROPERTIES OF THE DISCRIMINANT

> If $b^2 - 4ac$ is a perfect square (such as 16 or 25), the solution is two rational numbers. (See Example 2.) In fact, the equation can be solved by factoring over the integers.
>
> If $b^2 - 4ac = 0$, then the solution is a single number. The number that solves the equation has the form $\frac{-b}{2a}$. (See Example 4.)
>
> If $b^2 - 4ac > 0$, then the solution is two real numbers. (See Example 1.)
>
> If $b^2 - 4ac < 0$, then the solution is *no real numbers*. Both numbers that solve the equation are nonreal complex numbers. (See Example 3.)
>
> In each of the last two cases, the numbers are *conjugate pairs*.

In the following example we solve the equation $3x^2 - 2x = 4$ by the quadratic formula. Compare this with the arithmetic involved in solving the same equation by completing the square. (See Example 12 in Section 8.1.)

EXAMPLE 5 Solve: $3x^2 - 2x = 4$

Solution

Step 1 Standard form: $3x^2 - 2x - 4 = 0$

Step 2 Identify a, b, and c: $a = 3$, $b = -2$, $c = -4$

Step 3 Compute $b^2 - 4ac = (-2)^2 - 4(3)(-4) = 4 + 48 = 52$

Step 4 Write the solution: $\frac{2 + \sqrt{52}}{6}$ and $\frac{2 - \sqrt{52}}{6}$

Reduce: $\frac{2 + 2\sqrt{13}}{6}$ and $\frac{2 - 2\sqrt{13}}{6}$

$\frac{\cancel{2}(1 + \sqrt{13})}{\cancel{2}(3)}$ and $\frac{\cancel{2}(1 - \sqrt{13})}{\cancel{2}(3)}$

In reduced form the solution is $\frac{1 + \sqrt{13}}{3}$ and $\frac{1 - \sqrt{13}}{3}$. ■

As a another example of working with the quadratic formula, work through the next example.

EXAMPLE 6 Solve: $x^2 - 5x = 9$

Solution

Step 1 Write in standard form: $x^2 - 5x - 9 = 0$

Step 2 Identify a, b, and c: $a = 1$, $b = -5$, $c = -9$

Step 3 Compute $b^2 - 4ac$: $(-5)^2 - 4(1)(-9) = 61$

Step 4 Write the solution: $\frac{5 + \sqrt{61}}{2}$ and $\frac{5 - \sqrt{61}}{2}$ ■

Using the Graphics Calculator

To check our work in Example 6, store the number `(5+√61)/2` in memory location `X`. The number output is `6.405124838`, so this irrational number is between 6 and 7. Type `X²−5X` and then press the `ENTER` key. Is `9` output? If not, an error is present. Now, store the number `(5−√61)/2` in memory location `X`. The number `-1.405124838` is displayed. This irrational number is between -2 and -1. Type `X²−5X` and then press the `ENTER` key. Is `9` output? If not, an error is present.

To increase our understanding of the solution to Example 6, view the graph screen, using the Window 9.4 range settings. Adjust the `Ymin` value to `-16` (see Figure 1). Use the `TRACE` key to see that the x-values that result in $x^2 - 5x - 9$ evaluating to zero are approximately 6.4 and -1.4.

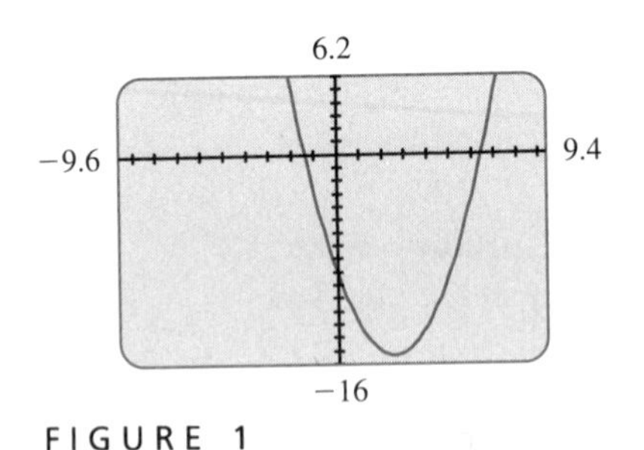

FIGURE 1

GIVE IT A TRY

1. Use the quadratic formula to solve the equation $x^2 - 5x - 3 = 0$:
   a. Report the values for a, b, and c.
   b. Report the discriminant.
   c. Report the solution (in completely reduced form).
2. Solve each equation using the quadratic formula:
   a. $x^2 - 5x = 7$
   b. $2x^2 - 3x = 7$
   c. $(x - 3)(x + 2) = 1$
3. Use the graphics calculator to check your work for Problem 2.

Special Cases We will now consider some special cases that occur when using the quadratic formula. Work through the following examples.

EXAMPLE 7 Solve: $-2x^2 + 3x = 1$

Solution Although we can certainly work with an x^2-coefficient of -2, it is wise to use a positive x^2-coefficient. We can accomplish this by multiplying each side of the equation by -1 to get:

$$2x^2 - 3x = -1$$

Step 1 Standard form: $2x^2 - 3x + 1 = 0$

Step 2 Identify a, b, c: $a = 2$, $b = -3$, $c = 1$

Step 3 Compute $b^2 - 4ac$: $9 - 4(2)(1) = 1$

Step 4 Write the solution: $\dfrac{3 + \sqrt{1}}{4}$ and $\dfrac{3 - \sqrt{1}}{4}$

Reduce the numbers to 1 and $\frac{1}{2}$. ■

EXAMPLE 8 Solve: $3x^2 - 5 = 0$

Solution For this equation, the x-term is missing. Thus, the x-coefficient, b, is 0.

Step 1 Standard form: $3x^2 - 5 = 0$

Step 2 Identify a, b, c: $a = 3, b = 0, c = -5$

Step 3 Compute $b^2 - 4ac$: $0 - 4(3)(-5) = 60$

Step 4 Write the solution: $\dfrac{0 + \sqrt{60}}{6}$ and $\dfrac{0 - \sqrt{60}}{6}$

Reduce the numbers to $\dfrac{\not{2}\sqrt{15}}{\not{2}(3)}$ and $\dfrac{\not{2}\sqrt{15}}{\not{2}(3)}$, so the solution is

$$\frac{\sqrt{15}}{3} \quad \text{and} \quad \frac{-\sqrt{15}}{3}$$

■

EXAMPLE 9 Solve: $2x^2 - 5x = 0$

Solution For this equation, the constant term is missing. Thus, c is zero.

Step 1 Standard form: $2x^2 - 5x = 0$

Step 2 Identify a, b, c: $a = 2, b = -5, c = 0$

Step 3 Compute $b^2 - 4ac$: $25 - 4(2)(0) = 25$

Step 4 Write the solution: $\dfrac{5 + \sqrt{25}}{4}$ and $\dfrac{5 - \sqrt{25}}{4}$

Reduce the numbers to $\dfrac{5}{2}$ and 0.

■

EXAMPLE 10 Solve: $x^2 - 3x = -5$

Solution

Step 1 Standard form: $x^2 - 3x + 5 = 0$

Step 2 Identify a, b, and c: $a = 1, b = -3, c = 5$

Step 3 Compute $b^2 - 4ac$: $9 - 4(1)(5) = -11$
Since this value is negative, the solution will be *no real number*. Both numbers will be nonreal complex numbers.

Step 4 Write the solution: $\dfrac{3 + i\sqrt{11}}{2}$ and $\dfrac{3 - i\sqrt{11}}{2}$

■

GIVE IT A TRY

Solve each equation.

4. $x^2 - 3x + 1 = 1$

5. $3x^2 = 2x + 3$

6. $x^2 + 5x + 7 = 0$

7. $(x - 2)(x + 1) = 2 - x$

8. $x^2 - 1 = x^2 + 2x + 5$

9. $x^2 - 5x + 8 = 0$

■ ***Quick Index***

■ SUMMARY

In this section, we have used reasoning by analogy from the completing-the-square technique to develop a general form that the solution to a second-degree equation in one variable must fit. For an equation $ax^2 + bx + c = 0$ (in standard form with $a \neq 0$), that general form (formula) for the solution is

$$\frac{-b + \sqrt{b^2 - 4ac}}{2a} \quad \text{and} \quad \frac{-b - \sqrt{b^2 - 4ac}}{2a}$$

This means that if we know the quadratic formula and the values for a, b, and c, then we can write the solution to a second-degree equation in one variable. This approach overcomes the arithmetic complexities inherent in the completing-the-square approach to solving second-degree equations in one variable.

To use the quadratic formula to solve a second-degree equation in one variable, we use four steps:

1. Rewrite the equation to standard form, $ax^2 + bx + c = 0$.
2. Identify a, b, and c—the x^2-coefficient, the x-coefficient, and the constant term.
3. Compute the discriminant, $b^2 - 4ac$.
4. Write the solution, and then completely reduce the radical expressions.

Step 3 of this procedure provides some insight into the solution. If the discriminant, $b^2 - 4ac$, is a perfect square (such as 9 or 64), then the equation can be solved by factoring. If the discriminant is greater than zero, the solution is two real numbers. If the discriminant is 0, the solution is exactly one number, with the form $-b/(2a)$. If the discriminant is less than zero, then no real number solves the equation. The solution is two nonreal complex numbers.

Special cases for using the quadratic formula include equations in which b is zero (the x-term is missing) or c is zero (the constant term is missing). Of course, if a is zero (the x^2-term is missing), the quadratic formula cannot be used (because we must divide by $2a$). However, with a equal to 0, the equation is a first-degree equation, which we already know how to solve (from Chapter 2).

8.2 ■ ■ ■ EXERCISES

Insert true *for a true statement, or* false *for a false statement.*

____ 1. The first step in using the quadratic formula is to rewrite the equation to standard form, which is $ax^2 + bx + c = 0$.

____ 2. The quadratic formula can be used to solve a first-degree equation in one variable.

____ 3. If a second-degree equation in one variable can be solved by factoring over the integers, then it cannot be solved by using the quadratic formula.

____ 4. The quadratic formula can be used to solve any second-degree equation in one variable.

____ 5. If the discriminant, $b^2 - 4ac$, is zero, then the second-degree equation in one variable will have exactly one number in its solution.

____ 6. If the x-term is missing, (b is zero), then the quadratic formula cannot be used.

____ 7. If the discriminant is a perfect square (such as 16 or 81), then the second-degree equation in one variable can be solved by factoring over the integers.

____ 8. If the discriminant is less than zero, then the solution to the second-degree equation in one variable contains no real number.

____ 9. For the solution to a second-degree equation in one variable to be a single number, the discriminant must be 0.

____ 10. The solution to a second-degree equation in one variable can be the three numbers: 5, 7, and 8.

For each second-degree equation in one variable, report the values for a, b, *and* c, *the discriminant, and the solution.*

11. $x^2 + 6x + 1 = 0$
12. $x^2 + 6x - 2 = 0$
13. $x^2 + 3x - 1 = 0$
14. $x^2 - 3x - 2 = 0$
15. $x^2 - 5x = 10$
16. $x^2 + 5x = 15$
17. $x^2 + 4x = 9$
18. $x^2 - 4x = 11$
19. $x^2 = 4x + 13$
20. $x^2 - 7 = 2x$
21. $2x^2 - x - 1 = 0$
22. $2x^2 - 3x - 5 = 0$
23. $(2x - 1)(x + 2) = 1$
24. $(x - 3)(2x) = 7$
25. $(x - 2)(x + 2) = x$
26. $3 - (x - 1)^2 = x$

The solution to each of the following quadratic equations is in the nonreal complex numbers. Report these number in a + bi *form (see Section 7.5).*

27. $x^2 - x + 5 = 0$
28. $x^2 + x + 7 = 0$
29. $x^2 = x - 7$
30. $x^2 = 2(x - 1)$
31. $(x - 2)^2 = -9$
32. $(x + 3)^2 = -16$
33. $2x^2 = x - 15$
34. $3x = 2x^2 + 12$
35. $6 = (x + 1)^2 + 7$
36. $3 - x^2 = 5$

Solve each second-degree equation in one variable.

37. $2x^2 - 3x = 1 - 2x$
38. $5 - x = 12 - x^2$
39. $-x^2 - 5x - 1 = 0$
40. $5x + 1 = x^2 + 2x - 1$
41. $1 = 3x^2 + 5x - 6$
42. $7 = x(x - 3) - 2$
43. $0.5x^2 + 1.2x - 3 = 0$
44. $0.1x^2 - 0.2x = 0.5$
45. $\frac{1}{2}x^2 - \frac{3}{2}x = \frac{7}{2}$
46. $x^2 + \frac{3}{2}x = \frac{5}{2}$
47. $x^2 - \frac{3}{2}x = \frac{7}{3}$
48. $\frac{x^2 - 7}{3} = \frac{x}{2}$
49. $x^2 - (2x + 1)(x + 1) = 0$
50. $x^3 = x^2(x + 1) - 4$
51. $(x + 2)(x - 1) = -x^2 - 2$
52. $2.5x^2 - x - 4 = 0$
53. $(x + 2)(x - 1) = x^2 - 2$
54. $(x - 3)(3 - x) = 6x - x^2 - 9$

8.3 ■ ■ ■ QUADRATIC FORMS AND OTHER EQUATIONS

Now that we know how to solve *any* first-degree equation in one variable and *any* second-degree equation in one variable, we will use this ability to solve some related equations. These equations, such as $x^4 - 3x^2 - 4 = 0$, have the basic form of a quadratic equation. Solving equations that have quadratic form makes use of the very important technique of substitution. Also, we will solve rational expression equations and radical expression equations that yield quadratic equations in the solution process. Our work in this section uses and combines several previous topics. (See Appendix B for a calculator program for the quadratic formula.)

Quadratic Forms An equation such as $x^4 - 3x^2 - 1 = 0$ is known as an **equation in quadratic form**. Although the equation is a fourth-degree equation in one variable, it has a *form* similar to a second-degree equation in one variable (a quadratic equation). If we let $u = x^2$, then the equation appears as

$$u^2 - 3u - 1 = 0$$

which is a quadratic equation in u. Some additional examples of quadratic forms are shown here along with the substitution used to write each one as a quadratic equation.

Quadratic form	*Substitution*	*Quadratic equation*
$2x^4 - 3x^2 - 3 = 0$	Let $u = x^2$	$2u^2 - 3u - 3 = 0$
$x^6 - x^3 - 2 = 0$	Let $u = x^3$	$u^2 - u - 2 = 0$
$x^{-2} + 2x^{-1} - 3 = 0$	Let $u = x^{-1}$	$u^2 + 2u - 3 = 0$
$x - 4x^{1/2} - 1 = 0$	Let $u = x^{1/2}$	$u^2 - 4u - 1 = 0$
$(x - 3)^2 + 2(x - 3) + 1 = 0$	Let $u = x - 3$	$u^2 + 2u + 1 = 0$

Notice that in each of these examples, after making the substitution we have a quadratic equation in u. Consider the equation $x^4 + x - 1 = 0$. Try a substitution, say $u = x^2$. The equation becomes $u^2 + \sqrt{u} - 1 = 0$, which is *not* a quadratic equation in u. No matter what substitution we try for u, this equation does not yield a quadratic equation in u. This equation is not a quadratic form. In all the examples of quadratic form equations we have listed, the *exponent* for the lead term is twice the exponent for the second term. We now use quadratic forms to extend the types of polynomial equations that we can solve. To learn how to solve polynomial equations that have quadratic form, work through the following examples.

EXAMPLE 1 Solve: $x^4 - 3x^2 - 4 = 0$

Solution Let $u = x^2$ and use substitution to get: $u^2 - 3u - 4 = 0$

Now, using factoring we get: $(u - 4)(u + 1) = 0$

$$u - 4 = 0 \quad \text{or} \quad u + 1 = 0$$
$$u = 4 \qquad\qquad u = -1$$

We now have the numbers that solve $u^2 - 3u - 4 = 0$. To find the numbers that solve the original equation $x^4 - 3x^2 - 4 = 0$, we use *back-substitution*:

$$u = 4 \quad \text{or} \quad u = -1$$
$$x^2 = 4 \qquad x^2 = -1 \qquad u \text{ represents } x^2.$$
$$x = 2 \text{ or } x = -2 \qquad x = i \text{ or } x = -i$$

The solution to $x^4 - 3x^2 - 4 = 0$ is the collection of numbers -2, 2, $-i$, and i. ■

The substitution approach produces only the *real* numbers that solve polynomial equations of degree $2n$, where $n = 3, 5, \ldots$. Consider the following examples.

EXAMPLE 2 Solve: $x^6 - x^3 - 2 = 0$

Solution Let $u = x^3$ to get: $u^2 - u - 2 = 0$

Using factoring, we get: $(u - 2)(u + 1) = 0$

$$u - 2 = 0 \quad \text{or} \quad u + 1 = 0$$
$$u = 2 \qquad\qquad u = -1$$

Using back-substitution, we get: $x^3 = 2$ or $x^3 = -1$

Taking the cube root, we get: $x = \sqrt[3]{2}$ $\quad x = \sqrt[3]{-1} = -1$

The real numbers in the solution to the equation $x^6 - x^3 - 2 = 0$ is the collection of numbers: $\sqrt[3]{2}$ and -1. ■

Using the Graphics Calculator

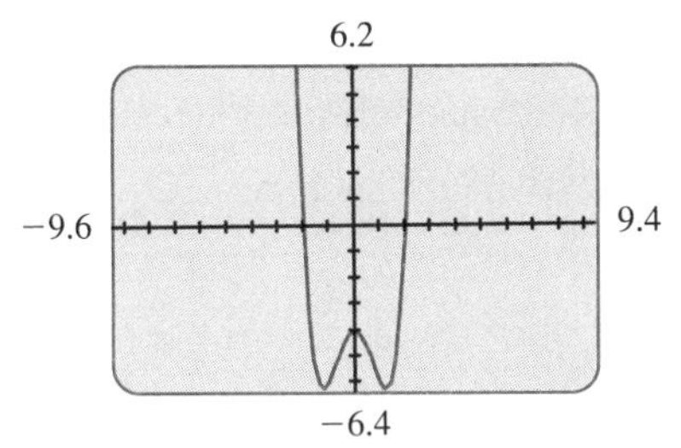

FIGURE 1

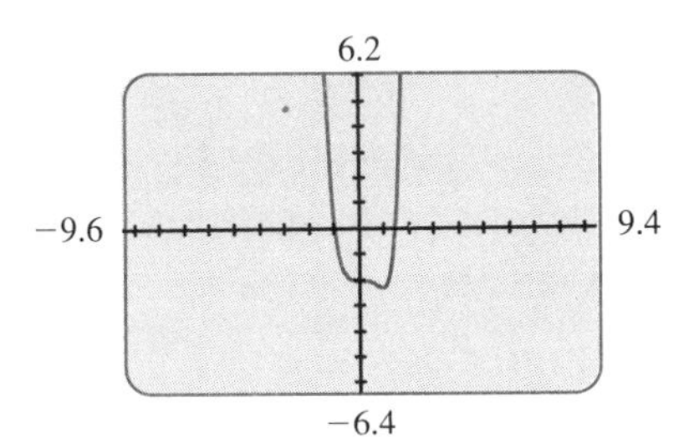

FIGURE 2

To increase our understanding of the solution to the equations in Examples 1 and 2, we can use the graph screen to display evaluations of the polynomials that are the left-hand members of the equations. For Y1, type X^4−3X²−4. View the graph screen using the Window 9.4 range settings. The evaluations of the polynomial $x^4 - 3x^2 - 4$ for various values of x are displayed in Figure 1. We are interested in x-values that result in an evaluation of 0. Use the TRACE feature and move the cursor right until the message reads X=2 Y=0. When x is replaced by 2, the polynomial $x^4 - 3x^2 - 4$ evaluates to 0. Press the left arrow key until the message reads X=⁻2 Y=0. When x is replaced by -2, the polynomial $x^4 - 3x^2 - 4$ evaluates to 0. Notice that only the real numbers that solve the equation are displayed.

For Y1, type X^6−X^3−2. Then view the graph screen. As Figure 2 shows, the evaluations of the polynomial $x^6 - x^3 - 2$ for various values of x are displayed. We are interested in x-values that result in an evaluation of 0. Use the TRACE feature and move the cursor right until the message reads X=1.2 Y=⁻.742016. Zoom in on the point. Press the TRACE key again and move the cursor until the Y-value in the message is near 0. Continue to zoom in. The x-value is approximately 1.26. The exact value is $\sqrt[3]{2}$. Enter the Window 9.4 range settings and then use the same process to see the other real number that results in $x^6 - x^3 - 2$ evaluating to 0.

In the next example the quadratic formula must be used to solve the equation in u after the substitution.

EXAMPLE 3 Solve: $x^4 - 4x^2 + 1 = 0$

Solution Let $u = x^2$ to get the quadratic form: $u^2 - 4u + 1 = 0$

Use the quadratic formula: $a = 1$, $b = -4$, $c = 1$

Compute the discriminant: $b^2 - 4ac = (-4)^2 - 4(1)(1) = 16 - 4 = 12$

Write the solution: $\dfrac{4 + \sqrt{12}}{2}$ and $\dfrac{4 - \sqrt{12}}{2}$

Reduce the numbers: $2 + \sqrt{3}$ and $2 - \sqrt{3}$

Use back-substitution:

$$x^2 = 2 + \sqrt{3} \qquad\qquad x^2 = 2 - \sqrt{3}$$

$$x = \sqrt{2 + \sqrt{3}} \quad x = -\sqrt{2 + \sqrt{3}} \quad x = \sqrt{2 - \sqrt{3}} \quad x = -\sqrt{2 - \sqrt{3}}$$

The solution to $x^4 - 4x^2 + 1 = 0$ is:

$$\sqrt{2 + \sqrt{3}}, \quad -\sqrt{2 + \sqrt{3}}, \quad \sqrt{2 - \sqrt{3}}, \quad \text{and} \quad -\sqrt{2 - \sqrt{3}}$$

■

GIVE IT A TRY

Solve each equation using the back-substitution approach.

1. $x^4 - 8x^2 + 15 = 0$ **2.** $x^4 - 5x^2 = -3$ **3.** $x^6 - 2x^3 = 3$

4. Use the graphics calculator to check your work in Problems 1–3.

Other Quadratic Forms Quadratic forms and the back-substitution technique can also be used with equations containing exponents such as -2 and $\frac{1}{3}$. Consider the following examples.

EXAMPLE 4 Solve: $x^{-2} - 3x^{-1} = 4$

Solution Although we could use the meaning of the negative exponents to write the equation as: $\frac{1}{x^2} - \frac{3}{x} = 4$, we will instead use the fact that we have a quadratic form.

Let $u = x^{-1}$ to get the quadratic equation:			$u^2 - 3u = 4$
Standard form:			$u^2 - 3u - 4 = 0$
Use factoring to get:			$(u - 4)(u + 1) = 0$
Use the zero property:	$u - 4 = 0$	or	$u + 1 = 0$
	$u = 4$		$u = -1$

The solution to $u^2 - 3u = 4$ is 4 and -1.

Use back-substitution:	$x^{-1} = 4$	and	$x^{-1} = -1$
So, we have:	$\frac{1}{x} = 4$		$\frac{1}{x} = -1$
Multiplying by x:	$1 = 4x$		$1 = -x$
So,	$\frac{1}{4} = x$		$-1 = x$

After checking that these numbers do not result in a denominator evaluating to 0, we conclude that the solution to $x^{-2} - 3x^{-1} = 4$ is $\frac{1}{4}$ and -1. ■

Solve the equation $\frac{1}{x^2} - \frac{3}{x} = 4$ to see that we get the same numbers in the solution as we found in Example 4. (Multiply each side by LCD, x^2.)

EXAMPLE 5 Solve: $x - x^{1/2} = 4$

Solution Although we could use the meaning of rational number exponents to write the equation as $x - \sqrt{x} = 4$, we will instead use the fact that we have a quadratic form.

Let $u = x^{1/2}$ to get the quadratic equation: $u^2 - u = 4$

Standard form: $u^2 - u - 4 = 0$

Use the quadratic formula: $a = 1, b = -1, c = -4$

Compute the discriminant: $b^2 - 4ac = (-1)^2 - 4(1)(-4) = 1 + 16 = 17$

Write the solution: $\frac{1 + \sqrt{17}}{2}$ and $\frac{1 - \sqrt{17}}{2}$

The solution to $u^2 - u = 4$ is $\frac{1 + \sqrt{17}}{2}$ and $\frac{1 - \sqrt{17}}{2}$.

Use back-substitution: $x^{1/2} = \frac{1 + \sqrt{17}}{2}$ and $x^{1/2} = \frac{1 - \sqrt{17}}{2}$

$$\sqrt{x} = \frac{1 + \sqrt{17}}{2} \qquad \sqrt{x} = \frac{1 - \sqrt{17}}{2}$$

Squaring, we get: $x = \left(\frac{1 + \sqrt{17}}{2}\right)^2 \qquad x = \left(\frac{1 - \sqrt{17}}{2}\right)^2$

$$x = \frac{9 + \sqrt{17}}{2} \qquad x = \frac{9 - \sqrt{17}}{2}$$

Checking for extraneous roots, we use the calculator to approximate these two numbers. We get

$$\frac{9 + \sqrt{17}}{2} \approx 6.56 \quad \text{and} \quad \frac{9 - \sqrt{17}}{2} \approx 2.44$$

Now, $6.56 - \sqrt{6.56} \approx 4$. So, we conclude that $\frac{9 + \sqrt{17}}{2}$ solves $x - x^{1/2} = 4$. Next, $2.44 - \sqrt{2.44} \approx 0.88$ (not the desired value of 4). So, we conclude that $\frac{9 - \sqrt{17}}{2}$ is extraneous. The solution to $x - x^{1/2} = 4$ is $\frac{9 + \sqrt{17}}{2}$. ■

Using the Graphics Calculator

To check our work in Example 5, type `X-X^(1/2)` for `Y1`. Store the number `(9+√17)/2` in memory location `X`. As you can see, this number is approximately 6.56. Display the resulting value in `Y1`. The number `4` is output, indicating that $\frac{9 + \sqrt{17}}{2}$ is in the solution. Now, store the number `(9-√17)/2` in memory location `X`. As you can see, this number is approximately 2.44. Display the resulting value in `Y1`. The number `0.8768943744` is output, not 4. This indicates that $\frac{9 - \sqrt{17}}{2}$ is not in the solution—it is an extraneous root.

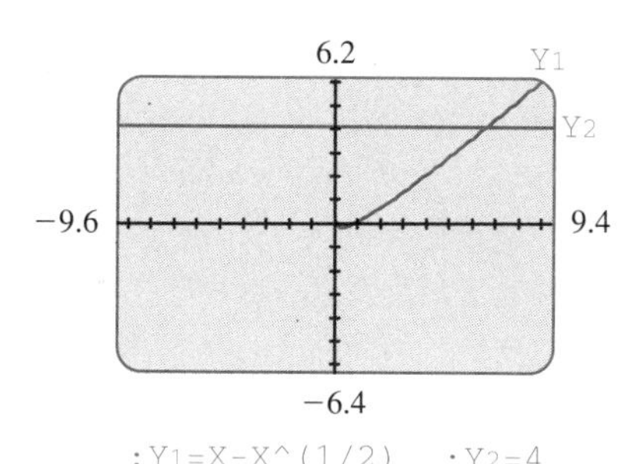

FIGURE 3

To increase our understanding of the solution in Example 5, for Y2 type 4. View the graph screen using the Window 9.4 range settings. Evaluations of the radical expression $x - x^{1/2}$ are displayed for various replacements of the x-variable (see Figure 3). Press the TRACE key, and then use the right arrow key. Notice that when the x-value is about 2.44, the evaluation is about 0.86, not 4. So, $\frac{9-\sqrt{17}}{2} \approx 2.44$ is not in the solution to the equation. Continue to move the cursor to the right. When the x-value is 6.6 $\left(\text{or, from algebra, exactly } \frac{9+\sqrt{17}}{2}\right)$, the radical expression evaluates to 4. This means that $\frac{9+\sqrt{17}}{2}$ is in the solution to the equation $x - \sqrt{x} = 4$.

GIVE IT A TRY

Solve each rational- or radical-expression equation using quadratic forms and substitution.

5. $x^{-2} = 2 - x^{-1}$ **6.** $x^{2/3} = x^{1/3} + 6$ **7.** $x + \sqrt{x} = 1$

8. Use the graphics calculator to check your work in Problems 5–7.

Rational Expression Equations In Section 6.4 we learned how to solve a rational expression equation by multiplying each side of the equation by the LCD (least common denominator). We now consider solving this type of equation, where a quadratic equation (requiring the quadratic formula) is produced in the equation-solving process.

EXAMPLE 6 Solve: $\frac{1}{x-1} + \frac{1}{x+2} = 1$

Solution We start by multiplying each side of the equation by the LCD, $(x-1)(x+2)$, to get:

$$\frac{\cancel{(x-1)}(x+2)}{\cancel{x-1}} + \frac{(x-1)\cancel{(x+2)}}{\cancel{x+2}} = 1(x-1)(x+2)$$

Reduce, to get: $x + 2 + x - 1 = 1(x-1)(x+2)$

$$2x + 1 = x^2 + x - 2$$

Standard form: $x^2 - x - 3 = 0$

Use the quadratic formula: $a = 1,\ b = -1,\ c = -3$

Compute the discriminant: $b^2 - 4ac = (-1)^2 - 4(1)(-3) = 1 + 12 = 13$

Write the solution: $\frac{1+\sqrt{13}}{2}$ and $\frac{1-\sqrt{13}}{2}$

Checking for extraneous roots, does either of these numbers result in a 0 denominator in the original equation? No. Thus, the solution to the equation $\frac{1}{x-1} + \frac{1}{x+2} = 1$ is the collection of numbers

$$\frac{1+\sqrt{13}}{2} \quad \text{and} \quad \frac{1-\sqrt{13}}{2}$$

■

Using the Graphics Calculator

To check our work in Example 3, for `Y1` type `1/(X−1)+1/(X+2)`. Store the number `(1+√13)/2` in memory location `X`. As you can see, this number is approximately 2.3. Display the resulting value in `Y1`. Is the number 1? If not, an error is present. Now, store the number `(1−√13)/2` in memory location `X`. As you can see, this number is approximately −1.3. Display the resulting value in `Y1`. Is the number 1? If not, an error is present.

Radical Expression Equations In Section 7.4 we learned how to solve a radical expression equation by raising each side of the equation to the *n*th power. We now consider solving this type of equation where a quadratic equation (requiring the quadratic formula) is produced in the equation-solving process.

EXAMPLE 7 Solve: $\sqrt{x+3} = x - 1$

Solution We start by squaring each side to get:

$$\left(\sqrt{x+3}\right)^2 = (x-1)^2$$
$$x + 3 = x^2 - 2x + 1$$

Standard form: $x^2 - 3x - 2 = 0$

Use the quadratic formula: $a = 1 \quad b = -3 \quad c = -2$

Compute the discriminant: $b^2 - 4ac = (-3)^2 - 4(1)(-2) = 9 + 8 = 17$

Write the solution: $\frac{3-\sqrt{17}}{2}$ and $\frac{3+\sqrt{17}}{2}$

Check Using the calculator to check for extraneous roots, we find that

$$\frac{3+\sqrt{17}}{2} \approx 3.56 \quad \text{and} \quad \frac{3-\sqrt{17}}{2} \approx -0.51$$

Replacing the variable by $\frac{3-\sqrt{17}}{2}$ yields a false arithmetic sentence, since we get a square root equal to a negative number. Thus, only $\frac{3+\sqrt{17}}{2}$ solves the original equation.

So, $\sqrt{x+3} = x - 1$ has as its solution the number $\frac{3+\sqrt{17}}{2}$. ■

EXAMPLE 8 Solve: $\sqrt{2x+1} + \sqrt{x} = 2$

Solution We start by isolating one of the radicals:

$$\begin{aligned} \sqrt{2x+1} + \sqrt{x} &= 2 \\ -\sqrt{x} \quad &\quad -\sqrt{x} \\ \hline \sqrt{2x+1} &= 2 - \sqrt{x} \end{aligned}$$

Square each side: $\left(\sqrt{2x+1}\right)^2 = \left(2 - \sqrt{x}\right)^2$

$$2x + 1 = 4 - 4\sqrt{x} + x$$

Isolate the radical: $\begin{aligned} -x - 4 \quad & -4 \quad -x \\ \hline x - 3 &= -4\sqrt{x} \end{aligned}$

Square each side: $(x-3)^2 = \left(-4\sqrt{x}\right)^2$

$$x^2 - 6x + 9 = 16x$$

Standard form: $x^2 - 22x + 9 = 0$

Use the quadratic formula: $a = 1 \quad b = -22 \quad c = 9$

Compute the discriminant: $b^2 - 4ac = (-22)^2 - 4(1)(9) = 484 - 36 = 448$

Write the solution: $\dfrac{22 + \sqrt{448}}{2}$ and $\dfrac{22 - \sqrt{448}}{2}$

Reducing, we get: $11 + 4\sqrt{7}$ and $11 - 4\sqrt{7}$.

Check Using the calculator to check for extraneous roots, we find that these numbers are approximately 0.42 and 21.58. Now, replacing x by 0.42 yields the arithmetic sentence $\sqrt{2(0.42)+1} + \sqrt{0.42} = 2$. Using the calculator, we get the approximations $1.36 + 0.65 = 2.01$. We conclude that $11 - 4\sqrt{7}$ solves the original equation.

Replacing x by 21.58, we get the arithmetic sentence

$$\sqrt{2(21.58)+1} + \sqrt{21.58} = 2$$

Using the calculator, we get the approximations $6.65 + 4.65 = 11.3$. We conclude that $11 + 4\sqrt{7}$ does not solve the original equation. Thus, $\sqrt{2x+1} + \sqrt{x} = 2$ has as its solution the number $11 - 4\sqrt{7}$. ■

Using the Graphics Calculator

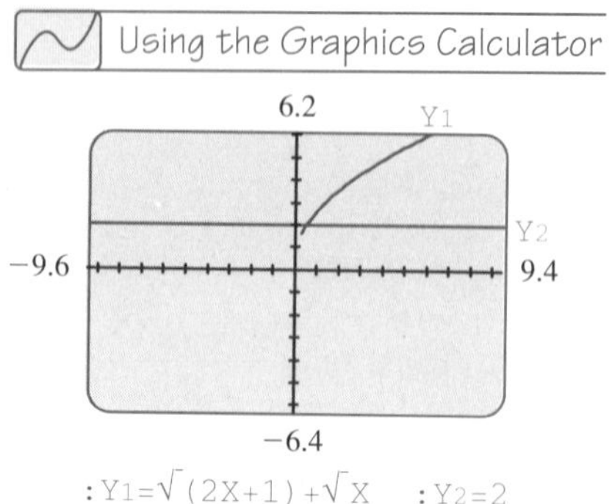

FIGURE 4

To increase our understanding of the solution found in Example 8, type `√(2X+1) + √X` for `Y1` and type `2` for `Y2`. View the graph screen using the Window 9.4 range settings. As the graphs in Figure 4 show, the radical expression $\sqrt{2x+1} + \sqrt{x}$ evaluates to 2 for an x-value just to the right of 0. That x-value is approximately 0.42 or exactly $11 - 4\sqrt{7}$. To see that $11 + 4\sqrt{7}$ (which is approximately 21.58) does *not* solve the equation, adjust the range settings to `Xmin=0`, `Xmax=30`, `Ymin=0`, `Ymax=15`, and view the graph screen. Press the `TRACE` key and use the right arrow key. When the x-value is about 21, the evaluation for the radical expression is about 11 (not 2).

GIVE IT A TRY

Solve each rational- or radical-expression equation.

9. $\frac{1}{x+1} = 2 - \frac{1}{x-1}$ **10.** $\sqrt{3-x} = x + 1$ **11.** $\sqrt{x-2} + \sqrt{x} = 2$ 

■ SUMMARY

In this section, we have used the quadratic formula in solving equations that have quadratic form. The process of solving equations that have quadratic form introduces the technique of substitution. Quadratic forms extend the types of polynomial equations in one variable that we can solve. Quadratic forms can also be found in selected rational-expression equations, such as

$$x^{-2} - x^{-1} - 2 = 0$$

and in selected radical expression equations, such as $x - x^{1/2} = 1$.

In solving rational expression equations and radical expression equations, the quadratic formula is often required in the process of solving the equations. With this type of equation we must check for extraneous roots (numbers generated in the solution process) that do not actually solve the original equation.

■ ***Quick Index***

8.3 ■ ■ ■ EXERCISES

Use the substitution technique to solve each polynomial equation. Report only the real numbers in the solution. Use the graphics calculator to check your work.

1. $x^4 + 3x^2 - 4 = 0$ [*Hint*: Let $u = x^2$.]
2. $x^6 - 3x^3 - 10 = 0$
3. $2x^4 - 3x^2 - 2 = 0$
4. $2x^6 - 3x^3 - 2 = 0$
5. $x^4 - x^2 - 1 = 0$
6. $2x^4 - x^2 = 4$

Use the substitution technique to solve each rational-expression or radical-expression equations. Report only the real numbers in the solution. Use the graphics calculator to check your work.

7. $x^{-2} - 3x^{-1} = 10$ [*Hint*: Let $u = x^{-1}$.]
8. $x^{-4} - x^{-2} = 2$ [*Hint*: Let $u = x^{-2}$.]
9. $x^{-2} - x^{-1} - 6 = 0$
10. $2x^{-2} + x^{-1} - 6 = 0$
11. $x = 3x^{1/2} + 10$ [*Hint*: Let $u = x^{1/2}$.]
12. $x^{1/2} = 3x^{1/4} + 10$ [*Hint*: Let $u = x^{1/4}$.]
13. $x - x^{1/2} = 3$
14. $x + x^{1/2} = 3$
15. $x^{2/3} - 3x^{1/3} - 4 = 0$
16. $\sqrt{x} + \sqrt[4]{x} = 5$ [*Hint*: Let $u = x^{1/4}$.]

Solve each rational-expression or radical-expression equation. Report only the real numbers in the solution. Use the graphics calculator to check your work.

17. $\frac{x-2}{x+3} = \frac{3}{x}$
18. $\frac{x+1}{2x+3} = \frac{x-2}{x-1}$
19. $\frac{1}{x+3} + \frac{1}{x-1} = 1$
20. $\frac{1}{x+3} - \frac{1}{x-1} = 1$
21. $\sqrt{x-3} = x + 1$
22. $\sqrt{2x+1} = x$
23. $\sqrt{x+1} - \sqrt{x} = 3$
24. $\sqrt{x+1} + \sqrt{x} = 3$
25. $\sqrt{3x+4} = \sqrt{x} + 2$
26. $\sqrt{3-y} + \sqrt{1-y} = -1$
27. The substitution approach can be used to solve an equation such as $(x-2)^4 - 3(x-2)^2 - 4 = 0$. Let $u = (x-2)^2$, to get $u^2 - 3u - 4 = 0$. Solve this equation in u and then use back-substitution to report the solution to the equation $(x-2)^4 - 3(x-2)^2 - 4 = 0$.
28. The substitution approach can be used to solve equations such as $(a - 2a^{-1})^2 - 5(a - 2a^{-1}) + 4 = 0$. Let $u = (a - 2a^{-1})$, to get $u^2 - 5u + 4 = 0$. Solve this equation in u and then use back-substitution to report the solution to $(a - 2a^{-1})^2 - 5(a - 2a^{-1}) + 4 = 0$.

DISCOVERY

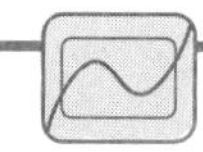

Key Points for Rational Expression Inequalities

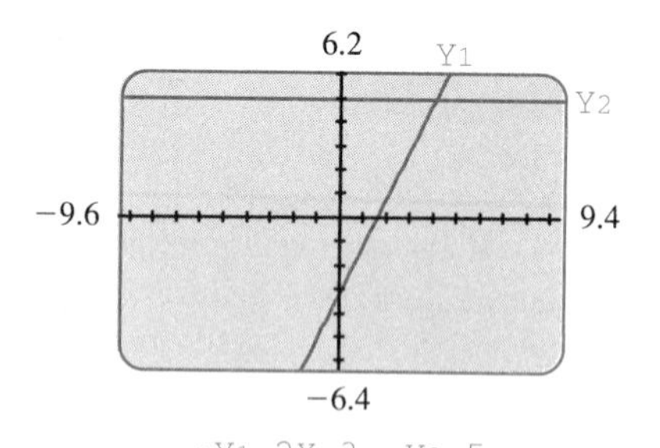

FIGURE 1

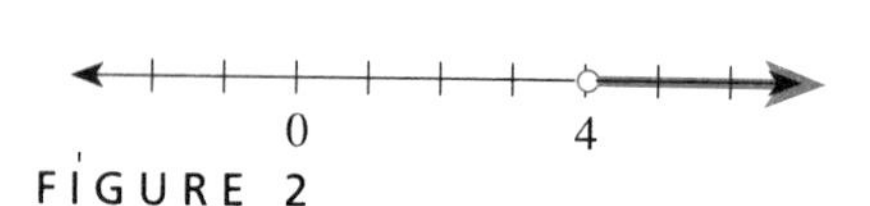

FIGURE 2

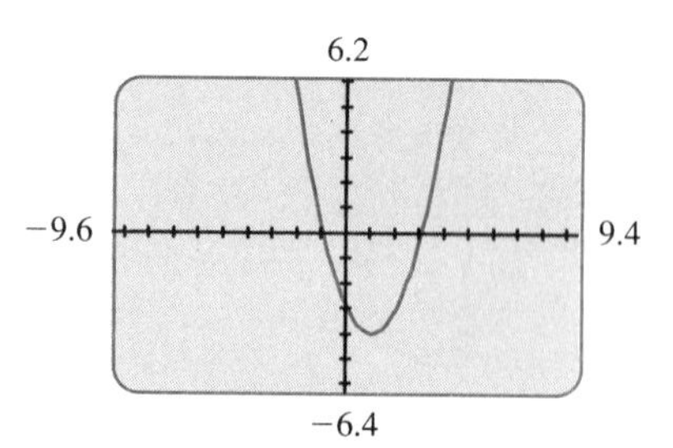

FIGURE 3

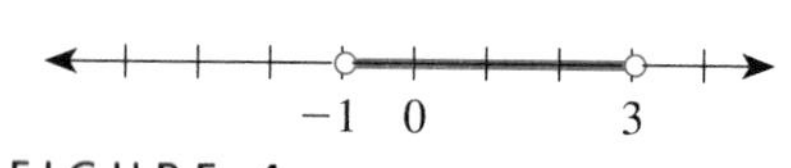

FIGURE 4

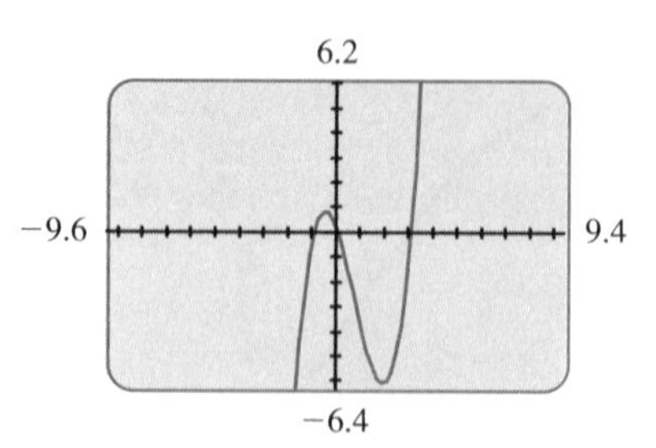

FIGURE 5

In the next section, we will solve quadratic inequalities whose key points are irrational numbers that contain radicals, such as $\frac{3 + \sqrt{5}}{2}$. In that section, we will also discuss rational expression inequalities. Here, we review information on solving inequalities in one variable. We also want to investigate the key points of a rational expression inequality. We start with the review.

First-degree inequalities in one variable, such as $2x - 3 > 5$, typically have a single key point—the solution to the equation $2x - 3 = 5$. Using the graphics calculator, we can see that this is indeed the case. For Y1, type 2X−3 and for Y2 type 5. Use the Window 9.4 range settings to view the graph screen. Figure 1 shows the evaluations of the polynomial $2x - 3$ that are displayed for various replacements of x. Also, a horizontal line representing 5 is drawn. Press the TRACE key, and then move the cursor right. Notice that all of these x-values result in the polynomial $2x - 3$ evaluating to a number less than 5. At the intersection point of the graph and the horizontal line, we have a key x-value. For x-values greater than (to the right of) this x-value, the polynomial $2x - 3$ evaluates to a number greater than 5. For x-values less than this key x-value, the polynomial $2x - 3$ evaluates to a number less than 5. Using algebra, and solving the equation $2x - 3 = 5$, we find that the key point is 4. We show the solution to the inequality on a number-line graph, as shown in Figure 2.

A second-degree inequality in one variable, such as $x^2 - 2x - 3 < 0$, usually has two key points—the solution to the equation $x^2 - 2x - 3 = 0$. Using the graphics calculator, we can see that this is indeed the case. For Y1, type X²−2X−3. View the graph screen as shown in Figure 3. The evaluations of the polynomial $x^2 - 2x - 3$ are displayed for various replacements of x. We want the x-values that result in the polynomial evaluating to a number less than 0. Press the TRACE key, and then move the cursor to the right. All of these x-values result in the polynomial $x^2 - 2x - 3$ evaluating to a number less than 0. At the intersection point of the graph and the x-axis, we have a key x-value. For x-values greater than (to the right of) this x-value, the polynomial $x^2 - 2x - 3$ evaluates to a number greater than 0. Now, move the cursor left to the leftmost intersection point with the x-axis. For x-values less than this key x-value, the polynomial $x^2 - 2x - 3$ evaluates to a number greater than 0. Using algebra, and solving the equation $x^2 - 2x - 3 = 0$, we find the two key points, -1 and 3. We show the solution to the inequality on a number line graph, as in Figure 4.

How many key points can a third-degree inequality have? Consider the inequality $x^3 - 2x^2 - 3x < 0$. For Y1, type X³−2X²−3X. View the graph screen (see Figure 5). The evaluations of the polynomial $x^3 - 2x^2 - 3x$ are displayed for various replacements of x. Here, we want the x-values that result in the polynomial evaluating to a number less than 0. Solving the equation (using factoring), we get a solution of -1, 0, and 3. As you can see, for the x-values less than -1, the evaluations of the polynomial are less than

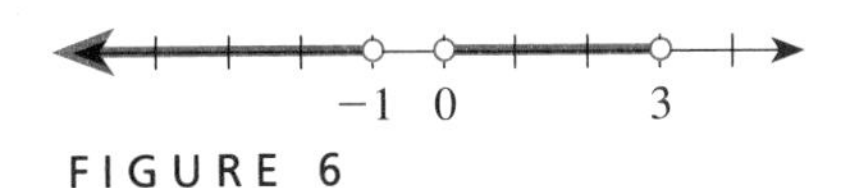

FIGURE 6

0. Also, the polynomial evaluates to a number less than 0 for x-values between 0 and 3. We show the solution to the inequality on a number-line graph, (see Figure 6).

In general, the key points for polynomial inequalities are the numbers that solve the related equation. Now, consider the rational expression inequality $\frac{1}{x} > 2$. Here, the solution to the equation $\frac{1}{x} = 2$ is $\frac{1}{2}$. For `Y1` type `1/X` and for `Y2` type `2`. View the graph screen. As Figure 7 shows, the evaluations of the rational expression $\frac{1}{x}$ are displayed, and a horizontal line is drawn for 2. We are interested in the x-values that result in $\frac{1}{x}$ evaluating to a number greater than 2. For x-values less than 0, the evaluations are less than 2. For x-values to the right of 0, the evaluations are greater than 2, up to a point—namely, the x-value $\frac{1}{2}$ $\left(\text{the solution to the equation } \frac{1}{x} = 2\right)$. There seems to be two key points. As expected, one key point is the solution to the related equation, namely, $\frac{1}{2}$. Do you know what the other key point is? It is the number that results in the denominator evaluating to 0, the restricted value for $\frac{1}{x}$. A number-line graph of the solution to the inequality, $\frac{1}{x} > 2$, is shown in Figure 8.

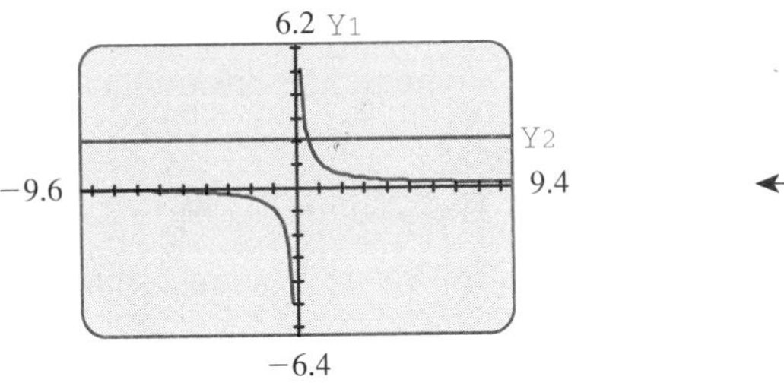

FIGURE 7

$\frac{1}{2}$

0 1

FIGURE 8

Exercises

1. In your own words, explain how to obtain the key points for a rational expression inequality. Test your conclusion with the inequality

$$\frac{1}{x-3} \geq 1$$

Use the graphics calculator to confirm your conclusion.

2. Draw a number-line graph of the solution to the inequality in Problem 1. Report the interval notation for the solution.

8.4 ■ ■ ■ QUADRATIC INEQUALITIES AND RATIONAL EXPRESSION INEQUALITIES

In this section, we will solve second-degree inequalities in one variable, whose key points are irrational numbers. This development extends our earlier work in Section 5.6, on solving second-degree inequalities in one variable. (You may want to review that section now.) We will also solve rational expression inequalities.

Irrational Key Points The ability to approximate irrational numbers containing radical expressions (that is, to find integer bounds for irrational numbers) is often required when solving second-degree inequalities. For example, we can use a calculator to approximate a number such as $3 - \sqrt{5}$. This number is approximately 0.76. On a number-line graph, we plot a point between 0 and 1 for the number $3 - \sqrt{5}$. This type of problem helps us get a *feel* for irrational numbers and for the real number line. Work through the following example.

EXAMPLE 1 Graph the solution to $x^2 - x < 3$.

Solution **Step 1** Solve the related equation: $x^2 - 3x = 3.$

Standard form: $x^2 - x - 3 = 0$

Using the quadratic formula, we start by identifying a, b, and c: $a = 1, b = -1, c = -3.$

Compute the discriminant: $b^2 - 4ac = (-1)^2 - 4(1)(-3) = 1 - (-12) = 13$

The solution is $\frac{1 - \sqrt{13}}{2}$ and $\frac{1 + \sqrt{13}}{2}$.

These two irrational numbers are the key points.

Step 2 Plot and label the key points. To plot these points, we must first estimate the two consecutive integers between which the key points lie. Using the calculator to approximate $\frac{1 + \sqrt{13}}{2}$, we obtain 2.302775638 as the output. Thus, this key point is between the integers 2 and 3. Plot and label the point on the number line. Repeating these steps for the number $\frac{1 - \sqrt{13}}{2}$, we get the approximation -1.302775638. So, we plot and label a point between -2 and -1. The key points are labeled in Figure 1.

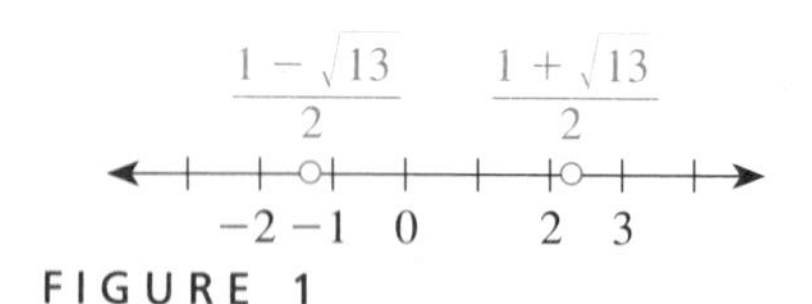

FIGURE 1

Step 3 Both dots remain open, because neither of the numbers solves $x^2 - x < 3$.

Step 4 Test a number in the leftmost interval, say -2: $(-2)^2 - (-2) < 3$. We get $4 + 2 < 3$, a false arithmetic sentence. So, we leave the interval unshaded.

Test a number in the middle interval, say 0: $(0)^2 - 0 < 3$. Here, we get a true arithmetic sentence, so we shade the middle interval. See Figure 2.

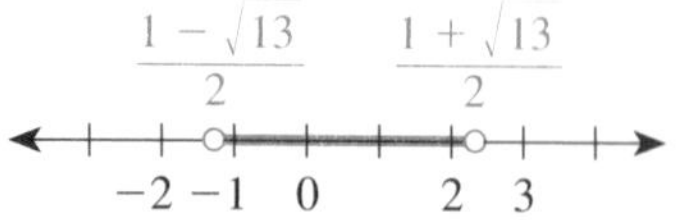

FIGURE 2

Test a number in the rightmost interval, say 3: $(3)^2 - 3 < 3$. We get $9 - 3 < 3$, a false arithmetic sentence. We leave the rightmost interval unshaded.

In interval notation, we report the solution as

$$\left(\frac{1 - \sqrt{13}}{2}, \frac{1 + \sqrt{13}}{2}\right)$$

■

REMEMBER

In interval notation, always report the smallest number first.

In Example 1 the points *between* the key points solve the inequality. Being able to find integer bounds for irrational numbers is important for finding the solution to this example. Finally, the example shows how mathematics grows, but that growth can be managed through using an organized approach: When we approached this problem, we knew how to solve second-degree equations in one variable and we knew the general approach to solving inequalities. To solve the problem, we combined these two skills.

Using the Graphics Calculator

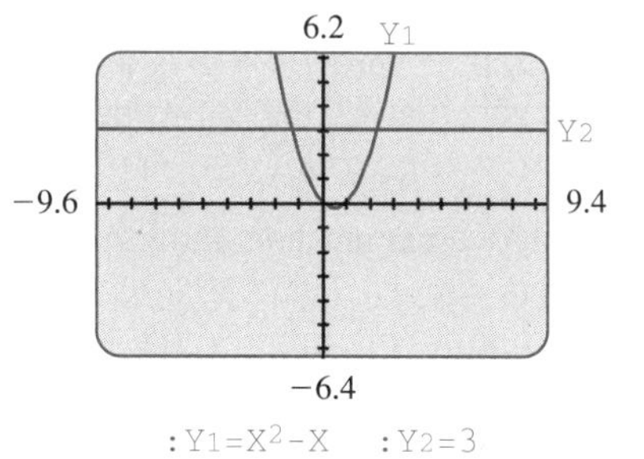

FIGURE 3

To increase our understanding of the solution to the inequality found in Example 2, for `Y1` type `X²−X`. For `Y2` type `3`. Use the Window 9.4 range settings to view the graph screen as shown in Figure 3. The solution to the inequality is the set of x-values for which the evaluations of the polynomial $x^2 - x$ are less than 3. Press the `TRACE` key and use the left arrow key until the message reads `X=-1.2 Y=2.64`. These x-values result in $x^2 - x$ evaluating to a number less than 3. Press the left arrow key again—this x-value, `-1.4`, results in $x^2 - x$ evaluating to a number larger than 3. Thus, we have a key point between -2 and -1. From algebra, we know that this key point is exactly $\frac{1 - \sqrt{13}}{2}$. Use the right arrow key to check that the other key point is between 2 and 3.

GIVE IT A TRY

1. Graph the solution to $x^2 + x - 4 \geq 0$. Report the solution in interval notation.

Complex Key Points As with first-degree inequalities in one variable, special cases occur when the related equation of a second-degree inequality produces a solution of *none* or *all*. That is, either no point is a key point or all the points are key points. With second-degree inequalities, we have yet another special case. The number-line graphs we construct are real number lines. What happens when the related second-degree equation for the inequality produces nonreal complex numbers? For example, consider the inequality $x^2 \geq -4$. Here, the related equation is $x^2 = -4$. The solution is two nonreal complex numbers $2i$

and $-2i$. The equation produces no real number key points; that is, no numbers divide the real number line into intervals. In fact, all real numbers solve the inequality $x^2 \geq -4$, because any real number squared is greater than or equal to -4 (in fact, all real numbers yield evaluations greater than or equal to 0). Consider the next example.

EXAMPLE 2 Graph the solution to $x^2 - 3x \geq -5$.

Solution

Step 1 Solve the related equation, $x^2 - 3x = -5$, to obtain the key points.

Standard form: $x^2 - 3x + 5 = 0$

Identify a, b, and c: $a = 1, b = -3, c = 5$

Compute the discriminant: $b^2 - 4ac = (-3)^2 - 4(1)(5)$
$= 9 - 20 = -11$

The solution is *no real number* [two complex numbers, $\left(3 + i\sqrt{11}\right)/2$ and $\left(3 - i\sqrt{11}\right)/2$, solve the equation].

Because our graph is the real number line, we have no key point to divide the real number line into intervals.

Step 2 Plot and label the key points. There is nothing to do in this step.

Step 3 See if the key points should be open or solid. There is nothing to do in this step.

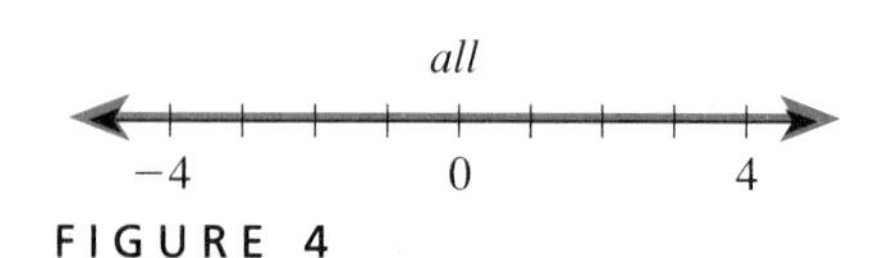

FIGURE 4

Step 4 Test a number in the one interval, say 0: $0^2 - 3(0) \geq -5$ is a true arithmetic sentence. So, shade the interval, that is, the entire number line. See Figure 4. In interval notation, we write $(-\infty, +\infty)$.

Just because the equation produces no key point, we cannot conclude the inequality has no solution. The lack of key points simply means that no number breaks up the interval (the real number line). This implies that the solution will be *all* or *none*. We must check a number to find out which of these is the solution. ■

Using the Graphics Calculator

To increase our understanding of the solution in Example 2, for `Y1` type `X²−3X`. For `Y2`, type `⁻5`. Using the Window 9.4 range settings, view the graph screen. Figure 5 shows that the evaluations of the polynomial $x^2 - 3x$ are always greater than or equal to -5.

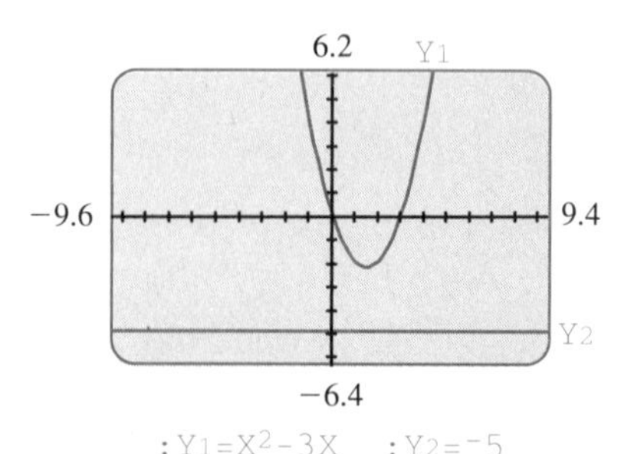

FIGURE 5

GIVE IT A TRY

2. Graph the solution to $x^2 - 5x + 7 \geq 0$.

Replacements for x in $\frac{x-2}{x+1} \geq 0$

x-value	*Inequality*
-2	$\frac{-2-2}{-2+1} \geq 0$ true
-1	$\frac{-1-2}{-1+1} \geq 0$ unknown
0	$\frac{0-2}{0+1} \geq 0$ false
1	$\frac{1-2}{1+1} \geq 0$ false
2	$\frac{2-2}{2+1} \geq 0$ true
3	$\frac{3-2}{3+1} \geq 0$ true

Rational Expression Inequalities We now consider solving an inequality in one variable that contains a rational expression. Such an inequality is $\frac{x-2}{x+1} \geq 0$. This rational expression inequality holds the place for arithmetic sentences, just as do inequalities containing polynomials—first-degree inequalities in one variable and second-degree inequalities in one variable, for example. To solve the inequality we find the numbers that replace the variable to produce true arithmetic sentences. A table of replacements for x and the resulting arithmetic sentence is shown in the table at the left. Some of these x-values result in true arithmetic sentences, while others produce false arithmetic sentences. As you may suspect by now, the numbers that solve the inequality form an interval, or intervals, when graphed on a number line. To find these intervals, we need to find the key points for the inequality. This particular inequality appears to have two key points. The related equation, $\frac{x-2}{x+1} = 0$ has only one number in its solution, namely, the number 2. This number is a key point. How do we find the other key point? It is the restricted value for the domain of definition, the number that results in the denominator evaluating to 0. In this case, it is the number -1. In general, we have the following property.

KEY POINTS FOR RATIONAL EXPRESSION INEQUALITIES

> The key points for a rational expression inequality are the numbers that solve the related equation and the numbers in the restricted set (the numbers that result in the denominator evaluating to 0).

To learn how to solve rational expression inequalities, work through the following examples.

EXAMPLE 3 Solve: $\frac{x-2}{x+1} \geq 0$

Solution

Step 1 Solve the related equation: $\frac{x-2}{x+1} = 0$

Multiply by the LCD, $x + 1$: $\frac{(x+1)(x-2)}{x+1} = 0(x+1)$

Reduce: $x - 2 = 0$

$x = 2$

From solving the equation, we get the key point 2.
The only restricted value is -1, so the other key point is -1.

Step 2 Plot and label the key points.

Step 3 Determine whether the key points should be open or solid. The key point 2 is a solid dot, because it solves the inequality. The key point -1 is an open dot, since it does not yield a true arithmetic sentence. These dots are shown in Figure 6.

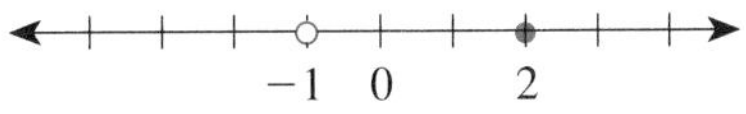

−1 0 2

FIGURE 6

Step 4 Test a number in the leftmost interval, say -2: $\frac{-2-2}{-2+1} \geq 0$ is a true arithmetic sentence. So, we shade the leftmost interval.

Test a number in the middle interval, say 0: $\frac{0-2}{0+1} \geq 0$ is a false arithmetic sentence. So, the interval is left unshaded.

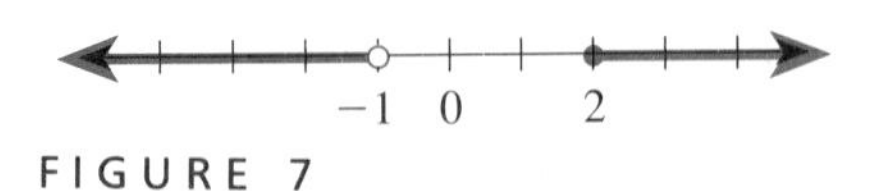

FIGURE 7

Test a number in the rightmost interval, say 3: $\frac{3-2}{3+1} \geq 0$ is a true arithmetic sentence. So, we shade the rightmost interval. Figure 7 shows the solution. In interval notation we write $(-\infty, -1) \cup [2, +\infty)$. ■

EXAMPLE 4 Solve: $\frac{3}{x+3} \geq \frac{1}{x-1}$

Solution

Step 1 Solve the related equation:

$$\frac{3}{x+3} = \frac{1}{x-1}$$

Multiply by the LCD, $(x+3)(x-1)$:

$$\frac{(x+3)(x-1)(3)}{x+3} = \frac{(x+3)(x-1)(1)}{x-1}$$

Reduce:

$$3x - 3 = x + 3$$
$$2x - 3 = 3$$
$$2x = 6$$
$$x = 3$$

From solving the equation, we get the key point 3.
The restricted values are -3 and 1.
So, the key points are -3, 1, and 3.

Step 2 Plot and label the key points.

Step 3 Determine whether the key points should be open or solid. The key point 3 is a solid dot, since it solves the inequality. The key points -3 and 1 are both open dots, because neither yields a true arithmetic sentence. As the points in Figure 8 shows, we have 4 intervals.

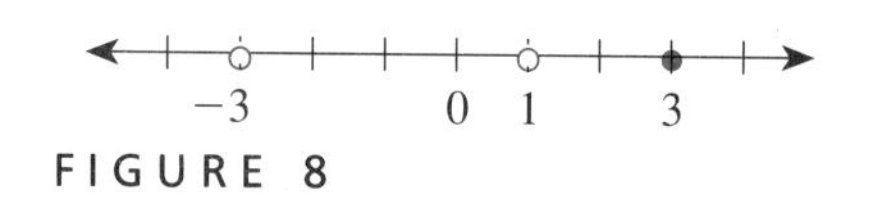

FIGURE 8

Step 4 Test a number in the leftmost interval, say -4: $\frac{3}{-4+3} \geq \frac{1}{-4-1}$, or $-3 \geq \frac{-1}{4}$ is a false arithmetic sentence. So, the leftmost interval is left unshaded.

Test a number in the next interval, say 0: $\frac{3}{0+3} \geq \frac{1}{0-1}$, or $1 \geq -1$ is a true arithmetic sentence. So, the interval is shaded.

Test a number in the next interval, say 2: $\frac{3}{2+3} \geq \frac{1}{2-1}$, or $\frac{3}{5} \geq 1$ is a false arithmetic sentence. So, this interval is left unshaded.

Test a number in the rightmost interval, say 4: $\frac{3}{4+3} \geq \frac{1}{4-1}$, or $\frac{3}{7} \geq \frac{1}{3}$ is a true arithmetic sentence. So, we shade the rightmost interval. The solution is shown in Figure 9. In interval notation we write $(-3, 1) \cup [3, +\infty)$. ■

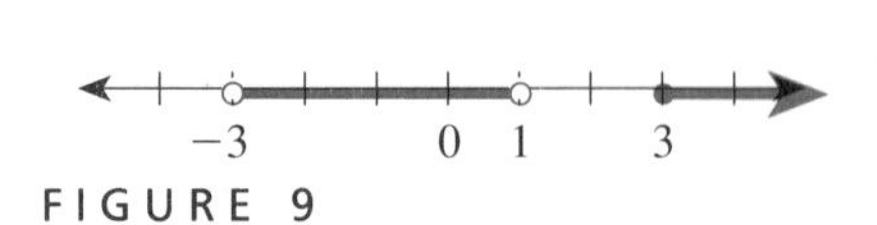

FIGURE 9

Viewing the x-values for which the evaluations of the rational expression $\frac{3}{x+3}$ are greater than the evaluations of the rational expression $\frac{1}{x-1}$ may be difficult. Using the fact that adding or subtracting an expression from an inequality yields an equivalent inequality (one with the same solution as the original), we can observe where the evaluations of $\frac{3}{x+3} - \frac{1}{x-1}$ are greater than or equal to 0. For Y1 type 3/(X+3)−1/(X−1). View the graph screen using the Window 9.4 range settings. As Figure 10 shows, the evaluations are greater than 0 when the x-values are between -3 and 1. Also, at 3 and to the right of 3, the evaluations are greater than or equal to 0.

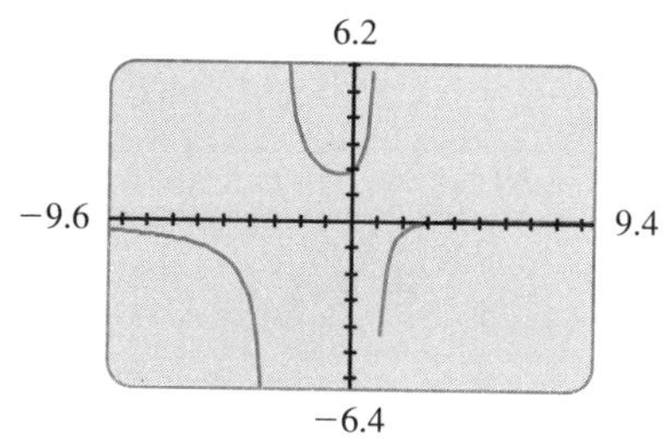

FIGURE 10

GIVE IT A TRY

3. Follow the work in Example 4 to solve the inequality

$$\frac{3}{x+3} \leq x$$

[*Hint*: Use the quadratic formula after multiplying each side by $x + 3$.]

■ SUMMARY

In this section, we have applied our knowledge of and skills in solving a second-degree equation in one variable to solving inequalities. The solution to a second-degree inequality in one variable is typically an infinite collection of numbers. To communicate these numbers, we use a one-dimensional number-line graph. To find the numbers that result is a true arithmetic sentence when each replaces the variable, we use the same procedure (process) that we used to solve first-degree inequalities. This procedure starts by solving the related equation (a second-degree equation in this case) to obtain the key points. Typically, a second-degree inequality has two key points. These points divide the number line into intervals. If one number in a interval solves the inequality, then every number in that interval solves the inequality. Thus, we select a number from each of the intervals to test the inequality. If the number from an interval produces a true arithmetic sentence, the interval is shaded. For two key points the solution will be either the numbers represented by the points outside the key points or the points inside the key points.

We also learned to solve rational expression inequalities in one variable. For these inequalities, the key points are the numbers that solve the related equation and the numbers in the restriction set (the numbers resulting in a denominator evaluating to 0). As with polynomial inequalities, the solution to a rational expression inequality is the set of numbers in the intervals determined by the key points.

■ ***Quick Index***

8.4 ■ ■ ■ EXERCISES

Draw a number-line graph of the solution to each inequality and report the interval notation for the solution. Check your work using the graphics calculator.

1. $x^2 - 5x \geq 0$
2. $x^2 - 2x - 3 \leq 0$
3. $(x - 2)(x + 3) > 0$
4. $(x + 4)(x - 1) < 0$
5. $(3x - 2)(2x + 3) \leq 0$
6. $4x^2 - 9 \geq 0$
7. $(x - 2)^2 > 9$
8. $(x - 2)^2 \leq 7$
9. $(x - 2)(x + 3) > 5$
10. $x(3 - x) \leq 5$
11. $x^2 < x^2 + 1$
12. $x^2 > x^2 + 1$
13. $2x^2 \geq x + 1$
14. $x^2 - 2x \geq 12$
15. $x^2 - 3x > -5$
16. $(x - 1)^2 < -2$
17. $2x^2 + 13x \geq -20$
18. $(x + 2)^2 > 2x + 1$
19. $x + 3 < x^2$
20. $2x - 1 \leq x^2 - x$
21. $\dfrac{3}{x + 1} \leq 2$
22. $\dfrac{x + 1}{x - 2} > 0$
23. $\dfrac{3x}{x - 4} \geq 1$
24. $\dfrac{-1}{2x + 1} \leq 2$
25. $\dfrac{3}{x - 2} + \dfrac{4}{x} > 0$
26. $\dfrac{3}{x - 2} - \dfrac{4}{x} \leq 1$
27. $\dfrac{-1}{x + 2} \leq \dfrac{2}{x - 1}$
28. $\dfrac{-5}{x + 3} > \dfrac{2}{x - 3}$
29. $\dfrac{3}{x + 2} \geq x$
30. $\dfrac{1}{x + 4} \geq \dfrac{x + 1}{x - 2}$

DISCOVERY

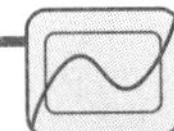

Points of View

At the beginning of Chapter 5, we discussed polynomials. A typical polynomial is $x^2 - 2x - 3$. In fact, this polynomial is a second-degree trinomial in one variable. This polynomial holds the place for (represents) number phrases of the form

$$2^2 - 2(5) - 3 \qquad \left(\frac{2}{3}\right)^2 - 3\left(\frac{2}{3}\right) - 3 \qquad \left(\sqrt{3}\right)^2 - 2\left(\sqrt{3}\right) - 3$$

and so forth. If we use the graphics calculator, we can see evaluations of this polynomial for different replacements of the variable, x. Type `X²−2X−3` for `Y1`. Press the `RANGE` key and enter the Window 9.4 range settings. View the graph screen. The curve drawn represents the evaluations of the polynomial $x^2 - 2x - 3$ for various replacements of the x-variable. Press the `TRACE` key. As shown in Figure 2, the message reads `X=0 Y=-3`.

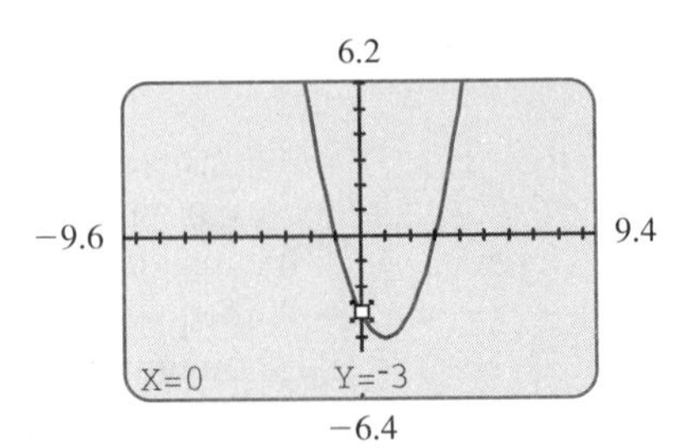

FIGURE 1

This means that replacing x by 0 results in the polynomial $x^2 - 2x - 3$ evaluating to the number -3. Use the arrow keys to complete the following table, which shows the evaluations of the polynomial $x^2 - 2x - 3$ for various replacements of the x-variable.

x	-1	0	1	2	3	4
$x^2 - 2x - 3$		-3				

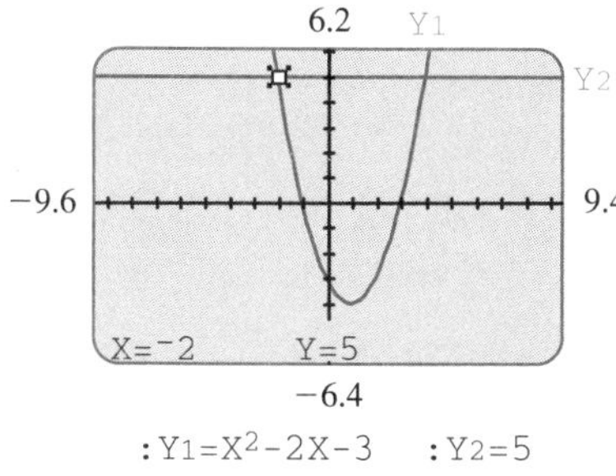

:Y1=X²-2X-3 :Y2=5

FIGURE 2

In Chapter 5, and in this chapter, we have solved second-degree equations in one variable. Such an equation is $x^2 - 2x - 3 = 5$. To solve this equation, we find the number replacements of the x-variable that produce a true arithmetic sentence. One way of viewing this problem is that we seek the x-values that result in the polynomial $x^2 - 2x - 3$ evaluating to the number 5. On the graphics calculator, press the `Y=` key and type `5` for `Y2`. View the graph screen. Use the `TRACE` key and the arrow keys to move to the left-most point of intersection of the curve representing the evaluations of the polynomial $x^2 - 2x - 3$ with the horizontal line representing the number 5. The message reads `X=-2 Y=5`. The point is shown in Figure 2. Thus, when the x-variable is replaced by -2, the polynomial $x^2 - 2x - 3$ evaluates to 5. That is, the number -2 solves the equation $x^2 - 2x - 3 = 5$. What is the other number that solves this equation?

At the end of Chapter 5, we also considered second-degree equations in two variables. An example of such an equation is $y = x^2 - 2x - 3$. The solution to this equation is an infinite set of ordered pairs of numbers. Using the graphics calculator, make sure `X²-2X-3` is entered for `Y1`, (and erase any entry for `Y2`, `Y3`, or `Y4`). View the graph screen, and press the `TRACE` key. As Figure 2 shows, the message `X=0 Y=-3` is displayed. That is, the ordered pair $(0, -3)$ solves the equation $y = x^2 - 2x - 3$. In fact, the ordered-pair name of every point on the curve drawn solves the equation $y = x^2 - 2x - 3$. Using the `TRACE` key, find the ordered pairs with y-coordinate 5 that solve the equation.

8.5 ■ ■ ■ APPLICATIONS

In this section, we consider situations for which a quadratic equation serves as the mathematical model. You may want to review the application problems presented in Section 5.5, where the model was a quadratic equation that was solved by factoring over the integers. Additionally, we will see how the quadratic formula can be used to rewrite a formula for a variable that has an exponent of 2 (for example, rewriting the formula $y = x^2 - 3x - 2$ for x).

EXAMPLE 1 The rectangular poster shown in Figure 1 has a perimeter of 28 feet. If the area measure is 42 square feet, find the width of the poster.

Solution The quantity requested is the width of the rectangle.
So, we let $x =$ the width of the rectangle.

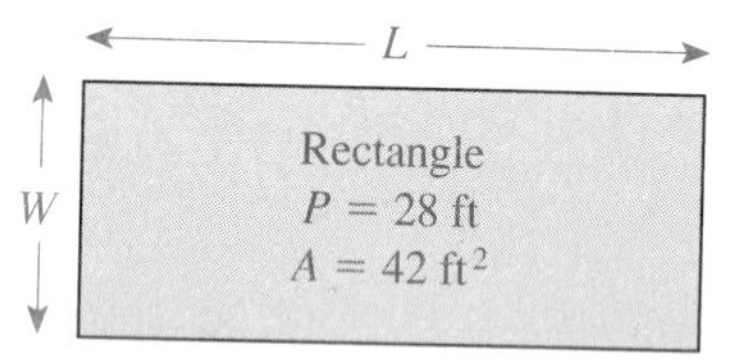

FIGURE 1

Because the perimeter is 28, we have: $2x + 2L = 28$

Multiply by $\frac{1}{2}$: $x + L = 14$

$$L = 14 - x$$

So, in terms of x, the length of the rectangle is $14 - x$.
Using $A = LW$, we have that the area of the rectangle is $(14 - x)x$.

Area is 42.

Equation: $(14 - x)(x) = 42$

Standard form: $-x^2 + 14x - 42 = 0$

Multiply by -1 to get: $x^2 - 14x + 42 = 0$

Identify a, b, and c: $a = 1, b = -14, c = 42$

Compute the discriminant: $b^2 - 4ac = (-14)^2 - 4(1)(42) = 28$

Write the solution: $\frac{14 + \sqrt{28}}{2}$ and $\frac{14 - \sqrt{28}}{2}$

Reducing the solutions, we get: $\frac{14 + \sqrt{28}}{2} = \frac{2(7 + \sqrt{7})}{2} = 7 + \sqrt{7}$

$$\frac{14 - \sqrt{28}}{2} = \frac{2(7 - \sqrt{7})}{2} = 7 - \sqrt{7}$$

The solution to $(14 - x)(x) = 42$ is $7 + \sqrt{7}$ and $7 - \sqrt{7}$. These numbers are approximately 9.65 and 4.35. Traditionally, we select the smaller number for the width, namely $7 - \sqrt{7}$. So, we report $7 - \sqrt{7}$ (≈ 4.35) for the width.

Check For a width of $7 - \sqrt{7}$ feet, the length is $14 - (7 - \sqrt{7})$, or $7 + \sqrt{7}$ feet.
The resulting perimeter is:

$$2(7 - \sqrt{7}) + 2(7 + \sqrt{7}) \quad \text{or} \quad 14 - 2\sqrt{7} + 14 + 2\sqrt{7} = 28$$

Also, the area is

$$(7 - \sqrt{7})(7 + \sqrt{7}) = 49 + 7\sqrt{7} - 7\sqrt{7} - \sqrt{49}$$
$$= 49 - 7 = 42$$

Using the approximation 4.35 for the width, we get:

Length is approximately 9.65
Perimeter is $2(4.35) + 2(9.65) = 8.7 + 19.3 = 28$
Area is $(4.35)(9.65) = 41.9775 \approx 42$ ■

Using the Graphics Calculator

To increase our understanding of Example 1, type `(14−X)X` for `Y1`. Evaluations of this polynomial represents area measures for different values for the width of the rectangle, `X`. Enter the range settings shown in Figure 2*, then view the graph screen (see Figure 3). Press the `TRACE` key and move the cursor until the message reads `X=12  Y=24`. This means that when the width is 12, the area measurement is 24. For `Y2`, type `42`. View the graph screen. Press the `TRACE` key and move the cursor until the message reads `X=4.4  Y=42.24`. Press the left arrow key. The message reads `X=4.2  Y=41.16`. The x-value that results in the polynomial $x(14 - x)$ evaluating to 42 is between 4.2 and 4.4. We know from the algebra work that it is exactly $7 - \sqrt{7}$, which is approximately 4.35.

Suppose the problem read: A rectangular poster has a perimeter of 28 feet with an area measure of 52 square feet. Find the width of the poster. To solve this problem, type `52` for `Y2`. View the graph screen, as shown in Figure 4, and use the `TRACE` feature. Are any of the evaluations of the polynomial $x(14 - x)$ as large as 52? What does this mean? What do you think is the solution to the equation $x(14 - x) = 52$? Solve the equation to see if your answer is correct!

```
RANGE
Xmin=0
Xmax=19
Xscl=5
Ymin=-5
Ymax=60
Yscl=10
Xres=1
```

FIGURE 2

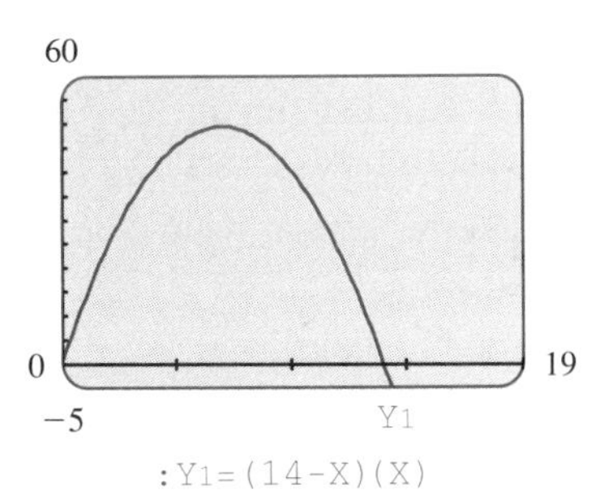

FIGURE 4

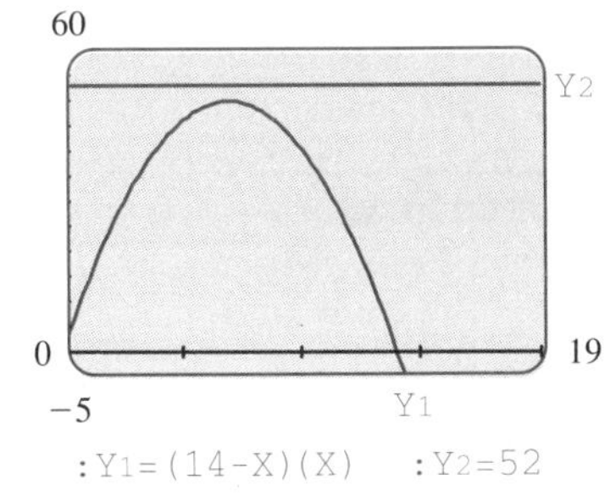

FIGURE 3

GIVE IT A TRY

1. A right triangle has a hypotenuse of 10 cm. The shortest side is 2 cm less than the other leg. Use the Pythagorean theorem, $c^2 = a^2 + b^2$, to find the length of the shortest side.

Projectiles One kind of physical situation that produces a quadratic equation is that of throwing, shooting, or dropping an object (a projectile) into the air. In the early 1600s, Galileo discovered that the distance traveled increases with the square of the elapsed time. Such motion was described (on the surface of the earth) by the general formula $h = -16t^2 + v_0t + h_0$, where v_0 is the velocity of the object at the start and h_0 is the height of the object at the start of the motion. Consider the next example.

TI-82 Note *Use `Xmax = 18.8`.

EXAMPLE 2 The height a ball reaches t seconds after being thrown straight up is given by the formula $h = -16t^2 + 64t + 5$, where t is measured in seconds and the height h is measured in feet. The physical situation is illustrated in Figure 5. How long will it take the ball to reach a height of 53 feet? How long will it take the ball to reach the ground (height zero)?

Solution Here the pattern is given by the formula: $h = -16t^2 + 64t + 5$.

To find the time the height is 53 feet, we replace h by 53: $53 = -16t^2 + 64t + 5$

We then solve this equation:

$$-16t^2 + 64t + 5 = 53 \qquad \text{If } a = b \text{ then } b = a.$$
$$-16t^2 + 64t - 48 = 0 \qquad \text{Standard form}$$
$$t^2 - 4t + 3 = 0 \qquad \text{Multiply by } \tfrac{-1}{16}.$$
$$(t - 3)(t - 1) = 0 \qquad \text{Factored form}$$

$$t - 3 = 0 \quad \text{or} \quad t - 1 = 0$$
$$t = 3 \qquad\qquad t = 1$$

Solution: 3 and 1

FIGURE 5

Replacing the t-variable by each of these numbers, we find two times at which when the height is 53 feet. The height is 53 feet at time 1 second (when the ball is going up) and again at 3 seconds (when the ball is coming down). These times are shown in Figure 6.

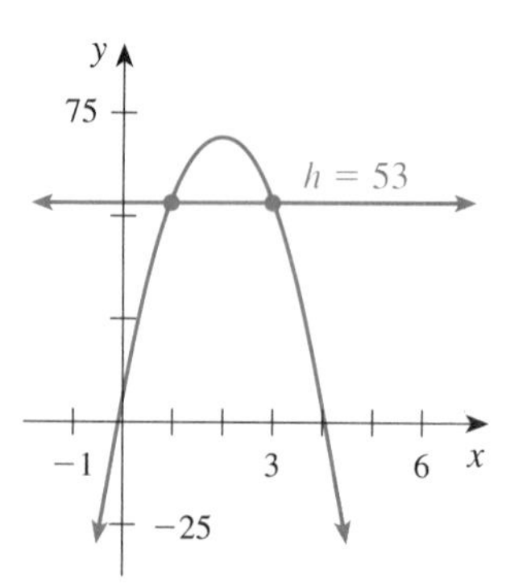

FIGURE 6

To find the time at which the ball hits the ground, we replace the height h with 0 to get $0 = -16t^2 + 64t + 5$. Solving this equation, we get:

$$0 = -16t^2 + 64t + 5$$
$$-16t^2 + 64t + 5 = 0$$
$$16t^2 - 64t - 5 = 0$$

$$a = 16 \quad b = -64 \quad c = -5$$

$$b^2 - 4ac = (-64)^2 - 4(16)(-5)$$
$$= 4416$$

Solution: $\dfrac{64 + \sqrt{4416}}{32}$ and $\dfrac{64 - \sqrt{4416}}{32}$

Reducing these numbers, we get:

$$\frac{64 + \sqrt{(8^2)(69)}}{32} = \frac{\not{8}(8 + \sqrt{69})}{(\not{8})(4)} = \frac{8 + \sqrt{69}}{4}$$

Similarly, $\dfrac{64 - \sqrt{4416}}{32}$ reduces to $\dfrac{8 - \sqrt{69}}{4}$.

Using a calculator to approximate these numbers, we get 4.08 and −0.08.

Thus, we conclude the ball reaches the ground at time $\frac{8 + \sqrt{69}}{4}$ seconds, or approximately 4.08 seconds after it is thrown upward. Here the negative value, $\frac{8 - \sqrt{69}}{4}$, is discarded. Why? ■

Using the Graphics Calculator

RANGE
Xmin=0
Xmax=4.75
Xscl=1
Ymin=0
Ymax=100
Yscl=10
Xres=1

FIGURE 7

We can increase our understanding of the work in Example 2 by typing `-16X²+64X+5` for `Y1`. Enter the range settings shown in Figure 7,* and view the graph screen. The curve represents the evaluations of the polynomial $-16x^2 + 64x + 5$ for replacements of the x-variable. In the physical situation of Example 2 these values represent the *height* of the ball at various times (not the *path* of the ball). Use the `TRACE` feature, and move the cursor until the message at the bottom of the screen reads `X=1.8 Y=68.36`. When the time is 1.8 seconds, the ball is 68.36 feet high.

Press the `Y=` key, and type `53` for `Y2`. View the graph, from Figure 8, we see that there are two times, x-values, at which the ball is 53 feet high. Use the `TRACE` key and the arrow keys to convince yourself that the times at which the height is 53 feet are 1 and 3. Move the cursor until the message reads `X=4.05 Y=1.76`. When the time is a little more than 4 seconds, the ball is about 2 feet high. Press the right arrow key again to get the message `X=4.1 Y=-1.56`. This means that the ball has already hit the ground (the height is less than 0). From our algebra work we know that the ball hit the ground at exactly $\frac{8 + \sqrt{69}}{4}$ seconds. Return to the home screen and store `(8+√69)/4` in memory location `X`. The number `4.076655966` is displayed. Display the resulting value stored in `Y1`. Is it 0? If not, an error is present.

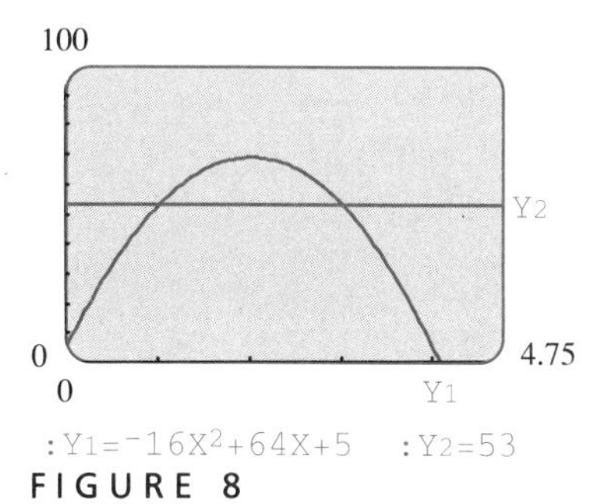

FIGURE 8

The next example involves a rational expression equation as a mathematical model. In solving this equation, a quadratic equation is produced.

EXAMPLE 3 Buffy sails 8 miles to the Redfish Point buoy, and then returns. The round-trip takes 3 hours. Her trip to the buoy is with a 10 mi/h wind, and her return is against the same wind. What is Buffy's speed in calm water?

Solution Let x = Buffy's speed in calm water.

Now, $x + 10$ = speed going to the buoy, with the wind of 10 mi/h

$x - 10$ = speed on the return trip, against the wind of 10 mi/h

$\frac{8}{x + 10}$ = time for trip to the buoy Use $t = \frac{d}{r}$.

$\frac{8}{x - 10}$ = time for the return trip Use $t = \frac{d}{r}$.

TI-82 Note *Type `4.7` for `Xmax`.

time going ↓ ↓ time returning

Equation: $\frac{8}{x+10} + \frac{8}{x-10} = 3$ ← total time

Multiply by the LCD:

$$\frac{8(x+10)(x-10)}{x+10} + \frac{8(x+10)(x-10)}{x-10} = 3(x+10)(x-10)$$

Reduce expressions: $8(x-10) + 8(x+10) = 3(x^2 - 100)$

Standard form: $3x^2 - 16x - 300 = 0$

Identify a, b, and c: $a = 3, b = -16, c = -300$

Compute the discriminant: $b^2 - 4ac = (-16)^2 - 4(3)(-300)$
$= 256 + 3600 = 3856$

Write the solution: $\frac{16 + \sqrt{3856}}{6}$ and $\frac{16 - \sqrt{3856}}{6}$

Reduce the numbers: $\frac{8 + 4\sqrt{241}}{3}$ and $\frac{8 - 4\sqrt{241}}{3}$

Approximations: 13.01611646 and −7.682783131

To answer the question, we discard the negative value and report that Buffy's speed in calm water is approximately 13.02 mi/h. ■

GIVE IT A TRY

2. Rework Example 2 if the height of the ball is given by the formula $h = -16t^2 + 96t + 6$.

Price and yield for pecans

Day	*Price per pound*	*Yield per tree*
0	\$0.62	50
1	\$0.61	51
2	\$0.60	52
x	$0.62 - 0.01x$	$50 + x$

3. Mr. Workman owns a pecan farm with 200 trees. The current selling price of pecans is \$0.62 per pound, and each tree contains 50 pounds of pecans. From experience, Mr. Workman knows that for each day he waits to sell his pecans, the yield per tree increases one pound, but the price per pound drops one cent. He has worked out the table shown in the margin. How many days from today will his pecan crop be worth \$6270?

4. Using the graphics calculator, for Y1 type the polynomial representing the income from pecan sales in Problem 3. For Y2 type the income desired, 6270. Use the TRACE key to verify your answer for Problem 3. Is this the maximum amount of income he can make from the pecans? Use the TRACE key to approximate how many days Mr. Workman should wait to get the maximum value from the crop. Also, report the approximation for the maximum amount. [*Hint*: Be sure to find appropriate range values for the graph screen.]

Rewriting Formulas In creating or writing mathematical models, we have worked with formulas in which one of the variables is to the second power. We now wish to consider rewriting such a formula. This process follows the same general pattern used in Section 2.8 with first-degree equations in one variable. Work through the following example.

EXAMPLE 4 Rewrite $y = x^2 - x$ for x.

Solution From our previous work with formulas, we know the meaning of rewriting a formula for x—we rearrange the formula so that only the variable x appears as the left-hand member, and this is the only place in the formula where x appears. To accomplish this, we will use the quadratic formula. As before (in Section 2.8), we can use an analogy as a guide. To create this analogy, we replace all the variables (only y in this case) with a number, except the one variable for which the formula is to be rewritten (x in this case).

Analogy (for $y = 10$)	*Formula*
Solve: $10 = x^2 - x$	Rewrite $y = x^2 - x$ for x.
Standard form: $x^2 - x - 10 = 0$	Standard form: $x^2 - x - y = 0$
Identify a, b, and c: $a = 1, b = -1, c = -10$	Identify a, b, and c: $a = 1, b = -1, c = -y$
Compute $b^2 - 4ac$: $(-1)^2 - 4(1)(-10) = 41$	Compute $b^2 - 4ac$: $(-1)^2 - 4(1)(-y) = 4y + 1$
Write solution: $\dfrac{1 + \sqrt{41}}{2}$ and $\dfrac{1 - \sqrt{41}}{2}$	Write formula for x: $x = \dfrac{1 + \sqrt{4y + 1}}{2}$ and $x = \dfrac{1 - \sqrt{4y + 1}}{2}$

As you can see, we get two formulas for x. Each of these formulas contains a radical expression. ■

GIVE IT A TRY

5. Rewrite $y = x^2 - 3x - 1$ for x.
6. Rewrite $d = -16t^2 + 10t$ for t.
7. Rewrite $x^2 + y^2 = 25$ for y. Two formulas are obtained for y. On the graphics calculator, enter one as the rule for Y1 and the other as the rule for Y2. Press the ZOOM key, select the Standard option, and then select the Square option. What shape is drawn?

■ SUMMARY

In this section, we have applied our knowledge of the quadratic formula to solve a second-degree equation in one variable. Many real world situations can be modeled by a second-degree equation in one variable. These situations often involve multiplication (such as problems dealing with area measure of geometric shapes). The procedure is to read the problem, produce an equation, solve the equation, and write the answer to the problem. Often, the model is provided directly in the story (as a formula).

We also learned to rewrite a formula for a variable whose exponent is 2. Here, the quadratic formula can be used as an analogy. That is, the steps in solving a second-degree equation in one variable can be used as a guide for rewriting the formula.

8.5 ■ ■ ■ EXERCISES

1. The area of a square is doubled when the side is increased by 3. What is the original length of the side of the square?

2. A rectangle is 5 feet longer than it is wide. If the area measure is 60 square feet, how wide is the rectangle?

3. A rectangle has a width 3 feet less than twice its length. If the area measure is 10 square feet, what is the length of the rectangle?

4. For the right triangle shown in the diagram, one leg is 3 cm longer than the other leg, and the hypotenuse is 7 cm. What is the length of the shorter leg?

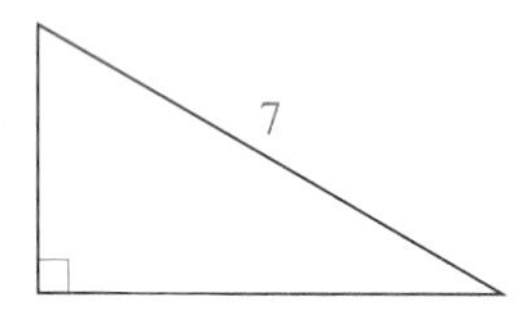

5. The right triangle sketched in the diagram has one leg that is 3 cm less than twice the length of the other leg. If the hypotenuse is 17 cm, what is perimeter of the right triangle?

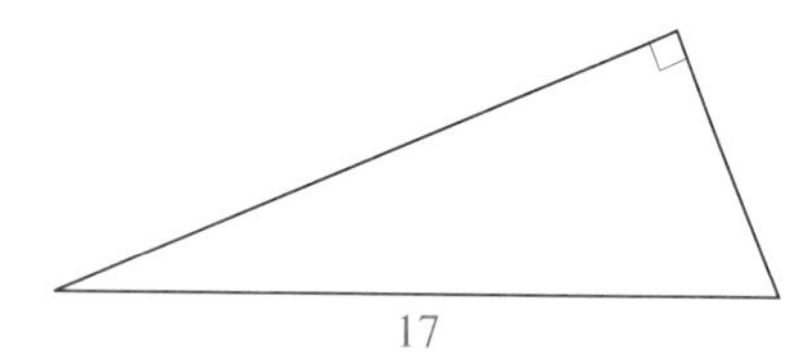

6. If the perimeter of a rectangle is 20 inches and its area measure is 10 square inches, what is the width of the rectangle?

7. One leg of a right triangle is 3 cm longer than the other leg. If the area measure of the right triangle is 40 cm^2, what is the length of the shorter leg?

8. A 16-foot ladder is leaning against a vertical wall. If the base of the ladder is 2 feet farther from the wall than the height reached by the top of the ladder, what is the height of the ladder on the vertical wall?

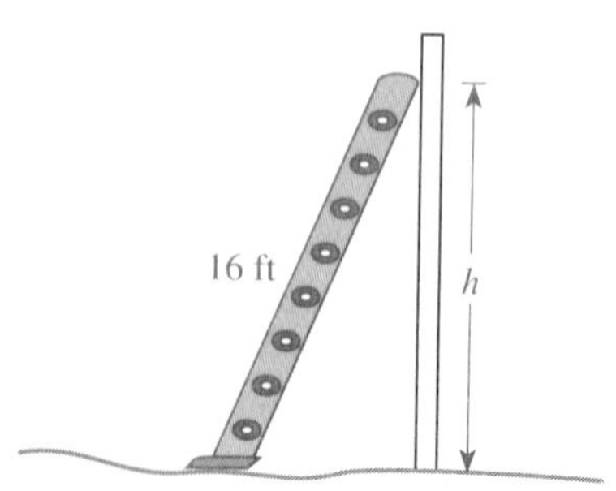

9. The distance from the point $(-2, 3)$ to a point on the horizontal line $y = 1$ is 5 (see the graph). What is the x-coordinate of the point on the line?

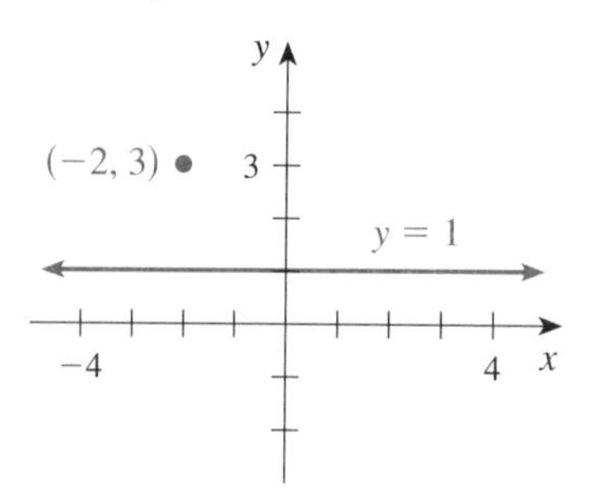

10. The distance from the point $(-2, 3)$ to a point on the vertical line $x = 1$ is 6 (see the graph). What is the y-coordinate of the point on the line?

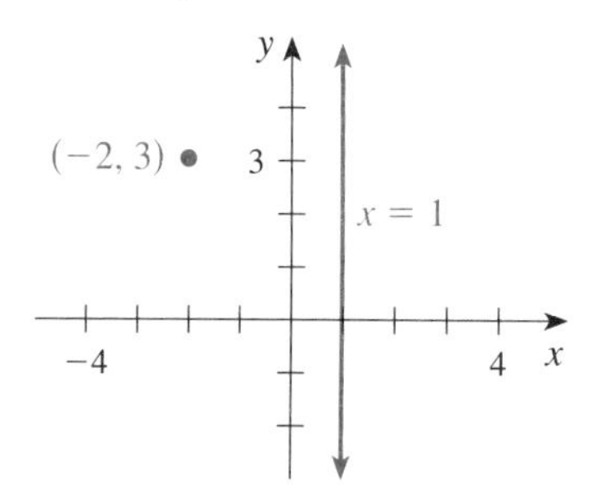

11. If a ball is thrown straight up, the height h that it reaches at a particular time t is given by $h = -16t^2 + 64t$. At what time will the ball be 50 feet high? At what time will the ball return to the ground (height zero)?

12. Devon flies his ultra-light plane 12 miles, against a 7 mi/hr wind, then turns and flies back with the same wind. If the trip takes one hour, what is the speed of the plane in calm air?

Rewrite each formula for the indicated variable.

13. $y = x^2 - x$ for x

14. $y = x^2 - 3x$ for x

15. $P = R(5 - R)$ for R

16. $n = 3r^2 - 2r$ for r

17. $R = W^2 - 3W$ for W

18. $t = gt^2 - 1$ for t

19. $x^2 - y^2 = 16$ for y

20. $x^2 - y^2 = 11$ for x

21. $x^2 + (y + 1)^2 = 9$ for y

22. $x^2 + (y - 2)^2 = 16$ for y

23. $\dfrac{1}{x - 2} = \dfrac{x}{y}$ for y

24. $\dfrac{1}{x - 2} = \dfrac{x}{y}$ for x

D I S C O V E R Y

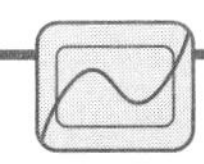

Translations and Reflections

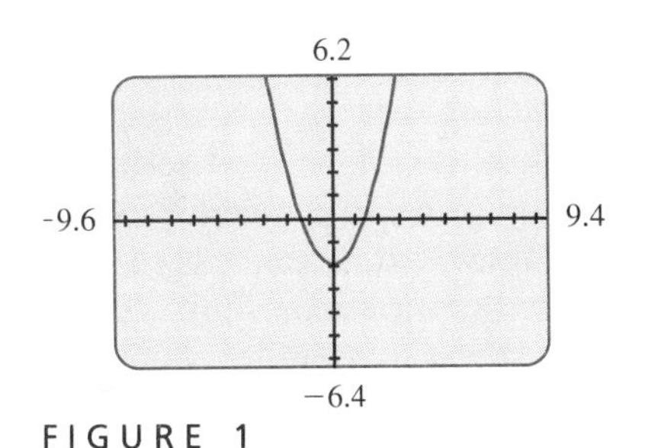

FIGURE 1

In the Discovery following Exercises 3.4, we considered the graphs of the equations $y = x^2$, $y = x^2 + A$, and $y = (x + A)^2$. You may wish to review that material now. Additionally, you may wish to review the material in Section 5.7 on parabolas. Now, type X² for Y1. Using the Window 9.4 range settings, view the graph screen. As you can see, a parabola is displayed. The vertex of this parabola is the origin (0, 0).

Edit the entry for Y1 to X²−2. Do you think you know how this alteration affects the graph? View the graph screen, as shown in Figure 1. The vertex is now at (0, −2). Edit the entry for Y1 to X²+1. What are the coordinates of the vertex? View the graph screen, and use the TRACE key. The vertex is now at (0, 1). Were you right? This shift of the parabola for $y = x^2$—either upward or downward—is known as a *vertical translation*.

GIVE IT A TRY

Predict the vertex of the parabola determined by the given equation. Use the graphics calculator to verify your prediction.

1. $y = x^2 - 3$ **2.** $y = x^2 + 2$ **3.** $y = x^2 - 5$

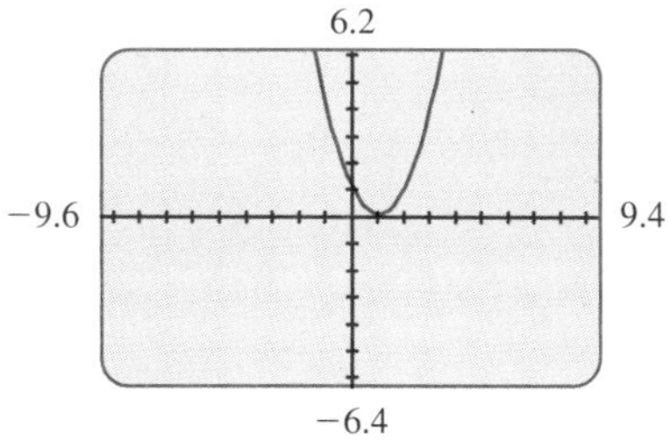

6.2
−9.6 9.4
−6.4

FIGURE 2

Type (X−1)² for Y1, and view the graph screen (see Figure 2). Use the TRACE key and the arrow keys to identify the coordinates of the vertex. As you can see, the vertex is now at (1, 0). Next, edit the entry for Y1 to (X−2)². What effect does this alteration have on the graph? View the graph screen. Were you right? The vertex now is at (2, 0). Edit the entry for Y1 to (X+2)². Where do you think the vertex is for this graph? View the graph screen. The vertex is now at (−2, 0). This shift, either right or left of the parabola for $y = x^2$ is known as a *horizontal translation*.

GIVE IT A TRY

Predict the vertex of the parabola determined by the given equation. Use the graphics calculator to verify your prediction.

4. $y = (x - 3)^2$ **5.** $y = (x + 3)^2$ **6.** $y = (x - 2)^2$

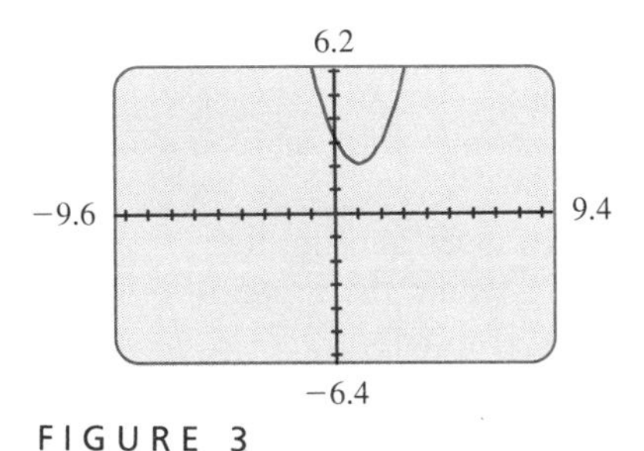

FIGURE 3

Type (X−1)²+2 for Y1. View the graph screen, as shown in Figure 3. Use the TRACE key and the arrow keys to identify the coordinates of the vertex. As you can see, the vertex is now at (1, 2). Next, edit the entry for Y1 to (X+2)²−3. Predict the coordinates of the vertex, then view the graph screen and confirm your prediction. For $y = (x + 2)^2 - 3$, we say that the parabola for $y = x^2$ has been translated downward 3 units and left 2 units.

GIVE IT A TRY

Predict the vertex of the parabola determined by the given equation. Use the graphics calculator to verify your prediction.

7. $y = (x - 3)^2 + 1$ **8.** $y = (x + 1)^2 - 2$ **9.** $y = (x - 2)^2 - 4$ 

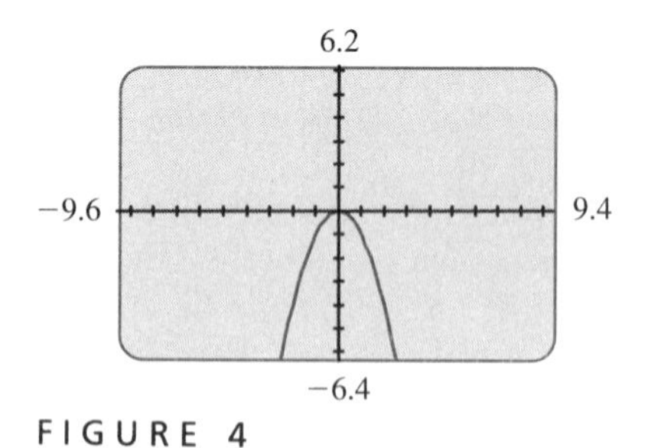

FIGURE 4

Type `-X²` for `Y1`. View the graph screen, as shown in Figure 4. Use the `TRACE` key and the arrow keys to identify the coordinates of the vertex. As you can see, the vertex is at (0, 0). The parabola drawn is the same as the one drawn for $y = x^2$, except that it is reflected across the x-axis. Edit the entry for `Y1` to `-X²+3`. What effect does this alteration have on the vertex of the parabola? View the graph screen (see Figure 5). The vertex is now at (0, 3). We say that the parabola for $y = x^2$ has been reflected across the x-axis and translated upward 3 units. Edit the entry for `Y1` to `-(X+2)²+3`. What effect does this alteration have on the vertex of the parabola? View the graph screen, as the graph in Figure 6 shows, the vertex is now at (−2, 3). We say that the parabola for $y = x^2$ has been reflected across the x-axis, translated left 2 units, and then translated upward 3 units.

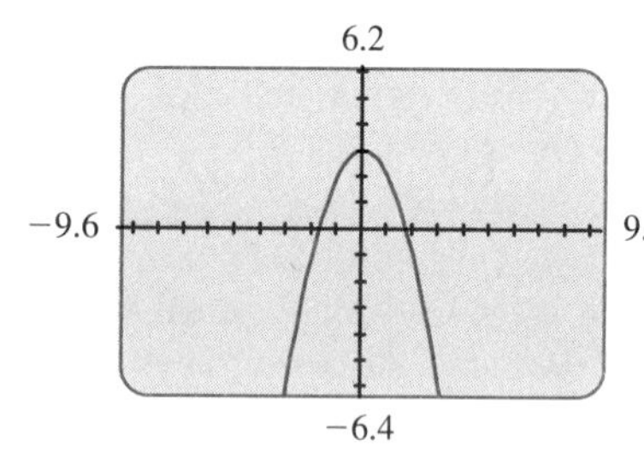

FIGURE 5

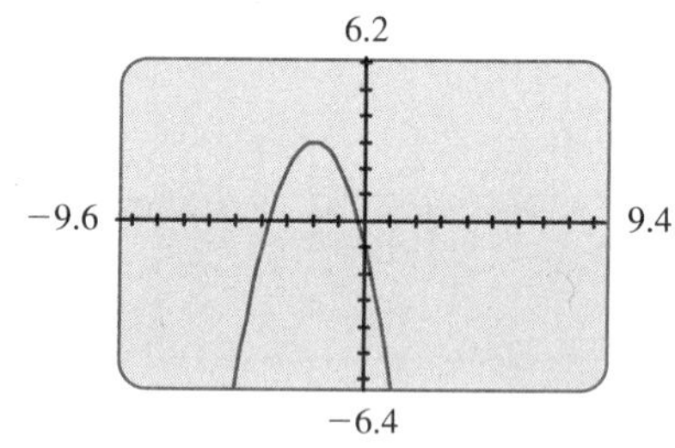

FIGURE 6

GIVE IT A TRY

Predict the vertex of the parabola determined by the given equation. Use the graphics calculator to verify your prediction.

10. $y = -x^2 - 2$ **11.** $y = -(x - 2)^2$ **12.** $y = -(x + 2)^2 + 4$ 

Y2 Y3 6.2 Y1
−9.6 9.4
−6.4
:Y1=X² :Y2=(1/2)X² :Y3=3X²

FIGURE 7

So far, we have considered translations (shifts up or down and right or left) and reflections (across the x-axis) of the parabola for $y = x^2$. Every graph produced has been congruent (identical in shape and size) to the parabola for $y = x^2$. For `Y1`, type `X²` and for `Y2`, type `(1/2)X²`. View the graph screen. A parabola is drawn for `Y2`; however, this parabola is a *distortion* of (is not congruent to) the parabola for $y = x^2$. For `Y3`, type `3X²`. View the graph screen, as shown in Figure 7. How does altering the x^2-coefficient affect the parabola produced? Where would the graph of $y = 2x^2$ fit on the current graph screen? Type `2X²` for `Y4`, and view the graph screen to verify your prediction.

Exercises

1. For $y = a(x - h)^2 + k$, describe how the values of a, h, and k affect the parabola determined by $y = x^2$.
2. Report the coordinates of the vertex of the parabola determined by each equation. Use the graphics calculator to check your work.
   a. $y = (x - 3)^2 - 2$ **b.** $y = -(x + 2)^2 - 1$ **c.** $y = -2(x - 1)^2 + 3$

8.6 ■ ■ ■ QUADRATIC FUNCTIONS

In this section we extend our work with quadratic functions. In Section 5.7 we stated that the graph of a quadratic function, $f(x) = ax^2 + bx + c$, is a curve known as a parabola and the parabola had a vertex with x-coordinate $\frac{-b}{2a}$. We begin this section, by discussing how altering a function's rule affects the graph of the function. These results, along with completing the square, will be used to develop the fact that the x-coordinate of the vertex is $\frac{-b}{2a}$.

Vertical Translations The graph of the ordered pairs that solve the second-degree equation in two variables, $y = x^2$, is a parabola with vertex at the origin. Likewise, the graph of the input-output pairs for the quadratic function $f(x) = x^2$ is a parabola with vertex at the origin. This graph is shown in Figure 1. We now discuss what happens to the graph as the rule for the function is altered.

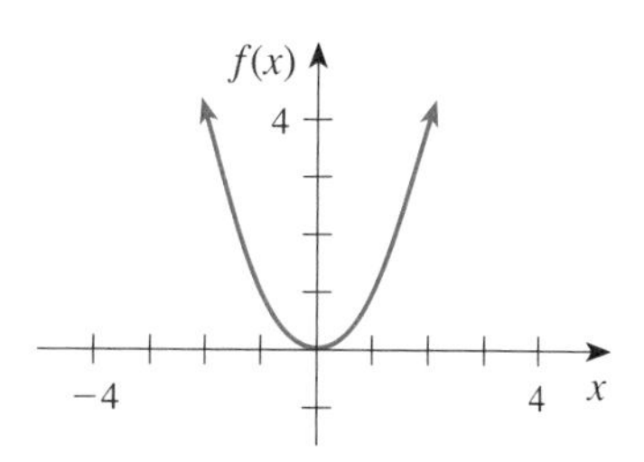

FIGURE 1

We start by giving a formal definition of a quadratic function.

DEFINITION OF A QUADRATIC FUNCTION

> A function that can be written in the form $f(x) = ax^2 + bx + c$, where a, b, and c are real numbers and $a \neq 0$, is known as a **quadratic function**.

EXAMPLE 1 Graph: $g(x) = x^2 - 2$

Solution Using a table of values (as shown below), and then plotting the input-output ordered pairs, we find that the graph for g is a parabola with vertex at $(0, -2)$. See Figure 2.

Input-output values for $g(x) = x^2 - 2$

Input, x	*Output, g(x)*
-2	$g(-2) = 2$
-1	$g(-1) = -1$
0	$g(0) = -2$
1	$g(1) = -1$
2	$g(2) = 2$

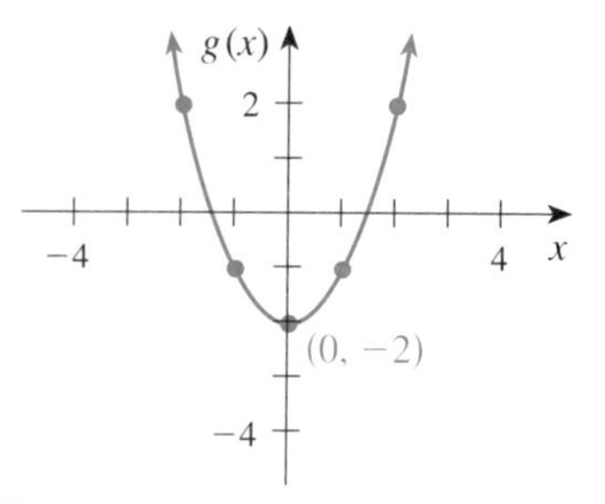

FIGURE 2 ■

The parabola produced for g in Example 1 is congruent to (has the same shape and size as) the parabola produced for $f(x) = x^2$. That is, the graph of g is the graph of f, translated down (shifted downward) 2 units. We have the following generalization.

VERTICAL TRANSLATION

For any function f, the graph of $g(x) = f(x) + c$ is the graph of f, translated vertically c units (upward if $c > 0$ or downward if $c < 0$).

Horizontal Translations We next consider the graph of a quadratic function of the form $h(x) = (x - b)^2$. Work through the following examples.

EXAMPLE 2 Graph: $h(x) = (x - 1)^2$

Solution Using a table of values (shown below), and then plotting the input-output ordered pairs, we find that the graph for h is a parabola with vertex at $(1, 0)$. The graph is shown in Figure 3.

Input-output values for $h(x) = (x - 1)^2$

Input, x	*Output, h(x)*
-1	$h(-1) = 4$
0	$h(0) = 1$
1	$h(1) = 0$
2	$h(2) = 1$
3	$h(3) = 4$

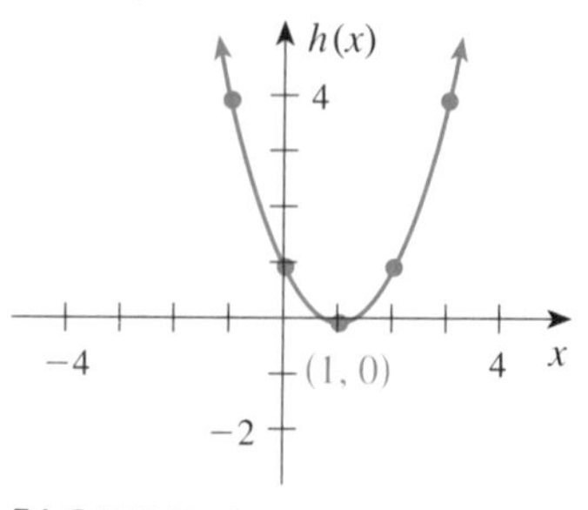

FIGURE 3 ■

The parabola produced for h in Example 2 is congruent to the parabola produced for $f(x) = x^2$. That is, the graph of h is the graph of f, translated horizontally (shifted right) 2 units. We have the following generalization.

HORIZONTAL TRANSLATION

> For any function f, the graph of $h(x) = f(x + b)$ is the graph of f, translated horizontally b units (to the left if $b > 0$ or to the right if $b < 0$).

The next example shows how the two generalizations on vertical and horizontal translations can be combined.

EXAMPLE 3 Graph: $d(x) = (x - 1)^2 - 2$

Solution We can view the graph of d as the graph of $f(x) = x^2$, translated to the right 1 unit [the $(x - 1)^2$ part of the rule for d], and down 2 units [the -2 part of the rule d]. That is, the vertex is shifted to $(1, -2)$. Notice that

$$d(1) = (1 - 1)^2 - 2 = -2$$

So, $(1, -2)$ is on the graph. Also, the least, or smallest, the variable part of the rule, $(x - 1)^2$, can be is 0, and this occurs when the x-variable is 1. These stages are shown in Figure 4.

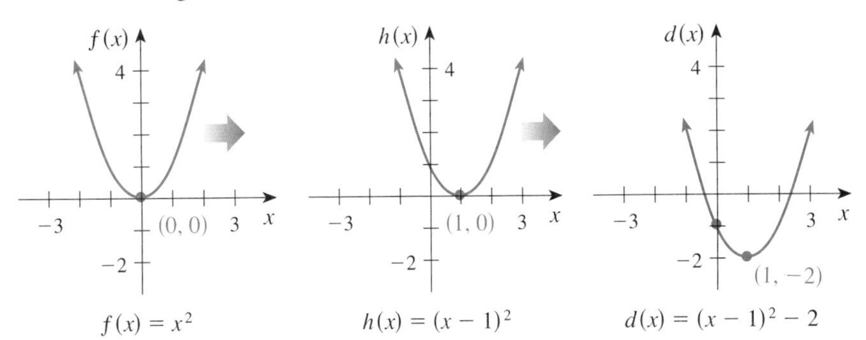

FIGURE 4 $f(x) = x^2$ $h(x) = (x - 1)^2$ $d(x) = (x - 1)^2 - 2$ ■

For quadratic functions, we have the following generalization.

THE GRAPH OF $g(x) = (x - h)^2 + k$

> The graph of $g(x) = (x - h)^2 + k$ is a parabola with the same shape as the parabola determined by $f(x) = x^2$, but with its vertex at (h, k).

Using the Graphics Calculator

To increase our understanding of the preceding examples, type X² for Y1. Press the RANGE key and enter the Window 4.7 range settings. View the graph screen. Now, type X²−2 for Y2, and view the graph screen. As you can see, the parabola $f(x) = x^2$ is drawn, and then the parabola $g(x) = x^2 - 2$ is drawn. This parabola is the first parabola, shifted down 2 units. Press the TRACE key and move the cursor to the point (1, 1) on the first parabola. Press the down arrow key to move

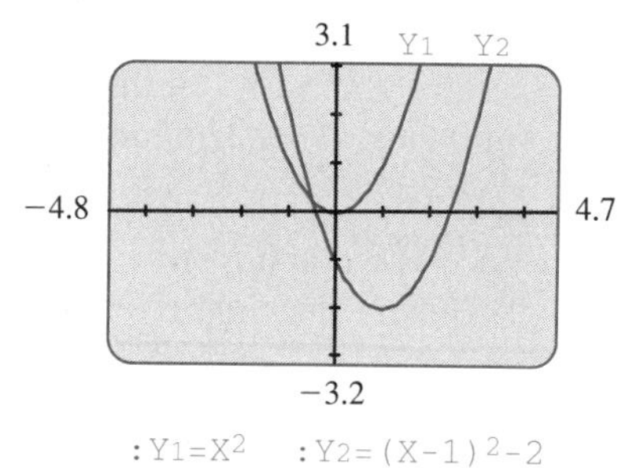

FIGURE 5

the cursor to the second parabola. The corresponding point is (1, −1). Move to the vertex of this parabola. Is it at (0, −2)? For Y2, type (X−1)². View the graph screen. This time the second parabola is shifted right 1 unit. Press the TRACE key, and then the down arrow key to move the cursor to the second parabola. Is the vertex at (1, 0)? Now, edit the rule for Y2 to (X−1)²−2. View the graph screen (see Figure 5). Press the TRACE key, and then the down arrow key to move the cursor to the second parabola. Is the vertex at (1, −2)?

GIVE IT A TRY

Report the vertex of the parabola determined by the given function.

1. $g(x) = x^2 + 1$ **2.** $h(x) = (x + 1)^2$ **3.** $d(x) = (x + 2)^2 - 3$

4. Use the graphics calculator to graph the functions in Problems 1–3, and then use the TRACE key to check your answers.

Distortions Before considering the general case of a quadratic function, we must consider a quadratic function of the form $f(x) = ax^2$, where $a > 0$. Work through the following examples.

EXAMPLE 4 Graph: $r(x) = \frac{1}{2}x^2$

Solution Using a table of values (as shown below) and then plotting the input-output ordered pairs, we find that the graph for r is a parabola with vertex at (0, 0), but its shape is distorted from that of the parabola determined by $f(x) = x^2$. See Figure 7.

Input-output values for
$r(x) = \frac{1}{2}x^2$

Input, x	*Output, r(x)*
−2	$h(-2) = 2$
−1	$h(-1) = \frac{1}{2}$
0	$h(0) = 0$
1	$h(1) = \frac{1}{2}$
2	$h(2) = 2$

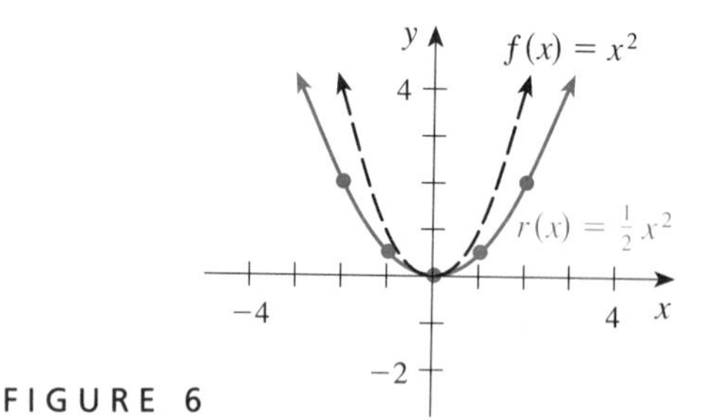

FIGURE 6

EXAMPLE 5 Graph: $s(x) = 2x^2$

Solution Using a table of values, and then plotting the input-output ordered pairs, we find that the graph for s is a parabola with vertex at (0, 0), but its shape is distorted from that of the parabola determined by $f(x) = x^2$. The graph is shown in Figure 7.

Input-output values for
$s(x) = 2x^2$

Input, x	*Output, s(x)*
-2	$h(-2) = 8$
-1	$h(-1) = 2$
0	$h(0) = 0$
1	$h(1) = 2$
2	$h(2) = 8$

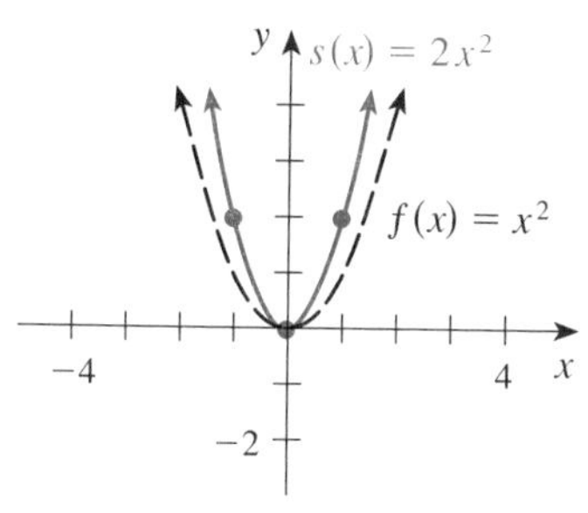

FIGURE 7

■

In general, we have the following.

DISTORTION OF THE PARABOLA FOR $f(x) = x^2$

For any function f, the graph of $g(x) = af(x)$, where $|a| \neq 1$, is a distortion of the graph of f.

The parabola determined by $g(x) = ax^2$, where $|a| \neq 1$, is a distortion of the parabola determined by $f(x) = x^2$. If $|a| > 1$, then the parabola is narrower, and if $0 < |a| < 1$, then the parabola is wider.

Reflections The last of the alterations to the function rule is where the x^2-coefficient, a, is less than 0. For $f(x) = x^2$, the graph of $g(x) = -x^2$ is a reflection of the graph of $f(x) = x^2$ across the x-axis. For the function g, the x^2-coefficient a is -1. Work through the next example.

EXAMPLE 6 Graph: $g(x) = -x^2$

Solution Using a table of values, and then plotting the input-output ordered pairs, we find that the graph for g is a parabola with vertex at (0, 0). From the graph shown in Figure 9, we see that its shape is congruent to the parabola determined by $f(x) = x^2$.

Input-output values for
$g(x) = -x^2$

Input, x	*Output, g(x)*
-2	$h(-2) = -4$
-1	$h(-1) = -1$
0	$h(0) = 0$
1	$h(1) = -1$
2	$h(2) = -4$

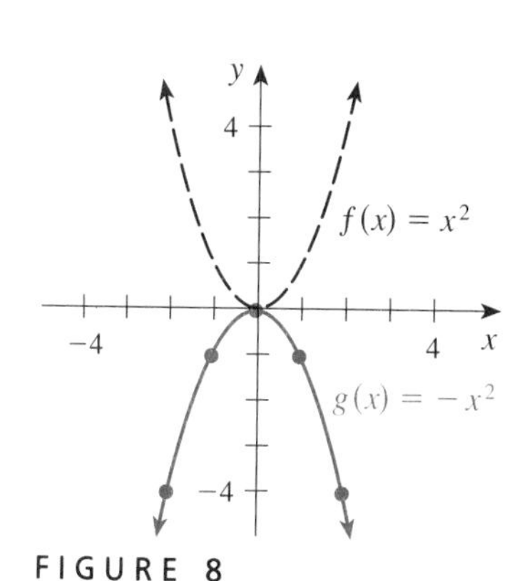

FIGURE 8

■

In general, we have the following.

REFLECTION OF THE GRAPH OF $f(x)$

> For any function f, the graph of $g(x) = -f(x)$ is the graph of f reflected across the x-axis.

Completing the Square Consider the function $f(x) = x^2 - 2x - 1$. As we did in Section 8.1, we can use the completing-the-square technique on the rule for this function to rewrite the rule to the form $f(x) = (x - h)^2 + k$. Once we have this form, we can draw the graph of the parabola. To learn how to do this, work through the next example.

EXAMPLE 7 Write $f(x) = x^2 - 2x - 1$ in the form $f(x) = (x - h)^2 + k$. Graph the function, and then report the vertex of the parabola.

Solution To complete the square for $x^2 - 2x$, we must add one-half the x-coefficient squared. To avoid altering the rule for the function, we also subtract this same amount. Now, $\left(\frac{1}{2} \text{ of } -2\right)^2$ is $(-1)^2$, or 1.

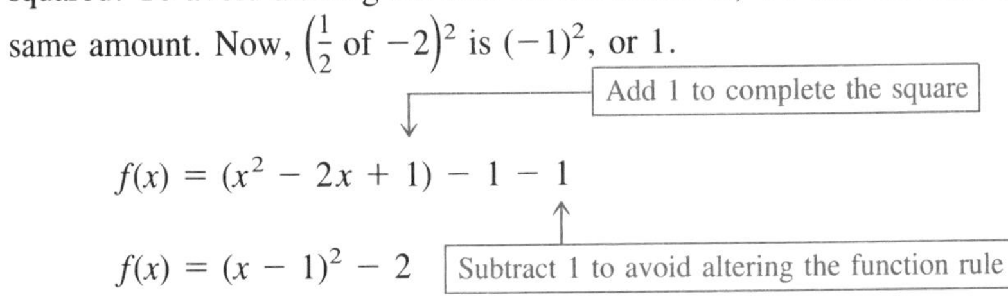

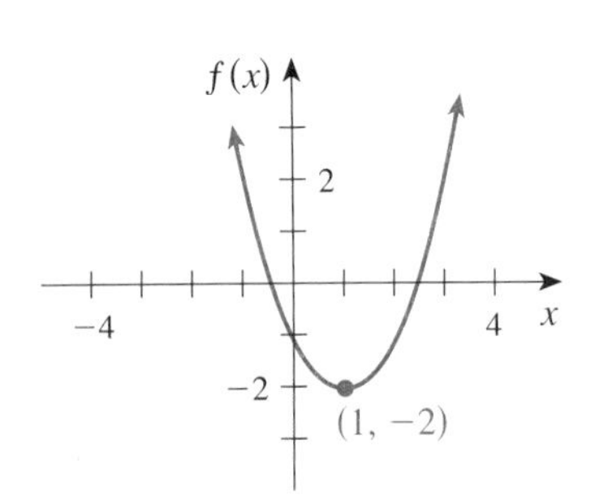

FIGURE 9

With the function rule written in this form, we know that the graph is a parabola, and that the parabola will have its vertex at $(1, -2)$. The graph is shown in Figure 9. ■

As is so common in algebra, we now consider an example that brings together many concepts. In the following example, we use translations, distortion, and completing the square.

EXAMPLE 8 Graph: $v(x) = 2x^2 + 4x - 1$

Solution Using the completing-the-square technique, we first factor 2 from the first two terms to get:

$$v(x) = 2(x^2 + 2x) - 1$$

To complete the square for $x^2 + 2x$, we must add $\left(\frac{1}{2} \text{ of } 2\right)^2$, or 1, to get the perfect square $x^2 + 2x + 1$. Because of the factor 2, we are actually adding 2(1). To compensate for this addition, we must subtract 2. We get the following:

$$v(x) = 2(x^2 + 2x + 1) - 1 - 2$$

$$v(x) = 2(x + 1)^2 - 3$$

narrower distortion (the factor 2); translations (the $+1$ and -3)

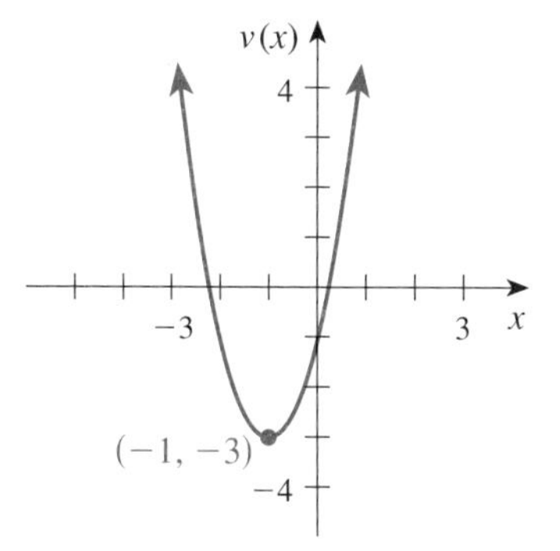

FIGURE 10

The graph is the narrower distortion caused by the 2 factor, and is translated left 1 unit and down 3 units. The vertex is $(-1, -3)$. Figure 10 shows the graph. ■

Vertex and Line of Symmetry The completing-the-square approach to find the graph of a quadratic function presents the same difficulty we found using this approach to solve second-degree equations in one variable—working with rational numbers. Just as we developed the quadratic formula, we now develop a general approach, or formula, to find the vertex of a parabola. To help clarify the following development, we will show a parallel problem, using numbers for a, b, and c. That is, we create an analogy using $f(x) = 2x^2 + 6x + 3$.

Analogy (for a = 2, b = 6, and c = 3)	*Formula*
$f(x) = 2x^2 + 6x + 3$	$f(x) = ax^2 + bx + c$
$f(x) = 2(x^2 + 3x) + 1$	$f(x) = a\left(x^2 + \frac{b}{a}x\right) + c$
$f(x) = 2\left(x^2 + 3x + \frac{9}{4}\right) + 3 - \frac{9}{2}$	$f(x) = a\left(x^2 + \frac{b}{a}x + \frac{b^2}{4a^2}\right) + c - \frac{b^2}{4a}$
$f(x) = 2\left(x + \frac{3}{2}\right)^2 + \frac{-3}{2}$	$f(x) = a\left(x + \frac{b}{2a}\right)^2 + \frac{4ac - b^2}{4a}$

The vertex occurs at the lowest point on the graph (this happens provided $a > 0$). From the form,

$$f(x) = a\left(x + \frac{b}{2a}\right)^2 + \frac{4ac - b^2}{4a}$$

We have the expression $\frac{4ac - b^2}{4a}$, which is constant for a given set of values for a, b, and c. Now, the smallest value the square expression, $\left(x + \frac{b}{2a}\right)^2$, can have is 0, and this happens when $x = \frac{-b}{2a}$. So, the vertex has x-coordinate $\frac{-b}{2a}$ and the line of symmetry is the vertical line through this point. So, the equation of the line of symmetry is $x = \frac{-b}{2a}$. We summarize this development as follows.

VERTEX AND LINE OF SYMMETRY

> For the parabola determined by $f(x) = ax^2 + bx + c$, with $a \neq 0$, the **vertex** is the point on the graph whose x-coordinate is $\frac{-b}{2a}$.
>
> The **line of symmetry** is the vertical line determined by $x = \frac{-b}{2a}$.

If we know the y-intercept, the x-intercepts, the line of symmetry, and the vertex, we can sketch the graph of most parabolas. To see that this is the case, work through the following example.

EXAMPLE 9 Consider the quadratic function $f(x) = x^2 + 3x - 1$.

a. Identify the y-intercept.
b. Identify the two x-intercepts.
c. Identify the vertex.
d. Find the equation of the line of symmetry.
e. Graph the parabola determined.

Solution

a. The y-intercept is $(0, -1)$.

b. To find the x-intercepts, we solve the equation $0 = x^2 + 3x - 1$.

Standard form: $x^2 + 3x - 1 = 0$

Use the quadratic formula: $a = 1, b = 3, c = -1$

Compute the discriminant: $b^2 - 4ac = 3^2 - 4(1)(-1) = 9 + 4 = 13$.

Write the solution: $\dfrac{-3 + \sqrt{13}}{2}$ and $\dfrac{-3 - \sqrt{13}}{2}$

So, the x-intercepts are: $\left(\dfrac{-3 + \sqrt{13}}{2}, 0\right)$ and $\left(\dfrac{-3 - \sqrt{13}}{2}, 0\right)$. Using the calculator, we approximate these points as $(0.3, 0)$ and $(-3.3, 0)$.

c. The x-coordinate of the vertex has the form $\dfrac{-b}{2a}$. For this example, $a = 1$ and $b = -2$. So, the x-coordinate of the vertex is $\dfrac{-(3)}{2(1)}$, or $\dfrac{-3}{2}$. So, we have $\left(\dfrac{-3}{2}, ?\right)$. To find the y-coordinate, we find $f\left(\dfrac{-3}{2}\right)$:

$$\begin{aligned} f\left(\frac{-3}{2}\right) &= \left(\frac{-3}{2}\right)^2 + 3\left(\frac{-3}{2}\right) - 1 \\ &= \frac{9}{4} - \frac{9}{2} - 1 \\ &= \frac{-9}{4} - \frac{-4}{4} = \frac{-13}{4} \end{aligned}$$

Thus, the vertex is $\left(\dfrac{-3}{2}, \dfrac{-13}{4}\right)$.

d. The line of symmetry is a vertical line through the vertex. Because the x-coordinate of the vertex is $\dfrac{-3}{2}$, the equation of line of symmetry is $x = \dfrac{-3}{2}$.

e. To draw the parabola determined by $f(x) = x^2 + 3x - 1$, we plot the y-intercepts, the x-intercepts, the vertex, and then draw a dotted or dashed line for the line of symmetry. Finally, we sketch in the parabola. These are all shown in Figure 11. ■

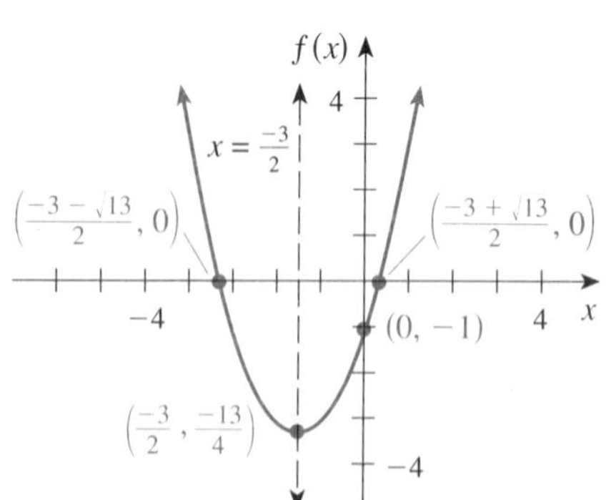

FIGURE 11

Using the Graphics Calculator

To check our work in Example 8, type X²+3X−1 for Y1. View the graph screen using the Window 9.4 range settings. Is the parabola drawn similar to the one shown in Figure 12? Press the TRACE key. As we can see, by the message X=0 Y=−1. The y-intercept is $(0, -1)$. Move the cursor to the rightmost x-intercept. Is the x-coordinate of this point approximately 0.3? Move the cursor to the leftmost x-intercept. Is the x-coordinate approximately -3.3? Finally, move the cursor as close to the vertex as possible. As you can see, the vertex is the minimum point on the curve. From the message presented, is it reasonable that the ordered-pair name of the vertex is $(-1.5, -3.25)$?

Because the x-intercepts of a parabola are determined by the solution to a second-degree equation in one variable, a parabola typically has two x-intercepts—it can also have exactly one, or it can have none. For Y1, type (X−2)². View the graph screen. The graph drawn is a plot of the input-output ordered pairs for $f(x) = (x - 2)^2$. See Figure 12. As you can see, the graph has one x-intercept—namely, $(2, 0)$. In fact, the x-intercept is also the vertex. Now, type X²+2X+3 for Y1, and view the graph. As you can see, this parabola has no x-intercept. In the next example we graph the parabola determined by $f(x) = x^2 + 2x + 3$ algebraically, in order to see why it has no x-intercept.

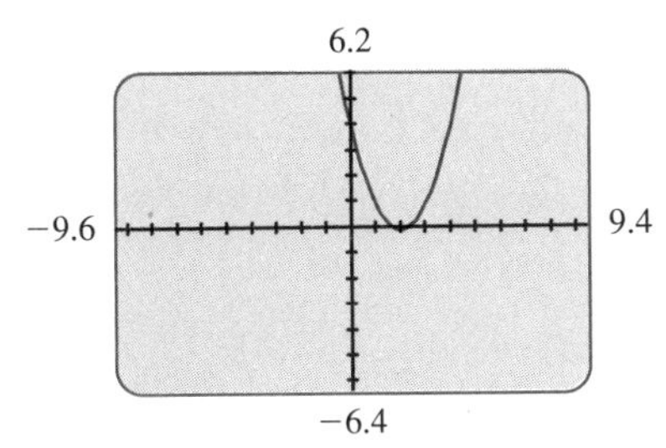

FIGURE 12

EXAMPLE 10 Consider the quadratic function $f(x) = x^2 + 2x + 3$.

a. Identify the y-intercept.
b. Identify the two x-intercepts.
c. Identify the vertex.
d. Find the equation of the line of symmetry.
e. Graph the parabola determined.

Solution **a.** The y-intercept is $(0, 3)$.

b. We solve the equation: $0 = x^2 + 2x + 3$

Standard form: $x^2 + 2x + 3 = 0$

Use the quadratic formula: $a = 1, b = 2, c = 3$

Compute the discriminant: $b^2 - 4ac = 2^2 - 4(1)(3) = 4 - 12 = -8$

Solution: *no real number*. The solution is two complex numbers,

$$-1 + i\sqrt{2} \quad \text{and} \quad -1 - i\sqrt{2}$$

Because we are graphing only ordered pairs with real number coordinates, the graph does not cross the x-axis—it has *no* x-intercept.

c. The x-coordinate of the vertex is $\frac{-b}{2a} = \frac{-2}{2} = -1$. To find the y-coordinate of the vertex, we replace the x-variable by -1 in the original function rule to get $f(-1) = (-1)^2 + 2(-1) + 3 = 1 - 2 + 3 = 2$.

So, the vertex is $(-1, 2)$.

d. The line of symmetry is $x = -1$.

e. We plot the points from parts (a)–(c), draw the line of symmetry, and then sketch in the parabola. See Figure 13. If needed, we can generate additional points:

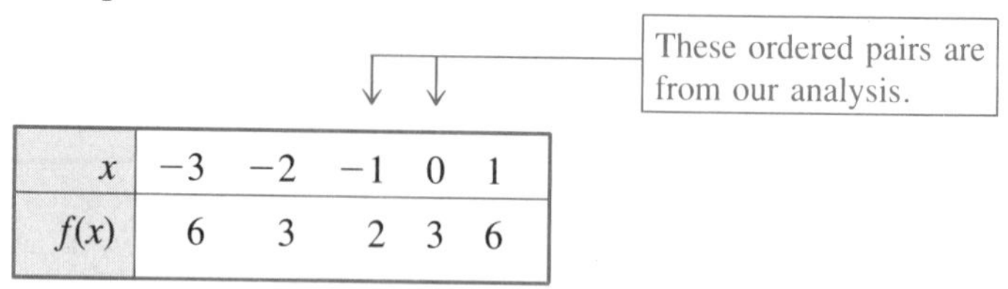

x	-3	-2	-1	0	1
$f(x)$	6	3	2	3	6

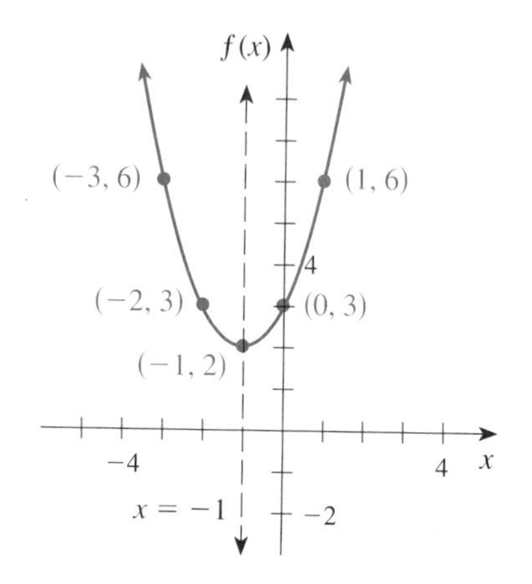

FIGURE 13

■

GIVE IT A TRY

A parabola is determined by each of the given quadratic functions.

a. Report the ordered-pair name of the y-intercept.
b. Report the ordered-pair name of the x-intercepts, if any.
c. Report the ordered-pair name of the vertex.
d. Report the equation of the vertical line that is the line of symmetry.
e. Use the information in parts (a)–(d) to graph the solution.

5. $f(x) = x^2 - 2x - 2$ **6.** $f(x) = x^2 + 2x + 3$ **7.** $f(x) = x^2 - 4x + 1$

The Discriminant In Chapter 5, we stated that for $f(x) = ax^2 + bx + c$, if $a < 0$, then the parabola turns downward and the vertex is a maximum point. From our earlier work with completing the square, we know that the function rule can be written as

$$f(x) = a\left(x + \frac{b}{2a}\right)^2 + \frac{4ac - b^2}{4a}$$

With $a < 0$, we argue that $a\left(x + \frac{b}{2a}\right)^2$ is negative, and it is as large as possible when $x = \frac{-b}{2a}$. So, the vertex is at $\left(\frac{-b}{2a}, f\left(\frac{-b}{2a}\right)\right)$, and this point is the maximum point for the parabola.

Because the x-intercepts are found by solving the equation $ax^2 + bx + c = 0$, the discriminant, $b^2 - 4ac$, determines the x-intercepts. In Figure 14, we use the discriminant to summarize the parabola determined by the quadratic function $f(x) = ax^2 + bx + c$.

If a > 0, the vertex is a minimum point:

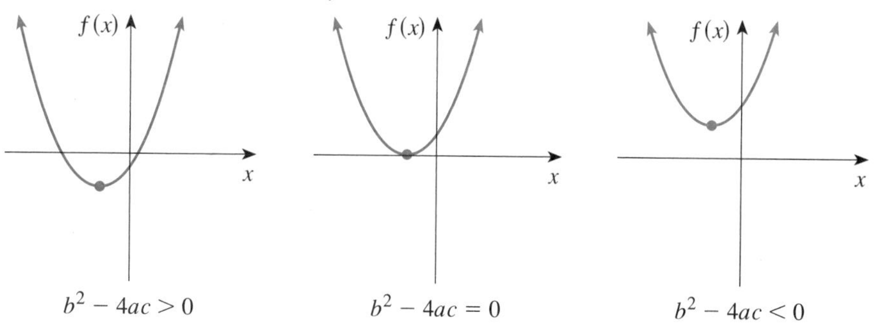

If a < 0, the vertex is a maximum point:

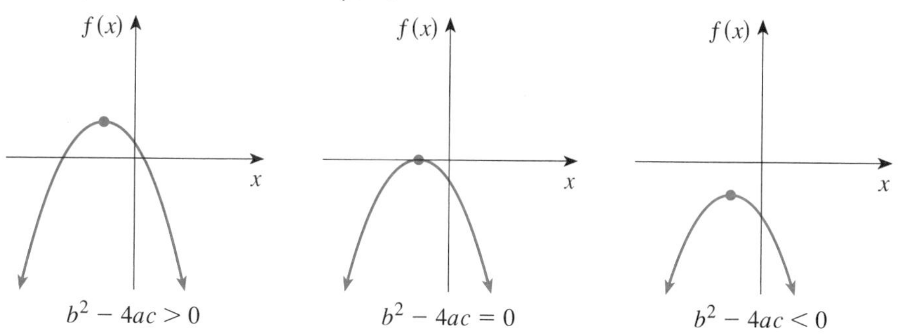

FIGURE 14

The following example summarizes our work in this section.

EXAMPLE 11 Consider the quadratic function $f(x) = -x^2 + 3x + 3$.

a. Find $f(-2)$.
b. What input value produces an output of 0?
c. Graph the quadratic function, and label the vertex with its ordered-pair name.
d. Is the vertex a maximum or a minimum point for the parabola?

Solution

a. $f(-2) = -(-2)^2 + 3(-2) + 3 = -4 - 6 + 3 = -10 + 3 = -7$

b. To find the input value x that produces an output $f(x)$ of 0, we replace $f(x)$ with 0 to get $0 = -x^2 + 3x + 3$, and solve this equation:

Standard form: $x^2 - 3x - 3 = 0$

Using the quadratic formula: $a = 1,\ b = -3,\ c = -3$

Compute the discriminant: $b^2 - 4ac = (-3)2 - 4(1)(-3) = 9 + 12 = 21$

Write the solution: $\dfrac{3 + \sqrt{21}}{2}$ and $\dfrac{3 - \sqrt{21}}{2}$

The x-intercepts are $\left(\dfrac{3 + \sqrt{21}}{2}, 0\right)$ and $\left(\dfrac{3 - \sqrt{21}}{2}, 0\right)$.

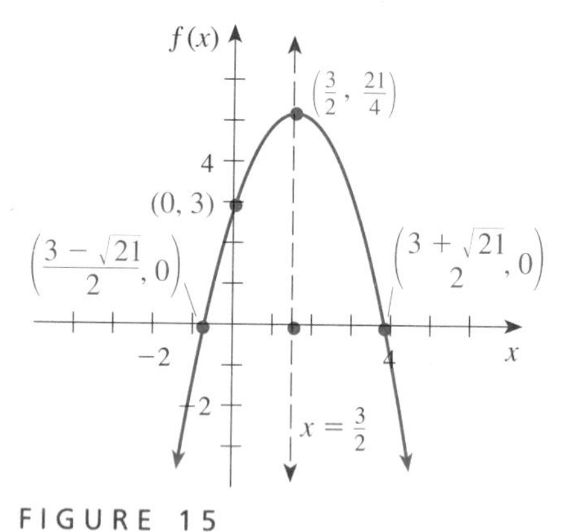

FIGURE 15

c. The vertex has x-coordinate $\frac{-b}{2a} = \frac{-3}{-2}$. The y-coordinate of the vertex is

$$f\left(\frac{3}{2}\right) = -\left(\frac{3}{2}\right)^2 + 3\left(\frac{3}{2}\right) + 3 = \frac{21}{4}$$

So, the vertex is $\left(\frac{3}{2}, \frac{21}{4}\right)$. The line of symmetry is $x = \frac{3}{2}$. Using this information and the information from parts (a) and (b), we draw the parabola shown in Figure 15.

d. Since $a < 0$, the vertex is a maximum point. ■

GIVE IT A TRY

For each of the given quadratic functions, do the following:

a. Report the ordered-pair name of the y-intercept.
b. Report the ordered-pair name for each of the two x-intercepts.
c. Report the ordered-pair name of the vertex.
d. Report the equation of the vertical line that is the line of symmetry.
e. Use the information in parts (a)–(d) to graph the function.

8. $f(x) = -x^2 + 5x + 6$ **9.** $g(x) = (3 - x)(2 + x)$ **10.** $h(x) = -x^2 + 2x - 2$

11. Use the graphics calculator to check your work in Problems 8–10.

■ ***Quick Index***

■ SUMMARY

In this section, we have learned that the graph of a function can be related to the graph of another function. These relationships are translations, distortions, and reflections. We have the following generalizations:

1. The graph of $g(x) = f(x) + c$ is a vertical translation of the graph of function f.
   The translation is upward if $c > 0$ or downward if $c < 0$.
2. The graph of $h(x) = f(x + c)$ is a horizontal translation of the graph of function f.
   The translation is to the left if $c > 0$ or to the right if $c < 0$.
3. The graph of $d(x) = ad(x)$, where $|a| \neq 1$, is a distortion of the graph of function f.
   For $f(x) = x^2$, if $|a| > 1$, then the graph of $d(x) = af(x)$ is narrower than the parabola determined by f, and if $0 < |a| < 1$, the graph is wider than the parabola determined by f.
4. The graph of $r(x) = -f(x)$ is a reflection across the x-axis of the graph of function f.

Using these generalizations and the completing-the-square approach, we found the graphs of various quadratic functions, such as $f(x) = x^2 - 3x - 4$, to be parabolas. Just as the major feature of a line is its slope, the major features of a parabola are its line of symmetry and its vertex. We developed a general procedure for graphing the quadratic function $f(x) = ax^2 + bx + c$:

1. To find the y-intercept, replace the x-variable by 0.
2. To find the x-intercepts, set the function rule equal to 0, and solve the quadratic equation produced.
3. Find the x-coordinate of the vertex $\left(\text{the } x\text{-coordinate is } \dfrac{-b}{2a}\right)$. The y-coordinate of the vertex is $f\left(\dfrac{-b}{2a}\right)$.
4. Find the equation of the line of symmetry: $x = \dfrac{-b}{2a}$.

In general, this information is sufficient to sketch in the parabola. If not, we can generate more ordered pairs by evaluating the function for selected input values. In addition to the given procedure for graphing a parabola, we learned that if $a > 0$, the parabola determined by $f(x) = ax^2 + bx + c$ turns upward, and the vertex is a minimum point. If $a < 0$, the parabola turns downward, and the vertex is a maximum point.

8.6 ■ ■ ■ EXERCISES

Report the vertex of the parabola determined by each function, and sketch its graph. Use the graphics calculator to check your answer.

1. $g(x) = x^2 + 1$
2. $h(x) = x^2 + 3$
3. $g(x) = (x - 4)^2$
4. $h(x) = (x + 3)^2$
5. $g(x) = (x - 5)^2$
6. $h(x) = (x + 4)^2$
7. $g(x) = (x - 2)^2 + 1$
8. $h(x) = (x - 4)^2 - 3$
9. $g(x) = (x + 3)^2 - 2$
10. $h(x) = (x + 5)^2 + 2$

Match graphs A–D with the functions given in Problems 11–14.

A.

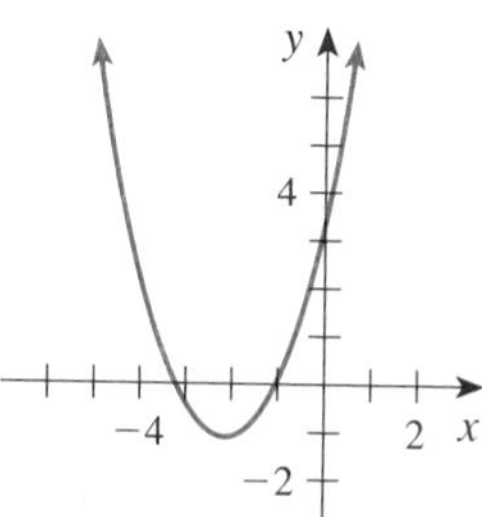

B.

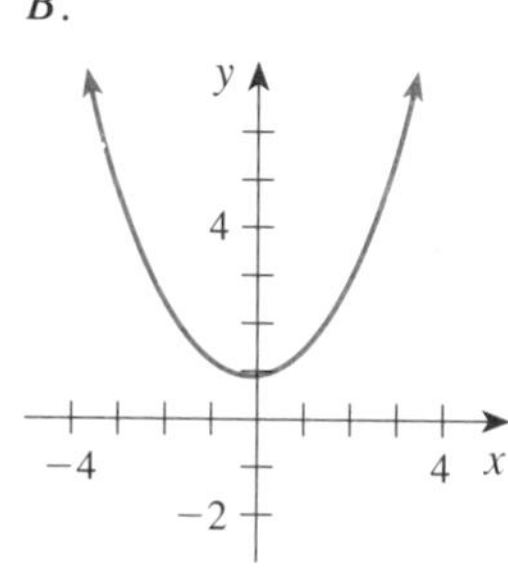

C.

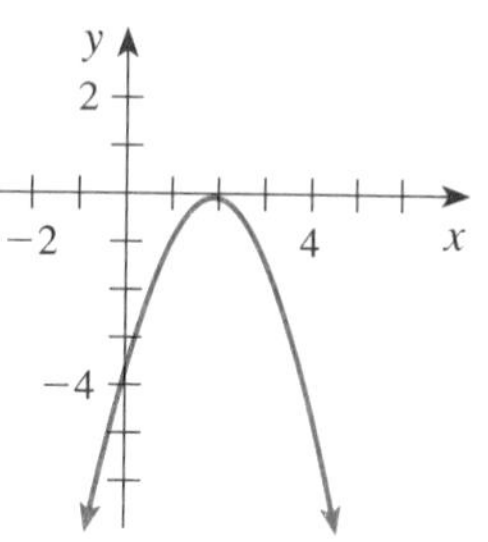

D.

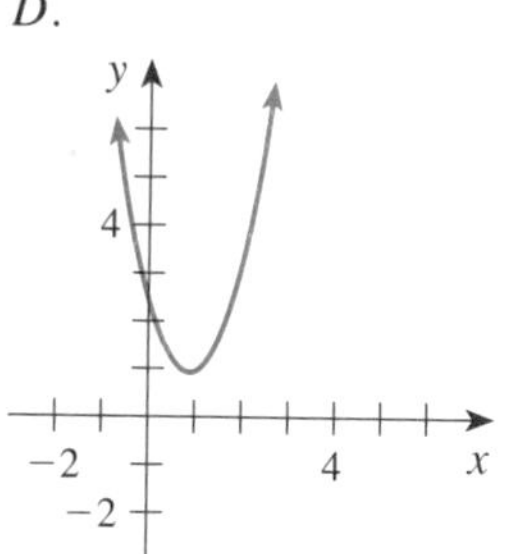

11. $g(x) = -(x - 2)^2$
12. $h(x) = (x + 2)^2 - 1$
13. $s(x) = \frac{1}{2}x^2 + 1$
14. $g(x) = 2(x - 1)^2 + 1$

Rewrite each equation in the form $f(x) = (x + h)^2 + k$. Report the vertex, and sketch a graph of the parabola determined. Use the graphics calculator to check your answer.

15. $d(x) = x^2 - 4x + 1$
16. $d(x) = x^2 + 4x - 2$
17. $d(x) = x^2 + 2x + 3$
18. $d(x) = x^2 - 2x - 5$
19. $d(x) = x^2 - 3x + 2$
20. $d(x) = x^2 + 3x + 1$

For the parabola determined by each quadratic function, complete the following:

a. Report the ordered-pair name of the y-intercept.
b. Report the ordered-pair name of each x-intercept.
c. Report the ordered-pair name of the vertex.
d. Report the equation of the vertical line that is the line of symmetry.
e. Use the information in parts (a)–(d) to graph the function.

21. $f(x) = x^2 + 2x - 2$
22. $g(x) = x^2 + 2x - 3$
23. $h(x) = x^2 - 3x - 1$
24. $r(x) = x^2 - 3x + 1$
25. $u(x) = x^2 - 4x - 4$
26. $v(x) = x^2 - 4x - 3$
27. $T(x) = x^2 - 6x - 3$
28. $R(x) = x^2 + 6x - 3$
29. $f(x) = 2x^2 - 6x - 1$
30. $h(x) = 2x^2 + x - 3$
31. $h(x) = (x - 2)(x + 1)$
32. $r(x) = (2x - 1)(x + 2)$
33. $f(x) = (2x - 3)(x + 1)$
34. $g(x) = (x - 1)(3x + 2)$
35. $T(x) = (3 - x)(x + 2)$
36. $S(x) = (2 + x)(1 - x)$
37. $n(x) = -x^2 + 3$
38. $p(x) = -2x^2 - 1$
39. $h(x) = -x^2 - 3x - 1$
40. $c(x) = -x^2 + 3x - 2$
41. $g(x) = -2x^2 - 5x - 2$
42. $d(x) = -3x^2 - 2x + 1$

8.7 ■ ■ ■ POLYNOMIAL, RATIONAL, AND RADICAL FUNCTIONS

So far, we have learned how to graph linear and quadratic functions. Now we introduce graphs of other polynomial functions, as well as graphs of rational functions and radical functions. We will make use of the ideas of distortions, reflections, and horizontal and vertical translations from Section 8.6. This section gives a global view of many of the objects of algebra.

Other Polynomial Functions From our earlier work, we know that the graph of a linear function, such as $f(x) = 2x - 1$, is a straight line. Also, we know that the graph of a quadratic function, such as $f(x) = x^2 - 3x - 4$, is a parabola. What about other polynomial functions, such as $g(x) = -x^4 - 3x^2 + 4$ and $h(x) = x^3 - x$? How do the graphs of these functions appear? We start by defining what is meant by a general polynomial function.

DEFINITION OF A POLYNOMIAL FUNCTION

An nth-degree polynomial function of x is any function of x that can be written in the form $f(x) = a_nx^n + a_{n-1}x^{n-1} + \cdots + a_1x^1 + a_0$, where $a_n \neq 0$. That is, a polynomial function is a function whose rule is a polynomial.

In general, the topic of graphing polynomial functions of degree higher than 2, using mathematics, is studied in calculus. However, we can get the general idea by building a table of points, plotting those points, and sketching in a smooth curve. Consider the next two examples.

EXAMPLE 1 Graph: $f(x) = x^3$

Solution First, we make a table of input-output ordered pairs:

x	−2	−1	0	1	2
$f(x)$	−8	−1	0	1	8

Plotting these ordered pairs and sketching in a smooth curve, we get the graph shown in Figure 1.

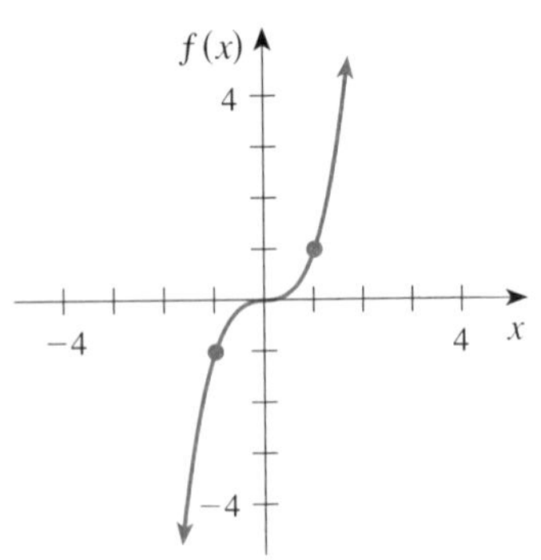

FIGURE 1 ■

EXAMPLE 2 Graph: $g(x) = -x^3$

Solution We first make a table of input-output ordered pairs:

x	-2	-1	0	1	2
$g(x)$	8	1	0	-1	-8

Plotting these ordered pairs and sketching in a smooth curve, we get the graph shown in Figure 2. ■

FIGURE 2

Notice from these graphs that just as the graph $g(x) = -x^2$ is a reflection of the graph of $f(x) = x^2$ across the x-axis, so is the graph of $g(x) = -x^3$ a reflection of the graph of $f(x) = x^3$ across the x-axis. Now, we consider two slightly more involved examples of polynomial functions.

EXAMPLE 3 Graph: $f(x) = x^3 + x^2 - 2x$

Solution We first construct a table of input-output ordered pairs:

x	-3	-2	-1	0	1	2
$f(x)$	-12	0	2	0	0	8

Plotting these ordered pairs and sketching in a smooth curve, we get the graph shown in Figure 3. For this graph, the y-intercept is $(0, 0)$ and the x-intercepts are $(-2, 0)$, $(0, 0)$, and $(1, 0)$. ■

FIGURE 3

EXAMPLE 4 Graph: $g(x) = -x^3 - x^2 + 2x$

Solution A table of input-output ordered pairs is given:

x	-3	-2	-1	0	1	2
$g(x)$	12	0	-2	0	0	-8

Plotting these ordered pairs and sketching in a smooth curve, we get the graph shown in Figure 4. ■

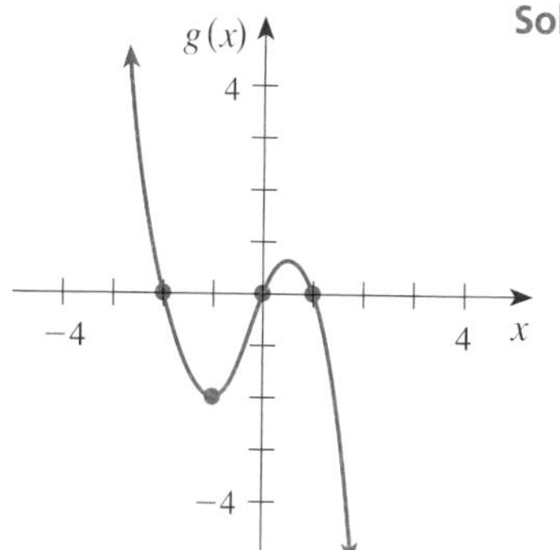

FIGURE 4

In Examples 3 and 4, we again see that multiplying the function rule by -1 [here $g(x) = -f(x)$] produces a reflection across the x-axis. These graphs have a y-intercept of $(0, 0)$ and x-intercepts $(-2, 0)$, $(0, 0)$, and $(1, 0)$.

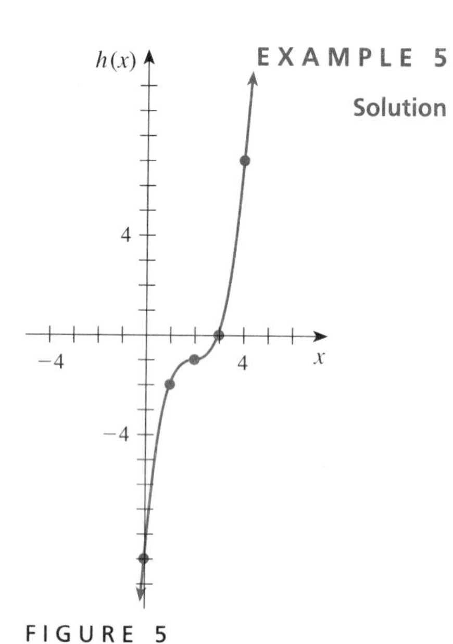

FIGURE 5

EXAMPLE 5 Graph: $h(x) = (x - 2)^3 - 1$

Solution A table of input-output ordered pairs is given:

x	−1	0	1	2	3	4
$h(x)$	−28	−9	−2	−1	0	7

Plotting these ordered pairs and sketching in a smooth curve, we get the graph shown in Figure 5. ■

Example 5 shows that the generalizations for vertical and horizontal translations hold for polynomial functions. The graph drawn is the graph of $f(x) = x^3$, translated (shifted) right 2 units and down 1 unit.

Using the Graphics Calculator

To increase our understanding of the graph of polynomial functions, type X³ for Y1. View the graph screen using the Window 4.7 range settings . For Y2 type X³−1. View the graph screen. As you can see, the second curve drawn is the graph of $f(x) = x^3$, translated downward 1 unit. For Y2 type (X−1)³. View the graph screen. The Y2 curve is the graph of $f(x) = x^3$, translated right 1 unit.

Erase the rule for Y2 and edit the rule for Y1 to X³+X². View the graph screen, as shown in Figure 6. Press the TRACE key. The y-intercept is (0, 0). Press the left arrow key. For any input value greater than −1, the output value is greater than or equal to 0. The x-intercepts are (−1, 0) and (0, 0). For any input value less than −1, the output value is negative. Notice that the function rule factors as $x^2(x + 1)$. Edit the rule for Y1 to X³+X²−2X. View the graph screen (see Figure 7). Compare this curve with the graph produced in Example 3. Use the TRACE feature to convince yourself that the y-intercept is the origin. Notice that the graph has three x-intercepts and that the function rule factors as $x(x + 2)(x - 1)$. What is the relationship be-

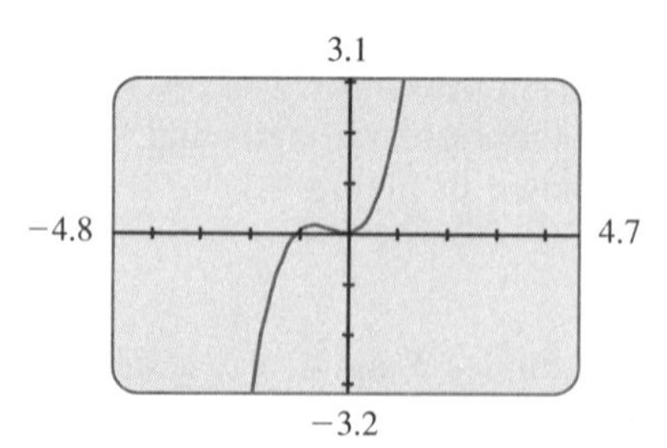

FIGURE 6

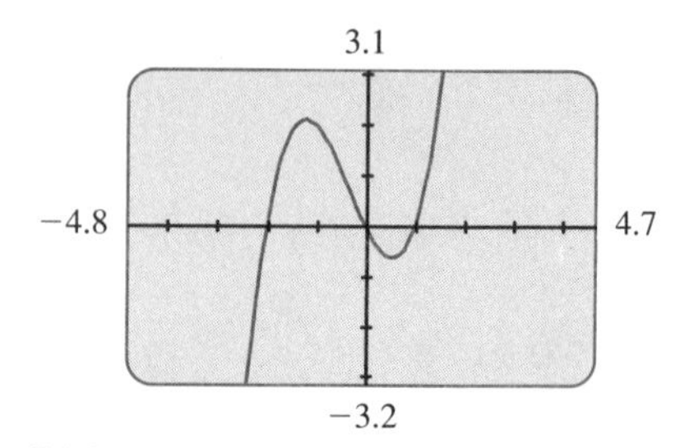

FIGURE 7

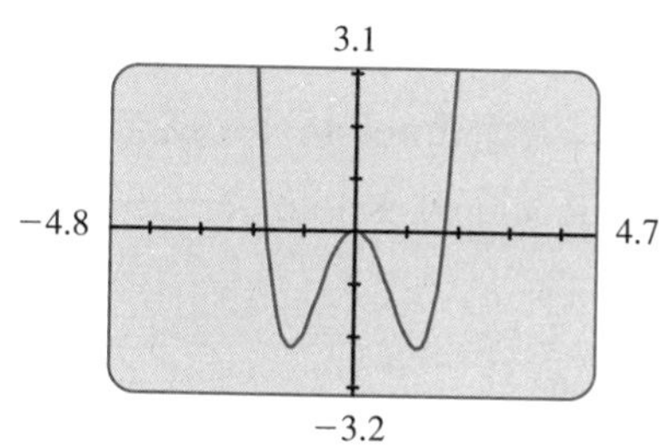

FIGURE 8

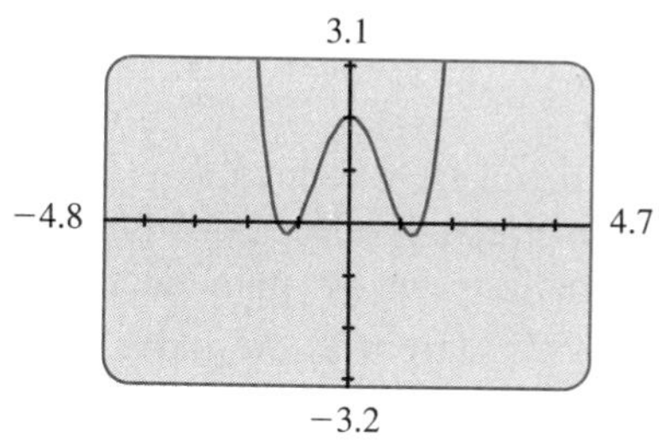

FIGURE 9

tween the x-intercepts and factors of the function rule? Edit the rule for Y1 to -X³−X²+2X. View the graph screen. How is the graph altered? Edit the function rule for Y1 to -X³−X²+2X+1. View the graph screen. How is the graph altered now?

For Y1 type X^4, and view the graph screen. Although this curve looks very much like a parabola, it is *not* a parabola. To see that this is the case, type X² for Y2. View the graph screen. The curve for $f(x) = x^4$ is much "flatter" than the parabola for input values between -1 and 1. Erase the entry for Y2, and edit the entry for Y1 to X^4−3X². View the graph screen, as shown in Figure 8. The y-intercept of the graph is the origin, and the graph has three x-intercepts: $\left(-\sqrt{3}, 0\right)$, $(0, 0)$, and $\left(\sqrt{3}, 0\right)$. Edit the rule for Y1 to X^4−3X²+2. View the graph screen (see Figure 9). What is the relationship between this graph and the graph for $f(x) = x^4 - 3x^2$? This graph has four x-intercepts. Can you find them?

GIVE IT A TRY

Match graphs *A–D* with the polynomial functions in Problems 1–4.

A.

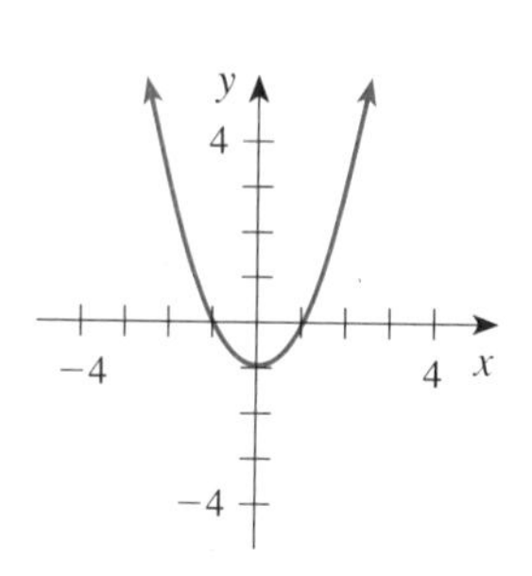

B.

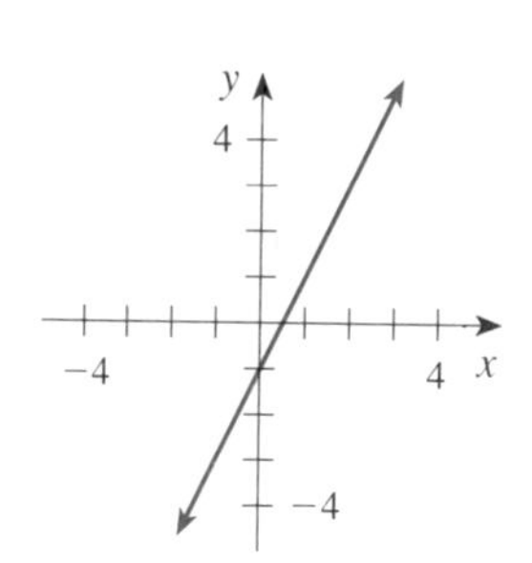

C.

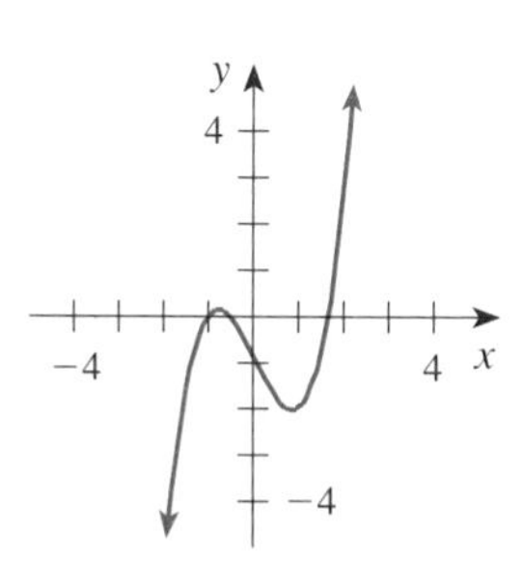

D.

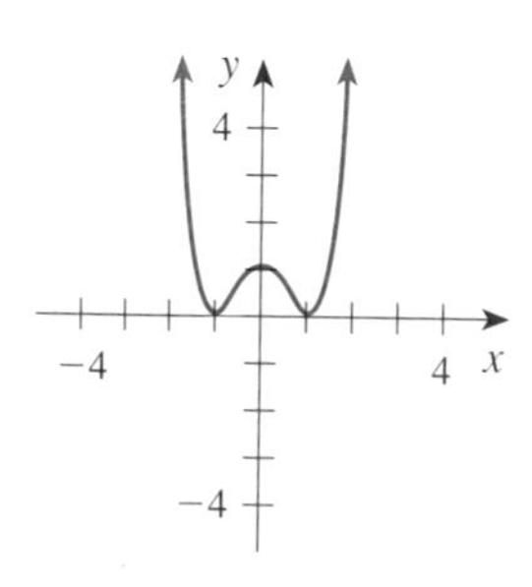

1. $f(x) = x^3 - 2x - 1$

2. $h(x) = x^2 - 1$

3. $g(x) = 2x - 1$

4. $r(x) = x^4 - 2x^2 + 1$

Rational Expression Functions A rational expression is a ratio of polynomials, for example, $\frac{x+2}{x-1}$. A **rational expression function** (or simply a rational function) is a function whose rule is a rational expression. For example,

$$r(x) = \frac{x+2}{x-1}$$

is a rational function. We can also define a rational function as a ratio of polynomials:

$$r(x) = \frac{f(x)}{g(x)}$$

where $f(x)$ and $g(x)$ are polynomials with no common factor and $g(x) \neq 0$. As you may suspect, there is great variety in the graph produced by rational functions. Here, we consider only rational functions whose numerator and denominator are polynomials of degree 1 or 0. We start with the most basic rational function, $f(x) = \frac{1}{x}$.

EXAMPLE 6 Graph: $f(x) = \frac{1}{x}$

Solution We start by producing a table of input-output ordered pairs. Plotting these ordered pairs and sketching a smooth curve, we get the graph shown in Figure 10.

Input-output values for $f(x) = \frac{1}{x}$

x	$f(x)$
-2	$\frac{-1}{2}$
-1	-1
0	undefined
1	1
2	$\frac{1}{2}$

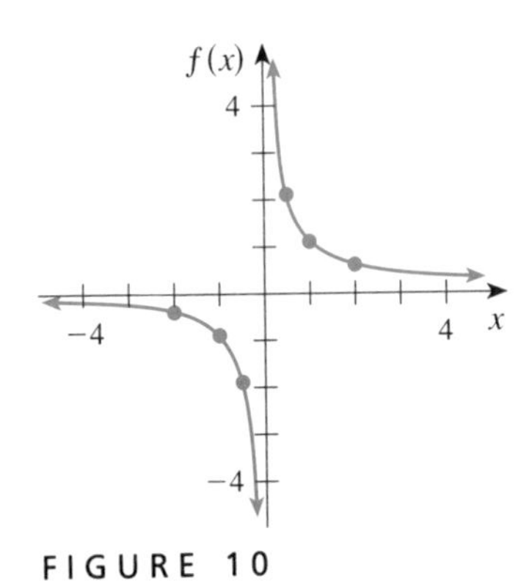

FIGURE 10

■

The curve produced in Example 6 is known as a **hyperbola**. A major difference between polynomial functions and rational expression functions involves the domain (the allowable input values) of the function. For a polynomial function, the domain is $\{x \mid x \text{ is a real number}\}$. For the rational function $f(x) = \frac{1}{x}$, the domain is $\{x \mid x \neq 0\}$. In general, the domain of a rational function is the domain of definition for the rational expression. The graph of $f(x) = \frac{1}{x}$ has neither a y-intercept nor an x-intercept.

EXAMPLE 7

Solution We first produce a table of input-output ordered pairs. Plotting these ordered pairs and sketching a smooth curve, we get the graph shown in Figure 11.

Input-output values for $g(x) = \frac{2}{x}$

x	$g(x)$
-2	-1
-1	-2
0	undefined
1	2
2	1

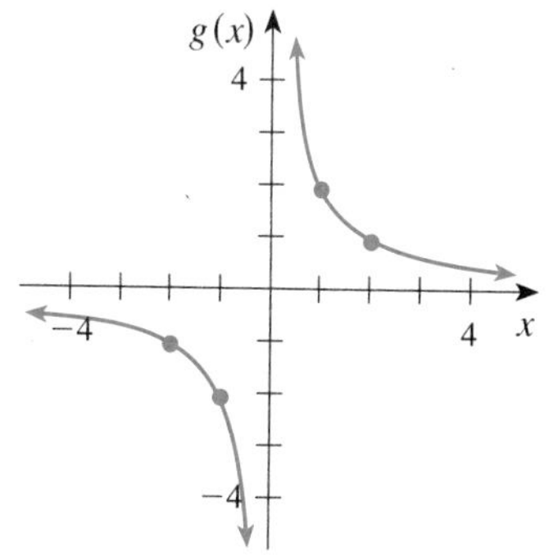

FIGURE 11

As you can see, the graph of this rational function is a hyperbola. It is a distortion of the graph of $f(x) = \frac{1}{x}$ because $g(x) = 2f(x)$. The domain is still $\{x \mid x \neq 0\}$. ■

Next, we consider the graph of $h(x) = \frac{-1}{x}$.

EXAMPLE 8 Graph: $h(x) = \frac{-1}{x}$

Solution A table of input-output ordered pairs is produced first. Plotting these ordered pairs and sketching a smooth curve, we get the graph shown in Figure 12.

Input-output values for $h(x) = \frac{-1}{x}$

x	$h(x)$
-2	$\frac{1}{2}$
-1	1
0	undefined
1	-1
2	$\frac{-1}{2}$

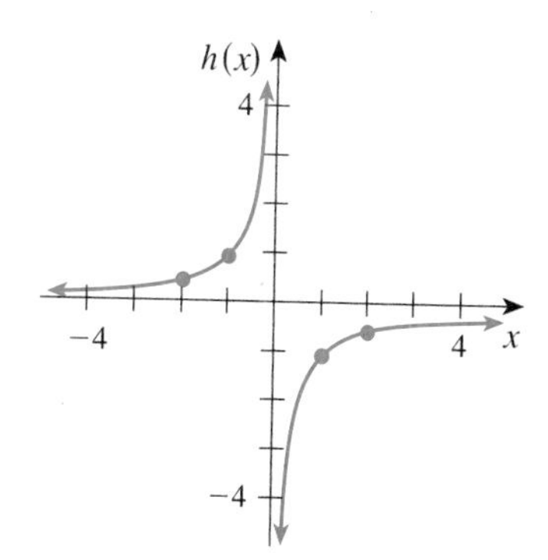

FIGURE 12

This graph is the reflection of the hyperbola for $f(x) = \frac{1}{x}$ across the x-axis because $h(x) = -f(x)$. The domain is still $\{x \mid x \neq 0\}$. ■

Next, we consider the graph of $k(x) = \frac{1}{x} + 1$. As with polynomial functions, vertical translations and horizontal translations hold for rational functions.

EXAMPLE 9 Graph: $k(x) = \dfrac{1}{x} + 1$

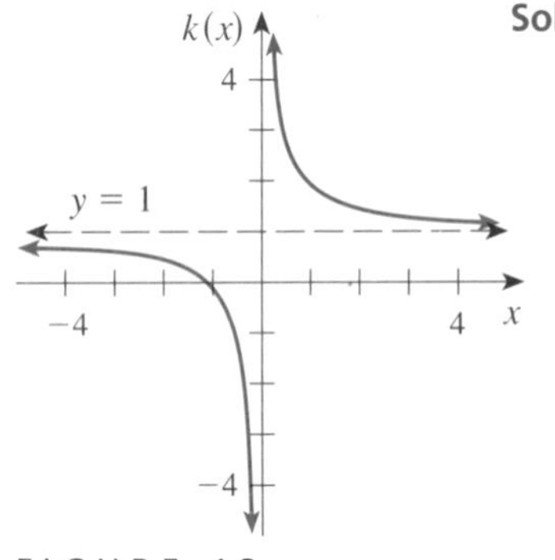

FIGURE 13

Solution Rather than using a set of table values, we produce this graph by translating the graph of $f(x) = \dfrac{1}{x}$ upward 1 unit. We use the dashed line $y = 1$ as a reference line. See Figure 13. ■

Just as the graph of $g(x) = (x - 2)^2$ is the graph of $f(x) = x^2$, translated right 2 units, we have a similar relationship for rational functions. The graph of $r(x) = \dfrac{1}{x - 2}$ is the graph of $f(x) = \dfrac{1}{x}$, translated right 2 units. To see this type of translation, consider the next example.

EXAMPLE 10 Graph: $r(x) = \dfrac{1}{x - 2}$

Solution Rather than using a set of table values, we produce this graph by translating the graph of $f(x) = \dfrac{1}{x}$ right 2 units. We use the dashed line $x = 2$ as a reference line. See Figure 14. Notice that this graph has a y-intercept, $\left(0, \dfrac{-1}{2}\right)$, but no x-intercept. The domain is $\{x \mid x \neq 2\}$. ■

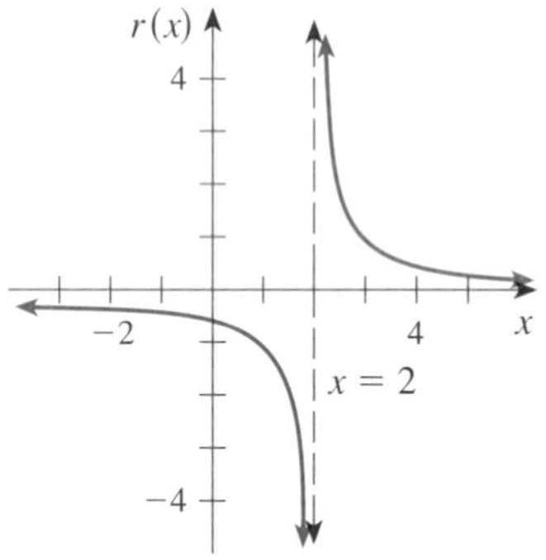

FIGURE 14

If the rule for the function in Example 9 is written as a completely reduced rational expression, we get $\dfrac{x + 1}{x}$. Of course, using long division (or synthetic division, in this case, because of the first-degree divisor), we could write $\dfrac{x + 1}{x}$ as:

$$x\overline{)\,x + 1\,}\ \text{quotient } 1;\quad \underline{x}\quad \text{remainder } 1 \qquad \text{or} \qquad \frac{x + 1}{x} = 1 + \frac{1}{x}$$

As our last example of this introduction to rational expression functions, we graph $h(x) = \dfrac{x + 2}{x - 1}$. Rather than constructing a table of values, we will use long division of polynomials to write the rule in a form for which we can use vertical and horizontal translations.

EXAMPLE 11 Graph: $h(x) = \frac{x + 2}{x - 1}$

Report the ordered-pair names for the x- and y-intercepts.

Solution Using long division on the function rule, we get:

$$\begin{array}{r} 1 \\ x - 1 \overline{)\, x + 2} \\ \underline{x - 1} \\ 3 \end{array}$$

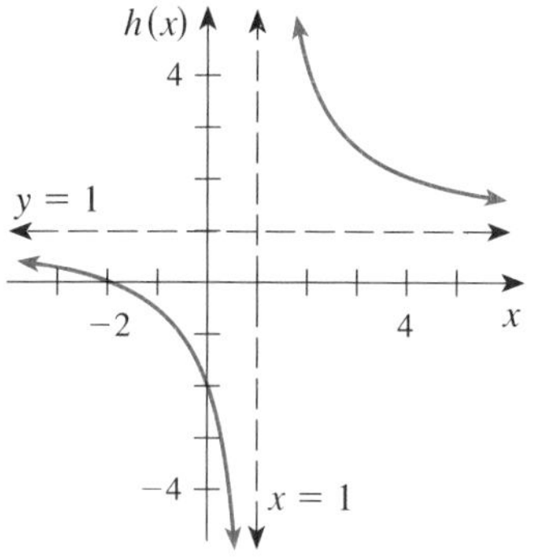

So, $h(x) = \frac{3}{x - 1} + 1$. Now, the graph is the graph of $f(x) = \frac{3}{x}$, translated right 1 unit $\left(\text{the } \frac{3}{x - 1} \text{ part of the rule}\right)$ and the translated up 1 unit (the $+ 1$ part of the rule). We get the hyperbola shown in Figure 15.

FIGURE 15

To find the y-intercept, we find $h(0)$. Here, we have $h(0) = \frac{0 + 2}{0 - 1} = -2$. So, the y-intercept is $(0, -2)$.

To find the x-intercept, we solve the equation

$$0 = \frac{x + 2}{x - 1}$$

Multiplying each side by $(x - 1)$, we get: $0 = x + 2$

$$x = -2$$

So, the x-intercept is $(-2, 0)$. ■

Using the Graphics Calculator

To increase our understanding of graphing rational functions, type `1/X` for `Y1`. View the graph screen using Window 4.7 range settings. Press the `TRACE` key. The message, `X=0 Y=` , means that for an input of 0, the output value is undefined. The domain is $\{x \mid x \neq 0\}$. Press the right arrow key until the message reads `X=.5 Y=2`. For an input of $\frac{1}{2}$ the output is 2 $\left(\text{the reciprocal of } \frac{1}{2}\right)$. Continue to press the right arrow key. As the input value gets larger and larger, the output value gets closer and closer to 0. However, the hyperbola determined by this function has no x-intercept (the output value never gets to zero). Edit the range settings to those shown in Figure 16.* Press the `TRACE` key and move the cursor until the message reads `X=.05 Y=20`. Now, move the cursor to the left and observe that as the input value gets closer to 0, the output value gets larger and larger. Continue until the message reads `X=-.01 Y=-100`. For negative input values, the reciprocal is negative.

Edit the function rule for `Y1` to `1/X+1`. Use the Window 4.7 range settings to view the graph screen. This graph in the hyperbola determined by $f(x) = \frac{1}{x}$ translated up 1 unit. It still has no y-intercept, but it

```
RANGE
Xmin=-.48
Xmax=.47
Xscl=1
Ymin=-100
Ymax=100
Yscl=10
Xres=1
```

FIGURE 16

TI-82 Note *Type `-.47` for `Xmin`.

does have an x-intercept of $(0, -1)$. Edit the function rule for `Y1` to `1/(X+1)`. Press the `TRACE` key. As Figure 17 shows, this graph is the hyperbola determined by $f(x) = \frac{1}{x}$, translated left 1 unit. This graph has no x-intercept, but it now has a y-intercept of $(0, 1)$.

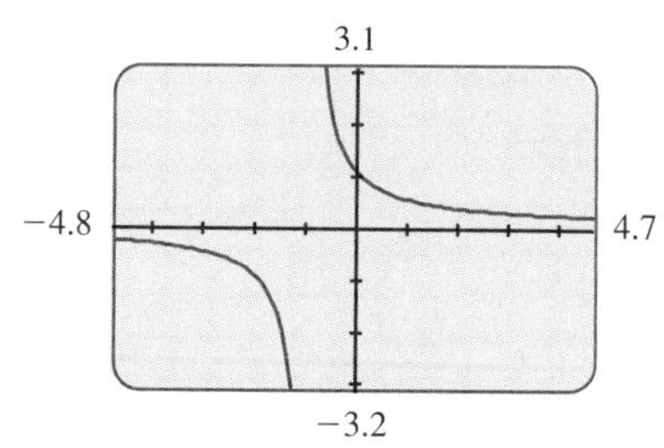

FIGURE 17

Dot Mode With the `TRACE` cursor on $(0, 1)$ and the message `X=0  Y=1` on the screen, select the `Zoom In` option and press the `ENTER` key. The screen appears as shown in Figure 18. The graphics calculator simply generates ordered pairs, plots them, and then connects the dots with a line segment (turns on the pixels between the plotted points). Often this produces a false vertical line, such as the one at `X=-1` shown in Figure 18. To avoid the display of such vertical lines, press the `MODE` key. Move the highlight to `Dot` on the line `Connected Dot`. Press the `ENTER` key. Press the `GRAPH` key to return to the graph screen. As Figure 19 shows, the false vertical line no longer appears. In this mode, the graphics calculator simply generates ordered pairs and plots them (it does not connect the points plotted). For this reason the `Dot` option is often selected when graphing rational functions. Set the mode back to `Connected`.

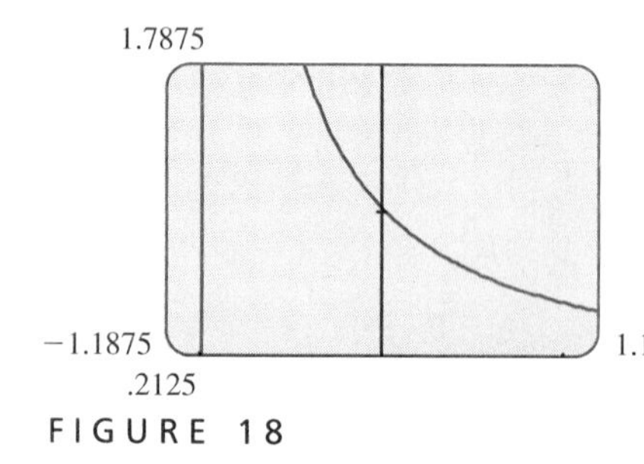

FIGURE 18

To view the graph of the function in Example 11, for `Y1` type `(X+2)/(X−1)`. Use the Window 9.4 range settings to view the graph screen, as shown in Figure 20. Use the `TRACE` key. Does the graph agree with the one we produced in Example 11?

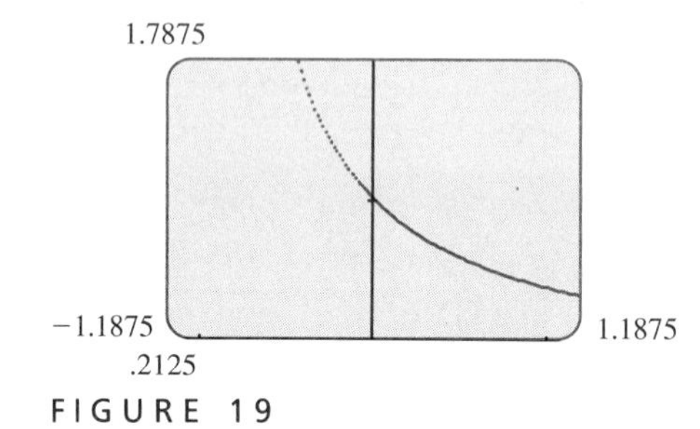

FIGURE 19

6.2
−9.6
9.4
−6.4

FIGURE 20

GIVE IT A TRY

5. For the rational expression function $h(x) = \frac{x-2}{x-1}$, complete the following table to generate ordered pairs on the graph. Plot the ordered pairs, and sketch in the hyperbola.

x	-2	-1	0	1	2
$h(x)$					

6. Use long division to rewrite the rule for $h(x) = \frac{x-2}{x-1}$. From the result of the long division, describe the function in terms of the hyperbola determined by $f(x) = \frac{1}{x}$ using the terms *translation* and *reflection*.

Radical Expression Functions The basic first-degree polynomial function is $f(x) = x$, the basic second-degree polynomial function is $f(x) = x^2$, and the basic third-degree polynomial function is $f(x) = x^3$. The basic rational expression function is $f(x) = \frac{1}{x}$, and the basic radical expression function is $f(x) = \sqrt{x}$. A **radical expression function** (or simply a radical function) is a function whose rule is a radical expression. As with rational functions, radical functions can produce a wide variety of graphs. In this introduction, we are concerned only with square roots and radicands that are first-degree polynomials, such as $g(x) = \sqrt{x-2}$. We start by considering the graph of the basic function $f(x) = \sqrt{x}$.

EXAMPLE 12 Graph: $f(x) = \sqrt{x}$

Solution We start by producing a table of input-output ordered pairs, as shown here. In graphing, we only consider real numbers. Plotting these real number ordered pairs and sketching a smooth curve, we get the graph shown in Figure 21.

Input-output values for $f(x) = \sqrt{x}$

x	$f(x)$
-2	undefined
-1	undefined
0	0
1	1
2	$\sqrt{2}$
3	$\sqrt{3}$

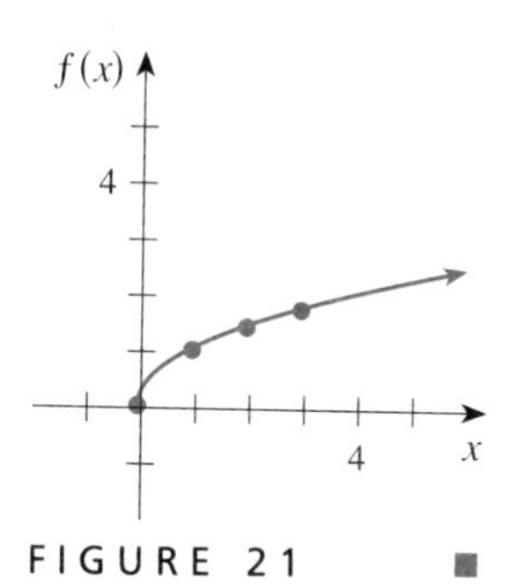

FIGURE 21

The curve produced in Example 12 is one-half of a parabola. The domain of a rational expression function excludes one or two numbers (values that result in its denominator evaluating to 0). Radical functions can exclude intervals of numbers from the domain. For $f(x) = \sqrt{x}$, the domain is $\{x \mid x \geq 0\}$. The range of this function—the output values produced—is $\{y \mid y \geq 0\}$. The y-intercept and the x-intercept for the graph is (0, 0).

EXAMPLE 13 Graph: $g(x) = -\sqrt{x}$

Solution The graph of this radical function is a reflection across the x-axis of the graph for $f(x) = \sqrt{x}$. Both graphs are shown in Figure 22. Here the domain is still $\{x \mid x \geq 0\}$, but the range is $\{y \mid y \leq 0\}$.

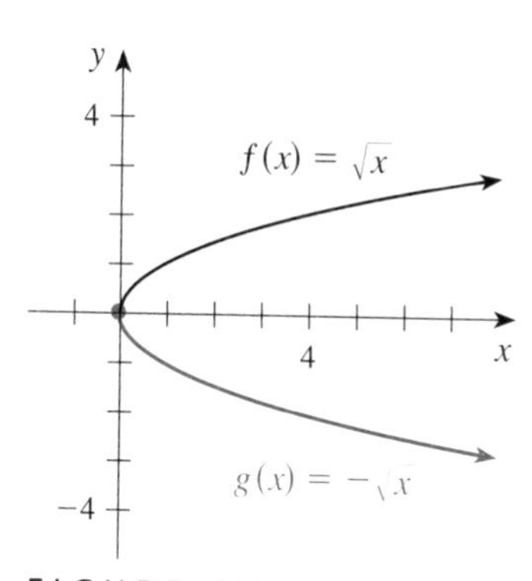

FIGURE 22

EXAMPLE 14 Graph: $h(x) = \sqrt{x + 2}$

Solution Just as the graph of $r(x) = (x + 2)^2$ is a translation, 2 units left, of the graph of $f(x) = x^2$, the graph of $h(x) = \sqrt{x + 2}$ is a translation, 2 units left, of the graph of $f(x) = \sqrt{x}$. See Figure 23. The domain is $\{x \mid x \geq -2\}$, and the range is $\{y \mid y \geq 0\}$. The x-intercept is $(-2, 0)$, and the y-intercept is $(0, \sqrt{2})$.

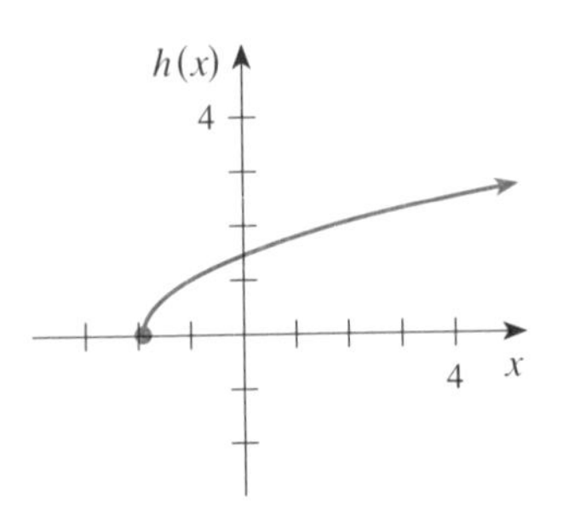

FIGURE 23

■

Using the Graphics Calculator

To increase our understanding of the graphs of radical functions, type √X for Y1. View the graph screen using the Window 9.4 range settings. Press the TRACE key and then the left arrow key until the x-value is negative. The domain of this function is $\{x \mid x \geq 0\}$. For Y2 type -√X and view the graph screen. From the graphs shown in Figure 24, we can see that the graph of $g(x) = -\sqrt{x}$ is the reflection of the graph of $f(x) = \sqrt{x}$. The two curves together form a horizontal parabola. (We will study such parabolas in Chapter 12.) Edit the entry for Y2 to √X+1. View the graph screen. The graph drawn is a vertical translation, up 1 unit, of the graph of $f(x) = \sqrt{x}$. Edit the entry for Y2 to √(X+1). View the graph screen. The new graph is a horizontal translation, left 1 unit, of the graph of $f(x) = \sqrt{x}$. Do you think the two curves drawn intersect at some point? Use the TRACE key to try to locate an input value at which both functions output the same value. Such an input value must solve the equation $\sqrt{x + 1} = \sqrt{x}$. What is the solution to this equation in one variable?

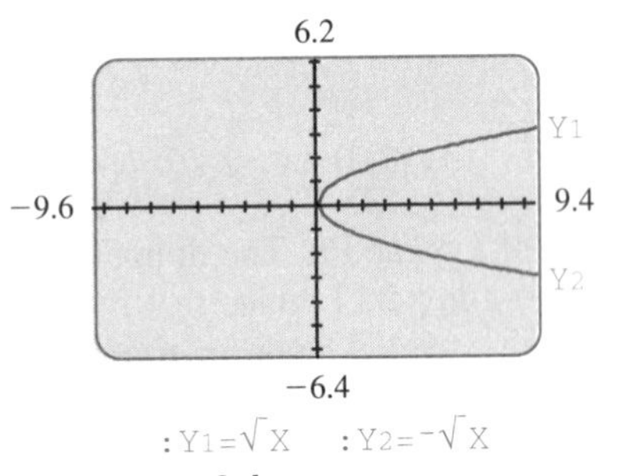

FIGURE 24

GIVE IT A TRY

For the radical function $g(x) = \sqrt{x} + 2$, complete the following.

7. Report the y-intercept.

8. Report the x-intercept.

9. Report the domain.

10. Report the range.

11. Graph the function.

12. Describe the graph in terms of a translation of the graph of $f(x) = \sqrt{x}$.

■ SUMMARY

■ ***Quick Index***

In this section, we have introduced the graphs of polynomial functions of degree greater than 2. We then introduced rational expression functions and their graphs, and finally, radical expression functions and their graphs.

The graphing techniques we discussed in Section 8.6 work with these functions. The basic polynomial functions have the rules x, x^2, x^3, and so on. The basic rational expression function has the rule $\frac{1}{x}$, and the basic radical function has the rule $\sqrt{x}$. Knowing the graph of these basic functions and using graphing techniques (distortions, reflections, and translations), we can find the the graphs of many related functions.

Unlike polynomial functions, rational and radical functions have a restricted domain. Every polynomial function has a domain of all real numbers. A rational function has a domain of all real numbers except the numbers that result in its denominator evaluating to 0. For example, $f(x) = \frac{x+2}{x-1}$ has a domain of $\{x \mid x \neq 1\}$. A radical function involving an even root, such as the square root, has a domain of intervals of real numbers. For example, $f(x) = \sqrt{x-1}$ has the domain $\{x \mid x \geq 1\}$.

8.7 ■ ■ ■ EXERCISES

Match graphs A–F with the polynomial functions in Exercises 1–6.

A.

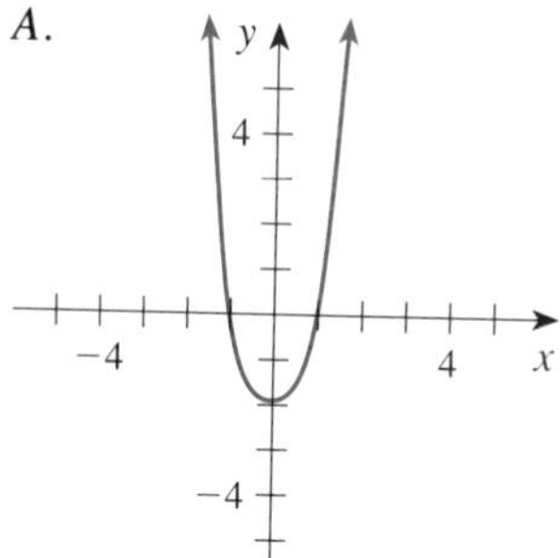

B.

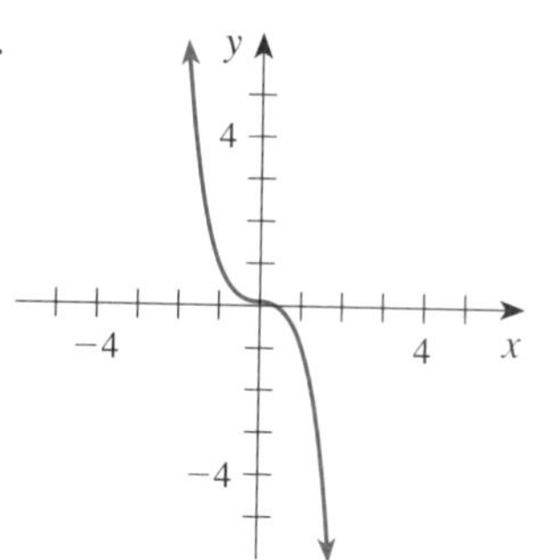

C.

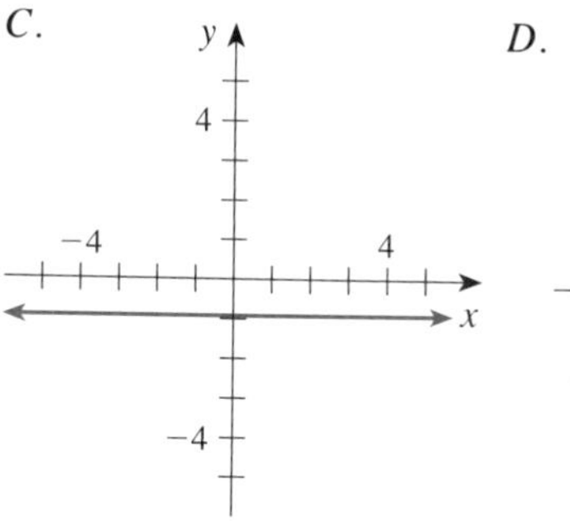

D.

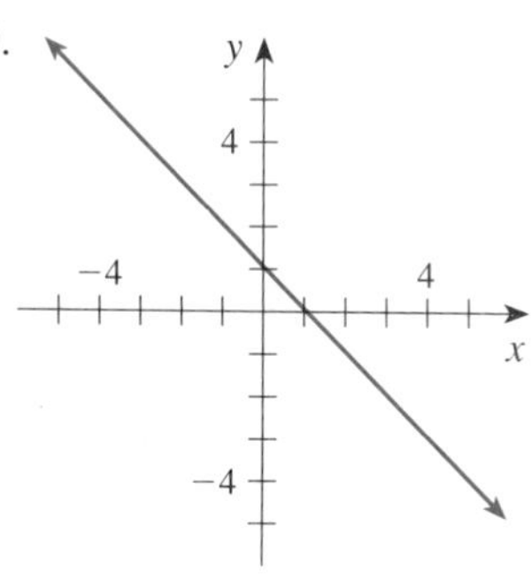

E.

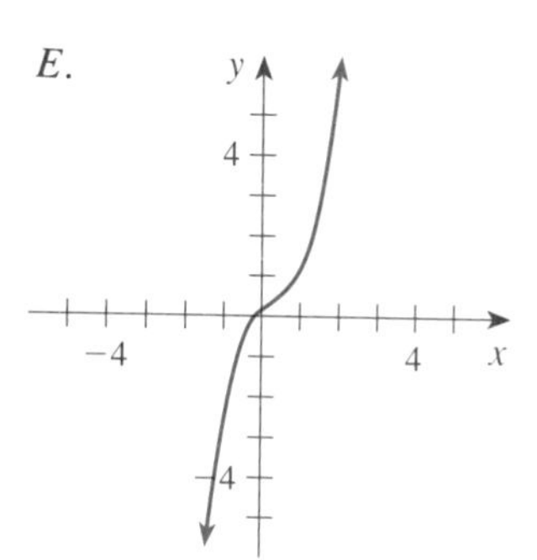

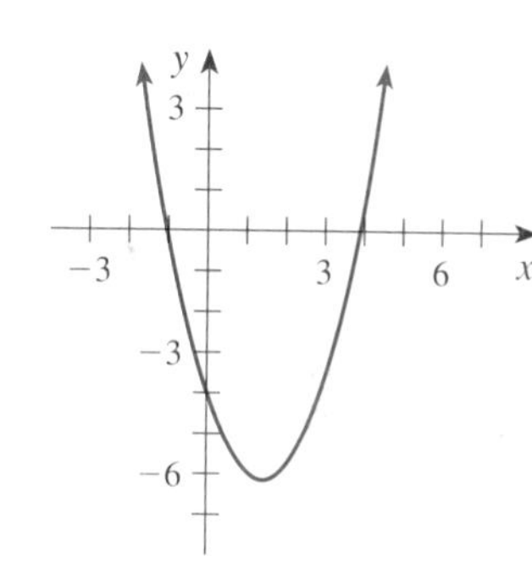

1. $f(x) = x^2 - 3x - 4$
2. $g(x) = x^3 - x^2 + x$
3. $h(x) = -x + 1$
4. $n(x) = -1$
5. $r(x) = x^4 + x^2 - 2$
6. $s(x) = -x^3$
7. **a.** For $f(x) = x^3 + 2x^2 - 3x$, complete the following table.

x	-3	-2	-1	0	1	2
$f(x)$						

b. Find the y-intercept for the graph.

c. Find the x-intercepts.

d. Plot the ordered pairs, and sketch a curve to fit the points.

8. **a.** Use the graph drawn in Exercise 7 to sketch the graph of $g(x) = x^3 + 2x^2 - 3x - 1$.
   b. Use the graphics calculator to approximate, to hundredths, the x-intercepts for the graph of g in part (a).

Use the graph of $f(x) = \frac{1}{x}$ to graph each rational expression function. Report the x- and y-intercepts and the domain of the function.

9. $g(x) = \frac{1}{x} - 1$
10. $h(x) = \frac{1}{x} + 2$
11. $d(x) = \frac{-1}{x} + 2$
12. $s(x) = \frac{-1}{x} - 1$
13. $k(x) = \frac{1}{x - 2}$
14. $n(x) = \frac{1}{x + 2}$
15. $h(x) = \frac{-1}{x + 2}$
16. $c(x) = \frac{-1}{x - 2}$

Use long division to rewrite the rational expression for each rational function. Graph the function and report the x- and y-intercepts.

17. $d(x) = \frac{2x + 1}{x}$
18. $s(x) = \frac{2x - 1}{x}$
19. $m(x) = \frac{2x + 1}{x - 1}$
20. $p(x) = \frac{2x + 1}{x + 1}$

Use the graph of $f(x) = \sqrt{x}$ to graph each radical expression functions. Report the x- and y-intercepts and the domain and range.

21. $g(x) = \sqrt{x} - 2$
22. $h(x) = \sqrt{x} + 1$
23. $k(x) = -\sqrt{x} + 1$
24. $n(x) = -\sqrt{x} - 1$
25. $c(x) = \sqrt{x + 1}$
26. $k(x) = \sqrt{x - 3}$
27. $d(x) = -\sqrt{x + 2}$
28. $p(x) = -\sqrt{x - 1}$
29. $h(x) = \sqrt{x + 1} + 1$
30. $v(x) = -\sqrt{x - 1} + 1$

CHAPTER 8 REVIEW EXERCISES

Add a number to the given polynomial to obtain a perfect square binomial. Report the resulting polynomial factored as a perfect square binomial.

1. $x^2 - 8x$
2. $x^2 - 6x$
3. $x^2 + 12x$
4. $x^2 - 14x$
5. $x^2 - 5x$
6. $x^2 + 3x$

Solve each equation by completing the square. Report the equation in $(x + a)^2 = b$ form, and report its solution.

7. $x^2 - 8x = 9$
8. $x^2 + 6x = 27$
9. $x^2 + 2x = 23$
10. $x^2 - 2x = 13$
11. $x^2 - 3x = 6$
12. $x^2 - 3x = 7$
13. $2x^2 - 4x = 3$
14. $2x^2 - 10x = 7$

Use the quadratic formula to solve each equation. Report the values for a, b, and c; the discriminant; and the solution.

15. $(2x - 3)(x + 1) = 0$
16. $2x^2 = 15 + 7x$
17. $x^2 - x = 13$
18. $x^2 = 11 - 2x$
19. $(2x - 1)^2 = x$
20. $(x + 1)(x - 3) = 5$
21. $2(x - 2)^2 - 3 = 5$
22. $3x^2 - x = 2x + 7$

Use quadratic forms and the substitution technique to solve each polynomial equation. Use the graphics calculator to check your work.

23. $x^4 - 4x^2 - 5 = 0$
24. $x^6 + 4x^3 - 5 = 0$
25. $x^4 - x^2 - 1 = 0$
26. $x^6 - x^3 - 1 = 0$
27. $x^{-2} - x^{-1} = 2$
28. $x^{-4} + 3x^{-2} = 2$
29. $x = x^{1/2} + 6$
30. $x^{1/2} = 2x^{1/4} + 8$

Solve each rational-expression or radical-expression equation. Report only the real numbers in the solution. Use the graphics calculator to check your work.

31. $\frac{2}{x} = \frac{x - 2}{x + 3}$
32. $\frac{x}{x + 3} = \frac{1}{x - 1}$
33. $\frac{1}{x + 3} + \frac{2}{x} = 1$
34. $\frac{1}{x - 1} - \frac{1}{x} = 1$
35. $\sqrt{x + 3} = x + 1$
36. $\sqrt{2x + 3} = x$
37. $\sqrt{x + 2} - \sqrt{x} = 1$
38. $\sqrt{x + 1} - \sqrt{x} = 2$

Draw a number-line graph of the solution to each inequality. Report the interval notation for the solution. Check your work using the graphics calculator.

39. $x^2 - 2x \geq 0$
40. $x^2 + 2x - 3 \leq 0$
41. $(x - 3)^2 > 2$
42. $(x + 2)^2 \leq 3$
43. $(x - 2)(x + 3) > 6$
44. $x(3 - x) \leq 4$
45. $x^2 \geq 2x^2 + 1$
46. $2x^2 \leq x^2 + 1$
47. $x^2 \geq x + 2$
48. $x^2 - 2x \geq 1$
49. $x^2 - 3x > -2$
50. $(x + 1)^2 < -1$
51. $\dfrac{3x}{x - 4} \geq 2$
52. $\dfrac{-1}{2x + 1} < -1$
53. $\dfrac{4}{x} < \dfrac{3}{x - 2}$
54. $\dfrac{1}{x + 4} \geq \dfrac{x + 1}{x - 2}$
55. The base of a triangle is 3 inches more than the altitude drawn to it. If the area measure is 30 square inches, what is the altitude of the triangle?
56. The height h (in feet) reached by a ball t seconds after it is thrown upward is given by the formula $h = -16t^2 + 640t$. When is the height of the ball 300 feet?
57. If the perimeter of a rectangle is 28 inches and its area measure is 10 square inches, what is the width of the rectangle?
58. A 20-foot ladder is leaning against a vertical wall. If the foot of the ladder is 3 feet further from the wall than the height reached by the top of the ladder, what is the height of the ladder on the vertical wall?

Rewrite the formula for the given variable.

59. $x^2 + (y + 1)^2 = 25$ for y
60. $h = -3t^2 + 6t + 3$ for t
61. $\dfrac{1}{x - 3} = \dfrac{x}{y}$ for x
62. $\dfrac{y}{x} = \dfrac{1}{y + 3}$ for y

Report the vertex of the parabola determined by each function, and sketch its graph. Use the graphics calculator to check your answer.

63. $g(x) = x^2 - 4$
64. $h(x) = x^2 + 2$
65. $g(x) = (x + 1)^2$
66. $h(x) = (x - 1)^2$
67. $g(x) = (x + 2)^2 + 1$
68. $h(x) = (x - 1)^2 + 3$

Rewrite each equation in the form $f(x) = (x + h)^2 + k$. Report the vertex, and sketch a graph of the parabola determined. Use the graphics calculator to check your answer.

69. $d(x) = x^2 - 4x + 2$
70. $d(x) = x^2 - 2x - 2$
71. $d(x) = x^2 + 4x + 2$
72. $d(x) = x^2 - 6x + 5$

Graph the given quadratic function. Report the y-intercept, x-intercept(s), the line of symmetry, and the vertex of the graph.

73. $g(x) = x^2 - 4x - 7$
74. $h(t) = 16 - t^2$
75. $r(n) = 5 - 2n - n^2$
76. $f(x) = 8 - 2x - x^2$
77. $h(x) = -x^2 - 3x + 1$
78. $c(x) = -x^2 + 3x$
79. $g(x) = -2x^2 - x - 2$
80. $d(x) = -x^2 - x + 1$

Match graphs A–D with the polynomial functions in Exercises 81–84.

A.

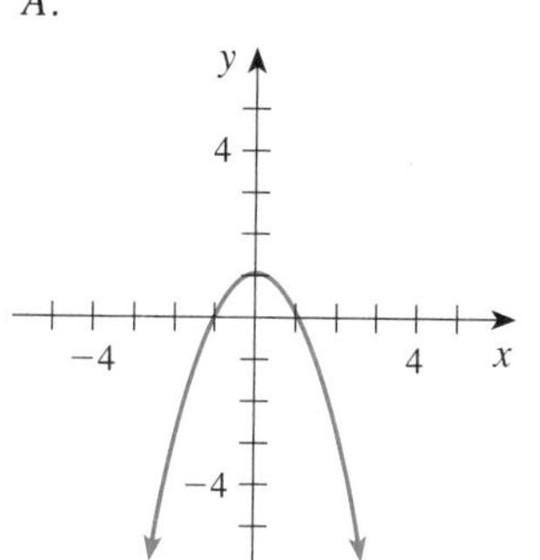

B.

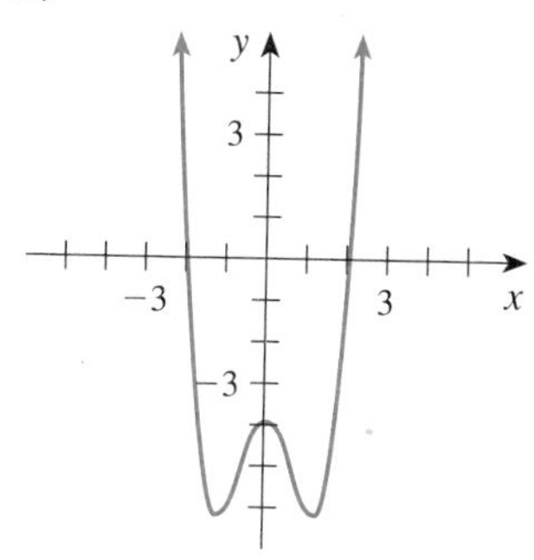

C.

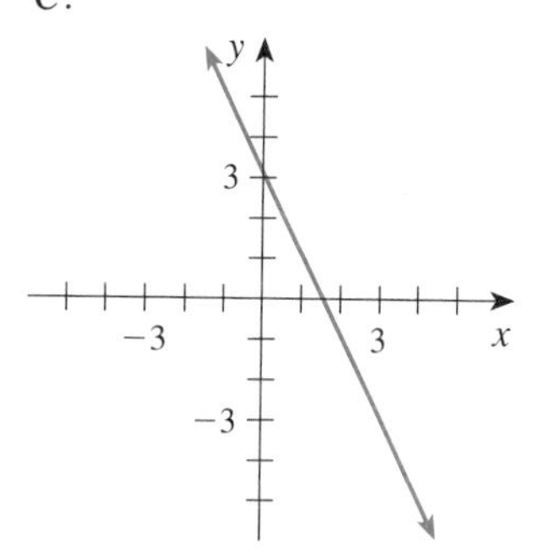

D.

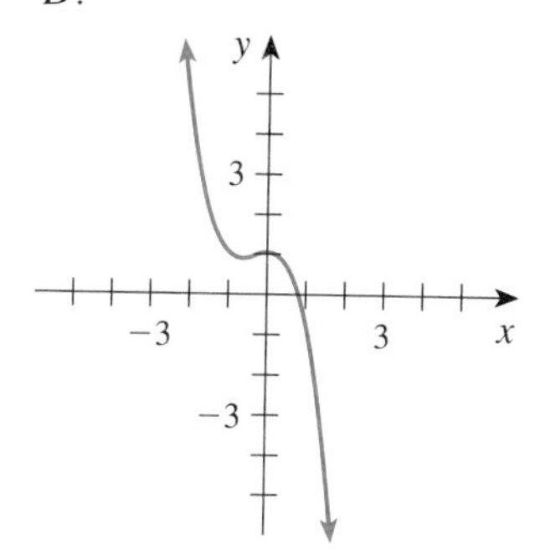

81. $f(x) = x^4 - 3x^2 - 4$
82. $g(x) = -x^3 - x^2 + 1$
83. $h(x) = -x^2 + 1$
84. $n(x) = -2x + 3$

Using the graphs of $f(x) = \dfrac{1}{x}$ and $r(x) = \sqrt{x}$, graph each rational- or radical-expression function. Report the x- and y-intercepts and the domain of the function.

85. $g(x) = \dfrac{1 - x}{x}$
86. $h(x) = \dfrac{x}{x + 1}$
87. $n(x) = \sqrt{3 - x}$
88. $t(x) = 3 + \sqrt{x}$

CHAPTER 8 TEST

1. For $x^2 - 4x = 2$, report the equation in $(x + a)^2 = b$ form, and report its solution.
2. For $2x^2 - x = 5$, report the standard form of the equation; the values of a, b, and c; the value of the discriminant; and the solution to the equation.
3. Use a substitution to solve each equation. Report the solution and the expression represented by u:

   a. $x^{-4} + x^{-2} = 2$ **b.** $x = x^{1/2} + 2$
4. Solve: $\dfrac{2x}{x - 4} = \dfrac{x}{x + 1}$
5. Report the number-line graph and the interval notation for the solution to each inequality:

   a. $x^2 - 3x \geq 1$ **b.** $\dfrac{x}{x - 1} \leq 2$
6. Report the vertex of the graph of $f(x) = (x - 3)^2 + 5$.
7. For $g(x) = -x^2 - 4x + 3$, graph the function, report the y-intercept, the x-intercept(s), the line of symmetry, and the vertex of the graph.
8. Use the graph of $f(x) = \dfrac{1}{x}$ to graph $h(x) = \dfrac{x}{x - 1}$. Report the x- and y-intercepts and the domain for function h.

9

FUNCTIONS

In this chapter, we will bring together some of the ideas of functions that we have discussed earlier. Also, we extend our knowledge of functions by producing new functions using the basic operations of addition, subtraction, multiplication, and division of functions. We then introduced another important operation that produces a new function—composition. Finally, we consider inverse functions and an algorithm for finding the inverse of a function.

9.1 ■ ■ ■ FUNCTION RULES

In this section, we review the basic idea of a function, including input values, the domain, the rule, the output values, and the range. Additionally, we review the evaluations of linear, absolute value, quadratic, rational, and radical functions.

Background Functions are a unifying concept in mathematics. That is, they are used in many areas of mathematics. In earlier chapters we have studied polynomial functions, rational functions, and radical functions. As you continue studying mathematics (either directly in courses such as trigonometry, calculus, and applications, or indirectly, in science and business courses), you will see many additional examples of functions.

In Section 3.4 we introduced a relation as a set of ordered pairs, and in Section 3.5 we defined a function as a *special type of relation* in which each first element of the ordered pairs is matched with exactly one second element. The collection (set) of first elements is known as the *domain* of the function, and the set of second elements is known as the *range* of the function. Consider the following examples.

EXAMPLE 1 Is the relation $\{(-1, 5), (0, 4), (2, 2), (4, 2)\}$ a function? If so, report its domain and range.

Solution This relation is a function. For each first element, there is exactly one second element. We can show this with the *mapping diagram* shown in Figure 1.

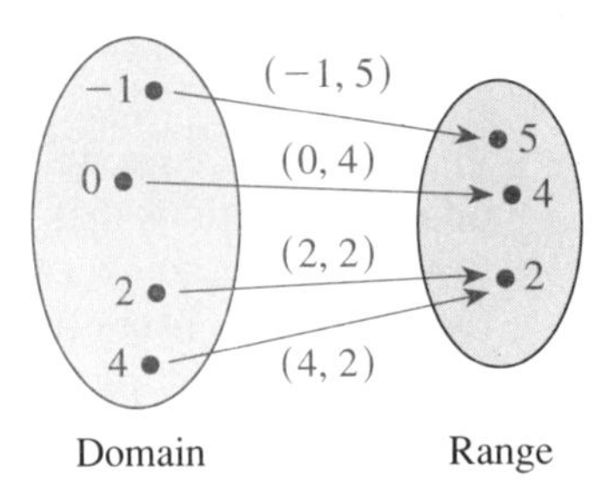

FIGURE 1
A function mapping

From this mapping, we can check that no first element is matched with two different second elements.

The domain $= \{-1, 0, 2, 4\}$. The range $= \{5, 4, 2\}$. ■

EXAMPLE 2 Determine if each mapping (correspondence) represents a function with domain set A. If so, report the range of the function. If not, report why it fails to be a function.

a.

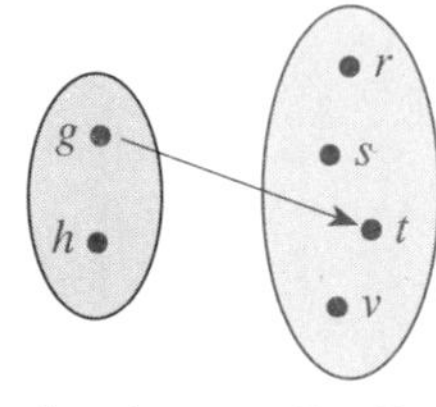

b.

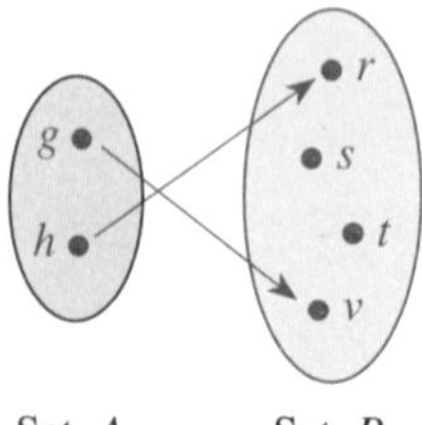

c.

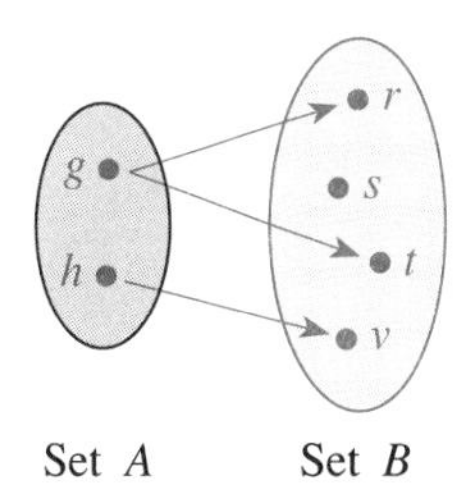

Solution a. This mapping does *not* represent a function with domain set A because it is not defined for each element in set A.

b. This mapping is a function. Its range is $\{r, v\}$.

c. This mapping is *not* a function, because the element g is matched with more than one element, namely, r and t. ■

GIVE IT A TRY

1. Draw a mapping diagram for each relation. Report which are functions.
   a. $\{(-2, 5), (-1, 4), (2, 5), (3, 0), (4, 0)\}$
   b. $\{(4, 0), (2, -1), (1, -1), (0, 0), (3, 0)\}$

2. Determine whether each mapping represents a function from set D:

a.

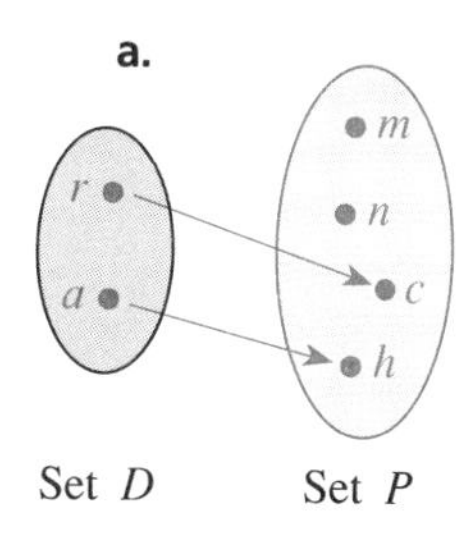

b.

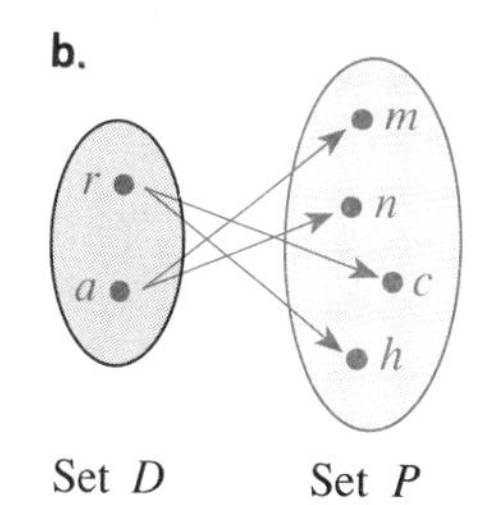

c.

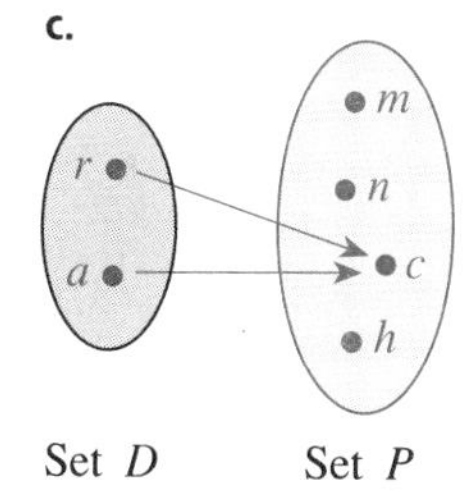

IPO Concept In Section 3.5, we discussed a function as a *rule* that assigns to each element of a set (usually the real numbers) a member of a second set (usually the real numbers). When we classify a function as a linear function, quadratic function, rational function, and so forth, we are referring to the *rule* being a first-degree polynomial, second-degree polynomial, rational expression, radical expression, and so on.

Using the rule idea, we can think of a function in terms of IPO (Input-Process-Output). That is, a function has input values (the domain) that it processes (operates on applying its rule), in order to produce a set of output values (the range). This last idea is very useful for understanding how to work with functions. Consider the following examples.

EXAMPLE 3 For the function $f(x) = 2x - 3$, find the indicated output:

a. $f(-3)$ b. $f(2)$ c. $f(a + 5)$

Solution a. $f(-3)$ means, "For an input of -3, what is the output?" To find the output, we apply the rule $2x - 3$ to the value -3. Thus, we get

$$\begin{aligned} f(-3) &= 2(-3) - 3 \\ &= -6 - 3 \\ &= -9 \end{aligned}$$

For an input of -3, the output is -9. That is, $f(-3) = -9$.

b.
$$\begin{aligned} f(2) &= 2(2) - 3 \\ &= 4 - 3 \\ &= 1 \end{aligned}$$

So, $f(2) = 1$.

c. We use the fact that this function operates on objects by doubling the object and then subtracting 3. This rule is applied to -3 in part (a) and to 2 in part (b). Applying this rule to $a + 5$, we get:

Double the input, $a + 5$, and subtract 3.

$$\begin{aligned} f(a + 5) &= 2(a + 5) - 3 \\ &= 2a + 10 - 3 \\ &= 2a + 7 \end{aligned}$$

So, for an input of $a + 5$, the output is $2a + 7$.
That is, $f(a + 5) = 2a + 7$. ■

WARNING

A common error many students make with the notation $f(a + 5)$ is to conclude that it is the same as $f(a) + f(5)$. This is an error! In general

$$f(a + 5) \neq f(a) + f(5)$$

In Example 3(c), $f(a + 5)$ is $2a + 7$. Now, $f(a) + f(5)$ is $(2a - 3) + 7$, or $2a + 4$. So,

$$f(a + 5) \neq f(a) + f(5)$$

EXAMPLE 4 For $g(x) = x^2 - 3x + 2$, find each output:

a. $g(-2)$ b. $g(3 + h)$

Solution a. We apply the rule $x^2 - 3x + 2$ to the number -2:

$$\begin{aligned} g(-2) &= (-2)^2 - 3(-2) + 2 \\ &= 4 + 6 + 2 \\ &= 12 \end{aligned}$$

So, $g(-2) = 12$.

b. We apply the rule $x^2 - 3x + 2$ to $3 + h$ to get

$$\begin{aligned} g(3 + h) &= (3 + h)^2 - 3(3 + h) + 2 \\ &= 9 + 6h + h^2 - 9 - 3h + 2 \\ &= h^2 + 3h + 2 \end{aligned}$$

So, $g(3 + h) = h^2 + 3h + 2$. ■

GIVE IT A TRY

3. For the function $h(x) = |3 - 2x|$, find $h(7)$.
4. For the function $d(x) = 3x^2 - x$, find $d(-2 + h)$.
5. For the function $f(x) = \frac{1}{x + 5}$, find $f(r - 1)$.

The Difference Quotient This next topic combines the skill of working with functions with the ability to do basic operations on polynomials, rational expressions, and radical expressions. Work through the next example.

EXAMPLE 5 For $f(x) = \frac{1}{x - 3}$, find $f(5 + h) - f(5)$.

Solution First, $f(5 + h) = \frac{1}{5 + h - 3} = \frac{1}{h + 2}$ and $f(5) = \frac{1}{5 - 3} = \frac{1}{2}$

Now, the difference is

$$\begin{aligned} f(5 + h) - f(5) &= \frac{1}{h + 2} - \frac{1}{2} \\ &= \frac{2}{2(h + 2)} - \frac{h + 2}{2(h + 2)} = \frac{2 - (h + 2)}{2(h + 2)} \\ &= \frac{-h}{2(h + 2)} \end{aligned}$$

So, $f(5 + h) - f(5) = \frac{-h}{2(h + 2)}$ ■

We now define an expression that uses function evaluations.

DEFINITION OF THE DIFFERENCE QUOTIENT

> For a function f, the expression $\frac{f(a + h) - f(a)}{h}$, where a and h are two real numbers, is known as the **difference quotient** at a.

The difference quotient is essential to the study of calculus. Here it is simply an application of our ability to work with functions.

EXAMPLE 6 For the rational function in Example 5, find the difference quotient at 5.

Solution The difference quotient at 5 for $f(x)$ is $\frac{f(5 + h) - f(5)}{h}$

From Example 5, we have the numerator, $\frac{-h}{2(h + 2)}$. Now, the difference quotient at 5 is

$$\begin{aligned} \frac{f(5 + h) - f(5)}{h} = \frac{\frac{-h}{2(h + 2)}}{h} &= \left(\frac{-h}{2(h + 2)}\right)\left(\frac{1}{h}\right) \\ &= \frac{-1}{2(h + 2)} \end{aligned}$$

■

EXAMPLE 7 For $f(x) = x^2 - x$, find the difference quotient at -2.

Solution We start by finding the numerator, $f(-2 + h) - f(-2)$:

$$\begin{aligned} f(-2 + h) - f(-2) &= [(-2 + h)^2 - (-2 + h)] - [(-2)^2 - (-2)] \\ &= [4 - 4h + h^2 + 2 - h] - [4 + 2] \\ &= 6 - 5h + h^2 - 6 \\ &= h^2 - 5h \end{aligned}$$

Now, the difference quotient at -2 is

$$\begin{aligned} \frac{f(-2 + h) - f(-2)}{h} &= \frac{h^2 - 5h}{h} \\ &= h - 5 \end{aligned}$$

■

6. For $g(x) = 2x - 1$, report the difference quotient at 3.
7. For $h(x) = \dfrac{-1}{x}$, report the difference quotient at -2.

Equations in Two Variables and Functions Consider the first-degree equation in two variables, $3x + y = 6$. The solution to this equation is a set of ordered pairs. In that set of ordered pairs, every first element (x-value) is matched with *exactly* one second element (y-value). The solution to this equation is a function. We say that the equation defines a function *implicitly*. The x-variable is thought of as the **independent variable** and the y-variable is then the **dependent variable**. If this equation is rewritten for y, the resulting equation is $y = -3x + 6$. In this form, the equation defines a function *explicitly*. In fact, we could write $y = f(x) = -3x + 6$.

Every equation in two variables has a solution that is a relation (a set of ordered pairs), but not every equation in two variables has a solution that is a function. The equation $x^2 + y^2 = 4$ has a solution that is a relation, but is not a function. Its solution contains the ordered pairs $(1, \sqrt{3})$ and $(1, -\sqrt{3})$. This violates the definition of function: "Every first element is matched with exactly one second element." If we rewrite the equation for y, we get two different equations: $y = \sqrt{-x^2 + 4}$ and $y = -\sqrt{-x^2 + 4}$.

■ ***Quick Index***

■ SUMMARY

In this section, we have reviewed some of the ideas of functions discussed in earlier chapters. These include the domain and range of a function. A new idea introduced was mapping the values in the domain to the values in the range. We discussed the ideas of IPO (input-process-output) and function rules as ways of understanding functions. It is often useful to evaluate a function for inputs that are expressions (such as $3 + h$ or $a - 4$). One area in which such evaluations are used is in finding the difference quotient at a particular value.

9.1 ■ ■ ■ EXERCISES

For each relation, draw a mapping diagram, and report whether or not the relation is a function. If so, report the domain and range. If not, report the reason the relation fails to be a function.

1. $\{(-1, 0), (0, -2), (0, -1), (2, 1)\}$
2. $\{(-3, 6), (-2, 4), (-2, 1)\}$
3. $\{(3, 1), (2, 1), (-1, 1), (-2, 1)\}$
4. $\{(a, b), (b, c), (c, d)\ (d, d)\}$

Determine whether each mapping (correspondence) represents a function with domain set A. If so, report the range of the function. If not, report why it fails to be a function.

5.

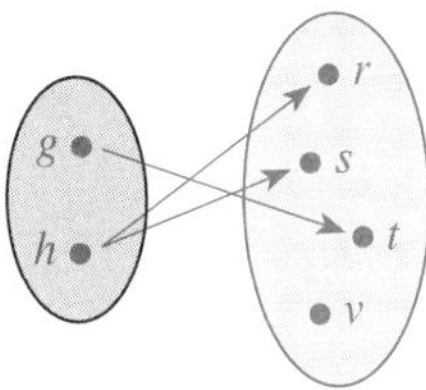

6.

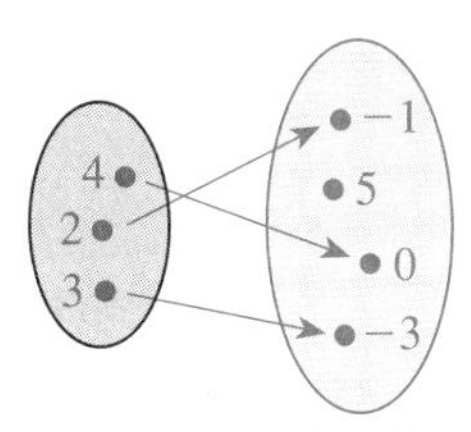

7.

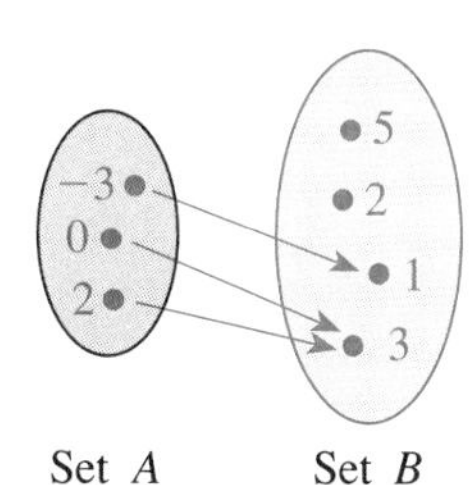

8. Report the domain for each function:

   a. $g(x) = \sqrt{3 - x}$ b. $f(x) = \dfrac{2x - 3}{x + 2}$

For the function $f(x) = -3x + 1$, find each output:

9. $f(-3)$
10. $f(2)$
11. $f(a + 5)$
12. $f(0)$
13. $f(2b)$
14. $f(d - 1)$
15. $f(2 + h) - f(2)$
16. $f(2a + 1) - f(2a)$

For the function $g(x) = x^2 - 2x - 3$, find each output:

17. $g(-1)$
18. $g(3)$
19. $g(r + 2)$
20. $g(0)$
21. $g(4)$
22. $g(s - 3)$
23. $g(-1 + h) - g(-1)$
24. $g(2 - k) - g(2k)$

For the function $h(x) = \dfrac{x}{x + 2}$, find each output:

25. $h(-1)$
26. $h(0)$
27. $h(2)$
28. $h(3)$
29. $h\left(\dfrac{1}{a}\right)$
30. $h(b^{-1})$
31. $h(1 + k) - h(k)$
32. $h(a + 3) - h(a)$

For the function $r(x) = \sqrt{x - 2}$, find each output:

33. $r(3)$
34. $r(5)$
35. $r(a^2)$
36. $r(0)$
37. $r(11)$
38. $r(2b)$
39. $r(2 + h) - r(h)$
40. $r(a^2 - 2a + 3)$

For each function, find the difference quotient at the indicated x-value.

41. $f(x) = 2 - 3x$ at $x = -2$
42. $d(x) = \dfrac{1}{2}x + 1$ at $x = 3$
43. $g(x) = x^2$ at $x = -1$
44. $h(x) = x^2 - 2x$ at $x = 2$
45. $f(x) = \dfrac{1}{x}$ at $x = 1$
46. $f(x) = \dfrac{x - 3}{x + 2}$ at $x = -2$
47. $f(x) = x^3$ at $x = -1$

9.2 ■ ■ ■ GRAPHS OF FUNCTIONS

In this section we review some of the ideas presented in earlier chapters about graphs of functions. We will learn to read graphs and to identify where a function is increasing and where it is decreasing. We then introduce piecewise-defined functions.

Background As we have seen in earlier chapters, we can plot the input-output pairs of a function on a (rectangular) coordinate system. In such a system, the horizontal number line (the x-axis) represents the input values and the vertical number line (the y-axis) represents the output values. Figure 1 gives a quick review of the functions we have graphed so far.

Constant function

$f(x) = 3$
Domain $= \{x \mid x \text{ is real}\}$
Range $= \{y \mid y = 3\}$

Linear function

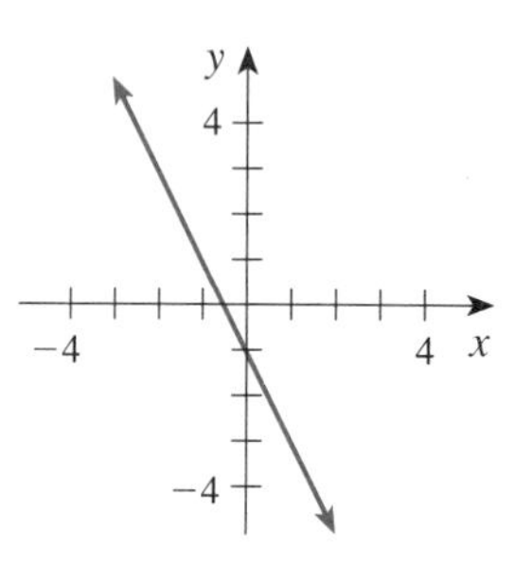

$f(x) = -2x - 1$
Domain $= \{x \mid x \text{ is real}\}$
Range $= \{y \mid y \text{ is real}\}$

Quadratic function

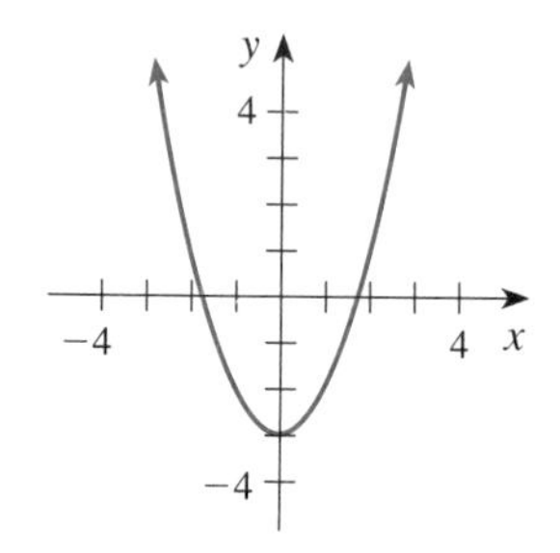

$f(x) = x^2 - 3$
Domain $= \{x \mid x \text{ is real}\}$
Range $= \{y \mid y \geq -3\}$

Rational function

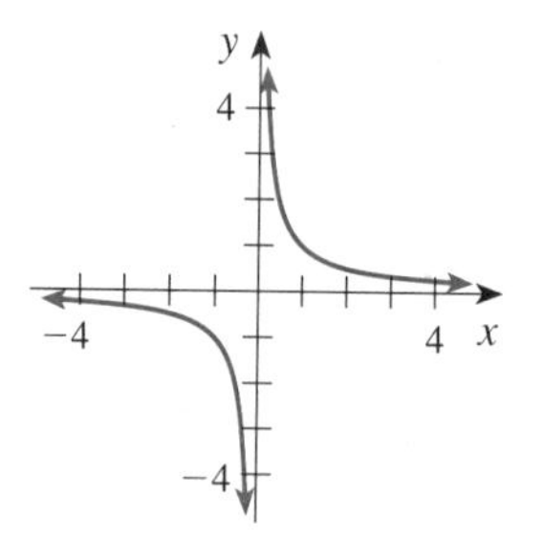

$f(x) = \dfrac{1}{x}$
Domain $= \{x \mid x \neq 0\}$
Range $= \{y \mid y \neq 0\}$

Radical function

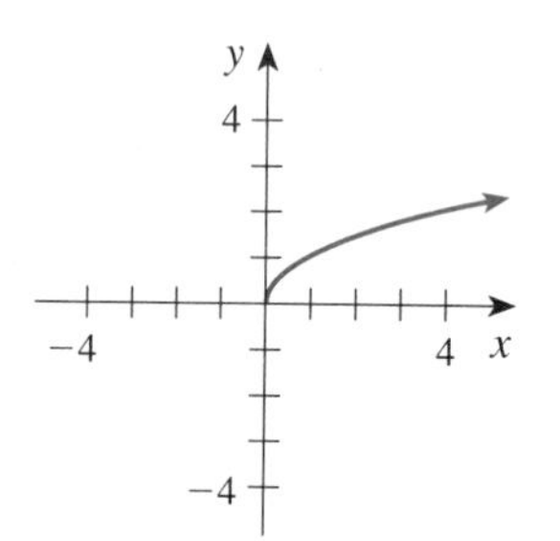

$f(x) = \sqrt{x}$
Domain $= \{x \mid x \geq 0\}$
Range $= \{y \mid y \geq 0\}$

Absolute value function

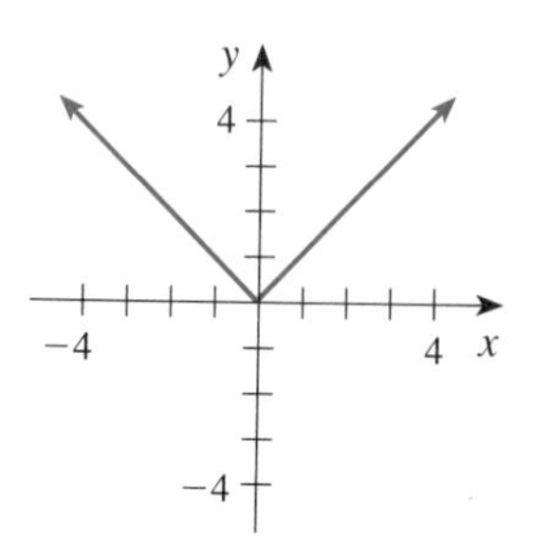

$f(x) = |x|$
Domain $= \{x \mid x \text{ is real}\}$
Range $= \{y \mid y \geq 0\}$

FIGURE 1

As we can see from the graphs, functions can have widely different domains (sets of allowable input values) and ranges (sets of output values produced). In Section 9.1 we introduced a mapping as a quick way to check whether a single value in the domain is matched with more than one value in the range. [If so, the set of ordered pairs (the relation) is not a function.] In a similar fashion, we can quickly view a graph and determine if the plotted points represent the ordered pairs of a function. This is accomplished by the vertical line test.

THE VERTICAL LINE TEST

> If any vertical line $x = a$ intersects a graph more than once, then the graph is *not* the graph of a function.

This test works because a vertical line $x = a$ that intersects the graph twice shows that (a, b) and (a, c) are both points on the graph. The ordered pairs represented by the graph cannot be a function, because the element a in the domain is matched with *more than one element* in the range, b and c. Consider the next example.

EXAMPLE 1 Which one of these graphs is the graph of a function?

a.

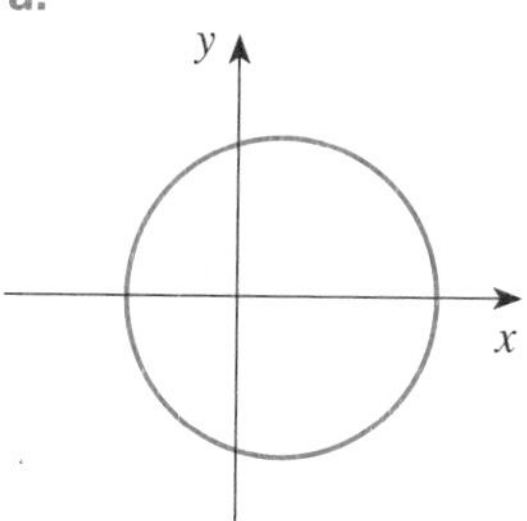

b.

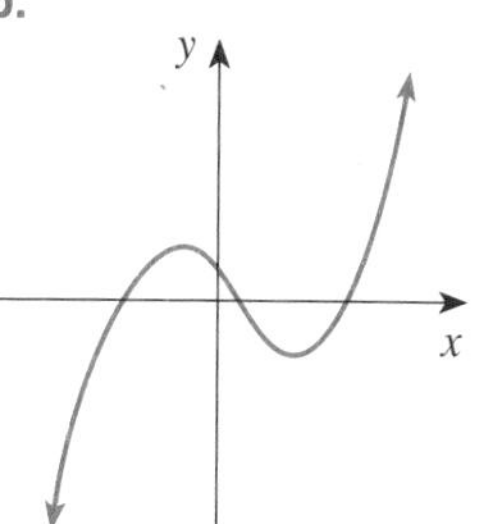

c.

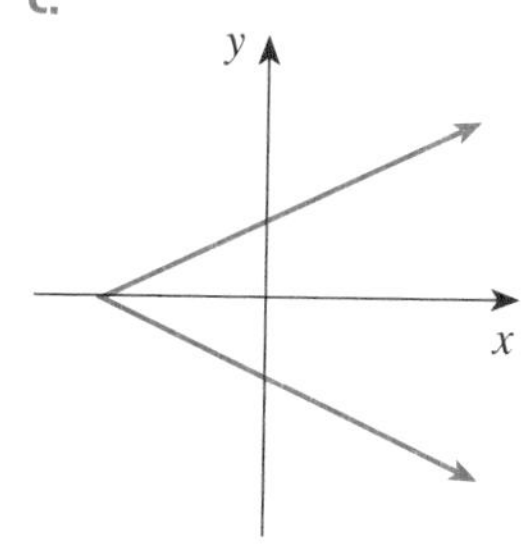

Solution a. This graph is *not* the graph of a function, because it fails the vertical line test. See Figure 2.

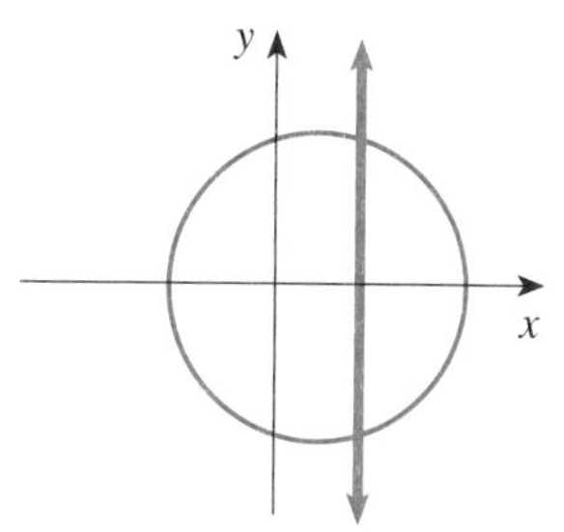

FIGURE 2

b. This graph is the graph of a function, because it passes the vertical line test.

c. The graph is *not* the graph of a function, because it fails the vertical line test. See Figure 3.

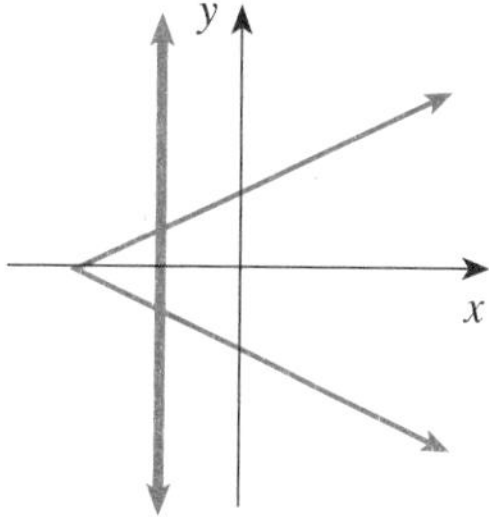

FIGURE 3

■

GIVE IT A TRY

Use the vertical line test to report whether each graph represents a function.

1.

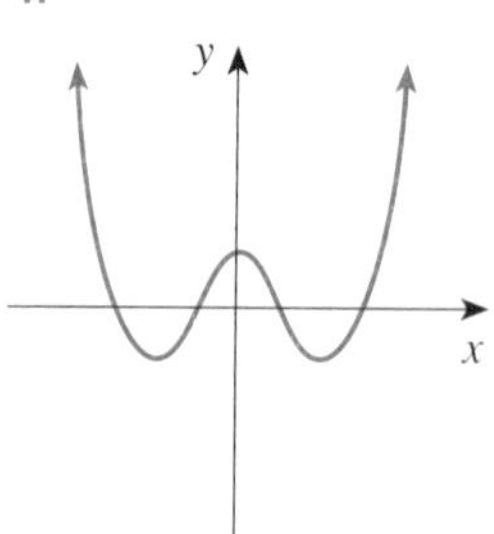

2.

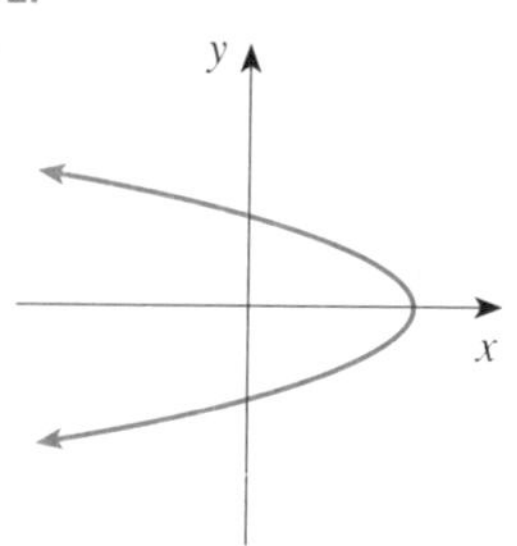

3.

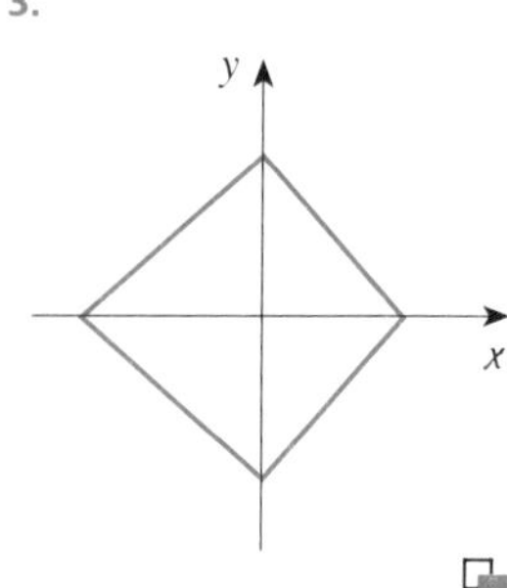

Reading a Graph Not only can a graph be used to identify a function, it can be *read* to identify input-output values. Consider the next example.

EXAMPLE 2 For the graph of function f shown in Figure 4, find each function value:
a. $f(2)$ b. $f(0)$ c. $f(-1)$
d. The input value whose output is 0.

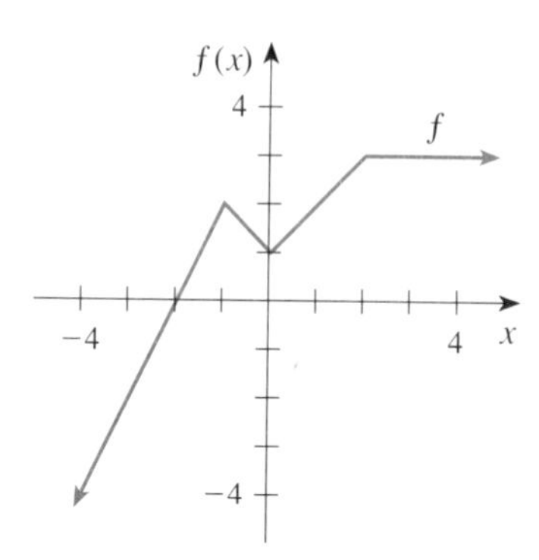

FIGURE 4

Solution a. We move along the x-axis to 2, then up to the graph, and then over to the y-axis, to get 3. Thus, $f(2) = 3$. The ordered pair is (2, 3).

b. We move from the origin, where x is 0, up to the graph, and read the value 1. Thus, $f(0) = 1$.

c. We move left to -1 on the x-axis, then up to the graph, and then over to the y-axis, to read 2. Thus, $f(-1) = 2$. The ordered pair is $(-1, 2)$.

d. We want the input value where the function outputs 0. Any such input is the x-coordinate of the x-intercept. Observing where the graph crosses the x-axis, we see that the x-value is -2. Thus, $f(-2) = 0$. ■

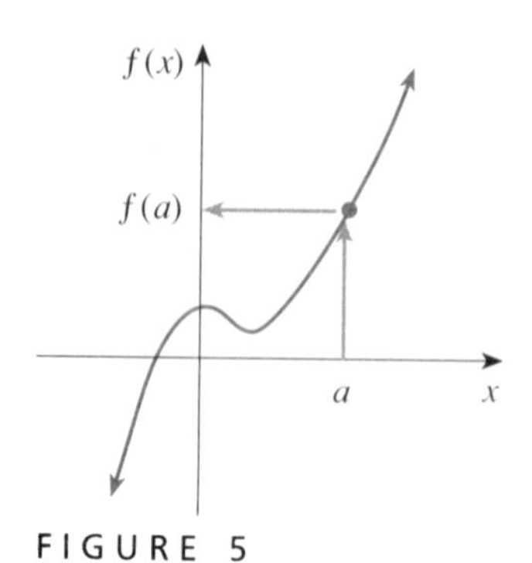

FIGURE 5

The process for reading a graph to find an output value is as shown in Figure 5. We think of this diagram as meaning, "Move to the input value a, read up to the graph, and then over to the y-axis, to get the output value $f(a)$." We often think of a as being *mapped* to $f(a)$.

Using the Graphics Calculator

We can use the graphics calculator to simulate reading the graph of a function. Type `X³+X²−2X` for `Y1`. Using the Window 4.7 range settings, view the graph screen. To find $f(-1)$ for the function $f(x) = x^3 + x^2 - 2x$, we can use the left arrow key (not the `TRACE` key) to move to -1 on the x-axis, and then use the up arrow key to move up to the curve. Now, we move the cursor over to the y-axis, and read the value, 2. That is, $f(-1) = 2$. Of course, this is exactly what the `TRACE` feature does.

GIVE IT A TRY

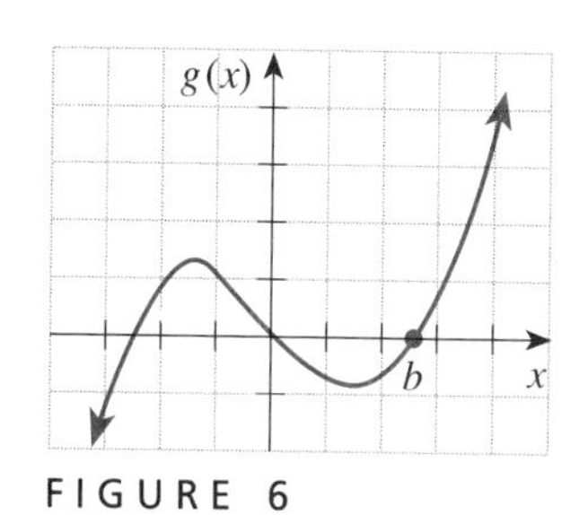

FIGURE 6

Read the graph in Figure 6 to find the function value.

4. $g(-2)$ **5.** $g(b)$

6. What input value(s) produce an output of 1?

Increasing and Decreasing Functions From our earlier work with linear functions, we know that if the slope measure is positive, then the function is increasing. (Its graph is going up, when read from left to right.) Likewise, we know that if the slope measure is negative, then the function is decreasing. (Its graph is going down, read from left to right.) We now want to formalize what we mean by *increasing* and *decreasing functions.*

DEFINITION OF INCREASING AND DECREASING FUNCTIONS

A function f is said to be **increasing on an interval** (a, b) if for values c and d in the interval, with $c < d$, we have $f(c) < f(d)$. A function f is said to be **decreasing on an interval** (a, b) if for values c and d in the interval, with $c < d$, we have $f(c) > f(d)$.

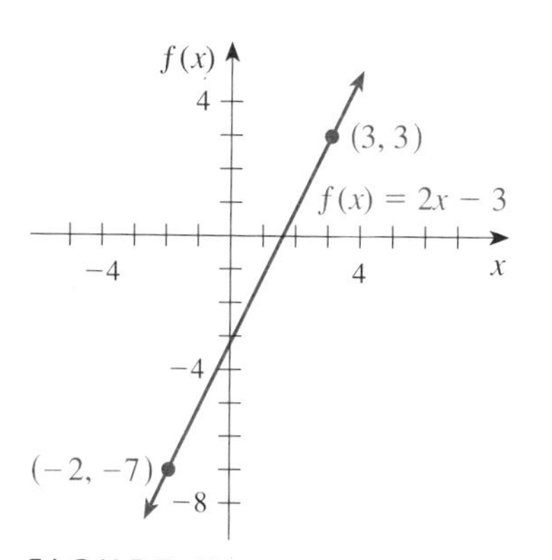

FIGURE 7

For the linear function, $f(x) = 2x - 3$, if we take two values, say -2 and 3, we have $-2 < 3$. Now, is $f(-2) < f(3)$? Because $f(-2) = -7$ and $f(3) = 3$, it is true that $f(-2) < f(3)$. So we conclude the function is increasing. The graph in Figure 7 confirms this fact.

Next, consider the function $g(x) = x^2 - 2x$. Viewing the graph in Figure 8, we can see that for the interval $(-\infty, 1)$, the function is decreasing. For the interval $(1, +\infty)$, the function is increasing. The point $(1, -1)$, the vertex of the

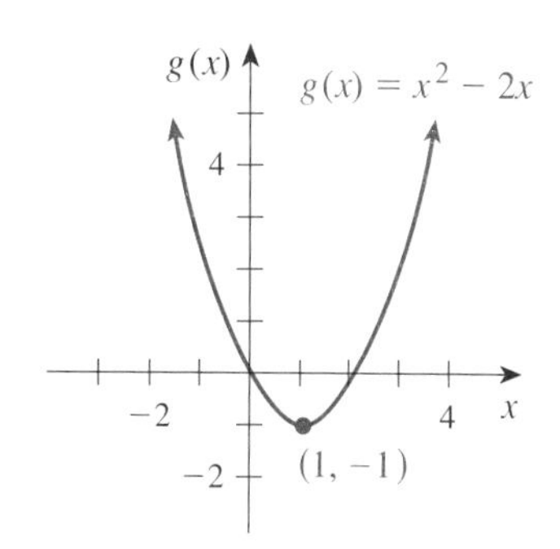

FIGURE 8

parabola in this case, is known as a turning point. A **turning point** is a point on the graph where the graph is increasing on one side of the point and decreasing on the other side of the point.

Using the Graphics Calculator

With the rule `X³+X²−2X` still entered for `Y1`, and using the Window 4.7 range settings, view the graph screen. Press the `TRACE` key. Now press the right arrow key. Notice that as the x-value increases, the function value decreases. We say this function is decreasing. Continue to press the right arrow key until the message reads `X=.5 Y=⁻.625`. This is a turning point for the graph. Press the right arrow key again. Now, as the x-value increases, the function value increases. The graph has another turning point at `X=⁻1.2 Y=2.112`. Use the `TRACE` feature to check the values close to this point.

In addition to increasing, decreasing, and turning points, another important feature of the graph of a function is that of a *zero of a function*. A zero of a function is an input value that results in the function producing an output of 0. Graphically, a zero occurs at the x-intercept. Thus, the linear function $f(x) = 2x - 3$ has a zero at $x = \frac{3}{2}$ (see Figure 7). The quadratic function $g(x) = x^2 - 2x$ has two zeros—one at $x = 0$ and one at $x = 2$ (see Figure 8). Notice that $g(0) = 0$ and $g(2) = 0$.

GIVE IT A TRY

7. For the function whose graph is shown in Figure 9, report the interval(s) where the function is increasing and those where the function is decreasing. Report the ordered-pair names of the turning points. Report the zeros of the function.
8. For $h(x) = x^2 - 3x - 4$, report the interval where the function is decreasing, the interval where the function is increasing, the turning point, and the location of the two zeros of the function.

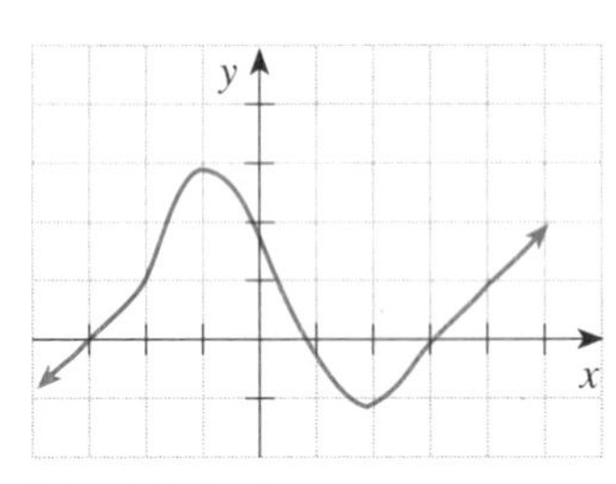

FIGURE 9

Explicit Domains The domain of a function is typically the set of input values where the function rule is defined (makes sense). This is the *understood domain*. We can override this understood domain by explicitly stating the domain or by having it implied by a graph. Consider the following example.

EXAMPLE 3 From the graph of each function, report the domain and the range of the function:

a.

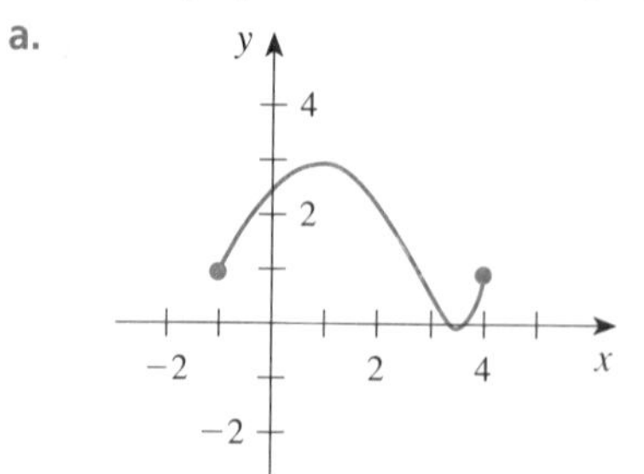

b.

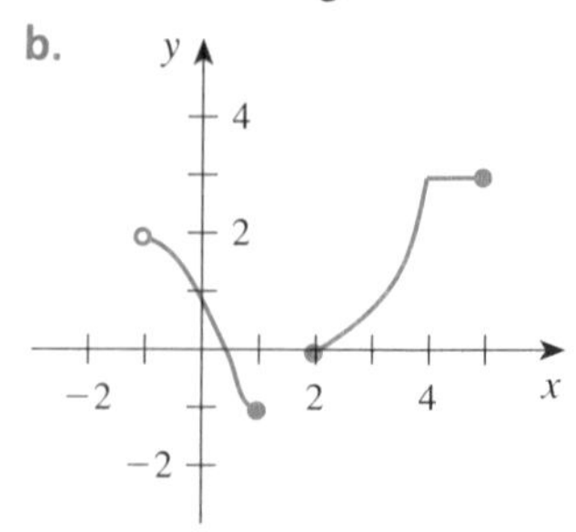

Solution a. The domain is the set of input values. Here, these values run from -1 to 4. In interval notation we write $[-1, 4]$. In set notation we write $\{x \mid -1 \leq x \leq 4\}$. The range is the set of output values. These values run from 0 to 3. We write this in interval notation as $[0, 3]$. In set notation we write $\{y \mid 0 \leq y \leq 3\}$.

b. The domain is the set of values from -1 to 1 (the open circle indicates that -1 is not included), and then from 2 to 5. In interval notation we write $(-1, 1] \cup [2, 5]$. In set notation we write

$$\{x \mid -1 < x \leq 1\} \cup \{x \mid 2 \leq x \leq 5\}$$

The range is $[-1, 3]$. In set notation, we write this as $\{y \mid -1 \leq y \leq 3\}$. ■

Piecewise-Defined Functions To write the rules for functions such as the ones in Example 3, we must write the rule in pieces. That is, a rule is given for each branch, and the function is **piecewise defined**. The most famous piecewise-defined function is that of absolute value:

$$f(x) = \begin{cases} x & \text{if } x \geq 0 \\ -x & \text{if } x < 0 \end{cases}$$

Here, the function has one rule, x, over the interval $[0, +\infty)$. It has a different rule, $-x$, over the interval $(-\infty, 0)$. The graph is shown in Figure 10.

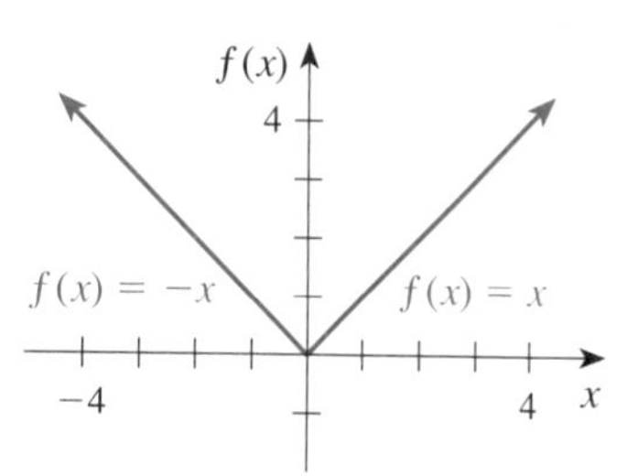

FIGURE 10

Consider the following definition:

$$f(x) = \begin{cases} x + 2 & \text{if } -5 \leq x \leq -1 \\ -2x + 1 & \text{if } x \geq 0 \end{cases}$$

This definition is known as a **piecewise definition** of a function. It means that the function follows the rule $x + 2$ when the input value is from -5 to -1. The function follows the rule $-2x + 1$ when the input value is greater than or equal to zero. A graph of this function appears in Figure 11. Notice that the domain of the function, the input values, are the set of reals from -5 to -1, and 0, and the positive reals. Using set notation, we write

$$\{x \mid -5 \leq x \leq -1\} \cup \{x \mid x \geq 0\}$$

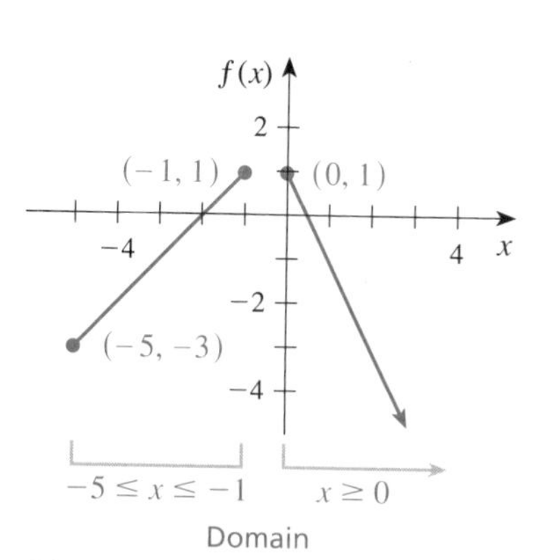

FIGURE 11

The range of this function is the set of values less than or equal to 1. That is, the function will never output a value larger than 1.

The function f has zeros at $x = -2$ and at $x = \frac{1}{2}$. That is, $f(-2) = -2 + 2$, or $f(-2) = 0$. Also, $f\left(\frac{1}{2}\right) = -2\left(\frac{1}{2}\right) + 1$. Thus, $f\left(\frac{1}{2}\right) = 0$.

GIVE IT A TRY

9. Graph the following piecewise-defined function and report the domain, range, and zeros of the function:

$$h(x) = \begin{cases} 2x - 1 & \text{if } x \geq 0 \\ x^2 & \text{if } x \leq -2 \end{cases}$$

■ SUMMARY

■ ***Quick Index***

In this section, we have seen the varieties of graphs of functions produced to this point, including the graphs of linear, quadratic, rational, radical, and absolute value functions. The vertical line test was discussed as a way of quickly judging whether a graph is the graph of a function. Also, we read graphs to obtain input and output values of the function.

We also introduced some important terminology concerning graphs of functions. We defined increasing and decreasing functions, turning points, and the zeros of a function. Finally, we discussed piecewised-defined functions.

9.2 ■ ■ ■ EXERCISES

Use the vertical line test to judge whether each graph is the graph of a function.

1.

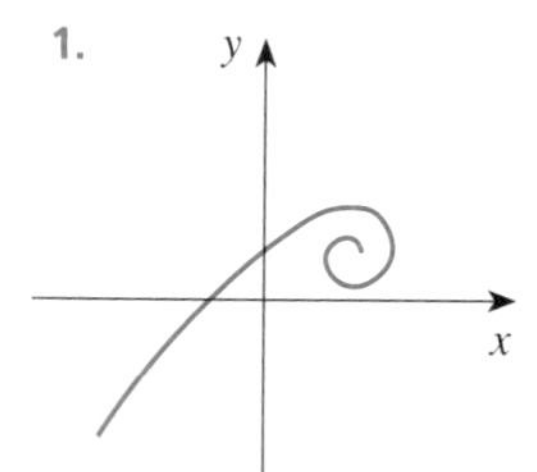

2.

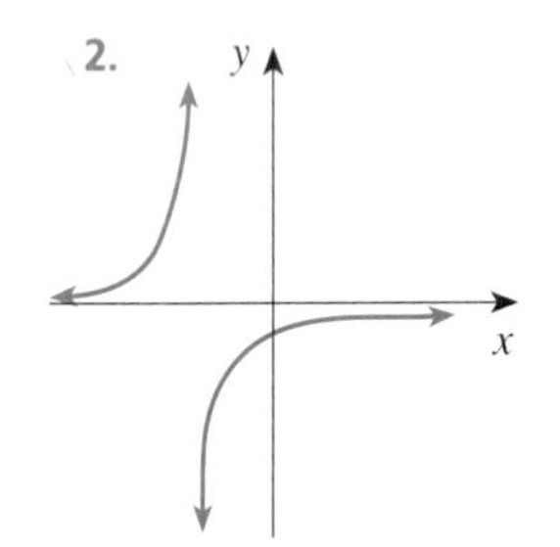

3.

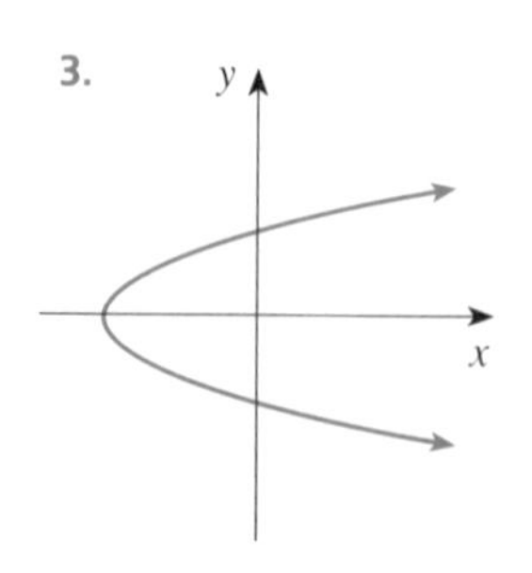

4.

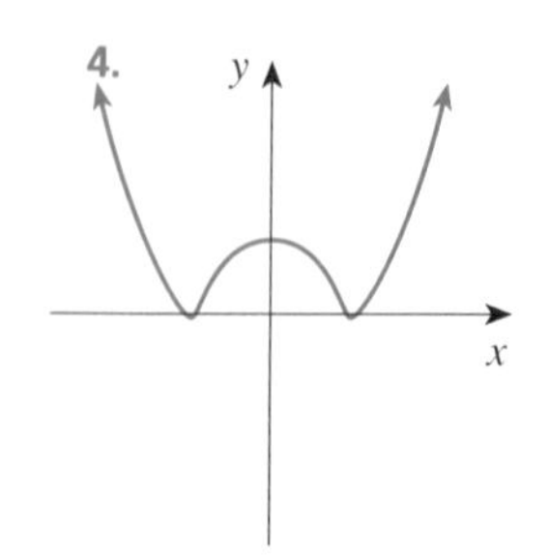

Use the following graph to answer Problems 5–10.

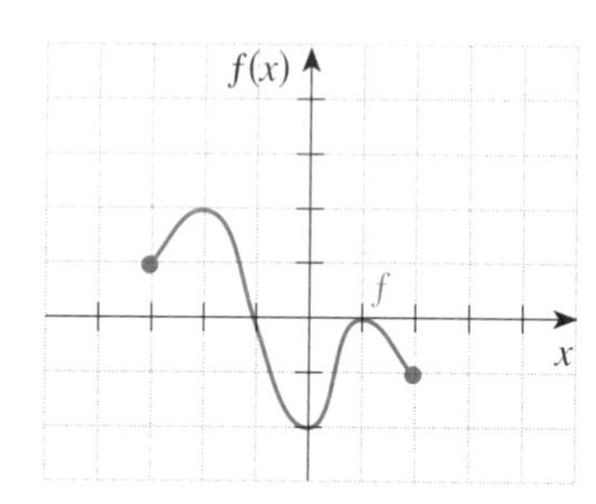

Read the graph of the function f *and report the indicated value.*

5. $f(-2)$
6. $f(0)$
7. $f(1)$
8. What input(s) produce an output of -2?
9. Report the zeros of function f.
10. Report the domain and range of function f.

Use the following graph to answer Problems 11–16.

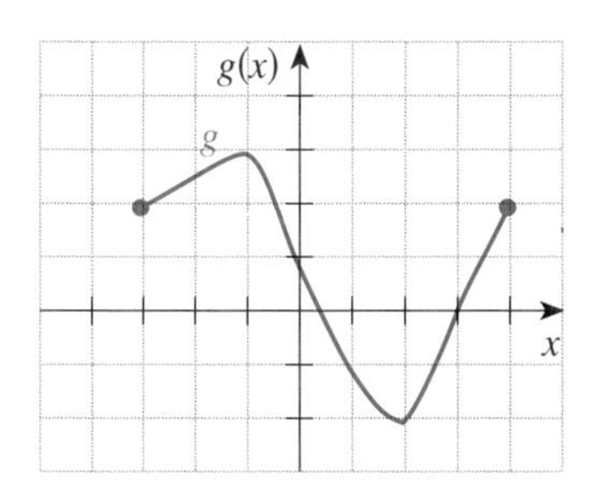

Read the graph of the function g *and report the indicated value.*

11. $g(-3)$
12. $g(0)$
13. $g(2)$
14. What input(s) produce an output of 3?
15. Report the zeros of function g.
16. Report the domain and range of function g.

Use the following graph to answer Problems 17–22.

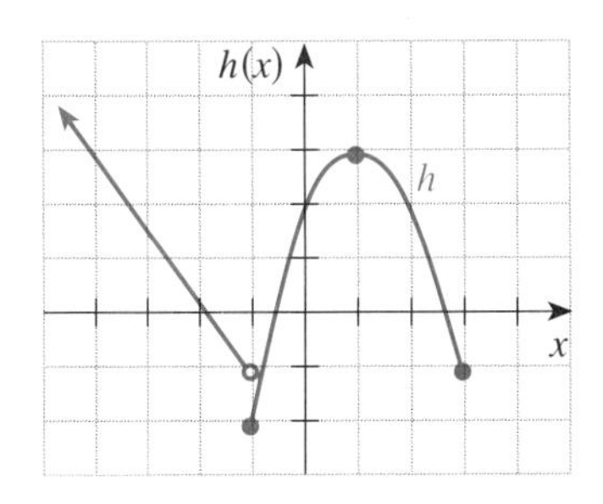

Read the graph of the function h *and report the indicated value.*

17. $h(-1)$
18. $h(0)$
19. $h(3)$
20. Report the interval(s) on which the function h is decreasing.
21. Report the interval(s) on which the function h is increasing.
22. Report the turning point(s) of the graph of h.

Use the following graph to answer Problems 23–28.

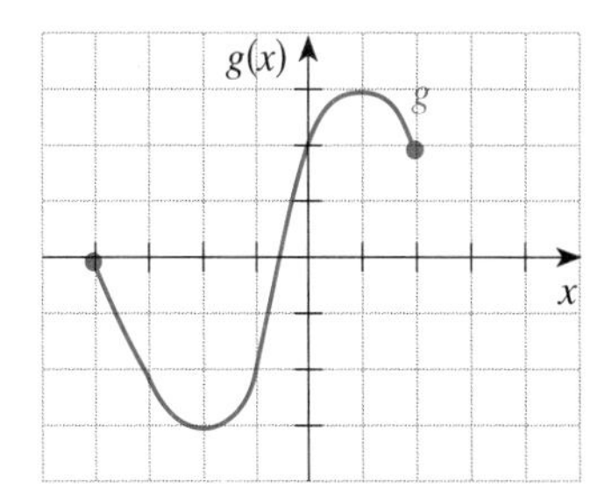

Read the graph of the function g *and report the indicated value.*

23. $g(-1)$
24. $g(0)$
25. $g(1)$
26. Report the interval(s) on which the function g is decreasing.
27. Report the interval(s) on which the function g is increasing.
28. Report the turning point(s) of the graph of g.

For the given piecewise-defined function, report the domain, the range, and the location of any turning point(s) and zero(s).

29. $f(x) = \begin{cases} 3x - 2 & \text{if } x < 1 \\ -x + 3 & \text{if } x \geq 1 \end{cases}$

30. $g(x) = \begin{cases} 4 - x^2 & \text{if } -3 \leq x < 1 \\ x^2 - x - 2 & \text{if } x \geq 2 \end{cases}$

31. $h(x) = \begin{cases} |x - 2| & \text{if } -4 \leq x < -1 \\ x^2 - 1 & \text{if } x \geq 0 \end{cases}$

32. $d(x) = \begin{cases} x & \text{if } x < 0 \\ 1 & \text{if } 0 \leq x < 2 \\ 4 - x^2 & \text{if } x \geq 2 \end{cases}$

33. $r(x) = \begin{cases} \sqrt{3 - x} & \text{if } -4 \leq x < 0 \\ \dfrac{1}{x + 1} & \text{if } 0 \leq x \leq 4 \end{cases}$

DISCOVERY

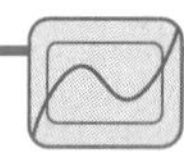

Piecewise-Defined Functions

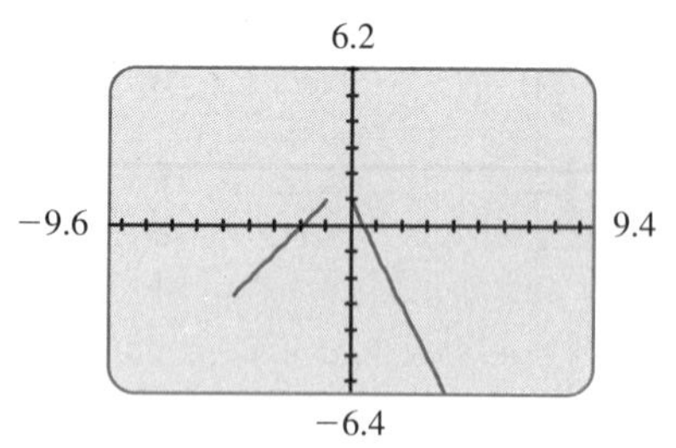

FIGURE 1

We can use the graphics calculator to graph piecewised-defined functions. Press the Y= key and erase any entries. For Y1, type (X+2)(X≥-5). (Use the TEST menu to type the relation, ≥.) Use the Window 9.4 range settings to view the graph screen. Entering a rule in parentheses, and then typing an interval in parentheses has the effect of graphing the function only for the input values in the given interval, in this case, for values greater than or equal to -5. Do you know why this works? Edit the rule for Y1 to (X+2)(X≥-5)(X<2). To eliminate the false vertical lines in the graph, set the mode to Dot. View the graph screen. Use the TRACE key and the arrow keys to explore the graph of the function.

Edit Y1 to read (X+2)(X≥-5)(X≤-1)+(-2X+1)(X≥0). View the graph screen, as shown in Figure 1. Press the TRACE key. As you can see, the y-intercept is (0, 1). Press the right arrow key. A zero occurs at $x = \frac{1}{2}$. Press the left arrow key until the message reads X=-.2 Y=0. This message is misleading, because the function is undefined for input values between 0 and -1. Continue to press the left arrow key until the message reads X=-1 Y=1. These values mean that for an input of -1, the function outputs 1 (it now follows the rule $x + 2$). Continue to press the left arrow key. The message X=-2 Y=0 means the function has a zero at $x = -2$.

Consider the following piecewise-defined function:

$$h(x) = \begin{cases} x^2 + 2x & x \leq 1 \\ -x - 1 & x > 2 \end{cases}$$

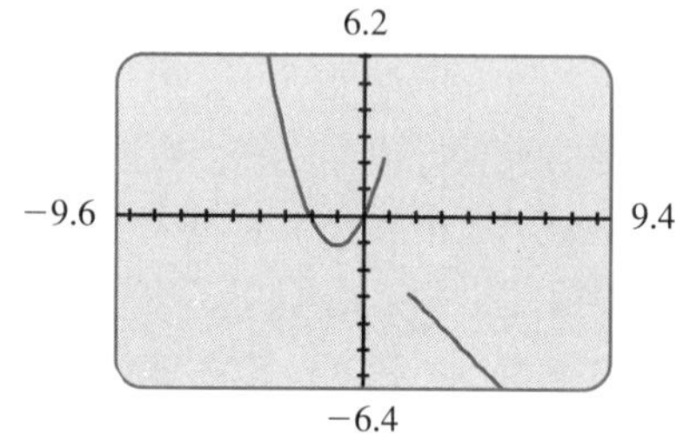

FIGURE 2

For Y1, type (X²+2X)(X≤1)+(-X−1)(X>2). View the graph screen (see Figure 2). Press the TRACE key. The function follows the rule $x^2 + 2x$ when x is 1 or less. To the right of 1, up to and including 2, the function is undefined (the calculator erroneously reports 0 for these values). For x greater than 2, the function follows the rule $-x - 1$. The domain of this function is $\{x \mid x \leq 1\} \cup \{x \mid x > 2\}$. The range of the function is

$$\{y \mid y \geq -1\} \cup \{y \mid y < -3\}$$

There are no input values that produce an output value in the interval $(-3, -1]$.

Exercises

Use this technique to check your work on Problems 29–33 in Exercises 9.2.

9.3 ■ ■ ■ OPERATIONS ON FUNCTIONS

In this section, we will use the basic operations of addition, subtraction, multiplication, and division to produce new functions. That is, we will learn about the algebra of functions.

Basic Operations Now that we have worked with several types of functions—linear, quadratic, rational, and radical functions—we can discuss operations on these objects. We first discuss the basic operations of addition, subtraction, multiplication, and division. Finally, we introduce the rather special operation of composition of functions.

Consider the two linear functions $f(x) = x + 3$ and $g(x) = x - 1$. The sum of the functions, denoted $f + g$, is

$$\begin{aligned}(f + g)(x) &= f(x) + g(x)\\ &= (x + 3) + (x - 1)\\ &= 2x + 2\end{aligned}$$

WARNING
Make sure you recognize the difference between

$$f(a + b) \neq f(a) + f(b)$$

and

$$(f + g)(a) = f(a) + g(a)$$

Thus, for an input of x, the *sum function* $f + g$ outputs the sum of the output for $f(x)$ and the output for $g(x)$. That is, the sum function has the rule $2x + 2$, which is the sum of the two rules for f and g.

Using the Graphics Calculator

Type X+3 for Y1 and X−1 for Y2. Return to the home screen. To find $(f + g)(-5)$, store ⁻5 in memory location X. Display the value for Y1. The value ⁻2 is output. That is, $f(-5) = -2$. Display the value for Y2. The value ⁻6 is output. That is, $g(-5) = -6$. Press the Y= key, and use the Y-VARS key to enter the rule Y1+Y2 for Y3. Return to the home screen. Display the value for Y3. The value ⁻8 is output (see Figure 1). Thus, $(f + g)(-5) = -8$. Note that $(f + g)(-5) = f(-5) + g(-5)$.

Using the Window 9.4 range settings, view the graph screen, and then press the TRACE key. Move the cursor until the message reads X=2 Y=5. This shows that $f(2) = 5$. Press the down arrow key. The message X=2 Y=1 shows that $g(2) = 1$. Press the down arrow key again. The message X=2 Y=6 means that $(f + g)(2) = 6$.

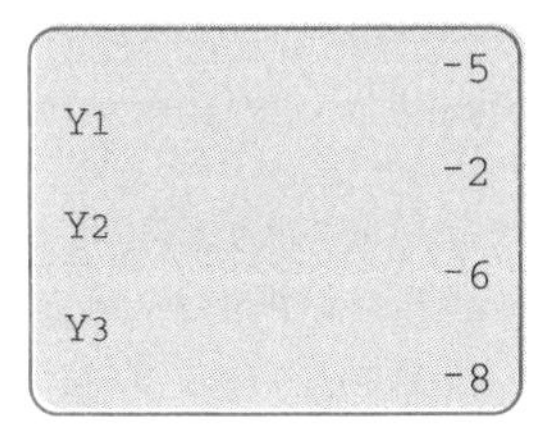

FIGURE 1

For $f(x) = x + 3$ and $g(x) = x - 1$, the difference function $f - g$ is

$$\begin{aligned}(f - g)(x) &= f(x) - g(x)\\ &= (x + 3) - (x - 1)\\ &= 4\end{aligned}$$

For an input of x, the *difference function* $f - g$ outputs the difference of $f(x)$ minus the output of $g(x)$. As you might suspect, $f - g \neq g - f$. The difference function $g - f$ is

$$\begin{aligned}(g - f)(x) &= g(x) - f(x)\\ &= (x - 1) - (x + 3)\\ &= -4\end{aligned}$$

The sum of two linear functions is a linear function. The difference of two linear functions is also a linear function.

Using the Graphics Calculator

Keeping the current rules X+3 for Y1 and X−1 for Y2, edit the rule for Y3 to Y1−Y2. Press the TRACE key. Is function Y3 the constant function 4? Press the Y= key, and enter the rule Y2−Y1 for Y4. View the graph. Does the graph of Y4 agree with the difference function we have found?

The *product* of the two linear functions, $f(x) = x + 3$ and $g(x) = x - 1$, is

$$\begin{aligned}(f \cdot g)(x) &= [f(x)][g(x)] \\ &= (x + 3)(x - 1) \\ &= x^2 + 2x - 3\end{aligned}$$

The *product function* of these two linear functions is a quadratic function.

Using the Graphics Calculator

Keeping the rules currently entered for Y1 and Y2, edit the rule for Y3 to Y1*Y2. Erase the rule for Y4. Press the TRACE key. Is the graph for function Y3 a parabola? What do you notice about the x-intercepts for the three functions?

We now define the addition, subtraction, and multiplication operations for functions.

DEFINITION OF THE SUM, DIFFERENCE, AND PRODUCT FUNCTIONS

Given two functions, f and g,

$$(f + g)(x) = f(x) + g(x)$$
$$(f - g)(x) = f(x) - g(x)$$

and

$$(f \cdot g)(x) = f(x) \cdot g(x)$$

The domain of the functions $f + g$, $f - g$, and $f \cdot g$ is the intersection of the domains for f and g.

The **sum function**, denoted $f + g$, is the function whose rule is the sum of the function rules for f and g. Likewise, the **difference function**, denoted $f - g$, is the function whose rule is the difference of the function rules for f and g. The **product function**, denoted $f \cdot g$, is the function whose rule is the product of the rules for f and g. The addition and multiplication operations are commutative, but the subtraction operation is not. To find $(f + g)(x)$ and $(f - g)(x)$ for a particular input, it is usually easier to find the new rule, and then evaluate. To find $(f \cdot g)(x)$ at a particular input, it is usually easier to evaluate $f(x)$ and $g(x)$, and then multiply the values. Consider the following example.

EXAMPLE 1 Given the functions $f(x) = x^2 + x$ and $g(x) = x - 3$, find each output value.

a. $(f + g)(7)$ b. $(f - g)(7)$ c. $(g - f)(7)$ d. $(f \cdot g)(7)$

Solution a. $f + g$ has the rule $(x^2 + x) + (x - 3)$, or

$$(f + g)(x) = x^2 + 2x - 3$$

So,

$$(f + g)(7) = 7^2 + 2(7) - 3 = 60$$

b. $f - g$ has the rule $(x^2 + x) - (x - 3)$, or $(f - g)(x) = x^2 + 3$. So,

$$(f - g)(7) = 7^2 + 3 = 52$$

c. $g - f$ has the rule $(x - 3) - (x^2 + x)$, or $(g - f)(x) = -x^2 - 3$. So,

$$(g - f)(7) = -7^2 - 3 = -52$$

d. $f \cdot g$ has the rule $(x^2 + x)(x - 3)$, or $(f \cdot g)(x) = x^3 - 2x^2 - 3x$. So,

$$(f \cdot g)(7) = 7^3 - 2(7^2) - 3(7) = 224$$

■

Using the Graphics Calculator

To check our work in Example 1, type X²+X as the rule for Y1. For Y2, type X−3 and for Y3, type Y1+Y2. Return to the home screen and store 7 in memory location X. Use the Y-VARS menu to display Y3. Press the ENTER key. Is 60 output?

Edit the rule for Y3 to Y1−Y2. Return to the home screen and display the resulting value in Y3. Is 52 output?

Edit the rule for Y3 to Y2−Y1. Return to the home screen and display the resulting value in Y3. Is ⁻52 output?

Edit the rule for Y3 to Y2*Y1. Return to the home screen and display the resulting value in Y3. Is 224 output?

GIVE IT A TRY

1. Consider the functions $f(x) = 2x - 1$ and $g(x) = x + 3$.
   - **a.** Report the rule for $f + g$.
   - **b.** Report the rule for $f - g$.
   - **c.** Report the rule for $f \cdot g$.
   - **d.** For an input of 10, report the output of the function $g - f$.
   - **e.** On the same set of axes, graph the functions f, g, and $f \cdot g$.
2. Consider the functions $f(x) = x^2$ and $h(x) = -x + 1$.
   - **a.** Report the rule for $f + h$.
   - **b.** Report the rule for $h - f$.
   - **c.** Report the rule for $f \cdot h$.
   - **d.** For an input of -1, report the output of the function $h - f$.
   - **e.** On the same set of axes, graph the functions f, h, and $f \cdot h$.

The Division Operation So far, we have learned that the operations of addition, subtraction, and multiplication can be defined for functions. We now discuss the division operation for functions. For the linear functions

$$f(x) = 2x - 1 \qquad \text{and} \qquad g(x) = x - 1$$

the quotient function $\frac{f}{g}$ is the function whose rule is $(2x - 1)/(x - 1)$. That is, we define the quotient of two functions as follows.

DEFINITION OF THE QUOTIENT FUNCTION

Given two functions, f and g,

$$\left(\frac{f}{g}\right)(x) = \frac{f(x)}{g(x)}$$

The domain of the function $\frac{f}{g}$ is the intersection of the domain of f with the domain of g, where $g(x) \neq 0$.

For an input of x, the function $\frac{f}{g}$ outputs the quotient of $f(x)$ and $g(x)$, provided $g(x) \neq 0$. Consider the following example.

EXAMPLE 2 For $f(x) = 2x - 1$ and $g(x) = x - 1$ find the indicated values.

a. $\left(\frac{f}{g}\right)(-3)$ b. $\left(\frac{g}{f}\right)(0)$ c. $\left(\frac{f}{g}\right)(1)$

Solution a. The quotient function is

$$\left(\frac{f}{g}\right)(x) = \frac{f(x)}{g(x)} = \frac{2x - 1}{x - 1}$$

So, we have

$$\left(\frac{f}{g}\right)(-3) = \frac{2(-3) - 1}{-3 - 1} = \frac{-7}{-4} = \frac{7}{4}$$

Notice that

$$\left(\frac{f}{g}\right)(-3) = \frac{f(-3)}{g(-3)} = \frac{-7}{-4} = \frac{7}{4}$$

In general, to evaluate the product or quotient of functions for a particular input value, it is easier to evaluate the component functions, and then find the product or quotient

b. $$\left(\frac{g}{f}\right)(0) = \frac{g(0)}{f(0)} = \frac{-1}{-1} = 1$$

c. $$\left(\frac{f}{g}\right)(1) = \frac{f(1)}{g(1)} = \frac{1}{0} \quad \text{undefined!}$$

For an input of 1, the quotient function (f/g) is undefined. This value violates the definition, because $g(1) = 0$. ■

Using the Graphics Calculator

A rational function rule is a ratio of polynomials. To see this using the graphics calculator, type X+2 for Y1. For Y2, type 2X−3. For Y3, edit the current entry to Y1/Y2. Use the Window 9.4 range settings to view the graph screen. Notice that the x-intercept for the rational function, Y3, is where the numerator, Y1, is 0. Also, notice that where the denominator, Y2, is 0, the rational function is undefined.

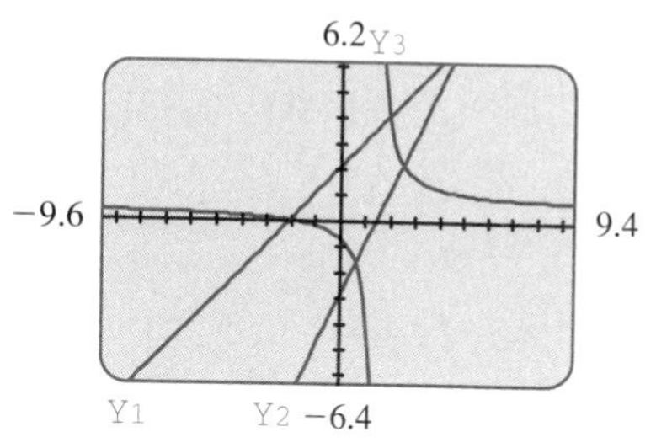

FIGURE 2

GIVE IT A TRY

3. For $f(x) = 2x - 5$ and $h(x) = 4x + 3$, find the indicated values.

   a. $\left(\frac{f}{h}\right)(0)$ **b.** $\left(\frac{h}{f}\right)\left(\frac{5}{2}\right)$ **c.** $\left(\frac{f}{h}\right)\left(\frac{5}{2}\right)$

4. For the function $\left(\frac{f}{g}\right)$, where $f(x) = 2x - 3$ and $g(x) = -x + 2$, report the function rule.

5. For the functions f and g from Problem 4, use the graphics calculator to enter 2X−3 for Y1, ⁻X+2 for Y2, and Y1/Y2 for Y3. Does the x-intercept of $\frac{f}{g}$ occur at the zero of f? Is the quotient function undefined at the zero of g?

The Composition of Functions So far, we have produced the sum, difference, product, and quotient of two functions. That is, given two functions f and g, we have the following operations.

Operations	*Examples*
	For $f(x) = x^2$ and $g(x) = 2x + 1$,
$(f + g)(x) = f(x) + g(x)$	$(f + g)(x) = x^2 + 2x + 1$
$(f - g)(x) = f(x) - g(x)$	$(f - g)(x) = x^2 - 2x - 1$
$(f \cdot g)(x) = f(x) \cdot g(x)$	$(f \cdot g)(x) = 2x^3 + x^2$
$\left(\frac{f}{g}\right)(x) = \frac{f(x)}{g(x)} \quad g(x) \neq 0$	$\left(\frac{f}{g}\right)(x) = \frac{x^2}{2x + 1} \quad x \neq \frac{-1}{2}$

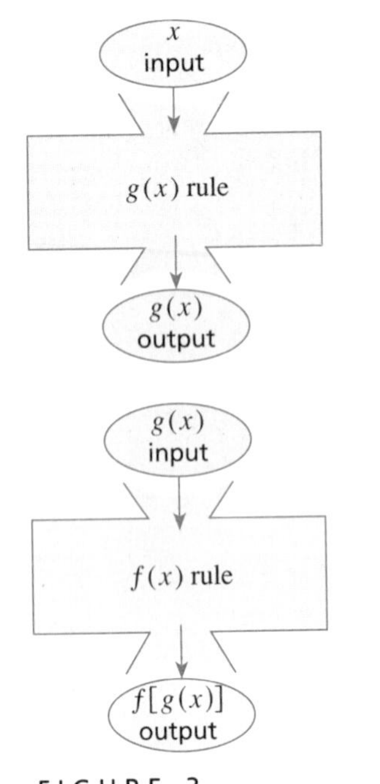

FIGURE 3
Composition of f on g, that is, $(f \circ g)(x) = f[g(x)]$

Each of these operations on two functions produces a third function. Suppose we have the radical function $f(x) = \sqrt{x}$ and the linear function $g(x) = 2x + 1$. None of these operations produce the radical function $h(x) = \sqrt{2x + 1}$. We now consider an operation that can be used to produce this function. This operation on two functions is known as the *composition operation*.

The composition of function f on g is denoted $f \circ g$, and is read *function f composed with function g*. It is the function whose input values are first acted on by the function g. The output produced, $g(x)$, is then input to the function f. See Figure 3. For $(f \circ g)(x)$, we have $f[g(x)]$. For example, for $f(x) = \sqrt{x}$ and $g(x) = 2x + 1$

$$\begin{aligned}(f \circ g)(x) &= f[g(x)] \\ &= f(2x + 1) \\ &= \sqrt{2x + 1}\end{aligned}$$

The domain of $f \circ g$ is the domain values of g whose range values are in the domain of function f.

Consider finding the output value of the composition for an input of 5:

First, the input value 5 is acted on by g, to output $2(5) + 1$ or 11.
Second, the value output by g, 11, is input to $f(x)$ to yield $\sqrt{11}$.

Thus, for an input of 5, the composition of f on g outputs $\sqrt{11}$. We write $(f \circ g)(5) = f[g(5)] = \sqrt{11}$.

EXAMPLE 3 For $f(x) = x^2$ and $g(x) = x - 3$, find the rule for each composition function:

a. $(f \circ g)(x)$ b. $(g \circ f)(x)$

c. Is the composition operation commutative?

Solution

a. $$\begin{aligned}(f \circ g)(x) &= f[g(x)] \\ &= f(x - 3) \\ &= (x - 3)^2 \\ &= x^2 - 6x + 9\end{aligned}$$

b. $$\begin{aligned}(g \circ f)(x) &= g[f(x)] \\ &= g(x^2) \\ &= x^2 - 3\end{aligned}$$

c. From the rules for $f \circ g$ and $g \circ f$, we see that $(f \circ g)(x) \neq (g \circ f)(x)$. Thus, composition is not a commutative operation. ■

Using the Graphics Calculator

We can use the graphics calculator to demonstrate the composition of functions. For Y1 type 2X+3. This is the function g. For Y2 type √Y1. This is the function $f(x) = \sqrt{x}$ composed on g, denoted $f \circ g$ and equal to $f[g(x)]$. View the graph screen with the Window 9.4 range settings. A straight line is drawn for Y1, the linear function, and the graph of √(2X+3) is drawn for Y2. To get a closer view of this, press the RANGE key and enter the Window 4.7 range settings. Press the GRAPH key to display the graph with this enlarged view. As Figure 4 shows, the x-intercept for both graphs is $\left(\frac{-3}{2}, 0\right)$. The y-intercept for Y1 is (0, 3) and the y-intercept for Y2, the composition, is $\left(0, \sqrt{3}\right)$.

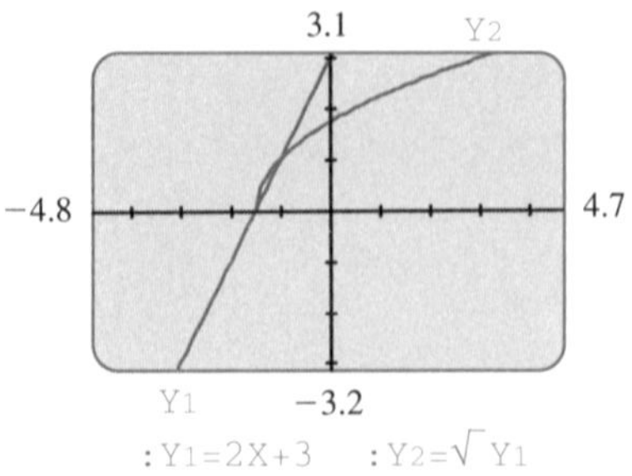

3.1 Y2
−4.8 4.7
Y1 −3.2
:Y1=2X+3 :Y2=√Y1

FIGURE 4

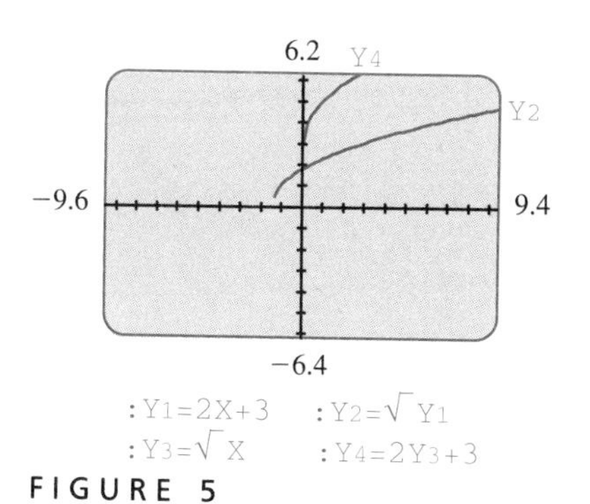

FIGURE 5

To see that the composition of functions is not a commutative operation, for Y3 type √X. This is the function f. For Y4 type 2Y3+3. This is the function g composed on the function f, denoted $g \circ f$ and equal to $g[f(x)]$. Move the cursor to the Y1 prompt, press the left arrow key and then the ENTER key to hide (prevent the display of) the graph of function Y1. Likewise, hide the graph of function Y3. View the graph using the Window 9.4 range settings. First the graph of $f \circ g$ is drawn. Next, the graph of $g \circ f$ is drawn. Figure 5 shows that the graphs are not the same. That is, $f \circ g \neq g \circ f$, so the composition operation is *not* commutative.

GIVE IT A TRY

6. For $f(x) = \sqrt{x}$ and $g(x) = 2x + 3$, find the indicated function values.
a. $f[g(-2)]$ **b.** $(f \circ g)(0)$ **c.** $(g \circ f)(-2)$ **d.** $g[f(0)]$ **e.** $(f \circ f)(16)$

7. For $f(x) = x^2 - 2x$ and $g(x) = \sqrt{x + 1}$, use the graphics calculator and the procedure from Using the graphics calculator to graph $f \circ g$, the composition of f on g. Use your algebra skills to write out the function rule for $f \circ g$. Enter this rule for function Y3 to check your work.

8. For $f(x) = x^2 - 2x$ and $g(x) = \sqrt{x + 1}$, use the graphics calculator and the procedure from Problem 7 to graph $g \circ f$, the composition of g on f. For a closer view, zoom in on the origin. What do you think is the rule for the composition? Use your algebra skills to write out the function rule for $(g \circ f)(x)$. Enter this rule for function Y3 to check your work.

A Set View Because you may have difficulty conceptualizing the operation of composition of functions, we now present a "set view" of this operation. Consider the following sets of ordered pairs:

$$A = \{(-2, 1), (-1, 0), (0, 2), (1, -1), (2, -2)\}$$

and

$$B = \{(-2, 1), (-1, 2), (0, -2), (1, 0), (2, -1)\}$$

These sets are functions, since every first element in the ordered pairs is matched with exactly one second element. So, we can construct the function $A \circ B$, the composition of function A on function B. The composition $A \circ B$ will be the set of ordered pairs where B acts on the first element to produce an output, then A takes the output as an input and acts on it. We construct a diagram to show this:

Function B		Function A		Composition $A \circ B$	
−2 →	1	1 →	−1	−2 →	−1
−1 →	2	2 →	−2	−1 →	−2
0 →	−2	−2 →	1	0 →	1
1 →	0	0 →	2	1 →	2
2 →	−1	−1 →	0	2 →	0

So, $A \circ B = \{(-2, -1), (-1, -2), (0, 1), (1, 2), (2, 0)\}$.

GIVE IT A TRY

9. For the following sets A and B, report $A \circ B$ and $B \circ A$.

$$A = \{(-2, 1), (-1, 0), (0, 2), (1, -1), (2, -2)\}$$
$$B = \{(-2, 0), (-1, -2), (0, 0), (1, 2), (2, -1)\}$$

Applications To see the operation of composition of functions in a real-world example, work through the next example.

EXAMPLE 4 An oil spill on a calm lake is spreading in a circular pattern. The area covered, A (in square feet), is a function of the radius r (in feet), given by $A(r) = \pi r^2$. The radius r is a function of time t (in minutes), given by $r(t) = 2t + 3$.

a. Write a function for the area covered by the spill, A, as a function of time t.

b. At what time is the area covered by the spill 300 square feet?

Solution a. Given that $A(r) = \pi r^2$ and that $r(t) = 2t + 3$, the desired function $A(t)$ is the composition of function A on function r:

$$\begin{aligned} A(t) &= (A \circ r)(t) \\ &= A[r(t)] \\ &= A(2t + 3) \\ &= \pi(2t + 3)^2 \end{aligned}$$

b. We solve $300 = \pi(2t + 3)^2$:

$$\begin{aligned} \pi(2t + 3)^2 &= 300 \\ (2t + 3)^2 &= \frac{300}{\pi} \end{aligned}$$

$$2t + 3 = \sqrt{\frac{300}{\pi}} \quad \text{and} \quad 2t + 3 = -\sqrt{\frac{300}{\pi}}$$

$$2t = -3 + \sqrt{\frac{300}{\pi}} \qquad 2t = -3 - \sqrt{\frac{300}{\pi}}$$

$$t = \frac{-3}{2} + \frac{1}{2}\sqrt{\frac{300}{\pi}} \qquad t = \frac{-3}{2} - \frac{1}{2}\sqrt{\frac{300}{\pi}}$$

Approximations of these numbers using a calculator yields 3.39 and −6.39. The negative value is discarded. Why? The oil spill will cover 300 square feet at time $t \approx 3.39$ minutes. ■

■ SUMMARY

In this section, we have introduced the idea of operations on functions. For functions f and g, we can find the sum, difference, product, and quotient of the functions:

$$(f + g)(x) = f(x) + g(x)$$
$$(f - g)(x) = f(x) - g(x)$$
$$(f \cdot g)(x) = f(x)g(x)$$
$$\left(\frac{f}{g}\right)(x) = \frac{f(x)}{g(x)} \quad \text{provided } g(x) \neq 0$$

We also introduced a new binary operation, the composition of two functions. The notation for composition is denoted $f \circ g$ and is read *function f composed on function g*. This operation produces a new function by one function operating (applying its rule) on a second function. For example, if $f(x) = 2x + 1$ and $g(x) = \sqrt{x}$, then $(f \circ g)(x) = 2\sqrt{x} + 1$. One important result is that the composition of functions is *not* commutative $\left[\text{for these functions } f \text{ and } g\text{, we have } (g \circ f)(x) = \sqrt{2x + 1}\right]$.

9.3 ■ ■ ■ EXERCISES

For $f(x) = 2x - 1$ and $h(x) = x + 2$, find the indicated function values.

1. $(f + h)(-1)$
2. $(f - h)(5)$
3. $(f \cdot h)(10)$
4. $(h - f)(5)$
5. $\left(\frac{f}{h}\right)(2)$
6. $\left(\frac{h}{f}\right)(2)$
7. $(f + h)(2a)$
8. $(f \cdot h)(a + 2)$
9. $\left(\frac{f}{h}\right)(-a)$
10. What input value produces an output of 0 for $(f - h)(x)$?

For $f(x) = \frac{1}{x}$ and $h(x) = x - 2$, find the indicated function values.

11. $(f + h)(-1)$
12. $(f - h)(5)$
13. $(f \cdot h)(10)$
14. $(h - f)(5)$
15. $\left(\frac{f}{h}\right)(2)$
16. $\left(\frac{h}{f}\right)(2)$
17. $(f + h)(2a)$
18. $(f \cdot h)(a + 2)$
19. $\left(\frac{f}{h}\right)(-a)$
20. What input value produces an output of 0 for $(f \cdot h)(x)$?

For $f(x) = \sqrt{x + 3}$ and $h(x) = -1$, find the indicated function values.

21. $(f + h)(-1)$
22. $(f - h)(5)$
23. $(f \cdot h)(10)$
24. $(h - f)(5)$
25. $\left(\frac{f}{h}\right)(2)$
26. $\left(\frac{h}{f}\right)(2)$
27. $(f + h)(2a)$
28. $(f \cdot h)(a + 2)$
29. $\left(\frac{f}{h}\right)(-a)$
30. Report the domain and range of the function $f + h$.

For $f(x) = \sqrt{x + 5}$ and $g(x) = x^2 + 2x$, find the indicated function values.

31. $f[g(0)]$
32. $g[f(0)]$
33. $f[g(-1)]$
34. $g[f(-1)]$
35. $(g \circ f)(2)$
36. $(f \circ g)(2)$

For $h(x) = 6 - 2x$ and $g(x) = \sqrt{-x}$, find the indicated function values.

37. $g[h(-1)]$
38. $h[g(-1)]$
39. $g[h(3)]$
40. $h[g(3)]$
41. $(g \circ h)(1)$
42. $(h \circ g)(1)$

43. For $f(x) = x^2 + 1$ and $g(x) = x - 1$, report the function rule for each composition:

a. $f \circ g$ b. $g \circ f$

Use the graphics calculator to check your work.

44. For $f(x) = x^2 - 1$ and $g(x) = 2x$, report the function rule for each composition:

a. $f \circ g$ b. $g \circ f$

Use the graphics calculator to check your work.

Use the graphics calculator to graph $f \circ g$ and $g \circ f$ for each pair of functions. State the domain, range, y-intercept, and x-intercept for both compositions.

45. $f(x) = x - 3$ and $g(x) = \sqrt{x}$

46. $f(x) = \sqrt{x} - 3$ and $g(x) = \dfrac{1}{x}$

47. $f(x) = \sqrt{x} + 1$ and $g(x) = x^2 - 2x + 1$ with $x \geq 1$

48. $f(x) = -\sqrt{x - 1}$ and $g(x) = x^2 + 1$ with $x \leq 0$

49. For the functions A and B given in set notation, report the set $B \circ A$:

$$A = \{(-2, 3), (-1, 2), (0, 1), (1, -1), (2, 0)\}$$
$$B = \{(-1, 4), (0, -3), (1, 2), (2, 3), (3, 0)\}$$

50. For the functions C and D given in set notation, report the set $D \circ C$:

$$C = \{(-2, 1), (-1, 1), (0, -2), (1, 0), (2, 2)\}$$
$$D = \{(-2, 0), (-1, 2), (0, -2), (1, -1), (2, 1)\}$$

Read the graphs of functions f and g to find the indicated function values.

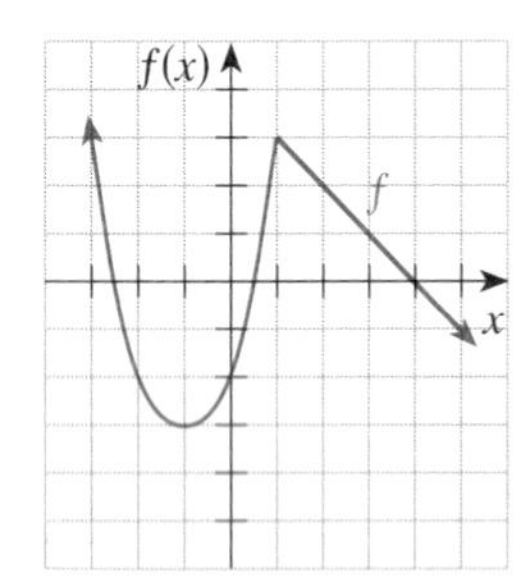

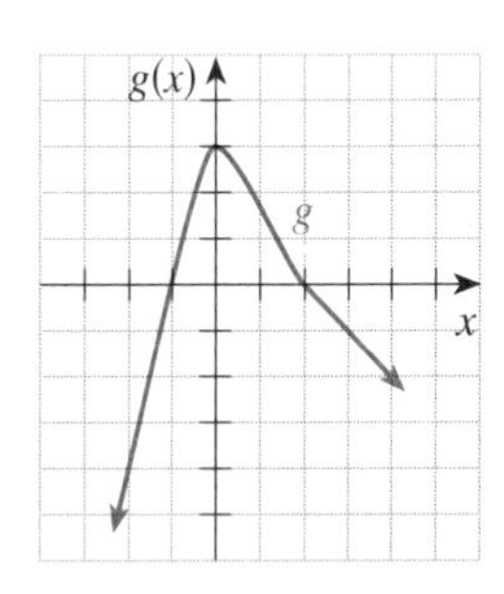

51. $(f \circ g)(2)$ 52. $(f \circ g)(-1)$ 53. $(g \circ f)(0)$ 54. $(g \circ f)(1)$

55. For $f(x) = x^2 - 2$, report the rule for $f \circ f$.

56. For $g(x) = \dfrac{1}{x - 2}$, report the rule for $g \circ g$.

57. For $h(x) = 3x - 4$ and $g(x) = x^2$, find the output $(h + g)(b - 1)$.

58. For $f(x) = \sqrt{x - 2}$ and $g(x) = x^{-1}$, report the rule for $g \circ f$.

59. An oil spill on a calm lake is spreading in a circular pattern. The area covered, A (in square feet), is a function of the radius r (in feet), given by $A(r) = \pi r^2$. The radius r is a function of time t (in minutes), given by $r(t) = t + 2$.

a. Write a function for the area covered by the spill, A, as a function of time t.

b. At what time is the area covered by the spill 200 square feet?

60. A 20-foot ladder is leaning against a building, as shown in the sketch. The height h of the top of the ladder as a function of the distance x from the foot of the ladder to the building is

$$h(x) = \sqrt{400 - x^2}$$

The distance from the foot of the ladder to the wall is a function of time t (in seconds), given by

$$x(t) = 2t + 1$$

a. Use composition of functions to write a function for the height h as a function of time t.

b. At what time is the height of the ladder 5 feet?

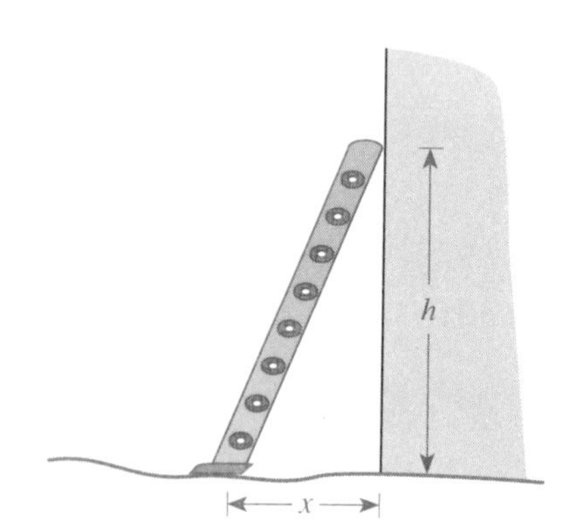

9.4 ■ ■ ■ FUNCTION INVERSES

We have discussed the basic binary operations of addition, subtraction, multiplication, division, and compositions for functions. The composition operation is a rather special binary operation. With this operation, we can produce the function $f(x) = \sqrt{2x + 3}$ from the functions $g(x) = \sqrt{x}$ and $h(x) = 2x + 3$. In this section, we use the composition operation in finding the inverse of a function.

1-to-1 Functions For any function, each input value is matched with exactly one output value. That is, if f is a function, we could not have a situation where $f(2) = 4$ and $f(2) = -4$, because the input value 2 would not be matched with exactly one output. Given the graph of a collection of ordered pairs, such as the graph in Figure 1, we can judge whether the ordered pairs represent the graph of a function by use of the vertical line test. That is, if any vertical line intersects the graph at most once, then each input value is matched with exactly one output value. The graph is then the graph of some function. Compare the graphs in Figure 1 and Figure 2.

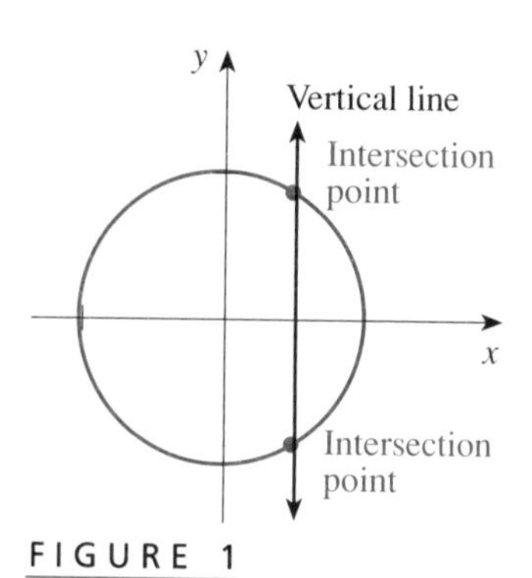

FIGURE 1

The graph of a relation that is *not* a function fails the vertical line test.

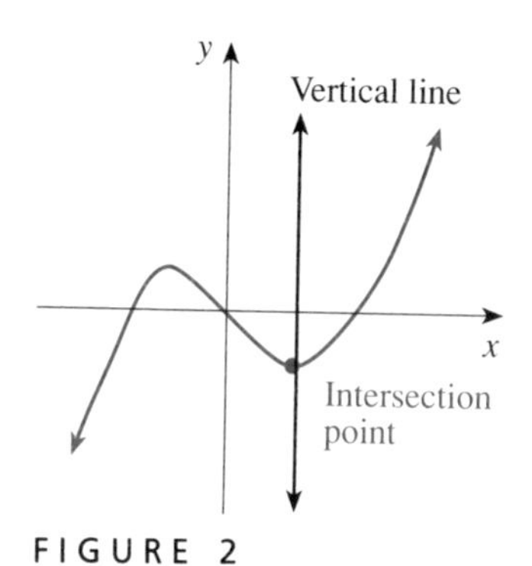

FIGURE 2

The graph of a function passes the vertical line test.

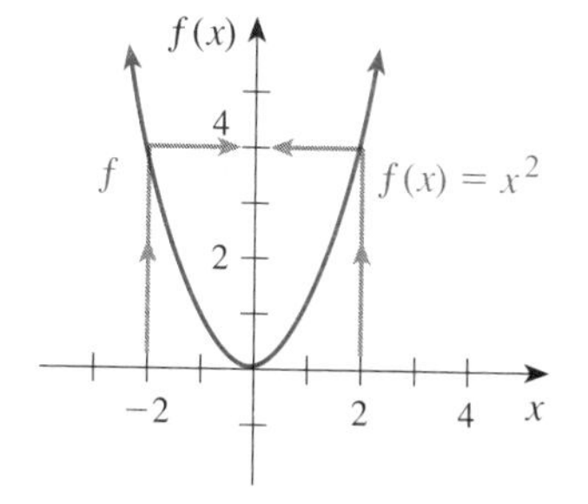

FIGURE 3

Of course, for many functions, two or more input values can be matched with the same output value. Consider the function $f(x) = x^2$. With this function, for an input of 2, the output is 4, and for an input of -2 the output is 4. That is, for $f(x) = x^2$, we have $f(2) = 4$ and $f(-2) = 4$. The graph of f is shown in Figure 3.

We now consider a class of function which eliminates the situation of two or more input values being matched with the same output value. This class of functions is known as 1-to-1 functions.

DEFINITION OF A 1-TO-1 FUNCTION

> A function is a 1-to-1 function if for each output value exactly one input value is matched with it.

Using this definition, the function $f(x) = 2x + 1$ is a 1-to-1 function. Replace the output, $f(x)$, with a value, say -3. Now, solve the equation $-3 = 2x + 1$. Its solution yields the input values that produce an output of -3. Because this equation has exactly one number in its solution, namely, -2, exactly one input value is matched with the output value, -3.

The function $f(x) = x^2$ is not a 1-to-1 function. Replace the output, $f(x)$, with a value, say 4. Now, solve the equation $4 = x^2$. Its solution yields the input values that produce an output of 4. Because this equation has two numbers in its solution, namely, 2 and -2, there is not exactly one input value matched with the output value, 4.

Horizontal Line Test Given the graph of a function, if any horizontal line intersects the graph at most once, then each output value is matched with exactly one input value. The graph is the graph of some 1-to-1 function. As you can see in Figure 4, the graph of $f(x) = 2x - 3$ passes the horizontal line test. The graph of $f(x) = x^2 - 2$ in Figure 5 fails the horizontal line test.

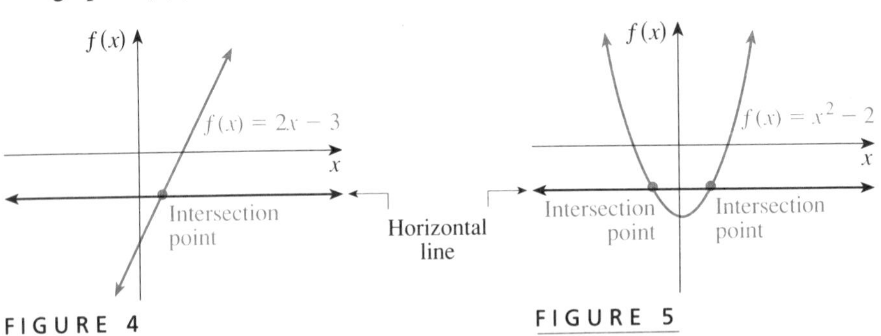

FIGURE 4
A 1-to-1 function passes the horizontal line test.

FIGURE 5
A function that fails the horizontal line test is *not* a 1-to-1 function.

Using the Graphics Calculator

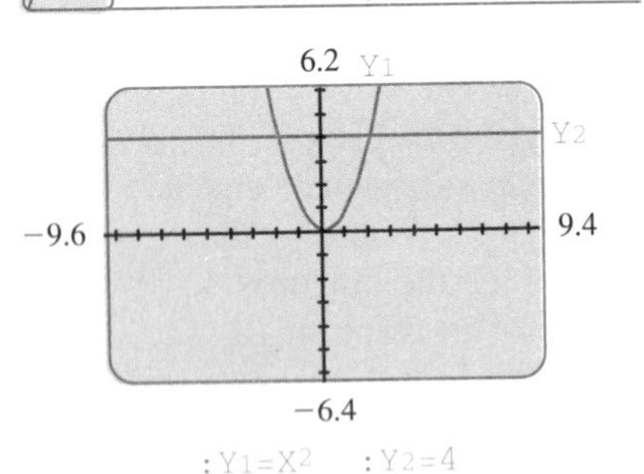

FIGURE 6

For Y1, type X². For Y2, type 4. Use the Window 9.4 range settings to view the graph. As Figure 6 shows, the horizontal line $y = 4$ intersects the graph of $f(x) = x^2$ two times. Press the TRACE key and then use the right arrow key to move the cursor until the message reads X=2 Y=4. Thus, for input of 2, the output is 4. Press the left arrow key until the message reads X=-2 Y=4. Thus, for an input of -2, the output is 4. We have an output value, 4, matched with two different input values, namely, 2 and -2. Thus, the function $f(x) = x^2$ is not a 1-to-1 function.

GIVE IT A TRY

Use the graphics calculator and the procedure from the preceding Using the Graphics Calculator to check whether each function is a 1-to-1 function.

1. $f(x) = \sqrt{2x + 5}$

2. $h(x) = x^2 - x - 2$

3. $g(x) = \dfrac{1}{x - 2}$

4. $f(x) = \dfrac{1}{(x - 2)^2}$

5. Use the horizontal line test to judge whether each graph represents the graph of a 1-to-1 function.

a.

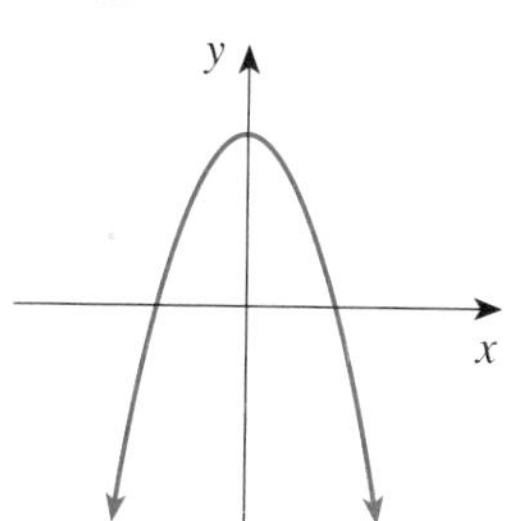

b.

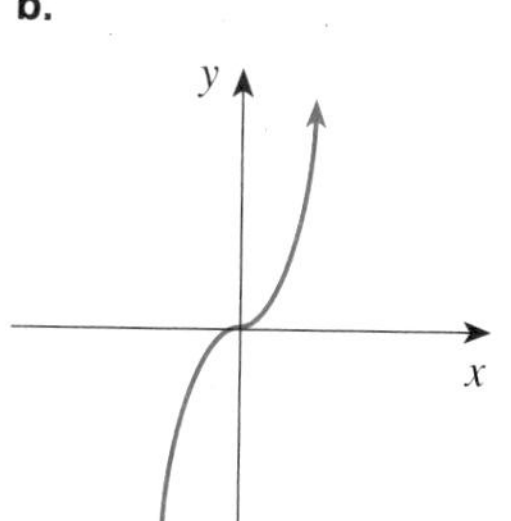

c.

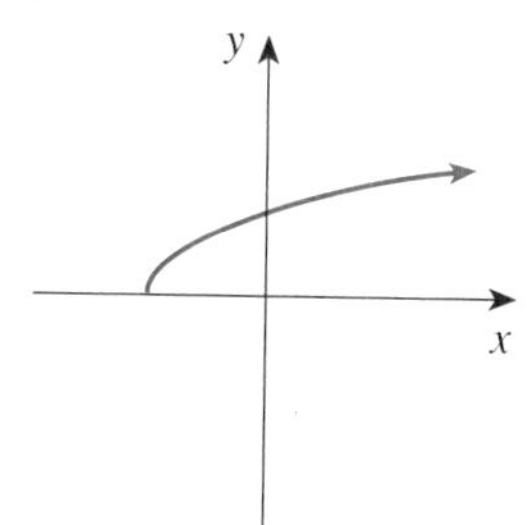

REMEMBER

Add 5 → $x + 5 - 5 = x$ ← Subtract 5

Multiply by 5 → $\dfrac{5x}{5} = x$ ← Divide by 5

Inverse Functions We know that the operations of addition of a number and subtraction of the same number are inverse operations. Likewise, the operation of multiplication by a nonzero number and the operation of division by that same number are inverse operations. (See the reminder in the margin.)

We define inverse functions as follows.

DEFINITION OF INVERSE FUNCTIONS

> For a 1-to-1 function containing elements (a, b), the inverse function is the set of ordered pairs containing the elements (b, a).

For the 1-to-1 function $A = \{(-2, 1), (-1, 3), (0, 4), (1, 5)\}$, the inverse function is $B = \{(1, -2), (3, -1), (4, 0), (5, 1)\}$. For B, we have simply "swapped" the components of the ordered pairs. A graph of the two functions (sets of ordered pairs) reveals that the function B is a reflection across the line $y = x$ of function A, as shown in Figure 7.

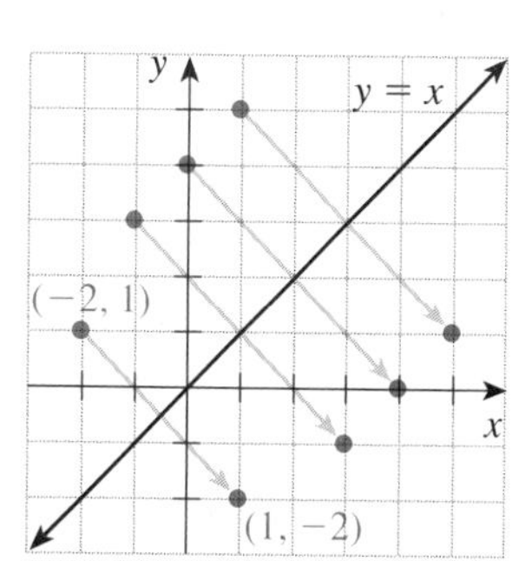

FIGURE 7

We can use composition of functions to state conditions for two functions to be inverse functions.

COMPOSITION OF FUNCTIONS AND INVERSE FUNCTIONS

A function f and a function g are inverses of each other if

1. their compositions are commutative, $(f \circ g)(x) = (g \circ f)(x)$, and
2. their compositions produce the identity function, $I(x) = x$, that is, $(f \circ g)(x) = x$ and $(g \circ f)(x) = x$.

Consider the linear functions $f(x) = 2x + 1$ and $g(x) = \frac{1}{2}x - \frac{1}{2}$. Their compositions are

$$\begin{aligned} f[g(x)] &= f\left[\tfrac{1}{2}x - \tfrac{1}{2}\right] \\ &= 2\left[\tfrac{1}{2}x - \tfrac{1}{2}\right] + 1 \\ &= x - 1 + 1 \\ &= x \end{aligned} \qquad \begin{aligned} g[f(x)] &= g[2x + 1] \\ &= \tfrac{1}{2}[2x + 1] - \tfrac{1}{2} \\ &= x + \tfrac{1}{2} - \tfrac{1}{2} \\ &= x \end{aligned}$$

Thus, $f[g(x)] = g[f(x)]$, and these compositions produce the identity function x. So f and g are inverse functions. That is, $g(x) = \frac{1}{2}x - \frac{1}{2}$ is the inverse of $f(x) = 2x + 1$. Likewise, $f(x) = 2x + 1$ is the inverse of the function $g(x) = \frac{1}{2}x - \frac{1}{2}$. For $f(x) = 2x + 1$, its inverse, $g(x) = \frac{1}{2}x - \frac{1}{2}$, is often denoted $f^{-1}(x) = \frac{1}{2}x - \frac{1}{2}$. Likewise, for $g(x) = \frac{1}{2}x - \frac{1}{2}$, its inverse, $f(x) = 2x + 1$, is often denoted $g^{-1}(x) = 2x + 1$. The notation $f^{-1}(x)$ is read *f inverse of x* (not "f to the negative one of x").

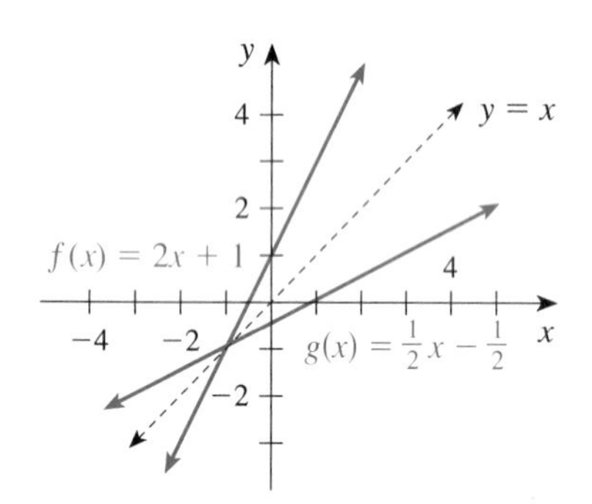

FIGURE 8

The importance of 1-to-1 functions is that every 1-to-1 function has an inverse function. Consider the function $f(x) = x^2$. This function is not 1-to-1. Thus, it has no inverse function. However, if we consider the function

$$f(x) = x^2, \; x \geq 0$$

[we limit the domain to get a 1-to-1 function], then the function has an inverse function, $g(x) = \sqrt{x}$.

$$\begin{aligned} f[g(x)] &= f\left(\sqrt{x}\right) \\ &= \left(\sqrt{x}\right)^2 \\ &= x \end{aligned} \qquad \begin{aligned} g[f(x)] &= g(x^2) \\ &= \sqrt{x^2} \\ &= x \qquad \text{provided } x \geq 0 \end{aligned}$$

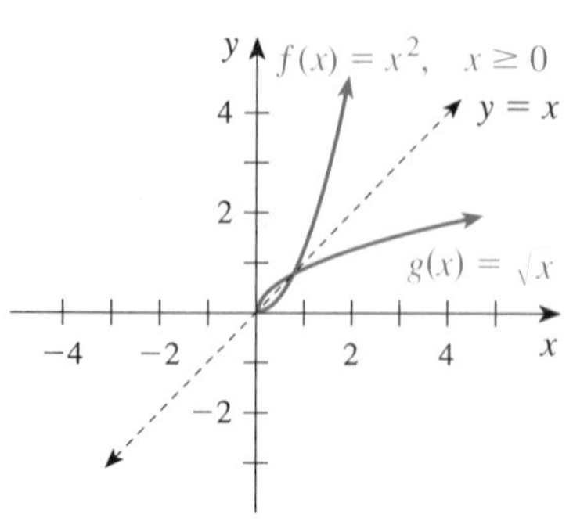

FIGURE 9

Thus, this *limited square function* composed with the square root function equals the square root function composed with the limited square function. Also, the compositions produce the identity function x.

A function and its inverse have graphs that are reflections about the line $y = x$. For example, consider the graph of $f(x) = 2x + 1$ and its inverse, $g(x) = \frac{1}{2}x - \frac{1}{2}$ shown in Figure 8.

Next, we consider the graph of $f(x) = x^2$, with $x \geq 0$, and its inverse, $g(x) = \sqrt{x}$, as shown in Figure 9. Visually, we think of folding the paper along the line $y = x$. The graph of the function and the graph of its inverse will match. We say the graphs are reflections of each other across the line determined by $y = x$. (This property holds true for the graphs in both figures.) Algebraically, for an ordered pair, say (a, b), that satisfies the function, the ordered pair (b, a) will satisfy the inverse function. That is, (2, 4) satisfies $f(x) = x^2$, with $x \geq 0$, and (4, 2) satisfies the inverse function, $g(x) = \sqrt{x}$.

Using the Graphics Calculator

```
RANGE
Xmin=-1
Xmax=3.75
Xscl=1
Ymin=-1
Ymax=4
Yscl=1
Xres=1
```

FIGURE 10

To judge whether the two functions, $f(x) = x^2$, with $x \geq 0$, and $g(x) = \sqrt{x}$, are inverses of each other, for `Y1`, type `X²(X≥0)`. (Use the `TEST` menu for the relation ≥.) For `Y2`, type `√X`. Use the settings in Figure 10* to view the graph screen. For `Y3`, type `√Y1`. This is the composition of the square root on the limited square function. View the graph. The graph of x^2, with $x \geq 0$, is drawn first, the graph of $\sqrt{x}$ is drawn next, and then the graph of the composition is drawn. See Figure 11.

To see that the composition is in fact the identity function x, type `X` as the function rule for `Y4`. View the graph, as shown in Figure 12. To verify that the composition the other way also yields the identity function, for `Y3` type `Y2²`. This is the composition of the limited square function on the square root function. Clear `Y4`. View the graph. As you can see, the composition is still the identity function x.

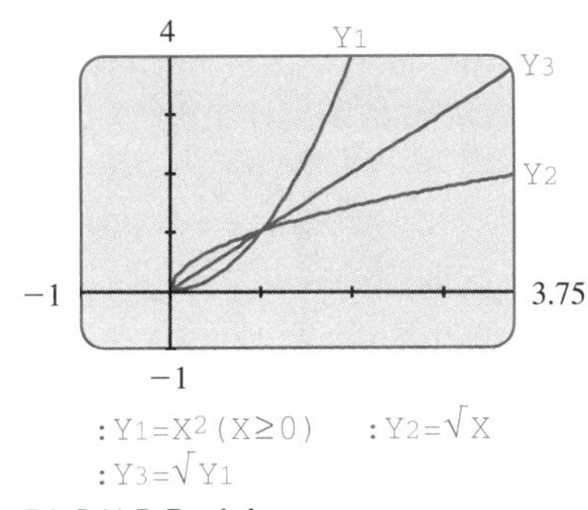

FIGURE 11

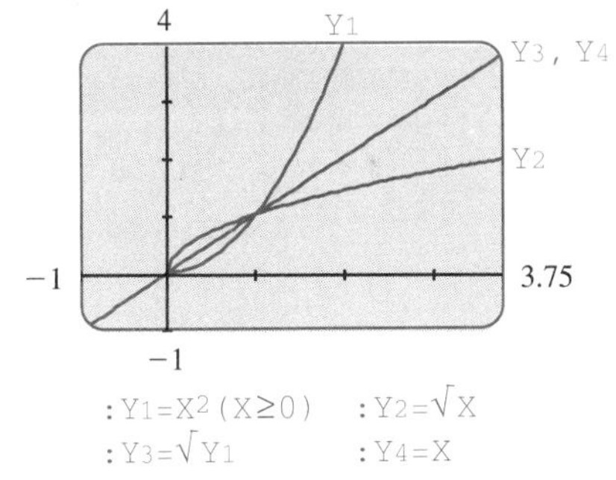

FIGURE 12

GIVE IT A TRY

6. For $A = \{(-3, 2), (-2, -1), (0, 3), (1, 4)\}$, find the inverse of A. Plot the two sets on the same set of axes. Are the plots reflections about the line $y = x$?
7. Use the graphics calculator to check that $f(x) = 2x + x^2$, with $x \geq -1$, and $g(x) = \sqrt{x + 1} - 1$ are inverse functions.

An Algorithm for Finding the Inverse Function Just as we have an algorithm (process) for finding the quotient of two polynomials, we also have an algorithm for finding the inverse of a function. Because the range of a function is the domain of its inverse function, we can follow these steps to find the rule

TI-82 Note *Type `3.7` for `Xmax`.

for the inverse function:

1. Set y equal to the function rule (the expression in x).
2. Interchange the x and y variables.
3. Rewrite the equation for y.
4. State the inverse.

This algorithm is used in the following examples.

EXAMPLE 1 Find the inverse of the function $f(x) = 3x - 2$.

Solution The algorithm involves the following steps:

Step 1 Write the function as the equation $y = 3x - 2$.

Step 2 Interchange x and y to get: $x = 3y - 2$.

Step 3 Rewrite this last equation for y.

$$x = 3y - 2$$

$$3y - 2 = x \qquad \text{Rewrite for } y.$$

$$\underline{\;\;2\;\;\;\;2\;\;} \qquad \text{Add 2 to each side.}$$

$$3y = x + 2 \qquad \text{Multiply each side by } \tfrac{1}{3}.$$

$$y = \tfrac{1}{3}x + \tfrac{2}{3}$$

Step 4 State the inverse function $f^{-1}(x) = \frac{1}{3}x + \frac{2}{3}$.

EXAMPLE 2 Find the inverse of $f(x) = x^2 - 2x$ for $x \leq 1$.

Solution We follow the steps of the inverse algorithm:

Step 1 $y = x^2 - 2x$ — Replace $f(x)$ with y.

Step 2 $x = y^2 - 2y$ — Interchange x and y.

Step 3 $y^2 - 2y = x$ — Rewrite for y.

$y^2 - 2y - x = 0$ — Standard form

$a = 1 \quad b = -2 \quad c = -x$ — Identify a, b, and c.

$b^2 - 4ac = 4 + 4x$ — Find the discriminant.

$$y = \frac{2 + \sqrt{4x + 4}}{2} \quad \text{or} \quad y = \frac{2 - \sqrt{4x + 4}}{2} \qquad \text{Quadratic formula}$$

The domain of function f is $\{x \mid x \leq 1\}$. That is, (0, 0) satisfies f, so (0, 0) must also satisfy f^{-1}. The ordered pair (0, 0) satisfies the second equation, but not the first, so we select the second equation. Reducing the right member of the equation, we get

$$y = \frac{2 - 2\sqrt{x + 1}}{2}$$

$$= 1 - \sqrt{x + 1}$$

Step 4 $f^{-1}(x) = 1 - \sqrt{x + 1}$ — Replace y with $f^{-1}(x)$. ■

EXAMPLE 3 Find the inverse of $f(x) = \sqrt{2x - 1}$.

Solution We follow the same steps used in Example 1:

Step 1 $y = \sqrt{2x - 1}$ Replace $g(x)$ with y.

Step 2 $x = \sqrt{2y - 1}$ Interchange x and y.

Step 3 $\sqrt{2y - 2} = x$ Rewrite for y.

$2y - 1 = x^2$ Square each side.

$2y = x^2 + 1$ Add 1 to each side.

$y = \frac{1}{2}x^2 + \frac{1}{2}$ Multiply each side by $\frac{1}{2}$.

Step 4 $f^{-1}(x) = \frac{1}{2}x^2 + \frac{1}{2}$ Replace y with $f^{-1}(x)$.

The range of function f is $\{f(x) \mid f(x) \geq 0\}$. Thus, the domain of f^{-1} is $\{x \mid x \geq 0\}$. Check the compositions to see that we do indeed have the inverse of f. ■

Using the Graphics Calculator

To check our work in Example 3, for `Y1` type `√(2X−1)` and for `Y2` type `(.5X²+.5)(X≥0)`. For `Y3` type `√(2Y2−1)`, the composition of f on the f^{-1}. Use the settings in Figure 10 to view the graph. Is the graph of `Y3` the graph of the identity function $I(x) = x$? If not, an error is present!

GIVE IT A TRY

8. Consider the functions $f(x) = \dfrac{1}{x - 1}$ and $g(x) = \dfrac{x + 1}{x}$.

a. Find $f[g(x)]$.

b. Find $g[f(x)]$.

c. Are f and g inverses of each other?

d. Graph f and g on the same set of axes. Are the curves symmetrical about the line $y = x$?

e. Report the range of f. Report the domain of g.

9. For $g(x) = \sqrt{x + 1}$, find g^{-1}. State the range of g. State the domain of g^{-1}.

■ ***Quick Index***

■ SUMMARY

A topic related to composition of functions is finding the inverse of a function. For a 1-to-1 function, each output value is matched with exactly one input value. Only 1-to-1 functions have function inverses. We found that a vertical line can be used to test to see if a graph has originated from a function, and a horizontal line can test whether a graph has originated from a 1-to-1 function. Functions f and g are inverses of each other if

1. the composition $(f \circ g)(x)$ equals the composition $(g \circ f)(x)$, and
2. the compositions $(f \circ g)(x)$ and $(g \circ f)(x)$ yield x.

The notation for the inverse of function f is f^{-1}. We also introduced an algorithm for finding the inverse of a function.

We discussed the relationship of the graph of a function and the graph of its inverse to the line determined by $y = x$. The graph of f^{-1} is the graph of f reflected across the line determined by $y = x$.

9.4 ■ ■ ■ EXERCISES

Use the vertical line test to determine whether each graph represents a function. Use the horizontal line test to determine which functions are 1-to-1 functions.

1.

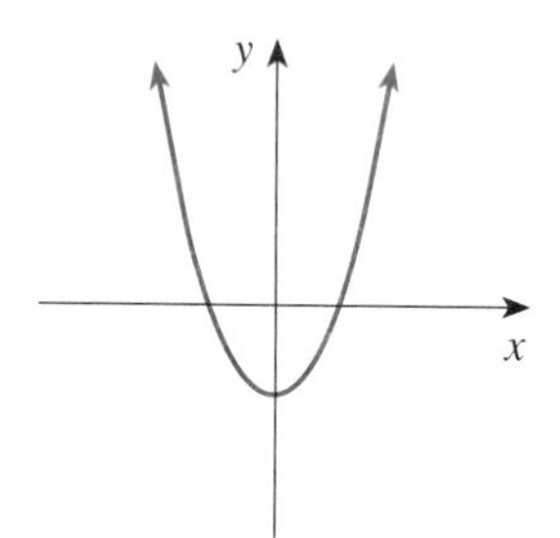

2.

3.

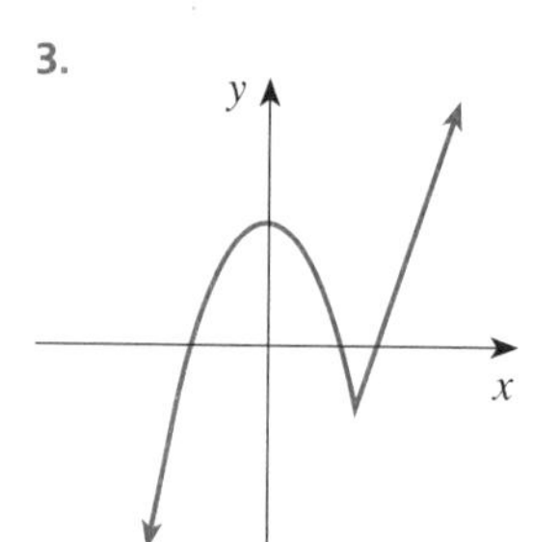

4.

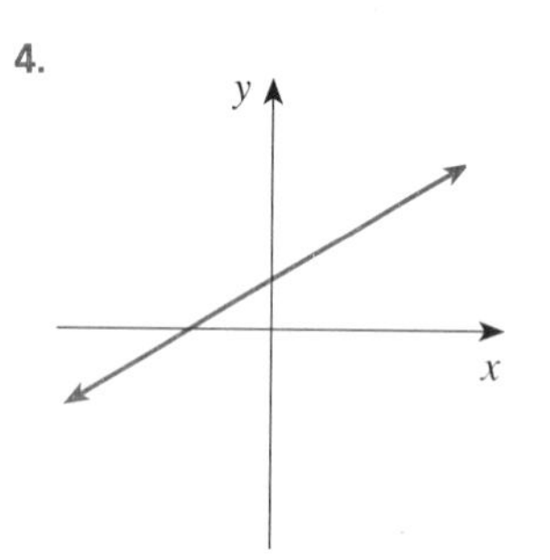

5.

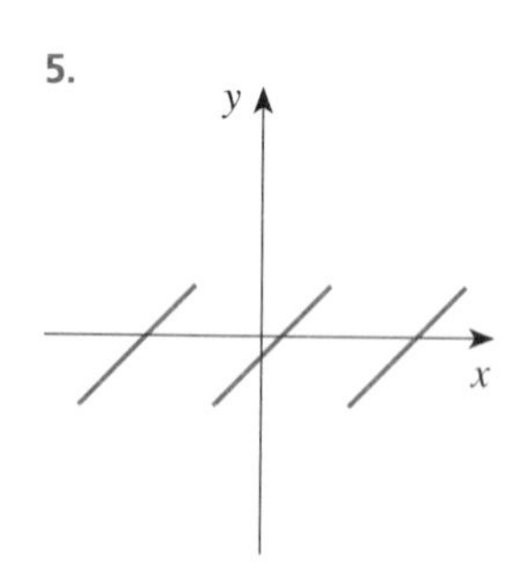

6.

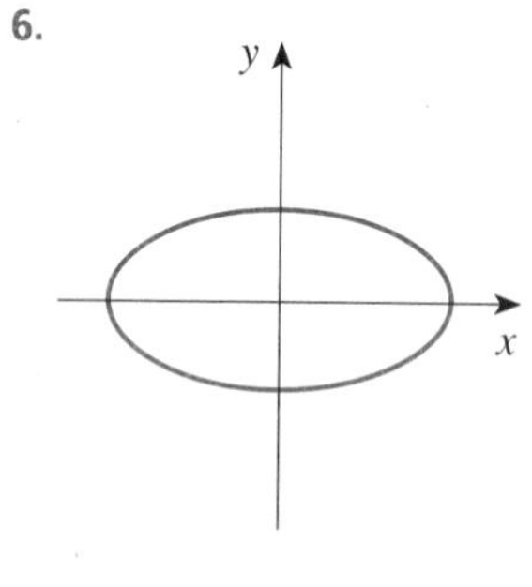

Report the inverse function of the given functions. Plot the function and its inverse on the same set of axes. Also, draw in the line determined by $y = x$.

7. $A = \{(-3, 2), (-2, 3), (-1, 5), (0, 6)\}$

8. $B = \{(3, 2), (2, -1), (1, 3), (0, 4)\}$

9. $C = \{(-4, -2), (-2, 2), (-1, 0), (0, -3)\}$

10. $D = \{(4, -1), (2, -3), (1, -1), (0, 1)\}$

Use the graphics calculator to graph $f \circ g$ and $g \circ f$ for each pair of functions. Report whether or not the functions are inverses of each other.

11. $f(x) = (x - 1)^{-1}$ and $g(x) = (x + 1)(x^{-1})$

12. $f(x) = \sqrt{x - 1}$ and $g(x) = x^2 + 1$, with $x \geq 0$

Use the given graph of a 1-to-1 function and the fact that if (a, b) is on the graph of a function, then (b, a) is on the graph of the inverse to draw the graph of the inverse function.

13.

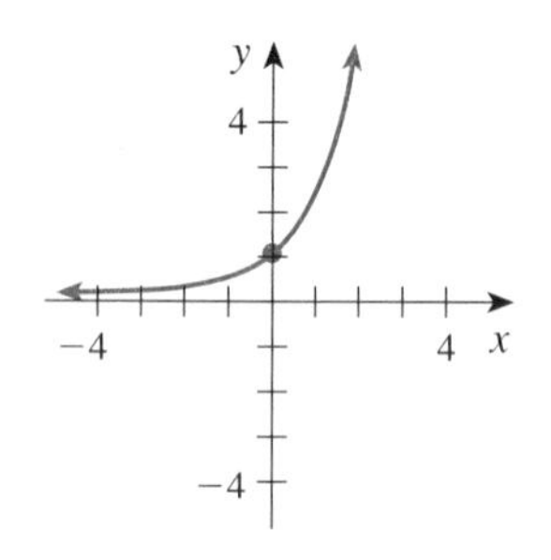

14.

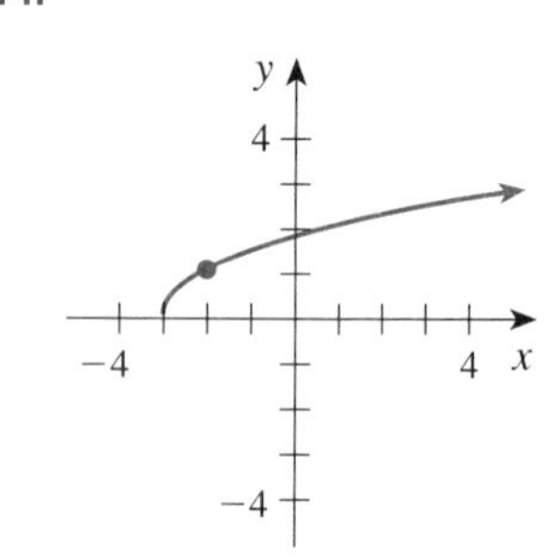

Find the inverse of each function.

15. $g(x) = 5 - 4x$

16. $h(x) = 2x - 3$

17. $f(x) = \frac{3}{2}x + \frac{1}{2}$

18. $c(x) = 2 - \frac{5}{3}x$

19. $f(x) = x^2 - 4x$, with $x \geq 2$

20. $h(x) = x^2 + 6x$, with $x \geq -3$

21. $r(x) = \dfrac{x + 3}{2 - x}$

22. $k(x) = \dfrac{x}{x + 2}$

23. $d(x) = \sqrt{x + 3}$

24. $c(x) = \sqrt{x - 2}$

25. $v(x) = 2x^{-1} + 1$

26. $w(x) = x^{-1} - 2$

27. $f(x) = x^{1/2} - 1$

28. $g(x) = x^{-1/2} - 2$

29. $j(x) = x^2 - 2x - 3$, with $x \geq 1$

30. $h(x) = 5 - 4x - x^2$, with $x \geq -2$

CHAPTER 9 REVIEW EXERCISES

For the given relation, draw a mapping diagram and report whether or not the relation is a function. If so, report the domain and range. If not, report why it fails to be a function.

1. $\{(5, 0), (0, -1), (2, -1), (-1, 1)\}$
2. $\{(-3, 0), (-2, 0), (-2, 3)\}$

Determine whether each mapping (correspondence) represents a function with domain set A. If so, report the range of the function. If not, report why it fails to be a function.

3.

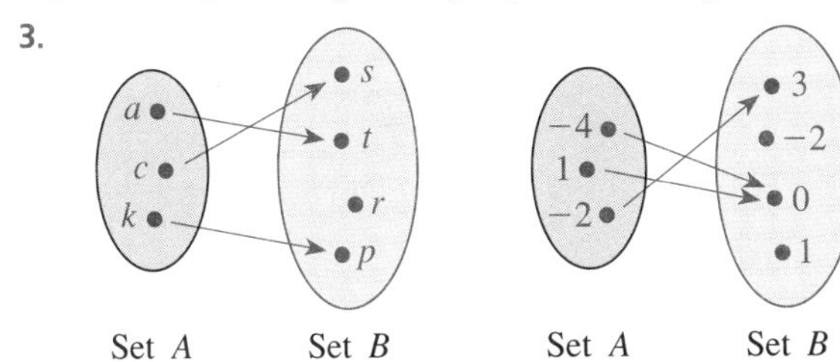

For the function $f(x) = 2x - 3$, find the indicated output.

5. $f(-3)$
6. $f(2)$
7. $f(a + 5) - f(5)$

For the function $g(x) = x^2 + 2x - 3$, find the indicated output.

8. $g(-1)$
9. $g(3)$
10. $g(h + 2) - g(2)$

For the function $h(x) = \dfrac{x - 1}{x}$, find the indicated output.

11. $h(-2)$
12. $h(c^{-1})$
13. $h(a - 1) - h(a)$

For the function $r(x) = \sqrt{x + 2}$, find the indicated output.

14. $r(2)$
15. $r(7)$
16. $r(3 + h) - r(h)$

Find the difference quotient of the function at the given x-value.

17. $f(x) = 1 - 2x$ at $x = -1$
18. $d(x) = \dfrac{-1}{x}$ at $x = 2$
19. $g(x) = -x^2$ at $x = 2$

Use the vertical line test to determine whether each graph represents a function.

20. 21.

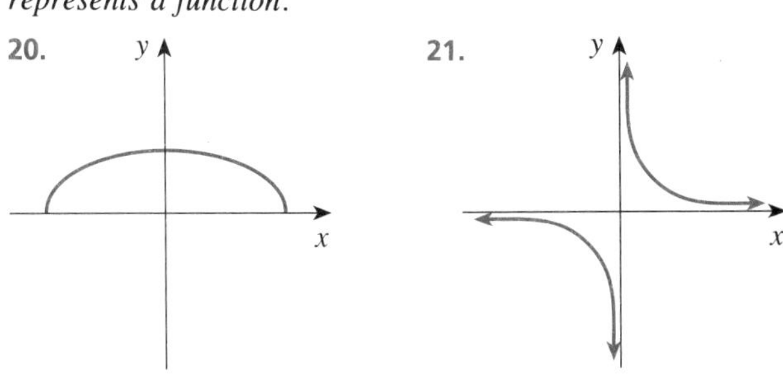

22.

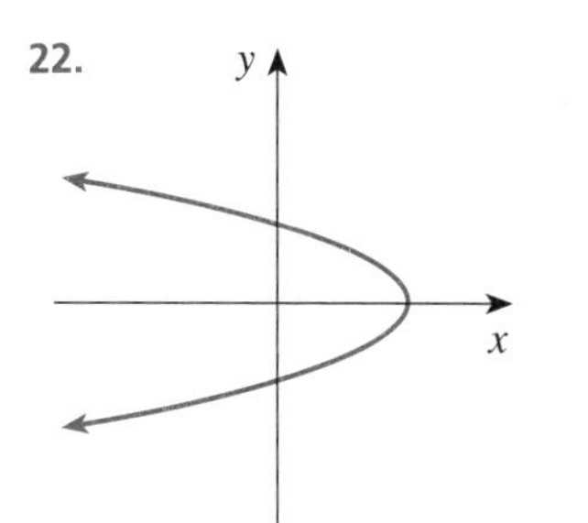

23.

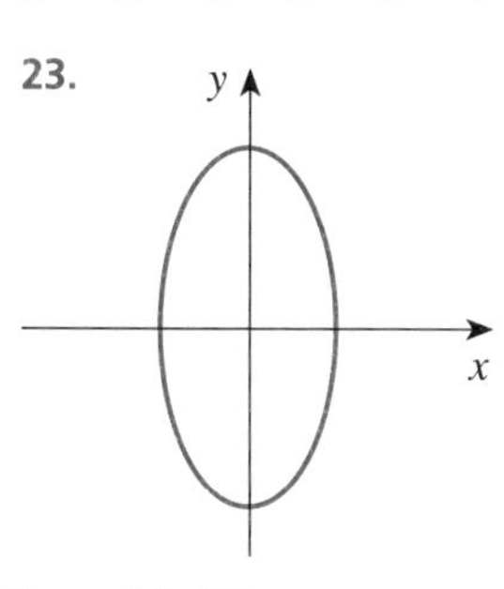

Use the given graph to answer Problems 24–32.

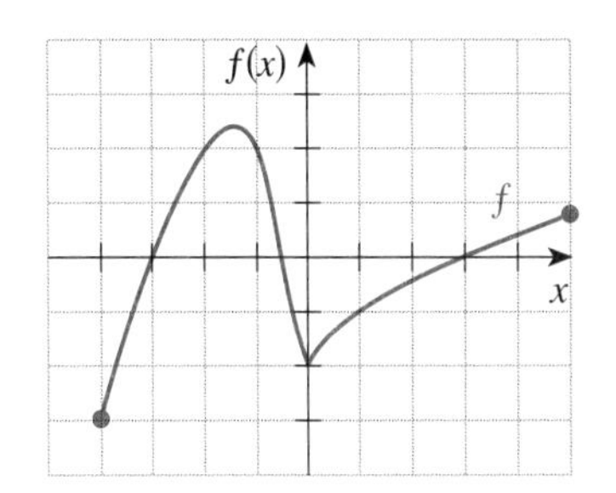

Read the graph of the function f and report the function value.

24. $f(-2)$
25. $f(0)$
26. $f(1)$
27. What input(s) produce an output of -3?
28. Report the zeros of the function.
29. Report the domain and range.
30. Report the interval(s) on which the function is increasing.
31. Report the interval(s) on which the function is decreasing.
32. Report any turning point(s).

For the given piecewise-defined function, report the domain and the range, and report the location of any turning point(s) and zero(s). Use the graphics calculator to check your work.

33. $f(x) = \begin{cases} -2x + 1 & \text{if } x < 0 \\ x + 3 & \text{if } x \geq 1 \end{cases}$

34. $g(x) = \begin{cases} \sqrt{4 - x} & \text{if } -5 \leq x < 0 \\ x^2 - x - 2 & \text{if } x \geq 1 \end{cases}$

For $f(x) = 3x - 2$ and $h(x) = -x + 2$, find the indicated output.

35. $(f + h)(-3)$
36. $(f - h)(2)$
37. $(f \cdot h)(2)$
38. $\left(\dfrac{f}{h}\right)(0)$
39. $\left(\dfrac{h}{f}\right)(0)$
40. $(f - h)(-b)$

For $f(x) = 2 - x$ and $h(x) = 2x^{-1}$, find the indicated output.

41. $(f \circ h)(2)$ **42.** $(f \circ h)(4)$ **43.** $(h \circ f)(2)$

44. $(h \circ h)(a + b)$ **45.** $(h \circ f)(c^2)$ **46.** $(f \circ f)(c)$

47. Use the given graph of a 1-to-1 function and the fact that if (a, b) is on the graph of a function, then (b, a) is on the graph of the inverse to draw the graph of the inverse function.

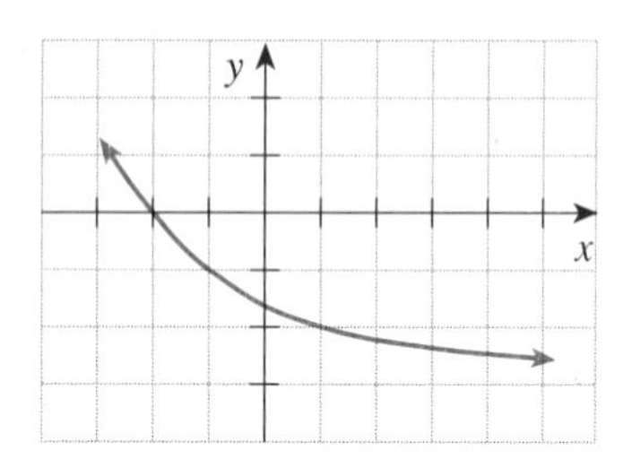

Find the inverse of each function.

48. $g(x) = -2x + 3$

49. $f(x) = \frac{-1}{3}x + \frac{4}{3}$

50. $f(x) = x^2 - 2x$, with $x \leq 1$

51. $r(x) = 4 - x^2$, with $x \geq 0$

52. $h(x) = x^{-1} - 3$

53. $d(x) = \frac{x + 3}{-2x}$

CHAPTER 9 TEST

1. For $f(x) = -3x - 4$ and $g(x) = x^2 - x$, find the indicated function value.

   a. $(g - f)(3)$ b. $(f \cdot g)(-2)$

2. For $f(x) = \frac{3}{x}$, find $\frac{f(x + h) - f(x)}{h}$ at $x = 2$.

3. Read the given graph of h to find each output.

   a. $h(-1)$
   b. $h(a)$
   c. Report the interval(s) on which the function is increasing.

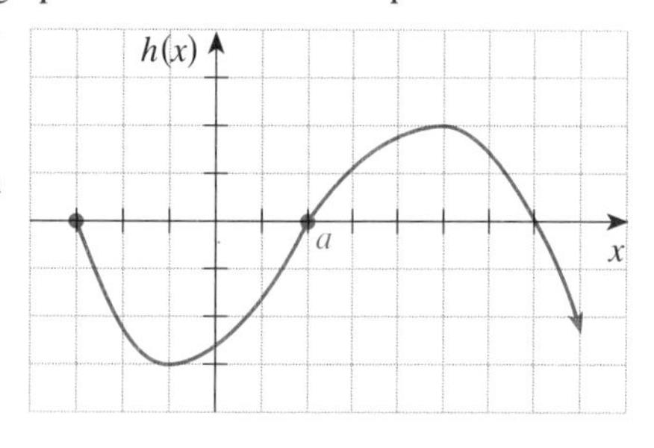

4. For $f(x) = 2x - 1$ and $g(x) = x^2$, find the given composition.

   a. $(g \circ f)(3)$ b. $(f \circ g)(-2)$

5. For $f(x) = x^{-2} - x^{-1}$ and $g(x) = x^{1/2}$, report the rule for $(f \circ g)(x)$ as a completely reduced radical expression.

6. For $f(x) = -2x + 5$, find $f^{-1}(x)$.

7. For $g(x) = \frac{x + 1}{x - 2}$, find $g^{-1}(x)$.

8. Graph $f(x) = 4x - x^2$, $x \geq 2$ and f^{-1} on the same set of axis. Also graph the line determined by $y = x$. Label the x- and y-intercepts with their ordered-pair names.

10

EXPONENTIAL AND LOGARITHMIC FUNCTIONS

In this chapter we will learn about two objects that are very different from the polynomials, rational expressions, and radicals expressions we have studied so far. These objects are exponential expressions and logarithmic expressions. We will also learn about the corresponding functions. In Chapter 6 we introduced integer exponents, and in Chapter 7, we considered rational number exponents. In this chapter we will use variables as exponents. Thus, we must consider exponents that are irrational numbers.

10.1 ■ ■ ■ EXPONENTIAL FUNCTIONS AND THEIR INVERSES

We begin with exponential functions such as $f(x) = 2^x$ and $g(x) = 10^x$. Using the ideas of Section 9.4, we will discuss the inverse of these functions, $f^{-1}(x) = \log_2 x$ and $g^{-1}(x) = \log_{10} x$. This discussion sets the background for our introduction to logarithmic equations and functions.

An Exponential Function All the functions we have discussed so far are algebraic functions—all have rules that are algebraic expressions, that is, a combination of numbers, variables, and the operations of addition, subtraction, multiplication, division, and nth roots. These algebraic functions have exponents that are rational numbers. We now consider a function in which the exponent is a variable, for example, $f(x) = 2^x$. Such a function is known as an **exponential function**. Consider the following definition.

DEFINITION OF AN EXPONENTIAL FUNCTION

For any real number a, with $a > 0$ and $a \neq 1$, $f(x) = a^x$ is an exponential function.

Input-output pairs for $f(x) = 2^x$

Input x	*Output* 2^x
-2	$\frac{1}{4}$
-1	$\frac{1}{2}$
0	1
1	2
2	4

As we know from our study of rational number exponents, such as $x^{a/b}$, if the denominator of the exponent, b, is an even number and $x < 0$, then $x^{a/b}$ is not a real number. Thus, we require that an exponential function have a base that is a positive number. Also, because 1 to any power is 1, we require that the base be a positive number other than 1. If $a = 1$, then $f(x) = a^x$ is just the constant function $f(x) = 1$.

A typical example of an exponential function is $f(x) = 2^x$. Compare $f(x) = 2^x$ with the linear function $g(x) = 2x$ and with the quadratic function $h(x) = x^2$. In $h(x) = x^2$, the variable is the base. For the exponential function $f(x) = 2^x$, the variable is an exponent.

We can quickly generate some input-output ordered pairs for the function $f(x) = 2^x$. Several pairs are shown in the table in the margin. Plotting these ordered pairs produces the graph shown in Figure 1. This shape is typical of all exponential functions (just as a parabola is typical of all quadratic functions). The y-intercept is (0, 1). (Remember that any number except 0 to the zero power is 1.) Also, the graph has no x-intercept. In fact, as the input value gets smaller and smaller (-10, -100, $-1{,}000$, and so on), the output value gets closer and closer to 0. The line $y = 0$, the x-axis, is a horizontal asymptote.

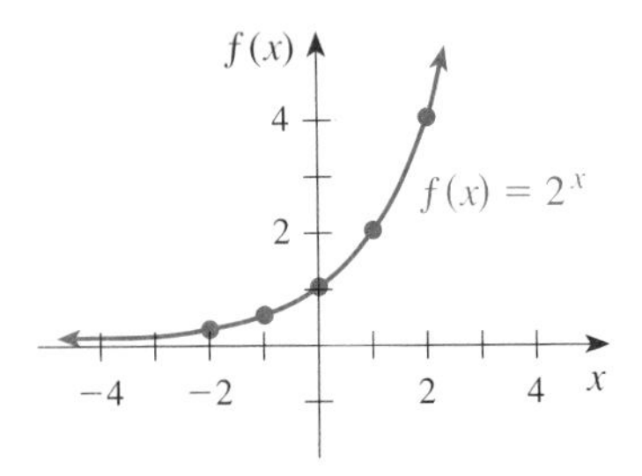

FIGURE 1

Using the Graphics Calculator

For `Y1`, type `2^X`. Enter the Window 9.4 range settings, and view the graph screen. The graph drawn is similar to the one shown in Figure 1. Press the `TRACE` key. The y-intercept is (0, 1), that is $2^0 = 1$. Now, press the right arrow key. The message `X=.2 Y=1.1486984` means $2^{0.2} \approx 1.1486984$. Now, $2^{0.2}$ is $2^{1/5}$, which means $\sqrt[5]{2}$. That is, the number to the fifth power that yields 2 is approximately 1.1486984. Continue to press the right arrow key and observe the messages. One of these messages reads `X=1 Y=2`. That is, $2^1 = 2$. Continue to press the right arrow key and observe the messages `X=2 Y=4` (that is, $2^2 = 4$) and `X=3 Y=8` (that is, $2^3 = 8$). As the right arrow key is pressed repeatedly, the function output increases rapidly. This is what is meant by an *exponential growth pattern.*

Press the left arrow key until the message `X=-1 Y=.5` appears. This message means that $2^{-1} = \frac{1}{2}$. Continue to press the left arrow key and observe the messages `X=-2 Y=.25` $\left(\text{that is, } 2^{-2} = \frac{1}{4}\right)$ and `Y=-3 Y=.125` $\left(\text{that is, } 2^{-3} = \frac{1}{8}\right)$. As the left arrow key is pressed repeatedly, the output value simply gets closer and closer to 0, but it always remains a positive number—the curve never touches or crosses the x-axis.

Irrational Exponents For the function $f(x) = 2^x$, we know the meaning of the exponent whenever x is a rational number. But, the graph in Figure 1 also includes irrational input values, such as $\sqrt{2}$. What does $2^{\sqrt{2}}$ mean? To assign a formal meaning to this irrational number exponent is beyond this course (because techniques of calculus must be used). We can, however, give this expression a meaning informally by approximating the number $\sqrt{2}$ and then observing the corresponding value of 2 raised to that power:

Approximation of $\sqrt{2}$	1.4	1.41	1.414
Meaning of 2^x	$2^{14/10}$	$2^{141/100}$	$2^{1414/1000}$
As a radical	$\sqrt[10]{2^{14}}$	$\sqrt[100]{2^{141}}$	$\sqrt[1000]{2^{1414}}$
Decimal approximation	2.6 . . .	2.65 . . .	2.664 . . .

As the sequence of approximations approaches $\sqrt{2}$, the sequence of radicals approach a real number that has a decimal name starting as 2.66. We name this real number $2^{\sqrt{2}}$. Informal meanings for other irrational number exponents can be developed similarly.

GIVE IT A TRY

For the given exponential function, construct a table of input-output values for the input values -2, -1, 0, 1, and 2. Plot the points, and then sketch a curve for the graph of the exponential function.

1. $g(x) = 3^x$ **2.** $h(x) = 4^x$ **3.** $d(x) = 2^{-x}$

4. On the graphics calculator, type √2 and press the ENTER key. Now, type 2^1.4 and press the ENTER key. Next, type 2^1.41, and press the ENTER key. Type 2^1.414 and press the ENTER key. Finally, type 2^√2 and press the ENTER key. What number is output for each entry?

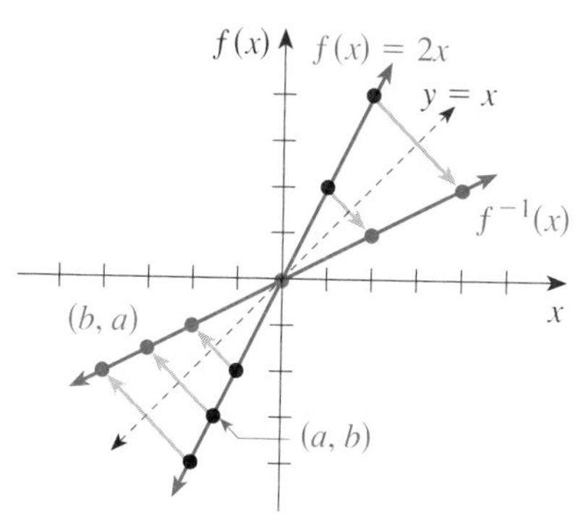

FIGURE 2

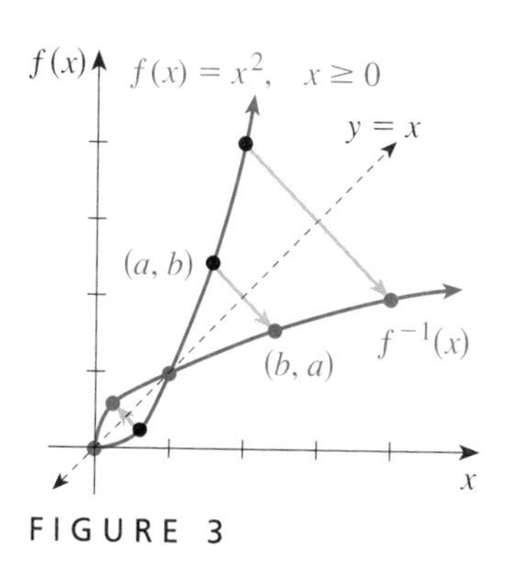

FIGURE 3

Inverse Functions Now that we have an idea of what is meant by an exponential function, we want to find the inverse of the function $f(x) = 2^x$. We start our argument with some simpler functions, namely $f(x) = 2x$ and $f(x) = x^2$.

From the graph of the *doubling function*, $f(x) = 2x$, we can obtain a graph of its inverse, by simply interchanging the ordered pairs. See Figure 2. We name this inverse function the *halving function*, and write $f^{-1}(x) = \frac{1}{2}x$. Next, consider the *squaring function*, $f(x) = x^2$ for $x \geq 0$. Again, from the graph of the squaring function we can draw the graph of its inverse function, as shown in Figure 3. We name the inverse function the *square root function* and write $f^{-1}(x) = \sqrt{x}$. To find an output for this inverse function, that is, to find $\sqrt{25}$, we must think *backwards*: "What number squared yields 25?" We write $\sqrt{25} = 5$. We will use a similar thought pattern to find the output for the inverse of an exponential function.

We are now ready to consider the function $f(x) = 2^x$ and its inverse. In the table with Figure 4 we have listed some ordered pairs from the function

$$f(x) = 2^x$$

and the corresponding ordered pairs for the inverse function. From the graph of the function $f(x) = 2^x$, we construct the graph of its inverse (see Figure 5).

Ordered pairs for the function f and its inverse function

$f(x) = 2^x$	f^{-1}
$\left(-2, \frac{1}{4}\right)$	$\left(\frac{1}{4}, -2\right)$
$\left(-1, \frac{1}{2}\right)$	$\left(\frac{1}{2}, -1\right)$
(0, 1)	(1, 0)
(1, 2)	(2, 1)
(2, 4)	(4, 2)

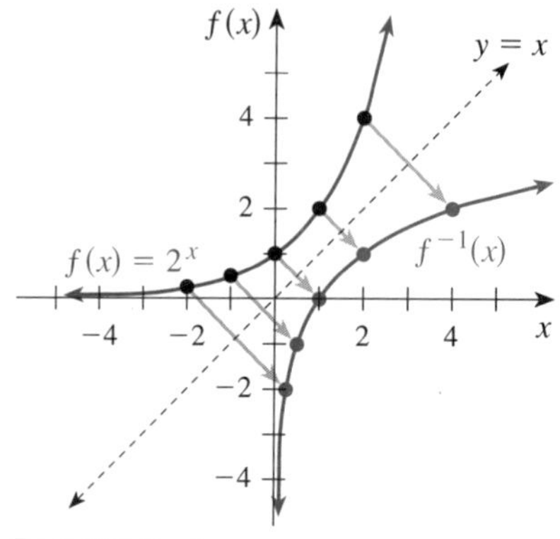

FIGURE 4

Notice that the graph of $f(x) = 2^x$ and its inverse are reflections across the line determined by $y = x$. This inverse function f^{-1}, is named *log base* 2 (*log* is short for *logarithm*), and we write $\log_2 x$. From the graph we see that the domain of this inverse function is $\{x \mid x > 0\}$ and the range is $\{y \mid y \text{ is real}\}$.

Just as we worked with $g(x) = \sqrt{x}$, we must think "backwards" from $h(x) = \log_2 x$ to $f(x) = 2^x$. We can view $\sqrt{x}$ as a set of instructions: *Find the nonnegative number such that it times itself yields x.* In a similar fashion, we can view $\log_2 x$ as a set of instructions: *Rewrite x as a power of* 2, *and then report the exponent.* That is, the $\log_2 x$ is an exponent. Consider the following example.

EXAMPLE 1 Find:

a. $\log_2 8$ b. $\log_2 \frac{1}{8}$ c. $\log_2 32$

Solution a. We rewrite 8 as a power of 2 to get 2^3. The exponent is 3, so

$$\log_2 8 = 3$$

b. We rewrite $\frac{1}{8}$ as a power of 2 to get $\frac{1}{8} = 2^{-3}$. The exponent is −3, so

$$\log_2 \frac{1}{8} = -3$$

c. $32 = 2^5$ So $\log_2 32 = 5$ ■

Soon, we will want to know the value of $\log_2 5$. That is, we will rename 5 as a power of 2 and ask the value of the exponent, $5 = 2^?$. Before considering this question, we must learn more about logarithms.

GIVE IT A TRY

5. Use the graph of the exponential $f(x) = 3^x$ to produce the graph of the inverse function, $f^{-1}(x) = \log_3 x$.

6. Use the result of Problem 5 to find $\log_3 1$ and $\log_3 \frac{1}{9}$.

Common Logarithm Just as we have found the inverse functions for exponential functions, such as $f(x) = 2^x$ and $f(x) = 3^x$, we can find the inverse of the function $f(x) = 10^x$. This inverse function is named *log base 10 of x* and is known as the **common logarithm**. That is, for $f(x) = 10^x$, we have $f^{-1}(x) = \log_{10} x$. Fortunately, values for this logarithmic function can be found on most calculators with a special key, the LOG key.

Using the Graphics Calculator

We have not viewed the inverse of $f(x) = 2^x$, that is, $f^{-1}(x) = \log_2 x$, on the graphics calculator because we have no function key for $\log_2 x$. (Later we will learn of a way to see the graph of this function.) Now, for Y1, type 10^X. Enter the Window 9.4 range values, and view the graph screen. The shape of this graph is the same as the graphs of other exponential functions. Of course, the rate of growth here is much greater than for $f(x) = 2^x$ or $f(x) = 3^x$.

Now, for Y2, use the LOG key to type log X. View the graph screen, which is shown in Figure 5. Press the TRACE key. The message X=0 Y=1 means that the y-intercept for $f(x) = 10^x$ is (0, 1). Press the down arrow key to move the cursor to the inverse function, $g(x) = \log_{10} x$. The message X=0 Y= means that the log function has no y-intercept. Its domain is $\{x \mid x > 0\}$. That is, $\log_{10} 0$ is undefined! Press the right arrow key. The message X=.2 Y=-.69897 means $\log_{10} 0.2 \approx -0.69897$. That is, $0.2 \approx 10^{-0.69897}$. Continue to press the right arrow key until the message reads X=1 Y=0. This message means $\log_{10} 1 = 0$. That is, $1 = 10^0$. Press the right arrow key until the message reads X=5 Y=.69897. This message means $\log_{10} 5 = 0.69897$. That is, $5 \approx 10^{0.69897}$. Continue to press the right arrow key (the screen will scroll) until the message reads X=10 Y=1. This message means $\log_{10} 10 = 1$. That is, $10 = 10^1$.

To see that the function entered for Y2, $\log_{10} x$, is indeed the inverse of the function entered for Y1, 10^x, type 10^Y2 for Y3. This is the composition of $f(x) = 10^x$ on the function $g(x) = \log_{10} x$. View the graph screen, as shown in Figure 6. The graph for $f(x) = 10^x$ is drawn, then the graph for $g(x) = \log_{10} x$ is drawn, and finally the composition is drawn. As you can see, this composition is the identity function, $I(x) = x$.

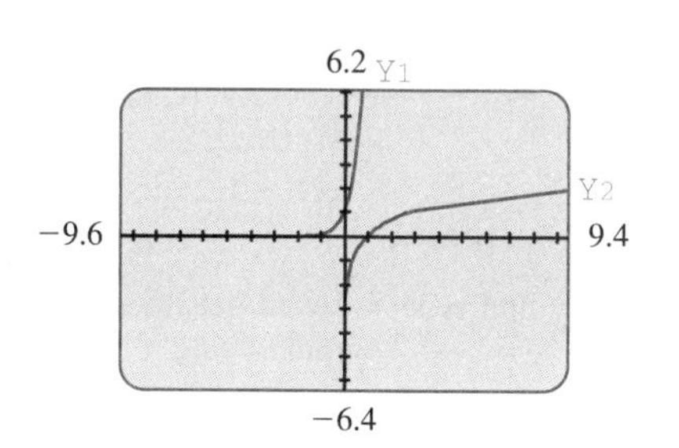

:Y1=10^X :Y2=log X

FIGURE 5

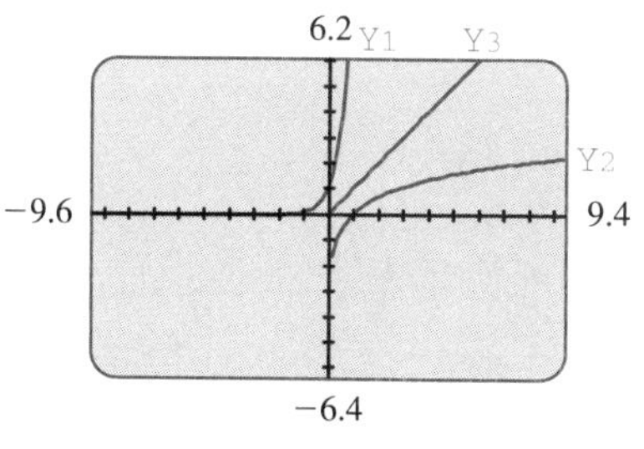

:Y1=10^X
:Y2=log X
:Y3=10^Y2

FIGURE 6

We define a general logarithmic function as follows.

DEFINITION OF A LOGARITHMIC FUNCTION

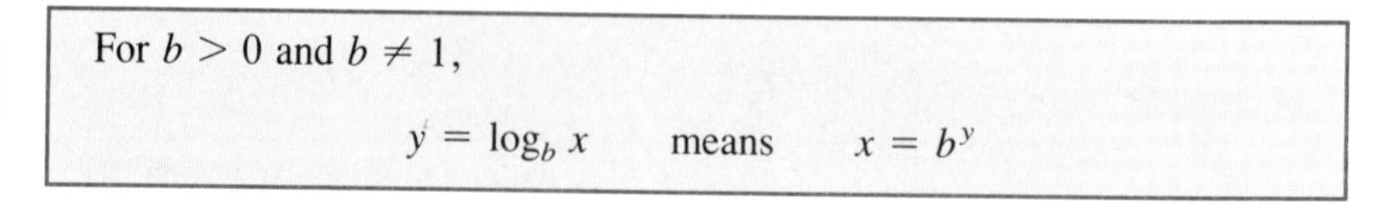

For $b > 0$ and $b \neq 1$,

$$y = \log_b x \quad \text{means} \quad x = b^y$$

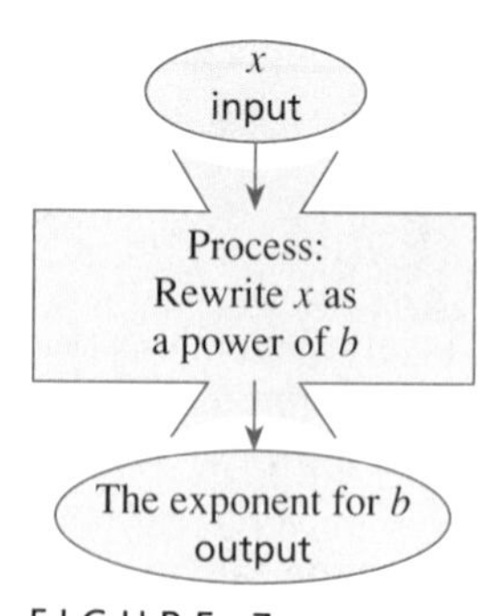

FIGURE 7

In words, $y = \log_b x$ means that if we rename x as a power of b, the exponent is y. That is, $\log_2 8 = 3$ means that when 8 is renamed as a power of 2, the exponent is 3, that is, $8 = 2^3$. Also, $\log_{10} 100 = 2$ means 100 renamed as a power of 10 has exponent 2, since $100 = 10^2$. Figure 7 shows an IPO diagram for the general logarithmic function.

From our work with exponents and the meaning of log, we now state two important properties of logarithms.

PROPERTIES OF LOGARITHMIC FUNCTIONS

For base b with $b > 0$ and $b \neq 1$,

1. $\log_b 1 = 0$ because $1 = b^0$

2. $\log_b b = 1$ because $b = b^1$

GIVE IT A TRY

7. Use the meaning of log to find the value of $\log_{10} \sqrt{10}$.

8. Using the graphics calculator, press the LOG key and type √10 to get the command log √10. Press the ENTER key. What number is output?

9. Use the properties of logarithmic functions to evaluate $\log_7 1$ and $\log_{18} 18$.

Evaluations To increase our understanding of logarithms, and to gain experience in working with log notation, we now evaluate some logarithmic expressions.

EXAMPLE 2 Evaluate each expression.

a. $\log_4 16$ b. $\log_5 \frac{1}{25}$ c. $\log_6 \frac{1}{6}$ d. $\log_7 0$

Solution a. We use the meaning of *log* to get $16 = 4^?$. Now, 4^2 is 16, so

$$\log_4 16 = 2$$

b. We use the meaning of *log* to get $\frac{1}{25} = 5^?$. Now, 5^{-2} is $\frac{1}{25}$, so

$$\log_5 \frac{1}{25} = -2$$

c. We use the meaning of *log* to get $\frac{1}{6} = 6^?$. Now, 6^{-1} is $\frac{1}{6}$, so

$$\log_6 \frac{1}{6} = -1$$

d. We use the meaning of *log* to get $0 = 7^?$. There is *no* exponent for 7 that will yield 0, so $\log_7 0$ is *undefined*. ■

Solving Equations We next consider solving logarithmic equations in one variable. A **logarithmic equation** is an equation in which the logarithm operation is applied to a variable expression. For example, $\log_3(x + 2) = 2$ is a logarithmic equation in one variable. To solve this equation means to find the numbers that replace the variable to yield true arithmetic sentences. The solution to $\log_3(x + 2) = 2$ is 7, because $\log_3 (7 + 2) = 2$ is a true arithmetic sentence. Remember,

$$\log_3(7 + 2) = 2 \quad \textit{means} \quad 7 + 2 = 3^2$$

In general, to solve this type of equation, we use the meaning of logarithms. Consider the following examples.

EXAMPLE 3 Solve: $\log_2(x + 1) = 3$

Solution

$$x + 1 = 2^3 \qquad \text{The meaning of } \log_b a$$

$$x + 1 = 8$$

$$x = 7$$

So, the solution to $\log_2(x + 1) = 3$ is 7. ■

EXAMPLE 4 Solve: $\log_3(x^2 + 2x) = 1$

Solution

$$x^2 + 2x = 3^1 \qquad \text{The meaning of } \log_b a$$

$$x^2 + 2x - 3 = 0$$

$$(x + 3)(x - 1) = 0$$

$$x + 3 = 0 \quad \text{or} \quad x - 1 = 0$$

$$x = -3 \qquad\qquad x = 1$$

So, the solution to $\log_3(x^2 + 2x) = 1$ is the numbers -3 and 1. ■

EXAMPLE 5 Solve: $\log_4\left(\frac{x+1}{x-2}\right) = -1$

Solution

$$\frac{x+1}{x-2} = 4^{-1} \qquad \text{The meaning of } \log_b a$$

$$\frac{x+1}{x-2} = \frac{1}{4}$$

$$\frac{4(x-2)(x+1)}{x-2} = \frac{4(x-2)}{4} \qquad \text{Multiply by the LCD.}$$

$$4(x+1) = x-2 \qquad \text{Reduce rational expressions.}$$

$$4x+4 = x-2$$

$$\underline{-x \qquad -x}$$

$$3x+4 = -2$$

$$\underline{-4 \qquad -4}$$

$$3x = -6$$

$$x = -2$$

So, the solution to $\log_4\left(\frac{x+1}{x-2}\right) = -1$ is the number -2. ■

EXAMPLE 6 Solve: $\log_5\left(\sqrt{x-2}\right) = 0$

Solution

$$\sqrt{x-2} = 5^0 \qquad \text{The meaning of } \log_b a$$

$$\sqrt{x-2} = 1$$

$$x-2 = 1 \qquad \text{Square each side.}$$

$$x = 3$$

So, the solution to $\log_5\left(\sqrt{x-2}\right) = 0$ is 3. ■

In Examples 3–6, after using the meaning of $\log_b a$, we produced an equation which we know how to solve.

GIVE IT A TRY

10. Evaluate: $\log_8 64$

Solve each logarithmic equation.

11. $\log_3(3x - 2) = 2$

12. $\log_4(x^2) = 2$

13. $\log_2\frac{1}{x} = 1$

14. $\log_6\sqrt{3 - x} = 1$

■ SUMMARY

In this section, we have learned about exponential functions, such as $f(x) = 2^x$. The inverse of such a function is a logarithmic function, such as $g(x) = \log_2 x$. The notation $\log_b a = c$ means that if a is written as a power of b, then the exponent is c, that is, $a = b^c$. Two important properties of logarithms were presented: for $b > 0$ and $b \neq 1$, $\log_b 1 = 0$, and $\log_b b = 1$. Using the meaning of logarithms, we solved equations involving logs, such as $\log_3(x - 2) = 2$.

■ ***Quick Index***

10.1 ■ ■ ■ EXERCISES

1. Graph the exponential function $f(x) = \left(\frac{1}{3}\right)^x$.
2. Use the graph from Problem 1 to construct the graph of $f^{-1}(x) = \log_{1/3} x$.
3. From the graphs in Problems 1 and 2, what is $\log_{1/3} 3$?
4. From the graphs in Problems 1 and 2, what is $\log_{1/3}(-1)$?
5. Graph the exponential function $f(x) = 5^x$.
6. Use the graph from Problem 5 to construct the graph of $f^{-1}(x) = \log_5 x$.
7. From the graphs in Problems 5 and 6, what is $\log_5(-2)$?
8. From the graphs in Problems 5 and 6, what is $\log_5 1$?

Using the meaning of the notation $\log_b a$ *and the properties of logarithmic functions to find the indicated value.*

9. $\log_6 36$
10. $\log_7 49$
11. $\log_4 64$
12. $\log_5 125$
13. $\log_{1/2} \frac{1}{4}$
14. $\log_{1/4} \frac{1}{2}$
15. $\log_{1/2} 8$
16. $\log_8 \frac{1}{2}$
17. $\log_{23} 23$
18. $\log_{89} 1$
19. $\log_3 \sqrt[5]{3}$
20. $\log_{10} \sqrt[4]{10}$
21. $\log_{10} 0.01$
22. $\log_{10}(0.1)^3$
23. $\log_7 0$
24. $\log_{1/2} 0$

Use the calculator and its LOG *key to compute the common log of each number. Use the calculator output to write the given number as an approximate power of* 10.

25. 9
26. 300
27. $\frac{1}{100}$
28. 1,000
29. 0.005
30. $(10000)^{30}$

Solve each logarithmic equation.

31. $\log_3 x = -2$
32. $\log_5 x = -1$
33. $\log_{1/2} x = -1$
34. $\log_{2/7} x = 0$
35. $\log_6(x - 3) = 2$
36. $\log_6(2x + 1) = 1$
37. $\log_3(x^2 - 8x) = 2$
38. $\log_6(x^2 - 9x) = 2$
39. $\log_2(x^2 - 2x) = 1$
40. $\log_4(x^2 - x) = 1$
41. $\log_3\left(\frac{x - 1}{x + 3}\right) = -1$
42. $\log_4\left(\frac{x + 2}{x}\right) = 1$
43. $\log_{1/3}\left(\frac{2x + 1}{x - 1}\right) = 2$
44. $\log_{1/4}\left(\frac{x - 3}{x + 1}\right) = -1$
45. $\log_5 \sqrt{2x - 1} = 1$
46. $\log_2 \sqrt{3x + 2} = 1$
47. $\log_7 \sqrt{3 - x} = 2$
48. $\log_3 \sqrt{5 - x} = 3$
49. $\log_2 \sqrt[3]{x} = 1$
50. $\log_3 \sqrt[6]{x} = 0$

DISCOVERY

Earthquakes and the Richter Scale

When an earthquake occurs, seismic (shock) waves radiate from the focus of the quake. The *focus* is located beneath the earth's surface. The quake's *epicenter* is a location on earth's surface directly above the focus. See Figure 1.

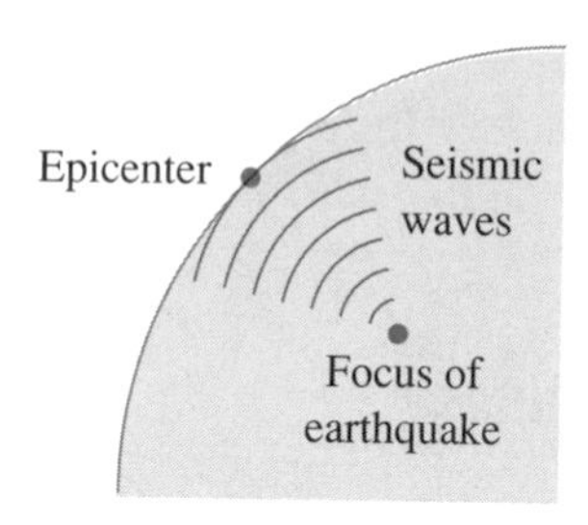

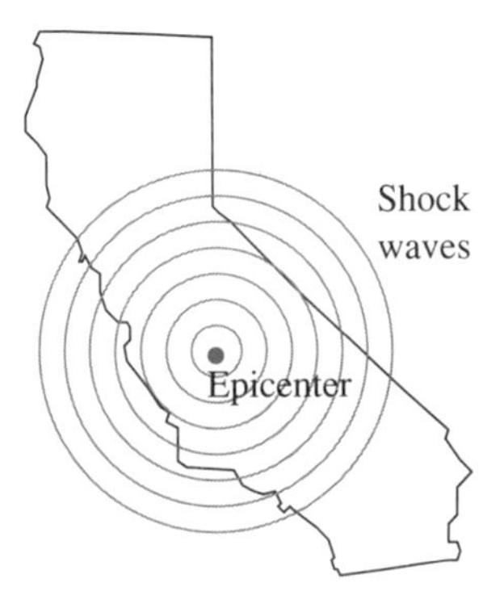

FIGURE 1

A seismograph is used to measure these seismic waves and, thus, the intensity of the earthquake. A simplified version of such a device is shown in Figure 2. The graph produced provides a measure of the vertical displace-

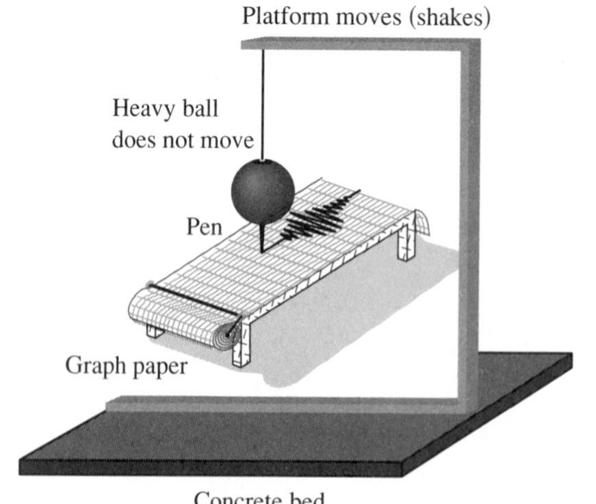

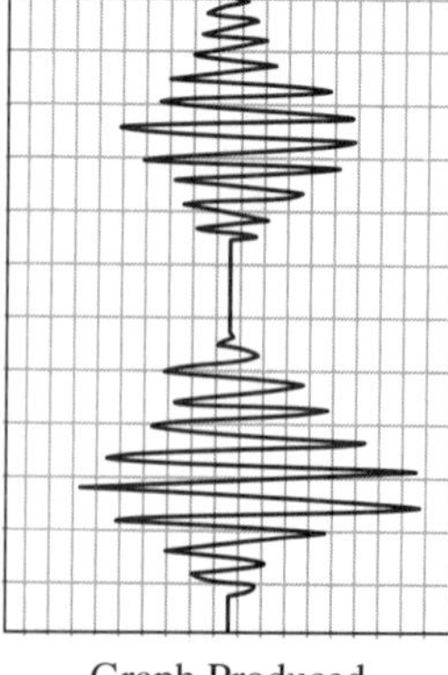

FIGURE 2

ment of the earth, the amplitude (in micrometers) and a measure of the time between the waves, the period (in seconds).

In 1935, at California Institute of Technology, Charles Richter devised a formula for assigning a value to a particular earthquake:

$$R = \log_{10} \frac{a}{T} + B$$

where a is the amplitude of the largest wave
T is the time between waves
and B is a constant based on the precision of the seismograph and its distance from the epicenter

The R-value, now used worldwide, is known as the Richter-scale value. It provides a method of comparing the intensity of a given earthquake with other earthquakes. Consider the following table of Richter values and effects.

Richter value	*Effect*
< 2.5	No observable effect
2.5 to 5.4	Felt, but no damage
5.5 to 6.0	Some damage to buildings
6.1 to 6.9	Damage to populated areas
7.0 to 7.9	Serious damage
> 8.0	Total destruction

The San Francisco earthquake of 1906 would have measured 8.25 on the Richter scale. The Tangshan, China, quake of 1976, which resulted in over 600,000 deaths, measured 7.6 on the Richter scale. The Loma Prieta quake during the 1989 Baseball World Series, measured 7.1 on the Richter scale.

Suppose the amplitude—the vertical earth movement—of an earthquake is measured to be 300 micrometers, the time interval between waves is 2 seconds, and the B-value for the particular seismograph is 4.25. What is the Richter scale value for this quake? Using the formula, we get

$$
\begin{aligned}
R &= \log_{10} \frac{a}{T} + B \\
&= \log_{10} \frac{300}{2} + 4.25 && a = 300,\ T = 2,\ B = 4.25 \\
&\approx 2.18 + 4.25 && \text{Use a calculator to evaluate } \log_{10} 150. \\
&\approx 6.4 && \text{Round to tenths.}
\end{aligned}
$$

We would say that the earthquake measures 6.4 on the Richter scale. Using the table of Richter values and effects, we would expect some damage to populated areas near the epicenter of the quake.

EXAMPLE 1 What is the amplitude measurement on the seismograph for a 8.2 Richter-scale earthquake? Assume the B-value for the seismograph is 4.2 and the time interval for a wave measurement is 2 seconds.

Solution Replacing R with 8.2, T with 2, and B with 4.2, we get the logarithmic equation

$$8.2 = \log_{10} \frac{a}{2} + 4.2$$

or

$$\begin{aligned} \log_{10} \frac{a}{2} + 4.2 &= 8.2 \\ -4.2 \quad & \quad -4.2 \\ \hline \log_{10} \frac{a}{2} &= 4 \\ \frac{a}{2} &= 10^4 && \text{Meaning of } \log_{10} \\ a &= 2 \times 10^4 && \text{Multiply by 2.} \end{aligned}$$

Thus, the amplitude is 2×10^4 micrometers. Since a micrometer is 1×10^{-6} meters, a is $(2 \times 10^4) \times 10^{-6}$ meters, or 2×10^{-2} meters, or 2 centimeters $\left(\approx \frac{3}{4} \text{ inch}\right)$.

Exercises

Using the graphics calculator, graph the Richter-scale value R as a function of the amplitude, a. Assume $B = 4.2$ and $T = 2$ seconds. Use the TRACE key to approximate the amplitude needed for an earthquake of the given Richter scale value.

1. 7.1 **2.** 7.8 **3.** 8.1 **4.** 8.5

10.2 ■ ■ ■ PROPERTIES OF LOGARITHMS

In Section 10.1 we introduced logarithmic functions. We now develop a collection of properties of logarithms, much like the properties of exponents. Using the properties of logarithms, we will solve some additional logarithmic equations in one variable.

Background In the previous section, we considered these properties:

> **REMEMBER**
>
> $\log_{10} 5 = \square$ ← same
>
> $5 = 10^{\square}$ ←

For $a > 0$ and $a \neq 1$,

1. $\log_a a = 1$ because $a = a^1$

Example: $\log_{10} 10 = 1$ because $10 = 10^1$

2. $\log_a 1 = 0$ because $1 = a^0$

Example: $\log_5 1 = 0$ because $1 = 5^0$

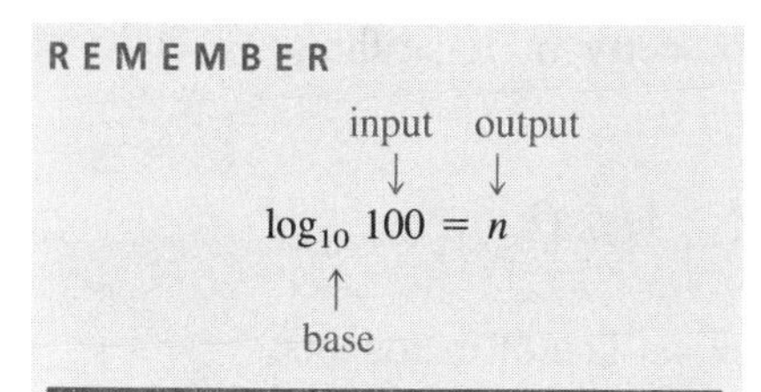

We also solved some logarithmic equations, such as $\log_3(2x + 1) = 2$. In this section, we extend this skill to logarithmic equations, such as

$$\log_2(x + 1) + \log_2(x - 2) = 2 \quad \text{and} \quad \log_5(2x - 1) - \log_5(x + 2) = -1$$

To solve such equations, we must know some additional properties of logarithms. Recall that the logarithm of a positive real number is an exponent. That is, $\log_3 9$ is the exponent for 3 that yields 9 (namely, 2). The properties of logarithms we now develop follows from these laws of exponents.

LAWS OF EXPONENTS

1. $(x^a)(x^b) = x^{a+b}$ — To multiply like bases, add the exponents.
2. $(x^a) \div (x^b) = x^{a-b}$ — To divide like bases, subtract the exponents.
3. $(x^a)^b = x^{ab}$ — To raise a power to a power, multiply the exponents.

In addition to these laws of exponents, we have the following property of equality of exponents.

EQUALITY PROPERTY OF EXPONENTS

If $a^x = a^y$, with $a \neq 0$ and $a \neq 1$, then $x = y$.

This statement says that if two powers are equal and their bases are equal, then their exponents are equal. Notice that $1^4 = 1^{10}$ and $0^4 = 0^{10}$, but $4 \neq 10$. Thus, we have the restriction that the base is neither 0 nor 1.

The Log of a Product The first property of logarithms matches the first law of exponents:

$$\log_b PQ = \log_b P + \log_b Q$$

In fact, we can establish this property by using the first law of exponents.

Let $x = \log_b P$. Then $P = b^x$

Let $y = \log_b Q$. Then $Q = b^y$.

Let $z = \log_b PQ$. Then $PQ = b^z$.

Now, $b^z = PQ = b^x b^y = b^{x+y}$.

Thus, $b^z = b^{x+y}$ and $z = x + y$ (property of equality of exponents). Using back-substitution, we get:

$$\begin{array}{ccccc} z & = & x & + & y \\ \downarrow & & \downarrow & & \downarrow \\ \log_b PQ & = & \log_b P & + & \log_b Q \end{array}$$

The Log of a Quotient The second property of logarithms we consider matches the second law of exponents:

$$\log_b \frac{P}{Q} = \log_b P - \log_b Q$$

We establish this property by using the second law of exponents.

Let $x = \log_b P$, then $P = b^x$.

Let $y = \log_b Q$, then $Q = b^y$.

Let $z = \log_b \frac{P}{Q}$, then $\frac{P}{Q} = b^z$.

Now, $b^z = \frac{P}{Q} = \frac{b^x}{b^y} = b^{x-y}$. Thus $z = x - y$. Using back-substitution,

$$\begin{array}{ccccc} z & = & x & - & y \\ \downarrow & & \downarrow & & \downarrow \\ \log_b \frac{P}{Q} & = & \log_b P & - & \log_b Q \end{array}$$

> **WARNING**
>
> Just as $(x + y)^2 \neq x^2 + y^2$ and $\sqrt{x + y} \neq \sqrt{x} + \sqrt{y}$,
>
> $$\log_b(x + y) \neq \log_b x + \log_b y$$
>
> Also, beware of making the following errors:
>
> $\log_b AB \neq (\log_b A)(\log_b B)$
>
> $\log_b \frac{A}{B} \neq \frac{\log_b A}{\log_b B}$

The Log of a Power The third property of logarithms matches the third law of exponents:

$$\log_b P^n = n \log_b P$$

We can establish this property by using the third law of exponents.

Let $x = \log_b P$. Then $P = b^x$.

Let $z = \log_b P^n$. Then $P^n = b^z$.

Now, $b^z = P^n = (b^x)^n = b^{xn}$. Thus, $z = nx$. Using back-substitution,

$$\begin{array}{ccc} z & = & nx \\ \downarrow & & \downarrow \\ \log_b P^n & = & n \log_b P \end{array}$$

We now summarize these properties of logarithms.

PROPERTIES OF LOGARITHMS

1. $\log_b PQ = \log_b P + \log_b Q$
2. $\log_b \frac{P}{Q} = \log_b P - \log_b Q$
3. $\log_b P^n = n \log_b P$
4. $\log_b b = 1$
5. $\log_b 1 = 0$

Using the Graphics Calculator

```
log AB
     1.653212514
log A+log B
     1.653212514
```

FIGURE 1

```
log (A/B)
    -.2552725051
log A-log B
    -.2552725051
```

FIGURE 2

```
log A^B
     6.290730039
Blog A
     6.290730039
```

FIGURE 3

We can use the graphics calculator to gain insight into these properties. Store the number `5` in memory location `A` and the number `9` in memory location `B`.

Evaluate `log (AB)`, and then evaluate `log A+log B`. See Figure 1. For these examples with $b = 10$,

$$\log_b PQ = \log_b P + \log_b Q$$

Evaluate `log (A/B)` and then evaluate `log A−log B`. See Figure 2. For these examples with $b = 10$,

$$\log_b \frac{P}{Q} = \log_b P - \log_b Q$$

Evaluate `log (A^B)`, and then evaluate `Blog A`. See Figure 3. For these examples with $b = 10$,

$$\log_b P^n = n \log_b P$$

For more familiarity with these properties, work through the following examples.

EXAMPLE 1 Given that the numeric value of the $\log_a x$ is 8 and the numeric value of $\log_a y$ is 10, find each numeric value.

a. $\log_a \sqrt{x}$ **b.** $\log_a(x^2y^5)$

Solution **a.** Using the properties of logarithms, we write

$$\begin{aligned} \log_a \sqrt{x} &= \frac{1}{2}\log_a x && \sqrt{x} = x^{1/2} \text{ and } \log_a x^n = n \log_a x \\ &= \frac{1}{2}(8) && \text{Value given for } \log_a x \\ &= 4 \end{aligned}$$

Thus, $\log_a \sqrt{x}$ has a numeric value of 4.

b. Using the properties of logarithms, we write:

$$\begin{aligned} \log_a(x^2y^5) &= \log_a x^2 + \log_a y^5 && \log_a xy = \log_a x + \log_a y \\ &= 2\log_a x + 5\log_a y && \log_a x^n = n \log_a x \\ &= 2(8) + 5(10) && \text{Values given in problem} \\ &= 66 \end{aligned}$$

Thus, the $\log_a(x^2y^5)$ has a numeric value of 66. ■

EXAMPLE 2 Given that the numeric value of the $\log_a x$ is -2 and the numeric value of $\log_a y$ is 6, find the numeric value of

$$\log_a \sqrt{\frac{x^2}{y}}$$

Solution Using the properties of logarithms and $\sqrt{x} = x^{1/2}$, we write:

$$\begin{aligned}
\log_a \sqrt{\frac{x^2}{y}} &= \frac{1}{2}\log_a \frac{x^2}{y} && \log_a x^n = n \log_a x \\
&= \frac{1}{2}(\log_a x^2 - \log_a y) && \log_a \frac{x}{y} = \log_a x - \log_a y \\
&= \frac{1}{2}(2\log_a x - \log_a y) && \log_a x^n = n \log_a x \\
&= \frac{1}{2}(2(-2) - 6) && \text{Values given in problem} \\
&= -5
\end{aligned}$$

Thus, the numeric value of $\log_a \sqrt{\frac{x^2}{y}}$ is -5. ■

GIVE IT A TRY

For $\log_a x = 5$ and $\log_a y = 7$, find the numeric value of each expression.

1. $\log_a\left(\sqrt{xy}\right)$ **2.** $\log_a\left[(x^2)\left(\sqrt[3]{y}\right)\right]$ **3.** $\log_a \sqrt[3]{\frac{x}{y^2}}$

A Change of Base We need to discuss one last property before considering solving equations using properties of logarithms. Using the fact that if $a = b$, then for any function f, $f(a) = f(b)$, we can state a property of equality.

EQUALITY PROPERTY OF LOGARITHMS

If $P > 0$, $Q > 0$, and $P = Q$, then $\log_b P = \log_b Q$.

Now, consider finding a numeric approximation for $\log_2 5$. That is, we want a number, say y, such that $y = \log_2 5$, or $5 = 2^y$. The number y should be between 2 and 3, because 2^2 is 4 and 2^3 is 8. The graphics calculator has a key for $\log_{10}$ (the LOG key), but has *no* key for $\log_2$. To approximate y such that $y = \log_2 5$, we use the following argument:

For $y = \log_2 5$, we have $5 = 2^y$. Now

$\log_{10} 5 = \log_{10} 2^y$ by the equality property for logs.

$\log_{10} 5 = y(\log_{10} 2)$ by the property $\log_b P^n = n \log_b P$.

So,

$$y = \frac{\log_{10} 5}{\log_{10} 2} \quad \text{Multiply by the nonzero number } \frac{1}{\log_{10} 2}.$$

Thus,

$$y = \log_2 5 = \frac{\log_{10} 5}{\log_{10} 2} \approx 2.32$$

That is, $5 \approx 2^{2.32}$. In general, we can use this procedure to rewrite (convert) a logarithm in one base to a logarithm in another base. We state this as follows.

CHANGE OF BASE PROPERTY

For $b > 0,\ b \neq 1,\ c > 0,\ c \neq 1,\quad \log_b P = \dfrac{\log_c P}{\log_c b}$

For $c = 10,\quad \log_b P = \dfrac{\log_{10} P}{\log_{10} b}$

Using the Graphics Calculator

We will now fulfill our promise of Section 10.1 to use the graphics calculator to graph the logarithmic function $f(x) = \log_2 x$. For `Y1` type `(log X)/(log 2)`. Use the Window 4.7 range settings to view the graph screen (see Figure 4). Press the `TRACE` key and use the arrow keys to convince yourself that this is the graph of $f(x) = \log_2 x$. Is $f(2) = 1$? Is $f(4) = 2$? For `Y2` type `2^X`. For `Y3`, use the `Y-VARS` menu to type the composition of $g(x) = 2^x$ on $f(x) = \log_2 x$ by typing `2^Y1`. View the graph screen. Notice that the composition does yield the identity function $I(x) = x$.

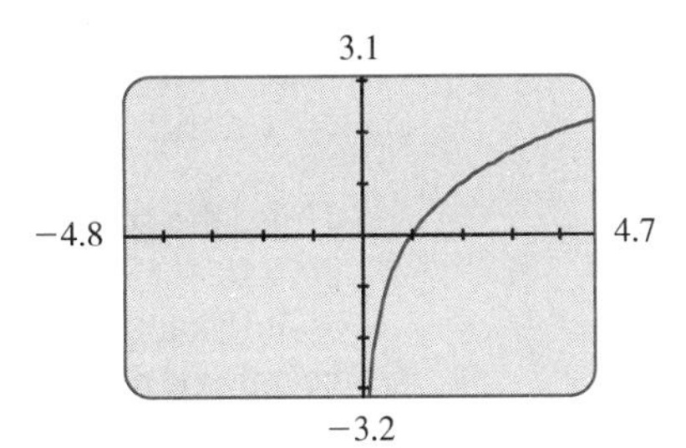

FIGURE 4

GIVE IT A TRY

Use the change of base property and the `LOG` key on the graphics calculator to report an approximate value (to three decimal places) for each number.

4. $\log_3 7$ **5.** $\log_2 8$ **6.** $\log_{17} 20$ 

Solving Logarithmic Equations Now that we are familiar with the properties of logarithms, we are ready to solve logarithmic equations such as

$$\log_3(x + 2) + \log_3 x = 1$$

Here, we will apply the properties of logarithms in order to get the basic form $\log_a b = c$. We can then use the meaning of "log" to obtain an equation that we know how to solve (such as a first-degree equation or a second-degree equation).

Consider the following examples.

EXAMPLE 3 Solve: $\log_3(x + 2) + \log_3 x = 1$

Solution We start by applying $\log_a x + \log_a y = \log_a(xy)$ to get

$$\log_3[(x + 2)(x)] = 1$$

The meaning of $\log_3[(x + 2)(x)] = 1$ is "write $(x + 2)(x)$ as a power of 3 and the exponent is 1."

Thus, we rewrite the log equation: $(x + 2)(x) = 3^1$

Doing the computation, we get: $(x + 2)(x) = 3$

Standard form: $x^2 + 2x - 3 = 0$

Factor the left-hand side: $(x + 3)(x - 1) = 0$

Solution: -3 and 1

Check

Replace x with -3: $\log_3(-3 + 2) + \log_3(-1) = 1$

Because the log of a negative number is undefined, -3 is extraneous and so is discarded.

Replace x with 1: $\log_3(1 + 2) + \log_3 1 = 1$

is a true arithmetic sentence, because $\log_3 3 = 1$ and $\log_3 1 = 0$.

Thus, the solution to $\log_3(x + 2) + \log_3 x = 1$ is 1. ■

In Example 3, the number -3 is discarded *not* because it is negative, but because it results in an attempt to find the *log* of negative number. The solution to $\log_5(3 - x) = 2$ is -22. Thus, the solution to a logarithmic equation *can* be a negative number.

EXAMPLE 4 Solve: $\log_2(x + 3) + \log_2(x - 1) = 3$

Solution We start by applying $\log_a x + \log_a y = \log_a(xy)$ to get

$$\log_2[(x + 3)(x - 1)] = 3$$

The meaning of $\log_2[(x + 2)(x - 1)] = 3$ is "write $(x + 3)(x - 1)$ as a power of 2 and the exponent is 3."

Thus, we rewrite the equation as $(x + 3)(x - 1) = 2^3$

Doing the computation, we get: $(x + 3)(x - 1) = 8$

Standard form: $x^2 + 2x - 11 = 0$

Identify a, b, c: $a = 1, b = 2, c = -11$

Compute discriminant: $b^3 - 4ac = 4 - (-44) = 48$

Solution: $-1 + 2\sqrt{3}$ and $-1 - 2\sqrt{3}$

Check Replacing x by $-1 - 2\sqrt{3}$ leads to our attempting to find the log of a negative number in $\log_2(x - 1)$. The log of a negative number is undefined, so $-1 - 2\sqrt{3}$ is discarded as an extraneous root.

Thus, the solution to $\log_2(x + 3) + \log_2(x - 1) = 3$ is $-1 + 2\sqrt{3}$. ■

Using the Graphics Calculator

To check our work in Example 4, store `-1+2√3` in memory location `X`. Now, we use the change of base property to evaluate the left-hand side of the equation. Type

```
log (X+3)/log 2+log (X-1)/log 2.
```

Press the `ENTER` key. Is the number 3 output? If not, an error is present! Running this same check on the number $-1 - 2\sqrt{3}$ produces the error screen! The log of a negative number does not exist.

GIVE IT A TRY

Solve each logarithmic equation in one variable.

7. $\log_2(x + 2) + \log_2(x - 2) = 2$ **8.** $\log_3(x - 3) + \log_3(x - 1) = 1$

EXAMPLE 5 Solve: $\log_3(x + 2) - \log_3 x = -2$

Solution We start by applying $\log_a x - \log_a y = \log_a \frac{x}{y}$ to get

$$\log_3\left(\frac{x + 2}{x}\right) = -2$$

Use definition of log: $$\frac{x + 2}{x} = 3^{-2}$$

or $$\frac{x + 2}{x} = \frac{1}{9}$$

Multiply by the LCD: $$\frac{9x(x + 2)}{x} = \frac{9x(1)}{9}$$

Reduce rational expressions: $$9(x + 2) = x(1)$$

$$9x + 18 = x$$

Add $-x$ to each side to get: $$8x + 18 = 0$$

Add -18 to each side to get: $$8x = -18$$

Solution: $$x = \frac{-18}{8} = \frac{-9}{4}$$

Check Replace x with $\frac{-9}{4}$: $\log_3\left(\frac{-9}{4} + 2\right) - \log_3\left(\frac{-9}{4}\right) = -2$

Because the log of a negative number is undefined, $\frac{-9}{4}$ is discarded.

Thus, the solution to $\log_3(x + 2) - \log_3 x = -2$ is *none*. ■

EXAMPLE 6 Solve: $\log_5(x+3)^3 = 6$

Solution We start by applying $\log_a x^n = n \log_a x$ to get

$$3 \log_5(x+3) = 6$$

Multiply by $\frac{1}{3}$ to get: $\log_5(x+3) = 2$

Use definition of log: $x + 3 = 5^2$

$$x + 3 = 25$$

$$x = 22$$

Check $\log_5(22+3)^3 = \log_5(25)^3$

$$= \log_5(5^2)^3$$

$$= \log_5 5^6$$

$$= 6$$

Thus, the solution to $\log_5(x+3)^3 = 6$ is 22. ■

Using the Graphics Calculator

To check our work in Example 6, store `22` in memory location `X`. Evaluate `log ((X+3)^3)/log 5`. Is the output `6`? If not, an error is present!

GIVE IT A TRY

Solve each logarithmic equation in one variable.

9. $\log_2(x+2) + \log_2 x = 3$

10. $\log_3(x-2) - \log_3 x = -1$

11. $\log_3(x-2) - 2\log_3 x = 0$

12. $\log_3(x-2) + \log_3 x = \log_3 3$

■ SUMMARY

■ ***Quick Index***

In this section, we have introduced a set of properties of logarithms that correspond to the laws of exponents:

$$\log_a AB = \log_a A + \log_a B$$

$$\log_a \frac{A}{B} = \log_a A - \log_a B$$

$$\log_a A^n = n \log_a A$$

These properties are used with the meaning of log to extend the types logarithmic equations in one variable that we can solve. Also, we learned how to change the base of a logarithm:

$$\log_a C = \frac{\log_b C}{\log_b a}$$

10.2 ■ ■ ■ EXERCISES

Given that $\log_a b = 2$ *and* $\log_a c = -3$, *use the properties of logarithms to find the numerical value of each logarithmic expression.*

1. $\log_a(bc)$
2. $3[\log_a (bc)]$
3. $\log_a b^3$
4. $\log_a\left(b\sqrt{c}\right)$
5. $\log_a \sqrt[5]{c}$
6. $\log_a(b^2c^3)$
7. $\log_a \dfrac{b^2}{c}$
8. $\log_a \dfrac{b}{\sqrt{c}}$
9. $\log_a \sqrt{\dfrac{b}{c}}$
10. $\log_a \dfrac{b}{c^3}$

Use the change of base property and the graphics calculator to report each evaluation as a decimal approximation (to the nearest ten-thousandth).

11. $\log_5 7$
12. $\log_3 8$
13. $\log_2 \sqrt{5}$
14. $\log_{12} 23$
15. $\log_6 9$
16. $\log_9 6$

Solve each logarithmic equation in one variable. Use the graphics calculator to check your work.

17. $\log_{10}(x - 3) + \log_{10} x = 1$
18. $\log_2(x - 3) + \log_2(x + 2) = 1$
19. $\log_4(x - 1) + \log_4(x + 2) = 2$
20. $\log_7(x - 3) + \log_7 x = 2$
21. $\log_2(x - 1) + \log_2(x + 1) = 3$
22. $\log_2(x + 3) + \log_2 x = \log_2 6$
    [*Hint*: If $\log A = \log B$, then $A = B$.]
23. $\log_7(x + 3) = \log_7(x - 1)$
24. $\log_2 x + 2 = \log_2(2x - 1)$
25. $\log_2(x - 3) - \log_2(x + 2) = -1$
26. $\log_5(x - 3) - \log_5 x^2 = 1$
27. $\log_5(x + 3) - \log_5 x = \log_5 6$
28. $\log_2(2x + 3) - \log_2 x = \log_2 5$
29. $\log_2(x - 3)^2 = 1$
30. $\log_2(x + 3) = \log_2(2x - 1)$
31. $\log_2(x + 3) - \log_2 x = \log_2 x$
32. $\log_2(3x + 3) = \log_2(2x - 1)$
33. $\log_5(x + 3) = \log_5(2x - 1)$
34. $\log_4(x - 2) = \log_4(3x + 2)$
35. $\log_3 x^2 = \log_3 x - 5$
36. $\log_7(2x^2) = \log_7 x - 5$
37. $\dfrac{\log_2(x - 1)}{\log_2 x} = \dfrac{1}{2}$
38. $\dfrac{\log_2(x + 1)}{\log_2 x} = \dfrac{1}{2}$

10.3 ■ ■ ■ SOLVING EXPONENTIAL EQUATIONS

We began this chapter by discussing exponential functions. The inverses of these functions lead us to discuss logarithmic functions and how to solve logarithmic equations in one variable. In this section, we will solve exponential equations in one variable, such as $2^x = 7$. Also, we will introduce a special exponential function, $f(x) = e^x$. The inverse of this function will lead to a discussion of natural logarithmic functions.

Exponential Equations An **exponential equation in one variable** is an equation in which a variable expression appears as an exponent. For example, $2^x = 16$ is an exponential equation. To solve such an equation, we must find the collection of numbers that replace the variable to produce true arithmetic sentences. The solution to $2^x = 16$ is 4, because $2^4 = 16$ is a true arithmetic sentence. In Section 10.2 we stated the following property.

EQUALITY PROPERTY OF EXPONENTS

> If $a^x = a^y$, with $a \neq 0$ and $a \neq 1$, then $x = y$.

We now use this property in solving exponential equations. Consider the following examples.

EXAMPLE 1 Solve: $5^x = 125$

Solution Because 125 is 5^3, we have: $5^x = 5^3$

Using the equality property, we have: $x = 3$

Thus, the solution to $5^x = 125$ is 3. ■

EXAMPLE 2 Solve: $2^{-x} = 8$

Solution Because $8 = 2^3$, we have: $2^{-x} = 2^3$

Use the equality property: $-x = 3$

$x = -3$

So, the solution to $2^{-x} = 8$ is -3. ■

To solve an exponential equation such as $2^x = 17$, we cannot use the preceding approach, because 17 is not a rational number power of 2. To solve this type of exponential equation, we make use of the equality property of logarithms, from Section 10.2.

EQUALITY PROPERTY OF LOGARITHMS

> If $A > 0$, $B > 0$, and $A = B$, then $\log_b A = \log_b B$.

That is, we can take the log of each side of an equation. Consider the following example.

EXAMPLE 3 Solve: $2^x = 17$

Solution We start by using the equality property of logarithms to take the log of each side:

$$\log_{10} 2^x = \log_{10} 17$$

$$x \log_{10} 2 = \log_{10} 17 \qquad \log_a P^n = n \log_a P$$

$$x = \frac{\log_{10} 17}{\log_{10} 2} \qquad \text{Multiply by } \frac{1}{\log_{10} 2}.$$

$$x \approx 4.087 \qquad \text{Use graphics calculator.}$$

Solution: $\dfrac{\log_{10} 17}{\log_{10} 2} \approx 4.087$

Check Using the calculator to check $x = 4.087$, we have $2^{4.087} \approx 16.99$, or 17. ■

WARNING

$$\frac{\log_{10} 17}{\log_{10} 2} \neq \log_{10} 8.5$$

That is,

$$\frac{\log_a A}{\log_a B} \neq \log_a \frac{A}{B}$$

Consider a second example of using the log operation to solve an exponential equation.

EXAMPLE 4 Solve: $100(1.01^n) = 300$

Solution We first multiply each side by $\frac{1}{100}$, to get:

$$1.01^n = 3$$

$$\log_{10} 1.01^n = \log_{10} 3 \qquad \text{Take the log of each side.}$$

$$n \log_{10} 1.01 = \log_{10} 3 \qquad \log_a x^n = n \log_a x$$

$$n = \frac{\log_{10} 3}{\log_{10} 1.01} \qquad \text{Multiply by } \frac{1}{\log_{10} 1.01}.$$

$$n \approx 110.41 \qquad \text{Use graphics calculator.}$$

Check Using the calculator to check 110.41, we have

$$100(1.01^{110.41}) \approx 300.00$$

■

EXAMPLE 5 Solve: $2^x = -3$

Solution This equation has no solution! The number 2 raised to any real number power x yields a positive number. (Recall the graph of $f(x) = 2^x$.)

We cannot use the equality property of logarithms because both members are not positive. Remember that the $\log_{10}(-3)$ is undefined. We conclude that the equation $2^x = -3$ has *no solution*. [Compare this reasoning with solving $\sqrt{x} = -2$.]

■

GIVE IT A TRY

Use either the equality property of exponents or the equality property of logarithms to solve each exponential equation.

1. $4^x = \frac{1}{2}$ **2.** $\left(\frac{1}{7}\right)^x = 49$ **3.** $17^x = 3$ **4.** $10(1.5)^x = 20$

Common and Natural Logarithms We have defined an exponential function as one of the form $f(x) = a^x$, where a is a real number with $a > 0$ and $a \neq 1$. We also defined a logarithmic function as the inverse of an exponential function. That is, $g(x) = \log_b x$ is the inverse of the function $f(x) = b^x$. Because the base of 10 appears so often, $\log_{10} x$ is referred to as the *common logarithm*. In fact, the notation, **log *x***, is assumed to be $\log_{10} x$.

Now we consider a rather special base, known as Euler's (pronounced Oiler's) constant, e. This number is an irrational number, like π, and is approximately 2.72. Using the calculator we can set an even better approximation of e. Press the `2nd` key followed by the `LN` key. The characters `e^` are displayed. Now, type a `1` and press the `ENTER` key. The number `2.718281828` is output. That is, e^1 is approximately 2.718281828.

Using the Graphics Calculator

(a)	(b)
RANGE	RANGE
Xmin=0	Xmin=0
Xmax=10	Xmax=100
Xscl=1	Xscl=10
Ymin=0	Ymin=2
Ymax=3	Ymax=3
Yscl=1	Yscl=1
Xres=1	Xres=1

FIGURE 1

To get an even better feel for Euler's constant, press the Y= key and erase any existing function rules. Euler's constant, e, is the value of the expression $\left(1 + \frac{1}{x}\right)^x$ as the x-value gets larger and larger (10, 100, 1,000, . . .). For Y1 enter the rule (1+1/X)^X. Use the range settings in Figure 1(a) to view the graph (see Figure 2). Press the TRACE key and follow the progression of this function as X increases. What are the outputs when the inputs are 6, 8, and 10? Press the RANGE key and edit the settings to those in Figure 1(b). Then press the TRACE key. When the input is 60, what is the output? To see that the curve is getting closer and a closer to e, press the Y= key and enter e^1 as the rule for Y2. View the graph as shown in Figure 3.

Now we define the natural exponential function.

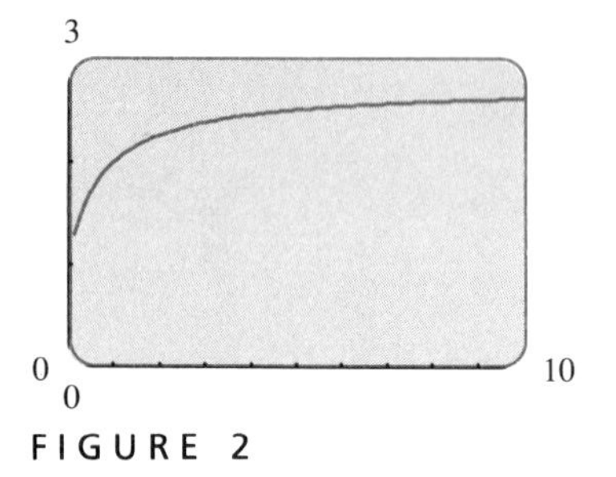

FIGURE 2

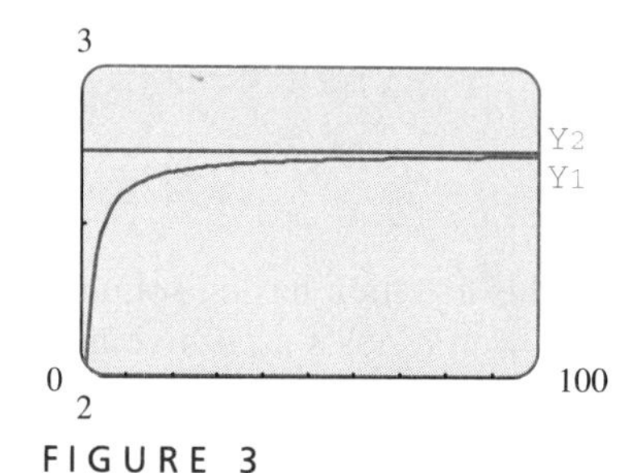

FIGURE 3

Now we define the natural exponential function.

DEFINITION OF THE NATURAL EXPONENTIAL FUNCTION

$f(x) = e^x$ where $e \approx 2.72$

Using the Graphics Calculator

To get a reference for the natural exponential function, type 2^X for Y1. View the graph using the Window 9.4 range settings. Zoom in on the origin. Next, for Y2 type e^X, and view the graph screen. Both exponential functions have y-intercept (0, 1). Any nonzero number to the 0 power is 1. Finally, for Y3 type 3^X, and view the graph screen. As Figure 4 shows, the graph of $f(x) = e^x$ lies between the graph of

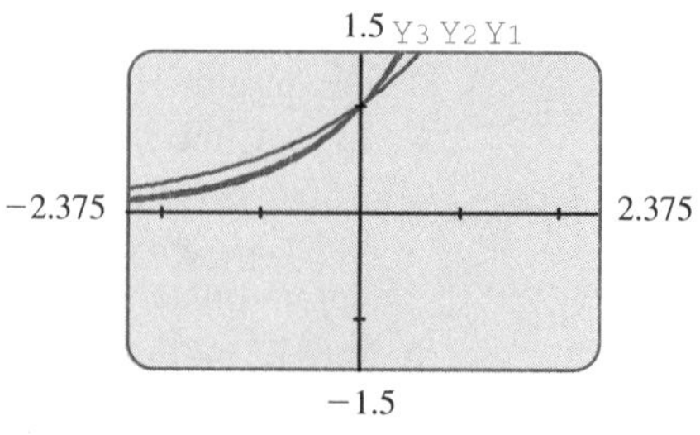

FIGURE 4

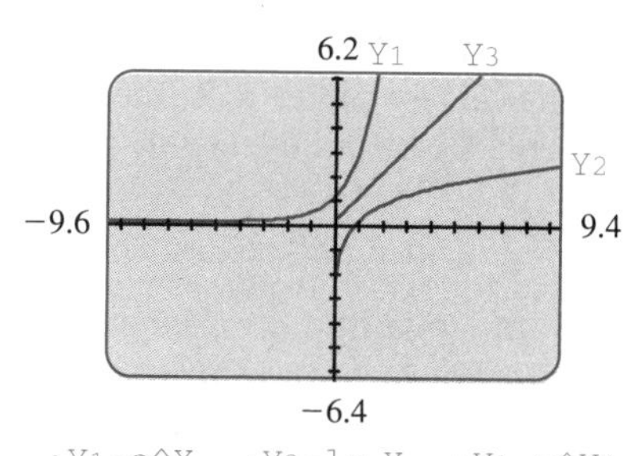

FIGURE 5

$g(x) = 2^x$ and the graph of $h(x) = 3^x$. This is to be expected, because $2 < e < 3$.

As with other exponential functions, the inverse of the function $f(x) = e^x$ is the logarithmic function $g(x) = \log_e x$. To see this, type e^X for Y1 and use the LN key to type ln X for Y2. (Erase the entry for Y3.) Use the Window 9.4 range settings to view the graph screen. To see that these are inverse functions, for Y3 type e^Y2. This is the composition of $f(x) = e^x$ on the function $g(x) = \log_e x$. View the graph screen. Figure 5 shows that the composition is the identity function $I(x) = x$.

The inverse function of the natural exponential function is known as the **natural logarithmic function**, and is defined as follows.

DEFINITION OF THE NATURAL LOGARITHMIC FUNCTION

$$g(x) = \log_e x = \ln x$$

This particular logarithmic function has many applications in business, the sciences, and in mathematics. We will consider some of these applications in the next section. The properties we have stated for other logarithmic functions still hold for this new function.

PROPERTIES OF THE NATURAL LOGARITHM

$\log_e e = 1$ because $e = e^1$. We write, $\ln e = 1$.

$\log_e 1 = 0$ because $1 = e^0$. We write, $\ln 1 = 0$.

Also, we have

$$\ln AB = \ln A + \ln B$$
$$\ln \frac{A}{B} = \ln A - \ln B$$
$$\ln A^n = n \ln A$$

The following examples show how to work with the natural logarithmic function.

EXAMPLE 6 Solve: $\ln(2x - 1) = 2$

Solution Using the meaning of ln, we know that $2x - 1$ is written as a power of e with an exponent of 2:

$$2x - 1 = e^2$$

Solve this equation:

$$\frac{1 \quad 1}{2x = 1 + e^2}$$

Multiply by $\frac{1}{2}$:

$$x = \frac{1 + e^2}{2} \approx 4.19$$

We can approximate e^2 on the graphics calculator. Press the `2nd` key followed by the `LN` key to display `e^`. Next, press the `2` key to get `e^2`, then press the `ENTER` key. The approximation `7.389056099` is displayed. ■

EXAMPLE 7 Solve: $e^{x-1} = 15$

Solution As with other exponential equations, we can take the logarithm of each side. Because the base of the exponential expression is e, we take the natural log of each side:

$$\begin{aligned} \ln e^{x-1} &= \ln 15 \\ (x-1)(\ln e) &= \ln 15 && \ln A^n = n \ln A \\ x - 1 &= \ln 15 && \ln e = 1 \\ \underline{\quad 1} &\quad \underline{1 \qquad\qquad} \\ x &= 1 + \ln 15 \approx 3.71 \end{aligned}$$

Check Using the calculator, $e^{3.71-1} = e^{2.71} \approx 15.03$. ■

GIVE IT A TRY

Solve each equation. Report the exact value and a two-decimal-place approximation of the solution.

5. $\ln(3x - 1) = 5$ **6.** $2e^n = 6$ **7.** $\ln x - \ln(x - 1) = -1$

■ SUMMARY

In this section, we have learned how to solve exponential equations. For exponential equations such as $2^x = 23$, we typically use logarithms in the equation-solving process. Using the fact that if $A > 0$, $B > 0$, and $A = B$, then $\log_b A = \log_b B$, we can take the log of each side of an equation. Carrying out this operation on exponential equations can result in a equation which we know how to solve (such as a first-degree equation).

Also, in this section we introduced a special base for exponential functions, $f(x) = e^x$. This base, known as Euler's constant, e, is approximately 2.72. The function is known as the natural exponential function. The inverse of this function is $g(x) = \ln x$. It is the logarithmic function whose base is e.

10.3 ■ ■ ■ EXERCISES

Use either the equality property of exponents or the equality property of logarithms to solve each exponential equation. Use the graphics calculator to check your work.

1. $6^x = 36$

2. $10^x = 1000$

3. $\left(\frac{1}{25}\right)^x = 25$

4. $\left(\frac{1}{3}\right)^x = \frac{1}{27}$

5. $5^{-x} = 5$

6. $4^{-x} = 2$

7. $36^x = 6$

8. $81^x = 3$

9. $\left(\frac{3}{5}\right)^x = \frac{5}{3}$

10. $\left(\frac{2}{3}\right)^{-x} = \frac{4}{9}$

11. $3^{x+2} = 9$

12. $5^{2x-1} = 125$

13. $3^x = 21$
14. $6^x = 40$
15. $17^x = 3$
16. $30^x = 3$
17. $10(1.5)^x = 20$
18. $5(2.1)^x = 20$
19. $6(5)^x = 30$
20. $8(2)^x = 64$

Solve each logarithmic equation. Report the exact value and a two-decimal-place approximation for each number in the solution. Use the graphics calculator to check your work.

21. $\ln x = 6$
22. $\ln 2x = 1$
23. $\ln(x + 1) = 1$
24. $\ln(2x - 1) = 2$
25. $\ln \frac{1}{x} = -1$
26. $\ln \frac{2}{x} = -2$
27. $\ln(x - 3) - \ln x = 1$
28. $\ln \sqrt{x + 1} = 2$

Solve each exponential equation. Report a two-decimal-place approximation for each number in the solution. Use the graphics calculator to check your work.

29. $e^x = 2$
30. $e^x = 8$
31. $e^{x-1} = 1$
32. $e^{2-x} = 2$
33. $3e^n = 5$
34. $2e^k = 10$
35. $200e^{0.05t} = 400$
36. $1000e^{0.9t} = 3000$
37. $500e^{2x} = 1000$
38. $300e^{0.5x} = 900$
39. $3e^{2t} = 18$
40. $300(1 + 0.1)^x = 500$

10.4 ■ ■ ■ APPLICATIONS

In this section, we consider real-world situations that lead to exponential and logarithmic equations. Our major concern will be situations where a quantity is growing at a rate that is a percentage of itself. The population of a city increasing at 10% each year is an example of a quantity increasing at a percentage of itself. For a present population of P, the population next year would be $P + 0.1P$. Another example is found in finance—compound interest.

Compound Interest Exponential equations find a wide range of applications in the sciences and in business. In a situation where a quantity is increasing (or decreasing) as a percentage of itself, an exponential equation often serves as the mathematical model.

Consider the idea of compound interest from finance. When you deposit money in a savings account in a bank, you are loaning the bank your money. For the use of the money, the banks pays rent. The rent (rather than being a fixed amount, such as $10) is usually computed based on a percentage of the money you have invested in the savings account. Suppose the bank agrees to pay you 6% interest per year. If you deposit $100 into the account, your money will grow as follows:

Time	*Deposit + Interest*	*or*	*Balance*
Beginning of year 1	$100		
End of year 1	$100 + 0.06($100)	or	$106.00
End of year 2	$106 + 0.06($106)	or	$112.36
End of year 3	$112.36 + 0.06($112.36)	or	$119.10
End of year 4	$119.10 + 0.06($119.10)	or	$126.25

Thus, the balance will continue to increase as long as you leave the money in the account.

Using the Graphics Calculator

```
100+0.06(100)
              106.00
Ans+0.06*Ans
              112.36
Ans+0.06*Ans
              119.10
```

FIGURE 1

To see this growth pattern, use the MODE key to set the number of decimal places to 2 on the second line, (Float 0123456789). Now, type 100+0.06(100), and then press the ENTER key. Type Ans+0.06*Ans (use the ANS key), and press the ENTER key. Use the replay feature and then press the ENTER key. See Figure 1. Continue to use the replay feature and watch the investment grow.

As you may suspect, there is a pattern to the process shown in Figure 1. Consider the following computations:

Time	*Computation*	*Pattern*
Beginning	100	$100(1 + 0.06)^0$
End of year 1	$100(1 + 0.06)$	$100(1 + 0.06)^1$
End of year 2	$100(1 + 0.06)^1 + 0.06[100(1 + 0.06)^1]$	$100(1 + 0.06)^2$
End of year 3	$100(1 + 0.06)^2 + 0.06[100(1 + 0.06)^2]$	$100(1 + 0.06)^3$
End of year 4	$100(1 + 0.06)^3 + 0.06[100(1 + 0.06)^3]$	$100(1 + 0.06)^4$

In general, for money invested in a savings account the formula is as follows.

FORMULA FOR YEARLY COMPOUND INTEREST

$$A = P(1 + r)^n$$

where A is the value of the investment
P is the principle (the amount invested)
r is the interest rate per year
n is the number of years

If interest (rent) is paid more often than yearly, such as each quarter (three months) or each month, then the formula becomes:

FORMULA FOR COMPOUND INTEREST

$$A = P\left(1 + \frac{r}{m}\right)^{nm}$$

where A is the value of the investment
P is the principle (the amount invested)
r is the interest rate per year
n is the number of years
m is the number of compounding periods per year

Suppose that \$100 is invested at 6%, compounded monthly for four years. The formula states that the value A of the investment is

$$A = 100\left(1 + \frac{0.06}{12}\right)^{4(12)}$$

Using the calculator to do the computation $100(1.005)^{48}$, we get 127.05. Thus, the value of the investment at the end of four years is \$127.05. (Compare with the \$126.25 when interest is paid only at the end of each year). As you can see, the interest rate r is the stated rate divided by 12. If the compounding period had been quarterly rather than monthly, the interest rate would have been divided by 4. Consider the following example.

EXAMPLE 1 Find the value of \$100 invested into a savings account that pays 8%, compounded quarterly, at the end of 20 years.

Solution The formula is $A = P\left(1 + \frac{r}{m}\right)^{nm}$. Before using the formula, we must identify the variables. Compounded quarterly means that the interest (rent) is paid on the account four times a year (every three months), so $m = 4$. For 20 years, we have $n = 20$. The interest rate is stated as 8%, compounded quarterly. This means the interest rate is $r = 0.08$. Now, we can use the formula:

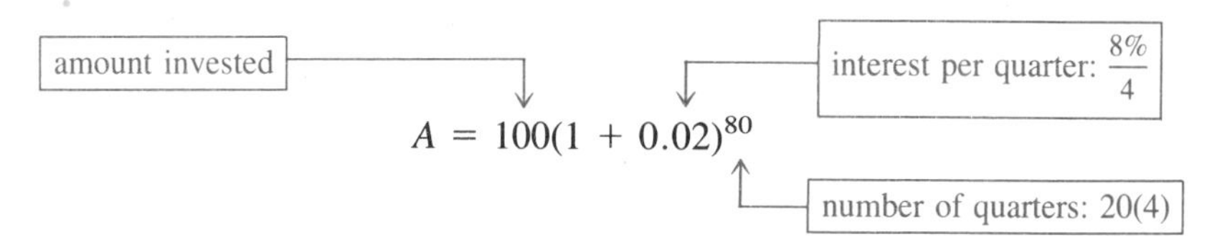

Doing the computation $100(1.02)^{80}$, we get 487.54. Thus, the \$100 investment is worth \$487.54 at the end of 20 years. ■

As an example of a compound interest problem leading to an exponential equation to be solved, consider the following.

EXAMPLE 2 How long will it take for \$1000 invested into a savings account that pays 6%, compounded monthly, to be worth \$3000?

Solution The formula is $A = P\left(1 + \frac{r}{m}\right)^{nm}$. We have $r = 6\%$, $m = 12$, and we want to find n, the number of interest payments. Also, we know that $P = 1000$ (the amount invested) and $A = 3000$ (the future value desired). Thus, we solve the equation

$$3000 = 1000\left(1 + \frac{0.06}{12}\right)^{12n}$$

$$3000 = 1000(1.005)^{12n}$$

Multiply each side by $\frac{1}{1000}$ to get: $$3 = 1.005^{12n}$$

Take $\log_{10}$ of each side: $$\log_{10} 3 = \log_{10}(1.005)^{12n}$$

Use $\log A = n \log A$ to get: $$\log_{10} 3 = 12n \log_{10} 1.005$$

$$n = \frac{\log_{10} 3}{12 \log_{10} 1.005}$$

Solution: $$\frac{\log_{10} 3}{12 \log_{10} 1.005} \approx 18.36$$

Thus, it will take approximately 18.36 years. ■

Continuous Compounding Suppose the money you have invested in the bank is compounded continuously (not monthly, not daily, not hourly, not minutely, not secondly, not microsecondly, not nanosecondly, but *continuously*). To compute this, we use the following progression.

$$\text{Monthly} \qquad A = P\left(1 + \frac{r}{12}\right)^{12t}$$

$$\text{Daily} \qquad A = P\left(1 + \frac{r}{365}\right)^{365t}$$

$$\text{Hourly} \qquad A = P\left(1 + \frac{r}{8760}\right)^{8760t}$$

$$\vdots$$

Using methods from calculus, it can be shown that as the compound period gets shorter and shorter, and the number of compounding periods gets larger and larger, the formula for compound interest becomes

$$A = Pe^{rt}$$

where e is Euler's constant, r is the interest rate, and t is time (in years). In the next example, the money in a savings account is compounded continuously.

EXAMPLE 3 If \$1000 is deposited into a savings account that pays 9% interest, compounded continuously, how much will the investment be worth at the end of 10 years?

Solution Using the formula for continuously compounded interest, $A = Pe^{rt}$, with $P = 1000$, $r = 0.09$, and $t = 10$, we have

$$A = 1000e^{(0.09)(10)}$$

Doing the computation (using the graphics calculator), we get $A = 2459.60$. Thus, at the end of 10 years, the \$1000 investment will be worth \$2,459.60. ■

The following table shows a comparison of various compounding periods for \$1,000 invested at 9% interest for 10 years.

Rate 9%, compounded	*Value of \$1,000 at end of 10 years:*	
Monthly	$1000\left(1 + \frac{0.09}{12}\right)^{120}$	≈ \$2,451.36
Daily (365 days)	$1000\left(1 + \frac{0.09}{365}\right)^{3650}$	≈ \$2,459.33
Continuously	$1000(e^{0.9})$	≈ \$2,459.60

GIVE IT A TRY

1. **a.** If $5,000 is invested in a savings account that pays 8%, compounded quarterly, what is the value of the money at the end of 10 years?
   b. How long will it take for the investment to be worth $25,000?
2. Rework Problem 1 for an interest rate of 8%, compounded continuously.
3. There is an old saying: If you could live forever, you would have to be rich. Suppose you invested 1 cent in an account paying 6%, compounded monthly. How much is the investment worth after 100 years? After 200 years? After 400 years? [Maybe a penny saved really is a *million* earned.]
4. Not only money can grow as a rate equal to a percentage of itself (in a compound fashion), but population is often viewed in this model. Suppose a small town has a population of 5,000 people. If the town is growing at 5% a year (compounded yearly), what is the expected population of the town in 10 years? In 20 years? In 40 years?

Radioactive Decay One of the reasons exponential functions and logarithmic functions are studied in algebra is that many scientific applications use these functions. One such application is radioactive decay. Radioactive substances (such as radium 226, carbon 14, and iodine 131) decay, or change into another substance, at various rates. For example, carbon 14 decays at a rate of 50% over 5,700 years. If you start with 10 grams of carbon 14, then after 5,700 years you will have 5 grams of carbon 14. The length of time for a substance to decay to one-half its original amount is known as the **half-life** of the substance. The formula for radioactive decay follows.

FORMULA FOR RADIOACTIVE DECAY

$$A = A_0(1 - r)^n$$

where A is the current amount
A_0 is the original amount
r is the decay rate (50% for half-life problems)
n is the number of periods

Suppose we start with 10 grams of carbon 14. How long would it take for the carbon 14 to decay to 7 grams? Using the formula, we get the equation

$$7 = 10(1 - 0.5)^n$$

or

$$7 = 10(0.5)^n$$

Now, we take the log of each side:

$$\log_{10} 7 = \log_{10} 10(0.5)^n$$

$$\log_{10} 7 = \log_{10} 10 + \log_{10}(0.5)^n \qquad \log AB = \log A + \log B$$

$$\log_{10} 7 = 1 + n \log_{10}(0.5) \qquad \log A^n = n \log A$$

$$\underline{\quad -1 \qquad -1 \quad}$$

$$\log_{10} 7 - 1 = \log_{10}(0.5)n$$

$$n = \frac{\log_{10} 7 - 1}{\log_{10} 0.5} \approx 0.5146$$

Thus, it will take 0.5146 of a decay period (for carbon 14, the decay period is 5,700 years). So, in 0.5146(5700), or 2933.22 years, the 10 grams of carbon 14 will decay to 7 grams.

Using knowledge of radioactive decay for carbon 14, archaeologists have developed a process known as *carbon dating* to determine the age of artifacts. The process is based on the fact that the ratio of carbon 14 to carbon 12 in living material is constant. Once the living material dies, the carbon 14 begins to decay, and the ratio of carbon 14 to carbon 12 begins to change. Consider the following example.

EXAMPLE 4 An archaeologist finds a fossil that contains 56 milligrams of carbon 14. She knows that the bone originally contained 100 milligrams of carbon 14. How old is the fossil?

Solution We use the formula for radioactive decay with our known values:

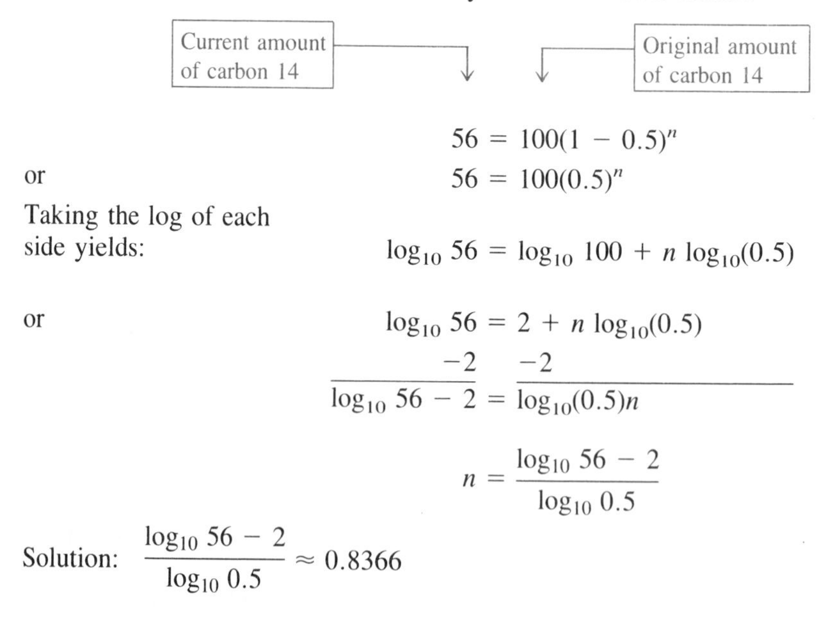

$$56 = 100(1 - 0.5)^n$$

or

$$56 = 100(0.5)^n$$

Taking the log of each side yields:

$$\log_{10} 56 = \log_{10} 100 + n \log_{10}(0.5)$$

or

$$\log_{10} 56 = 2 + n \log_{10}(0.5)$$

$$\underline{\quad -2 \qquad -2 \quad}$$

$$\log_{10} 56 - 2 = \log_{10}(0.5)n$$

$$n = \frac{\log_{10} 56 - 2}{\log_{10} 0.5}$$

Solution: $\dfrac{\log_{10} 56 - 2}{\log_{10} 0.5} \approx 0.8366$

Thus, the age of the fossil is 0.8366 of a half-life (5,700 years for carbon 14). So, the age of the fossil is 0.8366(5700), or 4768.62 years. ■

pH of a Solution Chemists classify a solution's acidity by use of the pH scale. That is, the pH of a solution is given by a logarithm.

DEFINITION OF THE pH OF A SOLUTION

$$\text{pH} = -\log[H^+]$$

where log is the log base 10 and H^+ is the concentration of hydrogen ions.

An *acid* is a solution which has a pH of less than 7 (the molar concentration of H^+ ions is greater than 1×10^{-7}). If the pH is greater than 7 (the molar concentration of H^+ ions is less than 1×10^{-7}), then the solution is classified as a *base*. A solution of pure water has a pH of 7, because pure water has a 1×10^{-7} molar concentration of H^+ ions [$\text{pH} = -\log_{10}(1 \times 10^{-7}) = -[0 + (-7)] = 7$]. Thus, pure water is neutral. For example, if a chemist finds that the concentration of H^+ ions in a solution is 2×10^{-3}, then the pH can be computed:

$$\begin{aligned} \text{pH} &= -\log_{10}(2 \times 10^{-3}) \\ &\approx -[0.301 + (-3)] \\ &\approx 2.699 \end{aligned}$$

The "p" in pH stands for $-\log$. That is, pH means $-\log$ of the molar concentration of H^+ ions. The molar concentrations of H^+ ions and corresponding pH levels of some common substances are listed below.

Substance	*Molar concentration of* H^+ *Ions*	pH
Ammonia cleaner	1×10^{-10}	10.0
Seawater	3.16×10^{-9}	8.5
Rainwater	1×10^{-6}	6.0
Coffee	1×10^{-5}	5.0
Wine	3.16×10^{-4}	3.5
Soft drinks	1×10^{-3}	3.0
Lemons	1×10^{-2}	2.0

As you can see, a pH number is easier to communicate than a molar concentration. Consider the following example of using the pH formula.

EXAMPLE 5 A chemist finds the H^+ ion concentration of a solution to be 5.2×10^{-8}. What is the pH of the solution?

Solution Using the pH formula, we get

$$\begin{aligned} \text{pH} &= -\log_{10}(5.2 \times 10^{-8}) \\ &\approx -[0.72 + (-8)] \\ &\approx -(-7.28) \\ &\approx 7.28 \end{aligned}$$

Thus, the pH of the solution is 7.28 (the solution is slightly basic). ■

Exponential Growth A model for the growth in population of a living organism (such as cats, bacteria, or humans) is given by the *Malthusian model* (named for English political economist Thomas Malthus, 1766–1834).

THE MALTHUSIAN GROWTH MODEL

$$N = Ie^{kt}$$

where N is the future population level
I is the initial population level
e is the Euler's constant (approximately 2.72)
k is the rate of growth
t is the time (in years)

As you can see, this population growth model is another example of continuous compounding. Consider the following example.

EXAMPLE 6 The population of Smallton is currently 2,000 people. How long will it take this population to double? Assume a growth rate of 2%.

Solution Using the Malthusian model, with $N = 4000$, $I = 2000$, and $k = 0.02$, we get

$$\begin{aligned} 4000 &= 2000e^{0.02t} \\ \ln 4000 &= \ln(2000e^{0.02t}) && \text{Take ln of each side.} \\ \ln 4000 &= \ln 2000 + 0.02t \ln e \\ \ln 4000 &= \ln 2000 + 0.02t(1) \\ -\ln 2000 & \quad -\ln 2000 \\ \hline \ln 2 &= 0.02t && \ln 4000 - \ln 2000 = \ln \frac{4000}{2000} \\ t &= \frac{\ln 2}{0.02} \end{aligned}$$

Solution: $\dfrac{\ln 2}{0.02} \approx 34.66$

Thus, it will take 34.66 years for the population of Smallton to double to 4,000 people. ■

A mathematical model of a situation is a representation consisting of equations or formulas that predict the behavior of the situation given a set of data. The value of the model is determined by the accuracy of its predictions. Often this depends not only on the model, but also on the quality of the data inserted in the model.

GIVE IT A TRY

5. Rework Example 4 using 65 milligrams as the amount of carbon 14 found in the fossil.
6. Rework Example 5 using 5.2×10^{-4} as the concentration of H^+ ions in the solution.
7. Rework Example 6 using 3,500 as the population of Smallton.

■ SUMMARY

In this section, we considered applications that use exponential and logarithmic equations as mathematical models. In general, applications of exponential equations involve a quantity that is growing at a percentage of itself, such as money, growing at a rate of 10% of itself per year.

The formulas developed, $A = P(1 + r)^n$ and $A = Pe^{rt}$, involve the exponential expressions $(1 + r)^n$ and e^{rt}. Similar applications discussed include carbon dating (radioactive decay and half-life of a substance), the pH of a solution, and population growth.

Carbon dating	*pH of a Solution*	*Exponential Growth*
where	where	where
$A = A_0(1 - r)^n$	$pH = -\log[H^+]$	$N = Ie^{kt}$
A is current amount of carbon 14	log is the log base 10	N is the number of organisms
A_0 is the initial amount of carbon 14	H^+ is concentration of hydrogen ions	I is the initial population,
r is the rate of decay (50%)		k is the rate of growth,
n is the number of decay periods		t is the length of time (in years)

10.4 ■ ■ ■ EXERCISES

Solve each exponential or logarithmic equation.

1. $e^{x-2} = 5$
2. $10e^{2x} = 300$
3. $\ln(2x - 3) = 1$
4. $\ln x - \ln(x + 1) = 2$
5. $4000 = 2000e^{0.01t}$
6. $7 = 5(0.03)^n$
7. a. For \$10,000 deposited into an account that pays 12% interest, compounded monthly, what is the value of the account after five years?
   b. How long will it take for the money in the account to double in value?
8. Lidwina has received \$30,000 as a gift. She can invest the money into an account at First City Bank, paying 10% interest, compounded quarterly, or she can invest the money into an account at Second City Bank, paying 9%, compounded monthly.
   a. How much interest will she earn on the money at each of the banks over a five-year period?
   b. Which bank is the best investment for the five years?
9. a. For a \$5,000 investment into an account that pays 10% interest, compounded continuously, what is the value of the investment after three years?
   b. How long will it take for the investment to be worth \$10,000?

The following formula is used to compute the monthly payments MP *on a loan* L *at a monthly rate of* r, *for* n *months:*

$$MP = \frac{rL(1 + r)^n}{(l + r)^n - 1}$$

Find the monthly payment of each loan. (Remember to adjust the stated rate and time period.)

10. \$2,000 loan at 18% interest for 1 year
11. \$50,000 loan at 9% for 10 years
12. \$12,000 loan at 6% for 3 years
13. \$200,000 loan at 9% for 30 years

14. An archaeologist finds a fossil containing 8 milligrams of carbon 14. She knows that the fossil originally contained 40 milligrams of carbon 14. How old is the fossil?

15. A chemist finds the H^+ ion concentration of a solution to be 7.2×10^{-3}. What is the pH of solution?

16. An archaeologist finds a piece of fabric containing 9 milligrams of carbon 14. He knows that the fabric is 530 years old. How many milligrams of carbon 14 did the fabric contain originally?

17. A chemist finds the pH of a solution to be 9.8. What is the H^+ ion concentration of the solution?

18. How long will it take the population of Maxville, currently 200,000 people, to triple? Assume a growth rate of 5% and use the Malthusian model.

19. The population of Big Valley in 1982 was 45,000 and in 1993 was 65,000. Assuming a Malthusien model, find the growth rate. If this rate of growth continues, when will the population reach 90,000?

20. Plutonium 241 has a half-life of 13 years. If you start with 0.1 gram of this substance, how long will it take for only 0.005 gram to remain?

21. Radium 226 has a half-life of 1620 years. If you start with 10 grams of this substance, how long will it take for only 1 gram to remain?

22. The Richter scale reading R is given by the formula $R = \log I$, where I is the size of the earthquake compared to the smallest measurable activity that a seismograph can measure. If an earthquake has a Richter scale reading of 5, what is the value for I?

23. The decibel scale for sound is given by the formula $d = 10 \log I$, where I is smallest sound audible. If a leaf blower has a decibel reading of 65, what is the value for I?

24. A local company finds that its share of the market can be increased by advertising on local TV stations. Their percentage P of the market is given by the formula $P = 1 - e^{-0.05n}$, where n is the number of minutes of advertisements run daily. How many minutes of ads must they run to capture 40% of the market?

CHAPTER 10 REVIEW EXERCISES

1. Graph the exponential function $f(x) = \left(\frac{5}{2}\right)^x$.

2. Use the graph from Problem 1 to construct the graph of $f^{-1}(x) = \log_{5/2} x$.

3. From the graphs in Problems 1 and 2, evaluate $\log_{5/2} 6.25$.

4. From the graphs in Problems 1 and 2, what is $\log_{5/2} \frac{2}{5}$?

5. If $a^x = c$, then write a logarithmic expression for x in terms of a and c.

6. If $\log_b x = c$, then write an exponential expression for x in terms of b and c.

Compute each value.

7. $\log_9 81$ **8.** $\log_8 64$

9. $\log_3 27$ **10.** $\log_2 64$

11. $\log_{1/2} \frac{1}{16}$ **12.** $\log_{1/5} \frac{1}{125}$

13. $\log_2 \sqrt{8}$ **14.** $\log_5 \sqrt[3]{25}$

15. $\log_2 \frac{1}{16}$ **16.** $\log_{1/2} 4$

Solve each equation using the LOG *key of the graphics calculator. Report your answers to the nearest hundredth.*

17. $2^x = 32$ **18.** $6^x = 36$

19. $\left(\frac{1}{27}\right)^x = 9$ **20.** $\left(\frac{1}{16}\right)^x = 4$

21. $5^x = 13$ **22.** $8^x = 17$

23. $7^x = 47$ **24.** $11^x = 121$

25. $19^x = 20$ **26.** $50^x = 25$

27. $3^{x+1} = 3^{2x-2}$ **28.** $3^{4x} = 9^{x-1}$

29. $7^{2-x} = 12^{x-3}$ **30.** $9^{2+x} = 4^{x-1}$

31. $\left(\frac{1}{9}\right)^{x-2} = 3$ **32.** $\left(\frac{1}{5}\right)^{1-x} = 5$

33. $(4^x)(16^{x-1}) = 16^2$ **34.** $(5^x)(25^{x-1}) = 5^2$

Solve each logarithmic equation. Use the graphics calculator to check your work.

35. $\log_6(2x - 5) = 2$ **36.** $\log_{13}(7x - 1) = 1$

37. $\log_{1/2}(4x^{-3}) = -2$ **38.** $\log_{1/3}(9x^{-1}) = -2$

39. $\log_{10}(0.1x + 0.3) = -2$ **40.** $\log_{10}(40 - 5x) = 2$

41. $\log_4(x + 4)(x - 4) = 2$

42. $\log_3\left(\dfrac{x + 1}{2x - 5}\right) = -1$

43. $\log_{1/3}\left(\dfrac{x + 1}{2x - 5}\right) = -2$

44. $\log_4 \sqrt{2x - 1} = 2$

Given that $\log_a m = 3$ and $\log_a n = -2$, use the properties of logarithms to find the numerical value of each logarithmic expression.

45. $\log_a(m^2 n)$

46. $\log_a\left(\dfrac{m^2}{n^4}\right)$

47. $\log_a n^5$

48. $\log_a\left(m^3\sqrt{n}\right)$

Solve each equation. Use the graphics calculator to check your work.

49. $\log_2(x + 3) + \log_2 x = 2$

50. $\log_3(x - 3) + \log_3 x = \log_3 6$

51. $\log_5(2x + 3) = \log_5(5x - 11)$

52. $\log_3 x - 2 = \log_3(2x - 1)$

53. $\log_5(x - 3) - \log_5(x + 2) = -1$

54. $\log_4(x - 3) - \log_4 x^2 = 1$

55. $e^{x-2} = 3$

56. $30e^{3x} = 90$

57. $\ln(3x - 1) = 2$

58. $\ln(3x - 2) - \ln(x + 1) = 1$

59. a. For \$20,000 deposited into an account that pays 8% interest, compounded quarterly, what is the value of the account after three years?
 b. How long will it take for the money in the account to double in value?

60. a. For a \$9,000 investment into an account that pays 7% interest, compounded continuously, what is the value of the investment after 20 years?
 b. How long will it take for the investment to be worth \$27,000?

61. An archaeologist finds a piece of fabric containing 20 milligrams of carbon 14. He knows that the fabric is 700 years old. How many milligrams of carbon 14 did the fabric contain originally?

62. A chemist finds the pH of a solution to be 8.7. What is the H^+ ion concentration of the solution?

63. How long will it take the population of Jay, currently 500 people, to triple? Assume a growth rate of 3%.

64. Plutonium 239 has a half-life of 25,000 years. If you start with 0.09 gram of this substance, how long will it take for only 0.02 gram to remain?

65. A local company finds that its share of the market can be increased by advertising on local TV stations. Their percentage P of the market is given by the formula $P = 1 - e^{-0.07n}$, where n is the number of minutes of advertisements run. How many minutes of ads must they run to capture 30% of the market?

66. A potato taken from an oven is 130°F. If it is placed in a room at 80°F, its temperature x minutes later is given by $T = 80 + (130 - 80)e^{-0.1x}$. How many minutes will it take for the potato to cool to 90°F?

CHAPTER 10 TEST

1. If $a^c = d$, then find $\log_a d$.

2. Graph $f(x) = 2^x$ and its inverse. Plot and label the x-intercept(s), the y-intercept(s), and show any asymptote(s).

3. Solve: $3^{x-2} = 9^{2x+1}$

Report the numeric value of each expression.

4. $\log_4 64$

5. $\log_3 \sqrt[3]{9}$

6. $\log_{23} 23$

Given that $\log_b h = -3$ and $\log_b k = 5$, find the numeric value of each expression.

7. $\log_b (kh)$

8. $\log_b \dfrac{h}{k^2}$

9. $\log_b \sqrt[3]{h^2k^4}$

Solve each equation. Report an approximate solution to the nearest hundredth.

10. $5^x = 23$

11. $\log_3(2x - 8) = 2$

12. $3^x = 2^{x-1}$

13. $\log_2(x - 2) + \log_2 x = 3$

14. $\log_5(2x - 3) - \log_5(x - 2) = -1$

15. Use the formula $A = Pe^{rt}$ to find the length of time it takes for an investment of \$5,000 invested at 10%, compounded continuously, to grow to a value of \$9,000.

11

LINEAR SYSTEMS AND MATRICES

In this chapter, we return to the linear systems of Chapter 4. This time we also consider first-degree equations in three variables. The solution to an equation in three variables is a collection of ordered triples. We will solve systems of these equations in this chapter. Such systems lead us to consider a rather special object in mathematics, the matrix. Learning how to work with matrices—and the closely related concept of determinants—provides us with additional approaches to solving systems of linear equations.

11.1 ■ ■ ■ LINEAR SYSTEMS IN THREE VARIABLES

We begin this section by reviewing the substitution method of solving a system of two linear equations in two variables. Next, we consider linear systems in three variables. With this type of system, we use the substitution method to reduce the problem to a linear system of two equations, which we can then solve with our usual approach.

The Substitution Method In Section 4.3 we used the substitution method to solve a system of equations. As a review, let's solve such a system:

$$\begin{cases} 3x + 5y = 15 \\ 2x - 3y = 8 \end{cases}$$

The solution to this system is an ordered pair of numbers. In fact, it is an ordered pair that is a solution to both the first equation and the second equation.

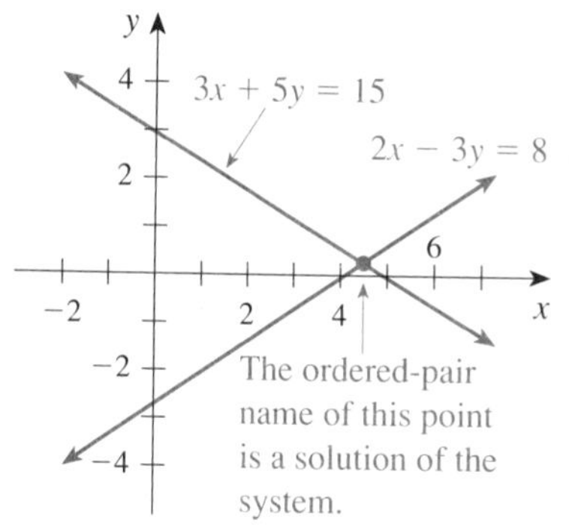

FIGURE 1

That is, it solves the equations simultaneously. A graph of the system is shown in Figure 1. We can solve this system by rewriting the first equation for y.

$$\begin{aligned} 3x + 5y &= 15 \\ -3x \qquad &\quad -3x \\ 5y &= -3x + 15 \\ y &= \frac{-3}{5}x + 3 \end{aligned}$$

Next, we substitute this expression for y into the second equation, and solve this equation in one variable.

Substitute for y

$$\begin{aligned} 2x - 3\left(\frac{-3}{5}x + 3\right) &= 8 \\ 2x + \frac{9}{5}x - 9 &= 8 \\ 10x + 9x - 45 &= 40 && \text{Multiply each side by 5.} \\ 19x - 45 &= 40 \\ 45 \quad &\quad 45 \\ 19x &= 85 \\ x &= \frac{85}{19} && \text{Multiply by } \tfrac{1}{19}. \end{aligned}$$

Thus, the x-coordinate of the ordered pair that is a solution to the system is $\frac{85}{19}$. Using this value $\frac{85}{19}$ for x in the first equation, $y = \frac{-3}{5}x + 3$, we get

$$y = \left(\frac{-3}{5}\right)\left(\frac{85}{19}\right) + 3$$

or

$$y = \frac{6}{19}$$

Thus, the solution to the system is the ordered pair $\left(\frac{85}{19}, \frac{6}{19}\right)$.

Using the Graphics Calculator

We can use the graphics calculator to check that $\left(\frac{85}{19}, \frac{6}{19}\right)$ does indeed solve the system

$$\begin{cases} 3x + 5y = 15 \\ 2x - 3y = \ \ 8 \end{cases}$$

Type `3(85/19)+5(6/19)` and then press the `ENTER` key. Is `15` output? If not, an error is present. Press the `2nd` key followed by the `ENTER` key, to replay the last command. Edit the command to read `2(85/19)-3(6/19)`. Press the `ENTER` key. Is `8` output? If not, an error is present.

GIVE IT A TRY

Use the substitution method to solve each linear system.

1. $\begin{cases} 3x - 2y = 8 \\ 2x + 5y = 10 \end{cases}$

2. $\begin{cases} 3x - (5 - 2y) = x + 2 \\ y = 3x - 1 \end{cases}$

A First-Degree Equation in Three Variables Learning to solve equations in algebra begins with solving a first-degree equation in one variable. Such an equation is $3x + 5 = 11$. The solution to this equation is the number 2, because replacing the variable by 2 produces the true arithmetic sentence $3(2) + 5 = 11$. The solution is shown on the number-line graph in Figure 2.

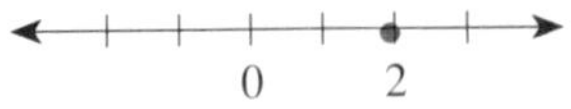

0 2

FIGURE 2
Solution to $3x + 5 = 11$

Next, we consider solving first-degree equations in two variables. Such an equation is $2x + y = 5$. The solution to this equation is an infinite collection of ordered pairs of numbers. One such ordered pair that solves the equation is (2, 1), because replacing x with 2 and y with 1 produces the true arithmetic sentence $2(2) + 1 = 5$. (Note that the ordered pair (1, 2) is *not* a solution to the equation—order is important.) If we graph the ordered pairs that solve $2x + y = 5$ on a set of axes, a straight line is produced. Figure 3 shows this graph.

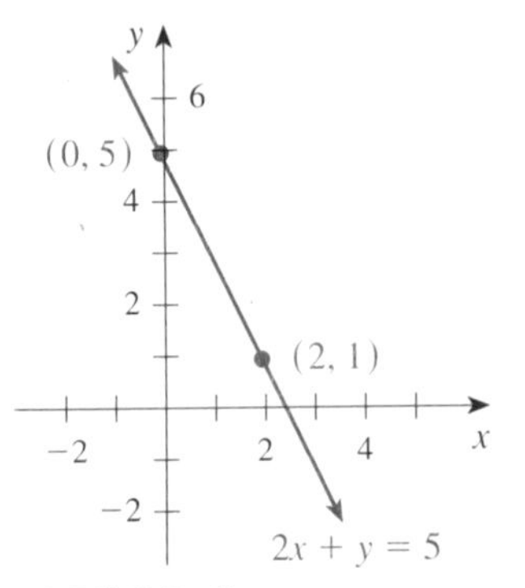

FIGURE 3
Solution to $2x + y = 5$

For a first-degree equation in three variables, say $2x + y - z = 6$, the solution is an infinite set of **ordered triples** of numbers. For example, the ordered triple (1, 4, 0) is a solution to this equation, because $2(1) + 4 - 0 = 6$ is a true arithmetic sentence. To graph ordered triples, we use three number lines, set at right angles to each other. On a two-dimensional page, we show such a three-axis coordinate system by drawing the y-axis horizontal, the z-axis vertical, and the x-axis is at an angle, as shown in Figure 4. The intersection of the three axes is the **origin**, (0, 0, 0). To plot a point, say (−2, 1, 3) on the xyz-coordinate system, we move along the x-axis 2 units in the negative direction, 1 unit along the positive y-axis, and then 3 units up the z-axis. Figure 5 illustrates how to plot this point.

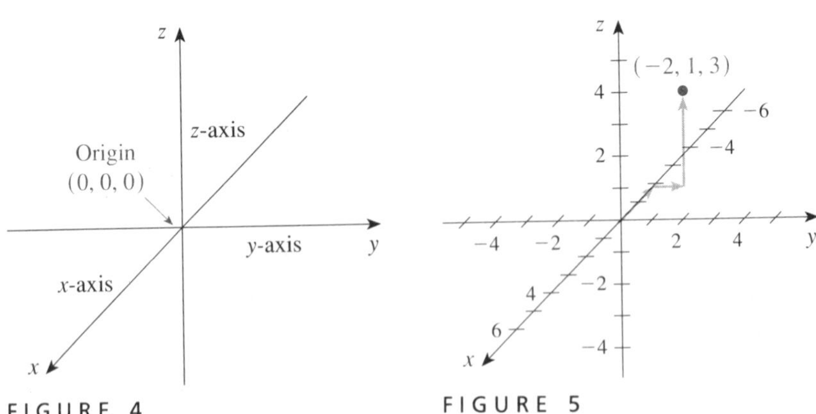

FIGURE 4
The xyz-coordinate system

FIGURE 5

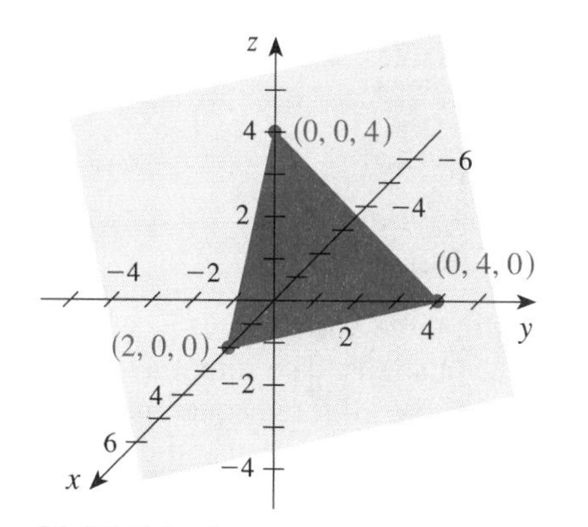

FIGURE 6

The graph of the solution to $2x + y + z = 4$

If we plot the ordered triples that solve a first-degree equation in three variables on an *xyz*-coordinate system, the graph is a **plane.** For example, the graph of the solution to the equation $2x + y + z = 4$ will intersect the x-axis at (2, 0, 0), the y-axis at (0, 4, 0) and the z-axis at (0, 0, 4). The plane determined is shown in Figure 6. Every ordered triple that solves this equation lies on this plane when plotted.

A System of Three Equations For a typical linear system of two equations in two variables, the solution is a single ordered pair. Graphically, two lines intersecting in a single point. However, the solution to such a system can be *none* if the two lines are parallel.

With a typical linear system of three equations in three variables, the solution is a single ordered triple. Graphically, the planes represented by the three equations intersect in a single point (see Figure 7). However, the solution can be either *none* or an infinite number of ordered triples. If the solution to the system is *none,* then (graphically) one of the situations shown in Figure 8 occurs. If the solution to the system is an infinite number of ordered triples, then the three planes intersect in a line, as shown in Figure 9.

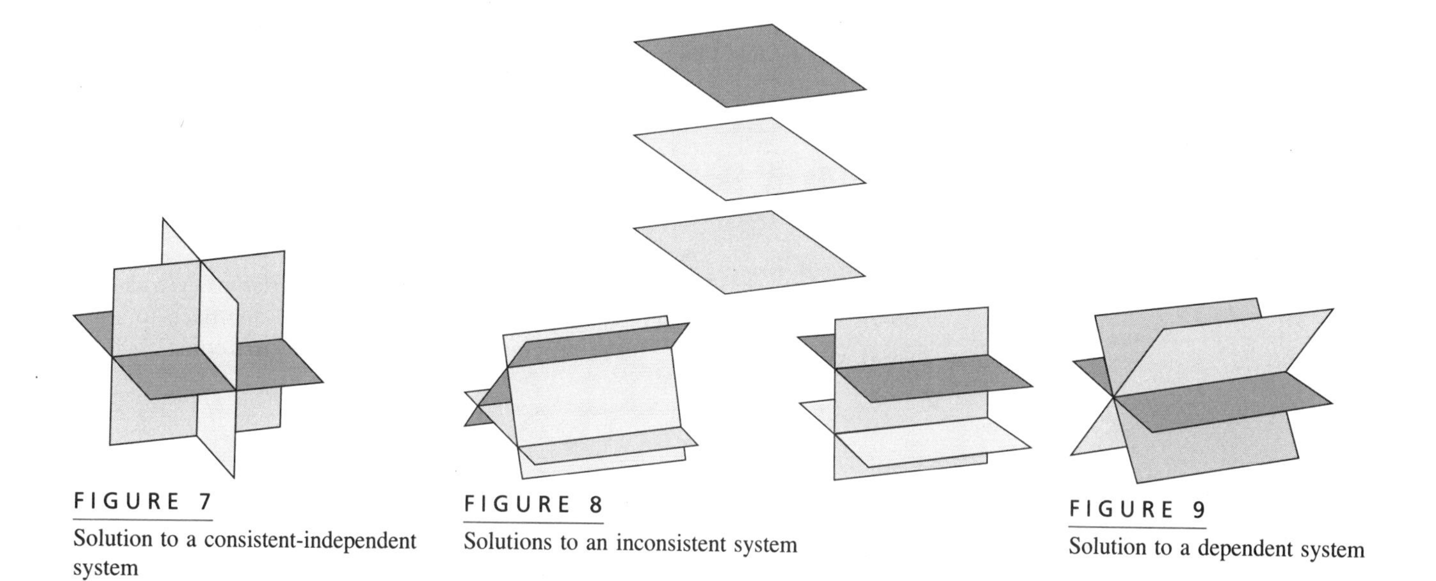

FIGURE 7

Solution to a consistent-independent system

FIGURE 8

Solutions to an inconsistent system

FIGURE 9

Solution to a dependent system

Solving a System of Three Equations For linear systems with two equations in two variables, using the substitution method essentially reduces the problem to solving a first-degree equation in one variable. (The addition method used in Chapter 4 also works with such linear systems.) We now consider a linear system of three equations in three variables, say x, y, and z. Using the same procedure that we used earlier (the substitution method), we can reduce the system to a system of two equations in two variables (which, of course, we know how to solve). Consider the linear system

$$\begin{cases} 2x + y - z = 6 \\ x - 2y + z = 8 \\ x + y - 2z = 4 \end{cases}$$

This is a system of three first-degree equations in three variables. It is a linear system because all the equations are first-degree equations. The solution to such a system is the collection of ordered triples that solve all three equations. The ordered triple (2, 3, 1) is not in the solution to the system, because it does not solve each of the equations in the system (it is only in the solution to the equation $2x + y - z = 6$). We now use the substitution approach to solve the system

$$\begin{cases} 2x + y - z = 6 \\ x - 2y + z = 8 \\ x + y - 2z = 4 \end{cases}$$

We start by rewriting one of the equations, say $2x + y - z = 6$, for one of the variables, say z. This yields

$$z = 2x + y - 6$$

Next, we substitute this expression for z into the remaining two equations to get

$$\begin{cases} x - 2y + (2x + y - 6) = 8 \\ x + y - 2(2x + y - 6) = 4 \end{cases}$$

or

$$\begin{cases} 3x - y = 14 \\ -3x - y = -8 \end{cases}$$

This last system is a linear system of two equations in two variables. We can use either the substitution method or the addition (elimination) approach to solve this system. We will use substitution here. Rewriting the first equation for y yields

$$y = 3x - 14$$

Substituting this expression for y in the second equation yields

$$\begin{aligned} -3x - (3x - 14) &= -8 \\ -6x + 14 &= -8 \\ -6x &= -22 \\ x &= \frac{-22}{-6} = \frac{11}{3} \end{aligned}$$

So, the x-coordinate of the ordered triple is $\frac{11}{3}$. Replacing the x-variable in the first equation, $y = 3x - 14$, with $\frac{11}{3}$ and doing the computation yields $y = -3$. The y-coordinate of the ordered triple is -3. Finally, replacing the x-variable and the y-variable with their values in the equation $z = 2x + y - 6$ yields $z = \frac{-5}{3}$. Thus, the ordered triple $\left(\frac{11}{3}, -3, \frac{-5}{3}\right)$ is the solution to the system

$$\begin{cases} 2x + y - z = 6 \\ x - 2y + z = 8 \\ x + y - 2z = 4 \end{cases}$$

Using the Graphics Calculator

To check that the ordered triple $\left(\frac{11}{3}, -3, \frac{-5}{3}\right)$ is the solution to the system, store `11/3` in memory location `X`, `-3` in memory location `Y`, and `-5/3` in memory location `Z`. Now, type `2X+Y-Z` and press the `ENTER` key. Is `6` output? Next, type `X-2Y+Z` and press the `ENTER` key. Is `8` output? Finally, type `X+Y-2Z` and press the `ENTER` key. Is `4` output? If any of the outputs are not obtained, then an error is present.

EXAMPLE 1 Use the substitution method to solve the system

$$\begin{cases} x - 2y + z = 4 \\ x + y - 2z = 21 \\ 5x - 2y - z = -3 \end{cases}$$

Solution We start by rewriting one of the equations, say $x - 2y + z = 4$, for z. This produces $z = -x + 2y + 4$.

Next, we substitute this expression for z in the other two equations to get:

$$\begin{cases} x + y - 2(-x + 2y + 4) = 21 \\ 5x - 2y - (-x + 2y + 4) = -3 \end{cases} \quad \text{or} \quad \begin{cases} 3x - 3y = 29 \\ 6x - 4y = 1 \end{cases}$$

Now, rewriting one of these equations, say $3x - 3y = 29$, for y, we get $y = x - \frac{29}{3}$. We then substitute this expression into the remaining equation and solve:

$$6x - 4\left(x - \frac{29}{3}\right) = 1$$

$$6x - 4x + \frac{116}{3} = 1$$

$$2x + \frac{116}{3} = 1$$

$$6x + 116 = 3$$

$$6x = -113$$

$$x = \frac{-113}{6}$$

Thus, the ordered triple has x-coordinate $\frac{-113}{6}$.

Now, using $x = \frac{-113}{6}$ and $y = x - \frac{29}{3}$, we get

$$y = \frac{-113}{6} - \frac{29}{3} = \frac{-171}{6} = \frac{-57}{2}$$

So, the y-coordinate is $\frac{-57}{2}$.

Finally, using $z = -x + 2y + 4$, we get

$$z = -\left(\frac{113}{6}\right) + 2\left(\frac{-57}{2}\right) + 4 = \frac{-205}{6}$$

The ordered triple that is the solution to the system is $\left(\frac{-113}{6}, \frac{-57}{2}, \frac{-205}{6}\right)$. ■

To check that the ordered triple for Example 1 actually is the solution to the system, store ⁻113/6 in memory location X, ⁻57/2 in memory location Y, and ⁻205/6 in memory location Z. Now, type X−2Y+Z and press the ENTER key. Is 4 output? Next, type X+Y−2Z and press the ENTER key. Is 21 output? Finally, type 5X−2Y−Z and press the ENTER key. Is ⁻3 output? If any outputs differ from these values, an error is present.

GIVE IT A TRY

3. Use the substitution approach to solve the given system. Use the graphics calculator to check your work.

$$\begin{cases} 2x - 3y + z = 12 \\ x + y - 2z = 21 \\ 5x - 2y - z = -3 \end{cases}$$

The Addition (Elimination) Approach We can also solve a system of three equations in three variables by use of the addition method used in Section 4.3. Here, the graph of the sum equation for two equations in the system will be a plane. The plane produced will still pass through the solution to the system.

Remember that we can always multiply each side of any equation by a nonzero number without altering the solution. Our strategy with the addition approach is to find two sum equations that will eliminate one preselected variable. This will produce a system of two equations with two variables, which we know how to solve.

EXAMPLE 2 Use the addition approach to solve the system

$$\begin{cases} x - 2y + z = 4 & \textbf{(1)} \\ x + y - 2z = 21 & \textbf{(2)} \\ 5x - 2y - z = -3 & \textbf{(3)} \end{cases}$$

Solution We start by selecting the variable to be eliminated. For this system, we will select z. Now, we multiply the first equation by 2 and the second equation by 1, and then form the sum equations.

$$\begin{cases} x - 2y + z = 4 \\ x + y - 2z = 21 \end{cases} \quad \begin{array}{l} \text{Multiply by 2} \\ \text{Multiply by 1} \\ \text{Sum equation} \end{array} \quad \begin{array}{rcl} 2x - 4y + 2z &=& 8 \\ x + y - 2z &=& 21 \\ \hline 3x - 3y &=& 29 \end{array}$$

Next, we find the sum equation for equations 1 and 3.

$$\begin{cases} x - 2y + z = 4 \\ 5x - 2y - z = -3 \end{cases} \quad \begin{array}{l} \text{Multiply by 1} \\ \text{Multiply by 1} \\ \text{Sum equation} \end{array} \quad \begin{array}{rcl} x - 2y + z &=& 4 \\ 5x - 2y - z &=& -3 \\ \hline 6x - 4y &=& 1 \end{array}$$

The two sum equations form a system in two variables.

$$\begin{cases} 3x - 3y = 29 & \textbf{(1)} \\ 6x - 4y = 1 & \textbf{(2)} \end{cases}$$

We now solve this system of two equations using the addition approach. Multiply equation (1) by -2 and equation (2) by 1:

$$\begin{array}{rcll} -6x + 6y &=& -58 & \textbf{(1)} \\ 6x - 4y &=& 1 & \textbf{(2)} \\ \hline 2y &=& -57 & \\ y &=& \dfrac{-57}{2} & \end{array}$$

Sum equation:

To eliminate y, we multiply equation (1) by 4: and multiply equation (2) by -3: Sum equation:

$$\begin{array}{rcl} 12x - 12y &=& 116 \\ -18x + 12y &=& -3 \\ \hline -6x &=& 113 \\ x &=& \dfrac{-113}{6} \end{array}$$

Although we could follow a similar approach to eliminate x and y using sum equations, it is easier to replace x and y with the values we have found, and then solve for z. We select the (original) first equation, $x - 2y + z = 4$. Replacing x with $\frac{-113}{6}$ and y with $\frac{-57}{2}$, we get

$$\frac{-113}{6} - 2\left(\frac{-57}{2}\right) + z = 4$$

Multiply by the LCD, 6:

$$\begin{aligned} -113 + 342 + 6z &= 24 \\ 229 + 6z &= 24 \\ 6z &= -205 \\ z &= \frac{-205}{6} \end{aligned}$$

So, the solution to the system is $\left(\frac{-113}{6}, \frac{-57}{2}, \frac{-205}{6}\right)$. ■

Compare the work required to produce the ordered triple in Example 2 with that required using the substitution method in Example 1.

GIVE IT A TRY

4. Use the addition approach to solve the system. Use the graphics calculator to check your work.

$$\begin{cases} 2x - y + z = 8 \\ x + y - z = 4 \\ 3x + y + z = 2 \end{cases}$$

Special Cases We have now solved a linear system whose solution is a single ordered triple. Next, consider the following linear system, whose solution is *none*, that is, the graphs of the planes determined by the first two equations are parallel. To solve the system

$$\begin{cases} 2x + 3y - z = 5 \\ 2x + 3y - z = 8 \\ x - y + z = 6 \end{cases}$$

we first rewrite the first equation for z to get $z = 2x + 3y - 5$. Substituting this expression into the second equation yields

$$2x + 3y - (2x + 3y - 5) = 8 \qquad \text{or} \qquad 5 = 8$$

The variables have "dropped out" (or are eliminated), and a false arithmetic sentence is produced. This indicates the system has solution *none*. The system is inconsistent.

Finally, we consider a system whose solution is an infinite collection of ordered triples:

$$\begin{cases} 2x + 6y + 4z = 8 \\ 4x + 12y + 10z = 20 \\ 3x + 9y + 6z = 12 \end{cases}$$

We begin by rewriting the first equation for x, to get $x = -3y - 2z + 4$. Now, substituting this expression in the remaining two equations, we get

$$\begin{cases} 4(-3y - 2z + 4) + 12y + 10z = 20 \\ 3(-3y - 2z + 4) + 9y + 6z = 12 \end{cases} \quad \text{or} \quad \begin{cases} 2z + 16 = 20 \\ 12 = 12 \end{cases} \quad \text{or} \quad \begin{matrix} z = 2 \\ \text{True} \end{matrix}$$

The equation $z = 2$ means that for all the ordered triples that solve the system, the z-coordinate must be 2. All the variables are eliminated in the second equation, and a true arithmetic sentence is produced. So, an infinite number of ordered triples solves the system, (all having z-coordinate 2). This system is dependent. Finally, returning to the equation we have rewritten for x, or $x = -3y - 2z + 4$, and replacing z with 2, we get $x = -3y$. So, the solution to the system is an infinite collection of ordered triples in the form $(-3a, a, 2)$, where a is any real number. Usually, we report a solution of this form using set notation. We write $\{(x, y, z) \mid z = 2 \text{ and } x = -3y\}$. From this set notation, we know $(3, -1, 2)$ and $(-15, 5, 2)$, as well as infinitely many other ordered triples of this form, are solutions to the system. But, $(-8, 4, 2)$ is not a solution to the system. Why?

GIVE IT A TRY

Solve each system. If the system is dependent (the solution is infinite), use set notation to report the ordered triples that are solutions to the system.

5. $\begin{cases} 3x - y + 2z = 8 \\ 5x + y - z = -5 \\ 6x - 2y + 4z = 3 \end{cases}$ **6.** $\begin{cases} x + 2y + z = 8 \\ x + 2z = 10 \\ y - 3z = 5 \end{cases}$ **7.** $\begin{cases} x + y = 12 \\ x + z = -9 \\ 2y + z = 5 \end{cases}$

Applications As an example of a situation that produces a system of three equations in three variables, we return to parabolas. In Chapter 4, we found the equation of a line, given *two points* on the line. A similar problem is to find the equation of a parabola, given *three points* on the parabola.

EXAMPLE 3 Find the equation of a parabola of the form $y = ax^2 + bx + c$ that passes through the points $(-2, 1)$, $(3, 0)$, and $(1, -2)$. The points are plotted in Figure 10.

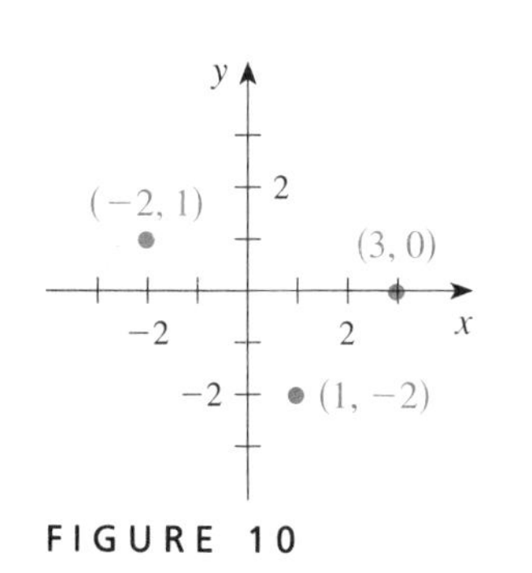

FIGURE 10

Solution Because the form is $y = ax^2 + bx + c$, we use the ordered pairs given to produce the following system.

$$\begin{cases} 1 = a(-2)^2 + b(-2) + c \\ 0 = a(3)^2 + b(3) + c \\ -2 = a(1)^2 + b(1) + c \end{cases} \quad \text{or} \quad \begin{cases} 4a - 2b + c = 1 \\ 9a + 3b + c = 0 \\ a + b + c = -2 \end{cases}$$

Using the substitution approach, we rewrite the first equation for c:

$$c = -4a + 2b + 1$$

Using this expression in the other two equations, we get

$$\begin{cases} 9a + 3b + (-4a + 2b + 1) = 0 \\ a + b + (-4a + 2b + 1) = -2 \end{cases} \quad \text{or} \quad \begin{cases} 5a + 5b = -1 \\ -3a + 3b = -3 \end{cases}$$

Repeating the same process with these equations, we rewrite the first equation for a:

$$a = -b - \frac{1}{5}$$

Using this expression in the other equation $-3a + 3b = -3$, we get

$$-3\left(-b - \frac{1}{5}\right) + 3b = -3$$

$$3b + \frac{3}{5} + 3b = -3$$

$$15b + 3 + 15b = -15$$

$$30b + 3 = -15$$

$$30b = -18$$

$$b = \frac{-18}{30} = \frac{-3}{5}$$

So, the value for b is $\frac{-3}{5}$.

Using this value for b and $a = -b - \frac{1}{5}$, we get $a = -\left(\frac{-3}{5}\right) - \frac{1}{5} = \frac{2}{5}$. Using $c = -4a + 2b + 1$, we get $c = -4\left(\frac{2}{5}\right) + 2\left(\frac{-3}{5}\right) + 1 = \frac{-9}{5}$.

Thus, the equation of the parabola is $y = \frac{2}{5}x^2 + \frac{-3}{5}x + \frac{-9}{5}$. ■

Using the Graphics Calculator

We can use the graphics calculator to check that the equation found in Example 3 actually does pass through the points $(-2, 1)$, $(3, 0)$, and $(1, -2)$. For `Y1`, type `(2/5)X²-(3/5)X-(9/5)`. Use the Window 4.7 range settings to view the graph screen. Use the `TRACE` key to verify that the given three points do lie on the parabola.

GIVE IT A TRY

8. Find the equation of the parabola in the form $y = ax^2 + bx + c$ that passes through the points $(-2, 1)$, $(1, -1)$, and $(2, 5)$.

■ SUMMARY

■ ***Quick Index***

In this section, we have learned that the solution to an equation in three variables, say x, y, and z, is a collection of ordered triples. That is, one member of the solution to the equation $x - y + z = 10$ is $(5, 1, 6)$. We also learned that the graph of the solution to a first-degree equation in three variables is a plane. The solution to a system of three first-degree equations in three variables is typically a single ordered triple. To find this ordered triple, we may use either the substitution method or the addition (elimination) approach. These approaches parallel the process for solving a system of first-degree equations in two variables.

11.1 ■ ■ ■ EXERCISES

Solve each linear system. Use the graphics calculator to check your work.

1. $\begin{cases} 3x - 2y = 90 \\ 2x + 5y = 82 \end{cases}$

2. $\begin{cases} 5x - 2y = 4 \\ 2x + 3y = 12 \end{cases}$

3. $\begin{cases} 3x - 2y + z = 9 \\ 2x + y - z = 8 \\ x + 2y + 2z = 2 \end{cases}$

4. $\begin{cases} x - y + 2z = 8 \\ 2x + 3y - z = 10 \\ 3x - 2y + z = -4 \end{cases}$

5. $\begin{cases} 3x - 2y + z = 9 \\ 2x + y - z = 8 \\ x - 2y + 2z = 10 \end{cases}$

6. $\begin{cases} z = 2x - y + 8 \\ z = 10 - 2x + y \\ z = 4y + 8 \end{cases}$

7. $\begin{cases} 2x + 3y + z = 10 \\ x + y - 2z = -3 \\ 3x - y + 2z = 11 \end{cases}$

8. $\begin{cases} 2x - y - z = -10 \\ x - 2y + z = -2 \\ 3x + y + 2z = -2 \end{cases}$

9. $\begin{cases} 2x + 2y + z = 1 \\ x - y + 6z = 21 \\ 3x + 2y - z = -4 \end{cases}$

10. $\begin{cases} 3x + y - 2z = 5 \\ z = 2y - x \\ y = 2z - 3x + 5 \end{cases}$

11. $\begin{cases} y = 3x - 2 \\ y + z = 7 \\ x = z + y - 1 \end{cases}$

12. $\begin{cases} 5x - y + 2z = 4 \\ 2x + 3y - z = 12 \\ 3z = 6 \end{cases}$

13. $\begin{cases} z = 3x - y + 10 \\ x = z + 3 \\ y = 2x + z \end{cases}$

14. $\begin{cases} x = 9 - 2x \\ y = 7 - x - z \\ z = 1 \end{cases}$

15. $\begin{cases} 2x - 2y + 3z = -3 \\ x + y + 2z = 1 \\ -x + 2y + z = 2 \end{cases}$

16. $\begin{cases} z = x - y + 1 \\ z = 8 - x + y \\ z = 2y - 3 \end{cases}$

17. $\begin{cases} x + z = 3 \\ y + 2z = 4 \\ x - y = -4 \end{cases}$

18. $\begin{cases} y - 3z = 3 \\ 4x + 5y = -2 \\ x - 3z = 3 \end{cases}$

19. $\begin{cases} 2x + y + z = 150 \\ x + y + 5z = 100 \\ 2x + 3y + z = 320 \end{cases}$

20. $\begin{cases} x + 2y = 2z - 9 \\ 2x + y = z - 3 \\ 3x - 2y + z = -6 \end{cases}$

21. Find the equation of the parabola in the form

$$y = ax^2 + bx + c$$

that passes through the points $(-3, 1)$, $(-1, -2)$, and $(2, 3)$.

22. Find the equation of the parabola that has a vertical line of symmetry and passes through the points $(1, -2)$, $(-1, 4)$, and $(3, 4)$.

23. Find the equation of the parabola in the form

$$y = ax^2 + bx + c$$

that passes through the points $(-2, -4)$, $(-3, -1)$, and $(-1, 0)$.

24. Find the equation of the parabola in the form

$$y = ax^2 + bx + c$$

that passes through the points $(-3, -1)$, $(1, 4)$, and $(-1, 6)$.

11.2 ■ ■ ■ MATRICES

In the last section, we solved systems of linear equations in three variables. The solution to such a system is the ordered triples of numbers that solve every equation in the system. So far, we have discussed two basic approaches to solving systems—the addition approach and the substitution method. In this section, we introduce a mathematical object known as a *matrix* (plural, matrices). We will use matrices in Section 11.3 to present another approach to solving systems of linear equations.

Introducing Matrices We start by defining a matrix. Informally, a *matrix* is simply a rectangular array of numbers. In fact, in working with computers and spreadsheets, matrices are often called *arrays*. We can think of a real number, such as -5, as being a 1×1 (read "one by one") matrix, and we can thus write $[-5]$. We can also think of an ordered pair of real numbers, such as $(3, 4)$, as a 1×2 matrix, and we write $[3 \quad 4]$. To state the *dimension* of a matrix, we always state the number of *rows,* followed by the number of *columns*. The ordered triple $(-2, 4, -1)$ can be thought of as the 1×3 matrix

$$[-2 \quad 4 \quad -1]$$

We now state a formal definition of a matrix.

DEFINITION OF A MATRIX

For positive integers m and n, an **m × n matrix** is a rectangular array of numbers with m rows and n columns:

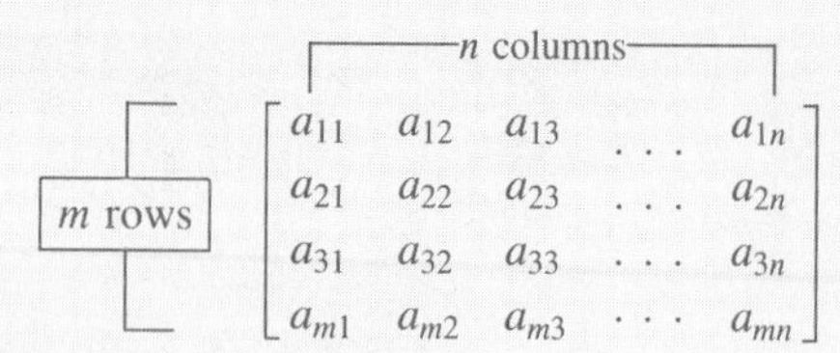

Each entry a_{ij} of the matrix is a number.

A matrix that has m rows and n columns (such as the one above) is known as an ***m* × *n* matrix**. We say that the matrix has *dimensions,* or **size** $m \times n$. If $m = n$, then the matrix is a **square matrix**. In the double-subscripted entry a_{ij}, the first reference i is to the row position of the entry. The second reference j is to the column position of the entry. Thus, a_{23} means the entry in the matrix that is located in row 2 and in column 3.

EXAMPLE 1 Report the size of the matrix $\begin{bmatrix} 2 & -3 & 4 \\ 5 & 0 & -1 \end{bmatrix}$. What is a_{21}?

Solution Since the matrix has 2 rows and 3 columns, its size is 2×3. The notation a_{21} means the entry at row 2, column 1. So, $a_{21} = 5$. ■

EXAMPLE 2 Construct a 3×2 matrix with -4 as entry a_{32}.

Solution We want a matrix with 3 rows and 2 columns:

2 columns

3 rows $\begin{bmatrix} \text{—} & \text{—} \\ \text{—} & \text{—} \\ \text{—} & \text{—} \end{bmatrix}$

Now, we can fill in the entries with any values we like, except the entry at row 3, column 2, which must be -4, that is, $a_{32} = -4$. So we have

$$\begin{bmatrix} 8 & -1 \\ -2 & 0 \\ 0 & -4 \end{bmatrix}$$

a_{32} ↑ Entry at row 3, column 2 is -4 ■

GIVE IT A TRY

Use this matrix to answer Problems 1–3:

$$\begin{bmatrix} -2 & 4 & 5 & -1 \\ 0 & 2 & 3 & -5 \\ -3 & 7 & 0 & -8 \end{bmatrix}$$

1. Report the size of the matrix.

2. Report the entry at row 1, column 3.

3. What is a_{24}?

Equality When discussing several different matrices, we refer to them by letters, A, B, C, and so on. We now define equal matrices.

DEFINITION OF EQUALITY BETWEEN TWO MATRICES

Matrix A is equal to matrix B provided that both matrices have the same size and that the matrix A entry a_{ij}, at row i, column j is equal to matrix B entry b_{ij}, at row i, and column j. For example, matrix A and matrix B are equal:

$$A = \begin{bmatrix} 3 & 8 & 7 \\ -2 & 1 & 0 \end{bmatrix} \qquad B = \begin{bmatrix} 3 & 8 & 7 \\ -2 & 1 & 0 \end{bmatrix}$$

We now consider an example that uses this definition of equality.

EXAMPLE 3 For $A = \begin{bmatrix} -2 & 3 + x \\ 4 & z - 2 \end{bmatrix}$ and $B = \begin{bmatrix} y & 2x - 1 \\ w & 6 \end{bmatrix}$

if $A = B$, report the values for x, y, z, and w.

Solution For these two matrices to be equal, their corresponding entries must be equal. Thus, $-2 = y$, $3 + x = 2x - 1$, $4 = w$, and $z - 2 = 6$. From these equations, we find that x is 4, y is -2, z is 8, and w is 4. ■

Scalar Multiplication Next, we consider an operation on matrices known as *scalar multiplication*. A typical example of a *scalar* is a real number. At this level of algebra, scalar multiplication will mean multiplication of a matrix by a real number. We now define scalar multiplication.

DEFINITION OF SCALAR MULTIPLICATION

For matrix A and scalar (real number) c, the scalar product of c times A, denoted cA, is the matrix whose entries are c times the entries of matrix A. For example,

if $\quad c = -2 \quad$ and $\quad A = \begin{bmatrix} -2 & 3 & 7 \\ 5 & 1 & 4 \end{bmatrix}$

then $\quad cA = \begin{bmatrix} 4 & -6 & -14 \\ -10 & -2 & -8 \end{bmatrix}$

EXAMPLE 4 For $A = \begin{bmatrix} -3 & 4 & 5 \\ 1 & 2 & -1 \\ 0 & 5 & 4 \end{bmatrix}$ and $c = 2$, find cA.

Solution

$$cA = \begin{bmatrix} 2(-3) & 2(4) & 2(5) \\ 2(1) & 2(2) & 2(-1) \\ 2(0) & 2(5) & 2(4) \end{bmatrix} = \begin{bmatrix} -6 & 8 & 10 \\ 2 & 4 & -2 \\ 0 & 10 & 8 \end{bmatrix}$$

■

Using the Graphics Calculator

```
MATRIX EDIT
1:RowSwap(
2:Row+(
3:*Row(
4:*Row+(
5:det
6:T
```

FIGURE 1

```
[A] 3×3
1,3↑5
2,1=1
2,2=2
2,3=-1
3,1=0
3,2=5
3,3=4
```

FIGURE 2

In addition to drawing graphs of functions, we can use a graphics calculator to work with matrices. To store an entry in a matrix, we use the MATRX key. We now rework Example 4 to demonstrate the use of the calculator. Press the MATRX key to display the MATRIX menu as shown in Figure 1. Press the right arrow key to display the EDIT menu. This displays the matrices [A], [B], and [C]. Select option 1 for matrix [A]. First, set the dimensions (the size) of the matrix. Press the 3 key, and then the ENTER key to set the number of rows to 3. Again press the 3 key, and then the ENTER key to set the number of columns to 3. Type -3 and press the ENTER key to enter -3 for a_{11}. Type in the remaining entries (4 and 5) for row 1 of matrix A (from Example 4). Next, type in the entries for row 2 (the values 1, 2, and -1), and then for row 3 (the values 0, 5, and 4). The screen should now match Figure 2*. Quit to the home screen.

To display matrix A, press the 2nd key followed by the 1 key (for the [A] key)**. The characters [A] are written to the screen. Press the ENTER key and matrix A is displayed, as shown in Figure 3. To have the graphics calculator carry out the scalar multiplication, type 2[A] and then press ENTER. The product of the scalar 2 times matrix A is displayed in Figure 4. Compare this with our answer for Example 4. If the matrices do not match, then an error is present.

```
[A]
[-3 4 5 ]
[ 1 2 -1]
[ 0 5 4 ]
```

FIGURE 3

```
[-3 4 5 ]
[ 1 2 -1]
[ 0 5 4 ]
2[A]
[-6 8  10]
[ 2 4  -2]
[ 0 10 8 ]
```

FIGURE 4

Addition of Matrices We now want to discuss the binary operation of addition for matrices. First, addition is defined only for matrices of the same size (same number of rows *and* the same number of columns).

DEFINITION OF MATRIX ADDITION

For matrices A and B of size $m \times n$, their sum $A + B$ is the matrix whose entries are the sum of the corresponding entries of A and B.

For example, for $A = \begin{bmatrix} -2 & 3 & 5 \\ 0 & 6 & 7 \end{bmatrix}$ and $B = \begin{bmatrix} 1 & 2 & 4 \\ 3 & 1 & 5 \end{bmatrix}$,

$$A + B = \begin{bmatrix} -2+1 & 3+2 & 5+4 \\ 0+3 & 6+1 & 7+5 \end{bmatrix} = \begin{bmatrix} -1 & 5 & 9 \\ 3 & 7 & 12 \end{bmatrix}$$

TI-82 Note

*The screen differs slightly.

**To display matrix A, press the MATRX key, followed by the 1 key.

EXAMPLE 5 For $A = \begin{bmatrix} -3 & 6 \\ 2 & 1 \\ 5 & 7 \end{bmatrix}$ and $B = \begin{bmatrix} -4 & 9 \\ 3 & -1 \\ -2 & -8 \end{bmatrix}$, find $A + B$.

Solution Using the definition of matrix addition, we have

$$A + B = \begin{bmatrix} -3 + (-4) & 6 + 9 \\ 2 + 3 & 1 + (-1) \\ 5 + (2) & 7 + (-8) \end{bmatrix} = \begin{bmatrix} -7 & 15 \\ 5 & 0 \\ 3 & -1 \end{bmatrix}$$ ■

Using the Graphics Calculator

To check our work in Example 5, use the MATRX key to edit the dimensions for [A] to 3×2 and then type the entries for matrix A. Next, use the MATRX key to select [B] from the EDIT menu. Set the dimensions for [B] to 3×2, then type the entries for matrix B. From the home screen, type [A]+[B] and press the ENTER key. Compare the display with the matrix we found in Example 5. If the matrices do not match, then an error is present.

Subtraction of Matrices We define subtraction of matrices in terms of addition and scalar multiplication.

DEFINITION OF MATRIX SUBTRACTION

For matrices A and B of size $m \times n$, their difference $A - B$ is the matrix $A + (-1B)$.

EXAMPLE 6 For $A = \begin{bmatrix} -3 & 6 \\ 2 & 1 \\ 5 & 7 \end{bmatrix}$ and $B = \begin{bmatrix} -4 & 9 \\ 3 & -1 \\ -2 & -8 \end{bmatrix}$, find $A - B$.

Solution Using the definition of matrix subtraction, we have

$$A - B = A + (-1B) = \begin{bmatrix} -3 & 6 \\ 2 & 1 \\ 5 & 7 \end{bmatrix} + \underbrace{\begin{bmatrix} 4 & -9 \\ -3 & 1 \\ 2 & 8 \end{bmatrix}}_{-1B}$$

$$= \begin{bmatrix} 1 & -3 \\ -1 & 2 \\ 7 & 15 \end{bmatrix}$$ ■

As you can see from our answer to Example 6, we can also find $A - B$ by simply subtracting corresponding entries in the matrices.

Using the Graphics Calculator

To check our work in Example 6, with entries from Example 6 stored in [A] and [B], type [A]+⁻1[B] and press the ENTER key. Compare the display with the matrix we found in Example 6. Now, type [A]−[B] and press the ENTER key. The outputs are the same.

GIVE IT A TRY

For $A = \begin{bmatrix} -4 & 5 & 2 \\ 7 & 8 & 4 \end{bmatrix}$ and $B = \begin{bmatrix} 4 & 6 & 9 \\ 6 & -3 & -5 \end{bmatrix}$, perform the indicated computation(s).

4. $A + B$ **5.** $B - A$ **6.** $2A + 4B$ **7.** $A - 2B$

8. Use the graphics calculator to check your work in Problems 4–7.

Properties We now consider some properties of addition of matrices. Compare these with the properties of numbers from Chapter 1.

PROPERTIES OF ADDITION

For matrices A, B, and C of size $m \times n$,

1. $A + B = B + A$ Matrix addition is commutative.
2. $A + (B + C) = (A + B) + C$ Matrix addition is associative.
3. There exists an $m \times n$ matrix O such that $A + O = O + A = A$. Matrix O is the $m \times n$ matrix with every entry the number 0 and is known as the **additive identity**.
4. For matrix A, there exists a matrix $-A$ such that $A + (-A) = O$. The matrix $-A$ is the matrix in which each entry is the additive opposite of the corresponding entry of matrix A. The matrix $-A$ is known as the **additive inverse** of matrix A.

GIVE IT A TRY

9. List the additive identity for matrices of size 2×3.

10. For the matrix $A = \begin{bmatrix} -3 & 5 & 7 \\ 2 & -3 & 1 \end{bmatrix}$, what is its additive inverse?

The Inner Product Before we consider multiplication of matrices, we need to introduce some additional terms. We have used the term *square matrix* to mean an $n \times n$ matrix. Now, a $1 \times n$ matrix, such as the 1×4 matrix $[1 \quad 5 \quad 7 \quad -3]$, is known as a **row matrix**, or a **row vector**. Next, an $n \times 1$ matrix, such as the 3×1 matrix

$$\begin{bmatrix} -3 \\ 2 \\ 5 \end{bmatrix}$$

is known as a **column matrix**, or a **column vector**. As you can see from our work so far, the operations of scalar multiplication, matrix addition, and matrix subtraction are fairly straightforward. This is not the case with multiplication. For matrix multiplication, we start by defining the *inner product* of a row matrix and a column matrix.

DEFINITION OF INNER PRODUCT

For A a row vector and B a column vector, with the number of columns in A equal to the number of rows in B, the **inner product** AB is the *number* that is the sum of the products of the corresponding entries a_{1i} of matrix A and b_{i1} of matrix B.

Example: $$\begin{bmatrix} 3 & 5 & 2 \end{bmatrix}\begin{bmatrix} -2 \\ 4 \\ 7 \end{bmatrix} = (3)(-2) + (5)(4) + (2)(7) = 28$$

The inner product is not defined if the number of columns in A is not equal to the number of rows in B. For example,

$$\begin{bmatrix} 3 & 7 & 1 & 4 \end{bmatrix}\begin{bmatrix} -5 \\ 8 \end{bmatrix} \quad \text{is undefined.}$$

GIVE IT A TRY

Find the inner product.

11. $\begin{bmatrix} -2 & 1 & 4 & 6 \end{bmatrix}\begin{bmatrix} 4 \\ -3 \\ 6 \\ 2 \end{bmatrix}$

12. $\begin{bmatrix} 6 & 7 \end{bmatrix}\begin{bmatrix} 10 \\ 5 \end{bmatrix}$

Matrix Multiplication Given two matrices, addition or subtraction is defined only if the matrices have the same size. For two matrices A and B, the product AB is defined only if the number of columns of matrix A equals the number of rows of matrix B:

$$\begin{array}{c} A \qquad B \\ \downarrow \qquad \downarrow \\ (m \times n)(n \times k) \end{array}$$

n same

size of product, $m \times k$

Now, we define multiplication of matrices.

DEFINITION OF MULTIPLICATION

For an $m \times n$ matrix A and an $n \times k$ matrix B, the product AB is the $m \times k$ matrix whose entry a_{ij} is the inner product of row i of matrix A and column j of matrix B. The product AB has the same number of rows as matrix A and the same number of columns as matrix B.

EXAMPLE 7 For $A = \begin{bmatrix} 2 & 3 \\ 1 & 0 \\ 7 & 9 \end{bmatrix}$ and $B = \begin{bmatrix} -5 & 6 & 7 & 2 \\ 4 & 2 & 3 & -1 \end{bmatrix}$, does AB exist?

Solution Matrix A is a 3×2 matrix and matrix B is a 2×4 matrix. So, AB does exist, and it will be a 3×4 matrix. ■

In Example 7, the product BA does not exist! Matrix B has 4 columns and matrix A has 3 rows. For square matrices (2×2 matrices, 3×3 matrices, and so on), the products AB and BA always exist.

EXAMPLE 8 For $A = \begin{bmatrix} 2 & 3 \\ 1 & 0 \\ 7 & 9 \end{bmatrix}$ and $B = \begin{bmatrix} -5 & 6 & 7 & 2 \\ 4 & 2 & 3 & 1 \end{bmatrix}$, find AB.

Solution We arrive at the entries of AB by finding inner products. Now, entry a_{11} is the inner product of row 1 of matrix A and column 1 of matrix B:

row 1 → $\begin{bmatrix} 2 & 3 \\ 1 & 0 \\ 7 & 9 \end{bmatrix} \begin{bmatrix} -5 & 6 & 7 & 2 \\ 4 & 2 & 3 & -1 \end{bmatrix}$ ↓ column 1

The inner product is the number 2. So, $a_{11} = -10 + 12 = 2$. We now list the entries for the inner products for row 1 of matrix A:

$$\begin{aligned} a_{11} &= (2)(-5) + (3)(4) = 2 \\ a_{12} &= (2)(6) + (3)(2) = 18 \\ a_{13} &= (2)(7) + (3)(3) = 23 \\ a_{14} &= (2)(2) + (3)(-1) = 1 \end{aligned}$$

row 1 entries — column 4 entries

So, row 1 of matrix AB is $[2 \quad 18 \quad 23 \quad 1]$.
Continuing this process for row 2 of matrix A, we get

$$\begin{aligned} a_{21} &= (1)(-5) + (0)(4) = -5 \\ a_{22} &= (1)(6) + (0)(2) = 6 \\ a_{23} &= (1)(7) + (0)(3) = 7 \\ a_{24} &= (1)(2) + (0)(-1) = 2 \end{aligned}$$

row 2 entries — column 4 entries

We now have row 2 of AB: $\begin{bmatrix} 2 & 18 & 23 & 1 \\ -5 & 6 & 7 & 2 \end{bmatrix}$

Continuing this process for row 3 of matrix A, we get

$$\begin{aligned} a_{31} &= (7)(-5) + (9)(4) = 1 \\ a_{32} &= (7)(6) + (9)(2) = 60 \\ a_{33} &= (7)(7) + (9)(3) = 76 \\ a_{34} &= (7)(2) + (9)(-1) = 5 \end{aligned}$$

row 3 entries column 4 entries

We now have row 3 of AB: $\begin{bmatrix} 2 & 18 & 23 & 1 \\ -5 & 6 & 7 & 2 \\ 1 & 60 & 76 & 5 \end{bmatrix}$ ■

To check our work in Example 8, edit the dimensions for [A] to 3×2 and then enter the values for matrix A. Set the dimensions for [B] to 2×4 and then enter the values for matrix B. From the home screen, type [A][B] and press the ENTER key. Compare the output with our result from Example 8.

GIVE IT A TRY

13. For $A = \begin{bmatrix} 2 & 8 \\ -4 & 5 \end{bmatrix}$ and $B = \begin{bmatrix} -3 & 6 \\ 2 & 3 \end{bmatrix}$, find AB and BA.

Properties of Multiplication From Give It a Try Problem 13, we can immediately see that multiplication of matrices is *not* a commutative operation. That is, for matrices A and B, the product AB is not always equal to BA. In fact, for *nonsquare matrices,* if AB is defined, then BA is often undefined. For square matrices, seldom does $AB = BA$. This noncommutativity of matrix multiplication is what originally intrigued mathematicians (such as Authur Cayley and James Joseph Sylvester, in the 1850s). Except for commutativity, many properties of binary operations do hold for matrix multiplication.

PROPERTIES OF MATRIX MULTIPLICATION

For matrices A, B, and C, assume their sizes are such that the following products and sums exists. Then

1. $A(BC) = (AB)C$ — Associative property
2. $A(B + C) = AB + AC$ — Distributive property for left-hand side
3. $(B + C)A = BA + CA$ — Distributive property for right-hand side
4. For an $n \times n$ matrix A, there exist a matrix I_n where $AI_n = I_nA = A$.

The matrix I_n is known as the **multiplicative identity**. It is the matrix which $a_{ij} = 1$ if $i = j$ and $a_{ij} = 0$ if $i \neq j$. For a 2×2 matrix, $I_2 = \begin{bmatrix} 1 & 0 \\ 0 & 1 \end{bmatrix}$.

For an $n \times n$ matrix A, the entries a_{ij} with $i = j$ form the **major diagonal**. For example,

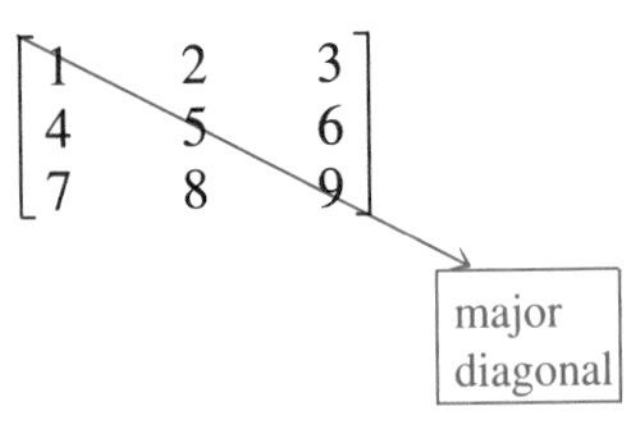

The identity matrix I_n has all 1s on its major diagonal and 0s elsewhere.

Solving a Matrix Equation Given matrices

$$A = \begin{bmatrix} 2 & 3 \\ -5 & 1 \end{bmatrix}$$

$$X = \begin{bmatrix} x \\ y \end{bmatrix}$$

$$C = \begin{bmatrix} 12 \\ -13 \end{bmatrix}$$

we can solve the matrix equation $AX = C$. Using the definition of multiplication of matrices, we get

$$AX = \begin{bmatrix} 2 & 3 \\ -5 & 1 \end{bmatrix} \begin{bmatrix} x \\ y \end{bmatrix} = \begin{bmatrix} 2x + 3y \\ -5x + 1y \end{bmatrix}$$

Next, we have

$$AX = C$$

$$\begin{bmatrix} 2x + 3y \\ -5x + y \end{bmatrix} = \begin{bmatrix} 12 \\ -13 \end{bmatrix}$$

and using the definition of equality, we have the system

$$\begin{cases} 2x + 3y = 12 \\ -5x + y = -13 \end{cases}$$

Solving this system of linear equations (using either the addition approach or the substitution approach), we get the ordered pair (3, 2). Thus, matrix X is $\begin{bmatrix} 3 \\ 2 \end{bmatrix}$.

GIVE IT A TRY

14. For $A = \begin{bmatrix} -5 & 6 \\ 2 & 3 \end{bmatrix}$, $X = \begin{bmatrix} x \\ y \end{bmatrix}$, and $C = \begin{bmatrix} -3 \\ 4 \end{bmatrix}$, solve $AX = C$.

■ SUMMARY

In this section, we have discussed operations on matrices. Two matrices are equal if their entries are equal. The first operation, scalar multiplication, is multiplication of a matrix by a number. To add matrices, we add corresponding entries. To subtract matrices, scalar multiplication and addition are combined, that is, $A - B = A + (-1B)$. As with real numbers, the addition operation for matrices is commutative and associative. Also, we have an identity matrix for addition, and each matrix has an additive inverse.

The operation of matrix multiplication is rather complicated. First, the inner product is defined for a row matrix and a column matrix. Next, multiplication for matrices is defined in terms of inner products. The product of two matrices, say A and B, exists if the number of columns in matrix A is equal to the number of rows in matrix B. An important result of the multiplication operation is that it is *not* commutative.

Finally, we discussed solving matrix equations, such as $AX = C$, where A, X, and C are matrices.

■ ***Quick Index***

11.2 ■ ■ ■ EXERCISES

Use the matrix

$$\begin{bmatrix} 2 & -3 & 5 & -2 \\ 3 & 4 & 1 & 8 \\ 6 & 7 & -4 & 9 \end{bmatrix}$$

to answer Problems 1–4.

1. Report the size of the matrix.
2. Report the entry whose location is row 2, column 3.
3. The entry 8 is in which row and which column?
4. Report each entry: a_{32} a_{21} a_{14}.

Find the values of the variables x, y, z, and w so that the given matrices are equal.

5. $A = \begin{bmatrix} -2 & x+1 \\ w & 2z-1 \end{bmatrix}$ and $B = \begin{bmatrix} y & 3x-1 \\ 2w & 6 \end{bmatrix}$
6. $C = \begin{bmatrix} -2 & 7 & 8 \\ 6 & 9 & 2z \end{bmatrix}$ and
$D = \begin{bmatrix} w-1 & 3x & 8 \\ y^2 - y & 9 & 4z \end{bmatrix}$

For $A = \begin{bmatrix} -4 & 5 \\ 7 & 8 \end{bmatrix}$ *and* $B = \begin{bmatrix} 4 & 6 \\ 6 & -3 \end{bmatrix}$, *perform the indicated operation.*

7. $-3B$
8. $B + A$
9. $A - B$
10. $B - A$
11. $3A - 2B$
12. $2B + 5B$

For

$$E = \begin{bmatrix} -2 & 3 \\ 5 & 8 \\ 10 & 4 \end{bmatrix} \quad \textit{and} \quad F = \begin{bmatrix} 9 & 1 \\ -5 & -6 \\ 7 & -3 \end{bmatrix},$$

perform the indicated operation.

13. $5F$
14. $F + E$
15. $F - E$
16. $E - F$
17. $E - 3E$
18. $F - 2E + 2F$
19. For 3×2 matrices, report the additive identity matrix O.
20. Report the additive inverse for matrix E used in Problems 13–18.
21. For 3×3 matrices, report the multiplicative identity matrix I_3.
22. For

$$A = [2 \quad -3 \quad 5] \quad \text{and} \quad B = \begin{bmatrix} 5 \\ -2 \\ -1 \end{bmatrix},$$

find the inner product.

23. For $A = \begin{bmatrix} 2 & -3 \\ 4 & 6 \end{bmatrix}$ and $B = \begin{bmatrix} 4 & 7 & -1 \\ 3 & -1 & 5 \end{bmatrix}$, find AB.

24. Use the graphics calculator and its matrix features with

$$A = \begin{bmatrix} 3 & 6 & 4 \\ -1 & 2 & 8 \end{bmatrix} \quad \text{and} \quad B = \begin{bmatrix} 4 & 7 \\ 3 & -1 \\ 6 & 2 \end{bmatrix}$$

to find AB.

For $C = \begin{bmatrix} -5 & 2 \\ 3 & 0 \end{bmatrix}$ *and* $D = \begin{bmatrix} 4 & -6 \\ 2 & -7 \end{bmatrix}$, *use the graphics calculator and its matrix features to perform the indicated operations.*

25. CD

26. DC

27. C^2 *Hint: Type* `[C]²`

28. D^2

29. $(C + D)(C + D)$

30. $C^2 + DC + CD + D^2$

31. $C^2 + 2CD + D^2$

32. $C(C + D)$

For

$$A = \begin{bmatrix} -5 & 2 & -1 \\ 3 & 0 & 5 \\ 8 & 1 & 0 \end{bmatrix} \quad \textit{and} \quad B = \begin{bmatrix} 4 & -6 & 8 \\ 1 & 0 & 2 \\ -2 & 0 & 1 \end{bmatrix}$$

use the graphics calculator to perform the indicated operations.

33. AB

34. BA

35. B^2

36. A^2

37. $(A - B)(A - B)$

38. $A^2 - AB - BA + B^2$

39. $A^2 - 2AB + B^2$

40. $B(A + B)$

For

$$A = \begin{bmatrix} -4 & 3 \\ 5 & 1 \end{bmatrix}, \quad X = \begin{bmatrix} x \\ y \end{bmatrix}, \quad \textit{and} \quad C = \begin{bmatrix} -3 \\ 2 \end{bmatrix},$$

solve the given equation.

41. $AX = C$

42. $-2AX = 3C$

For

$$C = \begin{bmatrix} 4 & -3 & 0 \\ -1 & 0 & 2 \\ -3 & 2 & 7 \end{bmatrix}$$

$$X = \begin{bmatrix} x \\ y \\ z \end{bmatrix},$$

and

$$D = \begin{bmatrix} 8 \\ -2 \\ 3 \end{bmatrix},$$

solve the given equation.

43. $CX = D$

44. $C^2X = D$

11.3 ■ ■ ■ INVERSE OF A MATRIX

In the previous section, we considered multiplication of matrices and we discussed properties of multiplication, such as the associative property and the fact that multiplication is not commutative. Although we listed the multiplicative identity for selected square matrices, we did not discuss multiplicative inverses nor division. As our work in this section will show, multiplicative inverses exists only for some square matrices. Also, division of matrices is never defined! In this section, we will use matrices to solve linear systems.

Systems and Inverses Figure 1 shows the identity for 2×2 matrices, I_2, and for 3×3 matrices, I_3. Now, a 2×2 square matrix A has a multiplicative inverse (if it exists) that is a matrix, denoted A^{-1}, such that $(A)(A^{-1}) = I_2$.

$$I_2 = \begin{bmatrix} 1 & 0 \\ 0 & 1 \end{bmatrix} \qquad I_3 = \begin{bmatrix} 1 & 0 & 0 \\ 0 & 1 & 0 \\ 0 & 0 & 1 \end{bmatrix}$$

FIGURE 1

Consider the matrix $A = \begin{bmatrix} 2 & 3 \\ -1 & 1 \end{bmatrix}$. Its multiplicative inverse A^{-1} is a matrix such that $(A)(A^{-1}) = \begin{bmatrix} 1 & 0 \\ 0 & 1 \end{bmatrix}$. To find A^{-1} we can use a system of equations:

$[A]$ → ← $[A]^{-1}$

$$\begin{bmatrix} 2 & 3 \\ -1 & 1 \end{bmatrix} \begin{bmatrix} a & b \\ c & d \end{bmatrix} = \begin{bmatrix} 1 & 0 \\ 0 & 1 \end{bmatrix}$$

We want to find a, b, c, and d. Using matrix multiplication, we get

$$\begin{bmatrix} 2a + 3c & 2b + 3d \\ -a + c & -b + d \end{bmatrix} = \begin{bmatrix} 1 & 0 \\ 0 & 1 \end{bmatrix}$$

Using equality of matrices, we get two systems of linear equations:

$$\begin{cases} 2a + 3c = 1 \\ -a + c = 0 \end{cases} \quad \text{and} \quad \begin{cases} 2b + 3d = 0 \\ -b + d = 1 \end{cases}$$

Solving the first system,

$$\begin{cases} 2a + 3c = 1 & \textbf{(1)} \\ -a + c = 0 & \textbf{(2)} \end{cases}$$

from equation (2) we get $a = c$. Substituting this in equation (1), we get

$$2c + 3c = 1$$
$$5c = 1$$
$$c = \frac{1}{5}$$

Thus, we have $a = \frac{1}{5}$ and $c = \frac{1}{5}$.

Solving the second system,

$$\begin{cases} 2b + 3d = 0 & \textbf{(3)} \\ -b + d = 1 & \textbf{(4)} \end{cases}$$

from equation (4) we get $d = b + 1$. Substituting this in equation (3), we get

$$2b + 3(b + 1) = 0$$
$$5b + 3 = 0$$
$$5b = -3$$
$$b = \frac{-3}{5}$$

Thus,

$$b = \frac{-3}{5} \quad \text{and} \quad d = \frac{-3}{5} + 1 = \frac{2}{5}$$

So we have $a = \frac{1}{5}$, $b = \frac{-3}{5}$, $c = \frac{1}{5}$, and $d = \frac{2}{5}$. Thus, A^{-1} is

$$\begin{bmatrix} \frac{1}{5} & \frac{-3}{5} \\ \frac{1}{5} & \frac{2}{5} \end{bmatrix}$$

Check To check that we do indeed have the inverse, we carry out the following multiplication:

$$\begin{bmatrix} 2 & 3 \\ -1 & 1 \end{bmatrix} \begin{bmatrix} \frac{1}{5} & \frac{-3}{5} \\ \frac{1}{5} & \frac{2}{5} \end{bmatrix} = \begin{bmatrix} \frac{2+3}{5} & \frac{-6+6}{5} \\ \frac{-1+1}{5} & \frac{3+2}{5} \end{bmatrix} = \begin{bmatrix} 1 & 0 \\ 0 & 1 \end{bmatrix}$$

EXAMPLE 1 Find the multiplicative inverse of $B = \begin{bmatrix} 2 & 1 \\ 3 & 2 \end{bmatrix}$.

Solution We start writing $B^{-1} = \begin{bmatrix} a & b \\ c & d \end{bmatrix}$. Now,

$$(B)(B^{-1}) = \begin{bmatrix} 2 & 1 \\ 3 & 2 \end{bmatrix} \begin{bmatrix} a & b \\ c & d \end{bmatrix} = \begin{bmatrix} 2a + c & 2b + d \\ 3a + 2c & 3b + 2d \end{bmatrix}$$

Because $(B)(B^{-1}) = I_2$, we have the systems

$$\begin{cases} 2a + c = 1 \\ 3a + 2c = 0 \end{cases} \quad \text{and} \quad \begin{cases} 2b + d = 0 \\ 3b + 2d = 1 \end{cases}$$

REMEMBER

$$I_2 = \begin{bmatrix} 1 & 0 \\ 0 & 1 \end{bmatrix}$$

Solving these systems of linear equations, we get $a = 2$, $b = -1$, $c = -3$, and $d = 2$. Thus,

$$B^{-1} = \begin{bmatrix} 2 & -1 \\ -3 & 2 \end{bmatrix}$$

■

Using the Graphics Calculator

To check B^{-1} in Example 1, set the size of [B] to 2×2 and then store the entries for matrix B. For matrix A, set the size to 2×2 and store in [A] the entries listed for B^{-1}. Next, compute the product BA. Is the matrix I_2 the output? If not, an error is present!

GIVE IT A TRY

1. For $A = \begin{bmatrix} -3 & -1 \\ 5 & 2 \end{bmatrix}$, find A^{-1}. Use the graphics calculator to check your work.

Definition of the Multiplicative Inverse We are now ready to define the multiplicative inverse of a square matrix.

DEFINITION OF THE MULTIPLICATIVE INVERSE

> An $n \times n$ matrix A has multiplicative inverse (if it exists) the matrix A^{-1} such that $(A)(A^{-1}) = (A^{-1})(A) = I_n$, where I_n is the multiplicative identity for $n \times n$ matrices.

Although multiplication of matrices is not generally commutative, it is so for a square matrix and its inverse. For matrix A, if A^{-1} does *not* exist, then matrix A is known as a **singular matrix**. If A^{-1} does exist, then matrix A is known as a **nonsingular matrix**.

The Calculator Approach As you might imagine, using the approach from Example 1 to find the inverse of a 3×3 matrix or a 4×4 matrix would be difficult. We can use the graphics calculator to find the inverse of a square matrix. We start by finding the inverse of matrix B from Example 1.

EXAMPLE 2 Use the graphics calculator to find B^{-1} for $B = \begin{bmatrix} 2 & 1 \\ 3 & 2 \end{bmatrix}$.

Solution Check that matrix B is still stored in [B]. Type [B] and then use the x^{-1} key to get the entry [B]$^{-1}$. Press the ENTER key, and the inverse of matrix B is displayed. As Figure 2 shows, we have $B^{-1} = \begin{bmatrix} 2 & -1 \\ -3 & 2 \end{bmatrix}$. ■

```
[B]-1
[ 2 -1]
[ -3 2]
```

FIGURE 2

Compare B^{-1} from Example 2 with our result in Example 1 to see that we have found the inverse of matrix B in Example 2. We can also use the graphics calculator to check that we have the inverse of matrix B. We start by storing the currently displayed matrix, B^{-1}, as matrix C. To do this, press the STO▶ key and then type [C]. Press the ENTER key and the matrix B^{-1} is stored in matrix C. Now, type [B]*[C] and press the ENTER key. As Figure 3 shows, the identity matrix is output (the entry -2E-12 is a calculator approximation of 0).

We now use the graphics calculator to find the inverse of a 3×3 matrix.

```
[ -3 2 ]
Ans→[C]
[ 2 -1 ]
[ -3 2 ]
[B]*[C]
[ 1 -2E-12]
[ 0 1      ]
```

FIGURE 3

EXAMPLE 3 Use the graphics calculator to find A^{-1} for $A = \begin{bmatrix} 2 & 2 & 5 \\ 1 & 1 & 2 \\ 3 & 2 & 6 \end{bmatrix}$.

Solution Set the size of [A] as 3×3. Now, store the entries for matrix A in [A]. From the home screen, type [A]$^{-1}$ and press the ENTER key. To see the entries in the rightmost column, use the right arrow key to scroll the screen (see Figure 4). The matrix output has some entries that are decimal approximations of the actual entries. For example, the entry 1E-12 at row 2, column 1 is the calculator's approximation of 0. We report A^{-1} as

$$A^{-1} = \begin{bmatrix} -2 & 2 & 1 \\ 0 & 3 & -1 \\ 1 & -2 & 0 \end{bmatrix}$$

■

```
[A]-1
...-2      2  1     ]
...1E-12 3   -1     ]
...1       -2 4E-13]
```

FIGURE 4

GIVE IT A TRY

2. Use the graphics calculator to find the inverse of $C = \begin{bmatrix} 2 & 1 \\ 5 & 3 \end{bmatrix}$.

3. Use the graphics calculator to find A^{-1} for $A = \begin{bmatrix} -2 & 2 & -2 & 4 \\ 0 & 1 & -1 & 1 \\ 1 & -2 & 3 & 0 \\ 0 & 2 & -3 & 1 \end{bmatrix}$.

Linear Systems We now return to linear systems and show a matrix approach to solving such systems. We first move the system

$$\begin{cases} -2x + 3y = 6 & \textbf{(1)} \\ 4x - 5y = 3 & \textbf{(2)} \end{cases}$$

to a matrix equation. By taking the coefficients on the variables of the system, and inserting them into a 2 × 2 matrix (in this case), we obtain the **coefficient matrix** for the system:

x-coefficients y-coefficients

$$\begin{matrix} \text{Equation (1)} \rightarrow \\ \text{Equation (2)} \rightarrow \end{matrix} \begin{bmatrix} -2 & 3 \\ 4 & -5 \end{bmatrix}$$

The coefficient matrix is the matrix formed by using the x-coefficients for column 1 and the y-coefficients for column 2. We will refer to this as matrix A.

Now, we let $X = \begin{bmatrix} x \\ y \end{bmatrix}$ and $C = \begin{bmatrix} 6 \\ 3 \end{bmatrix}$. The system can be written as the matrix equation $AX = C$. To solve this equation we can multiply each side by A^{-1}.

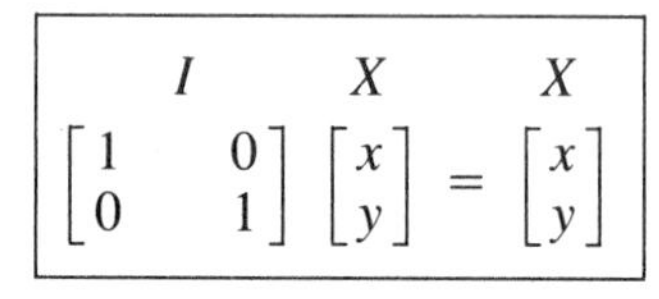

FIGURE 5

$$\begin{aligned} (A^{-1})(AX) &= (A^{-1})(C) && \text{Multiply each side by } A^{-1}. \\ (A^{-1}A)X &= A^{-1}C && \text{Matrix multiplication is associative.} \\ IX &= A^{-1}C && AA^{-1} = I \\ X &= A^{-1}C && IX = X. \text{ See Figure 5.} \end{aligned}$$

Thus, we start by finding A^{-1}. Using the graphics calculator, we get

$$A^{-1} = \begin{bmatrix} 2.5 & 1.5 \\ 2 & 1 \end{bmatrix}$$

Since $X = A^{-1}C$, we have

$$\begin{aligned} X &= \quad A^{-1} \qquad C \\ \begin{bmatrix} x \\ y \end{bmatrix} &= \begin{bmatrix} 2.5 & 1.5 \\ 2 & 1 \end{bmatrix} \begin{bmatrix} 6 \\ 3 \end{bmatrix} \\ \begin{bmatrix} x \\ y \end{bmatrix} &= \begin{bmatrix} 2.5(6) + 1.5(3) \\ 2(6) + 1(3) \end{bmatrix} \\ \begin{bmatrix} x \\ y \end{bmatrix} &= \begin{bmatrix} 19.5 \\ 15 \end{bmatrix} \end{aligned}$$

Thus, then we have $x = 19.5$ and $y = 15$, and the solution to the linear system is the ordered pair (19.5, 15).

Check To check that (19.5, 15) solves the system

$$\begin{cases} -2x + 3y = 6 \\ 4x - 5y = 3 \end{cases}$$

store `19.5` in memory location `X` and store `15` in memory location `Y`. Type `-2X+3Y` and then press the `ENTER` key. Is `6` output? Type `4X-5Y` and press the `ENTER` key. Is `3` output? If not, an error is present!

By using the fact that the solution to the matrix equation $AX = C$ is $X = A^{-1}C$, and using the calculator to find the inverse, we can solve the equation on the calculator. Check that the matrix $\begin{bmatrix} -2 & 3 \\ 4 & -5 \end{bmatrix}$ is stored in `[A]`. Store the matrix $\begin{bmatrix} 6 \\ 3 \end{bmatrix}$ in `[B]`. Now, type `[A]`$^{-1}$`[B]` and press the `ENTER` key. A 2×1 matrix is output. The entries yield the ordered pair that solves the system. The numbers with the solution we found by the algebraic approach.

GIVE IT A TRY

4. For $A = \begin{bmatrix} 3 & 8 \\ 4 & 11 \end{bmatrix}$, we have $A^{-1} = \begin{bmatrix} 11 & -8 \\ -4 & 3 \end{bmatrix}$. Use this to solve

$$\begin{bmatrix} 3 & 8 \\ 4 & 11 \end{bmatrix} \begin{bmatrix} x \\ y \end{bmatrix} = \begin{bmatrix} 2 \\ 7 \end{bmatrix}$$

5. Use the graphics calculator and matrices to solve the system

$$\begin{cases} x + y + 2z = 9 \\ 3x + 2y + 6z = 10 \\ 2x + 2y + 5z = 2 \end{cases}$$

Hint: The coefficient matrix is $A = \begin{bmatrix} 1 & 1 & 2 \\ 3 & 2 & 6 \\ 2 & 2 & 5 \end{bmatrix}$

Special Cases When we use the graphics calculator to find the inverse of a matrix, two problems may occur:

1. The entries computed by the calculator are decimal approximations.
2. The coefficient matrix has no inverse—it is a singular matrix.

Both of these problems are related to a special operation on matrices—the *determinant*. The determinant operates on a matrix to output a real number. We will discuss this operation in greater detail in Section 11.4, but here, we will use the operation as an aid to working around the problems we have just described.

```
[A]-1
[ .3846153846   ...
[ -.0769230769  ...
```

FIGURE 6

Suppose, using the graphics calculator, we enter the matrix $\begin{bmatrix} 2 & -3 \\ 1 & 5 \end{bmatrix}$ for matrix A. Now, from the home screen if we type `[A]`$^{-1}$, and press the `ENTER` key, the display appears as shown in Figure 6. These numbers are decimal approximations for the entries of the inverse matrix. Use the `MATRX` menu and select option 5, the `det`, to type `det [A]`. This command outputs a single

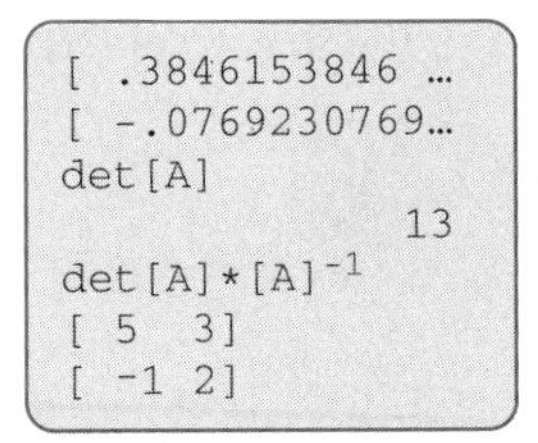

FIGURE 7

number, the determinant of matrix A. The determinant of matrix A is the denominator for all the entries in the inverse of matrix A. Press the ENTER key. The number 13 is output. To get the exact values in the inverse of matrix A, type det [A]*[A]$^{-1}$ and press the ENTER key. The entries displayed in Figure 7 are the numerators in the inverse. Thus, the inverse of matrix A is

$$A^{-1} = \begin{bmatrix} \frac{5}{13} & \frac{3}{13} \\ \frac{-1}{13} & \frac{2}{13} \end{bmatrix}$$

If the determinant of matrix A is 0, then matrix A is a singular matrix and has no multiplicative inverse. To see this, use the graphics calculator to find the inverse of $A = \begin{bmatrix} 2 & 6 \\ 1 & 3 \end{bmatrix}$. Use the MATRX key to set the dimensions for matrix A to 2×2, and then type the entries for [A]. From the home screen, type [A]$^{-1}$, and press the ENTER key. As you can see, the ERROR screen is presented. Now, type det [A] and press the ENTER key. The determinant of matrix A is 0. That is, this matrix is a singular matrix—it has no multiplicative inverse.

GIVE IT A TRY

6. For $A = \begin{bmatrix} 1 & 5 \\ 2 & 2 \end{bmatrix}$, use the graphics calculator and the determinant to find A^{-1}.

7. For $B = \begin{bmatrix} 1 & 1 & 1 & 0 \\ 1 & 0 & -1 & 2 \\ 2 & -1 & 1 & 1 \\ 1 & 2 & -1 & 2 \end{bmatrix}$, use the graphics calculator to find the determinant. What does this number tell you about B^{-1}?

Applications One application of linear systems of three equations in three variables is that of a mathematical model for physical situations that involve mixtures. Consider the following example.

EXAMPLE 4 A local company buys the parts and assembles three different sailboard models, A, B, and C. The model A sailboard costs \$200 for parts, requires 2 hours to assemble, and costs \$300 to market. The model B sailboard costs \$400 for parts, takes 1 hour to assemble, and costs \$200 to market. Model C costs \$500 for parts, requires 2 hours to assemble, and costs \$100 to market. If the accounting department reports \$11,900 spent for parts, 62 hours of labor for assembly, and \$8,500 spent for marketing for the previous month, how many of each type of sailboard were produced?

Solution Let x = number of model A sailboards produced
y = number of model B sailboards produced
z = number of model C sailboards produced

Model	Parts (in $)	Assembly	Marketing (in $)
A	200	2 hours	300
B	400	1 hour	200
C	500	2 hours	100
Total	11900	62 hours	8500

The table in the margin illustrates the given facts.

Using the money spent on parts, we get the equation

$$200x + 400y + 500z = 11900$$

Using the hours paid for labor, we get the equation

$$2x + 1y + 2z = 62$$

Finally, using the money spent on marketing, we get

$$300x + 200y + 100z = 8500$$

Thus, the model for this situation is a linear system of three equations in three variables:

$$\begin{cases} 200x + 400y + 500z = 11900 \\ 2x + y + 2z = 62 \\ 300x + 200y + 100z = 8500 \end{cases}$$

We can use matrices and the graphics calculator to solve this system. The coefficient matrix is

$$\begin{bmatrix} 200 & 400 & 500 \\ 2 & 1 & 2 \\ 300 & 200 & 100 \end{bmatrix}$$

So the matrix equation for the system is

$$\begin{bmatrix} 200 & 400 & 500 \\ 2 & 1 & 2 \\ 300 & 200 & 100 \end{bmatrix} \begin{bmatrix} x \\ y \\ z \end{bmatrix} = \begin{bmatrix} 11900 \\ 62 \\ 8500 \end{bmatrix}$$

```
[A]-1
det[A]
                0
[A]-1*[B]
[ 18]
[ 12]
[ 7 ]
```

FIGURE 8

Using the graphics calculator, select matrix [A], set the dimensions to 3×3, and type the entries from the coefficient matrix. Also, set the dimensions for matrix [B] to 3×1, and type the entries 11900, 62, and 8500. From the home screen, type $[A]^{-1}$*[B] and press the ENTER key. The matrix output shown in Figure 8 informs us that $x = 18$, $y = 12$, and $z = 7$. Thus, the solution to the system is the ordered triple (18, 12, 7).

The answer to the question is 18 sailboards of model A, 12 of model B, and 7 of model C were produced the previous month. ■

■ ***Quick Index***

■ SUMMARY

In this section, we have learned that only some square matrices have multiplicative inverses. These matrices are known as nonsingular matrices. One approach to finding the inverse of a matrix is to solve a system of linear equations, but a more efficient approach is using the graphics calculator. To find the inverse on the graphics calculator, use the x^{-1} key. We also found the determinant of a matrix yields the denominator for the entries in the matrix inverse. Multiplying the det [A] times A^{-1} yields the matrix whose entries are the numerators for the inverse matrix. In finding the inverse, if either the error screen is displayed or the determinant is 0, then the matrix has no inverse—it is a singular matrix.

Combining matrix equations, such as $AX = C$, and inverse matrices, we solved linear systems using matrices. In general, if A^{-1} exists, then the solution to the equation $AX = C$ is $X = A^{-1}C$. Also, we found that a linear system of three variables and three variables can be used as a model for a mixture problem.

11.3 ■ ■ ■ EXERCISES

Use multiplication and the graphics calculator to determine whether the given matrices are inverses of each other. If they are inverses, state this. If not, report their product.

1. $A = \begin{bmatrix} 4 & 5 \\ 7 & 9 \end{bmatrix}$ $B = \begin{bmatrix} 9 & -5 \\ -7 & 4 \end{bmatrix}$

2. $C = \begin{bmatrix} 2 & 1 \\ 7 & 3 \end{bmatrix}$ $D = \begin{bmatrix} -3 & 1 \\ 6 & -2 \end{bmatrix}$

3. $E = \begin{bmatrix} 2 & 2 & 1 \\ 7 & 9 & 3 \\ -4 & -7 & -1 \end{bmatrix}$ $F = \begin{bmatrix} 12 & -5 & -3 \\ -5 & 2 & 1 \\ -13 & 6 & 12 \end{bmatrix}$

4. $A = \begin{bmatrix} 2 & 2 & 1 \\ 7 & 9 & 3 \\ -4 & -7 & -1 \end{bmatrix}$ $T = \begin{bmatrix} 2 & 1 & -5 \\ -13 & -5 & 35 \\ -10 & -4 & 27 \end{bmatrix}$

5. Find a, b, c, and d:

$$\begin{bmatrix} -2 & 3 \\ 5 & 1 \end{bmatrix} \begin{bmatrix} a & b \\ c & d \end{bmatrix} = \begin{bmatrix} 1 & 0 \\ 0 & 1 \end{bmatrix}$$

6. Find a, b, c, and d:

$$\begin{bmatrix} 4 & -1 \\ 7 & -2 \end{bmatrix} \begin{bmatrix} a & b \\ c & d \end{bmatrix} = \begin{bmatrix} 1 & 0 \\ 0 & 1 \end{bmatrix}$$

Use the graphics calculator to find the inverse, if it exists, of each matrix. If the inverse does not exist, report singular.

7. $\begin{bmatrix} -5 & 2 \\ 3 & -1 \end{bmatrix}$

8. $\begin{bmatrix} 7 & -1 \\ 6 & -1 \end{bmatrix}$

9. $\begin{bmatrix} 3 & 2 \\ -1 & 1 \end{bmatrix}$

10. $\begin{bmatrix} 2 & -4 \\ 2 & 6 \end{bmatrix}$

11. $\begin{bmatrix} 3 & 6 \\ 2 & 4 \end{bmatrix}$

12. $\begin{bmatrix} -3 & 1 \\ 9 & 3 \end{bmatrix}$

13. $\begin{bmatrix} 5 & -4 \\ -1 & -2 \end{bmatrix}$

14. $\begin{bmatrix} -0.5 & 2 \\ -1 & 4 \end{bmatrix}$

15. $\begin{bmatrix} 3 & 2 & 1 \\ 5 & 4 & -1 \\ 4 & 3 & 1 \end{bmatrix}$

16. $\begin{bmatrix} -7 & 8 & 5 \\ 2 & -2 & -1 \\ 1 & -1 & -1 \end{bmatrix}$

17. $\begin{bmatrix} -1 & -1 & 2 \\ 1 & 1 & -1 \\ 0 & 1 & -1 \end{bmatrix}$

18. $\begin{bmatrix} 2 & 1 & 3 \\ 1 & 2 & 1 \\ 2 & 2 & 3 \end{bmatrix}$

19. $\begin{bmatrix} 0 & 2 & -1 \\ 3 & 1 & 1 \\ 2 & 0 & -2 \end{bmatrix}$

20. $\begin{bmatrix} -3 & 1 & 0 \\ 1 & 3 & 3 \\ 4 & 0 & -1 \end{bmatrix}$

Solve each matrix equation.

21. $\begin{bmatrix} -2 & 3 \\ 1 & -2 \end{bmatrix} \begin{bmatrix} x \\ y \end{bmatrix} = \begin{bmatrix} -2 \\ 5 \end{bmatrix}$

22. $\begin{bmatrix} 4 & -1 \\ 2 & 1 \end{bmatrix} \begin{bmatrix} x \\ y \end{bmatrix} = \begin{bmatrix} 1 \\ 5 \end{bmatrix}$

23. $\begin{bmatrix} 1 & 2 & 1 \\ 2 & -1 & 0 \\ 0 & 1 & 2 \end{bmatrix} \begin{bmatrix} x \\ y \\ z \end{bmatrix} = \begin{bmatrix} 0 \\ 8 \\ 0 \end{bmatrix}$

24. $\begin{bmatrix} 1 & 2 & 1 \\ 2 & -1 & 0 \\ 0 & 1 & 2 \end{bmatrix} \begin{bmatrix} x \\ y \\ z \end{bmatrix} = \begin{bmatrix} 3 \\ 8 \\ 1 \end{bmatrix}$

Solve the system of equations using a matrix equation. Let A *be the coefficient matrix. Report matrix A, A^{-1}, and the solution to the system.*

25. $\begin{cases} 3x - 2y = 6 \\ x + 2y = 2 \end{cases}$

26. $\begin{cases} 2x - 5y = 10 \\ 3x + 5y = 5 \end{cases}$

27. $\begin{cases} 4x + 3y = 8 \\ x - 2y = 4 \end{cases}$

28. $\begin{cases} 2x - 5y = 12 \\ 3x + y = 6 \end{cases}$

29. $\begin{cases} y = -2x + 3 \\ y = 3x + 1 \end{cases}$

30. $\begin{cases} y = 4x - 2 \\ y = -2x + 5 \end{cases}$

31. If $A = \begin{bmatrix} -3 & 4 \\ 2 & -3 \end{bmatrix}$, then $A^{-1} = \begin{bmatrix} -3 & -4 \\ -2 & -3 \end{bmatrix}$.

Use this fact to solve the system

$$\begin{cases} -3x + 4y = 9 \\ 2x - 3y = 1 \end{cases}$$

32. If $A = \begin{bmatrix} 3 & -5 \\ 1 & 2 \end{bmatrix}$, then $A^{-1} = \begin{bmatrix} \frac{2}{11} & \frac{5}{11} \\ \frac{-1}{11} & \frac{3}{11} \end{bmatrix}$.

Use this fact to solve the system

$$\begin{cases} 3x - 5y = 4 \\ x + 2y = 9 \end{cases}$$

33. If

$$A = \begin{bmatrix} 4 & 3 & -5 \\ -1 & 0 & 1 \\ -2 & -2 & 3 \end{bmatrix}, \text{ then } A^{-1} = \begin{bmatrix} 2 & 1 & 3 \\ 1 & 2 & 1 \\ 2 & 2 & 3 \end{bmatrix}$$

Use this fact to solve the system

$$\begin{cases} 4x + 3y - 5z = 8 \\ z - x = 2 \\ 3z = 2x + 2y + 1 \end{cases}$$

34. If

$$A = \begin{bmatrix} -2 & 2 & 1 \\ 0 & 3 & -1 \\ 1 & -2 & 0 \end{bmatrix}, \text{ then } A^{-1} = \begin{bmatrix} 2 & 2 & 5 \\ 1 & 1 & 2 \\ 3 & 2 & 6 \end{bmatrix}$$

Use this fact to solve the system

$$\begin{cases} -2x + 2y + z = 8 \\ 3y - z = 3 \\ x - 2y = 5 \end{cases}$$

Use the graphics calculator and the x^{-1} *key to solve each system.*

35. $$\begin{cases} 3x - 2y + z - 2w = 9 \\ x + y - z + 3w = 2 \\ y - 2z + w = 3 \\ 2x - y - w = 4 \end{cases}$$

36. $$\begin{cases} 2x - y - 2w = 6 \\ 3x + y - 2z + w = -4 \\ x + 2y - z + w = 2 \\ y + 2z - w = 4 \end{cases}$$

37. A company produces three types of calculators: the fx-70, the mx-80, and the rx-90. The fx-70 requires 3 hours to assemble, 1 hour to test, and 2 hours to package. The mx-80 requires 5 hours to assemble, 2 hours to test, and 1 hour to package. The rx-90 requires 4 hours to assemble and 2 hours to package (no testing is required). If, for a particular week, 895 hours of labor were used in assembling calculators, 176 hours were used in testing, and 403 hours were used in packaging calculators, how many of each type of calculator were produced?

38. A company wishes to blend three teas to produce 200 pounds of a blend to be sold as Mellow Meadow. Tea A costs \$0.23 per pound, tea B costs \$0.54 per pound, and tea C costs \$0.98 per pound. The company wants the cost per pound of the blend to be \$0.60. For taste, the amount of tea B used should equal the amount of tea C used. How many pounds of each tea should be used in the blend?

39. How many liters of 30% acid solution, 60% acid solution, and 90% acid solution must be mixed to produce 50 liters of 70% acid solution, provided that the amount of 60% acid solution used must be twice the amount of 30% acid solution used?

40. How many liters of 20% acid solution, 70% acid solution, and 80% acid solution must be mixed to produce 40 liters of 30% acid solution, if the amount of 80% acid solution used must be half the amount of 70% acid solution used?

41. A central computer receives calls from three remote locations. On a particular day, 500 calls are received, and the number of calls from remote location X equals the calls from both remote locations Y and Z. The combined number of calls from X and Z totals 420. How many calls are received from each of the three remote locations on the day in question.

11.4 ■ ■ ■ DETERMINANTS

In this section we will assign a numeric value to a square matrix. That value is known as the *determinant* of the matrix. As we will see, the determinant, used with an algorithm known as Cramer's Rule, can provide yet another approach to solving a linear system of equations. Determinants have applications beyond linear systems. For example, we will learn how to use determinants to find the area measure of a triangle and the equation of a line.

The Determinant of a 2 × 2 Matrix Determinants are defined only for square matrices, such as 2 × 2 matrices or 3 × 3 matrices. The approach we take to compute the determinant is a recursive approach, that is, each level is defined in terms of a previous level. The determinant of a 1 × 1 matrix is simply the single entry of the matrix. For example, the determinant of the 1 × 1 matrix $[-5]$ is the number -5. We now define the determinant of a 2 × 2 matrix.

DETERMINANT OF A 2 × 2 MATRIX

> For a 2 × 2 matrix $A = \begin{bmatrix} a & b \\ c & d \end{bmatrix}$, the **determinant** is denoted
>
> $$\begin{vmatrix} a & b \\ c & d \end{vmatrix} \quad \text{or} \quad \det(A) \quad \text{or} \quad |A|$$
>
> and is equal to the number $ad - bc$.

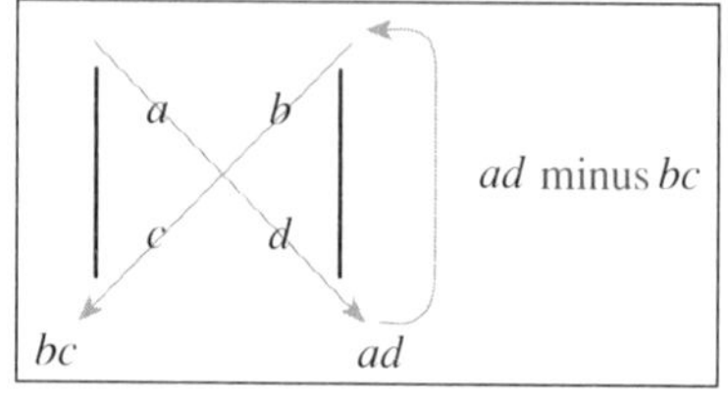

FIGURE 1

Figure 1 illustrates how to compute the determinant of a 2 × 2 matrix.
Consider the following examples.

$$\text{For } A = \begin{bmatrix} 2 & 1 \\ 4 & 3 \end{bmatrix}, \quad \begin{vmatrix} 2 & 1 \\ 4 & 3 \end{vmatrix} = (2)(3) - (1)(4) = 6 - 4 = 2$$

$$\text{For } B = \begin{bmatrix} -3 & 5 \\ 6 & 0 \end{bmatrix}, \quad \det(B) = (-3)(0) - (5)(6) = 0 - 30 = -30$$

$$\text{For } C = \begin{bmatrix} 8 & -2 \\ 1 & -1 \end{bmatrix}, \quad |C| = (8)(-1) - (-2)(1) = -8 - (-2) = -6$$

GIVE IT A TRY

Find the determinant of each 2 × 2 matrix.

1. $\begin{bmatrix} -6 & 2 \\ 2 & 6 \end{bmatrix}$ **2.** $\begin{bmatrix} 5 & 9 \\ 1 & 8 \end{bmatrix}$ **3.** $\begin{bmatrix} \sqrt{3} & \sqrt{5} \\ \sqrt{5} & \sqrt{6} \end{bmatrix}$

Minors and Cofactors To define the determinant of a 3 × 3 square matrix and matrices of greater dimensions, we need to define two terms—minor and cofactor. Consider the 3 × 3 square matrix A:

$$\begin{bmatrix} 2 & 3 & 1 \\ 4 & 5 & 6 \\ 7 & 8 & 9 \end{bmatrix}$$

(row 2 → the second row; column 3 → the third column; a_{23} → the entry 6)

Now, take any entry of the matrix, say a_{23}, which is 6. The minor for this particular entry is the determinant of the 2 × 2 matrix formed by eliminating row 2 and column 3 from matrix A:

$$\begin{bmatrix} 2 & 3 & 1 \\ 4 & 5 & 6 \\ 7 & 8 & 9 \end{bmatrix}$$

(column 3 and row 2 struck through)

Thus, the minor for a_{23} is $\begin{vmatrix} 2 & 3 \\ 7 & 8 \end{vmatrix}$, which is $(2)(8) - (3)(7) = 16 - 21 = -5$.

As another example, take entry a_{31}, which is 7. The minor for this entry, denoted M_{31}, is $\begin{vmatrix} 3 & 1 \\ 5 & 6 \end{vmatrix}$:

$$\begin{bmatrix} 2 & 3 & 1 \\ 4 & 5 & 6 \\ 7 & 8 & 9 \end{bmatrix} \text{ row 3}$$

column 1

Thus, $M_{31} = (3)(6) - (1)(5) = 13$.

Next, we consider the cofactor of an entry in matrix A. For a_{23}, the cofactor, denoted C_{23}, is

$$C_{23} = (-1)^{2+3}M_{23} = (-1)^5(-5) = 5$$

Also,

$$C_{31} = (-1)^{3+1}M_{31} = (-1)^4(13) = 13$$

That is, the cofactor always multiplies its minor by 1 or by -1.

We now define the minor and cofactor for an $n \times n$ square matrix.

DEFINITION OF MINOR AND COFACTOR

> For matrix A, an $n \times n$ matrix with $n > 1$, the **minor** for entry a_{ij} is denoted M_{ij} and is the determinant of the matrix formed from matrix A by deleting the ith row and the jth column.
>
> The **cofactor** for this entry, denoted C_{ij}, is $(-1)^{i+j}M_{ij}$.

GIVE IT A TRY

For $A = \begin{bmatrix} -3 & 6 & 7 \\ 2 & 0 & 4 \\ 3 & -1 & 5 \end{bmatrix}$ find the indicated quantity.

4. M_{11} **5.** C_{11} **6.** M_{32}

7. C_{32} **8.** M_{22} **9.** C_{22}

The Determinant of an $n \times n$ Matrix We are now ready to define the determinant of an $n \times n$ matrix.

DETERMINANT OF A MATRIX

> For A, an $n \times n$ matrix, the **determinant**, denoted $|A|$, is the sum of the product of the entries in the first row and their corresponding cofactors:
>
> $$|A| = a_{11}C_{11} + a_{12}C_{12} + a_{13}C_{13} + \cdots + a_{1n}C_{1n}$$

For example, if $A = \begin{bmatrix} 2 & 3 & 5 \\ 1 & 7 & 8 \\ 9 & 4 & 6 \end{bmatrix}$

then

entries from row 1

$$\begin{aligned} |A| &= 2C_{11} + 3C_{12} + 5C_{13} \\ &= 2(-1)^{1+1}M_{11} + 3(-1)^{1+2}M_{12} + 5(-1)^{1+3}M_{13} \\ &= 2(-1)^2M_{11} + 3(-1)^3M_{12} + 5(-1)^4M_{13} \\ &= 2M_{11} - 3M_{12} + 5M_{13} \\ &= 2\begin{vmatrix} 7 & 8 \\ 4 & 6 \end{vmatrix} - 3\begin{vmatrix} 1 & 8 \\ 9 & 6 \end{vmatrix} + 5\begin{vmatrix} 1 & 7 \\ 9 & 4 \end{vmatrix} \\ &= 2(42 - 32) - 3(6 - 72) + 5(4 - 63) \\ &= 2(10) - 3(-66) + 5(-59) \\ &= 20 - (-198) + (-295) \\ &= -77 \end{aligned}$$

This process is known as *expanding the determinant along row 1*. The following property of determinants states that the value of the determinant is not affected whether we expand it along any other row or, for that matter, any other column.

PROPERTY OF THE DETERMINANT

> The determinant of an $n \times n$ matrix A is the sum of the products of the entries in row i and their corresponding cofactors.
>
> The determinant of A is the sum of the products of the entries in column j and the corresponding cofactors.

$$A = \begin{bmatrix} 2 & 3 & 5 \\ 1 & 7 & 8 \\ 9 & 4 & 6 \end{bmatrix}$$

To demonstrate this property, we will find the determinant of matrix A (at the left) by expanding it about column 3.

Entries from column 3

$$|A| = 5C_{13} + 8C_{23} + 6C_{33}$$

$(-1)^{i+3}$

$$\begin{aligned} &= 5(1)\begin{vmatrix} 1 & 7 \\ 9 & 4 \end{vmatrix} + 8(-1)\begin{vmatrix} 2 & 3 \\ 9 & 4 \end{vmatrix} + 6(1)\begin{vmatrix} 2 & 3 \\ 1 & 7 \end{vmatrix} \\ &= 5(4 - 63) - 8(8 - 27) + 6(14 - 3) \\ &= 5(-59) - 8(-19) + 6(11) \\ &= -295 + 152 + 66 \\ &= -77 \end{aligned}$$

GIVE IT A TRY

10. For $B = \begin{bmatrix} 1 & 2 & 3 \\ 4 & 5 & 6 \\ 7 & 8 & 9 \end{bmatrix}$, find det (B).

Cramer's Rule We now consider an application of the determinant of a square matrix. Consider the general system of linear equations in two variables, x and y:

$$\begin{cases} ax + by = k_1 \\ cx + dy = k_2 \end{cases}$$

We can write a general form for the solution to this system, using the addition method (or any of the approaches discussed so far). For the system

$$\begin{cases} ax + by = k_1 & \textbf{(1)} \\ cx + dy = k_2 & \textbf{(2)} \end{cases}$$

we have

$$\begin{cases} acx + bcy = ck_1 \\ -acx - ady = -ak_2 \end{cases}$$ Multiply (1) by c. Multiply (2) by $-a$.

Sum equation: $$(bc - ad)y = ck_1 - ak_2$$

$$y = \frac{ck_1 - ak_2}{bc - ad}$$ Multiply by $\frac{1}{bc - ad}$.

or

$$y = \frac{ak_2 - ck_1}{ad - bc}$$ Multiply by $\frac{-1}{-1}$.

A similar development produces x. Thus, we have the following:

$$x = \frac{dk_1 - bk_2}{ad - bc} \qquad y = \frac{ak_2 - ck_1}{ad - bc}$$

Now, the denominator in these expressions is the determinant of the coefficient matrix for the system $\begin{vmatrix} a & b \\ c & d \end{vmatrix}$ and is denoted D. Also, the numerator for the x-expression is the determinant of the matrix $\begin{vmatrix} k_1 & b \\ k_2 & d \end{vmatrix}$ and is denoted D_x.

Finally, the numerator of the y-expression is the determinant of the matrix $\begin{vmatrix} a & k_1 \\ c & k_2 \end{vmatrix}$, denoted by D_y.

The solution to the system, in terms of determinants, is

$$\left(\frac{D_x}{D}, \frac{D_y}{D}\right)$$

This is Cramer's Rule (algorithm) for linear systems in two variables. To increase our understanding of this approach, work through the next example.

EXAMPLE 1 Use Cramer's Rule to solve the system $\begin{cases} 3x - 5y = 10 \\ 2x + y = 8 \end{cases}$

Solution We start by finding D, the determinant of the coefficient matrix.

$$D = \begin{vmatrix} 3 & -5 \\ 2 & 1 \end{vmatrix} = (3)(1) - (-5)(2) = 3 - (-10) = 13.$$

To find D_x, we form the matrix

constants

$$\begin{bmatrix} 10 & -5 \\ 8 & 1 \end{bmatrix}$$

by replacing the x-column with the constants. So,

$$D_x = \begin{vmatrix} 10 & -5 \\ 8 & 1 \end{vmatrix} = (10)(1) - (-5)(8) = 10 - (-40) = 50.$$

To find D_y, we replace the y-column with the constants.

constants

$$D_y = \begin{vmatrix} 3 & 10 \\ 2 & 8 \end{vmatrix} = (3)(8) - (10)(2) = 24 - 20 = 4.$$

The solution to the system is $\left(\dfrac{D_x}{D}, \dfrac{D_y}{D}\right) = \left(\dfrac{50}{13}, \dfrac{4}{13}\right)$. ■

We now state Cramer's Rule for a linear system in three variables. It can be extended to any linear system with n variables and n equations.

CRAMER'S RULE FOR A LINEAR SYSTEM OF THREE EQUATIONS IN THREE VARIABLES

For the linear system $\begin{cases} a_1x + b_1y + c_1z = k_1 \\ a_2x + b_2y + c_2z = k_2 \\ a_3x + b_3y + c_3z = k_3 \end{cases}$

let $D = \begin{vmatrix} a_1 & b_1 & c_1 \\ a_2 & b_2 & c_2 \\ a_3 & b_3 & c_3 \end{vmatrix}$ $\quad D_x = \begin{vmatrix} k_1 & b_1 & c_1 \\ k_2 & b_2 & c_2 \\ k_3 & b_3 & c_3 \end{vmatrix}$

$D_y = \begin{vmatrix} a_1 & k_1 & c_1 \\ a_2 & k_2 & c_2 \\ a_3 & k_3 & c_3 \end{vmatrix}$ $\quad D_z = \begin{vmatrix} a_1 & b_1 & k_1 \\ a_2 & b_2 & k_2 \\ a_3 & b_3 & k_3 \end{vmatrix}$

If $D \neq 0$, then the solution to the system is $\left(\dfrac{D_x}{D}, \dfrac{D_y}{D}, \dfrac{D_z}{D}\right)$.

If $D = 0$, then the system is either inconsistent or dependent.

We now use Cramer's Rule to solve a linear system of three equations in three variables.

EXAMPLE 2 Solve: $\begin{cases} 3x - y + z = 6 \\ x + y = 3 \\ 2x - 3z = 4 \end{cases}$

Solution We start by finding

$$D = \begin{vmatrix} 3 & -1 & 1 \\ 1 & 1 & 0 \\ 2 & 0 & -3 \end{vmatrix} = 3(-3) - (-1)(-3) + 1(-2) = -9 - 3 - 2 = -14$$

Next,

$$D_x = \begin{vmatrix} 6 & -1 & 1 \\ 3 & 1 & 0 \\ 4 & 0 & -3 \end{vmatrix} = 6(-3) - (-1)(-9) + 1(-4) = -18 - 9 - 4 = -31$$

$$D_y = \begin{vmatrix} 3 & 6 & 1 \\ 1 & 3 & 0 \\ 2 & 4 & -3 \end{vmatrix} = 3(-9) - (6)(-3) + 1(-2) = -27 - (-18) - 2 = -11$$

and, finally,

$$D_z = \begin{vmatrix} 3 & -1 & 6 \\ 1 & 1 & 3 \\ 2 & 0 & 4 \end{vmatrix} = 3(4) - (-1)(-2) + 6(-2) = 12 - 2 - 12 = -2$$

The solution is $\left(\frac{31}{14}, \frac{11}{14}, \frac{2}{14}\right)$. ■

Using the Graphics Calculator

We can use the graphics calculator to compute the determinant of a square matrix. To see this, enter the coefficient matrix of Example 2 for [A]. From the home screen, use the MATRX key select the det command. Complete the command to read det [A] and then press the ENTER key. The value -14 is output.

GIVE IT A TRY

11. For the linear system $\begin{cases} 2x - y + z = 1 \\ y - z = 4 \\ z = 2 \end{cases}$

report the values for D, D_x, D_y, and D_z. What is the solution to the system?

Coordinate Geometry We next consider some applications of determinants to coordinate geometry. We start by stating a method of using determinants to find the area measure of a triangle when given the coordinates of the vertices.

AREA OF A TRIANGLE

For a triangle with vertices (x_1, y_1), (x_2, y_2), and (x_3, y_3), and the matrix

$$A = \begin{bmatrix} x_1 & y_1 & 1 \\ x_2 & y_2 & 1 \\ x_3 & y_3 & 1 \end{bmatrix}$$

if D is the determinant of A, then the area measure is $\frac{1}{2}|D|$.

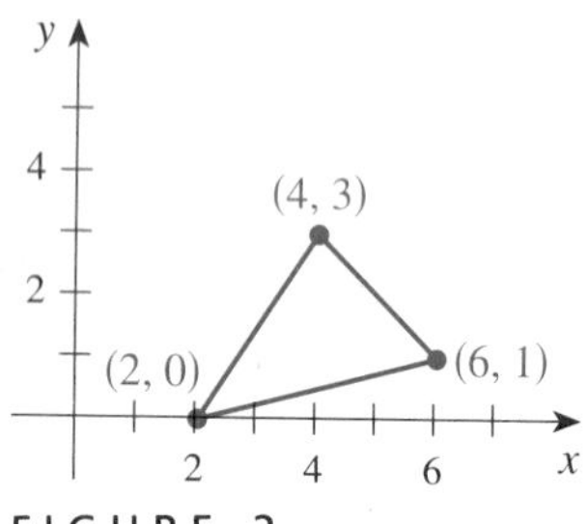

FIGURE 2

In this area property, the determinant can be either positive, negative, or zero. So, the absolute value is used to ensure a nonnegative area measure. If the determinant is zero, then the three points are collinear. Consider the three points shown in Figure 2. The area measure of the triangle formed is found as follows:

$$\begin{aligned} D = \begin{vmatrix} 2 & 0 & 1 \\ 4 & 3 & 1 \\ 6 & 1 & 1 \end{vmatrix} &= 2\begin{vmatrix} 3 & 1 \\ 1 & 1 \end{vmatrix} - 0\begin{vmatrix} 4 & 1 \\ 6 & 1 \end{vmatrix} + 1\begin{vmatrix} 4 & 3 \\ 6 & 1 \end{vmatrix} \\ &= 2(3 - 1) - 0(4 - 6) + 1(4 - 18) \\ &= 2(2) - 0 + 1(-14) \\ &= 4 - 14 \\ &= -10 \end{aligned}$$

So, the area A is $\frac{1}{2}|-10| = 5$ square units.

We next state a test to determine whether three points are collinear. (We know that any two points in the plane are collinear, that is, they lie on a line).

TEST FOR COLLINEAR POINTS

For the three points (x_1, y_1), (x_2, y_2), and (x_3, y_3) to be collinear, the determinant of

$$A = \begin{bmatrix} x_1 & y_1 & 1 \\ x_2 & y_2 & 1 \\ x_3 & y_3 & 1 \end{bmatrix}$$

must equal 0.

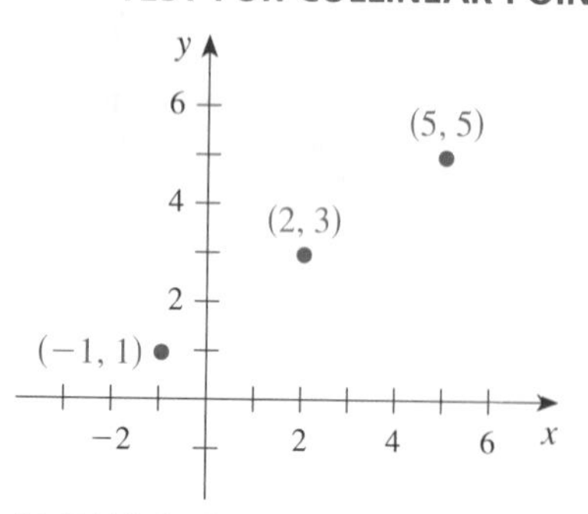

FIGURE 3

As an example of this test, consider the points with coordinates (2, 3), (−1, 1), and (5, 5), as shown in Figure 3. Do these points lie on one line?

To find out, we compute

$$\begin{vmatrix} 2 & 3 & 1 \\ -1 & 1 & 1 \\ 5 & 5 & 1 \end{vmatrix} = 2\begin{vmatrix} 1 & 1 \\ 5 & 1 \end{vmatrix} - 3\begin{vmatrix} -1 & 1 \\ 5 & 1 \end{vmatrix} + 1\begin{vmatrix} -1 & 1 \\ 5 & 5 \end{vmatrix}$$
$$= 2(1 - 5) - 3(-1 - 5) + 1(-5 - 5)$$
$$= 2(-4) - 3(-6) + 1(-10)$$
$$= -8 + 18 - 10$$
$$= 0$$

Thus, we conclude that the three points are collinear.

Of course, once we find that three points are collinear, it is natural to ask, "What is the equation of the line?" The next property provides a link between the determinant and the equation of a line.

DETERMINANTS AND THE EQUATION OF A LINE

For two points (x_1, y_1) and (x_2, y_2), the equation of the line that contains the points is

$$\begin{vmatrix} x & y & 1 \\ x_1 & y_1 & 1 \\ x_2 & y_2 & 1 \end{vmatrix} = 0$$

As an example of this property, consider finding the equation of the line through the points (2, 3) and (−1, 1). We first set up the determinant equation:

$$\begin{vmatrix} x & y & 1 \\ 2 & 3 & 1 \\ -1 & 1 & 1 \end{vmatrix} = 0$$

Finding the determinant, we get

$$D = x\begin{vmatrix} 3 & 1 \\ 1 & 1 \end{vmatrix} - y\begin{vmatrix} 2 & 1 \\ -1 & 1 \end{vmatrix} + 1\begin{vmatrix} 2 & 3 \\ -1 & 1 \end{vmatrix}$$
$$= x(3 - 1) - y(2 - (-1)) + 1(2 - (-3))$$
$$= 2x - 3y + 5$$

Thus, we get the equation $2x - 3y + 5 = 0$.

Using the Graphics Calculator

Rewrite the equation $2x - 3y + 5 = 0$ for y to get $y = \frac{2}{3}x + \frac{5}{3}$. Type `(2/3)X+(5/3)` for `Y1` and view the graph screen. Use the `TRACE` feature to convince yourself that the points (2, 3) and (−1, 1) are on the line. Also, check that the point (5, 5) is on this line.

GIVE IT A TRY

12. Find the area measure of the triangle with vertices $(2, -1)$, $(5, 6)$, and $(-1, 3)$.
13. Are the points $(-1, 0)$, $(3, -1)$, and $(7, -2)$ collinear?
14. Use determinants to find the equation of the line passing through the points $(3, 4)$ and $(-2, -1)$.

■ SUMMARY

■ Quick Index

In this section, we have introduced the determinant of a square matrix, which is always a number. To find the determinant, we use minors and cofactors. Using Cramer's Rule, determinants provide yet another approach to solving linear systems. For example, the solution to a linear system of two equations in two variables is $\left(\frac{D_x}{D}, \frac{D_y}{D}\right)$, where $D \neq 0$ and D is the determinant of the coefficient matrix. The number D_x is the determinant of the coefficient matrix, where the x-column has been replaced with the constant terms of the equations. Likewise, the number D_y is the determinant of the coefficient matrix, where the y-column has been replaced with the constant terms of the equations. If $D = 0$, then the system is either inconsistent or dependent.

Determinants find applications beyond solving linear systems. One such application is finding the area measure of a triangle when the coordinates of the vertices of the triangle are known. In a closely related area, determinants are used to test three points to determine whether they are collinear. Finally, determinants can be used to find the equation of a line through two points.

11.4 ■ ■ ■ EXERCISES

Find the determinant of each matrix. Use the graphics calculator to check your work in Problems 1–6.

1. $[-15]$
2. $[20]$
3. $\begin{bmatrix} -1 & 7 \\ 2 & -8 \end{bmatrix}$
4. $\begin{bmatrix} 4 & 5 \\ 9 & 2 \end{bmatrix}$
5. $\begin{bmatrix} -2 & 6 \\ 0 & 9 \end{bmatrix}$
6. $\begin{bmatrix} 10 & 3 \\ -7 & 3 \end{bmatrix}$
7. $\begin{bmatrix} 1 + \sqrt{7} & \sqrt{7} \\ \sqrt{7} & 1 - \sqrt{7} \end{bmatrix}$
8. $\begin{bmatrix} \sqrt{3} + \sqrt{2} & \sqrt{2} \\ \sqrt{3} & \sqrt{3} + \sqrt{2} \end{bmatrix}$

For $\begin{bmatrix} 5 & 1 & -3 \\ -4 & 2 & 2 \\ 3 & 0 & -1 \end{bmatrix}$, *find each quantity.*

9. M_{21}
10. C_{21}
11. M_{12}
12. C_{12}
13. M_{33}
14. C_{33}
15. For $A = \begin{bmatrix} 5 & 2 & 1 \\ -3 & 0 & 2 \\ 2 & 1 & 7 \end{bmatrix}$, find $\det(A)$.
16. For $A = \begin{bmatrix} -3 & 0 & 4 \\ -2 & 2 & 2 \\ -5 & 1 & 0 \end{bmatrix}$, find $|A|$.

Write each determinant as a standard polynomial.

17. $\begin{vmatrix} x-2 & x-2 \\ x & x+1 \end{vmatrix}$

18. $\begin{vmatrix} x^2 & -2 \\ x^2-x & x+1 \end{vmatrix}$

19. $\begin{vmatrix} x & y & 1 \\ -2 & 5 & 1 \\ 1 & -1 & 1 \end{vmatrix}$

20. $\begin{vmatrix} x & y & 1 \\ 3 & -2 & 1 \\ 0 & 1 & 1 \end{vmatrix}$

21. $\begin{vmatrix} x & y & 1 \\ 5 & 1 & 1 \\ 2 & -1 & 1 \end{vmatrix}$

22. $\begin{vmatrix} x & y & 1 \\ 2 & 3 & 1 \\ -1 & 5 & 1 \end{vmatrix}$

Use Cramer's Rule to solve the given linear system. In addition to the solution, report D, D_x, and D_y. Use the graphics calculator to check your work.

23. $\begin{cases} 3x - 5y = 9 \\ -x + 2y = 1 \end{cases}$

24. $\begin{cases} 4x - y = 6 \\ 7x + 2y = 8 \end{cases}$

25. $\begin{cases} 3x + 4y = 18 \\ 2x - 5y = 15 \end{cases}$

26. $\begin{cases} 2x - 9y = 16 \\ x - 5y = 12 \end{cases}$

27. $\begin{cases} y = -3x + 5 \\ y = \frac{-1}{2}x - 2 \end{cases}$

28. $\begin{cases} y = \frac{3}{5}x - 3 \\ y = -3x + 2 \end{cases}$

Use Cramer's Rule to solve the given linear system. In addition to the solution, report D, D_x, D_y, and D_z. Use the graphics calculator to check your work.

29. $\begin{cases} 3x - 5y + z = 9 \\ -x + 2y - z = 1 \\ 2x + 3z = 6 \end{cases}$

30. $\begin{cases} 4x - y - z = 6 \\ x + 2y + z = 8 \\ y - z = 3 \end{cases}$

31. $\begin{cases} x + 4y + 3z = 18 \\ 2x - 5y = 15 \\ z = x - y + 2 \end{cases}$

32. $\begin{cases} x - y + 2z = 6 \\ x - 2y - z = 10 \\ x = 5 - x - z \end{cases}$

33. $\begin{cases} y = -3x + z + 5 \\ y = x - 2z - 1 \\ z = 2x - y + 3 \end{cases}$

34. $\begin{cases} y = z - x - 3 \\ y = z - 3x + 2 \\ y = 2x - z - 4 \end{cases}$

Use determinants to find the area measure of the triangle with the given vertices.

35. $(-1, 5)$, $(3, 1)$, $(4, 3)$
36. $(6, -1)$, $(0, 0)$, $(-1, 3)$
37. $\left(3, \sqrt{2}\right)$, $\left(-1, \sqrt{3}\right)$, $\left(0, -\sqrt{6}\right)$
38. $\left(-1 + \sqrt{3}, 2\right)$, $\left(-1 - \sqrt{3}, 0\right)$, $\left(\sqrt{3}, 3\right)$

Use determinants to determine if the given points are collinear. If they are collinear, report the equation of the line. Use the graphics calculator to check your work.

39. $(-3, 5)$, $(2, 4)$, $(7, 3)$
40. $(6, -3)$, $(0, 1)$, $(3, -1)$
41. $(-2, 5)$, $(6, 0)$, $(1, 4)$
42. $(-3, -1)$, $(2, 3)$, $(5, 7)$

43. Use determinants to find the equation of the line that passes through points $(2, -3)$ and $(-3, 1)$.

44. The determinant of the following matrix has the same value, regardless of the values used for x, y, and z. What is the determinant?

$$\begin{bmatrix} 3 & x & y \\ 0 & -2 & z \\ 0 & 0 & 5 \end{bmatrix}$$

CHAPTER 11 REVIEW EXERCISES

Report true *for a true statement or* false *for a false statement.*

1. The ordered triple $(-2, 1, 5)$ is in the solution to the equation $2x - 5y + z = 4$.
2. The graph of the solution to $2x - y + z = 8$ has x-intercept $(0, -8, 0)$.
3. If the solution to a linear system of equations in three variables is *none*, the system is dependent.
4. A linear system can be both dependent and consistent.
5. To find the equation of the parabola through three points in the plane, a linear system in three equations and three variables is solved.
6. A square matrix is a matrix whose entries are perfect squares, such as 9, 4, and 25.
7. Scalar multiplication refers to finding the product of two 2×2 matrices.
8. The additive identity for $n \times n$ matrices is the $n \times n$ matrix with all entries 0.
9. The multiplicative identity for $n \times n$ matrices is the $n \times n$ matrix with all entries 1.
10. The product of a 3×5 matrix and a 5×6 matrix is a 5×5 matrix.

11. For all $n \times n$ matrices, the multiplication operation is commutative.

12. The determinant of a 2×2 matrix is a 2×2 matrix.

Use the substitution approach to solve each linear system. Use the graphics calculator to check your work.

13. $\begin{cases} 5x + 3y = 30 \\ 2x - 5y = 80 \end{cases}$

14. $\begin{cases} 2x - 3y = 8 \\ 3x + 4y = 12 \end{cases}$

15. $\begin{cases} x - 3y + z = 9 \\ 2x + y - 3z = 8 \\ 3x - y + 2z = 6 \end{cases}$

16. $\begin{cases} x - y + z = 8 \\ 2x + y - z = 5 \\ x + 2y + 3z = 2 \end{cases}$

17. Find the equation of the parabola in the form $y = ax^2 + bx + c$ that passes through the points $(2, -1)$, $(-3, -2)$, and $(-1, -2)$.

18. Find the equation of the parabola that has a vertical line of symmetry and passes through the points $(4, -2)$, $(-3, 4)$, and $(1, 4)$.

19. Find the equation of the parabola in the form $y = ax^2 + bx + c$ that passes through the points $(2, -1)$, $(0, 4)$, and $(-2, -2)$.

20. Find the equation of the parabola in the form $y = ax^2 + bx + c$ that passes through the points $(3, 1)$, $(1, 6)$, and $(-1, 1)$.

Use the matrix

$$\begin{bmatrix} 5 & -8 & 1 \\ 2 & 0 & 2 \\ 7 & 9 & -3 \\ 2 & -3 & 5 \end{bmatrix}$$

to answer Problems 21–24.

21. Report the size of the matrix.

22. Report the entry whose location is row 4, column 2.

23. The entry 9 is in which row and which column?

24. Report a_{42}, a_{23}, and a_{13}.

Find the values of the variables x, y, z, and w so that the following matrices are equal.

25. $A = \begin{bmatrix} -2 & x - 1 \\ w & 2z + 1 \end{bmatrix}$ and $B = \begin{bmatrix} y & 3x - 1 \\ 2w & 7 \end{bmatrix}$

26. $P = \begin{bmatrix} -6 & y^2 - 1 \\ 2z & 3w + 6 \\ -x & 20 \end{bmatrix}$ and

$$Q = \begin{bmatrix} -2y & 3y - 1 \\ z & -2 \\ 2x + 1 & 20 \end{bmatrix}$$

For $C = \begin{bmatrix} -2 & 3 & 10 \\ 5 & 8 & 4 \end{bmatrix}$ *and* $D = \begin{bmatrix} 9 & 1 & 7 \\ -5 & -6 & 3 \end{bmatrix}$, *use the graphics calculator to find each of the following.*

27. $5C$

28. $C + 2D$

29. $C - D$

30. $D - C$

31. $C - 3C$

32. $D - 2C + 2D$

33. For $A = [9 \quad -2 \quad -3]$ and $B = \begin{bmatrix} 1 \\ -2 \\ -5 \end{bmatrix}$, find the inner product.

For $A = \begin{bmatrix} 8 & -3 \\ 4 & 1 \end{bmatrix}$ and $B = \begin{bmatrix} 4 & 0 & -3 \\ 0 & -1 & 2 \end{bmatrix}$, find AB.

For $C = \begin{bmatrix} -1 & 2 \\ 3 & 1 \end{bmatrix}$ *and* $D = \begin{bmatrix} 2 & -5 \\ 2 & 0 \end{bmatrix}$, *use the graphics calculator to find each quantity.*

35. CD

36. DC

37. D^2

38. $D^2 + 2DC + C^2$

For $A = \begin{bmatrix} -1 & 0 & -1 \\ 2 & 1 & 5 \\ 8 & 3 & 0 \end{bmatrix}$ *and* $B = \begin{bmatrix} 0 & -6 & 3 \\ 1 & 3 & 2 \\ -1 & 0 & 1 \end{bmatrix}$

use the graphics calculator to find each quantity.

39. AB

40. BA

41. $A^2 - 2AB + B^2$

42. $B^2 - A^2$

Solve each system using the graphics calculator and matrix inverses.

43. $\begin{cases} 2x - 5y = 8 \\ 3x + y = 2 \end{cases}$

44. $\begin{cases} -3x + y = -2 \\ x + y = 3 \end{cases}$

45. $\begin{cases} x + 3y + z = 4 \\ x - 2y - z = 6 \\ 2x + z = 4 \end{cases}$

46. $\begin{cases} x + y + 2z = 8 \\ 3x - 2y = 5 \\ 3y - z = 3 \end{cases}$

47. A company produces three types of scanners, the V-35, the V-70, and the V-70X. The V-35 requires 2 hours to assemble, 3 hours to test, and 1 hour to package. The V-70 requires 3 hours to assemble, 1 hour to test, and 1 hour to package. The V-70X requires 4 hours to assemble and 2 hours to package (no testing is required). If, for a particular week, 329 hours of labor were used in assembling scanners, 156 hours were used in testing, and 148 hours were used in packaging scanners, how many of each type of scanner were produced?

48. How many liters of 10% acid solution, 90% acid solution, and 70% acid solution must be mixed to produce 100 liters of 65% acid solution if the amount of 90% acid solution used must be twice the amount of 10% acid solution used?

49. An answering service dispatches calls to three remote locations. For a particular day, 500 calls are dispatched. The number of calls sent to remote location Y equals the calls sent to both remote locations X and Z. The combined calls dispatched to Y and Z totals 320. How many calls were dispatched to each of the three remote locations?

50. The Blue Jays charge $10 for midfield seats, $7 for the 40-to-goal line seats, and $3 for end zone seats. After two days of ticket sales, 540 tickets have been sold, with total sales of $3,470. The number of 40-to-goal line seats sold equals the combined sales for the other two types of tickets. How many of each type of seat has been sold?

For $A = \begin{bmatrix} -2 & -3 \\ 2 & 2 \end{bmatrix}$, $X = \begin{bmatrix} x \\ y \end{bmatrix}$, *and* $C = \begin{bmatrix} -2 \\ 5 \end{bmatrix}$, *solve the matrix equation.*

51. $AX = C$

52. $-1AX = 2C$

For $C = \begin{bmatrix} 2 & 0 & 1 \\ -1 & 1 & 2 \\ 2 & 2 & 3 \end{bmatrix}$, $X = \begin{bmatrix} x \\ y \\ z \end{bmatrix}$ *and* $D = \begin{bmatrix} 8 \\ -1 \\ 4 \end{bmatrix}$, *solve the matrix equation.*

53. $CX = D$

54. $C^2X = CD$

Use multiplication to determine whether the given matrices are inverses. If they are inverse, state this. If not, report their product.

55. $A = \begin{bmatrix} 1 & 2 \\ 0 & 3 \end{bmatrix}$ $B = \begin{bmatrix} 9 & -5 \\ -4 & 0 \end{bmatrix}$

56. $E = \begin{bmatrix} 2 & -1 & 4 \\ -2 & 0 & -2 \\ 5 & 1 & 3 \end{bmatrix}$ $F = \begin{bmatrix} 1 & 1 & -2 \\ 2 & -1 & 2 \\ 1 & 0 & 1 \end{bmatrix}$

Use the graphics calculator to find inverse, if it exists, of each matrix. If the inverse does not exist, report that the matrix is singular.

57. $\begin{bmatrix} -1 & 2 \\ 3 & -4 \end{bmatrix}$

58. $\begin{bmatrix} 5 & -1 \\ 6 & -1 \end{bmatrix}$

59. $\begin{bmatrix} 2 & 0 & 1 \\ 1 & 3 & 0 \\ 0 & 2 & 1 \end{bmatrix}$

60. $\begin{bmatrix} 0 & 3 & -1 \\ 2 & 0 & 1 \\ 1 & 3 & 2 \end{bmatrix}$

61. Solve: $\begin{bmatrix} 4 & 1 & 2 \\ 1 & -2 & 0 \\ 0 & 2 & 1 \end{bmatrix} \begin{bmatrix} x \\ y \\ z \end{bmatrix} = \begin{bmatrix} 4 \\ 6 \\ 5 \end{bmatrix}$

Solve each system using a matrix equation. Let A be the coefficient matrix. Report matrix A, A^{-1}, *and the solution.*

62. $\begin{cases} 5x - 3y = 8 \\ 2x + y = 5 \end{cases}$

63. $\begin{cases} x - 2y = 6 \\ 3x + y = 8 \end{cases}$

64. If

$$A = \begin{bmatrix} 3 & 2 & 1 \\ -2 & 1 & 0 \\ 2 & 3 & 1 \end{bmatrix}, \text{ then } A^{-1} = \begin{bmatrix} -1 & -1 & 1 \\ -2 & -1 & 2 \\ 8 & 5 & -7 \end{bmatrix}$$

Use this fact to solve the system $\begin{cases} 3x + 2y + z = 5 \\ -2x + y = 4 \\ 2x + 3y + z = 10 \end{cases}$

Find the determinant of each matrix. Use the graphics calculator to check your work in Problems 65–68.

65. $[15]$

66. $[-3]$

67. $\begin{bmatrix} -8 & 4 \\ 2 & -1 \end{bmatrix}$

68. $\begin{bmatrix} 6 & 4 \\ 3 & 2 \end{bmatrix}$

69. $\begin{bmatrix} \sqrt{6} & 2 + \sqrt{3} \\ 1 - \sqrt{3} & \sqrt{2} \end{bmatrix}$

70. $\begin{bmatrix} 2 + \sqrt{3} & \sqrt{6} \\ \sqrt{3} & 2 + \sqrt{3} \end{bmatrix}$

Use determinants to solve the given linear system. In addition to the solution, report D, D_x, *and* D_y. *Use the graphics calculator to check your work.*

71. $\begin{cases} 2x - 3y = 9 \\ 5x + y = 2 \end{cases}$

72. $\begin{cases} 5x - 6y = 4 \\ 2x + 3y = 5 \end{cases}$

Use determinants to solve the given linear system. Along with the solution report D, D_x, D_y, *and* D_z. *Use the graphics calculator to check your work.*

73. $\begin{cases} x - 5y + 2z = 9 \\ -x + 4y - z = 5 \\ 3x + 3y = 6 \end{cases}$

74. $\begin{cases} x - 2y - 3z = 6 \\ y + 2z = 8 \\ 3x - y - z = 2 \end{cases}$

Use determinants to find the area measure of the triangle with the given vertices.

75. $(-3, 4)$, $(-2, 1)$, $(5, 3)$

76. $(3, \sqrt{3})$, $(1, \sqrt{2})$, $(2, -\sqrt{5})$

Use determinants to determine if the given points are collinear. If so, report the equation of the line.

77. $(-1, 8)$, $(3, -4)$, $(5, -10)$

78. $(-3, 13)$, $(1, 1)$, $(3, -5)$

79. Use determinants to find the equation of the line that passes through the points $(4, 5)$ and $(-2, -1)$.

80. Use determinants to find the equation of the line that passes through the points $(2, -1)$ and $(-5, -5)$.

CHAPTER 11 TEST

1. Use the substitution approach to solve the given linear system, and use the graphics calculator to check your work:

$$\begin{cases} x - 2y + z = 4 \\ 2x - y + 3z = 1 \\ 3x + y + z = 3 \end{cases}$$

2. Consider the matrix

$$\begin{bmatrix} -2 & -3 & 3 \\ 3 & 0 & 1 \\ 5 & 4 & -3 \\ -1 & -5 & -1 \end{bmatrix}$$

   a. Report the size of the matrix.

   b. Report the entry whose location is row 4, column 2.

   c. The entry 0 is in which row and which column?

   d. Report a_{32}, a_{13}, and a_{43}.

3. Solve the given system using the graphics calculator to find the inverse of the coefficient matrix. Report the coefficient matrix, its inverse, and the solution to the system.

$$\begin{cases} -3x + y = -2 \\ x - y = 4 \end{cases}$$

4. A solution of 100 liters of 60% acid is made by mixing 20% acid solution, 90% acid solution, and 40% acid solution. The amount of 90% acid solution used must be twice the amount of 20% acid solution used. List the variables for this problem, and what they represent. Report the system of linear equations needed to model the situation.

5. For $A = \begin{bmatrix} 8 & -3 \\ 4 & 1 \end{bmatrix}$ and $B = \begin{bmatrix} 4 & 0 & -3 \\ 0 & -1 & 2 \end{bmatrix}$, find AB.

6. If

$$A = \begin{bmatrix} 1 & 2 & -1 \\ 2 & 5 & 1 \\ 3 & 6 & -2 \end{bmatrix}, \text{ then } A^{-1} = \begin{bmatrix} -16 & -2 & 7 \\ 7 & 1 & -3 \\ -3 & 0 & 1 \end{bmatrix}$$

   Use this fact to solve the system

$$\begin{cases} x + 2y - z = 5 \\ 2x + 5y + z = 4 \\ 3x + 6y - 2z = 1 \end{cases}$$

7. Use determinants to solve the given linear system. In addition to the solution, report D, D_x, and D_y. Use the graphics calculator to check your work.

$$\begin{cases} 3x - y = 6 \\ 2x + 5y = 4 \end{cases}$$

12

CONIC SECTIONS AND NONLINEAR SYSTEMS

We begin this chapter by studying sections of the cone—conic sections. These sections are actually curves that are produced by intersecting (cutting) a cone with a plane. The figure shows such a cone.

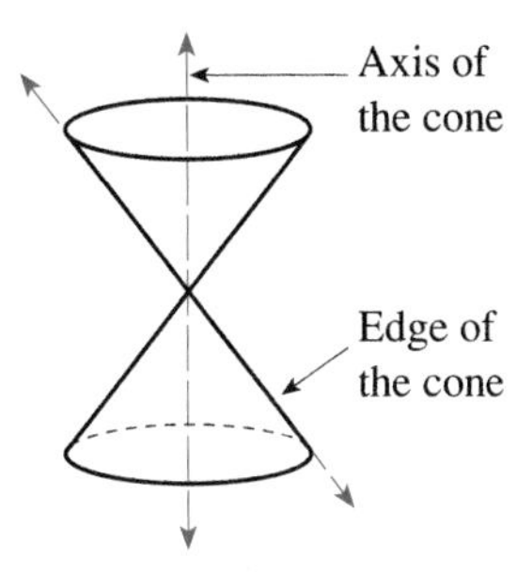

The parabola is one such curve formed by this method. We have become familiar with this curve from our earlier work with quadratic functions. Here, we will also study the parabola generated by the solution to an equation, such as $x = y^2 + 2$. In addition to the parabola, we will study circles, ellipses, and hyperbolas. These curves are generated by graphing the solutions to equations of the form $Ax^2 + By^2 = 1$. We will also extend our earlier work (from Chapter 4) on systems of equations. This time, one or more of the equations in the system will be a nonlinear equation, such as $y = x^2 - 2x + 1$. We start the chapter by reviewing the parabola.

12.1 ■ ■ ■ PARABOLAS

When a cone is intersected (cut) by a plane that is parallel to its edge, a *parabola* is generated. One such parabola is shown in Figure 1. Another definition of a parabola is that it is the set of points in the plane that are equidistant from a given point (known as the *focus*) and a given line (known as the *directrix*). These are shown in Figure 2.

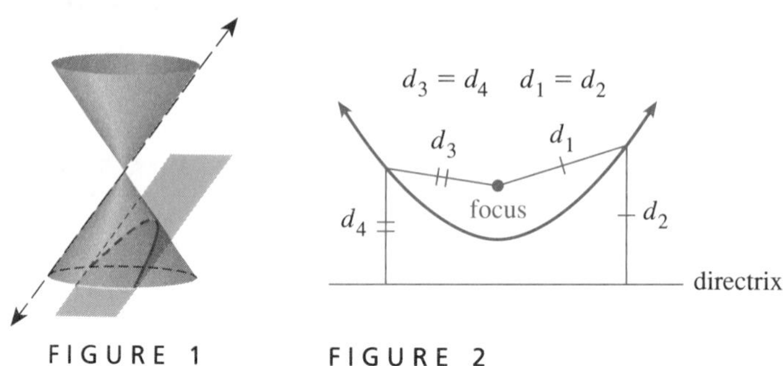

FIGURE 1 FIGURE 2

The definition we want to work with in this section involves equations in two variables and the graph of the solution. In Section 5.7, we learned that the solution to an equation such as $y = x^2 - 3x - 4$ is a set of ordered pairs. The curve generated when these ordered pairs are plotted on a set of axes is a parabola. In general, the solution to a second-degree equation in two variables is a *conic section*—a parabola, a circle, an ellipse, or a hyperbola. If the equation is of second degree in just one of the two variables, then the graph of its solution is a parabola.

We start by reviewing the conclusions from our earlier work on quadratic functions (in Sections 5.7 and 8.4). From this work we have the following properties.

THE PARABOLA FOR $y = ax^2 + bx + c$

The graph of the solution to an equation in the form $y = ax^2 + bx + c$ is a parabola.

If the equation is written in the form $y = a(x - h)^2 + k$, then the vertex of the parabola is (h, k). The line of symmetry is $x = h$.

For the equation in the form $y = ax^2 + bx + c$, the vertex has x-coordinate $\frac{-b}{2a}$ and the line of symmetry is the vertical line determined by $x = \frac{-b}{2a}$.

If $a > 0$, then the parabola opens upward and the vertex is a minimum point.

If $a < 0$, then the parabola opens downward and the vertex is a maximum point.

As a review of these properties, the following example involves graphing the parabola determined by an equation in two variables.

EXAMPLE 1 For $y = -x^2 + 2x + 3$, identify the intercepts, the vertex, and the line of symmetry for the graph. Also, sketch the graph.

Solution We start by finding the intercepts. To find the y-intercept, replace the x-variable with 0, to get $y = 3$. Thus, the y-intercept is (0, 3).

To find the x-intercept, replace the y-variable with 0, to get $0 = -x^2 + 2x + 3$. Solving this equation, we get

$$x^2 - 2x - 3 = 0$$
$$(x - 3)(x + 1) = 0$$

$$x - 3 = 0 \qquad \text{or} \qquad x + 1 = 0$$
$$x = 3 \qquad\qquad\qquad x = -1$$

The x-intercepts are (3, 0) and (−1, 0).

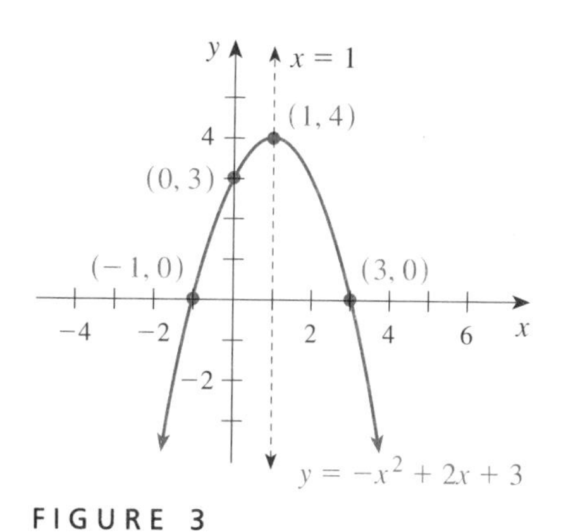

FIGURE 3

Knowing that the vertex has x-coordinate $\dfrac{-b}{2a}$, we get $\dfrac{-(2)}{2(-1)} = 1$ for the x-coordinate of the vertex. Using this x-value, we obtain

$$y = -(1)^2 + 2(1) + 3 \qquad \text{or} \qquad y = 4$$

So the vertex is (1, 4).

Knowing that the line of symmetry passes through the vertex, we can write the equation of this vertical line, $x = 1$.

Finally, observing that a (the x^2-coefficient) is less than 0, we determine that this parabola opens downward.

Plotting this information and sketching the parabola, we get the graph shown in Figure 3. ■

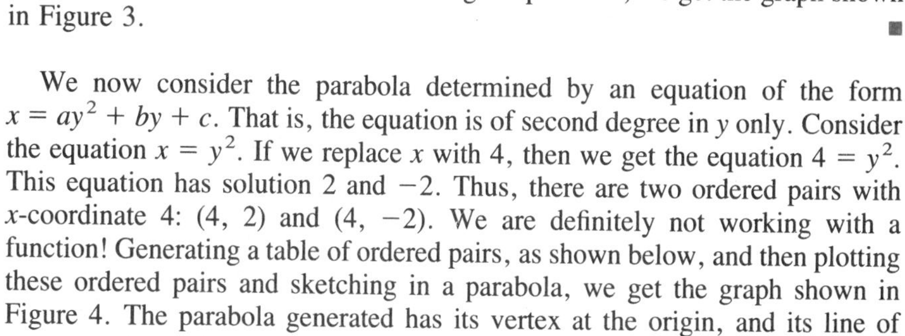

We now consider the parabola determined by an equation of the form $x = ay^2 + by + c$. That is, the equation is of second degree in y only. Consider the equation $x = y^2$. If we replace x with 4, then we get the equation $4 = y^2$. This equation has solution 2 and −2. Thus, there are two ordered pairs with x-coordinate 4: (4, 2) and (4, −2). We are definitely not working with a function! Generating a table of ordered pairs, as shown below, and then plotting these ordered pairs and sketching in a parabola, we get the graph shown in Figure 4. The parabola generated has its vertex at the origin, and its line of symmetry is the horizontal line $y = 0$.

Ordered pairs that are in the solution to $x = y^2$

x	y	(x, y)
4	2	(4, 2)
4	−2	(4, −2)
1	1	(1, 1)
1	−1	(1, −1)
0	0	(0, 0)

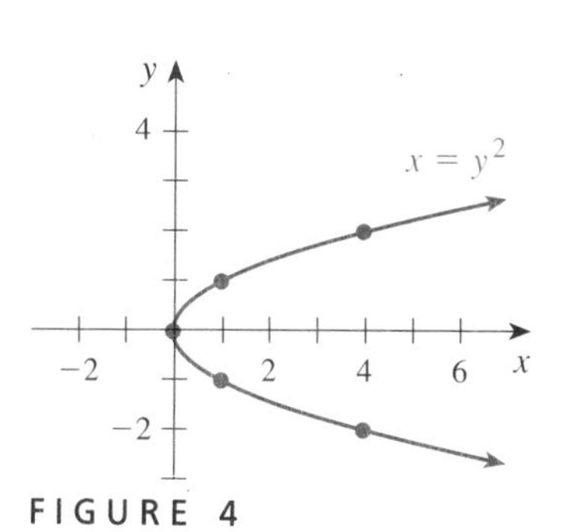

FIGURE 4

In general, an equation of the form $x = ay^2 + by + c$ has a graph that is a parabola with a horizontal line of symmetry. We have the following generalization.

THE PARABOLA FOR $x = ay^2 + by + c$

> The graph of the solution to an equation in the form $x = ay^2 + by + c$ is a parabola.
>
> If the equation is written in the form $x = a(y - k)^2 + h$, then the vertex of the parabola is (h, k). The line of symmetry is $y = k$.
>
> If the equation is written in the form $x = ay^2 + by + c$, then the vertex has y-coordinate $\frac{-b}{2a}$ and the line of symmetry is the horizontal line determined by $y = \frac{-b}{2a}$.
>
> If $a > 0$, then the parabola opens to the right.
> If $a < 0$, then the parabola opens to the left.

As an illustration of graphing the parabola determined by an equation that is second degree in y only, work through the following example.

EXAMPLE 2 For $x = y^2 + 2y - 3$, identify the intercepts, the vertex, and the line of symmetry for the graph. Also, sketch the graph.

Solution We start by finding the intercepts. To find the y-intercept, replace the x-variable with 0, to get $0 = y^2 + 2y - 3$. Solving this equation, we get

$$y^2 + 2y - 3 = 0$$
$$(y + 3)(y - 1) = 0$$

$$y + 3 = 0 \quad \text{or} \quad y - 1 = 0$$
$$y = -3 \qquad\qquad y = 1$$

The y-intercepts are $(0, -3)$ and $(0, 1)$.

To find the x-intercept, we replace the y-variable with 0, to get $x = -3$. Thus, the x-intercept is $(-3, 0)$.

Knowing that the vertex has y-coordinate $\frac{-b}{2a}$, we get $\frac{-2}{2(1)} = -1$ for the y-coordinate of the vertex. Using this y-value, we have

$$x = (-1)^2 + 2(-1) - 3 \quad \text{or} \quad x = -4$$

So, the vertex is $(-4, -1)$.

Knowing that the line of symmetry passes through the vertex, the equation of the horizontal line is $y = -1$.

Finally, observing that a (the y^2-coefficient) is greater than 0, we determine that the parabola opens to the right.

Plotting this information and sketching in the parabola, we get the graph shown in Figure 5. ■

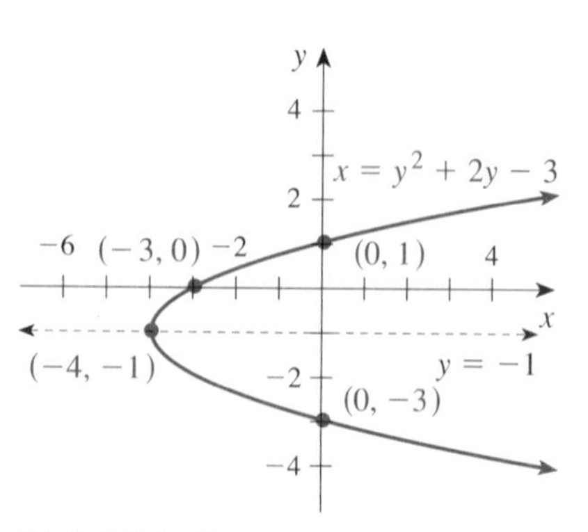

FIGURE 5

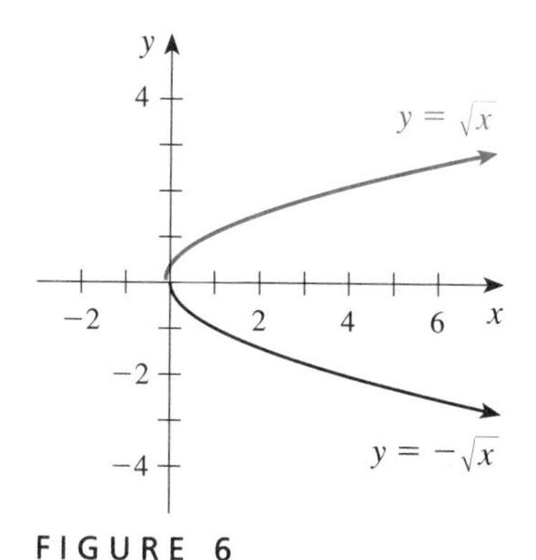

FIGURE 6

A Radical Approach Consider the equation $x = y^2$. Rewriting this equation for y, we get:

$$y = \sqrt{x} \qquad \text{and} \qquad y = -\sqrt{x}$$

Plotting the ordered pairs for these two radical functions, we have the graph shown in Figure 6.

Now, consider the equation $x = y^2 - 2y - 3$. We can rewrite this equation for y, using the completing-the-square technique:

$$\begin{aligned} y^2 - 2y - 3 &= x \\ \underline{\qquad\qquad 1 \quad} &\;\; \underline{1 \qquad} \qquad \text{Add the square of } \tfrac{1}{2}(-2) \text{ to each side.} \\ y^2 - 2y + 1 - 3 &= x + 1 \\ (y - 1)^2 - 3 &= x + 1 \\ \underline{\qquad\qquad 3 \quad} &\;\; \underline{3 \qquad} \qquad \text{Add 3 to each side.} \\ (y - 1)^2 &= x + 4 \end{aligned}$$

$$\begin{aligned} y - 1 &= \sqrt{x + 4} \qquad \text{or} \qquad & y - 1 &= -\sqrt{x + 4} \\ y &= 1 + \sqrt{x + 4} & y &= 1 - \sqrt{x + 4} \end{aligned}$$

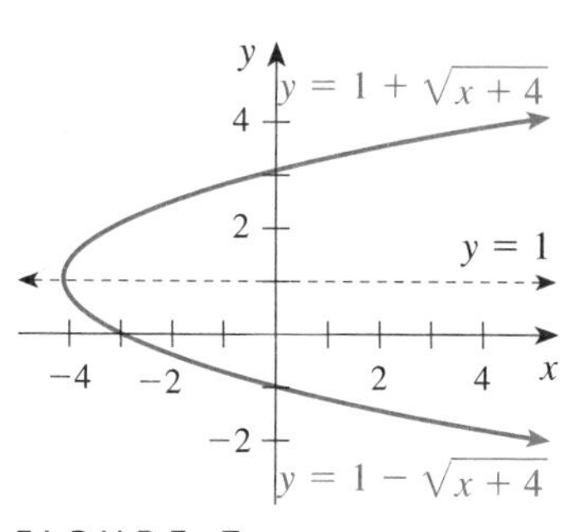

FIGURE 7

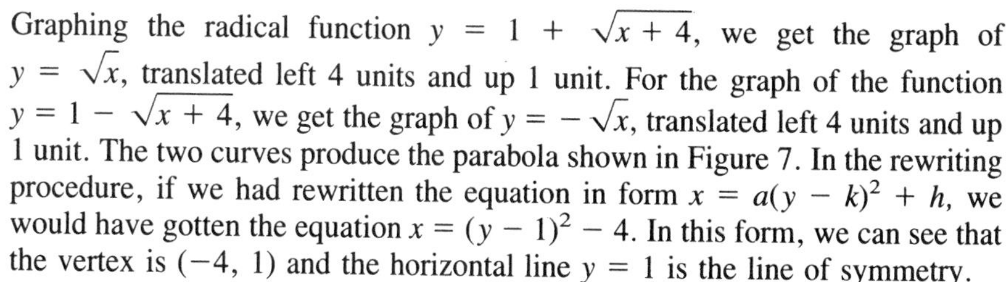
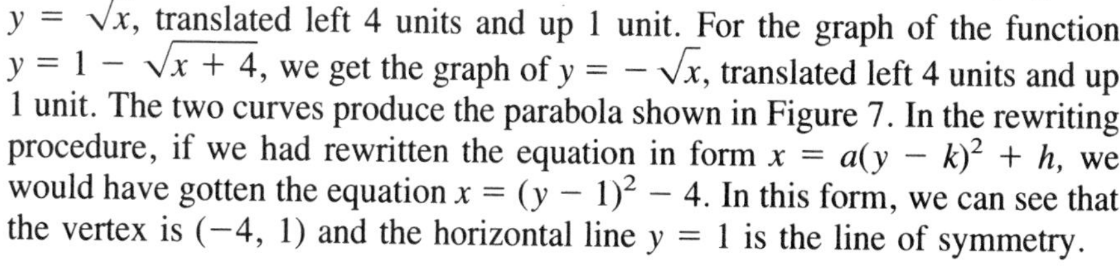

Graphing the radical function $y = 1 + \sqrt{x + 4}$, we get the graph of $y = \sqrt{x}$, translated left 4 units and up 1 unit. For the graph of the function $y = 1 - \sqrt{x + 4}$, we get the graph of $y = -\sqrt{x}$, translated left 4 units and up 1 unit. The two curves produce the parabola shown in Figure 7. In the rewriting procedure, if we had rewritten the equation in form $x = a(y - k)^2 + h$, we would have gotten the equation $x = (y - 1)^2 - 4$. In this form, we can see that the vertex is $(-4, 1)$ and the horizontal line $y = 1$ is the line of symmetry.

Relation to Functions The parabola determined by the equation

$$x = y^2 - 2y - 3$$

is *not* the graph of a function. Check the graph with the vertical line test. Rewriting the equation for y shows that the equation determines two radical functions. The graphs of these radical functions together produce the parabola.

In the next example, we graph the parabola determined by the equation

$$x = -y^2 + 3y + 4$$

Rather than completing the square to rewrite the equation for y (in function form), we will use the quadratic formula.

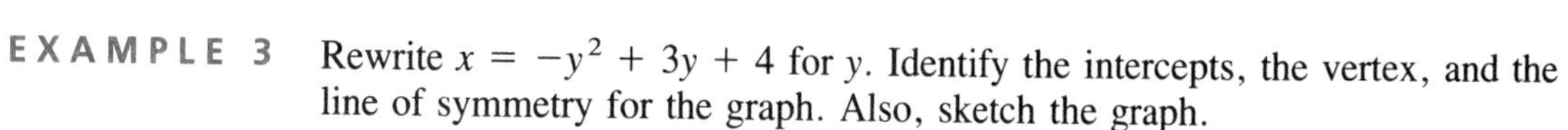

EXAMPLE 3 Rewrite $x = -y^2 + 3y + 4$ for y. Identify the intercepts, the vertex, and the line of symmetry for the graph. Also, sketch the graph.

Solution We start by rewriting the equation in standard form:

$$-y^2 + 3y + 4 = x$$

$$
\begin{array}{ll}
\text{Multiply by } -1: & y^2 - 3y - 4 = -x \\
\text{Add } x \text{ to each side:} & \underline{\, x \quad \; x} \\
\text{Standard form:} & y^2 - 3y - 4 + x = 0
\end{array}
$$

$$
\begin{array}{ll}
\text{Identify } a, b, \text{ and } c: & a = 1,\ b = -3,\ c = x - 4 \\
\text{Compute the discriminant:} & b^2 - 4ac = (-3)^2 - 4(1)(x - 4) \\
& = 9 - 4x + 16 \\
& = -4x + 25
\end{array}
$$

Write the formula: $y = \dfrac{3 + \sqrt{-4x + 25}}{2}$ and $y = \dfrac{3 - \sqrt{-4x + 25}}{2}$

To find the y-intercepts, replace x by 0 in the formulas to get

$$
y = \frac{3 + \sqrt{-4(0) + 25}}{2} = \frac{3 + \sqrt{25}}{2} = \frac{8}{2} = 4 \qquad \text{and} \qquad y = \frac{3 - \sqrt{-4(0) + 25}}{2} = \frac{3 - \sqrt{25}}{2} = \frac{-2}{2} = -1
$$

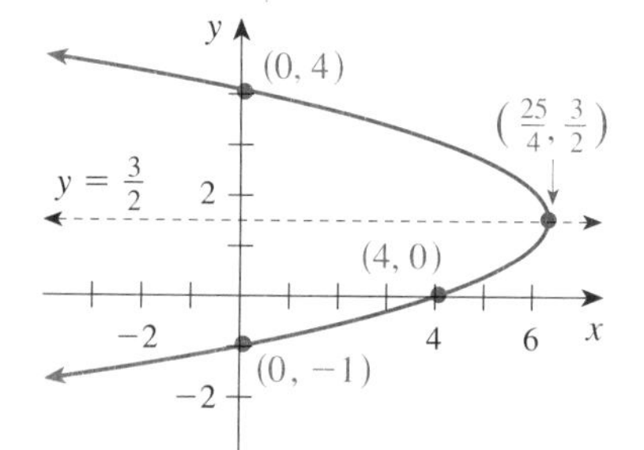

So, the y-intercepts are (0, 4) and (0, −1).

To find the x-intercept, replace y with 0 in the original equation to get $x = 4$. So, the x-intercept is (4, 0).

Using the original equation, $x = -y^2 + 3y + 4$, the y-coordinate of the vertex is $\dfrac{-b}{2a} = \dfrac{-3}{2(-1)} = \dfrac{3}{2}$. Do you see this value in the equations written for y? The x-coordinate of the vertex is $x = -\left(\dfrac{3}{2}\right)^2 + 3\left(\dfrac{3}{2}\right) + 4 = \dfrac{25}{4}$.

The vertex is $\left(\dfrac{25}{4}, \dfrac{3}{2}\right)$ and the line of symmetry is the horizontal line $y = \dfrac{3}{2}$. Using this information, we graph the parabola in Figure 8. ■

Using the Graphics Calculator

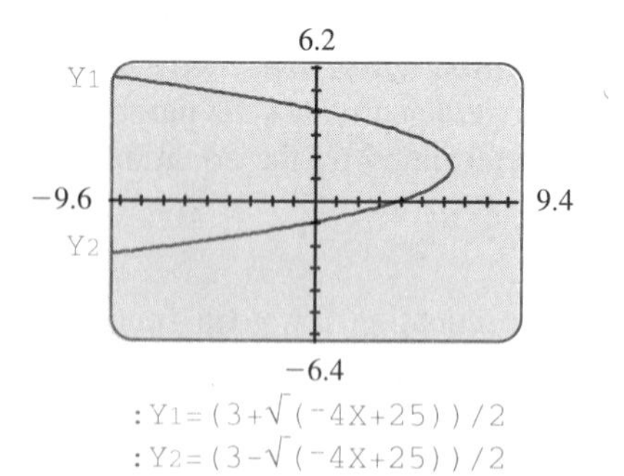

FIGURE 9

Graphics calculators graph only functions. That is, typing an expression for Y1, Y2, Y3, or Y4 is actually typing a function rule. To graph the ordered pairs that solve $x = -y^2 + 3y + 4$, we must first rewrite the equation for y. Using the equations for y produced in Example 3, type (3+√(-4X+25))/2 for Y1 and type (3−√(-4X+25))/2 for Y2. View the graph screen using the Window 9.4 settings. The graph of the radical function for Y1 is drawn, and then the graph of the radical function for Y2 is drawn, as shown in Figure 9. The curve produced is the parabola determined by the equation $x = -y^2 + 3y + 4$. Press the TRACE key. The message yields the y-intercept (0, 4). Press the down arrow key to move the cursor to the graph for the second function. The message yields the second y-intercept (0, −1). Press the right arrow key. The x-intercept is (4, 0). Continue to press the right arrow key. Is the vertex at (6.25, 1.5)? If you move the cursor to the right now, the Y= message is blank. For x greater than 6.25, the y-value is a complex number. In fact, for x greater than 6.25, there is *no real number* for y for either function.

GIVE IT A TRY

1. Graph the solution to each equation. Label the line of symmetry and the vertex.
   a. $x = (y + 1)^2 - 1$ b. $x = -(y - 3)^2 + 1$
2. Rewrite $x = 2y^2 - y + 1$ for y. Identify the intercepts, the vertex, and the line of symmetry for the graph. Also, sketch the graph.
3. Use the graphics calculator to check your work in Problems 1 and 2.

■ SUMMARY

In this section, we have extended our knowledge of parabolas. We studied the parabola as the graph of the solution to a second-degree equation in only one of the two variables. As shown in the following list, the key to drawing such a graph is to find the line of symmetry and the vertex of the parabola.

Equation form	*Features of the graph: Parabola*	*Line of Symmetry*	*Vertex*
$y = a(x - h)^2 + k,\ a > 0$	opens upward	$x = h$	(h, k)
$y = a(x - h)^2 + k,\ a < 0$	opens downward	$x = h$	(h, k)
$x = a(y - h)^2 + k,\ a > 0$	opens to the right	$y = h$	(k, h)
$x = a(y - h)^2 + k,\ a < 0$	opens to the left	$y = h$	(k, h)
$y = ax^2 + bx + c,\ a > 0$	opens upward	$x = \frac{-b}{2a}$	
$y = ax^2 + bx + c,\ a < 0$	opens downward	$x = \frac{-b}{2a}$	
$x = ay^2 + by + c,\ a > 0$	opens to right	$y = \frac{-b}{2a}$	
$x = ay^2 + by + c,\ a < 0$	opens to left	$y = \frac{-b}{2a}$	

12.1 ■ ■ ■ EXERCISES

Graph the solution to each equation. Label the line of symmetry and the vertex.

1. $y = (x - 2)^2 + 1$
2. $y = (x + 2)^2 - 1$
3. $x = (y - 2)^2 + 1$
4. $x = (y + 2)^2 - 1$
5. $x = -y^2 + 4$
6. $x = -y^2 + 3$
7. $y = -(x + 3)^2$
8. $y = -(x - 3)^2$
9. $y = -2(x - 1)^2 + 3$
10. $y = -2(x + 1)^2 - 3$
11. $x = -2(y - 1)^2 + 3$
12. $x = -2(y - 1)^2 + 3$
13. $y = \left(x - \frac{1}{2}\right)^2 + \frac{3}{2}$
14. $x = \left(y - \frac{4}{3}\right)^2 + \frac{1}{3}$

Rewrite each equation to the form

$$y = a(x - h)^2 + k \quad \text{or} \quad x = a(y - h)^2 + k$$

Then graph its solution. Label the line of symmetry and the vertex. Use the graphics calculator to check your work.

15. $y = x^2 - 4x + 4$
16. $x = y^2 - 6y + 9$
17. $x = y^2 - 2y$
18. $y = x^2 + 6y$
19. $2x = y - x^2 + 1$
20. $y^2 - 3x + 2 = 5y$
21. $3x + 4y = y^2 - 1$
22. $3x - 6y = y^2 + 2$
23. $9x^2 - 16y = 144$
24. $y^2 + x = 36$
25. $x = \frac{1}{2}y^2 + 2y$
26. $y = \frac{1}{2}x^2 - 4x$
27. $y = \frac{-1}{2}x^2 - 3x + 2$
28. $x = \frac{-1}{2}y^2 - 3y - 1$
29. $(2 + y)^2 = 2x - 3$
30. $3 - (x + 2)^2 = 2y$

12.2 ■ ■ ■ THE CIRCLE

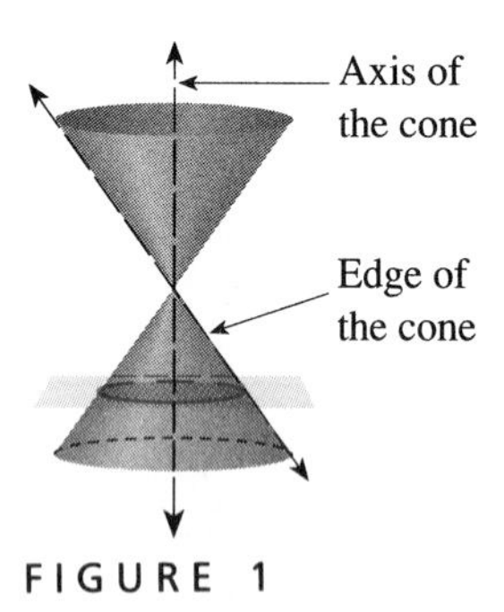

FIGURE 1

The second conic section we consider is the circle. For a plane that is perpendicular to the axis of the cone, the intersection of the plane with the cone is a circle, as shown in Figure 1. Another definition of a circle is that it is the set of points that are equidistant from a given point. Of course, the given point is the center of the circle, and the distance is the radius of the circle.

We begin this section by reviewing some of the basics of coordinate geometry, such as the distance between two points and the midpoint of a line segment. These topics are closely related to circles. We will develop the general equation of a circle (much as we learned the general equation of a line). Also, we will learn how to alter an equation of a circle to a form in which the center and the radius of the circle are obvious.

Background The set of points in a plane that are all equidistant from a given point form a shape known as a **circle**. A quick way to draw a circle is to use a thumb tack, a piece of string, and a pencil, as shown in Figure 2. To find an equation in two variables that describes this shape, we need the following ideas from coordinate geometry (see Section 4.2).

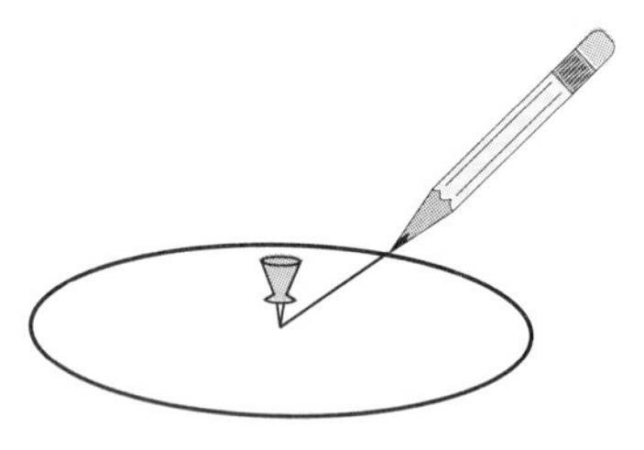

FIGURE 2

DISTANCE BETWEEN TWO POINTS

For two points (x_1, y_1) and (x_2, y_2), the distance between the two points is given by the formula

$$d = \sqrt{(x_2 - x_1)^2 + (y_2 - y_1)^2}$$

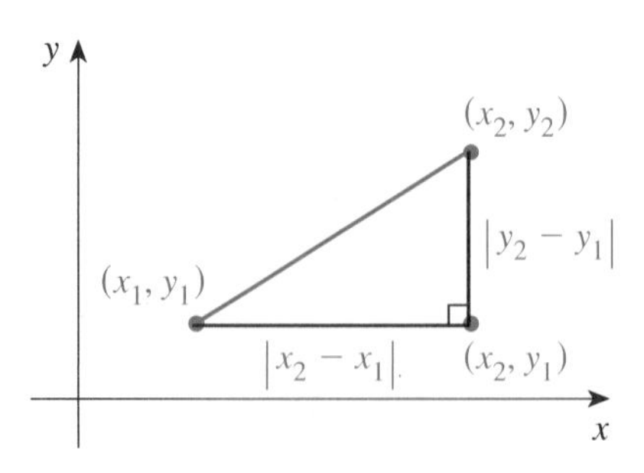

MIDPOINT OF A LINE SEGMENT

For two points (x_1, y_1) and (x_2, y_2), the midpoint of the line segment with the two points as endpoints is given by the formula

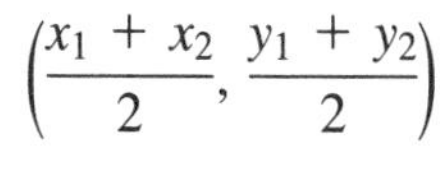

$$\left(\frac{x_1 + x_2}{2}, \frac{y_1 + y_2}{2}\right)$$

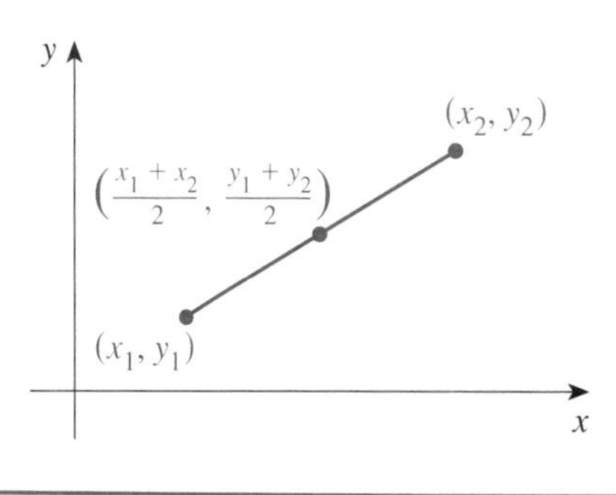

Now, the set of points, say (x, y), that are all the same distance from a fixed point must satisfy the distance formula. Suppose the fixed point is the origin, (0, 0), and the distance from this point is 3. Then, using the distance formula we have the following equation in two variables:

$$3 = \sqrt{(x - 0)^2 + (y - 0)^2}$$

Because each side of the equation always represents a positive value, we can square both sides, and write

$$9 = x^2 + y^2$$

Thus, the set of ordered pairs that solve $x^2 + y^2 = 9$ is a circle, centered at the origin, with radius $\sqrt{9} = 3$. In general, we have the following equation.

EQUATION OF A CIRCLE CENTERED AT THE ORIGIN

The equation of a circle centered at the origin with radius r is

$$x^2 + y^2 = r^2$$

Work through the following examples, which demonstrate how to graph a circle.

EXAMPLE 1 Graph the solution to the equation $x^2 + y^2 = 7$. Report the x- and y-intercepts.

Solution Because of the form of the equation, we know that the shape of the graph will be a circle centered at the origin. Also, we know that the radius of this circle is $\sqrt{7} \approx 2.65$. We find the x-intercepts by replacing the y-variable with 0, and solving the resulting equation:

$$x^2 + 0^2 = 7$$
$$x^2 = 7$$

Solution: $\sqrt{7}$ and $-\sqrt{7}$

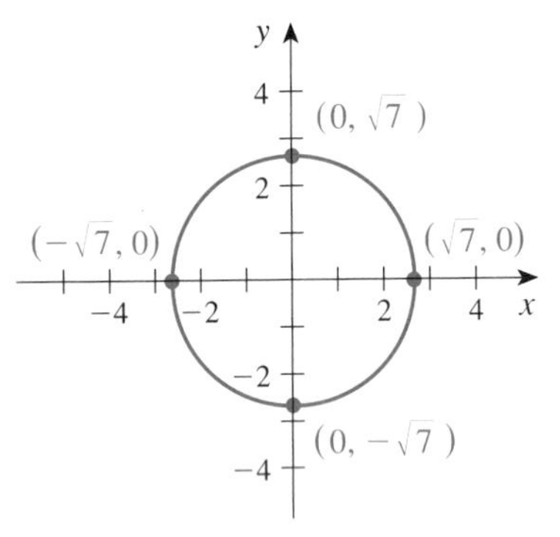

FIGURE 3
Graph of $x^2 + y^2 = 7$

Thus, there are two x-intercepts, $\left(\sqrt{7}, 0\right)$ and $\left(-\sqrt{7}, 0\right)$.

The y-intercepts are found by replacing the x-variable with 0, and then solving the equation:

$$0^2 + y^2 = 7$$
$$y^2 = 7$$

Solution: $\sqrt{7}$ and $-\sqrt{7}$

Thus, there are two y-intercepts, $\left(0, \sqrt{7}\right)$ and $\left(0, -\sqrt{7}\right)$.

Plotting these points and sketching the circle, we get the graph shown in Figure 3. ■

The graph in Figure 3 is *not* the graph of a function, because it fails the vertical line test. Actually, if we rewrite $x^2 + y^2 = 7$ for y, we get two radical functions:

$$y = \sqrt{7 - x^2} \quad \text{and} \quad y = -\sqrt{7 - x^2}.$$

The ordered pairs that satisfy $y = \sqrt{7 - x^2}$ are a function, and (graphically) form the top half of the circle (the top semicircle). The ordered pairs that satisfy $y = -\sqrt{7 - x^2}$ are a function and (graphically) form the bottom semicircle. Each of the individual graphs represents a function.

Using the Graphics Calculator

To graph the solution to $x^2 + y^2 = 7$, we must first rewrite the equation for y (in function form). Doing so, we get the radical expression equations listed above. For `Y1`, type `√(7−X²)`, and for `Y2` type `-√(7−X²)`. Use the Window 4.7 range settings to view the graph screen. Press the `TRACE` key. The message `X=0  Y=2.6457513` means that the y-intercept for the function $y = \sqrt{7 - x^2}$ is $\left(0, \sqrt{7}\right)$. Press the down arrow key to move the cursor to the second function. The y-intercept for the second function is $\left(0, -\sqrt{7}\right)$. Press the down arrow key to return to the first function. Now, press the right arrow key until the message reads `X=1  Y=2.4494897`. This is an approximation of the point $\left(1, \sqrt{6}\right)$.

To see that the point $\left(1, \sqrt{6}\right)$ is indeed $\sqrt{7}$ units from the origin, return to the home screen. Display the value in memory location `X` (it should be `1`). Now, use the `2nd` key and the `VARS` key to display the evaluation of the function `Y1` at $x = 1$ $\left(\text{the output should be } 2.449489743 \approx \sqrt{6}\right)$. Next, type `√(X²+Y1²)` and press the `ENTER` key. The approximation of `√7` is output. That is, the point $\left(1, \sqrt{6}\right)$ is $\sqrt{7}$ units from the origin. Press the `TRACE` key. For every point on the curve drawn, the distance from the origin is $\sqrt{7}$ units.

Next, we reverse the process to find the equation of a circle centered at the origin. The key is finding the radius of the circle.

EXAMPLE 2 Find the equation of a circle centered at the origin with an x-intercept at $(-2, 0)$.

Solution We know that the equation of such a circle has the form $x^2 + y^2 = r^2$. Knowing that an x-intercept is at $(-2, 0)$ means that the radius is 2, because $(-2, 0)$ is 2 units from the origin. Knowing that the radius is 2 yields the equation $x^2 + y^2 = 4$. ■

GIVE IT A TRY

1. Graph the solution to $x^2 + y^2 = 11$. Report the x- and y-intercepts.
2. Find the equation of the circle centered at the origin and passing through the point $(-4, 3)$. [*Hint*: How far from the origin is the point $(-4, 3)$?]

A Circle Centered at (h, k) Consider the equation $(x - 1)^2 + y^2 = 4$. From our study of the graphs of functions (in Chapter 9), we know that the graph of $y = (x - 1)^2$ is the graph of $y = x^2$, translated right 1 unit. In a similar manner, the graph of the solution to $(x - 1)^2 + y^2 = 4$ is the graph of the solution to $x^2 + y^2 = 4$, translated right 1 unit.

Using the Graphics Calculator

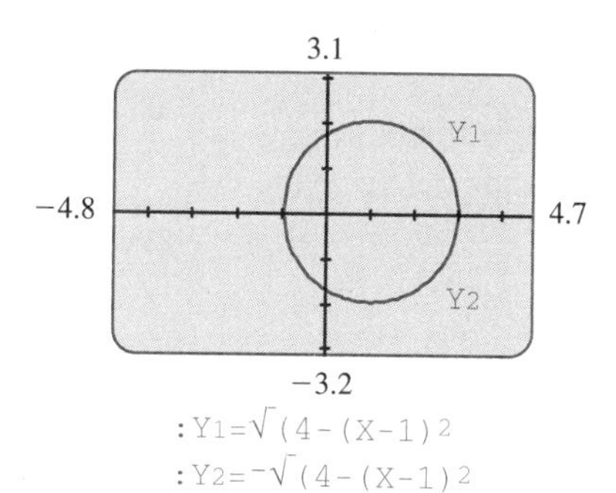

FIGURE 4

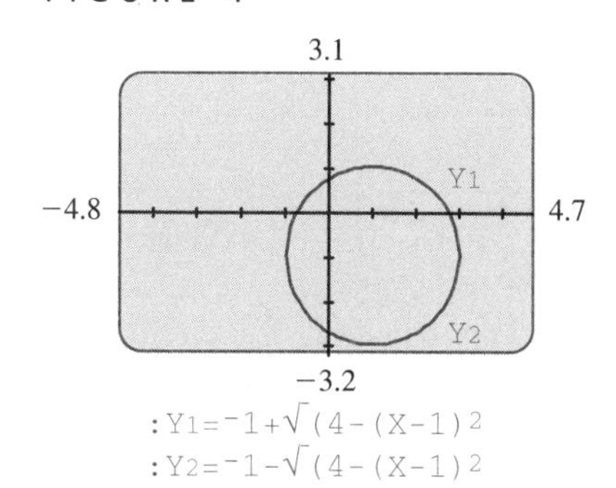

FIGURE 5

If we rewrite the equation $(x - 1)^2 + y^2 = 4$ for y (in function form), we get the two radical functions

$$y = \sqrt{4 - (x - 1)^2} \quad \text{and} \quad y = -\sqrt{4 - (x - 1)^2}$$

Type `√(4−(X−1)²)` for `Y1` and `-√(4−(X−1)²)` for `Y2`. Use the Window 4.7 range values to view the graph screen. As Figure 4 indicates, we still get a circle for a graph. However, the circle is not centered at the origin. This circle is centered at $(1, 0)$. Its radius is still 2 units.

Next, consider the graph of the solution to $(x - 1)^2 + (y + 1)^2 = 4$. The affect of the 1 in $(y + 1)^2$ in this equation is to translate the graph downward 1 unit. So, we get the graph of $x^2 + y^2 = 4$, translated right 1 unit and down 1 unit. If we rewrite the equation for y, we get the two radical functions

$$y = -1 + \sqrt{4 - (x - 1)^2} \quad \text{and} \quad y = -1 - \sqrt{4 - (x - 1)^2}$$

Edit the `Y1` rule to `-1+√(4−(X−1)²)` and the `Y2` rule to `-1−√(4−(X−1)²)`. View the graph screen, as shown in Figure 5. Again, the graph is a circle. This circle is centered at $(1, -1)$. Its radius is still 2 units.

In general, we have the following equation.

EQUATION OF A CIRCLE CENTERED AT (*h*, *k*)

> The equation of a circle centered at (h, k) with radius r is
>
> $$(x - h)^2 + (y - k)^2 = r^2$$

Work through the next example, which shows how to graph a circle centered at (h, k).

EXAMPLE 3 Graph the solution to the equation $(x - 2)^2 + (y + 1)^2 = 9$. Report the x- and y-intercepts.

Solution Because of the form of the equation, we know that the shape of the graph will be a circle centered at $(2, -1)$. We also know that the radius is $\sqrt{9}$, or 3 units.

To sketch the graph, first plot the center, $(2, -1)$. Convenient points on the circle to plot are the ones 3 units above, below, right, and left of the center. So, plot the four points $(2, 2)$, $(2, -4)$, $(5, -1)$, and $(-1, -1)$. These five points are shown in Figure 6.

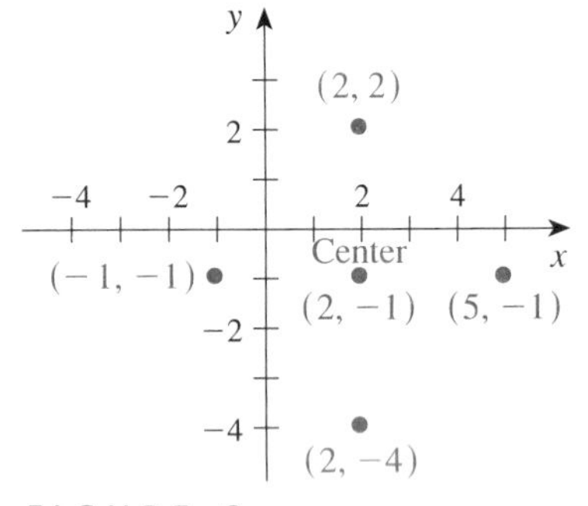

FIGURE 6

The x-intercepts are found by replacing y with 0 and solving the equation $(x - 2)^2 + (0 + 1)^2 = 9$. This second-degree equation has solution $2 + 2\sqrt{2}$ and $2 - 2\sqrt{2}$. Thus, the x-intercepts are $\left(2 + 2\sqrt{2}, 0\right)$ and $\left(2 - 2\sqrt{2}, 0\right)$.

The y-intercepts are found by replacing the x-variable with 0 and solving the equation $(0 - 2)^2 + (y + 1)^2 = 9$. This second-degree equation has solution $-1 + \sqrt{5}$ and $-1 - \sqrt{5}$. Thus, the y-intercepts are $\left(0, -1 + \sqrt{5}\right)$ and $\left(0, -1, - \sqrt{5}\right)$. Plotting these y-intercepts and sketching the circle, we get the graph shown in Figure 7. ■

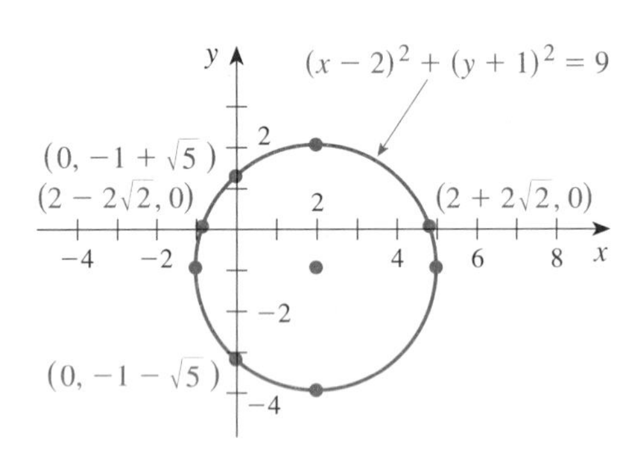

FIGURE 7

Next, we consider reversing this process to find the equation of a circle that satisfies selected conditions.

EXAMPLE 4 For the line segment with endpoints $A(-2, 3)$ and $B(4, 1)$, find the equation of the circle for which line segment AB is a diameter.

Solution To find the equation of a circle, we need the center and the radius. Because the circle has $\overline{AB}$ as a diameter, its center will be the midpoint of this diameter.

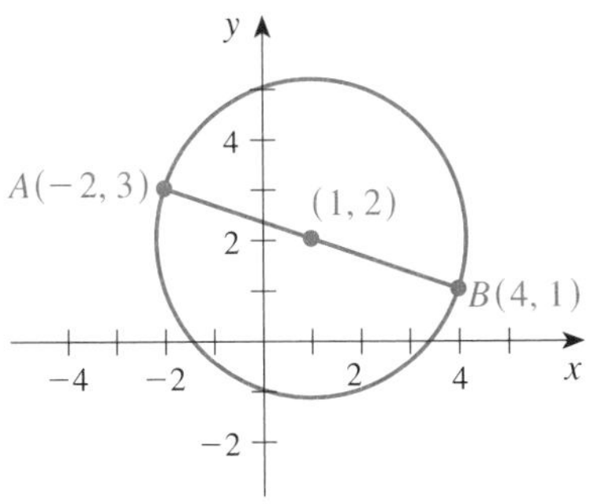

FIGURE 8

$$\text{center} = \text{midpoint of } \overline{AB} = \left(\frac{-2+4}{2}, \frac{3+1}{2}\right) = (1, 2)$$

The radius of the circle will be one-half the length of the diameter (the distance from A to B). See Figure 8.

$$\begin{aligned}\text{radius} &= \tfrac{1}{2}\left(\sqrt{(4-(-2))^2 + (1-3)^2}\right) \\ &= \tfrac{1}{2}\left(\sqrt{40}\right) \\ &= \tfrac{1}{2}\left(2\sqrt{10}\right) \\ &= \sqrt{10}\end{aligned}$$

Now, we can write the equation for the circle:

$$(x-1)^2 + (y-2)^2 = \left(\sqrt{10}\right)^2 \quad \text{or} \quad (x-1)^2 + (y-2)^2 = 10 \quad ■$$

GIVE IT A TRY

3. Graph the solution to $(x + 1)^2 + (y - 3)^2 = 4$. Label the x- and y-intercepts.
4. Find the equation of the circle that has center $(-1, 1)$ and passes through the point $(2, 4)$.

Completing the Square To graph the solution to a first-degree equation, such as $3x - 4(2 - y) = x - 2$, we write the equation in either standard form, $2x + 4y = 6$, or slope, y-intercept form, $y = \frac{1}{2}x + \frac{3}{2}$. With second-degree equations, we also must rewrite the equation to a particular form. To accomplish this, we use the completing-the-square technique from Section 5.3. Work through the following example.

EXAMPLE 5 Graph the solution to $x^2 + y^2 - 4x + 2y - 4 = 0$. Report the x- and y-intercepts.

Solution We start by rewriting the equation to the basic form for a circle. This will involve use of the completing-the-square technique. We start by grouping the x-variables and the y-variables, and we add 4 to each side of the equation to get

$$x^2 - 4x + y^2 + 2y = 4$$

We complete the square for the x-variable:

$$x^2 - 4x + \boxed{4} + y^2 + 2y = 4 + \boxed{4} \qquad \text{Add 4 to each side.}$$

$$\left(\tfrac{1}{2} \text{ of } -4\right)^2 = 4$$

Now, we complete the square for the y-variable:

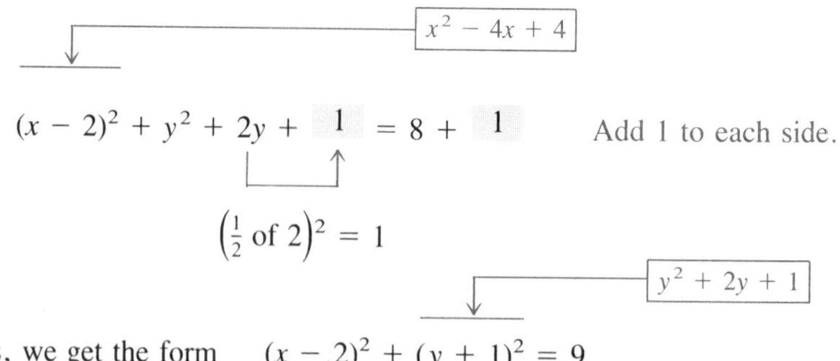

Thus, we get the form $\quad (x - 2)^2 + (y + 1)^2 = 9$

So, we have a circle centered at $(2, -1)$ with radius $\sqrt{9}$, or 3. The graph is given in Figure 9.

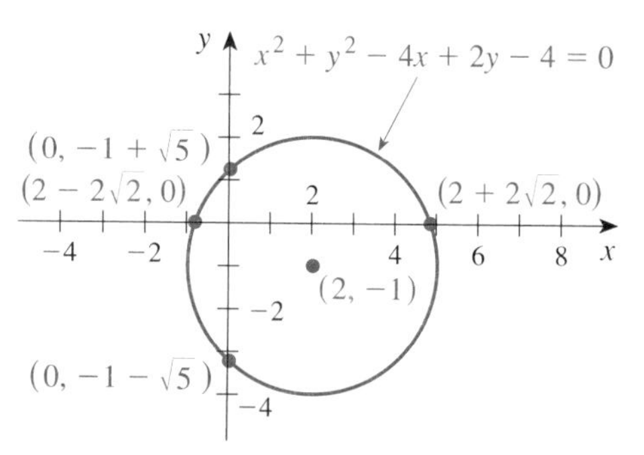

FIGURE 9

The x-intercepts are found by replacing the y-variable with 0 and solving $(x - 2)^2 + 1 = 9$. The x-intercepts are $\left(2 + 2\sqrt{2}, 0\right)$ and $\left(2 - 2\sqrt{2}, 0\right)$.

The y-intercepts are found by replacing the x-variable with 0 and solving $4 + (y + 1)^2 = 9$. The y-intercepts are $\left(0, -1 + \sqrt{5}\right)$ and $\left(0, -1 - \sqrt{5}\right)$. ■

GIVE IT A TRY

Graph the solution to each equation. Report the center, the radius, and the x- and y-intercepts.

5. $x^2 + y^2 + 4x - 2y - 3 = 0$ **6.** $x^2 + y^2 - 6x - 4y - 1 = 0$

■ SUMMARY

In this section, we have learned about another of the conic sections—the circle. The equation of a circle centered at (h, k) and with radius, r, is

$$(x - h)^2 + (y - k)^2 = r^2$$

We learned how to graph the solution to such an equation by plotting the center point and four points obtained from applying the radius. We also found the x- and y-intercepts. We learned how to reverse the process to find the equation of a circle given information about the circle. We used the completing-the-square technique to write equations of circles in standard form.

■ ***Quick Index***

12.2 ■ ■ ■ EXERCISES

Graph the solution to each equation. Label the center of the circle, the radius, and the x- and y-intercepts. Use the graphics calculator to check your work.

1. $x^2 + y^2 = 25$

   $x^2 + y^2 = 16$

3. $x^2 + y^2 = 6$

   $x^2 + y^2 = 14$

5. Report the equation of the circle that is centered at the origin and has radius 2.
6. Report the equation of the circle that is centered at the origin and has radius 3.
7. Report the equation of the circle centered at the origin, with y-intercept $(0, -3)$.
8. Report the equation of the circle centered at the origin, with y-intercept $(0, 2)$.
9. Report the equation of the circle that is centered at the origin and passes through the point $(-2, 3)$.
10. Report the equation of the circle that is centered at the origin and passes through the point $(2, -3)$.
11. Report the equation of the circle that is centered at the origin and passes through the point $(4, 1)$.
12. Report the equation of the circle that is centered at the origin and passes through the point $(-1, 4)$.

Graph the solution to each equation. Label the center of the circle, the radius, and the x- and y-intercepts. Use the graphics calculator to check your work.

13. $(x - 3)^2 + y^2 = 16$
14. $x^2 + (y - 4)^2 = 9$
15. $(x - 3)^2 + (y - 4)^2 = 4$
16. $(x - 2)^2 + y^2 = 7$
17. $x^2 + (y - 5)^2 = 13$
18. $(x - 2)^2 + (y - 5)^2 = 5$
19. Find the equation of the circle that is centered at $(-3, -2)$ and has radius 4.
20. Find the equation of the circle that is centered at $(3, -1)$ and has radius 5.
21. Find the equation of the circle that is centered at $(3, 1)$ and passes through the point $(3, 5)$.
22. Find the equation of the circle that is centered at $(4, -2)$ and passes through the point $(3, -2)$.
23. Find the equation of the circle that is centered at $(4, -1)$ and passes through the point $(3, 3)$.
24. Find the equation of the circle that is centered at $(-2, -2)$ and passes through the point $(3, 4)$.
25. For the line segment with endpoints $A(3, 5)$ and $B(1, -1)$, find the equation of the circle for which line segment AB is a diameter.
26. For the line segment with endpoints $A(4, -3)$ and $B(1, -1)$, find the equation of the circle for which line segment AB is a diameter.

Use the completing-the-square technique to write the equation in standard form. Graph the solution to the resulting equation. Report the center of the circle, the radius, and the x- and y-intercepts. Use the graphics calculator to check your work.

27. $x^2 + 4x + y^2 = 7$
28. $x^2 + y^2 - 6y - 1 = 0$
29. $x^2 + y^2 - 4x + 2y = 5$
30. $x^2 + y^2 + 8x - 4y - 2 = 0$
31. $x^2 + y^2 - 8x - 4y - 2 = 0$
32. $x^2 + y^2 - 2x + 6y = -1$
33. Find the area measure of the circle determined by the equation $x^2 + y^2 - 4y = 12$.
34. Find the circumference of the circle determined by the equation $x^2 + y^2 - 4x = 2y + 1$.

Read each graph and report the equation of the circle.

35.

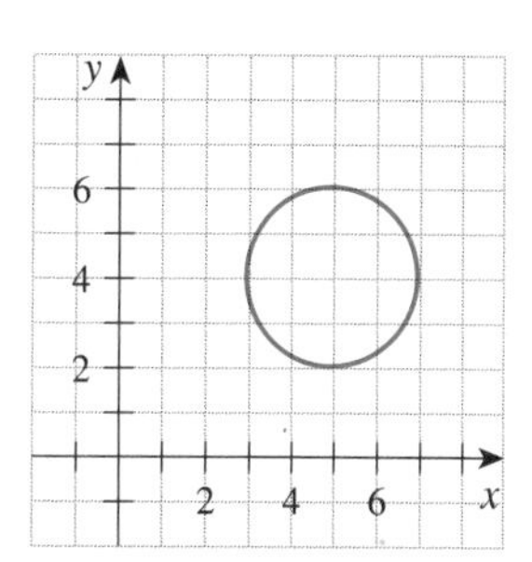

36.

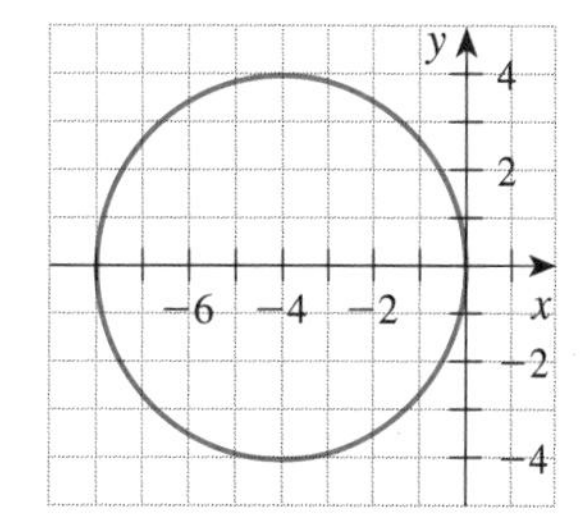

12.3 ■ ■ ■ THE ELLIPSE AND THE HYPERBOLA

In this section we continue our study of conic sections, introducing ellipses and hyperbolas. We will learn the general equations for an ellipse and a hyperbola (much as we learned the equation of a parabola in Section 12.1 and a circle in Section 12.2). Also, we will learn how to graph the solution to an equation that determines an ellipse or a hyperbola.

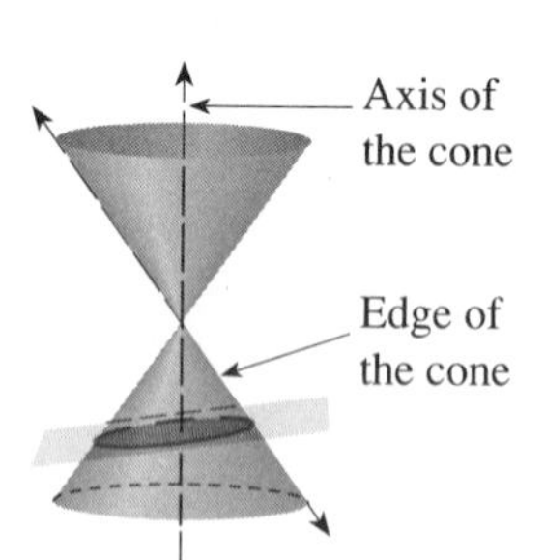

FIGURE 1

Background When a plane that is perpendicular to the axis of a cone intersects the cone, a circle is formed. If the plane is tilted slightly, an ellipse is formed, as shown in Figure 1. A circle is the set of points that are all the same distance from a fixed point. (The distance is the radius, and the fixed point is the center.) An **ellipse** is the set of points where the sum of the distance from two fixed points, known as the *foci,* is constant. Figure 2 identifies the foci.

An ellipse can be drawn using two thumb tacks, a piece of string, and a pencil, as shown in Figure 3. Notice that as the two foci are moved closer and closer together, the ellipse becomes closer and closer to being a circle. A circle is a special case of an ellipse. When the two foci are merged to a single point, the ellipse is a circle, and the single point is the center of the circle.

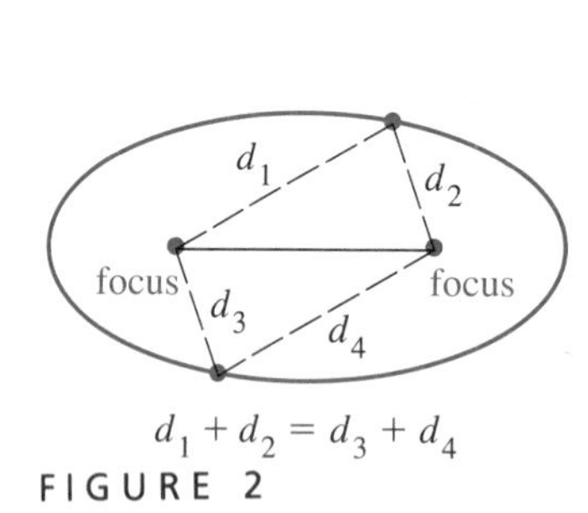

FIGURE 2

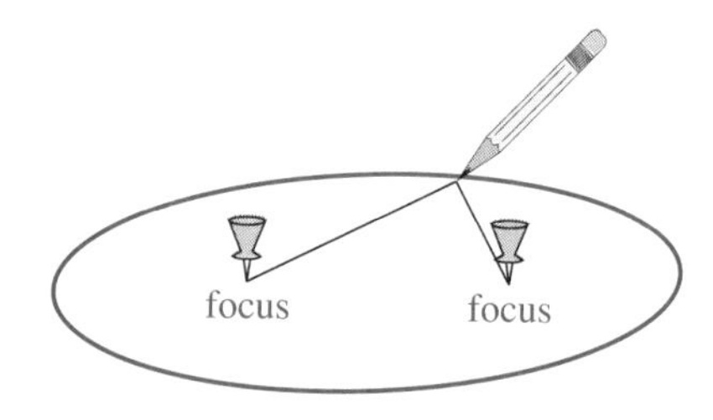

FIGURE 3

As you may suspect, the equation of an ellipse is similar to the equation of a circle. Consider the equations

$x^2 + y^2 = 16$	A circle centered at the origin with radius 4
$4x^2 + y^2 = 16$	An ellipse centered at the origin

As these equations show, if the x^2- and y^2-coefficients are not identical, then the shape formed when the solution is graphed is not a circle.

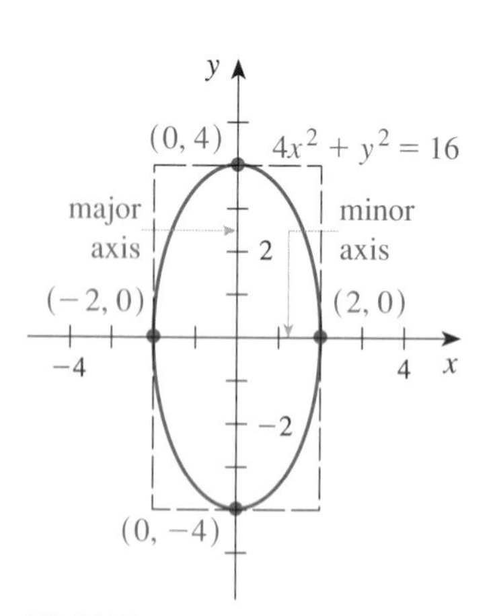

FIGURE 4

Graphing an Ellipse We now consider the graph of the solution to the equation $4x^2 + y^2 = 16$. First, to find the x-intercepts, we replace the y-variable with 0 and solve the equation $4x^2 + 0^2 = 16$. This equation has solution 2 and -2. Thus, the x-intercepts are $(2, 0)$ and $(-2, 0)$. Next, to find the y-intercepts, we replace the x-variable with 0 and solve the equation $4(0)^2 + y^2 = 16$. This equation has solution 4 and -4. Thus, the y-intercepts are $(0, 4)$ and $(0, -4)$. Now, we plot these intercepts and sketch in a box (to use as a guide), with the intercepts as midpoints of the sides. Finally, we sketch the ellipse shown in Figure 4. Here, the y-axis is known as the **major axis** of the ellipse and the x-axis is known as the **minor axis** of the ellipse.

For a circle centered at the origin, simply by observing the equation we can determine the radius of the circle and, thus, the x- and y-intercepts. For the equation of an ellipse centered at the origin, we can also quickly see the x- and y-intercepts by writing the equation in standard form:

$$4x^2 + y^2 = 16$$

Multiply by $\frac{1}{16}$:

$$\frac{4x^2}{16} + \frac{y^2}{16} = \frac{16}{16}$$

Standard form:

$$\frac{x^2}{4} + \frac{y^2}{16} = 1$$

In this standard form, the denominators now reveal the intercepts. Here, the denominator for y^2 is greater than the denominator for x^2. This results in the y-axis being the major axis. In the next example, the x-axis is the major axis.

EXAMPLE 1 Graph the solution to $4x^2 + 9y^2 = 36$. Report the x- and y-intercepts.

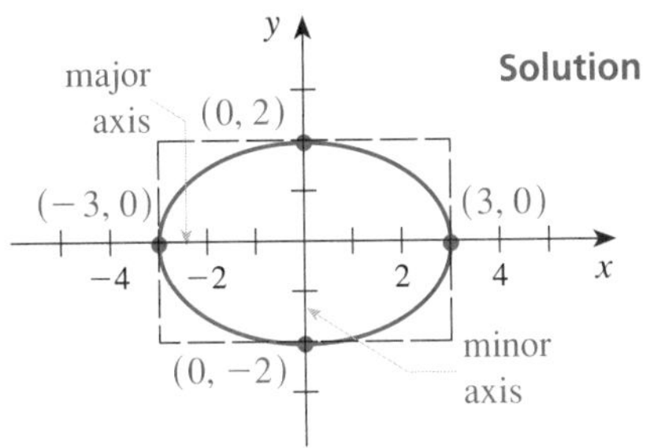

FIGURE 5

Solution Multiply by $\frac{1}{36}$:

$$\frac{x^2}{9} + \frac{y^2}{4} = 1$$

The x^2-denominator is 9, so the x-intercepts are (3, 0) and (−3, 0).
The y^2-denominator is 4, so the y-intercepts are (0, 2) and (0, −2).
After plotting these intercepts and the frame, we sketch the ellipse shown in Figure 5. ■

We have the following generalization.

STANDARD FORM FOR AN ELLIPSE CENTERED AT THE ORIGIN

An ellipse centered at the origin has an equation of the form

$$\frac{x^2}{a^2} + \frac{y^2}{b^2} = 1$$

The x-intercepts are $(a, 0)$ and $(-a, 0)$.
The y-intercepts are $(0, b)$ and $(0, -b)$.
If $a^2 > b^2$, then the x-axis is the major axis.
If $a^2 < b^2$, then the y-axis is the major axis.
If $a^2 = b^2$, then the ellipse is a circle.

To learn how to graph the solution to an equation of an ellipse written in the standard form, work through the following example.

EXAMPLE 2 Graph the solution to $\frac{x^2}{16} + \frac{y^2}{9} = 1$. Report the x- and the y-intercepts.

Solution First, because 16 is greater than 9, we know that the x-axis is the major axis. In fact, the x-intercepts are (4, 0) and (−4, 0).
The y-axis is the minor axis, and the y-intercepts are (0, 3) and (0, −3).

Now, we plot these points and sketch a box based on the points. Next, we sketch an ellipse, the graph shown in Figure 6.

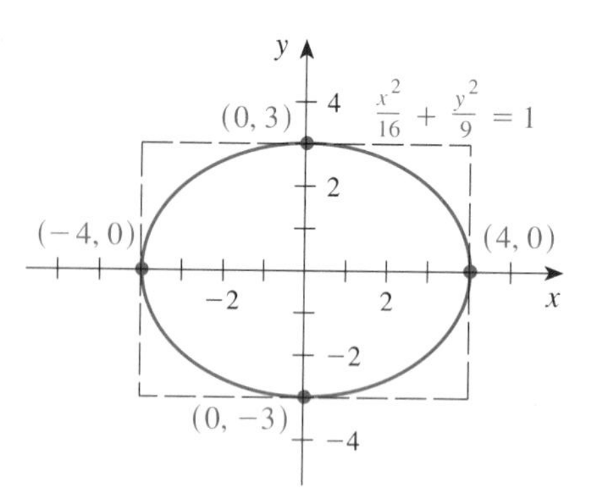

FIGURE 6

■

To graph the solution to an equation of an ellipse that is *not* given in standard form, work through the next example.

EXAMPLE 3 Graph the solution to $9x^2 + 4y^2 = 36$.

Solution We start by rewriting the equation in standard form. To accomplish this, we want the right-hand side to be the number 1. So, we multiply by $\frac{1}{36}$, to get

$$\frac{9x^2}{36} + \frac{4y^2}{36} = \frac{36}{36}$$

$$\frac{x^2}{4} + \frac{y^2}{9} = 1$$

Because 9 is greater than 4, the major axis is the y-axis.
The y-intercepts are (0, 3) and (0, −3). The x-intercepts are (2, 0) and (−2, 0).

Plotting these intercepts, sketching in the box, and then drawing the ellipse yields the graph in Figure 7. ■

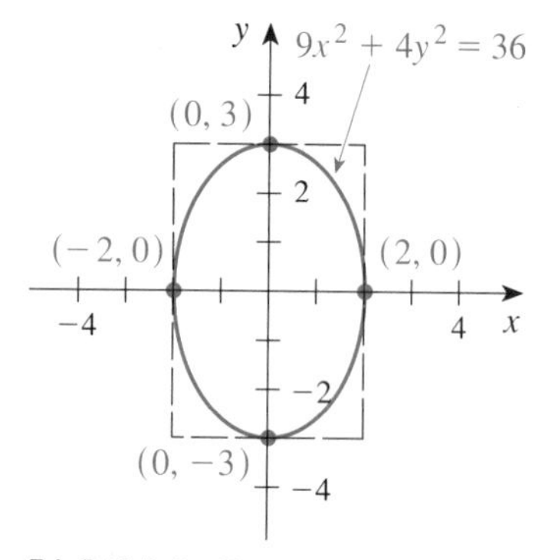

FIGURE 7

Using the Graphics Calculator

As with circles, we can use the graphics calculator to graph an ellipse. We must first rewrite the equation of y. For the equation in Example 3, we get

$$\begin{aligned} 9x^2 + 4y^2 &= 36 \\ -9x^2 \qquad &\quad -9x^2 \\ \hline 4y^2 &= 36 - 9x^2 \\ y^2 &= \frac{36 - 9x^2}{4} \end{aligned}$$

$$y = \sqrt{\frac{9(4 - x^2)}{4}} \quad \text{or} \quad y = -\sqrt{\frac{9(4 - x^2)}{4}}$$

$$y = \frac{3}{2}\sqrt{4 - x^2} \qquad y = \frac{-3}{2}\sqrt{4 - x^2}$$

Type `(3/2)√(4−X²)` for `Y1`, and type `(-3/2)√(4−X²)` for `Y2`. Use the Window 4.7 range settings to view the graph. As you can see, an ellipse is drawn, centered at the origin, with the y-axis as the major axis. Use the `TRACE` feature to verify that the y-intercepts are $(0, 3)$ and $(0, -3)$.

GIVE IT A TRY

Graph the solution to each equation. Label the x- and y-intercepts of the graph.

1. $3x^2 + 4y^2 = 12$ **2.** $4y^2 + x^2 = 36$

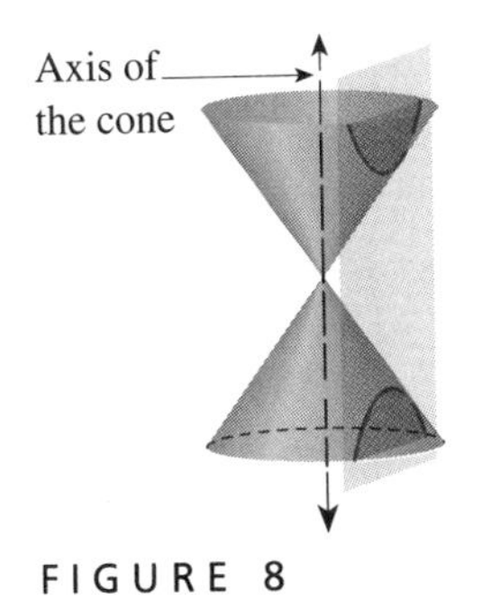

FIGURE 8

Hyperbolas When a plane that is perpendicular to the axis of a cone intersects the cone, a circle is formed. As we tilt the plane, an ellipse is formed, and then (tilting further), a parabola is formed. If we continue to tilt the plane still further, a hyperbola is formed. See Figure 8. A **hyperbola** is the set of points in the plane for which the *difference* of the distances between two fixed points (known as the *foci*) is constant. Figure 9 illustrates these distances from the foci for two points on a hyperbola.

conjugate axis conjugate axis

d_1 d_2 f_1 f_2 d_3 d_4 f_1 f_2

transverse axis transverse axis

$$|d_1 - d_2| = |d_3 - d_4|$$

FIGURE 9

The line segment with the foci as endpoints is known as the **transverse axis**, and its perpendicular bisector is known as the **conjugate axis**.

As you may suspect, the equation of a hyperbola is similar to the equation of an ellipse. Consider these equations:

$x^2 + y^2 = 16$	A circle centered at the origin with radius 4
$4x^2 + y^2 = 16$	An ellipse centered at the origin, y-axis major
$x^2 + 4y^2 = 16$	An ellipse centered at the origin, x-axis major
$4x^2 - y^2 = 16$	A hyperbola centered at the origin, x-axis transverse
$y^2 - 4x^2 = 16$	A hyperbola centered at the origin, y-axis transverse

If the sign of either the x^2- or the y^2-coefficient is negative (but not of both), the shape formed is a hyperbola. The transverse axis is on the x-axis if the x^2-coefficient is positive. The transverse axis is on the y-axis if the y^2-coefficient is positive.

Graphing a Hyperbola We now graph the solution to the equation

$$4x^2 - 3y^2 = 36$$

First, to find the x-intercepts, we replace the y-variable with 0 and solve the equation $4x^2 - 3(0)^2 = 36$. This equation has solution 3 and -3. Thus, the x-intercepts are $(3, 0)$ and $(-3, 0)$. Next, to find the y-intercepts, we replace the x-variable with 0 and solve the equation $4(0)^2 - 3y^2 = 36$. This equation has no real number in its solution. Thus, the graph has *no* y-intercept. We now generate some additional ordered pairs that solve the equation. A table of ordered pairs is shown in the margin. Plotting the x-intercepts and these points, then sketching the curve, we get the hyperbola shown in Figure 10. As the graph indicates, this is not the graph of a function. (Try the vertical line test).

Ordered pairs that solve $4x^2 - 3y^2 = 36$

x	y	(x, y)
4	$\frac{2\sqrt{21}}{3}$	$\left(4, \frac{2\sqrt{21}}{3}\right)$
	$\frac{-2\sqrt{21}}{3}$	$\left(4, \frac{-2\sqrt{21}}{3}\right)$
-4	$\frac{2\sqrt{21}}{3}$	$\left(-4, \frac{2\sqrt{21}}{3}\right)$
	$\frac{-2\sqrt{21}}{3}$	$\left(-4, \frac{-2\sqrt{21}}{3}\right)$
5	$\frac{8\sqrt{3}}{3}$	$\left(5, \frac{8\sqrt{3}}{3}\right)$
	$\frac{-8\sqrt{3}}{3}$	$\left(5, \frac{-8\sqrt{3}}{3}\right)$
-5	$\frac{8\sqrt{3}}{3}$	$\left(-5, \frac{8\sqrt{3}}{3}\right)$
	$\frac{-8\sqrt{3}}{3}$	$\left(-5, \frac{-8\sqrt{3}}{3}\right)$

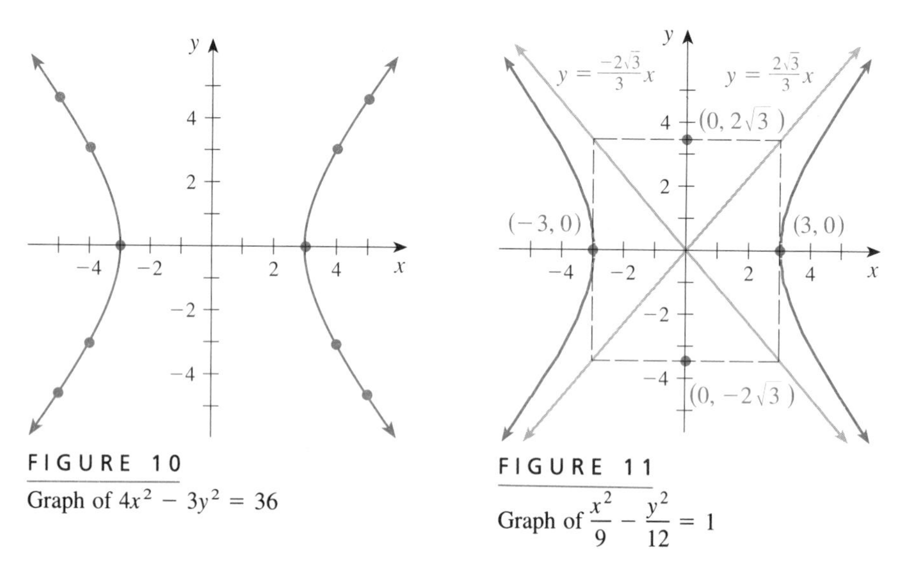

FIGURE 10
Graph of $4x^2 - 3y^2 = 36$

FIGURE 11
Graph of $\frac{x^2}{9} - \frac{y^2}{12} = 1$

An improved approach to drawing the hyperbola is to sketch a box, much like the one we used with an ellipse. To do this, first rewrite the equation in standard form. Multiplying by $\frac{1}{36}$, we get

$$\frac{x^2}{9} - \frac{y^2}{12} = 1$$

Next, we plot $(a, 0)$ and $(-a, 0)$, where a^2 is the denominator for x^2. We plot $(0, b)$ and $(0, -b)$, where b^2 is the denominator for y^2. We sketch in the rectangle that has these points as midpoints for its sides. Additionally, we sketch in the diagonals of the box. This framework is shown in Figure 11.

We can now sketch the hyperbola. The diagonals of the box are the lines determined by $y = \frac{2\sqrt{3}}{3}x$ and $y = \frac{-2\sqrt{3}}{3}x$. Notice that the line increasing from left to right, passes through the origin [has y-intercept (0, 0)], and has slope measure $\frac{b}{a}$. Likewise, the other diagonal has y-intercept (0, 0) and slope measure $\frac{-b}{a}$. These lines are known as **asymptotes** for the hyperbola. For a given x-value, the distance between a point on the hyperbola and the corresponding point on the asymptote becomes smaller and smaller as $|x|$ becomes larger and larger.

To use this technique, work through the following example.

EXAMPLE 4 Graph the solution to $\frac{x^2}{9} - \frac{y^2}{4} = 1$. Report the equations of the asymptotes and the x- and y-intercepts.

Solution Because the x^2-term is positive, we know that the transverse axis is on the x-axis. The x-intercepts are (3, 0) and (−3, 0). The graph has no y-intercept. Now, we sketch the box through (0, 2), (0, −2), (3, 0), and (−3, 0), as well as the diagonals of the box. These diagonals, $y = \frac{2}{3}x$ and $y = \frac{-2}{3}x$, are the asymptotes for the hyperbola. Finally, we sketch the hyperbola, as shown in Figure 12.

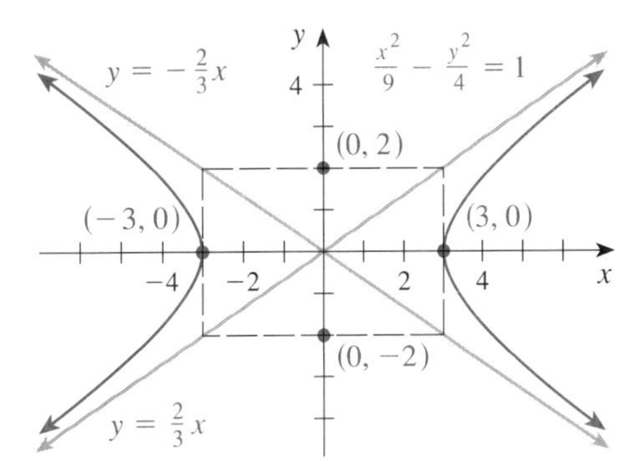

FIGURE 12
Graph of $\frac{x^2}{9} - \frac{y^2}{4} = 1$

■

We have the following generalization for hyperbolas.

STANDARD FORM FOR A HYPERBOLA CENTERED AT THE ORIGIN

The graph of the solution to $\frac{x^2}{a^2} - \frac{y^2}{b^2} = 1$ is a hyperbola centered at the origin with the transverse axis on the x-axis.

The graph of the solution to $\frac{y^2}{b^2} - \frac{x^2}{a^2} = 1$ is a hyperbola centered at the origin with the transverse axis on the y-axis.

In the next example, we graph a hyperbola whose equation is not given in standard form.

EXAMPLE 5 Graph the solution to $4y^2 - x^2 = 16$.

Solution As usual, we start by rewriting the equation in standard form. That is, we want the right-hand side to be a 1. To achieve this, we multiply each side by $\frac{1}{16}$, to get

$$\frac{y^2}{4} - \frac{x^2}{16} = 1$$

Observing that the y^2-term is positive, we conclude that the transverse axis is on the y-axis. The y-intercepts are (0, 2) and (0, −2). Next, we draw the box through the points (4, 0), (−4, 0), (0, 2), and (0, −2). Extending the diagonals of this box, we sketch the asymptotes for the hyperbola, $y = \frac{1}{2}x$ and $y = \frac{-1}{2}x$. Finally, sketching the hyperbola, we have the graph shown in Figure 13. ■

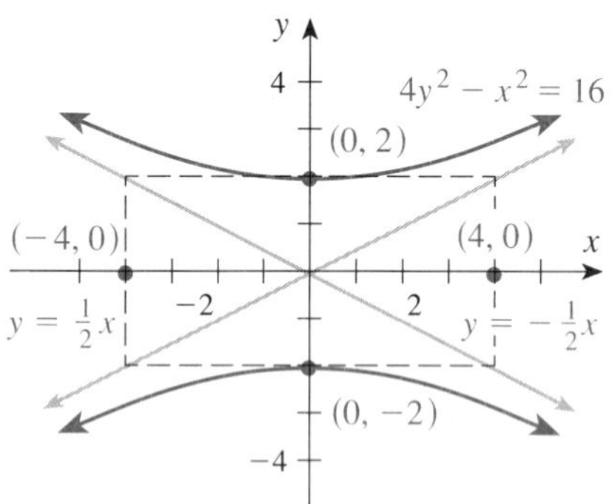

FIGURE 13
Graph of $4y^2 - x^2 = 16$

Using the Graphics Calculator

To use the graphics calculator to graph the hyperbola in Example 5, we must first rewrite the equation for y:

$$4y^2 - x^2 = 16$$
$$\underline{\quad x^2 \quad x^2}$$
$$4y^2 = 16 + x^2$$
$$y^2 = \frac{x^2 + 16}{4}$$

$$y = \sqrt{\frac{x^2 + 16}{4}} \quad \text{or} \quad y = -\sqrt{\frac{x^2 + 16}{4}}$$

$$y = \frac{1}{2}\sqrt{x^2 + 16} \qquad y = \frac{-1}{2}\sqrt{x^2 + 16}$$

Type `(1/2)√(X²+16)` for `Y1`, and type `(⁻1/2)√(X²+16)` for `Y2`. Use the Window 9.4 range settings to view the graph screen, then press the `TRACE` key. The message `X=0 Y=2` means that the y-intercept is (0, 2). Press the down arrow key to move the cursor to the other curve. The message `X=0 Y=⁻2` means that the other y-intercept is (0, −2). For `Y3` type `(1/2)X`, and for `Y4` type `(⁻1/2)X`, the equations for the asymptotes. View the graph screen. Now, zoom out from the origin. As $|x|$ gets larger, the hyperbola becomes indistinguishable from its asymptotes.

If the equation were $x^2 - 4y^2 = 16$, then the transverse axis is on the x-axis. Edit the rule for `Y1` to `(1/2)√(X²−16)` and the rule for `Y2` to `(⁻1/2)√(X²−16)`. Use the Window 9.4 range settings to view the graph screen, as shown in Figure 14. This time, the graph has no y-intercept. Notice that for x-values between −4 and 4, the functions are undefined. Why? The message `X=4 Y=0` means that one x-intercept is (4, 0).

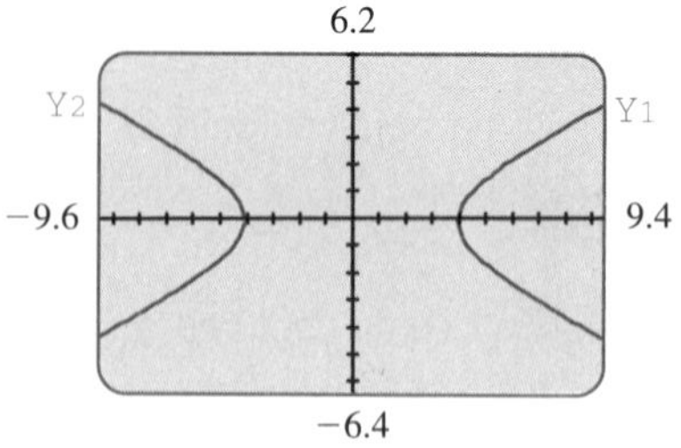

:Y1=(1/2)√(X²-16)
:Y2=(⁻1/2)√(X²-16)

FIGURE 14

GIVE IT A TRY

Graph the solution to each equation. Report the intercepts and asymptotes.

3. $16x^2 - 25y^2 = 400$

4. $25y^2 - 16x^2 = 400$

■ SUMMARY

In this section, we have learned about two more of the conic sections—the ellipse and the hyperbola. The equation of an ellipse centered at the origin is

$$\frac{x^2}{a^2} + \frac{y^2}{b^2} = 1$$

and the equation of a hyperbola centered at the origin is either

$$\frac{x^2}{a^2} - \frac{y^2}{b^2} = 1 \quad \text{or} \quad \frac{y^2}{b^2} - \frac{x^2}{a^2} = 1$$

We learned how to graph the solution to such an equation by sketching the box formed by the points $(a, 0)$, $(-a, 0)$, $(0, b)$, and $(0, -b)$. For the ellipse, we sketch the oval within the box. For a hyperbola, we sketch the diagonals of the box and extend them to the lines. These lines represent the asymptotes for the hyperbola. Also, with a hyperbola centered at the origin, we must determine whether the transverse axis is on the x-axis or the y-axis (by observing which variable has the positive coefficient). Once the asymptotes are drawn and the transverse axis is determined, we can sketch the hyperbola.

■ ***Quick Index***

12.3 ■ ■ ■ EXERCISES

Graph the solution to each equation. Report the x- and y-intercepts. If the graph is a hyperbola, also report the equation of the asymptotes.

1. $x^2 + 4y^2 = 16$
2. $4x^2 + y^2 = 16$
3. $4x^2 + 9y^2 = 36$
4. $9x^2 + 4y^2 = 36$
5. $\frac{x^2}{16} + \frac{y^2}{4} = 1$
6. $\frac{x^2}{25} + \frac{y^2}{9} = 1$
7. $\frac{x^2}{4} + \frac{y^2}{1} = 1$
8. $\frac{x^2}{1} + \frac{y^2}{9} = 1$
9. $x^2 - y^2 = 4$
10. $y^2 - x^2 = 4$
11. $4x^2 - 25y^2 = 100$
12. $25y^2 - 4x^2 = 100$
13. $9x^2 - 4y^2 = 36$
14. $9y^2 - 4x^2 = 36$
15. $\frac{x^2}{16} - \frac{y^2}{9} = 1$
16. $\frac{x^2}{9} - \frac{y^2}{16} = 1$
17. $\frac{y^2}{25} - \frac{x^2}{9} = 1$
18. $\frac{y^2}{25} - \frac{x^2}{16} = 1$
19. $2x^2 - 3y^2 = 6$
20. $2y^2 - 3x^2 = 12$
21. $16x^2 - 25y^2 = 100$
22. $9y^2 + 16x^2 = 144$
23. $9x^2 - 16y^2 = 144$
24. $y^2 + x^2 = 36$
25. $\frac{x^2}{2} + \frac{y^2}{3} = 1$
26. $\frac{x^2}{5} + \frac{y^2}{9} = 1$
27. $\frac{x^2}{4} - \frac{y^2}{3} = 1$
28. $\frac{y^2}{5} - \frac{x^2}{9} = 1$

12.4 ■ ■ ■ NONLINEAR SYSTEMS

In this section, we will solve a system of two equations in which at least one of the equations is not a first-degree equation. That is, one or more of the equations is a quadratic equation or an equation involving rational or radical expressions. Unlike the solution of a linear system, the solution of a nonlinear system is often more than one ordered pair. With this type of system, we will use substitution to reduce the problem to one of solving an equation in one variable.

Lines and Parabolas A nonlinear system is a collection of equations (in two variables, for this discussion). In such a system, at least one of the equations is not linear (first-degree). An example of a nonlinear system is

$$\begin{cases} y = x^2 - 3x - 4 \\ y = 2x - 4 \end{cases}$$

The solution to a nonlinear system is the collection of ordered pairs that are solutions to the first equation, $y = x^2 - 3x - 4$, *and* to the second equation, $y = 2x - 4$.

Using the Graphics Calculator

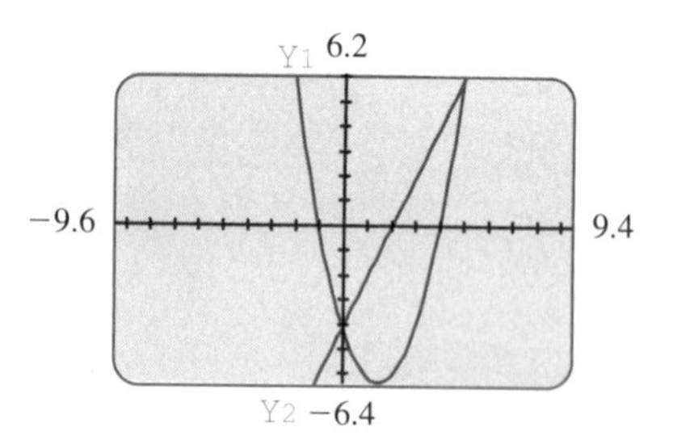

FIGURE 1

The ordered pairs that are solutions to the equation $y = x^2 - 3x - 4$ determine a parabola when plotted. For Y1, type the rule X²−3X−4. View the graph screen using the Window 9.4 range settings. The parabola is plotted. Type 2X−4 for Y2. Press the GRAPH key. For Y2, a line is drawn. The solution to the system will be the ordered pairs that name the intersection points of the parabola and the line. See Figure 1. Press the TRACE key. As you can see, the ordered-pair name of the leftmost intersection point is the y-intercept of the two curves, $(0, -4)$. Move the cursor to the rightmost intersection point. Its x-coordinate is 5 and its y-coordinate is 6. We will now use an algebraic approach to arrive at the names of these intersection points.

Substitution Because nonlinear systems are often written in function form, it is natural to use the substitution approach to solve the system. That is, for the given system, the ordered pairs in the solution will be those whose input value x results in the same output value for both the rule $2x - 4$ and the rule $x^2 - 3x - 4$. To find these input values, we equate the two rules:

$$x^2 - 3x - 4 = 2x - 4$$

This is a second-degree equation in one variable. (We known how to solve this type of equation—see Section 5.4.) Solving the equation, we proceed as follows:

Write in standard form: $x^2 - 5x = 0$

Factor the left-hand side: $x(x - 5) = 0$

Solution: 0 and 5

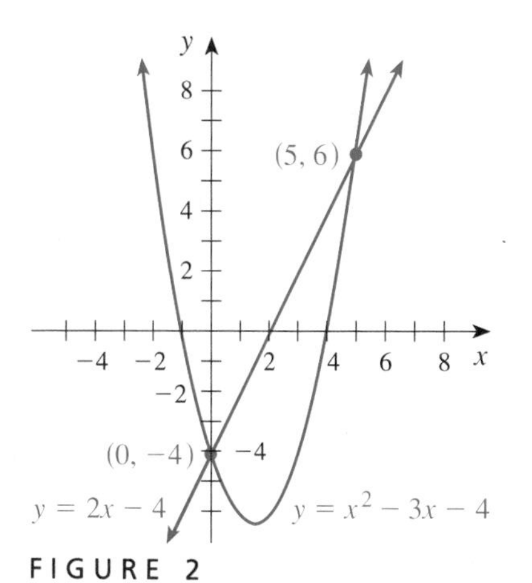

FIGURE 2

Thus, two input values, 0 and 5, produce the same output value for the two given function rules. Replacing the x-variable with 0 in either of the system equations yields a y-value of -4. Thus, the ordered pair $(0, -4)$ is a solution to the system. Replacing the x-variable with 5 in either of the equations yields a y-value of 6. Thus, the ordered pair $(5, 6)$ is also a solution to the system. The solution to the nonlinear system is the two ordered pairs $(0, -4)$ and $(5, 6)$. These points are shown in Figure 2.

Using the Graphics Calculator

> The input value 0 outputs -4 for both $y = x^2 - 3x - 4$ and $y = 2x - 4$. Store `0` in memory location `X`. Display the value stored in `Y1`—it should be `-4`. Display the value in `Y2`—it should be `-4`.
>
> The input value 5 outputs 6 for both $y = x^2 - 3x - 4$ and $y = 2x - 4$. Store `5` in memory location `X`. Display the values stored in `Y1` and `Y2`. The outputs should both be `6`.

We will now alter the given nonlinear system slightly.

EXAMPLE 1 Solve: $\begin{cases} y = x^2 - 3x - 4 \\ y = 2x - 1 \end{cases}$

Solution We start by sketching a graph, as shown in Figure 3.
Now, equating the function rules, we get $x^2 - 3x - 4 = 2x - 1$.
Next, we solve this equation in one-variable.

Write in standard form: $x^2 - 5x - 3 = 0$

Identify a, b, and c: $a = 1, b = -5, c = -3$

Compute $b^2 - 4ac$: $(-5)^2 - 4(1)(-3) = 25 + 12 = 37$

Solution: $\dfrac{5 + \sqrt{37}}{2}$ and $\dfrac{5 - \sqrt{37}}{2}$

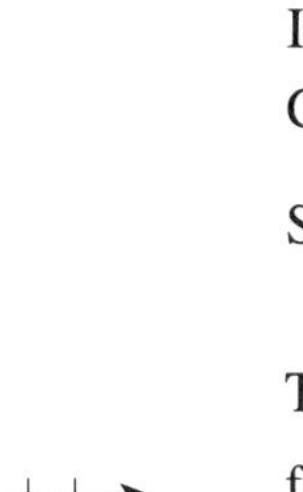

FIGURE 3

Thus, two input values, $\frac{5 + \sqrt{37}}{2}$ and $\frac{5 - \sqrt{37}}{2}$, produce the same output value for the given function rules.

The input value $\frac{5 + \sqrt{37}}{2}$, outputs $4 + \sqrt{37}$ for both $y = x^2 - 3x - 4$ and $y = 2x - 1$.

The input value $\frac{5 - \sqrt{37}}{2}$, outputs $4 - \sqrt{37}$ for both $y = x^2 - 3x - 4$ and $y = 2x - 1$.

Thus, the solution to the nonlinear system is the collection of ordered pairs

$$\left(\frac{5 + \sqrt{37}}{2}, 4 + \sqrt{37}\right) \quad \text{and} \quad \left(\frac{5 - \sqrt{37}}{2}, 4 - \sqrt{37}\right)$$

■

Using the Graphics Calculator

Check that `X²−3X−4` is still entered for `Y1` and that the Window 9.4 range settings are still set. Edit the rule for `Y2` to `2X−1`, and use the `TRACE` key. For an input near `⁻0.5`, both functions output a value near `⁻2.1`. Edit the `Ymax` setting to `12`, and press the `TRACE` key. At an input near `5.5`, both functions output a value near `10.4`.

As another example of solving a nonlinear system of equations, consider the following system, in which both functions are quadratic.

EXAMPLE 2 Solve: $\begin{cases} f(x) = x^2 - 3x - 4 \\ g(x) = -x^2 + 5 \end{cases}$

Solution Start by sketching the graph of the system, as shown in Figure 4. Equating the two function rules, we get

$$x^2 - 3x - 4 = -x^2 + 5$$

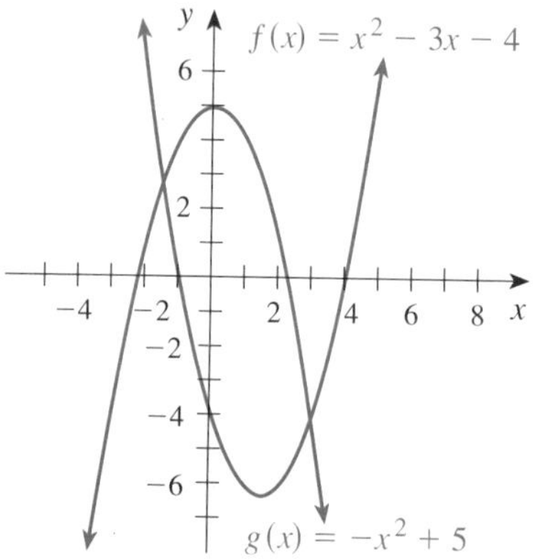

Solving this equation in one variable, we proceed as follows.

Standard form: $2x^2 - 3x - 9 = 0$

Factor the left-hand side: $(2x + 3)(x - 3) = 0$

$$2x + 3 = 0 \quad \text{or} \quad x - 3 = 0$$
$$2x = -3 \qquad\qquad x = 3$$
$$x = \frac{-3}{2}$$

FIGURE 4

The solution is $\frac{-3}{2}$ and 3.

Thus, two input values, $\frac{-3}{2}$ and 3, result in both function rules producing the same output.

For an input of $\frac{-3}{2}$, both $f(x) = x^2 - 3x - 4$ and $g(x) = -x^2 + 5$ output $\frac{11}{4}$.

For an input of 3, both $f(x) = x^2 - 3x - 4$ and $g(x) = -x^2 + 5$ output -4.

Thus, the solution to the system is $\left(\frac{-3}{2}, \frac{11}{4}\right)$ and $(3, -4)$. ■

GIVE IT A TRY

Solve each nonlinear system.

1. $\begin{cases} f(x) = -2x + 3 \\ h(x) = -x^2 + 3 \end{cases}$

2. $\begin{cases} y - 3x + 2 = 0 \\ y - (x - 2)^2 = 0 \end{cases}$

Systems Involving Conic Sections So far, we have considered systems involving functions. In the next example, we consider a system involving two of the conic sections discussed earlier in this chapter.

EXAMPLE 3 Solve: $\begin{cases} x^2 + y^2 = 9 \\ y = x^2 - 3 \end{cases}$

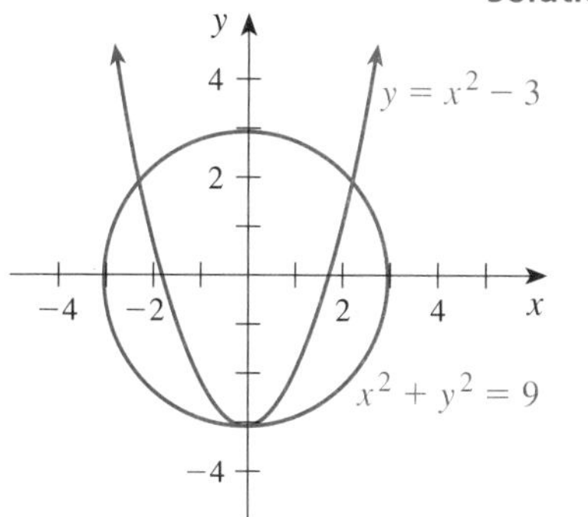

FIGURE 5

Solution A graph of this system is shown in Figure 5. Using the substitution technique, we get an equation in one variable:

$$y = x^2 - 3$$
$$\downarrow$$
$$x^2 + (x^2 - 3)^2 = 9$$

Now, we solve this equation in one variable.

Square the binomial: $x^2 + x^4 - 6x^2 + 9 = 9$

Standard form: $x^4 - 5x^2 = 0$

Factored form: $x^2(x^2 - 5) = 0$

$$x^2 = 0 \quad \text{or} \quad x^2 - 5 = 0$$

Solution: $0, \sqrt{5}$, and $-\sqrt{5}$

Thus, we have the x-coordinates of the three points where the curves intersect. Replacing the x-variable with these numbers in $y = x^2 - 3$ and computing the y-coordinates, we get the ordered pairs that solve the system: $(0, -3)$, $(\sqrt{5}, 2)$, and $(-\sqrt{5}, 2)$. ■

Using the Graphics Calculator

We first rewrite the equation $x^2 + y^2 = 9$ for y to get the functions $y = \sqrt{9 - x^2}$ and $y = -\sqrt{9 - x^2}$. Type √(9−X²) for Y1 and -√(9−X²) for Y2. Press the RANGE key and enter the Window 4.7 range settings. View the graph screen. A circle of radius 3 is drawn. For Y3, type the function rule X²−3. View the graph screen. After the circle is drawn, a parabola is drawn. As shown in Figure 6, the point $(0, -3)$ is a solution of the system. Press the TRACE key and approximate the other two ordered pairs that are solutions of this system.

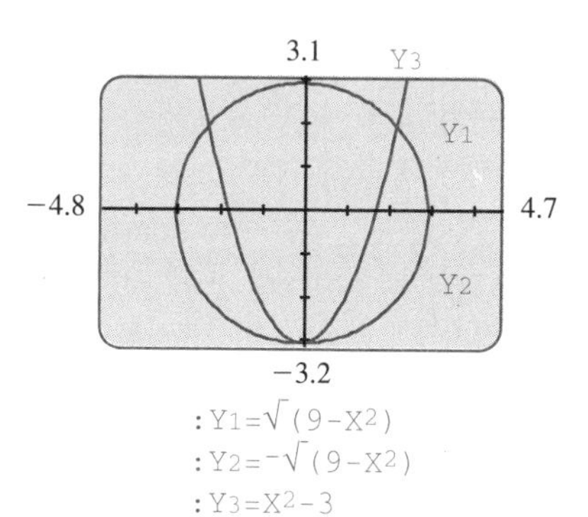

:Y1=√(9−X2)
:Y2=−√(9−X2)
:Y3=X2−3

FIGURE 6

Special Cases As we found for linear systems, for some systems of nonlinear equations in two variables, the solution is sometimes *none* or is an infinite set of ordered pairs. The following is an example of a system whose solution is *none*. Also, we can use the addition (elimination) approach to solve such a nonlinear system.

EXAMPLE 4 Solve: $\begin{cases} x^2 + y^2 = 25 & \textbf{(1)} \\ 9x^2 + 4y^2 = 36 & \textbf{(2)} \end{cases}$

Solution Here, we will use the addition (elimination) approach to solve the system. We multiply the first equation by -4 to get

$$\begin{cases} -4x^2 - 4y^2 = -100 & \text{Multiply equation (1) by } -4. \\ 9x^2 + 4y^2 = 36 \end{cases}$$
$$5x^2 = -64 \qquad \text{Sum equation}$$
$$x^2 = \frac{-64}{5} \qquad \text{Multiply by } \tfrac{1}{5}.$$

Here, the sum equation, $x^2 = -\frac{64}{5}$ has no real number in its solution. (Only the complex numbers $\frac{8i\sqrt{5}}{5}$ and $\frac{-8i\sqrt{5}}{5}$ are solutions to the equation.) Thus, we conclude that the solution to the system is *none*. ■

Using the Graphics Calculator

To check our work in Example 4, for Y1 type √(25−X²) and for Y2 type -√(25−X²). For Y3 type (1/2)√(36−9X²) and for Y4 type (-1/2)√(36−9X²). Use the Window 9.4 range settings to view the graph. The graph of the first equation is drawn, the circle. Next, the graph of the second equation is drawn, the ellipse. Figure 7 shows that the graph of this system has no intersection points and, thus, the system has no solution.

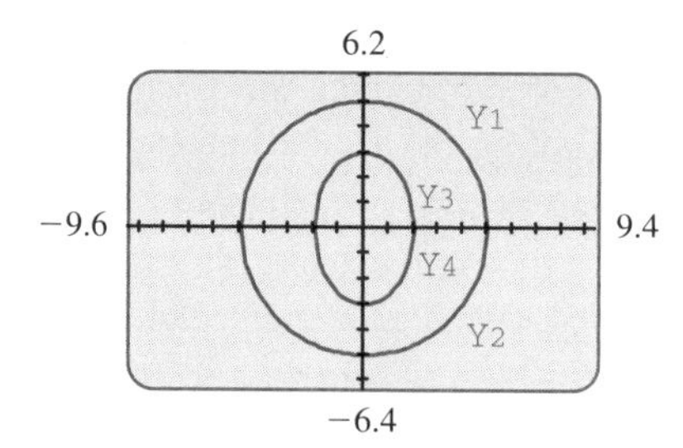

FIGURE 7

GIVE IT A TRY

Solve each nonlinear system involving conic sections.

3. $\begin{cases} 4x^2 + 3y^2 = 36 \\ x^2 - y^2 = 16 \end{cases}$

4. $\begin{cases} (x-2)^2 + y^2 = 4 \\ y = x - 1 \end{cases}$

Applications As an application of nonlinear systems, we now consider a physical situation for which a nonlinear system serves as a model. To learn how to find such a model, work through the next example.

EXAMPLE 5 The rectangular-shaped plate shown in Figure 8 must have an area measure of 70 cm². The perimeter of the plate is 34 cm. Find the measurements for the length and width of the plate.

Solution Because the requested quantities are the length and width, we define our variables as follows.

Let x = width of the rectangular plate
y = length of the rectangular plate

An expression for the perimeter is $2x + 2y$.
An expression for the area measure is xy.

Using the facts from the story, we write a nonlinear system as the mathematical model:

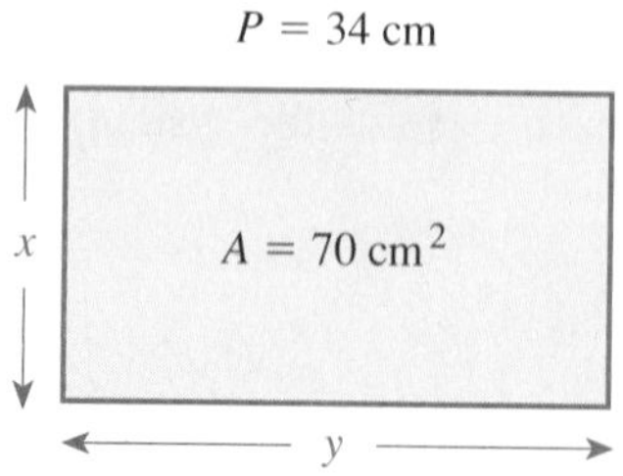

$P = 34$ cm

x $A = 70\text{ cm}^2$

y

FIGURE 8

Perimeter: $\begin{cases} 2x + 2y = 34 \\ xy = 70 \end{cases}$ Area: or $\begin{cases} x + y = 17 \\ y = \dfrac{70}{x} \quad x \neq 0 \end{cases}$

Using substitution, we get the equation: $x + \frac{70}{x} = 17$

Multiply each side by x: $x^2 + 70 = 17x$

Standard form: $x^2 - 17x + 70 = 0$

Factored form: $(x - 10)(x - 7) = 0$

Solution: 10 and 7

As is traditional, we will use the smaller value, 7, for the width. Replacing x by 7 in $y = \frac{70}{x}$, we find that the length is $\frac{70}{7}$, or 10 cm. ■

■ SUMMARY

In this section, we have learned how to solve a system of nonlinear equations in two variables. The solution to such a system is the set of ordered pairs that solve the equations in the system simultaneously. Unlike a linear system, where the solution is typically a single ordered pair, with nonlinear systems, the graphs of the solutions to the equations can intersect in more than one point. Thus, the solution can be several ordered pairs of numbers. Like linear systems, for some nonlinear systems the solution can be *none* (no ordered pairs of numbers solve each of the equations simultaneously) or the solution can be an infinite number of ordered pairs of numbers. We also learned that a nonlinear system can serve as a mathematical model for a particular physical situation.

12.4 ■ ■ ■ EXERCISES

Solve the given system. Use the graphics calculator to check your work.

1. $\begin{cases} y = 2x - 3 \\ y = -4x + 1 \end{cases}$

2. $\begin{cases} y = -3x + 5 \\ y = 3x + 1 \end{cases}$

3. $\begin{cases} y = x^2 - 1 \\ y = 2x - 2 \end{cases}$

4. $\begin{cases} y = 4 - x^2 \\ y = 5x - 2 \end{cases}$

5. $\begin{cases} y = x^2 - 3x - 2 \\ y = x^2 + 2x + 1 \end{cases}$

6. $\begin{cases} y = x^2 - x - 3 \\ y = 2x^2 - x + 7 \end{cases}$

7. $\begin{cases} y = x^2 \\ y = \sqrt{-x} \end{cases}$

8. $\begin{cases} y = -x^2 \\ y = -\sqrt{x} \end{cases}$

9. $\begin{cases} f(x) = x^2 - 3x - 2 \\ g(x) = x + 3 \end{cases}$

10. $\begin{cases} h(x) = x^2 - x \\ d(x) = 3x - 3 \end{cases}$

11. $\begin{cases} m(x) = |x - 2| \\ n(x) = |3 - x| \end{cases}$

12. $\begin{cases} h(x) = |x| - 3 \\ g(x) = |4 - x| \end{cases}$

13. $\begin{cases} d(x) = \sqrt{x + 2} \\ f(x) = x - 1 \end{cases}$

14. $\begin{cases} f(x) = \sqrt{2x - 1} \\ g(x) = x \end{cases}$

15. $\begin{cases} x^2 + y^2 = 4 \\ x + y = 1 \end{cases}$

16. $\begin{cases} x^2 + y^2 = 9 \\ y = x - 2 \end{cases}$

17. $\begin{cases} 4x^2 + y^2 = 16 \\ x^2 + y^2 = 4 \end{cases}$

18. $\begin{cases} x^2 + 3y^2 = 9 \\ x^2 + y^2 = 9 \end{cases}$

19. If the rectangular screen on a television set has a diagonal measure of 19 inches, and the area measure is 120 in^2, what are the measures of the length and the width (to the nearest hundredth inch)?

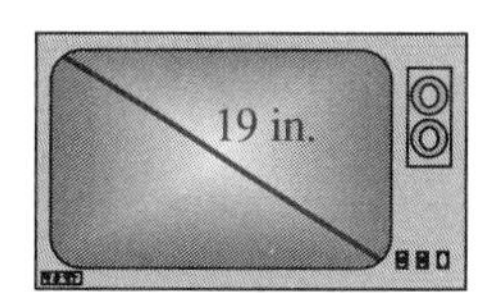

20. The diagonal of a rectangular flower garden must be 40 feet. If the area measure of the garden is to be 40 square feet, what are the length and width measures for the garden?

21. For the circle determined by $x^2 + y^2 = 4$, find the length of the chord determined by $y = x - 1$. [*Hint*: A chord of a circle is a line segment whose endpoints lie on the circle.]

22. For the circle determined by $x^2 + y^2 = 9$, find the length of the chord determined by $y = x - 2$.

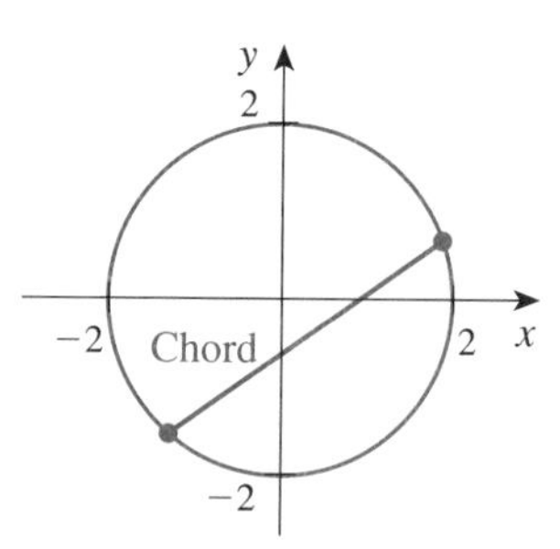

FIGURE FOR PROBLEMS 21 AND 22

12.5 ■ ■ ■ NONLINEAR INEQUALITIES AND SYSTEMS OF INEQUALITIES

In this section, we will learn to display the solution to a nonlinear inequality. Also, we will learn how to solve a system of inequalities in two variables. The solution to such a system is typically an infinite set of ordered pairs of numbers that form a region.

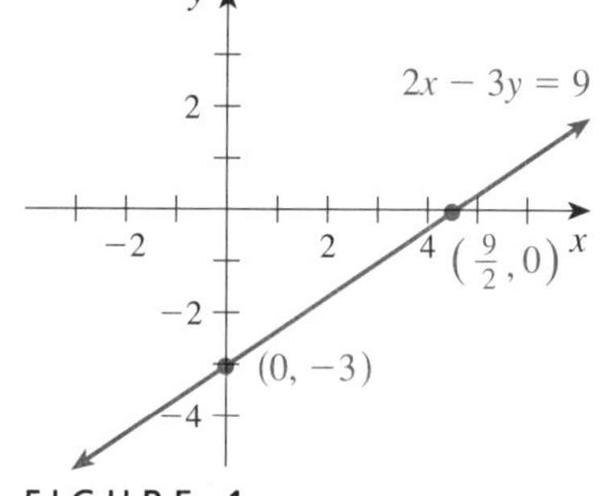

FIGURE 1

Nonlinear Inequalities In Section 3.3, we learned how to graph the solution to a linear inequality in two variables, such as $2x - 3y \leq 9$. The technique used involved the following steps:

1. Graph the solution to the related equation, $2x - 3y = 9$ in this case. This line is known as the key line. See Figure 1.
2. Determine if the key line drawn in step 1 should be solid (the ordered pairs represented by the key line solve the original inequality) or dashed (the ordered pairs represented by the key line do *not* solve the original inequality).
3. Test an ordered pair representing a point in the region above the key line. If it solves the original inequality, shade the region. If the ordered pair fails to solve the original inequality, leave the region unshaded. See Figure 2.

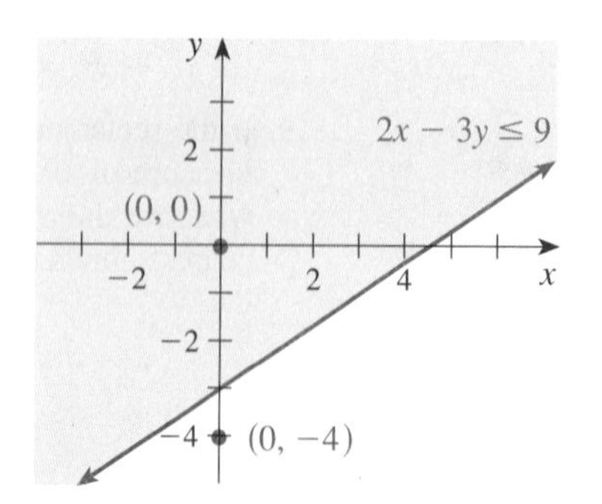

FIGURE 2

Test (0, 0): $2(0) - 3(0) \leq 9$ is true

Test (0, −4): $2(0) - 3(-4) \leq 9$ is false

4. Repeat step 3 for an ordered pair representing a point in the region below the key line.

We will now use this technique on a nonlinear inequality. Here, instead of a key line, we find a *key curve*. Work through the following example.

EXAMPLE 1 Graph the solution to $y \leq x^2 - 2x - 3$.

Solution We start with the equation. The graph of the solution to $y = x^2 - 2x - 3$ is a parabola. The y-intercept is $(0, -3)$.

Replacing y with 0 and solving the equation $0 = x^2 - 2x - 3$ yields a solution of 3 and -1. So, the x-intercepts are $(3, 0)$ and $(-1, 0)$.

The line of symmetry is $x = \dfrac{-b}{2a} = \dfrac{-(-2)}{(2)(1)} = 1$.

The vertex is on the line of symmetry, so its x-coordinate is 1. Replacing x by 1 in $y = x^2 - 2x - 3$ yields $y = (1)^2 - 2(1) - 3 = -4$. Thus, the vertex is $(1, -4)$. We plot this information, then draw the key curve (the parabola) shown in Figure 3.

The curve is solid (rather than dashed) because the points on the curve represent ordered pairs that solve the original inequality.

Trying a point in the region above the curve, say $(0, 0)$, we get a false arithmetic sentence, $0 \leq 0^2 - 2(0) - 3$. Thus, the region is left unshaded.

Trying a point in the region below the key curve, say $(0, -4)$, we get a true arithmetic sentence, $-4 \leq 0^2 - 2(0) - 3$. So, that region is shaded. The solution is graphed in Figure 4.

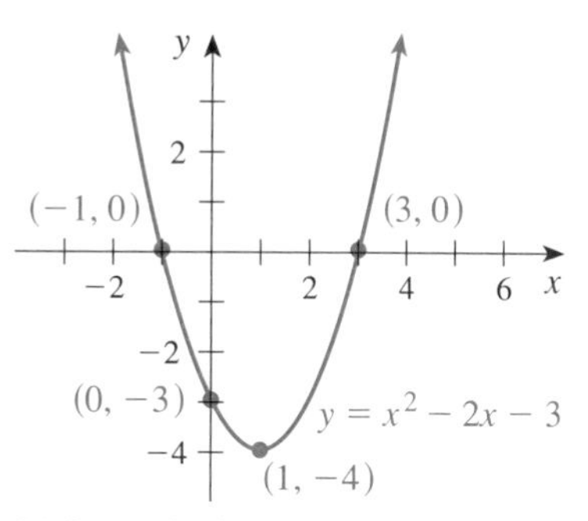

FIGURE 3

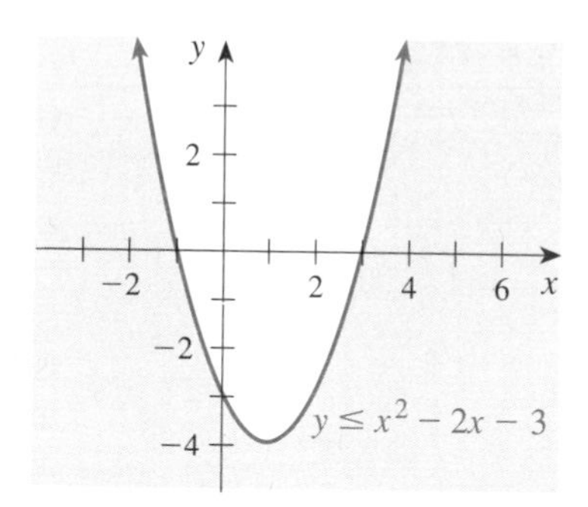

FIGURE 4

■

As an example of an inequality involving one of the conic sections, work through the following example.

EXAMPLE 2 Graph the solution to $x^2 + y^2 > 7$.

Solution The key curve is determined by the solution to $x^2 + y^2 = 7$. The graph of the solution to this equation is the circle centered at the origin with radius

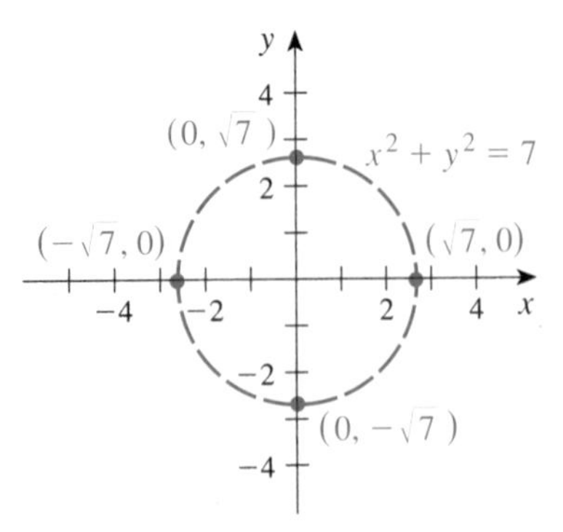

FIGURE 5

$\sqrt{7}$. To graph this key curve, we use a dashed line, because the ordered pairs represented by the points on the curve do not solve the original inequality, $x^2 + y^2 > 7$. See Figure 5.

Rather than dividing the plane into regions above and below the key curve as graphs of functions do, this conic section divides the plane into a region *inside* the key curve and a region *outside* the key curve.

Testing a point in the region inside the key curve, say (0, 0), we get $0^2 + 0^2 > 7$, a false arithmetic sentence. Thus, the region inside the curve is left unshaded.

Testing a point in the region outside the key curve, say (3, 0), we get $3^2 + 0^2 > 7$, a true arithmetic sentence. Thus, the region outside the key curve is shaded. See Figure 6. ■

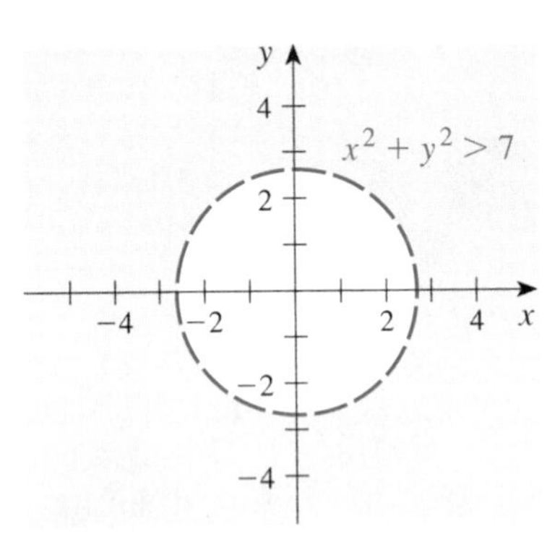

FIGURE 6

GIVE IT A TRY

Graph the solution to each nonlinear inequality.

1. $x > y^2 - 3y - 4$
2. $x^2 - y^2 \geq 4$ [*Hint:* The curve divides the plane into three regions.]

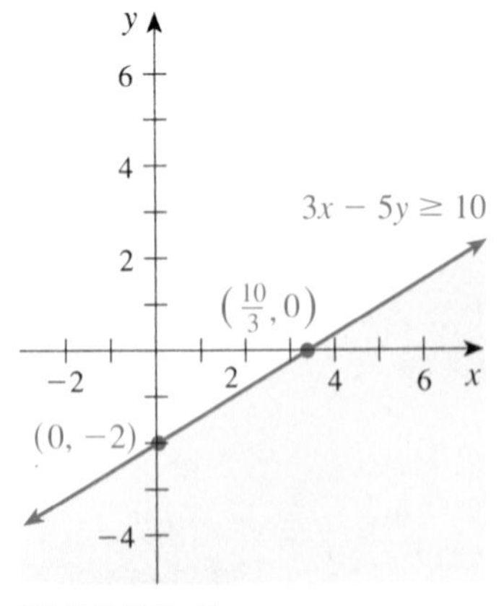

FIGURE 7
Solution to $3x - 5y \geq 10$

Systems of Linear Inequalities The solution to a system of linear *equations* is the ordered pairs (typically, a single ordered pair) that simultaneously solve the equations of the system. The solution to a linear system of *inequalities* is the ordered pairs that simultaneously solve the inequalities in the system. As with most inequalities, this collection of ordered pairs is infinite, and we resort to a graph to communicate the solution. Consider the following system of linear inequalities:

$$\begin{cases} 3x - 5y \geq 10 \\ 2x + y \leq 5 \end{cases}$$

The graph of the ordered pairs that are solutions to the first inequality, $3x - 5y \geq 10$, is the region below the key line (the line determined by $3x - 5y = 10$) and the points on the key line (because the relation is $\geq$). See Figure 7. The graph of the ordered pairs that are solutions to the second inequality, $2x + y \leq 5$, is the region below the key line (the line determined by

$2x + y = 5$) and the points on the key line (because the relation is $\leq$). See Figure 8.

The solution to the linear system of inequalities is the intersection (the overlap) of the two regions. See Figure 9. The point of intersection of the key lines is known as the **vertex** of the region. The ordered pair name of this point is $\left(\frac{35}{13}, \frac{-5}{13}\right)$, the solution to the linear system of equations.

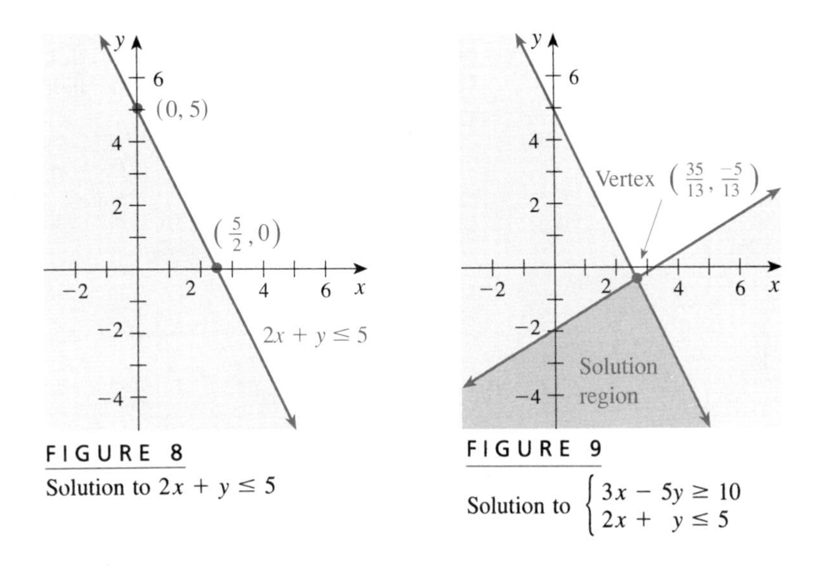

FIGURE 8

Solution to $2x + y \leq 5$

FIGURE 9

Solution to $\begin{cases} 3x - 5y \geq 10 \\ 2x + y \leq 5 \end{cases}$

A system of linear inequalities may have more than two inequalities in the system. Consider the following example.

EXAMPLE 3 Graph the solution to the system

$$\begin{cases} y \geq 3x - 2 \\ y < -2x + 6 \\ y \geq -2 \\ x > -1 \end{cases}$$

Name the vertices of the region.

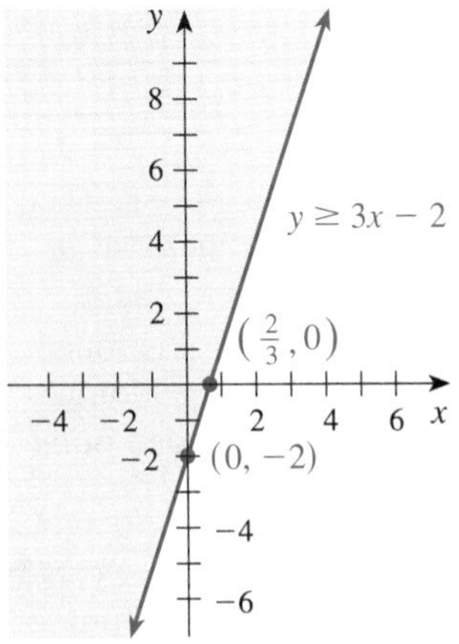

Solution We start with a graph of the solution to the key line determined by $y = 3x - 2$. Next, because an ordered pair in the region above the line, say (0, 0), solves the inequality, this region is shaded. Now, we have a graph of the solution to $y \geq 3x - 2$. See Figure 10.

Next, we graph the line $y = -2x + 6$ on the same set of axis. A dashed line is used because of the relation $<$. Since an ordered pair in the region below this line, say (0, 0), solves the inequality $y < -2x + 6$, the region is shaded. We now have the solution to

$$(y \geq 3x - 2) \quad \textit{and} \quad (y < -2x + 6).$$

FIGURE 10

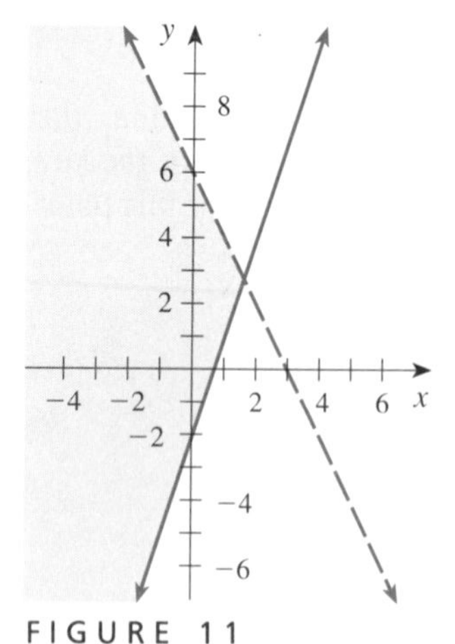

FIGURE 11

Solution to $\begin{cases} y \geq 3x - 2 \\ y < -2x + 6 \end{cases}$

This solution is shown in Figure 11.

Now, the key line determined by $y = -2$ is graphed, and a solid line is used because of the relation $\geq$. Since an ordered pair in the region above this key line, say (0, 0), solves the inequality $y \geq -2$, the region is shaded. We now have the solution to

$$(y \geq 3x - 2) \quad \textit{and} \quad (y < -2x + 6) \quad \textit{and} \quad (y \geq -2)$$

Figure 12 shows this solution.

Finally, the key line determined by $x = -1$ is graphed, and a dashed line is used because of the relation $>$. An ordered pair in the region right of this key line, say (0, 0), solves the inequality $x > -1$, so the region is shaded. We now have the solution to

$$(y \geq 3x - 2) \quad \textit{and} \quad (y < -2x + 6) \quad \textit{and} \quad (y \geq -2) \quad \textit{and} \quad (x > -1)$$

Figure 13 shows the graph of this solution.

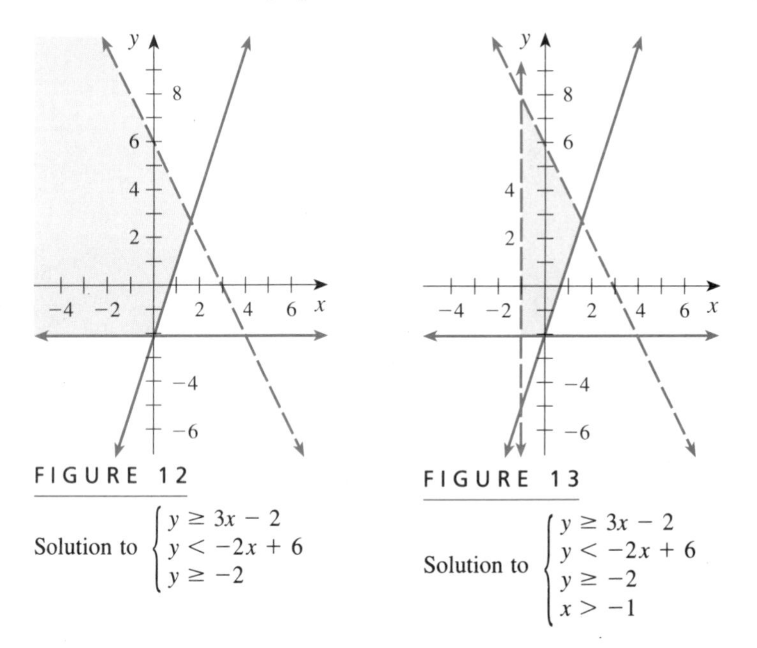

FIGURE 12

Solution to $\begin{cases} y \geq 3x - 2 \\ y < -2x + 6 \\ y \geq -2 \end{cases}$

FIGURE 13

Solution to $\begin{cases} y \geq 3x - 2 \\ y < -2x + 6 \\ y \geq -2 \\ x > -1 \end{cases}$

There are four vertices for this region. The easiest vertex to identify is the intersection of the horizontal line $y = -2$ and the vertical line $x = -1$. The ordered-pair name of this vertex is $(-1, -2)$. This vertex is not included in the solution region, so we use an open dot in the graph. See Figure 14.

The vertex determined by the key lines $y = -2$ and $y = 3x - 2$ is found by replacing y by -2 in $y = 3x - 2$ to get $-2 = 3x - 2$. This equation has solution 0. Thus, this vertex is named $(0, -2)$. This vertex is included in the solution region, so we use a solid dot.

The vertex determined by the key lines $x = -1$ and $y = -2x + 6$ is found by replacing x by -1 in $y = -2x + 6$ to get $y = -2(-1) + 6$. This equation has solution 8. Thus, this vertex is named $(-1, 8)$. This vertex is not in the solution region, so we graph it with an open dot.

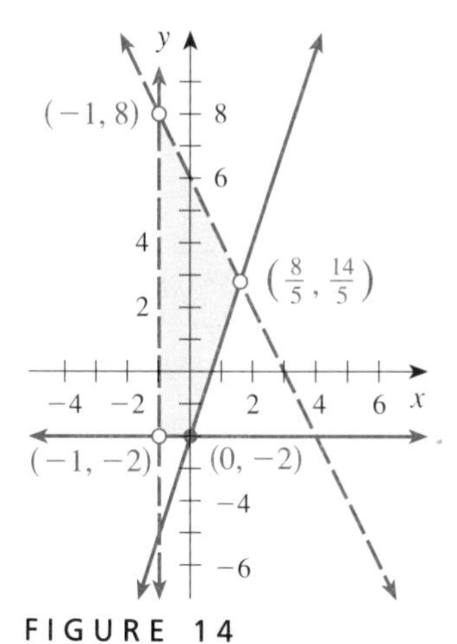

FIGURE 14

Vertices for solution region

Finally, the vertex determined by the key lines

$$y = 3x - 2 \qquad \text{and} \qquad y = -2x + 6$$

is found by solving the system

$$\begin{cases} y = 3x - 2 \\ y = -2x + 6 \end{cases}$$

The solution to this system is $\left(\frac{8}{5}, \frac{14}{5}\right)$. This vertex is not in the solution region, so we use an open dot.

Figure 14 shows the region with all the vertices labeled. ■

GIVE IT A TRY

3. Graph the solution to the given linear system of inequalities. Label each vertex of the region with its ordered-pair name.

$$\begin{cases} 3x - 2y \leq 6 \\ x \geq -1 \\ y \leq 2 \end{cases}$$

Nonlinear Systems Combining our skills in solving first- and second-degree equations in one variable, solving first- and second-degree inequalities in two variables, and solving linear and nonlinear systems, we are now ready to solve a system of nonlinear inequalities. The solution to such a system is a collection of ordered pairs that are solutions to each of the inequalities in the system. Typically, these ordered pairs form a region. Consider the following example.

EXAMPLE 4 Solve: $\begin{cases} y \leq 2x - 1 \\ y \geq x^2 - x - 6 \end{cases}$

Solution The graph of $y = 2x - 1$ is a line with y-intercept $(0, -1)$ and slope $m = 2$. The solution to $y \leq 2x - 1$ is the region above the line or the region below the line.

Trying $(0, 0)$, a point above the line, yields $0 \leq 2(0) - 1$, a false arithmetic sentence. Thus the region is left unshaded.

Trying $(0, -2)$, a point below the line, yields $-2 \leq 2(0) - 1$, a true arithmetic sentence. Thus the region below the line is shaded. See Figure 15.

The graph of $y = x^2 - x - 6$ is a parabola with y-intercept $(0, -6)$, x-intercepts $(3, 0)$ and $(-2, 0)$, line of symmetry $x = \frac{1}{2}$, and vertex $\left(\frac{1}{2}, \frac{-25}{4}\right)$. The solution to $y \geq x^2 - x - 6$ is a region above the parabola or the region below the parabola.

Trying $(0, 0)$, a point above the parabola, yields $0 \geq 0^2 - 0 - 6$, a true arithmetic sentence. Thus, the region above the parabola is shaded.

Trying $(0, -7)$, a point below the parabola, yields $-7 \geq 0^2 - 0 - 6$, a

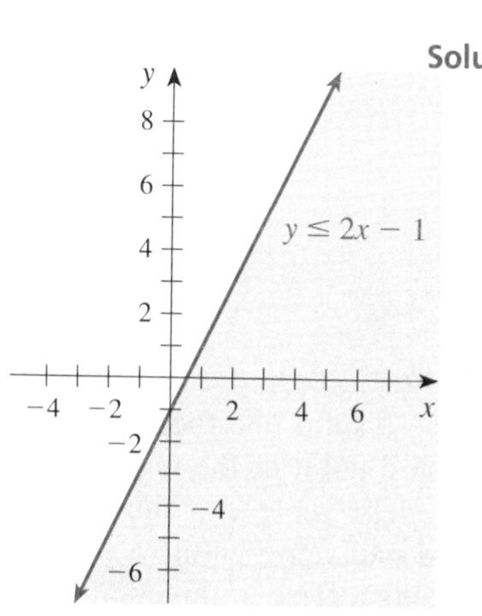

FIGURE 15

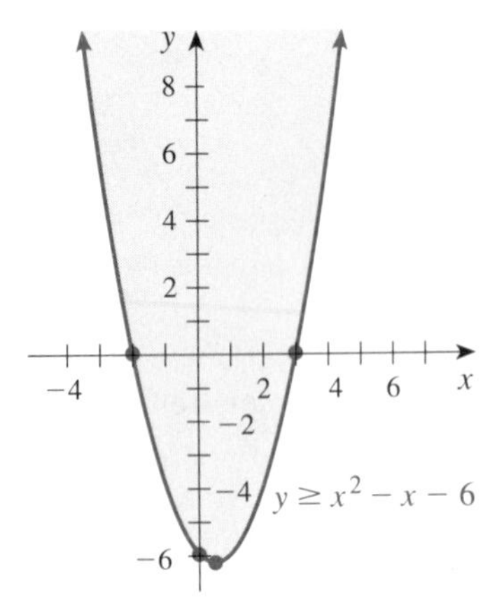

FIGURE 16

false arithmetic sentence. Thus, the region below the parabola is left unshaded. See Figure 16.

The solution to the system is the ordered pairs that solve

$$y \leq 2x - 1 \qquad \text{and} \qquad y \geq x^2 - x - 6$$

Graphically, the solution is the overlap, or intersection, of the two regions. See Figure 17.

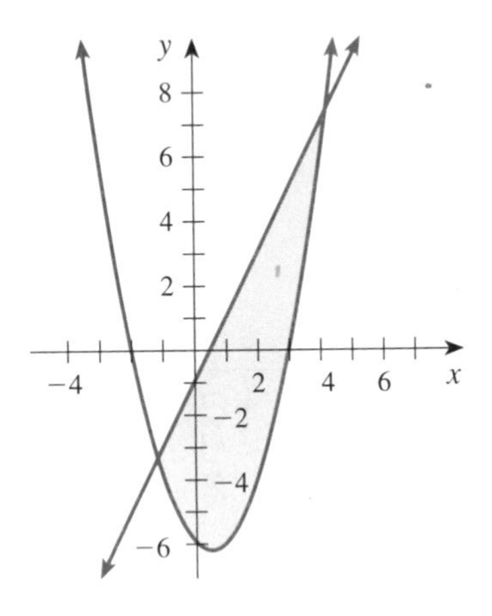

FIGURE 17

Solution to $\begin{cases} y \leq 2x - 1 \\ y \geq x^2 - x - 6 \end{cases}$

■

As a second example of solving a nonlinear system of inequalities, work through the next example, which involves the conic shapes of the ellipse and the hyperbola.

EXAMPLE 5 Graph the solution to $\begin{cases} 4x^2 + 9y^2 \leq 36 \\ x^2 - y^2 \geq 4 \end{cases}$

Solution Again, we start by graphing the solution to the first inequality,

$$4x^2 + 9x^2 \leq 36$$

To do this, we first graph the solution to the related equation,

$$4x^2 + 9x^2 = 36$$

or, in standard form,

$$\frac{x^2}{9} + \frac{y^2}{4} = 1 \qquad \text{Multiply by } \tfrac{1}{36}.$$

The graph is an ellipse with x-intercepts (3, 0) and (−3, 0) and y-intercepts (0, 2) and (0, −2). Drawing this key curve, we use a solid line (because of the relation ≥). Trying an ordered pair representing a point of the region inside the key curve, say (0, 0), yields a true arithmetic sentence. Thus, this region is shaded. See Figure 18. Trying an ordered pair representing a point of the region outside the key curve, say (0, 3), yields a false arithmetic sentence. Thus, that region is left unshaded.

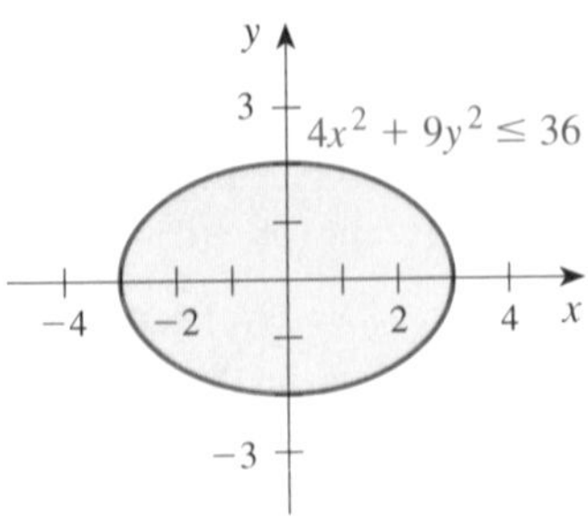

FIGURE 18

Solution to $4x^2 + 9y^2 \leq 36$

Next, we graph the solution to the second inequality in the system. The related equation is $x^2 - y^2 = 4$. This equation in standard form for a hyperbola is $\frac{x^2}{4} - \frac{y^2}{4} = 1$. We draw a box using the points (2, 0), (−2, 0), (0, 2), and (0, −2), and extend the diagonals for the asymptotes of the hyperbola. These asymptotes have equations $y = x$ and $y = -x$. Now, we draw the hyperbola (using solid lines, because of the relation $\geq$). Testing an ordered pair for a point from the right most region, say (3, 0), we get a true arithmetic sentence. This region is shaded. See Figure 19. Testing an ordered pair from the region between the key curves, say (0, 0), we get a false arithmetic sentence. Thus, the region between the key curves is left unshaded.

The solution to system of inequalities is the intersection (overlap) of the two regions. See Figure 20.

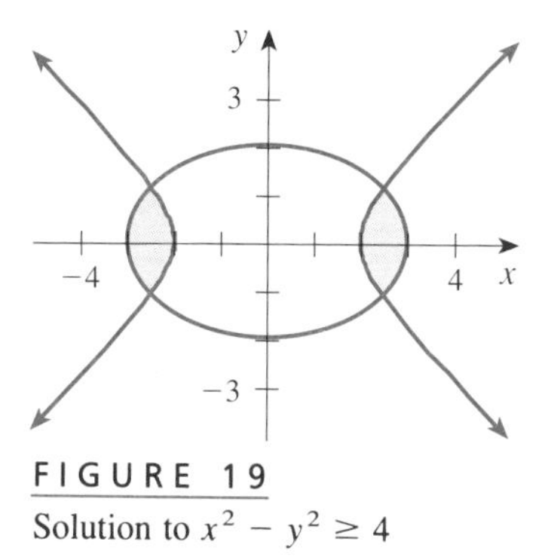

FIGURE 19

Solution to $x^2 - y^2 \geq 4$

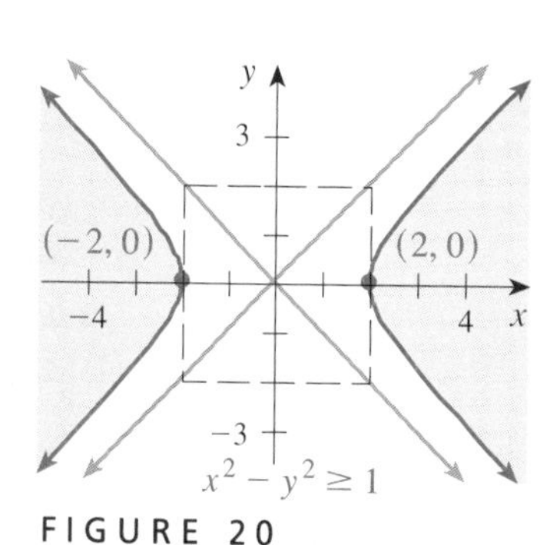

FIGURE 20

Solution to $\begin{cases} 4x^2 + 9y^2 \leq 36 \\ x^2 - y^2 \geq 4 \end{cases}$ ■

GIVE IT A TRY

4. Graph the solution: $\begin{cases} x^2 + y^2 < 4 \\ y \leq x^2 - 2 \end{cases}$

■ SUMMARY

In this section, we have learned to solve a nonlinear inequality, such as $x^2 - y^2 > 1$. The solution to such an inequality is the ordered pairs represented by the points of one or more of the regions created by the key curve.

We also solved systems of linear inequalities. The graph of such a system is the intersection of the regions determined by each of the inequalities in the system. A point where two sides of the region intersect is known as a vertex of the region. To find the ordered-pair name of a vertex, we solve a system of related linear equations.

As a final topic, we discussed solving nonlinear systems of inequalities. Here, the solution is the overlap (intersection) of the regions that solve each inequality in the system. To show this graphically, the region that represents the solution to the first inequality is shaded. The region that represents the solution to the second inequality is shaded. The region that represents the solution to the system is the intersection of these two shaded regions.

12.5 ■ ■ ■ EXERCISES

Solve the inequality. Use the graphics calculator to check your work.

1. $y \leq x^2 + 2x - 3$
2. $y \geq x^2 - 3x - 4$
3. $y > x^2 - 1$
4. $y < 4 - x^2$
5. $y < (x - 2)^2 - 2$
6. $y > (x + 2)^2 - 3$
7. $y < x^2$
8. $y \geq -x^2$
9. $x^2 + y^2 \geq 4$
10. $x^2 + y^2 > 9$
11. $(x - 1)^2 + y^2 < 9$
12. $x^2 + (y - 2)^2 < 7$

Graph the solution to the system of linear inequalities. For the solution region, use ordered pairs to name the vertex (or vertices).

13. $\begin{cases} x + y \geq 3 \\ 3x - y \leq 6 \end{cases}$
14. $\begin{cases} x - y \leq 5 \\ -x - 2y \geq 4 \end{cases}$
15. $\begin{cases} y \leq -2x + 5 \\ y > 4x - 1 \end{cases}$
16. $\begin{cases} y > 3x - 3 \\ y \leq 2x + 1 \end{cases}$
17. $\begin{cases} 4x + 2y \leq 4 \\ x + 4y \geq -2 \end{cases}$
18. $\begin{cases} 4x - y < 6 \\ -2x + 3y < 8 \end{cases}$
19. $\begin{cases} x - y > 4 \\ x + y \geq -3 \end{cases}$
20. $\begin{cases} 12x + 2y < 6 \\ 6x + y > 4 \end{cases}$
21. $\begin{cases} y \geq 2 \\ x < 3 \\ 2x + y \leq 8 \end{cases}$
22. $\begin{cases} x - 3y < 9 \\ x > -5 \\ y \leq 1 \end{cases}$
23. $\begin{cases} x + 3y \leq 8 \\ x - 2y \geq 3 \\ x - 2y \leq 6 \end{cases}$
24. $\begin{cases} 3x - 2y \leq 12 \\ 2x + 3y > 6 \\ y \leq 2 \end{cases}$
25. $\begin{cases} x + y \leq 4 \\ 2x - y \geq 3 \\ x \leq 3 \end{cases}$
26. $\begin{cases} x - y \leq 4 \\ y > 6 \\ x \leq 2 - y \end{cases}$

Graph the solution to each system. Use the graphics calculator to check your work.

27. $\begin{cases} y > 2x - 5 \\ y \geq x^2 - 3x - 4 \end{cases}$
28. $\begin{cases} y \leq x^2 + 6x - 7 \\ y \geq 6 - x - x^2 \end{cases}$
29. $\begin{cases} y > x^2 - 4 \\ y \leq 5 - x^2 \end{cases}$
30. $\begin{cases} y^2 > x^2 - 4 \\ y - x^2 \leq 5 \end{cases}$
31. $\begin{cases} y > 2x - 1 \\ x \geq y^2 - 3y - 4 \end{cases}$
32. $\begin{cases} x \leq y^2 + y - 2 \\ y \geq x^2 \end{cases}$
33. $\begin{cases} y > x^2 - 4 \\ x \leq 5 - y^2 \end{cases}$
34. $\begin{cases} y^2 > -x^2 + 4 \\ x - y^2 \geq 1 \end{cases}$

CHAPTER 12 REVIEW EXERCISES

Graph the solution to the given equations. Label the line of symmetry and the vertex.

1. $y = (x - 3)^2 + 1$
2. $y = (x + 3)^2 - 1$
3. $x = (y + 2)^2 - 1$
4. $x = (y - 2)^2 + 3$
5. $x = -(y + 1)^2$
6. $x = -(y + 1)^2 - 1$

Rewrite each equation to the form

$$y = a(x - h)^2 + k \qquad or \qquad x = a(y - h)^2 + k$$

Then graph the solution to the equation. Label the line of symmetry and the vertex. Use the graphics calculator to check your work.

7. $y = x^2 - 2x + 1$
8. $x = y^2 - 2y + 4$
9. $x = y^2 - 4y$
10. $y = x^2 + 4x$
11. $4y = x - y^2 + 1$
12. $x^2 - 3y + 2 = 5x$

Graph the solution to each equation. Label the center of the circle, the radius, and the x- and y-intercepts. Use the graphics calculator to check your work.

13. $x^2 + y^2 = 49$
14. $x^2 + y^2 = 81$
15. $x^2 + y^2 = 8$
16. $x^2 + y^2 = 15$
17. Report the equation of the circle that is centered at the origin and passes through the point $(-1, 4)$.
18. Report the equation of the circle that is centered at the origin and passes through the point $(-2, 3)$.

Graph the solution to each equation. Label the center of the circle, the radius, and the x- and y-intercepts. Use the graphics calculator to check your work.

19. $(x + 1)^2 + (y - 2)^2 = 9$
20. $(x - 2)^2 + (y + 4)^2 = 1$
21. Find the equation of the circle that is centered at $(-3, 2)$ and has radius 9.
22. Find the equation of the circle that is centered at $(-2, -1)$ and has radius 11.
23. For the line segment with endpoints $A(-2, 3)$ and $B(2, -1)$, find the equation of the circle for which line segment AB is a diameter.
24. For the line segment with endpoints $A(5, -3)$ and $B(-1, 1)$, find the equation of the circle for which line segment AB is a diameter.

Use the completing-the-square technique to write each equation in standard form for a circle. Graph the solution to the resulting equation. Label the center of the circle, the radius, and the x- and y-intercepts. Use the graphics calculator to check your work.

25. $x^2 + 6x + y^2 = 3$

26. $x^2 + y^2 + 4y - 1 = 0$

Read each graph and report the equation of the circle.

27.

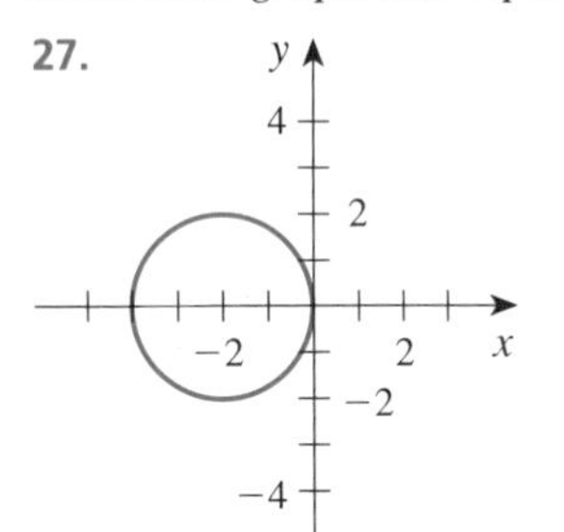

28.

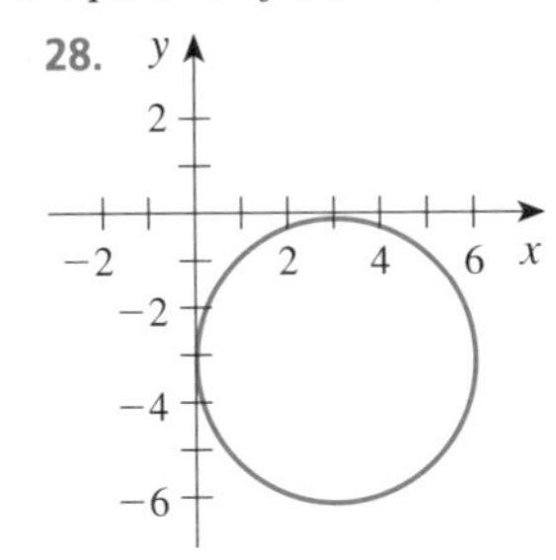

Graph the solution to the equation. Label the x- and y-intercepts. If the graph is a hyperbola, also report the equations of the asymptotes.

29. $3x^2 + y^2 = 9$

30. $x^2 - 9y^2 = 36$

31. $16x^2 - 9y^2 = 144$

32. $9x^2 + 16y^2 = 144$

33. $\dfrac{x^2}{16} + \dfrac{y^2}{4} = 1$

34. $\dfrac{x^2}{16} - \dfrac{y^2}{4} = 1$

35. $\dfrac{y^2}{1} - \dfrac{x^2}{16} = 1$

36. $\dfrac{x^2}{1} + \dfrac{y^2}{4} = 1$

Solve the given nonlinear system. Use the graphics calculator to check your work.

37. $\begin{cases} y = x^2 - 3 \\ y = 4x - 2 \end{cases}$

38. $\begin{cases} y = 4 - x^2 \\ y = x + 2 \end{cases}$

39. $\begin{cases} y = x^2 - 3x - 4 \\ y = x^2 - 2x + 1 \end{cases}$

40. $\begin{cases} y = x^2 - 2x - 3 \\ y = 2x^2 - x \end{cases}$

41. $\begin{cases} y = -x^2 \\ y = \sqrt{-x} \end{cases}$

42. $\begin{cases} y = -x^2 \\ y = \sqrt{x} \end{cases}$

43. $\begin{cases} x^2 + y^2 = 9 \\ x - y = 1 \end{cases}$

44. $\begin{cases} x^2 + y^2 = 1 \\ y = -x + 1 \end{cases}$

45. $\begin{cases} 4x^2 - y^2 = 16 \\ x^2 + y^2 = 9 \end{cases}$

46. $\begin{cases} x^2 - y^2 = 9 \\ x^2 + y^2 = 9 \end{cases}$

47. $\begin{cases} 4x^2 + y^2 = 16 \\ x^2 + y = 9 \end{cases}$

48. $\begin{cases} 4x^2 + y^2 = 16 \\ x + y^2 = 9 \end{cases}$

49. The area measure of a rectangle is 30 in^2 and its perimeter is 40 inches. Find the lengths of the sides of the rectangle.

50. For the circle determined by $x^2 + y^2 = 9$, find the length of the chord determined by $y = -x + 3$.

51. Find the intersection points of the circle where the line segment with endpoints at $(-1, 3)$ and $(1, -3)$ is a diameter, and the hyperbola determined by $y^2 = x^2 + 1$.

Graph the solution to each inequality.

52. $y \le 2 - x - x^2$

53. $4x^2 + y^2 \ge 16$

54. $x^2 - y^2 \ge 4$

Graph the solution to the system of inequalities.

55. $\begin{cases} 3x - 2y \le 6 \\ y \le 2 \\ x \ge 1 \end{cases}$

56. $\begin{cases} y \ge 2x - 3 \\ y < -x + 2 \\ x \ge -1 \end{cases}$

57. $\begin{cases} y \le -x^2 + 4 \\ y \ge 2x - 1 \end{cases}$

58. $\begin{cases} x^2 + y^2 \le 16 \\ y \ge x^2 - 1 \end{cases}$

CHAPTER 12 TEST

1. Graph the solution to the equation $x = -(y + 1)^2 - 1$. Label the vertex, the x- and y-intercepts, and the line of symmetry.

2. Report the equation of the circle that is centered at the origin and passes through the point $(-2, 3)$. Graph the circle.

3. For the line segment with endpoints $A(-3, 3)$ and $B(1, -3)$, find the equation of the circle for which line segment AB is a diameter. Graph the circle and draw the diameter $\overline{AB}$.

Graph the solution to the given equation. Label the x- and y-intercepts. If the graph is a hyperbola, report the equation of the asymptotes.

4. $x^2 + 3y^2 = 9$

5. $4x^2 - 9y^2 = 36$

6. Solve: $\begin{cases} y = x^2 + 2x - 3 \\ y = -x + 2 \end{cases}$

7. Using the solution found in Problem 6, graph the solution to the system

$$\begin{cases} y \ge x^2 + 2x - 3 \\ y \le -x + 2 \end{cases}$$

APPENDIXES

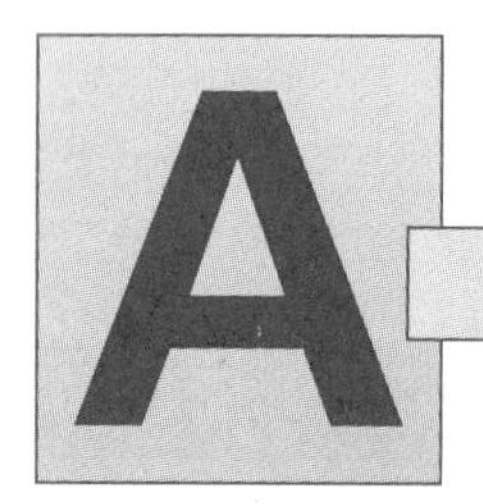

THE LCD

To add or subtract fractions, we first rename the fractions so that they have the same denominator. This same denominator is known as a *common denominator* for the two fractions. The product of the two denominators will always produce a common denominator. In this appendix we will find a *least common denominator*—a common denominator with fewer factors than any other common denominator. Consider the following example.

EXAMPLE 1 Write as a completely reduced rational number: $\frac{2}{3} - \frac{1}{5}$

Solution Because the denominators, 3 and 5, are unlike, we first rename the fractions so that they have the same denominator. That is, we find a common denominator. In this case, the product of the two denominators, 15, will be the common denominator. Renaming the fractions, we get

$$\frac{2}{3} \cdot \frac{5}{5} = \frac{10}{15} \quad \text{and} \quad \frac{1}{5} \cdot \frac{3}{3} = \frac{3}{15}$$

So,

$$\begin{aligned} \frac{2}{3} - \frac{1}{5} &= \frac{10}{15} - \frac{3}{15} \\ &= \frac{10 - 3}{15} \\ &= \frac{7}{15} \end{aligned}$$

■

When the denominators are unlike and have factors in common, we must find a **least common denominator**, an LCD. The LCD is the smallest number that both denominators will divide evenly (with no remainder). Suppose the unlike denominators are 12 and 10. Here, the product, 120, is a common denominator. Because the two numbers 12 and 10 have factors in common (namely, 2) we can find a common denominator smaller than the product of the denominators. We accomplish this by using the following steps:

Step 1 Completely (prime) factor each denominator.

Example $12 = (2)(2)(3)$ and $10 = (2)(5)$

Step 2 Form the least common denominator, the LCD, by taking each factor that appears in either factoring (2, 3, and 5, in our current example). If a factor appears more than once in a particular factoring, take the most number of times it appears in either factoring. So, in this example, the factor 2 will appear twice in the LCD. The LCD is (2)(2)(3)(5), or 60. Note that $60 < 120$.

EXAMPLE 2 Find the LCD for the fractions $\frac{2}{20}$ and $\frac{1}{18}$.

Solution Because the denominators, 20 and 18, have common factors, the LCD will be a number less than the product of 20 and 18. Following the preceding steps we get the following.

Step 1 Completely (prime) factor each denominator:

$$20 = (2)(2)(5) \quad \text{and} \quad 18 = (2)(3)(3)$$

Step 2 Taking each factor the most number of times it appears in either factoring, we get the LCD, (2)(2)(3)(3)(5), or 180. ■

We now show an example of adding two fractions, using an LCD.

EXAMPLE 3 Write as a completely reduced rational number: $\frac{5}{12} + \frac{3}{28}$

Solution Because the denominators, 12 and 28, are unlike, we first rename the fractions so that they have the same denominator. That is, we find a common denominator. In this case, the denominators have common factors, so we first find their LCD.

$$12 = (2)(2)(3) \quad \text{and} \quad 28 = (2)(2)(7)$$

So, the LCD is (2)(2)(3)(7), or 84. Renaming the fractions, we get

$$\frac{5}{12} \cdot \frac{7}{7} = \frac{35}{84} \quad \text{and} \quad \frac{3}{28} \cdot \frac{3}{3} = \frac{9}{84}$$

So,

$$\begin{aligned} \frac{5}{12} + \frac{3}{28} &= \frac{35}{84} + \frac{9}{84} \\ &= \frac{35 + 9}{84} \\ &= \frac{44}{84} \end{aligned}$$

Reducing the fraction $\frac{44}{84}$, we get $\frac{11}{21}$. ■

B

PROGRAMMING THE TI-81 AND THE TI-82

In this appendix we use the PRGM key to access a special memory area of the graphics calculator. We will enter program instructions and then have the calculator execute these stored instructions.

B.1 ■ ■ ■ PROGRAMMING THE TI-81

```
EXEC EDIT ERASE
1:Prgm1
2:Prgm2
3:Prgm3
4:Prgm4
5:Prgm5
6:Prgm6
7↓Prgm7
```

FIGURE 1

```
:(A+B)(A-B)=
:A²-B²
:
:(A+B)²=
:A²+2AB+B²
:
:(A-B)²=
:A²-2AB+B²
```

FIGURE 2

The PRGM Key As an introduction to the program area of memory on the TI-81, we first use the area to keep some notes. This application is not full usage of the area, and is intended only as an introduction.

Press the PRGM key. Figure 1 shows the PRGM menu, which has three headings: EXEC EDIT ERASE. Press the right arrow key to highlight the EDIT heading. Press the 1 key to select Prgm1 to edit. The screen changes to the prompt Prgm1: followed by the ALPHA cursor (the blinking A cursor). We can now type the name of the program. Type the word PRODUCT, and press the ENTER key. The cursor moves to the first line and changes to the overwrite cursor. To type A, press the ALPHA key and then the MATH key. Likewise, to type B, press the ALPHA key and then the MATRX key. To type the = symbol, press the 2nd key, the MATH key, and then the 1 key to select the = option from the TEST menu. Now, type (A+B)(A−B)= and press the ENTER key. The cursor moves to the second line. Type A^2-B^2 and press the ENTER key. Press the ENTER key again to leave a blank line. Now, on the fourth line, type $(A+B)^2=$ and press the ENTER key. On the fifth line, type $A^2+2AB+B^2$ and press the ENTER key. Press the ENTER key again to insert a blank line. Next, type $(A-B)^2=$ and press the ENTER key. [Notice that the top line scrolls off the screen. It is still in memory; however, its display is no longer on the screen.] Type $A^2-2AB+B^2$. The screen should now appear as shown in Figure 2. You can continue in this fashion entering lines of notes. To record and save these lines, quit to the home screen.

From the home screen, press the PRGM key. As you can see, the screen has changed. Now, after Prgm1 the name PRODUCT appears and the highlight is on the EXEC heading. Press the right arrow key to highlight the EDIT heading.

Press the 1 key and the contents of the PRODUCT program are presented. To see all the lines, press the down arrow key to move the cursor down the program lines. With the cursor on the last line, we could now add additional lines to the program. To move back up the program, press the up arrow key. Quit to the home screen.

```
ERROR 06 SYNTAX
1:Goto Error
2:Quit
```

FIGURE 3

The EXEC Heading Press the PRGM key. The highlight is on the EXEC heading. Press the 1 key to select program 1 to execute. The screen changes to the home screen with the characters Prgm1 displayed. Press the ENTER key to have the TI-81 execute the commands stored in the Prgm1 area of memory. In this case, the ERROR screen is presented, as shown in Figure 3, because the material we have entered for program 1 is just some notes, not commands that the calculator can execute. Press the 1 key to select the Goto Error option. The PRODUCT program is displayed, and the cursor is positioned at the end of the first line. This means the calculator did not understand this first line of the list of commands. Quit to the home screen. We will now insert some actual commands in the Prgm2 memory area.

THE RANGE PROGRAM

In our earlier work of viewing the graph screen, we have found that entering selected values for Xmin, Xmax, and so on, at the RANGE screen produces a friendly viewing screen (see page 92). That is, when using the TRACE feature, in the message at the bottom of the screen, the x-value changes in increments of 0.1, 0.2, and so forth. We will now enter instructions in program area 2 that will instruct the TI-81 to alter the range settings for us. These instructions are given in Figure 4.

```
Prgm2:RANGE
:Disp "CENTER"
:Disp "X="
:Input X
:Disp "Y="
:Input Y
:Disp "XFACTOR"
:Input A
:Disp "YFACTOR"
:Input B
:X-4.8A→Xmin
:X+4.7A→Xmax
:Y-3.2B→Ymin
:Y+3.1B→Ymax
:A→Xscl
:B→Yscl
```

FIGURE 4

Press the PRGM key and then use the right arrow key to highlight the EDIT heading. Press the 2 key to select the Prgm2 memory area. The screen changes to present the prompt Prgm2: with the ALPHA cursor waiting for the program name to be typed. Type RANGE and then press the ENTER key. As the first instruction, press the PRGM key. (Notice that the menu for this key is different than when the key is pressed from the home screen—see Figure 5.) The headings are now CTL I/O EXEC. Press the right arrow key to highlight the I/O (input/output) heading. The I/O menu is shown in Figure 6. Press the 1 key to select the Disp option. We are returned to the program entry area and the first line now contains the characters Disp. We must now tell the calculator *what* to display. We want to display the word CENTER. To do this, type a quotation mark (press the ALPHA key followed by the + key). Now, type the characters CENTER, and then type another quotation mark. Press the ENTER key to move to the next line.

```
CTL I/O EXEC
1:Lbl
2:Goto
3:If
4:IS>(
5:DS<(
6:Pause
7↓End
```

FIGURE 5

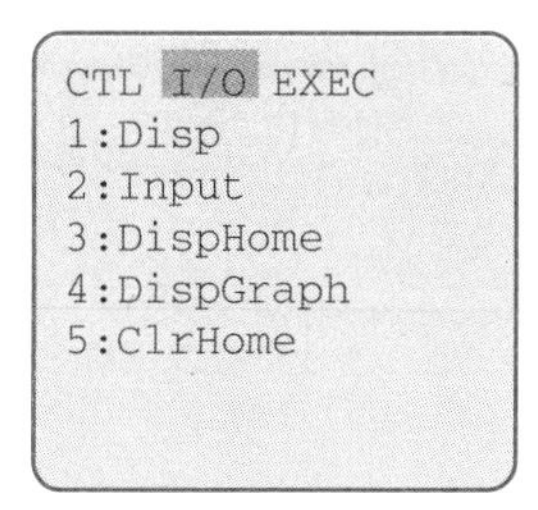

FIGURE 6

```
Prgm2:RANGE
:Disp "CENTER"
:Disp "X="
:Input X
:Disp "Y="
:Input Y
:Disp "XFACTOR"
:Input A
:Disp "YFACTOR"
:Input B
:X-4.8A→Xmin
:X+4.7A→Xmax
:Y-3.2B→Ymin
:Y+3.1B→Ymax
:A→Xscl
:B→Yscl
```

FIGURE 7

Repeat the preceding actions for the second line, DISP ''X=''. (Use the 2nd key, followed by the MATH key to access the = sign.) For the third line, press the PRGM key, the right arrow key, and then the 2 key for the Input option. The characters Input are written on line 3. We now need to select the variable for which to input a number. Press the X | T key to get the command Input X. Continue in this fashion to type in the other lines, through the line Input B (see Figure 7).

For the line X−4.8A→Xmin type X−4.8A and then press the STO▶ key. Next, press the VARS key and use the right arrow key to highlight the RNG heading. See Figure 8. Press the 1 key to select the Xmin option. Repeat these steps to enter the remaining instructions for the RANGE program.

Use the up and down arrow keys to scroll through the lines, checking each program command. If you find an entry error, use the INS and DEL keys to correct the command. When you have finished, quit to the home screen.

To run the program, press the PRGM key, the 2 key, and then the ENTER key. The message CENTER, followed by the prompt X=, is presented. Type 0 and press the ENTER key. The prompt Y= is presented. Type 0, and press the ENTER key. The prompt XFACTOR is presented. Type 1 and press the ENTER key. Likewise, enter 1 for the YFACTOR prompt. See Figure 9. The program ends its execution, and displays the characters Done. Press the RANGE key to see the alterations the program has made to the range settings.

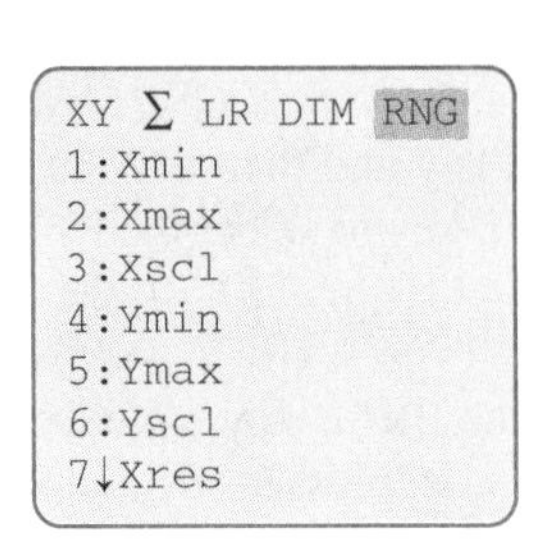

FIGURE 8

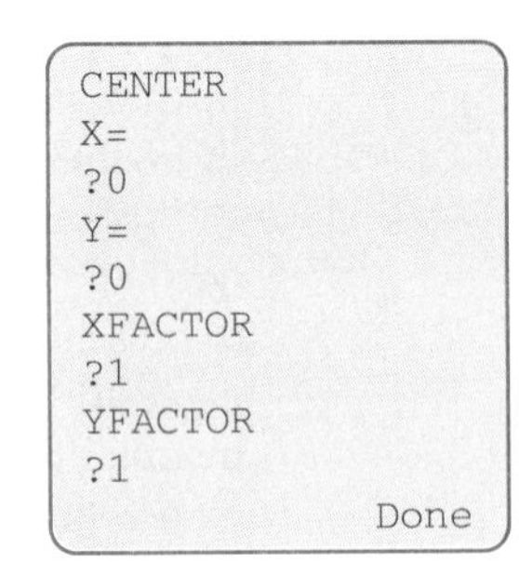

FIGURE 9

Press the Y= key and use the CLEAR key to erase any existing entries. For Y1, type 3X−1. Press the GRAPH key and use the TRACE key to see that these settings do indeed result in a friendly viewing screen.

Summary of the TI-81 Programming Techniques and Keys

1. Press the PRGM key, highlight the EDIT heading, and then select Prgm1, Prgm2, or other program. In this memory area, you can store material, such as formulas. Once you have entered one of these storage areas, the ALPHA cursor is presented, and you can type the name of the program (up to eight characters). Then press the ENTER key. The cursor moves to the first line and becomes the overwrite cursor. To leave this area, press the 2nd key, followed by the CLEAR key, to quite to the home screen.
2. Type the instructions in the program area for the calculator to execute (at a later time). Two important commands are the Disp com-

mand and the Input command (both found on the I/O menu, which is displayed by pressing the PRGM key while in the program entry area). The Disp command writes to the home screen any characters enclosed in quotation marks. The command Disp ``X='' will output X= on the home screen when the program is executed. The Input command specifies in which variable, X, Y, Xmax, and so forth, a value is to be stored. That is, Input X will output a ? on the home screen. When you type a number and press ENTER, the value is stored in memory location X.

3. Pressing the PRGM key and then a number, such as 2, results in the characters Prgm2 being written to the home screen. Pressing the ENTER key results in the instructions stored in that memory area being executed. If the calculator does not understand the instructions, the ERROR screen is presented.

EXERCISES

1. Press the Y= key, and use the CLEAR key to erase any existing entries. For Y1 type ⁻2X+3. Press the ZOOM key and select the Standard option. Press the TRACE key. As you can see by the message at the bottom of the screen, these numbers are not very friendly. Press the PRGM key, and then the 2 key, followed by the ENTER key, to run the RANGE program. For the X= and Y= prompts, type 0. [This sets the center of the graph screen to the origin, (0, 0).] For XFACTOR enter 5, and for YFACTOR enter 5. Press the TRACE key. When the x-value is replaced by 4, what is the value of the polynomial $-2x + 3$? When the x-value is replaced by -4, what is the value of the polynomial $-2x + 3$?
2. With ⁻2X+3 still entered for Y1, run the RANGE program. Enter 0 for the X= and Y= prompts. For the XFACTOR and YFACTOR prompts, enter .5. Press the TRACE key. When the x-variable is replaced by 1.35, what is the value of $-2x + 3$? When the x-value is replaced by 1.5, what is the value of $-2x + 3$?
3. Enter ⁻X²+3 for Y1, and run the RANGE program. Enter 0 for the X= prompt and 2 for the Y= prompt. For the XFACTOR and YFACTOR prompts, enter 1. Press the TRACE key. When the x-variable is replaced by -1, what is the value of $-x^2 + 3$? When the x-value is replaced by 1.5, what is the value of $-x^2 + 3$?
4. Display on the screen the contents of the RANGE program. Move to the last line, and press the ENTER key to add a line. Now, press the PRGM key, highlight the I/O heading, and then press the 4 key to select the DispGraph option. Quit to the home screen. Press the Y= key and erase any existing entries. For Y1, type X²−3X+2. Run the RANGE program and enter 0 for the X= and Y= prompts and 2 for the XFACTOR and YFACTOR prompts. When the x-variable is 2, what is the evaluation of $x^2 - 3x + 2$?
5. With X²−3X+2 entered for Y1, run the RANGE program and experiment with the XFACTOR and YFACTOR entries to find the x-value that produces the smallest evaluation for $x^2 - 3x + 2$. What is that x-value?

THE EQLINE PROGRAM

To write a program that instructs the graphics calculator to output the equation of a line, we can input the ordered-pair names of two points on the line, say (a, b) and (c, d). With these inputs, the TI-81 can compute the slope. From the value for the slope and one of the points, the TI-81 can compute the y-intercept. Finally, the TI-81 can output the value for the slope and the y-intercept, and we can write the equation of the line.

Press the PRGM key, then the right arrow key (for the Edit menu), and press the 3 key to select Prgm3. The program prompt is presented. Type the name EQLINE. (Remember, the cursor is the ALPHA cursor.) Press the ENTER key. Now type the commands listed in Figure 1. To type the Pause command, press the PRGM key, and then the 6 key, to select the Pause command from the CTL menu. When you are finished, quit to the home screen.

```
Prgm3:EQLINE
:Disp "X1"
:Input A
:Disp "Y1"
:Input B
:Disp "X2"
:Input C
:Disp "Y2"
:Input D
:(B-D)/(A-C)→M
:abs (A-C)→D
:Disp "SLOPE"
:M*D→F
:Disp F
:Disp "OVER"
:Disp D
:Pause
:Disp "Y-Int"
:(B-MA)*D→E
:Disp E
:Disp "OVER"
:Disp D
```

FIGURE 1

Running EQLINE We will now find the equation of the line that passes through the points $(-5, 3)$ and $(4, -1)$. Press the PRGM key, followed by the 3 key, to select the EQLINE line program. Press the ENTER key to execute the program. Type ⁻5 in response to the X1? prompt. Press the ENTER key. Continue, as shown in Figure 2, to enter the coordinates of the points. After the last coordinate is entered, the slope of the line is output, as shown in Figure 3. Press the ENTER key (for the Pause command in the program instructions). The y-coordinate of the y-intercept is then output, as shown in Figure 4. Thus, the equation of the line is through $(-5, 3)$ and $(4, -1)$ is

$$y = \frac{-4}{9}x + \frac{7}{9}$$

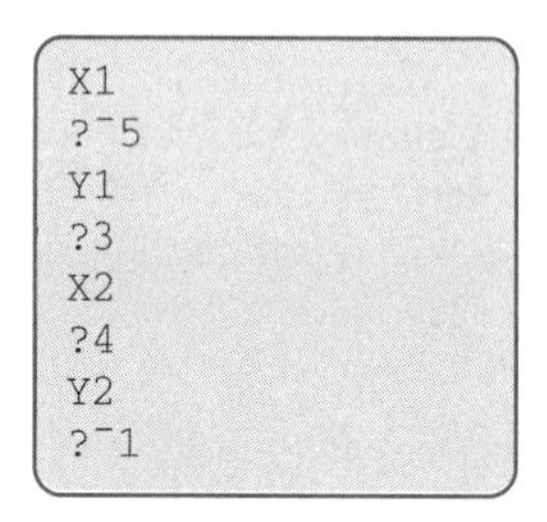

FIGURE 2

Entering $(-5, 3)$ and $(4, -1)$.

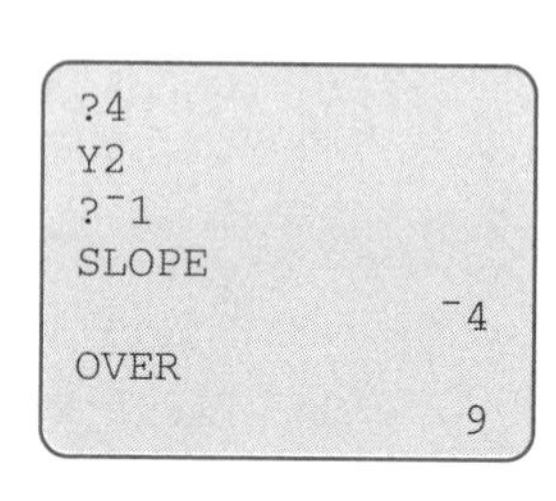

FIGURE 3

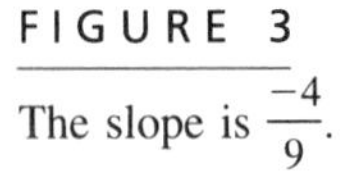

The slope is $\frac{-4}{9}$.

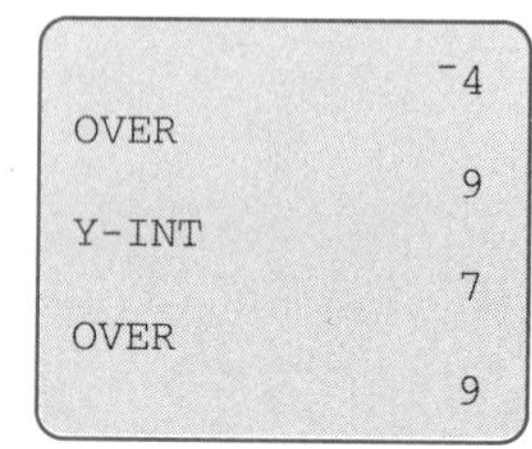

FIGURE 4

The y-intercept is $\left(0, \frac{7}{9}\right)$.

EXERCISES

1. Use the EQLINE program to find the equation of the line that passes through $(5, 7)$ and $(1, 1)$.
2. Use the EQLINE program to find the equation of the line that passes through $(2, 3)$ and has slope measure -3.
[*Hint*: Another point on the line is $(2 + \Delta x, 3 + \Delta y)$, with Δx and Δy from the slope, $m = \Delta y/\Delta x$.]

THE QUAD PROGRAM

The quadratic formula procedure is so routine that we can program it on the graphics calculator. That is, we will create a program which outputs the decimal approximations for the numbers that are solutions to a second-degree equation in one variable. Before writing this program based on the quadratic formula, we must make a decision considering the discriminant. This decision-making action is accomplished using a conditional command. We enter the conditional command on the calculator by selecting the `If` option from the `CTL` menu of the `PRGM` menu, as shown in Figure 1.

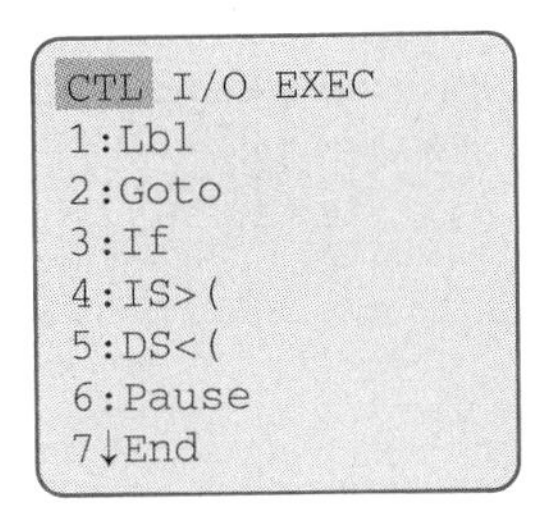

FIGURE 1

A typical conditional command is

```
:If B²-4AC≥0
:Disp ''TRUE''     ←Conditional B²-4AC≥0 is true.
:Disp ''FALSE''    ←Conditional B²-4AC≥0 is false.
```

Here, the expression `B²-4AC` is computed. If the result is larger than or equal to 0, the program executes the `Disp ''TRUE''` command. Otherwise, the program skips the `Disp ''TRUE''` command and executes the `Disp ''FALSE''` command. To develop a program to solve equations using the quadratic formula, we will use the following example as a model.

EXAMPLE 1 Solve: $x^2 - x - 3 = 0$

Solution Using the quadratic formula to solve the equation, we follow these steps:

Step 1 Write the equation in standard form: $x^2 - x - 3 = 0$

Step 2 Identify A, B, and C: $A = 1, B = -1, C = -3$

Step 3 Compute $B^2 - 4AC$: $(-1)^2 - 4(1)(-3) = 13$

Step 4 If $B^2 - 4AC$ is greater than or equal to 0, write the solution. Otherwise, the solution does not contain real numbers.

Solution: $\dfrac{1 + \sqrt{13}}{2}$ and $\dfrac{1 - \sqrt{13}}{2}$ ■

Based on Example 1, we can write a program, as shown in Figure 2.

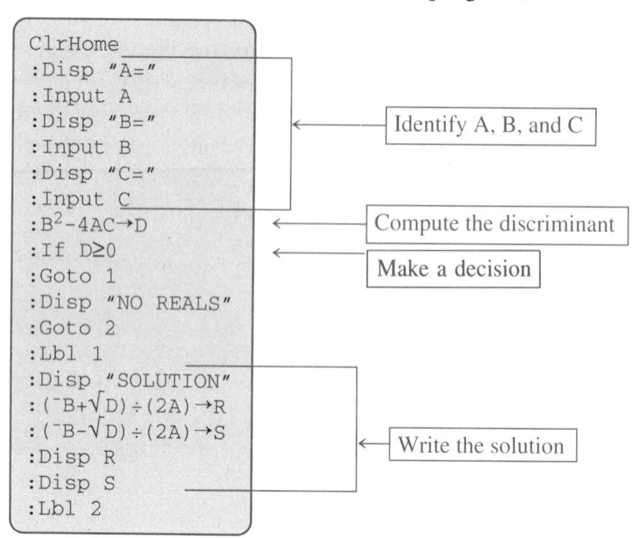

FIGURE 2
The QUAD program

Entering the Program Press the PRGM key, followed by the right arrow key, to select the EDIT option. Select the first unprogrammed slot. Type the name QUAD, and then press the ENTER key.

The A-LOCK (alpha-lock) feature is very useful for keying in words, such as ''NO REALS'' and ''SOLUTION''. Press the 2nd key, followed by the ALPHA key, to lock the cursor in ALPHA mode. The ALPHA key is pressed to return to the normal mode.

Type the commands, as listed in Figure 2. After typing the last line (Lbl 2), quit to the home screen.

Running the Program We now execute the program to solve the equation $x^2 - x - 3 = 0$. Press the PRGM key and the select the QUAD program. Press the ENTER key to run the program.

The program starts its execution by clearing the screen (the ClrHome command) and presenting the prompt A=. Type 1, and then press the ENTER key. Next, the prompt B= is presented. Type ⁻1, and then press the ENTER key. The prompt C= is presented. Type ⁻3, and then press the ENTER key. The word SOLUTION is written on the screen, followed by two outputs (from memory locations R and S). The output is shown in Figure 3. To run the program again, press the ENTER key. The screen will clear and the A= prompt will be displayed.

```
A=
?1
B=
?⁻1
C=
?⁻3
SOLUTION
      2.302775638
     -1.302775638
```

FIGURE 3

```
SOLUTION
                1
+√
               13
OVER
                2
```

FIGURE 4

EXERCISES

1. Use the QUAD program to solve each equation.

   a. $(x - 3)(x + 5) = 0$ **b.** $2x^2 - 3x = 5$

2. Edit the QUAD program to output the exact values of the solution to a second-degree equation in one variable. The screen display in Figure 4 represents the number $\dfrac{1 + \sqrt{13}}{2}$.

THE DIVPOLY PROGRAM

A plan (program design) for writing a program to do synthetic division of polynomials is presented below.

Get divisor, A.

Get degree of dividend, N (must be less than 6).

Set dimension of matrix A and matrix B to 1-by-N.

Get coefficients for dividend.

Do algorithm, store results in matrix B.

Report answer, matrix B.

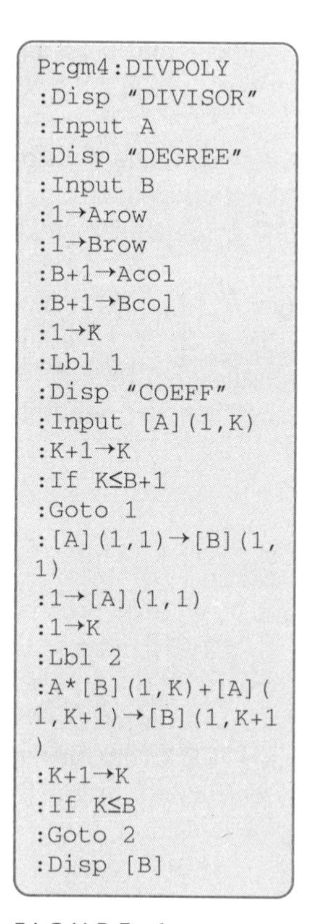

```
Prgm4:DIVPOLY
:Disp "DIVISOR"
:Input A
:Disp "DEGREE"
:Input B
:1→Arow
:1→Brow
:B+1→Acol
:B+1→Bcol
:1→K
:Lbl 1
:Disp "COEFF"
:Input [A](1,K)
:K+1→K
:If K≤B+1
:Goto 1
:[A](1,1)→[B](1,
1)
:1→[A](1,1)
:1→K
:Lbl 2
:A*[B](1,K)+[A](
1,K+1)→[B](1,K+1
)
:K+1→K
:If K≤B
:Goto 2
:Disp [B]
```

FIGURE 1

This design produces the DIVPOLY program displayed in Figure 1.

Press the PRGM key, then the right arrow key for EDIT, and the number of the first open program (probably 4). With the ALPHA cursor displayed, type the name DIVPOLY, and press the ENTER key. Now enter the commands listed in Figure 1. Use the following keystrokes to type specific commands:

Command	*Keystrokes*
Arow	VARS → → → 1
Brow	VARS → → → 3
Acol	VARS → → → 2
Bcol	VARS → → → 4
[A]	2nd 1
[B]	2nd 2

The symbol [A] or [B] refers to a *matrix,* which is introduced in Chapter 11. For now, simply think of a matrix as a list of memory locations where numbers can be stored.

After entering the commands, check the commands for typing errors. (Use your editing skills to correct any error.) Quit to the home screen.

Running the DIVPOLY Program To test the DIVPOLY program, we will use the example

$$(x^3 + 2x^2 - 4) \div (x - 3) \quad \text{is} \quad x^2 + 5x + 15, \text{ remainder } 41$$

From the home screen, press the PRGM key and select the DIVPOLY program. The DIVPOLY program starts its execution by presenting the prompt DIVISOR. Because the divisor in $(x^3 + 2x^2 - 4) \div (x - 3)$ is $x - 3$, type 3 and then press the ENTER key.

```
DIVISOR
?3
DEGREE
?3
COEFF
?1
COEFF
?
```

FIGURE 2

```
COEFF
?2
COEFF
?0
COEFF
?⁻4
[ 1 5 15 41 ]
```

FIGURE 3

Next, the prompt DEGREE is presented. Because the dividend,

$$x^3 + 2x^2 - 4$$

is a third-degree polynomial, type 3 and press the ENTER key. Now enter the coefficients for $x^3 + 2x^2 - 4$. For the prompt COEFF, type 1 and press the ENTER key. See Figure 2. Continue in this fashion, entering 2, 0, and ⁻4 for the coefficients of x^2, x, and the constant term.

After pressing the ENTER key for the last coefficient, the synthetic division algorithm is executed, and the matrix [1 5 15 41] is written to the screen. See Figure 3. Translating this matrix to a polynomial and a remainder yields $1x^2 + 5x + 15$, remainder 41.

As a second test of the program, consider the product $(2x + 3)(x - 5)$, which yields $2x^2 - 7x - 15$. Now, the division $(2x^2 - 7x - 15) \div (2x + 3)$ should produce $x - 5$, remainder 0. For synthetic division (and the DIVPOLY program) to work, the divisor must have x-coefficient 1. Thus, we multiply dividend and divisor by $\frac{1}{2}$ to get

$$\left(x^2 - \frac{7}{2}x - \frac{15}{2}\right) \div \left(x + \frac{3}{2}\right)$$

Execute the DIVPOLY program. Enter ⁻1.5 for the DIVISOR prompt. Enter 2 for the DEGREE prompt. Enter the coefficients 1, ⁻3.5, and ⁻7.5. The output is [1 ⁻5 0]. Thus, the quotient is $x - 5$, remainder 0.

If the last line output by DIVPOLY is too wide to be displayed, press the 2nd key followed by the 2 key for the [B] key. Press the ENTER key to display matrix B. Now, press the right arrow key to display any entries not originally visible.

EXERCISES

1. Use the DIVPOLY program to perform each division. (Remember to report the remainder.)
   a. $(x^3 + 2x^2 - 4x - 2) \div (x - 1)$
   b. $(x^4 + x^3 - x - 2) \div (x + 5)$
   c. $(3x^3 + x^2 - 4x) \div (x - 2)$
   d. $(32x^5 - 1) \div (2x - 1)$
2. If one factor of $2x^4 - 5x^3 + 2x^2 + 6x - 3$ is $2x - 1$, what is the other factor?
3. If one number that solves $x^3 + 5 = 4x^2 + 2x$ is 1, find the other numbers in the solution.

B.2 ■ ■ ■ PROGRAMMING THE TI-82

FIGURE 1

The PRGM Key As an introduction to the program area of memory on the TI-82, we first use the area to keep some notes. This application is not full usage of the area, and is intended only as an introduction.

Press the PRGM key. Figure 1 shows the PRGM menu, which has three headings: EXEC EDIT NEW. Press the right arrow key to highlight the NEW heading. Press the 1 key to select the Create New option. The screen changes to the prompt Name=, followed by the ALPHA cursor (the blinking A cursor). We can now type the name of the program. Type the word PRODUCT, and press the ENTER key. The cursor moves to the first line and changes to the overwrite cursor. To type A, press the ALPHA key and then the MATH key. Likewise, to type B, press the ALPHA key and then the MATRX key. To type the = symbol, press the 2nd key, the MATH key, and then the 1 key to select the = option from the TEST menu. Now, type (A+B)(A−B)= and press the ENTER key. The cursor moves to the second line. Type A^2-B^2 and press the ENTER key. Press the ENTER key again to leave a blank line. Now, on the fourth line, type $(A+B)^2=$ and press the ENTER key. On the fifth line, type $A^2+2AB+B^2$ and press the ENTER key. Press the ENTER key again to insert a blank line. Next, type $(A-B)^2=$ and press the ENTER key. [Notice that the top line scrolls off the screen. It is still in memory; however, its display is no longer on the screen.] Type $A^2-2AB+B^2$. The screen should now appear as shown in Figure 2. You can continue in this fashion entering lines of notes. To record and save these lines, quit to the home screen.

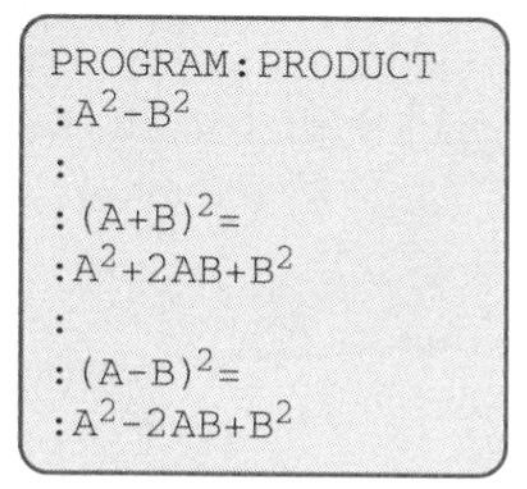

FIGURE 2

From the home screen, press the PRGM key. As you can see, the screen has changed. Now, the name PRODUCT appears and the highlight is on the EXEC heading. Press the right arrow key to highlight the EDIT heading. Press the 1 key and the contents of the PRODUCT program are presented. To see all the lines, press the down arrow key to move the cursor down the program lines. With the cursor on the last line, we could now add additional lines to the program. To move back up the program, press the up arrow key. Quit to the home screen.

The EXEC Heading Press the PRGM key. The highlight is on the EXEC heading. Press the 1 key to select program 1 to execute. The screen changes to the home screen with the characters PrgmPRODUCT displayed. Press the ENTER key to have the TI-82 execute the commands stored in the Prgm-PRODUCT area of memory. In this case, the ERROR screen is presented, as shown in Figure 3, because the material we have entered for program 1 is just some notes, not commands that the calculator can execute. Press the 1 key to select the Goto option. The PRODUCT program is displayed, and the cursor is positioned at the end of the first line. This means the calculator did not understand this first line of the list of commands. Quit to the home screen. We will now insert some actual commands in the program memory area.

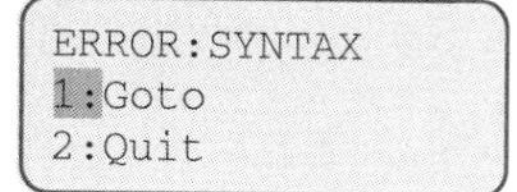

FIGURE 3

THE RANGE PROGRAM

In our earlier work of viewing the graph screen, we have found that entering selected values for Xmin, Xmax, and so on, at the WINDOW screen produces a friendly viewing screen (see page 92). That is, when using the TRACE feature,

```
PROGRAM:RANGE
:Disp "CENTER"
:Prompt X,Y
:Disp "XFACTOR"
:Prompt A
:Disp "YFACTOR"
:Prompt B
:X-4.7A→Xmin
:X+4.7A→Xmax
:Y-3.1B→Ymin
:Y+3.1B→Ymax
:A→Xscl
:B→Yscl
```

FIGURE 4

in the message at the bottom of the screen, the x-value changes in increments of 0.1, 0.2, and so forth. We will now enter instructions in program area 2 that will instruct the TI-82 to alter the range settings. These instructions are given in Figure 4.

Press the PRGM key and then use the right arrow key to highlight the NEW heading. Press the 1 key to select the Create New option. The screen changes to present the prompt Name= with the ALPHA cursor waiting for the program name to be typed. Type RANGE and then press the ENTER key. As the first instruction, press the PRGM key. (Notice that the menu for this key is different than when the key is pressed from the home screen—see Figure 5.) The headings are now CTL I/O EXEC. Press the right arrow key to highlight the I/O (input/output) heading. The I/O menu is shown in Figure 6. Press the 3 key to select the Disp option. We are returned to the program entry area and the first line now contains the characters Disp. We must now tell the calculator *what* to display. We want to display the word CENTER. To do this, type a quotation mark (press the ALPHA key followed by the + key). Now, type the characters CENTER, and then type another quotation mark. Press the ENTER key to move to the next line.

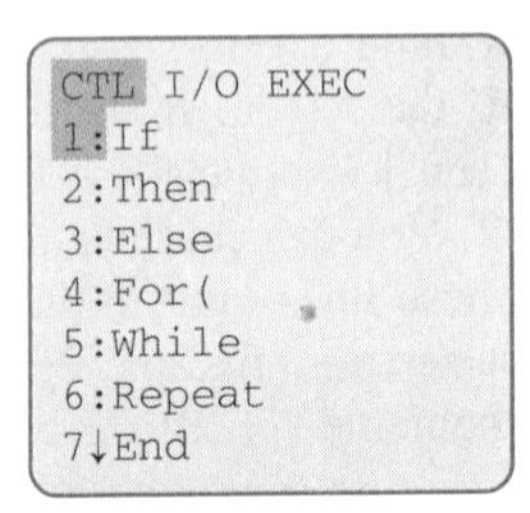

FIGURE 5

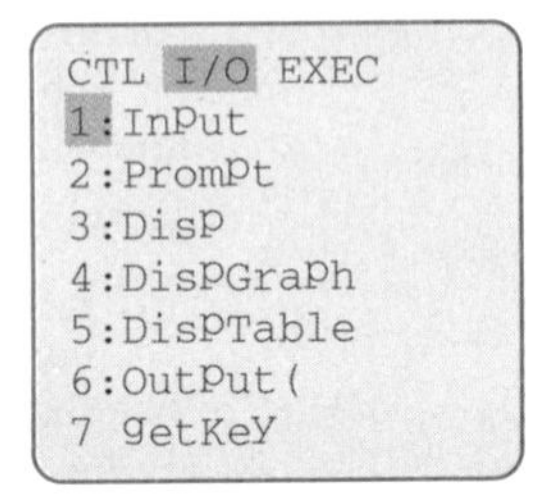

FIGURE 6

For the second line, select the Prompt option from the PRGM-I/O menu, then type X,Y to get the command Prompt X,Y. Continue in this fashion to type the other lines, through the line Prompt B (see Figure 4).

For the line X−4.7A→Xmin, type X−4.7A and then press the STO▶ key. Next, press the VARS key, followed by the 1 key to select the Window option. See Figure 7. Press the 1 key to select the Xmin option. Repeat these steps to enter the remaining instructions for the RANGE program.

```
X/Y T/θ U/V
1:Xmin
2:Xmax
3:Xscl
4:Ymin
5:Ymax
6:Yscl
7↓ΔX
```

FIGURE 7

Use the up and down arrow keys to scroll through the lines, checking each program command. If you find an entry error, use the INS and DEL keys to correct the command. When you have finished, quit to the home screen.

To run the program, press the PRGM key, select the program RANGE, and then press the ENTER key. The message CENTER, followed by the prompt X=?, is presented. Type 0 and press the ENTER key. The prompt Y=? is presented. Type 0, and press the ENTER key. The prompt XFACTOR is presented. Type 1 and press the ENTER key. Likewise, enter 1 for the YFACTOR prompt. See Figure 8. The program ends its execution. Press the WINDOW key to see the alterations the program has made to the range settings.

```
CENTER
X=?0
Y=?0
XFACTOR
A=?1
YFACTOR
B=?1
```

FIGURE 8

Press the Y= key and use the CLEAR key to erase any existing entries. For Y1, type 3X−1. Press the TRACE key to see that these settings do indeed result in a friendly viewing screen.

Summary of the TI-82 Programming Techniques and Keys

1. Press the PRGM key, highlight the NEW heading, and then type a name for the program memory area. In this memory area, you can store material, such as formulas. Once you have entered one of these storage areas, the cursor moves to the first line and becomes the overwrite cursor. To leave this area, press the 2nd key, followed by the MODE key, to quite to the home screen.

2. Type the instructions in the program area for the calculator to execute (at a later time). Two important commands are the Disp command and the Prompt command (both found on the I/O menu, which is displayed by pressing the PRGM key while in the program entry area). The Disp command writes to the home screen any characters enclosed in quotation marks. The command Disp ''A='' will output A= on the home screen when the program is executed. The Prompt command specifies in which variable, X, Y, Xmax, and so forth, a value is to be stored. That is, Prompt X will output X=? on the home screen. When you type a number and press ENTER, the value is stored in memory location X.

3. Pressing the PRGM key and then selecting the program RANGE results in the characters PrgmRANGE being written to the home screen. Pressing the ENTER key results in the instructions stored in the PrgmRANGE memory area being executed. If the calculator does not understand the instructions, the ERROR screen is presented.

EXERCISES

1. Press the Y= key, and use the CLEAR key to erase any existing entries. For Y1 type -2X+3. Press the ZOOM key and select the Standard option. Press the TRACE key, and then the right arrow key. As you can see by the message at the bottom of the screen, these numbers are not very friendly. Press the PRGM key, and select the program RANGE. For the X= and Y= prompts, type 0. [This sets the center of the graph screen to the origin, (0, 0).] For XFACTOR enter 5, and for YFACTOR enter 5. Press the TRACE key. When the x-value is replaced by 4, what is the value of the polynomial $-2x + 3$? When the x-value is replaced by -4, what is the value of the polynomial $-2x + 3$?

2. With -2X+3 still entered for Y1, run the RANGE program. Enter 0 for the X= and Y= prompts. For the XFACTOR and YFACTOR prompts, enter .5. Press the TRACE key. When the x-variable is replaced by 1.35, what is the value of $-2x + 3$? When the x-value is replaced by 1.5, what is the value of $-2x + 3$?

3. Enter -X²+3 for Y1, and run the RANGE program. Enter 0 for the X= prompt and 2 for the Y= prompt. For the XFACTOR and YFACTOR prompts, enter 1. Press the TRACE key. When the x-variable is replaced by -1, what is the value of $-x^2 + 3$? When the x-value is replaced by 1.5, what is the value of $-x^2 + 3$?

4. Press the PRGM key, highlight the EDIT heading, and select the PrgmRANGE option. Move to the last line, and press the ENTER key to add a line. Now, press the PRGM key, highlight the I/O heading, and then press the 4 key to select the DispGraph option. Quit to the home screen. Press the Y= key and erase any existing entries. For Y1, type X²−3X+2. Run the RANGE program and enter 0 for the X= and Y= prompts and 2 for the XFACTOR and YFACTOR prompts. When the x-variable is 2, what is the evaluation of $x^2 - 3x + 2$?
5. With X²−3X+2 entered for Y1, run the RANGE program and experiment with the XFACTOR and YFACTOR entries to find the x-value that produces the smallest evaluation for $x^2 - 3x + 2$. What is that x-value?

THE EQLINE PROGRAM

To write a program that instructs the graphics calculator to output the equation of a line, we can input the ordered-pair names of two points on the line, say (a, b) and (c, d). With these inputs, the calculator can compute the slope. From the value for the slope and one of the points, the calculator can compute the y-intercept. Finally, the calculator can output the value for the slope and the y-intercept, and we can write the equation of the line.

Press the PRGM key, then the right arrow key (for the New menu), and then the 1 key. The prompt Name= is presented. Type the name EQLINE. (Remember, the cursor is the ALPHA cursor.) Press the ENTER key. Now type the commands listed in Figure 1. To type the Pause command, press the PRGM key, and then the 8 key, to select the Pause command from the CTL menu. When you are finished, quit to the home screen.

```
PROGRAM:EQLINE
:Disp "FIRST
POINT"
:Prompt A,B
:Disp "SECOND
POINT"
:Prompt C,D
:(B-D)/(A-C)→M
:abs (A-C)→D
:Disp "SLOPE"
:M*D→F
:Disp F
:Disp "OVER"
:Disp D
:Pause
:Disp "Y-Int"
:(B-MA)*D→E
:Disp E
:Disp "OVER"
:Disp D
```

FIGURE 1

Running EQLINE We will now find the equation of the line that passes through the points $(-5, 3)$ and $(4, -1)$. Press the PRGM key, and select the EQLINE program. Press the ENTER key to execute the program. Type ⁻5 in response to the A=? prompt. Press the ENTER key. Continue, as shown in Figure 2, to enter the coordinates of the points.

```
FIRST POINT
A=?⁻5
B=?3
SECOND POINT
C=?4
D=?⁻1
```

FIGURE 2
Entering $(-5, 3)$ and $(4, -1)$.

After the last coordinate is entered, the slope of the line is output, as shown in Figure 3. Press the ENTER key (for the Pause command in the program instructions). The y-coordinate of the y-intercept is then output, as shown in Figure 4. Thus, the equation of the line is through $(-5, 3)$ and $(4, -1)$ is

$$y = \frac{-4}{9}x + \frac{7}{9}$$

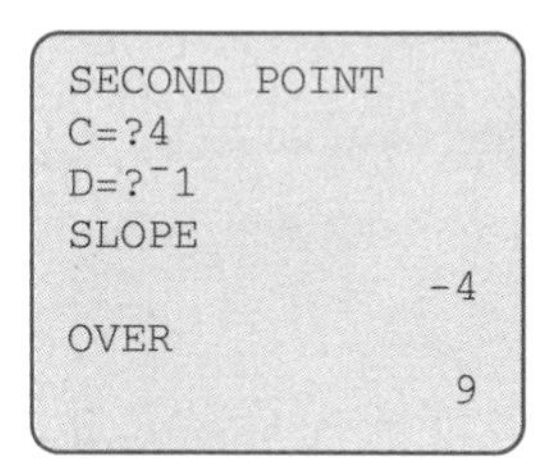

FIGURE 3

The slope is $\frac{-4}{9}$.

```
OVER
                9
Y-INT
                7
OVER
                9
             Done
```

FIGURE 4

The y-intercept is $\left(0, \frac{7}{9}\right)$.

EXERCISES

1. Use the EQLINE program to find the equation of the line that passes through (5, 7) and (1, 1).
2. Use the EQLINE program to find the equation of the line that passes through (2, 3) and has slope measure -3.
   [*Hint*: Another point on the line is $(2 + \Delta x, 3 + \Delta y)$, with Δx and Δy from the slope, $m = \Delta y/\Delta x$.]

THE QUAD PROGRAM

The quadratic formula procedure is so routine that we can program it on the graphics calculator. That is, we will create a program which outputs the decimal approximations for the numbers that are solutions to a second-degree equation in one variable. Before writing this program based on the quadratic formula, we must make a decision considering the discriminant. This decision-making action is accomplished using a conditional command. We enter the conditional command on the calculator by selecting the If option from the CTL menu of the PRGM menu, as shown in Figure 1. A typical conditional command is

```
CTL I/O EXEC
1:If
2:Then
3:Else
4:For(
5:While
6:Repeat
7↓End
```

FIGURE 1

```
:If B²−4AC≥0
:Then
:Disp ''TRUE''          ←Conditional B²−4AC≥0 is true.
:Else
:Disp ''FALSE''         ←Conditional B²−4AC≥0 is false.
:End
```

Here, the expression B^2-4AC is computed. If the result is larger than or equal to 0, the program executes the Disp ''TRUE'' command. Otherwise, Else, the program executes the Disp ''FALSE'' command, which follows the Else command. To develop a program to solve equations using the quadratic formula, we will use the following example as a model.

EXAMPLE 1 Solve: $x^2 - x - 3 = 0$

Solution Using the quadratic formula to solve the equation, we follow these steps:

Step 1 Write the equation in standard form: $x^2 - x - 3 = 0$

Step 2 Identify A, B, and C: $A = 1, B = -1, C = -3$

Step 3 Compute $B^2 - 4AC$: $(-1)^2 - 4(1)(-3) = 13$

Step 4 If $B^2 - 4AC$ is greater than or equal to 0, write the solution. Otherwise, the solution does not contain real numbers.

Solution: $\dfrac{1 + \sqrt{13}}{2}$ and $\dfrac{1 - \sqrt{13}}{2}$ ■

```
PROGRAM:QUAD
:ClrHome
:Prompt A,B,C
:B²-4AC→D
:If D≥0
:Then
:Disp "SOLUTION"

:(⁻B+ D)÷(2A)→R
:(⁻B- D)÷(2A)→S
:Disp R
:Disp S
:Else
:Disp "NO REALS"

:End
```

FIGURE 2

The QUAD program

Based on Example 1, we can write a program, as shown in Figure 2.

Entering the Program Press the PRGM key, followed by the right arrow key, to select the NEW option. Type the name QUAD, and then press the ENTER key. To enter the ClrHome command, press the PRGM key, highlight the I/O heading, then press the 8 key to select the ClrHome command.

The A-LOCK (alpha-lock) feature is very useful for keying in words, such as ''NO REALS'' and ''SOLUTION''. Press the 2nd key, followed by the ALPHA key, to lock the cursor in ALPHA mode. The ALPHA key is pressed to return to the normal mode.

Type the commands, as listed in Figure 2. After typing the last line (End), quit to the home screen.

Running the Program We now execute the program to solve the equation $x^2 - x - 3 = 0$. Press the PRGM key and the select the QUAD program. Press the ENTER key to run the program.

The program starts its execution by clearing the screen (the ClrHome command) and presenting the prompt A=?. Type 1, and then press the ENTER key. Next, the prompt B=? is presented. Type ⁻1, and then press the ENTER key. The prompt C=? is presented. Type ⁻3, and then press the ENTER key. The word SOLUTION is written on the screen, followed by two outputs (from memory locations R and S). The output is shown in Figure 3. To run the program again, press the ENTER key. The screen will clear and the A=? prompt will be displayed.

To terminate the execution of any program, press the ON key.

```
A=
?1
B=
?⁻1
C=
?⁻3
SOLUTION
      2.302775638
     -1.302775638
             Done
```

FIGURE 3

EXERCISES

1. Use the QUAD program to solve each equation.

   a. $(x - 3)(x + 5) = 0$ b. $2x^2 - 3x = 5$

2. Edit the QUAD program to output the exact values of the solution to a second-degree equation in one variable. The screen display in Figure 4 represents the number $\dfrac{1 + \sqrt{13}}{2}$.

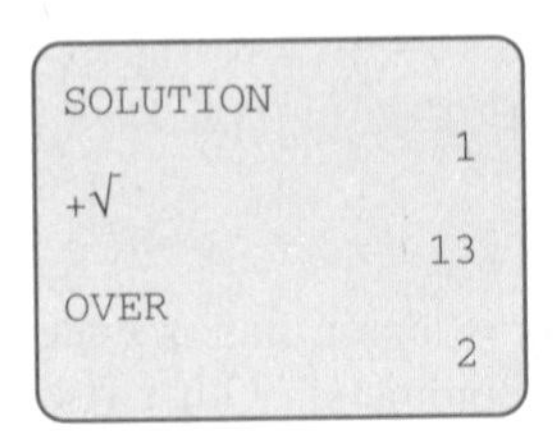

FIGURE 4

THE DIVPOLY PROGRAM

A plan (program design) for writing a program to do synthetic division of polynomials is presented below.

Get divisor, A.
Get degree of dividend, N.
Set dimension of list L1 and list L2 to N.
Get coefficients for dividend.
Do algorithm, store results in list L2.
Report answer, list L2.

```
PROGRAM:DIVPOLY
:Disp "DIVISOR"
:Input A
:Disp "DEGREE"
:Input B
:B+1→B
:B→dim L1
:B→dim L2
:For (N,1,B,1)
:Disp "COEFF",N
:Input K
:K→L1(N)
:End
:L1(1)→L2(1)
:1→L1(1)
:1→K
:For(J,1,B-1,1)
:A*L2(J)+L1(J+1)
→L2(J+1)
:End
:Disp L2
```

FIGURE 1

This design produces the DIVPOLY program displayed in Figure 1.

Press the PRGM key, then press the right arrow key to select the New option. Type the name DIVPOLY, and press the ENTER key. Now enter the commands listed in Figure 1. Use the following keystrokes to type specific commands:

Command	*Keystrokes*
dim	2nd STAT 4
L1	2nd 1
L2	2nd 2
For	Prgm 4
End	Prgm 7

The symbol L1 or L2 refers to a list. Think of a list as a collection of numbered (indexed) memory locations where numbers can be stored. In the DIVPOLY program, the commands between the For command and the End command are repeated a selected number of times. Such a structure is known as a **loop**.

After entering the commands, check the commands for typing errors. (Use your editing skills to correct any error.) Quit to the home screen.

Running the DIVPOLY Program To test the DIVPOLY program, we will use the example

$$(x^3 + 2x^2 - 4) \div (x - 3) \quad \text{is} \quad x^2 + 5x + 15, \text{ remainder } 41$$

From the home screen, press the PRGM key and select the DIVPOLY program. The DIVPOLY program starts its execution by presenting the prompt DIVISOR. Because the divisor in $(x^3 + 2x^2 - 4) \div (x - 3)$ is $x - 3$, type 3 and then press the ENTER key.

Next, the prompt DEGREE is presented. Because the dividend,

$$x^3 + 2x^2 - 4$$

is a third-degree polynomial, type 3 and press the ENTER key. Now enter the coefficients for $x^3 + 2x^2 - 4$. For the prompt COEFF, type 1 and press the

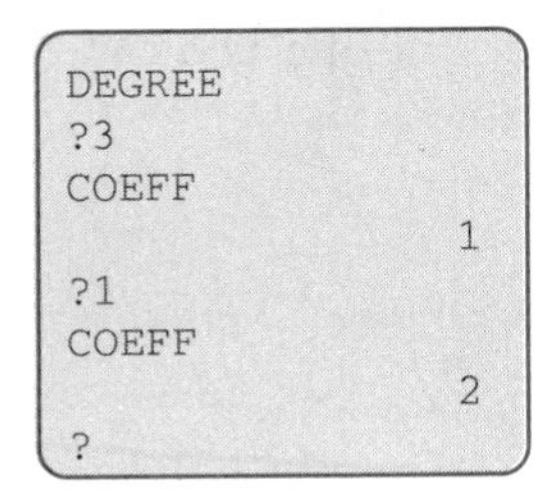

FIGURE 2

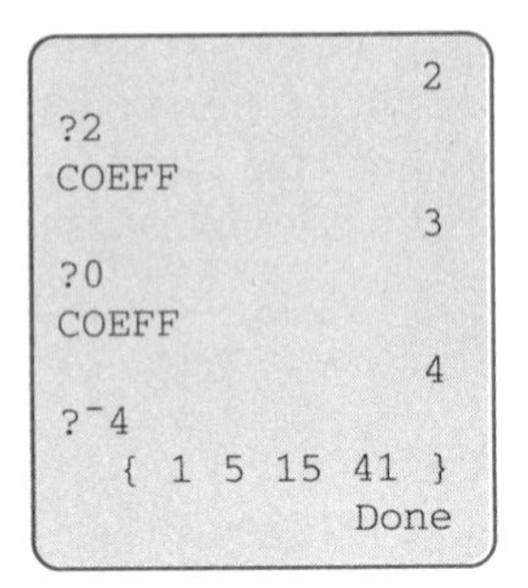

FIGURE 3

ENTER key. See Figure 2. Continue in this fashion, entering 2, 0, and ⁻4 for the coefficients of x^2, x, and the constant term.

After pressing the ENTER key for the last coefficient, the synthetic division algorithm is executed, and the list {1 5 15 41} is written to the screen. See Figure 3. Translating this matrix to a polynomial and a remainder yields $1x^2 + 5x + 15$, remainder 41.

As a second test of the program, consider the product $(2x + 3)(x - 5)$, which yields $2x^2 - 7x - 15$. Now, the division $(2x^2 - 7x - 15) \div (2x + 3)$ should produce $x - 5$, remainder 0. For synthetic division (and the DIVPOLY program) to work, the divisor must have x-coefficient 1. Thus, we multiply dividend and divisor by $\frac{1}{2}$ to get

$$\left(x^2 - \frac{7}{2}x - \frac{15}{2}\right) \div \left(x + \frac{3}{2}\right)$$

Execute the DIVPOLY program. Enter ⁻1.5 for the DIVISOR prompt. Enter 2 for the DEGREE prompt. Enter the coefficients 1, ⁻3.5, and ⁻7.5. The output is {1 ⁻5 0}. Thus, the quotient is $x - 5$, remainder 0.

If the last line output by DIVPOLY is too wide to be displayed, press the 2nd key followed by the 2 key for the L2 key. Press the ENTER key to display List 2. Now, press the right arrow key to display any entries not originally visible.

EXERCISES

1. Use the DIVPOLY program to perform each division. (Remember to report the remainder.)
   a. $(x^3 + 2x^2 - 4x - 2) \div (x - 1)$
   b. $(x^4 + x^3 - x - 2) \div (x + 5)$
   c. $(3x^3 + x^2 - 4x) \div (x - 2)$
   d. $(32x^5 - 1) \div (2x - 1)$
2. If one factor of $2x^4 - 5x^3 + 2x^2 + 6x - 3$ is $2x - 1$, what is the other factor?
3. If one number that solves $x^3 + 5 = 4x^2 + 2x$ is 1, find the other numbers in the solution.

ANSWERS TO GIVE IT A TRY PROBLEMS

CHAPTER 1

SECTION 1.1 ■ page 2

1. $A \cup B = \{1, 2, 3, 4, 5, 6, 7, 9\}$, $A \cap B = \{1, 3, 5\}$
2. Yes, yes **3.** No, no
4. -3 is greater than -1, (−3 −1 0) false
5. 0 is greater than or equal to -2, (−2 0) true
6. 8 is less than 23, (0 8 23) true
7. -8 is less than or equal to 2, (−8 0 2) true
8. 7 is greater than or equal to 7, (0 7) true
9. -3 is greater than or equal to -4, (−4 −3 0) true
10. a. -6 **b.** 3 **c.** 0 **d.** 147 **e.** $-35{,}678$
11. 9 **12. a.** 3 **b.** 17 **c.** 4 **d.** 0 **e.** 5
13. a. 3 **b.** 8 **c.** -5 **d.** 4
14. No, the absolute value of 0 is 0.
15. a. -15 **b.** -11 **c.** 14
16. a. 35 **b.** 6 **c.** -23
17. -25 **18.** 25 **19.** -8 **20.** -8 **21.** -1
22. $|-3 + 6| = |3| = 3$
23. $10 + 5 + (-1) = 14$ hours
24. $-10 + 20 = 10$ °C for the high;
$10 + (-12) = -2$ °C for the current temperature

DISCOVERY ■ page 14

1. a. -18 **b.** 18 **c.** 143 **2.** -8
3. a. 12 **b.** -300

SECTION 1.2 ■ page 19

1. $\frac{-29}{35}$ **2.** $\frac{5}{2}$ **3.** $\frac{1}{12}$ **4.** $\frac{-3}{2}$

SECTION 1.3 ■ page 28

1. a. 0.003 **b.** 132,800 **c.** 132,789.24
2. a. 23,500 **b.** 0.24
3. No, the number contains only 4 significant digits;
yes, 23.50 rounded to hundredths is 23.50;
yes, 23.50 rounded to hundreds is 0.
4. a. 0.030045 **b.** $\frac{469}{200}$ **c.** 0.2345%
5. 0.23 and 0.0023 **6. a.** 70% acid **b.** 390 miles
7. a. 2.34×10^{10} **b.** 3.4×10^{-5} **c.** 5×10^{0}
d. 1.02×10^{4}
8.

$\frac{23}{50}$	0.46	46%	4.6×10^{-1}
$\frac{23}{10000}$	0.0023	0.23%	2.3×10^{-3}
$\frac{461}{2000}$	0.2305	23.05%	2.305×10^{-1}
$\frac{81}{25000}$	0.00324	0.324%	3.24×10^{-3}

9. 3.6×10^{4} **10.** 3.5×10^{3} **11.** 2×10^{3}
12. 6.4×10^{7} **13.** -3.7×10^{2} **14.** 2.7×10^{16}
15. 3.84×10^{-7}

SECTION 1.4 ■ page 42

1. a. -6 **b.** -34 **c.** 12 **d.** -15
2. a. $\frac{1}{2}$ **b.** $\frac{305}{112}$ **c.** $\frac{25}{216}$ **d.** 6
3. $\frac{-17}{15}$ **4.** $\frac{-1}{21}$ **5.** $\frac{2089}{140}$ **6.** 9 **7.** $\frac{52}{5}$ or 10.4
8. 6 **9.** $\frac{13}{30}$ **10.** $\frac{1}{6}$

DISCOVERY ■ page 52

1. 2 **2.** 0.4 **3.** 9.5

4.

x	-3	-2	-1	0	1	2	3
y	20	12	6	2	0	0	2

SECTION 1.5 ■ page 54

1. 0.202 and 0.203 **2. a.** 6 **3.** -3 **c.** $\frac{2}{7}$ **d.** $\frac{1}{2}$

3. $\sqrt{29}$ **4.** 12 and 13 **5.** 5 and 6 **6.** -4 and -3

■ CHAPTER 2

SECTION 2.1 ■ page 65

1. 9 **2.** 27 **3.** 7 **4.** 1962

5. *Problem 1:* Store ⁻2 in memory location X then type 2X²−3X−5 and press the ENTER key. The number 9 is output.
Problem 2: Store 4 in memory location X and store ⁻3 in memory location Y. Type 3X−5Y and press the ENTER key. The number 27 is output.
Problem 3: Store 5 in memory location X and store 3 in memory location Y. Type X²−Y²−3Y and press the ENTER key. The number 7 is output.
Problem 4: Store 10 in memory location W then type 2W³−3W−8 and press the ENTER key. The number 1962 is output.

6. $4x^2 + 2xy - 2y^2 + y$

7. a. $-2x^3 - 2x^2 - 4x + 4$
b. $-x^3 - 7z^2 - 11xz + 3x - 2$

8. *Problem 6:* With ⁻5 stored in X and 3 in Y, the sum of the stated polynomials evaluates to 55 and the standard polynomial evaluates to 55.
Problem 7(a): Store a number, say ⁻3, in X. The original expression evaluates to 52 and the standard polynomial evaluates to 52 (if we store a different value in X, say 7, the original and the standard polynomials both evaluate to ⁻808).
Problem 7(b): Store values in memory locations X and Z, say 5 in X and ⁻2 in Z; the original expression evaluates to ⁻30 and the standard polynomial evaluates to ⁻30.

9. $6x^2 - 2x + 3$ **10.** $6x^2 + x - 2$

11. a. $-x^2 - 3x + 5$ **b.** $6x^2 - 4x - 7$

12. a. 3 **b.** 9

SECTION 2.2 ■ page 79

1. $\frac{-11}{3}$ **2.** $\frac{30}{7}$ **3.** $\frac{-7}{2}$ **4.** 8 **6.** -49 **7.** $\frac{-17}{6}$

8. $\frac{5}{4}$ **9.** $\frac{-3}{2}$

10. *Problem 8:* Type 15−X for Y1 and type 20−5X for Y2. Store 5/4 in X. Display the value for Y1 and Y2. In both cases, the output is 13.75. View the graph screen, zoom out, and use the TRACE key to see that the number which results in both 15−X and 20−5X evaluating to the same number is ≈1.25.

11. $\frac{22}{17}$ **12.** 5 **13.** *None* **14.** *All numbers*

SECTION 2.3 ■ page 94

1. 

2. 

3. 

4. $(3, +\infty)$ **5.** $[1, +\infty)$ **6.** $(-\infty, 0)$

7. With 1 stored in memory location X, both $-2x + 10$ and $3 + 5x$ evaluate to the same number, namely, 8. So, 1 is the key point. Storing 5, a number from the interval $[1, +\infty)$, and testing ⁻2X+10≥3+5X yields 0, indicating a false arithmetic sentence. So the interval $[1, +\infty)$ fails.

8. a. 

$(-\infty, 0]$

b. 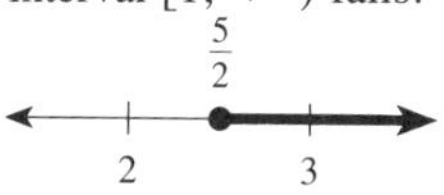

$[2.5, +\infty]$

9. 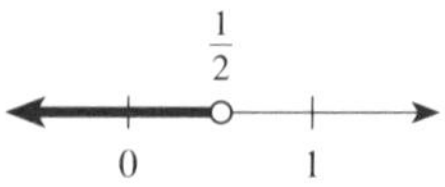

$\left(-\infty, \frac{1}{2}\right)$

10. 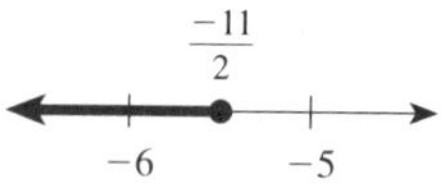

$\left(-\infty, \frac{-11}{2}\right]$

11. 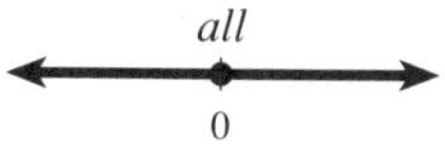

$(-\infty, +\infty)$

12. 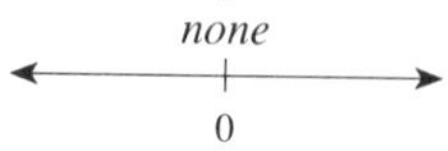

$\varnothing$

13.
$$\begin{aligned} -3x - 5 &\geq 2x + 7 \\ -2x \quad &\quad\ -2x \\ -5x - 5 &\geq 7 \\ 5 \quad &\quad 5 \\ -5x &\geq 12 \\ x &\leq \frac{-12}{5} \end{aligned}$$

Interval: $\left(-\infty, \frac{-12}{5}\right]$

14.
$$\begin{aligned} 6 - 3x &\geq -3 \\ -6 \quad &\quad -6 \\ -3x &\geq -9 \\ x &\leq 3 \end{aligned}$$

Interval: $(-\infty, 3]$

15.
$$\begin{aligned} 2x - (6 - x) &< 2 + 2x \\ 3x - 6 &< 2x + 2 \\ -2x \quad &\quad -2x \\ \hline x - 6 &< 2 \\ 6 \quad &\quad 6 \\ \hline x &< 8 \end{aligned}$$

Interval: $(-\infty, 8)$

16.
$$\begin{aligned} 2 - \tfrac{1}{2}x &> \tfrac{-1}{3}x + 1 \\ 12 - 3x &> -2x + 6 \\ 2x \quad &\quad 2x \\ \hline 12 - x &> 6 \\ -12 \quad &\quad -12 \\ \hline -x &> -6 \\ x &< 6 \end{aligned}$$

Interval: $(-\infty, 6)$

SECTION 2.4 ■ page 106

1. 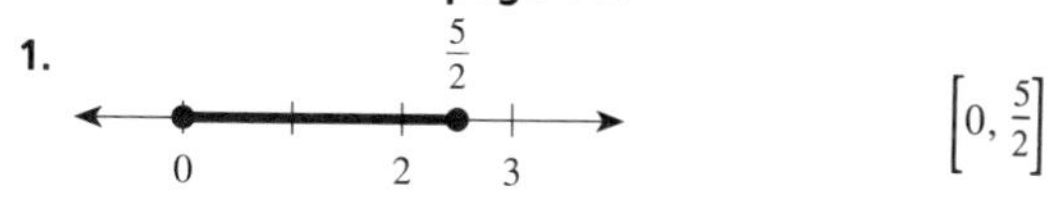

$\left[0, \frac{5}{2}\right]$

2. $\frac{1}{2}$; 0, 1, 3 $\qquad \left[\frac{1}{2}, 3\right]$

3. $\frac{5}{2}$; −1, 0, 2, 3 $\qquad \left(-1, \frac{5}{2}\right)$

4. $\frac{1}{2}$; 0, 1, 4 $\qquad \left[\frac{1}{2}, 4\right)$

5. $\frac{7}{3}$; 0, 1, 2, 3 $\qquad \left[1, \frac{7}{3}\right]$

6. $\frac{-7}{3}$; −3, −2, 0 $\qquad \left[\frac{-7}{3}, 0\right)$

8.
$$\begin{aligned} 0 < -3x - 5 &< 7 \\ 5 \qquad 5 \;&\; 5 \\ \hline 5 < -3x &< 12 \\ \tfrac{-5}{3} > x &> -4 \end{aligned}$$

Interval: $\left(-4, \frac{-5}{3}\right)$

9.
$$\begin{aligned} 8 \geq 6 - 3x &\geq -3 \\ -6 \quad -6 \;&\; -6 \\ \hline 2 \geq -3x &\geq -9 \\ \tfrac{-2}{3} \leq x &\leq 3 \end{aligned}$$

Interval: $\left[\frac{-2}{3}, 3\right]$

10. $2 \leq 2x - (6 - x) < 4 + 2x$; key points $\frac{8}{3}$ and 10

$\frac{8}{3}$; 0, 2, 4, 10 $\qquad \left[\frac{8}{3}, 10\right)$

11. $x - 1 < 2(3 - x) \leq x + 1$; key points $\frac{5}{3}$ and $\frac{7}{3}$

$\frac{5}{3}$, $\frac{7}{3}$; 0, 1, 2, 3 $\qquad \left[\frac{5}{3}, \frac{7}{3}\right)$

SECTION 2.5 ■ page 112

1. 3 and 2 **2.** $\frac{69}{5}$ and $\frac{81}{5}$ **3.** *None*

4. *Problem 1:* Type `abs (2X−5)` for `Y1`. Store `3` in memory location `X`. Display the resulting value in `Y1`. The number `1` is output. Store `2` in memory location `X`. Display the value in `Y1`. The number `1` is output.
Problem 2: Type `abs (5−(1/3)X)` for `Y1`. Store `69/5` in memory location `X`. Display the resulting value in `Y1`. The number `.4` is output. $\left[0.4 \text{ is the decimal name for } \frac{2}{5}\right]$. Store `81/5` in memory location `X`. Display the value in `Y1`. The number `.4` is output.
Problem 3: Type `abs (2X+3)` for `Y1` and `⁻2` for `Y2`. Display the graph screen. For every replacement of the x-variable, the expression $|2x + 3|$ evaluates to a number greater than -2. The evaluation is never equal to -2.

5. $[2, 3]$ **6.** $(-\infty, 1) \cup \left(\frac{7}{3}, +\infty\right)$ **7.** $(-\infty, +\infty)$

8.
$$\begin{aligned} 3x - 1 &> 8 & \text{or} \qquad 3x - 1 &< -8 \\ 3x &> 9 & 3x &< -7 \\ x &> 3 & x &< \tfrac{-7}{3} \end{aligned}$$

Solution: $\left(-\infty, \frac{-7}{3}\right) \cup (3, +\infty)$

9.
$$\begin{aligned} -7 < 2x + 3 &< 7 \\ -3 \qquad -3 \;&\; -3 \\ \hline -10 < 2x &< 4 \\ -5 < x &< 2 \end{aligned}$$

Solution: $(-5, 2)$

10.
$$\begin{aligned} 5 - 4x &\leq -3 & \text{or} \qquad 5 - 4x &\geq 3 \\ -5 \quad &\quad -5 & -5 \quad &\quad -5 \\ \hline -4x &\leq -8 & -4x &\geq -2 \\ x &\geq 2 & x &\leq \tfrac{1}{2} \end{aligned}$$

Solution: $\left(-\infty, \frac{1}{2}\right] \cup [2, +\infty)$

11. No, we get
$$\begin{aligned} -(x + 1) < 2x - 3 &< x + 1 \\ -x - 1 < 2x - 3 &< x + 1 \\ 3 \qquad 3 \;&\; 3 \\ \hline -x < 2x &< x + 4 \\ \tfrac{-1}{2}x + 1 < x &< \tfrac{1}{2}x + 2 \end{aligned}$$

The solution from the equivalent inequalities is not obvious. Using the key point approach, we get the number-line graph:

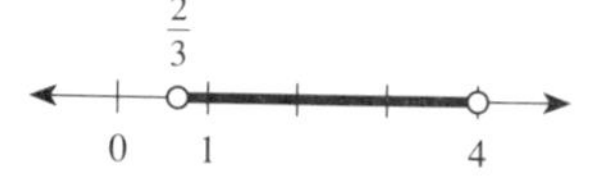

12. *Problem 8:* Type `abs (3X−1)` for `Y1`. Store a number from the interval $\left(-\infty, \frac{-7}{3}\right)$, say `⁻3`, in `X`. Display the resulting value in `Y1`. The value output is greater than 8.

Store a number from the interval $(3, +\infty)$, say 4, in X. Display the resulting value in Y1. The value output is greater than 8.

Problem 9: Type `abs (2X+3)` for Y1. Store a number from the interval $(-5, 2)$, say 0, in X. Display the resulting value in Y1. The value output is less than 7.

Problem 10: Type `abs (5-4X)` for Y1. Store a number from the interval $\left(-\infty, \frac{1}{2}\right]$, say 0, in X. Display the resulting value in Y1. The output is greater than 3. Store a number from the interval $[2, +\infty)$, say 4, in X. Display the resulting value in Y1. The value output is greater than 3.

Problem 11: Type `abs (2X-3)` for Y1 and type `X+1` for Y2. Store 0, a number from the leftmost interval, in X. Display the resulting value in Y1 and the value in Y2. The value output for Y1 is greater than the value for Y2. So the interval is left unshaded. Store a number from the middle interval, say 2, in X. Display the resulting value in Y1 and the value in Y2. The value output for Y1 is less than the value for Y2. So the interval is shaded. Store a number from the rightmost interval, say 5, in X. Display the resulting value in Y1 and the value in Y2. The value output for Y1 is greater than the value for Y2. So the interval is left unshaded.

SECTION 2.6 ■ page 120

1. Let x = the number. The polynomial is $(2x + 8) - 12$.
2. Let x = the number. The polynomial is $3(n - 7)$.
3. Let a = the number. The polynomial is $a - 0.3a$.
4. Let x = length of the hypotenuse (the longest side). Then

$$\frac{x}{2} = \text{length of one leg}$$

$$x - 6 = \text{length of other leg}$$

$$\text{So, perimeter} = x + \frac{x}{2} + (x - 6)$$

$$= \frac{5}{2}x - 6 \text{ as a standard polynomial}$$

SECTION 2.7 ■ page 128

1. $0.3x + 8 = 0.7(x + 10)$; add 2.5 liters of 30% acid solution
2. $6x + 8x = 200$; $14\frac{2}{7}$ hours, or 14 hours 17 minutes
3. $3.5x + 6(20) = 4(x + 20)$; 80 pounds of walnuts
4. ≈ 46.4 mi/h **5.** 400 miles

SECTION 2.8 ■ page 136

1. $T = \dfrac{C - 2R - 3}{R}$ **2.** $y = \frac{5}{2}x - \frac{13}{2}$

3. $x = \sigma s + \mu$ **4.** $R = \dfrac{-3V - 2}{V - T}$ or $R = \dfrac{3V + 2}{T - V}$

■ CHAPTER 3

SECTION 3.1 ■ page 144

1. a. $(-1, -3)$ **b.** $(9, 3)$ **c.** $(-1, -3)$

2.

x	-2	-1	0	$\frac{-8}{5}$	$\frac{-6}{5}$
y	0	$\frac{5}{2}$	5	1	2

3. a. $(0, 5)$ **b.** $(2, 0)$ **c.** $\left(3, \frac{-5}{2}\right)$

d.

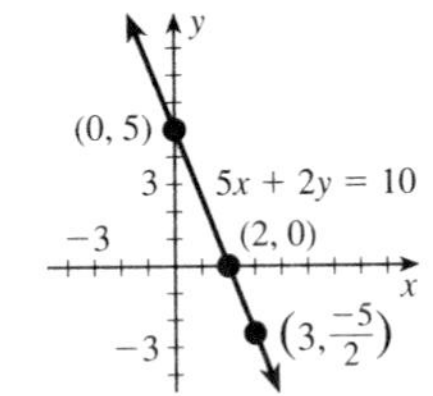

4. The x-intercept is $\left(\frac{8}{3}, 0\right)$ and the y-intercept is $\left(0, \frac{-8}{5}\right)$.

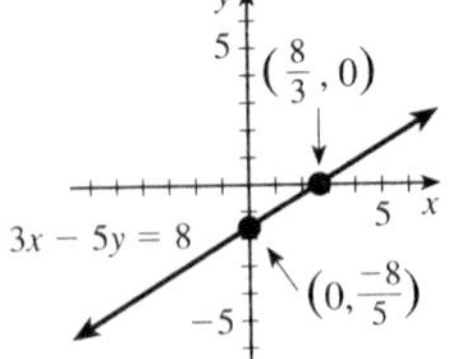

5. $(-2, -3.6), (-1, -3)$

6. y-intercept $(0, 2.4)$; x-intercept $(4, 0)$
The point with y-coordinate 2 is $\approx (.6, 2.04)$.
Using algebra, the point is $\left(\frac{2}{3}, 2\right)$.

7. The x-intercept is $(2.5, 0)$ and y-intercept is $(0, 5)$.

8. Type `(5/2)X-4` for Y1. The x-intercept is $(1.6, 0)$ and the y-intercept is $(0, -4)$.

SECTION 3.2 ■ page 152

1. a. $\Delta y = \frac{-12}{5}$ **b.** $\Delta x = -4$ **c.** $m = \left(\frac{-12}{5}\right) \div (-4)$, or $\frac{3}{5}$

2. a. $(0, 2)$ and $(3, 0)$ **b.** Negative
c. $\Delta y = -2, \Delta x = 3, m = \frac{-2}{3}$

3. The sign for the slope is negative.
$\Delta y = -6$
$\Delta x = 3$
$m = \frac{-6}{3} = -2$

4. $m = \frac{5}{7}$

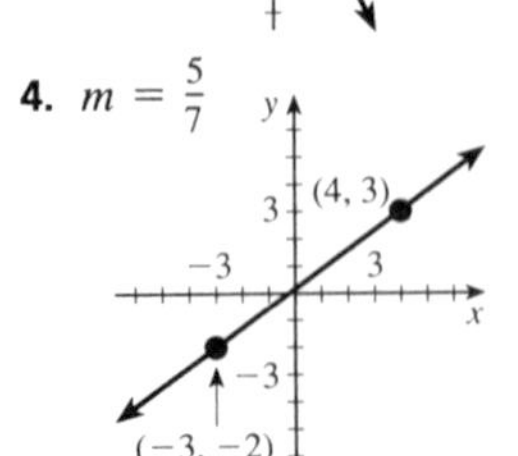

5. $m = \frac{-5}{3}$

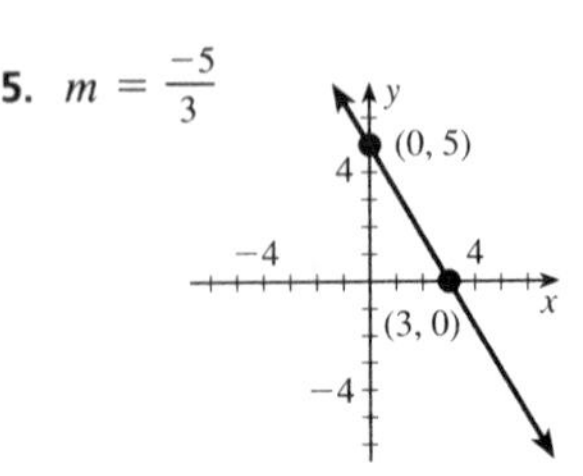

6. $m = \frac{2}{-4} = \frac{-1}{2}$

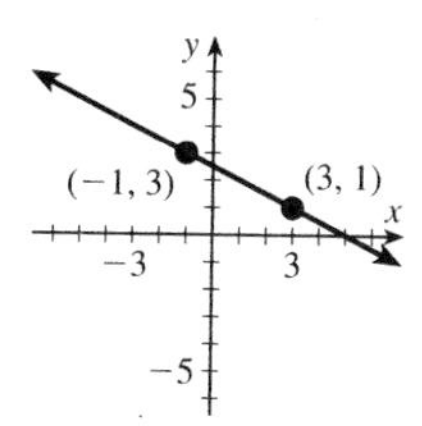

7. a. $y = \frac{-2}{3}x + 3$; $m = \frac{-2}{3}$; y-intercept $(0, 3)$

b. $(3, 1)$ **c.**

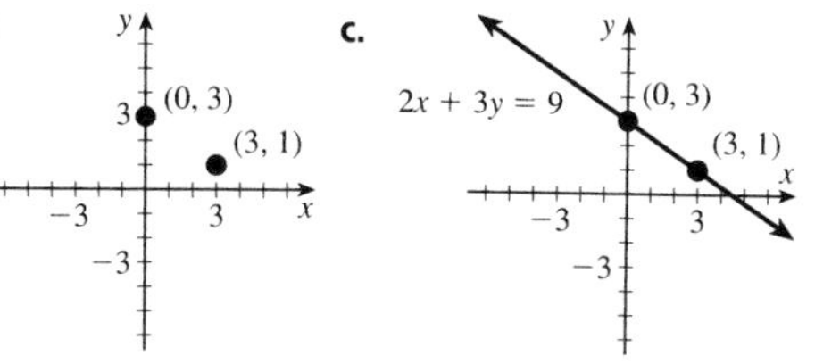

8. a. One possible choice is $(-1, 2)$, the point that is one unit right and 3 units up from the point $(-2, -1)$.

b.

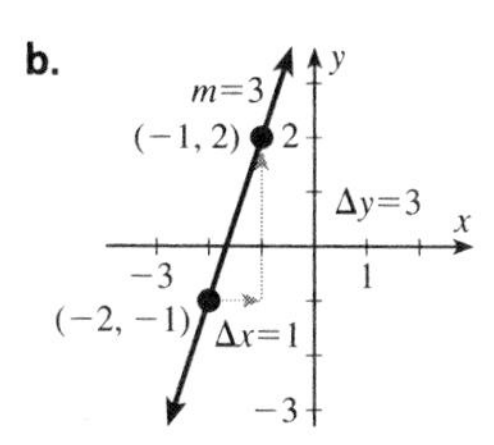

SECTION 3.3 ■ page 162

1. The x-intercept is $\left(\frac{7}{3}, 0\right)$.

The y-intercept is $\left(0, \frac{-7}{2}\right)$.

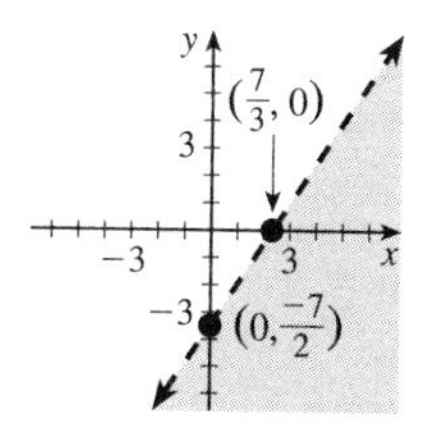

Try $(0, 0)$: $3(0) - 2(0) > 7$ is a false arithmetic sentence.
Try $(0, -4)$: $3(0) - 2(-4) > 7$ is a true arithmetic sentence.
So, shade the region below the key line.

2. The slope is $\frac{-3}{1}$.

The y-intercept is $(0, 4)$.
Another point on the line is $(1, 1)$.

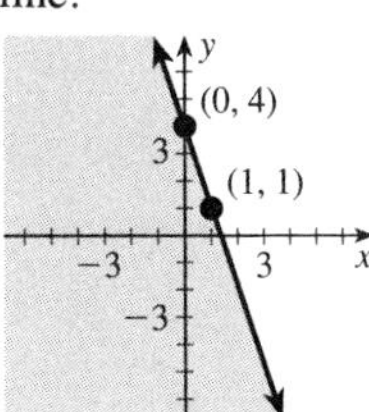

Try $(0, 5)$: $5 \le -3(0) + 4$ is a false arithmetic sentence.
Try $(0, 0)$: $0 \le -3(0) + 4$ is a true arithmetic sentence.
So, shade the region below the key line.

SECTION 3.4 ■ page 165

1. Not a relation **2.** A relation, but not linear
3. A linear relation **4.** Not a relation
5. Not a relation **6.** 3,333 games
7. In the year 2033 the predicted mile record is 3.5 minutes. The prediction is probably not realistic because there must be limitation on how fast a human can run one mile.

8. a.

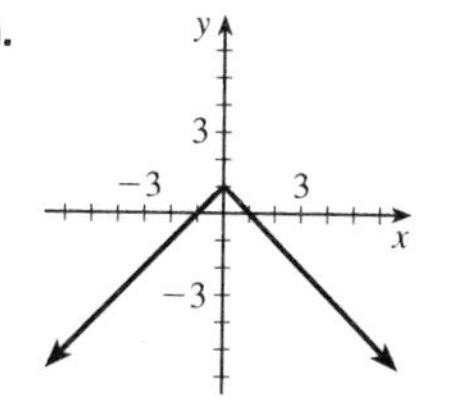

b. $(0, 1)$
c. $(-1, 0)$, $(1, 0)$

9. a.

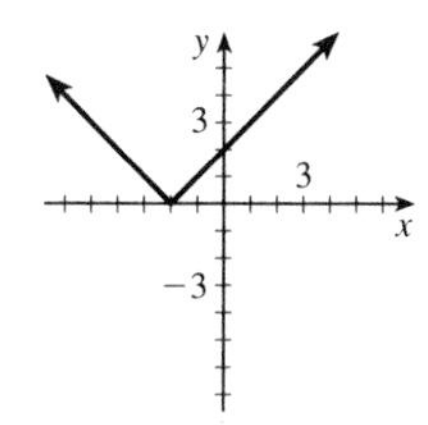

b. $(0, 2)$
c. $(-2, 0)$

10. Not a linear relation. The y-intercept is $(0, 1)$. The x-intercepts are $(1, 0)$ and $(-1, 0)$.
11. Not a linear relation. The y-intercept is $(0, -3)$. The x-intercepts are $(3, 0)$ and $(-1, 0)$.
12. Domain $= \{x \mid x \text{ is real}\}$; range $= \{y \mid y \text{ is real}\}$
13. Domain $= \{x \mid x \text{ is real}\}$; range $= \{y \mid y \le 2\}$

SECTION 3.5 ■ page 173

1. a. $f(-3) = -2(-3) + 5 = 11$
So, $f(-3)$ is 11.
b. $f(5) = -2(5) + 5 = -5$
So, output is -5.

2. a. Rule is $4x + 2$. **b.** $g(0) = 4(0) + 2 = 2$

3. a. $f(-2) = \frac{-7}{3}$
b. Solve $-3 = \frac{1}{3}x - \frac{5}{3}$. The solution is -4, so the input value x is -4.

4. a. $F(0) = 32$ **b.** $F(30) = 86\,°F$ **c.** $48\frac{8}{9}\,°C$
d.

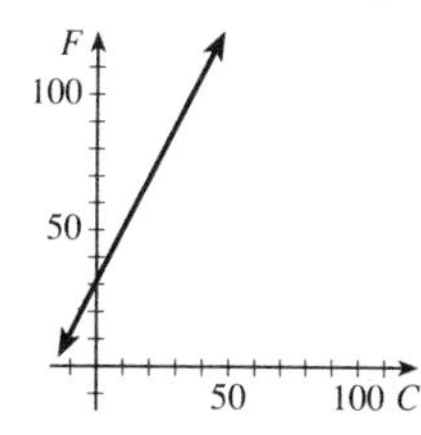

5. Domain $= \{x \mid x \text{ is real}\}$
Range $= \{h(x) \mid h(x) \le 2\}$
6. Domain $= \{x \mid x \le 3\}$
Range $= \{g(x) \mid g(x) \ge 0\}$
7. Domain $= \{x \mid x \text{ is real}\}$
Range $= \{r(x) \mid r(x) \ge -3\}$

■ CHAPTER 4

SECTION 4.1 ■ page 184

1. a.

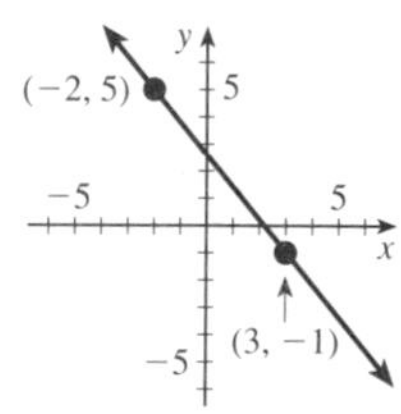

b. Line is decreasing, so slope is negative.

c. $m = \frac{5-(-1)}{-2-3} = \frac{-6}{5}$ **d.** Solve $5 = \frac{12}{5} + b$; $b = \frac{13}{5}$

e. $y = \frac{-6}{5}x + \frac{13}{5}$

f. Type `(-6/5)X+13/5` for `Y1`. Select `ZOOM` option 6, then press the `RANGE` key and type `-9` for `Xmin` (−8.8 on the TI-82). View the graph screen. Use `TRACE` to check that the points (−2, 5) and (3, −1) are on the line.

2. a. $y = 2x + 7$
b. $y = 3$

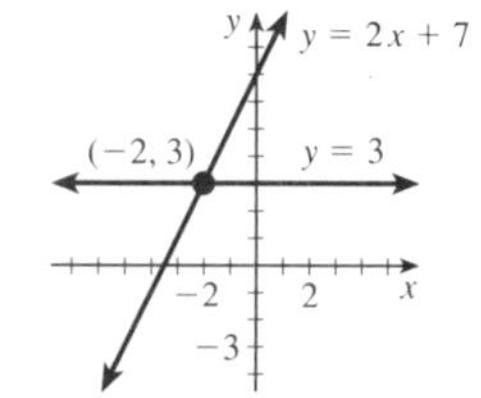

3. $y = -2x + 3$

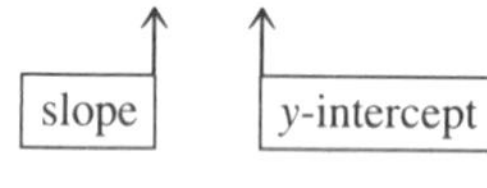

4. $m = \frac{5-1}{-3-4} = \frac{-4}{7}$

$f(x) = \frac{-4}{7}x + \frac{23}{7}$

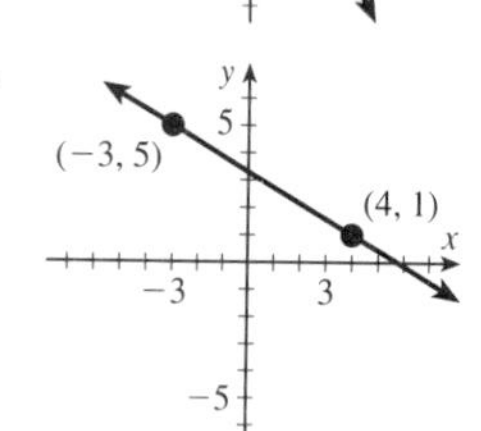

5.

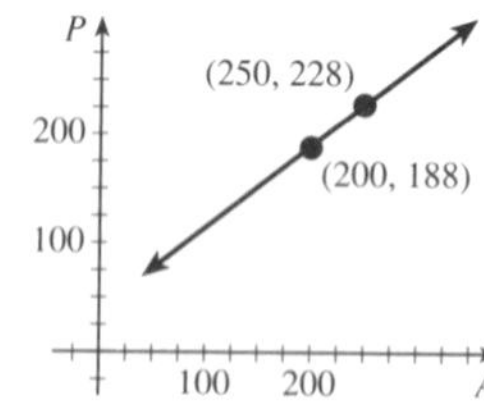

$m = \frac{228-188}{250-200} = \frac{40}{50} = \frac{4}{5}$, so we have $P = \frac{4}{5}A + b$
Now, use the data (200, 188) to find $b = 28$. The linear relationship is $P = \frac{4}{5}A + 28$.

6. a.

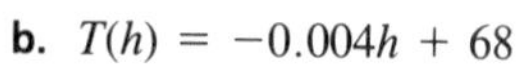

b. $T(h) = -0.004h + 68$
c. $T(4500)$, or 50°F
d. $h = 7500$ feet

7.

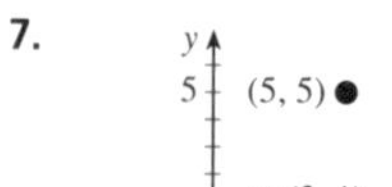

The equation of the line through (−1, −3) and (2, 1) is

$$y = \frac{4}{3}x - \frac{5}{3} \quad \text{or} \quad 4x - 3y = 5$$

Checking to see if the point (5, 5) is on this line, we get $4(5) - 3(5) = 5$, a true arithmetic sentence. So, the points are collinear.

SECTION 4.2 ■ page 198

1. a. $\sqrt{40}$, or $2\sqrt{10}$ **b.** (3, 2)

2. a. $\sqrt{41}$ **b.** $\left(\frac{-1}{2}, -1\right)$

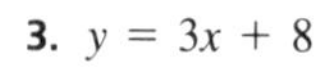

3. $y = 3x + 8$

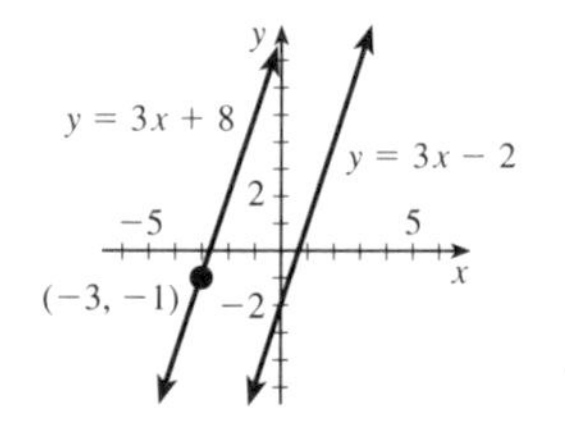

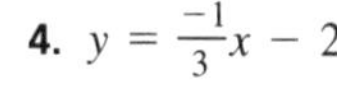

4. $y = \frac{-1}{3}x - 2$

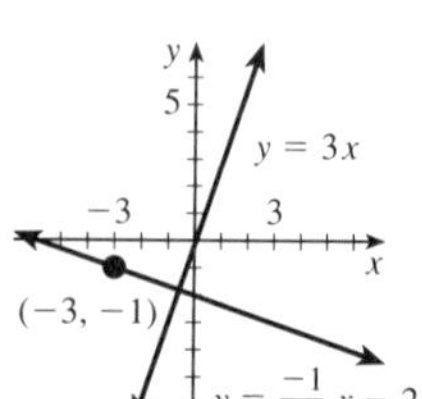

5.

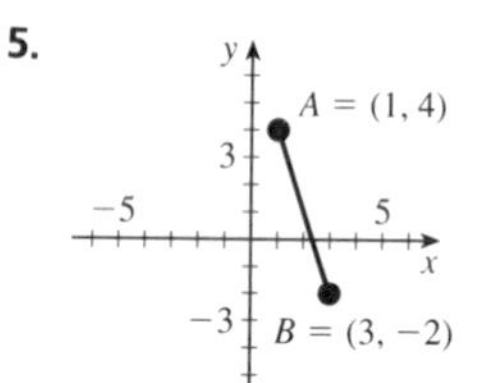

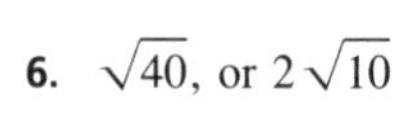

6. $\sqrt{40}$, or $2\sqrt{10}$

7. (2, 1)

8.

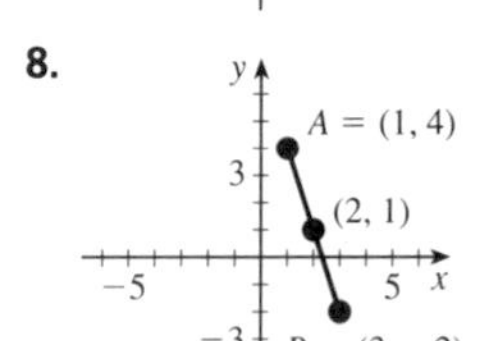

9. −3

10. $\frac{1}{3}$

11. $y = \frac{1}{3}x + \frac{1}{3}$

12.

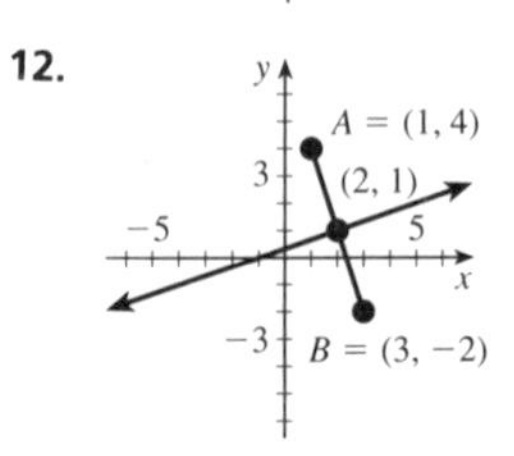

SECTION 4.3 ■ page 212

1. $\begin{cases} y = \dfrac{-3}{2}x + 4 \\ y = \dfrac{3}{2}x - 2 \end{cases}$

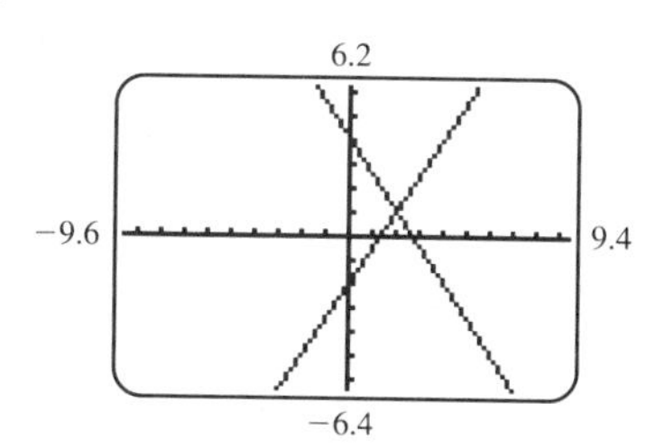

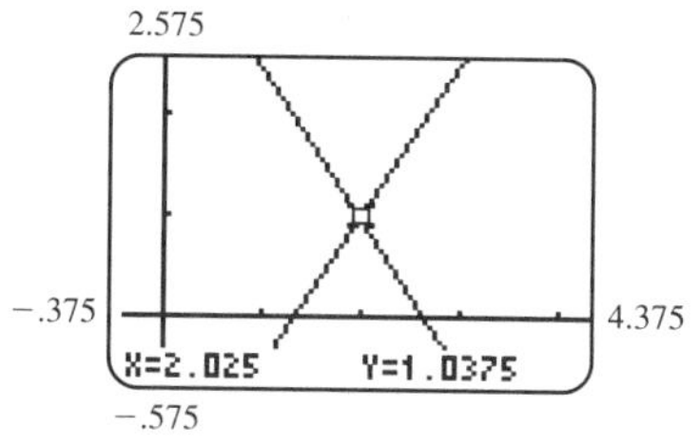

Solution ≈ (2, 1)

2. $\begin{cases} y = \dfrac{3}{5}x - \dfrac{9}{5} \\ y = \dfrac{-2}{3}x - 3 \end{cases}$

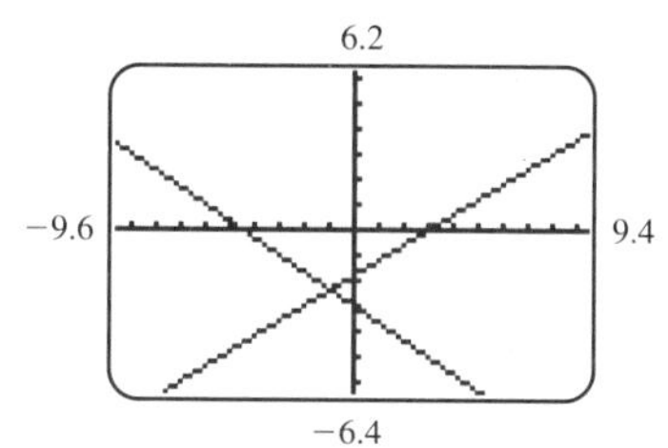

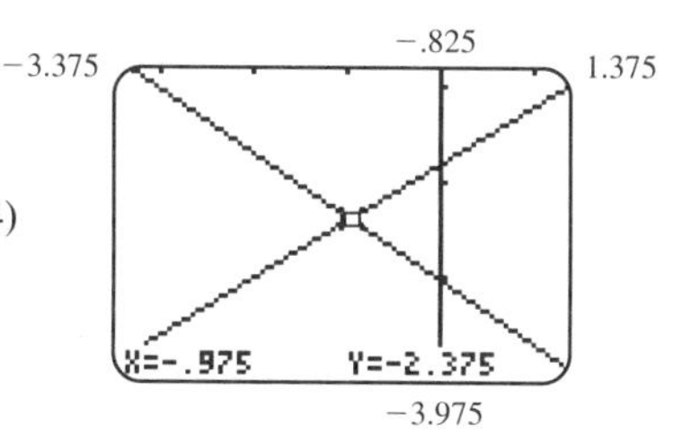

Solution ≈ (−1, −2.4)

3.

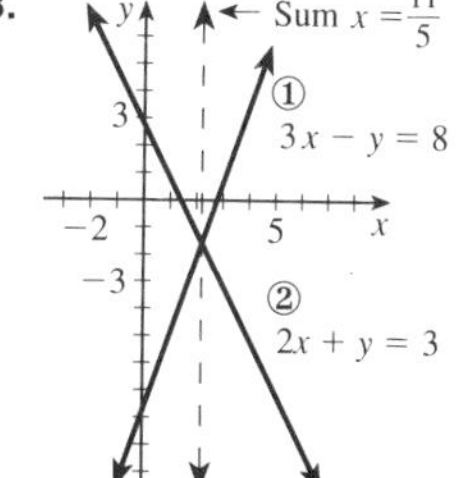

The x-coordinate of the intersection point is $\frac{11}{5}$. The solution to $3\left(\frac{11}{5}\right) - y = 8$ is $\frac{-7}{5}$, so the solution is $\left(\frac{11}{5}, \frac{-7}{5}\right)$.

4. a.

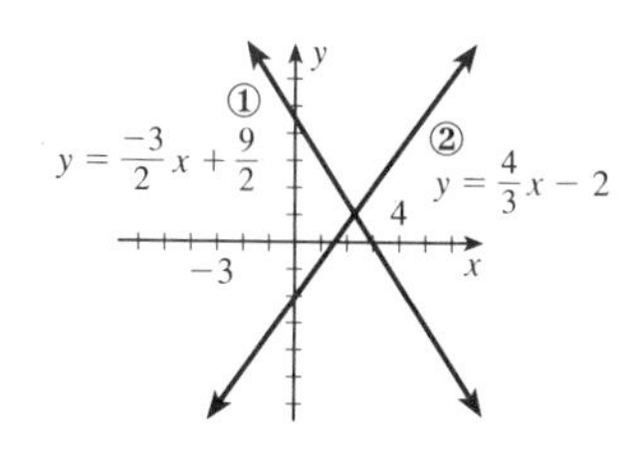

b. $\begin{cases} 12x + 8y = 36 \\ -12x + 9y = -18 \end{cases}$

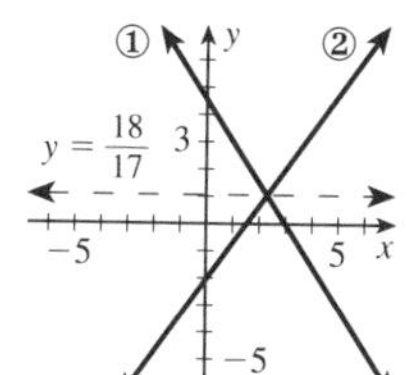

c. y-coordinate is $\frac{18}{17}$

d. $\begin{cases} 9x + 6y = 27 \\ 8x - 6y = 12 \end{cases}$

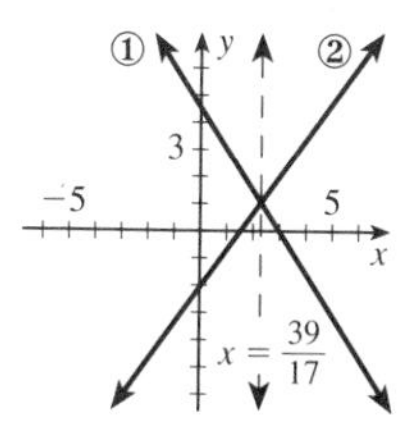

e. x-coordinate is $\frac{39}{17}$

f. $\left(\frac{39}{17}, \frac{18}{17}\right)$

5. ≈(2.875, 1.06875); $\left(\frac{39}{17}, \frac{18}{17}\right) \approx (2.9412, 1.0588)$

6. None **7.** $\left(\frac{8}{7}, \frac{-3}{7}\right)$ **8.** $\left(\frac{-19}{11}, \frac{-12}{11}\right)$

SECTION 4.4 ■ page 227

1. 25 liters of 30% acid solution and 125 liters of 90% acid solution
2. Let x = time traveled at 55 mi/h (in hours) and y = time traveled at 65 mi/h.
   The system is:
   $$\begin{cases} x + y = 10 & \text{(hours)} \\ 55x + 65y = 630 & \text{(miles)} \end{cases}$$
   The solution is (2, 8), so she travels 2 hours at 55 mi/h and 8 hours at 65 mi/h.
3. The system is:
   $$\begin{cases} C(x) = 0.2x + 30 \\ C(x) = 0.4x + 20 \end{cases}$$
   and the solution is (50, 40). Thus, when 50 miles are driven, the cost of the rentals will be the same, \$40.
4. To find the price where the units demanded will equal the units supplied, solve the system
   $$\begin{cases} P(x) = -3x + 100 \\ P(x) = 2x \end{cases}$$
   The solution is (20, 40). Thus, when the price is \$40, the units supplied will equal the units demanded, 20.

■ CHAPTER 5

SECTION 5.1 ■ page 241

1. $-80x^4y^6$ **2.** $-6p^5 - 3p^4 + 9p^2$
3. $-x^5 + 2x^4 + x^2$ **4.** $21x^7y^2$
5. $-6x^3 + 15x^2$ **6.** $x^2 + 15x + 56$
7. $6x^2 + x - 15$ **8.** $5x^2 + 3x - 8$
9. $x^3 - 3x^2 - 2x + 6$ **10.** $9x^2 - 4$ **11.** $9x^4 - 4$
12. $9x^2 + 12x + 4$ **13.** $9x^4 - 12x^2 + 4$
14. $9x^2 - 12x + 4$ **15.** $9x^4 + 12x^2 + 4$
16. $-2x^2 + 8x - 2$ **17.** $5x^2 - 14x$
18. $-4x^2 + 7x - 1$ **19.** $6x^2 - 6x - 2$
20. $-x^3 - x^2 + 4$

SECTION 5.2 ■ page 250

1. x **2.** 1 **3.** $10x^3y^2$ **4.** $10x^2(3x - 2)$
5. $-3wt(4w + 5t)$ **6.** prime **7.** $5x^2(x^2 + 3x - 2)$
8. $(x^2 + 5)(x + 1)$ **9.** $(x^2 - 5)(x + 1)$
10. $(a + b)(c - 2)$ **11.** $(2y)(y^2 + 1)(y + 3)$

SECTION 5.3 ■ page 256

1. $(x - 6)(x + 1)$ **2.** $4x(x - 3)(x - 2)$
3. $-1(x + 9)(x + 2)$ **4.** $(x - 7)(x - 2)$
5. $(5x + 3)(x - 1)$ **6.** $(x + 2y)(x - y)$
7. $2(3x + 2)(3x - 1)$ **8.** $(x^2 - 7)(x^2 - 2)$
9. $x^2(x - 3)(x + 1)$ **10.** $(x^2 + 2y^2)(x^2 + y^2)$
11. $(3x^2 + 2)(3x^2 - 1)$

SECTION 5.4 ■ page 266

1. $(x + 7)(x - 7)$ **2.** prime **3.** $x(x + 4)(x - 4)$
4. $(7x + 10)(7x - 10)$ **5.** $9(2x + 1)(2x - 1)$
6. $3x(2x + 3)(2x - 3)$ **7.** $(x - 2)(x^2 + 2x + 4)$
8. $(2x + 1)(4x^2 - 2x + 1)$ **9.** $(4y^2 + 1)(16y^4 - 4y^2 + 1)$
10. prime **11.** $3x(x - 3)(x^2 + 3x + 9)$
12. $(y + 2)(y - 2)(y^2 + 2y + 4)(y^2 - 2y + 4)$
13. $(2w - 3)(2w - 3)$ **14.** $(7y + 2)(7y + 2)$
15. $(x - 1)(x^2 + x + 1)(x - 1)(x^2 + x + 1)$
16. $x(x + 2)$ **17.** $2(x + 1)(2)(x - 4)$
18. $(x + 2)(x^2 + x + 1)$
19. *Problem 16:* Type (X+3)^2−4(X+3)+3 for Y1, and type X(X+2) for Y2. Store 3 in X, and display the values for Y1 and Y2. In each case, the output is 15.
20. $(x + 2)(x - 2)(x + 1)$
21. $(a + b)(c + 3)(c - 3)$
22. $(x + 2)(x^2 - 2x + 4)(x - 1)$
23. $2y(y + 1)(y - 1)(y + 3)$
24. *Problem 23:* Type 2Y^4+6Y^3−2Y^2−64 for Y1, and type 2Y(Y+1)(Y−1)(Y+3) for Y2. Store 5 in Y, and evaluate Y1 and Y2. In each case, the output is 1920.

SECTION 5.5 ■ page 276

1. Solution 0 and 6 **2.** Solution $\frac{3}{2}$ and -1
3. Solution 6 and -1 **4.** Solution 4 and -4
5. $x(2x - 5) = 0$ Solution 0 and $\frac{5}{2}$
6. $-1(2x + 1)(x - 3) = 0$ Solution $\frac{-1}{2}$ and 3
7. $(2x - 1)(x + 3) = 0$ Solution $\frac{1}{2}$ and -3
8. $(x - 7)(x + 1) = 0$ Solution 7 and -1
9. $(x - 4)(x + 2) = 0$ Solution 4 and -2
10. $(3x + 1)(x + 1) = 0$ Solution $\frac{-1}{3}$ and -1
11. *Problem 6:* Type ⁻2X^2+5X+3 for Y1. Store ⁻1/2 in X, and display the Y1 value. The output is 0. Store 3 in X, and display the Y1 value. The output is 0.
12. $\frac{19}{3}$ **13.** 6 and -1 **14.** 0 and 3 and -1
15. 2 and -2, the factor $x^2 + 4$ never evaluates to 0.

SECTION 5.6 ■ page 286

1. 1 2 $(1, 2)$

2. 

$(-\infty, -1] \cup [5, +\infty)$

SECTION 5.7 ■ page 290

1. a. y-intercept $(0, -6)$
b. x-intercepts $(6, 0)$ and $(-1, 0)$
c. Vertex $\left(\frac{5}{2}, \frac{-49}{4}\right)$
d. Line of symmetry $x = \frac{5}{2}$
e. See graph at right.

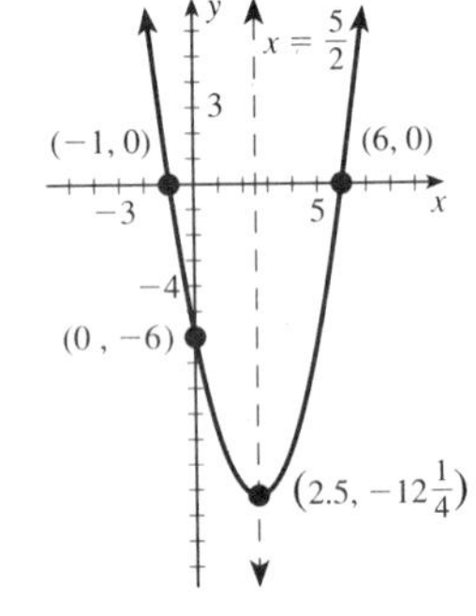

2. a. y-intercept $(0, 6)$
b. x-intercepts $(3, 0)$ and $(2, 0)$
c. Vertex $\left(\frac{5}{2}, \frac{-1}{4}\right)$
d. Line of symmetry $x = \frac{5}{2}$
e. See graph at right.

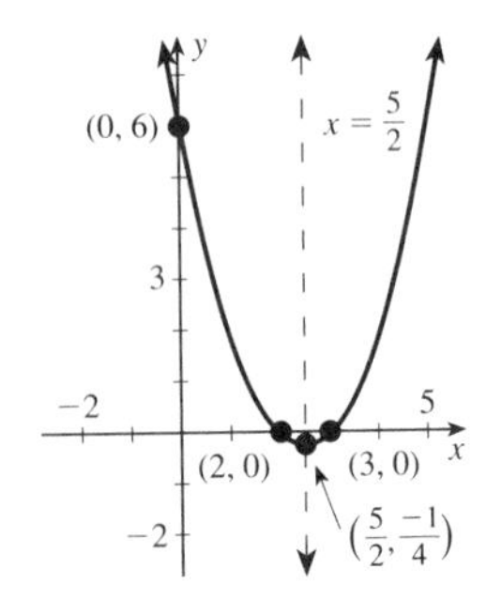

3. a. y-intercept $(0, 2)$
b. x-intercepts $(2, 0)$ and $(1, 0)$
c. Vertex $\left(\frac{3}{2}, \frac{-1}{4}\right)$
d. Line of symmetry $x = \frac{3}{2}$
e. See graph at right.

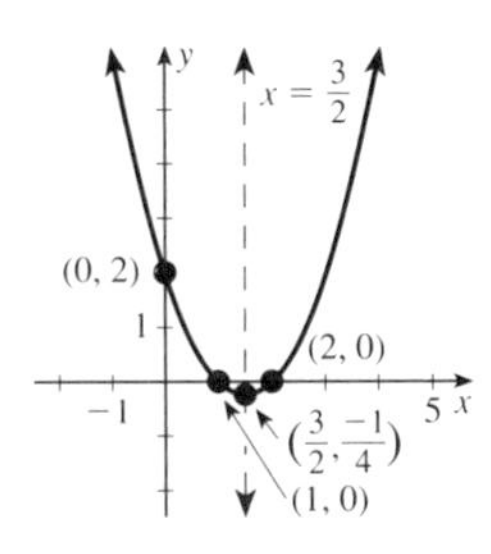

4. a. y-intercept $(0, 6)$
b. x-intercepts $(6, 0)$ and $(-1, 0)$
c. Vertex $\left(\frac{5}{2}, \frac{49}{4}\right)$
d. Line of symmetry $x = \frac{5}{2}$
e. See graph at right.

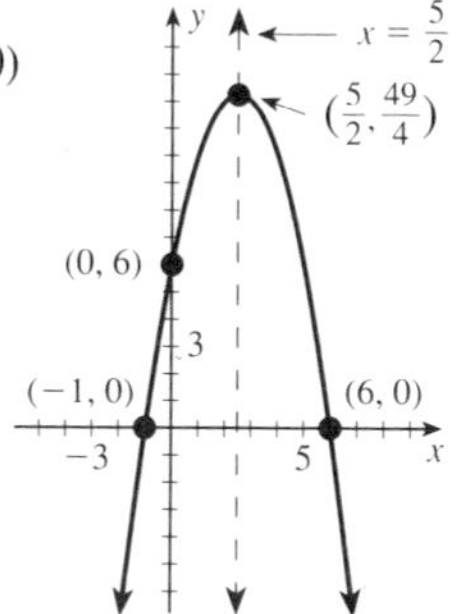

5. a. y-intercept $(0, 6)$
b. x-intercepts $(3, 0)$ and $(-2, 0)$
c. Vertex $\left(\frac{1}{2}, \frac{25}{4}\right)$
d. Line of symmetry $x = \frac{1}{2}$
e. See graph at right.

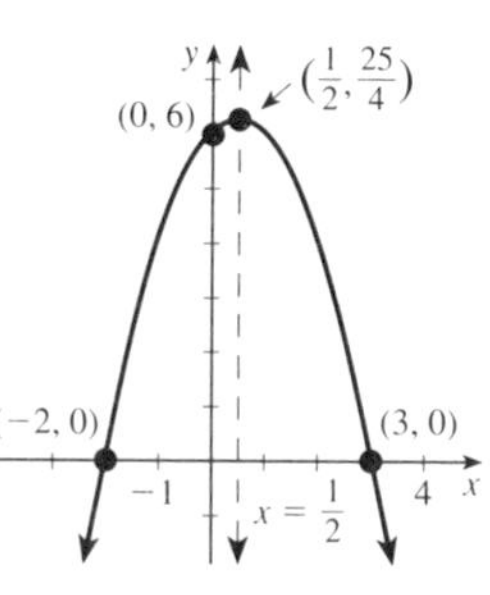

6. a. 0 **b.** -2 and 4 **c.** $(-3, 0)$ and $(5, 0)$
d. Domain $= \{x \mid x \text{ is real}\}$ and range $= \{y \mid y \leq 16\}$

SECTION 5.8 ■ page 299

1. $x^2 - \frac{1}{2}x$ remainder 2 **2.** x^2 remainder 3
3. $x^2 - x + 1$ remainder -2 **5.** $x^2 - x + 1$
6. $2x^2 - 2x + 1$
7. $(x^2 + x + 1)(x^2 - x + 1)(x + 1)(x - 1)$
8. Solution: $3, -2, 1$ **9.** Solution: $-2, 2, 1$
10.

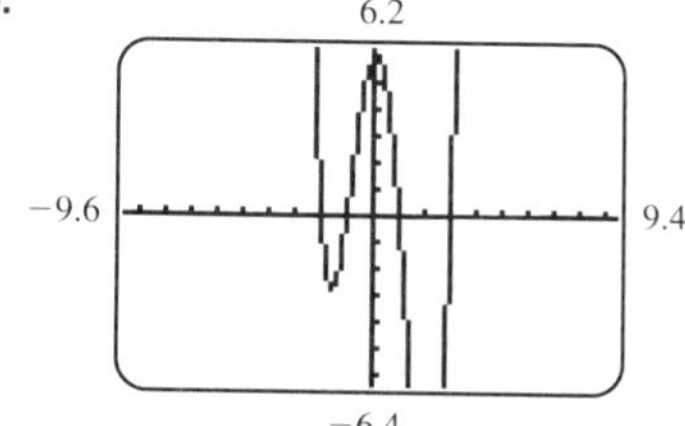

$(x + 2)(x + 1)(x - 1)(x - 3)$

11. $2x^2 - 6x + 17 + \frac{-55}{x + 3}$ **12.** $x^3 + x^2 + x + 1$
13. $x^2 + x + 1 + \frac{2}{x - 1}$

■ CHAPTER 6

SECTION 6.1 ■ page 314

1. $\{x \mid x \neq -4\}$ **2.** $\{x \mid x \neq 5, x \neq 1\}$
3. $\{x \mid x \neq 1, x \neq 2\}$
4. $\frac{2x^2 + 6x}{2x^2}$ **5.** $\frac{3x^2}{2x^2 + 3x}$ **6.** $\frac{x^3 - 5x^2 - 2x + 10}{x^2 - 25}$

7. *Problem 4:*
Store 4 in X.
Store (X+3)/X in Y1. Evaluation: 1.75
Store (2X²+6X)/(2X²) in Y2.
Evaluation: 1.75

Problem 5:
Store 10 in X.
Store (3X)/(2X+3) in Y1. Evaluation: 1.304347826
Store (3X²)/(2X²+3X) in Y2.
Evaluation: 1.304347826

Problem 6:
Store 8 in X.
Store (X²−2)/(X+5) in Y1.
Evaluation: 4.769230769
Store (X³−5X²−2X+10)/(X²−25) in Y2.
Evaluation: 4.769230769

8. $\frac{wy^4}{2x^2}$ **9.** $\frac{-x - 7}{x + 1}$ **10.** $\frac{4x^2 + 2x + 1}{x}$ **11.** $\frac{-2x^2}{5x + 1}$

SECTION 6.2 ■ page 323

1. $\frac{-4d^2y^2}{c}$ **2.** $\frac{x^2 - 5x}{x + 5}$ **3.** $\frac{-1}{6x(x + 5)}$ **4.** $\frac{x^2 + 5x - 6}{(x + 5)(x + 1)}$
5. $\frac{2y^2}{3m^2x}$ **6.** $\frac{x^3 - 2x^2 - 2x - 3}{2(x + 1)}$ **7.** $\frac{x^2 + 2x + 1}{2x}$
8. $\frac{x + 6}{x + 5}$ **9.** $\frac{-x + 2}{x^2 - 4x + 4}$ **10.** $\frac{4x^3 + 6x^2 + 9x}{8x^3 - 27}$

SECTION 6.3 ■ page 330

1. $x + 3$ **2.** $\frac{x^2 + 4x}{x - 4}$ **3.** $\frac{x - 1}{x - 5}$ **4.** $\frac{1}{x + 1}$
5. $\frac{-8x + 6}{(x + 3)(x - 2)}$ **6.** $\frac{x^3 + x^2 + 2x + 3}{(x - 1)(x^2 + 6)}$ **7.** $\frac{x^2 + y^2}{(x + y)(x - y)}$

8. *Problem 5:*

Store 8 in X.
Store (X−3)/(X+3)−X/(X−2) in Y1.
Evaluation: −.8787878788
Store (−8X+6)/((X+3)(X−2)) in Y2.
Evaluation: −.8787878788

9. $\dfrac{x^2 + 3x - 4}{(x + 2)(x - 2)(x + 1)}$ **10.** $\dfrac{5my - 4x}{6x^2y^2}$

11. $\dfrac{x^2 + 2x - 5}{(x - 5)(x - 5)(x + 1)}$

13. $\dfrac{x^4 - x - 1}{x^2}$ **14.** $\dfrac{x^3 - x - 1}{x^2}$

15. $\dfrac{x + 6}{x(x - 3)}$ **16.** $\dfrac{-3x + 1}{x}$

SECTION 6.4 ■ page 338

1. $\dfrac{x^2 + x - 8}{x + 3}$ **2.** $\dfrac{-1}{x - 3}$ **3.** $\dfrac{4x - 4}{(x - 2)(x - 2)(x^2)}$

4. $\dfrac{-1}{3x(x + 3)}$ **5.** $\dfrac{y}{x + y}$ **6.** $\dfrac{y}{x}$ **7.** $\dfrac{1}{xy(x - y)}$

8. $\dfrac{3x - 6}{4x - 9}$ **9. a.** $\dfrac{3}{xy}$ **b.** $\dfrac{1}{2xy}$ **c.** $\dfrac{y}{2x}$

SECTION 6.5 ■ page 346

1. $\frac{5}{2}$ **2.** -3 and 1 **3.** None **4.** 6 and -2 **5.** $\frac{8}{3}$

6. 2 **7.** $L = \dfrac{AF - A}{rF}$ **8.** $x = \dfrac{y + 2}{y - 1}$

9. $a = \dfrac{b}{c + bc}$

SECTION 6.6 ■ page 354

1. Let x = hours to plow the field

Worker	Rate of work per hour	Time	Fraction done
Farmer	$\frac{1}{6}$	x	$\frac{x}{6}$
Daughter	$\frac{1}{10}$	x	$\frac{x}{10}$

$\frac{x}{6} + \frac{x}{10} = 1$ Solution: $\frac{15}{4}$

The two, working together, can plow the field in $3\frac{3}{4}$ hours.

2. Let x = processing time for computer B

Computer	Rate of work per hour	Time	Fraction done
A	$\frac{1}{6}$	4	$\frac{4}{6}$
B	$\frac{1}{x}$	4	$\frac{4}{x}$

$\frac{4}{6} + \frac{4}{x} = 1$ Solution: 12

Computer B, working alone, can process the payroll in 12 hours.

3. Let x = number of hours helper works

Worker	Rate of work per hour	Time	Fraction done
Painter	$\frac{1}{10}$	4	$\frac{4}{10}$
Helper	$\frac{1}{15}$	x	$\frac{x}{15}$

$\frac{4}{10} + \frac{x}{15} = 1$ Solution: 9

Helper must work 9 hours to finish the job.

Cost of labor = (\$15)(4) + (\$8)(9) = \$132

4. Let x = miles driven at 50 mi/h

Distance	Rate	Time
x	50	$\frac{x}{50}$
$300 - x$	65	$\frac{300 - x}{65}$

$\dfrac{x}{50} + \dfrac{300 - x}{65} = 5$ Solution: $83\frac{1}{3}$

Jill drove $83\frac{1}{3}$ miles at 50 mi/h.

5. Let x = speed of plane in calm air.

	Distance	Rate	Time
With wind	250	$x + 10$	$\frac{250}{x + 10}$
Against wind	200	$x - 10$	$\frac{200}{x - 10}$

$\dfrac{250}{x + 10} = \dfrac{200}{x - 10}$ Solution: 90

Speed of the ultralight plane in calm air is 90 mi/h.

SECTION 6.7 ■ page 360

1. -3 **2.** $3x:(x - 1) = 2:5$

3. $V = \frac{kT}{P}$

$30 = \dfrac{k(60)}{50}$ So, $k = 25$.

$V = \dfrac{25T}{P}$ is the formula.

$V = \dfrac{25(70)}{50}$ So, volume is 35 in^3.

4. R varies directly with the square of T and inversely with V.

SECTION 6.8 ■ page 365

1. $\frac{26}{9x^2}$ **2.** $\frac{x^4}{9y^{10}}$ **3.** Solution: 3 **4.** Solution: 9

■ CHAPTER 7

SECTION 7.1 ■ page 372

1. 6 **2.** -3 **3.** Not a real number **4.** 2
5. $\frac{1}{9}$ **6.** $4x$ **7.** $\frac{-1}{5x}$

SECTION 7.2 ■ page 382

1. $2xy\sqrt{6x}$ **2.** $-2xy\sqrt[3]{2x^2}$ **3.** $wn\sqrt[4]{25w^2n}$
4. $\frac{4+\sqrt{7}}{3}$ **5.** $\frac{\sqrt{6xy}}{3y}$ **6.** $\frac{x\sqrt{2xy}}{2y}$ **7.** $\frac{2\sqrt[3]{3x^2y^2}}{3y}$
8. $\frac{x\sqrt[3]{4xy^2}}{2y}$ **9.** $\frac{\sqrt{3xy}}{x}$ **10.** $\frac{3}{x}$ **11.** $\frac{\sqrt[3]{3x^2y}}{x}$ **12.** $\frac{\sqrt[3]{9x}}{x}$

SECTION 7.3 ■ page 391

1. $14x\sqrt{2x} - 4x\sqrt{3x}$ **2.** $\frac{\sqrt{3x}}{x}$ **3.** $7\sqrt{x} - 18\sqrt{y}$
4. $\frac{4\sqrt{3x} - 3\sqrt{2x}}{6x}$ **5.** $4y\sqrt[3]{x^2}$
6. $x + 2\sqrt{x} - \sqrt{3x} - 2\sqrt{3}$ **7.** $x - 2\sqrt{2x} + 2$
9. $1 - \sqrt{3}$
10. $\left(2+\sqrt{3}\right)^2 - 4\left(2+\sqrt{3}\right) = 4 + 4\sqrt{3} + 3 - 8 - 4\sqrt{3}$
$= -1$
11. $\frac{3}{2}$ **12.** $\frac{\sqrt{x^2+2x} - \sqrt{2x+4}}{x+2}$ **13.** $\frac{x - 2\sqrt{2x} + 2}{x-2}$
14. $\frac{-3-2\sqrt{3}}{1}$ **15.** $\frac{3-\sqrt{5}}{4}$ **16.** $\frac{-x+5\sqrt{x}-6}{x-4}$
17. $\frac{y-3\sqrt{y}}{y-9}$

SECTION 7.4 ■ page 400

1. $\frac{-23}{3}$ **2.** 6 **3.** 1 (-4 is an extraneous root)
4. None $\left(\frac{49}{16}\text{ is an extraneous root}\right)$ **5.** $\frac{1}{4}$
6. Distance is $90\sqrt{2}$ ($\approx$127.28) feet **7.** $L = \frac{256T^2}{\pi^2}$

SECTION 7.5 ■ page 410

1. $-2\sqrt{5} + \left(3\sqrt{2}\right)i$ **2.** $5 + (-6)i$ **3.** $0 + \frac{-\sqrt{5}}{5}i$
4. $2 + 0i$ **5.** $5 + (-1)i$ **6.** $-9 + 11i$ **7.** $16 + 47i$
8. $-8 + (-6)i$ **9.** $10 + 0i$ **10.** $12 + 5i$
11. $\frac{5}{2} + \frac{-1}{2}i$
12. $(2+i)^2 - 4(2+i) + 5 = 4 + 4i + (-1) - 8 - 4i + 5$
$= 4 + (-1) - 8 + 5 + 4i - 4i$
$= 0$
13. $x^2 - 6x + 7 = 0$ **14.** $x^2 - 2x + 5 = 0$

■ CHAPTER 8

SECTION 8.1 ■ page 422

1. $x^2 - 10x + 25$; $(x-5)^2$ **2.** $x^2 + 2x + 1$; $(x+1)^2$
3. $x^2 + 5x + \frac{25}{4}$; $\left(x + \frac{5}{2}\right)^2$ **4.** 9 and -3
5. 4 and -14 **6.** $3 + \sqrt{23}$ and $3 - \sqrt{23}$
7. 8 and -2 **8.** $3 + \sqrt{17}$ and $3 - \sqrt{17}$
9. $-2 + \sqrt{11}$ and $-2 - \sqrt{11}$ **10.** $\frac{5+\sqrt{33}}{2}$ and $\frac{5-\sqrt{33}}{2}$
11. $\frac{5}{2}$ and -1 **12.** 2 and -1
13. *Problem 10:* Store (5+ √33)/2 in memory location X. Type X²−5X and press the ENTER key. The number 2 is output. Store (5− √33)/2 in memory location X. Type X²−5X and press the ENTER key. The number 2 is output.
14. 0 **15.** None **16.** All
17. No real number, nonreal complex numbers $i\sqrt{3}$ and $-i\sqrt{3}$

SECTION 8.2 ■ page 433

1. a. $a = 1, b = -5, c = -3$ **b.** $b^2 - 4ac = 37$
c. $\frac{5+\sqrt{37}}{2}$ and $\frac{5-\sqrt{37}}{2}$
2. a. $\frac{5+\sqrt{53}}{2}$ and $\frac{5-\sqrt{53}}{2}$ **b.** $\frac{3+\sqrt{65}}{4}$ and $\frac{3-\sqrt{65}}{4}$
c. $\frac{1+\sqrt{29}}{2}$ and $\frac{1-\sqrt{29}}{2}$
3. *Problem 2(a):* Type X²−5X for Y1. Store (5+√53)/2 in memory location X. Display the resulting value in Y1. The number 7 should be output. Store (5−√53)/2 in memory location Y1. Display the resulting value in Y1. The number 7 should be output.
4. 0 and 3 **5.** $\frac{1+\sqrt{10}}{3}$ and $\frac{1-\sqrt{10}}{3}$
6. $\frac{-5+i\sqrt{3}}{2}$ and $\frac{-5-i\sqrt{3}}{2}$ **7.** -2 and 2 **8.** -3
9. $\frac{5+i\sqrt{7}}{2}$ and $\frac{5-i\sqrt{7}}{2}$

SECTION 8.3 ■ page 441

1. Let $u = x^2$. Solution: $\sqrt{5}, -\sqrt{5}, \sqrt{3}, -\sqrt{3}$

2. Let $u = x^2$. Solution:

$$\frac{\sqrt{10 + 2\sqrt{13}}}{2}, \frac{-\sqrt{10 + 2\sqrt{13}}}{2}, \frac{\sqrt{10 - 2\sqrt{13}}}{2}, \frac{-\sqrt{10 - 2\sqrt{13}}}{2}$$

3. Let $u = x^3$. Solution: $\sqrt[3]{3}$ and -1
4. *Problem 1:* Type X^4−8X+15 for Y1. Store √5 in memory location X, and display the resulting value in Y1. The number −1E−11 is output, a value very close to 0. Store −√5 in X, and display Y1. Again, −1E − 11 is output. Store √3 in X, and display the resulting value in Y1. The output is 0. Store −√3 in X, and display the value in Y1. Again, 0 is output.
5. Let $u = x^{-1}$. Solution: $\frac{-1}{2}$ and 1
6. Let $u = x^{1/3}$. Solution: 27 and -8
7. Let $u = \sqrt{x}$. Solution: $\frac{3 - \sqrt{5}}{2}$, $\left[\frac{3 + \sqrt{5}}{2} \text{ is extraneous}\right]$
8. *Problem 5:* Type X^(−2) for Y1 and $2-\text{X}^{-1}$ for Y2. Store −1/2 in memory location X, and display the resulting values in Y1 and Y2. The number 4 is output for both. Store 1 in X, and display the resulting values in Y1 and Y2. The number 1 is output for both.
9. $\frac{1 + \sqrt{5}}{2}$ and $\frac{1 - \sqrt{5}}{2}$ 10. $\frac{-3 + \sqrt{17}}{2}$ 11. $\frac{9}{4}$

SECTION 8.4 ■ page 452

1. $\frac{-1-\sqrt{17}}{2}$ $\frac{-1+\sqrt{17}}{2}$

−3 −2 0 1 2

$$\left(-\infty, \frac{-1 - \sqrt{17}}{2}\right) \cup \left(\frac{-1 + \sqrt{17}}{2}, +\infty\right)$$

2. *all*

0

$(-\infty, +\infty)$

3. $\frac{-3-\sqrt{21}}{2}$ $\frac{-3+\sqrt{21}}{2}$

−4 −3 0 1

$$\left[\frac{-3 - \sqrt{21}}{2}, -3\right) \cup \left[\frac{-3 + \sqrt{21}}{2}, +\infty\right)$$

SECTION 8.5 ■ page 459

1.

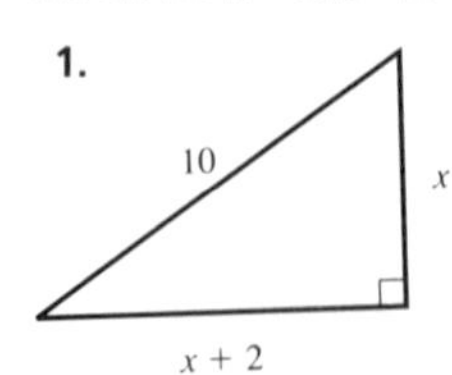

Let x = shortest side
$x + 2$ = other leg
Equation: $x^2 + (x + 2)^2 = 10^2$
Solution: -8, 6
Answer: shortest side has length 6 cm

2. Solve: $53 = -16t^2 + 96t + 6$

Solution: $\frac{12 + \sqrt{97}}{4}$ and $\frac{12 - \sqrt{97}}{4}$

Answer: The ball reaches a height of 53 feet at ≈0.538 sec and ≈5.46 sec after being thrown.
Solve: $16t^2 - 96t - 6 = 0$

Solution: $\frac{12 + 5\sqrt{6}}{4}$ and $\frac{12 - 5\sqrt{6}}{4}$

Answer: The ball reaches the ground at ≈6.06 sec.
3. Let x = days from today. Then V = value of the crop.
$V = 200(0.62 - 0.01x)(50 + x)$
Solve: $6270 = 200(0.62 - 0.01x)(50 + x)$
Solution: 5, 7
Answer: Mr. Workman's crop will be worth $6270 in 5 days and again in 7 days.
4. Type 200(0.62−0.01X)(50+X) for Y1. Type 6270 for Y2. Set Xmin to 0, Xmax to 19, Xscl to 5, Ymin to 6000, Ymax to 6500, and Yscl to 100. View the graph screen:

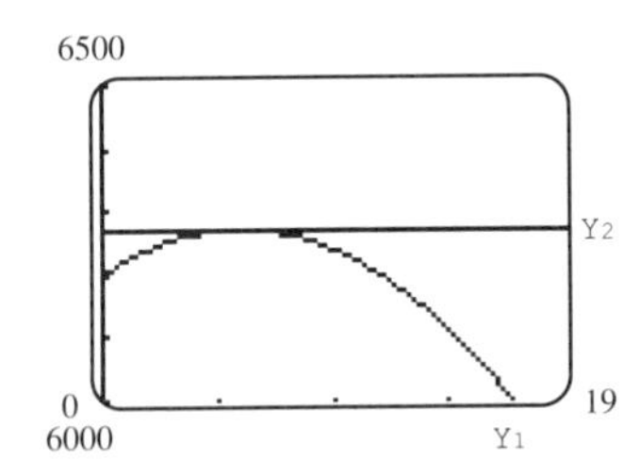

Press the TRACE key and move the cursor to X=6. Zoom in repeatedly. It appears that Mr. Workman should wait 6 days to achieve the maximum value of $6272 for his pecans.
5. Standard form: $x^2 - 3x - 1 - y = 0$
$a = 1$, $b = -3$, $c = -1 - y$; $b^2 - 4ac = 4y + 13$

Formula: $x = \frac{3 + \sqrt{4y + 13}}{2}$ and $x = \frac{3 - \sqrt{4y + 13}}{2}$
6. Standard form: $16t^2 - 10t + d = 0$
$a = 16$, $b = -10$, $c = d$; $b^2 - 4ac = -64d + 100$

Formula: $t = \frac{5 + \sqrt{-16d + 25}}{16}$ and $t = \frac{5 - \sqrt{-16d + 25}}{16}$
7. $y = \sqrt{25 - x^2}$ and $y = -\sqrt{25 - x^2}$

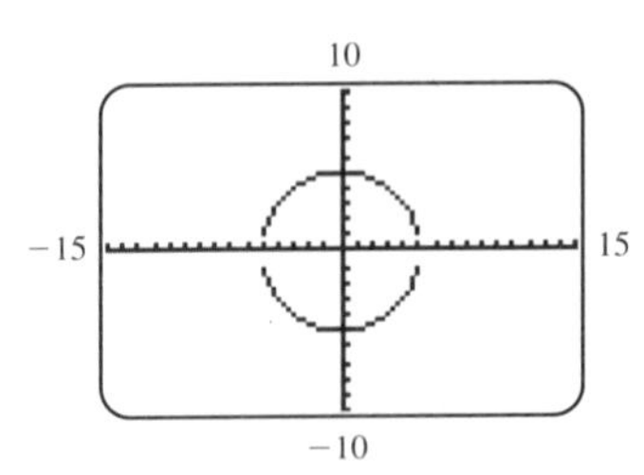

circle

DISCOVERY ■ page 467

1. (0, −3) **2.** (0, 2) **3.** (0, −5) **4.** (3, 0)
5. (−3, 0) **6.** (2, 0) **7.** (3, 1) **8.** (−1, −2)
9. (2, −4) **10.** (0, −2) **11.** (2, 0) **12.** (−2, 4)

SECTION 8.6 ■ page 469

1. (0, 1) **2.** (−1, 0) **3.** (−2, −3)

4. *Problem 3:* Type `(X+2)²-3` for `Y1` and use the settings given in the art to display the graph. Press the `TRACE` key and move the cursor to the lowest point. Its coordinates are (−2, −3).

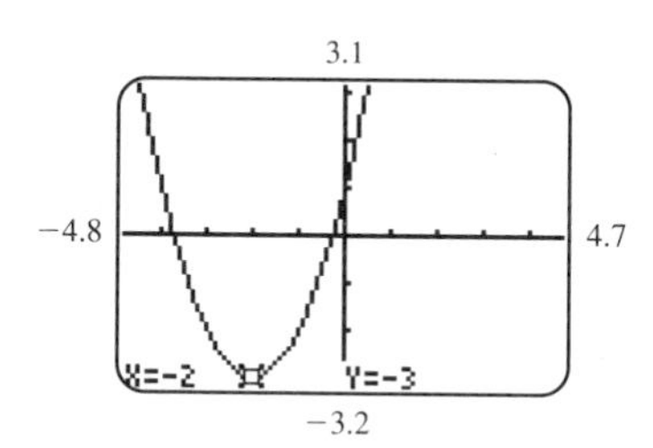

5. a. y-intercept (0, −2)
b. x-intercepts $\left(1 + \sqrt{3}, 0\right)$ and $\left(1 - \sqrt{3}, 0\right)$
c. Vertex (1, −3)
d. Line of symmetry $x = 1$
e. See graph at right

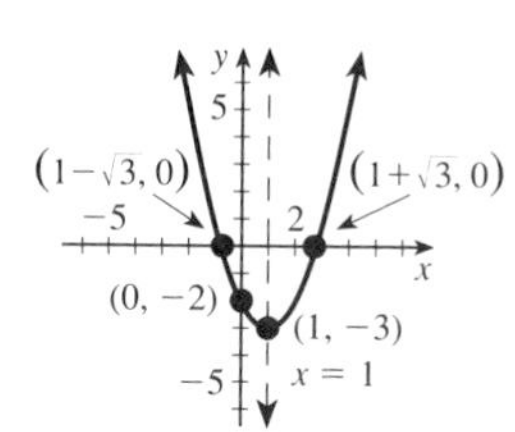

6. a. y-intercept (0, 3)
b. no x-intercept
c. Vertex (−1, 2)
d. Line of symmetry $x = -1$
e. See graph at right

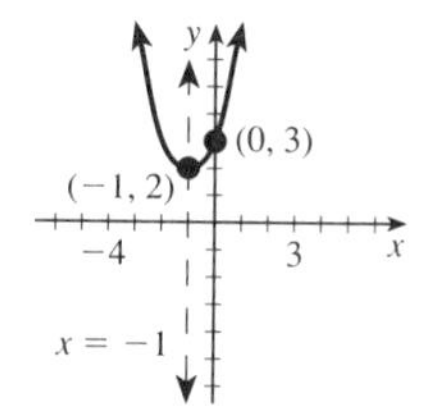

7. a. y-intercept (0, 1)
b. x-intercepts: $\left(2 + \sqrt{3}, 0\right)$ and $\left(2 - \sqrt{3}, 0\right)$
c. Vertex (2, −3)
d. Line of symmetry $x = 2$
e. See graph at right

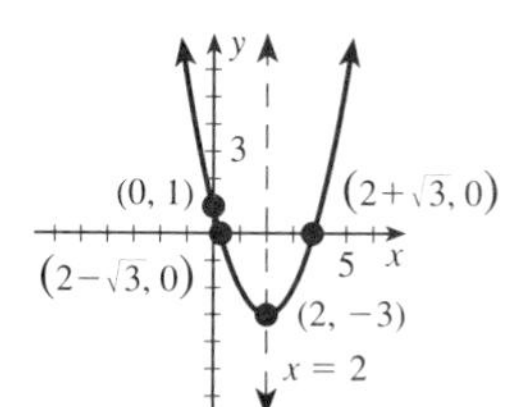

8. a. y-intercept (0, 6)
b. x-intercept (6, 0) and (−1, 0)
c. Vertex $\left(\frac{5}{2}, \frac{49}{4}\right)$
d. Line of symmetry $x = \frac{5}{2}$
e. See graph at right

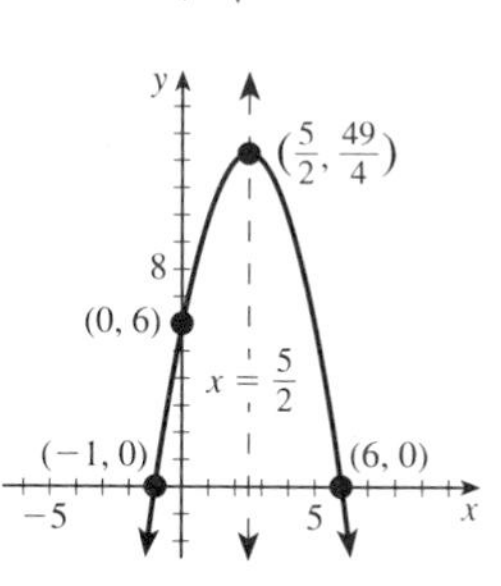

9. a. y-intercept (0, 6)
b. x-intercept (3, 0) and (−2, 0)
c. Vertex $\left(\frac{1}{2}, \frac{25}{4}\right)$
d. Line of symmetry $x = \frac{1}{2}$
e. See graph at right

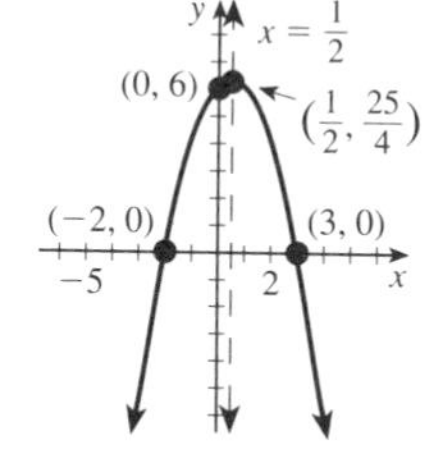

10. a. y-intercept (0, −2)
b. No x-intercept
c. Vertex: (1, −1)
d. Line of symmetry $x = 1$
e. See graph at right

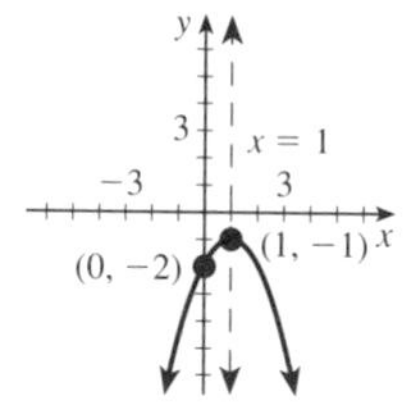

11. *Problem 9:* Type `(3-X)(2+X)` for `Y1` and use the settings given in the art to display the graph. Use the `TRACE` key to check the y-intercept and x-intercepts. Move the cursor until the message reads `X=.5  Y=6.25` to confirm that the vertex is at $\left(\frac{1}{2}, \frac{25}{4}\right)$.

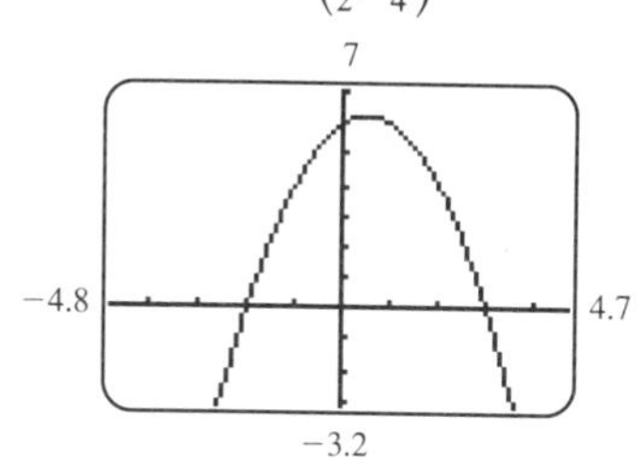

SECTION 8.7 ■ page 482

1. *C* **2.** *A* **3.** *B* **4.** *D*

5.

x	−2	−1	0	1	2
$h(x)$	$\frac{4}{3}$	$\frac{3}{2}$	2	undefined	0

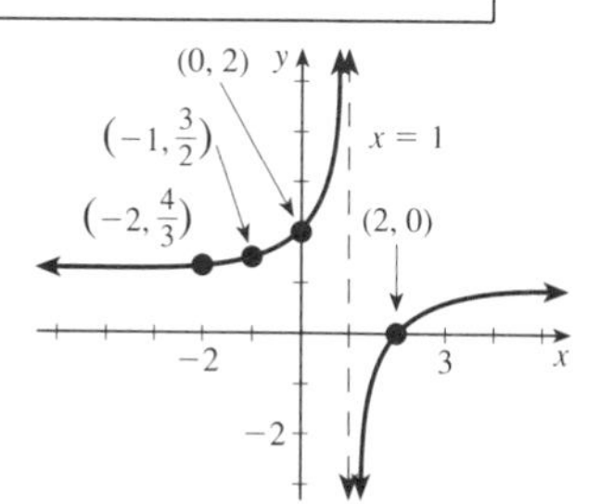

6. $(x - 2) \div (x - 1) = 1 + \frac{-1}{x - 1}$ so $h(x) = \frac{-1}{x - 1} + 1$

The graph of h is the graph of $f(x) = \frac{1}{x}$, reflected about the x-axis, translated right 1 unit, and translated up 1 unit.

7. y-intercept (0, 2) **8.** No x-intercept

9. Domain = $\{x \mid x \geq 0\}$ **10.** Range = $\{f(x) \mid f(x) \geq 2\}$

11.

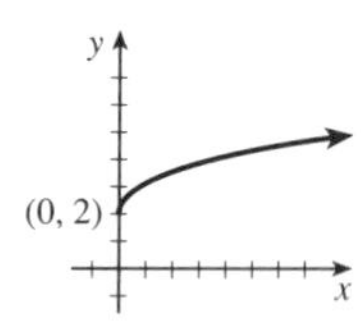

12. Graph of g is graph of $f(x) = \sqrt{x}$, translated up 2 units.

■ CHAPTER 9

SECTION 9.1 ■ page 497

1. a.

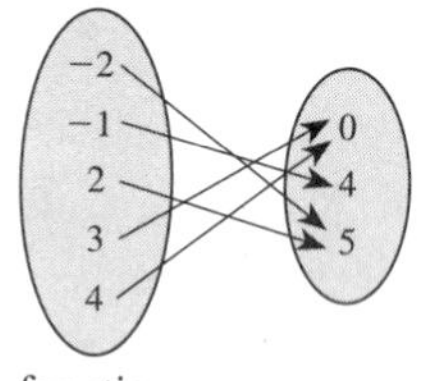

function

b.

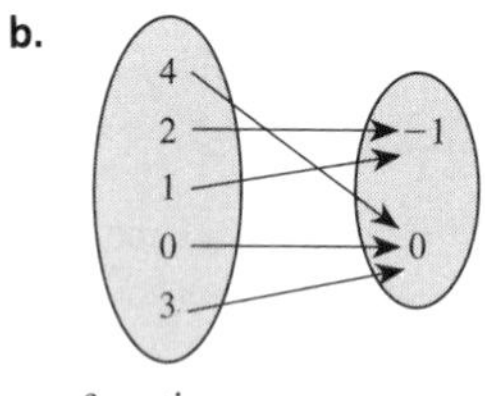

function

2. a. Function **b.** Not a function, because it contains (a, m) and (a, n) **c.** Function

3. $h(7) = |3 - 2(7)| = |-11| = 11$

4. $d(-2 + h) = 3(-2 + h)^2 - (-2 + h) = 3h^2 - 13h + 14$

5. $f(r - 1) = \dfrac{1}{(r - 1) + 5} = \dfrac{1}{r + 4}$

6. $\dfrac{g(3 + h) - g(3)}{h} = \dfrac{2h + 5 - 5}{h} = 2$

7. $\dfrac{h(-2 + h) - h(-2)}{h} = \dfrac{\dfrac{-1}{-2 + h} - \dfrac{-1}{-2}}{h} = \dfrac{-1}{2(h - 2)}$

SECTION 9.2 ■ page 504

1. Function **2.** Not a function **3.** Not a function

4. $g(-2) = 1$ **5.** $g(b) = 0$

6. Input values are -2, -1, and 3.

7. Increasing on $(-\infty, -1)$ and $(2, +\infty)$; decreasing on $(-1, 2)$; turning points $(-1, 3)$ and $(2, -1)$; zeros at $x = -3$, $x = 1$, $x = 3$

8. Increasing on $\left(\frac{3}{2}, +\infty\right)$; decreasing on $\left(-\infty, \frac{3}{2}\right)$; turning point $\left(\frac{3}{2}, \frac{-25}{4}\right)$; zeros at $x = -1$ and $x = 4$

9.

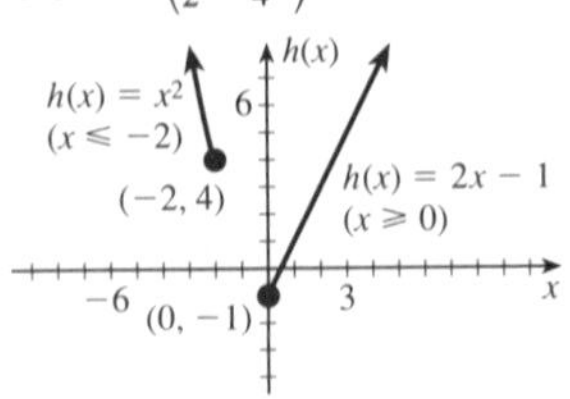

Domain $= (-\infty, -2] \cup [0, +\infty)$; range $= [-1, +\infty)$; zero at $x = \frac{1}{2}$

SECTION 9.3 ■ page 513

1. a. $(f + g)(x) = 3x + 2$
b. $(f - g)(x) = x - 4$
c. $(f \cdot g)(x) = 2x^2 + 5x - 3$
d. $(g - f)(10) = -6$
e. See graph at right.

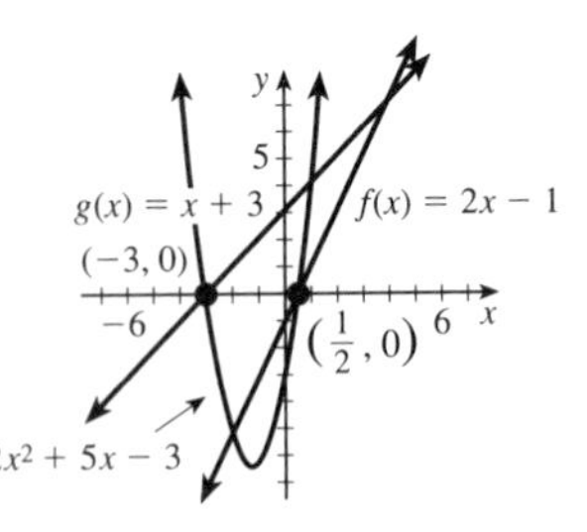

2. a. $(f + h)(x) = x^2 - x + 1$
b. $(h - f)(x) = -x^2 - x + 1$
c. $(f \cdot h)(x) = -x^3 + x^2$
d. $(h - f)(10) = 1$
e. See graph at right.

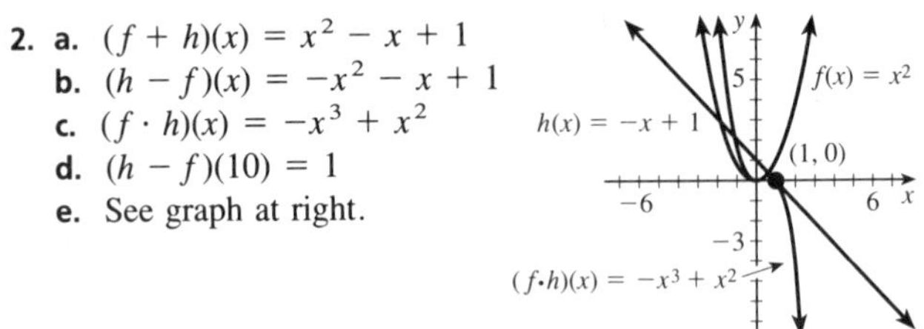

3. a. $\dfrac{f}{h}(0) = \dfrac{-5}{3}$ **b.** $\dfrac{h}{f}\left(\dfrac{5}{2}\right) = \dfrac{13}{0}$, undefined

c. $\dfrac{f}{h}\left(\dfrac{5}{2}\right) = \dfrac{0}{13} = 0$

4. $\dfrac{f}{g}(x) = \dfrac{2x - 3}{-x + 2}$ for $x \neq 2$

5. At the zero of f, $x = \frac{3}{2}$, a zero of f/g occurs. At the zero of g, $x = 2$, the function f/g is undefined.

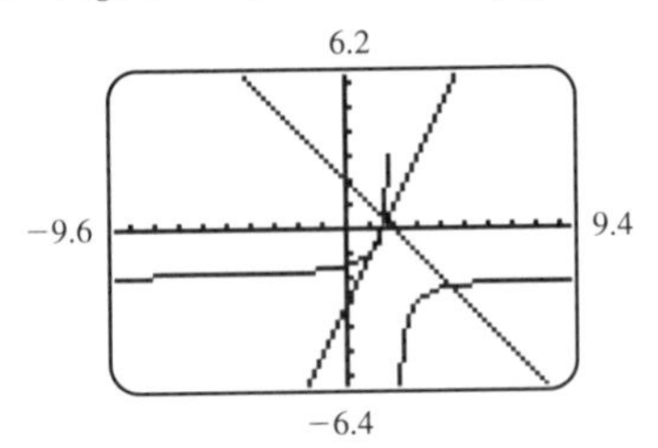

6. a. $f[g(-2)] = f(-1) = \sqrt{-1}$, not a real number
b. $(f \circ g)(0) = f[g(0)] = f(3) = \sqrt{3}$
c. $(g \circ f)(-2) = g[f(-2)] = g[f(\sqrt{-2})]$, not a real number
d. $g[f(0)] = g(0) = 3$
e. $(f \circ f)(16) = f[f(16)] = f(4) = 2$

7. $(f \circ g)(x) = f[g(x)] = f\left(\sqrt{x + 1}\right) = x + 1 - 2\sqrt{x + 1}$

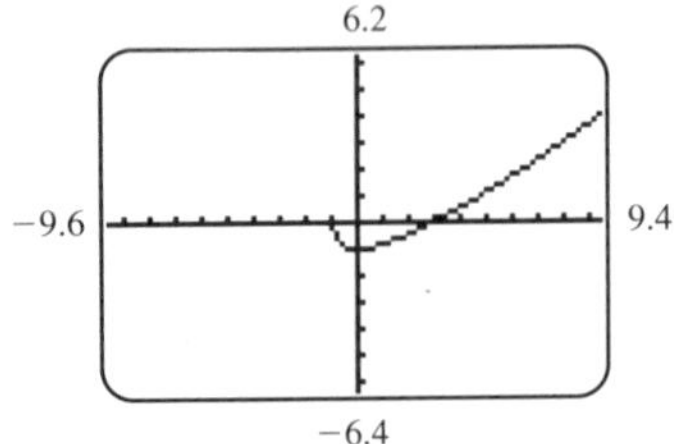

8. $(g \circ f)(x) = g(f(x)) = g(x^2 - 2x)$

$$= \sqrt{x^2 - 2x + 1} = |x - 1|$$

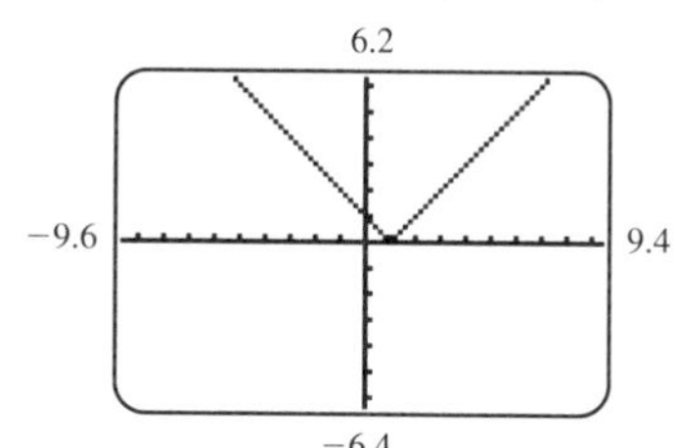

9. $A \circ B = \{(-2, 2), (-1, 1), (0, 2), (1, -2), (2, 0)\}$
$B \circ A = \{(-2, 2), (-1, 0), (0, -1), (1, -2), (2, 0)\}$

SECTION 9.4 ■ page 523

1.

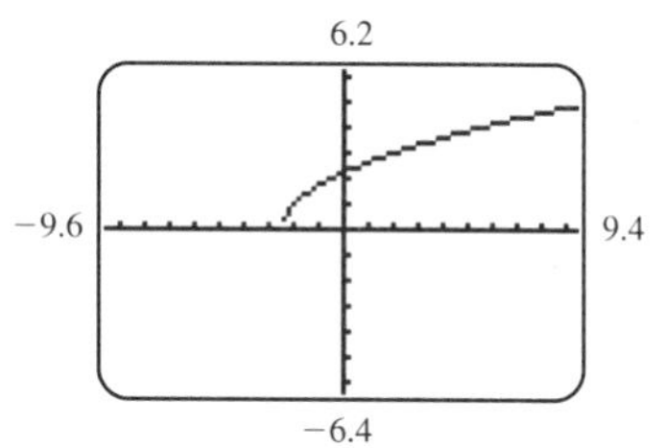

1-to-1

2.

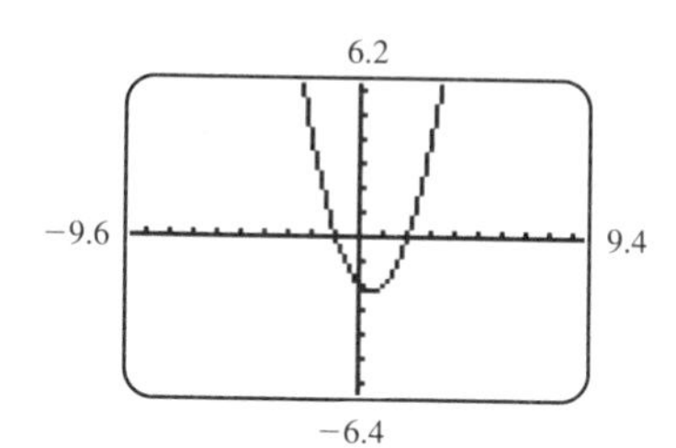

not 1-to-1

3.

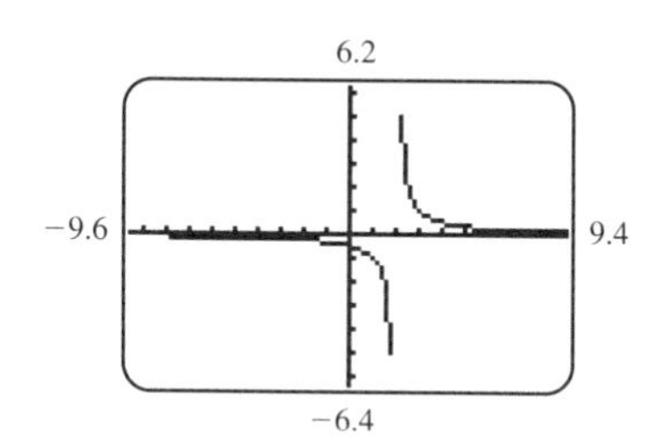

1-to-1

4.

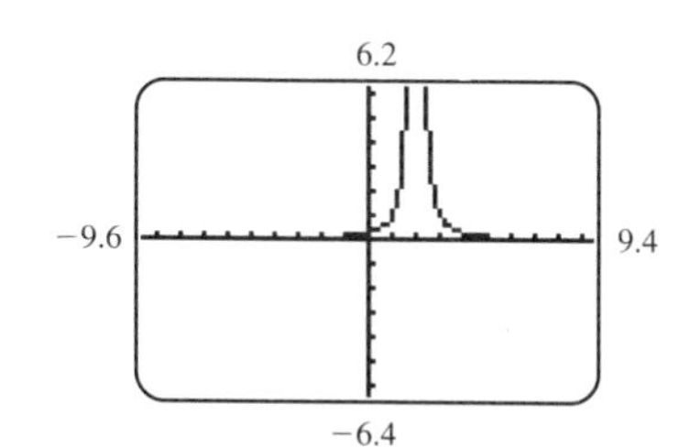

not 1-to-1

5. a. Not 1-to-1 **b.** 1-to-1 **c.** 1-to-1

6. $A^{-1} = \{(2, -3), (-1, -2), (3, 0), (4, 1)\}$

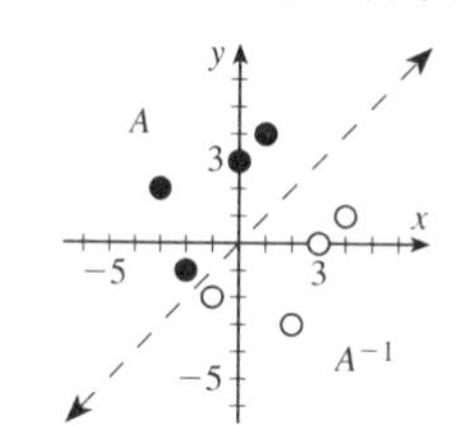

7.

3.1 Y1 Y3
Y2
−4.8 4.7
−3.2

:Y1=(2X+X²)(X≥−1)
:Y2=√(X+1)−1
:Y3=√(Y1+1)−1

Yes, the plots are reflections about the line $y = x$.

8. a. $f[g(x)] = f\left(\frac{x+1}{x}\right) = \frac{1}{\frac{x+1}{x} - 1} \cdot \frac{x}{x} = \frac{x}{x + 1 - x} = x$

b. $g[f(x)] = g\left(\frac{1}{x-1}\right) = \frac{\frac{1}{x-1} + 1}{\frac{1}{x-1}} \cdot \frac{x-1}{x-1} = \frac{1 + x - 1}{1} = x$

c. Yes; $(f \circ g)(x) = (g \circ f)(x)$, and both yield $I(x) = x$.

d.

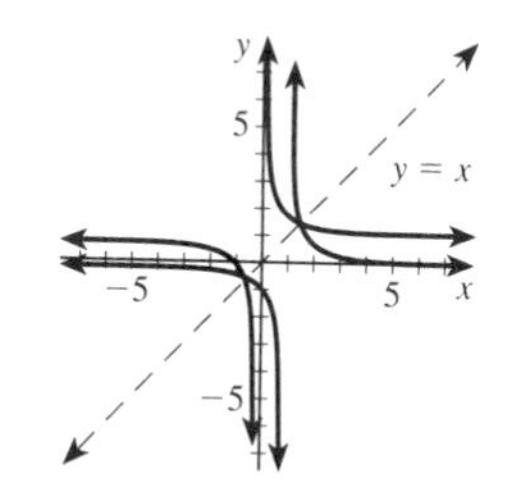

The graphs are symmetrical about the line $y = x$.

e. Range of $f = \{f(x) \mid f(x) \neq 0\}$
domain of $g = \{x \mid x \neq 0\}$

9. $g^{-1}(x) = x^2 - 1$ range of $g = \{g(x) \mid g(x) \geq 0\}$
domain of $g^{-1} = \{x \mid x \geq 0\}$

■ CHAPTER 10

SECTION 10.1 ■ page 533

1.

x	$g(x) = 3^x$
−2	$\frac{1}{9}$
−1	$\frac{1}{3}$
0	1
1	3
2	9

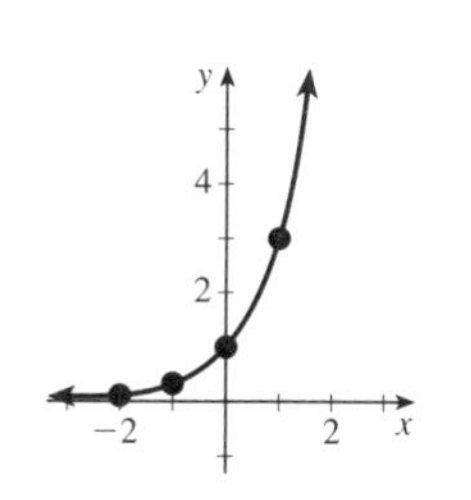

2.

x	$h(x) = 4^x$
-2	$\frac{1}{16}$
-1	$\frac{1}{4}$
0	1
1	4
2	16

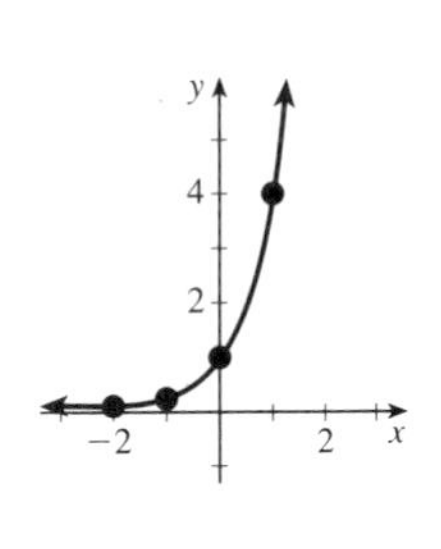

3.

x	$d(x) = 2^{-x}$
-2	4
-1	2
0	1
1	$\frac{1}{2}$
2	$\frac{1}{4}$

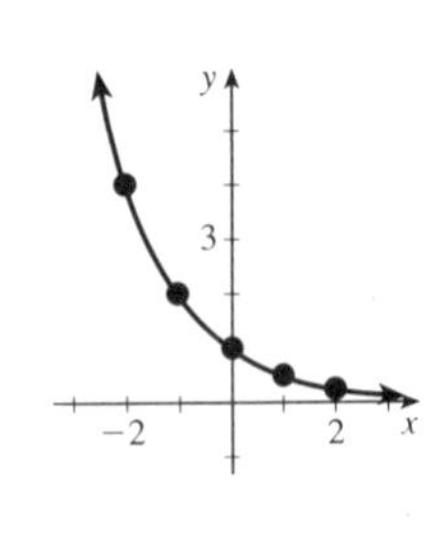

4.

$\sqrt{2}$	1.414213562
$2^{1.4}$	2.639015822
$2^{1.41}$	2.657371628
$2^{1.414}$	2.66474965
$2^{\sqrt{2}}$	2.665144143

5.

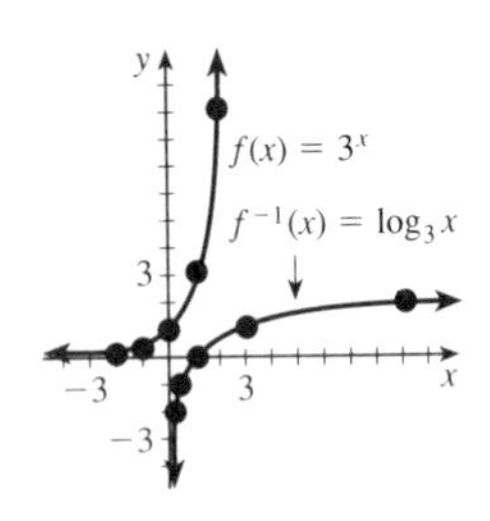

6. a. $\log_3 1 = 0$; $\log_3 \frac{1}{9} = -2$
7. $\log_{10} \sqrt{10} = \frac{1}{2}$ because $\sqrt{10} = 10^{1/2}$
8. LOG √10 outputs .5
9. $\log_7 1 = 0$ because $1 = 7^0$; $\log_{18} 18 = 1$ because $18 = 18^1$
10. $\log_8 64 = 2$ because $64 = 8^2$
11. $\log_3(3x - 2) = 2$ means $3x - 2 = 3^2$. Solution: $\frac{11}{3}$
12. $\log_4 x^2 = 2$ means $x^2 = 4^2$. Solution: 4 and -4
13. $\log_2 \frac{1}{x} = 1$ means $\frac{1}{x} = 2^1$. Solution: $\frac{1}{2}$
14. $\log_6 \sqrt{3 - x} = 1$ means $\sqrt{3 - x} = 6^1$. Solution: -33

SECTION 10.2 ■ page 544

1. 6 2. $\frac{37}{3}$ 3. -3

4. $\log_3 7 = \frac{\log_{10} 7}{\log_{10} 3} \approx 1.771$ 5. $\log_2 8 = \frac{\log_{10} 8}{\log_{10} 2} = 3.000$

6. $\log_{17} 20 = \frac{\log_{10} 20}{\log_{10} 17} \approx 1.057$

7. Solution $2\sqrt{2}$ ($-2\sqrt{2}$ is extraneous)
8. Solution 4 (0 is extraneous)
9. Solution 2 (-4 is extraneous)
10. Solution 3
11. Solution *no real numbers*
12. Solution 3 (-1 is extraneous)

SECTION 10.3 ■ page 553

1. $\frac{-1}{2}$ 2. -2 3. $\frac{\log_{10} 3}{\log_{10} 17} \approx 0.3878$

4. $\frac{\log_{10} 2}{\log_{10} 1.5} \approx 1.7095$

5. Solution $\frac{1 + e^5}{3} \approx 49.8044$ 6. Solution $\ln 3 \approx 1.0986$

7. Solution *none* $\left(\frac{-1}{e - 1} \approx -0.582 \text{ is an extraneous root}\right)$

SECTION 10.4 ■ page 559

1. a. $A = 5000(1.02)^{40} \approx \$11{,}040.20$
   b. Solve: $25000 = 5000(1.02)^{4n}$
   Solution: $\frac{\log_{10} 5}{4 \log_{10} 1.02} \approx 20.32$ years $\approx$ 20 years 4 months
2. a. $A = 5000e^{(0.08)(10)} \approx \$11{,}127.70$
   b. Solve: $25000 = 5000e^{0.08n}$
   Solution: $\frac{\ln 5}{0.08} \approx 20.12$ years
3. $A = 0.01(1.005)^{1200} \approx \$3.97$ after 100 years
   $A = 0.01(1.005)^{2400} \approx \$1579.60$ after 200 years
   $A = 0.01(1.005)^{4800} \approx \$249{,}514{,}869.10$ after 400 years
4. $P = 5000(1.05)^{10} \approx 8{,}144$ in 10 years
   $P = 5000(1.05)^{20} \approx 13{,}266$ in 20 years
   $P = 5000(1.05)^{40} \approx 35{,}200$ in 40 years
5. Solve: $65 = 100(1 - 0.5)^n$
   Solution: $\frac{\log_{10} 0.65}{\log_{10} 0.5} \approx 0.6215$
   Age $\approx 0.6215(5700) = 3542.55$ years
6. $\text{pH} = -\log_{10}(5.2 \times 10^{-4}) \approx 3.284$
7. Solve: $7000 = 3500e^{0.02t}$
   Solution: $\frac{\ln 2}{0.02} \approx 34.66$ years

■ CHAPTER 11

SECTION 11.1 ■ page 570

1. $\left(\frac{60}{19}, \frac{14}{19}\right)$ 2. $\left(\frac{9}{8}, \frac{19}{8}\right)$ 3. $(-18, -27, -33)$

4. $(4, -5, -5)$ 5. *None* (system is inconsistent)

6. $\left(\frac{74}{5}, \frac{-11}{5}, \frac{-12}{5}\right)$ 7. $\left(\frac{10}{3}, \frac{26}{3}, \frac{-37}{3}\right)$

8. Solve: $\begin{cases} 1 = a(-2)^2 + b(-2) + c \\ -1 = a(1)^2 + b(1) + c \\ 5 = a(2)^2 + b(2) + c \end{cases}$

   Equation of the parabola is $y = \frac{5}{3}x^2 + x - \frac{11}{3}$.

SECTION 11.2 ■ page 581

1. 3×4 2. 5 3. -5

4. $\begin{bmatrix} 0 & 11 & 11 \\ 13 & 5 & -1 \end{bmatrix}$ 5. $\begin{bmatrix} 8 & 1 & 7 \\ -1 & -11 & -9 \end{bmatrix}$

6. $\begin{bmatrix} 8 & 34 & 40 \\ 38 & 4 & -12 \end{bmatrix}$ 7. $\begin{bmatrix} -12 & -7 & -16 \\ -5 & 14 & 14 \end{bmatrix}$

8. *Problem 4:* Set the size for matrix [A] to 2 × 3. Type the entries for [A].
Set the size for matrix [B] to 2 × 3, and type the entries for [B].
From the home screen, type [A] + [B] and press enter. The output is

$$\begin{bmatrix} 0 & 11 & 11 \\ 13 & 5 & -1 \end{bmatrix}$$

9. $\begin{bmatrix} 0 & 0 & 0 \\ 0 & 0 & 0 \end{bmatrix}$ **10.** $\begin{bmatrix} 3 & -5 & -7 \\ -2 & 3 & -1 \end{bmatrix}$

11. $(-2)(4) + (1)(-3) + (4)(6) + (6)(2) = 25$

12. $(6)(10) + (7)(5) = 95$

13. $AB = \begin{bmatrix} 10 & 36 \\ 22 & -9 \end{bmatrix}$ $BA = \begin{bmatrix} -30 & 6 \\ -8 & 31 \end{bmatrix}$

14. System: $\begin{cases} -5x + 6y = -3 \\ 2x + 3y = 4 \end{cases}$

Solution: $\left(\frac{11}{9}, \frac{14}{27}\right)$, so $X = \begin{bmatrix} \frac{11}{9} \\ \frac{14}{27} \end{bmatrix}$

SECTION 11.3 ■ page 592

1. $A^{-1} = \begin{bmatrix} -2 & -1 \\ 5 & 3 \end{bmatrix}$ **2.** $C^{-1} = \begin{bmatrix} 3 & -1 \\ -5 & 2 \end{bmatrix}$

3. $A^{-1} = \begin{bmatrix} \frac{-1}{4} & \frac{1}{2} & \frac{1}{2} & \frac{1}{2} \\ \frac{-1}{2} & 4 & -1 & -2 \\ \frac{-1}{4} & \frac{5}{2} & \frac{-1}{2} & \frac{-3}{2} \\ \frac{1}{4} & \frac{-1}{2} & \frac{1}{2} & \frac{1}{2} \end{bmatrix}$

4. $\begin{bmatrix} 11 & -8 \\ -4 & 3 \end{bmatrix}\begin{bmatrix} 2 \\ 7 \end{bmatrix} = \begin{bmatrix} -34 \\ 13 \end{bmatrix}$ **5.** $(24, 17, -16)$

6. $\det A = {}^{-}8$ $A^{-1} = \begin{bmatrix} \frac{-1}{4} & \frac{5}{8} \\ \frac{1}{4} & \frac{-1}{8} \end{bmatrix}$

7. $\det B = 0$ B^{-1} does not exist

SECTION 11.4 ■ page 601

1. -40 **2.** 31 **3.** $-5 + 3\sqrt{2}$

4. $M_{11} = 4$ **5.** $C_{11} = (-1)^{1+1}(4) = 4$

6. $M_{32} = -26$ **7.** $C_{32} = (-1)^{3+2}(-26) = 26$

8. $M_{22} = -36$ **9.** $C_{22} = (-1)^{2+2}(-36) = -36$

10. $\det(B) = 0$

11. $D = 2, D_x = 5, D_y = 12, D_z = 4$

Solution: $\left(\frac{5}{2}, 6, 2\right)$

12. $D = 33; A = \frac{1}{2}|D| = \frac{33}{2} = 16\frac{1}{2}$

13. For $A = \begin{bmatrix} -1 & 0 & 1 \\ 3 & -1 & 1 \\ 7 & -2 & 1 \end{bmatrix}$, $\det(A) = 0$, so points are collinear.

14. $5x - 5y + 5 = 0$

■ CHAPTER 12

SECTION 12.1 ■ page 616

1. a.

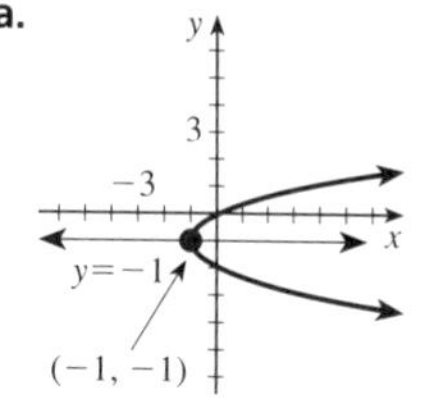

b.

2. Standard form: $2y^2 - y + 1 - x = 0$
$a = 2, b = -1, c = 1 - x$
$b^2 - 4ac = 1 - 4(2)(1 - x) = 8x - 7$

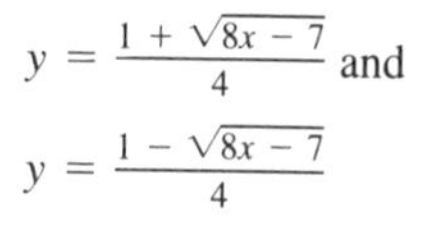

$y = \frac{1 + \sqrt{8x - 7}}{4}$ and

$y = \frac{1 - \sqrt{8x - 7}}{4}$

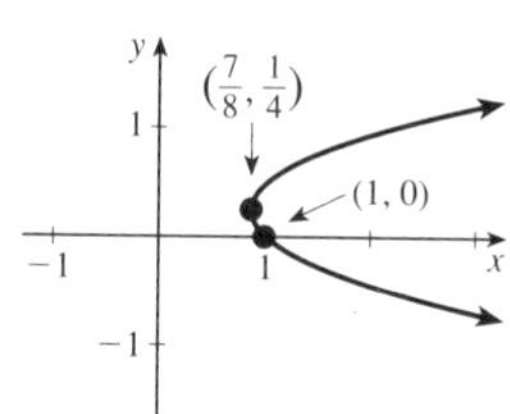

3. *Problem 1a*: Type −1+√(X+1) for Y1. Type −1−√(X+1) for Y2. View the graph screen. Use the TRACE feature to check the vertex and intercepts.

SECTION 12.2 ■ page 622

1.

2. $x^2 + y^2 = 25$

3.

4. $(x + 1)^2 + (y - 1)^2 = 18$

5. $x^2 + 4x + \underline{\ 4\ } + y^2 - 2y + \underline{\ 1\ } = 3 + 4 + 1$
or $(x + 2)^2 + (y - 1)^2 = 8$
center $(-2, 1)$; radius $2\sqrt{2}$

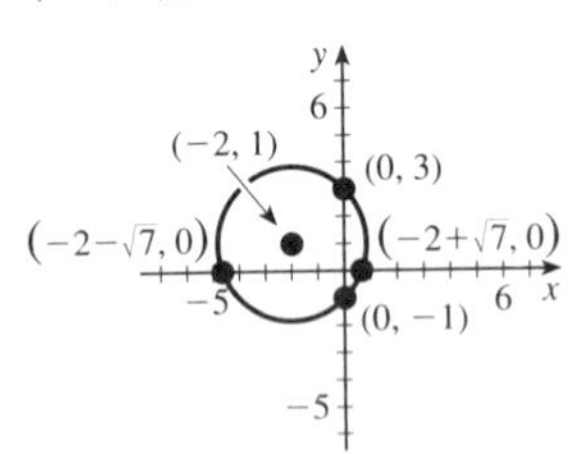

6. $(x - 3)^2 + (y - 2)^2 = 14$;
center $(3, 2)$; radius $\sqrt{14}$

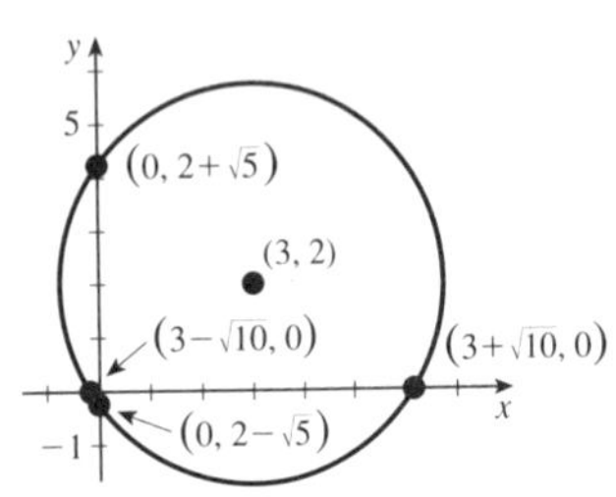

SECTION 12.3 ■ page 630

1.

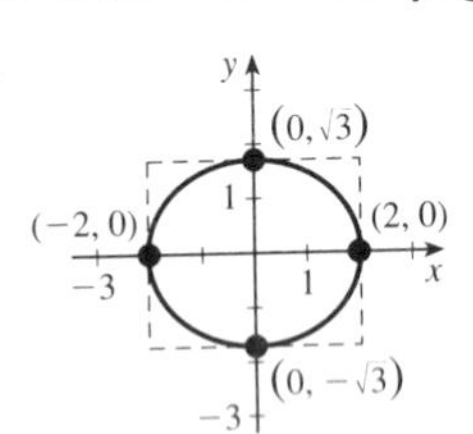

2.

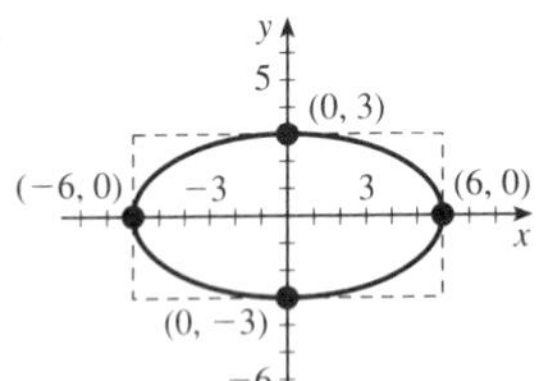

3. $\frac{x^2}{25} - \frac{y^2}{16} = 1$

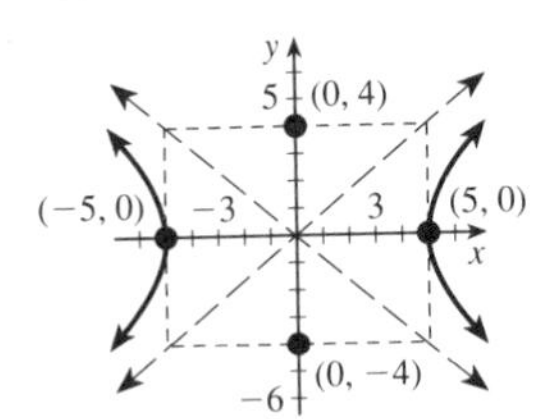

4. $\frac{y^2}{16} - \frac{x^2}{25} = 1$

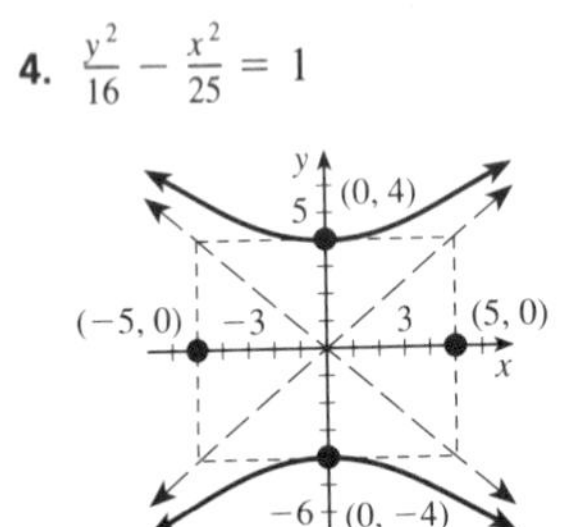

SECTION 12.4 ■ page 638

1. Solution $(0, 3)$ and $(2, -1)$
2. Solution $(6, 16)$ and $(1, 1)$
3. Solution *none*
4. Solution $\left(\frac{3 - \sqrt{7}}{2}, \frac{1 - \sqrt{7}}{2}\right) \approx (0.177, -0.823)$
and $\left(\frac{3 + \sqrt{7}}{2}, \frac{1 + \sqrt{7}}{2}\right) \approx (2.823, 1.823)$

SECTION 12.5 ■ page 644

1.

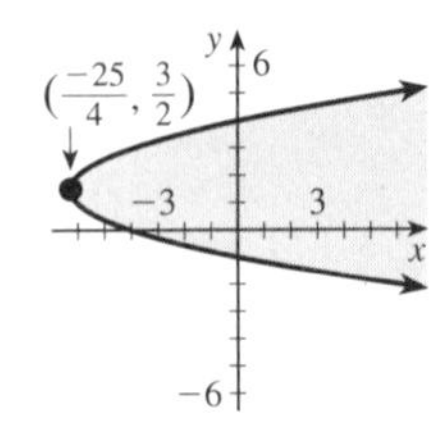

2.

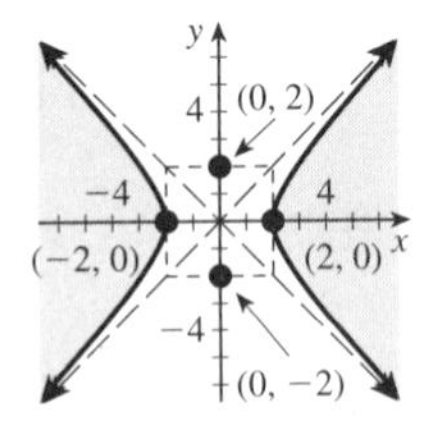

3.

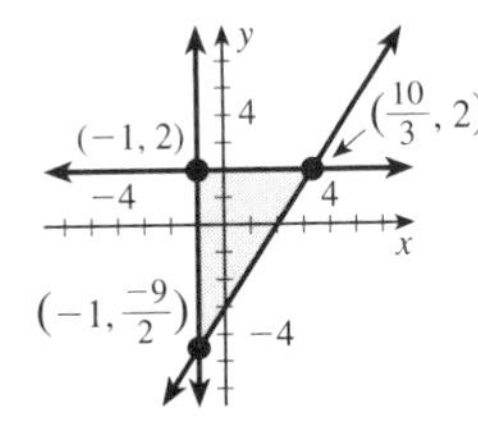

4.

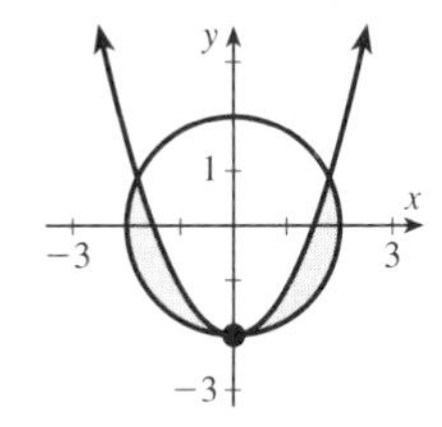

ANSWERS TO SELECTED EXERCISES

CHAPTER 1

EXERCISES 1.1 ■ page 12

1. True **3.** True **5.** True **7.** True **9.** True
11. Yes **13.** {1, 2, 3, 5, 6, 7, 8, 9, 11} **15.** Yes
17. W **19.** -36 **21.** 25 **23.** -100 **25.** 10
27. -65 **29.** 31 **31.** -196 **33.** -532 **35.** 69
37. -900 **39.** -75 **41.** 0 **43.** 11 **45.** -4
47. -9 **49.** -12 **51.** -200 **53.** -42 **55.** -18
57. -18 **59.** -23 **61.** -8
63. Addition, subtraction, multiplication **65.** 0
67. $1, -1, 2, -2, 4, -4, 8, -8$
69. $1, -1, 3, -3, 5, -5, 15, -15$
71. $1, -1, 2, -2, 3, -3, 6, -6, 9, -9, 18, -18$
73. $1, -1, 2, -2, 4, -4, 7, -7, 14, -14, 28, -28$
75. $1, -1, 13, -13$ **77.** -125 **79.** 256 **81.** -1[illegible]
83. 36 **85.** -8000 **87.** 1 **89.** 0
91. Commutative property of addition
93. Distributive property
95. Identity property of multiplication
97. 11 **99.** 44 **101.** $760 gain

DISCOVERY ■ page 18

1. a. Type: [(-)] [(] [37] [-] [2nd] [x^{-1}] [(] [(-)] [52] [+] [278] [)] [)] [ENTER]
Output: 189

b. Type: [2nd] [x^{-1}] [(] [(-)] [7] [-] [2nd] [x^{-1}] [(] [5] [-] [20] [)] [-] [8] [)] [-] [2] [ENTER]
Output: 28

3. Refer to the answers for Exercises 1.1. Keystrokes for selected problems follow.

19. Type: [(-)] [13] [+] [(-)] [23] [ENTER]
Output: -36

61. Type: [(-)] [800] [÷] [100] [ENTER]
Output: -8

83. Type: [(] [(-)] [6] [)] [x^2] [ENTER]
Output: 36

85. Type: [(-)] [20] [^] [3] [ENTER]
Output: -8000

97. Type: [2nd] [x^{-1}] [(] [(-)] [8] [-] [3] [)] [ENTER]
Output: 11

101. Type: [150] [×] [8] [+] [220] [×] [(-)] [2] [ENTER]
Output: 760

EXERCISES 1.2 ■ page 26

1. Addition, subtraction, multiplication, division **3.** 0
5. $\frac{-1}{3}$ **7.** $\frac{5}{9}$ **9.** $\frac{7}{9}$ **11.** $\frac{45}{14}$ **13.** $\frac{-5}{9}$ **15.** $\frac{-13}{15}$
17. $\frac{-3}{10}$ **19.** $\frac{9}{8}$ **21.** $\frac{-5}{6}$ **23.** $\frac{17}{20}$ **25.** $\frac{-53}{60}$ **27.** $\frac{1}{6}$
29. $\frac{-3}{8}$ **31.** $\frac{1}{3}$ **33.** $\frac{3}{35}$ **35.** -75 **37.** $\frac{25}{144}$
39. $\frac{27}{64}$ **41.** 1 **43.** $\frac{19}{24}$ **45.** $\frac{-36}{5}$ **47.** $\frac{1}{2}$
49. 1200 calls per day **51.** $5000
53. $56\frac{1}{4}$ inches remain.

55. Commutative property of addition
57. Distributive property
59. Associative property of multiplication
61. Commutative property of addition
63. Distributive property

DISCOVERY ■ page 28

Refer to the answers from Exercises 1.2. Keystrokes for selected problems follow.

13. Type: [(-)] [8] [÷] [36] [+] [(-)] [12] [÷] [36] [ENTER]
Output: -.5555555556
Type: [(-5)] [÷] [9] [ENTER]
Output: -.5555555556

21. Type: [(-)] [1] [÷] [2] [−] [1] [÷] [3] [ENTER]
Output: -.8333333333
Type: [(-)] [5] [÷] [6] [ENTER]
Output: -.8333333333

29. Type: [(-)] [1] [÷] [2] [×] [3] [÷] [4] [ENTER]
Output: -.375
Type: [(-)] [3] [÷] [8] [ENTER]
Output: -.375

31. Type: [(-)] [2] [÷] [5] [÷] [(] [(-)] [6] [÷] [5] [)] [ENTER]
Output: .333333333
Type: [1] [÷] [3] [ENTER]
Output: .333333333

53. Type: [100] [−] [5] [×] [(] [8] [+] [3] [÷] [4] [)] [ENTER]
Output: 56.25

EXERCISES 1.3 ■ page 39

1. 10^{-1} **3.** 10^{-3} **5.** 10^{0} **7.** 10^{-6} **9.** 50,600
11. −203 **13.** 0.056 **15.** −2.3 **17.** 32.53
19. 5 significant digits **21.** 5 significant digits
23. 4 significant digits **25.** 123,568,000 **27.** 1000.
29. 23,900 **31.** 0.00457 **33.** −0.9 **35.** 15,000
37. 0.003457 **39.** 350,000 **41.** 5.004 **43.** 0.003057
45. 23% **47.** 500% **49.** 37.5% **51.** 19.04
53. 24 **55.** 0.09 **57.** $\frac{140}{17}$ or $8\frac{4}{17}$
59. 24 liters of acid, 56 liters of pure water
61. 21 grams of silver, 42% silver **63.** 2.3×10^{13}
65. 6.782×10^{2} **67.** 3.98×10^{-7} **69.** 1.78×10^{5}
71. 1.89×10^{-10} **73.** 3.6×10^{3} **75.** 5.5225×10^{-14}
77. 2.8×10^{4} **79.** 1×10^{-4} **81.** 1.092×10^{8}
83. 6×10^{1} **85.** 2.05×10^{3} **87.** 3.238×10^{3}
89. 9.623×10^{-1} **91.** 8.5×10^{-1}

DISCOVERY ■ page 42

1. a. 2.577×10^{0} **b.** 2.706×10^{10}
3. a. $(5)^2(5)^3$ evaluates to 3125 and $(5)^5$ evaluates to 3125.
b. $((2)^2)^3$ evaluates to 64 and $(2)^6$ evaluates to 64.
c. $(2)^2 \div (2)^3$ evaluates to 0.5 and $(2)^{-1}$ evaluates to 0.5.

Refer to the answers from Exercises 1.3. Keystrokes for selected problems follow.

63. Set MODE to Sci.
Type: [23000000000000] [ENTER]
Output: 2.3E13
Scientific Notation: 2.3×10^{13}

71. Set MODE to Sci.
Type: [0.000000000189] [ENTER]
Output: 1.89E-10
Scientific Notation: 1.89×10^{-10}

79. Set MODE to Sci.
Type: [0.12] [÷] [(] [1.2] [EE] [3] [)] [ENTER]
Output: 1E-4
Scientific Notation: 1×10^{-4}

EXERCISES 1.4 ■ page 49

1. −20 **3.** −11 **5.** −11 **7.** −5 **9.** −69
11. −73 **13.** $\frac{3}{2}$ or 1.5 **15.** 9 **17.** $\frac{11}{12}$ **19.** $\frac{5}{12}$
21. $\frac{12}{5}$ or 2.4 **23.** 4 **25.** −135 **27.** −67
29. −14 **31.** 16 **33.** $\frac{15}{68}$ **35.** 6 **37.** $\frac{1}{15}$
39. $\frac{-28}{15}$ **41.** $\frac{-163}{42}$ **43.** $\frac{-39}{35}$ **45.** $\frac{-17}{5}$ **47.** $y = 20$
49. $y = 10$ **51.** $y = -6$ **53.** $y = \frac{-25}{42}$ **55.** $y = \frac{-1}{5}$

DISCOVERY ■ page 53

1. Type: [(] [5.2] [EE] [9] [)] [(] [1] [+] [0.02] [)] [^] [10] [ENTER]
Output: 6338770984
Answer: 6.3 billion people

3. Refer to the answers for Exercises 1.4. Keystrokes for selected problems are included.

49. Type: [(-)] [3] [÷] [5] [STO▶] [X|T] [ENTER]
Output: -.6
Type: [(-)] [5] [X|T] [+] [7] [ENTER]
Output: 10

53. Type: Y= CLEAR

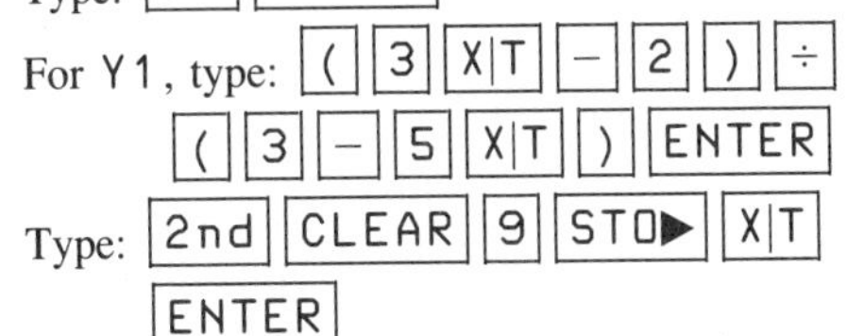

ENTER

Output: 9

Type: 2nd VARS 1 ENTER

Output: -.5952380952

Type: (−) 25 ÷ 42 ENTER

Output: -.5952380952

EXERCISES 1.5 ■ page 61

1. False **3.** True **5.** False **7.** False

9. $\sqrt{\frac{9}{25}}$ **11.** $\sqrt{\frac{49}{64}}$ **13.** $x = 15$ **15.** $c = 10$

17. One answer is $\frac{-7}{14}$, or $\frac{-1}{2}$ (Other answers are possible.)

19. Use the 2nd and x^2 keys to type √17.
Output: 4.123105626
Answer: 4 and 5

21. 5 and 6 **23.** 9 and 10 **25.** 14 and 15
27. 17 and 18 **29.** 22 and 23 **31.** 0 and 1
33. −1 and 0 **35.** −3 and −2 **37.** −3 and −2
39. −1 and 0 **41.** 8 **43.** 24 **45.** 1 **47.** −2
49. $\frac{4}{3}$

51. Type: π − 5
Output: -1.858407346
So, $|\pi - 5| = -(\pi - 5) = 5 - \pi$

53. $3\pi - 2e$ **55.** $5\pi - \sqrt{19}$ **57.** $\pi^2 - \sqrt{21}$

CHAPTER 1 REVIEW ■ page 63

1. a. $0, 13, \sqrt{16}$ **b.** $0, 13, -3, -1, \frac{-\sqrt{4}}{2}, \sqrt{16}$
c. $0, 13, -3, -1, \frac{-3}{7}, \frac{-\sqrt{4}}{2}, \sqrt{16}$ **d.** $-\sqrt{3}, \pi$

3. −5 **5.** 4 **7.** 3 **9.** 3 **11.** −8 **13.** 16
15. 8 **17.** $\frac{3}{2}$ or 1.5 **19.** −50 **21.** 48 **23.** $\frac{-10}{7}$
25. 15 **27.** −3 **29.** 2 **31.** $\frac{-17}{6}$ **33.** $\frac{-2}{7}$
35. $\frac{-2}{3}$ **37.** $\frac{-1}{20}$ **39.** 23.009 **41.** 23,800
43. 23,800 **45.** 3.5% **47.** 25% **49.** 500%
51. 33.3 **53.** 0.6 **55.** 0.0007 **57.** \$900
59. 77.1% acid **61.** \$448 **63.** 2.3×10^7
65. 4.056×10^{-4} **67.** 5.8×10^1 **69.** 9.0×10^6
71. 9.12×10^7 **73.** 350 **75.** 0.000567

77. 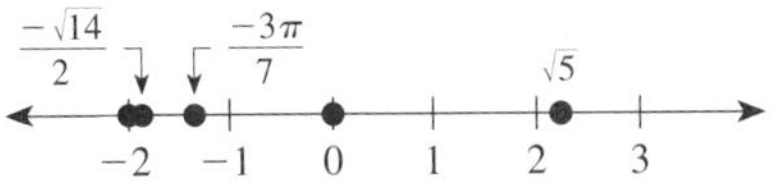 (number line from −2 to 3 with points marked at $\frac{-\sqrt{14}}{2}$, $\frac{-3\pi}{7}$, 0, $\sqrt{5}$) **79.** $\frac{-27}{5}$ **81.** $\frac{1}{4}$

83. −5 **85.** −4 **87.** 0 **89.** $\frac{1}{3}$ **91.** $\frac{9}{4}$

CHAPTER 1 TEST ■ page 64

1. a. $0, \sqrt{36}$ **b.** $-3, 0, \sqrt{36}$
c. $-3, 0, \sqrt{36}, \frac{8-\sqrt{25}}{6}$ **d.** $\frac{\sqrt{5}}{3}, e, \pi$

2. a. Distributive property
b. Associative property for multiplication

3. −12 **4.** −59 **5.** 10 **6.** −29 **7.** $\frac{-1}{12}$
8. $\frac{-10}{9}$ **9.** $\frac{17}{2}$ **10.** $\frac{-3}{4}$ **11.** 46.7%
12. 2.378×10^6 **13.** 4.56×10^{-5} **14.** 6.561×10^9
15. 2.81×10^{-2} **16.** 1.8×10^5 **17.** -5.41×10^4
18. 0.0035 **19. a.** 0 and 1 **b.** −2 and −1
c. 0 and 1 **20. a.** 9 **b.** 2 **c.** $\frac{4}{3}$

■ CHAPTER 2

EXERCISES 2.1 ■ page 76

1. True **3.** True **5.** False **7.** False
9. i. $3x + 2$ **11. ii.** −9
13. Third-degree trinomial in one variable
15. Zero-degree monomial
17. Seventh-degree quadrinomial in three variables
19. Answers will vary. Possible answers: $3x^2$ and $-3x^2 - 5$ or $7x^2 + 8$ and $-7x^2 - 2$
21. No **23.** −20 **25.** −45 **27.** 11300
29. $-9x^2 - x + 6$
To check for errors, store -7 in X. Type the original expression for Y1, and type the standard polynomial for Y2. Display the values of Y1 and Y2. In each case, the output is -428.
31. $2w^2 - 4w - 3$
To check for errors, store 9 in X. Type the original expression for Y1, and type the standard polynomial for Y2. Display the values of Y1 and Y2. In each case, the output is 123.
33. $5x^3 + 5x^2 - 3x - 8$ **35.** $3p^2 + 7p + 10$
37. $2x + 6$ **39.** $-x^2 + 7x$ **41.** $5x^3 + 3x^2 + 3x - 6$
43. $4s^2 + 2s^2 - 9$ **45.** $\frac{-x-5}{-5}$ or $\frac{x+5}{5}$ **47.** $-3x^2 + 7x$
49. $-5m^2 + 6m - 16$ **51.** $-5x^2 + 3x + 6$ **53.** 141
55. −2

DISCOVERY ■ page 78

1. a. 2.5 **b.** 7 **c.** −3.2 **d.** 10
3. a. −2 **b.** 2 **c.** −4.24 **d.** 2
5. a. −3.25 **b.** −4.09 **c.** 3.59 **d.** 8.71

EXERCISES 2.2 ■ page 91

1. True **3.** True **5.** False **7.** False **9.** False

11. a. $-2x + 2 = -x + 2$ **b.** $-x + 2 = 2$ **c.** $-x = 0$ **d.** $x = 0$ **e.** 0

13. -43
To check for errors, store ⁻43 in X, then type ⁻3X. The output is 129.

15. $\frac{16}{7}$ **17.** $\frac{15}{2}$ **19.** $\frac{-20}{21}$ **21.** -216 **23.** -2.54

25. 0.3 **27.** $\frac{5}{3}$ **29.** $\frac{11}{3}$ **31.** $\frac{-\sqrt{5}}{10}$ **33.** $\frac{-2}{5}$

35. -1 **37.** $\frac{-14}{5}$ **39.** 14

41. 1.1
To check for errors, store 1.1 in X. Type 3X−8 for Y1, and type ⁻7X+3 for Y2. Display the values of Y1 and Y2. In each case, the output is ⁻4.7.

43. $\frac{-2}{3}$ **45.** $\frac{15}{8}$ **47.** 2 **49.** $\frac{-10}{13}$ **51.** $\frac{-25}{4}$

53. $\frac{-15}{16}$ **55.** $\frac{-3}{5}$ **57.** 3.75 **59.** 80 **61.** $\frac{5}{8}$

63. $\frac{7}{11}$ **65.** $\frac{17}{5}$ **67.** 2 **69.** $\frac{2}{25}$ **71.** $\frac{1}{5}$

73. $\frac{5-\sqrt{3}}{6}$ **75.** 1 **77.** $\frac{3-\sqrt{5}}{3}$

DISCOVERY ■ page 94

1.

x-value	−2.1	−1.4	0.7	3.2	4.5
Evaluation of $-3x + 2$	8.3	6.2	−0.1	−7.6	−11.5

EXERCISES 2.3 ■ page 105

1. False **3.** True **5.** True **7.** False **9.** $\left(-\infty, \frac{3}{2}\right)$

11. $\left[\frac{-4}{3}, +\infty\right)$

13.
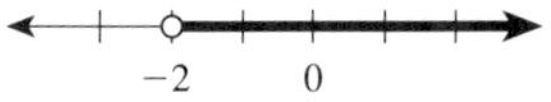

15.
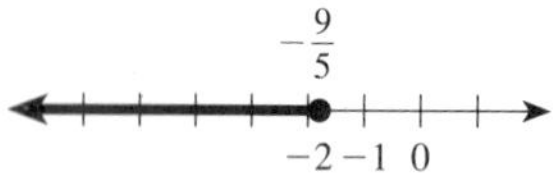

17. $(-4, +\infty)$

19. $(-\infty, -4]$

21. $\left(-\infty, \frac{15}{8}\right]$
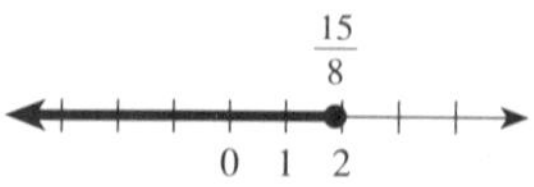

23. $\left(\frac{22}{5}, +\infty\right)$
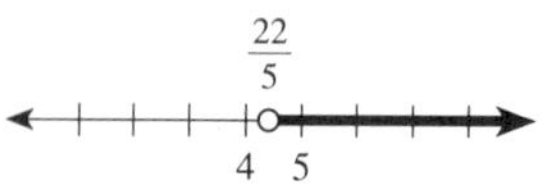

25. $\left[\frac{8}{3}, +\infty\right)$
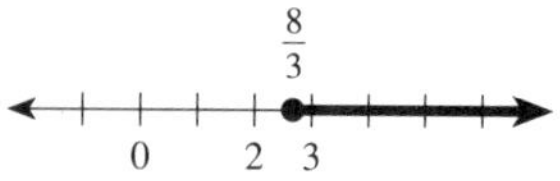

27. $[0, +\infty)$

29. $(-\infty, -6)$
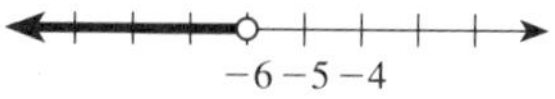

31. $[-5, +\infty)$
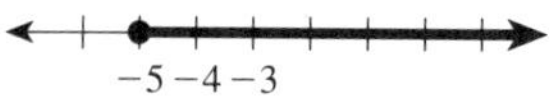

33. $\left(\frac{11}{2}, +\infty\right)$
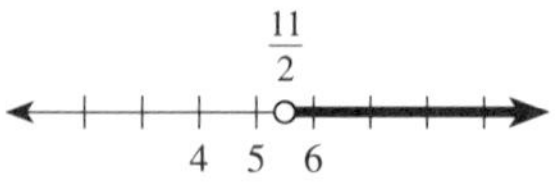

35. $\left[\frac{45}{14}, +\infty\right)$
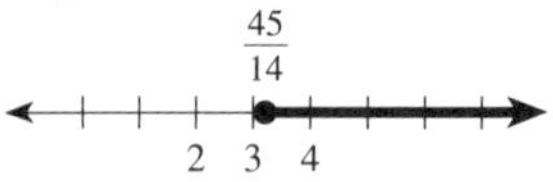

37. $\left[\frac{-3}{5}, +\infty\right)$
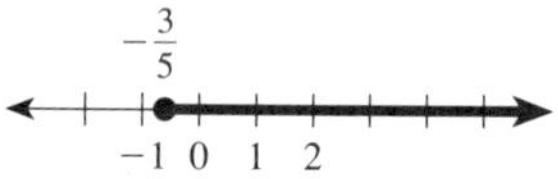

39. $\left(\frac{-1}{6}, +\infty\right)$
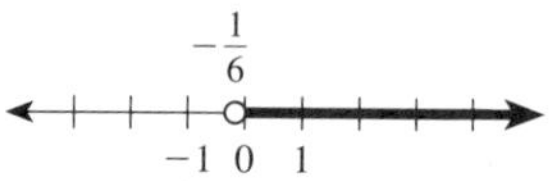

41. $(-\infty, +\infty)$
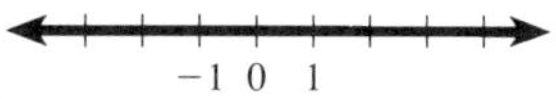

43. $(-\infty, 2]$

45. $(-\infty, 10]$

47. $[0, +\infty)$

49. $\left(-\infty, \frac{-3}{2}\right]$
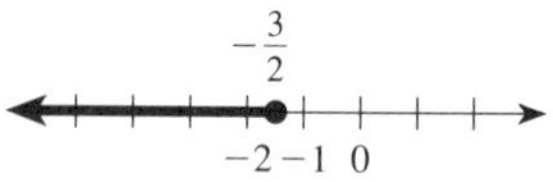

51. $\left(\frac{3}{2}, +\infty\right)$
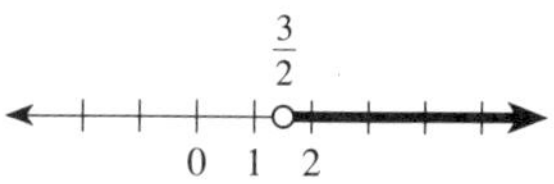

53. $\left(-\infty, \frac{-3}{2}\right]$
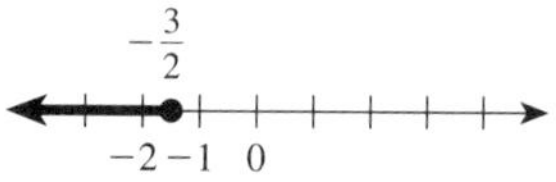

55. $[0, +\infty)$

57. $(2, +\infty)$
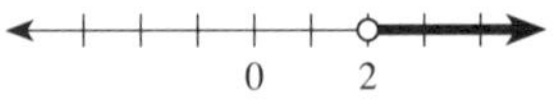

59. $\left[\frac{2}{25}, +\infty\right)$
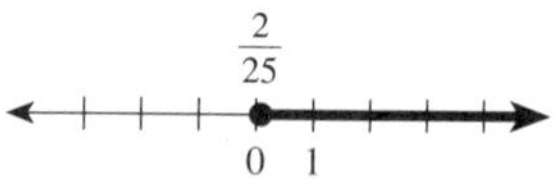

61. All **63.** None **65.** All **67.** $\left(\frac{6}{5}, +\infty\right)$

69. $(-\infty, 8]$

71. $[-1, +\infty)$
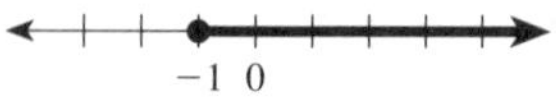

73. $(-\infty, -4)$ **75.** $(3, +\infty)$

77. $\varnothing$

none

EXERCISES 2.4 ■ page 111

1. $[-3, 3]$ **3.** $[2, +\infty)$

5.

7.

9.

11. $[-1, 4]$

13. $(-1, 2)$ **15.** $(-5, 2]$

17. $\left[1, \frac{8}{3}\right]$ **19.** $\left(\frac{1}{4}, \frac{13}{4}\right)$

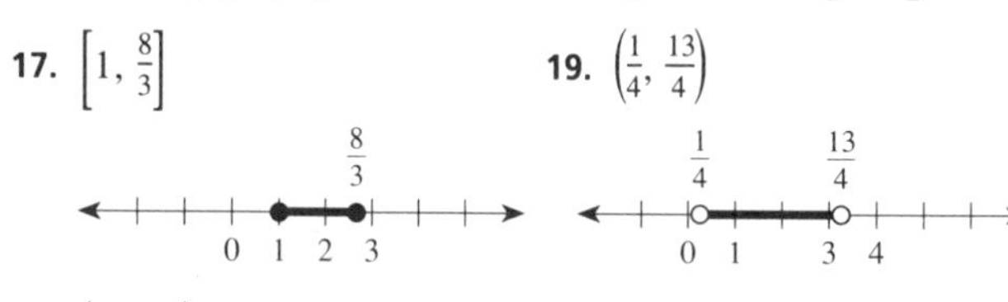

21. $\left(\frac{-4}{9}, \frac{13}{9}\right)$ **23.** $[-8, 4]$

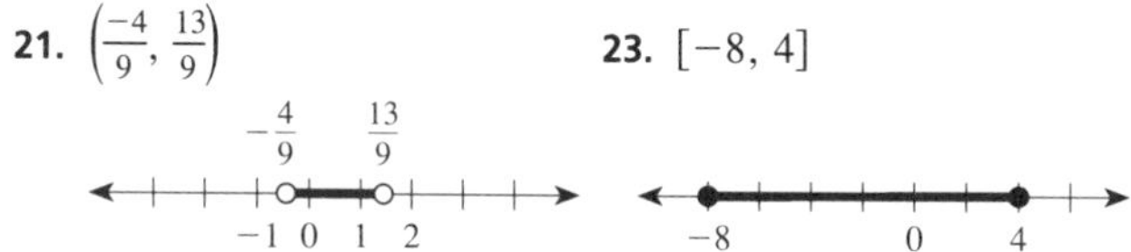

25. $\left(\frac{1}{4}, \frac{11}{6}\right)$ **27.** $\left[\frac{-1}{2}, \frac{9}{2}\right)$

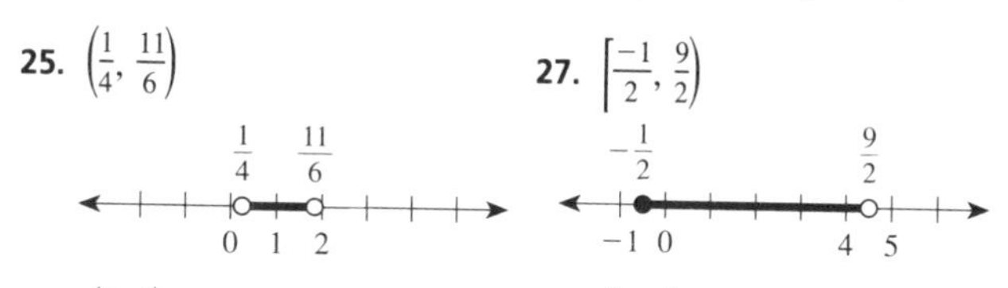

29. $\left(\frac{1}{3}, \frac{8}{3}\right)$ **31.** $\left(\frac{5}{4}, \frac{7}{4}\right)$

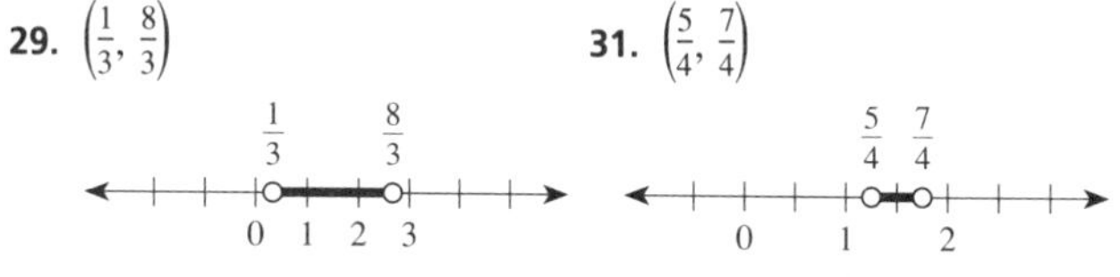

EXERCISES 2.5 ■ page 119

1. $[-2, 3)$ **3.** $(-\infty, -1] \cup [2, +\infty)$

5.

7.

9.

11. $4, -4$ **13.** $\varnothing$ **15.** $3, -13$ **17.** $\frac{5}{2}$ **19.** $\varnothing$

21. $2, \frac{-4}{3}$ **23.** $\frac{-1}{2}, \frac{-13}{2}$ **25.** $\frac{-16}{5}, 4$ **27.** $-8, 7$

29. $1, 4$ **31.** $\frac{1}{6}, \frac{3}{2}$ **33.** $1, 5$

35. $(-3, 3)$ **37.** $(-\infty, -6] \cup [6, +\infty)$

39. $[-2, 8]$ **41.** $(-\infty, 1) \cup (4, +\infty)$

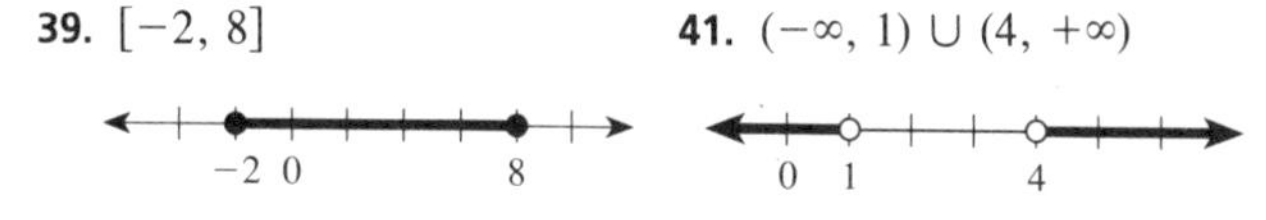

43. $\varnothing$ **45.** $(-\infty, -4] \cup [4, +\infty)$

none

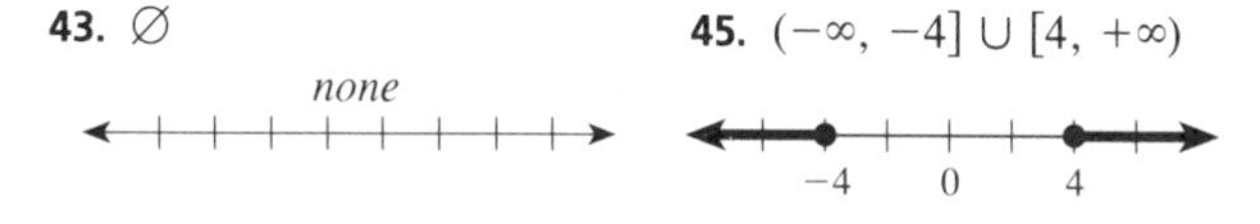

47. $(-\infty, +\infty)$ **49.** $[-7, 8]$

51. $\left(\frac{1}{4}, \frac{11}{4}\right)$ **53.** $\left(-\infty, \frac{4}{5}\right] \cup [2, +\infty)$

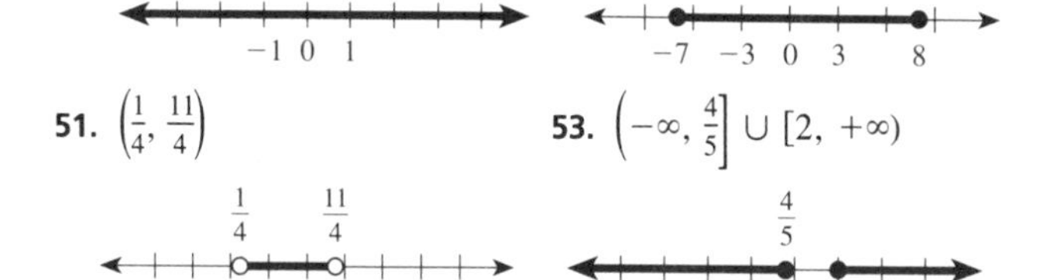

EXERCISES 2.6 ■ page 127

1. $3x + 6$ **3.** $\frac{x}{2} - 3$ **5.** $x + 0.6x$

7. $2x + 2(x + 20)$ **9.** $x + (x + 1)$

11. The difference of five from three times a number

13. The difference of two-thirds of a number from six

15. The sum of five-sevenths of a number and two-fifths of the number

17. $b - 0.3b + 0.05(b - 0.3b)$ **19.** $0.15m + 45$

21. a. n = the number **b.** $2n$ = twice the number; $20 - n$ = twenty decreased by the number
c. $2n = 20 - n$ **d.** The number is $\frac{20}{3}$, or $6\frac{2}{3}$.

23. a. x = Joey's age now
b. $0.5x$ = Joey's brother's age now
$x + 5$ = Joey's age in five years;
$0.5x + 5$ = Joey's brother's age in five years
c. $0.70(x + 5) = 0.5x + 5$
d. Joey is 7.5 years old now.

25. a. c = length of a side
b. $0.3c$ = length of second side
c. $42 + c + 0.3c = 172$
d. The unknown sides have lengths 100 and 30, so 30 meters is length of the shortest side.

27. a. s = sales in June **b.** $0.12s$ = commission on sales
c. $2000 + 0.12s = 2800$
d. \$6666.67 was the amount of Pete's sales in June

29. a. t = student tickets sold
b. $t + 0.4t$ = adult tickets sold;
$15(t + 0.4t)$ = income from adult tickets;
$6t$ = income from student tickets
c. $15(t + 0.4t) + 6t = 2700$
d. 100 student tickets were sold

31. a. p = sticker price **b.** $0.05p$ = tax
c. $p + 0.05p = 154$
d. \$146.67 is the sticker price of the CD player

33. a. x = the number
b. $0.3(x - 8)$ = 30% of difference of eight from the number
c. $0.3(x - 8) = \frac{2}{3}x$ **d.** $\frac{-72}{11}$ is the number

35. a. m = miles driven
b. $45 + 0.15m$ = daily rental charge
c. $45 + 0.15m = 90$ **d.** 300 miles driven by Jose

37. a. b = Bonnie's age today
b. $\frac{5}{2}b$ = Sister's age today;
$b + 10$ = Bonnie's age in ten years;
$\frac{5}{2}b + 10$ = sister's age in ten years
c. $b + 10 = 0.7\left(\frac{5}{2}b + 10\right)$
d. Bonnie is 4 years old today.

39. a. x = the number
b. $x + 3$ = three more than the number **c.** $\frac{x+3}{7} = \frac{x}{2}$
d. $\frac{6}{5}$ is the number

41. a. p = price of T-shirts
b. $100p$ = revenue;
$4(100) + 200 = 600$ is investment (or cost);
profit = 0.3 (investment)
c. $0.3(600) = 100p - 600$
d. \$7.80 should be the selling price of the T-shirts

EXERCISES 2.7 ■ page 135

1. a. 60 liters **b.** 48 liters **c.** 80% acid concentration
3. \$506 earned by the couple
5. a. $0.10x + 0.40(70)$ **b.** $0.30(x + 70)$ **c.** 35
d. 35 liters
7. a. $3n$ **b.** $3n + 4(n + 0.2n)$
c. $3n + 4(n + 0.2n) = 500$; $n = \frac{2500}{39}$
d. ≈ 64.1 miles per hour
9. 5.7 days **11.** 33.33 pounds of cashews
13. 54.6% acid concentration **15.** $\frac{20}{9}$ ($\approx$2.22 hours)
17. 70 games **19.** 0.6 hours **21.** 26.67 liters
23. 5:10 P.M.
25. Impossible—the rate would have to be infinite

EXERCISES 2.8 ■ page 140

1. $\frac{10}{3}$ **3.** $\frac{-11}{3}$ **5. a.** 13 **b.** $T = \frac{R+5}{3-C}$, $C \neq 3$
7. $y = \frac{-3x+21}{7}$ or $y = \frac{-3}{7}x + 3$
9. $x = \frac{5y+12}{2}$ or $x = \frac{5}{2}y + 6$
11. $V = \frac{10-2R}{P}$, $P \neq 0$ **13.** $W = 0$
15. $B = \frac{2A}{h} - b$ or $B = \frac{2A - bh}{h}$, $h \neq 0$
17. $g = \frac{s - s_0}{t^2}$, $t \neq 0$ **19.** $F = \frac{2+T}{C-R}$, $C \neq R$
21. $y = \frac{3x-5}{2}$ or $y = \frac{3}{2}x - \frac{5}{2}$
23. $T = \frac{3P-2}{3}$ or $T = P - \frac{2}{3}$
25. $x = \frac{-5y+3}{2}$ or $x = \frac{-5}{2}y + \frac{3}{2}$
27. $h = \frac{2A}{b_2 + b_1}$, $b_2 \neq -b_1$

CHAPTER 2 REVIEW ■ page 141

1. Fifth-degree binomial in two variables
3. Seventh-degree monomial in three variables
5. Fifth-degree quadrinomial in one variable
7. Second-degree binomial in three variables
9. Third-degree trinomial in two variables
11. ii $3x - 8$ **13.** $-2x^2 - 8x - 3$ **15.** $-p^2 - 8p + 3$
17. $2x^2 + 3x$ **19.** $-y^2 - 2y + 8$ **21.** $4y^2 - 2y + 1$
23. 25 **25.** -86 **27.** -45 **29.** $\frac{-1}{30}$ **31.** 6
33. 4 **35.** $\frac{-5}{2}$ **37.** $\frac{24}{5}$ **39.** $\frac{5}{6}$ **41.** $\frac{13}{11}$ **43.** $\frac{8}{9}$
45. $\frac{13}{4}$
47. $(-\infty, -3]$ **49.** $(0, +\infty)$ 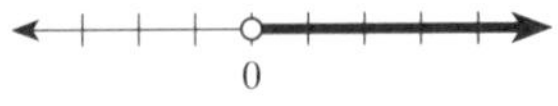
51. $\left[\frac{4}{3}, +\infty\right)$ **53.** $\left(-\infty, \frac{8}{3}\right)$
55. 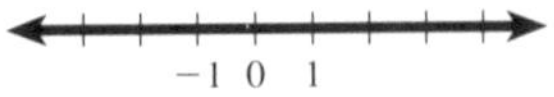 $(-\infty, +\infty)$ **57.** $\left[\frac{-7}{3}, +\infty\right)$
59. $\left[-2, \frac{1}{3}\right]$ 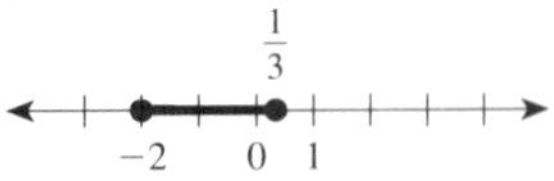 **61.** $\left[1, \frac{7}{2}\right]$

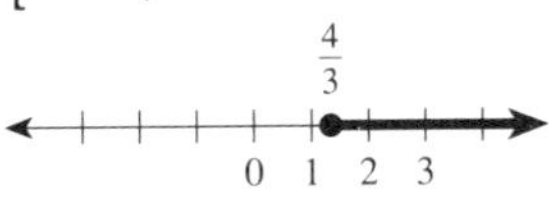

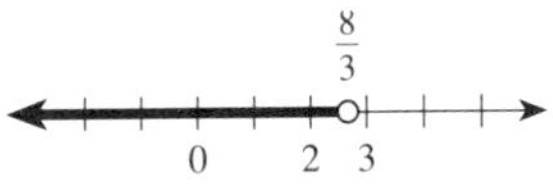

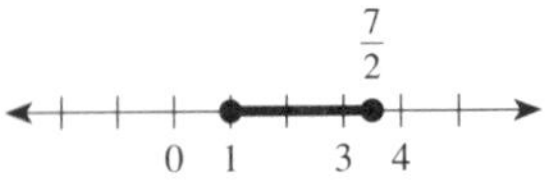

63. $\left[\frac{-7}{2}, -2\right]$

65. $\left(\frac{13}{6}, \frac{8}{3}\right)$

67. $\frac{-7}{2}, \frac{13}{2}$ **69.** $\varnothing$ **71.** $\frac{4}{3}, 4$ **73.** $\left[\frac{-7}{2}, \frac{13}{2}\right]$

75. $\left(-\infty, \frac{3}{2}\right) \cup \left(\frac{9}{2}, +\infty\right)$ **77.** $\left[\frac{4}{3}, 4\right]$ **79.** $2x - 3$

81. $2p + 5(12)$ **83.** $5x + 3(x + 0.2x)$

85. **a.** x = the number
b. $3x + 15$ = fifteen more than three times the number
c. $3x + 15 = 38$, $x = \frac{23}{3}$ **d.** $\frac{23}{3}$ is the number

87. **a.** n = the number
b. $n - 8$ = the number decreased by eight; $0.1(n + 8)$ = 10% of sum of the number and eight
c. $n - 8 = 0.1(n + 8)$ $n = \frac{88}{9}$
d. $\frac{88}{9}$ is the number

89. **a.** L = length of garden
b. $2L + 2(10)$ = perimeter of garden
c. $3[2L + 2(10)] = 300$; $L = 40$
d. length of the garden is 40 feet

91. **a.** x = number of liters of 6% alcohol solution
b. $0.06x + 0.1(3)$ = amount of alcohol in combination; $0.09(x + 3)$ = amount of alcohol in mixture
c. $0.06x + 0.1(3) = 0.09(x + 3)$; $x = 1$
d. 1 liter of 6% alcohol solution

93. **a.** s = original speed
b. $s - 15$ = decreased speed; $5s$ = distance at original speed; $1(s - 15)$ = distance at decreased speed
c. $5s + 1(s - 15) = 300$; $s = 52.5$
d. original speed was 52.5 mi/h

95. **a.** t = number of hours to assemble 170 packages
b. $7t$ = number of packages assembled by Becky; $10t$ = number of packages assembled by Holly
c. $7t + 10t = 170$; $t = 10$
d. 10 hours to assemble the packages

97. **a.** t = number of hours for both pumps to empty tank
b. $2500t$ = gallons emptied by pump A; $4000t$ = gallons emptied by pump B
c. $2500t + 4000t = 20000$; $t = 3.08$
d. 3.08 hours for both pumps to empty the tank

99. **a.** x = distance traveled by train A
b. $\frac{x}{80}$ = time traveled by train A;
$\frac{x}{100}$ = time traveled by train B
c. $\frac{x}{80} = \frac{x}{100} + 2$; $x = 800$
d. Train B overtakes train A 800 miles from Seatown.

101. $y = \frac{2x - 18}{9}$ or $y = \frac{2}{9}x - 2$ **103.** $x = \frac{-2y + 8}{7}$

105. $y = \frac{-10x + 15}{3}$ or $y = -\frac{10}{3}x + 5$

107. $V = \frac{W + C}{P}$, $P \neq 0$

109. $D = \frac{3C - T}{C}$ or $D = 3 - \frac{T}{C}$, $C \neq 0$ **111.** $r = \frac{C}{2\pi}$

113. $W = \frac{-P - C}{2}$ **115.** $R = \frac{8}{T + 2}$, $T \neq -2$

117. $y_1 = y - m(x - x_1)$ or $y_1 = y - mx + mx_1$

CHAPTER 2 TEST ■ page 143

1. Degree 3, -99
2. **a.** $2x^2 - x - 10$ **b.** $-4x^2 + 5x$ **c.** $p^2 - 5p - 6$
3. **a.** $\frac{-5}{27}$ **b.** $\frac{5}{2}$ **c.** -4
4. **a.** $(-\infty, 5]$

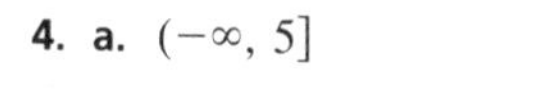

b. $\left(\frac{-1}{3}, \frac{8}{3}\right]$ **c.** $\left(\frac{-1}{3}, \frac{14}{3}\right)$

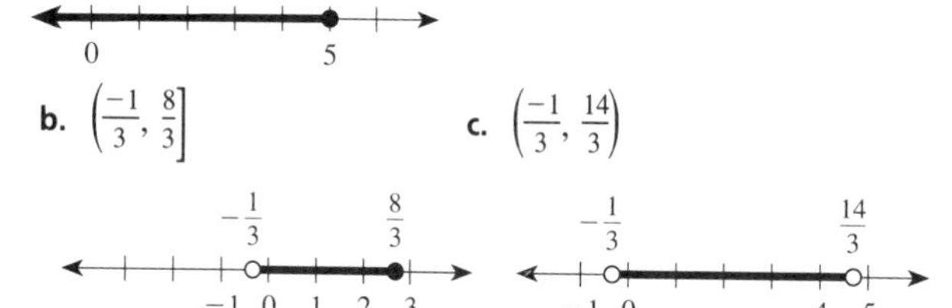

5. 3, 6
6. $(-\infty, 3) \cup (6, +\infty)$
7. **a.** x = amount of 20% alcohol solution
b. $0.2x + 0.6(15)$ = amount of alcohol in combination; $0.3(x + 15)$ = amount of alcohol in mixture
c. $0.2x + 0.6(15) = 0.3(x + 15)$; $x = 45$
d. 45 liters of 20% alcohol solution
8. **a.** s = original speed
b. $5s$ = distance traveled at original speed; $2(s - 10)$ = distance traveled at decreased speed
c. $5s + 2(s - 10) = 400$; $s = 60$
d. 60 mi/h was the original speed
9. $y = \frac{3x - 20}{8}$ or $y = \frac{3}{8}x - \frac{5}{2}$ **10.** $V = \frac{R}{P + 3}$, $P \neq -3$

■ CHAPTER 3

EXERCISES 3.1 ■ page 151

1. False **3.** True **5.** True **7.** False **9.** False
11. **iv.** $y = 3x$ **13.** B **15.** C

17. 

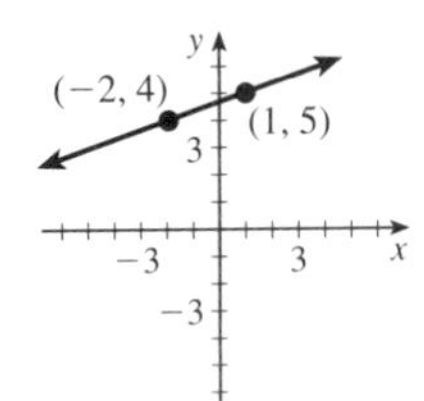

19.

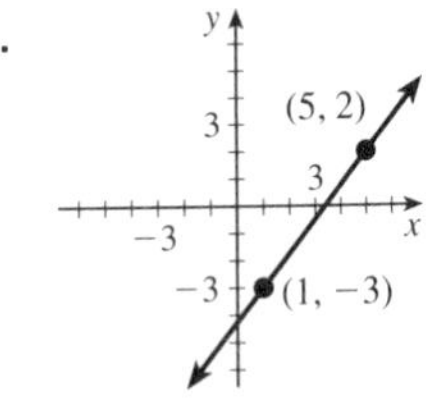

21.

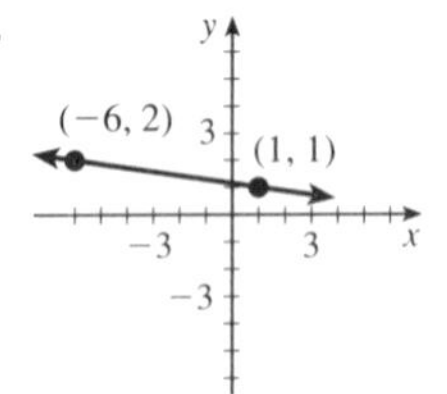

23.

x	-2	0	1	3	4
y	-4	$\frac{-12}{5}$	$\frac{-8}{5}$	0	$\frac{4}{5}$

25.

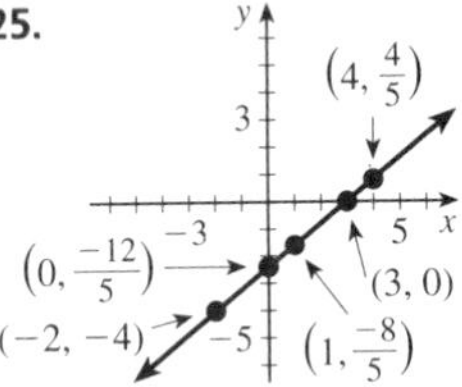

27. $y = -3x - 5$

calculator:
- x-intercept $(-1.6, -.2)$
- y-intercept $(0, -5)$

algebraic:
- x-intercept $\left(\frac{-5}{3}, 0\right)$
- y-intercept $(0, -5)$

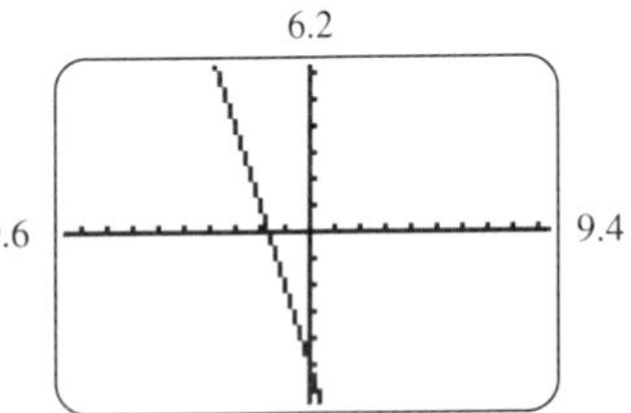

29. $y = \frac{-2}{3}x + \frac{7}{3}$

calculator:
- x-intercept $(3.5, 0)$
- y-intercept $(0, 2.3333333)$

algebraic:
- x-intercept $\left(\frac{7}{2}, 0\right)$
- y-intercept $\left(0, \frac{7}{3}\right)$

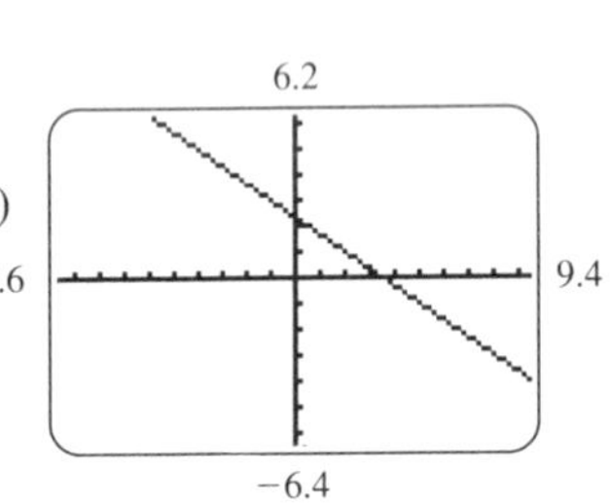

31. $y = \frac{6}{7}x - \frac{15}{7}$

calculator:
- x-intercept $(2.4, -.0857143)$
- y-intercept $(0, -2.142857)$

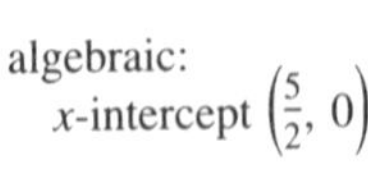

algebraic:
- x-intercept $\left(\frac{5}{2}, 0\right)$
- y-intercept $\left(0, \frac{-15}{7}\right)$

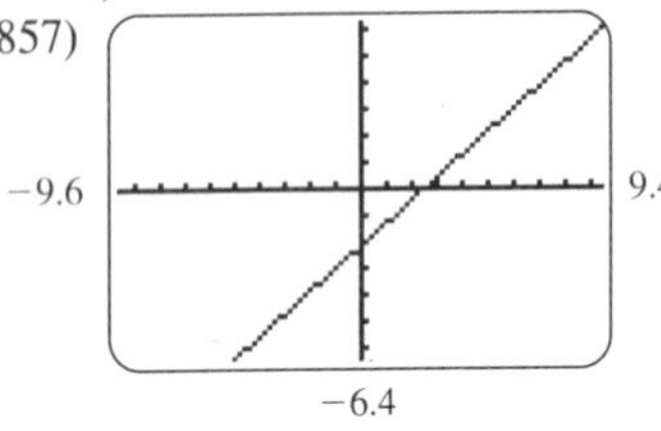

33. $y = \frac{-5}{2}x - 5$

calculator:
- x-intercept, $(-2, 0)$
- y-intercept $(0, -5)$

algebraic:
- x-intercept, $(-2, 0)$
- y-intercept $(0, -5)$

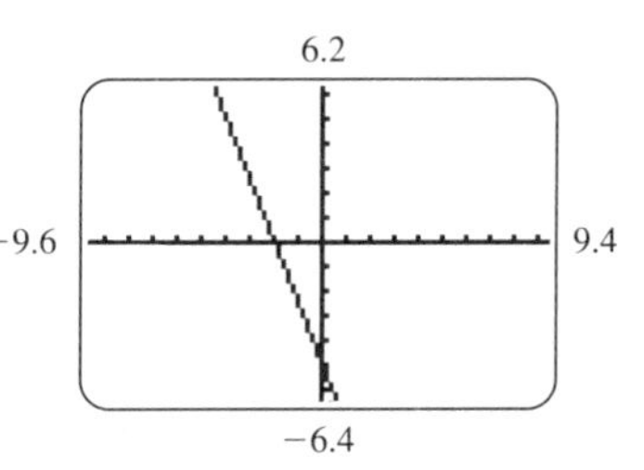

35. $y = \frac{7}{3}x - \frac{13}{3}$

calculator:
- x-intercept $(1.8, -.1333333)$
- y-intercept $(0, -4.333333)$

algebraic:
- x-intercept $\left(\frac{13}{7}, 0\right)$
- y-intercept $\left(0, \frac{-13}{3}\right)$

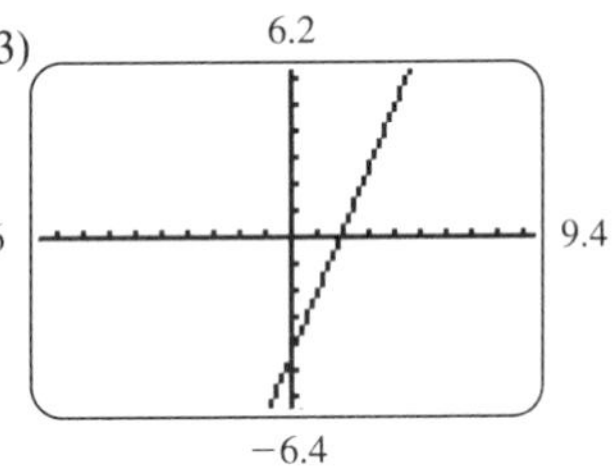

37. $y = \frac{-5}{2}x + \frac{13}{2}$

calculator:
- x-intercept $(2.6, 0)$
- y-intercept $(0, 6.5)$

algebraic:
- x-intercept $\left(\frac{13}{5}, 0\right)$
- y-intercept $\left(0, \frac{13}{2}\right)$

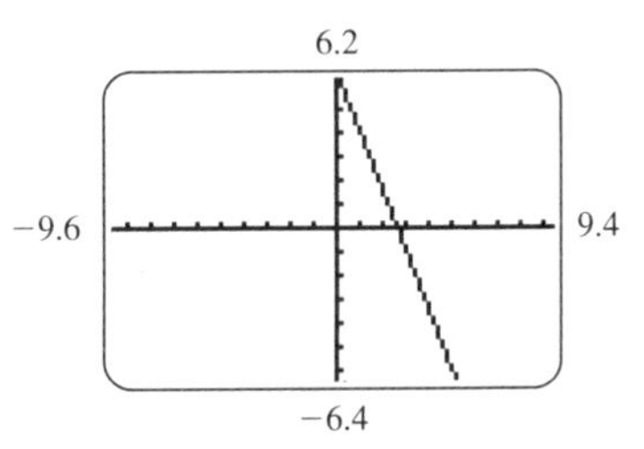

39. $y = \frac{-2}{5}x + 2$

calculator:
- x-intercept $(5, 0)$
- y-intercept $(0, 2)$

algebraic:
- x-intercept $(5, 0)$
- y-intercept $(0, 2)$

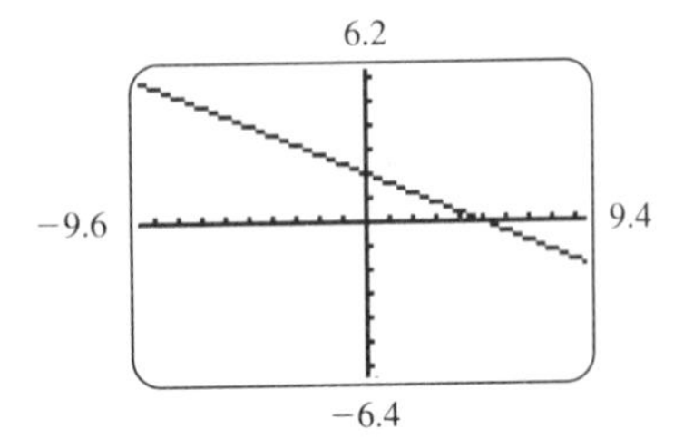

41. x-intercept $\left(\frac{-2}{3}, 0\right)$
y-intercept $(0, 2)$

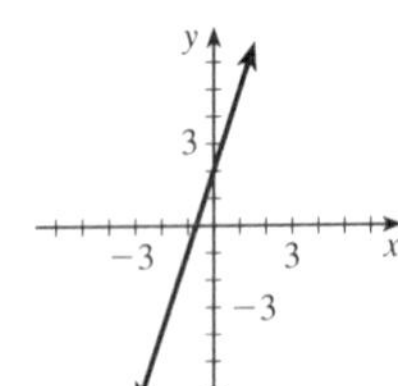

43. x-intercept $\left(\frac{-8}{3}, 0\right)$
y-intercept $(0, 4)$

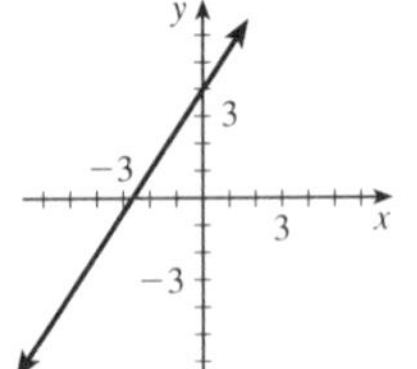

45. x-intercept $\left(\frac{5}{2}, 0\right)$
y-intercept $(0, 5)$

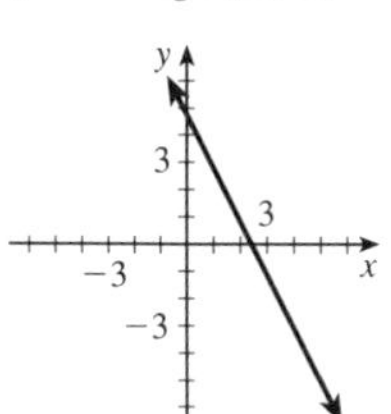

47. x-intercept $(2, 0)$
y-intercept $\left(0, \frac{-4}{5}\right)$

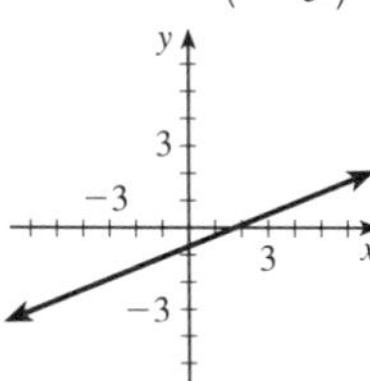

49. x-intercept $\left(\frac{-6}{5}, 0\right)$
y-intercept $(0, -1)$

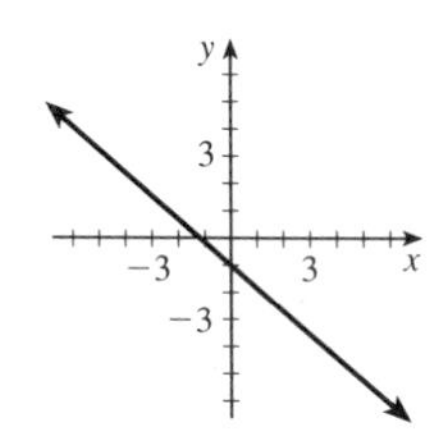

51. x-intercept $\left(\frac{-7}{3}, 0\right)$
y-intercept, $\left(0, \frac{-7}{2}\right)$

53. x-intercept $\left(\frac{13}{3}, 0\right)$
y-intercept $\left(0, \frac{-13}{5}\right)$

EXERCISES 3.2 ■ page 159

1. True **3.** True **5.** False **7.** False

9. **ii.** $5y = 10$ **11.** **iv.** $\frac{3}{2}$

13. **a.** $y = \frac{-3}{5}x + \frac{13}{5}$ **b.** $\frac{-3}{5}$ **c.** $\frac{-3}{5}$

15. $m_2 > m_1 > m_4 > m_3$

17. $(3, 1)$
(Other answers are possible.)

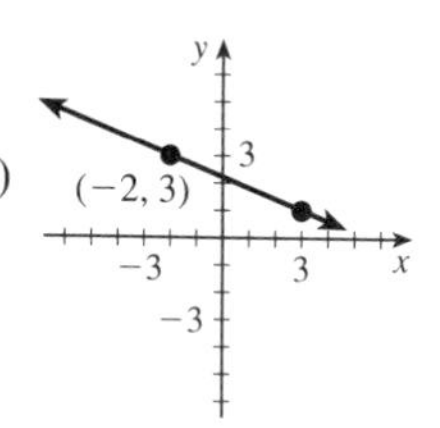

19. $\Delta x = 3$, $\Delta y = 1$,
$m = \frac{1}{3}$

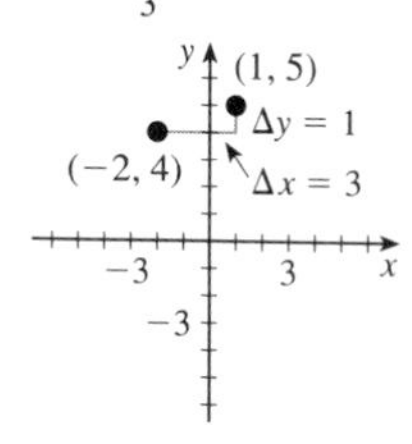

21. $\Delta x = -4$, $\Delta y = -5$,
$m = \frac{5}{4}$

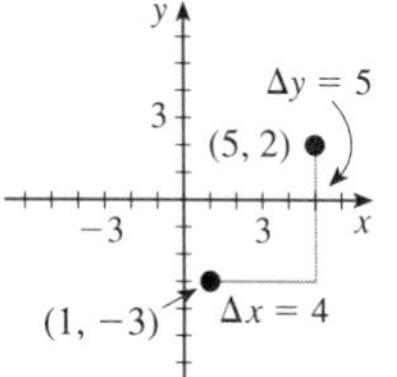

23. $\Delta x = 7$, $\Delta y = -1$,
$m = \frac{-1}{7}$

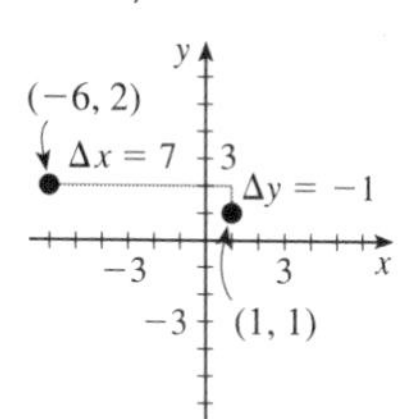

25. $\Delta x = \frac{-1}{2}$, $\Delta y = -1$,
$m = 2$

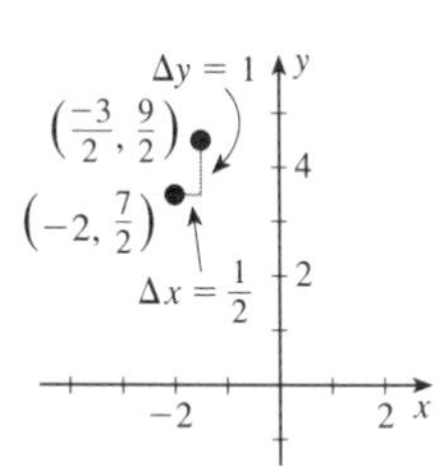

27. $y = \frac{3}{5}x - 4$, $m = \frac{3}{5}$,
y-intercept $(0, -4)$

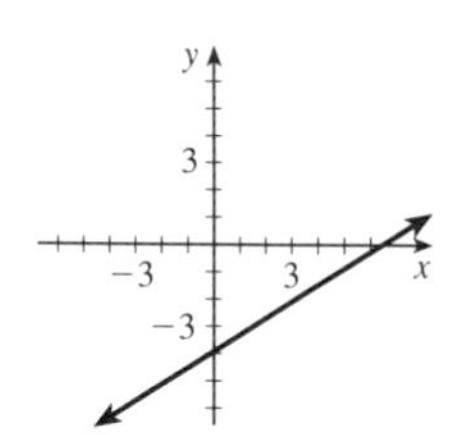

29. $y = \frac{1}{2}x + \frac{3}{2}$, $m = \frac{1}{2}$,
y-intercept $\left(0, \frac{3}{2}\right)$

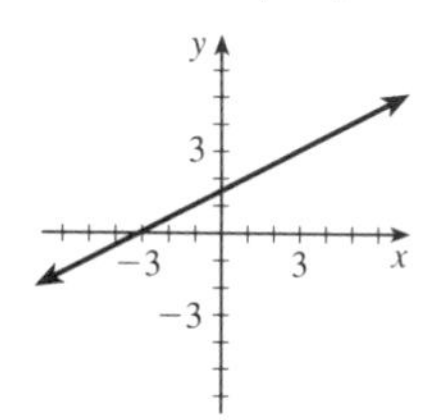

31. $y = \frac{5}{4}x - 2$, $m = \frac{5}{4}$,
y-intercept $(0, -2)$

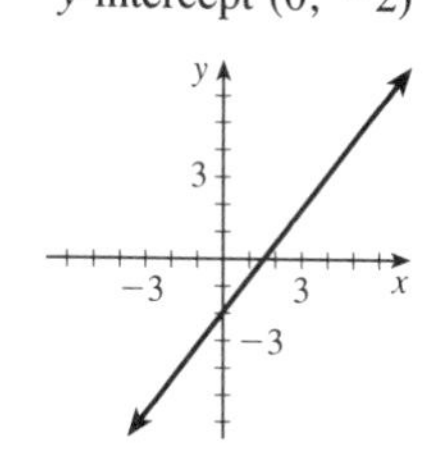

33. $m = 2$,
y-intercept $(0, 3)$

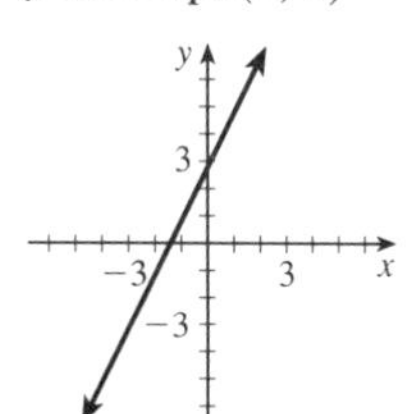

35. $m = \frac{2}{3}$,
y-intercept $\left(0, \frac{-8}{3}\right)$

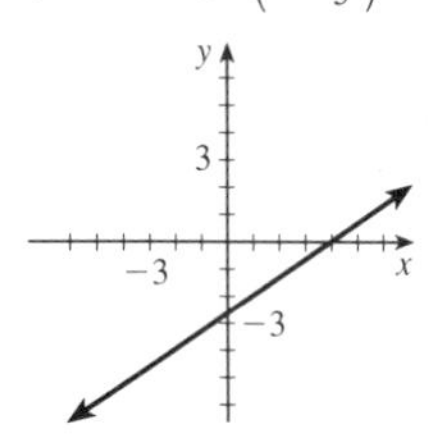

37. $m = -1$,
y-intercept $(0, -1)$

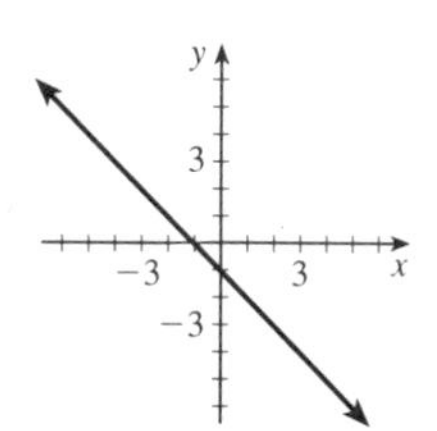

39. $m = \frac{2}{5}$,
y-intercept $\left(0, \frac{-4}{5}\right)$

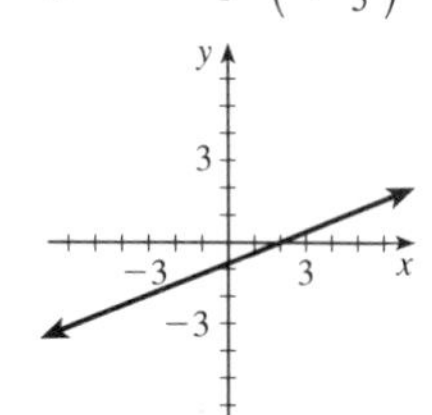

41. $m = \frac{-5}{6}$,
y-intercept $(0, -1)$

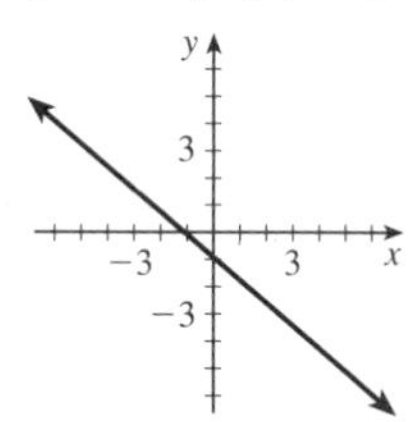

43. $m = \frac{3}{5}$, y-intercept $\left(0, \frac{-13}{5}\right)$

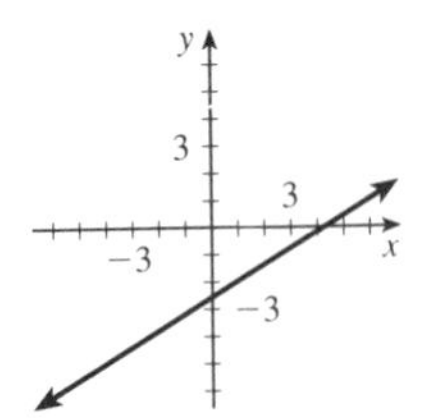

EXERCISES 3.3 ■ page 164

1. a. x-intercept $\left(\frac{10}{3}, 0\right)$, y-intercept $(0, -5)$, $m = \frac{3}{2}$

b.

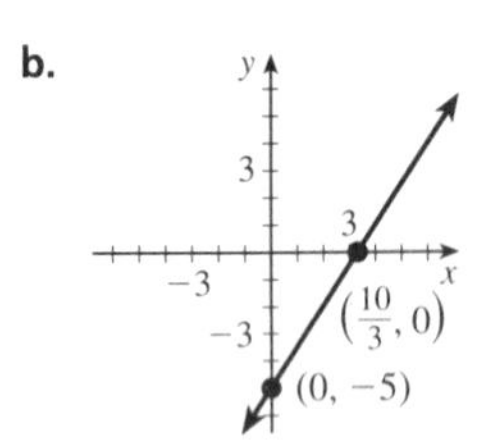

c. (0, 2), (−1, 2), (1, 2) (others are possible)

d. (−1, −7), (−1, −8), (−1, −9) (others are possible)

e.

3. a. y-intercept $(0, -3)$, $m = \frac{3}{5}$

b.

c. (0, 3), (2, 3), (−2, 3) (others are possible)

d. (−2, −5), (−2, −6), (−2, −7) (others are possible)

e.

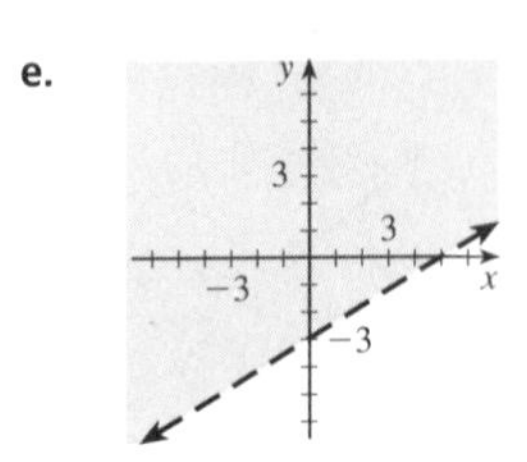

5.

7.

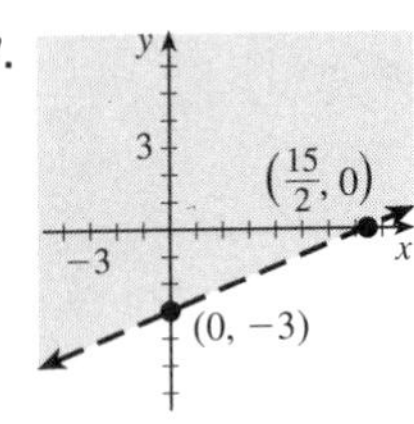

9.

11.

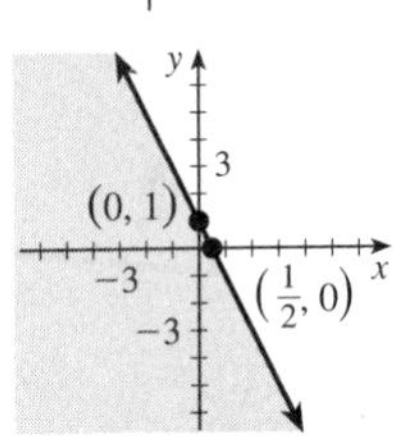

13.

15.

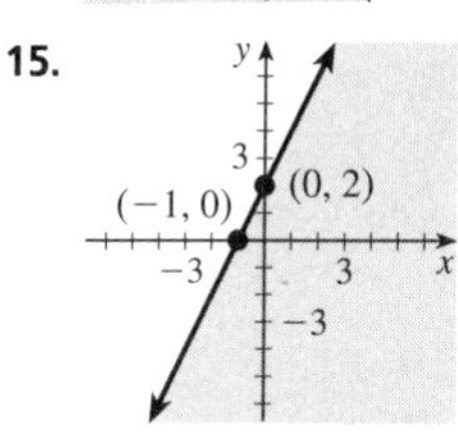

17.

19.

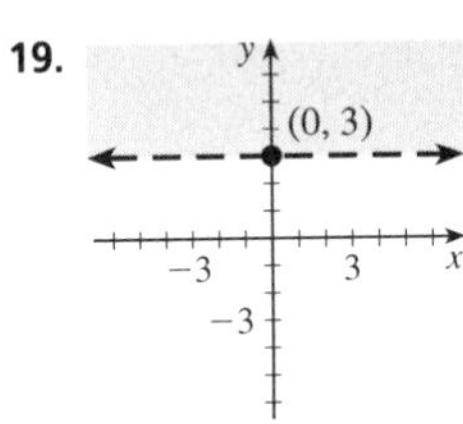

EXERCISES 3.4 ■ page 171

1. a.

b. \$38

c. 460 miles

3. a.

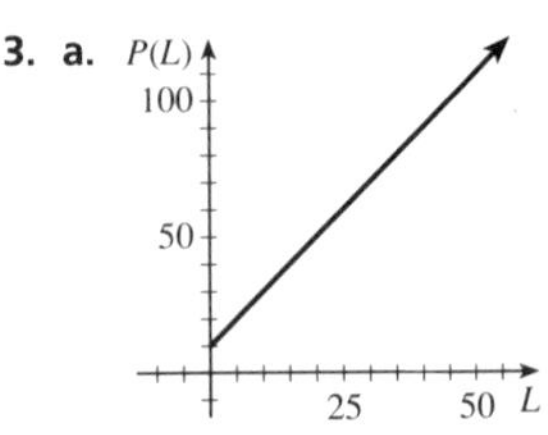

b. 30 meters

c. 22 meters

5. Not a linear relation, because the graph is not a straight line.

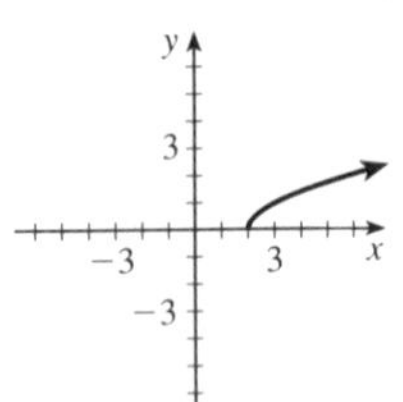

7. Domain: {−4, −2, 0, 1, 2, 3, 4};
range: {−2, −1, 0, 1, 2, 3}

9. a. (0, −3)
b. (3, 0), (−3, 0)
c. (0, −3)

11. a. (0, 1)
b. (−1, 0), (−5, 0)
c. (−3, −2)

13. a. (0, −1)
b. (1, 0), (3, 0)
c. (2, 1)

15. a. (0, 3)
b. (1, 0), (3, 0)
c. (2, −1)

17. a. (0, 4)
b. (2, 0), (−2, 0)
c. (0, 4)

19. Domain: $(-\infty, +\infty)$; range: $(-\infty, +\infty)$
21. Domain: $(-\infty, +\infty)$; range: $[0, +\infty)$
23. Domain: $(-\infty, +\infty)$; range: $(-\infty, 3]$

DISCOVERY ■ page 172

1.

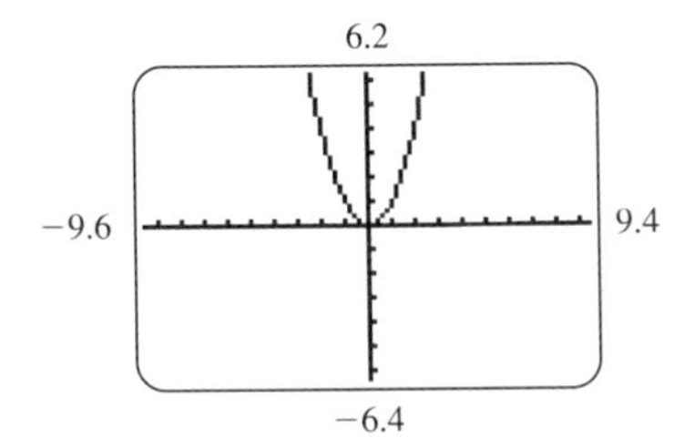

When $A > 0$, the graph is moved up A units from the location of the graph of $y = x^2$.
When $A < 0$, the graph is moved down A units from the location of the graph of $y = x^2$.

EXERCISES 3.5 ■ page 180

1. a. 5 **b.** 14 **c.** −7 **d.** 7 **3.** −5 **5.** $\frac{9}{2}$

7. $\frac{17}{2}$ **9.** $\frac{10}{7}$

11.

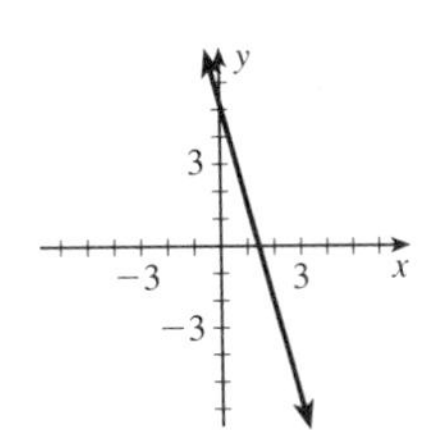

13.

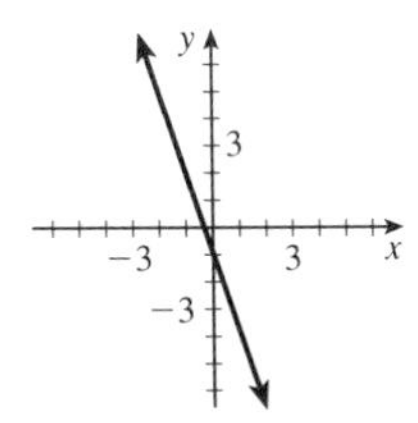

x-intercept $\left(\frac{-1}{3}, 0\right)$
y-intercept $(0, -1)$
slope -3

15.

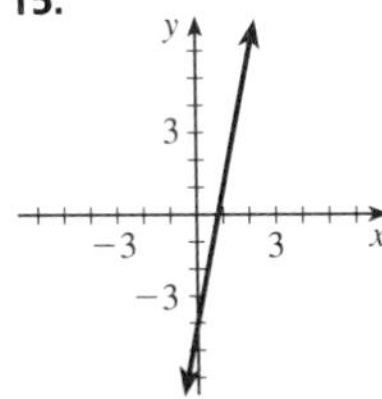

x-intercept $\left(\frac{4}{5}, 0\right)$
y-intercept $(0, -4)$
slope 5

17.

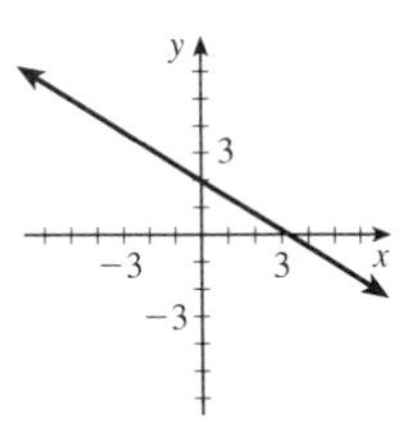

x-intercept $\left(\frac{10}{3}, 0\right)$
y-intercept $(0, 2)$
slope $\frac{-3}{5}$

19.

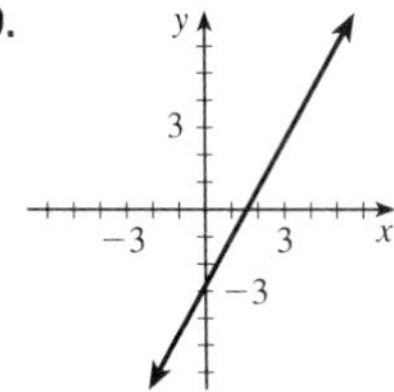

x-intercept $\left(\frac{11}{7}, 0\right)$
y-intercept $\left(0, \frac{-11}{4}\right)$
slope $\frac{7}{4}$

21.

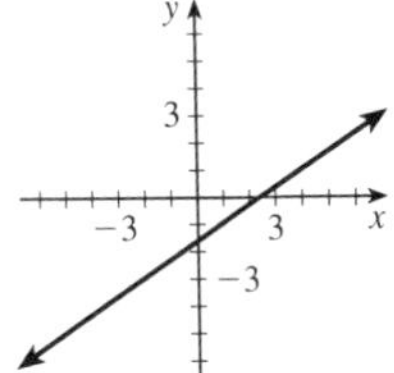

x-intercept $\left(\frac{5}{2}, 0\right)$
y-intercept $\left(0, \frac{-5}{3}\right)$
slope $\frac{2}{3}$

23.

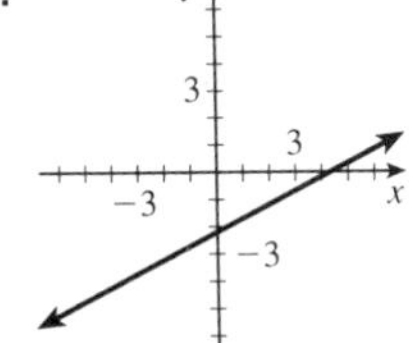

x-intercept (4.6, 0)
y-intercept (0, −2.3)
slope 0.5

25.

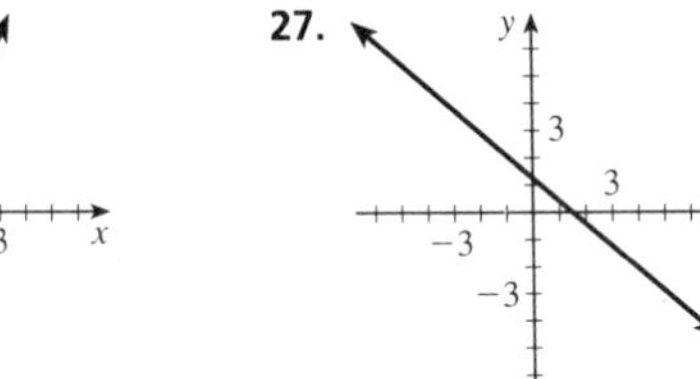

x-intercept (0, 0)
y-intercept (0, 0)
slope $\frac{17}{8}$

27.

x-intercept $\left(\frac{8}{5}, 0\right)$
y-intercept $\left(0, \frac{4}{3}\right)$
slope $\frac{-5}{6}$

29.

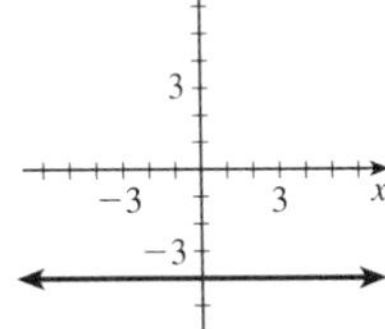

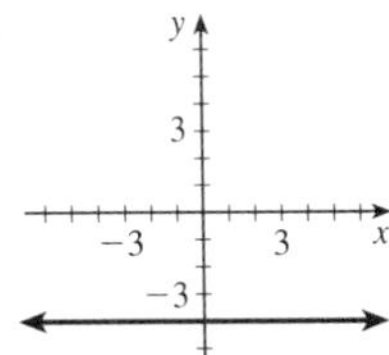

x-intercept none
y-intercept $(0, -4)$
slope 0

31. a. $\frac{-160}{9}$ **b.** 0 **c.** 68 **d.**

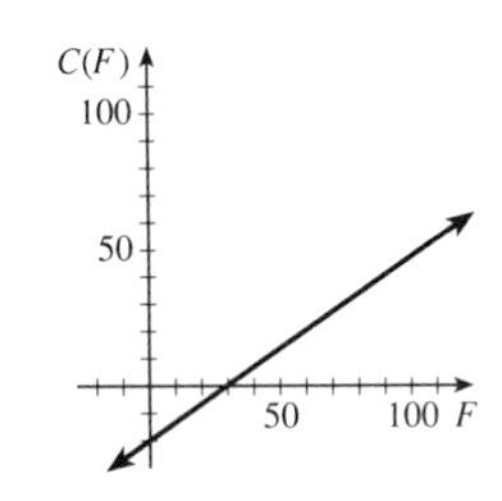

33. **a.** A loss of \$4000 **b.** 40,000 fish cakes

c.

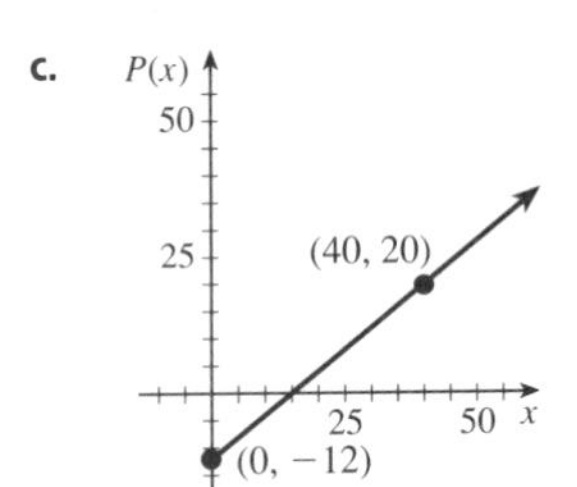

CHAPTER 3 REVIEW ■ page 180

1. **iv.** $y = -2x + 5$ **3.** **iii.** **5.** $(-3, -9)$

7.

x	-1	0	2	3.1	$\frac{5}{3}$	3	$\frac{11}{3}$
y	-6	$\frac{-9}{2}$	$\frac{-3}{2}$	0.15	-2	0	1

9. $y = \frac{3}{2}x - \frac{9}{2}$ **11.** $\left(\frac{9}{5}, -1\right)$

13. y-intercept $(0, -4)$ and x-intercept $\left(\frac{12}{5}, 0\right)$

15.

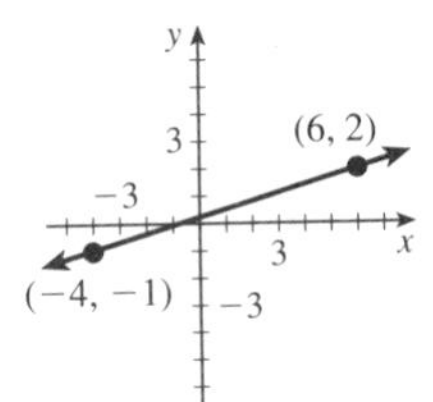

$m = \frac{3}{10}$

17.

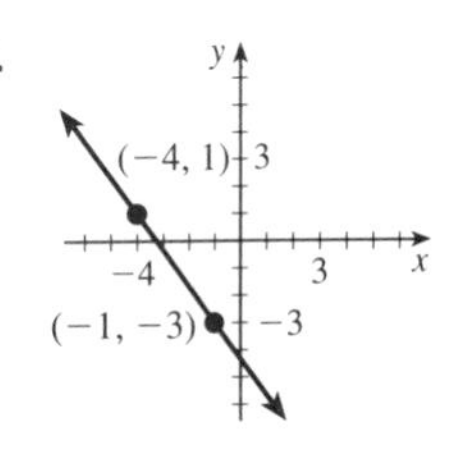

$m = \frac{-4}{3}$

19.

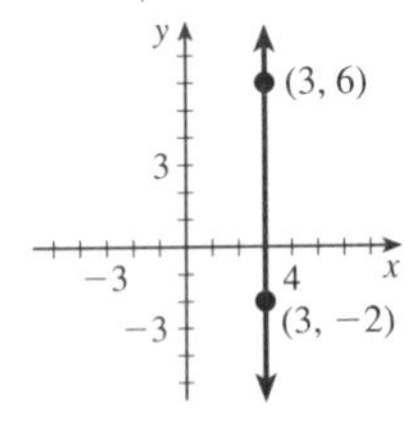

m = undefined (no slope)

21.

(2, 3) (others are possible)

23.

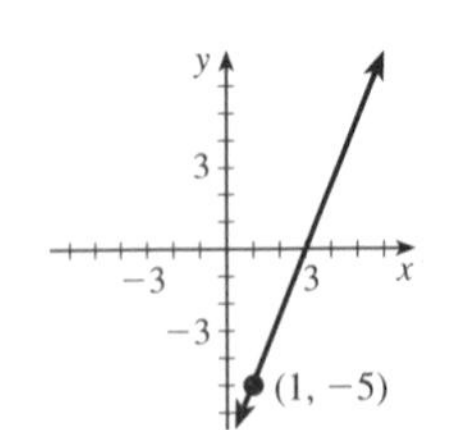

(4, 2) (others are possible)

25. D **27.** C **29.** **i.** Undefined **31.** $m = \frac{-7}{5}$

33.

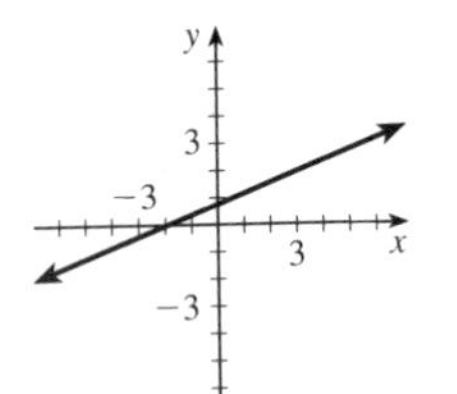

y-intercept $\left(0, \frac{4}{5}\right)$
x-intercept $(-2, 0)$
slope $\frac{2}{5}$

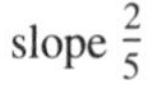

35.

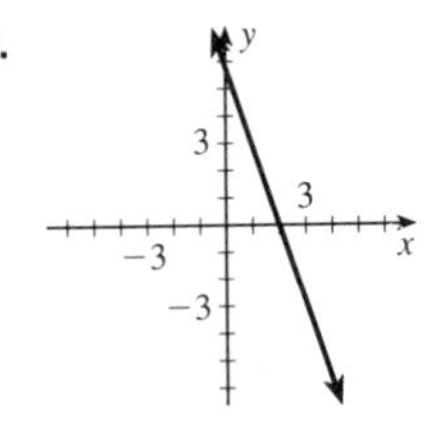

y-intercept $(0, 6)$
x-intercept $(2, 0)$
slope -3

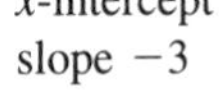

37.

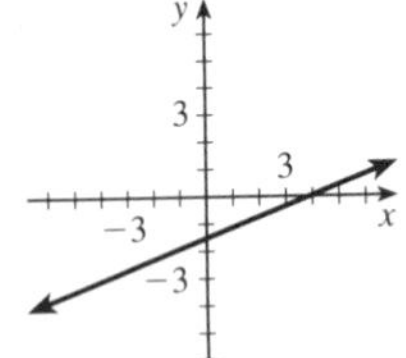

y-intercept $\left(0, \frac{-7}{5}\right)$
x-intercept $\left(\frac{7}{2}, 0\right)$
slope $\frac{2}{5}$

39. $y = \frac{3}{7}x - \frac{1}{7}$

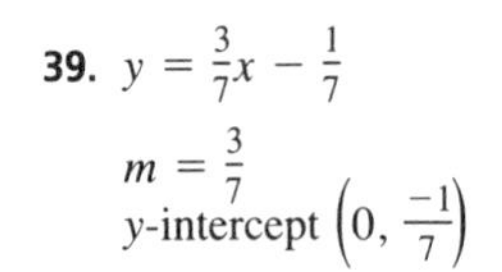

$m = \frac{3}{7}$
y-intercept $\left(0, \frac{-1}{7}\right)$

41. A. **43.** B.

45.

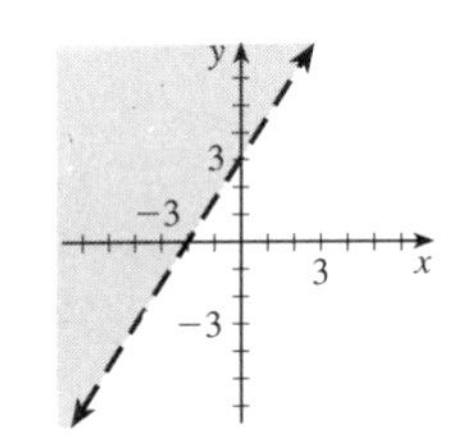

47.

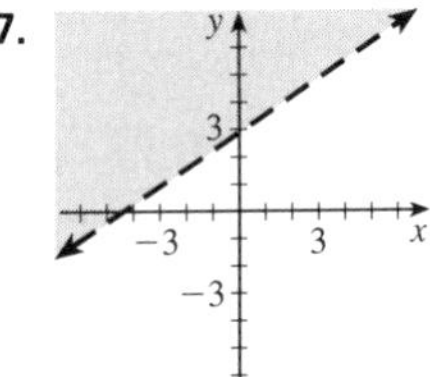

49. **ii.** **51.** m_3, m_1, m_4, m_2

53. **a.**

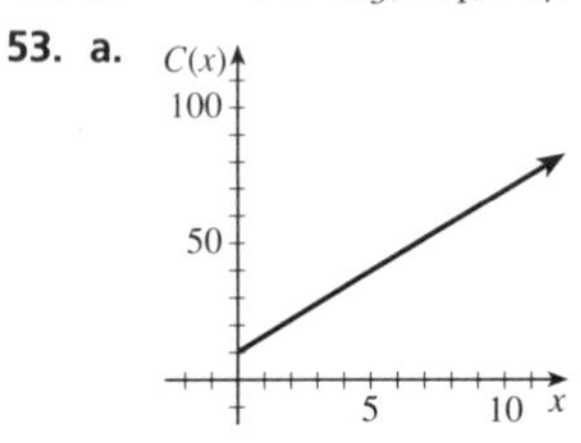

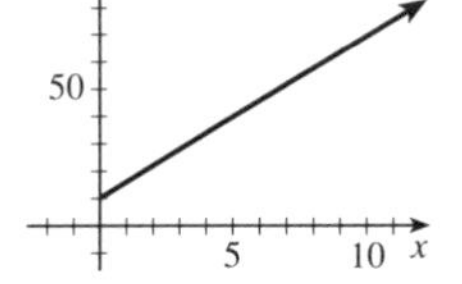

b. \$40
c. 1.5 hours
d. 10 to 94

55. **a.** -17 **b.** 4 **c.** -5 **d.** -4

57. **a.** 4
b. $\frac{9}{2}$
c.

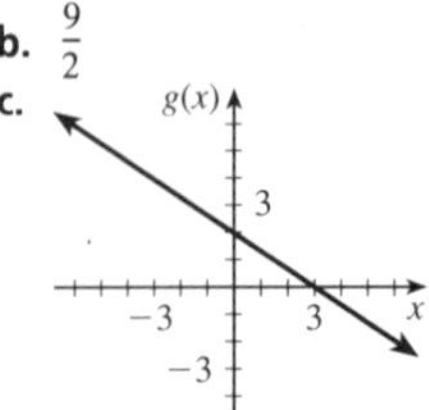

59. **a.** \$7.70
b. 21 ounces
c.

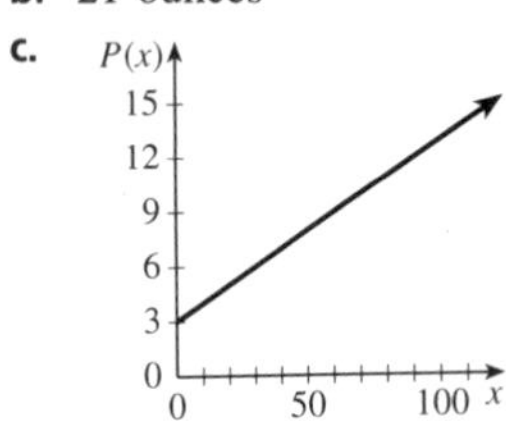

CHAPTER 3 TEST ■ page 183

1. a. **b.** $\left(0, \frac{-7}{2}\right)$ **c.** $\left(\frac{7}{5}, 0\right)$ **d.** $\frac{5}{2}$

2. a. $y = \frac{2}{3}x + 3$

b.

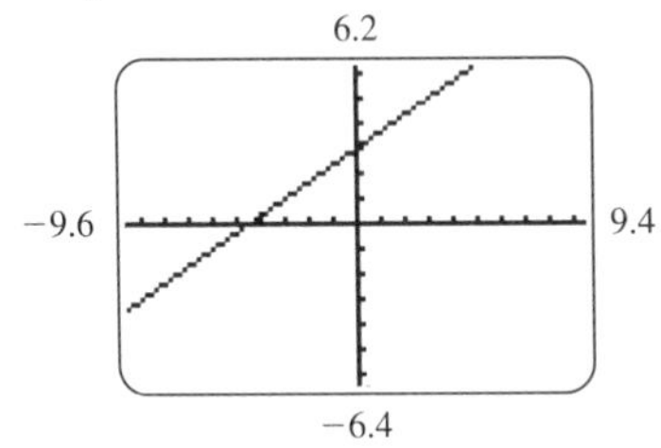

c.

x	−4.8	−1	2	5.2
y	−0.2	2.33	4.33	6.47

d.

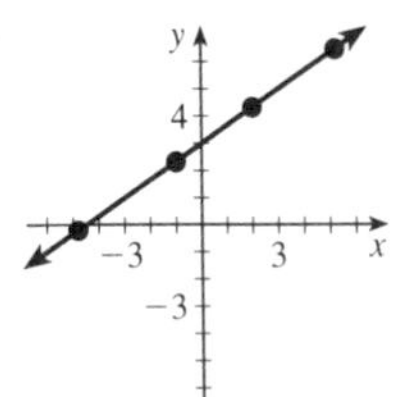

3. $\frac{-4}{3}$

4. a.

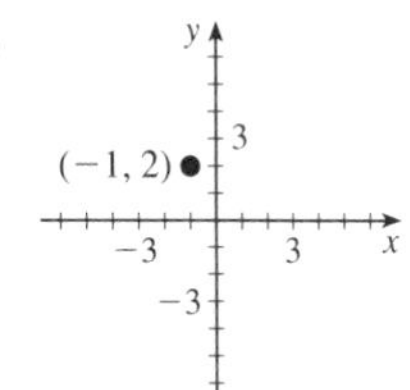

b.

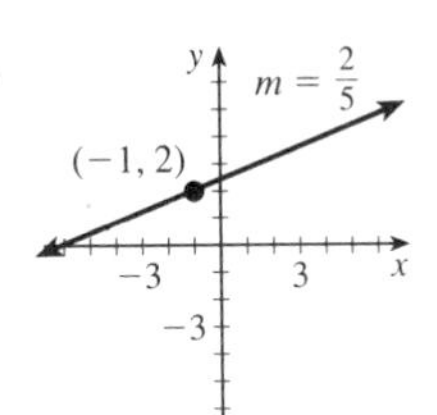

c. (4, 4) (Other answers are possible.)

5.

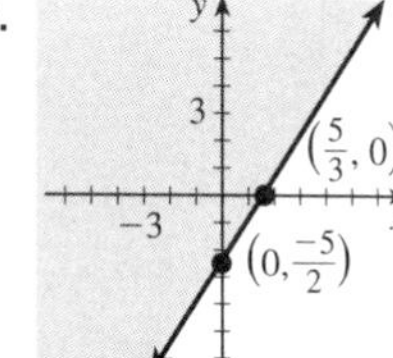

6. a. $P = -1$ **b.** $t = 1$ **c.** $\{P \mid -25 \le P \le 5\}$

7. a. $\{-2, 0, 2, 3\}$ **b.** $\{-1, 1, 3, 4\}$

8. a. $\frac{-27}{2}$ **b.** $\frac{27}{5}$ **9. a.** \$40 **b.** 300 miles

10. a. No **b.** $[-1, +\infty)$

■ CHAPTER 4

EXERCISES 4.1 ■ page 195

1. a.

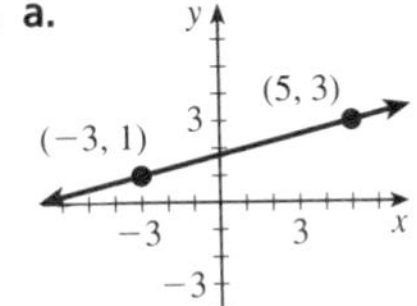

b. Increasing, positive slope
c. $\Delta y = 2$, $\Delta x = 8$, $m = \frac{1}{4}$
d. $y = \frac{1}{4}x + \frac{7}{4}$
e. $\left(3, \frac{5}{2}\right)$
f. (5, 3)

3.

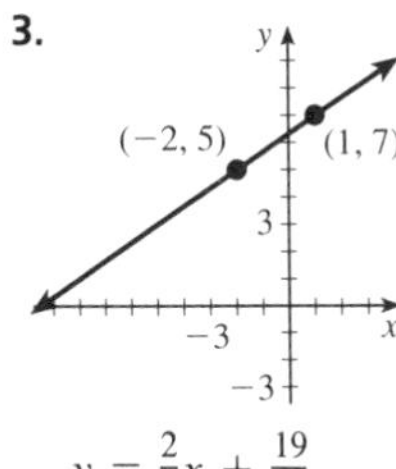

$y = \frac{2}{3}x + \frac{19}{3}$

(1, 7)

5. $y = x$

$(-3, -3)$

7. $y = -3x - 1$ **9.** $y = \frac{-2}{3}x + \frac{1}{3}$

11. $y = \frac{-1}{2}x - \frac{9}{2}$ **13.** $y = \frac{-2}{5}x + \frac{13}{5}$

15. $y = \frac{-4}{7}x + \frac{19}{7}$

17. a.

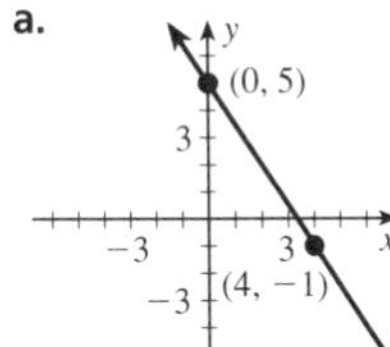

b. $f(x) = \frac{-3}{2}x + 5$
c. 8
d. 2
e. $\frac{5}{3}$

19.

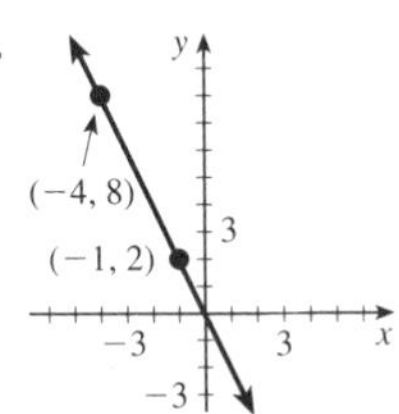

$f(x) = -2x$

21.

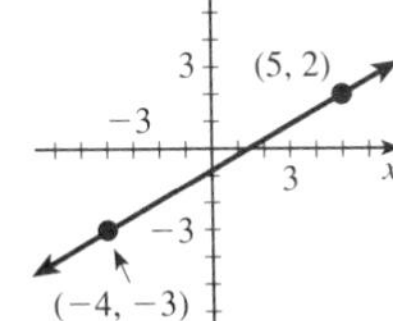

$g(x) = \frac{5}{9}x - \frac{7}{9}$

23.

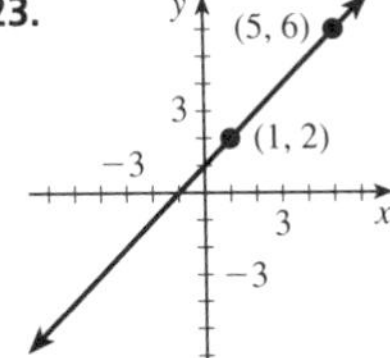

$f(x) = x + 1$

25. $f(x) = 6$

27. a.

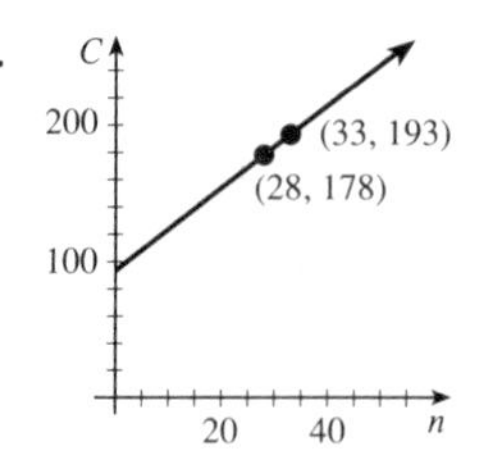

b. $C(n) = 3n + 94$

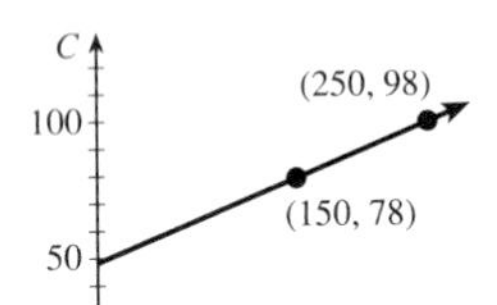

29. a.

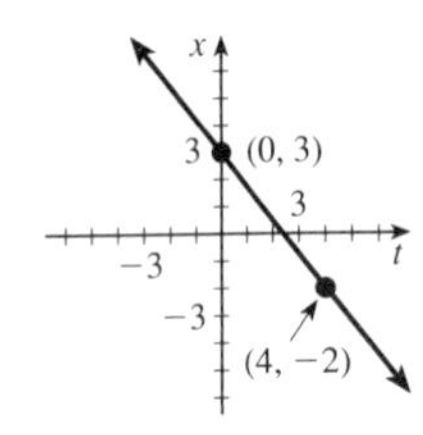

b. $x(t) = \frac{-5}{4}t + 3$

b. $C(x) = 0.20x + 48$
c. \$88
d. \$64

33. $S(x) = 10.6x + 79900$; expect \$92,620 in sales
35. $y = \frac{-3}{7}x + \frac{26}{7}$; $(11, -1)$ is on the line, and the points are collinear.
37. They are not collinear; $y = \frac{2}{3}x - 2$ and $y = \frac{3}{5}x - 2$

DISCOVERY ■ page 198

1. If the lines are to be perpendicular to each other, the slope of one line must be the negative reciprocal of the slope of the other line; $m = \frac{-1}{c}$ or $c = \frac{-1}{m}$
3. $y = \frac{-3}{5}x - 2$ (other answers are possible)

EXERCISES 4.2 ■ page 208

1. True **3.** True **5.** False **7.** False
9. iii. $(1, 3)$ **11.** i. $\frac{2}{3}$

13.

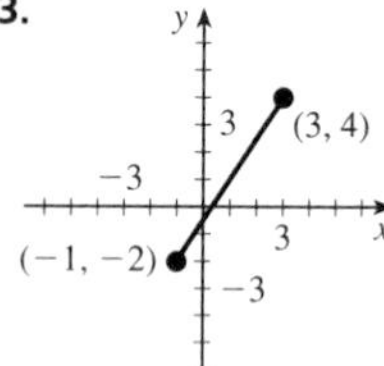

15. $(1, 1)$

17.

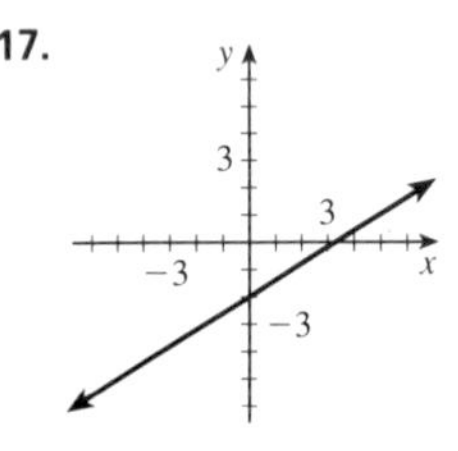

19. $y = \frac{3}{5}x + \frac{29}{5}$

21.

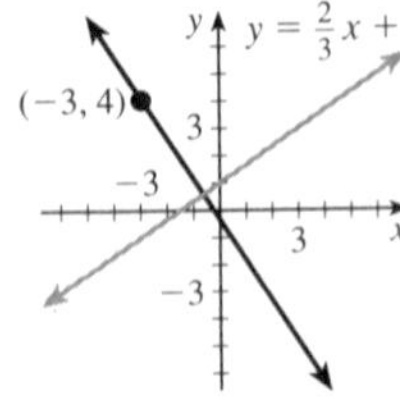

23.

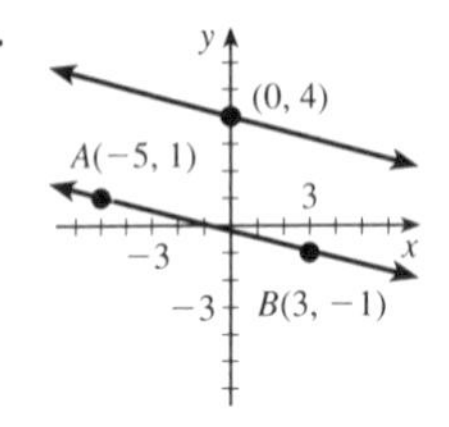

25. $y = \frac{-1}{4}x + 4$ **27.** $y = x + 1$

29.

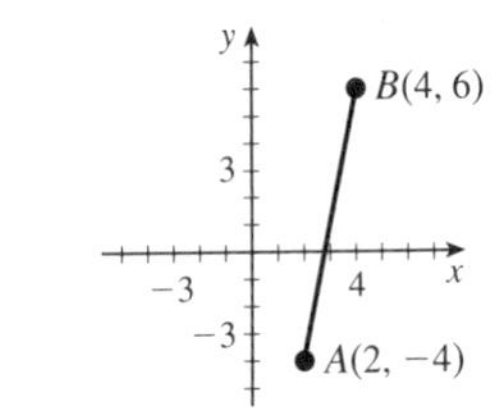

31. $y = \frac{-1}{5}x + \frac{8}{5}$ **33.** $(1, 0)$ **35.** $\sqrt{68}$ or $2\sqrt{17}$
37. $12 + \sqrt{72}$ **39.** $y = -x + 3$
41. Yes. By the Pythagorean theorem, $6^2 + 6^2 = \left(\sqrt{72}\right)^2$, or side AB is perpendicular to side BC.
43. midpoint $(1, 3)$, length $= \sqrt{52}$ or $2\sqrt{13}$
45. midpoint $\left(3, \frac{-1}{2}\right)$, length $= 5$
47. midpoint $\left(\frac{-1}{2}, \frac{3}{2}\right)$, length $= \sqrt{50}$, or $5\sqrt{2}$
49. midpoint $\left(\frac{-1}{6}, \frac{1}{3}\right)$, length $= \sqrt{\frac{221}{9}}$, or $\frac{\sqrt{221}}{3}$
51. a. $y = \frac{-3}{2}x$ **b.** $y = \frac{2}{3}x + \frac{13}{3}$

DISCOVERY ■ page 211

1. Delux: $C(x) = 0.45x + 50$; El Cheapo $C(x) = 0.50x + 40$

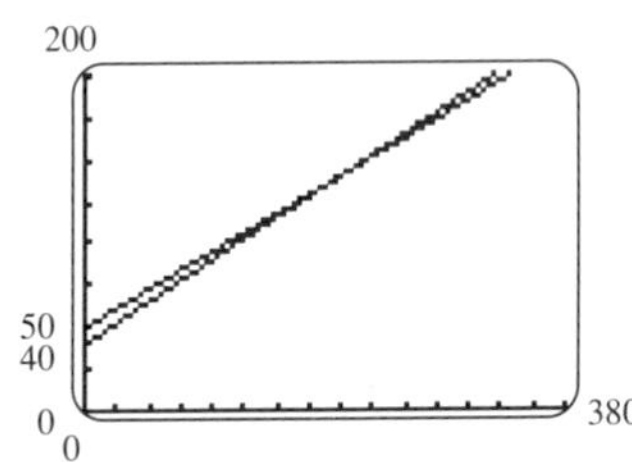

200.0 miles

3. Enter the two equations in Y1 and Y2. Then find the point of intersection of the two lines. The Celsius and Fahrenheit temperatures are equal at −40°.

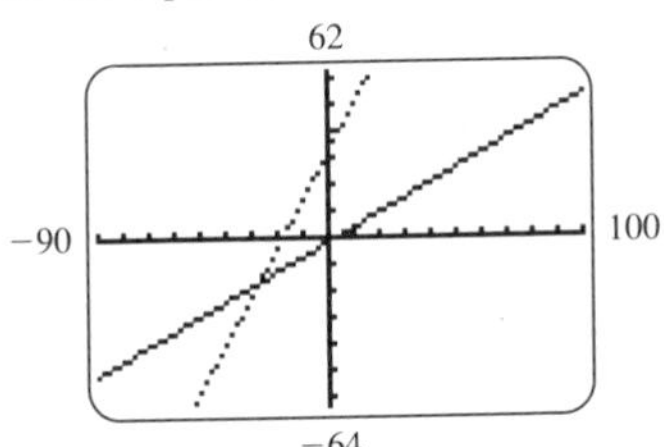

EXERCISES 4.3 ■ page 224

1.

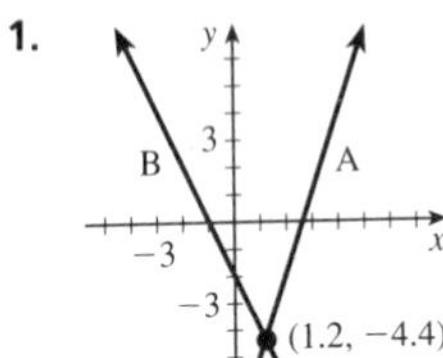

3.

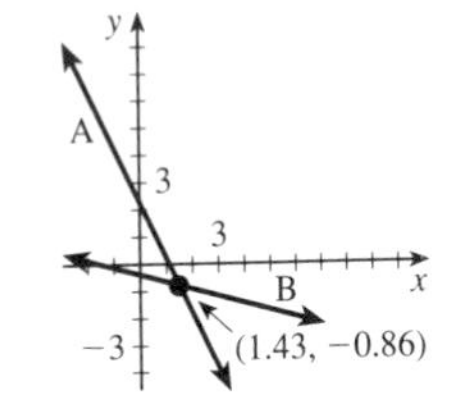

5. 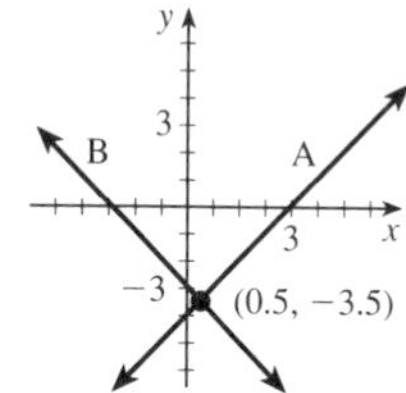

7. 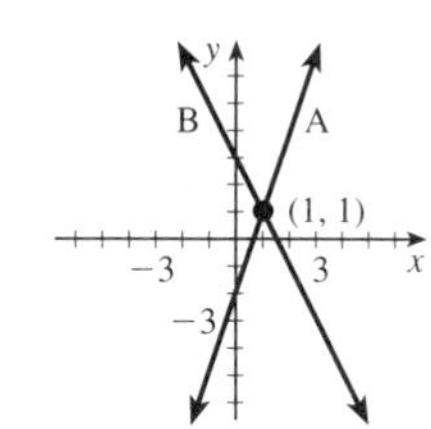

9. 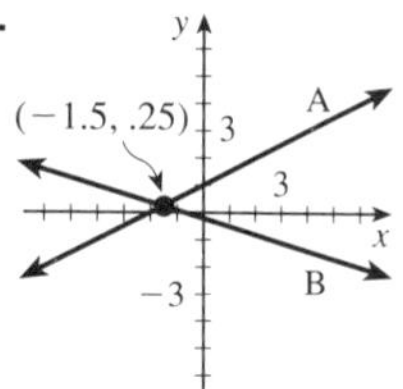

11. $\left(\frac{58}{33}, \frac{-13}{33}\right)$ 13. (3, 3) 15. $\left(\frac{16}{7}, \frac{38}{7}\right)$

17. (3, 2) 19. $\left(\frac{-1432}{99}, \frac{-452}{99}\right)$ 21. $\left(\frac{8}{5}, \frac{-1}{5}\right)$

23. $\left(\frac{10}{3}, \frac{14}{9}\right)$ 25. (14, 4)

27. 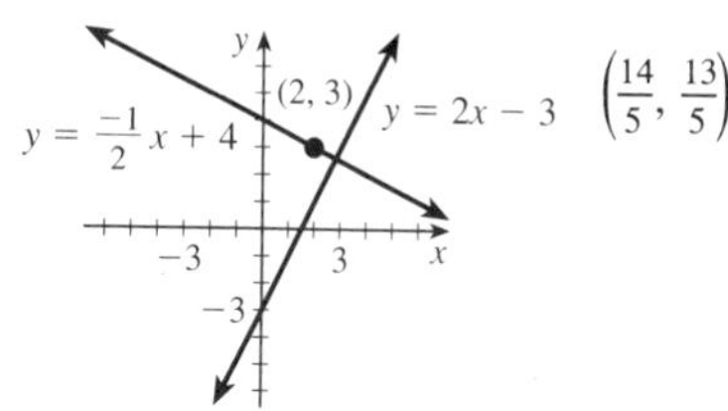

$\left(\frac{14}{5}, \frac{13}{5}\right)$

29. Consistent 31. Inconsistent 33. $\frac{9}{5}$

DISCOVERY ■ page 225

1. $C(x) = 2x + 3000$

3. 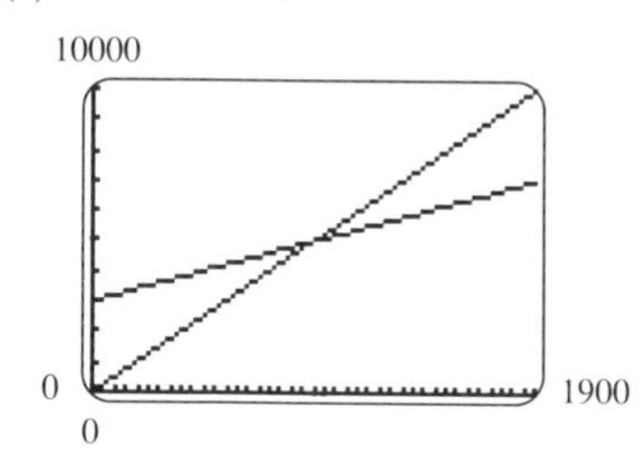

Break-even point (1000, 5000)

EXERCISES 4.4 ■ page 236

1. a. x = first number and y = second number
   b. $\begin{cases} x + y = 120 \\ x - y = 34 \end{cases}$
   c. Solution (77, 43); the two numbers are 77 and 43.
3. a. W = width of rectangle and L = length of rectangle
   b. $\begin{cases} W = L - 7 \\ 2L + 2W = 86 \end{cases}$
   c. Solution (18, 25); the width of the rectangle is 18 ft and the length is 25 ft.
5. a. x = speed of plane and y = speed of wind
   b. $\begin{cases} x + y = 300 \\ x - y = 240 \end{cases}$
   c. Solution (270, 30); the speed of the plane is 270 mi/h and the speed of the wind is 30 mi/h.
7. a. p = number of pounds of peanuts and a = number of pounds of almonds
   b. $\begin{cases} p + a = 10 \\ 1.20p + 4.50a = 20 \end{cases}$
   c. Solution (7.58, 2.42); 7.58 pounds of peanuts and 2.42 pounds of almonds are needed.
9. a. x = number of adult tickets and y = number of student tickets
   b. $\begin{cases} x + y = 300 \\ 15x + 6y = 2700 \end{cases}$
   c. Solution (100, 200); 100 adult and 200 student tickets were sold.
11. a. x = number of liters of 30% acid solution and y = number of liters of 70% acid solution
   b. $\begin{cases} x + y = 120 \\ .3x + .7y = .6(120) \end{cases}$
   c. Solution (30, 90); 30 liters of 30% acid solution must be mixed with 90 liters of 70% acid solution.
13. a. x = number of liters of 10% acid solution and y = number of liters of 60% acid solution
   b. $\begin{cases} x + y = 70 \\ .1x + .6y = .2(70) \end{cases}$
   c. Solution (56, 14); 56 liters of 10% acid solution must be mixed with 14 liters of 60% acid solution.
15. a. x = number of hours at 50 mi/h and y = number of hours at 65 mi/h
   b. $\begin{cases} x + y = 8 \\ 50x + 65y = 495 \end{cases}$
   c. Solution $\left(\frac{5}{3}, \frac{19}{3}\right)$; $1\frac{2}{3}$ hours at 50 mi/h and $6\frac{1}{3}$ hours at 65 mi/h.
17. a. x = Woo's rate and y = Tom's rate
   b. $\begin{cases} y = x + 10 \\ \frac{2}{3}x + \frac{2}{3}y = 100 \end{cases}$
   c. Solution (70, 80); Woo's rate is 70 mi/h and Tom's rate is 80 mi/h.
19. a. x = amount invested in stocks and y = amount invested in bonds
   b. $\begin{cases} x + y = 20000 \\ 0.15x + 0.10y = 2000 \end{cases}$
   c. Solution (0, 20000); \$0 is invested in stocks and \$20000 is invested in bonds.
21. a. x = number of hours Sally works and y = number of hours Jill works
   b. $\begin{cases} x + y = 8 \\ 20x + 15y = 150 \end{cases}$
   c. Solution (6, 2); Sally works 6 hours and Jill works 2 hours.

23. **a** m = slope and b = y-intercept

b. $\begin{cases} 6 = -3m + b \\ -1 = 5m + b \end{cases}$

c. Solution $\left(\frac{-7}{8}, \frac{27}{8}\right)$; $m = \frac{-7}{8}$ and $b = \frac{27}{8}$

25.

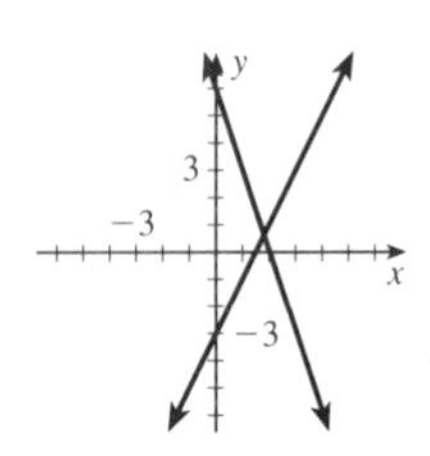

a. x = input value and y = output value

b. $\begin{cases} y = -3x + 6 \\ y = 2x - 3 \end{cases}$

c. Solution: $\left(\frac{9}{5}, \frac{3}{5}\right)$; the input is $\frac{9}{5}$ and the outut is $\frac{3}{5}$

27. **a.** x = number of miles driven and C = cost of car rental

b. $\begin{cases} C = 0.30x + 30 \\ C = 0.50x + 20 \end{cases}$

c. Solution (50, 45); when 50 miles are driven, the cost is the same, \$45.

29. **a.** t = time in years and $v(t)$ = value of item at time t

b. $\begin{cases} v(t) = 600 - 100t \\ v(t) = 800 - 150t \end{cases}$

c. Solution (4,200); at the end of 4 years, the values are equal, \$200.

CHAPTER 4 REVIEW ■ page 238

1. $y = \frac{4}{3}x + \frac{23}{3}$ **3.** $f(x) = \frac{-7}{5}x + \frac{9}{5}$

5. **a.** $C(t) = \frac{13}{2}t - \frac{19}{2}$ **b.** $\frac{57}{13}$ minutes **c.** 17.15°C

7. length $\sqrt{68}$, or $2\sqrt{17}$, and midpoint (1, 3)

9. length $\sqrt{85}$ and midpoint $\left(\frac{-3}{2}, -3\right)$

11. $y = \frac{4}{3}x + 2$ **13.** $y = -4x + 6$

15. $4x = 10$; point of intersection has x-coordinate $\frac{5}{2}$ and y-coordinate $\frac{-3}{10}$

17. $\left(1, \frac{-7}{5}\right)$ **19.** $\left(\frac{-2}{19}, \frac{-54}{19}\right)$ **21.** $\left(\frac{3}{2}, \frac{5}{2}\right)$

23. 54 and −6

25. 16.67 liters of 10% alcohol solution and 33.33 liters of 25% alcohol solution

27. Break-even point is 40 tubs; 25 tubs produce a loss of \$750.

CHAPTER 4 TEST ■ page 239

1. **a.** $y = \frac{1}{2}x - \frac{7}{2}$ **b.** (7, 0)

2. **a.**

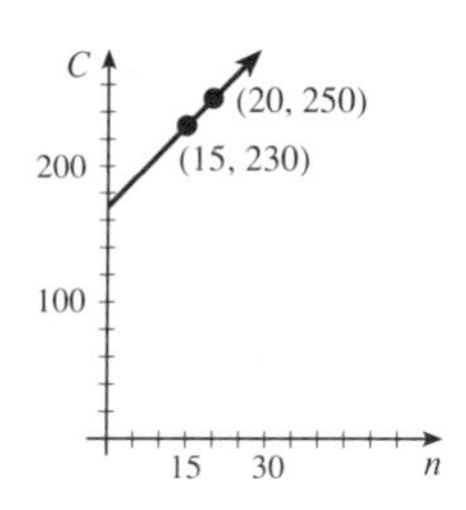

b. $C = 4x + 170$

c. 35 backpacks

3. The points are collinear; $y = -2x + 3$

4. **a.** $\sqrt{52}$, or $2\sqrt{13}$ **b.** (4, 2) **c.** $y = \frac{-3}{2}x + 8$ **d.** $\frac{2}{3}$

e. $y = \frac{2}{3}x - \frac{2}{3}$ **f.**

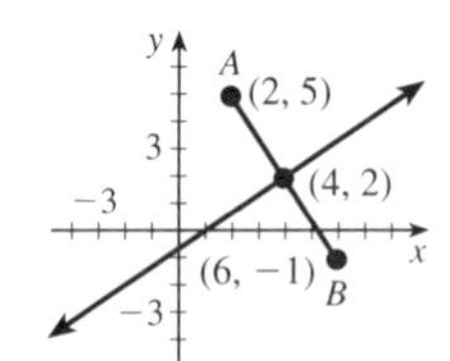

5. (15, 8) **6.** $\left(\frac{3}{2}, 0\right)$

7. 128 liters of 20% acid solution and 32 liters of 70% acid solution

8. 42 pigs and 27 chickens

■ CHAPTER 5

EXERCISES 5.1 ■ page 248

1. True **3.** False **5.** False **7.** False

9. **F** is $(3x)(x)$ or $3x^2$;
O + **I** is $(3x)(-5) + (2)(x)$ or $-13x$;
L is $(2)(-5)$ or -10;
standard polynomial $3x^2 - 13x - 10$

11. $-15x^4y^7$ Type `(3X²Y²)(-5X²Y^4)` for `Y1` and `-15X^4Y^7` for `Y2`. Store `7` in `X` and `2` in `Y`. Display the values in `Y1` and `Y2`. In each case, the output is `-4609920`.

13. $-8x^2y^3z^4$ **15.** $6x^2 - 10x$ **17.** $6x^3 - 10x^2$

19. $-3x^2 + 9x - 15$ **21.** $-12w^3 - 6w^2 + 18w$

23. $x^2 + 10x + 21$ **25.** $x^2 - 3x - 4$

27. $2a^2 + ab - b^2$ **29.** $4x^6 - 9$ **31.** $2x^2 - 7x - 4$

33. $x^4 - 14x^2 + 49$ **35.** $6p^2 + 7p - 20$

37. $2y^2 - 7y + 6$ **39.** $x^3 + 3x^2 - 3x - 9$

41. $x^2 - 10x + 25$ **43.** $9t^2 + 30t + 25$

45. $4y^2 - 4y + 1$ **47.** $x^2 - 9$ **49.** $9x^2 - 16$

51. $4a^2 - b^2$ **53.** $4x^6 - 12x^3 + 9$ **55.** $x^4 - x^3 + 1$

57. $-x^3 + x^2 + 5x + 12$ **59.** $6x^3 - 15x^2y^2 - 4xy + 10y^3$

61. $\frac{12}{25}x^2 - \frac{4}{15}x - \frac{8}{9}$ **63.** 145 **65.** 351

DISCOVERY ■ page 250

1. $27x^3 + 1$ **3.** $27x^3 - 18x^2 + 6x - 1$
5. $27n^3 - 8$ **7.** $-x^4 + 8$

EXERCISES 5.2 ■ page 255

1. $4x$ **3.** $2x$ **5.** 5 **7.** $15a^7$ **9.** 1
11. $11x(2x - 3)$ **13.** $3(2x^2 - x - 4)$ **15.** $(-10)(x - 2)$
17. $7(x^2 - 2x + 4)$ **19.** $2m^3x(5m^2x - 2)$
21. $3(2x^2 - x - 5)$ **23.** $3(2x - xy - 4y)$
25. $3(2x^2 - xy - 4y^2)$ **27.** $(x - 1)(x^2 - 3)$
29. $(x - 2)(x^2 - 5)$ **31.** $(x - 1)(x^2 + 5)$
33. $(x^3 + 2)(x^2 + 1)$ **35.** $(x - 2)(x^2 + 1)$
37. $(a + b)(x + 1)$ **39.** $(a + b)(a + 1)$
41. $(x - 2)(2x^2 + 1)$ **43.** $(x^2 + 2)(3x^3 - 1)$
45. $(a + b)(x - 1)$ **47.** $(a - b)(a - 1)$ **49.** Prime
51. Prime **53.** Prime **55.** $3(p^2 - 3)(p + 2)$

EXERCISES 5.3 ■ page 263

1. True **3.** True **5.** False **7.** True
9. $(x - 4)(x - 2)$ **11.** $(m - 2)(m - 1)$
13. $(m - 2)(m + 1)$ **15.** $(x - 8)(x - 1)$
17. $(x - 7)(x + 2)$ **19.** $(d - 8)(d + 5)$
21. $2(t - 4)(t + 1)$ **23.** Prime **25.** $7(p - 3)(p - 1)$
27. $2(t - 4)(t + 1)$ **29.** $(y - 9)(y + 7)$
31. $(2x - 1)(x - 1)$ **33.** $(2n - 3)(n - 2)$
35. $(2x + 1)(x - 6)$ **37.** $(7y - 3)(2y + 1)$ **39.** Prime
41. $(2t + 1)(t - 5)$ **43.** $(2x + 3)(2x + 3)$
45. $(3w + d)(w + 2d)$ **47.** $(5x + 1)(2x + 3)$
49. $(3x - 2)(x - 1)$ **51.** $(x - 4)(x - 2)$
53. $3x(x - 3)$ **55.** Prime **57.** $3t(t - 5)(t + 3)$
59. $x(x - 5)(x - 2)$ **61.** $y(y - 6)(y - 5)$
63. $(v - 8)(v + 3)$ **65.** Prime **67.** $2x(x^2 + 8x + 2)$
69. $18t^3(2t^2 + x)$ **71.** $(3x + 2)(x - 5)$
73. $2x(x + 5)(x - 4)$ **75.** $(3x^2 - 2)(2x^2 + 1)$

DISCOVERY ■ page 266

1. $(x - 12)(x + 7)$ **3.** $3(5y - 2)(2y - 7)$
5. Prime **7.** Prime **9.** $(6z - 1)(2z + 3)$

EXERCISES 5.4 ■ page 273

1. $(y - 9)(y + 9)$ **3.** $m(m - 1)(m + 1)$
5. $(4x - 1)(4x + 1)$ **7.** Prime **9.** $4(x - 2)(x + 2)$
11. $(5 - 3a)(5 + 3a)$ **13.** $(x + 5)(x^2 - 5x + 25)$
15. $(x - 1)(x^2 + x + 1)(x + 1)(x^2 - x + 1)$
17. $(10x^2 - 3)(10x^2 + 3)$ **19.** $(x - 2)(x + 2)(x^2 + 4)$
21. $(2x - 7)(2x + 7)$ **23.** $5(x^2 - 2)$
25. $5(x - 2)(x + 2)$ **27.** $(4x^2 - 3)(4x^2 + 3)$
29. $(2x - 3)(2x + 3)(4x^2 + 9)$ **31.** $(y - 9)^2$
33. $(m + 8)^2$ **35.** $(5x - 1)^2$ **37.** $(x^2 - 3)^2$
39. $(x - 2)^2(x + 2)^2$ **41.** $(10x - 3)^2$
43. $(x - 3)(x - 8)$ **45.** $(2x - 3)(2x + 5)$
47. $-2(x - 3)(4x^2 - 6x + 9)$ **49.** $(3x^2 - 2)(3x^2 + 1)$
51. $(a + b)(x + 1)(x - 1)$ **53.** $(a + b)(a^2 + 1)$
55. Prime **57.** $(c - 2b - a)(c - 2b + a)$
59. $(x + 1)(x - 2)(x + 2)$ **61.** $(y - 2)^2(y + 2)$
63. $(3x - 5)(3x + 5)$ **65.** $(3y + 5)(9y^2 - 15y + 25)$
67. $(2d - 1)(4d^2 + 2d + 1)$ **69.** Prime
71. $(x + 5)(3x + 13)$ **73.** $(x - 7)(x + 5)$
75. $(x - 3)(x^2 + 2)$ **77.** $(y + x)(y - 3x)$
79. $(x + 14)(x + 2)$ **81.** $(8d - 5t)(8d + 5t)$
83. $(b - 3c)(b + 3c)(b^2 + 9c^2)$ **85.** $(-x^2)(x + 5)(x - 5)$
87. $(4x - 1)(2x^2 - 3)$ **89.** $3x^2(3x - 1)(x + 5)$
91. $9x(x - 2)$ **93.** $y(y - 4)(y + 4)$ **95.** $8d(d^2 - 2)$
97. $2(2x - 1)(x - 2)$ **99.** $(3x - 14)(x - 6)$
101. $4(x - 4)(x + 2)$ **103.** $(x + 3)(x^2 - 2)$
105. $(y - 3x)(y - x)$ **107.** $(x - 14)(x - 2)$
109. $(8d^2 - 5t)(8d^2 + 5t)$ **111.** $(2b^2 - 9c^2)(2b^2 + 9c^2)$
113. $(-x^2)(4x + 5)(4x - 5)$

DISCOVERY ■ page 275

1. Press the Y= key and enter X² − 3X − 4 for Y1. Use the TRACE feature to find that the x-values that result in an evaluation of 0 are 4 and −1.
The factors are $(x - 4)\ (x + 1)$.

3. Press the Y= key and enter X² − 5X + 4 for Y1. Use the TRACE feature to find that the x-values that result in an evaluation of 0 are 4 and 1. The factors are $(x - 4)(x - 1)$.

5. Press the Y= key and enter X² − 5X − 6 for Y1. Use the TRACE feature to find that the x-values that result in an evaluation of 0 are 6 and −1.
The factors are $(x - 6)\ (x + 1)$.

7. Press the Y= key and enter X² − 1 for Y1. Use the TRACE feature to find that the x-values that result in an evaluation of 0 are 1 and −1.
The factors are $(x - 1)\ (x + 1)$.

9. Press the Y= key and enter X² + 1 for Y1. Use the TRACE feature to find that no x-values result in an evaluation of 0. The polynomial is prime, because the graph is always above the x-axis.

EXERCISES 5.5 ■ page 284

1. I **3.** D **5.** C **7.** B **9.** J **11.** 3, 5
13. $\frac{2}{3}, -1$ **15.** 2, −3, −5 **17.** 0, 7 **19.** $\frac{-3}{2}, 0$
21. 0, 5 **23.** 0, 3 **25.** 0, 13 **27.** −4, 3
29. −8, 3 **31.** −2, 5 **33.** $\frac{2}{3}, -2$ **35.** −5, 8
37. $\frac{13}{4}, 0$ **39.** $\frac{1}{2}, 1$ **41.** $\frac{10}{3}, \frac{-5}{2}$ **43.** 2, 7
45. 2, 4 **47.** −3, 0 **49.** 2, −3 **51.** −6, 1
53. $\frac{1}{2}$ **55.** 0 **57.** −1, 0, 3 **59.** $\frac{13}{6}$ **61.** −6, 1
63. None **65.** −3, −2, 2 **67.** 3 **69.** 6

71. 5 cm **73.** 10 meters **75.** 5 cm **77.** 2 cm
79. 8 feet **81.** 8 seconds

EXERCISES 5.6 ■ page 289

1. $(-\infty, 0] \cup [5, +\infty)$

3. $(-\infty, -3) \cup (2, +\infty)$

5. $\left[\frac{-3}{2}, \frac{2}{3}\right]$

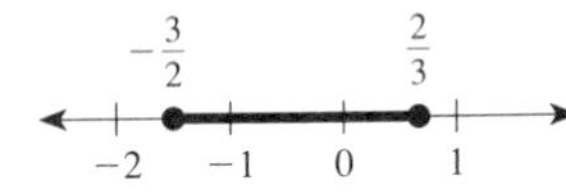

7. $(-\infty, -1) \cup (5, +\infty)$

9. $(-\infty, -4) \cup (3, +\infty)$

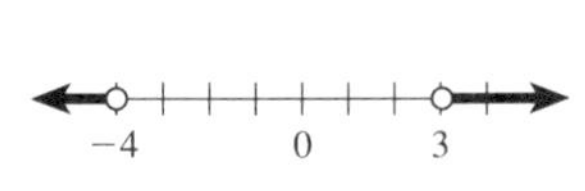

11. $(-\infty, 0) \cup \left(\frac{13}{4}, +\infty\right)$

13. $\left[\frac{1}{2}, 1\right]$

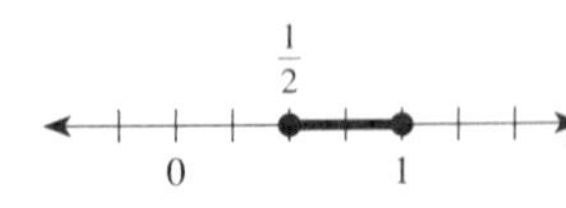

15. $\left(-\infty, \frac{-5}{2}\right] \cup \left[\frac{10}{3}, +\infty\right)$

17. $[2, 7]$

19. $(-\infty, 2) \cup (4, +\infty)$

21. $(-3, 0)$

23. $[-3, 2]$

25. None (value never is negative) $\varnothing$

27.

29. $(0, 8)$ (negative values are not allowed for the width)

EXERCISES 5.7 ■ page 298

1. False **3.** True

5. True **7.** False

9. False

11. a. $(0, -6)$
b. $(-3, 0)$ and $(2, 0)$
c. $\left(\frac{-1}{2}, \frac{-25}{4}\right)$
d. $x = \frac{-1}{2}$
e.

13. a. $(0, -3)$
b. $(-3, 0)$ and $(1, 0)$
c. $(-1, -4)$
d. $x = -1$
e.

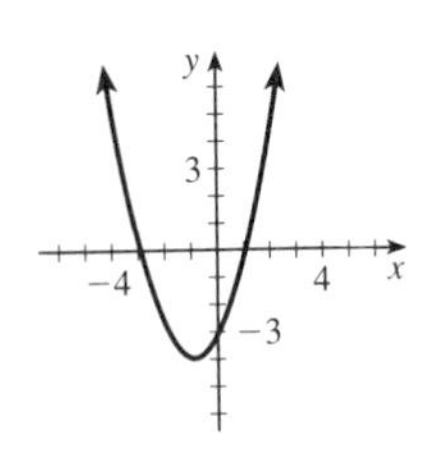

15. a. $(0, -6)$
b. $(-3, 0)$ and $(1, 0)$
c. $(-1, -8)$
d. $x = -1$
e.

17. a. $(0, -3)$
b. $(-3, 0)$ and $\left(\frac{1}{2}, 0\right)$
c. $\left(\frac{-5}{4}, \frac{-49}{8}\right)$
d. $x = \frac{-5}{4}$
e.

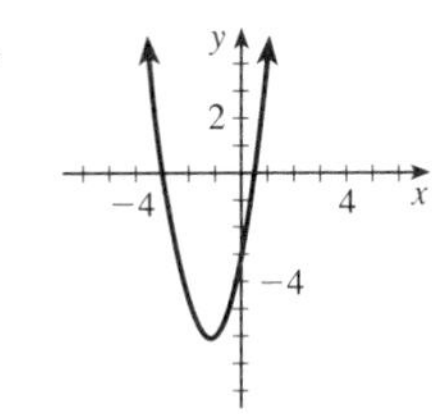

19. a. $(0, 3)$
b. $(-1, 0)$ and $(3, 0)$
c. $(1, 4)$
d. $x = 1$
e.

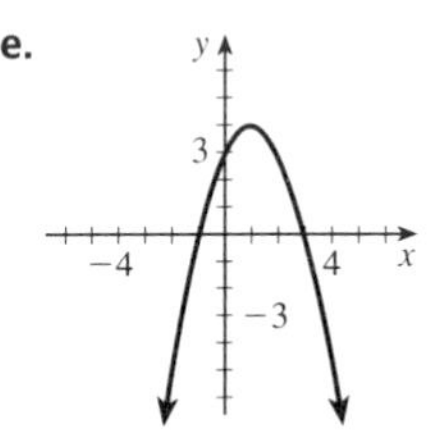

21. a. $(0, -5)$
b. $(-5, 0)$ and $(1, 0)$
c. $(-2, -9)$
d. $x = -2$
e.

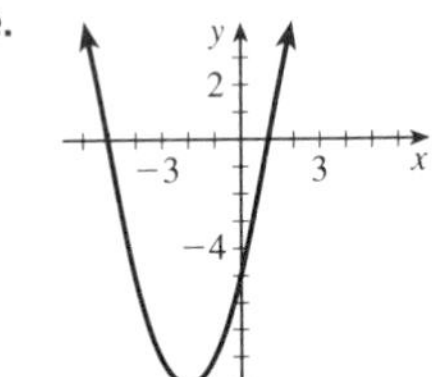

23. a. $(0, 5)$
b. $(-1, 0)$ and $\left(\frac{5}{2}, 0\right)$
c. $\left(\frac{3}{4}, \frac{49}{8}\right)$
d. $x = \frac{3}{4}$
e.

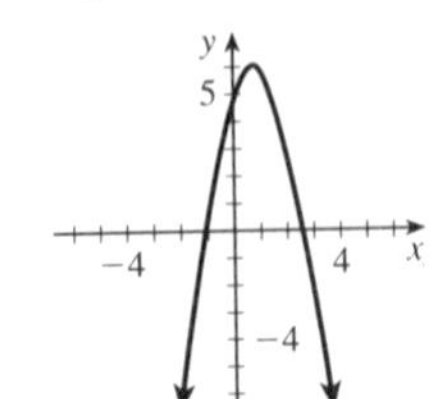

25. a. $(0, -8)$
b. $(-4, 0)$ and $(2, 0)$
c. $(-1, -9)$
d. $x = -1$
e.

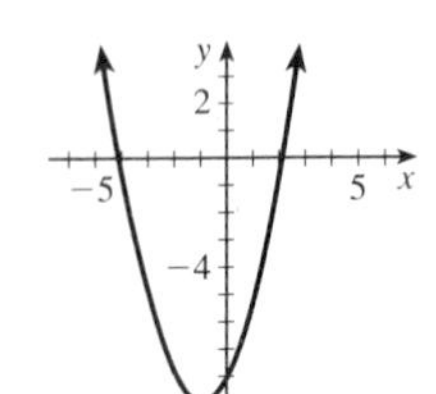

27. a. (0, 1)
b. (1, 0)
c. (1, 0)
d. $x = 1$
e.

29. a. $(0, -1)$
b. $(-1, 0)$
c. $(-1, 0)$
d. $x = -1$
e.

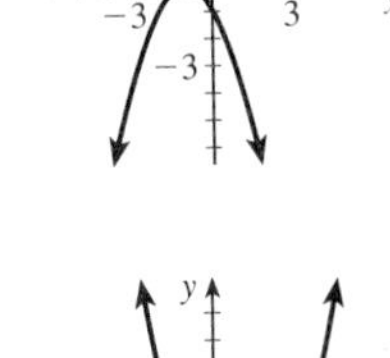

31. a. $(0, -3)$
b. (3, 0) and $(-1, 0)$
c. $(1, -4)$
d. $x = 1$
e.

33. a. (0, 3)
b. (3, 0) and $(-1, 0)$
c. (1, 4)
d. $x = 1$
e.

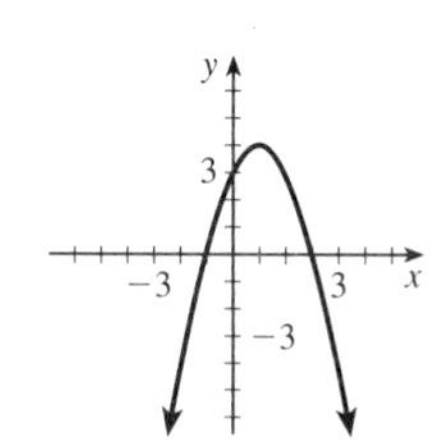

35. a. (0, 0)
b. (0, 0) and (8, 0)
c. (4, 16)
d. $x = 4$
e.

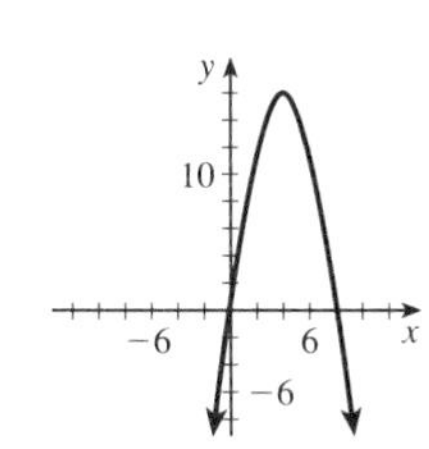

37.

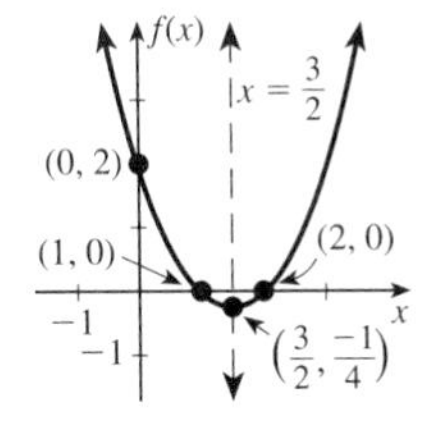

domain: $\{x \mid x \text{ is real}\}$;
range: $\left\{f(x) \mid f(x) \geq \frac{-1}{4}\right\}$

39.

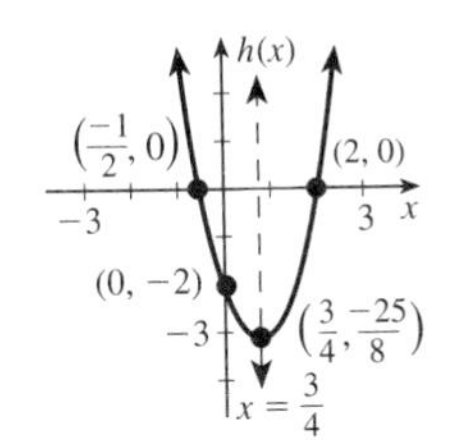

domain: $\{x \mid x \text{ is real}\}$;
range: $\left\{h(x) \mid h(x) \geq \frac{-25}{8}\right\}$

41.

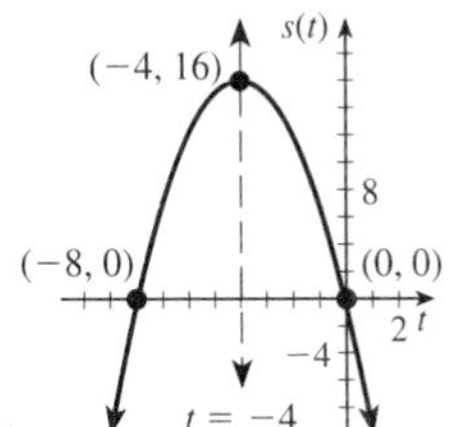

domain: $\{t \mid t \text{ is real}\}$;
range: $\{s(t) \mid s(t) \leq 16\}$

43.

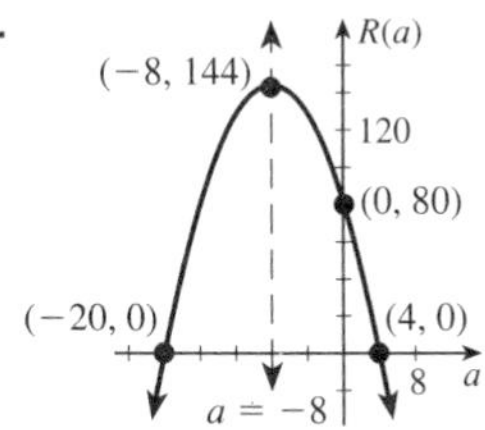

domain: $\{a \mid a \text{ is real}\}$;
range: $\{R(a) \mid R(a) \leq 144\}$

45. 40 chips or 60 chips

47. a.

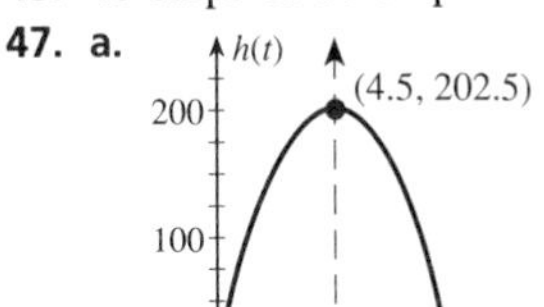

b. $h(2) = 140$
c. 4.5 seconds, 202.5 feet
d. 9 seconds

49. $P(x) = x(2x + 8)$; $x = -2$; $P(x) = -8$
51. $A(x) = x(40 - x)$; $x = 20$ meters

EXERCISES 5.8 ■ page 310

1. $-3(17) + 40 = -11$; correct remainder is 50
3. Quotient $x^2 - \frac{1}{2}x$; remainder 2
5. Quotient $x^2 + x + 1$; remainder 0
7. Quotient $2x^2 - 1$; remainder -5
9. Quotient $-3x^2$; remainder $-3x^2 - x + 2$
11. Quotient $x^2 - 5x + 23$; remainder -110
13. $x + 2$ **15.** $x^3 + 6$ **17.** $(x^2 - x - 1)(x + 4)$
19. $(2x + 3)(2x - 3)(x - 2)(x + 2)$ **21.** $-1, 2, 4$
23. $-3, -2, 1, 3$ **25.** $x - 7 + \dfrac{34}{x + 5}$
27. $x^2 - 2x + \dfrac{-1}{x - 1}$

CHAPTER 5 REVIEW ■ page 311

1. True **3.** True **5.** False **7.** False **9.** False
11. i. **13.** ii. **15.** $\frac{4}{9}x^6y^4$
17. $3x^4 - 9x^3 - 6x^2$ **19.** $6x^3 - 9x^2 + 2x - 3$
21. $8x^4 + 6x^2 - 5$ **23.** $x^2 + 4x - 5$
25. $3x^2 - 8x + 5$ **27.** $4x^2 - 9$ **29.** $x^4 - 25$
31. $4x^2 - 20x + 25$ **33.** $9x^2 + 6x + 1$ **35.** $x^3 + 27$
37. $5x^2 - 9x$ **39.** $-x^2 + 4x + 1$ **41.** $3x^2(3x + 1)$
43. $5xy(y - 4x)$ **45.** $4m^2x^3(2m^2x^2 - 3)$
47. $3x(2x^2 - 10x - 5)$ **49.** $x^2y^2(3y - 8x)$
51. $(x - 1)(x^2 - 5)$ **53.** $(x - 6)(x^2 - 2)$
55. $(x - 3)(x^2 + 3)$ **57.** Prime
59. $(x - 1)(x^3 - x - 1)$ **61.** $(x - 5)(x + 2)$
63. $(-1)(x + 6)(x - 1)$ **65.** $(5x - 1)(x - 3)$
67. $(3x + 1)(x - 3)$ **69.** $(7y + 3)(2y - 1)$

71. $(12x + 3)(x - 2)$ **73.** $(2t + 1)(t - 7)$
75. $(7x + 2)^2$ **77.** $(9x - 1)^2$ **79.** Prime
81. $(4x - 9)(4x + 9)$ **83.** $(2z - 3)(4z^2 + 6z + 9)$
85. $(x - 6)(x - 2)$ **87.** $(x^2 - 7)(x^2 + 1)$
89. $(x - 2)^2(x + 2)$ **91.** $-7, 3$ **93.** $0, 4$
95. $0, 2$ **97.** $0, \frac{3}{2}$ **99.** $-2, 5$ **101.** $-2, 9$
103. -8 and 6 **105.** -5 and $\frac{3}{2}$
107. $-3, -2, 2$ **109.** 5 **111.** -5 and 1

113. $[-2, 1]$ **115.** $\left[-1, \frac{3}{2}\right]$

117. The rocket is at $h = 1500$ at 2.5 sec and 37.5 sec

119.

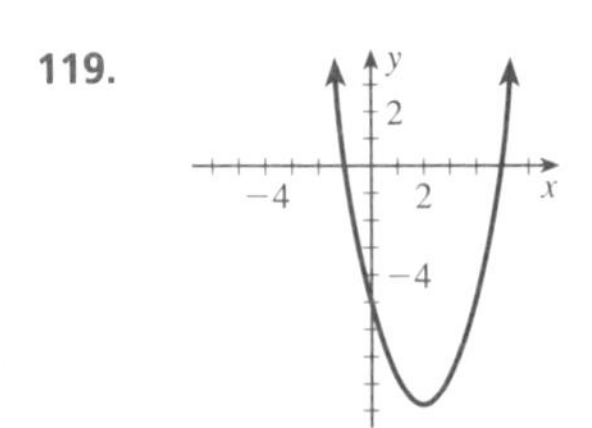

x-intercepts $(-1, 0)$ and $(5, 0)$
y-intercept $(0, -5)$
line of symmetry $x = 2$
vertex $(2, -9)$

121.

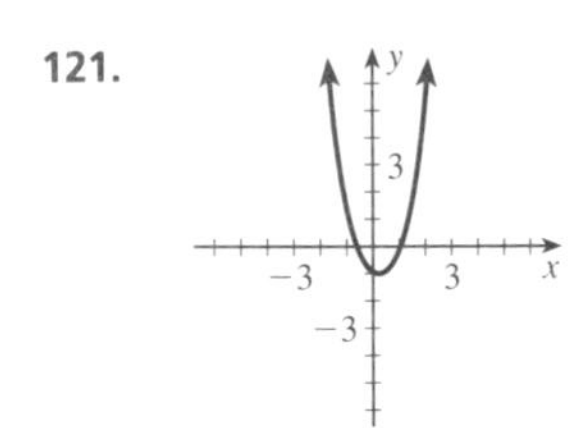

x-intercepts $\left(\frac{-1}{2}, 0\right)$ and $(1, 0)$
y-intercept $(0, -1)$
line of symmetry $x = \frac{1}{4}$
vertex $\left(\frac{1}{4}, \frac{-9}{8}\right)$

123.

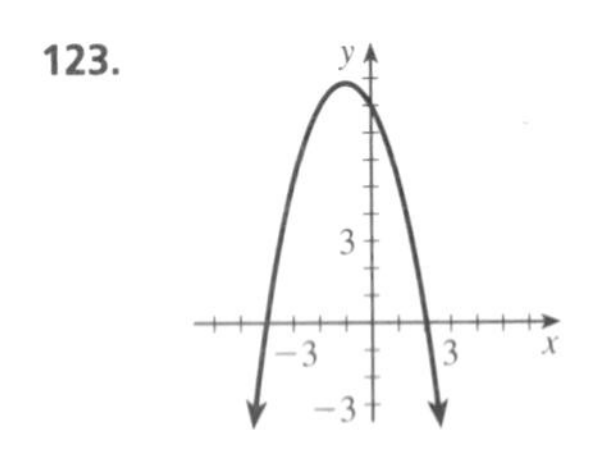

x-intercepts $(-4, 0)$ and $(2, 0)$
y-intercept $(0, 8)$
line of symmetry $x = -1$
vertex $(-1, 9)$

125.

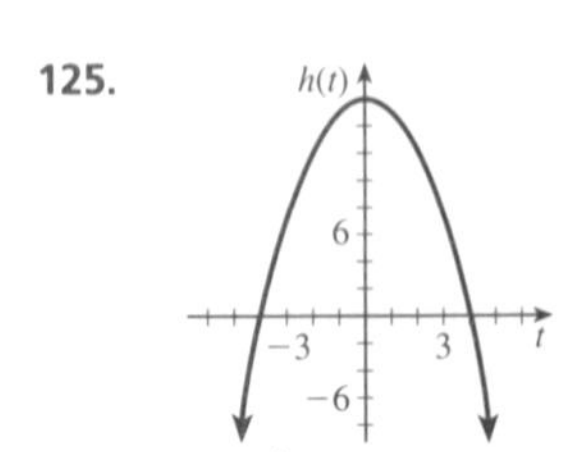

y-intercept $(0, 16)$
x-intercepts $(4, 0)$ and $(-4, 0)$
line of symmetry $x = 0$
vertex $(0, 16)$

127.

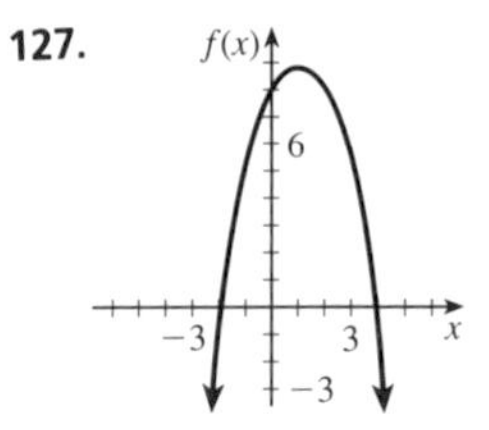

y-intercept $(0, 8)$
x-intercepts $(-2, 0)$ and $(4, 0)$
line of symmetry $x = 1$
vertex $(1, 9)$

129. $x^3 - 3x^2 + 8x - 24 + \frac{70}{x + 3}$

131. $(x + 2)$ and $(x^2 - 2x + 4)$ **133.** $x^2 + 2x + 1 + \frac{7}{x - 2}$

135. $x^4 + 4x^3 + 16x^2 + 61x + 244 + \frac{974}{x - 4}$

CHAPTER 5 TEST ■ page 313

1. a. $\frac{1}{4}x^4y^6$ **b.** $6x^4 - x^2 - 2$ **c.** $9x^2 - 25$
d. $8x^3 - 27$ **e.** $2x^2 + 4x - 10$

2. a. $(x - 5)(2x - 7)$ **b.** $(8x^2 + 3)(x + 2)$
c. $y(2y + 5)(y - 3)$ **d.** $(2x^2 - 9)(2x^2 + 9)$
e. $(3s - 5)(9s^2 + 15s + 25)$ **f.** $(3x - 5)(2x + 1)$
g. $(x + 3)(x - 2)(x + 2)$

3. a. $0, 5$ **b.** $\frac{5}{3}, 2$ **c.** $2, 3$ **d.** $\frac{3}{2}, -2$
e. $-3, -2, 2$

4. 36 feet

5. a. $(0, -15)$ **b.** $(-3, 0)$ and $\left(\frac{5}{2}, 0\right)$ **c.** $x = \frac{-1}{4}$
d. $\left(\frac{-1}{4}, \frac{-121}{8}\right)$ **e.**

6. a. -15 **b.** Inputs 0 and $\frac{1}{2}$ yield output 5.

7. $2x^2 + 3x + 6 + \frac{15}{x - 2}$

■ CHAPTER 6

EXERCISES 6.1 ■ page 322

1. $\{x \mid x \neq 0\}$ **3.** $\{s \mid s \neq -2, s \neq 2\}$
5. $\{r \mid r \neq -7\}$ **7.** $\{a \mid a \neq -3, a \neq 3\}$
9. $\left\{x \mid x \neq \frac{4}{3}, x \neq 2\right\}$ **11.** $\{v \mid v \neq 0, v \neq 1\}$

13. $\frac{x^2}{xy + x^2}, \frac{xy}{y^2 + xy}, \frac{x^2 + x}{x^2 + xy + x + y}$ (Other answers are possible.)

15. $\frac{3x}{y^2x - x^3}, \frac{3y}{y^3 - x^2y}, \frac{6}{2y^2 - 2x^2}$ (Other answers are possible.)

17. Store 5 in X; the first expression has an output of $2.3\overline{3}$ and the second expression has an output of $2.\overline{18}$. We conclude that the expressions are *not* equivalent.

19. $\frac{2x^2 - 10x + 12}{2x^2 - 4x}$ **21.** $\frac{y^3 - 4y^2}{y^2 - 2y - 8}$ **23.** $\frac{2x^2 + 7x + 6}{x^2 - 4}$

25. $\frac{-x^2 - 5x - 4}{16 - x^2}$ **27.** $\frac{x^3 - 4x^2 + 8x - 5}{x^3 - 1}$ **29.** $\frac{m^2}{2}$

31. $\frac{-2x^3}{d^2}$ **33.** $\frac{2x + 1}{x}$ **35.** $\frac{-h + 3}{4h^2}$ **37.** $\frac{2s + 3}{6u}$

39. $\frac{x + 2}{x - 2}$ **41.** Cannot be reduced further **43.** $\frac{xy}{x + y}$

EXERCISES 6.2 ■ page 329

1. $\frac{x - 1}{2}$ **3.** $\frac{1}{4mx^2}$ **5.** $\frac{2a}{7bmy}$ **7.** 1 **9.** $3x + 6$

11. $\frac{1}{x}$ **13.** $\frac{-8x^3 - 4x^2 + 2x + 1}{(2x^2 - x + 1)(x - 1)}$ **15.** $\frac{2d + 2c}{d + 1}$

17. $\frac{2x - 3}{(x + 1)(2x + 3)}$ **19.** $\frac{3x^3}{2m^2}$ **21.** $\frac{y - 1}{x + 1}$ **23.** $\frac{x + 1}{y - 1}$

25. y **27.** $\frac{x^3 - 6x^2 + 12x - 8}{3x}$ **29.** $\frac{x^3 + 2x^2 - 9x - 18}{x(x - 2)^2}$

31. $\frac{x^2 - y^2}{x^2y + xy^2}$ **33.** $\frac{2x^2 + 9x + 10}{2x^2 + x - 10}$

EXERCISES 6.3 ■ page 337

1. iii. $\frac{1}{x}$ **3.** $\frac{3x}{x - 3}$ **5.** $\frac{1}{x - 1}$ **7.** $\frac{5x + 1}{4}$

9. $\frac{-x^2 + 3x - 1}{x}$ **11.** $\frac{4x^2 - 6}{2x + 1}$ **13.** $\frac{x^3 - 3x^2 + 2x + 2}{(x + 1)(x - 3)}$

15. $\frac{-2x + 3}{6x^2}$ **17.** $\frac{x^2 - 2x + 6}{2(x - 2)(x + 2)}$ **19.** $\frac{x - 1}{(x - 2)(x + 1)}$

21. $\frac{-5x^2 + 7x - 7}{x^2(x - 1)^2}$ **23.** $\frac{3x - 3}{(x - 3)^2}$ **25.** $\frac{x + 2}{(2x)(x^2 + 2x + 4)}$

27. $\frac{-y^3 + xy^2 - x^3 + x^2}{xy(x - 1)(y - x)}$ **29.** $\frac{-4}{(x + 3)(x + 5)}$ **31.** $\frac{-s^3 + s^2 + 7s}{(s^2 - 9)(s - 2)}$

33. $\frac{-x^2 + 2x + 1}{(x - 1)(x + 1)}$ **35.** $\frac{x^2 + 1}{x}$ **37.** $\frac{x^2 + 3x + 1}{x(x + 1)}$

39. $\frac{2}{x + 1}$ **41.** $\frac{3y + 2x}{3x^2}$

EXERCISES 6.4 ■ page 345

1. $\frac{1}{x - 1}$ **3.** 1 **5.** $\frac{x^2 + 5x + 7}{x + 2}$ **7.** $\frac{x + 1}{x + 2}$

9. $\frac{2x^2 + x - 2}{2x^2 - 3}$ **11.** $\frac{(x - 1)^3}{(x + 2)^2(x - 2)}$ **13.** $\frac{-x^3 + 2x^2 + 28x - 47}{(x - 2)(x + 1)}$

15. $\frac{-1}{(x + 2)(x + 2 + h)}$ **17.** $\frac{r^2 - 2r - 3}{r^2 + r - 1}$ **19.** $\frac{r - 2}{r}$

21. $\frac{x - 1}{x - 2}$ **23.** $-s^2 + s$ **25.** $\frac{2a + 3}{a + 1}$ **27.** $\frac{n^3 - n^2 - 1}{n^2(n - 1)}$

29. $\frac{x^2 + x - 1}{x^2 - x}$ **31.** 1 **33.** $\frac{x - 2}{x - 1}$ **35.** $\frac{3x - 10}{x - 5}$

37. $x^2 - x + 1$ **39.** $\frac{x - 3}{x + 2}$ **41.** $\frac{-12x^2 - x - 6}{x(7x - 6)}$

43. $\frac{-1}{x + 1}$ **45.** $\frac{-1}{x^2 - y^2}$

EXERCISES 6.5 ■ page 354

1. iii. $\frac{2}{3x}$ **3. iv.** $7 - 3x = \frac{2}{3}$ **5.** False

7. Problem A requires only the addition of the rational expressions; problem B requires finding a value for x that makes the statement true.

9. 2 **11.** 5 **13.** $\frac{-21}{4}$ **15.** −4 and 3

17. −6, 1 **19.** $\frac{1}{2}$ **21.** $\frac{-2}{3}$ **23.** −4

25. $x = \frac{5}{y} + 2$ or $x = \frac{2y + 5}{y}$ **27.** $C = \frac{AB - 2}{A + 1}$

29. $x = \frac{2y + 10}{5}$ **31.** $R = \frac{R_1R_2}{R_1 + R_2}$ **33.** $y = \frac{-5x + 10}{x - 27}$

EXERCISES 6.6 ■ page 359

1. x = number of hours when both work; $\frac{x}{5} + \frac{x}{12} = 1$; $x = \frac{60}{17}$; it takes $\frac{60}{17}$ hours to plow the field when both work.

3. x = number of hours for pump B to empty the tank; $\frac{4}{6} + \frac{4}{x} = 1$; $x = 12$; pump B takes 12 hours to empty the tank when working alone.

5. x = number of days for both to paint 300 soldiers; $30x + 40x = 300$; $x = \frac{30}{7}$; it takes $\frac{30}{7}$ days to paint the soldiers when they work together.

7. x = number of hours for Walt to clean the apartment; $\frac{2}{3} + \frac{2}{x} = 1$; $x = 6$; Walt takes 6 hours to clean the apartment by himself.

9. x = number of hours for Ashley to mow the lawn by himself; $\frac{1}{4} + \frac{2}{4} + \frac{2}{x} = 1$; $x = 8$; It takes Ashley 8 hours to mow the lawn by himself.

11. x = rate of the express bus; $\frac{400}{x} = \frac{290}{x - 20}$; $x = 72.7$; the express bus travels at 72.7 mi/h.

13. x = rate of flying in calm air; $\frac{190}{x + 20} = \frac{150}{x - 20}$; $x = 170$; the plane's rate in calm air is 170 mi/h.

15. x = time for Thelma's husband to catch her; $50(x + 1) = 60x$; $x = 5$; It takes Thelma's husband 5 hours to catch her.

17. x = speed of the boat in calm water;
$\frac{8}{x+4} = \frac{5}{x-4}$; $x = \frac{52}{3}$;
The speed of the boat in calm water is $\frac{52}{3}$ mi/h.

19. x = speed John rides his bike;
$\frac{10}{x} = \frac{4}{x-10}$; $x = \frac{50}{3}$;
John rides his bike at $\frac{50}{3}$ mi/h.

21. x = number of miles from B-burgh to C-ville;
$80\left(\frac{x}{70} - 1\right) = 300$; $x = 332.5$;
The distance from B-burgh to C-ville is 332.5 miles.

23. x = positive ingeter
$x + 2 - 8\left(\frac{1}{x}\right) = 9$
$x = 8$ or $x = -1$
The positive integer is 8.

25. x = speed of the runner;
$\frac{20}{5x} + 3 = \frac{20}{x}$; $x = \frac{16}{3}$;
The runner's speed is 5.33 mi/h.

EXERCISES 6.7 ■ page 364

1. $\frac{48}{5}$ **3.** $\frac{5}{4}$ **5.** $\frac{-7}{5}$ **7.** $\frac{45}{14}$ **9.** 3 and -3
11. $k = \frac{6}{5}$; 9.6 inches **13.** $k = \pi$
15. $k = 3.5$; $\approx$\$2.2 million
17. V varies jointly as L and the square of r.
19. N varies inversely as d.
21. K varies jointly as m and the square of v.
23. I varies jointly as P and t.
25. v varies directly as d and inversely as t.

EXERCISES 6.8 ■ page 369

1. $\frac{x^5}{2y}$ **3.** $\frac{1}{18x}$ **5.** $-h^7$ **7.** $\frac{3}{2c^9}$ **9.** $\frac{16x^3}{y^4}$ **11.** $\frac{1}{2a^5}$
13. 1 **15.** $\frac{r}{9216}$ **17.** $\frac{4x+1}{(x-2)(x+1)}$ **19.** $\frac{1+18x}{9x^2}$
21. $\frac{x+5}{(x+1)(x-2)}$ **23.** $\frac{-3}{2}$ and 1 **25.** 8 **27.** -12
29. $\frac{-2}{5}$

CHAPTER 6 REVIEW ■ page 370

1. $\{x \mid x \neq 3\}$ **3.** $\{x \mid x \neq 0, x \neq 4\}$
5. $\left\{x \mid x \neq \frac{1}{2}, x \neq -2\right\}$ **7.** $\frac{-w^3}{2}$ **9.** $\frac{T+S}{-S}$
11. $\frac{-y}{y+5}$ **13.** $\frac{m}{6aby}$ **15.** $\frac{x^4+x^3}{y^3(y-1)}$ **17.** $x^2 + 2x + 1$
19. $\frac{x^2-2x+5}{(x+1)(x-3)}$ **21.** $\frac{-2y^2+3y-3}{y^2(y-1)^2}$ **23.** $\frac{x^2+x-1}{x(x-1)}$
25. $\frac{3x}{y^2}$ **27.** $\frac{x-1}{x-2}$ **29.** $\frac{5}{2}$ **31.** -11 **33.** -1 and 3
35. None **37.** Pump B takes $\frac{40}{3}$ hours to empty the tank.
39. $S = \frac{AR - R - 2}{A}$ **41.** $a = \frac{bc^2}{b-c}$ **43.** $k = \frac{1}{5}$; 5.6 inches
45. V varies jointly with h and the square of r.
47. $\frac{1}{18x}$ **49.** $\frac{4x+1}{(x-2)(x+1)}$
51. $\frac{2x}{x+2}$ **53.** 14 **55.** $\frac{-1}{4}$, 1

CHAPTER 6 TEST ■ page 371

1. $\left\{x \mid x \neq \frac{5}{3}\right\}$ **2.** $4x^2y^4$ **3.** $\frac{x-3}{x-2}$
4. $\frac{x^3+x^2-2x}{(x+3)(x-2)(2x+1)}$ **5.** $\frac{3x+5}{x-3}$ **6.** $\frac{x^2-4x-5}{(2x-5)(x-6)}$
7. $\frac{-x^3+x^2}{x+1}$ **8.** $\frac{1}{4}$ and -1 **9.** 0 **10.** 116.7 hours
11. 81 meters
12. H varies directly with the square of w and inversely as p. The constant of variation is $\frac{3}{2}$.

■ CHAPTER 7

EXERCISES 7.1 ■ page 382

1. $2x$, 2 **3.** 7 and -7, 7 **5.** 0 **7.** 10
9. Not a real number **11.** -1 **13.** 0
15. $\{x \mid x \text{ is real}\}$ **17.** $\{x \mid x \neq -3\}$ **19.** $\left\{x \mid x \geq \frac{5}{3}\right\}$
21. $\{x \mid x \text{ is real}\}$ **23.** $\{x \mid x \leq 0\} \cup \{x \mid x \geq 5\}$ **25.** 216
27. 16 **29.** 1000 **31.** 0 **33.** $\frac{1}{512}$ **35.** $\frac{1}{32}$
37. $\frac{1}{5}$ **39.** 0 **41.** $\frac{1}{64x^6}$ **43.** x **45.** x^{10}
47. $4x$ **49.** x **51.** $\frac{5}{x}$ **53.** $5x$ **55.** x^2
57. $2x^2$

EXERCISES 7.2 ■ page 390

1. False **3.** True **5.** False **7.** False
9. True **11.** $10y\sqrt{2x}$ **13.** $3x\sqrt[3]{y^2}$ **15.** $3mn\sqrt[4]{m^3n}$
17. $3x^2y\sqrt{2xy}$ **19.** $2dy\sqrt[5]{3x^3y}$ **21.** $-3m^2n\sqrt[3]{3mn^2}$

23. $5a^2b^2\sqrt{6b}$ **25.** $2y\sqrt[3]{6x^2y^2}$ **27.** $5\sqrt{x}$ **29.** $3\sqrt{2x}$

31. $\frac{2\sqrt{x}}{x}$ **33.** $3b\sqrt[3]{a^2}$ **35.** $\frac{2\sqrt[3]{3x^2}}{3}$ **37.** $\frac{\sqrt{5x}}{x}$

39. $\frac{\sqrt{3ay}}{3a}$ **41.** $\frac{\sqrt[3]{2x^2}}{x}$ **43.** $\frac{-\sqrt[3]{3a}}{3a}$ **45.** $\sqrt{y}$

EXERCISES 7.3 ■ page 398

1. $5x\sqrt{3x}$ **3.** $5x\sqrt[3]{2}$ **5.** $-3x\sqrt{2x}$

7. $-3\sqrt{2x}$ **9.** $x\sqrt{6} - 2\sqrt{xy}$ **11.** $x\sqrt[3]{x} - x$

13. $x + 2\sqrt{x} - 15$ **15.** $x + \sqrt{5x} - \sqrt{3x} - \sqrt{15}$

17. $x - 9$ **19.** -7 **21.** 7 **23.** $\frac{1}{4}$ **25.** $\frac{6 + 5\sqrt{5}}{9}$

27. $\left(1 - \sqrt{2}\right)^2 - 2\left(1 - \sqrt{2}\right) = 1 - 2\sqrt{2} + 2 - 2 + 2\sqrt{2} = 1$

29. $\left(-3 + \sqrt{5}\right)^2 + 6\left(-3 + \sqrt{5}\right) = 9 - 6\sqrt{5} + 5 - 18 + 6\sqrt{5} = -4$

31. $\frac{\sqrt{2xy}}{2xy}$ **33.** $\frac{1}{3}$ **35.** $\frac{x\sqrt{2x+1}}{2x+1}$

37. $\frac{x\sqrt{2x} - x}{2x - 1}$ **39.** $\frac{\sqrt{x^2 - 2x} + 2\sqrt{x - 2}}{x - 2}$

41. $5 + 2\sqrt{6}$ **43.** $\frac{x\sqrt{2} - \sqrt{6x}}{x - 3}$

45. $\frac{x\sqrt{3} + \sqrt{6x} - \sqrt{x} - \sqrt{2}}{x - 2}$ **47.** $\frac{\sqrt{x(x-1)} + \sqrt{x-1}}{x - 1}$

49. $\frac{x + 2\sqrt{x} + 1}{x - 1}$ **51.** $\frac{x\sqrt{2} - 3\sqrt{x} + \sqrt{2x} - 3}{2x - 9}$

53. a. $\frac{\sqrt[3]{x^2} - 2\sqrt[3]{x} + 4}{x + 8}$ **b.** $\frac{\sqrt[3]{x^2} + 3\sqrt[3]{x} + 9}{x - 27}$

EXERCISES 7.4 ■ page 409

1. ii. **3.** iii. **5.** 26 **7.** $\frac{23}{3}$ **9.** -2

11. 2 **13.** $\frac{5}{2}$ **15.** 32 **17.** 2 **19.** 1

21. $\frac{49}{16}$ **23.** None **25.** $\frac{33}{16}$ **27.** 19 **29.** 3 and 7

31. a. Two hours later, the automobiles are $40\sqrt{13}$ ($\approx$144.27) miles apart

b. In $\frac{5\sqrt{13}}{13}$ ($\approx$1.39) hours, they are 100 miles apart. The time is 11:23 A.M.

c. In $\frac{15\sqrt{13}}{13}$ ($\approx$4.16) hours, they are 300 miles apart. The time is 2:10 P.M.

33. $10\sqrt{91}$ ($\approx$95.4) feet

35. $200\sqrt{6}$ ($\approx$489.9) **37.** 9π ($\approx$28.3) in^2

39. $T = R^2 - 6R + 9$ **41.** $x = \frac{2}{z^2}$

43. $x = \sqrt{(y + 3)^2}$, $x = -\sqrt{(y + 3)^2}$, $y \geq -3$

EXERCISES 7.5 ■ page 418

1. $0 + 2i\sqrt{3}$ **3.** $0 + \left(3 - 3\sqrt{3} + 6\sqrt{2}\right)i$

5. $-5 - 2i$ **7.** $15 + 45i$ **9.** $-3 + 51i$

11. $-5 + 5i$ **13.** $2 + 0i$ **15.** $58 + 0i$

17. $-40 - 42i$ **19.** $-1 - 12i$ **21.** $\frac{2}{13} - \frac{3}{13}i$

23. $\frac{-7}{10} + \frac{1}{10}i$ **25.** $\frac{-8}{13} + \frac{1}{13}i$

27.
$$\begin{aligned}(2 + 2i)^2 - 4(2 + 2i) + 8 &= 4 + 8i + 4i^2 - 8 - 8i + 8\\ &= 4 - 4 - 8 + 8 + 8i - 8i\\ &= 0\end{aligned}$$

29.
$$\begin{aligned}(1 + i)^2 - 2(1 + i) + 2 &= 1 + 2i + i^2 - 2 - 2i + 2\\ &= 1 - 1 - 2 + 2 + 2i - 2i\\ &= 0\end{aligned}$$
$$\begin{aligned}(1 - i)^2 - 2(1 - i) + 2 &= 1 - 2i + i^2 - 2 + 2i + 2\\ &= 1 - 1 - 2 + 2 - 2i + 2i\\ &= 0\end{aligned}$$

31. $x^2 - 2x - 1 = 0$ **33.** $x^2 + 2x + 10 = 0$

35. $x^3 - 3x^2 + 3x - 3 = 0$

DISCOVERY ■ page 419

1. For a quadratic equation with integer coefficients, if $a + \sqrt{b}$ solves the equation, then its conjugate solves the equation.

CHAPTER 7 REVIEW ■ page 420

1. $3x^2y\sqrt{6y}$ **3.** $\frac{2\sqrt[3]{3x^2}}{3}$ **5.** $3xy\sqrt{3x}$ **7.** $\frac{\sqrt{5x}}{x}$

9. $xy\sqrt[3]{18x^2}$ **11.** $\frac{\sqrt[3]{45x^2}}{3x}$ **13.** $3xy^2\sqrt{2xy}$ **15.** $\frac{\sqrt{x}}{2x}$

17. $\frac{x\sqrt{2}}{2}$ **19.** $5x\sqrt{2x}$ **21.** $5x$ **23.** $-x\sqrt{x}$

25. $x - 8\sqrt{x} + 16$ **27.** $x - 3$

29. $x\sqrt{6} + 3\sqrt{2x} - \sqrt{6x} - 3\sqrt{2}$

31. $x + 3\sqrt[3]{x^2} - 3\sqrt[3]{x} - 9$ **33.** $\frac{\sqrt{6xy}}{3xy}$ **35.** 1

37. 1 and 2 **39.** None **41.** -22

43. None $\left(\frac{121}{36}\text{ is extraneous}\right)$ **45.** 9 (1 is extraneous)

47. 3 **49.** 17 **51.** 0 and 4 **53.** $\frac{72}{\pi^2}$ ($\approx$7.3) feet

55. $x = \frac{y^2 + 5}{3}$ **57.** $d = \frac{3}{g^2}$ **59.** $2\sqrt{3} - 6i\sqrt{5}$

61. $8 + 0i$ **63.** $-5 - 7i$ **65.** $-11 - 7i$

67. $34 + 0i$ **69.** $-21 - 20i$ **71.** $-15 - 8i$

73. $-1 - \frac{3}{2}i$ **75.** $3 + i$ **77.** $\frac{-12}{13} + \frac{5}{13}i$

79.
$$\begin{aligned}(1 - 2i)^2 - 2(1 - 2i) + 5 &= 1 - 4i + 4i^2 - 2 + 4i + 5\\ &= 1 - 4 - 2 + 5 - 4i + 4i\\ &= 0\end{aligned}$$

81. $(2+i)^2 - 4(2+i) + 5 = 4 + 4i + i^2 - 8 - 4i + 5$
$= 4 - 1 - 8 + 5 + 4i - 4i$
$= 0$
$(2-i)^2 - 4(2-i) + 5 = 4 - 4i + i^2 - 8 + 4i + 5$
$= 4 - 1 - 8 + 5 - 4i + 4i$
$= 0$

83. $x^2 - 6x + 7 = 0$

CHAPTER 7 TEST ■ page 421

1. $6xy^2z^2\sqrt{2xz}$ **2.** $\frac{\sqrt[3]{2x^2}}{2x^2}$

3. $x\sqrt{6} + 5\sqrt{3x} - 5\sqrt{2x} - 25$ **4.** $\frac{x+\sqrt{2x}}{x-2}$

5. $\frac{\sqrt{x}+\sqrt{y}}{x-y}$ **6.** $\frac{-59}{2}$ **7.** None

8. a. $-9 + 3i$ **b.** $-21 + i$ **c.** $-45 - 28i$
d. $58 + 0i$ **e.** $-1 + i$

9. $y = x + 2\sqrt{x} + 5$

10. $\frac{5\sqrt{41}}{41}$ ($\approx$0.78) hour

11. $(4 - 3i)^2 - 8(4 - 3i) + 25$
$= 16 - 24i + 9i^2 - 32 + 24i + 25$
$= 16 - 9 - 32 + 25 - 24i + 24i$
$= 0$
$(4 + 3i)^2 - 8(4 + 3i) + 25$
$= 16 + 24i + 9i^2 - 32 - 24i + 25$
$= 16 - 9 - 32 + 25 + 24i - 24i$
$= 0$

■ CHAPTER 8

EXERCISES 8.1 ■ page 432

1. c. **3.** c. **5.** b. **7.** 16; $(x-4)^2$

9. 25; $(x+5)^2$ **11.** $\frac{49}{4}$; $\left(x - \frac{7}{2}\right)^2$ **13.** $\frac{1}{9}$; $\left(x + \frac{1}{3}\right)^2$

15. -1 and 9 **17.** -5 and 3

19. $7 + \sqrt{19}$ and $7 - \sqrt{19}$

21. $-2 + 2\sqrt{5}$ and $-2 - 2\sqrt{5}$

23. $\frac{1+\sqrt{7}}{2}$ and $\frac{1-\sqrt{7}}{2}$ **25.** 1 and $\frac{-3}{2}$

27. $\frac{7+3\sqrt{3}}{2}$ and $\frac{7-3\sqrt{3}}{2}$ **29.** $3 + \sqrt{13}$ and $3 - \sqrt{13}$

31. $\frac{5+\sqrt{33}}{2}$ and $\frac{5-\sqrt{33}}{2}$ **33.** $2 + \sqrt{35}$ and $2 - \sqrt{35}$

35. $-5 + \sqrt{38}$ and $-5 - \sqrt{38}$ **37.** $1 + \sqrt{6}$ and $1 - \sqrt{6}$

39. $-1 + \sqrt{5}$ and $-1 - \sqrt{5}$ **41.** $\frac{3+\sqrt{29}}{2}$ and $\frac{3-\sqrt{29}}{2}$

43. $\frac{2+\sqrt{6}}{2}$ and $\frac{2-\sqrt{6}}{2}$ **45.** $\frac{6+\sqrt{42}}{2}$ and $\frac{6-\sqrt{42}}{2}$

47. $\frac{9+\sqrt{237}}{6}$ and $\frac{9-\sqrt{237}}{6}$ **49.** $\frac{5+\sqrt{13}}{2}$ and $\frac{5-\sqrt{13}}{2}$

51. $\frac{1+i\sqrt{14}}{3}$ and $\frac{1-i\sqrt{14}}{3}$

53. $-3 + 3\sqrt{2}$ and $-3 - 3\sqrt{2}$ **55.** All real numbers

57. 5 **59.** $4i$ and $-4i$ **61.** No solution

EXERCISES 8.2 ■ page 440

1. True **3.** False **5.** True **7.** True **9.** True

11. $a = 1, b = 6, c = 1$ discriminant $= 32$
$-3 + 2\sqrt{2}$ and $-3 - 2\sqrt{2}$

13. $a = 1, b = 3, c = -1$ discriminant $= 13$
$\frac{-3+\sqrt{13}}{2}$ and $\frac{-3-\sqrt{13}}{2}$

15. $a = 1, b = -5, c -10$ discriminant $= 65$
$\frac{5+\sqrt{65}}{2}$ and $\frac{5-\sqrt{65}}{2}$

17. $a = 1, b = 4, c = -9$ discriminant $= 52$
$-2 + \sqrt{13}$ and $-2 - \sqrt{13}$

19. $a = 1, b = -4, c = -13$ discriminant $= 68$
$2 + \sqrt{17}$ and $2 - \sqrt{17}$

21. $a = 2, b = -1, c = -1$ discriminant $= 9$
1 and $\frac{-1}{2}$

23. $a = 2, b = 3, c = -3$ discriminant $= 33$
$\frac{-3+\sqrt{33}}{4}$ and $\frac{-3-\sqrt{33}}{4}$

25. $a = 1, b = -1, c = -4$ discriminant $= 17$
$\frac{1+\sqrt{17}}{2}$ and $\frac{1-\sqrt{17}}{2}$

27. $\frac{1}{2} + \frac{i\sqrt{19}}{2}$ and $\frac{1}{2} - \frac{i\sqrt{19}}{2}$

29. $\frac{1}{2} + \frac{3i\sqrt{3}}{2}$ and $\frac{1}{2} - \frac{3i\sqrt{3}}{2}$ **31.** $2 + 3i$ and $2 - 3i$

33. $\frac{1}{4} + \frac{i\sqrt{119}}{4}$ and $\frac{1}{4} - \frac{i\sqrt{119}}{4}$ **35.** $-1 + i$ and $-1 - i$

37. $\frac{-1}{2}$ and 1 **39.** $\frac{-5-\sqrt{21}}{2}$ and $\frac{-5+\sqrt{21}}{2}$

41. $\frac{-5+\sqrt{109}}{6}$ and $\frac{-5-\sqrt{109}}{6}$

43. $\frac{-6+\sqrt{186}}{5}$ and $\frac{-6-\sqrt{186}}{5}$

45. $\frac{3+\sqrt{37}}{2}$ and $\frac{3-\sqrt{37}}{2}$ **47.** $\frac{9+\sqrt{417}}{12}$ and $\frac{9-\sqrt{417}}{12}$

49. $\frac{-3+\sqrt{5}}{2}$ and $\frac{-3-\sqrt{5}}{2}$ **51.** 0 and $\frac{-1}{2}$ **53.** 0

EXERCISES 8.3 ■ page 449

1. 1 and -1 **3.** $\sqrt{2}$ and $-\sqrt{2}$

5. $\sqrt{\frac{1+\sqrt{5}}{2}}$ and $-\sqrt{\frac{1+\sqrt{5}}{2}}$ **7.** $\frac{-1}{2}$ and $\frac{1}{5}$

9. $\frac{-1}{2}$ and $\frac{1}{3}$ **11.** 25 **13.** $\frac{7+\sqrt{13}}{2}$ **15.** -1 and 64

17. $\frac{5+\sqrt{61}}{2}$ and $\frac{5-\sqrt{61}}{2}$ **19.** $\sqrt{5}$ and $-\sqrt{5}$

21. No real number solutions **23.** No solution
25. 0 and 4 **27.** 0, 4, $2 + i$, and $2 - i$

DISCOVERY ■ page 451

1. The key points can be found by changing the inequality to its related equation, and solving. Key points also occur where the expression is undefined (the restricted values). For $\frac{1}{x-3} \geq 1$, the key points are 4 and 3. So the solution is the interval (3, 4]. On the graphics calculator type 1/(X−3) for Y1 and 1 for Y2. Viewing the graph screen, the evaluations of Y1 are greater than or equal to 1 for x values in the interval (3, 4].

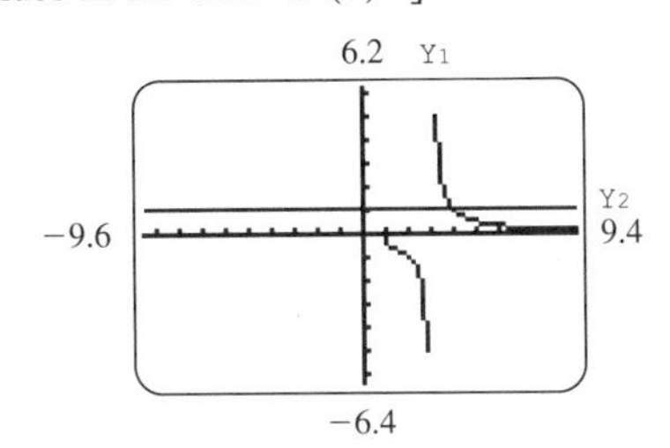

EXERCISES 8.4 ■ page 458

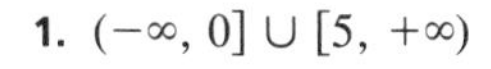

1. $(-\infty, 0] \cup [5, +\infty)$

3. $(-\infty, -3) \cup (2, +\infty)$

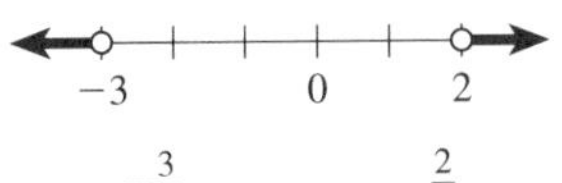

5. $\left[\frac{-3}{2}, \frac{2}{3}\right]$

7. $(-\infty, -1) \cup (5, +\infty)$

9. $\left(-\infty, \frac{-1-3\sqrt{5}}{2}\right) \cup \left(\frac{-1+3\sqrt{5}}{2}, +\infty\right)$

11. $(-\infty, +\infty)$

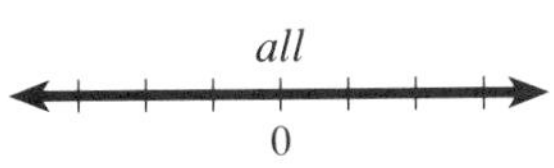

13. $\left(-\infty, \frac{-1}{2}\right] \cup [1, +\infty)$

15. $(-\infty, +\infty)$

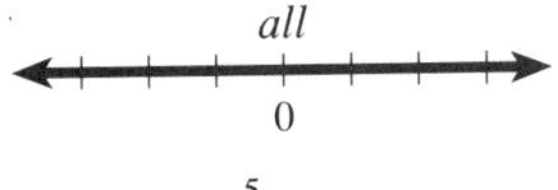

17. $(-\infty, -4] \cup \left[\frac{-5}{2}, +\infty\right)$

19. $\left(-\infty, \frac{1-\sqrt{13}}{2}\right) \cup \left(\frac{1+\sqrt{13}}{2}, +\infty\right)$

21. $(-\infty, -1) \cup \left[\frac{1}{2}, +\infty\right)$

23. $(-\infty, -2] \cup (4, +\infty)$

25. $\left(0, \frac{8}{7}\right) \cup (2, +\infty)$

27. $(-2, -1] \cup (1, +\infty)$

29. $(-\infty, -3] \cup (-2, 1]$

EXERCISES 8.5 ■ page 466

1. $3 + 3\sqrt{2}$ **3.** $\frac{3 + \sqrt{89}}{4}$ feet

5. $\frac{88 + 6\sqrt{359}}{5}$ (≈40.34) cm

7. $\frac{-3 + \sqrt{329}}{2}$ (≈7.57) cm

9. $-2 + \sqrt{21}$ or $-2 - \sqrt{21}$

11. $\frac{8 + \sqrt{14}}{4} \approx 2.94$ sec and $\frac{8 - \sqrt{14}}{4} \approx 1.06$ sec. The ball hit the ground in 4 sec.

13. $x = \frac{1 + \sqrt{1 + 4y}}{2}$ and $x = \frac{1 - \sqrt{1 + 4y}}{2}$

15. $R = \frac{5 + \sqrt{25 - 4P}}{2}$ and $R = \frac{5 - \sqrt{25 - 4P}}{2}$

17. $W = \frac{3 + \sqrt{9 + 4R}}{2}$ and $W = \frac{3 - \sqrt{9 + 4R}}{2}$

19. $y = \sqrt{x^2 - 16}$ and $y = -\sqrt{x^2 - 16}$

21. $y = -1 + \sqrt{9 - x^2}$ and $y = -1 - \sqrt{9 - x^2}$

23. $y = x(x - 2)$, $x \neq 2$, $x \neq 0$

DISCOVERY ■ page 469

1. The constant, a, determines whether the graph is opening up or down and also determines the distortion of the graph (how wide it is). The constant h determines the horizontal translation. The constant k determines the vertical translation. The vertex is at (h, k).

EXERCISES 8.6 ■ page 481

1. vertex $(0, 1)$

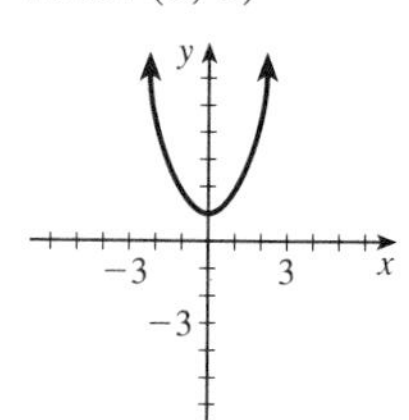

3. vertex $(4, 0)$

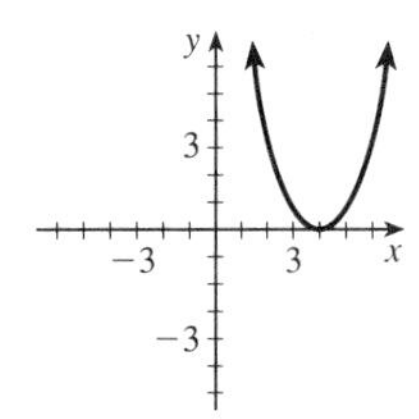

5. vertex $(5, 0)$

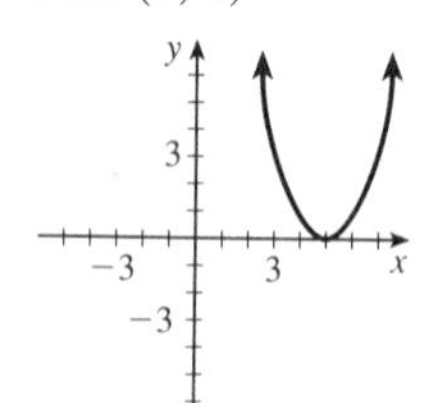

7. vertex $(2, 1)$

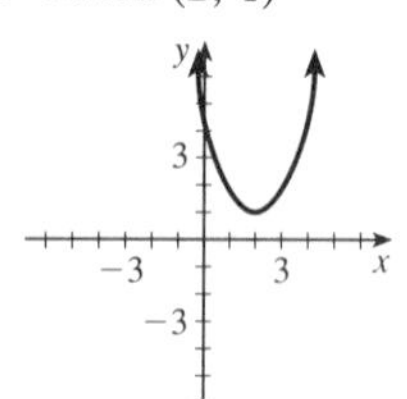

9. vertex $(-3, -2)$

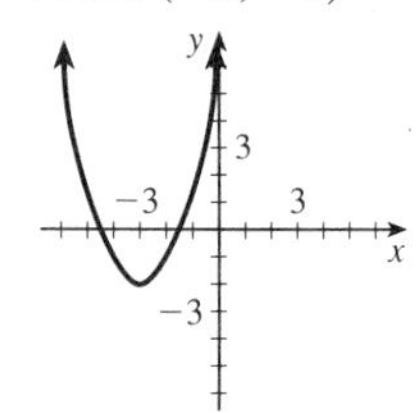

11. C **13.** B

15. $d(x) = (x - 2)^2 - 3$
vertex $(2, -3)$

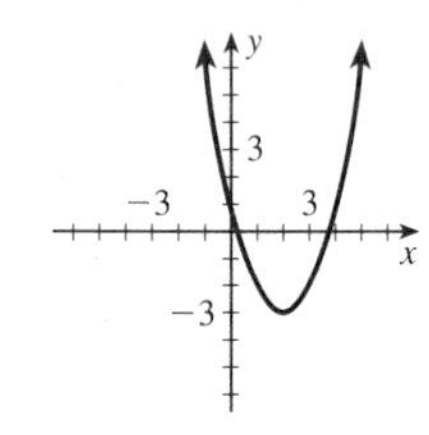

17. $d(x) = (x + 1)^2 + 2$
vertex $(-1, 2)$

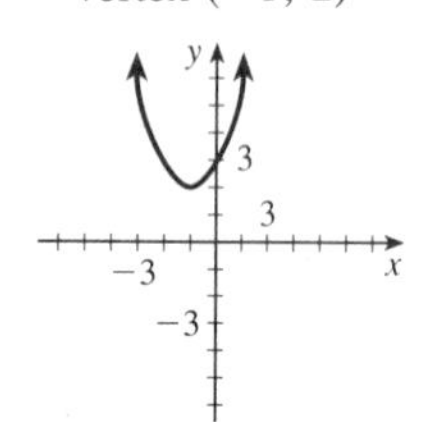

19. $d(x) = \left(x - \frac{3}{2}\right)^2 - \frac{1}{4}$
vertex $\left(\frac{3}{2}, \frac{-1}{4}\right)$

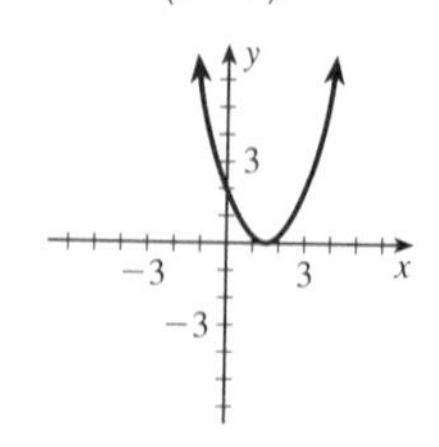

21. **a.** y-intercept $(0, -2)$
b. x-intercepts $\left(-1 + \sqrt{3}, 0\right), \left(-1 - \sqrt{3}, 0\right)$
c. vertex $(-1, -3)$
d. line of symmetry $x = -1$
e.

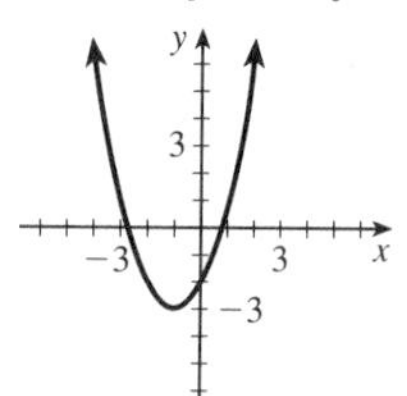

23. **a.** y-intercept $(0, -1)$
b. x-intercepts $\left(\frac{3 + \sqrt{13}}{2}, 0\right), \left(\frac{3 - \sqrt{13}}{2}, 0\right)$
c. vertex $\left(\frac{3}{2}, \frac{-13}{4}\right)$
d. line of symmetry $x = \frac{3}{2}$
e.

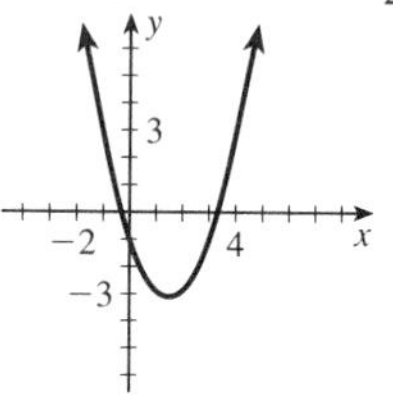

25. **a.** y-intercept $(0, -4)$
b. x-intercepts $\left(2 + 2\sqrt{2}, 0\right), \left(2 - 2\sqrt{2}, 0\right)$
c. vertex $(2, -8)$
d. line of symmetry $x = 2$
e.

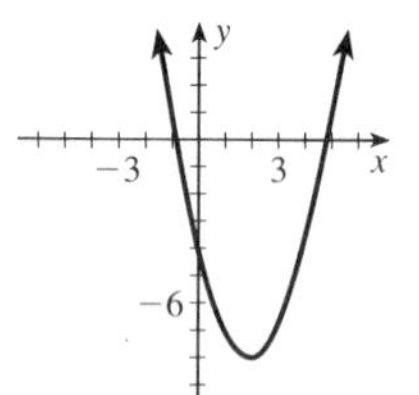

27. **a.** y-intercept $(0, -3)$
b. x-intercepts $\left(3 + 2\sqrt{3}, 0\right), \left(3 - 2\sqrt{3}, 0\right)$
c. vertex $(3, -12)$
d. line of symmetry $x = 3$
e.

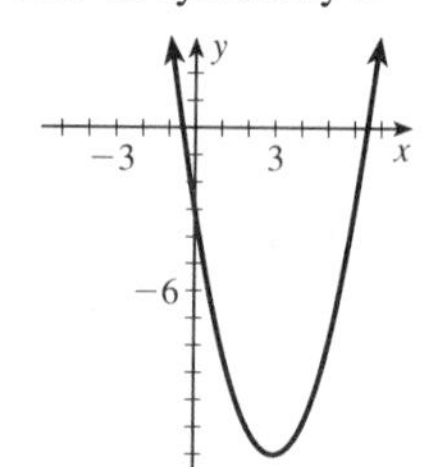

29. **a.** y-intercept $(0, -1)$
b. x-intercepts $\left(\frac{3+\sqrt{11}}{2}, 0\right), \left(\frac{3-\sqrt{11}}{2}, 0\right)$
c. vertex $\left(\frac{3}{2}, \frac{-11}{2}\right)$
d. line of symmetry $x = \frac{3}{2}$
e.

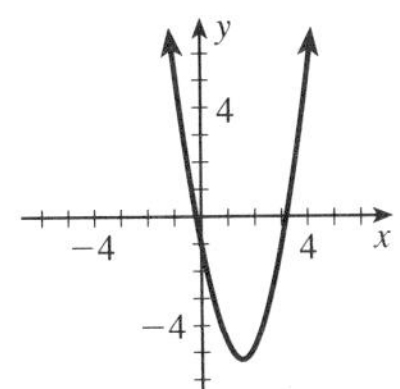

31. **a.** y-intercept $(0, -2)$
b. x-intercepts $(2, 0), (-1, 0)$
c. vertex $\left(\frac{1}{2}, \frac{-9}{4}\right)$
d. line of symmetry $x = \frac{1}{2}$
e.

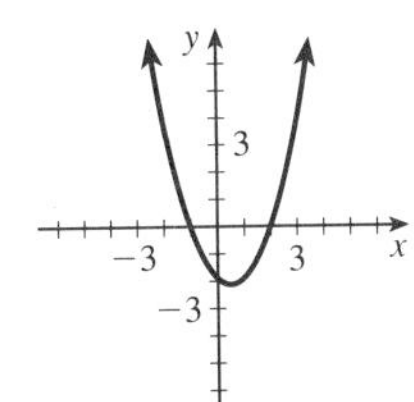

33. **a.** y-intercept $(0, -3)$
b. x-intercepts $\left(\frac{3}{2}, 0\right), (-1, 0)$
c. vertex $\left(\frac{1}{4}, \frac{-25}{8}\right)$
d. line of symmetry $x = \frac{1}{4}$
e.

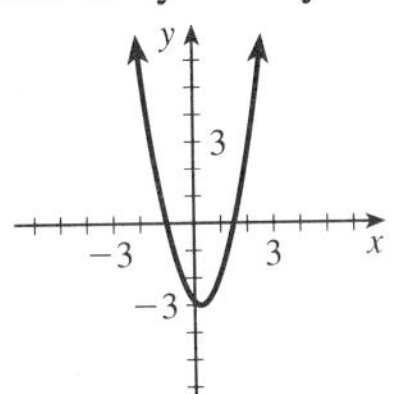

35. **a.** y-intercept $(0, 6)$
b. x-intercepts $(3, 0), (-2, 0)$
c. vertex $\left(\frac{1}{2}, \frac{25}{4}\right)$
d. line of symmetry $x = \frac{1}{2}$
e.

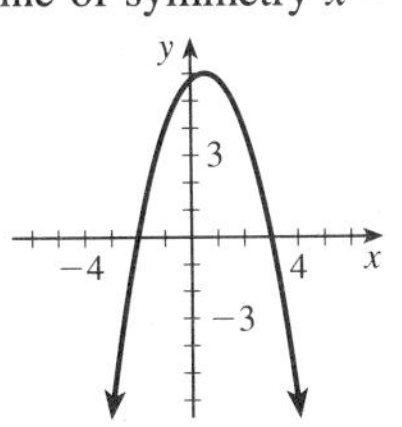

37. **a.** y-intercept $(0, 3)$
b. x-intercepts $(\sqrt{3}, 0), (-\sqrt{3}, 0)$
c. vertex $(0, 3)$
d. line of symmetry $x = 0$
e.

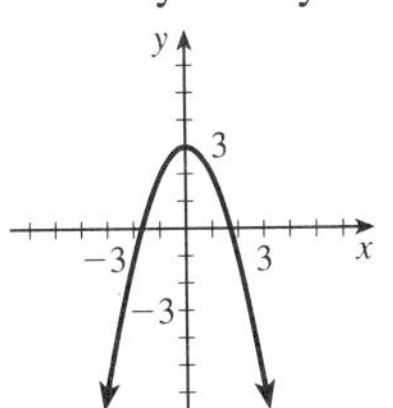

39. **a.** y-intercept $(0, -1)$
b. x-intercepts $\left(\frac{-3+\sqrt{5}}{2}, 0\right), \left(\frac{-3-\sqrt{5}}{2}, 0\right)$
c. vertex $\left(\frac{-3}{2}, \frac{5}{4}\right)$
d. line of symmetry $x = \frac{-3}{2}$
e.

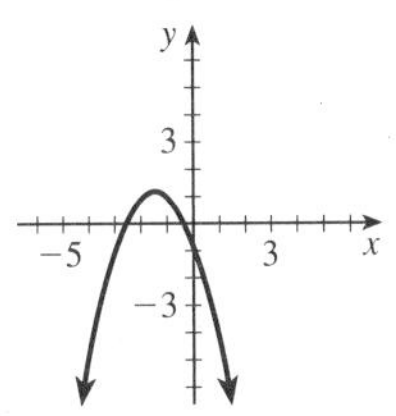

41. **a.** y-intercept $(0, -2)$
b. x-intercepts $\left(\frac{-1}{2}, 0\right), (-2, 0)$
c. vertex $\left(\frac{-5}{4}, \frac{9}{8}\right)$
d. line of symmetry $x = \frac{-5}{4}$
e.

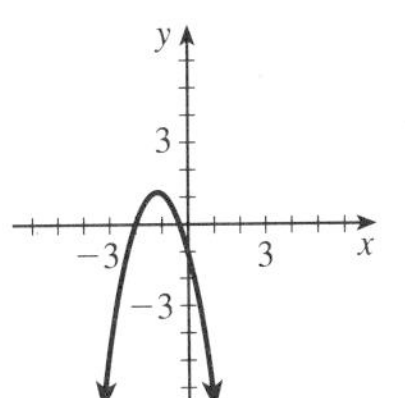

EXERCISES 8.7 ■ page 493

1. F **3.** D **5.** A

7. **a.**

x	−3	−2	−1	0	1	2
$f(x)$	0	6	4	0	0	10

b. y-intecept $(0, 0)$
c. x-intercepts
$(-3, 0), (0, 0), (1, 0)$
d. See graph at right.

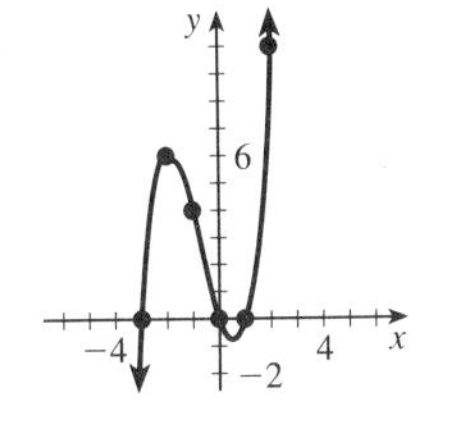

9.

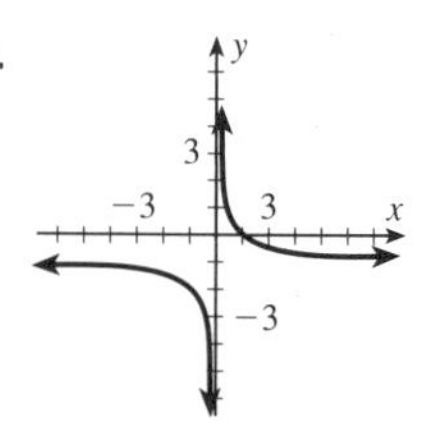

x-intercept (1, 0)
y-intercept none
domain: $\{x \mid x \neq 0\}$

11.

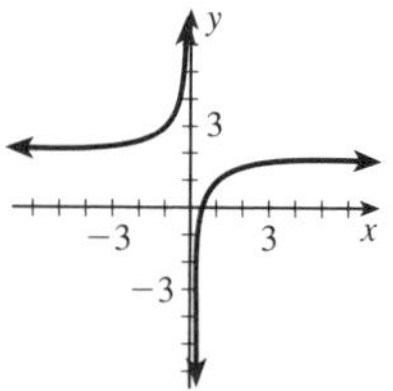

x-intercept $\left(\frac{1}{2}, 0\right)$
y-intercept none
domain: $\{x \mid x \neq 0\}$

13.

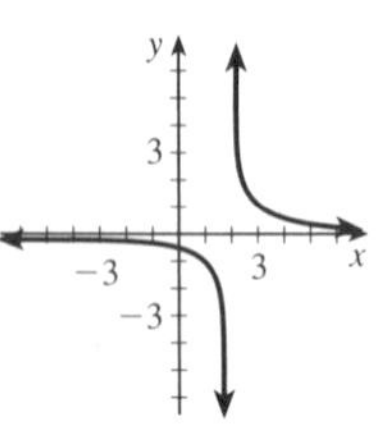

x-intercept none
y-intercept $\left(0, \frac{-1}{2}\right)$
domain: $\{x \mid x \neq 2\}$

15.

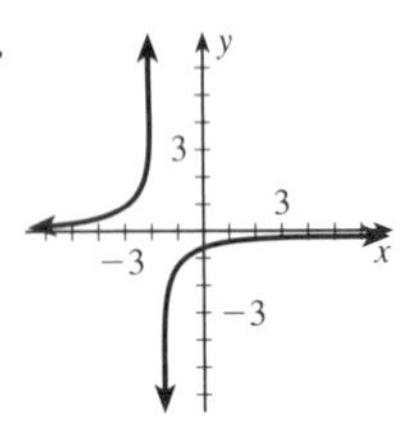

x-intercept none
y-intercept $\left(0, \frac{-1}{2}\right)$
domain: $\{x \mid x \neq -2\}$

17. $d(x) = 2 + \frac{1}{x}$

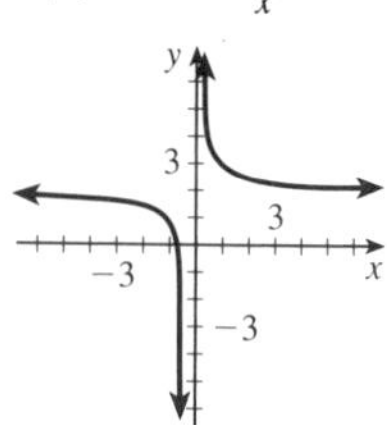

x-intercept $\left(\frac{-1}{2}, 0\right)$
y-intercept none

19. $m(x) = 2 + \frac{3}{x - 1}$

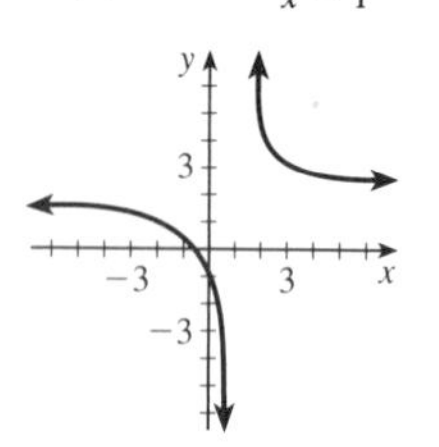

x-intercept $\left(\frac{-1}{2}, 0\right)$
y-intercept $(0, -1)$

21.

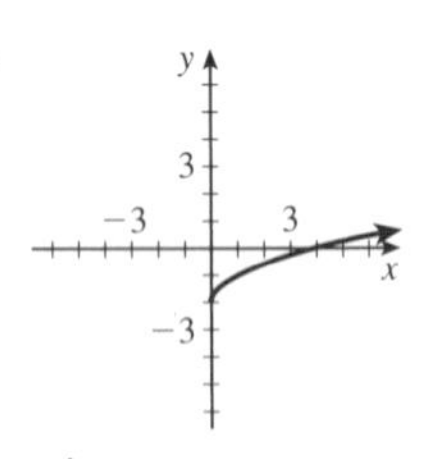

x-intercept (4, 0)
y-intercept $(0, -2)$
domain: $\{x \mid x \geq 0\}$
range: $\{y \mid y \geq -2\}$

23.

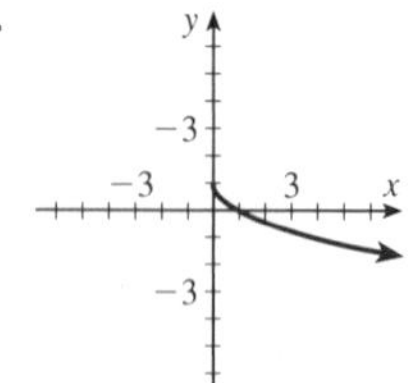

x-intercept (1, 0)
y-intercept (0, 1)
domain: $\{x \mid x \geq 0\}$
range: $\{y \mid y \leq 1\}$

25.

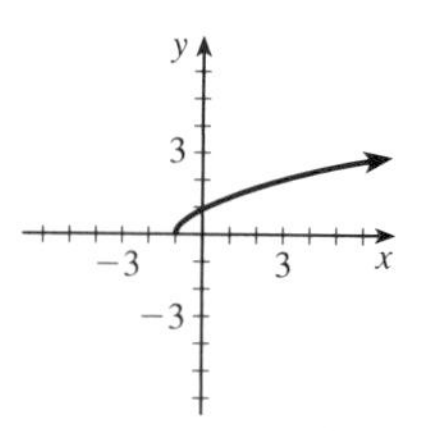

x-intercept $(-1, 0)$
y-intercept (0, 1)
domain: $\{x \mid x \geq -1\}$
range: $\{y \mid y \geq 0\}$

27.

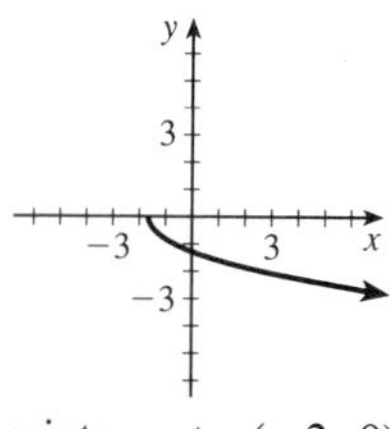

x-intercept $(-2, 0)$
y-intercept $\left(0, -\sqrt{2}\right)$
domain: $\{x \mid x \geq -2\}$
range: $\{y \mid y \leq 0\}$

29.

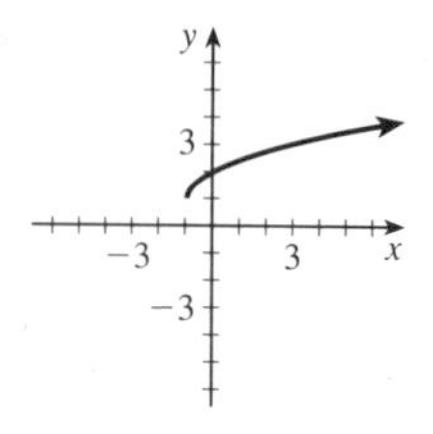

x-intercept none
y-intercept (0, 2)
domain: $\{x \mid x \geq -1\}$
range: $\{y \mid y \geq 1\}$

CHAPTER 8 REVIEW ■ page 494

1. $x^2 - 8x + 16$ or $(x - 4)^2$
3. $x^2 + 12x + 36$ or $(x + 6)^2$
5. $x^2 - 5x + \frac{25}{4}$ or $\left(x - \frac{5}{2}\right)^2$
7. $(x - 4)^2 = 25$ solution: $-1, 9$
9. $(x + 1)^2 = 24$ solution: $-1 + 2\sqrt{6}, -1 - 2\sqrt{6}$
11. $\left(x - \frac{3}{2}\right)^2 = \frac{33}{4}$ solution: $\frac{3 + \sqrt{33}}{2}, \frac{3 - \sqrt{33}}{2}$
13. $(x - 1)^2 = \frac{5}{2}$ solution: $1 + \frac{\sqrt{10}}{2}, 1 - \frac{\sqrt{10}}{2}$
15. $a = 2, b = -1, c = -3$
discriminant $= 25$ solution: $\frac{3}{2}$, and -1
17. $a = 1, b = -1, c = -13$
discriminant $= 53$ solution: $\frac{1 + \sqrt{53}}{2}$, and $\frac{1 - \sqrt{53}}{2}$
19. $a = 4, b = -5, c = 1$
discriminant $= 9$ solution: $\frac{1}{4}$, and 1
21. $a = 2, b = -8, c = 0$
discriminant $= 64$ solution: 0, and 4
23. $-\sqrt{5}, \sqrt{5}, -i$, and i
25. $\sqrt{\frac{1 + \sqrt{5}}{2}}, -\sqrt{\frac{1 + \sqrt{5}}{2}}, i\sqrt{\frac{\sqrt{5} - 1}{2}}, -i\sqrt{\frac{\sqrt{5} - 1}{2}}$
27. $\frac{1}{2}$ and -1 **29.** 9 (4 is extraneous)
31. $2 + \sqrt{10}$, and $2 - \sqrt{10}$ **33.** $-\sqrt{6}$, and $\sqrt{6}$
35. 1 **37.** $\frac{1}{4}$

39. $(-\infty, 0] \cup [2, +\infty)$

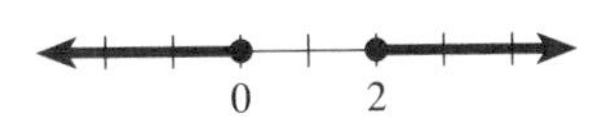

41. $\left(-\infty, 3 - \sqrt{2}\right) \cup \left(3 + \sqrt{2}, +\infty\right)$

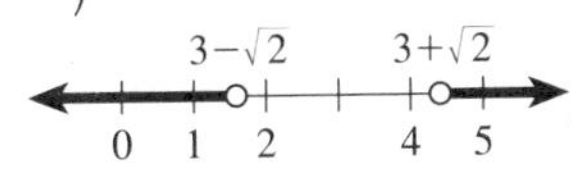

43. $(-\infty, -4) \cup (3, +\infty)$

45. $\varnothing$

none

47. $(-\infty, -1] \cup [2, +\infty)$

49. $(-\infty, 1) \cup (2, +\infty)$

51. $(-\infty, -8] \cup (4, +\infty)$

53. $(-\infty, 0) \cup (2, 8)$

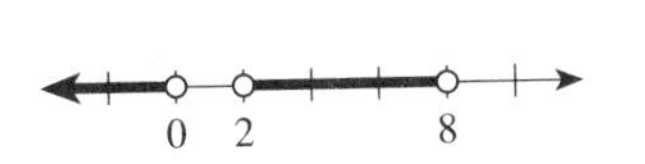

55. $\frac{-3 + \sqrt{249}}{2}$ (≈ 6.39) inches

57. $7 - \sqrt{39}$ (≈ 0.755) inch

59. $y = -1 + \sqrt{25 - x^2}$ and $y = -1 - \sqrt{25 - x^2}$

61. $x = \frac{3 + \sqrt{9 + 4y}}{2}$ and $x = \frac{3 - \sqrt{9 + 4y}}{2}$

63. vertex $(0, -4)$

65. vertex $(-1, 0)$

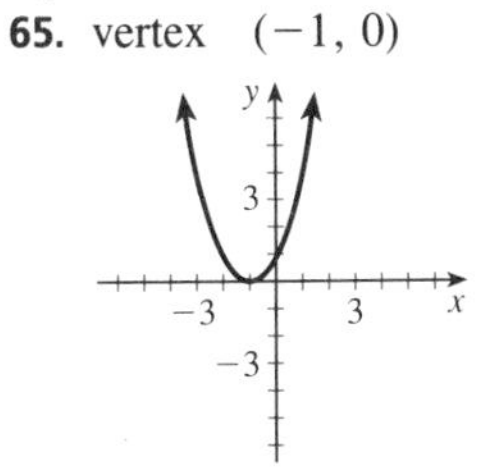

67. vertex $(-2, 1)$

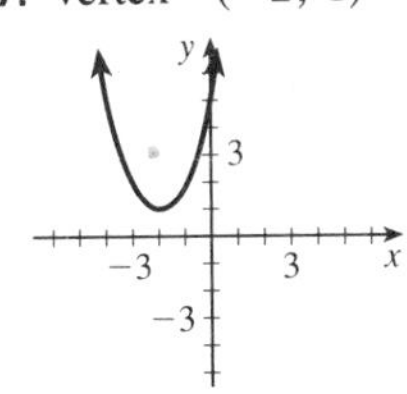

69. $d(x) = (x - 2)^2 - 2$
vertex $(2, -2)$

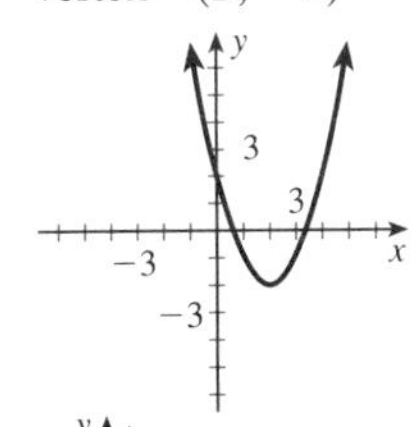

71. $d(x) = (x + 2)^2 - 2$
vertex $(-2, -2)$

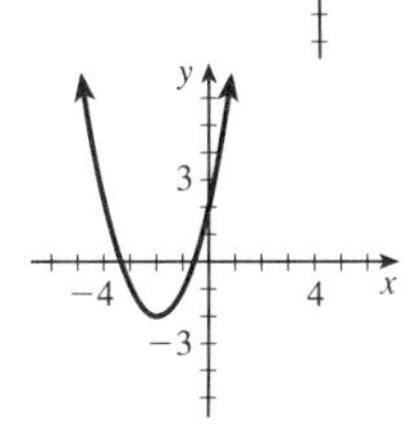

73.

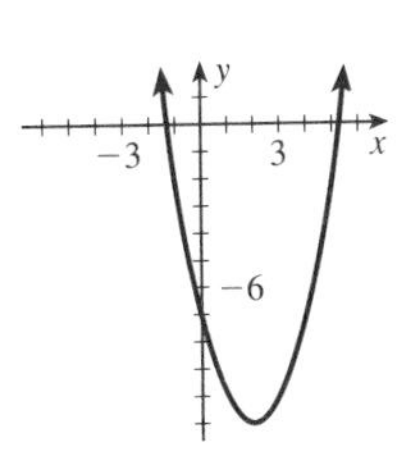

y-intercept $(0, -7)$
x-intercept
$\left(2 + \sqrt{11}, 0\right), \left(2 - \sqrt{11}, 0\right)$
line of symmetry $x = 2$
vertex $(2, -11)$

75.

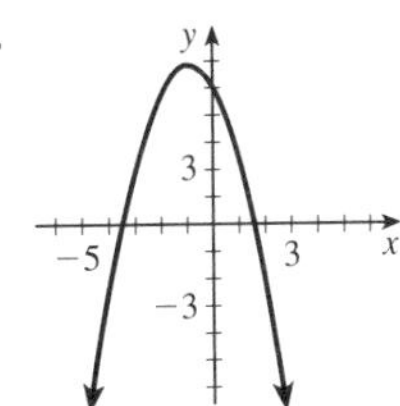

y-intercept $(0, 5)$
x-intercept
$\left(-1 + \sqrt{6}, 0\right)$ and $\left(-1 - \sqrt{6}, 0\right)$
line of symmetry $n = -1$
vertex $(-1, 6)$

77.

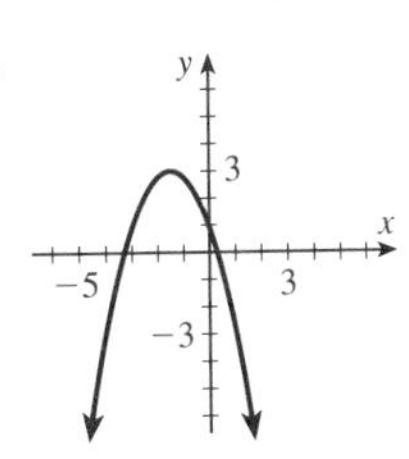

y-intercept $(0, 1)$
x-intercept
$\left(\frac{-3 + \sqrt{13}}{2}, 0\right)$ or $\left(\frac{-3 - \sqrt{13}}{2}, 0\right)$
line of symmetry $x = \frac{-3}{2}$
vertex $\left(\frac{-3}{2}, \frac{13}{4}\right)$

79.

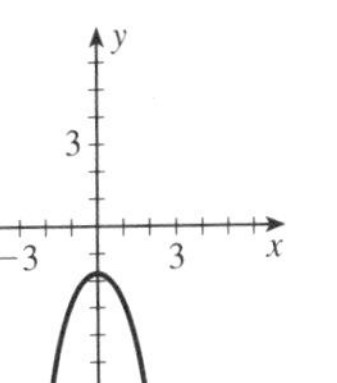

y-intercept $(0, -2)$
x-intercept none
line of symmetry $x = \frac{-1}{4}$
vertex $\left(\frac{-1}{4}, \frac{-15}{8}\right)$

81. B **83.** A

85.

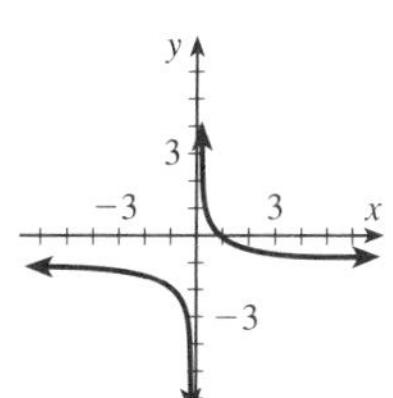

x-intercept $(1, 0)$
y-intercept none
domain: $\{x \mid x \neq 0\}$

87.

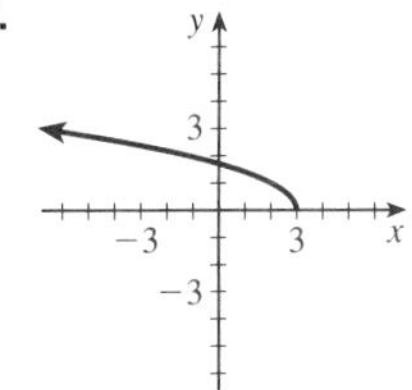

x-intercept $(3, 0)$
y-intercept $\left(0, \sqrt{3}\right)$
domain: $\{x \mid x \leq 3\}$

CHAPTER 8 TEST ■ page 496

1. $(x - 2)^2 = 6$
$2 + \sqrt{6}$ and $2 - \sqrt{6}$

2. Standard form: $2x^2 - x - 5 = 0$
$a = 2, b = -1, c = -5$
discriminant $= 41$ solution: $\frac{1 + \sqrt{41}}{4}$ and $\frac{1 - \sqrt{41}}{4}$

3. a. $u = x^{-2}$

solution: $1, -1, \frac{i\sqrt{2}}{2}, \frac{-i\sqrt{2}}{2}$

b. $u = x^{1/2}$ solution: 4

4. -6 and 0

5. a. $\left(-\infty, \frac{3 - \sqrt{13}}{2}\right] \cup \left[\frac{3 + \sqrt{13}}{2}, +\infty\right)$

b. $(-\infty, 1) \cup [2, +\infty)$

6. $(3, 5)$

7. y-intercept $(0, 3)$
x-intercept
$\left(-2 + \sqrt{7}, 0\right), \left(-2 - \sqrt{7}, 0\right)$
line of symmetry $x = -2$
vertex $(-2, 7)$

8.

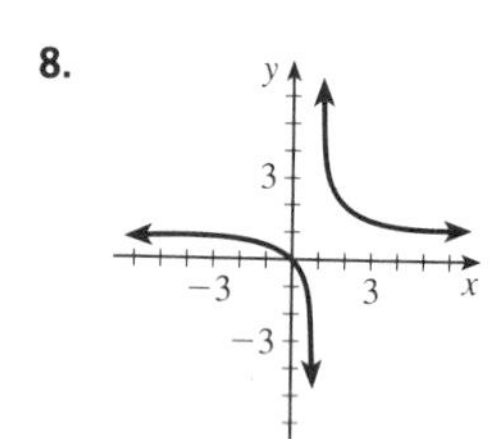

x-intercept $(0, 0)$
y-intercept $(0, 0)$
domain $\{x \mid x \neq 1\}$

■ CHAPTER 9

EXERCISES 9.1 ■ page 503

1. Not a function
0 maps to both -1 and -2

3.

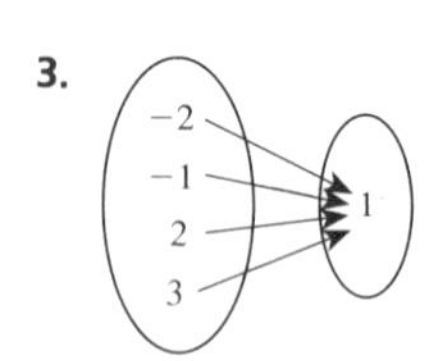

Is a function
domain: $\{-2, -1, 2, 3\}$
range: $\{1\}$

5. Not a function h maps to both r and s
7. Is a function range: $\{1, 3\}$

9. 10 **11.** $-3a - 14$ **13.** $-6b + 1$ **15.** $-3h$
17. 0 **19.** $r^2 + 2r - 3$ **21.** 5 **23.** $h^2 - 4h$
25. -1 **27.** $\frac{1}{2}$ **29.** $\frac{1}{2a + 1}$ **31.** $\frac{2}{(k + 3)(k + 2)}$
33. 1 **35.** $\sqrt{a^2 - 2}$ **37.** 3 **39.** $\sqrt{h} - \sqrt{h - 2}$
41. -3 **43.** $h - 2$ **45.** $\frac{-1}{h + 1}$ **47.** $h^2 - 3h + 3$

EXERCISES 9.2 ■ page 510

1. Not a function **3.** Not a function **5.** 2 **7.** 0
9. Zeros at $x = -1$ and $x = 1$ **11.** 2
13. -2 **15.** Zeros at $x = \frac{1}{2}$ and $x = 3$ **17.** -2
19. -1 **21.** $(-1, 1)$ **23.** -2 **25.** 3 **27.** $(-2, 1)$
29. Domain: $\{x \mid x \text{ is a real number}\}$
range: $\{y \mid y \leq 2\}$
turning points none
zeros at $x = \frac{2}{3}$ and $x = 3$
31. Domain: $\{x \mid -4 \leq x < -1\} \cup \{x|x \geq 0\}$
range: $\{y \mid y \geq -1\}$
turning points none
zeros at $x = 1$
33. Domain: $\{x \mid -4 \leq x \leq 4\}$
range: $\left\{y \mid \frac{1}{5} \leq y \leq 1\right\} \cup \left\{y \mid \sqrt{3} < y \leq \sqrt{7}\right\}$
turning points none
zeros none

EXERCISES 9.3 ■ page 521

1. -2 **3.** 228 **5.** $\frac{3}{4}$ **7.** $6a + 1$ **9.** $\frac{-2a - 1}{-a + 2}$
11. -4 **13.** $\frac{4}{5}$ **15.** Undefined **17.** $\frac{1}{2a} + 2a - 2$
19. $\frac{1}{a^2 + 2a}$ **21.** $\sqrt{2} - 1$ **23.** $-\sqrt{13}$ **25.** $-\sqrt{5}$
27. $\sqrt{2a + 3} - 1$ **29.** $-\sqrt{3 - a}$ **31.** $\sqrt{5}$ **33.** 2
35. $7 + 2\sqrt{7}$ **37.** Not a real number, $2i\sqrt{2}$ **39.** 0
41. Not a real number, $2i$
43. a. $x^2 - 2x + 2$ **b.** x^2
45. $(f \circ g)(x) = \sqrt{x} - 3$

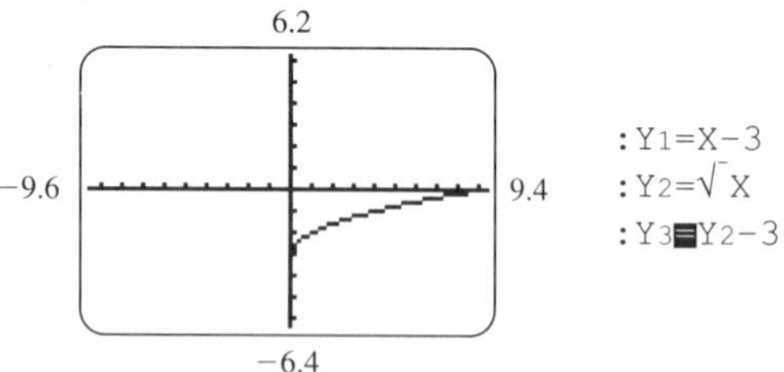

domain: $\{x \mid x \geq 0\}$
range: $\{y \mid y \geq -3\}$
y-intercept $(0, -3)$
x-intercept $(9, 0)$

$(g \circ f)(x) = \sqrt{x - 3}$

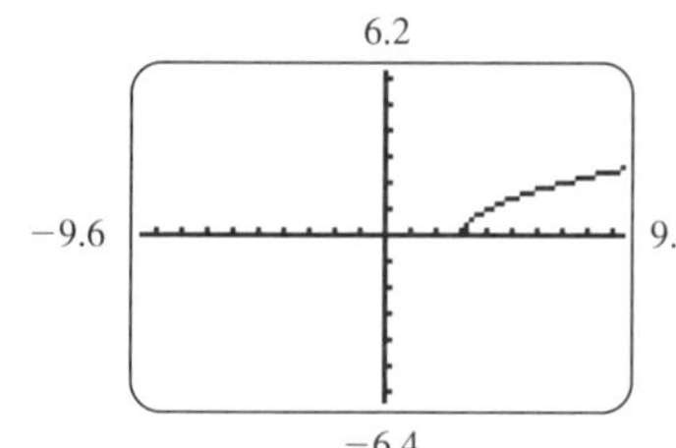

$:Y_1 = X - 3$
$:Y_2 = \sqrt{X}$
$:Y_3 = Y_2 - 3$
$:Y_4 \blacksquare \sqrt{Y_1}$

domain: $\{x \mid x \geq 3\}$
range: $\{y \mid y \geq 0\}$
y-intercept none
x-intercept $(3, 0)$

47. $(f \circ g)(x) = \sqrt{x^2 - 2x + 1} + 1$

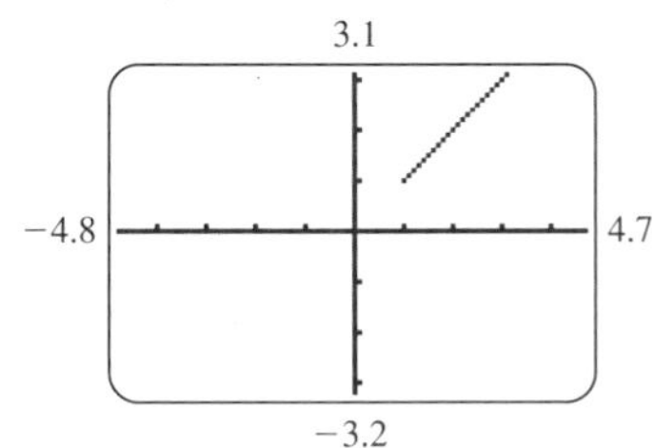

$:Y_1 = \sqrt{X} + 1$
$:Y_2 = (X^2 - 2X + 1)(X \geq 1)$
$:Y_3 \blacksquare (\sqrt{Y_2} + 1)(X \geq 1)$

domain: $\{x \mid x \geq 1\}$
range: $\{y \mid y \geq 1\}$
y-intercept none
x-intercept none

$(g \circ f)(x) = x$

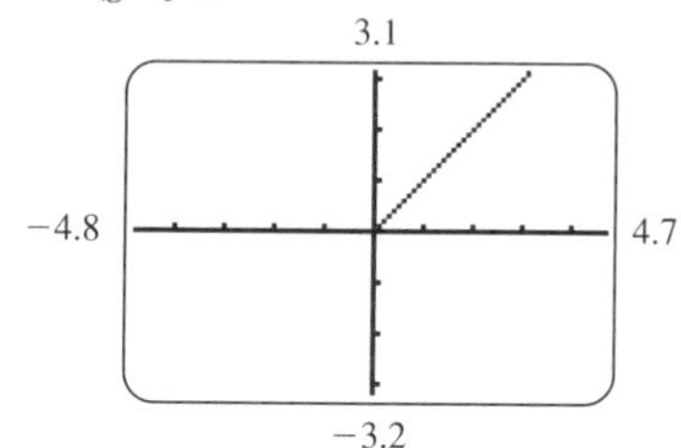

$:Y_1 = \sqrt{X} + 1$
$:Y_2 = (X^2 - 2X + 1)(X \geq 1)$
$:Y_3 = (\sqrt{Y_2} + 1)(X \geq 1)$
$:Y_4 \blacksquare Y_1{}^2 - 2Y_1 + 1$

domain: $\{x \mid x \geq 0\}$
range: $\{y \mid y \geq 0\}$
y-intercept $(0, 0)$
x-intercept $(0, 0)$

49. $\{(-2, 0), (-1, 3), (0, 2), (1, 4), (2, -3)\}$ **51.** -2
53. -4 **55.** $(f \circ f)(x) = f(x^2 - 2) = x^4 - 4x^2 + 2$
57. $b^2 + b - 6$
59. a. $A(t) = \pi(r + 2)^2$ **b.** $-2 + \dfrac{10\sqrt{2\pi}}{\pi} \approx 5.98$ minutes

EXERCISES 9.4 ■ page 530

1. Function, not a 1-1 function.
3. Function, not a 1-1 function.
5. Function, not a 1-1 function.
7. $\{(2, -3), (3, -2), (5, -1), (6, 0)\}$

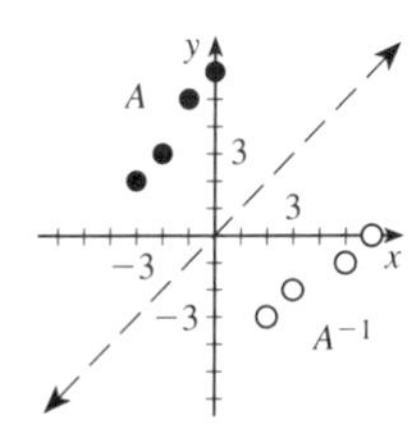

9. $\{(-2, -4), (2, -2), (0, -1), (-3, 0)\}$

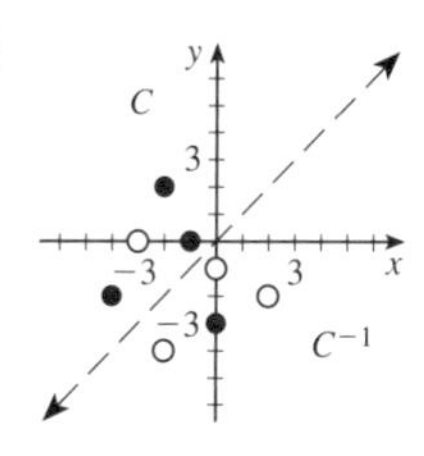

11.

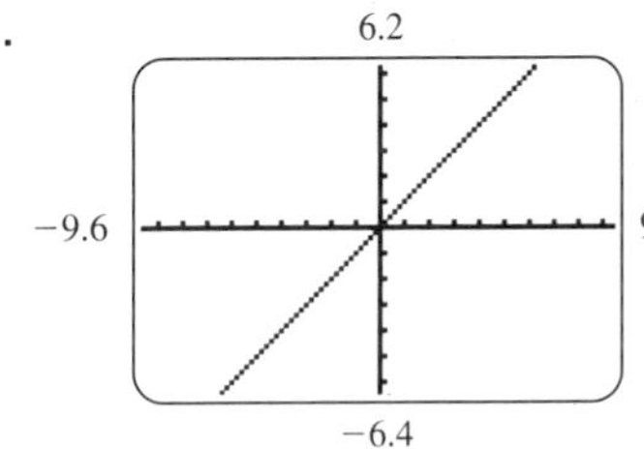

$:Y_1 = (X - 1)^{-1}$
$:Y_2 = (X + 1)(X^{-1})$
$:Y_3 \blacksquare (Y_2 - 1)^{-1}$

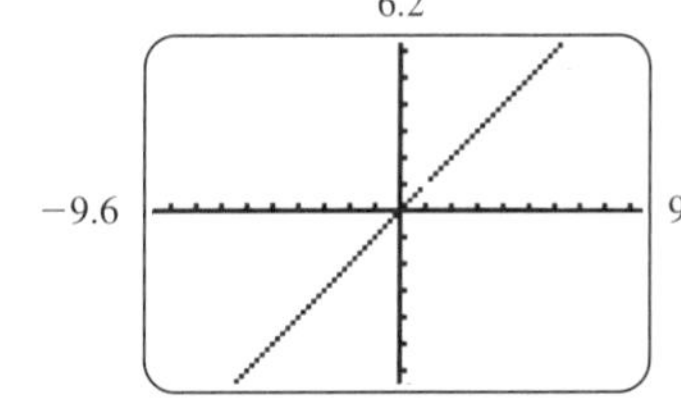

$:Y_1 = (X - 1)^{-1}$
$:Y_2 = (X + 1)(X^{-1})$
$:Y_3 = (Y_2 - 1)^{-1}$
$:Y_4 \blacksquare (Y_1 + 1)(Y_1{}^{-1})$

They are inverses.

13.

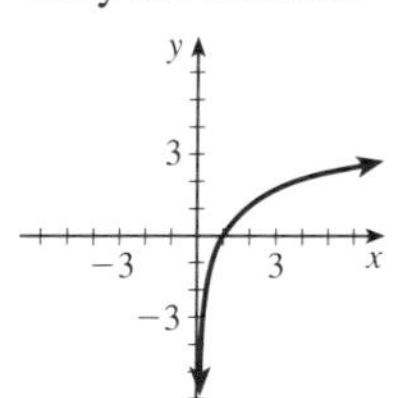

15. $g^{-1}(x) = \frac{-1}{4}x + \frac{5}{4}$
17. $f^{-1}(x) = \frac{2}{3}x - \frac{1}{3}$
19. $f^{-1}(x) = 2 + \sqrt{4 + x}, \quad x \geq -4$
21. $r^{-1}(x) = \dfrac{2x - 3}{x + 1}, \quad x \neq -1$
23. $d^{-1}(x) = x^2 - 3, \quad x \geq 0$
25. $v^{-1}(x) = \dfrac{2}{x - 1}, \quad x \neq 1$ **27.** $f^{-1}(x) = (x + 1)^2, \quad x \geq -1$
29. $j^{-1}(x) = 1 + \sqrt{4 + x}, \quad x \geq -4$

CHAPTER 9 REVIEW ■ page 531

1.

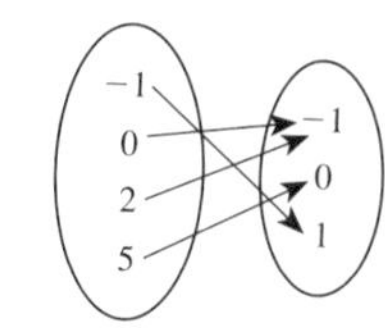

Is a function
domain: $\{-1, 0, 2, 5\}$
range: $\{-1, 0, 1\}$

3. Is a function Range: $\{p, s, t\}$

5. -9 **7.** $2a$ **9.** 12 **11.** $\frac{3}{2}$ **13.** $\frac{-1}{a(a-1)}$

15. 3 **17.** -2 **19.** $-4-h$ **21.** Is a function

23. Not a function **25.** -2 **27.** -4

29. Domain: $\{x \mid -4 \le x \le 5\}$
range: $\left\{y \mid -3 \le y \le \frac{7}{3}\right\}$

31. $\frac{-3}{2} < x < 0$

33. Domain: $\{x \mid x < 0\} \cup \{x \mid x \ge 1\}$
range: $\{y \mid y > 1\}$
turning points none
zeros none

35. -6 **37.** 0 **39.** -1 **41.** 1 **43.** Undefined

45. $\frac{2}{2-c^2}$

47.

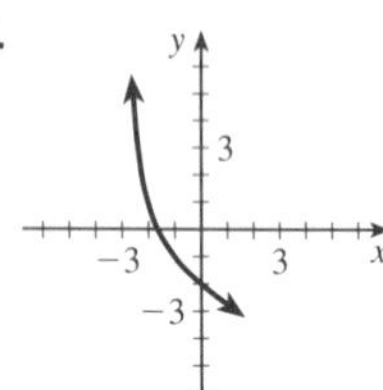

49. $f^{-1}(x) = -3x + 4$

51. $r^{-1}(x) = \sqrt{4-x},\ x \le 4$

53. $d^{-1}(x) = \frac{-3}{2x+1},\ x \ne \frac{-1}{2}$

CHAPTER 9 TEST ■ page 532

1. a. 19 **b.** 12 **2.** $\frac{-3}{2h+4}$ **3. a.** -3 **b.** 0

c. $-1 < x < 5$ **4. a.** 25 **b.** 7 **5.** $\frac{1-\sqrt{x}}{x}$

6. $f^{-1}(x) = \frac{-1}{2}x + \frac{5}{2}$ **7.** $g^{-1}(x) = \frac{2x+1}{x-1}$

8.

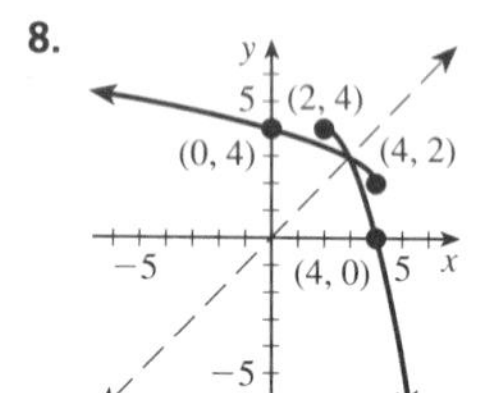

■ CHAPTER 10

EXERCISES 10.1 ■ page 541

1.

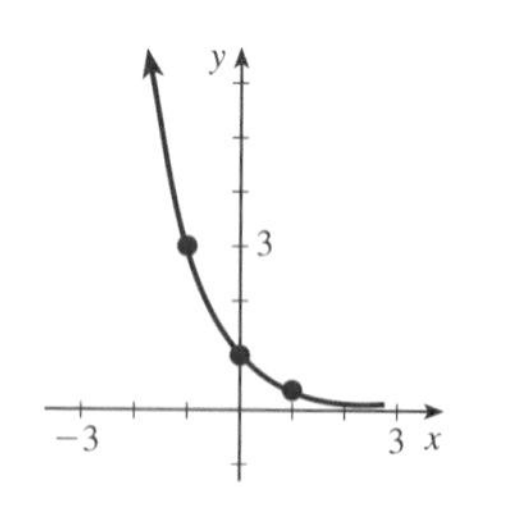

3. -1

5.

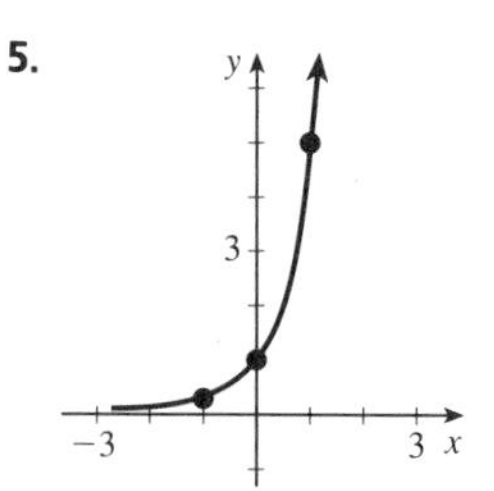

7. Undefined **9.** 2 **11.** 3 **13.** 2 **15.** -3

17. 1 **19.** $\frac{1}{5}$ **21.** -2 **23.** Undefined

25. 0.9542; $10^{0.9542} \approx 9$ **27.** -2; $10^{-2} = \frac{1}{100}$

29. -2.3010; $10^{-2.3010} \approx 0.005$ **31.** $\frac{1}{9}$ **33.** 2

35. 39 **37.** -1 and 9 **39.** $1+\sqrt{3}$ and $1-\sqrt{3}$

41. 3 **43.** $\frac{-10}{17}$ **45.** 13 **47.** -2398 **49.** 8

DISCOVERY ■ page 544

1. 1,590 micrometers **3.** 15,890 micrometers

EXERCISES 10.2 ■ page 553

1. -1 **3.** 6 **5.** $\frac{-3}{5}$ **7.** 7 **9.** $\frac{5}{2}$ **11.** 1.2091

13. 1.1610 **15.** 1.2263 **17.** 5 **19.** $\frac{-1+\sqrt{73}}{2}$

21. 3 **23.** No solution **25.** 8 **27.** $\frac{3}{5}$

29. $3+\sqrt{2}$ and $3-\sqrt{2}$ **31.** $\frac{1+\sqrt{13}}{2}$ **33.** 4

35. $\frac{1}{243}$ **37.** $\frac{3+\sqrt{5}}{2}$

EXERCISES 10.3 ■ page 558

1. 2 **3.** -1 **5.** -1 **7.** $\frac{1}{2}$ **9.** -1 **11.** 0

13. $\frac{\log 21}{\log 3} \approx 2.7712$ **15.** $\frac{\log 3}{\log 17} \approx 0.3878$

17. $\frac{\log 2}{\log 1.5} \approx 1.7095$ **19.** 1 **21.** $e^6 \approx 403.43$

23. $e - 1 \approx 1.72$ **25.** $e \approx 2.72$ **27.** None

29. $\ln 2 \approx 0.69$ **31.** 1 **33.** $\ln\left(\frac{5}{3}\right) \approx 0.51$

35. $\frac{\ln 2}{0.05} \approx 13.86$ **37.** $\frac{\ln 2}{2} \approx 0.35$ **39.** $\frac{\ln 6}{2} \approx 0.90$

EXERCISES 10.4 ■ page 567

1. $2 + \ln 5 \approx 3.61$ **3.** $\frac{e+3}{2} \approx 2.86$

5. $100 \ln 2 \approx 69.31$ **7. a.** $18,166.97 **b.** 5.8 years

9. a. $6,749.29 **b.** 6.93 years **11.** $633.38

13. $1,609.25 **15.** 2.14 **17.** 1.58×10^{-10}

19. Growth rate = 0.0368; In the year 2001, the population will reach 90,000.

21. 5381.52 years **23.** 3.16×10^6

CHAPTER 10 REVIEW ■ page 568

1.

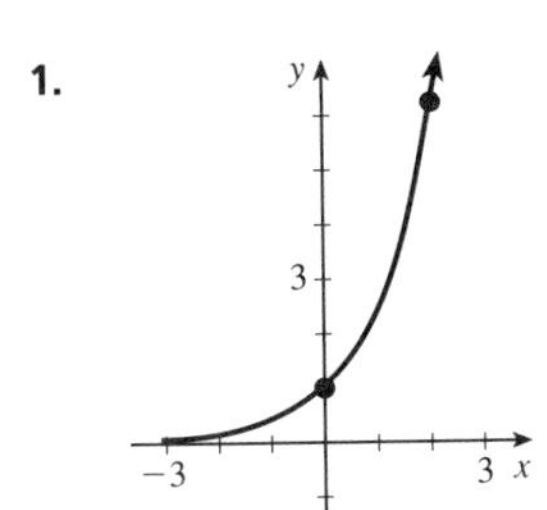

3. 2 **5.** $x = \log_a c$ **7.** 2 **9.** 3 **11.** 4 **13.** $\frac{3}{2}$

15. −4 **17.** 5 **19.** $\frac{-2}{3} \approx -0.67$ **21.** $\frac{\log 13}{\log 5} \approx 1.59$

23. $\frac{\log 47}{\log 7} \approx 1.98$ **25.** $\frac{\log 20}{\log 19} \approx 1.02$ **27.** 3

29. ≈ 2.56 **31.** 1.5 **33.** 2 **35.** $\frac{41}{2}$ **37.** 1

39. −2.9 **41.** $4\sqrt{2}$ **43.** $\frac{46}{17}$ **45.** 4 **47.** −10

49. 1 **51.** $\frac{14}{3}$ **53.** $\frac{17}{4}$ **55.** $2 + \ln 3 \approx 3.10$

57. $\frac{1 + e^2}{3} \approx 2.80$

59. a. $25,364.84 **b.** 8.75 years **61.** 21.777 milligrams

63. 36.62 years **65.** 5.10 minutes

CHAPTER 10 TEST ■ page 569

1. c

2.

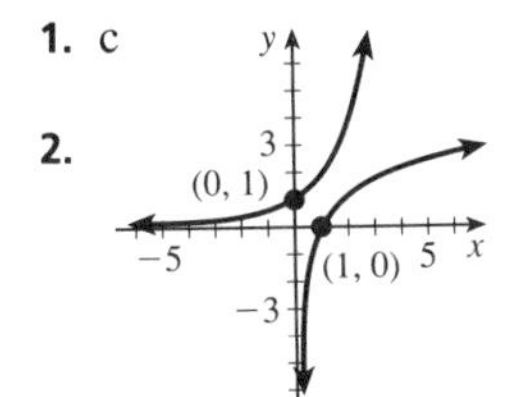

x-intercept (1, 0) for f^{-1}
y-intercept (0, 1) for f
vertical asymptote $x = 0$ for f^{-1}
horizontal asymptote $y = 0$ for f

3. $\frac{-4}{3}$ **4.** 3 **5.** $\frac{2}{3}$ **6.** 1 **7.** 2 **8.** −13

9. $\frac{14}{3}$ **10.** $\frac{\ln 23}{\ln 5} \approx 1.95$ **11.** $\frac{17}{2}$

12. $\frac{-\ln 2}{\ln 3 - \ln 2} \approx -1.71$ **13.** 4 **14.** $\frac{13}{9}$ **15.** 5.88 years

■ CHAPTER 11

EXERCISES 11.1 ■ page 580

1. $\left(\frac{614}{19}, \frac{66}{19}\right)$ **3.** $\left(\frac{86}{25}, \frac{1}{5}, \frac{-23}{25}\right)$ **5.** (5.2, 9, 11.4)

7. (2, 1, 3) **9.** (1, −2, 3) **11.** (6, 16, −9)

13. (16, 45, 13) **15.** $\left(\frac{1}{3}, \frac{4}{3}, \frac{-1}{3}\right)$ **17.** (6, 10, −3)

19. $\left(\frac{310}{9}, 85, \frac{-35}{9}\right)$ **21.** $y = \frac{19}{30}x^2 + \frac{31}{30}x - \frac{8}{5}$

23. $y = \frac{7}{2}x^2 + \frac{29}{2}x + 11$

EXERCISES 11.2 ■ page 591

1. 3×4 **3.** Row 2, column 4

5. $x = 1, y = -2, z = \frac{7}{2}, w = 0$

7. $\begin{bmatrix} -12 & -18 \\ -18 & 9 \end{bmatrix}$ **9.** $\begin{bmatrix} -8 & -1 \\ 1 & 11 \end{bmatrix}$ **11.** $\begin{bmatrix} -20 & 3 \\ 9 & 30 \end{bmatrix}$

13. $\begin{bmatrix} 45 & 5 \\ -25 & -30 \\ 35 & -15 \end{bmatrix}$ **15.** $\begin{bmatrix} 11 & -2 \\ -10 & -14 \\ -3 & -7 \end{bmatrix}$

17. $\begin{bmatrix} 4 & -6 \\ -10 & -16 \\ -20 & -8 \end{bmatrix}$ **19.** $\begin{bmatrix} 0 & 0 \\ 0 & 0 \\ 0 & 0 \end{bmatrix}$ **21.** $\begin{bmatrix} 1 & 0 & 0 \\ 0 & 1 & 0 \\ 0 & 0 & 1 \end{bmatrix}$

23. $\begin{bmatrix} -1 & 17 & -17 \\ 34 & 22 & 26 \end{bmatrix}$ **25.** $\begin{bmatrix} -16 & 16 \\ 12 & -18 \end{bmatrix}$

27. $\begin{bmatrix} 31 & -10 \\ -15 & 6 \end{bmatrix}$ **29.** $\begin{bmatrix} -19 & 32 \\ -40 & 29 \end{bmatrix}$ **31.** $\begin{bmatrix} 3 & 40 \\ 3 & 7 \end{bmatrix}$

33. $\begin{bmatrix} -16 & -30 & -37 \\ 2 & -18 & 29 \\ 33 & -48 & 66 \end{bmatrix}$ **35.** $\begin{bmatrix} -6 & -24 & 28 \\ 0 & -6 & 10 \\ -10 & 12 & -15 \end{bmatrix}$

37. $\begin{bmatrix} 7 & -81 & 114 \\ 12 & 19 & -21 \\ -98 & 79 & -86 \end{bmatrix}$ **39.** $\begin{bmatrix} 49 & -95 & 117 \\ 21 & 41 & -51 \\ -113 & 124 & -150 \end{bmatrix}$

41. $\begin{bmatrix} \frac{9}{19} \\ \frac{-7}{19} \end{bmatrix} \approx \begin{bmatrix} 0.47 \\ -0.37 \end{bmatrix}$ **43.** $\begin{bmatrix} \frac{92}{19} \\ \frac{72}{19} \\ \frac{27}{19} \end{bmatrix} \approx \begin{bmatrix} 4.84 \\ 3.79 \\ 1.42 \end{bmatrix}$

EXERCISES 11.3 ■ page 600

1. They are inverses of each other.

3. Not inverses of each other. $\begin{bmatrix} 1 & 0 & 8 \\ 0 & 1 & 24 \\ 0 & 0 & -7 \end{bmatrix}$

5. $a = \frac{-1}{17}, b = \frac{3}{17}, c = \frac{5}{17}, d = \frac{2}{17}$ **7.** $\begin{bmatrix} 1 & 2 \\ 3 & 5 \end{bmatrix}$

9. $\begin{bmatrix} 0.2 & -0.4 \\ 0.2 & 0.6 \end{bmatrix}$ **11.** Singular

13. $\begin{bmatrix} \frac{1}{7} & \frac{-2}{7} \\ \frac{-1}{14} & \frac{-5}{14} \end{bmatrix} \approx \begin{bmatrix} 0.14 & -0.29 \\ -0.07 & -0.36 \end{bmatrix}$

15. $\begin{bmatrix} 3.5 & 0.5 & -3 \\ -4.5 & -0.5 & 4 \\ -0.5 & -0.5 & 1 \end{bmatrix}$ **17.** $\begin{bmatrix} 0 & 1 & -1 \\ 1 & 1 & 1 \\ 1 & 1 & 0 \end{bmatrix}$

19. $\begin{bmatrix} \frac{-1}{9} & \frac{2}{9} & \frac{1}{6} \\ \frac{4}{9} & \frac{1}{9} & \frac{-1}{6} \\ \frac{-1}{9} & \frac{2}{9} & \frac{-1}{3} \end{bmatrix} \approx \begin{bmatrix} -0.11 & 0.22 & 0.17 \\ 0.44 & 0.11 & -0.17 \\ -0.11 & 0.22 & -0.33 \end{bmatrix}$

21. $\begin{bmatrix} -11 \\ -8 \end{bmatrix}$ 23. $\begin{bmatrix} 3 \\ -2 \\ 1 \end{bmatrix}$

25. $A = \begin{bmatrix} 3 & -2 \\ 1 & 2 \end{bmatrix}$ $A^{-1} = \begin{bmatrix} 0.25 & 0.25 \\ -0.125 & 0.375 \end{bmatrix}$ (2, 0)

27. $A = \begin{bmatrix} 4 & 3 \\ 1 & -2 \end{bmatrix}$ $A^{-1} = \begin{bmatrix} \frac{2}{11} & \frac{3}{11} \\ \frac{1}{11} & \frac{-4}{11} \end{bmatrix}$ $\left(\frac{28}{11}, \frac{-8}{11}\right)$

29. $A = \begin{bmatrix} 2 & 1 \\ -3 & 1 \end{bmatrix}$ $A^{-1} = \begin{bmatrix} 0.2 & -0.2 \\ 0.6 & 0.4 \end{bmatrix}$ (0.4, 2.2)

31. (−31, −21) 33. (21, 13, 23)
35. Singular
37. 70 *fx*-70, 53 *mx*-80, 105 *rx*-90
39. $8\frac{1}{3}$ liters of 30% acid solution
$16\frac{2}{3}$ liters of 60% acid solution
25 liters of 90% acid solution
41. 250 calls from location *X*
80 calls from location *Y*
170 calls from location *Z*

EXERCISES 11.4 ■ page 610

1. −15 3. −6 5. −18 7. −13 9. −1
11. −2 13. 14 15. 37 17. $x - 2$
19. $6x + 3y - 3$
21. $2x - 3y - 7$
23. $D = 1, D_x = 23, D_y = 12$; (23, 12)
25. $D = -23, D_x = -150, D_y = 9$; $\left(\frac{150}{23}, \frac{-9}{23}\right)$
27. $D = 2.5, D_x = 7, D_y = -8.5$; (2.8, −3.4)
29. $D = 9, D_x = 87, D_y = 28, D_z = -40$; $\left(\frac{29}{3}, \frac{28}{9}, \frac{-40}{9}\right)$
31. $D = -22, D_x = -75, D_y = 36, D_z = -155$; $\left(\frac{75}{22}, \frac{-18}{11}, \frac{155}{22}\right)$
33. $D = -7, D_x = 6, D_y = -31, D_z = 22$; $\left(\frac{-6}{7}, \frac{31}{7}, \frac{-22}{7}\right)$
35. 6 37. $\frac{3\sqrt{3} + 4\sqrt{6} + \sqrt{2}}{2} \approx 8.20$
39. Collinear $x + 5y - 22 = 0$ 41. Not collinear
43. $4x + 5y + 7 = 0$

CHAPTER 11 REVIEW ■ page 611

1. False 3. False 5. True 7. False 9. False
11. False 13. $\left(\frac{390}{31}, \frac{-340}{31}\right)$
15. $\left(\frac{79}{33}, \frac{-95}{33}, \frac{-67}{33}\right) \approx (2.39, -2.88, -2.03)$
17. $y = \frac{1}{15}x^2 + \frac{4}{15}x - \frac{9}{5}$ 19. $y = \frac{-11}{8}x^2 + \frac{1}{4}x + 4$
21. 4 × 3 23. Row 3, Column 2
25. $x = 0, y = -2, z = 3, w = 0$ 27. $\begin{bmatrix} -10 & 15 & 50 \\ 25 & 40 & 20 \end{bmatrix}$

29. $\begin{bmatrix} -11 & 2 & 3 \\ 10 & 14 & 1 \end{bmatrix}$ 31. $\begin{bmatrix} 4 & -6 & -20 \\ -10 & -16 & -8 \end{bmatrix}$

33. 28 35. $\begin{bmatrix} 2 & 5 \\ 8 & -15 \end{bmatrix}$ 37. $\begin{bmatrix} -6 & -10 \\ 4 & -10 \end{bmatrix}$

39. $\begin{bmatrix} 1 & 6 & -4 \\ -4 & -9 & 13 \\ 3 & -39 & 30 \end{bmatrix}$ 41. $\begin{bmatrix} -18 & -33 & 0 \\ 49 & 37 & -12 \\ -9 & 87 & -55 \end{bmatrix}$

43. $\left(\frac{18}{17}, \frac{-20}{17}\right) \approx (1.06, -1.18)$
45. $\left(\frac{30}{7}, \frac{10}{7}, \frac{-32}{7}\right) \approx (4.29, 1.43, -4.57)$
47. *V*-35 41
V-70 33
V-70*X* 37
49. 180 calls sent to location X
250 calls sent to location Y
70 calls sent to location Z

51. $\begin{bmatrix} \frac{11}{2} \\ -3 \end{bmatrix}$ 53. $\begin{bmatrix} \frac{7}{3} \\ \frac{-16}{3} \\ \frac{10}{3} \end{bmatrix}$

55. Not inverses of each other $\begin{bmatrix} 1 & -5 \\ -12 & 0 \end{bmatrix}$

57. $\begin{bmatrix} 2 & 1 \\ 1.5 & 0.5 \end{bmatrix}$ 59. $\begin{bmatrix} 0.375 & 0.25 & -0.375 \\ -0.125 & 0.25 & 0.125 \\ 0.25 & -0.50 & 0.75 \end{bmatrix}$

61. $\begin{bmatrix} -6 \\ -6 \\ 17 \end{bmatrix}$

63. $A = \begin{bmatrix} 1 & -2 \\ 3 & 1 \end{bmatrix}$ $A^{-1} = \begin{bmatrix} \frac{1}{7} & \frac{2}{7} \\ \frac{-3}{7} & \frac{1}{7} \end{bmatrix} \begin{bmatrix} 0.14 & 0.29 \\ -0.43 & 0.14 \end{bmatrix}$
$\left(\frac{22}{7}, \frac{-10}{7}\right) \approx (3.14, -1.43)$

65. 15 67. 0 69. $1 + 3\sqrt{3}$
71. $D = 17, D_x = 15, D_y = -41$; $\left(\frac{15}{17}, \frac{-41}{17}\right)$
73. $D = -12, D_x = 39, D_y = -63, D_z = -231$; $\left(\frac{-13}{4}, \frac{21}{4}, \frac{77}{4}\right)$
75. 11.5 77. Collinear $3x + y - 5 = 0$
79. $x - y + 1 = 0$

CHAPTER 11 TEST ■ page 614

1. $\left(\frac{28}{13}, \frac{-23}{13}, \frac{-22}{13}\right)$
2. a. 4 × 3 b. −5 c. Row 2 column 2
d. 4, 3, −1
3. $\begin{bmatrix} -3 & 1 \\ 1 & -1 \end{bmatrix}$, $\begin{bmatrix} -0.5 & -0.5 \\ -0.5 & -1.5 \end{bmatrix}$, (−1, −5)

4. x = number of liters of 20% acid solution
y = number of liters of 90% acid solution
z = number of liters of 40% acid solution

$$\begin{cases} x + y + z = 100 \\ 2x - y = 0 \\ 0.2x + 0.9y + 0.4z = 60 \end{cases}$$

5. $\begin{bmatrix} 32 & 3 & -30 \\ 16 & -1 & -10 \end{bmatrix}$ **6.** $(-81, 36, -14)$

7. $D = 17, D_x = 34, D_y = 0; (2, 0)$

■ CHAPTER 12

EXERCISES 12.1 ■ page 621

1.

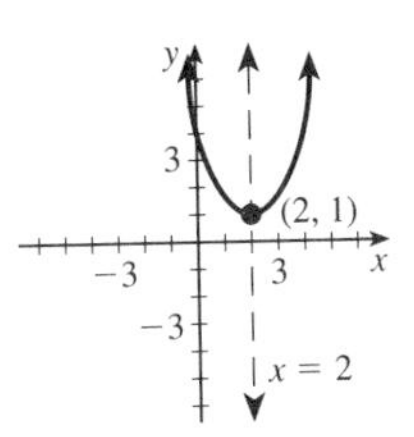

3.

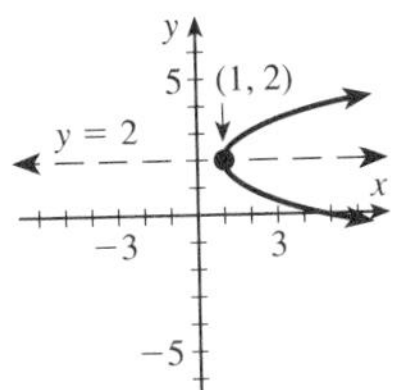

5.

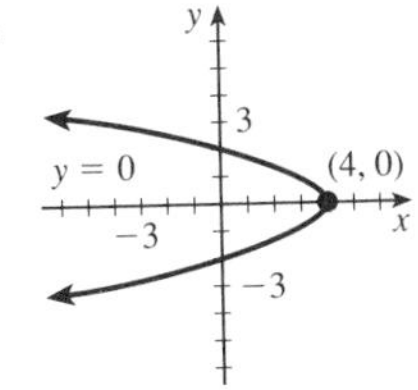

7.

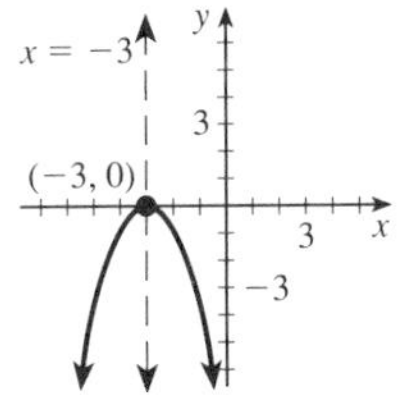

9.

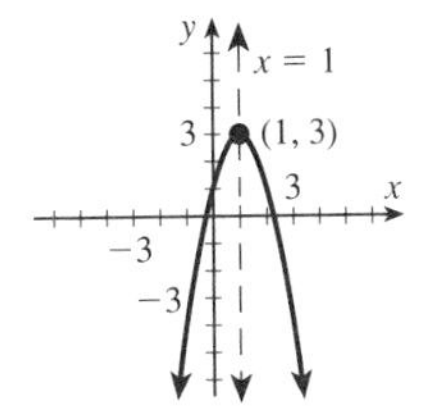

11.

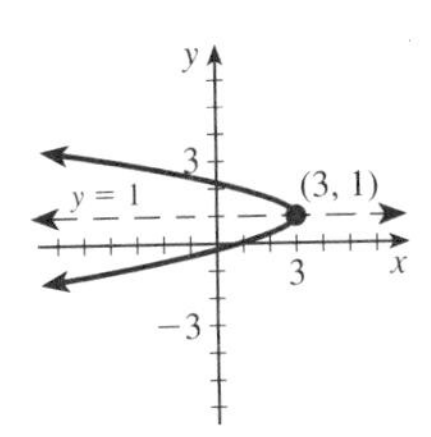

13.

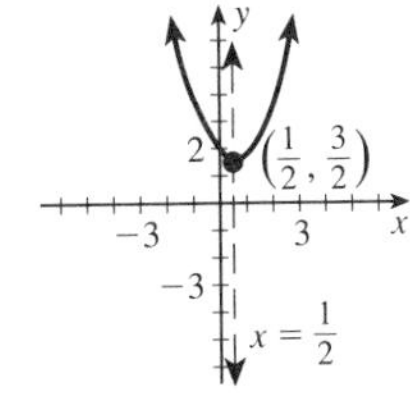

15. $y = (x - 2)^2$

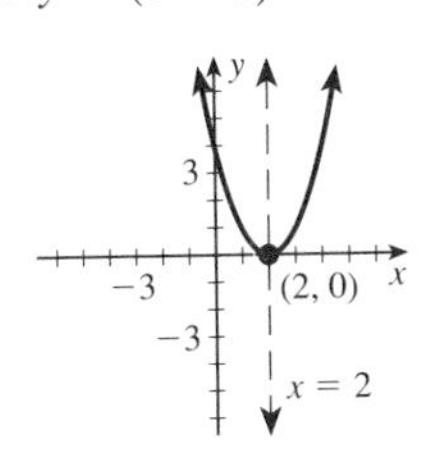

17. $x = (y - 1)^2 - 1$

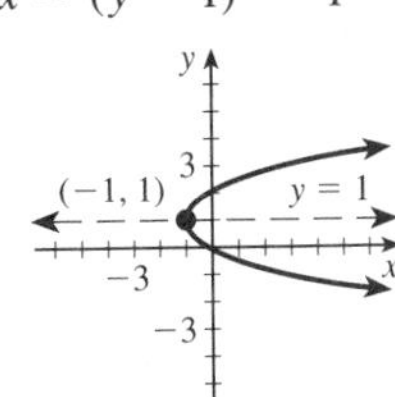

19. $y = (x + 1)^2 - 2$

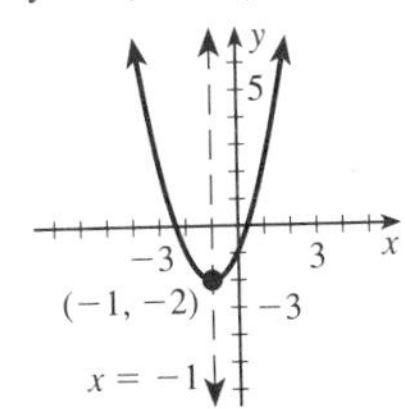

21. $x = \frac{1}{3}(y - 2)^2 - \frac{5}{3}$

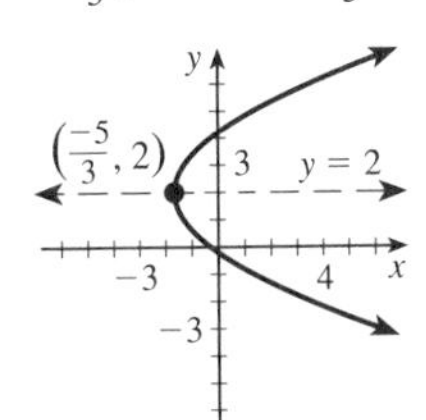

23. $y = \frac{9}{16}x^2 - 9$

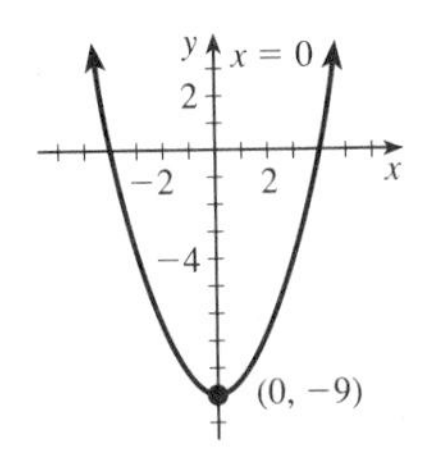

25. $x = \frac{1}{2}(y + 2)^2 - 2$

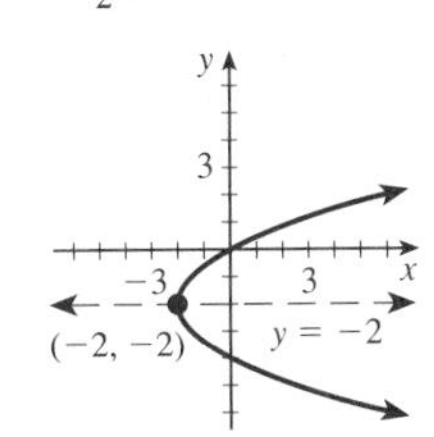

27. $y = \frac{-1}{2}(x + 3)^2 + \frac{13}{2}$

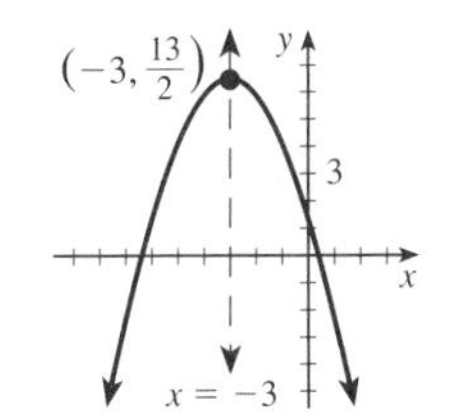

29. $x = \frac{1}{2}(y + 2)^2 + \frac{3}{2}$

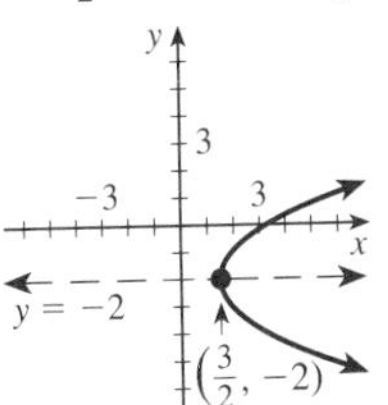

EXERCISES 12.2 ■ page 629

1.

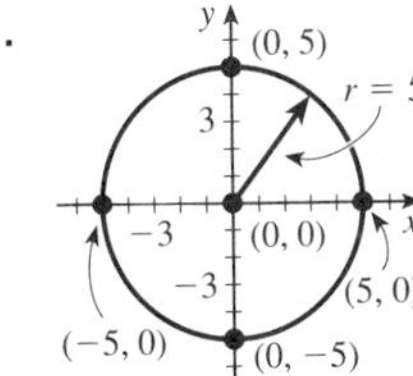

3. $(-\sqrt{6}, 0)$, $(0, \sqrt{6})$, $r = \sqrt{6}$, $(\sqrt{6}, 0)$, $(0, -\sqrt{6})$

5. $x^2 + y^2 = 4$ **7.** $x^2 + y^2 = 9$ **9.** $x^2 + y^2 = 13$
11. $x^2 + y^2 = 17$

13.

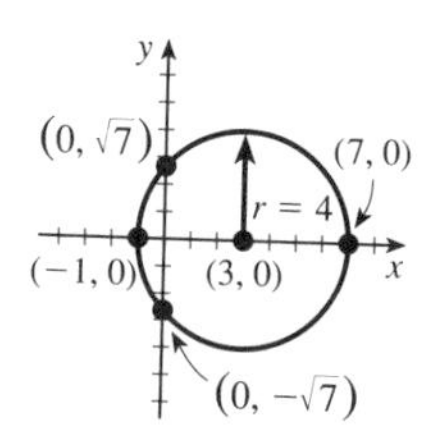

15.

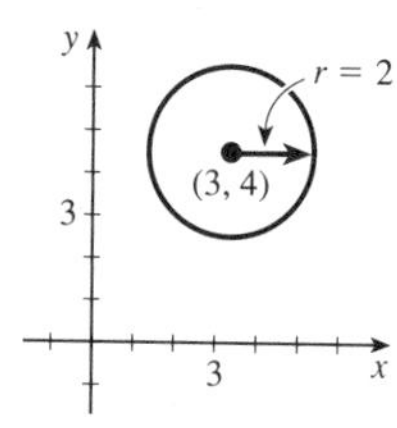

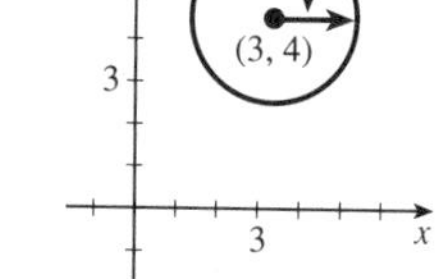

17.

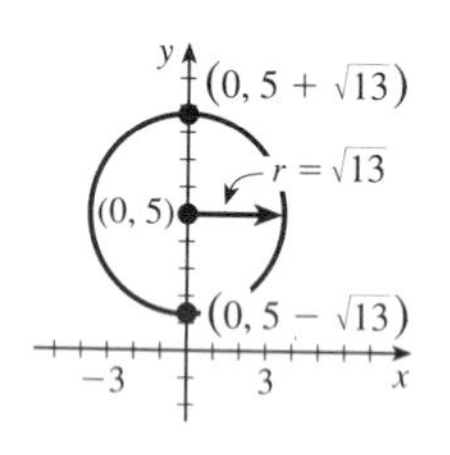

19. $(x + 3)^2 + (y + 2)^2 = 16$
21. $(x - 3)^2 + (y - 1)^2 = 16$
23. $(x - 4)^2 + (y + 1)^2 = 17$
25. $(x - 2)^2 + (y - 2)^2 = 10$

27. $(x + 2)^2 + y^2 = 11$

center $(-2, 0)$
x-intercepts $\left(-2 + \sqrt{11}, 0\right), \left(-2 - \sqrt{11}, 0\right)$
y-intercepts $\left(0, \sqrt{7}\right), \left(0, -\sqrt{7}\right)$
radius $= \sqrt{11}$

29. $(x - 2)^2 + (y + 1)^2 = 10$

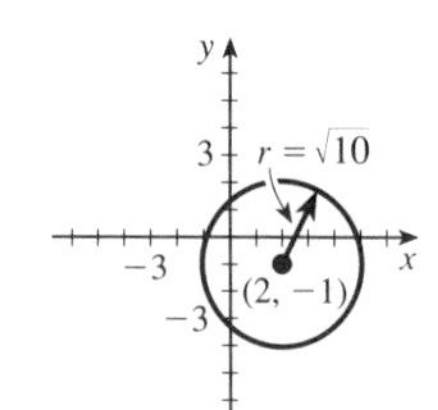

center $(2, -1)$
radius $= \sqrt{10}$
x-intercepts $(5, 0), (-1, 0)$
y-intercepts $\left(0, -1 + \sqrt{6}\right), \left(0, -1 - \sqrt{6}\right)$

31. $(x - 4)^2 + (y - 2)^2 = 22$

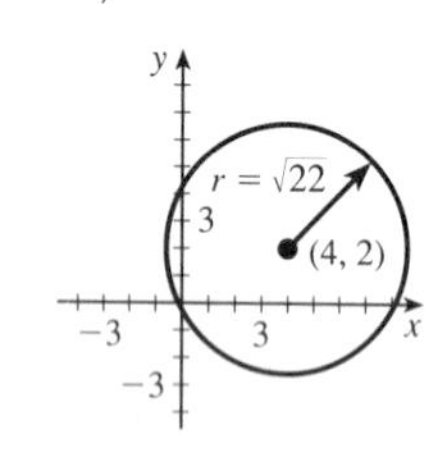

ccenter $(4, 2)$
radius $= \sqrt{22}$
x-intercepts $\left(4 + 3\sqrt{2}, 0\right), \left(4 - 3\sqrt{2}, 0\right)$
y-intercepts $\left(0, 2 + \sqrt{6}\right), \left(0, 2 - \sqrt{6}\right)$

33. $16\pi \approx 50.27$ **35.** $(x - 5)^2 + (y - 4)^2 = 4$

EXERCISES 12.3 ■ page 637

1.

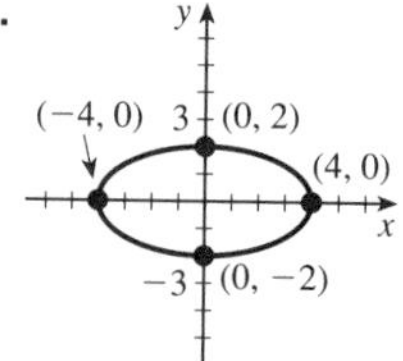

x-intercepts $(4, 0), (-4, 0)$
y-intercepts $(0, 2), (0, -2)$

3.

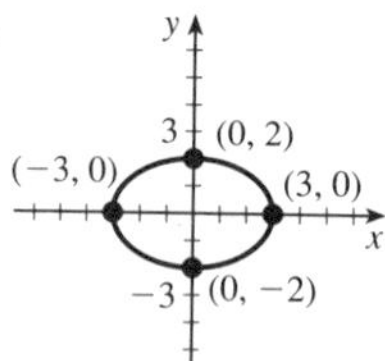

x-intercepts $(3, 0), (-3, 0)$
y-intercepts $(0, 2), (0, -2)$

5.

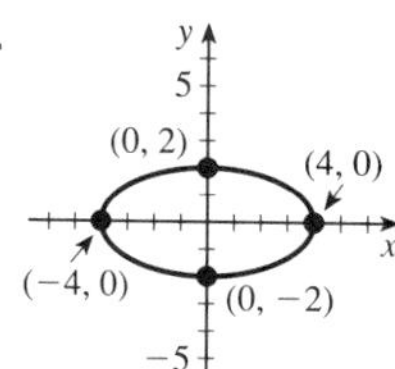

x-intercepts $(4, 0), (-4, 0)$
y-intercepts $(0, 2), (0, -2)$

7.

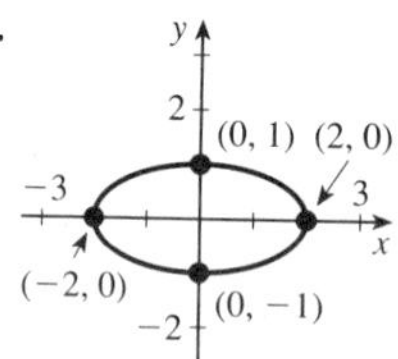

x-intercepts $(2, 0), (-2, 0)$
y-intercepts $(0, 1), (0, -1)$

9.

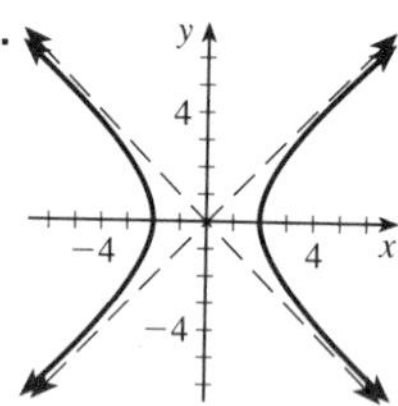

x-intercepts $(2, 0), (-2, 0)$
y-intercepts none
asymptotes $y = x, y = -x$

11.

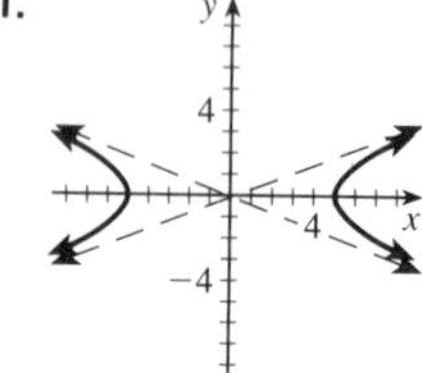

x-intercepts $(5, 0), (-5, 0)$
y-intercepts none
asymptotes $y = \frac{2}{5}x, y = \frac{-2}{5}x$

13.

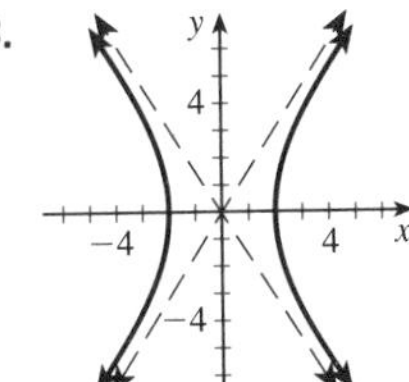

x-intercepts $(2, 0)$, $(-2, 0)$
y-intercepts none
asymptotes $y = \frac{3}{2}x$, $y = \frac{-3}{2}x$

15.

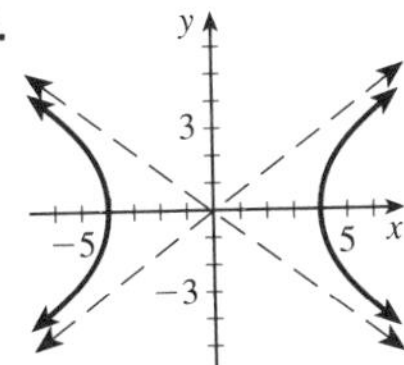

x-intercepts $(4, 0)$, $(-4, 0)$
y-intercepts none
asymptotes $y = \frac{3}{4}x$, $y = \frac{-3}{4}x$

17.

x-intercepts none
y-intercepts $(0, 5)$, $(0, -5)$
asymptotes $y = \frac{5}{3}x$, $y = \frac{-5}{3}x$

19.

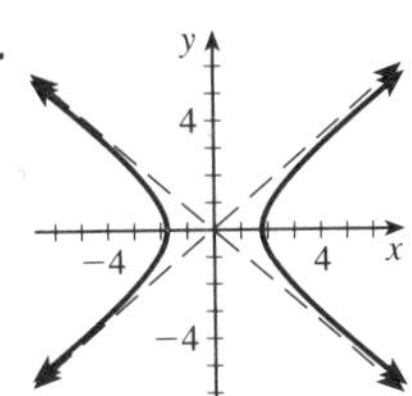

x-intercepts $\left(\sqrt{3}, 0\right)$, $\left(-\sqrt{3}, 0\right)$
y-intercepts none
asymptotes $y = \frac{\sqrt{6}}{3}x$, $y = \frac{-\sqrt{6}}{3}x$

21.

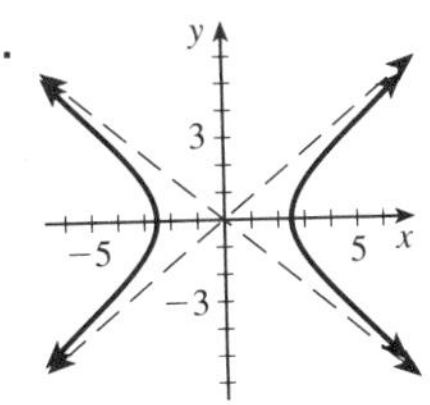

x-intercepts $\left(\frac{5}{2}, 0\right)$, $\left(\frac{-5}{2}, 0\right)$
y-intercepts none
asymptotes $y = \frac{4}{5}x$, $y = \frac{-4}{5}x$

23.

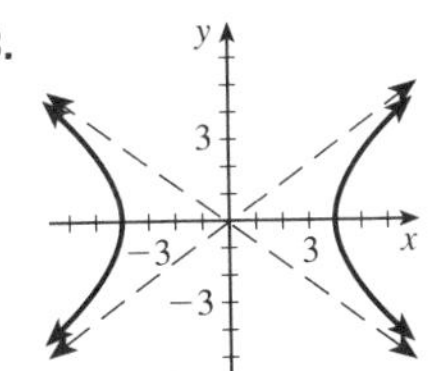

x-intercepts $(4, 0)$, $(-4, 0)$
y-intercepts none
asymptotes $y = \frac{3}{4}x$, $y = \frac{-3}{4}x$

25.

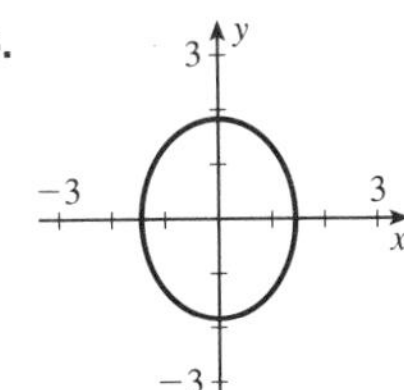

x-intercepts $\left(\sqrt{2}, 0\right)$, $\left(-\sqrt{2}, 0\right)$
y-intercepts $\left(0, \sqrt{3}\right)$, $\left(0, -\sqrt{3}\right)$

27.

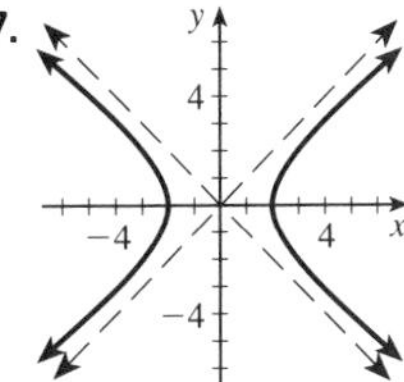
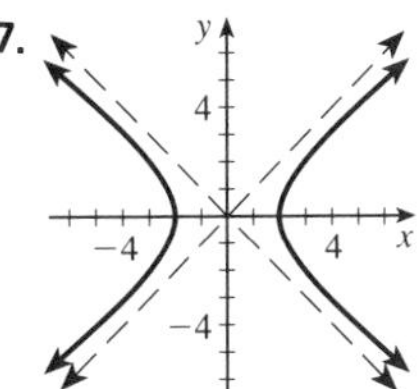

x-intercepts $(2, 0)$, $(-2, 0)$
y-intercepts none
asymptotes $y = \frac{\sqrt{3}}{2}x$, $y = \frac{-\sqrt{3}}{2}x$

EXERCISES 12.4 ■ page 643

1. $\left(\frac{2}{3}, \frac{-5}{3}\right)$ **3.** $(1, 0)$ **5.** $\left(\frac{-3}{5}, \frac{4}{25}\right)$

7. $(0, 0)$ and $(-1, 1)$ **9.** $(5, 8)$ and $(-1, 2)$ **11.** $\left(\frac{5}{2}, \frac{1}{2}\right)$

13. $\left(\frac{3 + \sqrt{13}}{2}, \frac{1 + \sqrt{13}}{2}\right)$ and $\left(\frac{3 - \sqrt{13}}{2}, \frac{1 - \sqrt{13}}{2}\right)$

15. $\left(\frac{1 + \sqrt{7}}{2}, \frac{1 - \sqrt{7}}{2}\right)$ and $\left(\frac{1 - \sqrt{7}}{2}, \frac{1 + \sqrt{7}}{2}\right)$

17. $(2, 0)$ and $(-2, 0)$

19. 6.76 inches by 17.76 inches **21.** $\sqrt{14}$

EXERCISES 12.5 ■ page 652

1.

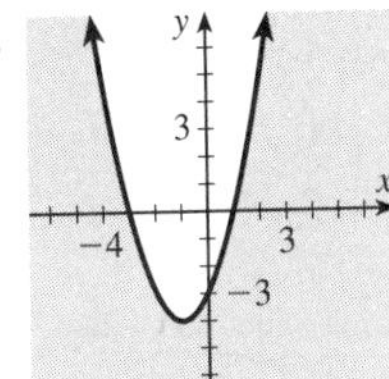

3.

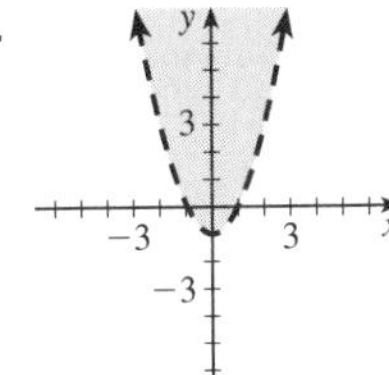

5.

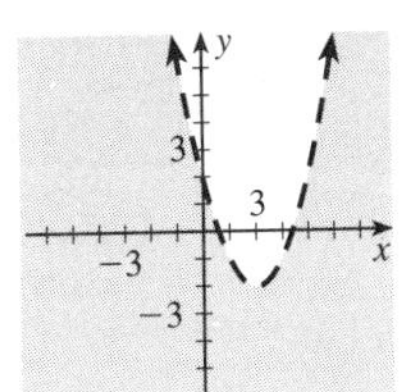

7.

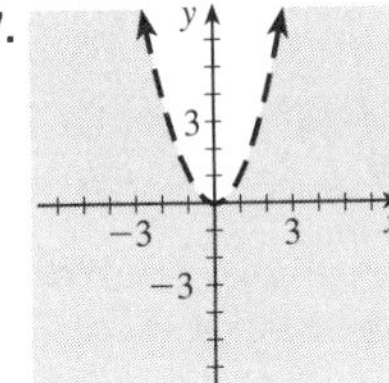

9.

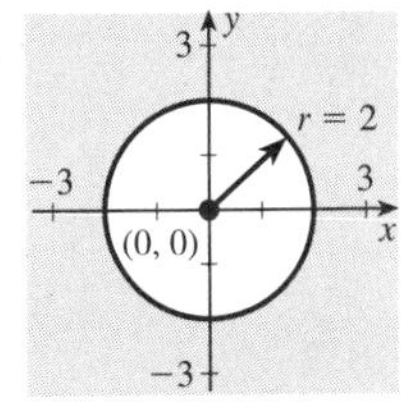

11.

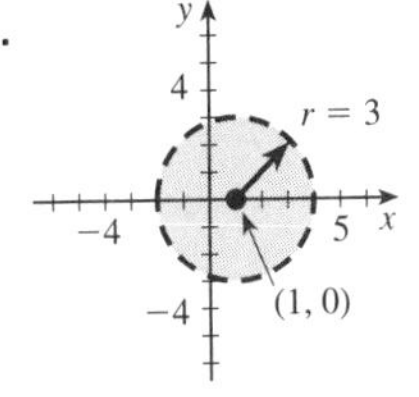

13.

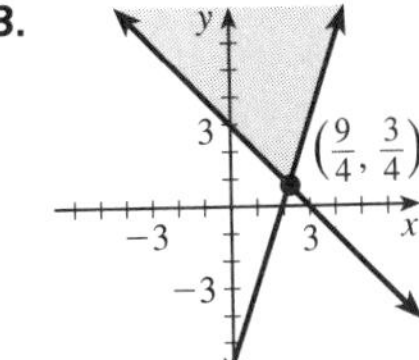

15.

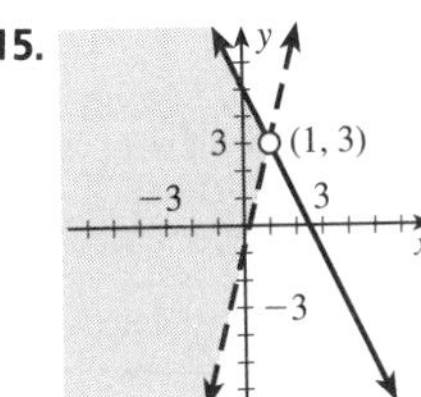

17.

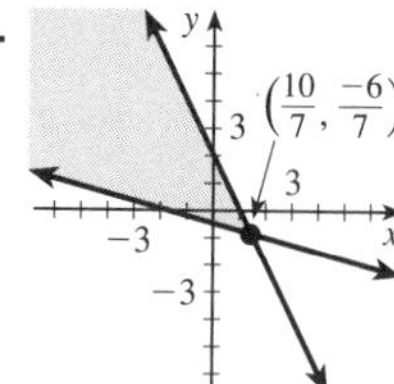

19.

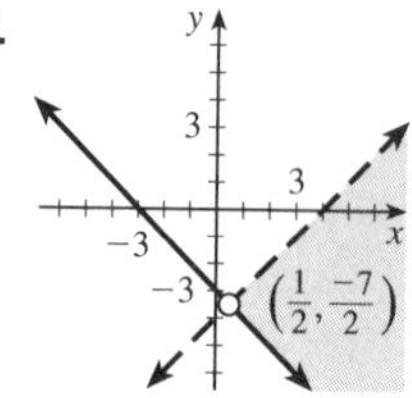

21.

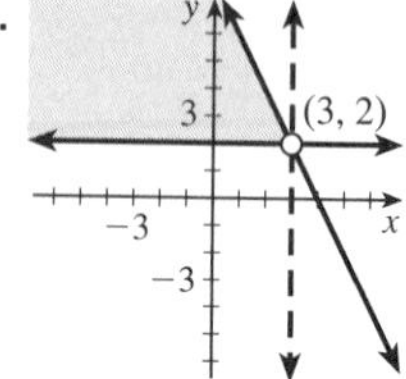

23.

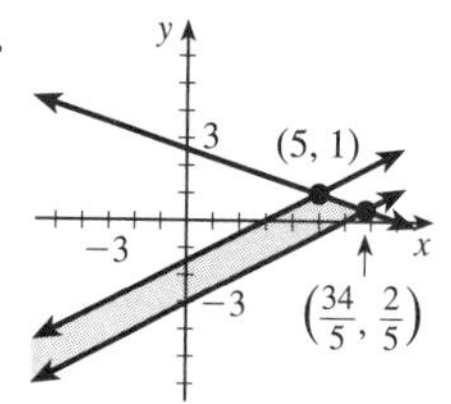

25.

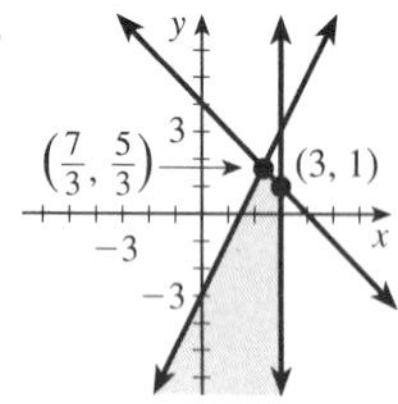

27.

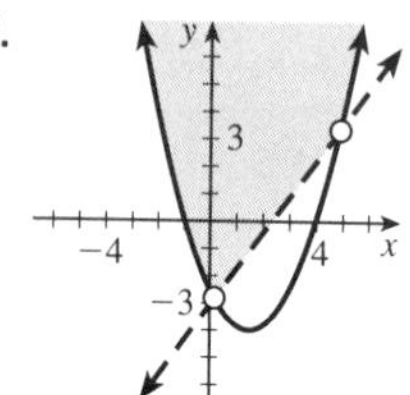

29.

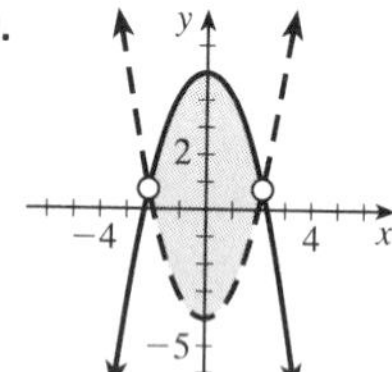

31.

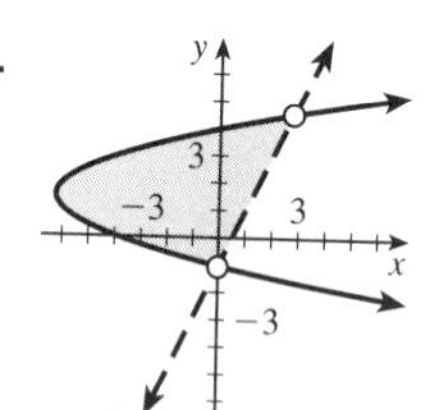

33.

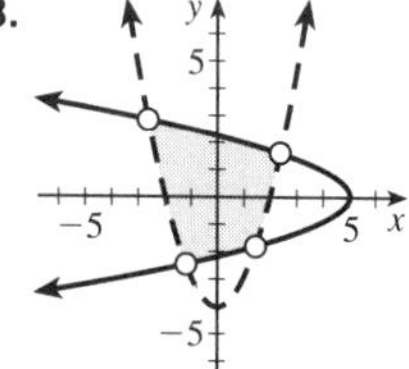

CHAPTER 12 REVIEW ■ page 652

1.

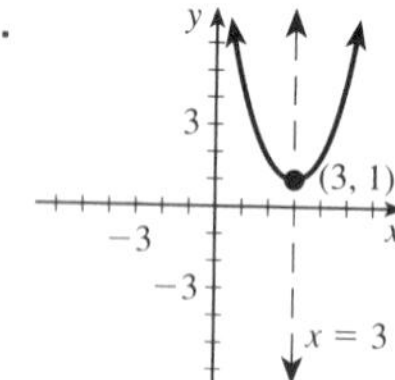

3.

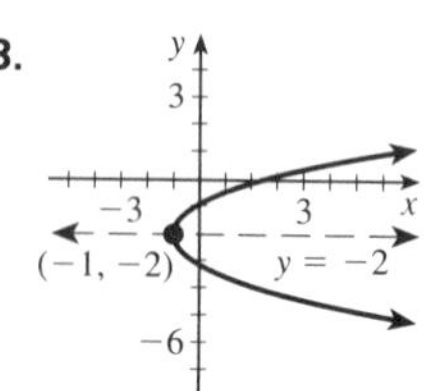

5.

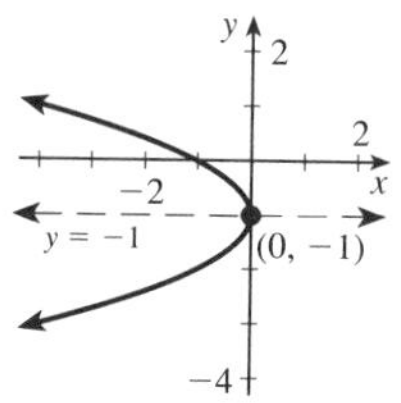

7. $y = (x - 1)^2$

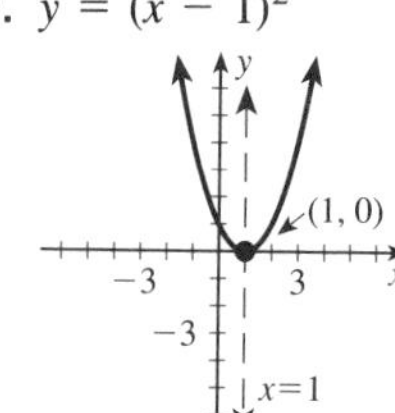

9. $x = (y - 2)^2 - 4$

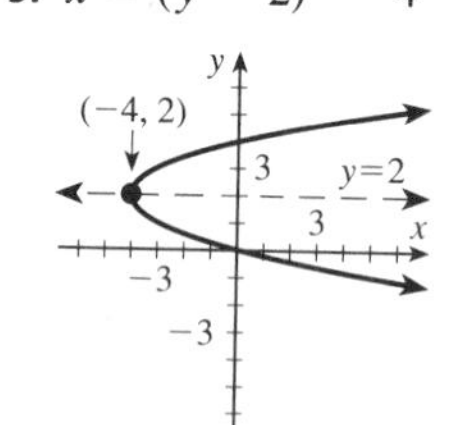

11. $x = (y + 2)^2 - 5$

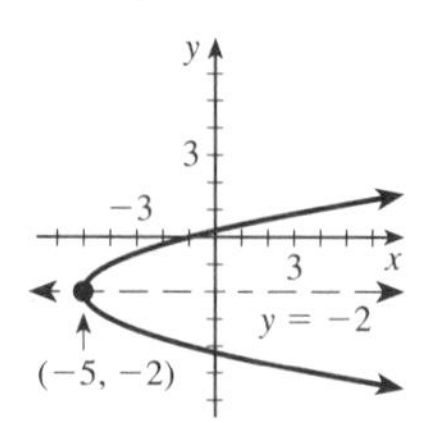

13.

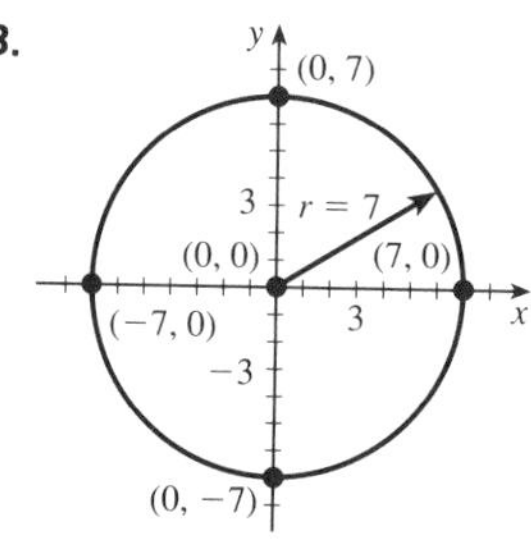

15.

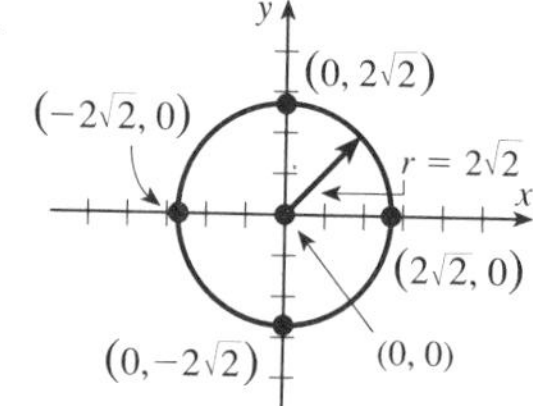

17. $x^2 + y^2 = 17$

19.

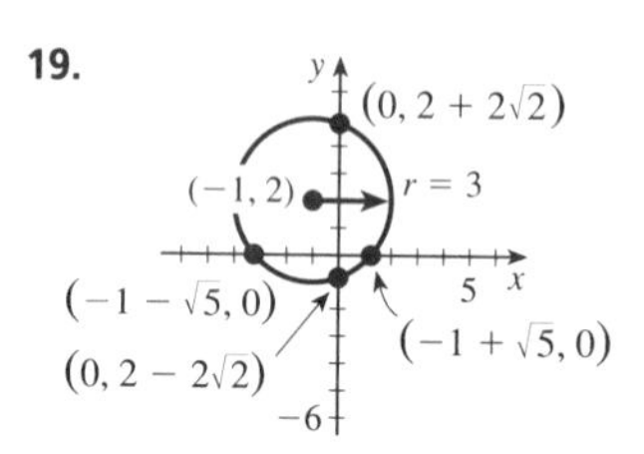

21. $(x + 3)^2 + (y - 2)^2 = 81$

23. $x^2 + (y - 1)^2 = 8$

25. $(x + 3)^2 + y^2 = 12$

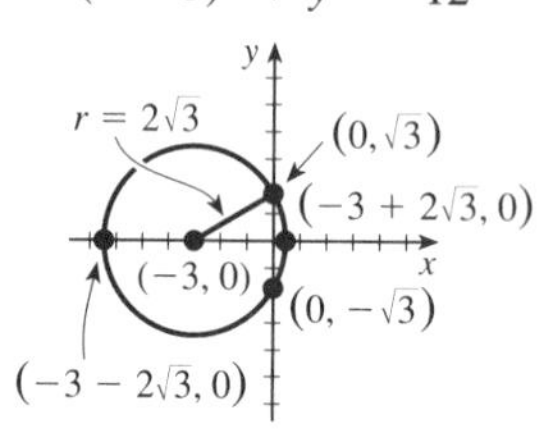

27. $(x + 2)^2 + y^2 = 4$

29.

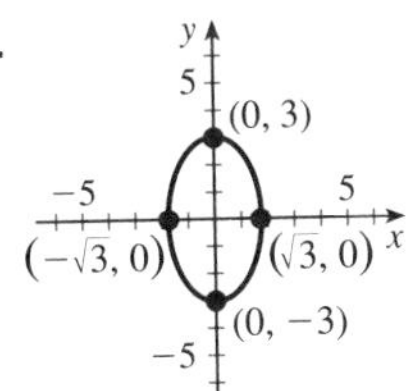

31.

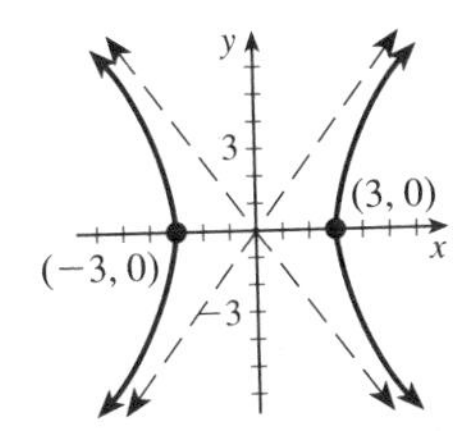

Asymptotes: $y = \frac{4}{3}x \quad y = -\frac{4}{3}x$

57.

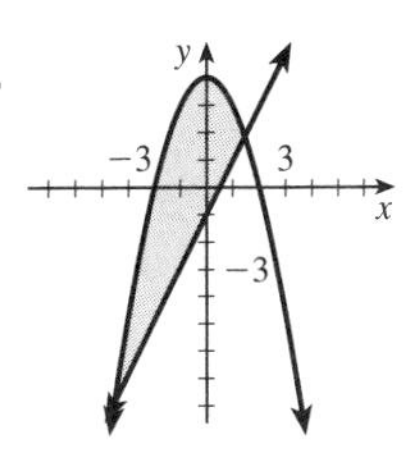

33.

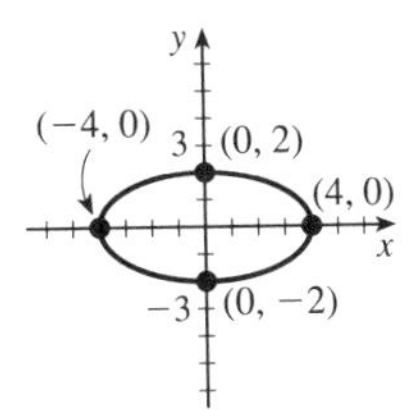

35.

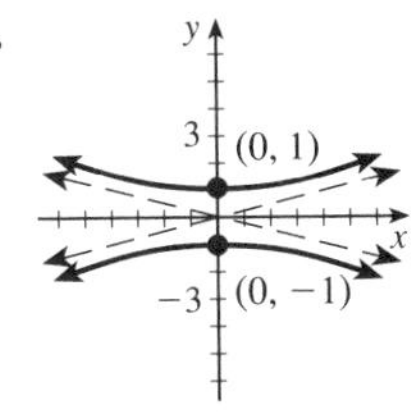

Asymptotes: $y = \frac{1}{4}x \quad y = -\frac{1}{4}x$

37. $\left(2 + \sqrt{5} \text{ and } 6 + 4\sqrt{5}\right)$ and $\left(2 - \sqrt{5}, 6 - 4\sqrt{5}\right)$

39. $(-5, 36)$ **41.** $(0, 0)$

43. $\left(\frac{1 + \sqrt{17}}{2}, \frac{-1 + \sqrt{17}}{2}\right)$ and $\left(\frac{1 - \sqrt{17}}{2}, \frac{-1 - \sqrt{17}}{2}\right)$

45. $\left(\sqrt{5}, 2\right), \left(\sqrt{5}, -2\right)$ and $\left(-\sqrt{5}, 2\right), \left(-\sqrt{5}, -2\right)$

47. No solution

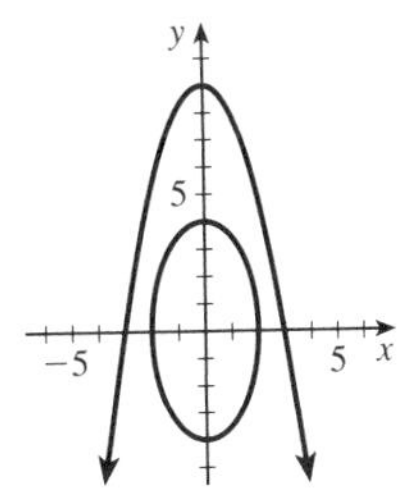

49. Width $10 - \sqrt{70}$ (≈ 1.63) inches and length $10 + \sqrt{70}$ (≈ 18.37) inches

51. $\left(\frac{3\sqrt{2}}{2}, \frac{\sqrt{22}}{2}\right), \left(\frac{-3\sqrt{2}}{2}, \frac{\sqrt{22}}{2}\right), \left(\frac{3\sqrt{2}}{2}, \frac{-\sqrt{22}}{2}\right), \left(\frac{-3\sqrt{2}}{2}, \frac{-\sqrt{22}}{2}\right)$

53.

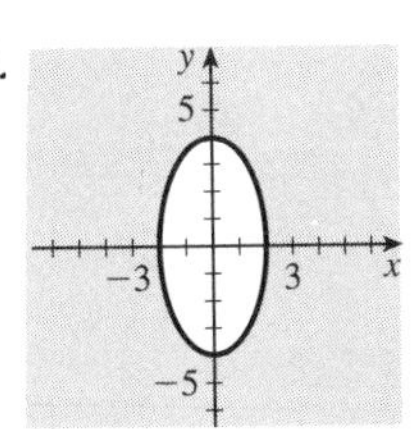

55.

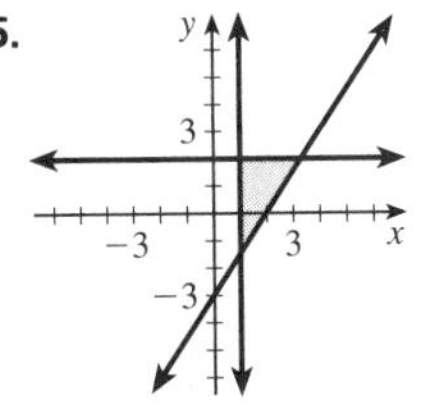

CHAPTER 12 TEST ■ page 653

1.

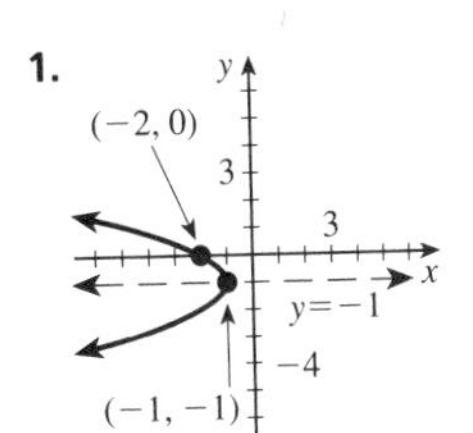

2. $x^2 + y^2 = 13$

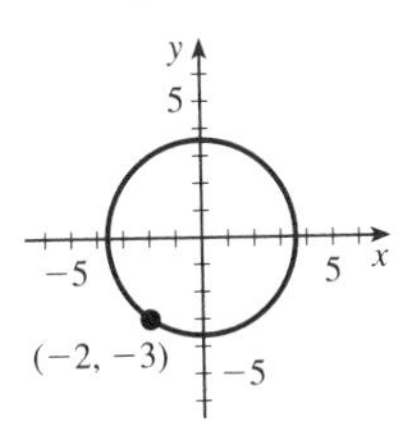

3. $(x + 1)^2 + y^2 = 13$

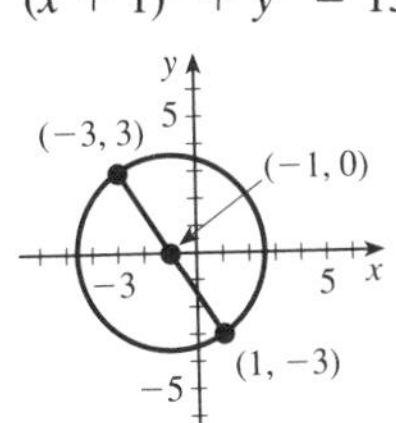

4.

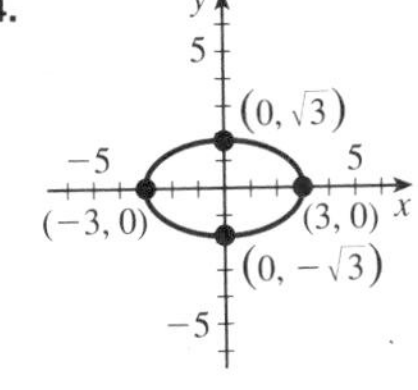

5.

Asymptotes: $y = \frac{2}{3}x, y = -\frac{2}{3}x$

6. $\left(\frac{-3 + \sqrt{29}}{2}, \frac{7 - \sqrt{29}}{2}\right)$ and $\left(\frac{-3 - \sqrt{29}}{2} \text{ and } \frac{7 + \sqrt{29}}{2}\right)$

7.

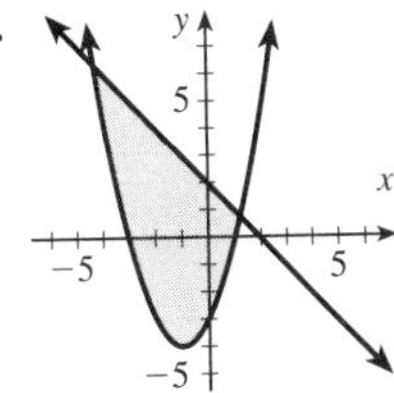

■ APPENDIX B

EXERCISES B.1

The RANGE Program (page A-7)

1. $-5 \quad 11$ **3.** $2 \quad 0.75$ **5.** 1.5

The EQLINE Program (page A-8)

1. $y = \frac{6}{4}x - \frac{2}{4}$ or $y = \frac{3}{2}x - \frac{1}{2}$

The QUAD Program (page A-10)

1. a. $3, -5$ **b.** $2.5, -1$

The DIVPOLY Program (page A-12)

1. a. $x^2 + 3x - 1$ remainder -3
b. $x^3 - 4x^2 + 20x - 101$ remainder 503
c. $3x^2 + 7x + 10$ remainder 20
d. $16x^4 + 8x^3 + 4x^2 + 2x + 1$ remainder 0

3. $\frac{3 + \sqrt{29}}{2}, \frac{3 - \sqrt{29}}{2}$

EXERCISES B.2

See answers to Exercises B.1.

INDEX

F

G

■ Q ■

■ R ■

■ S ■

■ T ■

■ U ■

■ V ■